TRANSPORT PHENOMENA

TRANSPORT PHENOMENA

R. BYRON BIRD
WARREN E. STEWART
EDWIN N. LIGHTFOOT

Department of Chemical Engineering
University of Wisconsin
Madison, Wisconsin

John Wiley & Sons, Inc.
New York · London · Sydney

14 15

LIBRARY OF CONGRESS CATALOG CARD NUMBER: 60-11717
PRINTED IN THE UNITED STATES OF AMERICA
ISBN 0 471 07392 X

Preface

This book is intended to be an introduction to the field of transport phenomena for students of engineering and applied science. Herein we present the subjects of momentum transport (viscous flow), energy transport (heat conduction, convection, and radiation), and mass transport (diffusion). In this treatment the media in which the transport phenomena are occurring are regarded as continua, and very little is said about the molecular explanation of these processes. Surely the continuum approach is of more immediate interest to engineering students, although it should be emphasized that both approaches are needed for complete mastery of the subject.

Because of the current demand in engineering education to put more emphasis on understanding basic physical principles than on the blind use of empiricism, we feel there is a very definite need for a book of this kind. Obviously the subject matter is sufficiently basic that it cuts across traditional departmental lines. Our thought has been that the subject of transport phenomena should rank along with thermodynamics, mechanics, and electromagnetism as one of the key "engineering sciences." Knowledge of the basic laws of mass, momentum, and energy transport has certainly become important, if not indispensable, in engineering analysis. In addition, the material in this text may be of interest to some who are working in physical chemistry, soil physics, meteorology, and biology.

Since the field of transport phenomena has not heretofore been recognized as a distinct engineering subject, it seems worthwhile for us to tell the reader how we have organized the material. Diverse methods of organization were studied, and, with the help of our departmental colleagues, we settled on the outline shown in Table I. Each topic has been assigned a pigeonhole in a two-dimensional array in order to emphasize the relation of each subject to other subjects in the same row or column. Division of the material into columns labeled mass, momentum, and energy transport allows for one method of classification, based on the entity being transported. In the various rows another mode of classification, based on the type of transport, is indicated. Clearly, on the basis of this chart, one can organize a course on transport phenomena in one of two ways: by working down the columns (Chapters 1, 2, 3, 4, 5, etc.) or by working across the rows (Chapters 1, 8, 16, 2, 9, 17, 3, etc.). Actually, the text material is arranged in such a way that either method may be used. The "column" approach is probably better for beginners, whereas the "row" approach may be more suited to advanced students.

Each chapter is provided with illustrative examples which show how to use various techniques or which give further elaboration on the text. Discussion questions at the end of the chapter are included in an effort to catalyze thinking about the material from several different viewpoints. The problems at the end of each chapter have been grouped into four classes (designated by a subscript after the problem number):

Class 1: Problems that illustrate direct numerical applications of the formulas in the text.

Class 2: Problems that require elementary analysis of physical situations, based on the subject material in the chapter.

Class 3: Problems that require somewhat more mature analysis, sometimes involving information from several chapters or material not specifically covered in the text.

Class 4: Problems that require mathematical analysis involving Bessel functions, partial differential equations, Laplace transforms, complex variable, and tensor analysis.

Of these four classes of problems the first three should be appropriate for junior and senior courses in transport phenomena; none of the problems in these classes involves mathematics beyond ordinary differential equations.

Obviously there is more material in this book than can be conveniently used in an introductory course. As a guide to prospective teachers of transport phenomena, we have indicated with an asterisk (*) those sections that we feel are suitable for a well-balanced three- or four-credit undergraduate course. Having some additional material in the book

TABLE I. SCHEMATIC DIAGRAM OF THE ORGANIZATION OF TRANSPORT PHENOMENA

Entity Being Transported → / Type of Transport ↓	Momentum	Energy	Mass
TRANSPORT BY MOLECULAR MOTION	1 VISCOSITY μ Newton's law of viscosity Temperature, pressure, and composition dependence of μ Kinetic theory of μ	8 THERMAL CONDUCTIVITY k Fourier's law of heat conduction Temperature, pressure, and composition dependence of k Kinetic theory of k	16 DIFFUSIVITY $\mathfrak{D}_{AB}$ Fick's law of diffusion Temperature, pressure, and composition dependence of $\mathfrak{D}_{AB}$ Kinetic theory of $\mathfrak{D}_{AB}$
TRANSPORT IN LAMINAR FLOW OR IN SOLIDS, IN ONE DIMENSION	2 SHELL MOMENTUM BALANCES Velocity profiles Average velocity Momentum flux at surfaces	9 SHELL ENERGY BALANCES Temperature profiles Average temperature Energy flux at surfaces	17 SHELL MASS BALANCES Concentration profiles Average concentration Mass flux at surfaces
TRANSPORT IN AN ARBITRARY CONTINUUM	3 EQUATIONS OF CHANGE (ISOTHERMAL) Equation of continuity Equation of motion Equation of energy (isothermal)	10 EQUATIONS OF CHANGE (NONISOTHERMAL) Equation of continuity Equation of motion for forced and free convection Equation of energy (nonisothermal)	18 EQUATIONS OF CHANGE (MULTICOMPONENT) Equations of continuity for each species Equation of motion for forced and free convection Equation of energy (multicomponent)
TRANSPORT IN LAMINAR FLOW OR IN SOLIDS, WITH TWO INDEPENDENT VARIABLES	4 MOMENTUM TRANSPORT WITH TWO INDEPENDENT VARIABLES Unsteady viscous flow Two-dimensional viscous flow Ideal two-dimensional flow Boundary-layer momentum transport	11 ENERGY TRANSPORT WITH TWO INDEPENDENT VARIABLES Unsteady heat conduction Heat conduction in viscous flow Two-dimensional heat conduction in solids Boundary-layer energy transport	19 MASS TRANSPORT WITH TWO INDEPENDENT VARIABLES Unsteady diffusion Diffusion in viscous flow Two-dimensional diffusion in solids Boundary-layer mass transport
TRANSPORT IN TURBULENT FLOW	5 TURBULENT MOMENTUM TRANSPORT Time-smoothing of equations of change Eddy viscosity Turbulent velocity profiles	12 TURBULENT ENERGY TRANSPORT Time-smoothing of equations of change Eddy thermal conductivity Turbulent temperature profiles	20 TURBULENT MASS TRANSPORT Time-smoothing of equations of change Eddy diffusivity Turbulent concentration profiles
TRANSPORT BETWEEN TWO PHASES	6 INTERPHASE MOMENTUM TRANSPORT Friction factor f Dimensionless correlations	13 INTERPHASE ENERGY TRANSPORT Heat-transfer coefficient h Dimensionless correlations (forced and free convection)	21 INTERPHASE MASS TRANSPORT Mass-transfer coefficient k_x Dimensionless correlations (forced and free convection)
TRANSPORT BY RADIATION	*Numbers refer to the chapters in this book*	14 RADIANT ENERGY TRANSPORT Planck's radiation law Stefan-Boltzmann law Geometrical problems Radiation through absorbing media	*This book may be studied either by "columns" or by "rows"*
TRANSPORT IN LARGE FLOW SYSTEMS	7 MACROSCOPIC BALANCES (ISOTHERMAL) Mass balance Momentum balance Mechanical energy balance (Bernoulli equation)	15 MACROSCOPIC BALANCES (NONISOTHERMAL) Mass balance Momentum balance Mechanical and total energy balance	22 MACROSCOPIC BALANCES (MULTICOMPONENT) Mass balances for each species Momentum balance Mechanical and total energy balance

will be helpful to instructors and advanced students and will, in addition, serve as a warning to the undergraduate that the "boundaries of the course" do not coincide with the "boundaries of the subject."

Our notation is uniform throughout the text, and a table of notation has been appended for the readers' convenience. Unfortunately, it is not possible to adopt notation in agreement with that used by all our readers, inasmuch as the subject material includes several fields that have developed independently. Generally, our notation represents a compromise between that used by physicists and that used by engineers.

Early in 1957 the Chemical Engineering Department of the University of Wisconsin decided, after considerable deliberation, to inaugurate a required one-semester junior course in transport phenomena. No textbook was available; hence mimeographed notes were prepared for the students' use and in the fall of 1958 were published as *Notes on Transport Phenomena*. These notes have also been used at several other universities, and we have benefited immensely from the comments sent to us by both students and teachers.

This book represents the result of an exhaustive revision of the *Notes on Transport Phenomena*. The text has been completely rewritten, several chapters have been entirely reorganized, and numerous problems and examples have been added. Most of the changes were made in an effort to provide a better text for beginning students.

R. BYRON BIRD
WARREN E. STEWART
EDWIN N. LIGHTFOOT

Madison, Wisconsin
June 1960

Acknowledgments

Many persons have contributed either directly or indirectly to this book; we should like to mention some of them by name:

Professor O. A. Hougen and Dean W. R. Marshall, Jr., of the University of Wisconsin deserve many thanks for their continued interest in the field of transport phenomena and their enthusiasm for promoting increased instruction in this area.

Professor R. A. Ragatz, Chairman, of the Chemical Engineering Department, University of Wisconsin, assisted us by handling the administrative problems associated with the introduction of this course into the new chemical engineering curriculum and by providing us with some additional time for preparation of the manuscript.

Our colleagues, Professors R. J. Altpeter, C. C. Watson, W. K. Neill, and E. J. Crosby, who have worked with us in developing the undergraduate course in transport phenomena, have given us many useful suggestions.

Professor J. E. Powers and his undergraduate class at the University of Oklahoma and Professor J. Dranoff and his graduate class at Northwestern University supplied us with detailed reviews of the *Notes on Transport Phenomena*.

Professor Eric Weger (Johns Hopkins University) and Professor K. M. Watson (Illinois Institute of Technology) also offered their comments after experiences in teaching with *Notes on Transport Phenomena*.

A number of our students have read certain chapters of the manuscript of this book and have contributed materially to the accuracy of the final text: Donald R. Woods, Allyn J. Ziegenhàgen, David O. Edwards, Paul F. Korbach, Donald W. McEachern, Rosendo J. Sánchez Palma, James P. Hutchins, Raffi M. Turian, Davis W. Hubbard, Boudewijn van Nederveen, Willam A. Hunt, John P. Lawler. In addition the following students have checked the statements and solutions to all of the Class 1 and Class 2 problems: Vipin D. Shah, Thomas J. Sadowski, Richard H. Weaver, Gary F. Kuether.

Professors J. O. Hirschfelder and C. F. Curtiss of the University of Wisconsin, with whom we have had many years of pleasant association, first introduced our chemical engineering department to the subject of transport phenomena some ten years ago via a graduate course; our present course is in a sense a direct descendent of theirs.

Professor H. Kramers (Technische Hogeschool, Delft, Holland) in 1956 prepared a set of lecture notes entitled *Physische Transportverschijnselen*, which represented the first attempt that we know of to teach transport phenomena to engineering students; one of us (R.B.B.) had the pleasure of spending a semester at Professor Kramers' laboratory as a Fulbright Lecturer and Guggenheim Fellow, during which period he profited very much from discussions related to the teaching of transport phenomena.

Miss Jeanne O. Lippert deserves our warmest thanks for typing the bulk of the manuscript and some parts of it several times. We are deeply indebted to Mr. Stuart E. Schreiber for his tireless efforts in mimeographing and assembling the original set of notes. Also we wish to thank Miss Ellen Gunderson for her part in assisting us with the preparation of the manuscript.

R. B. B.
W. E. S.
E. N. L.

Contents

PART I MOMENTUM TRANSPORT

PART III MASS TRANSPORT

PART I

MOMENTUM TRANSPORT

CHAPTER **1**

Viscosity and the Mechanism of Momentum Transport

The first part of this book deals with the flow of viscous fluids. The physical property that characterizes the flow resistance of simple fluids is the *viscosity*. Anyone who has purchased motor oil for his automobile is aware of the fact that some oils are more "viscous" than others and that viscosity is a function of the temperature. It is the purpose of this chapter to discuss the viscosities of gases and liquids in a quantitative way. This information will be needed immediately in Chapter 2 for the solution of viscous flow problems.

We begin in §1.1 by stating Newton's law of viscosity and then present a few numerical values to show how the viscosity varies with the conditions and the nature of the fluid. In §1.2 we touch briefly on the subject of non-Newtonian fluids, for which Newton's law of viscosity is inadequate. The effects of temperature and pressure on viscosities of gases and liquids are summarized in §1.3. Finally, in §§1.4 and 1.5, viscosity is discussed in terms of molecular processes, and the mechanisms of momentum transport in gases and liquids are contrasted.

§1.1 NEWTON'S LAW OF VISCOSITY

Consider a fluid—either a gas or a liquid—contained between two large parallel plates of area A, which are everywhere separated by a very small distance Y. (See Fig. 1.1–1.) We imagine that the system is initially at rest but that at time $t = 0$ the lower plate is set in motion in the x-direction at a constant velocity V. As time proceeds, the fluid gains momentum, and finally the steady-state velocity profile shown in Fig. 1.1–1 is established. When this final state of steady motion has been attained, a constant force

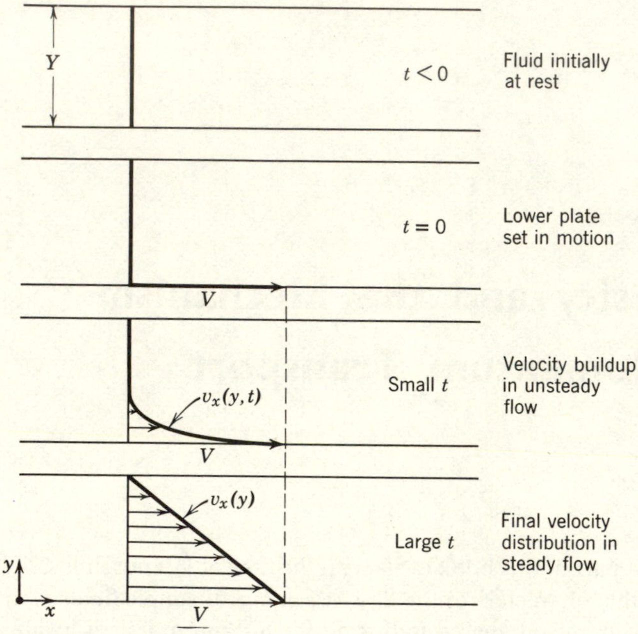

Fig. 1.1–1. Buildup to steady laminar velocity profile for fluid contained between two plates.

F is required to maintain the motion of the lower plate. This force may be expressed as follows (provided the flow is laminar):

$$\frac{F}{A} = \mu \frac{V}{Y} \tag{1.1–1}$$

That is, the force per unit area is proportional to the velocity decrease in the distance Y; the constant of proportionality μ is called the *viscosity* of the fluid.

We shall find it useful in the treatment that follows to rewrite Eq. 1.1–1 in a somewhat more explicit form. The shear stress exerted in the x-direction on a fluid surface of constant y by the fluid in the region of lesser y is designated as τ_{yx}, and the x-component of the fluid velocity vector is designated as v_x. Note that v_x is *not* equal to $\partial v/\partial x$; in this text we do not use subscripts to indicate differentiation of velocity components. Then, in terms of these symbols, Eq. 1.1–1 is rewritten as[1]

$$\tau_{yx} = -\mu \frac{dv_x}{dy} \qquad (1.1\text{–}2)$$

This states that the shear force per unit area is proportional to the negative of the local velocity gradient; this is known as *Newton's law of viscosity,* and fluids that behave in this fashion are termed *Newtonian fluids.* All gases and most simple liquids are described by Eq. 1.1–2; fluids that do not obey this simple law (primarily pastes, slurries, and high polymers) are discussed in §1.2.

Equation 1.1–2 may be interpreted usefully in another fashion. In the very neighborhood of the moving surface at $y = 0$ the fluid acquires a certain amount of x-momentum. This fluid, in turn, imparts some of its momentum to the adjacent "layer" of liquid causing it to remain in motion in the x-direction. Hence x-momentum is transmitted through the fluid in the y-direction. Consequently, τ_{yx} may also be interpreted as the viscous flux[2] of x-momentum in the y-direction. This interpretation ties in better with the molecular nature of the momentum transport process and corresponds to the treatment given later for energy and mass transport. Furthermore, the convention of sign for τ_{yx} seems easier to visualize in terms of momentum flux.

It may be seen from Eq. 1.1–2 that the viscous momentum flux is in the direction of the negative velocity gradient;[3] that is, the momentum tends to go in the direction of decreasing velocity. In other words, the momentum goes "downhill" in the sense that it "coasts" from a region of high velocity to a region of low velocity—just as a sled goes downhill from a region of high elevation to a region of low elevation or heat flows from a hot region toward a colder one. A velocity gradient can thus be thought of as a "driving force" for momentum transport.

We shall in the ensuing paragraphs refer to Newton's law in Eq. 1.1–2 sometimes in terms of forces (this brings out the essentially mechanical nature of the subject with which we are dealing) and sometimes in terms of momentum transport (this brings out the analogies with energy and mass

[1] The correspondence between these two equations is clearer if we note that Eq. 1.1–1 is equivalent to

$$\frac{F}{A} = -\mu \frac{0 - V}{Y - 0} \qquad (1.1\text{–}1a)$$

[2] By flux is meant "rate of flow per unit area." Momentum flux then has units of momentum per unit area per unit time. The student should verify that this is equivalent to force per unit area.

[3] A glance at §8.1 will show that this is the same behavior that one experiences with heat conduction, in which the heat flux is proportional to the negative temperature gradient.

transport). This dual viewpoint should cause no special difficulty, and indeed it should actually prove helpful in some instances.

In some formulas that appear in later chapters it will be useful to have a symbol to represent the viscosity divided by the mass density (mass per unit volume) of the fluid. Hence at this point we define the quantity ν by

$$\nu = \mu/\rho \tag{1.1-3}$$

which is called the *kinematic viscosity.*

A few words deserve to be said about the units of some of the quantities we have already defined. The situation is the simplest in the cgs system for which[4]

$$\begin{aligned} \tau_{yx} &[=] \text{ dyne cm}^{-2} \\ v_x &[=] \text{ cm sec}^{-1} \\ y &[=] \text{ cm} \end{aligned} \tag{1.1-4}$$

Since the two sides of Eq. 1.1–2 must agree in units as well as in numerical value, we may solve for the units of μ in the cgs system as follows:

$$\begin{aligned} \mu = -\tau_{yx}\left(\frac{dv_x}{dy}\right)^{-1} &[=] (\text{g cm}^{-1}\text{ sec}^{-2})(\text{cm sec}^{-1}\text{ cm}^{-1})^{-1} \\ &[=] \text{ g cm}^{-1}\text{ sec}^{-1} \end{aligned} \tag{1.1-5}$$

Correspondingly,

$$\nu = \mu/\rho \, [=] \text{ cm}^2\text{ sec}^{-1} \tag{1.1-6}$$

The cgs unit of g cm^{-1} sec^{-1} is called the poise; most viscosity data are reported either in this unit or in centipoises (1 cp = 0.01 poise). The analogous set of units in the English system is

$$\begin{aligned} \tau_{yx} &[=] \text{ poundals ft}^{-2} \\ v_x &[=] \text{ ft sec}^{-1} \\ y &[=] \text{ ft} \\ \mu &[=] \text{ lb}_m\text{ ft}^{-1}\text{ sec}^{-1} \\ \nu &[=] \text{ ft}^2\text{ sec}^{-1} \end{aligned} \tag{1.1-7}$$

These units are consistent with Eq. 1.1–2. Because it is not common to work in terms of poundals of force, many people prefer to rewrite Eq. 1.1–2 thus:

$$g_c\tau_{yx} = -\mu \frac{dv_x}{dy} \tag{1.1-8}$$

[4] Read [=] as "has units of."

in which

$$\begin{aligned} \tau_{yx} &[=] \text{lb}_f \text{ ft}^{-2} \\ v_x &[=] \text{ft sec}^{-1} \\ y &[=] \text{ft} \\ \mu &[=] \text{lb}_m \text{ ft}^{-1} \text{ sec}^{-1} \\ g_c &[=] (\text{lb}_m/\text{lb}_f)(\text{ft sec}^{-2}) \quad \text{or} \quad \text{poundals/lb}_f \end{aligned} \tag{1.1-9}$$

In these units the numerical value of g_c, the "gravitational conversion factor," is 32.174. Note that $g_c\tau_{yx}$ in Eq. 1.1–8 has units of poundals ft^{-2} and that division by g_c gives τ_{yx} in lb$_f$ ft^{-2}.

In this book we shall consistently employ Eq. 1.1–2 and understand that the units given in Eqs. 1.1–4 or 1.1–7 are used. The student should, however, be able to use formulas in either system because both systems are in current use in the technical literature. Careful checks for dimensional consistency are needed in all practical calculations.

Example 1.1–1. Calculation of Momentum Flux

Referring to Fig. 1.1–1, compute the steady-state momentum flux τ_{yx} in lb$_f$ ft^{-2} when the lower plate velocity V is 1 ft/sec in the positive x-direction, the plate separation Y is 0.001 ft, and the fluid viscosity μ is 0.7 cp.

Solution. Since τ_{yx} is desired in lb$_f$ ft^{-2}, we first convert all data to lb$_f$-ft-sec units. Thus, making use of Appendix C, Table C.3–4, we find

$$\begin{aligned} \mu &= (0.7 \text{ cp})(2.0886 \times 10^{-5}) \\ &= 1.46 \times 10^{-5} \text{ lb}_f \text{ sec ft}^{-2} \end{aligned}$$

The velocity profile is linear; hence

$$\frac{dv_x}{dy} = \frac{\Delta v_x}{\Delta y} = \frac{-1.0 \text{ ft sec}^{-1}}{0.001 \text{ ft}} = -1000 \text{ sec}^{-1}$$

Substitution in Eq. 1.1–2 then gives

$$\begin{aligned} \tau_{yx} = -\mu \frac{dv_x}{dy} &= -(1.46 \times 10^{-5})(-1000) \\ &= 1.46 \times 10^{-2} \text{ lb}_f \text{ ft}^{-2} \end{aligned}$$

If Eq. 1.1–8 were used, one would first convert μ as follows:

$$\mu = (0.7 \text{ cp})(6.7197 \times 10^{-4}) = 4.70 \times 10^{-4} \text{ lb}_m \text{ ft}^{-1} \text{ sec}^{-1}$$

Then substitution into Eq. 1.1–8 gives

$$\begin{aligned} \tau_{yx} &= -\frac{\mu}{g_c}\frac{dv_x}{dy} = -\frac{(4.70 \times 10^{-4})}{(32.174)}(-1000) \\ &= 1.46 \times 10^{-2} \text{ lb}_f \text{ ft}^{-2} \end{aligned}$$

This agrees with the result obtained from Eq. 1.1–2.

In Tables 1.1–1, 1.1–2, and 1.1–3 some experimental viscosity data are given for pure fluids at 1 atm pressure.[5] Note that at room temperature μ is about 1 cp for water and about 0.02 cp for air. Note also that for *gases* at low density the viscosity *increases* with increasing temperature, whereas for

TABLE 1.1–1

VISCOSITY OF WATER AND AIR AT 1 ATM PRESSURE

	Water (liq.)[a]		Air[b]	
Temperature T (°C)	Viscosity μ (cp)	Kinematic Viscosity $\nu \times 10^2$ (cm² sec⁻¹)	Viscosity μ (cp)	Kinematic Viscosity $\nu \times 10^2$ (cm² sec⁻¹)
0	1.787	1.787	0.01716	13.27
20	1.0019	1.0037	0.01813	15.05
40	0.6530	0.6581	0.01908	16.92
60	0.4665	0.4744	0.01999	18.86
80	0.3548	0.3651	0.02087	20.88
100	0.2821	0.2944	0.02173	22.98

[a] Calculated from the results of R. C. Hardy and R. L. Cottington, *J. Research Nat. Bur. Standards*, **42,** 573–578 (1949), and J. F. Swindells, J. R. Coe, Jr., and T. B. Godfrey, *J. Research Nat. Bur. Standards*, **48,** 1–31 (1952).

[b] Calculated from "Tables of Thermal Properties of Gases," *Nat. Bur. Standards Circ.* **464** (1955), Chapter 2.

TABLE 1.1–2

VISCOSITIES OF SOME GASES AND LIQUIDS AT ATMOSPHERIC PRESSURE[a]

Substance	Temperature T (°C)	Viscosity μ (cp)	Substance	Temperature T (°C)	Viscosity μ (cp)
Gases			Liquids		
i-C_4H_{10}	23	0.0076	$(C_2H_5)_2O$	20	0.245
CH_4	20	0.0109[b]	C_6H_6	20	0.647
H_2O	100	0.0127	Br_2	26	0.946
CO_2	20	0.0146[b]	C_2H_5OH	20	1.194
N_2	20	0.0175[b]	Hg	20	1.547
O_2	20	0.0203[b]	H_2SO_4	25	19.15
Hg	380	0.0654	Glycerol	20	1069.

[a] Values taken from N. A. Lange, *Handbook of Chemistry*, McGraw-Hill, New York (1956), Ninth Edition, pp. 1658–1664.

[b] H. L. Johnston and K. E. McKloskey, *J. Phys. Chem.*, **44,** 1038 (1940).

[5] Very complete data may be found in the Landolt-Börnstein *Physikochemische Tabellen.*

liquids the viscosity usually *decreases* with increasing temperature. This difference in temperature dependence is discussed from a molecular viewpoint in §1.4 and §1.5; we simply mention here that in gases (in which the molecules travel long distances between collisions) the momentum is transported primarily by the molecules in free flight, whereas in liquids (in which the molecules travel only very short distances between collisions) the principal mechanism for momentum transfer is the actual colliding of the molecules.

In §§1.3, 1.4, and 1.5 we take up the matter of calculating viscosities of gases and liquids. Before doing so, however, we digress briefly to indicate the kinds of deviations from Eq. 1.1–2 that are known to exist.

TABLE 1.1–3

VISCOSITIES OF SOME LIQUID METALS[a]

Metal	Temperature T (°C)	Viscosity μ (cp)
Li	183.4	0.5918
	216.0	0.5406
	285.5	0.4548
Na	103.7	0.686
	250	0.381
	700	0.182
K	69.6	0.515
	250	0.258
	700	0.136
Na-K alloy	103.7	0.546
56% Na by wt.	250	0.316
44% K by wt.	700	0.161
Hg	−20	1.85
	20	1.55
	100	1.21
	200	1.01
Pb	441	2.116
	551	1.700
	844	1.185

[a] Data taken from *The Reactor Handbook*, Vol. 2, Atomic Energy Commission AECD-3646, U.S. Government Printing Office, Washington D.C. (May 1955), pp. 258 *et seq.*

§1.2 NON-NEWTONIAN FLUIDS[1]

According to Newton's law of viscosity in Eq. 1.1–2, a plot of τ_{yx} versus $-(dv_x/dy)$ for a given fluid should give a straight line through the origin, and the slope of this line is the viscosity of the fluid at the given temperature and pressure. (See Fig. 1.2–1.) Experiments have shown that τ_{yx} is indeed

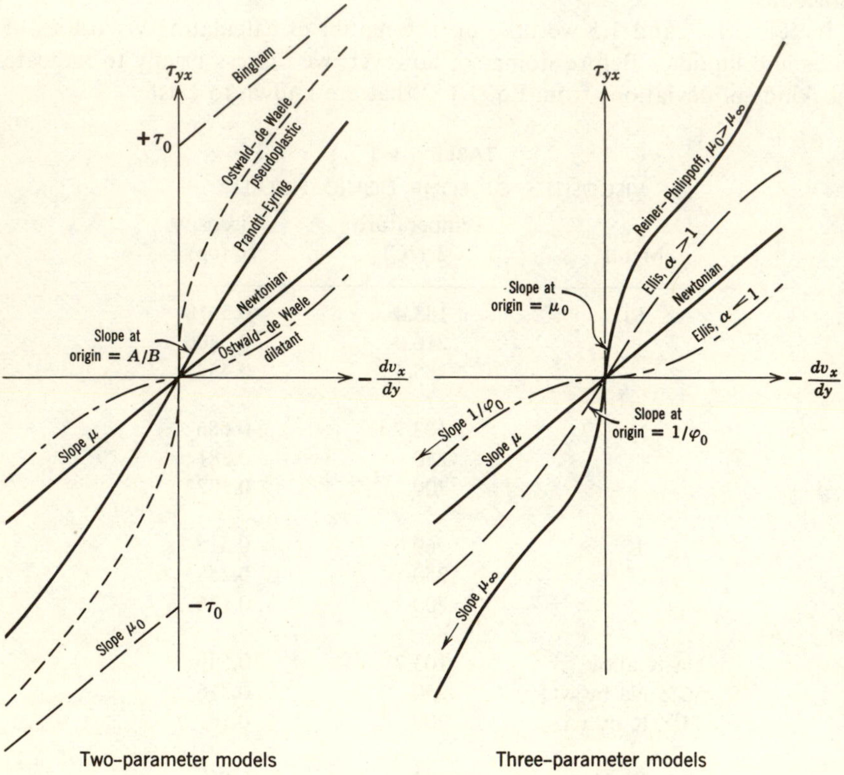

Fig. 1.2–1. Summary of steady-state non-Newtonian models (the Newtonian model is shown for reference).

proportional to $-(dv_x/dy)$ for all gases and for homogeneous nonpolymeric liquids. There are, however, quite a few industrially important materials that are not described by Eq. 1.1–2, and they are referred to as *non-Newtonian* fluids.

The subject of non-Newtonian flow is actually a subdivision of the larger science of *rheology*. This is the "science of deformation and flow"

[1] M. Reiner, *Deformation, Strain, and Flow*, Interscience, New York, (1960); W. Philippoff, *Viskosität der Kolloïde*, Steinkopff, Leipzig (1942).

and includes the study of the mechanical properties of gases, liquids, plastics, asphalts, and crystalline materials. Hence rheology includes Newtonian fluid mechanics at one end of the spectrum of subject material, and Hookean elasticity at the other. The region in between concerns the deformation and flow of all sorts of gunky and gooey materials.

The steady-state rheological behavior of most fluids in the situation of Fig. 1.1–1 can be expressed by a generalized[2] form of Eq. 1.1–2:

$$\tau_{yx} = -\eta \frac{dv_x}{dy} \tag{1.2–1}$$

where η may be expressed as a function of either dv_x/dy or τ_{yx}. In regions in which η decreases with increasing rate of shear $(-dv_x/dy)$ the behavior is termed *pseudoplastic*; in regions in which η increases with increasing rate of shear the behavior is termed *dilatant*. If η is independent of the rate of shear, the behavior is Newtonian, with $\eta = \mu$. (See Eq. 1.1–2.)

Numerous empirical equations, or "models," have been proposed to express the steady-state relation between τ_{yx} and dv_x/dy. Five representative models are summarized below. Each of these equations contains empirical positive parameters, which can be evaluated numerically to fit data on τ_{yx} versus dv_x/dy at constant temperature and pressure.

The Bingham Model

$$\tau_{yx} = -\mu_0 \frac{dv_x}{dy} \pm \tau_0 \quad \text{if } |\tau_{yx}| > \tau_0 \tag{1.2–2a}$$

$$\frac{dv_x}{dy} = 0 \quad \text{if } |\tau_{yx}| < \tau_0 \tag{1.2–2b}$$

The positive sign is used in Eq. 1.2–2*a* when τ_{yx} is positive, and the negative sign is used when τ_{yx} is negative. A substance that follows this two-parameter model is called a *Bingham plastic*; it remains rigid when the shear stress is of smaller magnitude than the yield stress τ_0 but flows somewhat like a Newtonian fluid when the shear stress exceeds τ_0. This model has been found reasonably accurate for many fine suspensions and pastes. Bingham parameters for suspensions of nuclear fuel particles in heavy water are given in Table 1.2–1.

The Ostwald-de Waele Model

$$\tau_{yx} = -m \left| \frac{dv_x}{dy} \right|^{n-1} \frac{dv_x}{dy} \tag{1.2–3}$$

This two-parameter equation is also known as the *power law*. For $n = 1$, it reduces to Newton's law of viscosity with $m = \mu$; thus the deviation of n

[2] For a more complete discussion, see §3.6.

TABLE 1.2–1
BINGHAM PLASTIC PARAMETERS FOR AQUEOUS NUCLEAR FUEL SLURRIES[a]

Suspended Material	Particle Size Distribution: D (microns)	σ (dimensionless)	$k_1 = \frac{\tau_0}{\phi^4}$ ($lb_f\ ft^{-2}$)	$k_2 = \frac{\ln \mu_0/\mu_w}{\phi}$ (dimensionless)
UO_2	1.4	1.7	150	18
U_3O_8	1.3	2.0	230	22
$UO_3 \cdot H_2O$	1.2	1.9	430	22
			$k_3 = \frac{\tau_0}{\phi^3}$ ($lb_f\ ft^{-2}$)	$k_4 = \frac{\ln \mu_0/\mu_w}{\phi}$ (dimensionless)
ThO_2	0.030	2.7	1100	24
	0.75	2.8	550	14
	1.6	1.5	100	12
	2.4	1.7	33	12

D = mass-median particle diameter
σ = standard deviation of ln (particle diameter) from ln (D)
ϕ = volume fraction solids in the suspension
μ_0, τ_0 = Bingham parameters for the suspension
μ_w = viscosity of water at the same temperature and pressure

[a] Taken from *Fluid Fuel Reactors*, edited by J. A. Lane, H. G. MacPherson, and F. Maslan, Addison-Wesley, Reading, Mass. (1958), §4.4 contributed by D. G. Thomas. Corrected values of k_2 are given here by courtesy of Dr. Thomas.

from unity indicates the degree of deviation from Newtonian behavior. For values of n less than unity, the behavior is pseudoplastic, whereas for n greater than unity the behavior is dilatant. Approximate values of m and n for various fluids are given in Table 1.2–2.

The Eyring Model

$$\tau_{yx} = A \operatorname{arcsinh}\left(-\frac{1}{B}\frac{dv_x}{dy}\right) \tag{1.2–4}$$

This two-parameter model is derivable by the Eyring kinetic theory[3] of

[3] See Eq. 1.5–7; for more complex models see F. H. Ree, T. Ree, and H. Eyring, *Ind. Eng. Chem.*, **50**, 1036–1040 (1958).

TABLE 1.2–2

POWER MODEL PARAMETERS FOR VARIOUS FLUIDS AT ROOM TEMPERATURE[a]

Fluid Composition (weight %)	m ($lb_f\ sec^n\ ft^{-2}$)	n (dimensionless)
23.3 % Illinois yellow clay in water	0.116	0.229
0.67 % CMC[b] in water	0.00634	0.716
1.5 % CMC in water	0.0653	0.554
3.0 % CMC in water	0.194	0.566
33 % lime in water	0.150	0.171
10 % napalm in kerosene	0.0893	0.520
4 % paper pulp in water	0.418	0.575
54.3 % cement rock in water	0.0524	0.153

[a] A. B. Metzner, *Advances in Chemical Engineering*, Vol. I, Academic Press, New York (1956), p. 103.
[b] Carboxymethylcellulose.

TABLE 1.2–3

ELLIS PARAMETERS FOR SOLUTIONS OF CARBOXYMETHYLCELLULOSE IN WATER[a]

Solution Concentration (weight %)	Temperature T (°F)	Parameters in Eq. 1.2–5: α (dimensionless)	φ_0 ($cm^2\ sec^{-1}\ dyne^{-1}$)	φ_1 ($cm^{2\alpha}\ sec^{-1}\ dyne^{-\alpha}$)	Experimental Range of Shear Stress ($dyne\ cm^{-2}$)
4.0% CMC low[b]	85.0	1.170	0.1377	0.3211	8 to 440
5.0% CMC low	85.0	1.337	0.0000	0.0521	8 to 1010
1.5% CMC medium	85.0	1.185	0.4210	0.2724	6 to 300
2.5% CMC medium	85.0	1.412	0.0383	0.0181	17 to 720
0.6% CMC high	85.0	1.707	0.2891	0.0280	8 to 270

[a] J. C. Slattery, doctoral thesis, University of Wisconsin (1959), p. 79a. Corrected values of φ_0 and φ_1 are given here by courtesy of Dr. Slattery.
[b] CMC low, medium, and high are commercial carboxymethylcellulose preparations of low, medium, and high molecular weight.

liquids, as shown in §1.5. It predicts pseudoplastic behavior at finite values of τ_{yx} but reduces asymptotically to Newton's law of viscosity with $\mu = A/B$ as τ_{yx} approaches zero.

The Ellis Model

$$-\frac{dv_x}{dy} = (\varphi_0 + \varphi_1 |\tau_{yx}|^{\alpha-1})\tau_{yx} \tag{1.2-5}$$

This model contains three adjustable positive parameters: φ_0, φ_1, and α If α is chosen greater than unity, the model approaches Newton's law for small τ_{yx}; on the other hand, if α is chosen less than unity, Newton's law is approached for large τ_{yx}. This model is extremely flexible and includes Newton's law ($\varphi_1 = 0$) and the power law ($\varphi_0 = 0$) as special cases. Ellis parameters are given for several fluids in Table 1.2–3.

The Reiner-Philippoff Model

$$-\frac{dv_x}{dy} = \left(\frac{1}{\mu_\infty + \dfrac{\mu_0 - \mu_\infty}{1 + (\tau_{yx}/\tau_s)^2}}\right)\tau_{yx} \tag{1.2-6}$$

This model contains three adjustable positive parameters: μ_0, μ_∞, and τ_s. Since Newtonian behavior has often been observed both at very low and very high shearing rates, Eq. 1.2–6 has been set up to reduce to Newton's law of viscosity with $\mu = \mu_0$ and $\mu = \mu_\infty$, respectively, in these two limiting

TABLE 1.2–4

REINER-PHILIPPOFF PARAMETERS FOR VARIOUS FLUIDS[a]

Substance	Temperature T (°C)	Parameters in Eq. 1.2–6: μ_0 (poise)	μ_∞ (poise)	τ_s (dyne cm^{-2})	Experimental Range of Shear Stress (dyne cm^{-2})
Molten sulfur	120	0.215	0.0105	0.073	0.2 to 10
Cholesterol butyrate	100	2.4	0.35	1.05	0.8 to 20
30.4% methanol in hexane[b]	34	0.035	0.0035	0.05	0.1 to 4
0.4% polystyrene[c] in tetralin	20	4.0	1.0	500	500 to 4000

[a] W. Philippoff, *Kolloid Z.*, **71**, 1–16 (1935). Verlag Dr. Dietrich Steinkopff, Darmstadt. [b] Mixture of two liquid phases. [c] Polystyrene with molecular weight of 600,000.

cases. The plot of τ_{yx} versus dv_x/dy has inflection points located at $\tau_{yx} = \pm\tau_s\sqrt{3\mu_0/\mu_\infty}$. Table 1.2–4 gives values of μ_0, μ_∞, and τ_s for several fluids.

The rheological behavior of the foregoing models is sketched in Fig. 1.2–1. Keep in mind that these equations are nothing more than empirical curve-fitting formulas, and it is hazardous to apply them beyond the range of available data. Note also that the parameters of any of these models are functions of temperature, pressure, composition, and, usually, of the range of dv_x/dy over which the equation is fitted; therefore, the conditions of measurement must be carefully specified in reporting rheological parameters.

Under unsteady-state conditions, a number of additional types of non-Newtonian behavior are possible. For example, fluids that show a limited *decrease* in η (see Eq. 1.2–1) with time under a suddenly applied constant stress τ_{yx} are called *thixotropic*, whereas those that show an *increase* in η with time are called *rheopectic*. Fluids that partially return to their original form when the applied stress is released are called *visco-elastic*. The quantitative study of these and other types of time-dependent behavior is an important and largely undeveloped area of fluid mechanics.

§1.3 PRESSURE AND TEMPERATURE DEPENDENCE OF VISCOSITY

Extensive data on viscosities of pure gases and liquids are available in a number of references, notably the Landolt-Börnstein *Physikochemische Tabellen*. When experimental data are lacking and there is not time to obtain them, one can estimate the viscosity by empirical methods, making use of other data on the given substance. We present here two correlations to facilitate such estimates and to illustrate the general trends of viscosity with temperature and pressure for ordinary fluids. These correlations represent two independent analyses of a large number of data on various fluids by corresponding-states approaches.[1]

In Fig. 1.3–1 a plot is shown of the reduced viscosity $\mu_r = \mu/\mu_c$, which is the viscosity at a given temperature and pressure divided by the viscosity at the critical point. This quantity is plotted as a function of the reduced temperature $T_r = T/T_c$ and the reduced pressure $p_r = p/p_c$. This chart shows that the viscosity of a gas approaches a definite limit (the low-density limit on Fig. 1.3–1) as the pressure approaches zero at a given temperature; for most gases, this limit is essentially reached at 1 atm pressure. The viscosity of a *gas* at low density *increases* with increasing temperature, whereas the viscosity of a *liquid decreases* with increasing temperature.

Experimental values of μ_c are seldom available. However, μ_c may be estimated in one of the following ways: (i) if a value of viscosity is known

[1] For an introductory discussion of the principle of corresponding states, see F. Daniels and R. A. Alberty, *Physical Chemistry*, Wiley, New York (1955), pp. 21–23.

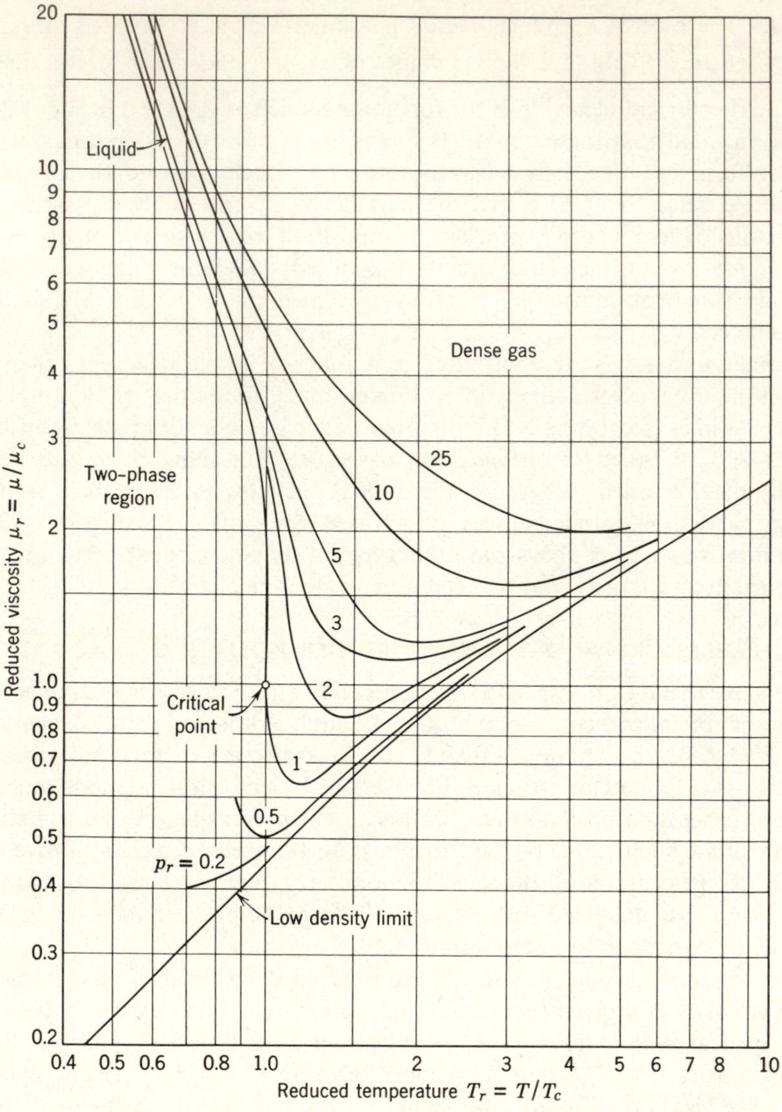

Fig. 1.3–1. Reduced viscosity $\mu_r = \mu/\mu_c$ as a function of temperature for several values of the reduced pressure $p_r = p/p_c$. [O. A. Uyehara and K. M. Watson, *Nat. Petroleum News, Tech. Section 36*, 764 (Oct. 4, 1944.); revised by K. M. Watson (1960). A large-scale version of this graph is available in O. A. Hougen, K. M. Watson, and R. A. Ragatz, *C.P.P. Charts*, Wiley, New York (1960), Second Edition.]

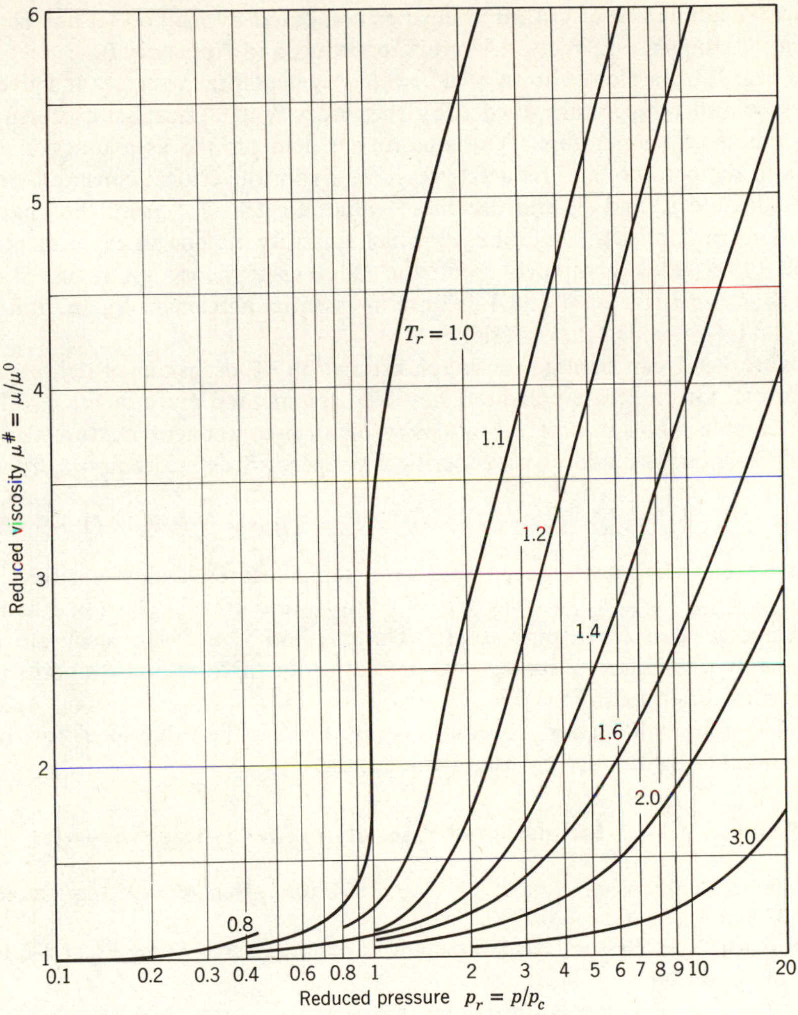

Fig. 1.3–2. Reduced viscosity $\mu^{\#} = \mu/\mu^{\circ}$ as function of reduced pressure $p_r = p/p_c$ and reduced temperature $T_r = T/T_c$. [N. L. Carr, R. Kobayashi, and D. B. Burroughs, *Am. Inst. Min. & Met. Engrs., Petroleum Tech.*, **6**, 47 (1954).]

at a given reduced pressure and temperature, preferably at conditions as near as possible to those of interest, then μ_c can be calculated from $\mu_c = \mu/\mu_r$; or (ii) if only critical *p-V-T* data are available then μ_c may be estimated from

$$\mu_c = 61.6(MT_c)^{1/2}(\tilde{V}_c)^{-2/3} \qquad \text{or} \qquad \mu_c = 7.70M^{1/2}p_c^{2/3}T_c^{-1/6} \qquad (1.3\text{–}1,2)$$

in which μ_c is in micropoises, p_c in atm, T_c in °K, and $\tilde{V}_c$ in cc per gram mole. A tabulation of critical viscosities computed by method (i) has been given by Hougen and Watson,[2] and is reproduced in Appendix B.

In Fig. 1.3–2 a plot is shown of $\mu^{\#} = \mu/\mu^0$, which is the viscosity at a given pressure and temperature divided by the viscosity at atmospheric pressure and the same temperature. This quantity is also plotted as a function of reduced temperature and reduced pressure. From the critical constants one can calculate p_r and T_r and use these values to get μ/μ^0 from the chart. Then one multiplies the latter by μ^0, which quantity may be an experimental value or a value computed from the dilute-gas theory given in §1.4. The charts in Figs. 1.3–1 and 1.3–2 are in essential agreement for the range of p_r and T_r in which they overlap.

Figure 1.3–1 can be used for rough estimation of viscosities of dense gas mixtures; for moderate pressures, however, the method given in Eq. 1.4–19 is preferable. To estimate the viscosity of an n-component mixture from Fig. 1.3–1, use is made of pseudocritical properties[3] defined empirically as

$$p_c' = \sum_{i=1}^{n} x_i p_{ci} \qquad T_c' = \sum_{i=1}^{n} x_i T_{ci} \qquad \mu_c' = \sum_{i=1}^{n} x_i \mu_{ci} \qquad (1.3\text{–}3,4,5)$$

That is, one computes reduced pressures, temperatures, and viscosities in terms of p_c', T_c', and μ_c', instead of p_c, T_c, and μ_c, and uses the chart in exactly the same way as for a pure fluid. This method is not very accurate if chemically dissimilar substances are present in the mixture or if the critical properties differ greatly.[4]

One can apply a similar procedure to Fig. 1.3–2. The value of μ^0 for the mixture can be estimated by use of Eq. 1.4–19.

Example 1.3–1. Estimation of Viscosity From Critical Properties

Estimate the viscosity of N_2 at 50° C and 854 atm, given $M = 28.0$ g/g-mole, $p_c = 33.5$ atm, and $T_c = 126.2°$ K.

Solution. The Watson-Uyehara method is required here. Using Eq. 1.3–2 to estimate μ_c, we get

$$\begin{aligned}\mu_c &= 7.70(28.0)^{1/2}(33.5)^{2/3}(126.2)^{-1/6}\\ &= 189 \text{ micropoises} = 189 \times 10^{-6} \text{ g cm}^{-1}\text{ sec}^{-1}\end{aligned}$$

The reduced temperature and pressure are

$$T_r = \frac{273.2 + 50}{126.2} = 2.56$$

$$p_r = \frac{854}{33.5} = 25.5$$

[2] O. A. Hougen and K. M. Watson, *Chemical Process Principles*, Part III, Wiley, New York (1947), p. 873.

[3] *Ibid.*, Part II, p. 604.

[4] E. W. Comings, *High Pressure Technology*, McGraw-Hill, New York (1956), p. 281.

From Fig. 1.3–1, we read $\mu/\mu_c = 2.39$. Hence the predicted viscosity is

$$\mu = \mu_c(\mu/\mu_c) = 189 \times 10^{-6} \times 2.39$$
$$= 452 \times 10^{-6} \text{ g cm}^{-1} \text{ sec}^{-1}$$

The measured value[5] is 455×10^{-6} g cm^{-1} sec^{-1}. This is unusually good agreement.

Example 1.3–2. Effect of Pressure on Gas Viscosity

The viscosity of CO_2 is reported[6] as 1800×10^{-7} poise at 45.3 atm and 40.3° C. Estimate the viscosity at 114.6 atm and 40.3° C, using Fig. 1.3–2.

Solution. The reduced temperature and pressure corresponding to the given viscosity value are

$$T_r = \frac{273.2 + 40.3}{304.2} = 1.03; \qquad p_r = \frac{45.3}{72.9} = 0.622$$

From Fig. 1.3–2, we find $\mu^{\#} = 1.12$ at this condition; hence $\mu^0 = \mu/\mu^{\#} = 1610 \times 10^{-7}$ g cm^{-1} sec^{-1}.

The reduced pressure at the condition of interest is

$$p_r = \frac{114.6}{72.9} = 1.57$$

and T_r is the same as above. From Fig. 1.3–2, we then read $\mu^{\#} = 3.7$; hence the predicted viscosity at 114.6 atm and 40.3° C is $\mu = 3.7 \times 1610 \times 10^{-7} = 6000 \times 10^{-7}$ g cm^{-1} sec^{-1}. The observed value[6] is 5800×10^{-7} g cm^{-1} sec^{-1}.

§1.4 THEORY OF VISCOSITY OF GASES AT LOW DENSITY

The viscosities of gases at low density have been extensively studied, both experimentally and theoretically. To illustrate the momentum transport mechanism, we start with a simplified derivation of the viscosity from a molecular point of view.

Consider a pure gas composed of rigid, nonattracting spherical molecules of diameter d and mass m, present in a concentration of n molecules per unit volume. We consider n to be small enough so that the average distance between molecules is many times their diameter d. At equilibrium in such a gas it is known from kinetic theory[1] that the molecular velocities, relative to the fluid velocity[2] $\boldsymbol{v}$, are randomly directed and have an average magnitude

[5] A. M. J. F. Michels and R. E. Gibson, *Proc. Roy. Soc.* (London), **A134**, 288 (1931).

[6] E. Warburg and L. v. Babo, *Wied Ann.*, **17**, 390 (1882).

[1] The first four equations here are given without proof. Detailed proofs and discussion are given in most texts on kinetic theory, e.g., E. H. Kennard, *Kinetic Theory of Gases*, McGraw-Hill, New York (1938), Chapters II and III.

[2] Here $\boldsymbol{v}$ is the fluid velocity vector representing the number average of the molecular velocities in the neighborhood of a given point.

$\bar{u}$ given by

$$\bar{u} = \sqrt{\frac{8\kappa T}{\pi m}} \tag{1.4–1}$$

in which κ is the Boltzmann constant. The frequency of molecular bombardment on one side of any stationary surface exposed to the gas, per unit area, is

$$Z = \tfrac{1}{4}n\bar{u} \tag{1.4–2}$$

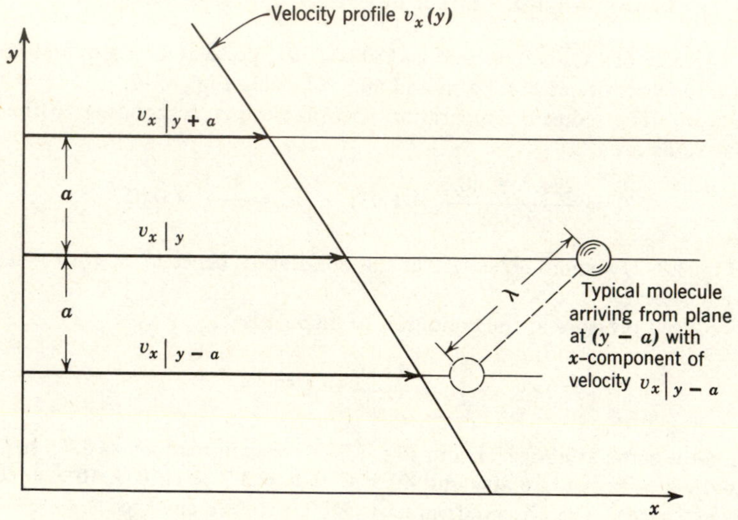

Fig. 1.4–1. Molecular transport of x-momentum from plane at $(y - a)$ to plane at y.

The average distance traveled by a molecule between successive collisions is the mean free path[3] λ, given by

$$\lambda = \frac{1}{\sqrt{2}\pi d^2 n} \tag{1.4–3}$$

The molecules reaching a plane have, on the average, had their last collision at a distance a from the plane, where

$$a = \tfrac{2}{3}\lambda \tag{1.4–4}$$

To determine the viscosity of such a gas in terms of the molecular properties, we consider the behavior of the gas when it flows parallel to the x-axis with a velocity gradient dv_x/dy. (See Fig. 1.4–1.) We assume that Eqs. 1.4–1,2,3,4 will remain valid in this nonequilibrium situation, provided that all molecular velocities are calculated relative to the average velocity $\boldsymbol{v}$ in the region in which the given molecule had its last collision. The flux of x-momentum

[3] The concept of the mean free path is intuitively appealing but is applicable only to molecules that exert no long-range forces on one another. The concept is easily applied to the "billiard-ball" molecules considered here and in early kinetic theory but does not appear in the modern kinetic theory discussed later in this section.

across any plane of constant y is found by summing the x-momenta of the molecules that cross in the positive y-direction and subtracting the x-momenta of those that cross in the opposite direction. Thus

$$\tau_{yx} = Zmv_x|_{y-a} - Zmv_x|_{y+a} \tag{1.4–5}$$

In writing this equation, we have assumed that all molecules have velocities representative of the region in which they last collided and that the velocity profile $v_x(y)$ is essentially linear for a distance of several mean free paths. In view of the latter assumption, we may further write

$$\begin{aligned} v_x|_{y-a} &= v_x|_y - \frac{2}{3}\lambda\frac{dv_x}{dy} \\ v_x|_{y+a} &= v_x|_y + \frac{2}{3}\lambda\frac{dv_x}{dy} \end{aligned} \tag{1.4–6}$$

By combining Eqs. 1.4–2, 5, and 6, we get

$$\tau_{yx} = -\frac{1}{3}nm\bar{u}\lambda\frac{dv_x}{dy} \tag{1.4–7}$$

This result corresponds to Newton's law of viscosity (Eq. 1.1–2), with the viscosity given by

$$\mu = \tfrac{1}{3}nm\bar{u}\lambda = \tfrac{1}{3}\rho\bar{u}\lambda \tag{1.4–8}$$

Equation 1.4–8 was obtained by Maxwell in 1860.

By combining Eqs. 1.4–1, 3, and 8, one obtains

$$\mu = \frac{2}{3\pi^{3/2}}\frac{\sqrt{m\kappa T}}{d^2} \tag{1.4–9}$$

which represents the viscosity of a gas composed of hard spheres at low density. Note that an experimental value of μ is needed to determine the collision diameter d; one can then predict μ at other conditions.

The above derivation gives a qualitatively correct picture of momentum transfer in a gas at low density. The prediction of Eq. 1.4–9 that μ is independent of pressure agrees well with experimental data up to about 10 atm. (See Figs. 1.3–1,2.) The predicted temperature dependence is less satisfactory; data for various gases indicate that μ varies more rapidly than $T^{0.5}$ and is not accurately represented by any power function of T. (See Fig. 1.3–1.) To predict the temperature dependence of μ accurately one has to replace the rigid-sphere model by a more realistic molecular force field; also one has to analyze the departure from equilibrium behavior more carefully. The results of a detailed analysis along these lines are given below.[4]

[4] The remainder of this section summarizes results given in J. O. Hirschfelder, C. F. Curtiss, and R. B. Bird, *Molecular Theory of Gases and Liquids*, Wiley, New York (1954), Chapters 8 and 9. A review article on the same subjects highlights the results of interest in engineering calculations; see R. B. Bird, J. O. Hirschfelder, and C. F. Curtiss, *Trans. ASME*, **76**, 1011–1038 (1954).

A rigorous kinetic theory of monatomic gases at low density was developed before World War I by Chapman in England and independently by Enskog in Sweden.[5] The Chapman-Enskog theory gives expressions for the transport coefficients in terms of the *potential energy of interaction* between a pair of molecules in the gas. This potential energy φ is related to the force of interaction F by the relation $F = -d\varphi/dr$, in which r is the distance between the molecules. Now, if one knew exactly how the forces between molecules vary as a function of the distance between them, then one could substitute this information into the Chapman-Enskog formulas and calculate the transport coefficients.

The exact functional form of $\varphi(r)$ is not known; fortunately, however, a considerable amount of research has shown that a fairly good empirical potential energy function is the *Lennard-Jones (6–12) potential*:

$$\varphi(r) = 4\epsilon\left[\left(\frac{\sigma}{r}\right)^{12} - \left(\frac{\sigma}{r}\right)^{6}\right] \tag{1.4–10}$$

in which σ is a characteristic diameter[6] of the molecule (the "collision diameter") and ϵ is a characteristic energy of interaction between the molecules (the maximum energy of attraction between a pair of molecules). This function is shown in Fig. 1.4–2; note that it displays the characteristic features of molecular interactions—weak attraction at large separations (very nearly proportional to r^{-6}) and strong repulsion at small separations (roughly proportional to r^{-12}). Equation 1.4–10 has been shown to be quite useful for many nonpolar molecules. Values of σ and ϵ are known for many substances; a partial list is given in Table B–1, and a more extensive list is available elsewhere.[4] When values of σ and ϵ are not known, they may be estimated from the properties of the fluid at the critical point (c), liquid at the normal boiling point (b), or the solid at the melting point (m), by means of the following empirical relations.[4]

$$\epsilon/\kappa = 0.77T_c \qquad \sigma = 0.841\tilde{V}_c^{1/3} \quad \text{or} \quad 2.44\left(\frac{T_c}{p_c}\right)^{1/3} \tag{1.4–11,12,13}$$

$$\epsilon/\kappa = 1.15T_b \qquad \sigma = 1.166\tilde{V}_{b,\,\text{liq.}}^{1/3} \tag{1.4–14,15}$$

$$\epsilon/\kappa = 1.92T_m \qquad \sigma = 1.222\tilde{V}_{m,\,\text{sol.}}^{1/3} \tag{1.4–16,17}$$

in which ϵ/κ and T are in ° K, σ is in Ångström units,[7] $\tilde{V}$ is in cm³ g–mole⁻¹, and p_c is in atmospheres.

[5] This theory is given in detail in the treatise by S. Chapman and T. G. Cowling, *Mathematical Theory of Non-Uniform Gases*, Cambridge University Press (1951), Second Edition.

[6] Note that σ is different from the molecular diameter d previously used in the simple kinetic theory. The quantities are of the same order of magnitude, but there is no simple relation between them.

[7] 1 Ångström unit = 10^{-8} cm = 10^{-10} m. The Ångström unit (abbr. Å) is the standard "yardstick" for giving molecular dimensions.

The coefficient of viscosity at absolute temperature T of a pure monatomic gas of molecular weight M may be written in terms of the parameters σ and ϵ as

$$\mu = 2.6693 \times 10^{-5} \frac{\sqrt{MT}}{\sigma^2 \Omega_\mu} \qquad (1.4\text{–}18)$$

in which μ [=] g cm^{-1} sec^{-1}, T [=] ° K, σ [=] Å, and Ω_μ is a slowly varying

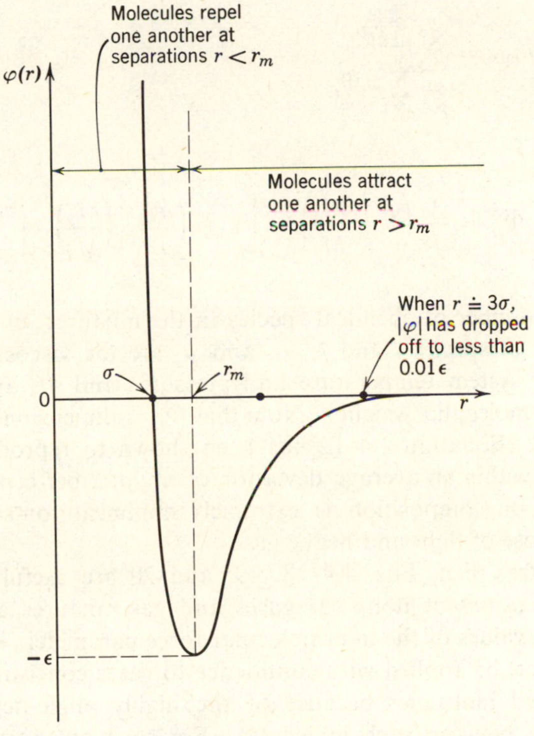

Fig. 1.4–2. Potential energy function describing the interaction of two spherical, nonpolar molecules. Equation 1.4–10 is one of the many empirical equations proposed for fitting this curve.

function of the dimensionless temperature $\kappa T/\epsilon$ given in Table B–2. Although this formula was derived for monatomic gases, it has been found to be remarkably good for polyatomic gases as well; the predicted temperature dependence of μ is in good agreement with the low density line on the reduced viscosity chart, Fig. 1.3–1. Note again that the *viscosity of gases at low density increases with temperature*, roughly as the 0.6 to 1.0 power of the temperature. Note also that there is no dependence on pressure in the low-density range.

If the gas were made up of rigid spheres of diameter σ (instead of real molecules with attractive and repulsive forces), then Ω_μ would be unity. Hence the function Ω_μ may be interpreted as giving the deviation from rigid sphere behavior.

The Chapman-Enskog theory has been extended to include multicomponent gas mixtures at low density by Curtiss and Hirschfelder.[8] For most purposes, the semiempirical formula of Wilke[9] is quite adequate:

$$\mu_{\text{mix}} = \sum_{i=1}^{n} \frac{x_i \mu_i}{\sum_{j=1}^{n} x_j \Phi_{ij}} \tag{1.4–19}$$

in which

$$\Phi_{ij} = \frac{1}{\sqrt{8}}\left(1 + \frac{M_i}{M_j}\right)^{-1/2}\left[1 + \left(\frac{\mu_i}{\mu_j}\right)^{1/2}\left(\frac{M_j}{M_i}\right)^{1/4}\right]^2 \tag{1.4–20}$$

Here n is the number of chemical species in the mixture; x_i and x_j are the mole fractions of species i and j; μ_i and μ_j are the viscosities of species i and j at the system temperature and pressure; and M_i and M_j are the corresponding molecular weights. Note that Φ_{ij} is dimensionless and, when $i = j$, $\Phi_{ij} = 1$. Equation 1.4–19 has been shown to reproduce measured values of μ_{mix} within an average deviation of about 2 per cent. The dependence of μ_{mix} on composition is extremely nonlinear for some mixtures, particularly those of light and heavy gases.

To summarize, then, Eqs. 1.4–18, 19, and 20 are useful formulas for computing viscosities of nonpolar gases and gas mixtures at low density from tabulated values of the intermolecular force parameters σ and ϵ. They cannot, however, be applied with confidence to gases consisting of polar or highly elongated molecules because of the highly angle-dependent force fields that exist between such molecules. For such polar vapors as H_2O, NH_3, CH_3OH, and NOCl, an angle-dependent modification of Eq. 1.4–10 has given good results.[10] A further, usually unimportant, limitation is that these equations require modification in the temperature range below 100° K to take account of quantum effects.[11] For gases heavier than H_2 and He, the

[8] C. F. Curtiss and J. O. Hirschfelder, *J. Chem. Phys.*, **17**, 550–555 (1949).

[9] C. R. Wilke, *J. Chem. Phys.*, **18**, 517–519 (1950). See also J. W. Buddenberg and C. R. Wilke, *Ind. Eng. Chem.*, **41**, 1345–1347 (1949).

[10] E. A. Mason and L. Monchick, *J. Chem. Phys.*, **35** (1961).

[11] J. O. Hirschfelder, C. F. Curtiss, and R. B. Bird, *op. cit.*, Chapter 10.

quantum effects on viscosity may be neglected down to even lower temperatures.

Example 1.4–1. Computation of the Viscosity of a Gas at Low Density

Compute the viscosity of CO_2 at 200, 300, and 800° K and 1 atm.

Solution. Use Eq. 1.4–18. From Table B–1, we find the Lennard-Jones constants for CO_2 to be $\epsilon/\kappa = 190°$ K, $\sigma = 3.996$ Å. The molecular weight of CO_2 is 44.010. Substitution of the constant factors M and σ into Eq. 1.4–18 gives

$$\mu = 2.6693 \times 10^{-5} \frac{\sqrt{44.010T}}{(3.996)^2\Omega_\mu} = 1.109 \times 10^{-5} \frac{\sqrt{T}}{\Omega_\mu} \tag{1.4–21}$$

in which μ [=] g cm^{-1} sec^{-1} and T [=] °K. The remaining calculations are conveniently done in tabular form:

T (°K)	$\frac{\kappa T}{\epsilon} = \frac{T}{190}$	Ω_μ from Table B–2	$\sqrt{T}$	Predicted Viscosity (Eq. 1.4–21) (g cm^{-1} sec^{-1})	Observed Viscosity[12] (g cm^{-1} sec^{-1})
200	1.053	1.548	14.14	1.013×10^{-4}	1.015×10^{-4}
300	1.58	1.286	17.32	1.494×10^{-4}	1.495×10^{-4}
800	4.21	0.9595	28.28	3.269×10^{-4}	. . .

Experimental data are shown in the last column for comparison. This good agreement is to be expected, since the Lennard-Jones constants of Table B–1 are derived from viscosity data.

Example 1.4–2. Prediction of the Viscosity of a Gas Mixture at Low Density

Predict the viscosity of the following gas mixture at 1 atm and 293° K from the given data on the pure components at 1 atm and 293° K:

Species	Mole Fraction, x	Molecular Weight, M	Viscosity, μ (g cm^{-1} sec^{-1})
1: CO_2	0.133	44.010	1462×10^{-7}
2: O_2	0.039	32.000	2031×10^{-7}
3: N_2	0.828	28.016	1754×10^{-7}

[12] H. L. Johnston and K. E. McCloskey, *J. Phys. Chem.*, **44**, 1038 (1940).

Solution. Use Eqs. 1.4–20 and 1.4–19, in that order. The following table shows the main steps in the calculation:

i	j	M_i/M_j	μ_i/μ_j	Φ_{ij} (Eq. 1.4–20)	$\sum_{j=1}^{3} x_j\Phi_{ij}$
1	1	1.000	1.000	1.000	
	2	1.375	0.720	0.730	0.763
	3	1.571	0.834	0.727	
2	1	0.727	1.389	1.394	
	2	1.000	1.000	1.000	1.057
	3	1.142	1.158	1.006	
3	1	0.637	1.200	1.370	
	2	0.876	0.864	0.993	1.049
	3	1.000	1.000	1.000	

Equation 1.4–19 gives

$$\mu_{\text{mix}} = \sum_{i=1}^{3} \frac{x_i\mu_i}{\sum_{j=1}^{3} x_j\Phi_{ij}}$$

$$= \frac{(0.133)(1462)(10^{-7})}{0.763} + \frac{(0.039)(2031)(10^{-7})}{1.057} + \frac{(0.828)(1754)(10^{-7})}{1.049}$$

$$= 1714 \times 10^{-7} \text{ g cm}^{-1} \text{ sec}^{-1}$$

The observed value[13] is 1793×10^{-7} g cm^{-1} sec^{-1}.

§1.5 THEORY OF VISCOSITY OF LIQUIDS

Our knowledge about the viscosity of liquids is largely empirical, since the kinetic theory of liquids is only partly developed. It is of interest, however, to consider the approximate theory developed by Eyring and co-workers,[1] which illustrates the mechanisms involved and does permit rough estimation of the viscosity from other physical properties.

In a pure liquid at rest the individual molecules are constantly in motion; however, because of the close packing, the motion is largely confined to vibration of each molecule within a "cage" formed by its nearest neighbors. This cage is represented by the energy barrier of height $\Delta\tilde{G}_0^{\ddagger}/\tilde{N}$, sketched in Fig. 1.5–1. Eyring has suggested that a liquid at rest continually

[13] F. Herning and L. Zipperer, *Gas- und Wasserfach*, **79**, 49–54, 69–73 (1936).

[1] S. Glasstone, K. J. Laidler, and H. Eyring, *Theory of Rate Processes*, McGraw-Hill, New York (1941), Chapter 9.

undergoes rearrangements in which one molecule at a time escapes from its "cage" into an adjoining "hole," as shown in Fig. 1.5–1, and that the molecules thus move in each of the cartesian coordinate directions in jumps of length a at a frequency k per molecule, where k is given by the rate equation

$$k = \frac{\kappa T}{h} e^{-\Delta \tilde{G}_0^{\ddagger}/RT} \tag{1.5–1}$$

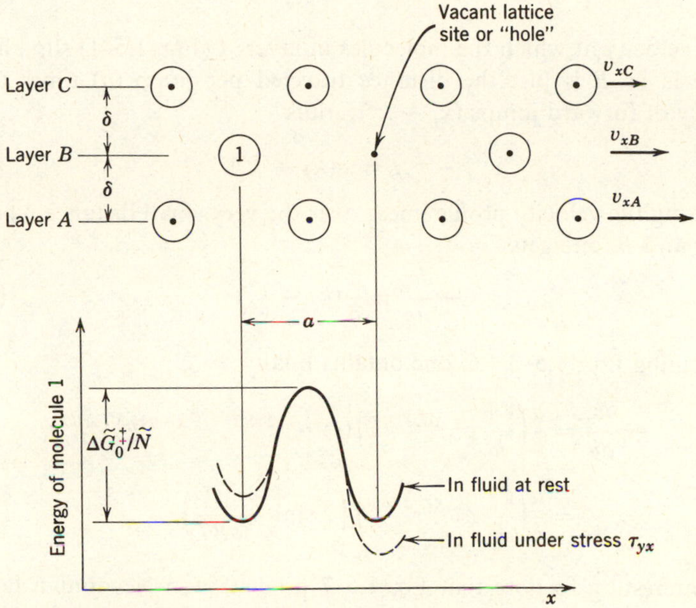

Fig. 1.5–1. Illustration of an escape process in flow of a liquid. Molecule 1 must pass through a "bottleneck" to reach the vacant site.

Here κ and h are the Boltzmann and Planck constants (see Appendix C), R is the molar gas constant, and $\Delta \tilde{G}_0^{\ddagger}$ is the molar "free energy of activation" in the stationary fluid.

In a fluid that flows in the x-direction with a velocity gradient dv_x/dy, the frequency of molecular rearrangements is increased. This effect can be explained by considering the potential energy barrier as distorted under the applied stress τ_{yx} (see Fig. 1.5–1), so that

$$-\Delta \tilde{G}^{\ddagger} = -\Delta \tilde{G}_0^{\ddagger} \pm \left(\frac{a}{\delta}\right)\left(\frac{\tau_{yx}\tilde{V}}{2}\right) \tag{1.5–2}$$

where $\tilde{V}$ is the volume of a mole of liquid, and $\pm(a/\delta)(\tau_{yx}\tilde{V}/2)$ is an approximation to the work done on the molecules as they move to the top of the energy barriers, moving with the applied shear stress (plus sign) or against the applied shear stress (minus sign). Defining k_f as the frequency of forward jumps and k_b as the frequency of backward jumps, one finds from Eqs. 1.5–1 and 1.5–2 that

$$k_f = \frac{\kappa T}{h} e^{-\Delta \tilde{G}_0^{\ddagger}/RT} e^{a\tau_{yx}\tilde{V}/2\delta RT} \tag{1.5–3}$$

$$k_b = \frac{\kappa T}{h} e^{-\Delta \tilde{G}_0^{\ddagger}/RT} e^{-a\tau_{yx}\tilde{V}/2\delta RT} \tag{1.5–4}$$

The net velocity at which the molecules in layer A (Fig. 1.5–1) slip ahead of those in layer B is just the distance traveled per jump (a) times the *net* frequency of forward jumps ($k_f - k_b$); thus

$$v_{xA} - v_{xB} = a(k_f - k_b) \tag{1.5–5}$$

Considering the velocity profile linear over the very small distance δ between layers A and B, one gets

$$-\frac{dv_x}{dy} = \frac{a}{\delta}(k_f - k_b) \tag{1.5–6}$$

By combining Eqs. 1.5–3,4,6, one obtains finally

$$\begin{aligned} -\frac{dv_x}{dy} &= \frac{a}{\delta}\left(\frac{\kappa T}{h} e^{-\Delta \tilde{G}_0^{\ddagger}/RT}\right)(e^{a\tau_{yx}\tilde{V}/2\delta RT} - e^{-a\tau_{yx}\tilde{V}/2\delta RT}) \\ &= \frac{a}{\delta}\left(\frac{\kappa T}{h} e^{-\Delta \tilde{G}_0^{\ddagger}/RT}\right)\left(2 \sinh \frac{a\tau_{yx}\tilde{V}}{2\delta RT}\right) \end{aligned} \tag{1.5–7}$$

It is interesting to note that Eq. 1.5–7 predicts non-Newtonian flow for liquids in general; in fact, this equation has the same general form as the Eyring model given in Eq. 1.2–4. If $a\tau_{yx}\tilde{V}/2\delta RT$ is small compared to unity, however, Eq. 1.5–7 becomes consistent with Newton's law of viscosity, with

$$\mu = \left(\frac{\delta}{a}\right)^2 \frac{\tilde{N}h}{\tilde{V}} e^{\Delta \tilde{G}_0^{\ddagger}/RT} \tag{1.5–8}$$

in which $\tilde{N}$ is Avogadro's number. In most applications δ/a is taken as unity; this simplification involves no loss of accuracy, since $\Delta \tilde{G}_0^{\ddagger}$ is determined empirically to make the equation agree with experimental data.

It has been found that free energies of activation, $\Delta \tilde{G}_0^{\ddagger}$, determined by fitting Eq. 1.5–7 to experimental data on viscosity versus temperature, are almost constant for a given fluid and correlate well with the internal energy of vaporization at the normal boiling point.[2]

$$\Delta \tilde{G}_0^{\ddagger} = 0.408 \Delta \tilde{U}_{\text{vap}} \tag{1.5–9}$$

[2] J. F. Kincaid, H. Eyring, and A. E. Stearn, *Chem. Revs.*, **28**, 301 (1941).

By using this empirical information and setting $\delta/a = 1$ (this is consistent with the calculations that led to Eq. 1.5–9), Eq. 1.5–8 becomes

$$\mu = \frac{\tilde{N}h}{\tilde{V}} e^{0.408\Delta\tilde{U}_{\text{vap}}/RT} \tag{1.5–10}$$

The energy of vaporization at the normal boiling point can be roughly estimated from Trouton's rule:

$$\Delta\tilde{U}_{\text{vap}} \doteq \Delta\tilde{H}_{\text{vap}} - RT_b \doteq 9.4RT_b \tag{1.5–11}$$

With this further approximation, Eq. 1.5–10 becomes

$$\mu \doteq \frac{\tilde{N}h}{\tilde{V}} e^{3.8T_b/T} \tag{1.5–12}$$

Equations 1.5–10 and 1.5–12 indicate an exponential *decrease* in viscosity with temperature, which agrees with the observed behavior of most liquids. These two equations are not highly accurate, errors of as much as 30 per cent being common; they are useful primarily for rough estimation and as guides for interpolation or extrapolation of fragmentary viscosity data. In particular, they should not be used for very long slender molecules, such as n-$C_{20}H_{42}$, which deviate markedly from Eq. 1.5–9. Other empirical formulas are given in the treatise by Partington.[3]

Example 1.5–1. Estimation of the Viscosity of a Pure Liquid

Estimate the viscosity of liquid benzene (C_6H_6) at 20° C.

Solution. Use Eq. 1.5–12 with the following information:

$$\begin{aligned}
\tilde{N} &= 6.023 \times 10^{23}\ (\text{g-mole})^{-1} \\
h &= 6.624 \times 10^{-27}\ \text{erg sec or g cm}^2\ \text{sec}^{-1} \\
\tilde{V} &= 89.0\ \text{cm}^3\ (\text{g-mole})^{-1}\ \text{at } 20^\circ\ \text{C} \\
T_b &= 273.2 + 80.1 = 353.3^\circ\ \text{K} \\
T &= 273.2 + 20 = 293.2^\circ\ \text{K}
\end{aligned}$$

Substitution into Eq. 1.5–12 gives

$$\begin{aligned}
\mu &\doteq (\tilde{N}h/\tilde{V}) \exp(3.8T_b/T) \\
&= \frac{(6.023 \times 10^{23})(6.624 \times 10^{-27})}{(89.0)} \exp\left(\frac{3.8 \times 353.2}{293.2}\right) \\
&= 4.46 \times 10^{-3}\ \text{g cm}^{-1}\ \text{sec}^{-1} \\
&= 0.45\ \text{cp}
\end{aligned}$$

The observed viscosity given in Table 1.1–2 is 0.647 cp. This disagreement emphasizes the approximate nature of the theory.

[3] J. R. Partington, *Treatise on Physical Chemistry*, Longmans, Green (1949). This multivolume work contains an unusually complete survey of the physicochemical literature.

QUESTIONS FOR DISCUSSION

1. Compare Newton's law of viscosity with Hooke's law of elasticity.

2. Show that the dimensions of "momentum per unit area per unit time" are the same as those of "force per unit area."

3. What are the units of viscosity and kinematic viscosity?

4. In what reference books may viscosity data be found?

5. Compare the magnitude of the viscosity and kinematic viscosity of air, water, and mercury at 1 atm and 20° C.

6. What is meant by the term "non-Newtonian"? What types of substance exhibit this behavior?

7. What is the science of "rheology"?

8. How does the viscosity vary with temperature and pressure for (*a*) dilute gases (*b*) liquids?

9. What methods are available for estimating the viscosity of dense gases and liquids?

10. What is the Lennard-Jones potential and what does it represent?

11. Determine the force between two molecules as a function of distance, as given by the Lennard-Jones potential. Determine the value of r_m in Fig. 1.4–2 in terms of σ and ϵ.

12. What is the significance of the parameters σ and ϵ? Illustrate with the aid of a sketch.

13. How can σ and ϵ be obtained from viscosity data?

14. The Lennard-Jones potential energy function describes a spherically symmetrical field. Name two molecular characteristics that give rise to potential energy fields that do not have spherical symmetry.

15. Sketch the potential energy function $\varphi(r)$ for rigid, nonattracting spherical molecules.

16. What is the physical significance of the Ω_μ function in Eq. 1.4–18?

17. Molecules containing different atomic isotopes have the same values of σ and ϵ; knowing this, how would you compute the viscosity of CD_4 at 20° C and 1 atm from the viscosity of CH_4 at 20° C?

18. How much difference in dilute gas viscosities would one anticipate for $U^{235}F_6$ and $U^{238}F_6$?

19. How does the Wilke equation simplify when $\Phi_{ij} = 1$ for all i and j in a gas mixture? For what sort of gas mixtures would this situation arise?

20. Can the momentum flux in a liquid be expressed by Eq. 1.4–5?

21. Fluid A has a viscosity that is twice that of fluid B; which fluid would you expect to flow more rapidly through a horizontal tube of length L under the same pressure drop?

PROBLEMS[1]

1.A$_1$ Computation of Viscosities of Gases at Low Density

Predict the viscosities of molecular oxygen, nitrogen, and methane at 20° C and atmospheric pressure. Express all results in centipoises. Compare your results with the experimental values given in this chapter. *Answers:* 0.0203, 0.0175, 0.0109 cp

[1] See the Preface regarding Classes of Problems.

1.B$_1$ Calculation of Viscosities of Gas Mixtures at Low Density

The following data[2] are available on the viscosity of mixtures of hydrogen and Freon-12 (dichlorodifluoromethane) at 25° C and 1 atm:

x_1 Mole Fraction H_2	$\mu \times 10^6$ (g cm^{-1} sec^{-1})
0.00	124.0
0.25	128.1
0.50	131.9
0.75	135.1
1.00	88.4

Compute and compare the values given by Eqs. 1.4–19 and 20 at the three intermediate compositions, using the observed viscosities for the pure components.

Sample answer: At $x_1 = 0.50$, $\mu = 0.01315$ cp

1.C$_1$ Estimation of Dense Gas Viscosity

Estimate the viscosity of N_2 at 68° F and 1000 psig by means of (*a*) Fig. 1.3–1, using μ_c from Table B–1; (*b*) Fig. 1.3–2, using μ^0 from Table 1.1–2. Give your results in $\text{lb}_m\ \text{ft}^{-1}\ \text{sec}^{-1}$. (For meaning of psig, see Table C.3–2.) *Answer: a.* $1.30 \times 10^{-5}\ \text{lb}_m\ \text{ft}^{-1}\ \text{sec}^{-1}$

1.D$_1$ Estimation of Liquid Viscosity

Estimate the viscosity of saturated liquid water at 0 and 100° C by means of (*a*) Eq. 1.5–10, with $\Delta\hat{U}_{\text{vap}} = 897.5$ Btu lb_m^{-1} at 100° C; (*b*) Eq. 1.5–12. Compare the results with the values in Table 1.1–1. *Answer: b.* 4.0 cp, 0.95 cp

1.E$_1$ Molecular Velocity and Mean Free Path

Compute the mean molecular velocity $\bar{u}$, cm sec^{-1}, and the mean free path λ, cm, in O_2 at 1 atm and 273.2° K. Assume $d = 3.0$ Å. What is the ratio of the mean free path to the molecular diameter in this situation? What would be the order of magnitude of the corresponding ratio in the liquid state?

Answers: $\bar{u} = 4.25 \times 10^4$ cm sec^{-1}, $\lambda = 9.3 \times 10^{-6}$ cm

1.F$_2$ Comparison of the Uyehara-Watson Chart with Kinetic Theory

a. Use Eqs. 1.4–11,18 and Table B–2 to construct a logarithmic plot of $\mu\sigma^2 M^{-1/2} T_c^{-1/2}$ versus T_r up to $T_r = 10$. On the same graph plot μ_r versus T_r, as given by Fig. 1.3–1 for gases at low density. What do you conclude from a comparison of the shapes of the two curves?

b. Determine μ_c in terms of M, T_c, and σ by comparison of the results plotted in (*a*).

c. Combine the result of (*b*) with Eqs. 1.4–12,13 to obtain two alternate expressions for μ_c in terms of the critical properties. Compare your results with the corresponding equations given by Uyehara and Watson.

1.G$_2$ Comparison of the Simple Kinetic Theory with the Exact Theory for Rigid Spheres

Convert Eq. 1.4–9 to the form of Eq. 1.4–18 and compare the two equations numerically for rigid spherical molecules with $d = \sigma$. What percentage of error is incurred by using the simple kinetic theory for such molecules?

[2] J. W. Buddenberg and C. R. Wilke, *Ind. Eng. Chem.*, **41**, 1345–1347 (1949).

1.H$_3$ Mean Distance of Free Flight in a Prescribed Direction

The mean free path λ is so defined that $e^{-s/\lambda}$ is the probability that a molecule will travel a distance greater than s between successive collisions. By making use of this definition and the fact that the molecular velocities in a gas at equilibrium are randomly directed, determine the mean distance b traveled in the y-direction between successive collisions by molecules that have positive v_y.

1.I$_3$ Calculation of Average Velocities from the Maxwell-Boltzmann Distribution

In the simplified kinetic theory derivation given in §1.4, several statements concerning the equilibrium behavior of a gas were made without proof. In this problem, and the next, some of these statements are shown to be exact consequences of the Maxwell-Boltzmann velocity distribution.

The Maxwell-Boltzmann distribution of molecular velocities in a gas at rest is as follows:

$$f(u_x, u_y, u_z) = f(\mathbf{u}) = \left(\frac{m}{2\pi \kappa T}\right)^{3/2} e^{-mu^2/2\kappa T} \tag{1.I–1}$$

in which $\mathbf{u}$ and its components denote individual molecular velocities, and $f(u_x, u_y, u_z)\, du_x\, du_y\, du_z$ is the fraction of the total molecules that is expected to have velocities between u_x and $u_x + du_x$, u_y and $u_y + du_y$, u_z and $u_z + du_z$ at any instant. It follows from Eq. 1.I–1 that the distribution of molecular speed u is

$$f(u) = 4\pi u^2 \left(\frac{m}{2\pi \kappa T}\right)^{3/2} e^{-mu^2/2\kappa T} \tag{1.I–2}$$

a. Make a two-dimensional sketch of the distribution $f(\mathbf{u})$ in velocity space, by holding u_z constant and placing dots on the u_x-, u_y-plane with a concentration proportional to $f(u_x, u_y, u_z)$.

b. Compute the *mean speed* $\bar{u}$ by integrating $uf(u)$ with respect to u:

$$\bar{u} = \frac{\int_0^\infty uf(u)\, du}{\int_0^\infty f(u)\, du} \tag{1.I–3}$$

The correct result is given in Eq. 1.4–1.

c. Compute the components $\bar{u}_x$, $\bar{u}_y$, and $\bar{u}_z$ of the *mean velocity* $\bar{\mathbf{u}}$ by integration of the components of $\mathbf{u}f(\mathbf{u})$ throughout velocity space. The integral which determines $\bar{u}_x$ is

$$\bar{u}_x = \frac{\int_{-\infty}^{\infty}\int_{-\infty}^{\infty}\int_{-\infty}^{\infty} u_x f(u_x, u_y, u_z)\, du_x\, du_y\, du_z}{\int_{-\infty}^{\infty}\int_{-\infty}^{\infty}\int_{-\infty}^{\infty} f(u_x, u_y, u_z)\, du_x\, du_y\, du_z} \tag{1.I–4}$$

and the integrals for $\bar{u}_y$ and $\bar{u}_z$ are analogous. The results should show that the gas is at rest.

d. Compute the mean kinetic energy per molecule by integration of $\frac{1}{2}mu^2 f(u)$ with respect to u:

$$\tfrac{1}{2}m\overline{u^2} = \frac{\int_0^\infty \frac{1}{2}mu^2 f(u)\, du}{\int_0^\infty f(u)\, du} \tag{1.I–5}$$

The correct result is $\frac{1}{2}m\overline{u^2} = \frac{3}{2}\kappa T$.

1.J$_3$ Calculation of Wall Collision Frequency from the Maxwell-Boltzmann Distribution

It is desired to know the frequency Z with which the molecules in an ideal gas strike unit area of a wall from one side only. The gas is at rest and at equilibrium at temperature T and contains n identical molecules per cm^3 of mass m. By considering the wall to be the surface $y = c$ and the fluid to be in the region $y < c$, one can show that Z is simply one-half the number-concentration n, times the average of u_y for all molecules moving toward the wall. Thus

$$Z = \frac{n}{2} \frac{\displaystyle\int_{-\infty}^{\infty} \int_{0}^{\infty} \int_{-\infty}^{\infty} u_y f(u_x, u_y, u_z)\, du_x\, du_y\, du_z}{\displaystyle\int_{-\infty}^{\infty} \int_{0}^{\infty} \int_{-\infty}^{\infty} f(u_x, u_y, u_z)\, du_x\, du_y\, du_z} \qquad \text{(1.J-1)}$$

Evaluate Z, using $f(u_x, u_y, u_z)$ as given in Eq. 1.I–1. Is your result consistent with Eqs. 1.4–1, 2?

CHAPTER 2

Velocity Distributions in Laminar Flow

In this chapter we show how one may calculate the laminar velocity profiles for some flow systems of simple geometry. These calculations make use of the definition of viscosity and the concept of a momentum balance. Actually, a knowledge of the complete velocity distributions is not usually needed in engineering problems. Rather, we need to know the maximum velocity, the average velocity, or the shear stress at a surface. These quantities can be obtained easily once the velocity profiles are known.

In the first section we make a few general remarks about differential momentum balances. Then, in subsequent sections, we work out in detail several classical examples of viscous flow patterns. These examples should be thoroughly understood, inasmuch as we shall have frequent occasion to refer to them in subsequent chapters. The reader may feel that these systems are too simple to be of engineering interest; it is certainly true that they represent highly idealized situations, but the results find considerable use in the development of numerous topics in engineering fluid mechanics.

The systems studied in this chapter are so arranged that the reader is gradually introduced to various factors that are considered in the solution of viscous flow patterns. In §2.2 the falling film problem illustrates the role of gravity forces and the use of cartesian coordinates; it is also shown how the solution is obtained when viscosity may be a function of position. In §2.3

the flow in a circular tube illustrates the role of pressure and gravity forces and the use of cylindrical coordinates; it is also shown how non-Newtonian flow problems are handled. In §2.4 the flow in a cylindrical annulus emphasizes the role played by the boundary conditions in obtaining the final solution. In §2.5 the question of boundary conditions is pursued further in the discussion of the flow of two adjacent immiscible fluids. Finally, in §2.6, the flow around a sphere is discussed briefly in order to illustrate a problem in spherical coordinates and also to indicate how both normal and tangential forces are handled.

The methods and problems given in this chapter apply to steady-state flow only. By "steady state" one means that the conditions at each point in the stream do not change with time. That is, a photograph of the flow system at time t looks exactly like a photograph taken at some later time, $t + \Delta t$. The general equations for unsteady flow are given in Chapter 3.

§2.1 SHELL MOMENTUM BALANCES: BOUNDARY CONDITIONS

The problems discussed in §§2.2 through 2.5 are approached by setting up momentum balances over a thin "shell" of fluid. For the *steady-state* flow, the momentum balance is

$$\begin{Bmatrix}\text{rate of} \\ \text{momentum in}\end{Bmatrix} - \begin{Bmatrix}\text{rate of} \\ \text{momentum out}\end{Bmatrix} + \begin{Bmatrix}\text{sum of forces} \\ \text{acting on system}\end{Bmatrix} = 0 \qquad (2.1\text{–}1)$$

Momentum may enter the system by momentum transfer according to the Newtonian (or non-Newtonian) expression for the momentum flux. Momentum may also enter by virtue of the over-all fluid motion. The forces we are concerned with are pressure forces (acting on *surfaces*) and gravity forces (acting on the *volume* as a whole).

The momentum balance in Eq. 2.1–1 is easy to apply *only when the streamlines of the system are straight lines* (i.e., in rectilinear flow). The method of handling systems with curved streamlines is discussed in the form of examples in §3.5. Actually the origin of Eq. 2.1–1 cannot be understood until one has studied the derivation in Chapter 3 and specifically Eq. 3.2–10. At this point we simply ask the reader to accept the principle set forth in Eq. 2.1–1 and to learn to use it in solving simple steady-state viscous flow problems.

Generally, the procedure for setting up and solving viscous flow problems is as follows: first we write a momentum balance of the form of Eq. 2.1–1 for a shell of finite thickness; then we let this thickness approach zero and make use of the mathematical definition of the first derivative to obtain the corresponding differential equation describing the momentum flux distribution. At this juncture one may insert the appropriate Newtonian or non-Newtonian expression for the momentum flux to obtain a differential equation

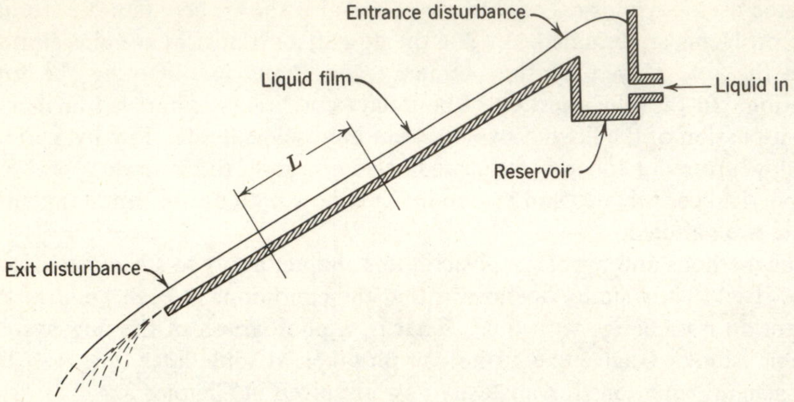

Fig. 2.2-1. Schematic diagram of falling film experiment, illustrating end effects. In the region of length L the velocity distribution is fully developed.

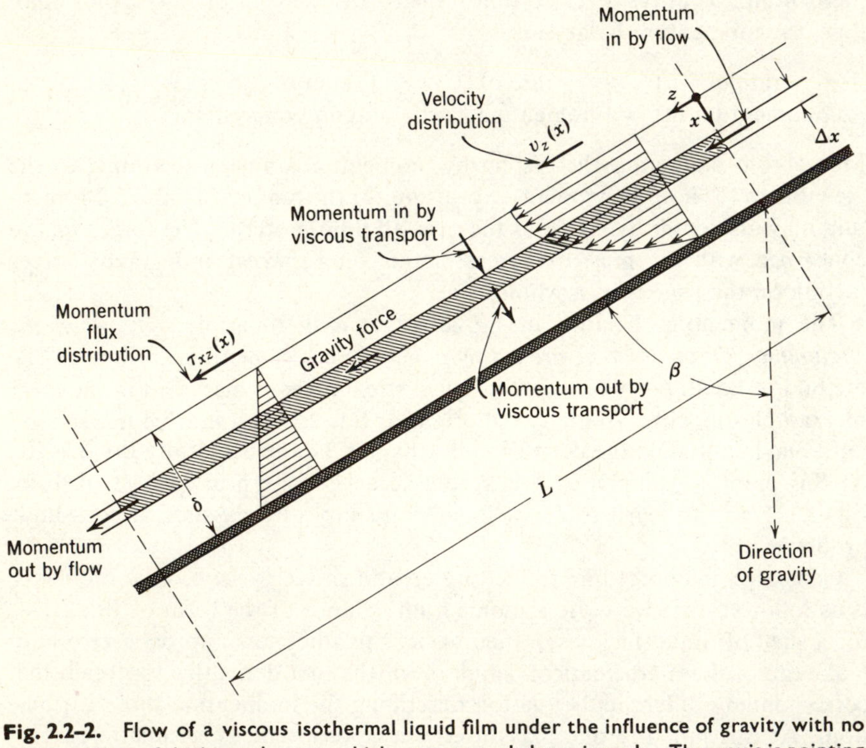

Fig. 2.2-2. Flow of a viscous isothermal liquid film under the influence of gravity with no rippling. Slice of thickness Δx over which momentum balance is made. The y-axis is pointing outward from the plane of paper.

for the velocity distribution. The integration of these two differential equations yields, respectively, the momentum flux and velocity distributions for the system. This information can then be used to calculate various other quantities, such as average velocity, maximum velocity, volume rate of flow, pressure drop, and forces on boundaries.

In the integrations mentioned above, several constants of integration appear, which are evaluated by the use of "boundary conditions," that is, statements of physical facts at specified values of the independent variable. The following are the most used boundary conditions:

a. At solid-fluid interfaces the fluid velocity equals the velocity with which the surface itself is moving; that is, the fluid is assumed to cling to any solid surfaces with which it is in contact.

b. At liquid-gas interfaces the momentum flux (hence the velocity gradient) in the liquid phase is very nearly zero and can be assumed to be zero in most calculations.

c. At liquid-liquid interfaces the momentum flux perpendicular to the interface, and the velocity, are continuous across the interface. (In the notation of §A.5, $\boldsymbol{v}$ and $\boldsymbol{n}p + [\boldsymbol{n} \cdot \boldsymbol{\tau}]$ are continuous for planar interfaces).

All three types of boundary conditions are encountered in the sections that follow.

In this section we have endeavored to present some general rules for solving elementary viscous flow problems. We now proceed to illustrate the application of these rules to a number of simple flow systems.

§2.2 FLOW OF A FALLING FILM

As our first example, we consider the flow of a fluid along an inclined flat surface, as shown in Fig. 2.2–1. Such films have been studied in connection with wetted-wall towers, evaporation and gas absorption experiments, and application of coatings to paper rolls. We consider the viscosity and density of the fluid to be constant. We focus our attention on a region of length L, sufficiently far from the ends of the wall that the entrance and exit disturbances are not included in L, that is, in this region the velocity component v_z does not depend on z.

We begin by setting up a z-momentum balance over a system of thickness Δx, bounded by the planes $z = 0$ and $z = L$, and extending a distance W in the y-direction. (See Fig. 2.2–2.) The various components of the momentum balance are then

rate of z-momentum in across surface at x	$(LW)(\tau_{xz})\|_x$	(2.2–1)

rate of z-momentum out across surface at $x + \Delta x$	$(LW)(\tau_{xz})\|_{x+\Delta x}$	(2.2–2)
rate of z-momentum in across surface at $z = 0$	$(W\Delta x\, v_z)(\rho v_z)\|_{z=0}$	(2.2–3)
rate of z-momentum out across surface at $z = L$	$(W\Delta x\, v_z)(\rho v_z)\|_{z=L}$	(2.2–4)
gravity force acting on fluid	$(LW\Delta x)(\rho g \cos\beta)$	(2.2–5)

Note that we always take the "in" and "out" directions in the direction of the positive x- and z-axes (in this problem these happen to coincide with the direction of momentum transport). The notation $|_{x+\Delta x}$ means "evaluated at $x + \Delta x$."

When these terms are substituted into the momentum balance of Eq. 2.1–1, we get

$$LW\tau_{xz}|_x - LW\tau_{xz}|_{x+\Delta x} + W\Delta x\, \rho v_z^2|_{z=0} - W\Delta x\, \rho v_z^2|_{z=L} + LW\Delta x\, \rho g \cos\beta = 0 \quad (2.2\text{–}6)$$

Because v_z is the same at $z = 0$ as it is at $z = L$ for each value of x, the third and fourth terms just cancel one another. We now divide Eq. 2.2–6 by $LW\,\Delta x$ and take the limit as Δx approaches zero:

$$\lim_{\Delta x \to 0} \left(\frac{\tau_{xz}|_{x+\Delta x} - \tau_{xz}|_x}{\Delta x} \right) = \rho g \cos\beta \quad (2.2\text{–}7)$$

The quantity on the left side may be recognized as the definition of the first derivative of τ_{xz} with respect to x. Therefore, Eq. 2.2–7 may be rewritten as

$$\frac{d}{dx}\tau_{xz} = \rho g \cos\beta \quad (2.2\text{–}8)$$

This is the differential equation for the momentum flux τ_{xz}. It may be integrated to give

$$\tau_{xz} = \rho g x \cos\beta + C_1 \quad (2.2\text{–}9)$$

The constant of integration may be evaluated by making use of the boundary condition at the liquid-gas interface (see §2.1):

B.C. 1: $\quad$ at $\quad x = 0, \quad \tau_{xz} = 0 \quad$ (2.2–10)

Substitution of this boundary condition into Eq. 2.2–9 reveals that $C_1 = 0$. Hence the momentum-flux distribution is

$$\boxed{\tau_{xz} = \rho g x \cos \beta} \tag{2.2–11}$$

as shown in Fig. 2.2–2.

If the fluid is Newtonian, then we know that the momentum flux is related to the velocity gradient according to

$$\tau_{xz} = -\mu \frac{dv_z}{dx} \tag{2.2–12}$$

Substitution of this expression for τ_{xz} into Eq. 2.2–11 gives the following differential equation for the velocity distribution:

$$\frac{dv_z}{dx} = -\left(\frac{\rho g \cos \beta}{\mu}\right) x \tag{2.2–13}$$

This equation is easily integrated to give

$$v_z = -\left(\frac{\rho g \cos \beta}{2\mu}\right) x^2 + C_2 \tag{2.2–14}$$

The constant of integration is evaluated by using the boundary condition

B.C. 2: $\quad$ at $\quad x = \delta, \quad v_z = 0$ $\quad$ (2.2–15)

Substitution of this boundary condition into Eq. 2.2–14 shows that $C_2 = (\rho g \cos \beta/2\mu)\delta^2$. Therefore, the velocity distribution is

$$\boxed{v_z = \frac{\rho g \delta^2 \cos \beta}{2\mu}\left[1 - \left(\frac{x}{\delta}\right)^2\right]} \tag{2.2–16}$$

Hence the velocity profile is parabolic. (See Fig. 2.2–2.)

Once the velocity profile has been found, a number of quantities may be calculated:

(i) The *maximum velocity* $v_{z,\max}$ is clearly the velocity at $x = 0$; that is

$$v_{z,\max} = \frac{\rho g \delta^2 \cos \beta}{2\mu} \tag{2.2–17}$$

(ii) The *average velocity* $\langle v_z \rangle$ over a cross section of the film is obtained by the following calculation:

$$\langle v_z \rangle = \frac{\int_0^W \int_0^\delta v_z \, dx \, dy}{\int_0^W \int_0^\delta dx \, dy}$$

$$= \frac{1}{\delta} \int_0^\delta v_z \, dx$$

$$= \frac{\rho g \, \delta^2 \cos \beta}{2\mu} \int_0^1 \left[1 - \left(\frac{x}{\delta} \right)^2 \right] d\left(\frac{x}{\delta} \right)$$

$$= \frac{\rho g \, \delta^2 \cos \beta}{3\mu} \tag{2.2–18}$$

(iii) The *volume rate of flow* Q is obtained from the average velocity or by integration of the velocity distribution:

$$Q = \int_0^W \int_0^\delta v_z \, dx \, dy = W\delta \langle v_z \rangle = \frac{\rho g W \delta^3 \cos \beta}{3\mu} \tag{2.2–19}$$

(iv) The *film thickness* δ may be given in terms of the average velocity, the volume rate of flow, or the mass rate of flow per unit width of wall ($\Gamma = \rho \, \delta \langle v_z \rangle$):

$$\delta = \sqrt{\frac{3\mu \langle v_z \rangle}{\rho g \cos \beta}} = \sqrt[3]{\frac{3\mu Q}{\rho g W \cos \beta}} = \sqrt[3]{\frac{3\mu \Gamma}{\rho^2 g \cos \beta}} \tag{2.2–20}$$

(v) The z-component of the *force* $\mathbf{F}$ *of the fluid on the surface* is given by integrating the momentum flux over the fluid-solid interface:

$$F_z = \int_0^L \int_0^W \tau_{xz}|_{x=\delta} \, dy \, dz$$

$$= \int_0^L \int_0^W -\mu \left. \frac{dv_z}{dx} \right|_{x=\delta} dy \, dz$$

$$= (LW)(-\mu) \left(-\frac{\rho g \, \delta \cos \beta}{\mu} \right)$$

$$= \rho g \, \delta L W \cos \beta \tag{2.2–21}$$

This is clearly just the z-component of the weight of the entire fluid in the film.

The foregoing analytical results are valid only when the film is falling in laminar flow with straight streamlines. For the slow flow of thin viscous films, these conditions are satisfied. It has been found experimentally that as the film velocity $\langle v_z \rangle$ increases, as the thickness of the film δ increases, and as the kinematic viscosity $\nu = \mu/\rho$ decreases, the nature of the flow gradually changes; in this gradual change three more or less distinct stable types of flow can be observed: (*a*) laminar flow with straight streamlines, (*b*) laminar flow with rippling, and (*c*) turbulent flow. Quantitative information concerning the type of flow that can be expected under a given set of physical conditions seems to be only fragmentary. For vertical walls, the following information may be given:[1,2]

laminar flow without rippling	$\text{Re} < 4$ to 25
laminar flow with rippling	4 to $25 < \text{Re} < 1000$ to 2000
turbulent flow	$\text{Re} > 1000$ to 2000

in which $\text{Re} = 4\,\delta\langle v_z\rangle\rho/\mu = 4\Gamma/\mu$ is the *Reynolds number* for this system. Why this dimensionless group is used as a criterion for flow patterns is discussed further in subsequent chapters.

Example 2.2–1. Calculation of Film Velocity

An oil has a kinematic viscosity of 2×10^{-4} m^2 sec^{-1} and a density of 0.8×10^3 kg m^{-3}. What should the mass rate of flow of this film down a vertical wall be in order to have a film thickness of 2.5 mm?

Solution. According to Eq. 2.2–20, the mass rate of flow per unit width of wall is (all numerical values given in mks units)

$$\Gamma = \frac{\delta^3\rho g}{3\nu} = \frac{(2.5 \times 10^{-3})^3(0.8 \times 10^3)(9.80)}{3(2 \times 10^{-4})}$$

$$= 0.204 \text{ kg m}^{-1}\text{ sec}^{-1}$$

This is the desired result if and only if the flow is indeed laminar. To ascertain the nature of the flow, we calculate a Reynolds number based on the mass rate of flow just found:

$$\text{Re} = \frac{4\Gamma}{\mu} = \frac{4\Gamma}{\rho\nu} = \frac{4(0.204)}{(0.8 \times 10^3)(2 \times 10^{-4})} = 5.1 \qquad \text{(dimensionless)}$$

This Reynolds number is below the observed upper limit for laminar flow stated above, and therefore the calculated value of Γ is valid.

Example 2.2–2. Falling Film with Variable Viscosity

Rework the falling film problem for the situation in which the viscosity depends upon position in the following manner:

$$\mu = \mu_0 e^{-\alpha(x/\delta)} \qquad (2.2\text{–}22)$$

[1] T. K. Sherwood and R. L. Pigford, *Absorption and Extraction*, McGraw-Hill, New York (1952), p. 265.

[2] S. S. Grimley, *Trans. Inst. Chem. Engrs.* (London), **23**, 228–235 (1948).

in which μ_0 is the viscosity at the surface of the film and α is a constant that tells how rapidly μ changes as x increases. Such a variation arises in the flow of condensate down a wall with a linear temperature gradient through the film. Show how the results of this problem simplify to the results already obtained for the special limit that $\alpha = 0$ (film of constant viscosity).

Solution. The setting up of the momentum balance and the calculation of the momentum flux distribution in Eq. 2.2–11 proceeds as before. Then, substituting Newton's law (with the variable viscosity of Eq. 2.2–22) in Eq. 2.2–11 gives

$$-\mu_0 e^{-\alpha(x/\delta)} \frac{dv_z}{dx} = \rho g x \cos \beta \tag{2.2–23}$$

This differential equation can be integrated, and use of the boundary condition in Eq. 2.2–15 allows for the evaluation of the integration constant. The velocity profile is then

$$v_z = \frac{\rho g \,\delta^2 \cos \beta}{\mu_0}\left[e^{\alpha}\left(\frac{1}{\alpha} - \frac{1}{\alpha^2}\right) - e^{\alpha x/\delta}\left(\frac{x}{\alpha\,\delta} - \frac{1}{\alpha^2}\right)\right] \tag{2.2–24}$$

In order to get the velocity for a constant-viscosity film, we put α equal to zero in Eq. 2.2–24. When we do this, we find we get into difficulty mathematically. The difficulty is met by expanding exp α and exp $(\alpha x/\delta)$ in a Taylor series:

$$\begin{aligned}(v_z)_{\alpha=0} &= \frac{\rho g\,\delta^2 \cos \beta}{\mu_0} \lim_{\alpha\to 0}\left[\left(1 + \alpha + \frac{\alpha^2}{2!} + \frac{\alpha^3}{3!} + \cdots\right)\left(\frac{1}{\alpha} - \frac{1}{\alpha^2}\right)\right.\\ &\quad \left. - \left(1 + \frac{\alpha x}{\delta} + \frac{\alpha^2 x^2}{2!\delta^2} + \frac{\alpha^3 x^3}{3!\delta^3} + \cdots\right)\left(\frac{x}{\alpha\,\delta} - \frac{1}{\alpha^2}\right)\right]\\ &= \frac{\rho g\,\delta^2 \cos \beta}{\mu_0} \lim_{\alpha\to 0}\left[\left(\frac{1}{2} + \frac{1}{3}\alpha + \cdots\right) - \left(\frac{1}{2}\frac{x^2}{\delta^2} + \frac{1}{3}\frac{x^3}{\delta^3}\alpha + \cdots\right)\right]\\ &= \frac{\rho g\,\delta^2 \cos \beta}{2\mu_0}\left[1 - \left(\frac{x}{\delta}\right)^2\right]\end{aligned} \tag{2.2–25}$$

which is in agreement with Eq. 2.2–16.

From Eq. 2.2–24 it may be shown that the average velocity is

$$\langle v_z \rangle = \frac{\rho g\,\delta^2 \cos \beta}{\mu_0}\left[e^{\alpha}\left(\frac{1}{\alpha} - \frac{2}{\alpha^2} + \frac{2}{\alpha^3}\right) - \frac{2}{\alpha^3}\right] \tag{2.2–26}$$

The reader may verify that this result simplifies to Eq. 2.2–18 when $\alpha = 0$.

§2.3 FLOW THROUGH A CIRCULAR TUBE

The flow of fluids in circular tubes is encountered frequently in physics, chemistry, biology, and engineering. The laminar flow of fluids in circular tubes may be analyzed by means of the momentum balance described in §2.1. The only new feature introduced here is the use of cylindrical coordinates, which are the natural coordinates to describe positions in a circular pipe.

We consider then the steady laminar flow of a fluid of constant density ρ in a "very long" tube of length L and radius R; we specify that the tube be "very long" because we want to assume that there are no "end effects"; that is, we ignore the fact that at the tube entrance and exit the flow will not necessarily be parallel everywhere to the tube surface.

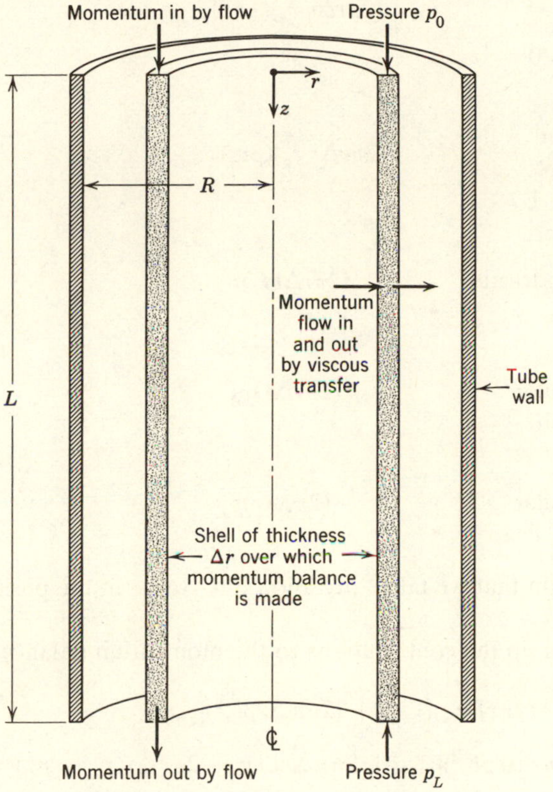

Fig. 2.3-1. Cylindrical shell of fluid over which momentum balance is made to get the velocity profile and the Hagen-Poiseuille formula for the volume rate of flow.

We select as our system a cylindrical shell of thickness Δr and length L (see Fig. 2.3–1), and we begin by listing the various contributions to the momentum balance in the z–direction:

rate of momentum in across cylindrical surface at r

$$(2\pi r L\tau_{rz})|_r \tag{2.3–1}$$

rate of momentum out across cylindrical surface at $r + \Delta r$

$$(2\pi r L \tau_{rz})|_{r+\Delta r} \tag{2.3–2}$$

rate of momentum in across annular surface at $z = 0$

$$(2\pi r \Delta r\, v_z)(\rho v_z)|_{z=0} \tag{2.3–3}$$

rate of momentum out across annular surface at $z = L$

$$(2\pi r \Delta r\, v_z)(\rho v_z)|_{z=L} \tag{2.3–4}$$

gravity force acting on cylindrical shell

$$(2\pi r \Delta r L)\rho g \tag{2.3–5}$$

pressure force acting on annular surface at $z = 0$

$$(2\pi r \Delta r)p_0 \tag{2.3–6}$$

pressure force acting on annular surface at $z = L$

$$-(2\pi r \Delta r)p_L \tag{2.3–7}$$

Note once again that we take "in" and "out" to be in the positive direction of the axes.

We now add up the contributions to the momentum balance:

$$(2\pi r L \tau_{rz})|_r - (2\pi r L \tau_{rz})|_{r+\Delta r} + (2\pi r\, \Delta r \rho v_z^2)|_{z=0} - (2\pi r\, \Delta r \rho v_z^2)|_{z=L} + 2\pi r\, \Delta r L \rho g + 2\pi r\, \Delta r(p_0 - p_L) = 0 \tag{2.3–8}$$

Because the fluid is assumed to be incompressible, v_z is the same at $z = 0$ and $z = L$, hence the third and fourth terms cancel one another. We now divide Eq. 2.3–8 by $2\pi L\, \Delta r$ and take the limit as Δr goes to zero; this gives

$$\lim_{\Delta r \to 0} \left(\frac{(r\tau_{rz})|_{r+\Delta r} - (r\tau_{rz})|_r}{\Delta r} \right) = \left(\frac{p_0 - p_L}{L} + \rho g \right) r \tag{2.3–9}$$

The expression on the left side is the definition of the first derivative. Hence Eq. 2.3–9 may be written as

$$\frac{d}{dr}(r\tau_{rz}) = \left(\frac{\mathscr{P}_0 - \mathscr{P}_L}{L} \right) r \tag{2.3–10}$$

in which[1] $\mathscr{P} = p - \rho gz$. Eq. 2.3–10 may be integrated to give:

$$\tau_{rz} = \left(\frac{\mathscr{P}_0 - \mathscr{P}_L}{2L}\right) r + \frac{C_1}{r} \tag{2.3–11}$$

The constant C_1 must be zero if the momentum flux is not to be infinite at $r = 0$. Hence the momentum flux distribution is

$$\boxed{\tau_{rz} = \left(\frac{\mathscr{P}_0 - \mathscr{P}_L}{2L}\right) r} \tag{2.3–12}$$

This distribution is shown in Fig. 2.3–2.

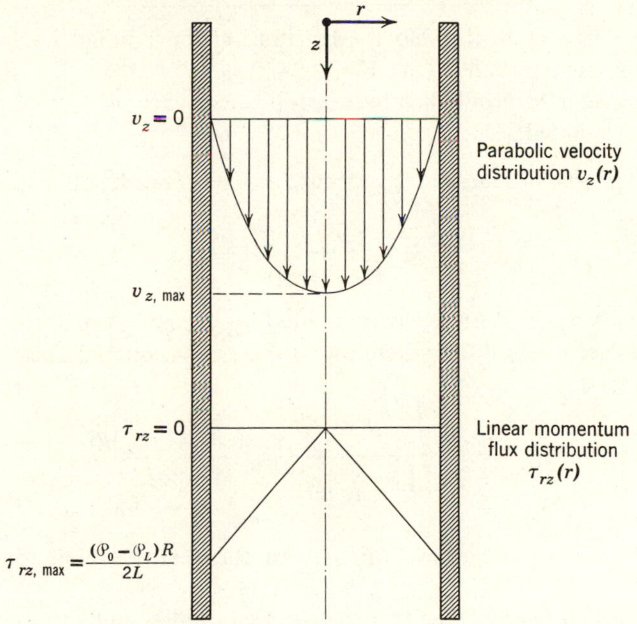

Fig. 2.3–2. Momentum flux and velocity distributions in flow in cylindrical tubes.

Newton's law of viscosity for this situation is

$$\tau_{rz} = -\mu \frac{dv_z}{dr} \tag{2.3–13}$$

[1] The quantity $\mathscr{P}$ represents the combined effect of static pressure and gravitational force. To allow for other flow orientations, $\mathscr{P}$ may be defined more generally as $\mathscr{P} = p + \rho gh$, where h is the distance *upward* (that is, in the direction opposed to gravity) from any chosen reference plane.

Substitution of this relation into Eq. 2.3–12 then gives the following differential equation for the velocity:

$$\frac{dv_z}{dr} = -\left(\frac{\mathscr{P}_0 - \mathscr{P}_L}{2\mu L}\right) r \tag{2.3–14}$$

Integration of this gives

$$v_z = -\left(\frac{\mathscr{P}_0 - \mathscr{P}_L}{4\mu L}\right) r^2 + C_2 \tag{2.3–15}$$

Because of the boundary condition that v_z be zero at $r = R$, the constant C_2 has the value $(\mathscr{P}_0 - \mathscr{P}_L)R^2/4\mu L$. Hence the velocity distribution is

$$\boxed{v_z = \frac{(\mathscr{P}_0 - \mathscr{P}_L)R^2}{4\mu L}\left[1 - \left(\frac{r}{R}\right)^2\right]} \tag{2.3–16}$$

This result tells us that the velocity distribution for laminar, incompressible flow in a tube is parabolic. (See Fig. 2.3–2.)

Once the velocity profile has been established, various derived quantities are easily calculated:

(i) The *maximum velocity* $v_{z,\max}$ occurs at $r = 0$ and has the value

$$v_{z,\max} = \frac{(\mathscr{P}_0 - \mathscr{P}_L)R^2}{4\mu L} \tag{2.3–17}$$

(ii) The *average velocity* $\langle v_z \rangle$ is calculated by summing up all the velocities over a cross section and then dividing by the cross-sectional area:

$$\langle v_z \rangle = \frac{\int_0^{2\pi}\int_0^R v_z r\, dr\, d\theta}{\int_0^{2\pi}\int_0^R r\, dr\, d\theta} = \frac{(\mathscr{P}_0 - \mathscr{P}_L)R^2}{8\mu L} \tag{2.3–18}$$

The details of the integration are left to the reader. Note that $\langle v_z \rangle = \frac{1}{2}v_{z,\max}$.

(iii) The *volume rate of flow* Q is the product of area and average velocity; thus

$$Q = \frac{\pi(\mathscr{P}_0 - \mathscr{P}_L)R^4}{8\mu L} \tag{2.3–19}$$

This rather famous result is called the Hagen-Poiseuille[2] law in honor of the two scientists[3,4] credited with its formulation. It gives the relationship

[2] Pronounce Poiseuille as "Pwah-zø'-yuh," in which ø is roughly the same as the "oo" in the American pronunciation of "book".

[3] G. Hagen, *Ann. Phys. Chem.*, **46**, 423–442 (1839).

[4] J. L. Poiseuille, *Compte Rendus*, **11**, 961 and 1041 (1840); **12**, 112 (1841).

between the volume rate of flow and the forces causing the flow—the forces associated with the pressure drop and the gravitational acceleration.

(iv) The z-component of the *force of the fluid on the wetted surface of* the pipe, F_z, is just the momentum flux integrated over the wetted area:

$$F_z = (2\pi RL)\left(-\mu \frac{dv_z}{dr}\right)\bigg|_{r=R} = \pi R^2(\mathscr{P}_0 - \mathscr{P}_L)$$

$$= \pi R^2(p_0 - p_L) + \pi R^2 L\rho g \tag{2.3–20}$$

This result is not surprising—all it says is that the net force acting downstream on the cylinder of fluid by virtue of the pressure difference and gravitational acceleration is just counterbalanced by the viscous force F_z, which tends to resist the fluid motion.

The results of this section are valid only for values of Reynolds numbers less than about 2100, for which the flow is laminar. For this system, it is customary to define the Reynolds number by $\text{Re} = D\langle v_z\rangle\rho/\mu$, where $D = 2R$ is the tube diameter.

We may now summarize all the assumptions that are implied in the development of the Hagen-Poiseuille law:

a. The flow is laminar—Re less than about 2100.

b. The density ρ is constant ("incompressible flow").

c. The flow is independent of time ("steady state")—the corresponding unsteady-state problem is discussed in §4.1.

d. The fluid is Newtonian—that is, $\tau_{rz} = -\mu(dv_z/dr)$.

e. End effects are neglected—actually an "entrance length" (beyond the tube entrance) on the order of $L_e = 0.035D$ Re is required for build-up to the parabolic profiles; if the section of pipe of interest includes the entrance region, a correction must be applied.[5] The fractional correction introduced in either $\Delta\mathscr{P}$ or Q never exceeds L_e/L if $L > L_e$.

f. The fluid behaves as a continuum—this assumption is valid except for very dilute gases or very narrow capillary tubes, in which the molecular mean free path is comparable to the tube diameter ("slip flow" regime) or much greater than the tube diameter ("Knudsen flow" or "free molecule flow" regime).[6]

g. There is no slip at the wall—this is an excellent assumption for pure fluids under the conditions assumed in (*f*).

[5] J. H. Perry, *Chemical Engineers Handbook*, McGraw-Hill, New York (1950), Third Edition, pp. 388–389. W. M. Kays and A. L. London, *Compact Heat Exchangers*, McGraw-Hill, New York (1958), p. 49.

[6] M. Knudsen, *The Kinetic Theory of Gases*, Methuen, London (1934). E. H. Kennard, *Kinetic Theory of Gases*, McGraw-Hill, New York (1938). G. N. Patterson, *Molecular Flow of Gases*, Wiley, New York (1956).

Example 2.3–1. Determination of Viscosity from Capillary Flow Data

Glycerine ($CH_2OH{\cdot}CHOH{\cdot}CH_2OH$) at 26.5° C is flowing through a horizontal tube 1 ft long and 0.1 in. inside diameter. For a pressure drop of 40 psi, the flow rate is 0.00398 ft³ min⁻¹. The density of glycerine at 26.5° C is 1.261 g cm⁻³. From the flow data, find the viscosity of glycerine in centipoises. (Measurement of flow in circular tubes is one of the common methods for determining viscosity; such devices are referred to as "capillary viscometers").

Solution. From the Hagen-Poiseuille law (Eq. 2.3–19)

$$\mu = \frac{\pi \Delta p R^4}{8QL}$$

$$= \frac{\pi\left(40\,\frac{\text{lb}_f}{\text{in.}^2}\right)\left(6.8947 \times 10^4\,\frac{\text{dyne cm}^{-2}}{\text{lb}_f\,\text{in.}^{-2}}\right)\left(0.05\text{ in.} \times \frac{1}{12}\frac{\text{ft}}{\text{in.}}\right)^4}{8\left(0.00398\text{ ft}^3\text{ min}^{-1} \times \frac{1}{60}\frac{\text{min}}{\text{sec}}\right)(1\text{ ft})}$$

$$= 4.92\text{ g cm}^{-1}\text{ sec}^{-1} = 492\text{ cp} \tag{2.3–21}$$

We must now check to be sure that the flow is laminar. The Reynolds number for this flow is

$$\text{Re} = \frac{D\langle v_z\rangle\rho}{\mu} = \frac{4}{\pi}\frac{Q\rho}{D\mu}$$

$$= \frac{4}{\pi}\,\frac{(0.00398\text{ ft}^3\text{ min}^{-1})(2.54\text{ cm in.}^{-1} \times 12\text{ in. ft}^{-1})^3\left(\frac{1}{60}\frac{\text{min}}{\text{sec}}\right)\left(1.261\,\frac{\text{g}}{\text{cm}^3}\right)}{(0.1\text{ in.} \times 2.54\text{ cm in.}^{-1})(4.92\text{ g cm}^{-1}\text{ sec}^{-1})}$$

$$= 2.41\text{ (dimensionless)} \tag{2.3–22}$$

Hence the flow is indeed laminar. Furthermore, the entrance length L_e is

$$L_e \doteq 0.035D\ \text{Re} = (0.035)(0.1/12)(2.41)$$
$$= 0.0007\text{ ft}$$

Hence entrance effects are not important, and the value of viscosity given above has been calculated properly.

Example 2.3–2. Bingham Flow in a Circular Tube

A fluid that is very nearly described by the Bingham model (see Eq. 1.2–2) is flowing through a vertical tube as the result of a pressure gradient and/or gravitational acceleration. The radius and length of the tube are R and L, respectively. It is desired to obtain a relation between the volume rate of flow Q and the combined pressure and gravity forces acting on the fluid.

Solution. The momentum flux distribution for flow of any kind of fluid through a circular tube is given by Eq. 2.3–12. According to Fig. 1.2–1, for a Bingham fluid the velocity gradient is zero when the momentum flux is less than the value τ_0. Hence one expects a "plug flow" region in the central part of the tube, as sketched in Fig. 2.3–3. Outside the plug-flow region the momentum flux and the velocity

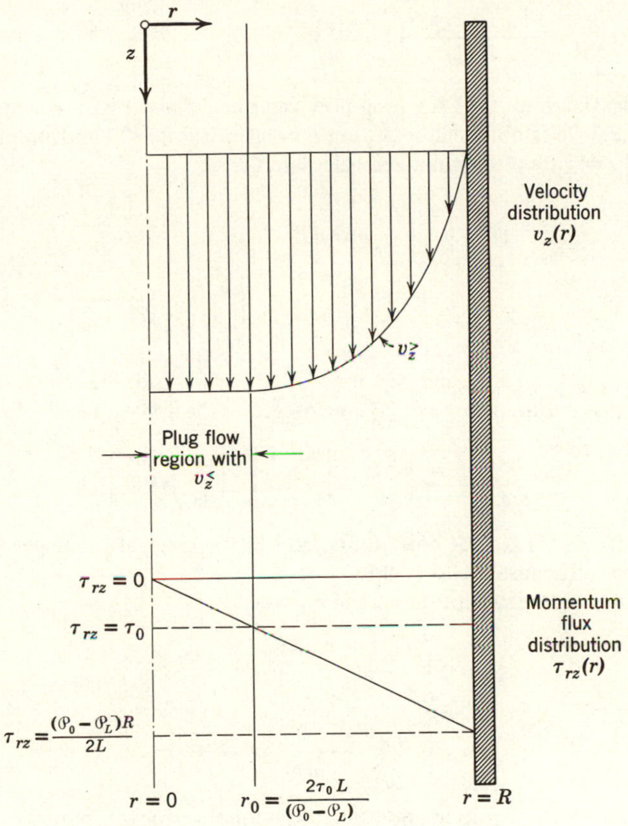

Fig. 2.3–3. Flow of a Bingham fluid in a circular tube.

gradient are related according to Eq. 1.2–2*a*. Substitution of the cylindrical coordinate version of Eq. 1.2–2*a* into Eq. 2.3–12 gives

$$\tau_0 - \mu_0 \frac{dv_z}{dr} = \left(\frac{\mathcal{P}_0 - \mathcal{P}_L}{2L}\right) r \tag{2.3–23}$$

This separable first-order differential equation may be integrated to give

$$v_z = -\left(\frac{\mathcal{P}_0 - \mathcal{P}_L}{4\mu_0 L}\right) r^2 + \frac{\tau_0}{\mu_0} r + C_2 \tag{2.3–24}$$

The constant C_2 is evaluated by making use of the boundary condition that $v_z = 0$ at $r = R$. Then the velocity distribution becomes finally

$$v_z^{>} = \frac{(\mathscr{P}_0 - \mathscr{P}_L)R^2}{4\mu_0 L}\left[1 - \left(\frac{r}{R}\right)^2\right] - \frac{\tau_0 R}{\mu_0}\left[1 - \left(\frac{r}{R}\right)\right] \qquad r \geqslant r_0 \qquad (2.3\text{–}25)$$

$$v_z^{<} = \frac{(\mathscr{P}_0 - \mathscr{P}_L)R^2}{4\mu_0 L}\left(1 - \frac{r_0}{R}\right)^2 \qquad r \leqslant r_0 \qquad (2.3\text{–}26)$$

Here r_0 is the radius of the plug-flow region, defined by $\tau_0 = (\mathscr{P}_0 - \mathscr{P}_L)r_0/2L$. Equation 2.3–26 is obtained by setting $r = r_0$ in Eq. 2.3–25 and simplifying.

The volume rate of flow may be calculated from

$$\begin{aligned} Q &= \int_0^{2\pi}\int_0^R v_z r\,dr\,d\theta \\ &= 2\pi\int_0^{r_0} v_z^{<}\,r\,dr + 2\pi\int_{r_0}^R v_z^{>}\,r\,dr \end{aligned} \qquad (2.3\text{–}27)$$

The expressions for $v_z^{<}$ and $v_z^{>}$ may be inserted and the integrals evaluated. Less algebra is required, however, if one integrates the first expression for Q by parts:

$$Q = 2\pi\left[\frac{1}{2}r^2 v_z\Big|_0^R - \frac{1}{2}\int_0^R \left(\frac{dv_z}{dr}\right) r^2\,dr\right] \qquad (2.3\text{–}28)$$

The quantity $r^2 v_z$ is zero at both limits, and in the integral the lower limit may be replaced by r_0 because $dv_z/dr = 0$ for $r \leqslant r_0$.

Hence the volume rate of flow is, if $\tau_R > \tau_0$

$$\begin{aligned} Q &= \pi\int_{r_0}^R (-dv_z^{>}/dr) r^2\,dr \\ &= \pi\int_{r_0}^R \left[\left(\frac{\mathscr{P}_0 - \mathscr{P}_L}{2\mu_0 L}\right) r - \frac{\tau_0}{\mu_0}\right] r^2\,dr \end{aligned} \qquad (2.3\text{–}29)$$

By performing the integration and using the symbol τ_R for the momentum flux at the wall, $\dfrac{(\mathscr{P}_0 - \mathscr{P}_L)R}{2L}$, one obtains when $\tau_R > \tau_0$:

$$Q = \frac{\pi(\mathscr{P}_0 - \mathscr{P}_L)R^4}{8\mu_0 L}\left[1 - \frac{4}{3}\left(\frac{\tau_0}{\tau_R}\right) + \frac{1}{3}\left(\frac{\tau_0}{\tau_R}\right)^4\right] \qquad (2.3\text{–}30)$$

which is known as the *Buckingham-Reiner Equation.*[7,8,9] When τ_0 is zero, the Bingham model reduces to the Newtonian model and Eq. 2.3–30 reduces appropriately to the Hagen-Poiseuille equation.

[7] E. Buckingham, *Proc. ASTM*, **21**, 1154–1161 (1921).

[8] M. Reiner, *Deformation and Flow*, Lewis, London (1949).

[9] M. Reiner, "Phenomenological Macrorheology," Chapter 2 in *Rheology*, F. R. Eirich (Ed.), Academic Press, New York (1956), Vol. 1, p. 45.

§2.4 FLOW THROUGH AN ANNULUS

Let us now consider another viscous flow problem in cylindrical coordinates, but one for which the boundary conditions are different. An incompressible fluid is flowing in steady state in the annular region between two coaxial circular cylinders of radii κR and R. (See Fig. 2.4–1.) We begin by setting up

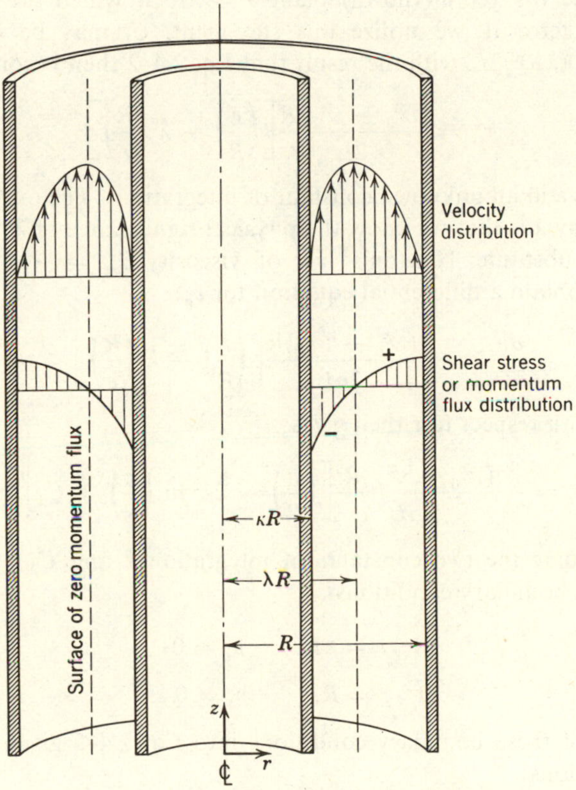

Fig. 2.4–1. Upward flow through a cylindrical annulus.

a momentum balance over a thin cylindrical shell and arrive at the same differential equation that we obtained before for tube flow (see Eq. 2.3–10):

$$\frac{d}{dr}(r\tau_{rz}) = \left(\frac{\mathscr{P}_0 - \mathscr{P}_L}{L}\right)r \tag{2.4–1}$$

Note that for this problem $\mathscr{P} = p + \rho g z$, inasmuch as the pressure and gravity forces are acting in opposite directions (that is, z is the same as h in fn. 1 on p. 45). This may be integrated as before (see Eq. 2.3–11) to give

$$\tau_{rz} = \left(\frac{\mathscr{P}_0 - \mathscr{P}_L}{2L}\right)r + \frac{C_1}{r} \tag{2.4–2}$$

The constant C_1 cannot be determined immediately, since we have no information on the momentum flux at either of the fixed surfaces $r = \kappa R$ or $r = R$. The most we can say is that there will be a maximum in the velocity curve at some (as yet unknown) plane $r = \lambda R$ at which the momentum flux will be zero. If we utilize this statement, C_1 may be replaced by $-(\mathscr{P}_0 - \mathscr{P}_L)(\lambda R)^2/2L$, with the result that Eq. 2.4–2 then becomes

$$\tau_{rz} = \frac{(\mathscr{P}_0 - \mathscr{P}_L)R}{2L}\left[\left(\frac{r}{R}\right) - \lambda^2\left(\frac{R}{r}\right)\right] \tag{2.4–3}$$

Note that λ is still an unknown constant of integration. The only reason for replacing C_1 by λ is that we know the physical significance of λ.

We now substitute Newton's law of viscosity $\tau_{rz} = -\mu(dv_z/dr)$ into Eq. 2.4–3 to obtain a differential equation for v_z:

$$\frac{dv_z}{dr} = -\frac{(\mathscr{P}_0 - \mathscr{P}_L)R}{2\mu L}\left[\left(\frac{r}{R}\right) - \lambda^2\left(\frac{R}{r}\right)\right] \tag{2.4–4}$$

Integration with respect to r then gives

$$v_z = -\frac{(\mathscr{P}_0 - \mathscr{P}_L)R^2}{4\mu L}\left[\left(\frac{r}{R}\right)^2 - 2\lambda^2 \ln\left(\frac{r}{R}\right) + C_2\right] \tag{2.4–5}$$

We now evaluate the two constants of integration λ and C_2 by using the following two boundary conditions:

B.C. 1: at $r = \kappa R, \quad v_z = 0$ (2.4–6)

B.C. 2: at $r = R, \quad v_z = 0$ (2.4–7)

Substitution of these boundary conditions into Eq. 2.4–5 gives two simultaneous equations:

$$0 = -\frac{(\mathscr{P}_0 - \mathscr{P}_L)R^2}{4\mu L}(\kappa^2 - 2\lambda^2 \ln \kappa + C_2) \tag{2.4–8}$$

$$0 = -\frac{(\mathscr{P}_0 - \mathscr{P}_L)R^2}{4\mu L}(1 + C_2) \tag{2.4–9}$$

Therefore the constants C_2 and λ are given by

$$C_2 = -1 \qquad 2\lambda^2 = \frac{1 - \kappa^2}{\ln(1/\kappa)} \tag{2.4–10,11}$$

Substitution of these values into Eqs. 2.4–3 and 2.4–5 then gives, respectively, the momentum flux distribution[1] and the velocity distribution for steady, incompressible flow in an annulus:

$$\tau_{rz} = \frac{(\mathscr{P}_0 - \mathscr{P}_L)R}{2L}\left[\left(\frac{r}{R}\right) - \left(\frac{1-\kappa^2}{2\ln(1/\kappa)}\right)\left(\frac{R}{r}\right)\right] \tag{2.4-12}$$

$$v_z = \frac{(\mathscr{P}_0 - \mathscr{P}_L)R^2}{4\mu L}\left[1 - \left(\frac{r}{R}\right)^2 + \left(\frac{1-\kappa^2}{\ln(1/\kappa)}\right)\ln\left(\frac{r}{R}\right)\right] \tag{2.4-13}$$

Note that when κ goes to zero these results reduce to the corresponding results for flow in circular tubes. (See Eqs. 2.3–12 and 2.3–16.) The reader should make it a habit to check analytical results in this way to ascertain whether they describe "limiting cases" properly.

Once we have the momentum flux and velocity distributions, it is straightforward to get other information of interest:

(i) The *maximum velocity* is

$$v_{z,\max} = v_z\Big|_{r=\lambda R} = \frac{(\mathscr{P}_0 - \mathscr{P}_L)R^2}{4\mu L}\left\{1 - \left(\frac{1-\kappa^2}{2\ln(1/\kappa)}\right)\left[1 - \ln\left(\frac{1-\kappa^2}{2\ln(1/\kappa)}\right)\right]\right\} \tag{2.4-14}$$

(ii) The *average velocity* is

$$\langle v_z\rangle = \frac{\displaystyle\int_0^{2\pi}\int_{\kappa R}^{R} v_z r\,dr\,d\theta}{\displaystyle\int_0^{2\pi}\int_{\kappa R}^{R} r\,dr\,d\theta} = \frac{(\mathscr{P}_0 - \mathscr{P}_L)R^2}{8\mu L}\left(\frac{1-\kappa^4}{1-\kappa^2} - \frac{1-\kappa^2}{\ln(1/\kappa)}\right) \tag{2.4-15}$$

(iii) The *volume rate of flow* is

$$Q = \pi R^2(1-\kappa^2)\langle v_z\rangle = \frac{\pi(\mathscr{P}_0 - \mathscr{P}_L)R^4}{8\mu L}\left((1-\kappa^4) - \frac{(1-\kappa^2)^2}{\ln(1/\kappa)}\right) \tag{2.4-16}$$

(iv) The *force exerted by the fluid on the solid* is obtained by summing the forces acting on the inner cylinder and outer cylinder, respectively:

$$\begin{aligned} F_z &= -\tau_{rz}|_{r=\kappa R}\cdot 2\pi\kappa RL + \tau_{rz}|_{r=R}\cdot 2\pi RL \\ &= \pi R^2(1-\kappa^2)(\mathscr{P}_0 - \mathscr{P}_L) \end{aligned} \tag{2.4-17}$$

The reader should verify and interpret these results as an exercise.

The foregoing simplify to the formulas for flow through circular tubes in the limit as $\kappa \to 0$. Also they go smoothly over to the formulas for flow in

[1] The momentum flux distribution is *not* given by Eq. 2.4–12 for non-Newtonian flow. See problem 2.Q.

plane slits in the limit as κ becomes very nearly 1. The limiting processes needed to obtain the latter are discussed in problems 2.F (for the volume rate of flow) and 2.P (for the velocity profiles).

The solution given above is valid only for laminar flow. The laminar-turbulent transition occurs in the neighborhood of Re = 2000, where the Reynolds number is defined as $\text{Re} = 2R(1 - \kappa)\langle v_z \rangle \rho / \mu$. Actually, before the transition the stable laminar flow appears to exhibit a sinuous motion.[2]

§2.5 ADJACENT FLOW OF TWO IMMISCIBLE FLUIDS[1]

Thus far we have considered flow situations with solid-fluid and liquid-gas boundaries. We now give one example of a flow problem with a liquid-liquid boundary. (See Fig. 2.5–1.)

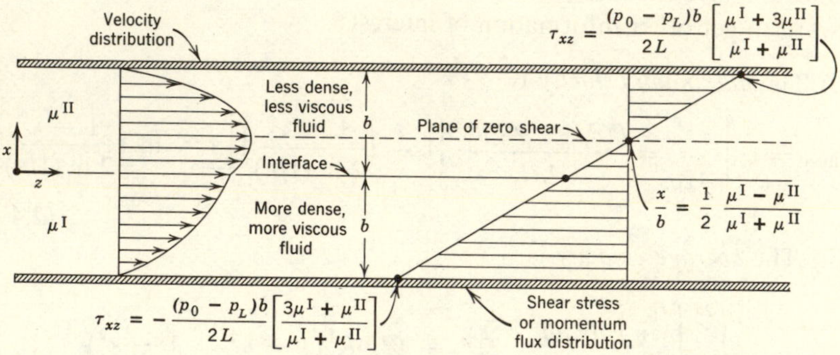

Fig. 2.5–1. Flow of two immiscible fluids between two flat plates under the influence of a pressure gradient.

Two immiscible incompressible fluids are flowing in the z-direction in a horizontal thin slit of length L and width W under the influence of a pressure gradient. The fluid rates are so adjusted that the slit is half filled with Fluid I (the more dense phase) and half filled with Fluid II (the less dense phase). It is desired to analyze the system in terms of the distribution of velocity and momentum flux.

A differential momentum balance leads to the following differential equation for τ_{xz}:

$$\frac{d\tau_{xz}}{dx} = \frac{p_0 - p_L}{L} \tag{2.5–1}$$

[2] R. S. Prengle and R. R. Rothfus, *Ind. Eng. Chem.*, **47**, 379–386 (1955).

[1] The adjacent flow of a turbulent and a laminar layer has been studied by T. J. Hanratty and J. M. Engen, *A.I.Ch.E. Journal*, **3**, 299–304 (1957). Annular gas-liquid flow in tubes has been discussed by A. D. K. Laird, *Trans. ASME.* **76**, 1005–1010 (1954), and S. Calvert and B. Williams, *A.I.Ch.E. Journal*, **1**, 78–86 (1955).

This equation is obtained for either phase I or phase II. Integration of Eq. 2.5–1 gives for the two regions

$$\tau_{xz}{}^{\mathrm{I}} = \left(\frac{p_0 - p_L}{L}\right) x + C_1{}^{\mathrm{I}} \tag{2.5–2}$$

$$\tau_{xz}{}^{\mathrm{II}} = \left(\frac{p_0 - p_L}{L}\right) x + C_1{}^{\mathrm{II}} \tag{2.5–3}$$

We may immediately make use of one of the boundary conditions, namely, that the momentum transport is continuous through the interface between the two fluids:

B.C. 1: $\quad$ at $x = 0, \quad \tau_{xz}{}^{\mathrm{I}} = \tau_{xz}{}^{\mathrm{II}}$ (2.5–4)

This tells us that $C_1{}^{\mathrm{I}} = C_1{}^{\mathrm{II}}$; hence we drop the superscript and call the integration constant simply C_1.

If Newton's law of viscosity is substituted into Eq. 2.5–2 and 2.5–3, we then get

$$-\mu^{\mathrm{I}} \frac{dv_z{}^{\mathrm{I}}}{dx} = \left(\frac{p_0 - p_L}{L}\right) x + C_1 \tag{2.5–5}$$

$$-\mu^{\mathrm{II}} \frac{dv_z{}^{\mathrm{II}}}{dx} = \left(\frac{p_0 - p_L}{L}\right) x + C_1 \tag{2.5–6}$$

Integration of these equations gives

$$v_z{}^{\mathrm{I}} = -\frac{(p_0 - p_L)x^2}{2\mu^{\mathrm{I}} L} - \frac{C_1}{\mu^{\mathrm{I}}} x + C_2{}^{\mathrm{I}} \tag{2.5–7}$$

$$v_z{}^{\mathrm{II}} = -\frac{(p_0 - p_L)x^2}{2\mu^{\mathrm{II}} L} - \frac{C_1}{\mu^{\mathrm{II}}} x + C_2{}^{\mathrm{II}} \tag{2.5–8}$$

We now have three integration constants to determine by using these three additional boundary conditions:

B.C. 2: $\quad$ at $x = 0, \quad v_z{}^{\mathrm{I}} = v_z{}^{\mathrm{II}}$ (2.5–9)

B.C. 3: $\quad$ at $x = -b, \quad v_z{}^{\mathrm{I}} = 0$ (2.5–10)

B.C. 4: $\quad$ at $x = +b, \quad v_z{}^{\mathrm{II}} = 0$ (2.5–11)

The mathematical statement of these boundary conditions is then

B.C. 2: $\quad C_2{}^{\mathrm{I}} = C_2{}^{\mathrm{II}}$ (2.5–12)

B.C. 3: $$0 = -\frac{(p_0 - p_L)b^2}{2\mu^{\mathrm{I}} L} + \frac{C_1 b}{\mu^{\mathrm{I}}} + C_2{}^{\mathrm{I}} \tag{2.5–13}$$

B.C. 4: $$0 = -\frac{(p_0 - p_L)b^2}{2\mu^{\mathrm{II}} L} - \frac{C_1 b}{\mu^{\mathrm{II}}} + C_2{}^{\mathrm{II}} \tag{2.5–14}$$

From these equations we find that

$$C_1 = -\frac{(p_0 - p_L)b}{2L}\left(\frac{\mu^{\mathrm{I}} - \mu^{\mathrm{II}}}{\mu^{\mathrm{I}} + \mu^{\mathrm{II}}}\right) \tag{2.5-15}$$

$$C_2^{\mathrm{I}} = +\frac{(p_0 - p_L)b^2}{2\mu^{\mathrm{I}}L}\left(\frac{2\mu^{\mathrm{I}}}{\mu^{\mathrm{I}} + \mu^{\mathrm{II}}}\right) = C_2^{\mathrm{II}} \tag{2.5-16}$$

Hence the momentum flux and velocity profiles are

$$\boxed{\tau_{xz} = \frac{(p_0 - p_L)b}{L}\left[\left(\frac{x}{b}\right) - \frac{1}{2}\left(\frac{\mu^{\mathrm{I}} - \mu^{\mathrm{II}}}{\mu^{\mathrm{I}} + \mu^{\mathrm{II}}}\right)\right]} \tag{2.5-17}$$

$$\boxed{v_z^{\mathrm{I}} = \frac{(p_0 - p_L)b^2}{2\mu^{\mathrm{I}}L}\left[\left(\frac{2\mu^{\mathrm{I}}}{\mu^{\mathrm{I}} + \mu^{\mathrm{II}}}\right) + \left(\frac{\mu^{\mathrm{I}} - \mu^{\mathrm{II}}}{\mu^{\mathrm{I}} + \mu^{\mathrm{II}}}\right)\left(\frac{x}{b}\right) - \left(\frac{x}{b}\right)^2\right]} \tag{2.5-18}$$

$$\boxed{v_z^{\mathrm{II}} = \frac{(p_0 - p_L)b^2}{2\mu^{\mathrm{II}}L}\left[\left(\frac{2\mu^{\mathrm{II}}}{\mu^{\mathrm{I}} + \mu^{\mathrm{II}}}\right) + \left(\frac{\mu^{\mathrm{I}} - \mu^{\mathrm{II}}}{\mu^{\mathrm{I}} + \mu^{\mathrm{II}}}\right)\left(\frac{x}{b}\right) - \left(\frac{x}{b}\right)^2\right]} \tag{2.5-19}$$

These distributions are shown in Fig. 2.5–1. Note that if $\mu^{\mathrm{I}} = \mu^{\mathrm{II}}$ both velocity distributions are the same and the results simplify to the parabolic velocity profile for laminar flow of a pure fluid in a slit.

The *average velocity* in each layer may now be calculated:

$$\langle v_z^{\mathrm{I}} \rangle = \frac{1}{b}\int_{-b}^{0} v_z^{\mathrm{I}}\, dx = \frac{(p_0 - p_L)b^2}{12\mu^{\mathrm{I}}L}\left(\frac{7\mu^{\mathrm{I}} + \mu^{\mathrm{II}}}{\mu^{\mathrm{I}} + \mu^{\mathrm{II}}}\right) \tag{2.5-20}$$

$$\langle v_z^{\mathrm{II}} \rangle = \frac{1}{b}\int_{0}^{b} v_z^{\mathrm{II}}\, dx = \frac{(p_0 - p_L)b^2}{12\mu^{\mathrm{II}}L}\left(\frac{\mu^{\mathrm{I}} + 7\mu^{\mathrm{II}}}{\mu^{\mathrm{I}} + \mu^{\mathrm{II}}}\right) \tag{2.5-21}$$

From the velocity and momentum-flux distributions given above, one can in addition calculate the maximum velocity, the velocity at the interface, the plane of zero shear stress, and the drag on the walls of the slit.

§2.6 CREEPING FLOW AROUND A SOLID SPHERE[1,2]

In the preceding sections several elementary viscous flow problems have been solved by making differential momentum balances. In the introductory section of this chapter it was emphasized that this method of analysis is

[1] See C. G. Stokes, *Trans. Cambridge Phil. Soc.*, **9**, 8 (1850). Also H. Lamb, *Hydrodynamics*, Dover, New York (1945), First American Edition, §338, pp. 602 *et seq.*; V. L. Streeter, *Fluid Dynamics*, McGraw-Hill, New York (1948), pp. 235–240. A more detailed discussion is to be found in H. Villat, *Leçons sur les fluides visqueux*, Gauthier-Villars, Paris (1943), Chapter 7, in which the unsteady motion of a sphere is considered.

[2] Non-Newtonian flow around spheres has been studied by J. C. Slattery, doctoral thesis, University of Wisconsin (1959).

restricted to flow systems with straight streamlines. Because the problem of flow around a sphere involves curved streamlines, it cannot be solved by the techniques introduced in this chapter. Nevertheless, a brief discussion is appropriate here because of the importance of flow around submerged objects in engineering. No attempt is made to derive the expressions for the distributions of momentum flux, pressure, and velocity. We simply state these results and then proceed to use them to derive some important relations, which we shall need in later discussions.

Let us consider the very slow flow of an incompressible fluid about a solid sphere, as shown in Fig. 2.6–1. The sphere has a radius R and a diameter D. The fluid has a viscosity μ and density ρ and approaches the sphere vertically upward along the negative z-axis with uniform velocity v_∞. For very slow flow, the momentum flux distribution, pressure distribution, and velocity components in spherical coordinates have been found analytically to be

$$\tau_{r\theta} = \frac{3}{2}\frac{\mu v_\infty}{R}\left(\frac{R}{r}\right)^4 \sin\theta \tag{2.6–1}$$

$$p = p_0 - \rho g z - \frac{3}{2}\frac{\mu v_\infty}{R}\left(\frac{R}{r}\right)^2 \cos\theta \tag{2.6–2}$$

$$v_r = v_\infty\left[1 - \frac{3}{2}\left(\frac{R}{r}\right) + \frac{1}{2}\left(\frac{R}{r}\right)^3\right]\cos\theta \tag{2.6–3}$$

$$v_\theta = -v_\infty\left[1 - \frac{3}{4}\left(\frac{R}{r}\right) - \frac{1}{4}\left(\frac{R}{r}\right)^3\right]\sin\theta \tag{2.6–4}$$

In Eq. 2.6–2 the quantity p_0 is the pressure in the plane $z = 0$ far away from the sphere, $-\rho g z$ is the contribution of the fluid weight (hydrostatic effect), and the term containing v_∞ results from the flow of the fluid around the sphere. These equations are valid only for "creeping flow," which for this system occurs when the Reynolds number $Dv_\infty\rho/\mu$ is less than about 0.1. This region is characterized by the virtual absence of eddying downstream from the sphere.

Note that the velocity distribution satisfies the condition that $v_r = v_\theta = 0$ at the surface of the sphere. Furthermore, it may be shown that v_z approaches v_∞ for distances far from the sphere. In addition, the pressure distribution clearly reduces to the hydrostatic equation $p = p_0 - \rho g z$ far from the sphere surface. Hence the expressions do satisfy the boundary conditions at $r = R$ and $r = \infty$.

Let us now calculate the net force exerted by the fluid on the sphere. This force is computed by integrating the normal force and tangential force over the sphere surface.

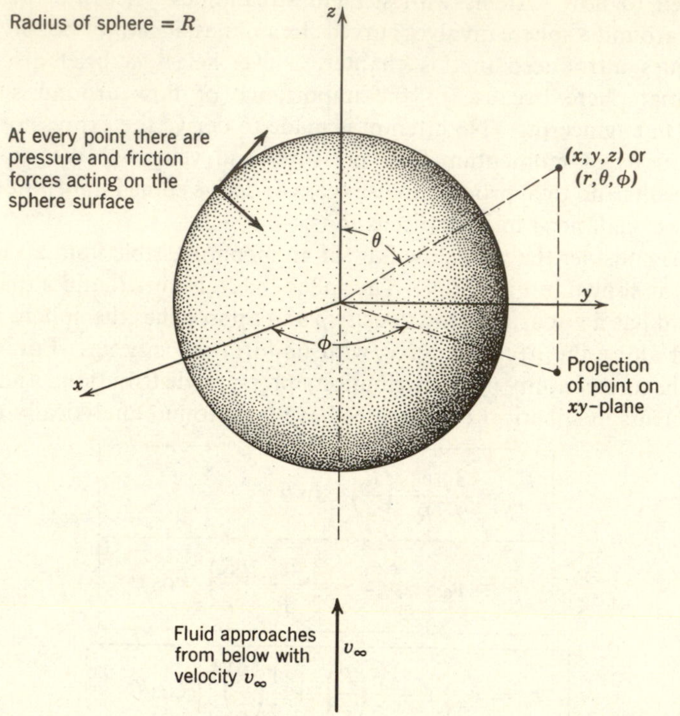

Fig. 2.6–1. Coordinate system used in describing the flow of a fluid about a rigid sphere.

Integration of the Normal Force

At each point on the surface of the sphere there is a force per unit area, p, on the solid acting perpendicularly to the surface. The z-component of this force is $-p \cos \theta$. We now multiply this local force per unit area by the surface area on which it acts, $R^2 \sin \theta \, d\theta \, d\phi$, and integrate over the surface of the sphere to get the resultant force in the z-direction:

$$F_n = \int_0^{2\pi} \int_0^{\pi} (-p|_{r=R} \cos \theta) R^2 \sin \theta \, d\theta \, d\phi \tag{2.6–5}$$

The pressure distribution at the surface of the sphere is

$$p|_{r=R} = p_0 - \rho g R \cos \theta - \frac{3}{2} \frac{\mu v_\infty}{R} \cos \theta \tag{2.6–6}$$

We now substitute this expression into the integral in Eq. 2.6–5. The integral involving p_0 vanishes identically, the integral involving $-\rho g R \cos \theta$ gives the buoyant force of the fluid on the solid, and the integral involving the velocity gives the "form drag"; hence we get finally

$$F_n = \tfrac{4}{3}\pi R^3 \rho g + 2\pi\mu R v_\infty \tag{2.6–7}$$

Integration of the Tangential Force

At each point on the surface there is also a shear stress acting tangentially. This stress, $-\tau_{r\theta}$, is the force in the θ-direction acting on a unit area of sphere surface. The z-component of this force per unit area is $(-\tau_{r\theta})(-\sin\theta)$. We now multiply it by $R^2 \sin\theta\, d\theta\, d\phi$ and integrate over the sphere surface to get the resultant force in the z-direction:

$$F_t = \int_0^{2\pi}\int_0^{\pi} (+\tau_{r\theta}|_{r=R} \sin\theta) R^2 \sin\theta\, d\theta\, d\phi \tag{2.6–8}$$

The shear stress distribution at the surface of the sphere from Eq. 2.6–1 is

$$\tau_{r\theta}|_{r=R} = \frac{3}{2}\frac{\mu v_\infty}{R}\sin\theta \tag{2.6–9}$$

Substitution of this expression into the integral in Eq. 2.6–8 gives the "friction drag"

$$F_t = 4\pi\mu R v_\infty \tag{2.6–10}$$

Hence the total force F of the fluid on the sphere is given by the sum of Eqs. 2.6–7 and 2.6–10:

$$F = \underset{\substack{\text{(buoyant}\\ \text{force)}}}{\tfrac{4}{3}\pi R^3 \rho g} + \underset{\text{(form drag)}}{2\pi\mu R v_\infty} + \underset{\text{(friction drag)}}{4\pi\mu R v_\infty} \tag{2.6–11}$$

or

$$F = \tfrac{4}{3}\pi R^3 \rho g + 6\pi\mu R v_\infty \tag{2.6–12}$$

The first term on the right side of Eq. 2.6–12 represents the buoyant force and the second results from the fluid motion about the sphere. It is convenient for later discussions to designate these two terms as F_s (the force exerted even if the fluid is stationary) and F_k (the force associated with the fluid movement, i.e., the "kinetic" contribution); for the problem at hand these forces are

$$F_s = \tfrac{4}{3}\pi R^3 \rho g \tag{2.6–13}$$

$$F_k = 6\pi\mu R v_\infty \tag{2.6–14}$$

Equation 2.6–14 is known as *Stokes's law*. It finds use in the motion of colloidal particles under the influence of an electric field, in the theory of sedimentation, and in the study of the movement of aerosol particles. Keep in mind that Stokes's law is valid up to a Reynolds number (based on the diameter of the sphere) of about 0.1; at Re = 1 Stokes's law predicts a drag force that is about 10 per cent too low. The behavior of the same system for higher Reynolds numbers is discussed in Chapter 6.

This problem points out the need for developing a more general formulation of fluid mechanics problems to cope with curved streamlines. Such a formulation is given in Chapter 3.

Example 2.6–1. Determination of Viscosity from Terminal Velocity of a Falling Sphere

Derive a relation that enables one to get the viscosity of a fluid by measuring the steady-state rate of fall of a sphere in the fluid.

Solution. If a sphere is allowed to fall from rest in a viscous fluid, it will accelerate until it reaches a constant ("terminal") velocity. When this state has been reached, the sum of all the forces acting on the sphere must be zero. The force of gravity on the solid acts in the direction of fall, and the buoyancy force and the force due to fluid motion act in the opposing direction:

$$\frac{4}{3}\pi R^3 \rho_s g = \frac{4}{3}\pi R^3 \rho g + 6\pi\mu v_t R \qquad (2.6\text{–}15)$$

Here R is the radius of the sphere, ρ_s is the density of the sphere, ρ is the density of fluid, and v_t is the "terminal velocity." By solving Eq. 2.6–15 for μ, we get

$$\mu = 2R^2(\rho_s - \rho)g/9v_t \qquad (2.6\text{–}16)$$

This result is valid only when $Dv_t\rho/\mu$ is less than about 0.1.

QUESTIONS FOR DISCUSSION

1. What is the definition of the first derivative and how is this definition used in connection with shell balances?

2. Contrast the dependence of τ_{rz} on r for the laminar flow of a viscous fluid in a tube with that in an annulus.

3. What is the Hagen-Poiseuille law and why is it important? Show that it is dimensionally consistent.

4. What is the Reynolds number? What are its dimensions?

5. For annular flow, is the surface of zero momentum flux closer to the inner or the outer wall?

6. What are the limitations placed on the derivation of the formula for the thickness of a falling film?

7. In §2.5 what is the physical significance of the four boundary conditions needed to determine the four integration constants?

8. What is the range of applicability of Stokes's law?

9. Would Stokes's law be expected to be valid for liquid droplets of A falling in an immiscible liquid medium made up of B?

10. Would Stokes's law be expected to hold for very tiny particles falling in air if the diameter of the particles is of the order of the mean free path of the air molecules?

11. In §2.6 F_s and F_k should really be treated as vector quantities. How should Eq. 2.6–12 be modified if the direction of fluid flow is not directly opposed to the direction of gravitational acceleration?

12. How does one select the form and orientation of volume element used to make a shell balance?

13. At what point in the derivation in §2.3 would one have to begin to make changes if (*a*) the coefficient of viscosity were a function of r (because of nonisothermal conditions, for example), (*b*) the fluid were non-Newtonian?

14. Discuss the difficulties of measurement involved in determining absolute viscosities with the aid of the Hagen-Poiseuille formula. Consider the relative effects on the result of a 1 per cent error in the various measurements.

15. Two immiscible liquids A and B are flowing in laminar flow between two parallel plates. Would there ever be the possibility that the velocity profiles would be of the following form? (Explain briefly the reasons for your answer.)

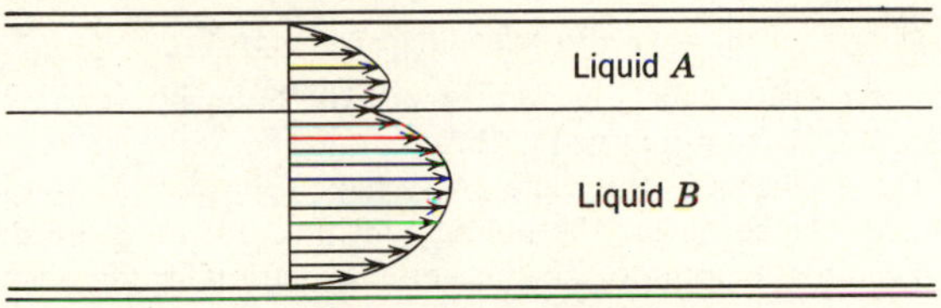

16. What is the terminal velocity of a spherical colloidal particle having a charge e in an electric field of strength $\mathscr{E}$? (Assume Stokes's law.)

PROBLEMS

2.A₁ Determination of Capillary Radius by Flow Measurements

One method of determining the radius of a capillary tube is to measure the rate of flow of a viscous fluid through the tube. Find the radius of a capillary from the following flow data:

Length of capillary	$= 50.02$ cm
Kinematic viscosity of fluid	$= 4.03 \times 10^{-5}$ m² sec⁻¹
Density of fluid	$= 0.9552 \times 10^{3}$ kg m⁻³
Pressure drop across (horizontal) capillary tube	$= 4.829 \times 10^{5}$ newtons m⁻²
	$= 4.766$ atm
Mass rate of flow through tube	$= 2.997 \times 10^{-3}$ kg sec⁻¹

What is a major drawback to this method? Suggest some other means for determining the radii of capillary tubes. *Answer:* 0.7512 mm

2.B₁ Volume Rate of Flow Through an Annulus

A horizontal annulus is 27 ft long. The outside radius of the inner cylinder is 0.495 in.; the inside radius of the outer cylinder is 1.1 in. A 60 per cent aqueous solution of sucrose ($C_{12}H_{22}O_{11}$) is to be pumped through the annulus at 20° C. At this temperature the fluid density ρ is 80.3 lb ft^{-3} and its viscosity is 136.8 lb$_m$ ft^{-1} hr^{-1}. What is the volume rate of flow when the impressed pressure drop is 5.39 psi? *Answer:* 0.108 ft³ sec⁻¹

2.C₁ Loss of Catalyst Particles in a Stack Gas

a. Estimate the maximum diameter of a microspherical catalyst that could be lost in the stack gas of a fluid cracking unit under the following conditions:

Gas velocity at axis of stack	= 1.0 ft sec^{-1} (vertically upward)
Gas viscosity	= 0.026 cp
Gas density	= 0.045 lb ft^{-3}
Density of single catalyst particle	= 1.2 gm cm^{-3}

Express the result in microns (1 micron = 10^{-6} m).

b. Is it permissible to use Stokes's law in (*a*)?

Answer: D_{max} = 110 microns; Re = 0.93

2.D₂ Flow of Falling Film—Alternate Derivations

a. Perform the derivation of the velocity profile and the average velocity by placing the center of coordinates in such a way that $\bar{x}$ is measured away from the wall (that is $\bar{x} = 0$ is at the wall, $\bar{x} = \delta$ is at the edge of the film). Show that the velocity distribution is then given by

$$v_z = (\rho g \delta^2/\mu) \cos \beta[(\bar{x}/\delta) - \tfrac{1}{2}(\bar{x}/\delta)^2] \tag{2.D-1}$$

and that the average velocity is that given in Eq. 2.2–18. Show how one can get the velocity distribution in Eq. 2.D–1 from that in Eq. 2.2–16.

b. In each of the problems in this chapter we have followed the following procedure: (i) derive a first-order equation for the momentum flux, (ii) integrate this equation, (iii) insert Newton's law into the result to get a first-order equation for the velocity, (iv) integrate the latter to get the velocity distribution. An alternative procedure is (i) derive a first-order equation for the momentum flux, (ii) insert Newton's law into this equation to get a second-order differential equation for the velocity, (iii) integrate the latter to get the velocity distribution. Apply this alternative procedure by substituting Eq. 2.2–12 into Eq. 2.2–8 and continuing as directed until the velocity distribution has been obtained.

2.E₂ Laminar Flow in a Narrow Slit

A viscous fluid is in laminar flow in a slit formed by two parallel walls a distance $2B$ apart. Make a differential momentum balance and obtain the expressions for the distributions of momentum flux and velocity (see Fig. 2. E):

$$\tau_{xz} = \left(\frac{\mathscr{P}_0 - \mathscr{P}_L}{L}\right)x \tag{2.E-1}$$

$$v_z = \frac{(\mathscr{P}_0 - \mathscr{P}_L)B^2}{2\mu L}\left[1 - \left(\frac{x}{B}\right)^2\right] \tag{2.E-2}$$

in which $\mathscr{P} = p + \rho g h = p - \rho g z$. What is the ratio of average to maximum velocity in the slit? Obtain the analog of the Hagen-Poiseuille law for the slit.

Answer: $\langle v_z \rangle = \frac{2}{3} v_{z,\max}$; $\quad Q = \dfrac{2}{3}\dfrac{(\mathscr{P}_0 - \mathscr{P}_L)B^3 W}{\mu L}$

2.F₂ Interrelation of Slit and Annulus Formulas

When an annulus is very thin, it may to a good approximation be considered as a thin slit. Then the results of Problem 2.E can be applied. For example, the volume rate of flow in an annulus with outer wall of radius R and inner wall of radius $(1 - \epsilon)R$, where ϵ is small, may be obtained from Problem 2.E by setting $2B$ equal to ϵR and W equal to $2\pi(1 - \frac{1}{2}\epsilon)R$ and thereby obtaining

$$Q = \frac{\pi(\mathscr{P}_0 - \mathscr{P}_L)R^4\epsilon^3}{6\mu L}(1 - \tfrac{1}{2}\epsilon) \tag{2.F-1}$$

Show that this same result may be obtained from Eq. 2.4–16 by setting κ equal to $1 - \epsilon$

everywhere in the formula and expanding the expression for Q in powers of ϵ. This operation involves using the Taylor series

$$\ln(1 - \epsilon) = -\epsilon - \tfrac{1}{2}\epsilon^2 - \tfrac{1}{3}\epsilon^3 - \tfrac{1}{4}\epsilon^4 - \cdots \tag{2.F-2}$$

and then performing a long division. The first term in the resulting series will be Eq. 2.F–1. (*Hint:* In the derivation use the first *four* terms of the Taylor series in Eq. 2.F–2.)

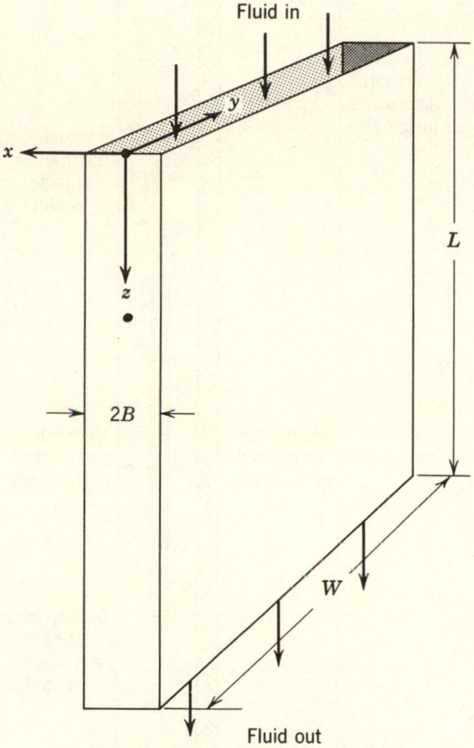

Fig. 2.E. Flow through a slit.

2.G₂ Laminar Flow of Falling Film on Outside of a Circular Tube

In a gas absorption experiment a viscous fluid flows upward through a small circular tube and then downward on the outside. (See Fig. 2.G.) Set up a momentum balance over a shell of thickness Δr in the film, as shown in the figure. Note that the "Momentum in" and "Momentum out" arrows are always taken in the positive r-direction in setting up the balance, even though in this case the momentum turns out to be flowing in the negative r-direction.

a. Show that the velocity distribution in the falling film (neglecting end effects) is

$$v_z = \frac{\rho g R^2}{4\mu}\left[1 - \left(\frac{r}{R}\right)^2 + 2a^2 \ln\left(\frac{r}{R}\right)\right] \tag{2.G-1}$$

b. Obtain an expression for the volume rate of flow in the film.

c. Show that the result in (*b*) simplifies to Eq. 2.2–19 if the film thickness is very small.

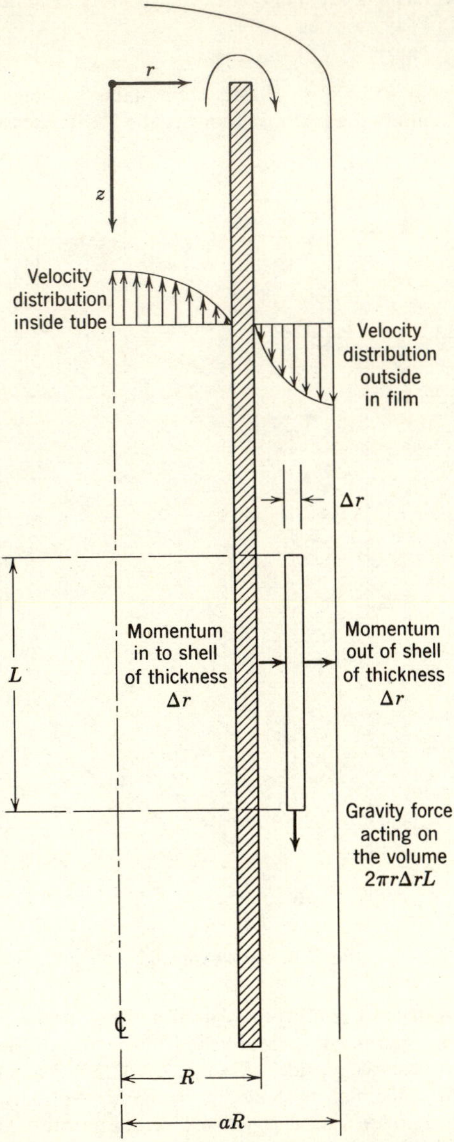

Fig. 2.G. Velocity distributions and momentum balance for falling film on outside of a circular tube.

2.H₂ Non-Newtonian Flow in a Tube

a. Derive the analog of the Hagen-Poiseuille formula for the Ostwald-de Waele (power-law) model. In making the derivation, one must first get rid of the absolute value sign. Because in tube flow dv_z/dr is everywhere negative the power law for this problem becomes

$$\tau_{rz} = -m\left|\frac{dv_z}{dr}\right|^{n-1}\frac{dv_z}{dr} = m\left|-\frac{dv_z}{dr}\right|^{n-1}\left(-\frac{dv_z}{dr}\right) = m\left(-\frac{dv_z}{dr}\right)^n \tag{2.H-1}$$

Explain carefully the manipulations in Eq. 2.H–1.

b. Derive an expression for the volume rate of flow through a tube for an Ellis fluid (see Eq. 1.2–5):

$$-\frac{dv_z}{dr} = \varphi_0\tau_{rz} + \varphi_1|\tau_{rz}|^{\alpha-1}\tau_{rz} \tag{2.H-2}$$

2.I₂ Flow of a Bingham Fluid from a Circular Tube[1]

A vertical tube is filled with a Bingham fluid and a plate is held over the lower end. (See Fig. 2.I.) When the plate is removed, the fluid may or may not flow out of the tube by gravity. Explain this and establish a criterion for flow in such an experiment.

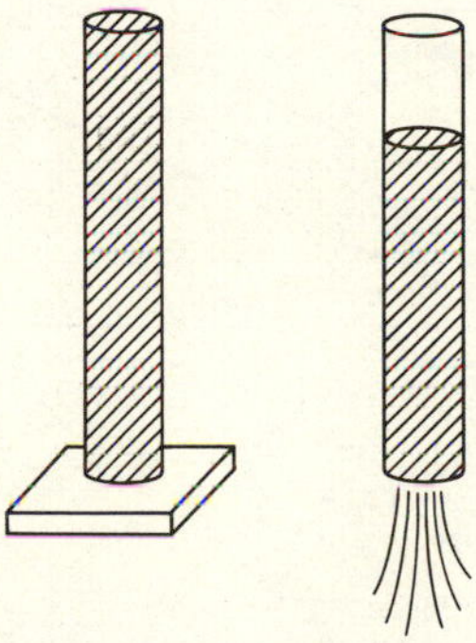

Fig. 2.I. Flow of a Bingham fluid from a circular tube.

2.J₂ Annular Flow with Inner Cylinder Moving Axially

Consider the system pictured in Fig. 2.J, in which the cylindrical rod is being moved with a velocity V. The rod and the cylinder are coaxial. Find the steady-state velocity

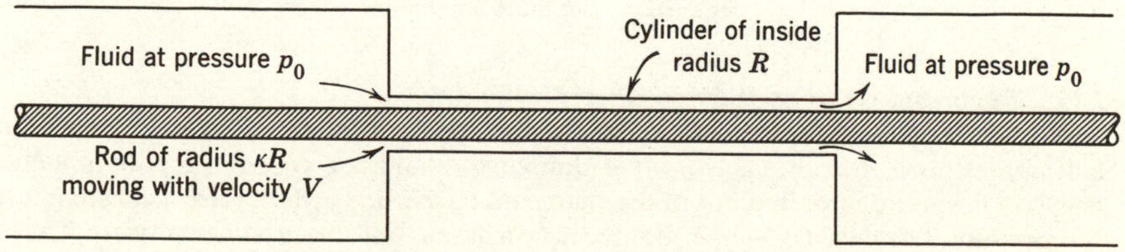

Fig. 2.J. Annular flow with inner cylinder moving axially.

[1] Suggested by Prof. H. Kramers, Technische Hogeschool (Delft).

distribution and the volume rate of flow. Problems of this kind arise in describing the performance of wire-coating dies.[2]

$$\textit{Answer:}\quad \frac{v_z}{V} = \frac{\ln (r/R)}{\ln \kappa}; \qquad Q = \frac{\pi R^2 V}{2}\left(\frac{(1-\kappa^2)}{\ln (1/\kappa)} - 2\kappa^2\right)$$

2.K$_2$ Non-Newtonian Film Flow

Derive a formula for the thickness of a film of a Bingham fluid falling down a vertical flat surface at a rate Γ (g sec^{-1} per unit width of wall).

2.L$_3$ Analysis of Capillary Flowmeter

Determine the rate of flow (in lb$_m$ hr^{-1}) through the capillary flowmeter shown in Fig. 2.L. The fluid flowing in the inclined tube is water at 20° C, and the manometer fluid is carbon tetrachloride (CCl_4) with density 1.594 gm cm^{-3}. The capillary diameter is 0.010 in. (Note that measurements of H and L are sufficient to calculate the flow rate; that is, θ need not be measured. Why?)

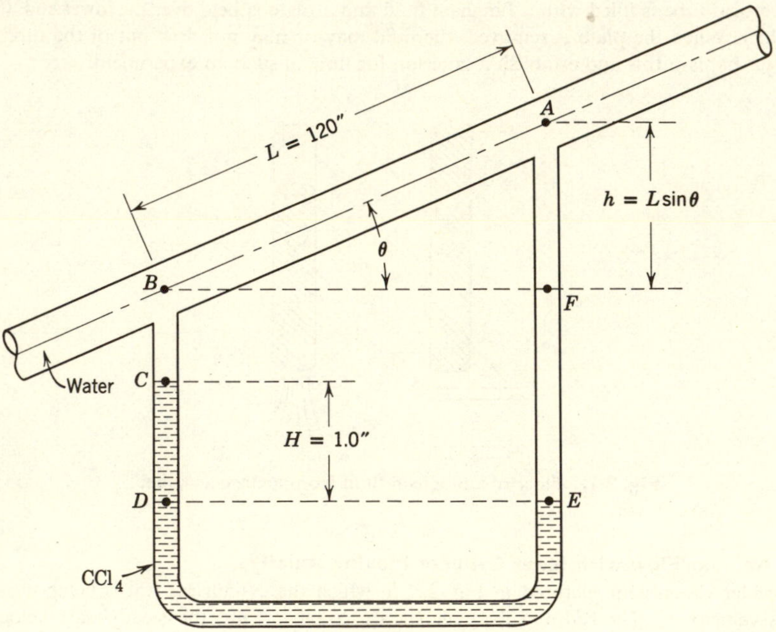

Fig. 2.L. Capillary flowmeter.

2.M$_3$ Performance of an Electric Dust Collector

A dust precipitator consists of a pair of oppositely charged plates between which dust-laden gases flow. (See Fig. 2.M.) It is desired to establish a criterion for the minimum length of the precipitator in terms of the charge on the particle e, the electric field strength $\mathscr{E}$, the pressure difference $(p_0 - p_L)$, the particle mass m, and the gas viscosity μ. That is,

[2] J. B. Paton, P. H. Squires, W. H. Darnell, F. M. Cash, and J. F. Carley, *Processing of Thermoplastic Materials*, E. C. Bernhardt (Ed.), Reinhold, New York (1959), Chapter 4, pp. 209–301.

for what length L will the smallest particle present (mass m) reach the bottom plate just before it has a chance to be swept out of the channel? Assume that the flow between the two plates is laminar so that Eq. 2.E–2 describes the velocity distribution. Assume also that the particle velocity in the z-direction is the same as the fluid velocity in the z-direction; assume further that the Stokes drag on the sphere as well as the gravity force acting on the particle as it is accelerated in the $-x$-direction can be neglected. Discuss the probable error in the neglect of the Stokes drag. *Answer:* $L_{\min} = [64(p_0 - p_L)^2 B^5 m / 225 \mu^2 e \mathscr{E}]^{1/4}$

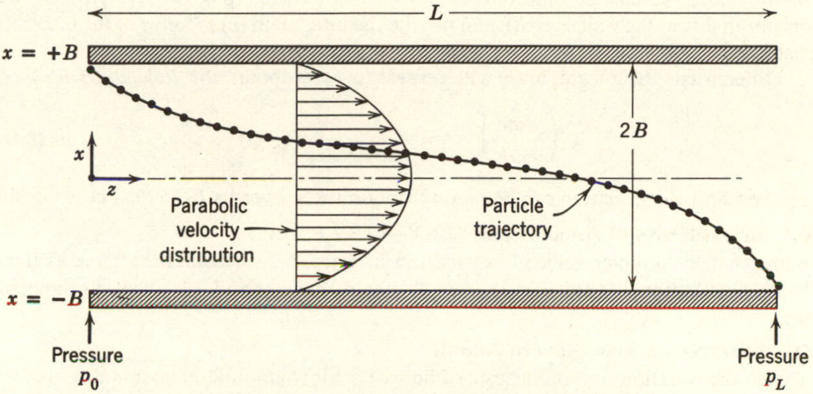

Fig. 2.M. Particle trajectory in electric dust collector.

2.N$_3$ Tube Flow with Slip at Wall

Obtain a modification of the Hagen-Poiseuille law by assuming fluid slip at the cylinder wall. That is, instead of assuming that $v_z = 0$ at $r = R$, use the boundary condition that

$$\beta v_z = -\mu \frac{dv_z}{dr} \qquad \text{at } r = R \tag{2.N–1}$$

in which β is the "coefficient of sliding friction." (What is the physical significance of $\beta = \infty$?)

In most fluid-flow problems slip is not important.[3] However, the solution for flow around spheres with slip[4] has been used in connection with the "hydrodynamic theories" of diffusion.[5] In addition, slip appears to be important in some non-Newtonian flow problems.[6]

2.O$_4$ Derivation of the Rabinowitsch Equation

By means of the Rabinowitsch equation,[7,8] one can derive a plot of momentum flux versus velocity gradient (see Fig. 1.2–1) for any fluid from experimental data on pressure

[3] H. Lamb, *Hydrodynamics*, Dover, New York (1945), p. 576.

[4] *Ibid.*, pp. 601 *et seq.*

[5] R. B. Bird, "Theory of Diffusion." In *Advances in Chemical Engineering*, T. B. Drew and J. W. Hoopes, Jr. (Eds.), Academic Press, New York (1956), Vol. 1, pp. 195–196.

[6] J. G. Oldroyd, "Non-Newtonian Flow of Liquids and Solids." In *Rheology*, F. R. Eirich (Ed.), Academic Press, New York (1956), Vol. 1, pp. 663–664.

[7] B. Rabinowitsch, *Z. physik. Chemie*, **A145**, 1–26 (1929).

[8] J. G. Oldroyd, "Non-Newtonian Flow of Liquids and Solids." In *Rheology*, F. R. Eirich (Ed.), Academic Press, New York (1956), Chapter 16, Vol. 1, pp. 662–666.

drop versus flow rate for laminar, isothermal flow through circular tubes. The only assumptions that need to be made are that the fluid is homogeneous throughout and that the fluid does not slip at the tube wall.

a. Show that the integral for the volume rate of flow may be integrated by parts to give

$$Q = -\pi \int_0^R \frac{dv_z}{dr} r^2 \, dr \tag{2.O-1}$$

b. Introduce the change of variable $r/R = \tau_{rz}/\tau_R$ (where $\tau_R = (\mathscr{P}_0 - \mathscr{P}_L)R/2L$ is the momentum flux at the wall, $r = R$), and rewrite the integral in (*a*) in terms of the integration variable τ_{rz}.

c. Differentiate the integral in (*b*) with respect to τ_R to obtain[9] the *Rabinowitsch equation*

$$\left(-\frac{dv_z}{dr}\right)_{r=R} = \frac{1}{\pi R^3 \tau_R^2} \frac{d}{d\tau_R} (\tau_R^3 Q) \tag{2.O-2}$$

Explain how this equation can be used to obtain the τ_{rz} versus $(-dv_z/dr)$ curve desired.

2.P$_4$ Interrelation of Annular and Slit Velocity Profiles

Show that the annular velocity distribution in Eq. 2.4–13 becomes the same as the slit velocity distribution (see Eq. 2.E–2) when R is held constant and κ is allowed to approach unity.

2.Q$_4$ Non-Newtonian Flow in Annuli

Derive the equation for volume rate of flow of a Bingham fluid in an annulus, using the notation introduced in §2.4 and in Example 2.3–2. Introduce the following dimensionless quantities:

$$T = 2\tau_{rz}L/(\mathscr{P}_0 - \mathscr{P}_L)R \quad = \text{dimensionless momentum flux}$$
$$T_0 = 2\tau_0 L/(\mathscr{P}_0 - \mathscr{P}_L)R \quad = \text{dimensionless rheological parameter}$$
$$\phi = (2\mu_0 L/(\mathscr{P}_0 - \mathscr{P}_L)R^2)v_z = \text{dimensionless velocity}$$
$$\xi = r/R \quad = \text{dimensionless radial coordinate}$$

Show that the momentum-flux distribution and the rheological law may then be written

$$T = \xi - \lambda^2\xi^{-1} \tag{2.Q-1}$$

$$T = \pm T_0 - \frac{d\phi}{d\xi} \tag{2.Q-2}$$

Show that the bounds on the plug flow region λ_+ and λ_- are given by

$$\pm T_0 = \lambda_\pm - \frac{\lambda^2}{\lambda_\pm} \tag{2.Q-3}$$

and that λ is just the geometric mean of λ_+ and λ_-. Obtain the velocity distribution for the three ranges:

ϕ_- for the range $\kappa \leqslant \xi \leqslant \lambda_-$

ϕ_0 for the range $\lambda_- \leqslant \xi \leqslant \lambda_+$

ϕ_+ for the range $\lambda_+ \leqslant \xi \leqslant 1$

[9] Use the "Leibnitz formula" for differentiating an integral:

$$\frac{d}{dt}\int_{a_1(t)}^{a_2(t)} f(x, t)\, dx = \int_{a_1(t)}^{a_2(t)} \frac{\partial f}{\partial t}\, dx + \left(f(a_2, t)\frac{da_2}{dt} - f(a_1, t)\frac{da_1}{dt}\right)$$

Then integrate over the velocity distribution to obtain the volume rate of flow:[10]

$$Q = \frac{\pi(\mathscr{P}_0 - \mathscr{P}_L)R^4}{8\mu_0 L}\left[(1-\kappa^4) - 2\lambda^2(1-\kappa^2) - \tfrac{4}{3}(1+\kappa^3)T_0 + \tfrac{1}{3}(4\lambda^2 + T_0^2)^{3/2}T_0\right] \tag{2.Q-4}$$

How is λ determined? Is the momentum-flux distribution for non-Newtonian flow in annuli the same as that for Newtonian flow?

2.R$_4$ Drainage of Liquids[11]

Let it be desired to find out how much liquid clings to the inside surface of a large vessel when it is drained. The situation near the wall of the vessel is shown in Fig. 2.R. The local film thickness is a function of both z and t.

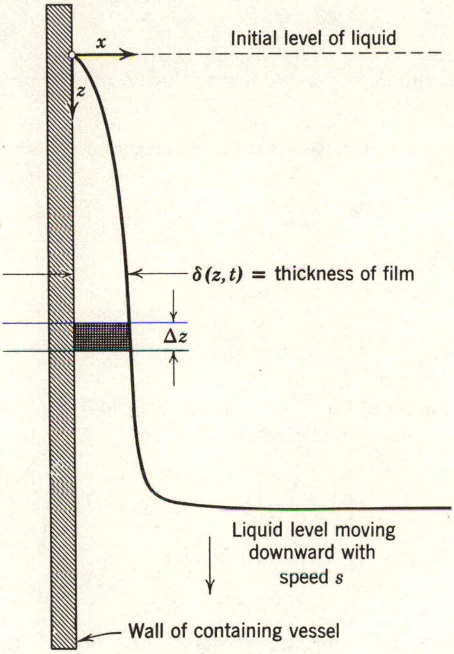

Fig. 2.R. Clinging of a viscous fluid to wall of vessel during draining.

a. Make an unsteady-state mass balance on a portion of the film between z and $z + \Delta z$ and show that

$$\frac{\partial}{\partial z}\langle v_z\rangle\delta = -\frac{\partial \delta}{\partial t} \tag{2.R-1}$$

b. Use Eq. 2.2-18 to obtain the following first-order partial differential equation for $\delta(z, t)$:

$$\frac{\partial \delta}{\partial t} + \frac{\rho g}{\mu}\delta^2\frac{\partial \delta}{\partial z} = 0 \tag{2.R-2}$$

What assumptions are embodied in Eq. 2.R-2?

[10] A. G. Fredrickson and R. B. Bird, *Ind. Eng. Chem.*, **50**, 347–352 (1958); a solution for the power-law model is also given in this article.

[11] For a further discussion of this problem, see J. J. van Rossum, *Appl. Sci. Research*, **A7**, 121–144 (1958).

c. Show that Eq. 2.R–2 may be solved to give

$$\delta = \sqrt{\frac{\mu}{\rho g}\frac{z}{t}} \tag{2.R–3}$$

What restrictions must be placed on this result?

2.S$_3$ Flow of Non-Newtonian Greases in Tubes

In a study of the flow of non-Newtonian greases in circular tubes the following model has been proposed:[12]

$$\tau_{rz} = a\left(-\frac{dv_z}{dr}\right) + b\left(-\frac{dv_z}{dr}\right)^n$$

$$= a\gamma + b\gamma^n \tag{2.S–1}$$

Show that the expression for the volume rate of flow Q may be obtained in the following way:

a. Show that the general expression for the volume rate of flow may be integrated by parts twice to obtain

$$Q = \frac{\pi R^3 \gamma_R}{3} - \frac{\pi}{3}\int_0^{\gamma_R} r^3\, d\gamma \tag{2.S–2}$$

in which $\gamma_R = (-dv_z/dr)|_{r=R}$.

b. Show that r may be written as a function of γ:

$$r = \frac{R}{\tau_R}(a\gamma + b\gamma^n) \tag{2.S–3}$$

c. Show that substitution of Eq. 2.S–3 into the integral in Eq. 2.S–2 and application of the appropriate boundary condition gives finally

$$Q = \frac{\pi R^3 a^3 \gamma_R^{\,4}}{4\tau_R^{\,3}}\left\{1 + 4\left[\left(\frac{n+2}{n+3}\right)X + \left(\frac{2n+1}{2n+2}\right)X^2 + \left(\frac{n}{3n+1}\right)X^3\right]\right\} \tag{2.S–4}$$

where $X = (b/a)(\gamma_R^{n-1})$ and γ_R is an implicit function of τ_R:

$$\tau_R = a\gamma_R + b\gamma_R^{\,n} \tag{2.S–5}$$

2.T$_3$ An Integral Expression for the Velocity Distribution in a Tube

You have been sent a paper to review by the editor of a professional journal. This paper concerns heat transfer in tube flow. The authors state that because they are dealing with nonisothermal flow they must have a "general" expression for the velocity distribution, which may be used even when the viscosity of the fluid is a function of temperature (hence of position). The authors then state that a "general expression for velocity distribution for flow in a tube" is

$$\frac{v_z}{\langle v_z\rangle} = \int_y^1 \frac{y}{\mu}\,dy \bigg/ \int_0^1 \frac{y^3}{\mu}\,dy \tag{2.T–1}$$

in which $\langle v_z\rangle$ is the average flow velocity, y is the reduced radial coordinate (that is, r/R), and μ is the viscosity (which may vary with y). The authors give no derivation of their formula, nor do they state where it may be found in the literature. As reviewer of their paper you feel obliged to (*a*) derive the formula they have given, (*b*) state the restrictions that have been implicitly placed on their formula.

[12] A. W. Sisko, *Ind. Eng. Chem.*, **50**, 1789–1792 (1958).

CHAPTER 3

The Equations of Change for Isothermal Systems

In Chapter 2 velocity distributions were determined for several simple flow systems by the application of shell momentum balances. These velocity distributions were then used to calculate other quantities, such as average velocity and drag force. The shell balance approach was used to acquaint the beginner with the application of the principle of conservation of momentum to viscous flow problems. It is not, however, necessary to formulate a momentum balance whenever one begins to work on a new flow problem. In fact, it is seldom desirable to do so. It is quicker, easier, and safer to start with the equations of conservation of mass and momentum in general form and to simplify these equations to fit the problem at hand. These two equations describe all problems of the viscous flow of a pure isothermal fluid. For nonisothermal fluids, and for multicomponent fluid mixtures, additional equations are needed to describe the conservation of energy (Chapter 10) and the conservation of individual chemical species (Chapter 18). These various conservation equations are sometimes called the "equations of change," inasmuch as they describe the change of velocity, temperature, and concentration with respect to time and position in the system.

In §3.1 the "equation of continuity" is developed by applying the law of conservation of mass to a small volume element within a flowing fluid.

The principle of conservation of mass has already been tacitly used throughout Chapter 2; for example, it justified the assumption that velocity is independent of axial distance in the Hagen-Poiseuille problem. We shall see that for complex systems a more general statement of this principle is of value.

In §3.2 we introduce the second equation of change, the "equation of motion," which is a generalization of the momentum balance of Chapter 2. This is a most important equation. Once we have developed it, we can, with some help from the equation of continuity, solve all of the problems of Chapter 2 and many that are much more complicated.

In §3.3 we use the equation of motion to derive an expression describing the interconversion of the various forms of mechanical energy in a moving fluid. This equation is particularly useful for describing the irreversible degradation of mechanical energy into thermal energy, which accompanies all real flow processes. It is also the basis for the important macroscopic mechanical energy balance, or Bernoulli equation, discussed in Chapter 7.

In the first three sections the derivations are given in rectangular coordinates. For many problems, cylindrical or spherical coordinates are more useful. In §3.4 the question of curvilinear coordinates is discussed briefly, and many important relations are summarized in all three coordinate systems. The reader will find that this tabulation reduces the art of setting-up viscous flow problems to what almost amounts to a simple "cookbook" procedure. The examples in §3.5 illustrate the method as applied to the solution of some laminar flow problems.

In §3.6 the material in the preceding five sections is extended to non-Newtonian flow. The principal point discussed here is the proper way of writing the momentum flux for the various non-Newtonian models so that it can be transformed into curvilinear coordinates.

In §3.7, the last in the chapter, we present the equations of change in terms of dimensionless variables. By writing the equations in this manner, we can collect the "scale factors" (e.g., system size, average fluid velocity, and fluid properties) into a small number of dimensionless ratios that are useful for characterizing fluid systems. In this chapter we show how these dimensionless ratios can be used to prepare empirical small-scale model studies of systems too complex to permit exact analysis. In Chapter 6 we extend these ideas to the empirical correlation of drag force in complex systems.

In this chapter vector and tensor notation are used occasionally, primarily for the purpose of abbreviating otherwise lengthy expressions.[1] The beginner

[1] An introduction to vector operations can be found in some undergraduate calculus texts, such as G. B. Thomas *Calculus and Analytic Geometry*, Addison-Wesley, Reading, Mass. (1953), Second Edition. In our text we use *lightface italic* symbols for scalars, ***boldface italic*** symbols for vectors, and **boldface Greek** for tensors. In addition, dot-product operations enclosed in ( ) are scalars and operations enclosed in [ ] are vectors.

will find that a knowledge of vector and tensor mathematics is not needed to utilize the summary in §3.4 for problem-solving purposes. The advanced student will find Appendix A helpful toward an understanding of some of the details of vector and tensor manipulations.

In many ways Chapter 3 is the most important chapter in the book, for it paves the way for everything that follows. The reader who masters its contents before proceeding will be rewarded for his efforts.

Before taking up the main business of the chapter, we pause briefly to make a few comments regarding three kinds of time derivatives used in the text. We might illustrate them with a homely example—namely the problem of reporting the concentration of fish in the Kickapoo River. Because the fish are moving, the fish concentration c will be a function of position (x, y, z) and time (t).

The Partial Time Derivative, $\partial c/\partial t$

Suppose we stand on a bridge and note how the concentration of fish just below us changes with time. We are observing then how the concentration changes with time at a *fixed* position in space. Hence by $\partial c/\partial t$ we mean the "partial of c with respect to t, holding x, y, z constant."

Total Time Derivative, dc/dt

Suppose now that instead of standing on the bridge we get in a motorboat and speed around on the river, sometimes going upstream, sometimes across the current, and perhaps sometimes downstream. If we report the change of fish concentration with respect to time, the numbers we report must also reflect the motion of the boat. The total time derivative is given by

$$\frac{dc}{dt} = \frac{\partial c}{\partial t} + \frac{\partial c}{\partial x}\frac{dx}{dt} + \frac{\partial c}{\partial y}\frac{dy}{dt} + \frac{\partial c}{\partial z}\frac{dz}{dt} \tag{3.0–1}$$

in which dx/dt, dy/dt, and dz/dt are the components of the velocity of the boat.

Substantial Time Derivative, Dc/Dt

Suppose that we get into a canoe, and, not feeling energetic, we simply float along counting fish. Now the velocity of the observer is just the same as the velocity of the stream $\boldsymbol{v}$. When we report the change of fish concentration with respect to time, the numbers depend on the local stream velocity. This derivative is a special kind of total time derivative and is called the "substantial derivative" or sometimes (more logically) the "derivative following the motion." It is related to the partial time derivative as follows:

$$\frac{Dc}{Dt} = \frac{\partial c}{\partial t} + v_x\frac{\partial c}{\partial x} + v_y\frac{\partial c}{\partial y} + v_z\frac{\partial c}{\partial z} \tag{3.0–2}$$

in which v_x, v_y, and v_z are the components of the local fluid velocity $\boldsymbol{v}$.

The reader should thoroughly master the physical meaning of these three derivatives. Remember that $\partial c/\partial t$ is the derivative at a fixed point in space and Dc/Dt is a derivative computed by an observer floating downstream with the fluid.

§3.1 THE EQUATION OF CONTINUITY

This equation is developed by writing a mass balance over a *stationary* volume element $\Delta x\ \Delta y\ \Delta z$ through which the fluid is flowing (see Fig. 3.1–1):

$$\begin{Bmatrix} \text{rate of} \\ \text{mass} \\ \text{accumulation} \end{Bmatrix} = \begin{Bmatrix} \text{rate of} \\ \text{mass} \\ \text{in} \end{Bmatrix} - \begin{Bmatrix} \text{rate of} \\ \text{mass} \\ \text{out} \end{Bmatrix} \tag{3.1–1}$$

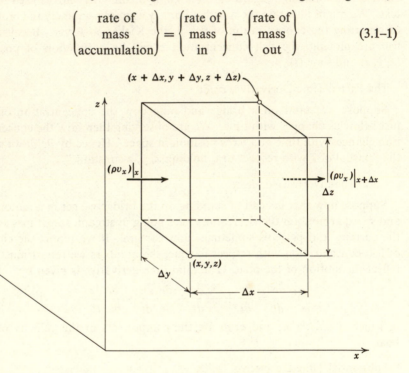

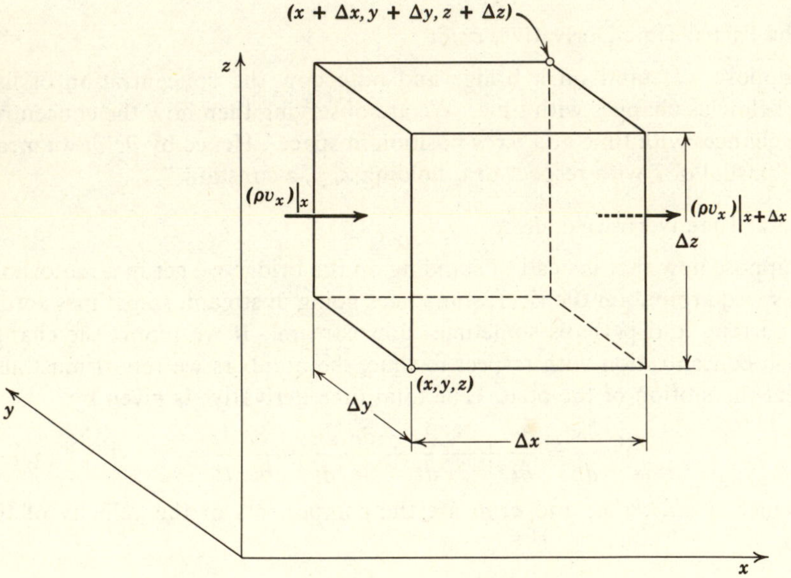

Fig. 3.1–1. Region of volume $\Delta x\ \Delta y\ \Delta z$ *fixed in space* through which a fluid is flowing.

We begin by considering the pair of faces perpendicular to the x-axis. The rate of mass in through the face at x is $(\rho v_x)|_x\ \Delta y\ \Delta z$, and the rate of mass out through the face at $x + \Delta x$ is $(\rho v_x)|_{x+\Delta x}\ \Delta y\ \Delta z$. Similar expressions may be written for the other two pairs of faces. The rate of mass accumulation within the volume element is $(\Delta x\ \Delta y\ \Delta z)(\partial \rho/\partial t)$. The mass balance then becomes

$$\begin{aligned} \Delta x\,\Delta y\,\Delta z\,\frac{\partial \rho}{\partial t} &= \Delta y\,\Delta z[(\rho v_x)|_x - (\rho v_x)|_{x+\Delta x}] + \Delta x\,\Delta z[(\rho v_y)|_y - (\rho v_y)|_{y+\Delta y}] \\ &\quad + \Delta x\,\Delta y[(\rho v_z)|_z - (\rho v_z)|_{z+\Delta z}] \end{aligned} \tag{3.1–2}$$

By dividing this entire equation by $(\Delta x\ \Delta y\ \Delta z)$ and taking the limit as these dimensions approach zero, we get

$$\frac{\partial \rho}{\partial t} = -\left(\frac{\partial}{\partial x}\rho v_x + \frac{\partial}{\partial y}\rho v_y + \frac{\partial}{\partial z}\rho v_z\right) \tag{3.1–3}$$

This is the *equation of continuity*, which describes the rate of change of density at a fixed point resulting from the changes in the mass velocity vector $\rho \boldsymbol{v}$. We may write Eq. 3.1–3 more conveniently in vector symbolism:[1]

$$\frac{\partial \rho}{\partial t} = -(\nabla \cdot \rho \boldsymbol{v}) \tag{3.1–4}$$

Here $(\nabla \cdot \rho \boldsymbol{v})$ is called the "divergence" of $\rho \boldsymbol{v}$, sometimes written as div $\rho \boldsymbol{v}$. Note that the vector $\rho \boldsymbol{v}$ is the mass flux, and its divergence has a simple significance: it is the net rate of mass efflux per unit volume. Then Eq. 3.1–4 simply states that the rate of increase of the density within a small volume element fixed in space is equal to the net rate of mass influx to the element divided by its volume.

It is frequently desirable to modify Eq. 3.1–3 by performing the indicated differentiation and collecting all derivatives of ρ on the left side:

$$\frac{\partial \rho}{\partial t} + v_x \frac{\partial \rho}{\partial x} + v_y \frac{\partial \rho}{\partial y} + v_z \frac{\partial \rho}{\partial z} = -\rho\left(\frac{\partial v_x}{\partial x} + \frac{\partial v_y}{\partial y} + \frac{\partial v_z}{\partial z}\right) \tag{3.1–5}$$

The left side of Eq. 3.1–5 is the substantial derivative of density, that is, the time derivative *for a path following the fluid motion*. Hence Eq. 3.1–5 may be abbreviated thus:

$$\frac{D\rho}{Dt} = -\rho(\nabla \cdot \boldsymbol{v}) \tag{3.1–6}$$

in which the operator D/Dt is that defined in Eq. 3.0–2. The equation of continuity in this form describes the rate of change of density as seen by an observer "floating along" with the fluid.

Remember that this equation in any form is simply a statement of conservation of mass. It should also be pointed out that the derivation can be performed for a volume element of any arbitrary shape; it is not necessary to restrict ourselves to a rectangular volume element as we have done here.

A very important special form of the equation of continuity, which we shall use subsequently, is that for a fluid of constant density for which

(incompressible fluid) $$(\nabla \cdot \boldsymbol{v}) = 0 \tag{3.1–7}$$

Of course, no fluid is truly incompressible, but very frequently in engineering

[1] Note that ∇ has the dimensions of reciprocal length!

practice the assumption of constant density results in considerable simplification and almost no error. Note that for Eq. 3.1–7 to be valid it is necessary only that ρ remain constant for a fluid element as it moves along a streamline, that is, that $D\rho/Dt$ be zero.

§3.2 THE EQUATION OF MOTION

For a volume element $\Delta x\ \Delta y\ \Delta z$, such as that used in the previous section, we write a momentum balance in this form:

$$\begin{Bmatrix}\text{rate of}\\ \text{momentum}\\ \text{accumulation}\end{Bmatrix} = \begin{Bmatrix}\text{rate of}\\ \text{momentum}\\ \text{in}\end{Bmatrix} - \begin{Bmatrix}\text{rate of}\\ \text{momentum}\\ \text{out}\end{Bmatrix} + \begin{Bmatrix}\text{sum of forces}\\ \text{acting on}\\ \text{system}\end{Bmatrix} \qquad (3.2\text{–}1)$$

Note that Eq. 3.2–1 is just an extension of Eq. 2.1–1 to unsteady-state systems. We may proceed then in much the same way as in Chapter 2.

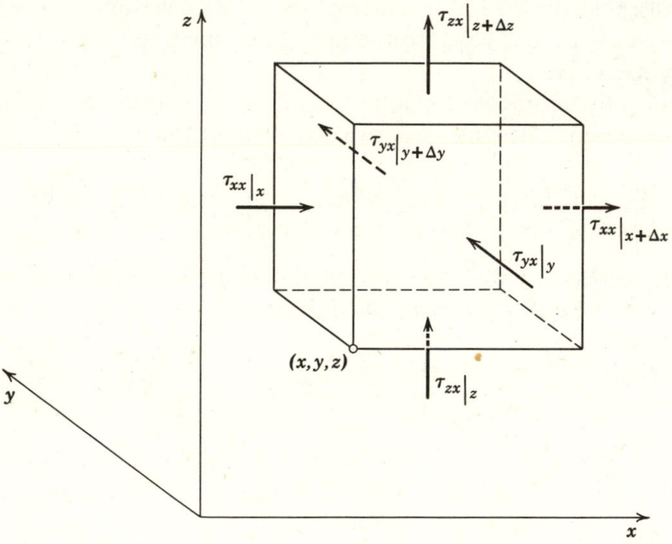

Fig. 3.2–1. Volume element $\Delta x\ \Delta y\ \Delta z$ with arrows indicating the direction in which the x-component of momentum is transported through the surfaces.

However, in addition to considering unsteady-state behavior, we will allow the fluid to move through all six faces of our volume element in any arbitrary direction, as in §3.1. It should be emphasized that Eq. 3.2–1 is a *vector* equation with components in each of the three coordinate directions x, y, and z. For simplicity, we begin by considering the x-component of each term in Eq. 3.2–1; the y- and z-components may be handled analogously.

First let us consider the rates of flow of the x-component of momentum into and out of the volume element shown in Fig. 3.2–1. Momentum flows into and out of the volume element by two mechanisms: by convection (that is, by virtue of the bulk fluid flow) and by molecular transfer (that is, by virtue of the velocity gradients).

The rate at which the x-component of momentum enters the face at x by convection is $\rho v_x v_x|_x \, \Delta y \, \Delta z$, and the rate at which it leaves at $x + \Delta x$ is $\rho v_x v_x|_{x+\Delta x} \, \Delta y \, \Delta z$. The rate at which it enters at y is $\rho v_y v_x|_y \, \Delta x \, \Delta z$. Similar expressions may be written for the other three faces. We see, then, that convective flow of x-momentum must be considered across all six faces and that the net convective x-momentum flow into the volume element is

$$\Delta y \, \Delta z(\rho v_x v_x|_x - \rho v_x v_x|_{x+\Delta x}) + \Delta x \, \Delta z(\rho v_y v_x|_y - \rho v_y v_x|_{y+\Delta y}) + \Delta x \, \Delta y(\rho v_z v_x|_z - \rho v_z v_x|_{z+\Delta z}) \quad (3.2\text{–}2)$$

Similarly, the rate at which the x-component of momentum enters the face at x by molecular transport is $\tau_{xx}|_x \, \Delta y \, \Delta z$, and the rate at which it leaves at $x + \Delta x$ is $\tau_{xx}|_{x+\Delta x} \, \Delta y \, \Delta z$. The rate at which it enters at y is $\tau_{yx}|_y \, \Delta x \, \Delta z$; similar expressions can be written for the remaining three faces (see Fig. 3.2–1). Keep in mind that τ_{yx} is the flux of x-momentum through a face perpendicular to the y-axis. When these six contributions are summed up, we get

$$\Delta y \, \Delta z(\tau_{xx}|_x - \tau_{xx}|_{x+\Delta x}) + \Delta x \, \Delta z(\tau_{yx}|_y - \tau_{yx}|_{y+\Delta y}) + \Delta x \, \Delta y(\tau_{zx}|_z - \tau_{zx}|_{z+\Delta z}) \quad (3.2\text{–}3)$$

Note that, as before, these momentum fluxes may be considered stresses. Thus τ_{xx} is the *normal* stress on the x-face, and τ_{yx} the x-directed *tangential* (or *shear*) stress on the y-face resulting from viscous forces.

In most cases the only important forces will be those arising from the fluid pressure p and the gravitational force per unit mass g. Clearly the resultant of these forces in the x-direction will be

$$\Delta y \, \Delta z(p|_x - p|_{x+\Delta x}) + \rho g_x \, \Delta x \, \Delta y \, \Delta z \quad (3.2\text{–}4)$$

Pressure in a moving fluid is defined by the equation of state $p = p(\rho, T)$ and is a scalar quantity.

Finally, the rate of accumulation of x-momentum within the element is $\Delta x \, \Delta y \, \Delta z \, (\partial \rho v_x / \partial t)$. We now substitute the foregoing expressions into Eq. 3.2–1. By dividing the entire resulting equation by $\Delta x \, \Delta y \, \Delta z$ and taking the limit as Δx, Δy, and Δz approach zero, we obtain the x-component of the equation of motion:

$$\frac{\partial}{\partial t} \rho v_x = -\left(\frac{\partial}{\partial x} \rho v_x v_x + \frac{\partial}{\partial y} \rho v_y v_x + \frac{\partial}{\partial z} \rho v_z v_x\right) - \left(\frac{\partial}{\partial x} \tau_{xx} + \frac{\partial}{\partial y} \tau_{yx} + \frac{\partial}{\partial z} \tau_{zx}\right) - \frac{\partial p}{\partial x} + \rho g_x \quad (3.2\text{–}5)$$

The y- and z-components, which may be obtained similarly, are

$$\frac{\partial}{\partial t}\rho v_y = -\left(\frac{\partial}{\partial x}\rho v_x v_y + \frac{\partial}{\partial y}\rho v_y v_y + \frac{\partial}{\partial z}\rho v_z v_y\right) - \left(\frac{\partial}{\partial x}\tau_{xy} + \frac{\partial}{\partial y}\tau_{yy} + \frac{\partial}{\partial z}\tau_{zy}\right) - \frac{\partial p}{\partial y} + \rho g_y \tag{3.2–6}$$

$$\frac{\partial}{\partial t}\rho v_z = -\left(\frac{\partial}{\partial x}\rho v_x v_z + \frac{\partial}{\partial y}\rho v_y v_z + \frac{\partial}{\partial z}\rho v_z v_z\right) - \left(\frac{\partial}{\partial x}\tau_{xz} + \frac{\partial}{\partial y}\tau_{yz} + \frac{\partial}{\partial z}\tau_{zz}\right) - \frac{\partial p}{\partial z} + \rho g_z \tag{3.2–7}$$

The quantities ρv_x, ρv_y, ρv_z are the components of the mass velocity vector $\rho \boldsymbol{v}$; similarly g_x, g_y, g_z are the components of the gravitational acceleration $\boldsymbol{g}$. Furthermore, $\partial p/\partial x$, $\partial p/\partial y$, $\partial p/\partial z$ are the components of a vector $\boldsymbol{\nabla} p$, known as the "gradient of p" (sometimes written grad p). The terms $\rho v_x v_x$, $\rho v_x v_y$, $\rho v_x v_z$, $\rho v_y v_z$, etc., are the nine components of the convective momentum flux $\rho \boldsymbol{vv}$, which is the "dyadic product" of $\rho \boldsymbol{v}$ and $\boldsymbol{v}$. Similarly, τ_{xx}, τ_{xy}, τ_{xz}, τ_{yx}, etc., are the nine components of $\boldsymbol{\tau}$, known as the "stress tensor." Because Eqs. 3.2–5, 6, 7 take up so much space, it is convenient to combine them to give the single vector equation:

$$\underbrace{\frac{\partial}{\partial t}\rho \boldsymbol{v}}_{\substack{\text{rate of increase}\\ \text{of momentum per}\\ \text{unit volume}}} = \underbrace{-[\boldsymbol{\nabla} \cdot \rho \boldsymbol{vv}]}_{\substack{\text{rate of momentum}\\ \text{gain by convection}\\ \text{per unit volume}}} \underbrace{- \boldsymbol{\nabla} p}_{\substack{\text{pressure force}\\ \text{on element per}\\ \text{unit volume}}} \underbrace{- [\boldsymbol{\nabla} \cdot \boldsymbol{\tau}]}_{\substack{\text{rate of momentum}\\ \text{gain by viscous}\\ \text{transfer per unit}\\ \text{volume}}} \underbrace{+ \rho \boldsymbol{g}}_{\substack{\text{gravitational}\\ \text{force on element}\\ \text{per unit}\\ \text{volume}}} \tag{3.2–8}$$

The reader should be cautioned that $[\boldsymbol{\nabla} \cdot \rho \boldsymbol{vv}]$ and $[\boldsymbol{\nabla} \cdot \boldsymbol{\tau}]$ are *not* simple divergences because of the tensorial nature of $\rho \boldsymbol{vv}$ and $\boldsymbol{\tau}$. The physical interpretation is, however, analogous to that of $(\boldsymbol{\nabla} \cdot \rho \boldsymbol{v})$ in §3.1; whereas $(\boldsymbol{\nabla} \cdot \rho \boldsymbol{v})$ represents the rate of loss of mass (a scalar) per unit volume by fluid flow, the quantity $[\boldsymbol{\nabla} \cdot \rho \boldsymbol{vv}]$ represents the rate of loss of momentum (a vector) per unit volume by fluid flow. Equation 3.2–5 may be rearranged, with the help of the equation of continuity, to give

$$\rho \frac{Dv_x}{Dt} = -\frac{\partial p}{\partial x} - \left(\frac{\partial \tau_{xx}}{\partial x} + \frac{\partial \tau_{yx}}{\partial y} + \frac{\partial \tau_{zx}}{\partial z}\right) + \rho g_x \tag{3.2–9}$$

Similar rearrangements can be made for the y- and z-components. When all three components are added together vectorially, we get

$$\underset{\substack{\text{mass per unit}\\ \text{volume times}\\ \text{acceleration}}}{\rho\frac{D\boldsymbol{v}}{Dt}} = \underset{\substack{\text{pressure}\\ \text{force on}\\ \text{element}\\ \text{per unit}\\ \text{volume}}}{-\nabla p} \underset{\substack{\text{viscous force}\\ \text{on element}\\ \text{per unit}\\ \text{volume}}}{-[\nabla\cdot\boldsymbol{\tau}]} \underset{\substack{\text{gravitational}\\ \text{force on element}\\ \text{per unit volume}}}{+\rho\boldsymbol{g}} \tag{3.2–10}$$

In this form the equation of motion states that a small volume element moving with the fluid is accelerated because of the forces acting upon it. In other words, this is a statement of Newton's second law in the form *mass* × *acceleration* = *sum of forces*. We see then that the momentum balance is completely equivalent to Newton's second law of motion. Note that the two forms of the equation of motion given in Eqs. 3.2–8 and 3.2–10 correspond to the two forms of the equation of continuity, Eqs. 3.1–4 and 3.1–6. In each case the first form represents a balance over a volume element fixed in space and the second is a description of the changes taking place in an element following the fluid motion. It should be kept in mind that Eqs. 3.2–5 through 10 are valid for any continuous medium.

In order to use these equations to determine velocity distributions, however, we must insert expressions for the various stresses in terms of velocity gradients and fluid properties. For Newtonian fluids, these expressions are[1]

$$\tau_{xx} = -2\mu\frac{\partial v_x}{\partial x} + \tfrac{2}{3}\mu(\nabla\cdot\boldsymbol{v}) \tag{3.2–11}$$

$$\tau_{yy} = -2\mu\frac{\partial v_y}{\partial y} + \tfrac{2}{3}\mu(\nabla\cdot\boldsymbol{v}) \tag{3.2–12}$$

$$\tau_{zz} = -2\mu\frac{\partial v_z}{\partial z} + \tfrac{2}{3}\mu(\nabla\cdot\boldsymbol{v}) \tag{3.2–13}$$

$$\tau_{xy} = \tau_{yx} = -\mu\left(\frac{\partial v_x}{\partial y} + \frac{\partial v_y}{\partial x}\right) \tag{3.2–14}$$

$$\tau_{yz} = \tau_{zy} = -\mu\left(\frac{\partial v_y}{\partial z} + \frac{\partial v_z}{\partial y}\right) \tag{3.2–15}$$

$$\tau_{zx} = \tau_{xz} = -\mu\left(\frac{\partial v_z}{\partial x} + \frac{\partial v_x}{\partial z}\right) \tag{3.2–16}$$

[1] Actually, the normal stresses should contain one additional term; for example, Eq. 3.2–11 should be

$$\tau_{xx} = -2\mu\frac{\partial v_x}{\partial x} + (\tfrac{2}{3}\mu - \kappa)(\nabla\cdot\boldsymbol{v}) \tag{3.2–11a}$$

in which κ is the "bulk viscosity." The bulk viscosity is identically zero for low density monatomic gases and is probably not too important in dense gases and liquids. Formulas to estimate the value of κ may be found in J. O. Hirschfelder, C. F. Curtiss, and R. B. Bird, *The Molecular Theory of Gases and Liquids*, Wiley, New York (1954), p. 503 (dilute polyatomic gases), p. 647 (dense gases). Experimental measurements have been discussed by S. M. Karim and L. Rosenhead, *Revs. Mod. Phys.*, **24**, 108–116 (1952).

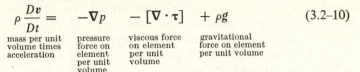

Equations 3.2–11 through 16 have been presented here without proof because the arguments involved are quite lengthy.[2] These equations, which constitute a more general statement of Newton's law of viscosity than that given in Eq. 1.1–2, apply to complex flow situations with the fluid flowing in all directions. When the fluid flows in the x-direction between two plates perpendicular to the y-direction (as depicted in Fig. 1.1–1) so that v_x is a function of y alone, then this set of six equations gives $\tau_{xx} = \tau_{yy} = \tau_{zz} = \tau_{yz} = \tau_{xz} = 0$ and $\tau_{yx} = -\mu(dv_x/dy)$, which is the same as Eq. 1.1–2. Hence the definition of viscosity in Chapter 1 is consistent with the more general definition here.

Substitution of Eqs. 3.2–11 through 16 into Eq. 3.2–9 and the corresponding y- and z-equations gives the general equations of motion for a Newtonian fluid with varying density and viscosity:

$$\rho \frac{Dv_x}{Dt} = -\frac{\partial p}{\partial x} + \frac{\partial}{\partial x}\left[2\mu \frac{\partial v_x}{\partial x} - \tfrac{2}{3}\mu(\nabla \cdot v)\right] + \frac{\partial}{\partial y}\left[\mu\left(\frac{\partial v_x}{\partial y} + \frac{\partial v_y}{\partial x}\right)\right] + \frac{\partial}{\partial z}\left[\mu\left(\frac{\partial v_z}{\partial x} + \frac{\partial v_x}{\partial z}\right)\right] + \rho g_x \quad (3.2\text{–}17)$$

$$\rho \frac{Dv_y}{Dt} = -\frac{\partial p}{\partial y} + \frac{\partial}{\partial x}\left[\mu\left(\frac{\partial v_y}{\partial x} + \frac{\partial v_x}{\partial y}\right)\right] + \frac{\partial}{\partial y}\left[2\mu \frac{\partial v_y}{\partial y} - \tfrac{2}{3}\mu(\nabla \cdot v)\right] + \frac{\partial}{\partial z}\left[\mu\left(\frac{\partial v_z}{\partial y} + \frac{\partial v_y}{\partial z}\right)\right] + \rho g_y \quad (3.2\text{–}18)$$

$$\rho \frac{Dv_z}{Dt} = -\frac{\partial p}{\partial z} + \frac{\partial}{\partial x}\left[\mu\left(\frac{\partial v_z}{\partial x} + \frac{\partial v_x}{\partial z}\right)\right] + \frac{\partial}{\partial y}\left[\mu\left(\frac{\partial v_z}{\partial y} + \frac{\partial v_y}{\partial z}\right)\right] + \frac{\partial}{\partial z}\left[2\mu \frac{\partial v_z}{\partial z} - \tfrac{2}{3}\mu(\nabla \cdot v)\right] + \rho g_z \quad (3.2\text{–}19)$$

These equations, along with the equation of continuity, the equation of state $p = p(\rho)$, the density dependence of viscosity $\mu = \mu(\rho)$, and the boundary and initial conditions, determine completely the pressure, density, and velocity components in a flowing isothermal fluid.

These equations in their complete form are seldom used to set up flow problems. Usually restricted forms of the equations of motion are used for convenience.

(i) *For constant ρ and constant μ*, Eqs. 3.2–17, 18, and 19 may be simplified by means of the equation of continuity $[(\nabla \cdot v) = 0]$ to give[3]

$$\rho \frac{Dv}{Dt} = -\nabla p + \mu \nabla^2 v + \rho g \quad (3.2\text{–}20)$$

[2] H. Lamb, *Hydrodynamics*, Dover, New York (1945), Sixth Edition, pp. 571–575; see also E. U. Condon, *Handbook of Physics*, E. U. Condon and H. Odishaw (Eds.), McGraw-Hill, New York (1958), pp. 3–10 to 3–13.

[3] The operator $\nabla^2 = \partial^2/\partial x^2 + \partial^2/\partial y^2 + \partial^2/\partial z^2$ is called the Laplacian operator (see §A.3 for further details).

The cartesian components of this equation are given in Eqs. (*D*), (*E*), and (*F*) in Table 3.4–2. Equation 3.2–20 is the celebrated *Navier-Stokes equation*, first developed by Navier[4] in France in 1822 by molecular arguments.[5]

(ii) *For* $[\nabla \cdot \tau] = 0$, Eq. 3.2–10 reduces to

$$\rho \frac{Dv}{Dt} = -\nabla p + \rho g \tag{3.2–21}$$

Equation 3.2–21 is the famous *Euler*[6] *equation*, first derived in 1755. It has been widely used for describing flow systems in which viscous effects are relatively unimportant.

§3.3 THE EQUATION OF MECHANICAL ENERGY

In this section we show how the equation of motion can be used to obtain a description of the energy interconversions that take place in a flowing fluid. We begin by forming the scalar product of the local velocity v with the equation of motion as given in Eq. 3.2–10:

$$\rho \frac{D}{Dt}(\tfrac{1}{2}v^2) = -(v \cdot \nabla p) - (v \cdot [\nabla \cdot \tau]) + \rho(v \cdot g) \tag{3.3–1}$$

This scalar equation describes the rate of change of kinetic energy per unit mass ($\frac{1}{2}v^2$) for an element of fluid moving downstream.

For convenience in subsequent discussions, we rewrite this equation in terms of $\partial/\partial t$ by using the equation of continuity; we also split up the pressure and viscous contributions into two terms each. The terms in the resulting equation can be interpreted in terms of a stationary volume element through which the fluid flows.

$$\underbrace{\frac{\partial}{\partial t}(\tfrac{1}{2}\rho v^2)}_{\substack{\text{rate of increase}\\ \text{in kinetic energy}\\ \text{per unit volume}}} = \underbrace{-(\nabla \cdot \tfrac{1}{2}\rho v^2 v)}_{\substack{\text{net rate of input of}\\ \text{kinetic energy by}\\ \text{virtue of bulk flow}}}$$

$$\underbrace{-(\nabla \cdot pv)}_{\substack{\text{rate of work done by}\\ \text{pressure of surround-}\\ \text{ings on volume element}}} \qquad \underbrace{-p(-\nabla \cdot v)}_{\substack{\text{rate of \textit{reversible}}\\ \text{conversion to}\\ \text{internal energy}}}$$

$$\underbrace{-(\nabla \cdot [\tau \cdot v])}_{\substack{\text{rate of work done by}\\ \text{viscous forces on}\\ \text{volume element}}} \qquad \underbrace{-(-\tau : \nabla v)}_{\substack{\text{rate of \textit{irreversible}}\\ \text{conversion to}\\ \text{internal energy}}}$$

$$\underbrace{+\rho(v \cdot g)}_{\substack{\text{rate of work done}\\ \text{by gravity force on}\\ \text{volume element}}} \tag{3.3–2}$$

[4] Pronounced "Nah-vyay'."

[5] For an interesting discussion of the history of these and other famous relations in fluid mechanics, see H. Rouse and S. Ince, *History of Hydraulics*, Iowa Institute of Hydraulics, Iowa City (1959).

[6] Pronounced "Oiler."

At this point it is not clear why we have ascribed the indicated physical significance to the terms $p(\nabla \cdot \boldsymbol{v})$ and $(\boldsymbol{\tau} : \nabla \boldsymbol{v})$; their significance cannot be properly appreciated until one has studied the energy balance in Chapter 10. There it will be seen that these same two terms enter with opposite sign into the equation for the internal energy.

It should be emphasized that for Newtonian fluids $(-\boldsymbol{\tau} : \nabla \boldsymbol{v})$ is always positive because it may be written as a sum of squared terms:

$$(-\boldsymbol{\tau} : \nabla \boldsymbol{v}) = \mu \Phi_v = \tfrac{1}{2}\mu \Sigma_i \Sigma_j \left[\left(\frac{\partial v_i}{\partial x_j} + \frac{\partial v_j}{\partial x_i}\right) - \tfrac{2}{3}(\nabla \cdot \boldsymbol{v})\delta_{ij}\right]^2 \qquad (3.3\text{–}3)$$

in which i and j take on the values x, y, z, and $\delta_{ij} = 1$ for $i = j$ and $\delta_{ij} = 0$ for $i \neq j$. (In Table 3.4–8 Φ_v is given for several coordinate systems.) This means that in all flow systems there is a degradation of mechanical to thermal energy and that therefore no real processes are reversible. In the absence of the $(\boldsymbol{\tau} : \nabla \boldsymbol{v})$ term, all forms of energy included in Eq. 3.3–2—kinetic, internal, and potential—would be freely interconvertible.

Because of the terms $p(\nabla \cdot \boldsymbol{v})$ and $(\boldsymbol{\tau} : \nabla \boldsymbol{v})$, the fluid may be heated (or cooled) internally. Hence when we speak of an "isothermal system," we really mean one in which the heat thus generated (or absorbed) does not cause appreciable temperature change. The $p(\nabla \cdot \boldsymbol{v})$ term causes appreciable temperature change for gases undergoing sudden expansion or compression, as in compressors, turbines, and shock tubes. The $(\boldsymbol{\tau} : \nabla \boldsymbol{v})$ term causes measurable temperature changes only in high-speed flow systems in which the velocity gradients are large, as in high-speed flight, rapid extrusion, and lubrication; one illustration of this kind of heating is discussed in §9.4.

Equation 3.3–2 is used in Chapter 7 as the starting point for developing the mechanical energy balance, or engineering Bernoulli equation.

§3.4 THE EQUATIONS OF CHANGE IN CURVILINEAR COORDINATES

It may be noted that all of the foregoing derivations have for simplicity been made in rectangular coordinates. However, rectangular coordinates are not always the most useful for solving problems. We have already seen in Chapter 2 that curvilinear coordinates are sometimes more convenient. For example, in the Hagen-Poiseuille problem the axial velocity v_z was found to be a function of only a single variable, r, when cylindrical coordinates were used. If rectangular coordinates had been employed, v_z would have been a function of the two variables x and y. Similarly, statement of the boundary condition at the tube wall would have been more difficult. In the analysis of the flow around a sphere the use of spherical coordinates allowed us to describe the velocity in terms of the two components v_r and v_θ, rather than v_x, v_y, and v_z, and again resulted in a simplification of the boundary conditions. Similar

TABLE 3.4–1
THE EQUATION OF CONTINUITY IN SEVERAL COORDINATE SYSTEMS

Rectangular coordinates (x, y, z):

$$\frac{\partial \rho}{\partial t} + \frac{\partial}{\partial x}(\rho v_x) + \frac{\partial}{\partial y}(\rho v_y) + \frac{\partial}{\partial z}(\rho v_z) = 0 \qquad (A)$$

Cylindrical coordinates (r, θ, z):

$$\frac{\partial \rho}{\partial t} + \frac{1}{r}\frac{\partial}{\partial r}(\rho r v_r) + \frac{1}{r}\frac{\partial}{\partial \theta}(\rho v_\theta) + \frac{\partial}{\partial z}(\rho v_z) = 0 \qquad (B)$$

Spherical coordinates (r, θ, ϕ):

$$\frac{\partial \rho}{\partial t} + \frac{1}{r^2}\frac{\partial}{\partial r}(\rho r^2 v_r) + \frac{1}{r \sin\theta}\frac{\partial}{\partial \theta}(\rho v_\theta \sin\theta) + \frac{1}{r \sin\theta}\frac{\partial}{\partial \phi}(\rho v_\phi) = 0 \qquad (C)$$

advantages will be found for curvilinear coordinates when setting up flow problems by simplification of the equations of change.

The equations of continuity and motion, as we derived them in §§3.1 and 3.2, are given in terms of the coordinates x, y, z, the velocity components v_x, v_y, v_z, and the shear components τ_{xx}, τ_{xy}, etc. If we wish to rewrite these equations in spherical coordinates, we must know the following: (*a*) the relations between x, y, z and r, θ, ϕ (see Fig. A.6–1), (*b*) the relation between v_x, v_y, v_z and the corresponding components v_r, v_θ, v_ϕ, and (*c*) the relations between τ_{xx}, τ_{xy}, etc., and $\tau_{r\theta}$, $\tau_{r\phi}$, etc. (The relations among vector and tensor components are summarized in §A.6.) The transformation from rectangular to spherical coordinates can then be performed by a straightforward but tedious procedure. The reader will not have to wade through the details of this process, for in Tables 3.4–1, 2, 3, and 4 (and elsewhere in the book) important equations are tabulated in rectangular, cylindrical, and spherical coordinates.

It is only fair to warn the beginner at this point that although the equation of continuity may be readily obtained in curvilinear coordinates by a shell balance the same is not true of the equation of motion. In general, it may be said that the shell balance method is very difficult to apply in systems with curved streamlines, and the use of this method in such systems is not recommended. Rather, one should always start with the general equations as illustrated in §3.5.

TABLE 3.4-2

THE EQUATION OF MOTION IN RECTANGULAR COORDINATES (x, y, z)

In terms of $\boldsymbol{\tau}$:

x-component
$$\rho\left(\frac{\partial v_x}{\partial t} + v_x\frac{\partial v_x}{\partial x} + v_y\frac{\partial v_x}{\partial y} + v_z\frac{\partial v_x}{\partial z}\right) = -\frac{\partial p}{\partial x} - \left(\frac{\partial \tau_{xx}}{\partial x} + \frac{\partial \tau_{yx}}{\partial y} + \frac{\partial \tau_{zx}}{\partial z}\right) + \rho g_x \quad (A)$$

y-component
$$\rho\left(\frac{\partial v_y}{\partial t} + v_x\frac{\partial v_y}{\partial x} + v_y\frac{\partial v_y}{\partial y} + v_z\frac{\partial v_y}{\partial z}\right) = -\frac{\partial p}{\partial y} - \left(\frac{\partial \tau_{xy}}{\partial x} + \frac{\partial \tau_{yy}}{\partial y} + \frac{\partial \tau_{zy}}{\partial z}\right) + \rho g_y \quad (B)$$

z-component
$$\rho\left(\frac{\partial v_z}{\partial t} + v_x\frac{\partial v_z}{\partial x} + v_y\frac{\partial v_z}{\partial y} + v_z\frac{\partial v_z}{\partial z}\right) = -\frac{\partial p}{\partial z} - \left(\frac{\partial \tau_{xz}}{\partial x} + \frac{\partial \tau_{yz}}{\partial y} + \frac{\partial \tau_{zz}}{\partial z}\right) + \rho g_z \quad (C)$$

In terms of velocity gradients for a Newtonian fluid with constant ρ and μ:

x-component
$$\rho\left(\frac{\partial v_x}{\partial t} + v_x\frac{\partial v_x}{\partial x} + v_y\frac{\partial v_x}{\partial y} + v_z\frac{\partial v_x}{\partial z}\right) = -\frac{\partial p}{\partial x} + \mu\left(\frac{\partial^2 v_x}{\partial x^2} + \frac{\partial^2 v_x}{\partial y^2} + \frac{\partial^2 v_x}{\partial z^2}\right) + \rho g_x \quad (D)$$

y-component
$$\rho\left(\frac{\partial v_y}{\partial t} + v_x\frac{\partial v_y}{\partial x} + v_y\frac{\partial v_y}{\partial y} + v_z\frac{\partial v_y}{\partial z}\right) = -\frac{\partial p}{\partial y} + \mu\left(\frac{\partial^2 v_y}{\partial x^2} + \frac{\partial^2 v_y}{\partial y^2} + \frac{\partial^2 v_y}{\partial z^2}\right) + \rho g_y \quad (E)$$

z-component
$$\rho\left(\frac{\partial v_z}{\partial t} + v_x\frac{\partial v_z}{\partial x} + v_y\frac{\partial v_z}{\partial y} + v_z\frac{\partial v_z}{\partial z}\right) = -\frac{\partial p}{\partial z} + \mu\left(\frac{\partial^2 v_z}{\partial x^2} + \frac{\partial^2 v_z}{\partial y^2} + \frac{\partial^2 v_z}{\partial z^2}\right) + \rho g_z \quad (F)$$

TABLE 3.4–3

THE EQUATION OF MOTION IN CYLINDRICAL COORDINATES (r, θ, z)

In terms of $\boldsymbol{\tau}$:

r-component[a]
$$\rho\left(\frac{\partial v_r}{\partial t} + v_r\frac{\partial v_r}{\partial r} + \frac{v_\theta}{r}\frac{\partial v_r}{\partial \theta} - \frac{v_\theta^2}{r} + v_z\frac{\partial v_r}{\partial z}\right) = -\frac{\partial p}{\partial r} - \left(\frac{1}{r}\frac{\partial}{\partial r}(r\tau_{rr}) + \frac{1}{r}\frac{\partial \tau_{r\theta}}{\partial \theta} - \frac{\tau_{\theta\theta}}{r} + \frac{\partial \tau_{rz}}{\partial z}\right) + \rho g_r \quad (A)$$

θ-component[b]
$$\rho\left(\frac{\partial v_\theta}{\partial t} + v_r\frac{\partial v_\theta}{\partial r} + \frac{v_\theta}{r}\frac{\partial v_\theta}{\partial \theta} + \frac{v_r v_\theta}{r} + v_z\frac{\partial v_\theta}{\partial z}\right) = -\frac{1}{r}\frac{\partial p}{\partial \theta} - \left(\frac{1}{r^2}\frac{\partial}{\partial r}(r^2\tau_{r\theta}) + \frac{1}{r}\frac{\partial \tau_{\theta\theta}}{\partial \theta} + \frac{\partial \tau_{\theta z}}{\partial z}\right) + \rho g_\theta \quad (B)$$

z-component
$$\rho\left(\frac{\partial v_z}{\partial t} + v_r\frac{\partial v_z}{\partial r} + \frac{v_\theta}{r}\frac{\partial v_z}{\partial \theta} + v_z\frac{\partial v_z}{\partial z}\right) = -\frac{\partial p}{\partial z} - \left(\frac{1}{r}\frac{\partial}{\partial r}(r\tau_{rz}) + \frac{1}{r}\frac{\partial \tau_{\theta z}}{\partial \theta} + \frac{\partial \tau_{zz}}{\partial z}\right) + \rho g_z \quad (C)$$

In terms of velocity gradients for a Newtonian fluid with constant ρ and μ:

r-component[a]
$$\rho\left(\frac{\partial v_r}{\partial t} + v_r\frac{\partial v_r}{\partial r} + \frac{v_\theta}{r}\frac{\partial v_r}{\partial \theta} - \frac{v_\theta^2}{r} + v_z\frac{\partial v_r}{\partial z}\right) = -\frac{\partial p}{\partial r} + \mu\left[\frac{\partial}{\partial r}\left(\frac{1}{r}\frac{\partial}{\partial r}(rv_r)\right) + \frac{1}{r^2}\frac{\partial^2 v_r}{\partial \theta^2} - \frac{2}{r^2}\frac{\partial v_\theta}{\partial \theta} + \frac{\partial^2 v_r}{\partial z^2}\right] + \rho g_r \quad (D)$$

θ-component[b]
$$\rho\left(\frac{\partial v_\theta}{\partial t} + v_r\frac{\partial v_\theta}{\partial r} + \frac{v_\theta}{r}\frac{\partial v_\theta}{\partial \theta} + \frac{v_r v_\theta}{r} + v_z\frac{\partial v_\theta}{\partial z}\right) = -\frac{1}{r}\frac{\partial p}{\partial \theta} + \mu\left[\frac{\partial}{\partial r}\left(\frac{1}{r}\frac{\partial}{\partial r}(rv_\theta)\right) + \frac{1}{r^2}\frac{\partial^2 v_\theta}{\partial \theta^2} + \frac{2}{r^2}\frac{\partial v_r}{\partial \theta} + \frac{\partial^2 v_\theta}{\partial z^2}\right] + \rho g_\theta \quad (E)$$

z-component
$$\rho\left(\frac{\partial v_z}{\partial t} + v_r\frac{\partial v_z}{\partial r} + \frac{v_\theta}{r}\frac{\partial v_z}{\partial \theta} + v_z\frac{\partial v_z}{\partial z}\right) = -\frac{\partial p}{\partial z} + \mu\left[\frac{1}{r}\frac{\partial}{\partial r}\left(r\frac{\partial v_z}{\partial r}\right) + \frac{1}{r^2}\frac{\partial^2 v_z}{\partial \theta^2} + \frac{\partial^2 v_z}{\partial z^2}\right] + \rho g_z \quad (F)$$

[a] The term $\rho v_\theta^2/r$ is the *centrifugal force*. It gives the effective force in the r-direction resulting from fluid motion in the θ-direction. This term arises automatically on transformation from rectangular to cylindrical coordinates; it does not have to be added on physical grounds. Two problems in which this term arises are discussed in Examples 3.5–1 and 3.5–2.

[b] The term $\rho v_r v_\theta/r$ is the *Coriolis force*. It is an effective force in the θ-direction when there is flow in both the r- and θ-directions. This term also arises automatically in the coordinate transformation. The Coriolis force arises in the problem of flow near a rotating disk (see, for example, H. Schlichting, *Boundary-Layer Theory*, McGraw-Hill, New York (1955), Chapter 5, §10.

TABLE 3.4–4

THE EQUATION OF MOTION IN SPHERICAL COORDINATES (r, θ, ϕ)

In terms of $\boldsymbol{\tau}$:

r-component

$$\rho\left(\frac{\partial v_r}{\partial t} + v_r\frac{\partial v_r}{\partial r} + \frac{v_\theta}{r}\frac{\partial v_r}{\partial \theta} + \frac{v_\phi}{r\sin\theta}\frac{\partial v_r}{\partial \phi} - \frac{v_\theta^2 + v_\phi^2}{r}\right)$$

$$= -\frac{\partial p}{\partial r} - \left(\frac{1}{r^2}\frac{\partial}{\partial r}(r^2\tau_{rr}) + \frac{1}{r\sin\theta}\frac{\partial}{\partial\theta}(\tau_{r\theta}\sin\theta) + \frac{1}{r\sin\theta}\frac{\partial\tau_{r\phi}}{\partial\phi} - \frac{\tau_{\theta\theta} + \tau_{\phi\phi}}{r}\right) + \rho g_r \qquad (A)$$

θ-component

$$\rho\left(\frac{\partial v_\theta}{\partial t} + v_r\frac{\partial v_\theta}{\partial r} + \frac{v_\theta}{r}\frac{\partial v_\theta}{\partial \theta} + \frac{v_\phi}{r\sin\theta}\frac{\partial v_\theta}{\partial \phi} + \frac{v_r v_\theta}{r} - \frac{v_\phi^2\cot\theta}{r}\right)$$

$$= -\frac{1}{r}\frac{\partial p}{\partial \theta} - \left(\frac{1}{r^2}\frac{\partial}{\partial r}(r^2\tau_{r\theta}) + \frac{1}{r\sin\theta}\frac{\partial}{\partial\theta}(\tau_{\theta\theta}\sin\theta) + \frac{1}{r\sin\theta}\frac{\partial\tau_{\theta\phi}}{\partial\phi} + \frac{\tau_{r\theta}}{r} - \frac{\cot\theta}{r}\tau_{\phi\phi}\right) + \rho g_\theta \qquad (B)$$

φ-component

$$\rho\left(\frac{\partial v_\phi}{\partial t} + v_r\frac{\partial v_\phi}{\partial r} + \frac{v_\theta}{r}\frac{\partial v_\phi}{\partial \theta} + \frac{v_\phi}{r\sin\theta}\frac{\partial v_\phi}{\partial \phi} + \frac{v_\phi v_r}{r} + \frac{v_\theta v_\phi}{r}\cot\theta\right)$$

$$= -\frac{1}{r\sin\theta}\frac{\partial p}{\partial \phi} - \left(\frac{1}{r^2}\frac{\partial}{\partial r}(r^2\tau_{r\phi}) + \frac{1}{r}\frac{\partial\tau_{\theta\phi}}{\partial\theta} + \frac{1}{r\sin\theta}\frac{\partial\tau_{\phi\phi}}{\partial\phi} + \frac{\tau_{r\phi}}{r} + \frac{2\cot\theta}{r}\tau_{\theta\phi}\right) + \rho g_\phi \qquad (C)$$

TABLE 3.4–4 (contd.)

In terms of velocity gradients for a Newtonian fluid with constant ρ and μ:[a]

r-component
$$\rho\left(\frac{\partial v_r}{\partial t} + v_r\frac{\partial v_r}{\partial r} + \frac{v_\theta}{r}\frac{\partial v_r}{\partial \theta} + \frac{v_\phi}{r\sin\theta}\frac{\partial v_r}{\partial \phi} - \frac{v_\theta^2 + v_\phi^2}{r}\right)$$
$$= -\frac{\partial p}{\partial r} + \mu\left(\nabla^2 v_r - \frac{2}{r^2}v_r - \frac{2}{r^2}\frac{\partial v_\theta}{\partial \theta} - \frac{2}{r^2}v_\theta\cot\theta - \frac{2}{r^2\sin\theta}\frac{\partial v_\phi}{\partial \phi}\right) + \rho g_r \tag{D}$$

θ-component
$$\rho\left(\frac{\partial v_\theta}{\partial t} + v_r\frac{\partial v_\theta}{\partial r} + \frac{v_\theta}{r}\frac{\partial v_\theta}{\partial \theta} + \frac{v_\phi}{r\sin\theta}\frac{\partial v_\theta}{\partial \phi} + \frac{v_r v_\theta}{r} - \frac{v_\phi^2\cot\theta}{r}\right)$$
$$= -\frac{1}{r}\frac{\partial p}{\partial \theta} + \mu\left(\nabla^2 v_\theta + \frac{2}{r^2}\frac{\partial v_r}{\partial \theta} - \frac{v_\theta}{r^2\sin^2\theta} - \frac{2\cos\theta}{r^2\sin^2\theta}\frac{\partial v_\phi}{\partial \phi}\right) + \rho g_\theta \tag{E}$$

φ-component
$$\rho\left(\frac{\partial v_\phi}{\partial t} + v_r\frac{\partial v_\phi}{\partial r} + \frac{v_\theta}{r}\frac{\partial v_\phi}{\partial \theta} + \frac{v_\phi}{r\sin\theta}\frac{\partial v_\phi}{\partial \phi} + \frac{v_\phi v_r}{r} + \frac{v_\theta v_\phi}{r}\cot\theta\right)$$
$$= -\frac{1}{r\sin\theta}\frac{\partial p}{\partial \phi} + \mu\left(\nabla^2 v_\phi - \frac{v_\phi}{r^2\sin^2\theta} + \frac{2}{r^2\sin\theta}\frac{\partial v_r}{\partial \phi} + \frac{2\cos\theta}{r^2\sin^2\theta}\frac{\partial v_\theta}{\partial \phi}\right) + \rho g_\phi \tag{F}$$

---

[a] In these equations

$$\nabla^2 = \frac{1}{r^2}\frac{\partial}{\partial r}\left(r^2\frac{\partial}{\partial r}\right) + \frac{1}{r^2\sin\theta}\frac{\partial}{\partial \theta}\left(\sin\theta\frac{\partial}{\partial \theta}\right) + \frac{1}{r^2\sin^2\theta}\left(\frac{\partial^2}{\partial \phi^2}\right)$$

TABLE 3.4–5

COMPONENTS OF THE STRESS TENSOR FOR NEWTONIAN FLUIDS IN RECTANGULAR COORDINATES (x, y, z)

$$\tau_{xx} = -\mu\left[2\frac{\partial v_x}{\partial x} - \tfrac{2}{3}(\nabla \cdot \boldsymbol{v})\right] \quad (A)$$

$$\tau_{yy} = -\mu\left[2\frac{\partial v_y}{\partial y} - \tfrac{2}{3}(\nabla \cdot \boldsymbol{v})\right] \quad (B)$$

$$\tau_{zz} = -\mu\left[2\frac{\partial v_z}{\partial z} - \tfrac{2}{3}(\nabla \cdot \boldsymbol{v})\right] \quad (C)$$

$$\tau_{xy} = \tau_{yx} = -\mu\left[\frac{\partial v_x}{\partial y} + \frac{\partial v_y}{\partial x}\right] \quad (D)$$

$$\tau_{yz} = \tau_{zy} = -\mu\left[\frac{\partial v_y}{\partial z} + \frac{\partial v_z}{\partial y}\right] \quad (E)$$

$$\tau_{zx} = \tau_{xz} = -\mu\left[\frac{\partial v_z}{\partial x} + \frac{\partial v_x}{\partial z}\right] \quad (F)$$

$$(\nabla \cdot \boldsymbol{v}) = \frac{\partial v_x}{\partial x} + \frac{\partial v_y}{\partial y} + \frac{\partial v_z}{\partial z} \quad (G)$$

TABLE 3.4-6

COMPONENTS OF THE STRESS TENSOR FOR NEWTONIAN FLUIDS IN CYLINDRICAL COORDINATES (r, θ, z)

$$\tau_{rr} = -\mu\left[2\frac{\partial v_r}{\partial r} - \tfrac{2}{3}(\nabla \cdot \boldsymbol{v})\right] \tag{A}$$

$$\tau_{\theta\theta} = -\mu\left[2\left(\frac{1}{r}\frac{\partial v_\theta}{\partial \theta} + \frac{v_r}{r}\right) - \tfrac{2}{3}(\nabla \cdot \boldsymbol{v})\right] \tag{B}$$

$$\tau_{zz} = -\mu\left[2\frac{\partial v_z}{\partial z} - \tfrac{2}{3}(\nabla \cdot \boldsymbol{v})\right] \tag{C}$$

$$\tau_{r\theta} = \tau_{\theta r} = -\mu\left[r\frac{\partial}{\partial r}\left(\frac{v_\theta}{r}\right) + \frac{1}{r}\frac{\partial v_r}{\partial \theta}\right] \tag{D}$$

$$\tau_{\theta z} = \tau_{z\theta} = -\mu\left[\frac{\partial v_\theta}{\partial z} + \frac{1}{r}\frac{\partial v_z}{\partial \theta}\right] \tag{E}$$

$$\tau_{zr} = \tau_{rz} = -\mu\left[\frac{\partial v_z}{\partial r} + \frac{\partial v_r}{\partial z}\right] \tag{F}$$

$$(\nabla \cdot \boldsymbol{v}) = \frac{1}{r}\frac{\partial}{\partial r}(r v_r) + \frac{1}{r}\frac{\partial v_\theta}{\partial \theta} + \frac{\partial v_z}{\partial z} \tag{G}$$

TABLE 3.4-7

COMPONENTS OF THE STRESS TENSOR FOR NEWTONIAN FLUIDS IN SPHERICAL COORDINATES (r, θ, ϕ)

$$\tau_{rr} = -\mu\left[2\frac{\partial v_r}{\partial r} - \tfrac{2}{3}(\nabla \cdot \boldsymbol{v})\right] \tag{A}$$

$$\tau_{\theta\theta} = -\mu\left[2\left(\frac{1}{r}\frac{\partial v_\theta}{\partial \theta} + \frac{v_r}{r}\right) - \tfrac{2}{3}(\nabla \cdot \boldsymbol{v})\right] \tag{B}$$

$$\tau_{\phi\phi} = -\mu\left[2\left(\frac{1}{r\sin\theta}\frac{\partial v_\phi}{\partial \phi} + \frac{v_r}{r} + \frac{v_\theta \cot\theta}{r}\right) - \tfrac{2}{3}(\nabla \cdot \boldsymbol{v})\right] \tag{C}$$

$$\tau_{r\theta} = \tau_{\theta r} = -\mu\left[r\frac{\partial}{\partial r}\left(\frac{v_\theta}{r}\right) + \frac{1}{r}\frac{\partial v_r}{\partial \theta}\right] \tag{D}$$

$$\tau_{\theta\phi} = \tau_{\phi\theta} = -\mu\left[\frac{\sin\theta}{r}\frac{\partial}{\partial \theta}\left(\frac{v_\phi}{\sin\theta}\right) + \frac{1}{r\sin\theta}\frac{\partial v_\theta}{\partial \phi}\right] \tag{E}$$

$$\tau_{\phi r} = \tau_{r\phi} = -\mu\left[\frac{1}{r\sin\theta}\frac{\partial v_r}{\partial \phi} + r\frac{\partial}{\partial r}\left(\frac{v_\phi}{r}\right)\right] \tag{F}$$

$$(\nabla \cdot \boldsymbol{v}) = \frac{1}{r^2}\frac{\partial}{\partial r}(r^2 v_r) + \frac{1}{r\sin\theta}\frac{\partial}{\partial \theta}(v_\theta \sin\theta) + \frac{1}{r\sin\theta}\frac{\partial v_\phi}{\partial \phi} \tag{G}$$

TABLE 3.4–8
THE FUNCTION $-(\boldsymbol{\tau} : \nabla \boldsymbol{v}) = \mu \Phi_v$ FOR NEWTONIAN FLUIDS[a]

Rectangular

$$\Phi_v = 2\left[\left(\frac{\partial v_x}{\partial x}\right)^2 + \left(\frac{\partial v_y}{\partial y}\right)^2 + \left(\frac{\partial v_z}{\partial z}\right)^2\right] + \left[\frac{\partial v_y}{\partial x} + \frac{\partial v_x}{\partial y}\right]^2 + \left[\frac{\partial v_z}{\partial y} + \frac{\partial v_y}{\partial z}\right]^2 + \left[\frac{\partial v_x}{\partial z} + \frac{\partial v_z}{\partial x}\right]^2 - \frac{2}{3}\left[\frac{\partial v_x}{\partial x} + \frac{\partial v_y}{\partial y} + \frac{\partial v_z}{\partial z}\right]^2 \qquad (A)$$

Cylindrical

$$\Phi_v = 2\left[\left(\frac{\partial v_r}{\partial r}\right)^2 + \left(\frac{1}{r}\frac{\partial v_\theta}{\partial \theta} + \frac{v_r}{r}\right)^2 + \left(\frac{\partial v_z}{\partial z}\right)^2\right] + \left[r\frac{\partial}{\partial r}\left(\frac{v_\theta}{r}\right) + \frac{1}{r}\frac{\partial v_r}{\partial \theta}\right]^2 + \left[\frac{1}{r}\frac{\partial v_z}{\partial \theta} + \frac{\partial v_\theta}{\partial z}\right]^2 + \left[\frac{\partial v_r}{\partial z} + \frac{\partial v_z}{\partial r}\right]^2 - \frac{2}{3}\left[\frac{1}{r}\frac{\partial}{\partial r}(rv_r) + \frac{1}{r}\frac{\partial v_\theta}{\partial \theta} + \frac{\partial v_z}{\partial z}\right]^2 \qquad (B)$$

Spherical

$$\Phi_v = 2\left[\left(\frac{\partial v_r}{\partial r}\right)^2 + \left(\frac{1}{r}\frac{\partial v_\theta}{\partial \theta} + \frac{v_r}{r}\right)^2 + \left(\frac{1}{r\sin\theta}\frac{\partial v_\phi}{\partial \phi} + \frac{v_r}{r} + \frac{v_\theta \cot\theta}{r}\right)^2\right] + \left[r\frac{\partial}{\partial r}\left(\frac{v_\theta}{r}\right) + \frac{1}{r}\frac{\partial v_r}{\partial \theta}\right]^2 + \left[\frac{\sin\theta}{r}\frac{\partial}{\partial \theta}\left(\frac{v_\phi}{\sin\theta}\right) + \frac{1}{r\sin\theta}\frac{\partial v_\theta}{\partial \phi}\right]^2 + \left[\frac{1}{r\sin\theta}\frac{\partial v_r}{\partial \phi} + r\frac{\partial}{\partial r}\left(\frac{v_\phi}{r}\right)\right]^2 - \frac{2}{3}\left[\frac{1}{r^2}\frac{\partial}{\partial r}(r^2 v_r) + \frac{1}{r\sin\theta}\frac{\partial}{\partial \theta}(v_\theta \sin\theta) + \frac{1}{r\sin\theta}\frac{\partial v_\phi}{\partial \phi}\right]^2 \qquad (C)$$

[a] These expressions are obtained by inserting the components of $\boldsymbol{\tau}$ from Tables 3.4–5,6,7 into the expression for $(\boldsymbol{\tau}:\nabla \boldsymbol{v})$ given in Appendix A. (See Tables A.7–1, 2, and 3.)

§3.5 USE OF THE EQUATIONS OF CHANGE TO SET UP STEADY FLOW PROBLEMS

To set up constant-density, constant-viscosity flow problems, one needs

equation of continuity..........Table 3.4–1 (with constant ρ)
equation of motion ....Table 3.4–2, 3, or 4 (Eqs. *D*, *E*, *F*)

and initial and boundary conditions. From these two equations one then obtains the pressure and velocity distributions.

To set up problems involving isothermal flow with variable density and viscosity, one needs

equation of continuity ....Table 3.4–1
equation of motion ....Tables 3.4–2, 3, or 4 (Eqs. *A*, *B*, *C*, with expressions for the components of $\boldsymbol{\tau}$ in Tables 3.4–5, 6, or 7)
equation of state ....$\rho = \rho(p)$
equation for viscosity ....$\mu = \mu(p)$

along with the initial and boundary conditions. From these four relations we get the distributions of velocity, pressure, density, and viscosity for a given flow system.

In this section we show how to set up viscous flow problems by simplifying the foregoing equations. We do this by discarding those terms in the general equations that are zero (or nearly zero) in the situation being studied. In determining the terms to discard, we are aided by our intuitive feeling as to the behavior of the system: the flow patterns, the pressure distribution, etc. One advantage of this procedure is that by the time one has finished the "discarding process" one automatically has a complete listing of the assumptions that have been made. We introduce this method by setting up two problems from Chapter 2. We then further illustrate the usefulness of the method by working out some slightly more complicated examples.

For the *axial flow of an incompressible fluid in a circular tube*, we set up a momentum balance in §2.3 and solved for the velocity distribution. Now let us see how the same result may be obtained by simplification of the equations of change. Clearly cylindrical coordinates are the most appropriate for this problem. Again we will consider a long tube and set v_θ and v_r equal to zero. The remaining velocity component v_z will not be a function of θ because of cylindrical symmetry. The z-component of the equation of motion for constant ρ and μ (see Table 3.4–3) may then be written

$$\rho v_z \frac{\partial v_z}{\partial z} = -\frac{\partial \mathscr{P}}{\partial z} + \mu\left[\frac{1}{r}\frac{\partial}{\partial r}\left(r\frac{\partial v_z}{\partial r}\right) + \frac{\partial^2 v_z}{\partial z^2}\right] \tag{3.5–1}$$

This equation may be further simplified by taking advantage of the equation of continuity, which reduces here to

$$\frac{\partial v_z}{\partial z} = 0 \tag{3.5–2}$$

Clearly then $\partial^2 v_z/\partial z^2$ also is zero and Eq. 3.5–1 becomes

$$0 = -\frac{d\mathscr{P}}{dz} + \mu\frac{1}{r}\frac{d}{dr}\left(r\frac{dv_z}{dr}\right) \tag{3.5–3}$$

Integration twice with respect to r and use of the boundary conditions $v_z = 0$ at $r = R$ and v_z = finite at $r = 0$ gives

$$v_z = \frac{(\mathscr{P}_0 - \mathscr{P}_L)R^2}{4\mu L}\left[1 - \left(\frac{r}{R}\right)^2\right] \tag{3.5–4}$$

This is the same as Eq. 2.3–16.

For the *falling film of variable viscosity* discussed in Example 2.2–2, we find from Eq. C of Table 3.4–2 that for steady state (and neglecting end effects)

$$0 = -\frac{d\tau_{xz}}{dx} + \rho g \cos\beta \tag{3.5–5}$$

Then we insert the expression for τ_{xz} (Eq. F in Table 3.4–5) to get

$$0 = +\frac{d}{dx}\left(\mu\frac{dv_z}{dx}\right) + \rho g \cos\beta \tag{3.5–6}$$

Insertion of Eq. 2.2–22 then gives

$$0 = \mu_0\frac{d}{dx}\left(e^{-\alpha(x/\delta)}\frac{dv_z}{dx}\right) + \rho g \cos\beta \tag{3.5–7}$$

This equation may then be integrated with the boundary conditions that $v_z = 0$ at $x = \delta$ and $dv_z/dx = 0$ at $x = 0$ to get the result in Eq. 2.2–24.

Many other problems can be set up by these methods. Some additional steady-state examples are given below; a few unsteady-state problems are solved in Chapter 4. Many more analytical solutions are available in standard references.[1]

[1] H. Schlichting, *Boundary Layer Theory*, McGraw-Hill, New York (1955), Chapter 4; H. Lamb, *Hydrodynamics*, Dover, New York (1945).

Example 3.5–1. Tangential Annular Flow of a Newtonian Fluid

Determine the velocity and shear stress distributions for the tangential laminar flow of an incompressible fluid between two vertical coaxial cylinders, the outer one of which is rotating with an angular velocity Ω_o. (See Fig. 3.5–1.) End effects may be neglected.

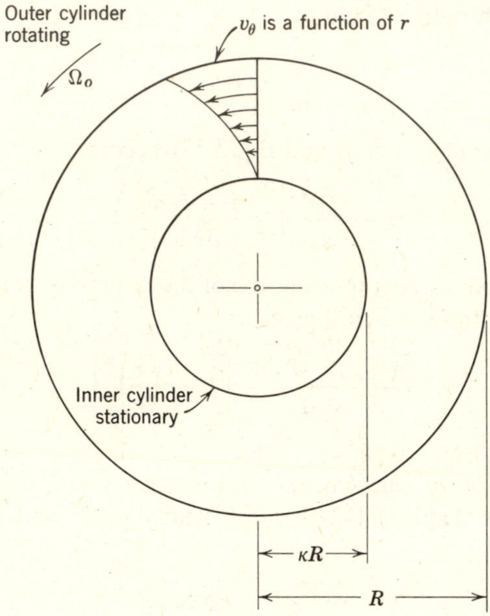

Fig. 3.5–1. Laminar flow of an incompressible fluid in the space between two coaxial cylinders, the outer one of which is rotating with an angular velocity Ω_o.

Solution. In steady-state laminar flow the fluid moves in a circular pattern, and the velocity components v_r and v_z are zero. There is no pressure gradient in the θ-direction. These statements are made on physical grounds. For this system all terms of the equation of continuity as written in cylindrical coordinates (Eq. *B* of Table 3.4–1) are zero, and the equations of motion (Eqs. *D*, *E*, *F* in Table 3.4–3) reduce to

r-component
$$-\rho \frac{v_\theta^2}{r} = -\frac{\partial p}{\partial r} \tag{3.5–8}$$

θ-component
$$0 = \frac{d}{dr}\left(\frac{1}{r}\frac{d}{dr}(rv_\theta)\right) \tag{3.5–9}$$

z-component
$$0 = -\frac{\partial p}{\partial z} + \rho g_z \tag{3.5–10}$$

Equation 3.5–9 may be integrated with respect to r with the boundary conditions: at $r = \kappa R$, $v_\theta = 0$; at $r = R$, $v_\theta = \Omega_o R$. The result is

$$v_\theta = \Omega_o R \frac{\left(\dfrac{\kappa R}{r} - \dfrac{r}{\kappa R}\right)}{\left(\kappa - \dfrac{1}{\kappa}\right)} \tag{3.5–11}$$

The shear stress distribution $\tau_{r\theta}(r)$ may now be obtained with the help of Table 3.4–6. Thus

$$\begin{aligned}\tau_{r\theta} &= -\mu \left\{ r \frac{d}{dr} \left[\frac{\Omega_o R}{r} \frac{\left(\dfrac{\kappa R}{r} - \dfrac{r}{\kappa R}\right)}{\left(\kappa - \dfrac{1}{\kappa}\right)} \right] \right\} \\ &= -2\mu\Omega_o R^2 \left(\frac{1}{r^2}\right)\left(\frac{\kappa^2}{1-\kappa^2}\right) \end{aligned} \tag{3.5–12}$$

The torque $\mathcal{T}$ required to turn the outer shaft may also be easily calculated as the product of the force times the lever arm:

$$\begin{aligned}\mathcal{T} &= 2\pi RL \cdot (-\tau_{r\theta})\big|_{r=R} \cdot R \\ &= 4\pi\mu L\Omega_o R^2 \left(\frac{\kappa^2}{1-\kappa^2}\right)\end{aligned} \tag{3.5–13}$$

The system considered here is a reasonably good model for certain kinds of friction bearings. Such systems are also frequently used to determine fluid viscosities from observations of torque and angular velocities. Viscometers of this kind are known as Couette-Hatschek viscometers.

Once the velocity distribution has been calculated, then the radial pressure distribution may be calculated from Eq. 3.5–8. For a discussion of the integration of this equation the reader is referred to the literature.[2]

Laminar flow in this system is strongly stabilized by centrifugal forces. Thus a fluid particle from an outer layer opposes being moved inwards because the centrifugal force on it is greater than on particles nearer the axis of rotation. At the same time, its outward movement is resisted by the higher centrifugal force on particles it would have to replace. As a result, the transition to turbulent flow takes place at a much higher Reynolds number here than in the corresponding system in which the inner cylinder is rotating and in which the centrifugal force tends to introduce instability. Both systems have been investigated,[3] and their

[2] R. B. Bird, C. F. Curtiss, and W. E. Stewart, *Chem. Eng. Sci.*, **11,** 114–117 (1959).

[3] These two systems are briefly discussed by H. Schlichting, *op. cit.*, pp. 355–357 and 359–360. Systems with the outer cylinder rotating have been investigated by H. Schlichting, *Nach. Ges. Wiss., Göttingen, Math. Physik. Kl.*, 160 (1932). Systems with rotating inner cylinders have been very thoroughly studied by G. I. Taylor: *Phil. Trans.*, **A223,** 289 (1923); *Proc. Roy. Soc.* (London), **A151,** 494 (1935); *ibid.*, **A157,** 546 and 565 (1936).

transition Reynolds numbers are found to be strongly dependent upon the ratio of the annulus thickness to the radius of the outer cylinder $(1 - \kappa)$. When the *outer* cylinder is rotating, the transition Reynolds number, defined as $(\Omega_o R^2 \rho/\mu)_{\text{trans}}$, reaches a minimum of about 50,000 when $(1 - \kappa)$ is about 0.05, as shown in Fig. 3.5–2. When the *inner* cylinder is rotating at an angular velocity of Ω_i (and the outer one is stationary), the transition Reynolds number may be expressed approximately as

$$(\Omega_i \kappa R^2 \rho/\mu)_{\text{trans}} \doteq 41.3/(1 - \kappa)^{3/2} \tag{3.5-14}$$

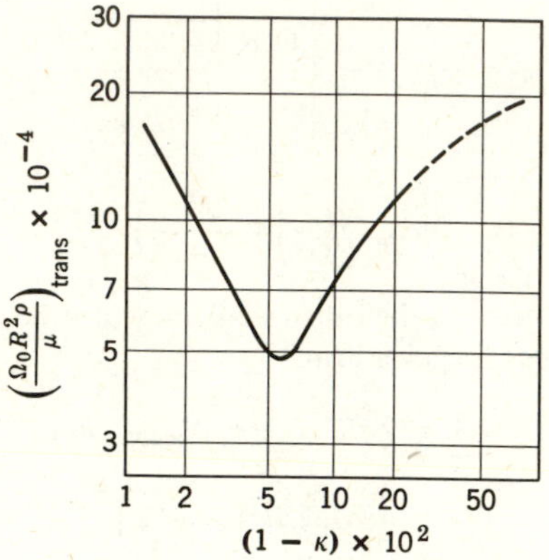

Fig. 3.5–2. Critical Reynolds number for tangential flow in annulus; outer cylinder rotating and inner cylinder stationary. [H. Schlichting, *Boundary Layer Theory*, McGraw-Hill, New York (1955), p. 357.]

The angular velocity distributions for these two systems are quite similar in the stable laminar range.[4]

Example 3.5–2. Shape of the Surface of a Rotating Liquid

A fluid of constant density and viscosity is in a cylindrical container of radius R, as shown in Fig. 3.5–3. The container is caused to rotate about its own axis at an angular velocity Ω. The cylinder axis is vertical so that $g_r = g_\theta = 0$ and $g_z = -g$. Find the shape of the free surface when steady state has been established.

Solution. This system is described most easily in cylindrical coordinates, and we therefore use the equations of change in this coordinate system, given in Table 3.4–3. At steady state we know that $v_z = v_r = 0$ and that v_θ is a function of r alone. We also know that pressure will depend upon r because of the centrifugal force and upon z because of gravitational force.

[4] H. Schlichting, *op. cit.*, p. 64.

As in Example 3.5–1, there is no contribution from the equation of continuity, and the equation of motion reduces to

r-component
$$\rho \frac{v_\theta^2}{r} = \frac{\partial p}{\partial r} \tag{3.5–15}$$

θ-component
$$0 = \mu \frac{\partial}{\partial r}\left(\frac{1}{r}\frac{\partial}{\partial r}(r v_\theta)\right) \tag{3.5–16}$$

z-component
$$0 = -\frac{\partial p}{\partial z} - \rho g \tag{3.5–17}$$

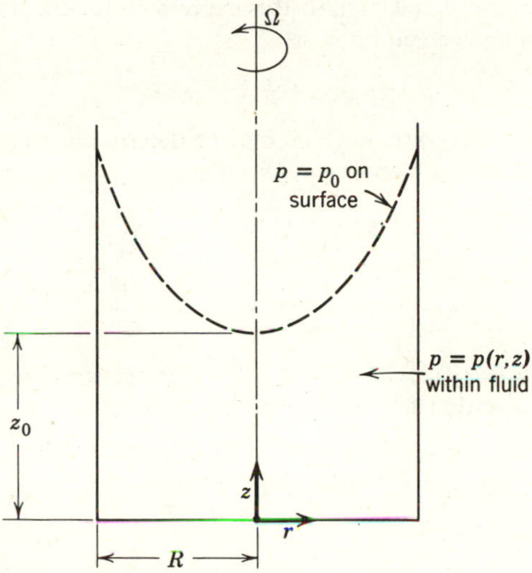

Fig. 3.5–3. Rotating liquid with a free surface, the shape of which is a paraboloid of revolution.

The θ-equation may be integrated immediately to give

$$v_\theta = \frac{1}{2} C_1 r + \frac{C_2}{r} \tag{3.5–18}$$

in which C_1 and C_2 are constants of integration. Because v_θ cannot be infinite at $r = 0$, the constant C_2 is set equal to zero. We know further that at $r = R$ the velocity v_θ is $R\Omega$. Hence C_1 may be evaluated and we get

$$v_\theta = \Omega r \tag{3.5–19}$$

This states that each element of the rotating fluid moves as the elements of a rigid body. This result may be substituted into the r-component of the equation of motion.

Hence we have two expressions for the pressure gradients:

$$\frac{\partial p}{\partial r} = \rho\Omega^2 r \tag{3.5–20}$$

$$\frac{\partial p}{\partial z} = -\rho g \tag{3.5–21}$$

Since p is an analytic function of position, we may write

$$dp = \frac{\partial p}{\partial r}\,dr + \frac{\partial p}{\partial z}\,dz \tag{3.5–22}$$

By substituting Eqs. 3.5–20 and 21 into this expression for the total differential of pressure and integrating, we get

$$p = -\rho g z + \tfrac{1}{2}\rho\Omega^2 r^2 + C \tag{3.5–23}$$

Here C is a constant of integration, which may be determined by making use of the statement that $p = p_0$ at $r = 0$ and $z = z_0$ so that

$$p_0 = -\rho g z_0 + C \tag{3.5–24}$$

Hence

$$p - p_0 = -\rho g(z - z_0) + \frac{\rho\Omega^2 r^2}{2} \tag{3.5–25}$$

The locus of the free surface consists of *all* points on the free surface—that is, where $p = p_0$. The equation for the free surface is thus

$$0 = -\rho g(z - z_0) + \frac{\rho\Omega^2 r^2}{2} \tag{3.5–26}$$

or

$$z - z_0 = \left(\frac{\Omega^2}{2g}\right) r^2 \tag{3.5–27}$$

This is the equation of a parabola. The reader might like to verify that the free surface of a fluid in a rotating annular container obeys a similar relation.

Example 3.5–3. Torque Relationships and Velocity Distribution in the Cone-and-Plate Viscometer[5]

The cone-and-plate viscometer, shown schematically in Fig. 3.5–4, consists essentially of a stationary flat plate, upon which is placed a puddle of the liquid or paste to be tested, and an inverted cone, which is lowered into the puddle until its apex just contacts the plate. The cone is rotated at a known angular velocity Ω, and the viscosity of the fluid is determined by measuring the torque required to turn the cone. In practice, the angle θ_0 between the conical and flat surfaces is kept

[5] The authors wish to acknowledge discussions with Professor A. G. Fredrickson (University of Minnesota) and Professor J. C. Slattery (Northwestern University) in connection with this problem.

small, say, about one half of a degree. This kind of instrument offers some important advantages, particularly in the case of non-Newtonian fluids:

a. Only one stress component $\tau_{\theta\phi}$ is important.
b. The magnitude of $\tau_{\theta\phi}$ is very nearly constant throughout the fluid.
c. End effects can be almost completely eliminated.

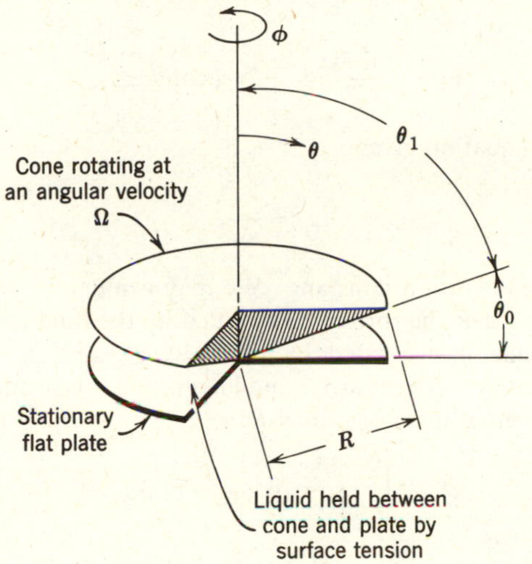

Fig. 3.5–4. Cutaway view of the cone-and-plate viscometer.

Analyze the system in the following way: (i) show that only $\tau_{\theta\phi}$ is important; (ii) determine $\tau_{\theta\phi}$ as a function of r and θ; (iii) use the results of (i) and (ii) to determine $v_\phi(r, \theta)$ for steady flow of a Newtonian fluid with constant μ and ρ—this expression will contain the torque $\mathscr{T}$; (iv) obtain an alternate expression for $v_\phi(r, \theta)$ containing the angular velocity rather than the torque.

Solution. *a.* If it is assumed that the flow is entirely tangential, then v_ϕ is a function of r and θ and $v_r = v_\theta = 0$. From Table 3.4–7 it can be seen that the only possible nonvanishing components of $\boldsymbol{\tau}$ are $\tau_{r\phi}$ and $\tau_{\theta\phi}$. Therefore, the three components of the equation of motion (see Eqs. *A*, *B*, and *C* of Table 3.4–4) are

r-component
$$-\rho \frac{v_\phi^2}{r} = -\frac{\partial p}{\partial r} \tag{3.5–28a}$$

θ-component
$$-\rho \cot\theta \frac{v_\phi^2}{r} = -\frac{1}{r}\frac{\partial p}{\partial \theta} \tag{3.5–28b}$$

φ-component
$$0 = -\left(\frac{1}{r^2}\frac{\partial}{\partial r}(r^2\tau_{r\phi}) + \frac{1}{r}\frac{\partial \tau_{\theta\phi}}{\partial \theta} + \frac{\tau_{r\phi}}{r} + 2\cot\theta \frac{\tau_{\theta\phi}}{r}\right) \tag{3.5–28c}$$

Next we assume "creeping flow"; that is, we assume that the flow is sufficiently slow that terms containing v_ϕ^2 can be set equal to zero. This leaves us with only the ϕ-component of the equation of motion.

Then we postulate[6] that the velocity distribution will be of the form $v_\phi(r, \theta) = rf(\theta)$. We are led to try this functional form because it satisfies the boundary conditions at both $\theta = \theta_1$ and $\theta = \pi/2$. By this hypothesis that the angular velocity v_ϕ/r is independent of r, we find from Eq. F of Table 3.4–7 that $\tau_{r\phi} = 0$.

b. When $\tau_{r\phi}$ is set equal to zero, we get the following ordinary differential equation for $\tau_{\theta\phi}$:

$$\frac{d\tau_{\theta\phi}}{d\theta} = -2\tau_{\theta\phi} \cot \theta \tag{3.5–29}$$

Integration of this equation gives

$$\tau_{\theta\phi} = \frac{C_1}{\sin^2 \theta} \tag{3.5–30}$$

in which C_1 is an integration constant. We may evaluate C_1 from the boundary condition at $\theta = \pi/2$ that the torque transmitted by the fluid to the plate is known to be $\mathscr{T}$. The torque is evaluated by multiplying $\tau_{\theta\phi}|_{\theta=\pi/2}$ by the differential area $r\,dr\,d\phi$ and then by the lever arm r and integrating this product over the surface of the completely wetted plate of radius R:

$$\begin{aligned} \mathscr{T} &= \int_0^{2\pi} \int_0^R \tau_{\theta\phi}|_{\theta=\pi/2}\, r^2\, dr\, d\phi \\ &= (2\pi)\left(\frac{R^3}{3}\right)\left(\frac{C_1}{\sin^2 \frac{\pi}{2}}\right) \end{aligned} \tag{3.5–31}$$

Combination of Eqs. 3.5–30 and 31 then gives

$$\tau_{\theta\phi} = \frac{3\mathscr{T}}{2\pi R^3 \sin^2 \theta} \tag{3.5–32}$$

If θ_0 is small, $\sin^2 \theta$ will be very nearly unity, and $\tau_{\theta\phi}$ will be very nearly independent of position.

c. From Eq. E of Table 3.4–7 we obtain the relationship between $\tau_{\theta\phi}$ and the gradient of v_ϕ/r. When this expression is inserted into Eq. 3.5–32, we obtain the following ordinary differential equation for the local angular velocity v_ϕ/r:

$$-\mu \sin \theta \frac{d}{d\theta}\left(\frac{v_\phi/r}{\sin \theta}\right) = \frac{3\mathscr{T}}{2\pi R^3 \sin^2 \theta} \tag{3.5–33}$$

By separating variables and integrating, we get the angular velocity distribution:

$$\frac{v_\phi}{r} = \frac{3\mathscr{T}}{4\pi R^3 \mu}\left[\cot \theta + \tfrac{1}{2}\left(\ln \frac{1 + \cos \theta}{1 - \cos \theta}\right) \sin \theta\right] + C_2 \tag{3.5–34}$$

The constant of integration C_2 is zero, inasmuch as $v_\phi = 0$ at $\theta = \pi/2$.

[6] See Problem 3.T for an alternate viewpoint.

d. We may now write Eq. 3.5–34 for the special case that $\theta = \theta_1 = \frac{\pi}{2} - \theta_0$ and $v_\phi = \Omega r \sin \theta_1$.

$$\Omega \sin \theta_1 = \frac{3\mathscr{T}}{4\pi R^3 \mu}\left[\cot \theta_1 + \frac{1}{2}\left(\ln \frac{1 + \cos \theta_1}{1 - \cos \theta_1}\right) \sin \theta_1\right] \tag{3.5–35}$$

(Note that this expression enables one to determine μ from measurements of $\mathscr{T}$ and Ω.)

Division of Eq. 3.5–34 by Eq. 3.5–35 then causes $\mathscr{T}$ to be eliminated, and thereby we get v_ϕ in terms of Ω:

$$\frac{v_\phi}{r} = \Omega \sin \theta_1 \left[\frac{\cot \theta + \frac{1}{2}\left(\ln \frac{1 + \cos \theta}{1 - \cos \theta}\right) \sin \theta}{\cot \theta_1 + \frac{1}{2}\left(\ln \frac{1 + \cos \theta_1}{1 - \cos \theta_1}\right) \sin \theta_1}\right] \tag{3.5–36}$$

For θ and θ_1 very nearly equal to $\pi/2$, this expression may be very closely approximated by

$$\frac{v_\phi}{r} \doteq \Omega \frac{\cos \theta}{\cos \theta_1} \doteq \Omega \left(\frac{\frac{\pi}{2} - \theta}{\frac{\pi}{2} - \theta_1}\right) \tag{3.5–37}$$

Equation 3.5–36 can also be derived by an alternate method described in Problem 3.T.

§3.6 THE EQUATIONS OF CHANGE FOR INCOMPRESSIBLE NON-NEWTONIAN FLOW

The equations of motion given in terms of $\boldsymbol{\tau}$ (see Eqs. *A*, *B*, and *C* in Tables 3.4–2, 3, and 4) may be used for describing non-Newtonian flow. In order to use them, however, we need relations between the components of $\boldsymbol{\tau}$ and the various velocity gradients; in other words, we have to replace the expressions given in Tables 3.4–5, 6, and 7 by other relations appropriate for the non-Newtonian fluid of interest. Therefore most of this section is devoted to the expressions for $\boldsymbol{\tau}$ for non-Newtonian models.

In §1.1 the coefficient of viscosity is defined in terms of Newton's law of viscosity. Then in §1.2 it is pointed out that certain classes of fluids are not described by Newton's law; it is further pointed out that some empirical "non-Newtonian models" are useful for describing the rheological behavior of some of the simpler non-Newtonian materials.

In §3.2 we have given Newton's law of viscosity in somewhat more general form; in Eqs. 3.2–11 through 16 we find the expressions in cartesian coordinates, and the analogous expressions for cylindrical and spherical coordinates are also given. In this section we wish to perform a similar generalization of the non-Newtonian models. We restrict this discussion, however, to *incompressible* flow.

First we recall from §3.2 that Newton's law of viscosity for an incompressible fluid is

$$\boldsymbol{\tau} = -\mu\boldsymbol{\Delta} \tag{3.6–1}$$

in which $\boldsymbol{\Delta}$ is the symmetrical "rate of deformation tensor" with cartesian components $\Delta_{ij} = (\partial v_i/\partial x_j) + (\partial v_j/\partial x_i)$. The coefficient of viscosity depends on the local pressure and temperature but *not* on $\boldsymbol{\tau}$ or $\boldsymbol{\Delta}$.

For non-Newtonian materials, the relation between $\boldsymbol{\tau}$ and $\boldsymbol{\Delta}$ is not the simple proportionality given in Eq. 3.6–1. We can, however, write for certain simple types of non-Newtonian fluids

$$\boldsymbol{\tau} = -\eta\boldsymbol{\Delta} \tag{3.6–2}$$

in which the non-Newtonian viscosity η, a scalar, is a function of $\boldsymbol{\Delta}$ (or a function of $\boldsymbol{\tau}$) as well as of temperature and pressure. The assumption of various empirical functions to describe the dependence of η on $\boldsymbol{\Delta}$ (or on $\boldsymbol{\tau}$) corresponds to the assumption of the various "models" given in §1.2.

In order for η to be a scalar function of the tensor $\boldsymbol{\Delta}$, it must depend only on the "invariants" of $\boldsymbol{\Delta}$. The invariants are those special combinations of the components of $\boldsymbol{\Delta}$ that transform as scalars under a rotation of the coordinate system:

$$I_1 = (\boldsymbol{\Delta} : \boldsymbol{\delta}) = \Sigma_i \Delta_{ii} \tag{3.6–3}$$

$$I_2 = (\boldsymbol{\Delta} : \boldsymbol{\Delta}) = \Sigma_i \Sigma_j \Delta_{ij} \Delta_{ji} \tag{3.6–4}$$

$$I_3 = \det \boldsymbol{\Delta} = \Sigma_i \Sigma_j \Sigma_k \epsilon_{ijk} \Delta_{1i} \Delta_{2j} \Delta_{3k} \tag{3.6–5}$$

The first invariant can easily be shown to be $2(\nabla \cdot \boldsymbol{v})$, which is zero for an incompressible fluid. Hence

$$\eta = \eta(I_2, I_3) \tag{3.6–6}$$

For "viscometric flows" (e.g., axial tube flow, axial and tangential annular flow, flow in a film) the third invariant vanishes identically. It is commonly assumed that I_3 is not very important in other flows. Hence it is customary to assume[1] that η can be taken to be some function of $(\boldsymbol{\Delta} : \boldsymbol{\Delta})$. For some materials models have been proposed for which η depends on the momentum flux; in that case η can be taken to be some function of $(\boldsymbol{\tau} : \boldsymbol{\tau})$.

With the foregoing development in mind, we are now in a position to

[1] This was apparently first suggested by K. Hohenemser and W. Prager, *Z. angew. Math. Mech.*, **12**, 216–226 (1932).

rewrite the non-Newtonian models of §1.2 in a form that allows us to describe flow in complex geometries:

Bingham model $$\boldsymbol{\tau} = -\left\{\mu_0 + \frac{\tau_0}{\left|\sqrt{\frac{1}{2}(\boldsymbol{\Delta}:\boldsymbol{\Delta})}\right|}\right\}\boldsymbol{\Delta} \quad \text{for } \tfrac{1}{2}(\boldsymbol{\tau}:\boldsymbol{\tau}) > \tau_0^2 \tag{3.6–7}$$

$$\boldsymbol{\Delta} = 0 \qquad \text{for } \tfrac{1}{2}(\boldsymbol{\tau}:\boldsymbol{\tau}) < \tau_0^2 \tag{3.6–8}$$

Ostwald-de Waele model $$\boldsymbol{\tau} = -\{m\,|\sqrt{\tfrac{1}{2}(\boldsymbol{\Delta}:\boldsymbol{\Delta})}|^{n-1}\}\boldsymbol{\Delta} \tag{3.6–9}$$

Reiner-Philippoff model $$\boldsymbol{\tau} = -\left\{\mu_\infty + \frac{\mu_0 - \mu_\infty}{1 + \dfrac{\frac{1}{2}(\boldsymbol{\tau}:\boldsymbol{\tau})}{\tau_0^2}}\right\}\boldsymbol{\Delta} \tag{3.6–10}$$

The utility of these expressions is that one can write down the components of $\boldsymbol{\tau}$ for non-Newtonian materials in cylindrical and spherical coordinates by the following procedure: (*a*) look up the corresponding Newtonian relations in Tables 3.4–5, 6, and 7; (*b*) replace μ in the Newtonian expression by the quantity in the braces in the desired model; and (*c*) insert the relation for $(\boldsymbol{\Delta}:\boldsymbol{\Delta})$ or $(\boldsymbol{\tau}:\boldsymbol{\tau})$ in the proper coordinates. Actually $\frac{1}{2}(\boldsymbol{\Delta}:\boldsymbol{\Delta})$ turns out to be exactly the function Φ_v in Table 3.4–8 with the $-\frac{2}{3}(\nabla \cdot \boldsymbol{v})^2$ term omitted; hence it is readily available in rectangular, cylindrical, and spherical coordinates. We illustrate this procedure presently with an example.

Before concluding this section, several points should be made. Expressions such as those in Eqs. 3.6–7, 8, 9, and 10 are *empirical* expressions designed to approximate the actual behavior of various materials; the only thing we have done here is to write the expressions in such a way that they transform properly in going from one coordinate system to another. Otherwise, we could not be assured that the parameters—*m* and *n*, for example—would be the same when determined in various geometrical arrangements. It should be emphasized that we have neglected the effect of I_3 largely because of lack of experimental information of its importance. Furthermore, we have restricted the discussion even more by the assumption that Eq. 3.6–2 can be used to describe non-Newtonian fluid behavior.

Equation 3.6–2 can describe non-Newtonian viscosity, but it is not capable of describing the "normal stresses" (stresses normal to the shearing surfaces in viscometric flows[2]). An equation which includes Eq. 3.6–2 but which can describe properly the normal stresses in steady-state viscometric

[2] B. D. Coleman, H. Markovitz, and W. Noll, *Viscometric Flows of Non-Newtonian Fluids*, Springer, Berlin (1966); all the specific flow examples discussed herein can be described exactly by Eq. 3.6–11.

flows is a special case of the Rivlin-Erickson equation due to Erickson:[3]

$$\boldsymbol{\tau} = -\eta\boldsymbol{\Delta} + \tfrac{1}{2}\theta \frac{\mathscr{D}}{\mathscr{D}t}\boldsymbol{\Delta} - \tfrac{1}{2}(\theta + 2\beta)\{\boldsymbol{\Delta}\cdot\boldsymbol{\Delta}\} \tag{3.6–11}$$

in which $\mathscr{D}/\mathscr{D}t$ is the corotational derivative defined by:

$$\frac{\mathscr{D}}{\mathscr{D}t}\boldsymbol{\Delta} = \frac{\partial}{\partial t}\boldsymbol{\Delta} + \{\boldsymbol{v}\cdot\nabla\boldsymbol{\Delta}\} + \tfrac{1}{2}(\{\boldsymbol{\Omega}\cdot\boldsymbol{\Delta}\} - \{\boldsymbol{\Delta}\cdot\boldsymbol{\Omega}\}) \tag{3.6–12}$$

Here $\boldsymbol{\Omega}$ is the vorticity tensor with cartesian components $\Omega_{ij} = \partial v_j/\partial x_i - \partial v_i/\partial x_j$. To describe time-dependent viscoelastic effects and non-viscometric flows, as are often encountered in polymer processing, still more complicated expressions for $\boldsymbol{\tau}$ have to be used.[4]

Example 3.6–1. Tangential Annular Flow of a Bingham Plastic

Determine the velocity and stress distributions for the flow of a Bingham plastic in the apparatus of Example 3.5–1, as functions of the torque $\mathscr{T}$ applied to the outer cylinder. Again assume incompressible laminar flow and ignore end effects.

Solution. For this system $v_r = v_z = 0$ and $v_\theta = v_\theta(r)$. Hence the only nonvanishing component of $\boldsymbol{\tau}$ is $\tau_{r\theta}$, and the equation of motion for steady state is

$$0 = \frac{1}{r^2}\frac{d}{dr}(r^2\tau_{r\theta}) \tag{3.6–12}$$

This may be integrated to give

$$\tau_{r\theta} = \frac{C_1}{r^2} \tag{3.6–13}$$

If the torque at the outer cylinder is known to be $\mathscr{T}$, then

$$\mathscr{T} = -\tau_{r\theta}\big|_{r=R}\cdot 2\pi RL\cdot R \tag{3.6–14}$$

where the minus sign is chosen because the θ-momentum flux is in the $-r$ direction. Hence $C_1 = -\mathscr{T}/2\pi L$ and

$$\tau_{r\theta} = -\frac{\mathscr{T}}{2\pi Lr^2} \tag{3.6–15}$$

This result, which is good for any kind of fluid, can also be obtained by recognizing that angular momentum must be transmitted undiminished from the outer cylinder to the inner one.

For the Bingham model, the analytical expression to be used depends on the value of $\frac{1}{2}(\boldsymbol{\tau}:\boldsymbol{\tau}) = \frac{1}{2}\Sigma_i\Sigma_j\tau_{ij}^2$. Because $\tau_{r\theta}$ is the only nonvanishing component of $\boldsymbol{\tau}$, we have

$$\tfrac{1}{2}(\boldsymbol{\tau}:\boldsymbol{\tau}) = \tau_{r\theta}^2 \tag{3.6–16}$$

[3] R. S. Rivlin and J. L. Ericksen, J. Rat. Mech. and Anal., **4**, 323–425 (1955), Eq. 37.10; J. L. Ericksen, in *Viscoelasticity* (ed. J. T. Bergen), Academic Press, New York (1960), Eq. 2.12.

[4] Introductions to the fluid dynamics of viscoelastic fluids have been given by J. G. Oldroyd, Proc. Roy. Soc., **A200,** 523–541 (1950); **A245,** 278–297 (1958); A. S. Lodge, *Elastic Liquids*, Academic Press, New York (1964); A. G. Fredrickson *Principles and Applications of Rheology*, Prentice-Hall, New York (1963); and R. B. Bird, *Chem. Engr. Prog. Symp. Series, No.* 58, **61** (1965), Chapter 6.

Hence we use Eq. 3.6–7 when $|\tau_{r\theta}| > \tau_0$ (that is, when the critical shear stress is exceeded) and Eq. 3.6–8 when $|\tau_{r\theta}| < \tau_0$. From Eq. 3.6–15, we find that we can define a quantity r_0, which is that value of r for which $|\tau_{r\theta}| = \tau_0$:

$$r_0 = \sqrt{\left|\frac{\mathscr{T}}{2\pi\tau_0 L}\right|} \tag{3.6–17}$$

We can then distinguish three situations:

a. If $r_0 \leqslant \kappa R$, then there will be no fluid motion at all.

b. If $R > r_0 \geqslant \kappa R$ then there will be viscous flow in the region $\kappa R < r < r_0$ and "plug flow" for $r_0 < r < R$.

c. If $r_0 \geqslant R$, then there is flow throughout.

Next we write Eq. 3.6–7 for the system at hand. Inasmuch as

$$\tfrac{1}{2}(\boldsymbol{\Delta}:\boldsymbol{\Delta}) = \left[r\frac{d}{dr}\left(\frac{v_\theta}{r}\right)\right]^2$$

we get

$$\tau_{r\theta} = -\left\{\mu_0 + \frac{\tau_0}{r\dfrac{d}{dr}\left(\dfrac{v_\theta}{r}\right)}\right\} r\frac{d}{dr}\left(\frac{v_\theta}{r}\right)$$

$$= -\tau_0 - \mu_0 r\frac{d}{dr}\left(\frac{v_\theta}{r}\right) \tag{3.6–18}$$

Since v_θ/r does not decrease with increasing r, $\tau_{r\theta}$ is negative. That is, the θ-component of momentum flows in the $-r$-direction.

By substituting Eq. 3.6–18 into Eq. 3.6–15 and integrating, we get for Case *b*

$$\frac{v_\theta}{r} = \Omega + \frac{\mathscr{T}}{4\pi L\mu_0 r_0^2}\left[1 - \left(\frac{r_0}{r}\right)^2\right] - \frac{\tau_0}{\mu_0}\ln\left(\frac{r}{r_0}\right) \qquad \text{for } \kappa R \leqslant r \leqslant r_0 \tag{3.6–19}$$

$$\frac{v_\theta}{r} = \Omega \qquad \text{for } r_0 \leqslant r \leqslant R \tag{3.6–20}$$

in which the boundary condition $v_\theta/r = \Omega$ at $r = r_0$ has been used. For Case *c*, we use the boundary condition that $v_\theta/r = \Omega$ at $r = R$ and get

$$\frac{v_\theta}{r} = \Omega + \frac{\mathscr{T}}{4\pi L\mu_0 R^2}\left[1 - \left(\frac{R}{r}\right)^2\right] - \frac{\tau_0}{\mu_0}\ln\left(\frac{r}{R}\right) \qquad \kappa R \leqslant r \leqslant R \tag{3.6–21}$$

Now for Case *c*, if we utilize the condition that at $r = \kappa R$, $v_\theta = 0$, we get

$$\Omega = \frac{\mathscr{T}}{4\pi L\mu_0 R^2}\left(\frac{1}{\kappa^2} - 1\right) + \frac{\tau_0}{\mu_0}\ln\kappa \tag{3.6–22}$$

This relation between Ω and $\mathscr{T}$—known as the Reiner-Riwlin equation[6]—gives a means for determining μ_0 and τ_0 from viscometric data.

[6] M. Reiner and R. Riwlin, *Kolloid-Z.*, **43,** 1 (1927).

Example 3.6–2. Components of the Momentum Flux Tensor for Non-Newtonian Radial Flow between Two Parallel Disks

Consider the radial creeping flow of a power-model fluid between two circular disks shown in Fig. 3.6–1. Neglect entrance and exit effects.

a. Does I_3 vanish for this system?
b. What are the components of $\boldsymbol{\tau}$ according to Eq. 3.6–9?

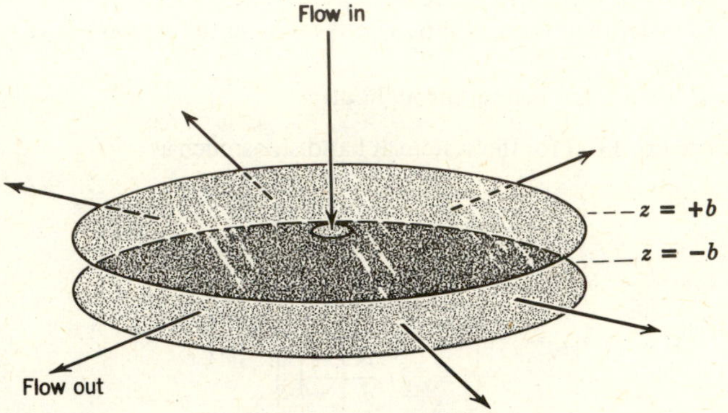

Fig. 3.6–1. Outward radial flow between two parallel disks of radius r_2. The radius of the entrance hole is r_1. The plates are a distance $2b$ apart.

Solution. *a.* For this system we use cylindrical coordinates, and, for creeping flow, we note that $v_z = v_\theta = 0$ and also that $v_r = v_r(r, z)$. Hence $\boldsymbol{\Delta}$ is

$$\boldsymbol{\Delta} = \begin{pmatrix} 2(\partial v_r/\partial r) & 0 & (\partial v_r/\partial z) \\ 0 & 2(v_r/r) & 0 \\ (\partial v_r/\partial z) & 0 & 0 \end{pmatrix} \tag{3.6–23}$$

We see that det $\boldsymbol{\Delta} \neq 0$ and that the third invariant does *not* vanish.

b. For cylindrical coordinates and the statements in (*a*) regarding the velocity components,

$$\tfrac{1}{2}(\boldsymbol{\Delta}:\boldsymbol{\Delta}) = 2\left(\frac{\partial v_r}{\partial r}\right)^2 + 2\left(\frac{v_r}{r}\right)^2 + \left(\frac{\partial v_r}{\partial z}\right)^2 \tag{3.6–24}$$

This result may be obtained either by forming the double-dot product of $\boldsymbol{\Delta}$ with itself or by using Eq. *B* of Table 3.4–8 along with the relation $\Phi_v = \frac{1}{2}(\boldsymbol{\Delta}:\boldsymbol{\Delta})$. Hence the nonvanishing components of $\boldsymbol{\tau}$ are

$$\tau_{rr} = -2\{m\left|\sqrt{\tfrac{1}{2}(\boldsymbol{\Delta}:\boldsymbol{\Delta})}\right|^{n-1}\}\frac{\partial v_r}{\partial r} \tag{3.6–25}$$

$$\tau_{\theta\theta} = -2\{m\left|\sqrt{\tfrac{1}{2}(\boldsymbol{\Delta}:\boldsymbol{\Delta})}\right|^{n-1}\}\frac{v_r}{r} \tag{3.6–26}$$

$$\tau_{rz} = \tau_{zr} = -\{m\left|\sqrt{\tfrac{1}{2}(\boldsymbol{\Delta}:\boldsymbol{\Delta})}\right|^{n-1}\}\frac{\partial v_r}{\partial z} \tag{3.6–27}$$

§3.7 DIMENSIONAL ANALYSIS OF THE EQUATIONS OF CHANGE[1]

Before leaving the subject of the general formulation of these partial differential equations and their significance, we discuss them from the standpoint of dimensional analysis. For simplicity, the development below is limited to systems of constant density and viscosity; however, it can easily be extended to situations in which these quantities vary.[2]

For many flow systems one can select a characteristic length D and a characteristic velocity V. Thus, for flow in a circular tube, D is usually taken to be the tube diameter and V, the average velocity of flow. For flow around a sphere, D is usually taken to be the diameter of the sphere and V, the velocity of the fluid far from the sphere, relative to that of the sphere, the so-called "approach velocity." The choice is an arbitrary one, but in any case it should be carefully specified. Once this choice has been made, we may define the following dimensionless variables and differential operations:

$$\boldsymbol{v}^* = \frac{\boldsymbol{v}}{V}; \qquad p^* = \frac{(p - p_0)}{\rho V^2}; \qquad t^* = \frac{tV}{D} \tag{3.7–1,2,3}$$

$$x^* = \frac{x}{D}; \qquad y^* = \frac{y}{D}; \qquad z^* = \frac{z}{D} \tag{3.7–4}$$

$$\nabla^* = D\nabla = \left(\boldsymbol{\delta}_1 \frac{\partial}{\partial x^*} + \boldsymbol{\delta}_2 \frac{\partial}{\partial y^*} + \boldsymbol{\delta}_3 \frac{\partial}{\partial z^*}\right) \tag{3.7–5}$$ [3]

$$\nabla^{*2} = D^2\nabla^2 = \frac{\partial^2}{\partial x^{*2}} + \frac{\partial^2}{\partial y^{*2}} + \frac{\partial^2}{\partial z^{*2}} \tag{3.7–6}$$

$$\frac{D}{Dt^*} = \left(\frac{D}{V}\right)\frac{D}{Dt} \tag{3.7–7}$$

In Eq. 3.7–2 p_0 is some convenient reference pressure.

Recall that the equations of continuity and motion for Newtonian fluids of constant density and viscosity are

(equation of continuity)
$$(\nabla \cdot \boldsymbol{v}) = 0 \tag{3.7–8}$$

(equation of motion)
$$\rho \frac{D\boldsymbol{v}}{Dt} = -\nabla p + \mu\nabla^2\boldsymbol{v} + \rho\boldsymbol{g} \tag{3.7–9}$$

[1] A recent book on the problems of scale-up is R. E. Johnstone and M. W. Thring, *Pilot Plants, Models, and Scale-up Methods in Chemical Engineering*, McGraw-Hill, New York (1957). The dimensional analysis of the Navier-Stokes equation is discussed here on pp. 50 *et seq.* Somewhat more detailed analyses are given by H. Schlichting, *Boundary Layer Theory*, McGraw-Hill, New York (1955).

[2] Compressibility effects are considered by H. Schlichting, *op. cit.*, pp. 248 *et seq.*

[3] The $\boldsymbol{\delta}_i$ are "unit vectors." (See §A.2.)

We may rewrite these two equations in terms of the foregoing dimensionless variables by setting $\boldsymbol{v} = \boldsymbol{v}^* V$, $(p - p_0) = p^* \rho V^2$, etc.:

$$\left(\frac{1}{D} \nabla^* \cdot \boldsymbol{v}^* V\right) = 0 \qquad (3.7\text{–}10)$$

$$\rho\left(\frac{V}{D}\right) \frac{D}{Dt^*} (\boldsymbol{v}^* V) = -\left(\frac{1}{D} \nabla^* \cdot p^* \rho V^2\right) + \mu \frac{1}{D^2} \nabla^{*2}(\boldsymbol{v}^* V) + \rho \boldsymbol{g} \qquad (3.7\text{–}11)$$

Multiplication of Eq. 3.7–10 by D/V and Eq. 3.7–11 by $D/\rho V^2$ gives

$$(\nabla^* \cdot \boldsymbol{v}^*) = 0 \qquad (3.7\text{–}12)$$

$$\frac{D\boldsymbol{v}^*}{Dt^*} = -\nabla^* p^* + \left[\frac{\mu}{DV\rho}\right] \nabla^{*2} \boldsymbol{v}^* + \left[\frac{gD}{V^2}\right] \frac{\boldsymbol{g}}{g} \qquad (3.7\text{–}13)[4]$$

Note that in these dimensionless forms of the equations of change the "scale factors," that is, those variables describing the over-all size and speed of the system and its physical properties, are concentrated in two dimensionless groups. These groups occur so often in engineering studies that they have been given names in honor of two of the pioneers in fluid mechanics:

$$\text{Re} = \left[\frac{DV\rho}{\mu}\right] = \text{Reynolds number} \qquad (3.7\text{–}14)$$

$$\text{Fr} = \left[\frac{V^2}{gD}\right] = \text{Froude number} \qquad (3.7\text{–}15)$$

Now, if in two different systems the scale factors are such that the Froude and Reynolds numbers are the same for both, then both systems are described by identical dimensionless differential equations. If, in addition, the dimensionless initial and boundary conditions are the same (this is possible only if the two different systems are geometrically similar), then the two systems are mathematically identical; that is, the dimensionless velocity distribution $\boldsymbol{v}^*(x^*, y^*, z^*, t^*)$ and the dimensionless pressure distribution $p^*(x^*, y^*, z^*, t^*)$ are the same in each. Such systems are said to be "dynamically similar." In the scale-up of processes that are not well understood it is often desirable to maintain dynamic similarity, as indicated in the example below.

Example 3.7–1. Prediction of Vortex Depth in an Agitated Tank

It is desired to predict vortex depth for steady-state flow in a large unbaffled tank of oil, shown in Fig. 3.7–1, as a function of agitator speed. We propose to do this

[4] Note that $\boldsymbol{g}/g$ is just a unit vector in the direction of gravity.

by means of a model study in a smaller geometrically similar tank. Determine the conditions under which the model study must be carried out to provide a valid means of prediction.

Solution. The flow pattern in this system is too complex to permit exact calculation; hence the methods of dimensional analysis will be used. The vortex shape will be the same for each tank if the dimensionless differential equations and dimensionless boundary conditions describing the flow are the same. Clearly the

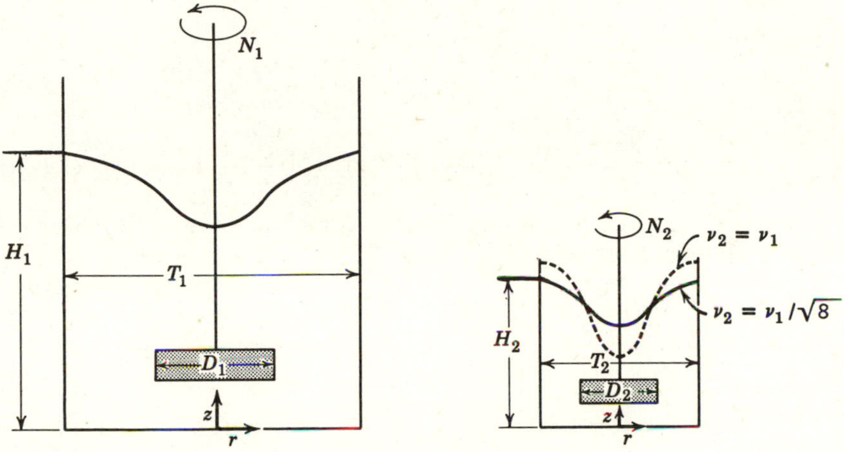

Fig. 3.7-1. Vortex depth in an agitated tank.

differential equations are the equations of continuity and motion; the boundary conditions are

	Large Tank	*Small Tank*	
$\mathbf{v} = 0$	at $z = 0$	at $z = 0$	
	for $0 < r < T_1/2$	for $0 < r < T_2/2$	(3.7–16)
	at $r = T_1/2$	at $r = T_2/2$	
	for $0 < z < H_1$	for $0 < z < H_2$	(3.7–17)
$p = p_0$	at $S_1(r, z)$	at $S_2(r, z)$	(3.7–18)

Here S_1 and S_2 are the vortex surfaces in the large and small tanks, and p_0 is the atmospheric pressure. We may also write that velocity relative to all moving solid surfaces is zero at these surfaces, but the foregoing conditions are sufficient to demonstrate the technique being used. It will be assumed that this is a steady-state operation, hence that no initial condition is necessary. We choose for this system the impeller diameter D as our reference length, and DN, the impeller diameter times the rate of impeller rotation in revolutions per unit time, as our reference velocity. The Reynolds and Froude numbers for this system then become $[D^2N\rho/\mu]$ and $[DN^2/g]$, respectively. The boundary conditions in terms of our dimensionless variables are

	Large Tank	*Small Tank*	
$v^* = 0$	at $z^* = 0$	at $z^* = 0$	
	for $0 < r^* < \dfrac{T_1}{2D_1}$	for $0 < r^* < \dfrac{T_2}{2D_2}$	(3.7–19)
$v^* = 0$	at $r^* = \dfrac{T_1}{2D_1}$	at $r^* = \dfrac{T_2}{2D_2}$	
	for $0 < z^* < \dfrac{H_1}{D_1}$	for $0 < z^* < \dfrac{H_2}{D_2}$	(3.7–20)
$p^* = 0$	at $S_1^*\left(\dfrac{r}{D_1}, \dfrac{z}{D_1}\right)$	at $S_2^*\left(\dfrac{r}{D_2}, \dfrac{z}{D_2}\right)$	(3.7–21)

It can be seen then that for the flow patterns in the two tanks to be similar the following equalities must exist:

$$\frac{T_1}{D_1} = \frac{T_2}{D_2} \tag{3.7–22}$$

$$\frac{H_1}{D_1} = \frac{H_2}{D_2} \tag{3.7–23}$$

$$S_1^*\left(\frac{r}{D_1}, \frac{z}{D_1}\right) = S_2^*\left(\frac{r}{D_2}, \frac{z}{D_2}\right) \tag{3.7–24}$$

$$\frac{D_1^2N_1\rho_1}{\mu_1} = \frac{D_2^2N_2\rho_2}{\mu_2} \tag{3.7–25}$$

$$\frac{D_1N_1^2}{g} = \frac{D_2N_2^2}{g} \tag{3.7–26}$$

Equations 3.7–22 and 23 show the requirement of geometric similarity. Clearly, the more detailed our description of the surfaces of zero velocity, the more of these size ratios we would have. In practice, even the relative smoothness of the tank surfaces and the sizes of bolt heads may be important. Equation 3.7–24 will obviously be satisfied if the vortices are of the same shape, since $S_1^*(r/D_1, z/D_1)$ and $S_2^*(r/D_2, z/D_2)$ are the dimensionless vortex shapes.

Equations 3.7–25 and 26 state the required relationships between the scale factors and are the most interesting from our standpoint. Since it is hardly practical to change the gravitational field, it follows from Eq. 3.7–26 that

$$\frac{N_2}{N_1} = \sqrt{\frac{D_1}{D_2}} \tag{3.7–27}$$

Substituting this expression into Eq. 3.7–25,

$$\frac{\mu_2}{\rho_2} = \frac{\mu_1}{\rho_1}\left(\frac{D_2}{D_1}\right)^{3/2} \tag{3.7–28}$$

We come then to the rather interesting result that dynamic similarity, that is, similar vortices in this example, cannot be achieved if the same fluid is used in the two tanks. Rather, a less viscous fluid must be used in the smaller tank. If the smaller tank is to have half the linear dimensions of the larger, then the kinematic viscosity of the liquid in it must be $1/\sqrt{8}$ times that of the oil in the large tank. If the same fluid were used in both and at the same Reynolds number, then the Froude number in the small tank would be greater and the vortex would be proportionately deeper. This is the situation indicated by the dotted curve in Fig. 3.7–1.

QUESTIONS FOR DISCUSSION

1. Give the physical significance of the three derivatives $\partial T/\partial t$, dT/dt, and DT/Dt, in which T is the local fluid temperature.
2. In a flowing fluid can Dc/Dt be zero when $\partial c/\partial t$ is nonzero? Can Dc/Dt be nonzero when $\partial c/\partial t$ is zero? Explain.
3. What is the physical meaning of the equation of continuity?
4. How may the equation of continuity be simplified for steady-state flow?
5. What form does the equation of continuity take for an incompressible fluid?
6. What is meant by the divergence of a vector and the gradient of a scalar?
7. What are the dimensions of ∇p, $\nabla^2 \boldsymbol{v}$, $[\nabla \cdot \boldsymbol{\tau}]$?
8. What is the physical law underlying the equation of motion?
9. What is the Laplacian operator?
10. Compare Eqs. 3.2–8 and 3.2–10 in regard to physical interpretation.
11. What is the origin of the differential mechanical energy balance?
12. Can the rate of mechanical energy dissipation ever be negative? What is the physical significance of your answer?
13. Give the physical significance of the terms $\rho v_\theta^2/r$ and $\rho v_\theta v_r/r$ in the equation of motion in cylindrical coordinates.
14. Do Eqs. 3.6–7 and 3.6–9 necessarily follow from the definitions of power-law fluids and Bingham plastics given in Chapter 1? Explain.
15. How may the equations given in §3.4 be used to solve viscous flow problems?
16. How can one obtain the basic differential equation for hydrostatics from Eq. 3.2–8?
17. Use the result of Question 16 to get the density distribution in an isothermal column of an ideal gas.
18. Show that the surfaces of two immiscible liquids contained in a glass rotating at constant angular velocity assume parabolic shapes.
19. How do the expressions in Eqs. 3.2–17, 18, 19 simplify for constant ρ and μ?
20. What information does one obtain by writing the equations of motion and continuity in dimensionless form?
21. Write the differential equations and boundary conditions for the flow system in §2.6.
22. Discuss some consequences of the Coriolis force in meteorology.
23. Is Eq. 3.5–27 valid if $z < z_0$?
24. Show how Eq. 3.5–37 can be obtained quite simply by a modification of the expression for the velocity profile for the flow between two parallel planes, the upper one of which is moving with a uniform velocity and the lower one of which is stationary.

25. Show that the solution in Eq. 3.5–36 does satisfy Eq. 3.5–28*c* but that it does *not* satisfy the equation for v_ϕ obtained by eliminating p between Eqs. 3.5–28*a* and *b*. Discuss.

PROBLEMS

3.A_1 Torque Required to Turn a Friction Bearing

Calculate the required torque in lb_f-ft and power consumption in horsepower to turn the shaft in the friction bearing shown in Fig. 3.A. You may assume that the length of the bearing surface on the shaft is 2 in., that the shaft is turning at 200 rpm, that the viscosity of the lubricant is 200 cp, and that the fluid density is 50 lb_m ft^{-3}.

Answer: 0.32 lb_f-ft
0.012 hp

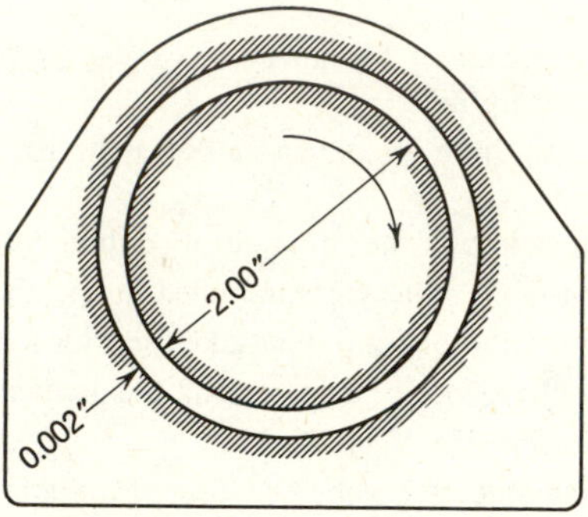

Fig. 3.A. Friction bearing.

3.B_1 The Cone-and-Plate Viscometer

A cone-and-plate viscometer, similar to that shown in Fig. 3.5–4, with cone radius R of 10 cm and of aperture angle θ_0 of 0.5° is to be used to measure the viscosity of Newtonian solutions. What torque, in dyne-centimeters, will be required to turn the cone at an angular velocity of 10 rad/min if the fluid viscosity is 100 cp?

Suggestion. Use the limiting velocity distribution for small θ_0.

Answer: 4.0×10^4 dyne-cm

3.C_1 The Effect of Altitude on Air Pressure

When standing at the mouth of the Ontonagon River on the south shore of Lake Superior (602 ft above mean sea level), your portable barometer indicates a pressure of 750 mm Hg. Use the equation of motion to estimate the barometric pressure at the top of Government Peak (2023 ft above mean sea level) in the nearby Porcupine Mountains. Assume the temperature at lake level is 70° F and that temperature decreases with increasing altitude at a steady rate of 3° F/1000 ft. The gravitational acceleration at the south shore of Lake Superior is about 32.19 ft sec^{-2}, and its variation with altitude may be neglected for the purposes of this problem. *Answer:* 712 mm Hg

3.D_2 Viscosity Determination with a Couette-Hatschek, or MacMichael, Viscometer

It is desired to measure the viscosities of sucrose solutions of about 60 per cent concentration by weight at about 20° C with a Couette-Hatschek viscometer. This instrument consists essentially of a stationary inner cylinder 4.000 cm in diameter and 4.00 cm long surrounded by a rotating concentric cylinder 4.500 cm in diameter and of 4.00 cm effective length. The outer cylinder is rotated by application of a known torque, and the viscosity of the test solution is determined from measurements of the resulting angular velocity. The viscosity of 60 per cent sucrose at 20° C is about 57 cp, and its density is about 1.29 g cm^{-3}.

On the basis of past experience it seems possible that "end effects" will be important, and it is therefore decided to calibrate the viscometer by measurements on some known solutions of approximately the same viscosity as those of the unknown sucrose solutions.

Determine a reasonable value for the applied torque to be used in calibration if the torque measurements are reliable within 100 dyne-cm and the angular velocity can be measured within 0.5 per cent. What will be the resultant angular velocity?

3.E$_2$ Use of the Navier-Stokes Equations to Set up Simple Problems

Use the Navier-Stokes equations for constant density to obtain the differential equations for velocity distribution: (*a*) for the flow of an isothermal film, as in §2.2, (*b*) for two-phase flow in a horizontal slit, as in §2.5, and (*c*) for axial flow in an annulus with a moving boundary as in Problem 2.J.

3.F$_2$ Velocity Distribution in a Stormer Viscometer

A Stormer viscometer consists essentially of two concentric cylinders, the inner of which rotates while the outer is held stationary. Viscosity is determined by measuring the rate of rotation of the inner cylinder under the application of a known torque. It is thus quite similar to the Couette-Hatschek viscometer discussed in §3.5.

Develop an expression for the velocity distribution in this kind of apparatus, as a function of applied torque, for laminar flow of a Newtonian fluid. Neglect end effects.

$$\textit{Answer: } \frac{v_\theta}{r} = \frac{\mathscr{T}}{4\pi\mu L}\left(\frac{1}{r^2} - \frac{1}{R^2}\right)$$

3.G$_2$ Shear Stresses in Terms of Fluid Motion and Properties

Show how one goes from the x-component of the equation of motion in terms of the components of $\boldsymbol{\tau}$ (Eq. 3.2–9) to the analogous equation in terms of μ (Eq. *D* of Table 3.4–2 for the *incompressible* fluid). Note that for the incompressible fluid $(\nabla \cdot \boldsymbol{v}) = 0$, according to the equation of continuity, which must be used (*a*) in simplifying the expression for τ_{xx} in Table 3.4–5,

$$\tau_{xx} = -2\mu \frac{\partial v_x}{\partial x} + \frac{2}{3}\mu(\nabla \cdot \boldsymbol{v})$$

and (*b*) in simplifying the collection of second derivatives obtained when the $\boldsymbol{\tau}$ terms are differentiated:

$$-\left(\frac{\partial \tau_{xx}}{\partial x} + \frac{\partial \tau_{yx}}{\partial y} + \frac{\partial \tau_{zx}}{\partial z}\right) = \mu\nabla^2 v_x + \mu\frac{\partial}{\partial x}(\nabla \cdot \boldsymbol{v})$$

3.H$_2$ Velocity Distribution between Two Rotating Cylinders

Determine $v_\theta(r)$ between two coaxial cylinders of radii R and κR rotating at angular velocities Ω_o and Ω_i, respectively. Assume that the space between cylinders is filled with an incompressible isothermal fluid in laminar flow.

$$\textit{Answer: } v_\theta = \frac{1}{R^2(1-\kappa^2)}\left(r(\Omega_o R^2 - \Omega_i \kappa^2 R^2) - \frac{\kappa^2 R^4}{r}(\Omega_o - \Omega_i)\right)$$

3.I$_2$ Changing the Form of the Equation of Motion

Show that the two forms of the equation of motion as written for a stationary volume element and one moving with the fluid motion, Eqs. 3.2–8 and 3.2–10, respectively, are equivalent.

3.J$_2$ The Equation of Continuity in Cylindrical Coordinates

a. Derive the equation of continuity in cylindrical coordinates by means of a mass balance over a stationary volume element $r\,\Delta r\,\Delta\theta\,\Delta z$.

b. Develop the equation of continuity in cylindrical coordinates from the rectangular form by change of variables. In this development the following relations may be assumed without proof:

$$x = r\cos\theta \qquad r = \sqrt{x^2 + y^2} \qquad v_x = v_r\cos\theta - v_\theta\sin\theta$$
$$y = r\sin\theta \qquad \theta = \arctan(y/x) \qquad v_y = v_r\sin\theta + v_\theta\cos\theta$$
$$z = z \qquad z = z \qquad v_z = v_z$$

c. Demonstrate the validity of the expressions in (*b*) for v_x and v_y in terms of v_r and v_θ.

3.K$_2$ Radial Flow between Two Parallel Disks

A part of a lubrication system consists of two circular disks between which a lubricant flows radially. The flow takes place because of a pressure drop Δp between the inner and outer radii r_1 and r_2. The system is sketched in Fig. 3.6–1.

a. Write the equations of continuity and motion for the flow system, assuming steady, laminar, incompressible Newtonian flow. Consider only the region $r_1 < r < r_2$ and assume that the flow is directed radially so that $v_\theta = v_z = 0$.

b. Show how the equation of continuity enables one to simplify the equation of motion to give

$$-\rho\frac{\phi^2}{r^3} = -\frac{dp}{dr} + \frac{\mu}{r}\frac{d^2\phi}{dz^2} \tag{3.K–1}$$

in which $\phi = rv_r$ is a function of z only. Why is ϕ independent of r?

c. It can be shown that no solution exists for Eq. 3.K–1 unless the nonlinear term (that is, the term containing ϕ^2) is neglected; for the proof, see Problem 3.Y. Omission of the ϕ^2 term corresponds to the "creeping flow" assumption made in deriving Stokes's law for flow around a sphere in §2.6. Show that when the nonlinear term is discarded, the equation of motion can be integrated to give:

$$0 = \Delta p + \left(\mu \ln\frac{r_2}{r_1}\right)\frac{d^2\phi}{dz^2} \tag{3.K–2}$$

d. Show that Eq. 3.K–2 leads to the velocity profile

$$v_r(r, z) = \frac{b^2\Delta p}{2\mu r\ln\dfrac{r_2}{r_1}}\left[1 - \left(\frac{z}{b}\right)^2\right] \tag{3.K–3}$$

e. Finally show that the volume rate of flow Q (in ft^3 sec^{-1}) through the slit is given by

$$Q = \frac{4\pi b^3\Delta p}{3\mu\ln\dfrac{r_2}{r_1}} \tag{3.K–4}$$

f. Draw meaningful sketches to show the pressure distribution $p(r)$ and the velocity distribution $v_r(r, z)$.

3.L$_2$ Symmetry of the Tensor $\boldsymbol{\tau}$

Consider a rectangular volume element within the fluid, as shown in Fig. 3.L, and note that the four shear stresses shown perpendicular to the z-axis form a pair of opposed couples. Write an expression for the angular acceleration of the volume element resulting from the action of these couples and show, by letting the dimensions of the element approach zero, that

$$\tau_{xy} = \tau_{yx} \tag{3.L-1}$$

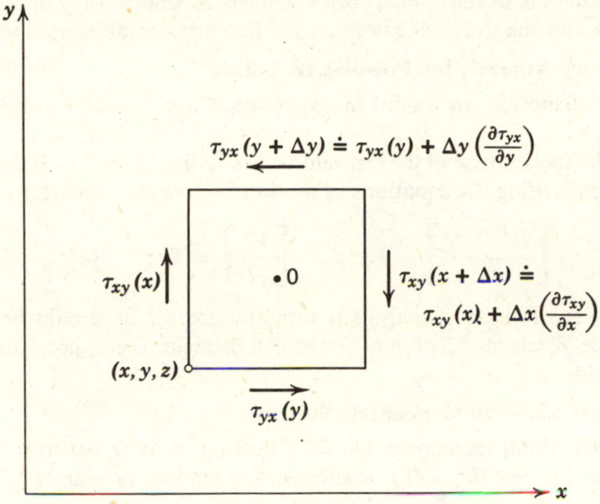

Fig. 3.L. Element of volume of size $\Delta x\ \Delta y\ \Delta z$ on which a moment balance is made. Moments are computed about an axis through 0.

(Note that if a tensor is symmetric in the rectangular coordinate system it is also symmetric in any other orthogonal coordinate system.)

3.M$_2$ Air Entrainment in a Draining Tank

A molasses storage tank 60 ft in diameter is to be built with a draw-off line 1 ft in diameter, 4 ft from the sidewall of the tank and extending vertically upward 1 ft from the tank

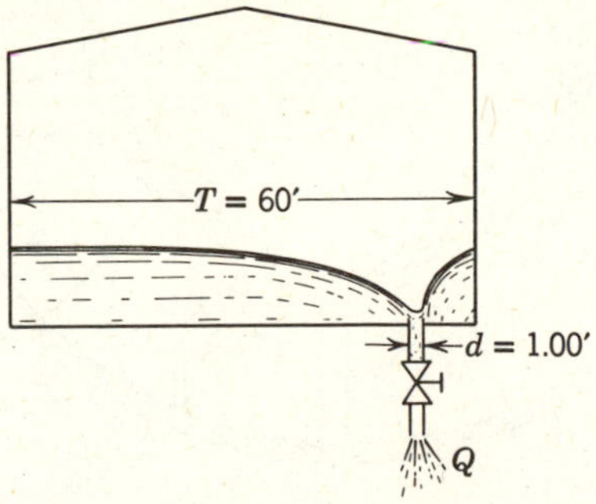

Fig. 3.M. A draining molasses tank.

bottom. (See Fig. 3.M.) It is known from experience that as molasses is withdrawn from the tank a vortex will form, and, as the liquid level drops, this vortex will ultimately reach the draw-off pipe, allowing air to be sucked into the molasses. This is to be avoided.

It is proposed to predict the minimum liquid level at which this entrainment can be avoided, at a draw-off rate of 800 gal/min, by model study in a smaller tank. For convenience, water at 68° F is used for the fluid in this model study.

Determine the proper tank dimensions and operating conditions for the model if the density of the molasses is 1.286 g cm^{-3} and its viscosity is 56.7 cp. It may be assumed that, in either the full-size tank or the model, vortex shape is dependent only upon the amount of liquid in the tank and the draw-off rate, that is, the vortex establishes itself very rapidly.

3.N_3 Dimensional Analysis for Power-Law Fluids

The equations of motion are useful in dimensional analysis of the behavior of non-Newtonian fluids.

Show that for the specific case of the Ostwald-de Waele fluid (Eq. 3.6–9) the dimensionless groups obtained by writing the equations of motion in dimensionless form are

$$\left[\frac{D^n V^{2-n} \rho}{m}\right] = \text{Re}_n; \qquad \left[\frac{V^2}{gD}\right] = \text{Fr}; \qquad [n] \tag{3.N-1,2,3}$$

N.B.: Whereas the foregoing analysis is formally correct, it should be kept in mind that the Ostwald-de Waele model of non-Newtonian behavior is empirical and not followed exactly by real fluids.

3.O_3 Radial Flow between Concentric Spheres

Consider an isothermal, incompressible fluid flowing radially between two concentric porous spherical shells. (See Fig. 3.O.) Assume steady laminar flow and neglect end effects.

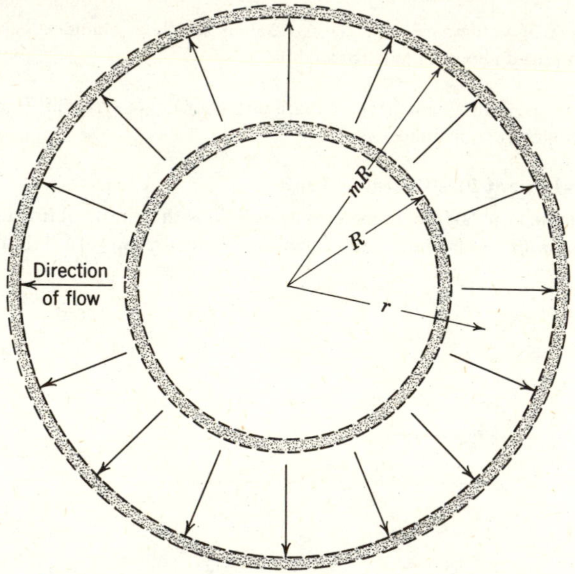

Fig. 3.O. Radial flow between concentric porous spheres.

Note that here the velocity is not assumed zero at the solid surfaces.

a. Show by use of the equation of continuity that

$$r^2 v_r = \psi \tag{3.O-1}$$

where ψ is a constant.

b. Show by use of the equations of motion that the pressure distribution in this system is described by the relations

$$\frac{1}{r \sin\theta}\frac{\partial p}{\partial \phi} = \rho g_\phi \tag{3.O-2}$$

$$\frac{1}{r}\frac{\partial p}{\partial \theta} = \rho g_\theta \tag{3.O-3}$$

$$\frac{\partial p}{\partial r} = \mu\left[\nabla^2 v_r - \frac{2v_r}{r^2}\right] - \rho v_r \frac{\partial v_r}{\partial r} + \rho g_r \tag{3.O-4}$$

c. With the aid of footnote a of Table 3.4-4, show that the radial pressure distribution may be expressed in terms of the quantity $\mathscr{P}$ as

$$\mathscr{P} - \mathscr{P}_R = \tfrac{1}{2}\rho v_r^2\big|_{r=R}\left[1 - \left(\frac{R}{r}\right)^4\right] \tag{3.O-5}$$

d. Write expressions for τ_{rr}, $\tau_{\theta\theta}$, $\tau_{\phi\phi}$, $\tau_{r\theta}$, $\tau_{r\phi}$, and $\tau_{\theta\phi}$ for this system. Which of the normal stresses are compressive and which tensile? What is the physical significance of your answer?

e. Are the compressive and tensile stresses exerted by the fluid on the inner and outer shells equal to $p|_R$ and $p|_{mR}$, respectively? Why?

3.P$_3$ Radial Flow between Coaxial Cylinders

Consider an isothermal, incompressible fluid flowing radially between two porous cylindrical shells. Assume steady laminar flow and neglect end effects.

a. Show by means of the equation of continuity that

$$r v_r = \phi \tag{3.P-1}$$

where ϕ is a constant.

b. Simplify the equations of motion to obtain the following expressions for pressure distribution in terms of the quantity $\mathscr{P}$:

$$\frac{d\mathscr{P}}{dr} = -\rho v_r \frac{dv_r}{dr} \tag{3.P-2}$$

$$\frac{d\mathscr{P}}{d\theta} = \frac{d\mathscr{P}}{dz} = 0 \tag{3.P-3}$$

c. Integrate the expression for $\dfrac{d\mathscr{P}}{dr}$ in (*b*) to obtain

$$\mathscr{P} - \mathscr{P}_R = \tfrac{1}{2}\rho v_r^2\big|_{r=R}\left[1 - \left(\frac{R}{r}\right)^2\right] \tag{3.P-4}$$

3.Q$_3$ Creeping Flow between Two Concentric Spheres

A very viscous fluid flows in the space between two concentric spheres, as shown in Fig. 3.Q. It is desired to find the rate of flow in the system as a function of the pressure drop imposed. Neglect end effects and assume $v_\theta = v_\theta(r, \theta)$ and $v_r = v_\phi = 0$.

a. Show by use of the equation of continuity that $v_\theta \sin\theta = u(r)$, where $u(r)$ is a function of r to be determined.

b. Write the θ-component of the equation of motion for this system, assuming flow rates low enough that the entire left side of the equation may be neglected. Show that this equation reduces to

$$0 = -\frac{1}{r}\frac{\partial p}{\partial \theta} + \mu\left[\frac{1}{\sin\theta}\frac{1}{r^2}\frac{d}{dr}\left(r^2\frac{du}{dr}\right)\right] \tag{3.Q-1}$$

c. Separate the foregoing relation into the two equations,

$$\sin\theta \frac{dp}{d\theta} = B \tag{3.Q–2}$$

$$\frac{\mu}{r}\frac{d}{dr}\left(r^2 \frac{du}{dr}\right) = B \tag{3.Q–3}$$

where B is a separation constant.

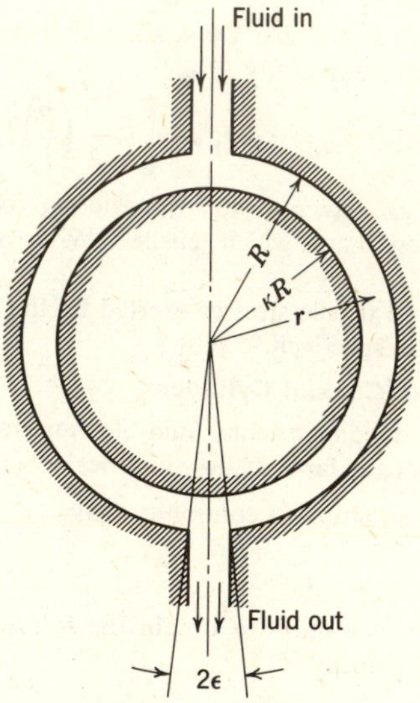

Fig. 3.Q. Creeping flow between two stationary concentric spheres.

d. Show that

$$\Delta p = B \ln\left(\frac{1 - \cos\epsilon}{1 + \cos\epsilon}\right) = -BE(\epsilon) \tag{3.Q–4}$$

$$u = \frac{R\,\Delta p}{2\mu E(\epsilon)}\left[\left(1 - \frac{r}{R}\right) + \kappa\left(1 - \frac{R}{r}\right)\right] \tag{3.Q–5}$$

where Δp is the imposed pressure drop.

e. Use the foregoing result to show that the volumetric flow rate is

$$Q = \frac{\pi R^3\,\Delta p}{6\mu E(\epsilon)}(1 - \kappa)^3 \tag{3.Q–6}$$

3.R$_3$ Stress Components in Terms of Fluid Motion and Properties in non-Newtonian Systems

a. Starting with Eq. 3.6–7, justify Eq. 2.3–23.

b. Starting with Eq. 3.6–9, obtain the expression for $\tau_{r\theta}$ corresponding to Eq. 3.6–18 for a power-law fluid.

c. Use the result in (*b*) to derive the power-law analog of the Reiner-Riwlin equation.

3.S$_4$ Derivation of the Equation of Motion through Application of the Divergence Theorem

Consider any stationary volume element G of finite volume V_G and surface S_G. Write Newton's second law of motion for the whole element in the form of Eq. 3.2–1. Since conditions cannot be considered uniform over the element, it will be necessary to integrate the rate of momentum gain and the body forces over the volume of the element and the convective flux and surface forces over the surface. Now convert the surface integrals into volume integrals by use of the divergence theorem and show that the resulting expression can be reduced to Eq. 3.2–8.

3.T$_4$ Velocity Distribution in the Plate-and-Cone Viscometer by Simplification of the Equation of Motion[1]

Develop the expression given for $v_\phi(r, \theta)$ in Eq. 3.5–36 by the following alternative method:

a. Show that Eq. *F* of Table 3.4–4 gives

$$\frac{\partial}{\partial r}\left(r^2 \frac{\partial v_\phi}{\partial r}\right) + \frac{1}{\sin\theta}\frac{\partial}{\partial\theta}\left(\sin\theta \frac{\partial v_\phi}{\partial\theta}\right) - \frac{v_\phi}{\sin^2\theta} = 0 \tag{3.T–1}$$

with these boundary conditions:

$$\text{at } \theta = \frac{\pi}{2}, \qquad v_\phi = 0 \tag{3.T–2}$$

$$\text{at } \theta = \theta_1, \qquad v_\phi = r\Omega\sin\theta_1 \tag{3.T–3}$$

$$\text{at } r = 0, \qquad v_\phi = 0 \tag{3.T–4}$$

b. Assume a solution of the form $v_\phi = R(r)\cdot\Theta(\theta)$ and show that the equation of motion may be split into the two ordinary differential equations:

$$\frac{d}{dr}\left(r^2\frac{dR}{dr}\right) = \beta R \tag{3.T–5}$$

$$\frac{d}{dx}\left[(1 - x^2)\frac{d\Theta}{dx}\right] = \Theta\left(\frac{1}{1 - x^2} - \beta\right) \tag{3.T–6}$$

Here β is a separation constant and $x = \cos\theta$.

c. Solve Eqs. 3.T–5 and 3.T–6 to obtain the desired velocity distribution. Note (i) Eq. 3.T–5 is a form of Cauchy's linear equation, (ii) Eq. 3.T–6 is a form of Legendre's equation, which gives rise to an associated Legendre function of the first order and nth degree, in which n is the positive root of $n(n + 1) = \beta$.

3.U$_4$ Alternate Forms of the Equation of Motion

a. Show that for a compressible fluid of constant μ the equation of motion may be written as

$$\rho\frac{Dv}{Dt} = -\nabla p - \mu[\nabla \times [\nabla \times v]] + \tfrac{4}{3}\mu\nabla(\nabla\cdot v) + \rho g \tag{3.U–1}$$

b. Show that if the fluid is incompressible, then

$$\rho\left[\frac{\partial v}{\partial t} - [v \times [\nabla \times v]] + \nabla\tfrac{1}{2}v^2\right] = -\nabla p + \mu\nabla^2 v + \rho g \tag{3.U–2}$$

[1] The authors are indebted to Professors A. G. Fredrickson and J. C. Slattery for this problem.

3.V₄ Periodic Laminar Flow of a Viscoelastic Fluid in a Tube

The simplest model of an incompressible, viscoelastic fluid is the Maxwell model,[2] which is a superposition of the Hookean solid and the Newtonian liquid:

$$\tau_{ij} + t_0 \frac{\partial \tau_{ij}}{\partial t} = -\mu\left(\frac{\partial v_i}{\partial x_j} + \frac{\partial v_j}{\partial x_i}\right) \tag{3.V-1}$$

Here $t_0 = \mu/G$ is a characteristic time for the fluid, and G is the modulus of shear rigidity. When Eq. 3.V–1 is used for the components of τ in the equations of change, hydrodynamic problems for *small-amplitude* motions may be solved.

In this problem we consider an example of a linearized, periodic solution to the equation of motion. By "linearized" we mean that the term $[\boldsymbol{v} \cdot \nabla \boldsymbol{v}]$ in the equation of motion may be neglected. By "periodic" we mean here that the local fluid motion is a sinusoidal function of the time.

Let us now consider the oscillations of a viscoelastic fluid in a tube of radius R about a mean position when an oscillating pressure gradient is imposed on the system. Then the two equations describing the system are

linearized equation of motion
$$\rho \frac{\partial v_z}{\partial t} = -\frac{\partial p}{\partial z} - \frac{1}{r}\frac{\partial}{\partial r}(r\tau_{rz}) \tag{3.V-2}$$

linearized Maxwell model
$$\tau_{rz} + t_0 \frac{\partial \tau_{rz}}{\partial t} = -\mu \frac{\partial v_z}{\partial r} \tag{3.V-3}$$

We use the fact that the momentum flux, the pressure gradient, the local velocity, and the volume flow rate are all periodic in time with frequency ω:

$$-\partial p/\partial z = b_0\, \mathcal{R}e\{e^{i\omega t}\} \tag{3.V-4}$$

$$\tau_{rz} = \mathcal{R}e\{\tau_0(r)e^{i\omega t}\} \tag{3.V-5}$$

$$v_z = \mathcal{R}e\{v_0(r)e^{i\omega t}\} \tag{3.V-6}$$

$$Q = \mathcal{R}e\{Q_0 e^{i\omega t}\} \tag{3.V-7}$$

in which b_0 is real, whereas τ_0, v_0, and Q_0 are complex. When the first three of these quantities are inserted into Eqs. 3.V–2,3, a second-order equation for $v_0(r)$ is obtained. Show that its solution is (with $k^2 = -i\omega\rho(1 + i\omega t_0)/\mu$)

$$v_0(r) = \frac{b_0}{i\omega\rho}\left(1 - \frac{J_0(kr)}{J_0(kR)}\right) \tag{3.V-8}$$

Show further that the amplitude of the volume rate of flow is

$$Q_0 = \frac{\pi R^2 b_0}{i\omega\rho}\left(1 - \frac{2J_1(kR)}{kRJ_0(kR)}\right) \tag{3.V-9}$$

[2] L. J. F. Broer, *Appl. Sci. Research*, **A6**, 226–236 (1957); see also J. G. Oldroyd, *Proc. Roy. Soc.*, **A200**, 523–541 (1950), for a more general approach to viscoelastic fluids.

Expand the Bessel functions in Eq. 3.V–9, retaining terms through k^4, thereby obtaining:

$$Q_0 = \frac{\pi R^2 b_0}{8i\omega\rho}\left[-(kR)^2 - \tfrac{1}{6}(kR)^4 - \cdots\right] \tag{3.V–10}$$

Using Eq. 3.V–7 and the definition of k gives for $\omega t_0 \ll 1$ and $R^2\omega\rho/6\mu \ll 1$:

$$Q(t) = \frac{\pi R^4 b_0}{8\mu}\left[\cos\omega t - \left(\omega t_0 - \frac{\omega\rho R^2}{6\mu}\right)\sin\omega t + \cdots\right] \tag{3.V–11}$$

Interpret the result physically.

3.W$_4$ Falling Film on a Cone

In connection with gas-absorption studies, it is decided to explore several arrangements for gas–liquid contacting in systems of simple geometry. One system that seems attractive

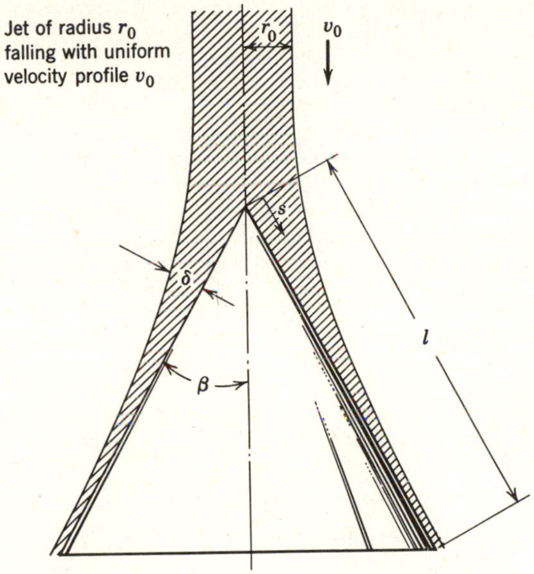

Fig. 3.W. Fluid jet impinging on a cone.

is that in which a liquid jet impinges upon the point of a cone and flows down along the cone as a liquid film, shown in Fig. 3.W. Obtain an approximate expression for the change of film thickness with s, the distance along the conical surface. List carefully any assumptions you make in your development and indicate any restrictions that may have to be placed on the final result.

An answer: $\delta = \sqrt[3]{\dfrac{3\mu r_0^2 v_0}{\rho g s \sin 2\beta}}$

3.X$_4$ Variational Principle for Non-Newtonian Flow[3]

Consider the following class of flows of an incompressible non-Newtonian fluid described in Eq. 3.6–2, in which η is taken to be a function of $I_2 = (\boldsymbol{\Delta} : \boldsymbol{\Delta})$ only:

a. The flow is steady ($\partial v/\partial t = 0$).

[3] R. B. Bird, *Physics of Fluids*, **3**, 539–541 (1960).

b. The inertial term $[\boldsymbol{v} \cdot \nabla \boldsymbol{v}]$ is either identically zero or negligibly small, as for "creeping flow."

c. The external force $\boldsymbol{g}$ is derivable from a potential $\hat{\Phi}$ (that is, $\boldsymbol{g} = -\nabla\hat{\Phi}$).

d. The velocity $\boldsymbol{v}$ is known to be $\boldsymbol{v}_0(x, y, z)$ on the surface S of the flow system of volume V.

For these flows, the integral

$$B = \underset{\substack{\text{volume of flow} \\ \text{system with fixed} \\ \text{boundaries}}}{\iiint} F\, dx\, dy\, dz \tag{3.X–1}$$

is a minimum, in which

$$F = -(p + \rho\hat{\Phi})I_1 + \tfrac{1}{2}\int_0^{I_2} \eta(I_2)\, dI_2 \tag{3.X–2}$$

Here I_1 and I_2 are the first and second invariants of $\boldsymbol{\Delta}$. (See Eqs. 3.6–3 and 4.) That is, of all the functions $\boldsymbol{v}(x, y, z)$ which reduce to the given function $\boldsymbol{v}_0(x, y, z)$ on S, the true velocity distribution will be that which gives the minimum value of B. In using the foregoing variational principle, one would normally select a trial function $\boldsymbol{v}(x, y, z)$ such that the equation of continuity is automatically satisfied. Then I_1 computed from the trial velocity distribution vanishes, and it is not necessary to select a trial function $p(x, y, z)$ for the pressure.

a. Show that the Euler-Lagrange equations[4] corresponding to the foregoing variational principle are just the equation of continuity and the three components of the equation of motion.

b. Verify that this variational principle simplifies to Helmholtz's principle[5] for a Newtonian fluid.

3.Y$_4$ Radial Flow between Parallel Disks

In solving the equations of motion one usually postulates the form of the solution for the velocity profiles. These postulates are then used to discard many terms in the equations so that one finally obtains simple starting equations for the problem at hand. Such a procedure is clearly intuitive in nature and can occasionally lead to difficulties.

To illustrate this point, consider the radial flow between a pair of circular disks. In solving this problem (see Problem 3.K) it was *postulated* that $v_r = v_r(r, z)$, and that v_θ and v_z are both zero, and that $p = p(r)$. These postulates lead to Eq. 3.K–1.

Differentiate this equation with respect to z and show that it leads to an inconsistency, except in the limit of creeping flow. Discuss.

[4] See, for example, E. Madelung, *Die mathematischen Hilfsmittel des Physikers*, Dover, New York (1943), p. 216, Eq. 4.

[5] H. Lamb, *Hydrodynamics*, Dover, New York (1945), pp. 617–619; see also D. F. Hays, *Journal of Basic Engineering*, **81**, 13–23 (1959) for an interesting application to lubrication problems.

CHAPTER 4

Velocity Distributions with More Than One Independent Variable

In Chapter 2 it was shown that a variety of steady-state viscous flow problems with straight streamlines may be solved by shell momentum balances. In Chapter 3 more general methods were developed, which enable us to describe more complicated flow systems. In §3.5 a few applications of the general equations were given, but in order to emphasize the physics and minimize the mathematics these applications were restricted for the most part to problems leading to ordinary rather than to partial differential equations. In this chapter we discuss several classes of applications that involve the solutions of partial differential equations: unsteady flow, viscous flow in more than one direction, two-dimensional ideal flow, and viscous flow in boundary layers. The purpose of this chapter is to provide an introduction to each of these topics along with some of the standard mathematical techniques. None of these topics is discussed in detail, inasmuch as all are treated extensively in fluid mechanics treatises.

§4.1 UNSTEADY VISCOUS FLOW

The study of transient phenomena of all types is becoming important in connection with current interest in problems of start-up and control. The equations given in Chapter 3 provide the starting point for describing unsteady flow. We give here two illustrations of how such problems are set up and solved.[1]

In the first we examine the flow of an infinite body of fluid near a wall suddenly set in motion. This problem will illustrate the use of the *method of combination of variables*, which enables one to reduce the partial differential equation to a single ordinary differential equation; such a procedure is possible only when two of the boundary conditions can be combined into one. It should be noted that this system never approaches a limiting steady state.

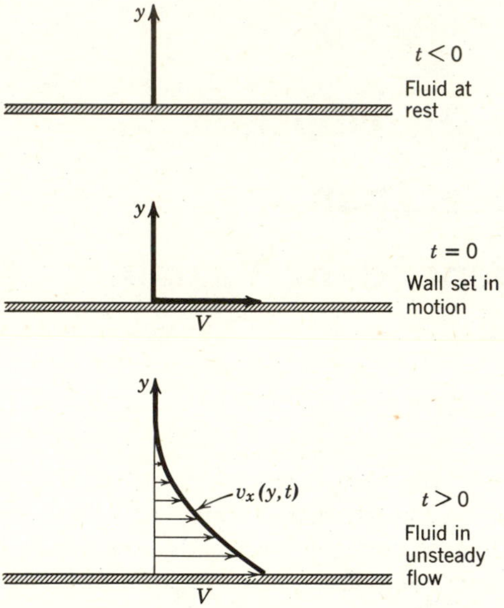

Fig. 4.1–1. Viscous flow of a fluid near a wall suddenly set in motion.

In the second illustration we consider the transient behavior of a viscous fluid in a tube. This problem is handled by the *method of separation of variables*, which enables one to reduce the partial differential equation to two ordinary differential equations. This system does have a limiting velocity profile as time goes to infinity—that is, the system approaches a steady state.

Example 4.1–1. Flow Near a Wall Suddenly Set in Motion

A semi-infinite body of liquid with constant ρ and μ is bounded on one side by a flat surface (the xz-plane). Initially, the fluid and the solid surface are at rest; but at time $t = 0$ the solid surface is set in motion in the positive x-direction with

[1] Other examples may be found in H. Lamb, *Hydrodynamics*, Dover, New York (1945), Chapter 11.

a velocity V, as shown in Fig. 4.1–1. It is desired to know the velocity as a function of y and t. There is no pressure gradient or gravity force in the x-direction, and the flow is assumed to be laminar.

Solution. For this system, $v_y = v_z = 0$ and $v_x = v_x(y, t)$. Then from Eq. A of Table 3.4–1 (continuity) and Eq. D of Table 3.4–2 (motion) we find that

$$\frac{\partial v_x}{\partial t} = \nu \frac{\partial^2 v_x}{\partial y^2} \tag{4.1–1}$$

in which $\nu = \mu/\rho$. The initial and boundary conditions are

I.C.:	at	$t \leqslant 0,$	$v_x = 0$	for all y	(4.1–2)
B.C. 1:	at	$y = 0,$	$v_x = V$	for all $t > 0$	(4.1–3)
B.C. 2:	at	$y = \infty,$	$v_x = 0$	for all $t > 0$	(4.1–4)

We now anticipate that a solution can be found of the form $v_x/V = \phi(\eta)$, in which η is a dimensionless variable $y/\sqrt{4\nu t}$. In terms of the new variables,

$$\frac{\partial(v_x/V)}{\partial t} = -\frac{1}{2}\frac{\eta}{t}\phi'; \qquad \frac{\partial^2(v_x/V)}{\partial y^2} = +\frac{\eta^2}{y^2}\phi'' \tag{4.1–5,6}$$

in which primes indicate differentiation with respect to η. Then Eq. 4.1–1 becomes

$$\phi'' + 2\eta\phi' = 0 \tag{4.1–7}$$

with the boundary conditions

B.C. 1:	at	$\eta = 0,$	$\phi = 1$	(4.1–8)
I.C. + B.C. 2:	at	$\eta = \infty,$	$\phi = 0$	(4.1–9)

If ϕ' is replaced by ψ, then we get a first-order separable equation, which may be solved to give

$$\psi \equiv \phi' = C_1 e^{-\eta^2} \tag{4.1–10}$$

A second integration then gives

$$\phi = C_1 \int_0^\eta e^{-\eta^2} d\eta + C_2 \tag{4.1–11}$$

where we have arbitrarily selected $\eta = 0$ for the lower limit of the indefinite integral, which we cannot evaluate in closed form; changing the lower limit $\eta = 0$ to a different limit would simply change the value of constant C_2, still undetermined. Application of the two boundary conditions then gives

$$\phi = 1 - \frac{\int_0^\eta e^{-\eta^2} d\eta}{\int_0^\infty e^{-\eta^2} d\eta} = 1 - \frac{2}{\sqrt{\pi}} \int_0^\eta e^{-\eta^2} d\eta \tag{4.1–12}$$

The ratio of the integrals appearing here is called the "error function," abbreviated erf η; it is a well-known function, and tables of it are readily available.[2] Hence the

[2] See, for example, H. B. Dwight *Tables of Integrals and Other Mathematical Data*, Macmillan, New York (1957), Third Edition, p. 275; 1 − erf u is sometimes written as erfc u, the "complementary error function."

solution may be written

$$\frac{v_x}{V} = 1 - \operatorname{erf} \frac{y}{\sqrt{4\nu t}} \tag{4.1-13}$$

The instantaneous velocity profiles are shown in Fig. 4.1–2; we also show the analogous profiles for some power-model non-Newtonian fluids.[3]

The error function is a monotone increasing function ranging from 0 to 1 and reaches a value of 0.99 when η is about 2.0. We can use this fact to define a "boundary-layer thickness" δ as that distance y for which v_x has dropped to a value $0.01V$. That is, we find that $\delta = 4\sqrt{\nu t}$. This distance is a measure of the extent to which momentum has "penetrated" the body of the fluid. Note that this thickness is proportional to the square root of the time.

Example 4.1–2. Unsteady Laminar Flow in a Circular Tube

A fluid of constant ρ and μ is contained in a very long horizontal pipe of length L and radius R. Initially the fluid is at rest. At $t = 0$, a pressure gradient $(p_0 - p_L)/L$ is impressed on the system. Determine how the velocity profiles change with time.

Solution. For this problem, cylindrical coordinates are used; it is assumed that $v_r = v_\theta = 0$ and that $v_z = v_z(r, t)$. Then Eq. *B* of Table 3.4–1 (continuity) and Eq. *F* of Table 3.4–3 (motion) may be combined to give

$$\rho \frac{\partial v_z}{\partial t} = \frac{p_0 - p_L}{L} + \mu \frac{1}{r} \frac{\partial}{\partial r}\left(r \frac{\partial v_z}{\partial r}\right) \tag{4.1-14}$$

The initial and boundary conditions are

I.C.: at $t = 0$, $v_z = 0$ for $0 \leqslant r \leqslant R$ (4.1–15)

B.C. 1: at $r = 0$, $v_z = \text{finite}$ (4.1–16)

B.C. 2: at $r = R$, $v_z = 0$ (4.1–17)

First we introduce the following dimensionless variables:

$$\phi = \frac{v_z}{(p_0 - p_L)R^2/4\mu L}; \qquad \xi = \frac{r}{R}; \quad \tau = \frac{\mu t}{\rho R^2} \tag{4.1-18,19,20}$$

The velocity is made dimensionless by dividing by the maximum velocity at $t = \infty$. When Eq. 4.1–14 is multiplied by $4L/(p_0 - p_L)$, we get in dimensionless variables

$$\frac{\partial \phi}{\partial \tau} = 4 + \frac{1}{\xi} \frac{\partial}{\partial \xi}\left(\xi \frac{\partial \phi}{\partial \xi}\right) \tag{4.1-21}$$

This is to be solved with the conditions that at $\tau = 0$, $\phi = 0$, at $\xi = 1$, $\phi = 0$, and at $\xi = 0$, ϕ is finite.

Now we make use of the fact that the system will attain a steady state at $\tau = \infty$. Hence a solution of the following form must be sought:

$$\phi(\xi, \tau) = \phi_\infty(\xi) - \phi_t(\xi, \tau) \tag{4.1-22}$$

[3] R. B. Bird, *A.I.Ch.E. Journal*, **5**, 565 (1959).

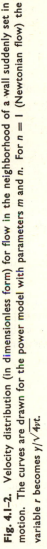

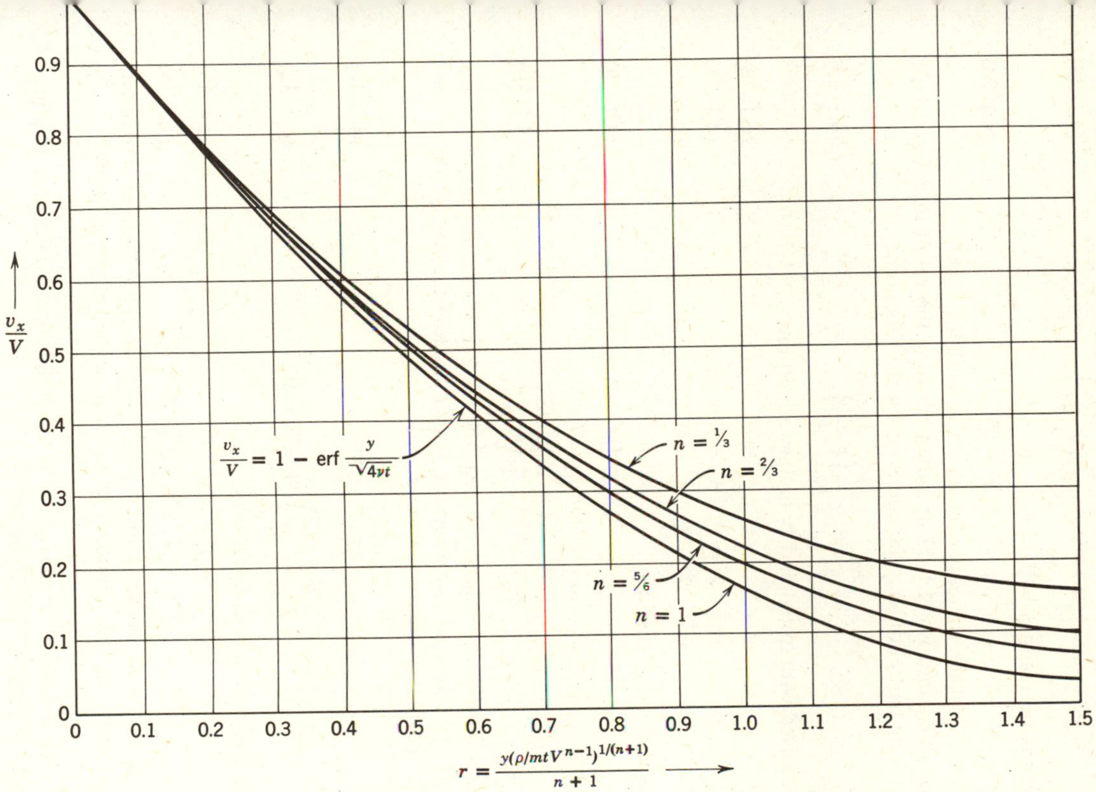

Fig. 4.1–2. Velocity distribution (in dimensionless form) for flow in the neighborhood of a wall suddenly set in motion. The curves are drawn for the power model with parameters *m* and *n*. For $n = 1$ (Newtonian flow) the variable *r* becomes $y/\sqrt{4\nu t}$.

That is, we split up the solution into a steady-state limiting solution $\phi_\infty(\xi)$ and a transient function $\phi_t(\xi, \tau)$. The steady-state solution is obtained from Eq. 4.1–21 by setting $\partial\phi/\partial\tau = 0$ at $\tau = \infty$:

$$0 = 4 + \frac{1}{\xi}\frac{d}{d\xi}\left(\xi\frac{d\phi_\infty}{d\xi}\right) \tag{4.1–23}$$

for which the solution with $\phi = 0$ at $\xi = 1$ is

$$\phi_\infty = 1 - \xi^2 \tag{4.1–24}$$

This is, of course, just the Poiseuille velocity distribution.

Substitution of ϕ_∞ into Eq. 4.1–22 and then inserting the result into Eq. 4.1–21 gives the following partial differential equation for the function ϕ_t:

$$\frac{\partial\phi_t}{\partial\tau} = \frac{1}{\xi}\frac{\partial}{\partial\xi}\left(\xi\frac{\partial\phi_t}{\partial\xi}\right) \tag{4.1–25}$$

This must be solved with initial and boundary conditions:

I.C.:	at	$\tau = 0, \quad \phi_t = \phi_\infty$	(4.1–26)
B.C. 1:	at	$\xi = 0, \quad \phi_t = \text{finite}$	(4.1–27)
B.C. 2:	at	$\xi = 1, \quad \phi_t = 0$	(4.1–28)

We now try a solution of the form

$$\phi_t(\xi, \tau) = \Xi(\xi)T(\tau) \tag{4.1–29}$$

Substitution into Eq. 4.1–25 and division by ΞT gives

$$\frac{1}{T}\frac{dT}{d\tau} = \frac{1}{\Xi}\frac{1}{\xi}\frac{d}{d\xi}\left(\xi\frac{d\Xi}{d\xi}\right) \tag{4.1–30}$$

The left side is a function of τ alone, whereas the right side is a function of ξ alone. Hence both may be set equal to a constant, which we choose to designate as $-\alpha^2$. We then get two ordinary differential equations:

$$\frac{dT}{d\tau} = -\alpha^2 T \tag{4.1–31}$$

$$\frac{1}{\xi}\frac{d}{d\xi}\left(\xi\frac{d\Xi}{d\xi}\right) + \alpha^2\Xi = 0 \tag{4.1–32}$$

These equations have solutions as follows:

$$T = C_0 e^{-\alpha^2\tau} \tag{4.1–33}$$

$$\Xi = C_1 J_0(\alpha\xi) + C_2 Y_0(\alpha\xi) \tag{4.1–34}$$

in which C_0, C_1, and C_2 are constants and J_0 and Y_0 are Bessel functions of zero order. We now apply the initial and boundary conditions in the following way:

B.C. 1:

Because ϕ_t is finite, Ξ must therefore also be finite. Because $Y_0(\alpha\xi)$ tends to minus infinity as ξ goes to zero, C_2 has to be zero.

B.C. 2:

Because $\phi_t = 0$ at $\xi = 1$, so must Ξ also vanish at $\xi = 1$. This is true only if $J_0(\alpha)$ is zero. But $J_0(\alpha)$ is an oscillating function that crosses the α-axis at $\alpha_1 = 2.405$, $\alpha_2 = 5.520$, $\alpha_3 = 8.654$, etc.[4] Hence there are many solutions $\Xi_n = C_{1n}J_0(\alpha_n)$ with $n = 1,2,3, \ldots \infty$, which will satisfy the differential equation and the boundary conditions.

[4] E. Jahnke and F. Emde, *Tables of Functions*, Dover, New York (1945), p. 166.

I.C.:

We know that ϕ_t must equal ϕ_∞ at $\tau = 0$. But because of the arguments under B.C. 2, we know that a superposition of the form

$$\phi_t(\xi, \tau) = \sum_{n=1}^{\infty} B_n e^{-\alpha_n^2 \tau} J_0(\alpha_n \xi) \tag{4.1–35}$$

satisfies the boundary conditions and the partial differential equation. Application of the initial condition then gives

$$(1 - \xi^2) = \sum_{n=1}^{\infty} B_n J_0(\alpha_n \xi) \tag{4.1–36}$$

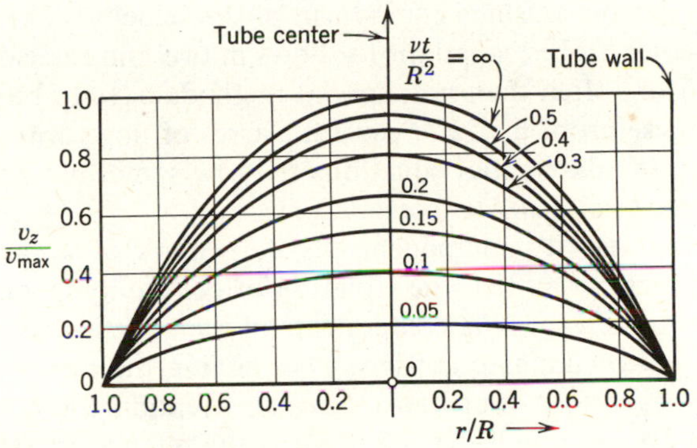

Fig. 4.1–3. Velocity distribution for unsteady state "start-up" flow in a circular tube. [P. Szymanski, *J. Math. pures appl.*, Series 9, **11**, 67–107 (1932).]

To get the B_n (which are just $C_0 C_{1n}$), we multiply both sides by $J_0(\alpha_m \xi)\xi\, d\xi$ and then integrate from 0 to 1:

$$\int_0^1 J_0(\alpha_m \xi)(1 - \xi^2)\xi\, d\xi = \sum_{n=1}^{\infty} B_n \int_0^1 J_0(\alpha_n \xi) J_0(\alpha_m \xi)\xi\, d\xi \tag{4.1–37}$$

Because of the orthogonality properties of Bessel functions, the only term on the right side that contributes is that for which $m = n$. The integral on the left side may be integrated by making use of some standard relations for Bessel functions.[5] We obtain finally

$$\frac{4J_1(\alpha_m)}{\alpha_m^{\ 3}} = B_m \cdot \tfrac{1}{2}[J_1(\alpha_m)]^2 \tag{4.1–38}$$

from which

$$B_m = \frac{8}{\alpha_m^{\ 3} J_1(\alpha_m)} \tag{4.1–39}$$

Hence the final expression for the reduced velocity distribution is

$$\phi = (1 - \xi^2) - 8 \sum_{n=1}^{\infty} \frac{J_0(\alpha_n \xi)}{\alpha_n^{\ 3} J_1(\alpha_n)} e^{-\alpha_n^2 \tau} \tag{4.1–40}$$

[5] *Ibid.*, pp. 145–146.

The profiles computed from Eq. 4.1–40 are shown in Fig. 4.1–3, in which one can see that the maximum (center-line) velocity is within 10 per cent of the steady-state value when τ is about 0.45.

The transient profiles for axial annular flow[6] and tangential annular flow[7] have also been calculated.

§4.2 STEADY VISCOUS FLOW WITH TWO NONVANISHING VELOCITY COMPONENTS: THE STREAM FUNCTION

In Chapters 2 and 3 and in §4.1 the problems were carefully selected so that there was only one nonvanishing component of the velocity. The solution to the complete Navier-Stokes equations for flows in two and three directions is very difficult. More often than not, special methods have to be invoked in order to solve these problems. For several classes of flows with constant ρ and μ, however, the differential equations may be somewhat simplified by formulation in terms of a *stream function* ψ.

We express the velocity components as derivatives of ψ (as shown in Table 4.2–1) in such a way that the equation of continuity is automatically satisfied. The two nonvanishing components of the equation of motion can then be combined to eliminate the terms containing the components of p. This procedure leads to a fourth-order scalar equation for ψ. The most important cases are listed in Table 4.2–1 for ready reference.[1] The physical significance of the stream function is that lines of $\psi = \text{constant}$ are *streamlines;* in steady flow these are the curves actually traced out by the particles of the fluid.

In this section we discuss as an illustrative example the steady flow of a viscous fluid about a sphere. This solution is valid only for *creeping flow* for which the inertial term $[\boldsymbol{v} \cdot \nabla \boldsymbol{v}]$ in the equation of motion may be neglected. This treatment leads to *Stokes's law*, which has been discussed briefly in §2.6.

This problem has been chosen for discussion because of the current interest in particle technology and droplet mechanics. The problem of ideal flow around a sphere is discussed in problem 4.H. That solution and the one given here form the two important limiting cases on which many analyses of complex systems are based.

[6] W. Müller, *Z. Angew. Math. Mech.*, **16,** 227–238 (1936).

[7] R. B. Bird and C. F. Curtiss, *Chem. Eng. Sci.*, **11,** 108–113 (1959).

[1] To get the differential equation for ψ for rectangular coordinates with $v_z = 0$, we start with the x- and y-components of the equation of motion for constant ρ and μ. We differentiate the x-equation with respect to y and the y-equation with respect to x and subtract; in this process the pressure and gravity terms are eliminated. Insertion of $v_x = -\partial\psi/\partial y$ and $v_y = +\partial\psi/\partial x$ leads directly to the fourth-order differential equation shown in the table.

EQUATIONS FOR THE STREAM FUNCTION[a]

Type of Motion	Coordinate System	Velocity Components	Differential Equations for ψ Which Are Equivalent to the Navier-Stokes Equation[b]	Expressions for Operators
Two-dimensional (Planar)	Rectangular with $v_z = 0$ and no z-dependence	$v_x = -\frac{\partial\psi}{\partial y}$ $v_y = +\frac{\partial\psi}{\partial x}$	$\frac{\partial}{\partial t}(\nabla^2\psi) + \frac{\partial(\psi, \nabla^2\psi)}{\partial(x, y)} = \nu\nabla^4\psi$ (*A*)	$\nabla^2 \equiv \frac{\partial^2}{\partial x^2} + \frac{\partial^2}{\partial y^2}$ $\nabla^4\psi \equiv \nabla^2(\nabla^2\psi) \equiv \left(\frac{\partial^4}{\partial x^4} + 2\frac{\partial^4}{\partial x^2\,\partial y^2} + \frac{\partial^4}{\partial y^4}\right)\psi$
	Cylindrical with $v_z = 0$ and no z-dependence	$v_r = -\frac{1}{r}\frac{\partial\psi}{\partial\theta}$ $v_\theta = +\frac{\partial\psi}{\partial r}$	$\frac{\partial}{\partial t}(\nabla^2\psi) + \frac{1}{r}\frac{\partial(\psi, \nabla^2\psi)}{\partial(r, \theta)} = \nu\nabla^4\psi$ (*B*)	$\nabla^2 \equiv \frac{\partial^2}{\partial r^2} + \frac{1}{r}\frac{\partial}{\partial r} + \frac{1}{r^2}\frac{\partial^2}{\partial\theta^2}$
Axisymmetrical	Cylindrical with $v_\theta = 0$ and no θ-dependence	$v_z = -\frac{1}{r}\frac{\partial\psi}{\partial r}$ $v_r = +\frac{1}{r}\frac{\partial\psi}{\partial z}$	$\frac{\partial}{\partial t}(E^2\psi) - \frac{1}{r}\frac{\partial(\psi, E^2\psi)}{\partial(r, z)} - \frac{2}{r^2}\frac{\partial\psi}{\partial z}E^2\psi = \nu E^4\psi$ (*C*)	$E^2 \equiv \frac{\partial^2}{\partial r^2} - \frac{1}{r}\frac{\partial}{\partial r} + \frac{\partial^2}{\partial z^2}$ $E^4\psi \equiv E^2(E^2\psi)$
	Spherical with $v_\phi = 0$ and no ϕ-dependence	$v_r = -\frac{1}{r^2\sin\theta}\frac{\partial\psi}{\partial\theta}$ $v_\theta = +\frac{1}{r\sin\theta}\frac{\partial\psi}{\partial r}$	$\frac{\partial}{\partial t}(E^2\psi) + \frac{1}{r^2\sin\theta}\frac{\partial(\psi, E^2\psi)}{\partial(r, \theta)} - \frac{2E^2\psi}{r^2\sin^2\theta}\left(\frac{\partial\psi}{\partial r}\cos\theta - \frac{1}{r}\frac{\partial\psi}{\partial\theta}\sin\theta\right) = \nu E^4\psi$ (*D*)	$E^2 \equiv \frac{\partial^2}{\partial r^2} + \frac{\sin\theta}{r^2}\frac{\partial}{\partial\theta}\left(\frac{1}{\sin\theta}\frac{\partial}{\partial\theta}\right)$

[a] Similar relations in general orthogonal coordinates may be found in S. Goldstein, *Modern Developments in Fluid Dynamics*, Oxford University Press (1938), pp. 114–115; in this reference formulas are also given for axisymmetrical flows with a nonzero component of the velocity around the axis.

[b] Here the Jacobians are designated by

$$\frac{\partial(f, g)}{\partial(x, y)} = \begin{vmatrix} \partial f/\partial x & \partial f/\partial y \\ \partial g/\partial x & \partial g/\partial y \end{vmatrix}$$

Example 4.2–1. "Creeping Flow" Around a Sphere[2]

Use Table 4.2–1 to set up the differential equation for the stream function for the flow of a Newtonian fluid around a nonrotating sphere of radius R. Solve and get the velocity distribution when the fluid approaches the sphere in the $+z$-direction, as in §2.6.

Solution. For steady, creeping flow, the entire left side of Eq. D in Table 4.2–1 vanishes; hence the ψ equation for axisymmetric flow is simply

$$E^4\psi = 0 \tag{4.2–1}$$

or, in spherical coordinates,

$$\left[\frac{\partial^2}{\partial r^2} + \frac{\sin\theta}{r^2}\frac{\partial}{\partial\theta}\left(\frac{1}{\sin\theta}\frac{\partial}{\partial\theta}\right)\right]^2 \psi = 0 \tag{4.2–2}$$

This has to be solved with the boundary conditions:

B.C. 1: $$v_r = -\frac{1}{r^2\sin\theta}\frac{\partial\psi}{\partial\theta} = 0 \qquad \text{at } r = R \tag{4.2–3}$$

B.C. 2: $$v_\theta = +\frac{1}{r\sin\theta}\frac{\partial\psi}{\partial r} = 0 \qquad \text{at } r = R \tag{4.2–4}$$

B.C. 3: $$\psi \to -\tfrac{1}{2}v_\infty r^2 \sin^2\theta \qquad \text{as } r \to \infty \tag{4.2–5}$$

The first two boundary conditions describe the clinging of the fluid to the solid sphere surface, and the third implies that $v_z = v_\infty$ far from the sphere.

This last boundary condition suggests that $\psi(r, \theta)$ is of the form

$$\psi = f(r)\sin^2\theta \tag{4.2–6}$$

When this function is substituted into Eq. 4.2–2, we get

$$\left(\frac{d^2}{dr^2} - \frac{2}{r^2}\right)\left(\frac{d^2}{dr^2} - \frac{2}{r^2}\right) f(r) = 0 \tag{4.2–7}$$

This is a linear, homogeneous fourth-order equation. A trial solution of the form $f(r) = Cr^n$ reveals that n may have the values $-1, 1, 2$, and 4. Therefore, the function $f(r)$ has the form

$$f(r) = \frac{A}{r} + Br + Cr^2 + Dr^4 \tag{4.2–8}$$

In order for the third boundary condition to be satisfied (Eq. 4.2–5), D must be zero and C has to be $-\frac{1}{2}v_\infty$. Hence the stream function is

$$\psi(r, \theta) = \left(\frac{A}{r} + Br - \tfrac{1}{2}v_\infty r^2\right)\sin^2\theta \tag{4.2–9}$$

[2] The solution given here follows that given by L. M. Milne-Thomson, *Theoretical Hydrodynamics*, Macmillan, New York (1955), Third Edition, pp. 555–557. For other approaches, see H. Lamb, *Hydrodynamics*, Dover, New York (1945), §§337, 338.

The velocity components (from Table 4.2–1) are

$$v_r = -\frac{1}{r^2 \sin\theta}\frac{\partial\psi}{\partial\theta} = \left(+v_\infty - 2\frac{A}{r^3} - 2\frac{B}{r}\right)\cos\theta \tag{4.2-10}$$

$$v_\theta = +\frac{1}{r\sin\theta}\frac{\partial\psi}{\partial r} = \left(-v_\infty - \frac{A}{r^3} + \frac{B}{r}\right)\sin\theta \tag{4.2-11}$$

We now use the first two boundary conditions to get $A = -\frac{1}{4}v_\infty R^3$ and $B = \frac{3}{4}v_\infty R$, so that

$$\frac{v_r}{v_\infty} = \left[1 - \frac{3}{2}\left(\frac{R}{r}\right) + \frac{1}{2}\left(\frac{R}{r}\right)^3\right]\cos\theta \tag{4.2-12}$$

$$\frac{v_\theta}{v_\infty} = -\left[1 - \frac{3}{4}\left(\frac{R}{r}\right) - \frac{1}{4}\left(\frac{R}{r}\right)^3\right]\sin\theta \tag{4.2-13}$$

These are the velocity distributions given in §2.6 without proof.

In §2.6 we showed how one can integrate the pressure and velocity distributions over the sphere surface to get the drag force; that method for getting the force of the fluid on the solid is general. Here we evaluate F_k by integrating the rate of energy dissipation:[3]

$$v_\infty F_k = -\int_0^{2\pi}\int_0^{\pi}\int_R^{\infty}(\boldsymbol{\tau}:\nabla \boldsymbol{v})r^2\,dr\,\sin\theta\,d\theta\,d\phi \tag{4.2-14}$$

Insertion of the function $-(\boldsymbol{\tau}:\nabla\boldsymbol{v})$ in spherical coordinates from Table 3.4–8 gives

$$v_\infty F_k = 2\pi\mu\int_0^{\pi}\int_R^{\infty}\left[2\left(\frac{\partial v_r}{\partial r}\right)^2 + 2\left(\frac{1}{r}\frac{\partial v_\theta}{\partial\theta} + \frac{v_r}{r}\right)^2 + 2\left(\frac{v_r}{r} + \frac{v_\theta\cot\theta}{r}\right)^2 + \left(r\frac{\partial}{\partial r}\left(\frac{v_\theta}{r}\right) + \frac{1}{r}\frac{\partial v_r}{\partial\theta}\right)^2\right] r^2\sin\theta\,dr\,d\theta \tag{4.2-15}$$

Insertion of Eqs. 4.2–12,13 into this expression and integration yields

$$F_k = 6\pi\mu v_\infty R \tag{4.2-16}$$

which is *Stokes's Law.*

§4.3 STEADY TWO-DIMENSIONAL POTENTIAL FLOW

A good approximation of the flow patterns can frequently be obtained by solving the equations of change for "potential flow," that is, with the assumptions that the fluid is *ideal* (ρ = constant, $\mu = 0$) and that the flow is *irrotational* ($[\nabla \times \boldsymbol{v}] = 0$). These assumptions are good except in the neighborhood

[3] The method of Eq. 4.2–14 has been used for a non-Newtonian flow study by J. C. Slattery, doctoral dissertation, University of Wisconsin (1959).

of the surfaces of the containing conduit or the surfaces of submerged objects. Near such surfaces viscous effects are of great importance, and in the proximity of these surfaces an alternative set of approximations can be used, which lead to the *boundary-layer* equations. In this section we discuss the ideal irrotational flow, and in the next section we treat the boundary-layer flow. The two topics are complementary.

For an ideal fluid, the equations of continuity and motion for steady flow are[1]

(continuity) $$(\nabla \cdot \boldsymbol{v}) = 0 \tag{4.3-1}$$

(motion) $$\rho \nabla \tfrac{1}{2} v^2 - \rho[\boldsymbol{v} \times [\nabla \times \boldsymbol{v}]] = -\nabla \mathscr{P} \tag{4.3-2}$$

in which $-\nabla \mathscr{P} = -\nabla p + \rho \boldsymbol{g}$. For two-dimensional irrotational flow the statement that curl $\boldsymbol{v} = 0$ becomes

(irrotational) $$\frac{\partial v_x}{\partial y} - \frac{\partial v_y}{\partial x} = 0 \tag{4.3-3}$$

and the equations of continuity and motion become

(continuity) $$\frac{\partial v_x}{\partial x} + \frac{\partial v_y}{\partial y} = 0 \tag{4.3-4}$$

(motion) $$\tfrac{1}{2}\rho(v_x^2 + v_y^2) + \mathscr{P} = \text{constant} \tag{4.3-5}$$

These three equations are to be used to determine v_x, v_y, and $\mathscr{P}$ as functions of x and y.

Actually it turns out to be easier to work in terms of a *stream function* $\psi(x, y)$ and a *velocity potential* $\phi(x, y)$ rather than in terms of the velocity components v_x and v_y. Clearly, if ψ and ϕ are defined thus,[2]

$$v_x = -\frac{\partial \psi}{\partial y} \qquad v_y = +\frac{\partial \psi}{\partial x} \tag{4.3-6,7}$$

$$v_x = -\frac{\partial \phi}{\partial x} \qquad v_y = -\frac{\partial \phi}{\partial y} \tag{4.3-8,9}$$

then Eqs. 4.3-3 and 4.3-4 will automatically be satisfied. From the foregoing set of relations we immediately get

$$\frac{\partial \phi}{\partial x} = \frac{\partial \psi}{\partial y}; \qquad \frac{\partial \phi}{\partial y} = -\frac{\partial \psi}{\partial x} \tag{4.3-10,11}$$

[1] Here we have used the identity $[\boldsymbol{v} \cdot \nabla \boldsymbol{v}] = \nabla \tfrac{1}{2} v^2 - [\boldsymbol{v} \times [\nabla \times \boldsymbol{v}]]$.

[2] For three-dimensional flow, ϕ is defined by $\boldsymbol{v} = -\nabla \phi$. Inasmuch as the curl of the gradient of any scalar function is zero, setting $\boldsymbol{v} = -\nabla \phi$ insures that curl $\boldsymbol{v} = 0$.

These are the *Cauchy–Riemann* equations, which are the relations that must be satisfied by the real and imaginary parts of any analytic function[3] $w(z) = \phi(x, y) + i\psi(x, y)$. The quantity $w(z)$ is called the *complex potential*. Differentiation of Eq. 4.3–10 with respect to x and Eq. 4.3–11 with respect to y and adding shows that $\nabla^2\phi = 0$; that is, ϕ satisfies the two-dimensional Laplace equation. It may be similarly shown that $\nabla^2\psi = 0$ also.

As a consequence of the foregoing development, it appears that *any* analytic function $w(z)$ yields a pair of functions $\phi(x, y)$ and $\psi(x, y)$, which are the velocity potential and the stream function to *some* flow problem. Furthermore, the curves $\phi(x, y) =$ constant and $\psi(x, y) =$ constant are then the *equipotential lines* and *streamlines* for the problem. The velocity components can be obtained from Eqs. 4.3–6 through 9 or from

$$\frac{dw}{dz} = -v_x + iv_y \tag{4.3–12}$$

in which dw/dz is called the *complex velocity*. Once the velocity is known, the pressure can be found from Eq. 4.3–5.

An alternative method of generating equipotential lines and streamlines can be set forth in terms of the *inverse function* $z(w) = x(\phi, \psi) + iy(\phi, \psi)$, in which $z(w)$ is *any* analytic function of w. Between the functions $x = x(\phi, \psi)$ and $y = y(\phi, \psi)$, we can eliminate ψ and get

$$F(x, y, \phi) = 0 \tag{4.3–13}$$

Similar elimination of ϕ gives

$$G(x, y, \psi) = 0 \tag{4.3–14}$$

Setting $\phi =$ a constant in Eq. 4.3–13 gives the equations for the equipotential lines, and setting $\psi =$ a constant in Eq. 4.3–14 gives equations for the streamlines for *some* flow problem. The velocity components can be obtained from

$$-\frac{dz}{dw} = \frac{1}{v^2}(v_x + iv_y) \tag{4.3–15}$$

in which $v^2 = v_x^2 + v_y^2$.

Thus far all we have said is: given an analytic function $w = w(z)$ [or alternatively $z = z(w)$], a flow net with streamlines $\psi(x, y) =$ constant and equipotential lines $\phi(x, y) =$ constant can be constructed. This flow net will describe *some* ideal flow problem. The inverse problem [given the flow problem, find the analytic function $w(z)$] is considerably more difficult and is not discussed here. Suffice it to say only that some special methods are

[3] Some knowledge of the analytic functions of a complex variable is assumed here. Helpful introductions to this subject can be found in V. L. Streeter, *Fluid Dynamics*, McGraw-Hill, New York (1948), Chapter 5, and in C. R. Wylie, *Advanced Engineering Mathematics*, McGraw-Hill, New York (1951), Chapters 9, 13, 14.

available[3] but that recourse to tables of conformal mappings frequently provides quicker solutions.[4] Here we discuss the ideal flow around a cylinder as an example of use of the complex potential $w = w(z)$ and the flow out of a channel as an example of the use of an inverse function $z = z(w)$. A few general comments should be kept in mind:

a. The streamlines are everywhere perpendicular to the equipotential lines.

b. Streamlines and equipotential lines can be interchanged to get the solution of another flow problem.

c. Any streamline may be replaced by a solid surface. (Note that this implies that the fluid does not cling to the solid surfaces.)

Example 4.3-1. Ideal Flow Around a Cylinder

a. Show that the complex potential

$$w(z) = v_\infty\left(z + \frac{R^2}{z}\right) \tag{4.3-16}$$

describes the ideal flow around a circular cylinder of radius R, when the approach velocity is v_∞. (See Fig. 4.3-1.)

b. Find the velocity distribution.

c. Find the pressure distribution on the cylinder surface when the pressure far from the cylinder is p_∞.

Solution. *a.* To find the stream function and the velocity potential, we first split up $w(z)$ into its real and imaginary parts:

$$w(z) = \underbrace{v_\infty x\left(1 + \frac{R^2}{x^2 + y^2}\right)}_{\text{potential function}} + i\,\underbrace{v_\infty y\left(1 - \frac{R^2}{x^2 + y^2}\right)}_{\text{stream function}} \tag{4.3-17}$$

The stream function is given by

$$\psi(x, y) = v_\infty y\left(1 - \frac{R^2}{x^2 + y^2}\right) \tag{4.3-18}$$

or in terms of dimensionless quantities

$$\Psi(X, Y) = Y\left(1 - \frac{1}{X^2 + Y^2}\right) \tag{4.3-19}$$

in which $\Psi = \psi/v_\infty R$, $X = x/R$, and $Y = y/R$.

[4] H. Kober, *Dictionary of Conformal Representations*, Dover, New York (1957), Second Edition.

In Fig. 4.3–1 the curves $\Psi = \text{constant}$, plotted according to Eq. 4.3–19, are shown. The streamline $\Psi = 0$ gives a unit circle, which is taken to represent the cylinder surface. The streamline $\Psi = \frac{3}{2}$ goes through the point $X = 0$, $Y = 2$, etc. A simple method of plotting the streamlines is suggested in Problem 4.C.

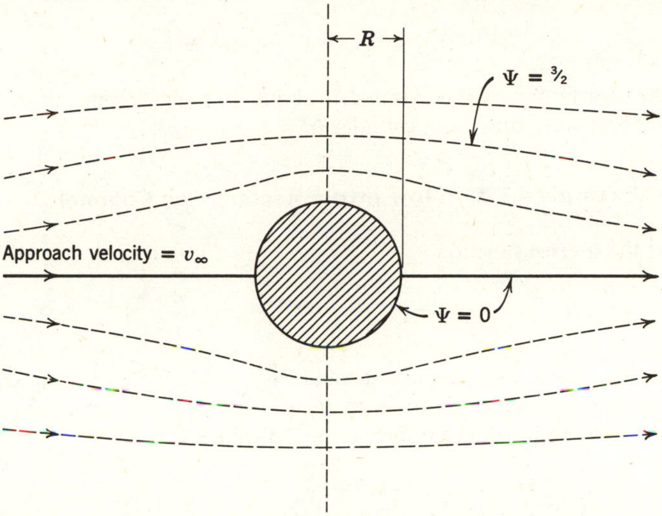

Fig. 4.3–1. Streamlines for ideal flow around a cylinder according to Eq. 4.3–19.

b. The velocity components are obtainable from the stream function by using Eqs. 4.3–6 and 7. They may also be obtained from the complex velocity according to Eq. 4.3–12:

$$\frac{dw}{dz} = v_\infty\left(1 - \frac{R^2}{z^2}\right)$$

$$= v_\infty\left(1 - \frac{R^2}{r^2}\, e^{-2i\theta}\right)$$

$$= v_\infty\left(1 - \frac{R^2}{r^2}(\cos 2\theta - i \sin 2\theta)\right) \tag{4.3–20}$$

Hence

$$-v_x = v_\infty\left(1 - \frac{R^2}{r^2}\cos 2\theta\right) \tag{4.3–21}$$

$$+v_y = v_\infty\left(\frac{R^2}{r^2}\sin 2\theta\right) \tag{4.3–22}$$

c. On the cylinder surface $r = R$ and

$$v^2 = v_x^2 + v_y^2$$

$$= v_\infty^2[(1 - \cos 2\theta)^2 + (\sin 2\theta)^2]$$

$$= 4v_\infty^2 \sin^2\theta \tag{4.3–23}$$

When $\theta = 0$ or $\theta = \pi$, the speed v equals zero; such points are known as *stagnation points*. Then from Eq. 4.3–5

$$\tfrac{1}{2}\rho v^2 + \mathscr{P} = \tfrac{1}{2}\rho v_\infty^2 + \mathscr{P}_\infty \tag{4.3–24}$$

From Eqs. 4.3–23 and 24, we get the pressure distribution on the cylinder:

$$(\mathscr{P} - \mathscr{P}_\infty) = \tfrac{1}{2}\rho v_\infty^2(1 - 4\sin^2\theta) \tag{4.3–25}$$

Note that the distribution of $\mathscr{P}$ is symmetric about the y-axis; hence the ideal fluid theory predicts no form drag on a cylinder (*d'Alembert's paradox*[5]).

Example 4.3–2. Flow into a Rectangular Channel

Show that the inverse function

$$z = \frac{w}{v_\infty} + \frac{b}{\pi} e^{\pi w/bv_\infty} \tag{4.3–26}$$

describes the flow into a rectangular channel of half-width b. The quantity v_∞ is the magnitude of the velocity far downstream from the entrance to the channel.

Solution. First we introduce dimensionless distance variables

$$X = \frac{\pi x}{b}; \qquad Y = \frac{\pi y}{b}; \qquad Z = \frac{\pi z}{b} = X + iY \tag{4.3–27}$$

and the dimensionless quantities

$$\Phi = \frac{\pi\phi}{bv_\infty}; \qquad \Psi = \frac{\pi\psi}{bv_\infty}; \qquad W = \frac{\pi w}{bv_\infty} = \Phi + i\Psi \tag{4.3–28}$$

These dimensionless quantities are similar to those used in Example 4.3–1. The function in Eq. 4.3–26, written in terms of the dimensionless quantities, may be split up into real and imaginary parts:

$$Z = W + e^W = (\Phi + e^\Phi \cos\Psi) + i(\Psi + e^\Phi \sin\Psi) \tag{4.3–29}$$

Hence

$$X = \Phi + e^\Phi \cos\Psi \tag{4.3–30}$$

$$Y = \Psi + e^\Phi \sin\Psi \tag{4.3–31}$$

We can now set Ψ equal to a constant, and the streamline $Y = Y(X)$ is expressed parametrically in Φ. For example, the streamline $\Psi = 0$ is given by

$$X = \Phi + e^\Phi \tag{4.3–32}$$

$$Y = 0 \tag{4.3–33}$$

[5] For a discussion of hydrodynamic paradoxes, see G. Birkhoff, *Hydrodynamics*, Dover, New York (1955).

As Φ goes from $-\infty$ to $+\infty$, X goes from $-\infty$ to $+\infty$; hence the X-axis is a streamline. Next consider $\Psi = \pi$ for which

$$X = \Phi - e^{\Phi} \tag{4.3–34}$$

$$Y = \pi \tag{4.3–35}$$

As Φ goes from ∞ to $-\infty$, X goes from $-\infty$ to -1 and then back to $-\infty$; that is, the streamline doubles back on itself. We select this streamline to be one of the solid walls of the reactangular channel. Similarly, the streamline $\Psi = -\pi$ is the other wall. The streamlines $\Psi = c$, where $-\pi < c < +\pi$, then give the flow pattern for the flow into the rectangular channel as shown in Fig. 4.3–2.

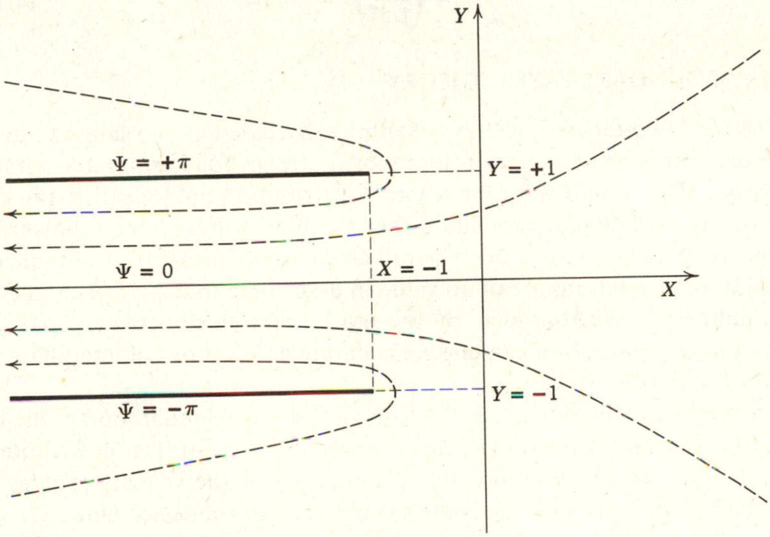

Fig. 4.3–2. Flow into a rectangular channel.

From Eq. 4.3–29 the derivative $-dz/dw$ may be found:

$$-\frac{dz}{dw} = -\frac{1}{v_\infty}\frac{dZ}{dW} = -\frac{1}{v_\infty}(1 + e^{W}) = -\frac{1}{v_\infty}(1 + e^{\Phi}\cos\Psi + ie^{\Phi}\sin\Psi) \tag{4.3–36}$$

Comparison of this expression with Eq. 4.3–15 gives the expressions for the velocity components:

$$\frac{v_x v_\infty}{v^2} = -(1 + e^{\Phi}\cos\Psi) \tag{4.3–37}$$

$$\frac{v_y v_\infty}{v^2} = -(e^{\Phi}\sin\Psi) \tag{4.3–38}$$

These equations have to be used in conjunction with Eqs. 4.3–30 and 31 in order to get the velocity components as a function of position.

To illustrate the procedure, let us find the velocity at the point $x = 0$, $y = 0$. When X and Y are set equal to zero in Eqs. 4.3–30 and 31, we find that $\Psi = 0$ and $\Phi = -0.57$. Insertion of these values into Eq. 4.3–38 gives $v_y = 0$; insertion of these same values into Eq. 4.3–37 gives

$$\frac{v_x v_\infty}{v^2} = -(1 + e^{\Phi})$$

$$= -(1 + 0.57) \tag{4.3–39}$$

But since $v_y = 0$, we have $v = v_x$, so that

$$\frac{v_x}{v_\infty} = -\left(\frac{1}{1.57}\right) = -0.64 \tag{4.3–40}$$

§4.4 BOUNDARY-LAYER THEORY

The ideal two-dimensional flow examples discussed in §4.3 showed how the flow net can be represented in terms of a stream function and a potential function. The solutions for the velocity distribution do not satisfy the usual hydrodynamic boundary condition that the fluid clings to the solid surfaces in the system (i.e., $v_x = v_y = 0$ on all fixed solid surfaces). Consequently, the ideal fluid solutions are of no value in describing transport phenomena in the immediate neighborhood of the wall. Specifically, the viscous drag cannot be computed, nor can one get accurate descriptions of interphase heat and mass transfer processes.

To obtain information about the behavior in the neighborhood of the wall, we take recourse to *boundary-layer* approaches. For the description of viscous flow, we obtain approximate solutions for the velocity profiles in a thin boundary layer near the wall, taking viscosity into account. Then we "splice" this solution onto the ideal flow solution that describes the flow outside the boundary layer.

We begin by giving two very simple examples of boundary-layer flow rather than by attempting to present a formal introduction to the subject. The first example considers the boundary-layer development as a function of time [with boundary-layer thickness $\delta = \delta(t)$], and the second concerns itself with boundary-layer development as a function of position in steady flow [with boundary-layer thickness $\delta = \delta(x)$].

Example 4.4–1. Flow Near a Wall Suddenly Set in Motion

Solve Example 4.1–1 by the following approximate procedure. Intuitively we suspect that the true solution to this problem will look something like the curves sketched in Fig. 4.4–1*a*, but we will assume that the solution can be satisfactorily represented by the sketch in Fig. 4.4–1*b*. That is, we assume that at any time t, there is a boundary-layer thickness $\delta(t)$ beyond which there is no flow, so that the viscosity effects are confined to the region $0 \leqslant y \leqslant \delta(t)$. We further assume that the dimensionless velocity profiles remain similar as time proceeds; analytically, this idea is expressed thus:

$$\frac{v_x}{V} = \phi(\eta) \qquad \text{where} \qquad \eta = \frac{y}{\delta(t)} \tag{4.4–1}$$

in which $\phi(\eta)$ is any "reasonable" function that one selects.

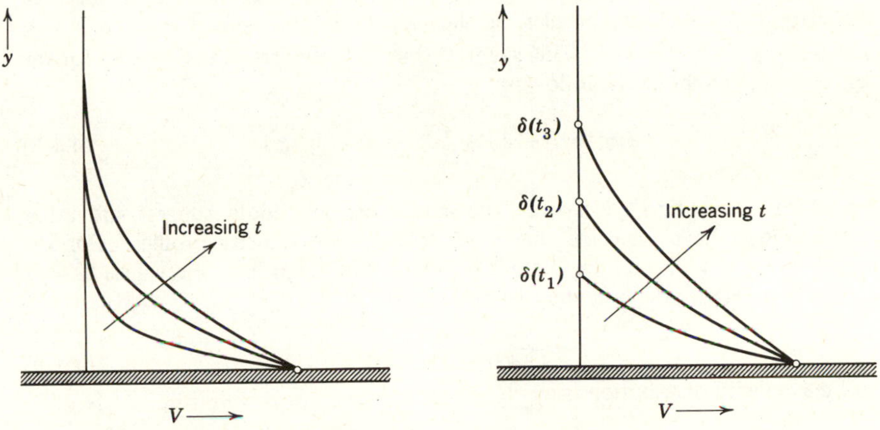

(a) Postulated true solution (b) Boundary-layer approximation

Fig. 4.4–1. Comparison of "true" and approximate unsteady velocity profiles near a wall suddenly set in motion with velocity *V*.

Solution. The flow problem is described by Eq. 4.1–1, a partial differential equation. We evaluate the partial derivatives appearing therein by using the assumed "similar" profiles of Eq. 4.4–1:

$$\frac{\partial v_x}{\partial t} = V\phi'\left(-\frac{\eta}{\delta}\right)\frac{d\delta}{dt} \tag{4.4–2}$$

$$\frac{\partial^2 v_x}{\partial y^2} = V\phi''\left(\frac{1}{\delta^2}\right) \tag{4.4–3}$$

where primes denote differentiation with respect to η. By substituting the expressions into Eq. 4.1–1 and then integrating over η, we get

$$M\delta\,\frac{d\delta}{dt} = \nu N \tag{4.4–4}$$

where

$$M = \left(-\int_0^1 \phi'\eta\, d\eta\right) = +\int_0^1 \phi\, d\eta \tag{4.4–5}$$

$$N = \left(+\int_0^1 \phi'' d\eta\right) = \phi'\Big|_0^1 \tag{4.4–6}$$

Integration of Eq. 4.4–4 with respect to time (using the initial condition that $\delta = 0$ when $t = 0$) gives

$$\delta(t) = \sqrt{2(N/M)\nu t} \tag{4.4–7}$$

Hence we find that the boundary-layer thickness increases as the square root of the time elapsed after the wall is set in motion.

We have thus far developed the expression for $\delta(t)$ in terms of any function $\phi(\eta)$. Let us now choose a specific function $\phi(\eta)$ that gives a reasonable description of the shape of the velocity profiles, as sketched in Fig. 4.4–1*b*. Clearly $\phi = 1$ at $\eta = 0$ and $\phi = 0$ at $\eta = 1$. This suggests that an appropriate assumption for the function $\phi(\eta)$ might be the following:

$$\phi(\eta) = 1 - \tfrac{3}{2}\eta + \tfrac{1}{2}\eta^3 \qquad 0 < \eta < 1 \tag{4.4–8}$$

for which in addition $\phi'(1) = 0$. The reader can no doubt suggest still other functions that have plausible shapes to describe the velocity profiles. For the profile in Eq. 4.4–8, the integrals in Eqs. 4.4–5 and 4.4–6 have the values $M = \frac{3}{8}$ and $N = \frac{3}{2}$. Hence the boundary-layer thickness is

$$\delta(t) = 2\sqrt{2\nu t} \tag{4.4–9}$$

and the velocity distribution is

$$\frac{v_x}{V} = 1 - \frac{3}{2}\left(\frac{y}{2\sqrt{2\nu t}}\right) + \frac{1}{2}\left(\frac{y}{2\sqrt{2\nu t}}\right)^3; \qquad 0 < y < \delta(t) \tag{4.4–10a}$$

$$\frac{v_x}{V} = 0; \qquad y > \delta(t) \tag{4.4–10b}$$

This result is to be compared with the exact solution in Eq. 4.1–13, which has the series expansion

$$\frac{v_x}{V} = 1 - 2\sqrt{\frac{2}{\pi}}\left(\frac{y}{2\sqrt{2\nu t}}\right) + \frac{2}{3}\sqrt{\frac{2^3}{\pi}}\left(\frac{y}{2\sqrt{2\nu t}}\right)^3 - \cdots \tag{4.4–11}$$

Hence the use of the function $\phi(\eta)$, chosen in Eq. 4.4–8, has led to the first three terms of the exact solution and modified the constants to get the "best" choice consistent with the restrictions inherent in the approximate procedure.

Example 4.4–2. Flow Near the Leading Edge of a Flat Plate

Obtain a description of the incompressible flow pattern near the leading edge of a flat plate immersed in a fluid stream, as shown in Fig. 4.4–2. Use a procedure similar to that developed in Example 4.4–1.

Solution. First we must write down the differential equations that describe the flow field. These are the two-dimensional, steady-state equation of continuity and the x-component of the equation of motion:

(continuity)
$$\frac{\partial v_x}{\partial x} + \frac{\partial v_y}{\partial y} = 0 \tag{4.4–12}$$

(motion)
$$v_x \frac{\partial v_x}{\partial x} + v_y \frac{\partial v_x}{\partial y} = \nu \left(\frac{\partial^2 v_x}{\partial x^2} + \frac{\partial^2 v_x}{\partial y^2} \right) \tag{4.4–13}$$

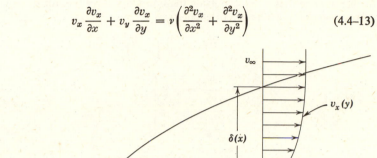

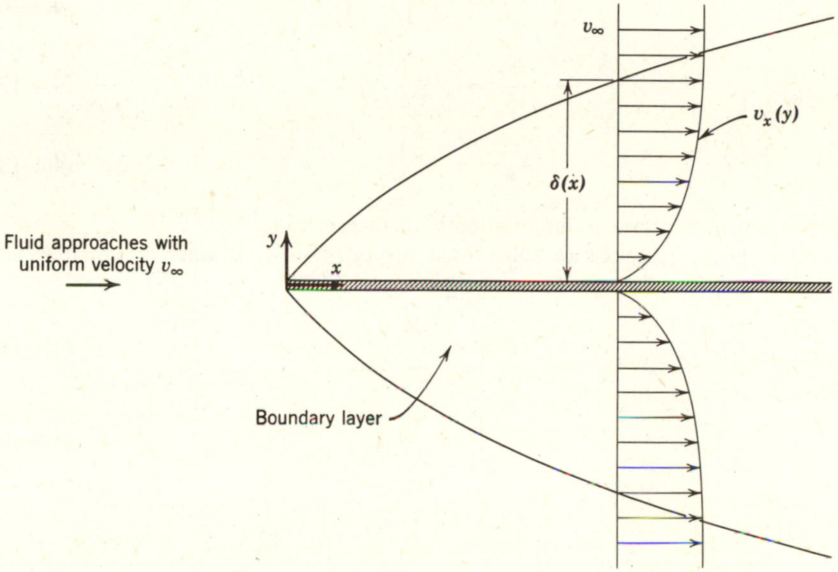

Fig. 4.4–2. Boundary-layer development near a flat plate of negligible thickness.

We do not consider the y-component of the equation of motion because the flow is not great in that direction. In Eq. 4.4–13 we can neglect the term $\partial^2 v_x/\partial x^2$ on the ground that it will be small[1] in comparison with $v_x(\partial v_x/\partial x)$. We then solve Eq. 4.4–12 for v_y (using the boundary condition that $v_y = 0$ at $y = 0$) and substitute the result into Eq. 4.4–13 to get

$$v_x \frac{\partial v_x}{\partial x} - \left(\int_0^y \frac{\partial v_x}{\partial x}\, dy \right) \frac{\partial v_x}{\partial y} = \nu \frac{\partial^2 v_x}{\partial y^2} \tag{4.4–14}$$

This then is our starting equation to determine $v_x(x, y)$; it is clearly nonlinear. It is to be solved with the boundary conditions $v_x = 0$ at $y = 0$, $v_x = v_\infty$ at $y = \infty$, and $v_x = v_\infty$ at $x = 0$ for all y.

[1] For more detailed arguments for determining which terms can be neglected, see, for example, H. Schlichting, *Boundary Layer Theory*, McGraw-Hill, New York (1955), Chapter 7.

As in Example 4.4–1, we assume that the velocity profiles at various values of x have the same shape:

$$\frac{v_x}{v_\infty} = \phi(\eta) \qquad \text{where} \qquad \eta = \frac{y}{\delta(x)} \tag{4.4–15}$$

in which $\delta(x)$ is the boundary-layer thickness a distance x down the plate. We further assume that $v_x = v_\infty$ outside the boundary layer (i.e. for $y > \delta$). Then the derivatives in Eq. 4.4–14 are calculated:

$$\frac{\partial v_x}{\partial x} = v_\infty \phi' \left(-\frac{\eta}{\delta} \right) \frac{d\delta}{dx} \tag{4.4–16}$$

$$\frac{\partial v_x}{\partial y} = v_\infty \phi' \frac{1}{\delta} \tag{4.4–17}$$

$$\frac{\partial^2 v_x}{\partial y^2} = v_\infty \phi'' \frac{1}{\delta^2} \tag{4.4–18}$$

in which primes denote differentiation with respect to η.

When these expressions are substituted into Eq. 4.4–14 and when the equation has been integrated with respect to η, we get

$$(B - A)\delta \frac{d\delta}{dx} = \frac{\nu}{v_\infty} C \tag{4.4–19}$$

where

$$A = \left(\int_0^1 \phi\phi'\eta \, d\eta \right) \tag{4.4–20}$$

$$B = \left[\int_0^1 \phi' \left(\int_0^{\bar{\eta}} \phi'\eta \, d\eta \right) d\bar{\eta} \right]$$

$$= -A + \left(\int_0^1 \phi'\eta \, d\eta \right) \tag{4.4–21}$$

$$C = \left(\int_0^1 \phi'' \, d\eta \right)$$

$$= \phi' \Big|_0^1 \tag{4.4–22}$$

Integration of Eq. 4.4–19 then gives

$$\delta(x) = \sqrt{2 \left(\frac{C}{B - A} \right) \nu x / v_\infty} \tag{4.4–23}$$

where we have used the boundary condition that $\delta = 0$ at $x = 0$. Hence we find that the boundary-layer thickness is proportional to the square root of the distance down the plate.

We now select as a reasonable velocity profile

$$\phi(\eta) = \tfrac{3}{2}\eta - \tfrac{1}{2}\eta^3 \tag{4.4–24}$$

because it has approximately the correct shape. For this choice, we find that $A = \frac{9}{35}$, $B = \frac{33}{280}$, and $C = -\frac{3}{2}$. Hence the boundary-layer thickness is

$$\delta(x) = \sqrt{\frac{280}{13}}\sqrt{\frac{\nu x}{v_\infty}} = 4.64\sqrt{\frac{\nu x}{v_\infty}} \tag{4.4–25}$$

and the velocity distribution is

$$\frac{v_x}{v_\infty} = \frac{3}{2}\left(\frac{y}{4.64\sqrt{\nu x/v_\infty}}\right) - \frac{1}{2}\left(\frac{y}{4.64\sqrt{\nu x/v_\infty}}\right)^3; \qquad 0 \leqslant y \leqslant \delta(x) \tag{4.4–26}$$

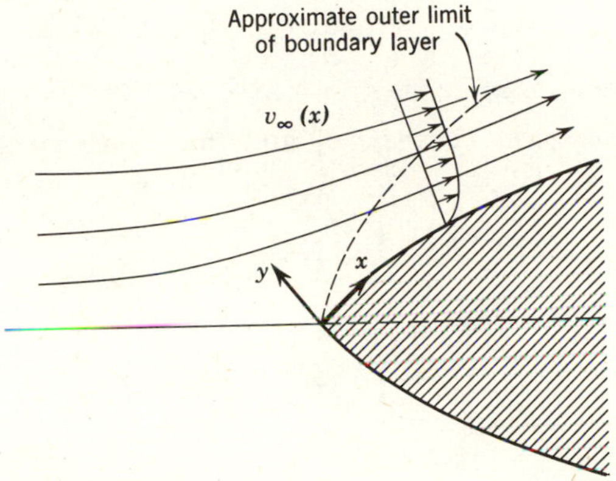

Fig. 4.4–3. Coordinate system for two-dimensional flow around a submerged object.

From this result we can get the drag force by integrating the momentum flux over the two wetted surfaces:

$$\begin{aligned} F_x &= 2\int_0^W \int_0^L \left(+\mu\frac{\partial v_x}{\partial y}\right)\bigg|_{y=0} dx\,dz \\ &= 1.292\sqrt{\rho\mu L W^2 v_\infty{}^3} \end{aligned} \tag{4.4–27}$$

An exact numerical solution to Eq. 4.4–14 has been obtained, and from that solution the drag force has also been calculated.[2] The result is of the same form as Eq. 4.4–27, but the constant is 1.328. This comparison gives some idea of the quantitative accomplishment of the approximate boundary-layer method used here.

The preceding examples represent the very simplest types of boundary-layer problems. For the steady two-dimensional incompressible flow around an object more complicated than a flat plate (see Fig. 4.4–3), one uses the following generalization of Eq. 4.4–14:

$$v_x\frac{\partial v_x}{\partial x} - \left(\int_0^y \frac{\partial v_x}{\partial x}\,dy\right)\frac{\partial v_x}{\partial y} = v_\infty\frac{dv_\infty}{dx} + \nu\frac{\partial^2 v_x}{\partial y^2} \tag{4.4–28}$$

[2] The numerical solution was obtained by H. Blasius, *Z. Math. Phys.*, **56**, 1–37 (1908) or NACA Tech. Memo. No. 1256; see also H. Schlichting, *Boundary Layer Theory*, Pergamon Press, London (1955), §7f.

in which $v_\infty(x)$ describes the change in the "free-stream velocity" with x at the outer edge of the boundary layer. This external flow velocity is found with the potential flow theory discussed in §4.3. Equation 4.4–28 is valid for surfaces that are not too curved; x is the coordinate along the surface and y is the coordinate perpendicular to the surface.

Equation 4.4–28 may be integrated with respect to y; in this integration use is made of the fact that v_x approaches $v_\infty(x)$ for large y. The result is

$$-\frac{1}{\rho}\tau_{yx}\Big|_{y=0} = \frac{d}{dx}(\delta_2 v_\infty{}^2) + \delta_1 v_\infty \frac{dv_\infty}{dx} \tag{4.4–29}$$

where the *displacement thickness* δ_1 and the *momentum thickness* δ_2 are defined by

$$\delta_1 = \int_0^\infty \left(1 - \frac{v_x}{v_\infty}\right) dy \tag{4.4–30}$$

$$\delta_2 = \int_0^\infty \frac{v_x}{v_\infty}\left(1 - \frac{v_x}{v_\infty}\right) dy \tag{4.4–31}$$

Equations 4.4–28 and 29 provide the basis for the solution of many two-dimensional, incompressible, steady-state boundary-layer problems. Similar equations are available for unsteady flows, compressible fluids, and three-dimensional boundary layers.[3] Equation 4.4–28 is the *Prandtl boundary-layer equation*. This equation has been solved numerically for quite a few problems;[3,4] such solutions are referred to as the "exact" solutions to boundary-layer problems.[5] "Approximate" solutions, on the other hand, are those obtained by assuming similar profiles for v_x/v_∞ in Eqs. 4.4–29, 30, and 31 in terms of a boundary-layer thickness very much as we have done in the illustrative examples. Equation 4.4–29 is usually referred to as the *von Kármán boundary-layer momentum balance*.

Although many exact and approximate boundary-layer solutions have been obtained, considerable work remains to be done in applying the methods to systems of engineering interest. For example, the flow on the back side of submerged objects is not amenable to boundary-layer calculation, for in that region "backward flow" occurs and the Prandtl boundary-layer equations are not valid.

[3] H. Schlichting, *op. cit.*

[4] S. I. Pai, *Viscous Flow Theory*, Van Nostrand, Princeton (1957).

[5] See §19.3 for a discussion of exact solutions for simultaneous mass, momentum, and energy transfer.

QUESTIONS FOR DISCUSSION

1. Are "steady" flow equations obtained from "unsteady" flow equations by setting $\partial/\partial t = 0$ or $D/Dt = 0$?

2. Check Eq. 4.1–1 for dimensional correctness.

3. How is the integral in the denominator of Eq. 4.1–12 evaluated?

4. What happens if one tries to solve Eq. 4.1–21 by the method of separation of variables without first recognizing that the solution can be written as the sum of a steady-state solution and a transient solution?

5. On a carefully labeled sketch, show the meaning of the $J_1(\alpha_n)$, which appear in Eq. 4.1–40.

6. Verify that B.C. 3 in Eq. 4.2–5 has been written properly.

7. Is the parabolic Newtonian flow in a circular tube irrotational? The creeping flow around a sphere?

8. Show that for two-dimensional ideal flow the stream function satisfies Laplace's equation, $\nabla^2\psi = 0$. Compare with heat transfer where $\nabla^2 T = 0$.

9. What is the significance of "d'Alembert's paradox"?

10. Suggest some other functions with suitable properties as the function $\phi(\eta)$ in Eq. 4.4–8.

11. Suggest some viscous flow problems in which boundary-layer approaches would be expected not to work.

12. Verify Eq. 4.4–21.

13. Show how the von Kármán momentum balance (Eq. 4.4–29) is obtained from the Prandtl boundary-layer equation (Eq. 4.4–28).

PROBLEMS

4.A$_1$ Time for Attainment of Steady State in Tube Flow

a. A heavy oil, with a kinematic viscosity of 3.45×10^{-4} m^2 sec^{-1}, is at rest in a long vertical tube with a radius of 0.7 cm. The fluid is suddenly allowed to flow from the bottom of the tube by virtue of gravity. After what time will the velocity at the tube center be within 10 per cent of its final value?

b. What is the result if water at 68° F is used? *Answer: a.* 6.4×10^{-2} sec
b. 22 sec

4.B$_1$ Velocity Near a Moving Sphere

A sphere of radius R is falling in creeping flow through a stationary fluid of viscosity μ with a terminal velocity v_∞. At what horizontal distance from the sphere in a plane perpendicular to the direction of fall does the velocity of the fluid fall to 0.01 times the terminal velocity of the sphere? *Answer:* About 37 diameters from the sphere

4.C$_1$ Construction of Streamlines for Ideal Flow Around a Cylinder

Plot the streamlines for flow around a cylinder by use of the information in Example 4.3–1 and with the help of the following procedure:

a. Select a value of $\Psi = C$ (that is, choose a streamline).

b. Plot $Y = C + K$ (lines parallel to X-axis) and $Y = K(X^2 + Y^2)$ (circles tangent to X-axis at origin with radii $1/2K$).

c. Plot the intersections of lines and circles that have the same K values.

d. Join these points to get the streamline for $\Psi = C$.

a'. Select another value of C and repeat the foregoing process to generate a new streamline.

4.D₁ Comparison of Exact Result and Boundary-Layer Result for Flow Near Suddenly Moved Wall

Compare the values of v_x/V obtained from Eq. 4.1–13 with those from Eq. 4.4–10 for the following values of $y/\sqrt{4\nu t}$: (*a*) 0.2, (*b*) 0.5, (*c*) 1.0. Express results as per cent error in the local velocity.

Answer: *a.* +1.5 per cent
b. +2.5 per cent
c. −26 per cent

4.E₂ Unsteady Pseudoplastic Flow Near a Moving Wall

Extend Example 4.4–1 to describe the unsteady flow of a pseudoplastic fluid near a flat surface suddenly set in motion. Use the Ostwald-de Waele (power-law) model of Eq. 1.2–3 to describe the fluid. Show that the boundary layer thickness is

$$\delta(t) = \left[\frac{8}{3}\left(\frac{3}{2}\right)^n \frac{m(n+1)V^{n-1}}{\rho}\, t\right]^{1/(n+1)} \tag{4.E-1}$$

when the same profile $\phi(\eta)$ is used here as in Example 4.4–1.

4.F₂ Use of von Kármán Momentum Balance

For the velocity profile in Eq. 4.4–24, use Eq. 4.4–29 to obtain (*a*) the boundary-layer thickness given in Eq. 4.4–25 and (*b*) the drag force on the flat plate given in Eq. 4.4–27.

4.G₂ Ideal Flow Near a Stagnation Point

a. Show that the complex potential $w = -v_0 z^2$ can describe the flow near a plane stagnation point. (See Fig. 4.G.)

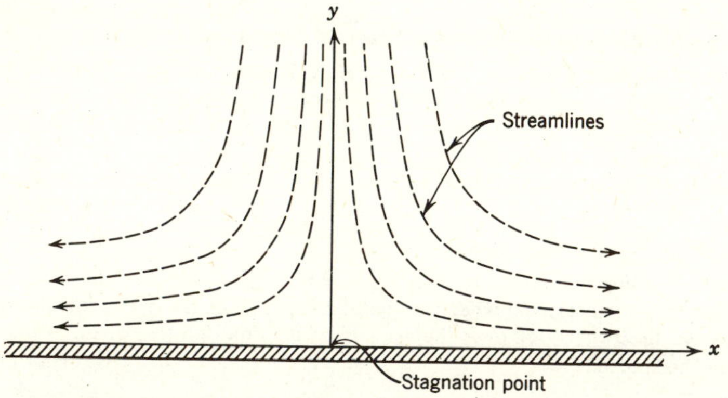

Fig. 4.G. Ideal flow near a stagnation point—two-dimensional flow.

b. What are the velocity components $v_x(x, y)$ and $v_y(x, y)$?

c. What is the physical significance of v_0?

4.H$_2$ Ideal Flow Around a Sphere

For the steady potential flow around a sphere with approach velocity v_∞ (see Fig. 2.6–1), the stream function and velocity potential are

$$\psi = +\frac{v_\infty R^3}{2r}\sin^2\theta - \frac{v_\infty r^2}{2}\sin^2\theta \tag{4.H–1}$$

$$\phi = -\frac{v_\infty R^3}{2r^2}\cos\theta - v_\infty r\cos\theta \tag{4.H–2}$$

The stream function is related to the velocity components, as indicated in Table 4.2–1, and the velocity potential is related to the velocity components by $\boldsymbol{v} = -\nabla\phi$, or specifically in spherical coordinates

$$v_r = -\frac{\partial\phi}{\partial r} \tag{4.H–3}$$

$$v_\theta = -\frac{1}{r}\frac{\partial\phi}{\partial\theta} \tag{4.H–4}$$

a. Show that Eqs. 4.H–1,2 give $v_z = v_\infty$ far from the sphere. (*Hint:* $v_z = v_r\cos\theta - v_\theta\sin\theta$.)

b. Show that the velocity at any point on the surface of the sphere is $v_\theta = -\frac{3}{2}v_\infty\sin\theta$.

c. Show that the pressure distribution on the surface of the sphere is $p - p_\infty = (\rho v_\infty^2/2)(1 - \frac{9}{4}\sin^2\theta)$.

4.I$_2$ Vortex Flow[1]

a. Show that the complex potential

$$w = \frac{i\Gamma}{2\pi}\ln z \tag{4.I–1}$$

describes the flow in a vortex. Show that the tangential velocity is given by

$$v_\theta = \frac{\Gamma}{2\pi r} \tag{4.I–2}$$

and that $v_r = 0$. This type of flow is sometimes called a *free vortex.*

b. Compare the functional dependence of v_θ on r in Eq. 4.I–2 with the function $v_\theta(r)$ which arose in Example 3.5–2. That kind of flow is sometimes called a *forced vortex.* Actual vortices, such as those that occur in an agitated tank, have a behavior intermediate between these two idealizations.

4.J$_3$ Basic Equations for Flow Through Porous Media[2,3,4]

For the flow of a fluid through a porous medium the equations of continuity and motion

[1] Vortices in agitated tanks have been experimentally investigated by S. Nagata, N. Yoshioka, and T. Yokoyama, *Memoirs of the Faculty of Engineering*, Kyoto University, Vol. XVII, No. III, July 1955.

[2] M. Muskat, *Flow of Homogeneous Fluids Through Porous Media*, McGraw-Hill (1937).

[3] A.Houpeurt, *Éléments de mécanique des fluides dans les milieux poreux*, Institut Français du Pétrole, Paris (1957). This book unfortunately contains a number of typographical errors.

[4] P. C. Carman, *Flow of Gases Through Porous Media*, Butterworths, London (1956).

may be replaced by

modified equation of continuity $\epsilon \frac{\partial \rho}{\partial t} = -(\nabla \cdot \rho \boldsymbol{v}_0)$ (4.J–1)

Darcy's law:[5] $\boldsymbol{v}_0 = -\frac{\kappa}{\mu}(\nabla p - \rho \boldsymbol{g})$ (4.J–2)

in which ϵ is the *porosity* (that is, the ratio of pore volume to total volume), and κ is the *permeability* of the porous medium. The velocity $\boldsymbol{v}_0$ in these equations is the *superficial velocity* (volume rate of flow through a unit cross-sectional area of the solid plus fluid) averaged over a small region of space—small with respect to macroscopic dimensions in the flow system but large with respect to the pore size. The quantities ρ and p are averaged over a region available to flow that is large with respect to the pore size. When Eqs. 4.J–1 and 2 are combined and the symbol $\mathscr{P}$ introduced (defined by $\nabla \mathscr{P} = \nabla p - \rho \boldsymbol{g}$), we get

$$\left(\frac{\epsilon \mu}{\kappa}\right) \frac{\partial \rho}{\partial t} = (\nabla \cdot [\rho \nabla \mathscr{P}]) \tag{4.J–3}$$

for constant viscosity μ and permeability κ. Equation 4.J–3 and the equation of state describe the motion of a fluid in a porous medium. For most purposes, we may write the *equation of state* as

$$\rho = \rho_0 p^m \exp \beta p \tag{4.J–4}$$

in which ρ_0 is the fluid density at unit pressure, and

For liquids	$m = 0$
Case 1: incompressible	$\beta = 0$
Case 2: compressible	$\beta \neq 0$
For gases	$\beta = 0$
Case 3: isothermal expansion	$m = 1$
Case 4: adiabatic expansion	$m = 1/\gamma = C_v/C_p$

Show that Eqs. 4.J–3 and 4.J–4 can be combined and simplified for these four cases to give

Case 1: $$\nabla^2 \mathscr{P} = 0 \quad \text{or} \quad \nabla^2 p = 0 \tag{4.J–5}$$

Case 2: $$\frac{\epsilon \mu \beta}{\kappa} \frac{\partial \rho}{\partial t} = \nabla^2 \rho - (\nabla \cdot \rho^2 \beta \boldsymbol{g}) \tag{4.J–6}$$

Case 3: $$\frac{2\epsilon \mu \rho_0}{\kappa} \frac{\partial \rho}{\partial t} = \nabla^2 \rho^2 \tag{4.J–7}$$

Case 4: $$\frac{(m+1)\epsilon \mu \rho_0^{1/m}}{\kappa} \frac{\partial \rho}{\partial t} = \nabla^2 \rho^{(1+m)/m} \tag{4.J–8}$$

It is customary to neglect the gravity terms in Cases 3 and 4, since they are small for gases with respect to the pressure terms. Note that Case 1 leads to *Laplace's equation*, Case 2 without the gravity term leads to the *heat conduction equation* or *diffusion equation*, and Cases 3 and 4 lead to nonlinear equations.

[5] An empirical modification of Darcy's law has been suggested by H. C. Brinkman, *Appl. Sci. Research*, **A1**, 27–34, 81–86 (1947):

$$0 = -\nabla p - \frac{\mu}{\kappa} \boldsymbol{v}_0 + \mu \nabla^2 \boldsymbol{v}_0 + \rho \boldsymbol{g} \tag{4.J–2a}$$

The term $\mu \nabla^2 \boldsymbol{v}_0$ is intended to account for distortion of the velocity profiles near containing walls.

4.K₃ Radial Flow Through a Porous Medium

A fluid flows through a porous cylindrical shell with inner and outer radii r_1 and r_2. At these surfaces, the pressures are known to be p_1 and p_2, respectively. The length of the cylindrical shell is h. See Problem 4.J for further notation.

a. Find the pressure distribution, radial flow velocity, and volume rate of flow for an incompressible fluid.

b. Rework for a compressible liquid and an ideal gas.

Answer: a.
$$\frac{p - p_1}{p_2 - p_1} = \frac{\ln r/r_1}{\ln r_2/r_1}$$

$$v_{0r} = -\frac{\kappa}{\mu r}\frac{p_2 - p_1}{\ln p_2/p_1}$$

$$Q = \frac{2\pi\kappa h(p_2 - p_1)}{\mu \ln (r_2/r_1)}$$

b.
$$Q = \frac{2\pi\kappa h(\rho_2 - \rho_1)}{\beta\mu\rho_1 \ln (r_2/r_1)}$$

$$Q = \frac{\pi\kappa h(p_2^2 - p_1^2)}{\mu p_1 \ln (r_2/r_1)}$$

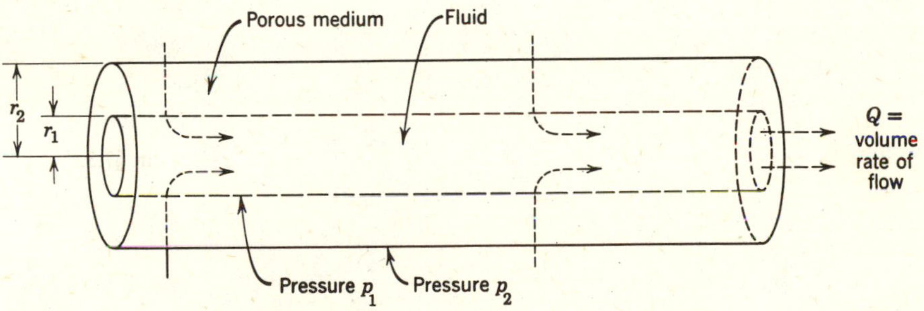

Fig. 4.K. Radial flow through a porous medium.

4.L₄ Unsteady Flow in an Annulus

a. By using Example 4.1–2 as a guide, derive the analogous equation for the unsteady *axial* flow in a coaxial annulus. Check your results with the published solution.[6]

b. Obtain a solution for the unsteady *tangential* flow in a coaxial annulus, when the fluid is at rest for $t < 0$ and the outer cylinder rotates with angular velocity Ω to cause laminar viscous flow for $t > 0$. Check your results with the published solution.[7]

4.M₄ Flow Through a Porous Medium in a Tube

a. A tube of radius R is filled with a porous material of uniform permeability κ. Use Eq. 4.J–2*a* to find the velocity distribution in the porous medium, taking into account that the velocity is zero at the tube surface.

b. Write the volume rate of flow as the flow rate given by Darcy's law multiplied by a correction factor accounting for the "wall effects."

[6] W. Müller, *Z. angew. Math. Mech.*, **16**, 227–228 (1936).

[7] R. B. Bird and C. F. Curtiss, *Chem. Eng. Sci.*, **11**, 108–113 (1959).

c. Show that the result in (*b*) simplifies to the Hagen-Poiseuille law when κ approaches ∞.

d. Look up sample values of μ and κ and determine whether or not the correction factor in (*b*) is normally important in industrial calculations.

e. Would the correction factor in (*b*) be larger or smaller than the correction needed to account for the nonrandomness of packing near the tube wall?

Answer: b. $$Q = \pi R^2 \left(\frac{\Delta p}{L}\frac{\kappa}{\mu}\right)\left(1 - 2\frac{\sqrt{\kappa}}{R}\frac{I_1(R/\sqrt{\kappa})}{I_0(R/\sqrt{\kappa})}\right)$$

4.N$_4$ The Bernoulli Equation for Inviscid Flow

Starting with Eq. 3.U–2, show that for *steady, irrotational, inviscid* flow with the external force derivable from a potential ($\boldsymbol{g} = -\nabla\hat{\Phi}$), the equation of motion may be integrated between any two points "1" and "2" in a flow field to get

$$\tfrac{1}{2}(v_2^2 - v_1^2) + \int_1^2 \frac{dp}{\rho} + (\hat{\Phi}_2 - \hat{\Phi}_1) = 0 \qquad (4.\text{N–1})$$

This is the *Bernoulli equation.*

It can be further shown[8] that the same equation results when the *steady, inviscid* (but not necessarily irrotational) equation of motion is integrated between two points "1" and "2" on the same streamline.

The generalization of the Bernoulli equation to viscous fluids is discussed in §7.3.

[8] H. Lamb, *Hydrodynamics*, Dover, New York (1945), pp. 20–21.

CHAPTER 5

Velocity Distributions in Turbulent Flow

In the preceding chapters only laminar flow problems have been discussed. We have seen that the differential equations describing laminar flow are known and that for a number of simple systems the velocity distribution and various derived quantities can be calculated in a straightforward way. The limiting factor in applying the equations of change is the mathematical complexity that one encounters in laminar flow systems involving several nonvanishing velocity components.

In this chapter we shall see that clean-cut methods for calculating turbulent velocity profiles are not at our disposal. It is true that the equations of continuity and motion *do* apply to turbulent flow; this seems plausible because the average size of the turbulent eddies is large in comparison with the mean free path of the molecules of the fluid. These equations, if they could be solved, would give the instantaneous values of the velocity and pressure, which in turbulent flow are fluctuating wildly about their mean values. (See §5.1.) Because such solutions cannot be found, other approaches must be developed.

In §5.2 we show how the equations of change can be averaged over a short time interval to get the "time-smoothed" equations of change. These equations describe the time-smoothed velocity and pressure distributions;

these profiles would then be the velocity measured by a pitot tube and the pressure measured by pressure gauges. The only difficulty is that the time-smoothed equation of motion contains "turbulent momentum flux" components $\tau_{xy}^{(t)}$. Whereas in laminar flow the components τ_{xy} could be expressed according to Newton's law of viscosity, the quantities $\tau_{xy}^{(t)}$ are generally handled empirically.

In §5.3 we summarize the empirical expressions for $\tau_{xy}^{(t)}$ that have been popular in the last few decades. Then several examples are given to show how they have been used to get information about the time-smoothed velocity distributions. Note that in each solution obtained at least one free constant has to be determined from experimental measurements. Therefore, the semi-empirical theories do not enable us to compute velocity profiles entirely a priori. Nevertheless, these theories have found worthwhile applications in engineering.

Because of the chaotic motion in turbulence, it seems evident that progress will ultimately be made by methods that capitalize on random motions and events. To date there is no entirely satisfactory self-contained statistical theory of turbulence. However, much has been done in the way of theoretical and experimental explorations into the behavior of various statistical quantities that must ultimately be explained by any successful statistical theory. In §5.4 we give the reader a glimpse of the statistical approach by discussing the decay of turbulence behind a grid.

In this brief introduction to turbulence we concern ourselves primarily with the description of the fully developed turbulent flow of an incompressible fluid. We do not consider at all the theoretical methods for predicting the inception of turbulence,[1] nor do we consider the experimental methods devised for probing into the structure of turbulent flows.[2,3] We hope the reader will realize that the literature on turbulence has grown to tremendous dimensions. All that we can hope to do in an elementary text is to present a few topics, which provide the background needed for further reading.

§5.1 FLUCTUATIONS AND TIME-SMOOTHED QUANTITIES

Before discussing the description of turbulent flow from an analytical viewpoint, we make a few remarks about the physical nature of turbulent flow. We use the flow in a circular tube as a basis for most of the discussion.

In §2.3 it was shown that for *laminar* flow in a circular tube the velocity distribution and the average velocity are given by

$$\frac{v_z}{v_{z,\max}} = \left[1 - \left(\frac{r}{R}\right)^2\right]; \qquad \frac{\langle v_z \rangle}{v_{z,\max}} = \frac{1}{2} \tag{5.1-1,2}$$

[1] H. Schlichting, *Boundary Layer Theory*, McGraw-Hill, New York (1955), Chapter 16.
[2] A. A. Townsend, *The Structure of Turbulent Shear Flow*, Cambridge University Press (1956).
[3] S. Goldstein, *Modern Developments in Fluid Dynamics*, Oxford University Press (1939).

Furthermore, it was shown that the pressure drop is exactly proportional to the volume rate of flow.

For *turbulent* flow, it has been shown experimentally that the time-smoothed quantities $\bar{v}_z$ and $\langle \bar{v}_z \rangle$ are given roughly by

$$\frac{\bar{v}_z}{\bar{v}_{z,\max}} \doteq \left(1 - \frac{r}{R}\right)^{1/7}; \qquad \frac{\langle \bar{v}_z \rangle}{\bar{v}_{z,\max}} \doteq \frac{4}{5} \tag{5.1-3,4}$$

These expressions are pretty good for the Reynolds number range 10^4 to 10^5; more accurate expressions are given later. In this Reynolds number range the

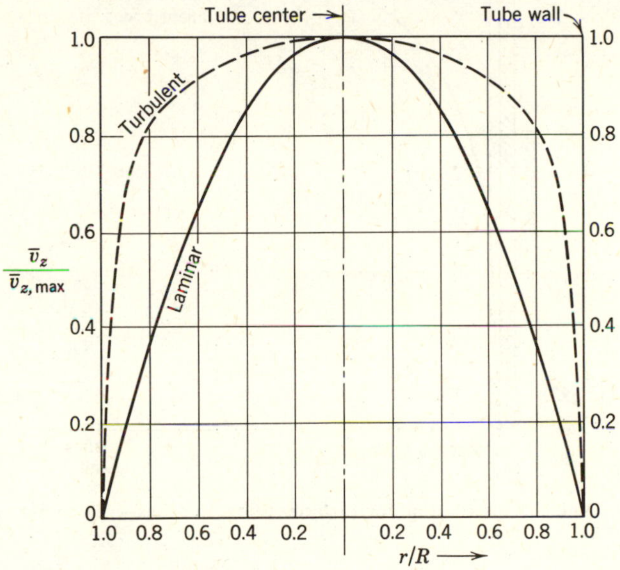

Fig. 5.1-1. Qualitative comparison of laminar and turbulent velocity distributions.

pressure drop is approximately proportional to the 7/4 power of the volume rate of flow. The laminar and turbulent velocity profiles are compared in Fig. 5.1-1.

For circular tubes the flow is laminar when $\text{Re} = D\langle v_z \rangle \rho/\mu$ is less than about 2.1×10^3, although a stable sinuous motion[1] sets in at a Reynolds number of about 1225. Above 2.1×10^3 laminar motion may be maintained temporarily if the tubes are very smooth and free from vibrations, but if the system is disturbed or if there is any appreciable surface roughness the laminar motion will give way to the random motion that characterizes turbulent flow.

[1] R. S. Prengle and R. R. Rothfus, *Ind. Eng. Chem.*, **47**, 379–386 (1955).

Actually, this statement is an oversimplification because the motion is not entirely random throughout the tube. At the center of the tube the velocity fluctuations are almost completely random. However, in the immediate neighborhood of the wall the fluctuations in the axial direction are greater than those in the radial direction, and all fluctuations approach zero at the wall itself. Hence it is apparent that there is a marked change in the physical behavior with radial distance. Although this change is continuous, it has

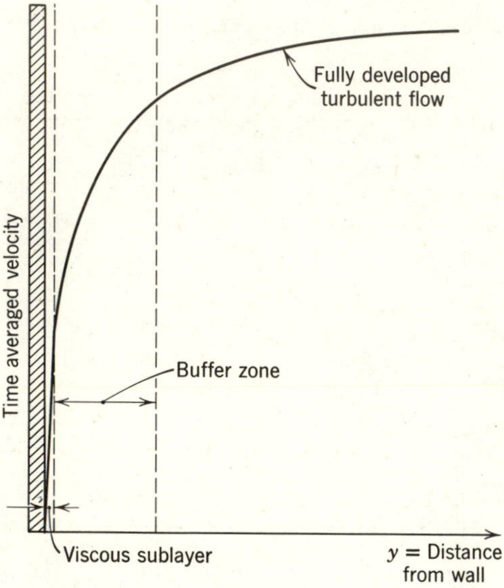

Fig. 5.1–2. Velocity distribution for turbulent flow in tubes—region near tube wall.

become customary to think of three arbitrary zones within the tube: the *viscous sublayer*, in which Newton's law of viscosity is used to describe the flow; the *buffer zone*, in which the laminar and turbulent effects are both important; and the *region of fully developed turbulence*, in which purely viscous effects are of negligible importance. These regions, which should not be interpreted too literally, are shown in Fig. 5.1–2.

So far we have been discussing primarily the time-smoothed velocity; let us now turn to the velocity fluctuations. We focus our attention on the fluid behavior at one point in the tube, where turbulent flow exists. We imagine that while we are watching this one spot the pressure drop causing the flow is increased slowly so that the mean velocity also increases slowly, in which case the axial component of velocity will behave as shown in Fig. 5.1–3. The instantaneous velocity v_z is an irregularly oscillating function. We define the time-smoothed velocity $\bar{v}_z$ by taking a time average of v_z over a time interval

t_0 large with respect to the time of turbulent oscillation but small with respect to the time-changes in the impressed pressure drop causing the flow:

$$\bar{v}_z = \frac{1}{t_0}\int_t^{t+t_0} v_z \, dt \tag{5.1–5}$$

The instantaneous velocity may then be written as the sum of the time-smoothed velocity $\bar{v}_z$ and a velocity fluctuation v_z':

$$v_z = \bar{v}_z + v_z' \tag{5.1–6}$$

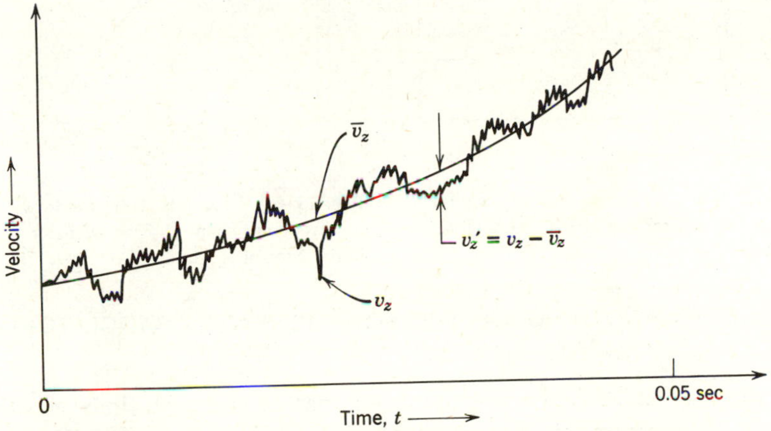

Fig. 5.1–3. Oscillation of velocity component about a mean value.

A similar expression can be written for the pressure, which is also fluctuating. Clearly $\overline{v_z'} = 0$ by the foregoing definitions. But $\overline{v_z'^2}$ will not be zero, and in fact $\sqrt{\overline{v_z'^2}}/\langle\bar{v}_z\rangle$ is a measure of the magnitude of the turbulent disturbance. This quantity, known as the "intensity of turbulence," may have values in the neighborhood of 1 to 10 per cent in tube flow.

It is helpful to have some idea of the orders of magnitude of the turbulent fluctuations; some sample curves for turbulent flow between two flat plates are given in Figs. 5.1–4 and 5.1–5. From Fig. 5.1–4 we see that the fluctuations in the direction of flow are greater than those in the direction perpendicular to the flow; there is a tendency for the two to be very nearly the same at the center of the duct (i.e. tendency toward *isotropy*). From Fig. 5.1–5 we see that the momentum flux over the greater part of the channel results almost entirely from the turbulent transfer. Only near the wall is molecular momentum transport important (this point is amplified in Example 5.3–3). The curves in Figs. 5.1–4 and 5 should be borne in mind in the discussion which follows.

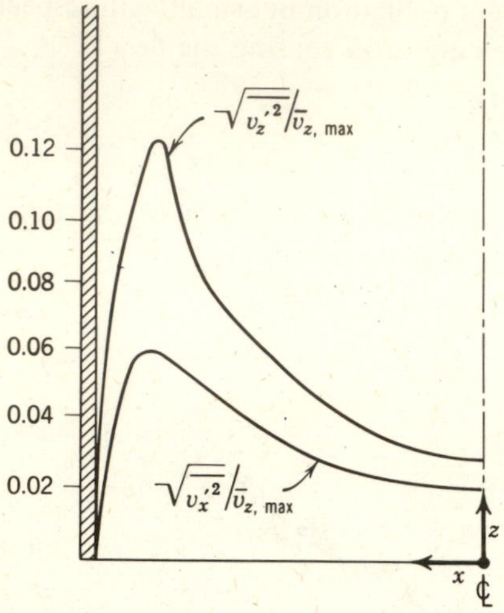

Fig. 5.1–4. Turbulent fluctuations for flow in z-direction in a rectangular channel.

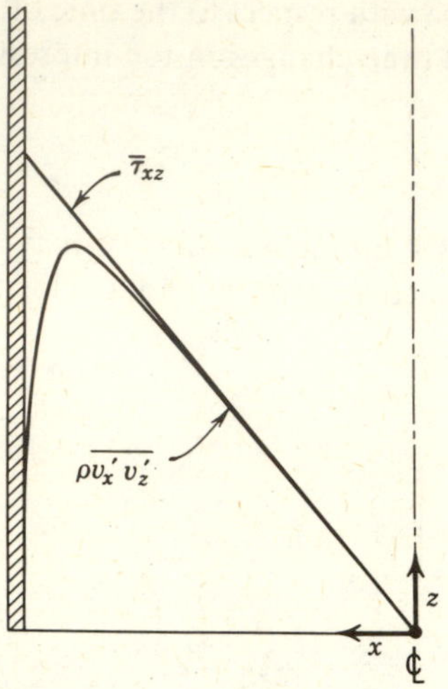

Fig. 5.1–5. Reynolds stress for flow in a rectangular channel. The quantity $\bar{\tau}_{xz}$ is the sum of $\bar{\tau}_{xz}^{(l)}$ and $\bar{\tau}_{xz}^{(t)}$.

§5.2 TIME-SMOOTHING OF THE EQUATIONS OF CHANGE FOR AN INCOMPRESSIBLE FLUID

In this section we derive the equations that describe the time-smoothed velocity and pressure for an incompressible fluid. To do this, we rewrite the equations of continuity and motion by replacing v_x by $\bar{v}_x + v_x'$ and p by $\bar{p} + p'$ everywhere they occur:

equation of continuity:
$$\frac{\partial}{\partial x}(\bar{v}_x + v_x') + \frac{\partial}{\partial y}(\bar{v}_y + v_y') + \frac{\partial}{\partial z}(\bar{v}_z + v_z') = 0 \qquad (5.2\text{–}1)$$

equation of motion:
$$\begin{aligned}\frac{\partial}{\partial t}\rho(\bar{v}_x + v_x') = &-\frac{\partial}{\partial x}(\bar{p} + p') \\ &-\left(\frac{\partial}{\partial x}\rho(\bar{v}_x + v_x')(\bar{v}_x + v_x')\right. \\ &+\frac{\partial}{\partial y}\rho(\bar{v}_y + v_y')(\bar{v}_x + v_x') \\ &\left.+\frac{\partial}{\partial z}\rho(\bar{v}_z + v_z')(\bar{v}_x + v_x')\right) \\ &+\mu\nabla^2(\bar{v}_x + v_x') \\ &+\rho g_x \end{aligned} \qquad (5.2\text{–}2)$$

Here we have used the equation of motion in the form of Eq. 3.2–5, with the stress components of a Newtonian fluid with constant ρ. The y- and z-components of the equation of motion may be similarly written.

We may now take the time average of Eqs. 5.2–1 and 2 according to Eq. 5.1–5, to get the following:

time-smoothed equation of continuity

$$\frac{\partial \bar{v}_x}{\partial x} + \frac{\partial \bar{v}_y}{\partial y} + \frac{\partial \bar{v}_z}{\partial z} = 0 \tag{5.2-3}$$

time-smoothed equation of motion

$$\frac{\partial}{\partial t} \rho \bar{v}_x = -\frac{\partial \bar{p}}{\partial x} - \left(\frac{\partial}{\partial x} \rho \bar{v}_x \bar{v}_x + \frac{\partial}{\partial y} \rho \bar{v}_y \bar{v}_x + \frac{\partial}{\partial z} \rho \bar{v}_z \bar{v}_x\right) - \underline{\left(\frac{\partial}{\partial x} \rho \overline{v_x' v_x'} + \frac{\partial}{\partial y} \rho \overline{v_y' v_x'} + \frac{\partial}{\partial z} \rho \overline{v_z' v_x'}\right)} + \mu \nabla^2 \bar{v}_x + \rho g_x \tag{5.2-4}$$

These are the time-smoothed equation of continuity and x-component of the equation of motion. By comparing them with the equations of §§3.1, 2, we see that the equation of continuity is the same as we had previously, except that time-smoothed velocity components replace the instantaneous velocity components. In the equation of motion, too, v_x, v_y, v_z, and p are everywhere replaced by $\bar{v}_x$, $\bar{v}_y$, $\bar{v}_z$, and $\bar{p}$; but, in addition, new terms arise (dashed underline), which are associated with the turbulent velocity fluctuations. For convenience, we introduce the notation

$$\bar{\tau}_{xx}^{(t)} = \rho \overline{v_x' v_x'}; \qquad \bar{\tau}_{xy}^{(t)} = \rho \overline{v_x' v_y'}; \quad \text{etc.} \tag{5.2-5,6}$$

These terms are the components of the *turbulent momentum flux* $\bar{\boldsymbol{\tau}}^{(t)}$; they are usually referred to as the *Reynolds stresses.*

In summary, we write the time-smoothed equations of continuity and motion in vector notation:

time-smoothed equation of continuity

$$(\nabla \cdot \bar{\boldsymbol{v}}) = 0 \tag{5.2-7}$$

time-smoothed equation of motion

$$\rho \frac{D\bar{\boldsymbol{v}}}{Dt} = -\nabla \bar{p} - [\nabla \cdot \bar{\boldsymbol{\tau}}^{(l)}] - \underline{[\nabla \cdot \bar{\boldsymbol{\tau}}^{(t)}]} + \rho \boldsymbol{g} \tag{5.2-8}$$

Here again we have indicated with a dashed underline the additional term that arises as a result of time-smoothing. The quantity $\bar{\boldsymbol{\tau}}^{(l)}$ is given by the expressions in Tables 3.4–5, 6 and 7, except that $\boldsymbol{v}$ is everywhere replaced by $\bar{\boldsymbol{v}}$.

The principal result in this section is that the equations summarized in Tables 3.4–2, 3 and 4 can be used for setting up turbulent flow problems provided that one changes all v_i to $\bar{v}_i$, p to $\bar{p}$, and τ_{ij} to $\bar{\tau}_{ij}^{(l)} + \bar{\tau}_{ij}^{(t)}$.

§5.3 SEMIEMPIRICAL EXPRESSIONS FOR THE REYNOLDS STRESSES

In order to use Eq. 5.2–8 to get velocity profiles, some expression for $\bar{\boldsymbol{\tau}}^{(t)}$ has to be inserted. Several semiempirical relations have been widely used, and we simply list them here. The reader desirous of finding out how these various expressions were "derived" should consult the original references.

Boussinesq's Eddy Viscosity[1]

One of the earliest proposals was that one write

$$\bar{\tau}_{yx}^{(t)} = -\mu^{(t)} \frac{d\bar{v}_x}{dy} \tag{5.3–1}$$

by analogy with Newton's law of viscosity; $\mu^{(t)}$ is a "turbulent coefficient of viscosity" or "eddy viscosity" and usually depends strongly on position.

Prandtl's Mixing Length[2,6]

By assuming that eddies move around in a fluid very much as molecules move about in a gas (actually a *very* poor analogy), Prandtl developed an expression for momentum transfer in a fluid in which the mixing length plays a role roughly analogous to that of the mean free path in gas kinetic theory. (See §1.4.) This way of thinking led Prandtl to the relation

$$\bar{\tau}_{yx}^{(t)} = -\rho l^2 \left|\frac{d\bar{v}_x}{dy}\right| \frac{d\bar{v}_x}{dy} \tag{5.3–2}$$

The *mixing length* l is also a function of position; Prandtl achieved some success by letting l be proportional to the distance y from the solid surface, that is, $l = \kappa_1 y$. A result similar to Eq. 5.3–2 was also obtained by G. I. Taylor's vorticity transport theory.[3]

von Kármán's Similarity Hypothesis[4,5,6]

On the basis of dimensional considerations, von Kármán suggested that the Reynolds' stresses have the form

$$\bar{\tau}_{yx}^{(t)} = -\rho {\kappa_2}^2 \left|\frac{(d\bar{v}_x/dy)^3}{(d^2\bar{v}_x/dy^2)^2}\right| \frac{d\bar{v}_x}{dy} \tag{5.3–3}$$

in which κ_2 is a "universal" constant whose value is given as 0.40 by some investigators and as 0.36 by others (these values having been determined from tube flow velocity profile data).

[1] T. V. Boussinesq, *Mém. prés. Acad. Sci.*, Third Edition, *Paris*, XXIII, 46 (1877).

[2] L. Prandtl, *Z. angew. Math. Mech:*, **5**, 136 (1925).

[3] G. I. Taylor, *Proc. Roy. Soc.* (*London*), **A135**, 685–701 (1932).

[4] T. von Kármán, *Nachr. Ges. Wiss. Göttingen, Math-physik. Kl.* (1930); also *NACA, TM 611.*

[5] R. G. Deissler, *NACA Report 1210* (1955).

[6] H. Schlichting, *Boundary Layer Theory*, McGraw-Hill, New York (1955), Chapter **XIX**.

Deissler's Empirical Formula for the Region Near the Wall[5]

Deissler proposed the following empirical expression for use in the neighborhood of the solid surfaces where the von Kármán equation and the Prandtl equation are inadequate:

$$\bar{\tau}_{yx}^{(t)} = -\rho n^2 \bar{v}_x y(1 - \exp\{-n^2 \bar{v}_x y/\nu\}) \frac{d\bar{v}_x}{dy} \tag{5.3–4}$$

in which n is a constant determined empirically by Deissler as 0.124 (from tube flow velocity distributions).

In the illustrative examples that follow we indicate how the foregoing relations are used to get some information about turbulent flows. In applying these relations to curvilinear systems, workers have frequently neglected the curvature effects for simplicity.[7]

Example 5.3–1. Derivation of the Logarithmic Distribution Law for Tube Flow (Far from Wall)

Find the time-smoothed velocity distribution for turbulent flow in a long tube, using the Prandtl mixing-length relation. The radius and length of the tube are R and L.

[7] In order to write appropriate expressions in curvilinear coordinates we must have $\bar{\boldsymbol{\tau}}^{(t)}$ expressed in tensor form. For example, the form for the von Kármán similarity hypothesis expression that is consistent with von Kármán's original derivation is

$$\bar{\boldsymbol{\tau}}^{(t)} = -\rho\kappa_2^2 \left| \frac{[\frac{1}{2}(\boldsymbol{\Delta}:\boldsymbol{\Delta})]^{3/2}}{[\frac{1}{2}(\boldsymbol{\Omega}:\boldsymbol{\Omega})]} \right| \boldsymbol{\Delta} \tag{5.3–3a}$$

in which $\Delta_{ij} = (\partial\bar{v}_i/\partial x_j) + (\partial\bar{v}_j/\partial x_i)$ and $\Omega_{ij} = (\partial\bar{w}_i/\partial x_j) + (\partial\bar{w}_j/\partial x_i)$. Here $\bar{w}_i$ is the ith component of the vorticity $\bar{\boldsymbol{w}} = [\nabla \times \bar{\boldsymbol{v}}]$, and the double-dot product of the tensors $\boldsymbol{\Delta}$ and $\boldsymbol{\Omega}$ is the operation defined in §A.4. For plane shear flow, Eq. 5.3–3a simplifies to Eq. 5.3–3. For symmetrical axial tube flow with $\bar{v}_z = \bar{v}_z(r)$ and $\bar{v}_\theta = \bar{v}_r = 0$, Eq. 5.3–3$a$ simplifies to

$$\bar{\tau}_{rz}^{(t)} = -\rho\kappa_2^2 \left| \frac{\left(\dfrac{d\bar{v}_z}{dr}\right)^3}{\left(\dfrac{d^2\bar{v}_z}{dr^2} - \dfrac{1}{r}\dfrac{d\bar{v}_z}{dr}\right)^2} \right| \frac{d\bar{v}_z}{dr} \tag{5.3–3b}$$

For the tangential flow between two rotating cylinders with $\bar{v}_\theta = \bar{v}_\theta(r)$ and $\bar{v}_r = \bar{v}_z = 0$, Eq. 5.3–3$a$ becomes

$$\bar{\tau}_{r\theta}^{(t)} = -\rho\kappa_2^2 \left| \frac{\left(\dfrac{d\bar{v}_\theta}{dr} - \dfrac{\bar{v}_\theta}{r}\right)^3}{\left[\dfrac{d}{dr}\left(\dfrac{d\bar{v}_\theta}{dr} + \dfrac{\bar{v}_\theta}{r}\right)\right]^2} \right| \left(\frac{d\bar{v}_\theta}{dr} - \frac{\bar{v}_\theta}{r}\right) \tag{5.3–3c}$$

which is in agreement with T. von Kármán, *Collected Works*, Butterworths, London (1956), p. 459. (The authors wish to acknowledge helpful correspondence with Dr. J. B. Opfell and Professor W. H. Corcoran in connection with the foregoing discussion.)

Solution. We let $s = R - r$ be the distance from the tube wall and assume that $l = \kappa_1 s$. For axial flow in tubes, Eq. 5.3–2 then becomes,

$$\begin{aligned}\bar{\tau}_{rz}^{(t)} &= +\rho\kappa_1^2(R-r)^2\left(-\frac{d\bar{v}_z}{dr}\right)^2 \\ &= +\rho\kappa_1^2 s^2\left(\frac{d\bar{v}_z}{ds}\right)^2\end{aligned} \tag{5.3–5}$$

The equation of motion is obtained from Eq. 5.2–8. For the situation in which $\bar{v}_z = \bar{v}_z(r)$ and the fluid is incompressible, we get

$$0 = \frac{\mathscr{P}_0 - \mathscr{P}_L}{L} - \frac{1}{r}\frac{d}{dr}(r\bar{\tau}_{rz}) \tag{5.3–6}$$

in which $\bar{\tau}_{rz} = \bar{\tau}_{rz}^{(l)} + \bar{\tau}_{rz}^{(t)}$. This may be integrated with the boundary condition that $\bar{\tau}_{rz} = 0$ at $r = 0$ to give

$$\bar{\tau}_{rz} = \frac{(\mathscr{P}_0 - \mathscr{P}_L)R}{2L}\cdot\frac{r}{R} = \tau_0\left(1 - \frac{s}{R}\right) \tag{5.3–7}$$

in which τ_0 is the wall shear stress (at $s = 0$). In the turbulent core (i.e., for most values of s) $\bar{\tau}_{rz}^{(l)}$ will be negligible with respect to $\bar{\tau}_{rz}^{(t)}$; in other words, momentum transport by molecules is small in comparison with that by eddy motion. Then Eqs. 5.3–5 and 5.3–7 may be combined to give

$$\rho\kappa_1^2 s^2\left(\frac{d\bar{v}_z}{ds}\right)^2 = \tau_0\left(1 - \frac{s}{R}\right) \tag{5.3–8}$$

Prandtl made a mathematical simplification (physically indefensible) at this point by setting the right side equal to τ_0. This simplifies the mathematics somewhat, and it has been shown that the result differs very little from that obtained by integrating Eq. 5.3–8. Hence after taking the square root we get

$$\frac{d\bar{v}_z}{ds} = \pm\frac{1}{\kappa_1}v_*\frac{1}{s} \tag{5.3–9}$$

in which the plus sign will have to be used. Here v_* is $\sqrt{\tau_0/\rho}$, which has dimensions of a velocity. If Eq. 5.3–9 is integrated from the outer edge of the buffer layer $s = s_1$ to any position s, we get

$$\bar{v}_z - \bar{v}_{z,1} = \frac{1}{\kappa_1}v_*\ln\frac{s}{s_1} \qquad s \geqslant s_1 \tag{5.3–10}$$

or in dimensionless variables ($v^+ = \bar{v}_z/v_*$, $s^+ = sv_*\rho/\mu$):

$$v^+ - v_1^+ = \frac{1}{\kappa_1}\ln\frac{s^+}{s_1^+} \qquad s^+ \geqslant s_1^+ \tag{5.3–11}$$

Deissler[5] has found from experimental velocity distribution data that the best value of κ_1 is 0.36; he has also found that the outer edge of the buffer zone can be conveniently chosen as $s_1{}^+ = 26$ (where $v_1{}^+ = 12.85$). With these values, recommended by Deissler, Eq. 5.3–11 becomes

$$v^+ = \frac{1}{0.36} \ln s^+ + 3.8 \qquad s^+ \geqslant 26 \tag{5.3–12}$$

This *logarithmic distribution* has been found[8] to give a rather good description of velocity profiles in turbulent flow at Re > 20,000 (except, of course, near the tube walls).

Example 5.3–2. Velocity Distribution for Tube Flow (Near Wall)

Use Deissler's empirical formula to calculate the velocity profile in the viscous sublayer and in the buffer region.

Solution. In the region near the wall we add Newton's law of viscosity and Deissler's expression to get $\bar{\tau}_{rz} = \bar{\tau}_{rz}^{(l)} + \bar{\tau}_{rz}^{(t)}$:

$$\bar{\tau}_{rz} = -\mu \frac{d\bar{v}_z}{dr} - \rho n^2 \bar{v}_z (R - r)\left(1 - \exp\{-n^2 \bar{v}_z (R - r)/\nu\}\right) \frac{d\bar{v}_z}{dr} \tag{5.3–13}$$

This is substituted into Eq. 5.3–7 and once again $1 - (s/R)$ is replaced by 1, which is a good approximation near the wall; this gives

$$\tau_0 = +\mu \frac{d\bar{v}_z}{ds} + \rho n^2 \bar{v}_z s\left(1 - \exp\{-n^2 \bar{v}_z s/\nu\}\right) \frac{d\bar{v}_z}{ds} \tag{5.3–14}$$

This is integrated from $s = 0$ to s; in dimensionless variables the result is

$$v^+ = \int_0^{s^+} \frac{ds^+}{1 + n^2 v^+ s^+ (1 - \exp\{-n^2 v^+ s^+\})} \qquad 0 \leqslant s^+ \leqslant 26 \tag{5.3–15}$$

in which n has been found empirically to be 0.124 for long, smooth tubes. Iterative procedures had to be used to solve for v^+ in terms of s^+; the result is given graphically in Fig. 5.3–1. For small values of s^+, Eq. 5.3–15 simplifies to

$$v^+ = s^+ \qquad 0 \leqslant s^+ \leqslant 5 \tag{5.3–16}$$

This result may also be obtained by integrating Newton's law of viscosity over the viscous sublayer.

Experimental velocity profile data for Re > 20,000 are in good agreement with the curve shown in Fig. 5.3–1. For a comparison of the work of several experimenters, see Corcoran, Opfell, and Sage.[8]

The effect of roughness on velocity distributions in tube flow has been studied, and a nice summary of this work may be found in Schlichting.[9] In later chapters

[8] A critical survey of various empirical results has been made by W. H. Corcoran. J. B. Opfell, and B. H. Sage, *Momentum Transport in Fluids*, Academic Press, New York (1956), Chapters III and IV.

[9] H. Schlichting, *op. cit.*, pp. 416–426.

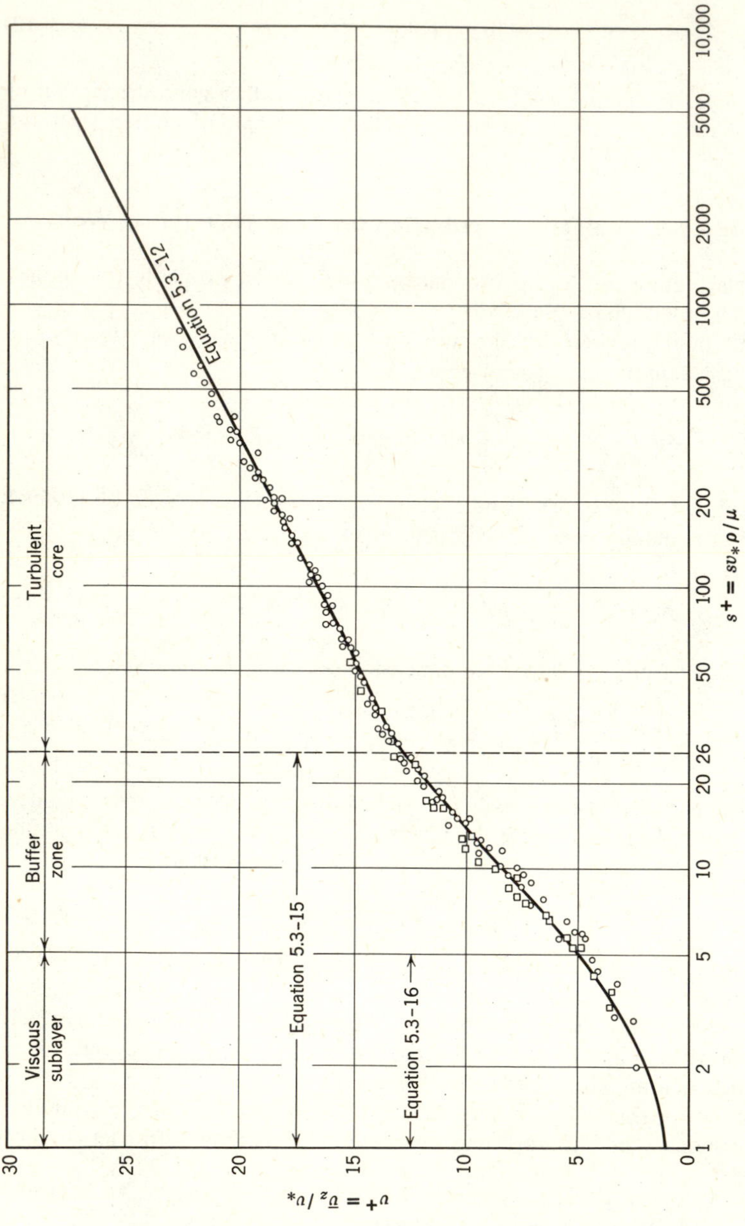

Fig. 5.3-1. Velocity distribution for turbulent, isothermal flow in tubes. [R. G. Deissler, *NACA Rept 1210* (1955).] Experimental data are those of R. G. Deissler [*NACA Tech. Note 3016* (1953)] and J. Laufer [*NACA Tech. Note 2954* (1953)], circles and squares, respectively.

of the same book one may also find summaries of theoretical studies on other turbulent flow systems, such as flow in jets, flow between rotating cylinders, flow near rotating disks, and flow around stationary cylinders. A discussion of this material lies outside the scope of this book.

Example 5.3–3. Relative Magnitude of Molecular and Eddy Viscosity

Determine the ratio $\mu^{(t)}/\mu$ at $s = R/2$ for water flowing at a steady rate in a long smooth round tube under the following conditions:

$$R = \text{tube radius} = 3 \text{ in.}$$
$$\tau_0 = \text{wall shear stress} = 2.36 \times 10^{-5} \text{ lb}_f \text{ in}^{-2}$$
$$\rho = \text{density} = 62.4 \text{ lb}_m \text{ ft}^{-3}$$
$$\nu = \text{kinematic viscosity} = 1.1 \times 10^{-5} \text{ ft}^2 \text{ sec}^{-1}$$

Solution. The eddy viscosity is defined by

$$\bar{\tau}_{rz} = -\mu \frac{d\bar{v}_z}{dr} - \mu^{(t)} \frac{d\bar{v}_z}{dr}$$
$$= +(\mu + \mu^{(t)}) \frac{d\bar{v}_z}{ds} \tag{5.3–17}$$

This relation may be solved for $\mu^{(t)}/\mu$ and the result expressed in terms of the dimensionless variables of the preceding example:

$$\frac{\mu^{(t)}}{\mu} = \frac{1}{\mu} \frac{\bar{\tau}_{rz}}{d\bar{v}_z/ds} - 1$$
$$= \frac{1}{\mu} \frac{\tau_0(1 - s/R)}{d\bar{v}_z/ds} - 1$$
$$= \frac{[1 - (s/R)]}{dv^+/ds^+} - 1 \tag{5.3–18}$$

When $s = R/2$, the value of s^+ is

$$s^+ = \frac{s v_* \rho}{\mu} = \frac{(R/2)\sqrt{\tau_0/\rho}\,\rho}{\mu} = 485 \tag{5.3–19}$$

For this value of s^+ we can use Eq. 5.3–12 to compute dv^+/ds^+:

$$\frac{dv^+}{ds^+} = (1/0.36)(1/485) \tag{5.3–20}$$

By substituting this into Eq. 5.3–18, we then get

$$\mu^{(t)}/\mu = \tfrac{1}{2}(0.36)(485) - 1 = 86 \tag{5.3–21}$$

This result emphasizes that far from the tube wall molecular momentum transport is negligible in comparison with eddy transport.

§5.4 THE SECOND-ORDER CORRELATION TENSOR AND ITS PROPAGATION (THE VON KÁRMÁN-HOWARTH EQUATION)[1,2]

In §5.2 it was found that the time-smoothing of the equation of motion leads to the Reynolds stress tensor, which has components $\rho\overline{v_i'v_j'}$. The quantities v_i' and v_j' are the velocity fluctuations in the i- and j-directions at *one* point, and $\overline{v_i'v_j'}$ is a measure of the "correlation" between the fluctuations.

In the last two decades considerable effort has been devoted to the understanding of the correlations between fluctuations at *two* points. These quantities, which are measurable, give the size and orientation of eddies. In the following development we consider the relations among three correlations, in which the fluctuations at point B' are indicated by a single prime and those at point B'' by a double prime:

$\overline{p'v_i''}$ = the ith component of a vector describing the correlation of the pressure fluctuation at B' with the velocity fluctuation in the i-direction at B''

$\overline{v_i'v_j''}$ = the ijth component of a second-order tensor describing the correlation of the velocity fluctuations at B' and B''

$\overline{v_i'v_j'v_k''}$ = the ijkth component of a third-order tensor describing the correlation of two velocity fluctuations at B' with one at B''

The behavior of these correlations has been studied extensively for isotropic homogeneous turbulence in a fluid for which $\bar{v} = 0$; this simplest of all cases is the only one to be discussed here. *Isotropic turbulence* is defined as that condition under which the time-smoothed value of any function of the velocity components and their space derivatives at a particular point, defined with respect to a given set of axes, is unaltered if the axes are rotated or reflected in any plane through the origin; in particular, $\overline{v_1'^2} = \overline{v_2'^2} = \overline{v_3'^2}$ $(= \overline{v'^2}$, say) and $\overline{v_1'v_2'} = \overline{v_2'v_3'} = \overline{v_1'v_3'} = 0$. *Homogeneous turbulence* is defined as that state for which the various time-smoothed quantities are independent of position; in particular, $\overline{v'^2} = \overline{v''^2}$. It has been found[3] that locally homogeneous isotropic turbulence with $\bar{v} = 0$ can be approximately realized in the flow behind a uniform grid, if the coordinate system is taken to move with the stream velocity.

[1] T. von Kármán and L. Howarth, *Proc. Roy. Soc. (London)*, **A164**, 192–215 (1938).
[2] G. I. Taylor, *Proc. Roy. Soc. (London)*, **A151**, 421–478 (1935).
[3] A. A. Townsend, *The Structure of Turbulent Shear Flow*, Cambridge University Press (1956), pp. 33 *et seq.*

We now define dimensionless correlations for isotropic turbulence:

$$P_i = \frac{\overline{p'v_i''}}{\sqrt{\overline{p'^2}\,\overline{v'^2}}} \tag{5.4–1}$$

$$R_{ij} = \frac{\overline{v_i'v_j''}}{\overline{v'^2}} \tag{5.4–2}$$

$$T_{ijk} = \frac{\overline{v_i'v_j'v_k''}}{(\overline{v'^2})^{3/2}} \tag{5.4–3}$$

We want to derive an equation of change for the second-order correlation tensor R_{ij}. We do this by the following procedure:

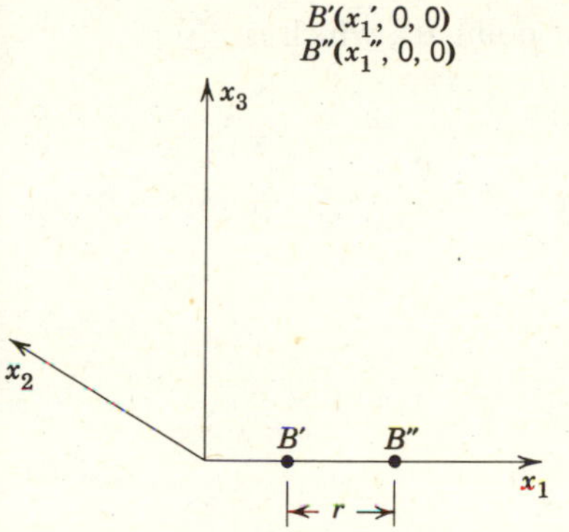

Fig. 5.4–1. B' and B'' both on x_1-axis, with $\xi_1 = r$, $\xi_2 = \xi_3 = 0$.

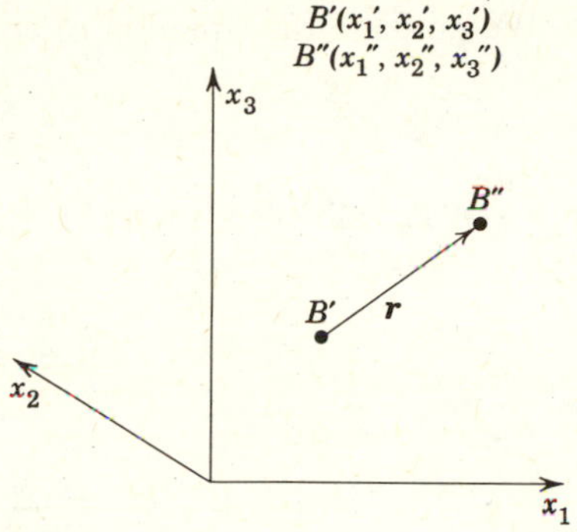

Fig. 5.4–2. Arbitrary location of B' and B'' with $\boldsymbol{r}$ having components ξ_1, ξ_2, ξ_3.

a. For each of the three correlations P_i, R_{ij}, and T_{ijk} we (i) find the nonvanishing components by use of the condition of isotropy, when B' and B'' are both on the x_1-axis (see Fig. 5.4–1); (ii) establish the general form of the correlation that will insure that the correlation is endowed with the proper transformation properties; (iii) use the equation of continuity to obtain relations among the nonvanishing components.

b. We then use the Navier-Stokes equations to obtain the von Kármán-Howarth equation describing the second-order correlation.

In the development we indicate the location of B' by coordinates x_i' and that of B'' by coordinates x_i''. The distance from B' to B'' is designated by the vector $\boldsymbol{r}$ with components $\xi_i = x_i'' - x_i'$. (See Fig. 5.4–2.)

The Correlation P_i

Consider first the special case in which B' and B'' are both on the x_1-axis. Because of isotropy, $\overline{p'v_2''} = \overline{p'v_3''} = 0$, since these quantities change sign when the axes are rotated 180° about the x_1-axis. The remaining component will at most be a function of r and t because of isotropy; hence we write for the positions of B' and B'' shown in Fig. 5.4–1

$$P_1 = \frac{\overline{p'v_1''}}{\sqrt{\overline{p'^2}\,\overline{v'^2}}} = s(r, t) \tag{5.4–4}$$

A vector expression which simplifies to Eq. 5.4–4 for $i = 1$ is

$$P_i = \frac{\overline{p'v_i''}}{\sqrt{\overline{p'^2}\,\overline{v'^2}}} = s(r, t)\,\frac{\xi_i}{r} \tag{5.4–5}$$

and it also gives $P_2 = P_3 = 0$. Because this expression transforms like a vector, it is the general expression for P_i when B' and B'' are arbitrarily located.

We now use the equation of continuity at point B'', which is

$$\sum_{i=1}^{3} \frac{\partial v_i''}{\partial x_i''} = 0 \tag{5.4–6}$$

Because p' depends only on the x_i' and t, we can insert p' in Eq. 5.4–6 to get (after time-smoothing)

$$\sum_{i=1}^{3} \frac{\partial}{\partial x_i''}\,\overline{p'v_i''} = 0 \tag{5.4–7}$$

or

$$\sum_{i=1}^{3} \frac{\partial}{\partial \xi_i}\,\overline{p'v_i''} = 0 \tag{5.4–8}$$

Insertion of Eq. 5.4–5 into Eq. 5.4–8 then leads to the following differential equation for $s(r, t)$:

$$\frac{ds}{dr} + 2\,\frac{s}{r} = 0 \tag{5.4–9}$$

for which the solution is $s = C/r^2$. Because $s(r)$ does not go to infinity as $r = 0$, the integration constant C must be zero. Therefore, $s(r) = 0$, and we conclude that

$$P_i = 0 \qquad \text{or} \qquad \overline{p'v_i''} = 0 \qquad \text{for } i = 1, 2, 3 \tag{5.4–10}$$

Hence there is no correlation between the pressure fluctuation at one point and the velocity fluctuation at a nearby point.

The Correlation R_{ij}

Consider again the special case in which B' and B'' are both on the x_1-axis. Because of isotropy;

$\overline{v_1'v_2''} = \overline{v_1'v_3''} = 0$ because these quantities would change sign under rotation through 180° about the x_1-axis

$\overline{v_2'v_3''} = \overline{v_3'v_2''} = 0$ because these quantities would change sign under reflection in the x_1x_3-plane

The remaining nonvanishing components are functions of r and t; hence we write

$$R_{11} = \frac{\overline{v_1'v_1''}}{\overline{v'^2}} = f(r, t) \tag{5.4-11}$$

$$R_{jj} = \frac{\overline{v_j'v_j''}}{\overline{v'^2}} = g(r, t) \qquad j = 2, 3 \tag{5.4-12}$$

When B' and B'' are both on the x_1-axis, the matrix of the R_{ij} has the following form:

$$\|R_{ij}\| = \begin{Vmatrix} f & 0 & 0 \\ 0 & g & 0 \\ 0 & 0 & g \end{Vmatrix} \tag{5.4-13}$$

This matrix may be split up into two parts:

$$\|R_{ij}\| = \begin{Vmatrix} g & 0 & 0 \\ 0 & g & 0 \\ 0 & 0 & g \end{Vmatrix} + \begin{Vmatrix} f - g & 0 & 0 \\ 0 & 0 & 0 \\ 0 & 0 & 0 \end{Vmatrix} \tag{5.4-14}$$

The second-order tensor expression that simplifies to Eqs. 5.4–11,12 for R_{11} R_{22}, and R_{33} and also gives $R_{ij} = 0$ for $i \neq j$ is

$$R_{ij} = g(r, t)\delta_{ij} + [f(r, t) - g(r, t)]\frac{\xi_i\xi_j}{r^2} \tag{5.4-15}$$

Because this expression transforms as a second-order tensor, it is the general expression for R_{ij} when B' and B'' are arbitrarily located.

We now use the equation of continuity at B'' (Eq. 5.4–6) to write (after multiplication by v_j')

$$\sum_{i=1}^{3} \frac{\partial}{\partial x_i''} v_j'v_i'' = 0 \tag{5.4-16}$$

Division by $\overline{v'^2}$ and time-smoothing gives

$$\sum_{i=1}^{3} \frac{\partial}{\partial x_i''} R_{ji} = 0 \tag{5.4-17}$$

or

$$\sum_{i=1}^{3} \frac{\partial}{\partial \xi_i} R_{ji} = 0 \tag{5.4-18}$$

Substitution of Eq. 5.4–15 into this last result leads to

$$f - g = -\frac{1}{2} r \frac{\partial f}{\partial r} \tag{5.4-19}$$

which allows us to rewrite Eq. 5.4–15 entirely in terms of $f(r, t)$:

$$R_{ij} = \left(f + \frac{1}{2} r \frac{\partial f}{\partial r}\right)\delta_{ij} + \left(-\frac{1}{2} r \frac{\partial f}{\partial r}\right)\frac{\xi_i\xi_j}{r^2} \tag{5.4-20}$$

Hence we have expressed all the components R_{ij} in terms of the single scalar function $f(r, t)$.

The Correlation T_{ijk}

We consider once again the special case in which B' and B'' are both on the x_1-axis. By isotropy arguments similar to those used before, it is possible to show[1] that there are seven nonvanishing components which may be expressed in terms of three scalar functions:

$$T_{111} = \frac{\overline{v_1'^2 v_1''}}{\left(\overline{v'^2}\right)^{3/2}} = n(r, t) \tag{5.4–21}$$

$$T_{122} = T_{133} = T_{212} = T_{313} = m(r, t) \tag{5.4–22}$$

$$T_{221} = T_{331} = h(r, t) \tag{5.4–23}$$

Hence when B' and B'' are both on the x_1-axis, the matrix of the T_{ijk} has the following form: (5.4–24)

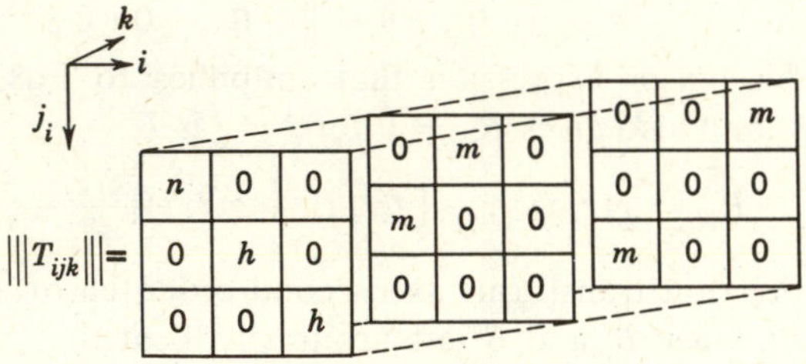

This matrix may be split up into four parts: (5.4–25)

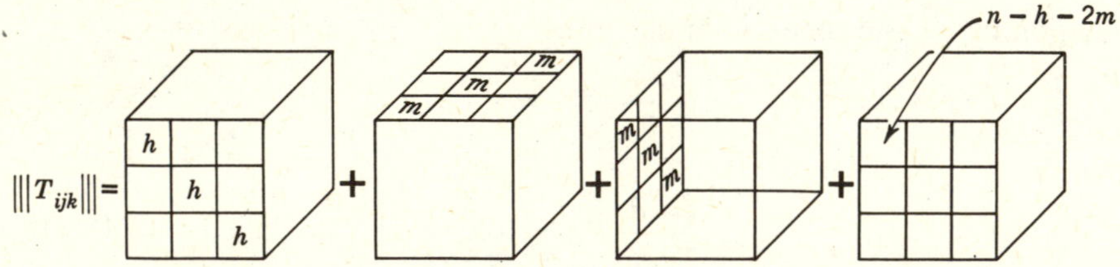

All unspecified cubes are zeros. Hence an appropriate third-order tensor expression, which reproduces Eq. 5.4–25, is

$$T_{ijk} = h\delta_{ij}\frac{\xi_k}{r} + m\delta_{jk}\frac{\xi_i}{r} + m\delta_{ik}\frac{\xi_j}{r} + (n - h - 2m)\frac{\xi_i\xi_j\xi_k}{r^3} \tag{5.4–26}$$

Because this transforms as a third-order tensor, it is the general expression for T_{ijk} when B' and B'' are arbitrarily located.

Application of the equation of continuity in a manner analogous to that in the preceding two developments leads to the fact that $n = -2h$ and

$m = -h - \frac{1}{2}r(\partial h/\partial r)$. This allows Eq. 5.4–26 to be written entirely in terms of h:

$$T_{ijk} = h\delta_{ij}\frac{\xi_k}{r} + \left(-h - \frac{1}{2}r\frac{\partial h}{\partial r}\right)\left(\delta_{jk}\frac{\xi_i}{r} + \delta_{ik}\frac{\xi_j}{r}\right) + \left(-h + r\frac{\partial h}{\partial r}\right)\frac{\xi_i\xi_j\xi_k}{r^3} \tag{5.4–27}$$

Let us now derive an equation for the R_{ij} from the Navier-Stokes equation for the situation in which $\bar{v} = 0$; at point B' the i-component of the equation of motion is

$$\frac{\partial v_i'}{\partial t} + \Sigma_j v_j' \frac{\partial v_i'}{\partial x_j'} = -\frac{1}{\rho}\frac{\partial p'}{\partial x_i'} + \nu\Sigma_j \frac{\partial^2}{\partial x_j'^2} v_i' \tag{5.4–28}$$

When this equation is multiplied by v_k'' and time-smoothed (making use of the fact that $\overline{p'v_k''} = 0$), one obtains

$$\overline{v_k''\frac{\partial v_i'}{\partial t}} - \Sigma_j \frac{\partial}{\partial \xi_j}\overline{v_i'v_j'v_k''} = \nu\Sigma_j \frac{\partial^2}{\partial \xi_j^2}\overline{v_i'v_k''} \tag{5.4–29}$$

Here the equation of continuity $\Sigma_j\, \partial v_j/\partial x_j = 0$ has been used.

Similar multiplication of the k-component of the equation of motion at B'' by v_i' and time-smoothing leads to

$$\overline{v_i'\frac{\partial v_k''}{\partial t}} - \Sigma_j \frac{\partial}{\partial \xi_j}\overline{v_j'v_k'v_i''} = \nu\Sigma_j \frac{\partial^2}{\partial \xi_j^2}\overline{v_i'v_k''} \tag{5.4–30}$$

Addition of these last two results then gives

$$\frac{\partial}{\partial t}\overline{v'^2}R_{ik} - (\overline{v'^2})^{3/2}\Sigma_j \frac{\partial}{\partial \xi_j}(T_{ijk} + T_{kji}) = 2\nu\overline{v'^2}\Sigma_j \frac{\partial^2}{\partial \xi_j^2}R_{ik} \tag{5.4–31}$$

This equation for R_{ik} also contains the third-order correlation T_{ijk}. Similarly, an equation for T_{ijk} can be given only in terms of the fourth-order correlation and so on. That is, a whole "hierarchy" of equations for the correlation functions results; hence it is impossible to solve exactly to get the R_{ik}.

Next we perform the operation of "contraction" on Eq. 5.4–31 (set $i = k$ and sum on i):

$$\frac{\partial}{\partial t}(\overline{v'^2}\Sigma_i R_{ii}) - 2(\overline{v'^2})^{3/2}\Sigma_j \frac{\partial}{\partial \xi_j}\Sigma_i T_{iji} = 2\nu\overline{v'^2}\Sigma_j \frac{\partial^2}{\partial \xi_j^2}\Sigma_i R_{ii} \tag{5.4–32}$$

The expressions for $\Sigma_i R_{ii}$ and $\Sigma_i T_{iji}$ are, from Eqs. 5.4–20 and 27,

$$\Sigma R_{ii} = \left(3 + r\frac{\partial}{\partial r}\right)f \tag{5.4–33}$$

$$\Sigma T_{iji} = -\frac{\xi_j}{r}\left(4 + r\frac{\partial}{\partial r}\right)h \tag{5.4–34}$$

We now substitute these expressions into Eq. 5.4–32 [in which $\Sigma_j \partial^2/\partial\xi_j^2$ has been replaced by its equivalent in spherical coordinates: $\partial^2/\partial r^2 + (2/r)\,\partial/\partial r$]. After some rearrangement this gives

$$\left(3 + r\frac{\partial}{\partial r}\right)\left[\frac{\partial}{\partial t}\overline{v'^2}f + 2(\overline{v'^2})^{3/2}\left(\frac{4}{r} + \frac{\partial}{\partial r}\right)h - 2\nu\overline{v'^2}\left(\frac{\partial^2}{\partial r^2} + \frac{4}{r}\frac{\partial}{\partial r}\right)f\right] = 0 \quad (5.4\text{–}35)$$

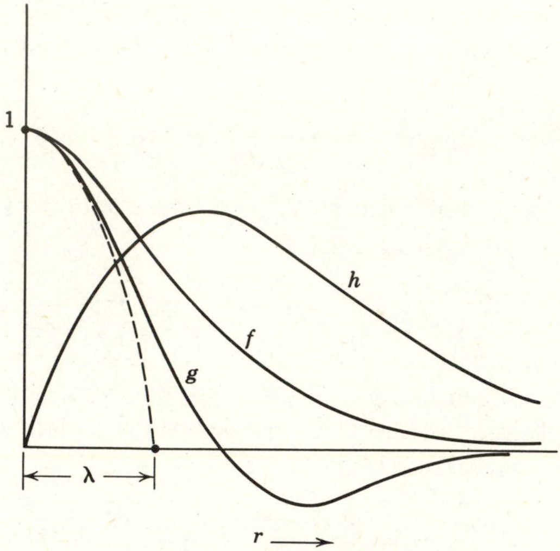

Fig. 5.4–3. Sketch showing the shape of the functions f, g, h. Taylor's microscale of turbulence is the intercept on the r-axis of a parabola that fits the curve $g(r)$ at small r.

Integration of this equation indicates that the expression within brackets equals C/r^3 where C is a constant of integration; inasmuch as [ ] does not become infinite as $r \to 0$, C must be zero. Therefore, we finally obtain

$$\frac{\partial}{\partial t}(\overline{v'^2}f) = 2\nu\overline{v'^2}\left(\frac{\partial}{\partial r} + \frac{4}{r}\right)\left(\frac{\partial f}{\partial r} - \frac{\sqrt{\overline{v'^2}}}{\nu}h\right) \quad (5.4\text{–}36)$$

which is the *von Kármán-Howarth equation* for the second-order correlation function $f(r, t)$. Hot-wire anemometry measurements of f, h, and $\overline{v'^2}$ have shown this equation to be satisfied within experimental error.[4] This agreement between theory and measurement has been taken by some[5] as a justification of the use of the Navier-Stokes equation as the starting point for the description of turbulent flow. The general shapes of the functions f, g, h are sketched in Fig. 5.4–3. The integral under the $f(r)$ curve, that is, $\int_0^\infty f(r)\,dr$,

[4] R. W. Stewart, *Proc. Camb. Phil. Soc.*, **47**, 146–157 (1951).

[5] A. A. Townsend, *op. cit.*, p. 35.

is a measure of the eddy size; it is called the *scale of turbulence*. Considerable effort has been devoted to the study of Eq. 5.4–36 or its Fourier transform,[1,6] and a number of limiting solutions have contributed to an improved knowledge of the structure of turbulence. For further information the reader is referred to the original references and to several books that have been devoted to turbulence.[3,7,8,9]

Example 5.4–1. Decay of Turbulence Behind a Grid

Set $r = 0$ in the von Kármán-Howarth equation and obtain an equation for the change of $\overline{v'^2}$ with time (i.e., the "decay" of the intensity of the turbulence). Use Taylor's expression for the microscale of turbulence[2] to perform an integration of the decay equation.

Solution. When $r = 0$, we have $f = 1$ and $h = 0$. Because $f(r, t)$ is an even function of r,

$$f(r, t) = 1 + \frac{1}{2!} f_0'' r^2 + \frac{1}{4!} f_0'''' r^4 + \cdots \tag{5.4–37}$$

and

$$\left[\left(\frac{\partial}{\partial r} + \frac{4}{r}\right)\frac{\partial f}{\partial r}\right]\bigg|_{r=0} = 5 f_0'' \tag{5.4–38}$$

Hence, at $r = 0$, Eq. 5.4–36 becomes

$$\frac{d}{dt}\overline{v'^2} = 10\nu f_0'' \overline{v'^2} = -10\nu \frac{\overline{v'^2}}{\lambda^2} \tag{5.4–39}$$

in which we have introduced the length λ known as Taylor's "microscale of turbulence"; its geometrical meaning is given in Fig. 5.4–3.

On the basis of dimensional arguments and a study of data on turbulence decay, Taylor[2] proposed that λ is related to the mesh spacing M of the grid producing the turbulence according to the following equation:

$$\frac{\lambda}{M} = A\sqrt{\frac{\nu}{M\sqrt{\overline{v'^2}}}} \tag{5.4–40}$$

in which A is a constant. When this is substituted into Eq. 5.4–39 and d/dt is replaced by $V(d/dx)$, where V is the (uniform) time-smoothed velocity of the fluid, we get

$$V\frac{d}{dx}\overline{v'^2} = -\frac{10\nu\overline{v'^2}}{M^2A^2}\left(\frac{M(\overline{v'^2})^{1/2}}{\nu}\right) \tag{5.4–41}$$

[6] T. von Kármán and C. C. Lin, *Advanced Applied Mechanics*, Academic Press, Vol. II (1951).

[7] G. K. Batchelor, *The Theory of Homogeneous Turbulence*, Cambridge University Press (1953).

[8] J. O. Hinze, *Turbulence*, McGraw-Hill, New York (1959).

[9] L. Landau and E. M. Lifshitz, *Fluid Mechanics*, Addison-Wesley, Reading (1960), Chapter 3.

Integration then gives

$$\frac{V}{\sqrt{\overline{v'^2}}} = \frac{5}{MA^2}x + \text{constant} \tag{5.4–42}$$

This behavior has been observed experimentally for the decay of the turbulent intensity with distance behind a grid; the quantity A is found to have a value of 1.95 to 2.20. The adequacy of Eq. 5.4–42 leads one to conclude that Taylor's proposal in Eq. 5.4–40 has given useful information regarding the relation between eddy size and mesh spacing.

QUESTIONS FOR DISCUSSION

1. Why can Eq. 5.1–3 *not* be used to get the drag force on the tube wall by evaluating the velocity gradient at the wall?
2. Define instantaneous pressure, time-smoothed pressure, and pressure fluctuation.
3. Discuss the physical significance of the curves in Fig. 5.1–5 in terms of the terms $[\nabla \cdot \bar{\tau}^{(l)}]$ and $[\nabla \cdot \bar{\tau}^{(t)}]$ in Eq. 5.2–8.
4. What is the connection between the R_{ij} tensor of §5.4 and the Reynolds stresses of §5.2?
5. Is $\text{Re} = 2.1 \times 10^3$ always the value of the Reynolds number at which laminar-turbulent transition occurs?
6. Compare laminar and turbulent tube flow in regard to (*a*) velocity profile, (*b*) ratio of average to maximum velocity, (*c*) dependence of flow rate on pressure drop.
7. Summarize all the assumptions that led to Eq. 5.3–12 (frequently referred to as the "universal velocity profile").
8. Derive Eq. 5.3–16 by the two methods suggested in the text.
9. Why is the von Kármán-Howarth f-function expected to go to zero for large r?
10. Is the turbulence in tube flow homogeneous or isotropic? (Use Figs. 5.1–4 and 5 and the definitions in §5.4.)
11. Summarize the assumptions that are inherent in the von Kármán-Howarth equation.
12. Explain how one gets rid of the absolute value sign in going from Eq. 5.3–2 to Eq. 5.3–5.

PROBLEMS

5.A$_1$ Pressure Drop Required for Laminar-Turbulent Transition

A fluid with viscosity 18.3 cp and density 1.32 g cm^{-3} is flowing in a horizontal tube of radius 0.21 in. For what pressure gradient (expressed in psi/ft) will the flow become turbulent?

Answer: 0.62 psi/ft

5.B$_1$ Velocity Distribution in Turbulent Pipe Flow

Water is flowing through a long, straight, level run of smooth 6.00-in. i.d. pipe, at a temperature of 68° F. The pressure gradient along the length of the pipe is 1.0 psi/mile.

a. Determine the wall shear stress τ_0 in psi (pounds force per square inch).

b. Assuming the flow to be turbulent, determine the radial distances from the pipe wall at which $\bar{v}/\bar{v}_{\max} = 0.0, 0.1, 0.2, 0.4, 0.7, 0.85$, and 1.0. Use Fig. 5.3–1 for your calculations.

c. Plot the complete velocity profile, $\bar{v}/\bar{v}_{\max}$ versus $s = R - r$.

d. Is the assumption of turbulent flow justified?

e. What is the volume rate of flow?

5.C$_2$ Average Flow Velocity in Turbulent Tube Flow

a. For the turbulent flow in smooth circular tubes the curve-fit function

$$\frac{\bar{v}_z}{\bar{v}_{z,\max}} = \left(1 - \frac{r}{R}\right)^{1/n} \tag{5.C–1}$$

is sometimes useful:[1] near Re $= 4 \times 10^3$, $n = 6$; near Re $= 1.1 \times 10^5$, $n = 7$; and near 3.2×10^6, $n = 10$. Show that the ratio of average to maximum velocity is

$$\frac{\langle \bar{v}_z \rangle}{\bar{v}_{z,\max}} = \frac{2n^2}{(n+1)(2n+1)} \tag{5.C–2}$$

and verify the result in Eq. 5.1–4.

b. Obtain the ratio of average to maximum velocity by integrating Eq. 5.3–12 over the entire tube cross section (that is, ignore the error introduced by not taking into account the laminar sublayer). In what respect does the form of this result differ from that obtained in (*a*).

5.D$_2$ Velocity Distribution in a Rectangular Channel

A fluid is flowing in the z-direction in fully developed turbulent flow in a rectangular channel of half-width h (in the y-direction). The channel is assumed to be very wide in the x-direction so that $\bar{v}_z = \bar{v}_z(y)$.

a. Show that the equation describing the turbulent core according to the von Kármán similarity hypothesis is

$$\tau_0 \frac{y}{h} = \rho \kappa_2^2 \frac{(d\bar{v}_z/dy)^4}{(d^2\bar{v}_z/dy^2)^2} \qquad 0 \leqslant y \leqslant h \tag{5.D–1}$$

where τ_0 is the wall shear stress and y is the distance from the mid-plane of the channel.

b. Integrate Eq. 5.D–1 to show that the velocity distribution is given by

$$\frac{\bar{v}_{z,\max} - \bar{v}_z}{v_*} = -\frac{1}{\kappa_2}\left[\ln\left(1 - \sqrt{\frac{y}{h}}\right) + \sqrt{\frac{y}{h}}\right] \qquad 0 \leqslant y \leqslant h \tag{5.D–2}$$

in which $v_* = \sqrt{\tau_0/\rho}$. (*Note:* In performing the integration, the fictitious boundary condition that $d\bar{v}_z/dy = -\infty$ at $y = h$ will have to be used. Explain.)

c. Is the velocity profile $\bar{v}_z(y)$ flat at the center of the channel ($y = 0$)?

5.E$_3$ Turbulent Velocity Profile in a Tube (Pai's Method)[2]

In §5.3 we showed that some information can be obtained about the time-smoothed velocity distribution by employing empirical relations between the turbulent momentum flux and the time-smoothed velocity gradient. Another approach is to assume a relation between the turbulent momentum flux and the distance. Such an assumption usually leads to much simpler integrations. As an example, let us consider the flow in a circular tube.

a. For turbulent flow in a tube, show that the time-smoothed equations of motion reduce to

$$-\rho \frac{\overline{v_\theta'^2}}{r} = -\frac{\partial \bar{p}}{\partial r} - \frac{1}{r}\frac{\partial}{\partial r} r\bar{\tau}_{rr}^{(t)} \tag{5.E–1}$$

$$0 = -\frac{\partial \bar{p}}{\partial z} + \mu \frac{1}{r}\frac{\partial}{\partial r}\left(r\frac{\partial \bar{v}_z}{\partial r}\right) - \frac{1}{r}\frac{\partial}{\partial r}(r\bar{\tau}_{rz}^{(t)}) \tag{5.E–2}$$

[1] H. Schlichting, *Grenzschichttheorie*, Braun, Karlsruhe (1951), pp. 364–366.
[2] S. I. Pai, *Viscous Flow Theory*, Vol. II, Van Nostrand, Princeton (1957), pp. 41–44.

b. Justify the statement that $\overline{v_\theta'^2}$, $\bar{\tau}_{rr}^{(t)}$, $\bar{v}_z$, and $\bar{\tau}_{rz}^{(t)}$ are functions of r (but not of θ and z). Show that this leads to the result that $-\partial\bar{p}/\partial z = A$, a constant.

c. Introduce the following dimensionless quantities:

$$\phi = \frac{\bar{v}_z}{v_*}; \qquad \xi = \frac{r}{R}; \qquad \tau = \frac{\bar{\tau}_{rz}^{(t)}}{\tau_0}; \qquad \text{Re}_* = \frac{Rv_*\rho}{\mu} \tag{5.E-3,4,5,6}$$

where $v_* = \sqrt{\tau_0/\rho}$ and τ_0 is the wall shear stress. Show that Eq. 5.E–2 may then be rewritten thus:

$$0 = B + \frac{1}{\text{Re}_*}\frac{1}{\xi}\frac{d}{d\xi}\left(\xi\frac{d\phi}{d\xi}\right) - \frac{1}{\xi}\frac{d}{d\xi}(\xi\tau) \tag{5.E-7}$$

in which B is a constant simply related to A. Show that this equation may be integrated once to give

$$\frac{C}{\xi} = \frac{1}{2}B\xi + \frac{1}{\text{Re}_*}\frac{d\phi}{d\xi} - \tau \tag{5.E-8}$$

d. Determine the constants B and C by using the following boundary conditions (explain their physical significance):

$$\text{At } \xi = 0: \quad \tau = \text{finite}, \quad d\phi/d\xi = \text{finite} \tag{5.E-9}$$

$$\text{At } \xi = 1: \quad \tau = 0, \quad d\phi/d\xi = -\text{Re}_* \tag{5.E-10}$$

and show that Eq. 5.E–8 finally becomes

$$0 = \xi + \frac{1}{\text{Re}_*}\frac{d\phi}{d\xi} - \tau \tag{5.E-11}$$

e. Equation 5.E–11 can be integrated if τ is given as a function of ξ. From Fig. 5.1–5, we can suggest the following function:

$$\tau = (1 - \epsilon)(\xi - \xi^{2n-1}) \tag{5.E-12}$$

where $\epsilon \lll 1$ and n is a very large positive integer. Why is this a satisfactory choice? Show that Eq. 5.E–12 leads to the following velocity distribution:

$$\frac{\bar{v}_z}{v_{\max}} = 1 - \left(\frac{n\epsilon}{(n-1)\epsilon + 1}\right)\xi^2 - \left(\frac{1-\epsilon}{(n-1)\epsilon + 1}\right)\xi^{2n} \tag{5.E-13}$$

where the boundary conditions at the wall have been satisfied.

f. Show that the average velocity is

$$\frac{\langle\bar{v}_z\rangle}{\bar{v}_{z,\max}} = \frac{n[(n-1)\epsilon + 2]}{2(n+1)[(n-1)\epsilon + 1]} \tag{5.E-14}$$

The constants n and ϵ are functions of the Reynolds number. For a Reynolds number of $\bar{v}_{z,\max}R\rho/\mu = 25{,}000$, Pai has found that $\epsilon = 0.0161$ and $n = 33$; he claims a good fit of the experimental data even near the wall.

5.F$_3$ Turbulent Velocity Distribution in Annuli[3]

Experimental profile measurements[4,5] have suggested that λ (see Fig. 5.F) for turbulent flow is (at least at high Reynolds numbers) roughly the same as λ for laminar flow. Hence

[3] D. M. Meter and R. B. Bird, *A.I.Ch.E. Journal*, **7**, 41–45 (1961).

[4] J. G. Knudsen and D. L. Katz, *Fluid Dynamics and Heat Transfer*, McGraw-Hill, New York (1958).

[5] R. R. Rothfus, doctoral thesis, Carnegie Institute of Technology (1948); J. E. Walker, doctoral thesis, Carnegie Institute of Technology, (1957); G. A. Whan, doctoral thesis, Carnegie Institute of Technology (1956).

assume that in turbulent flow

$$\lambda = \sqrt{\frac{1-\kappa}{2 \ln (1/\kappa)}} \tag{5.F-1}$$

(See Eq. 2.4–11.) By paralleling the development in Example 5.3–1, show that the velocity distribution in the annulus is, to the same degree of approximation,

$$\bar{v}_{z,\max} - \bar{v}_z = \frac{1}{\kappa_1}\sqrt{\frac{\tau_0}{\rho}}\sqrt{\frac{\lambda^2-\kappa^2}{\kappa}} \ln \frac{(\lambda-\kappa)R}{r-\kappa R} \qquad r < \lambda R \tag{5.F-2}$$

$$\bar{v}_{z,\max} - \bar{v}_z = \frac{1}{\kappa_1}\sqrt{\frac{\tau_0}{\rho}}\sqrt{1-\lambda^2} \ln \frac{R-\lambda R}{R-r} \qquad r > \lambda R \tag{5.F-3}$$

in which $\tau_0 = \Delta p(R/2L)$ (Δp = pressure drop, L = annulus length).

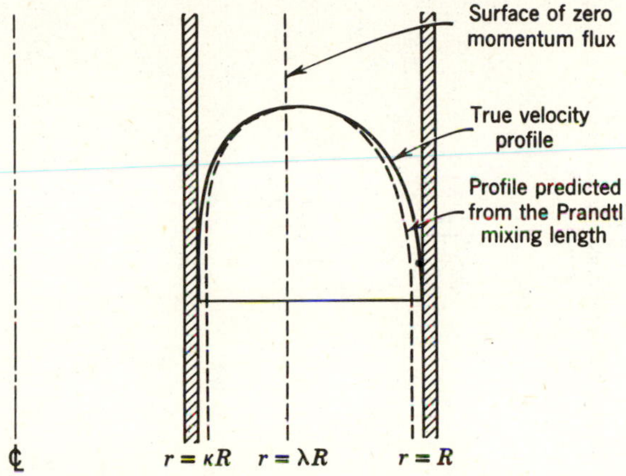

Fig. 5.F. Turbulent velocity profile in an annulus.

5.G$_3$ Alternate Derivation of the Taylor Decay Equation

Eq. 5.4–39 was originally derived by Taylor[6] in a way different from that given in Example 5.4–1. Here we work through a modified version of Taylor's derivation, due to von Kármán and Howarth:

a. First, show that

$$\overline{\frac{\partial v_k'}{\partial x_i'} v_l''} = \frac{\partial}{\partial x_i'} \overline{v_k' v_l''} = -\overline{v'^2} \frac{\partial R_{kl}}{\partial \xi_i} \tag{5.G-1}$$

b. Then take the derivative of this expression with respect to x_j'' to get

$$\overline{\frac{\partial v_k'}{\partial x_i'} \frac{\partial v_l''}{\partial x_j''}} = \frac{\partial}{\partial x_j''} \overline{\frac{\partial v_k'}{\partial x_i'} v_l''} = -\overline{v'^2} \frac{\partial^2 R_{kl}}{\partial \xi_j \, \partial \xi_i} \tag{5.G-2}$$

c. From (*b*) show that when B' and B'' coincide

$$\overline{\frac{\partial v_k'}{\partial x_i'} \frac{\partial v_l'}{\partial x_j'}} = -\overline{v'^2} \left(\frac{\partial^2 R_{kl}}{\partial \xi_i \, \partial \xi_j}\right)_{\xi_1=\xi_2=\xi_3=0} \tag{5.G-3}$$

[6] G. I. Taylor, *Proc. Roy. Soc.* (London), **A151,** 421–512 (1935).

d. Show that for homogeneous, isotropic turbulence, equating the decrease in kinetic energy behind a grid to the mechanical energy "loss" by viscous dissipation gives

$$-\frac{3}{2}\rho V\frac{d}{dx}\overline{v'^2} = 6\mu\left[\overline{\left(\frac{\partial v_1'}{\partial x_1}\right)^2} + \overline{\left(\frac{\partial v_1'}{\partial x_2}\right)^2} + \overline{\left(\frac{\partial v_1'}{\partial x_2}\frac{\partial v_2'}{\partial x_1}\right)}\right] \tag{5.G-4}$$

e. Use the result in (*c*) and the expression for the R_{kl} in terms of f to evaluate the derivatives in (*d*) and obtain Eq. 5.4–39.

5.H$_3$ Velocity Distribution in a Turbulent Plane Jet

Turbulent flow without a containing surface is referred to as "free turbulence." One of the simplest problems in free turbulent shear flow is the flow in a plane jet.(See Fig. 5.H.)

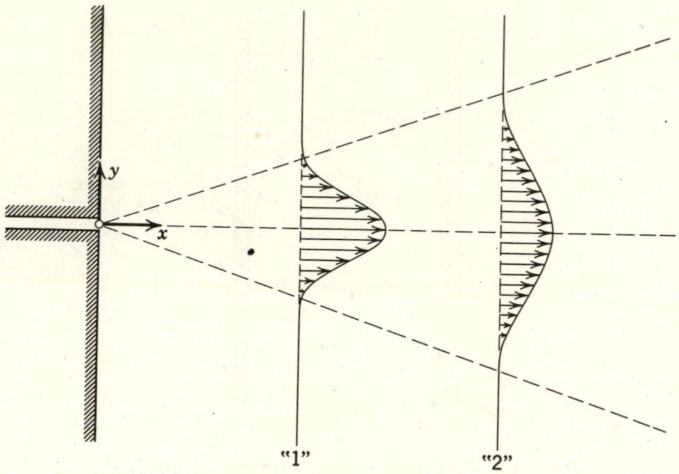

Fig. 5.H. Plane turbulent jet.

The first description of this system was one based on the Prandtl mixing length theory[7] with l taken as proportional to x. In this problem we work though a somewhat simpler theory based on the use of the eddy viscosity.[8,9]

a. Show that the following form for the velocity components is consistent with the equation of continuity:

$$\bar{v}_x/V = x^{-1/2}F'(\eta) \tag{5.H-1}$$

$$\bar{v}_y/V = [-\tfrac{1}{2}x^{-1/2}F(\eta) + \eta x^{-1/2}F'(\eta)]\sigma^{-1} \tag{5.H-2}$$

in which $\eta = \sigma(y/x)$ where σ is an arbitrary constant and V is a characteristic velocity, as yet unspecified.

b. Show that Eq. 5.H–1 leads to the fact that the rate of flow of x-momentum across any plane (such as "1" or "2") does not change with x.

[7] W. Tollmien, *Z. angew. Math. Mech.*, **6**, 468–478 (1926).
[8] H. Görtler, *Z. angew. Math. Mech.*, **22**, 244–254 (1942).
[9] H. Schlichting, *Boundary Layer Theory*, McGraw-Hill, New York (1955), Chapter 23, §5.

c. Show that the time-smoothed equation of motion with $\nu^{(t)} = \mu^{(t)}/\rho$ taken to be proportional to $x^{1/2}$ (i.e., $\nu^{(t)} = kx^{1/2}$) is

$$\bar{v}_x \frac{\partial \bar{v}_x}{\partial x} + \bar{v}_y \frac{\partial \bar{v}_x}{\partial y} = k\sqrt{x}\, \frac{\partial^2 \bar{v}_x}{\partial y^2} \tag{5.H-3}$$

d. Show that by (*a*) Eq. 5.H–3 becomes

$$\tfrac{1}{2}(F')^2 + \tfrac{1}{2}FF'' + \frac{k\sigma^2}{V} F''' = 0 \tag{5.H-4}$$

Because two arbitrary constants k and σ appear in a product, we can arbitrarily let $\dfrac{k\sigma^2}{V}$ be $\frac{1}{4}$, so that Eq. 5.H–4 simplifies to

$$\frac{1}{2}\frac{d}{d\eta}(FF') + \frac{1}{4}\frac{d}{d\eta}(F'') = 0 \tag{5.H-5}$$

e. Show that this equation may be integrated three times to give

$$F = \tanh \eta \tag{5.H-6}$$

f. If J is defined as

$$J = \rho \int_{-\infty}^{+\infty} \bar{v}_x^2 \, dy \tag{5.H-7}$$

then show that

$$\bar{v}_x = \frac{\sqrt{3}}{2}\sqrt{\frac{J\sigma}{\rho x}}\,(1 - \tanh^2 \eta) \tag{5.H-8}$$

When σ is chosen to be 7.67, this result does a pretty good job of fitting the data up to about $\eta = 1.5$. Above $\eta = 1.5$ it gives values of $\bar{v}_x$ that are too high.

CHAPTER 6

Interphase Transport in Isothermal Systems

In the preceding chapters we have shown how laminar flow problems may be formulated and solved, and we have indicated that the solution of turbulent flow problems depends on the use of some semiempirical relations between the momentum flux and the gradients of the time-smoothed velocity. In all of our work thus far we have considered systems with relatively simple geometry, inasmuch as geometrically complex problems are soluble only at the expense of considerable effort, usually with the help of high-speed computers. Even then rigorous determination of the velocity profiles for a packed column by analytical or numerical procedures is entirely out of the question.

Many engineering flow problems fall into one of two broad categories: flow in channels and flow around submerged objects. Examples of flow in channels are pumping of oil in pipes, flow of water in open channels, extrusion of plastics through dies, and flow of a fluid through a filter. Examples of flow around submerged objects are the motion of air around an airplane wing, motion of fluid around particles undergoing sedimentation, and flow across tube banks in a heat exchanger.

In the problems of flow in channels one is generally concerned with obtaining the relationship between the pressure drop and the volume rate of flow. In problems of flow around submerged objects one generally wants to know the relation between the velocity of the approaching fluid and the drag force. We have seen in the preceding chapters that if one knows the velocity and pressure distributions in the system then the desired relationships for these two cases may be obtained. (See the discussions in §§2.3 and 2.6 for examples of the two categories we are discussing here.)

Because for many systems the velocity and pressure profiles cannot be calculated, we are led to seek other ways of getting pressure drop versus throughput and drag force versus velocity for systems of engineering interest. This is accomplished by using some experimental data on these quantities to construct charts or "correlations" that can be used to estimate the flow behavior in geometrically similar systems. In the establishment of these correlations we shall find it logical and useful to employ dimensionless quantities and in doing so to make use of the discussion in §3.7.

This chapter is devoted to the preparation and use of "friction factor" charts. The discussion that follows should be regarded as an introduction to the subject; much more information concerning friction factors is to be found in the handbooks of chemical, mechanical, and civil engineering. Many important practical problems can be solved by means of friction factor charts, and students are urged to learn this topic well.

§6.1 DEFINITION OF FRICTION FACTORS

Let us consider the steady flow of a fluid of constant ρ in one of two systems: (*a*) the fluid flows in a straight conduit of uniform cross section; (*b*) the fluid flows around a submerged object which has either an axis or a plane of symmetry parallel to the direction of the velocity of the approaching fluid. The fluid will exert a force $\boldsymbol{F}$ on the solid surfaces. This force may be conveniently split into two parts: $\boldsymbol{F}_s$, that force which would be exerted by the fluid even if it were stationary, and $\boldsymbol{F}_k$, that additional force associated with the kinetic behavior of the fluid (see §2.6 for further discussion of this notation). In systems of type (*a*) $\boldsymbol{F}_k$ points in the same direction as the average velocity $\langle \boldsymbol{v} \rangle$ in the conduit, and in systems of type (*b*) $\boldsymbol{F}_k$ points in the same direction as the approach velocity, $\boldsymbol{v}_\infty$.

For both systems the magnitude of the force F_k may be arbitrarily expressed as the product of a characteristic area A, a characteristic kinetic energy per unit volume K, and a dimensionless quantity f, known as the *friction factor*:

$$F_k = AKf \tag{6.1-1}$$

Note that Eq. 6.1-1 is *not* a law of fluid mechanics but just a definition for f; clearly, for any given flow system f is not defined until A and K are specified. This is a useful definition because the dimensionless quantity f can be given as a relatively simple function of the Reynolds number and the system shape. Before presenting these dimensionless correlations of f, we give several relations to show how f is obtained from experimental data for two specific systems.

Usually, for *flow in conduits*, A is taken to be the wetted surface and K is taken to be the quantity $\frac{1}{2}\rho\langle v\rangle^2$. Specifically, for circular tubes of radius R and length L, we define f by

$$F_k = (2\pi RL)(\tfrac{1}{2}\rho\langle v\rangle^2)f \tag{6.1-2}$$

Generally, the quantity measured is not F_k but rather the pressure drop $p_0 - p_L$ and the elevation difference $h_0 - h_L$. A force balance on the fluid between 0 and L in the direction of flow gives for fully developed flow

$$\begin{aligned} F_k &= [(p_0 - p_L) + \rho g(h_0 - h_L)]\pi R^2 \\ &= (\mathscr{P}_0 - \mathscr{P}_L)\pi R^2 \end{aligned} \tag{6.1-3}$$

Elimination of F_k between Eqs. 6.1-2 and 6.1-3 then gives (with $D = 2R$)

$$f = \frac{1}{4}\left(\frac{D}{L}\right)\left(\frac{\mathscr{P}_0 - \mathscr{P}_L}{\frac{1}{2}\rho\langle v\rangle^2}\right) \tag{6.1-4}$$

This equation shows explicitly how f is calculated from experimental data. The quantity f is sometimes called the *Fanning friction factor*.[1]

For *flow around submerged objects*, the characteristic area A is usually taken to be the area obtained by projecting the solid onto a plane perpendicular to the velocity of approach of the fluid; and K is taken to be $\frac{1}{2}\rho v_\infty^2$, where v_∞ is the approach velocity of the fluid at a large distance from the object. For example, for flow around spheres of radius R, we define f by

$$F_k = (\pi R^2)(\tfrac{1}{2}\rho v_\infty^2)f \tag{6.1-5}$$

Usually, the quantity measured is not F_k, but rather the terminal velocity of the object when it falls through the fluid (this terminal velocity is then v_∞). For the steady-state fall of a sphere in a fluid, the force F_k is just counterbalanced by the gravitational force on the sphere less the buoyancy force (cf. Eq. 2.6-11):

$$F_k = \tfrac{4}{3}\pi R^3\rho_{\text{sph}}g - \tfrac{4}{3}\pi R^3\rho g \tag{6.1-6}$$

Elimination of F_k between Eqs. 6.1-5 and 6.1-6 then gives

$$f = \frac{4}{3}\frac{gD}{v_\infty^2}\left(\frac{\rho_{\text{sph}} - \rho}{\rho}\right) \tag{6.1-7}$$

[1] The definition of f varies from text to text; hence extreme caution must be exercised in the use of tables and formulas involving friction factors.

This expression may be used to calculate f from terminal-velocity data. The friction factor defined in Eqs. 6.1–5 and 7 is sometimes called the *drag coefficient* and given the symbol c_D. We have seen that drag coefficients for submerged objects and friction factors for flow in channels are defined in the same way; hence we prefer to use one symbol and one name to designate both of them.

§6.2 FRICTION FACTORS FOR FLOW IN TUBES

We now combine the definition of f in Eq. 6.1–2 with the dimensional analysis of §3.7 to show what f must depend on. As our system we consider a length L of smooth horizontal pipe, shown in Fig. 6.2–1. We restrict the

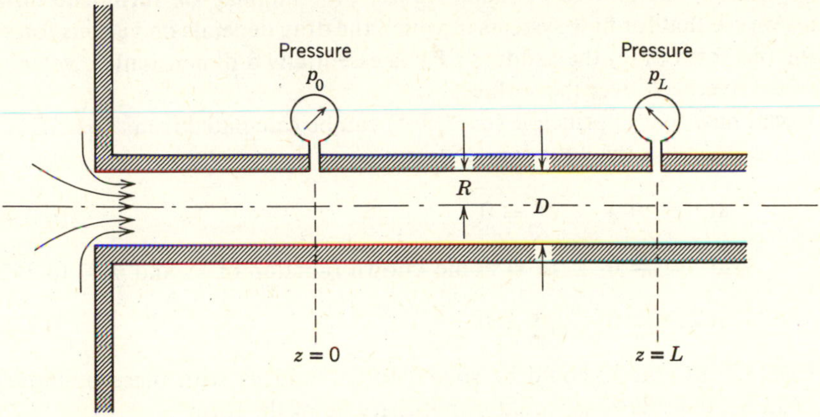

Fig. 6.2–1. Section of circular pipe for discussion of dimensional analysis.

discussion to the steady flow[1] of a fluid with constant ρ and μ, flowing with an average velocity $\langle v \rangle$. It is presumed that the pressure p_0 at $r = 0$ and $z = 0$ is known and that the velocity distribution at the plane $z = 0$ is also known. Clearly, the velocity distribution depends upon the nature of the flow system for $z < 0$. If that part of the pipe is very long, then v_z at $z = 0$ will be the fully established velocity profile, independent of z for $z > 0$. If the portion of the pipe for $z < 0$ is very short, or nonexistent, then v_z will depend on z for $z > 0$.

The force of the fluid on the inner pipe wall for either laminar or turbulent flow is

$$F_k = \int_0^L \int_0^{2\pi} \left(-\mu \frac{\partial v_z}{\partial r} \right) \bigg|_{r=R} R \, d\theta \, dz \tag{6.2–1}$$

[1] In the following development we consider the pressure and velocity to be time-smoothed. For brevity, the overlines (see §5.1) are omitted.

If we equate Eqs. 6.2–1 and 6.1–2, we get the following expression for the friction factor:

$$f = \frac{\displaystyle\int_0^L \int_0^{2\pi} \left(-\mu \frac{\partial v_z}{\partial r}\right)\bigg|_{r=R} R\, d\theta\, dz}{(2\pi RL)(\frac{1}{2}\rho\langle v\rangle^2)} \tag{6.2–2}$$

Next we introduce the dimensionless quantities from §3.7: $v_z^* = v_z/\langle v\rangle$, $r^* = r/D$, $p^* = (p - p_0)/\rho\langle v\rangle^2$, and $\mathrm{Re} = D\langle v\rangle\rho/\mu$. Then Eq. 6.2–2 may be written

$$f = \frac{1}{\pi}\frac{D}{L}\frac{1}{\mathrm{Re}} \int_0^{L/D} \int_0^{2\pi} \left(-\frac{\partial v_z^*}{\partial r^*}\right)\bigg|_{r^*=\frac{1}{2}} d\theta\, dz^* \tag{6.2–3}$$

This relation is valid for circular tubes with laminar or turbulent flow. Thus we see that for flow systems in which the drag depends on viscous forces alone (no form drag) the product $f\,\mathrm{Re}$ is essentially a dimensionless velocity gradient averaged over the surface.

Recall now that in principle $(\partial v_z^*/\partial r^*)$ can be calculated from Eqs. 3.7–12 and 13 along with the boundary conditions

$$\text{at} \quad r^* = \tfrac{1}{2} \qquad \boldsymbol{v}^* = 0 \tag{6.2–4}$$

$$\text{at} \quad z^* = 0 \qquad \boldsymbol{v}^* = \text{some known function of } r^* \text{ and } \theta \tag{6.2–5}$$

$$\text{at} \quad z^* = 0 \qquad p^* = 0 \tag{6.2–6}$$

If Eqs. 3.7–12 and 13 could be solved for p^* and v^* with these boundary conditions, the solutions would of necessity be of the form[2]

$$\boldsymbol{v}^* = \boldsymbol{v}^*(r^*, \theta, z^*; \mathrm{Re}) \tag{6.2–7}$$

$$p^* = p^*(r^*, \theta, z^*; \mathrm{Re}) \tag{6.2–8}$$

That is, the functional dependence will include all the reduced variables and the one dimensionless group appearing in the differential equations. No additional dimensionless groups enter via the foregoing boundary conditions. As a consequence, $\partial v_z^*/\partial r^*$ must also depend only on r^*, θ, z^*, and Re. When the gradient is evaluated at $r^* = \frac{1}{2}$ and integrated over z^* and θ, the result depends only on Re and L/D (which appears in the upper limit in the integration over z^*). Therefore, we are led to the conclusion that

$$f = f(\mathrm{Re}, L/D) \tag{6.2–9}$$

[2] Since this discussion deals with the flow of a fluid of constant ρ in a completely filled pipe (that is, no free surface), gravity does not influence the dimensionless velocity distribution. Hence the Froude number, Fr, can be disregarded in this development.

That is, the friction factor depends only on the Reynolds number and the length-to-diameter ratio.[3]

If the velocity profile is fully developed at plane $z = 0$, then $\partial v^*/\partial r^*$ is independent of z^*. The integration over z^* in Eq. 6.2–3 can be performed to give L/D; this factor just cancels the factor D/L appearing before the integral. Hence, for this situation, f is independent of L/D and

$$f = f(\text{Re}) \tag{6.2–10}$$

If the velocity profile is not fully developed at plane $z = 0$ but the entrance length is very small with respect to L, then the integral over z^* will yield very nearly L/D, and Eq. 6.2–10 is a very good approximation. Hence we conclude that Eq. 6.2–10 is valid when (*a*) the velocity profile is fully developed or (*b*) $L/D \gg 1$.

Equations 6.2–9 and 10 are useful results in that they provide a guide to the systematic presentation of data on volume rate of flow versus pressure drop for laminar and turbulent flow in circular tubes. That is, for *long* tubes we need only a single curve of f plotted against the particular combination $D\langle v\rangle\rho/\mu$. Just think how much simpler this is than the plotting of pressure drop versus volume rate of flow for separate values of D, ρ, and μ, which is what the uninitiated might do.

There is much experimental information for tubes on pressure drop versus average velocity (or volume rate of flow); hence one can calculate f from experimental data by Eq. 6.1–4. Then one can plot f versus Re for smooth tubes to obtain the *solid curves* shown in Fig. 6.2–2. These curves reflect

[3] The same result may be obtained without reference to the basic differential equations at all, provided one can write down a set of variables just sufficient to specify the physical situation. According to "Buckingham's Pi theorem," *the functional relationship among q quantities, whose units may be given in terms of u fundamental units, may be written as a function of q − u dimensionless groups* (*the* Π'*s*). If we seek the functional form relating $p_0 - p_L$, ρ, $\langle v\rangle$, μ, D, and L, we note that

$$p_0 - p_L \,[=]\, ml^{-1}t^{-2} \qquad \langle v\rangle \,[=]\, lt^{-1} \qquad D \,[=]\, l$$
$$\rho \,[=]\, ml^{-3} \qquad \mu \,[=]\, ml^{-1}t^{-1} \qquad L \,[=]\, l$$

Hence $q = 6$ and $u = 3$. One may then select as three independent dimensionless quantities

$$\Pi_1 = \frac{p_0 - p_L}{\frac{1}{2}\rho\langle v\rangle^2} \qquad \Pi_2 = \frac{L}{D} \qquad \Pi_3 = \frac{D\langle v\rangle\rho}{\mu}$$

The choice of these quantities is arbitrary; any one of them may be replaced by the product of that one (raised to any power) by any of the others raised to any power. The Pi theorem then tells us that the general relation must be of the form

$$F(\Pi_1, \Pi_2, \Pi_3) = 0 \qquad \text{or} \qquad \Pi_1 = G(\Pi_2, \Pi_3)$$

This Pi theorem has the disadvantage that it does not select the variables nor does it determine their relative significance. The technique used in the text is to be preferred. For further information about the Pi theorem, see W. H. McAdams, *Heat Transmission*, McGraw-Hill, New York (1954), Third Edition, Chapter 5.

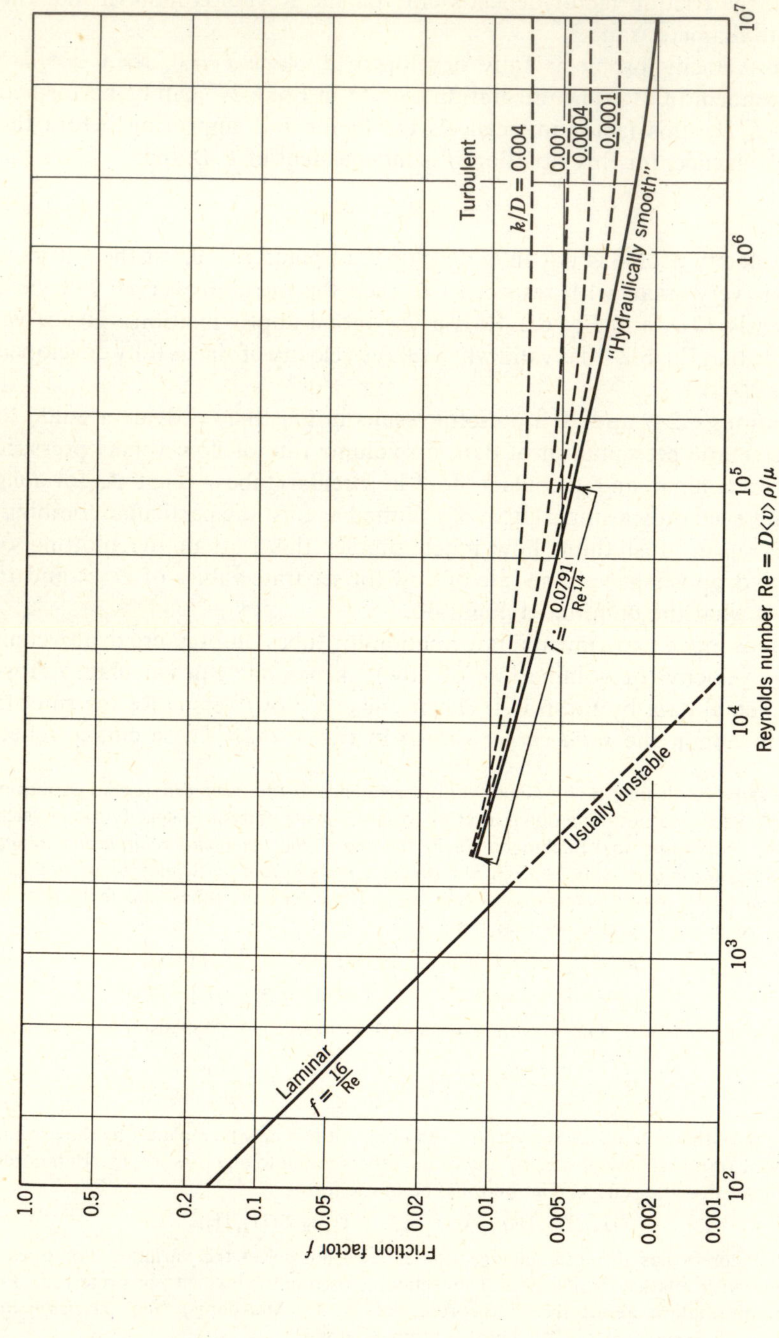

Fig. 6.2-2. Friction factors for tube flow (see definition of f in Eqs. 6.1-2 and 6.1-3. [Curves of L. F. Moody, *Trans. ASME*, **66**, 671 (1944) as presented in W. L. McCabe and J. C. Smith, *Unit Operations of Chemical Engineering*, McGraw-Hill, New York (1954).]

the laminar and turbulent behavior for flows of fluids in *long, smooth, circular* tubes.

Note that the *laminar* curve on the friction factor chart is nothing more than a plot of the *Hagen-Poiseuille* law given in Eq. 2.3–18. This can be seen by substituting the expression for $\langle v \rangle$ from Eq. 2.3–18 for *one* of the $\langle v \rangle$'s in the denominator of Eq. 6.1–4. For laminar flow in long tubes this gives

$$f = \frac{16}{\text{Re}} \qquad \begin{array}{ll} \text{Re} < 2.1 \times 10^3 & \text{stable} \\ \text{Re} > 2.1 \times 10^3 & \text{usually unstable} \end{array} \tag{6.2–11}$$

in which $\text{Re} = D\langle v \rangle \rho / \mu$; this is exactly the laminar line in Fig. 6.2–2.

The analogous *turbulent* curve on the friction factor chart has been established by *experimental data.* One can also calculate an f versus Re curve from some of the turbulent velocity distributions given in Chapter 5; this subject is discussed at great length by Schlichting.[4] We consider here only the simplest turbulent velocity distribution, namely the ($\frac{1}{7}$)-power law:

$$\frac{\bar{v}_z}{v_*} = 8.56 \left(\frac{s v_* \rho}{\mu} \right)^{1/7} \tag{6.2–12}$$

in which $v_* = \sqrt{\tau_0 / \rho} = \sqrt{(\mathscr{P}_0 - \mathscr{P}_L) R / 2L\rho}$ and $s = R - r$. The reader may show that the average velocity is

$$\frac{\langle \bar{v}_z \rangle}{v_*} = (0.817)(8.56) \left(\frac{R v_* \rho}{\mu} \right)^{1/7} \tag{6.2–13}$$

By forcing Eq. 6.2–13 into the form of Eq. 6.1–4, we get for turbulent tube flow

$$f = 2 \left(\frac{v_*}{\langle \bar{v}_z \rangle} \right)^2 = \left(\frac{2^5}{(8.56)^7 (0.817)^7} \right)^{1/4} \left(\frac{D \langle v \rangle \rho}{\mu} \right)^{-1/4} \tag{6.2–14}$$

or

$$f = \frac{0.0791}{\text{Re}^{1/4}} \qquad 2.1 \times 10^3 < \text{Re} < 10^5 \tag{6.2–15}$$

This equation, known as the *Blasius formula*, is good up to about $\text{Re} = 10^5$; it is useful for making estimates. Somewhat more accurate relations between f and Re may be obtained by using more accurate expressions for the velocity distribution. (See Problem 6.J.) For most engineering calculations involving turbulent flow, one should simply use the chart in Fig. 6.2–2, since it summarizes the experimental data available on tubes.

If the circular tubes are *rough*, then in the turbulent region higher pressure drops are required for a given flow rate than would be indicated by the solid line on Fig. 6.2–2. If k is the height of the protuberances, then the "relative roughness" k/D would be expected to enter the correlation. The dotted lines indicate the f versus Re curves for various values of k/D. Note that the roughness tends to make f approach a constant value at high Reynolds

[4] H. Schlichting, *Boundary Layer Theory*, McGraw-Hill, New York (1955), Chapter XX,

numbers. The parameter k/D is not really sufficient to define the shape and distribution of the roughness.

If the tubes are not circular, then one may use an empirical "mean hydraulic radius" R_h, which is defined as

$$R_h = S/Z \tag{6.2-16}$$

in which S is the cross section of the stream and Z is the wetted perimeter, For *turbulent flow*, as an approximation one may use Eq. 6.1–4 and Fig. 6.2–2, with the diameter D of the circular tube everywhere replaced by $4R_h$. That is, we calculate pressure drops by using the formula

$$\frac{\mathscr{P}_0 - \mathscr{P}_L}{\frac{1}{2}\rho\langle v\rangle^2} = \left(\frac{L}{R_h}\right)f \tag{6.2-17}$$

and we get f from Fig. 6.2–2 by using a Reynolds number defined as

$$\mathrm{Re}_h = \frac{4R_h\langle v\rangle\rho}{\mu} \tag{6.2-18}$$

This empiricism is *not* recommended for *laminar flow*. (See Problem 6.L.)

Example 6.2–1. Pressure Drop Required for a Given Flow Rate

What pressure gradient is required to cause N,N-diethylaniline ($C_6H_5N(C_2H_5)_2$ to flow in a horizontal smooth circular tube of inside diameter $D = 3$ cm at a volume rate of $Q = 1.1$ liter/sec at 20° C? At this temperature the density of diethylaniline is $\rho = 0.935$ g cm^{-3} and its viscosity is $\mu = 1.95$ cp (or 1.95×10^{-2} g cm^{-1} sec^{-1}).

Solution. First we determine the Reynolds number for the flow

$$\begin{aligned}\mathrm{Re} &= \frac{D\langle v\rangle\rho}{\mu} = \frac{DQ\rho}{(\pi D^2/4)\mu} = \frac{4Q\rho}{\pi D\mu} \\ &= \frac{4(1100\ \mathrm{cm^3\,sec^{-1}})(0.935\ \mathrm{g\,cm^{-3}})}{\pi(3\ \mathrm{cm})(1.95\times10^{-2}\ \mathrm{g\,cm^{-1}\,sec^{-1}})} = 2.24\times10^4\end{aligned} \tag{6.2-19}$$

From Fig. 6.2–2, we find that for this Reynolds number the friction factor f has a value of 0.0063 for smooth tubes. Hence the pressure gradient required to sustain the flow, according to Eq. 6.1–4, is

$$\begin{aligned}\frac{p_0 - p_L}{L} &= \frac{4}{D}\cdot\frac{1}{2}\rho\langle v\rangle^2\cdot f \\ &= \frac{4}{D}\cdot\frac{1}{2}\rho\left(\frac{4Q}{\pi D^2}\right)^2 f \\ &= \frac{32\rho Q^2 f}{\pi^2 D^5} \\ &= \frac{(32)(0.935)(1100)^2(0.0063)}{\pi^2(3.0)^5} \\ &= 95\ (\mathrm{dynes\ cm^{-2}})/\mathrm{cm} \\ &= 0.071\ (\mathrm{mm\ Hg})/\mathrm{cm}\end{aligned} \tag{6.2-20}$$

Example 6.2–2. Flow Rate for a Given Pressure Drop

Determine the flow rate, in pounds per hour, of water at 68° F. through a 1000 ft length of horizontal 8-in., schedule 40 steel pipe (internal diameter 7.981 in.) under a pressure difference of 3.00 psi. For such a pipe use Fig. 6.2–2 and assume that $k/D = 2.3 \times 10^{-4}$.

Solution. We want to use Eq. 6.1–4 and Fig. 6.2–2 to solve for $\langle v \rangle$ when $p_0 - p_L$ is known. But the quantity $\langle v \rangle$ appears explicitly on the left side of the equation and implicitly on the right side in f (which depends on $\text{Re} = D\langle v \rangle \rho/\mu$). Clearly, a trial-and-error solution could be found. However, if one has to make more than a few calculations of $\langle v \rangle$, it pays to use a more systematic approach; two methods are suggested here. Because of the fact that experimental data are frequently presented in graphical form, it behooves the engineering student to use his originality in developing special methods such as those described here.

Method A. Figure 6.2–2 may be used to construct a plot[5] of Re versus the group $\text{Re}\sqrt{f}$, which does not contain $\langle v \rangle$:

$$\text{Re}\sqrt{f} = \frac{D\langle v \rangle \rho}{\mu}\sqrt{\frac{(p_0 - p_L)D}{2L\rho\langle v \rangle^2}} = \frac{D\rho}{\mu}\sqrt{\frac{(p_0 - p_L)D}{2L\rho}} \tag{6.2–21}$$

The quantity $\text{Re}\sqrt{f}$ can be computed for this problem, and a value of the Reynolds number can be read off the Re versus $\text{Re}\sqrt{f}$ plot. From Re the average velocity and flow rate can be calculated.

Method B. Figure 6.2–2 may be used directly without any replotting, by devising a scheme which is tantamount to the graphical solution of two simultaneous equations. The two equations are

$$f = f(\text{Re}, k/D) \qquad \text{curve given in Fig. 6.2–2} \tag{6.2–22}$$

$$f = \frac{(\text{Re}\sqrt{f})^2}{\text{Re}^2} \qquad \text{straight line of slope } -2 \text{ on log-log plot} \tag{6.2–23}$$

The procedure is then to compute $\text{Re}\sqrt{f}$ according to Eq. 6.2–21 (one does not have to know $\langle v \rangle$ to do this) and then to plot Eq. 6.2–23 on the log-log plot of f versus Re in Fig. 6.2–2. The intersection point gives us the Reynolds number of the flow from which $\langle v \rangle$ can be computed.

For the problem at hand we have

$$\begin{aligned}
p_0 - p_L &= 3.00\ (\text{lb}_f\ \text{in.}^{-2}) \times 32.17\ (\text{lb}_m\ \text{ft}\ \text{lb}_f^{-1}\ \text{sec}^{-2}) \times 144(\text{in.}^2\ \text{ft}^{-2}) \\
&= 1.39 \times 10^4\ (\text{lb}_m\ \text{ft}^{-1}\ \text{sec}^{-2}) \\
D &= 7.981\ (\text{in.}) \times \tfrac{1}{12}\ (\text{ft/in.}) = 0.665\ \text{ft} \\
L &= 1000\ \text{ft} \\
\rho &= 62.3\ (\text{lb}_m\ \text{ft}^{-3}) \\
\mu &= 1.03\ (\text{cp}) \times 6.72 \times 10^{-4}\ (\text{lb}_m\ \text{ft}^{-1}\ \text{sec}^{-1})/(\text{cp}) \\
&= 6.93 \times 10^{-4}\ (\text{lb}_m\ \text{ft}^{-1}\ \text{sec}^{-1})
\end{aligned}$$

[5] A related plot was proposed by T. von Kármán, *Nachr. Ges. Wiss. Göttingen, Fachgruppen*, I, **5**, 58–76 (1930); see, for example, W. L. McCabe and J. C. Smith, *Unit Operations of Chemical Engineering*, McGraw-Hill, New York (1956), p. 72.

Then, according to Eq. 6.2–21,

$$\mathrm{Re}\sqrt{f} = \frac{D\rho}{\mu}\sqrt{\frac{(p_0 - p_L)D}{2L\rho}} = \frac{(0.665)(62.3)}{(6.93 \times 10^{-4})}\sqrt{\frac{(1.39 \times 10^4)(0.665)}{2(10^3)(62.3)}}$$
$$= 1.63 \times 10^4, \text{ dimensionless} \tag{6.2–24}$$

The line of Eq. 6.2–23 for this value of $\mathrm{Re}\sqrt{f}$ passes through $f = 1.0$ at $\mathrm{Re} = 1.63 \times 10^4$ and through $f = 0.01$ at $\mathrm{Re} = 1.63 \times 10^5$. Extension of the straight line through these points to the curve of Eq. 6.2–22 for $k/D = 0.00023$ gives the solution to the two simultaneous equations:

$$\mathrm{Re} = \frac{D\langle v\rangle\rho}{\mu} = 2.4 \times 10^5 \tag{6.2–25}$$

Now for circular pipes the mass rate of flow is $w = \rho\langle v\rangle\pi D^2/4$; hence Eq. 6.2–25 becomes

$$\mathrm{Re} = \frac{4w}{\pi D\mu} = 2.4 \times 10^5 \tag{6.2–26}$$

Solving for w gives

$$w = \frac{\pi}{4} D\mu\,\mathrm{Re}$$
$$= 0.7854 \times 0.665 \times 6.93 \times 10^{-4} \times 3600 \times 2.4 \times 10^5$$
$$= 3.12 \times 10^5\ \mathrm{lb}_m\ \mathrm{hr}^{-1} \tag{6.2–27}$$

§6.3 FRICTION FACTORS FOR FLOW AROUND SPHERES

In this section we use the definition of f in Eq. 6.1–5 along with the dimensional analysis of §3.7 to determine the dependence of f. Once again we restrict the development to fluids of constant ρ; the coordinate system is that used in §2.6.

The contribution F_k to the force acting on a sphere in a fluid flowing in the $+z$-direction is the total force $F = F_n + F_t$ minus the force F_s:

$$F_k = (F_n - F_s) + F_t$$
$$= F_{\text{form drag}} + F_{\text{friction drag}} \tag{6.3–1}$$

in which

$$F_n = \int_0^{2\pi}\int_0^{\pi}\{-p|_{r=R}\cos\theta\}R^2\sin\theta\,d\theta\,d\phi \tag{6.3–2}$$

$$F_s = \int_0^{2\pi}\int_0^{\pi}\{-(p_0 - \rho g z)|_{r=R}\cos\theta\}R^2\sin\theta\,d\theta\,d\phi \tag{6.3–3}$$

$$F_t = \int_0^{2\pi}\int_0^{\pi}\left\{-\mu\left[r\frac{\partial}{\partial r}\left(\frac{v_\theta}{r}\right) + \frac{1}{r}\frac{\partial v_r}{\partial\theta}\right]\bigg|_{r=R}\sin\theta\right\}R^2\sin\theta\,d\theta\,d\phi \tag{6.3–4}$$

Here p_0 is the pressure at the plane $z = 0$, which goes through the equator of the sphere.

If we now split f into two contributions analogous to $F_{\text{form drag}}$ and $F_{\text{friction drag}}$ and if we use the definition of f given in Eq. 6.1–5, then

$$f = f_{\text{form}} + f_{\text{friction}} \tag{6.3–5}$$

$$f_{\text{form}} = \frac{2}{\pi}\int_0^{2\pi}\int_0^{\pi}\{-\mathscr{P}^* \cos\theta\} \sin\theta \, d\theta \, d\phi \tag{6.3–6}$$

$$f_{\text{friction}} = \frac{4}{\pi}\frac{1}{\text{Re}}\int_0^{2\pi}\int_0^{\pi}\left\{-\left[r^*\frac{\partial}{\partial r^*}\left(\frac{v_\theta^*}{r^*}\right) + \frac{1}{r^*}\frac{\partial v_r^*}{\partial \theta}\right]\right\}\Bigg|_{r^*=1} \sin^2\theta \, d\theta \, d\phi \tag{6.3–7}$$

The friction factor has thus been expressed in terms of dimensionless variables:

$$\mathscr{P}^* = \frac{p - p_0 + \rho g z}{\rho v_\infty^2}; \qquad v_\theta^* = \frac{v_\theta}{v_\infty}; \qquad v_r^* = \frac{v_r}{v_\infty}; \qquad r^* = \frac{r}{R} \tag{6.3–8,9,10,11}$$

and a Reynolds number defined as

$$\text{Re} = \frac{D v_\infty \rho}{\mu} = \frac{2 R v_\infty \rho}{\mu} \tag{6.3–12}$$

Clearly, in order to evaluate f one has to know $\mathscr{P}^*$, v_r^*, and v_θ^* as functions of r^*, θ, and ϕ. In §§2.6 and 4.2 we showed how one could derive the expression for Stokes's law for the condition of "creeping flow," that is, very slow flow with Re less than about 0.1.

For $\text{Re} > 0.1$, very little is known quantitatively about the reduced pressure and velocity distributions; but we do know that for incompressible flow these distributions can *in principle* be obtained from Eqs. 3.7–12 and 3.7–13.[1] These equations have to be solved with the boundary conditions that

$$\text{at} \quad r^* = 1, \qquad v_r^* = v_\theta^* = 0 \tag{6.3–13}$$

$$\text{at} \quad r^* = \infty, \qquad v_z^* = 1 \tag{6.3–14}$$

$$\text{at} \quad r^* = \infty, \qquad \mathscr{P}^* = 0 \tag{6.3–15}$$

Because no additional groups enter in via the boundary conditions, we see that

$$\mathscr{P}^* = \mathscr{P}^*(x^*, y^*, z^*; \text{Re}) \tag{6.3–16}$$

$$\mathbf{v}^* = \mathbf{v}^*(x^*, y^*, z^*; \text{Re}) \tag{6.3–17}$$

and that

$$f = f(\text{Re}) \tag{6.3–18}$$

[1] For the system under consideration, Eq. 3.7–13 may be rewritten

$$\frac{D\mathbf{v}^*}{Dt^*} = -\nabla^*\mathscr{P}^* + \left[\frac{\mu}{R v_\infty \rho}\right]\nabla^{*2}\mathbf{v}^* \tag{6.3–12a}$$

We see that the Froude number appears neither in the differential equations nor in the boundary conditions.

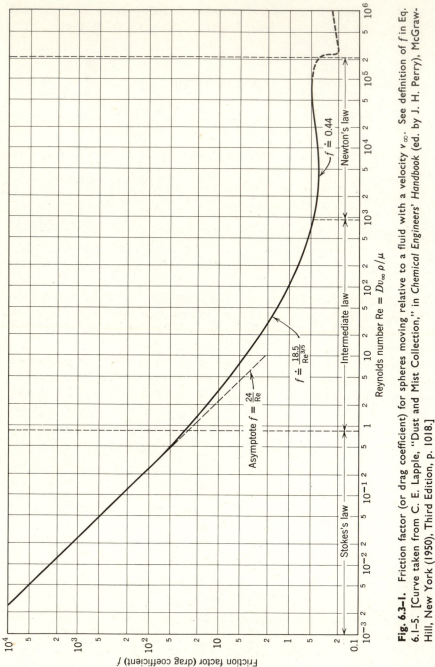

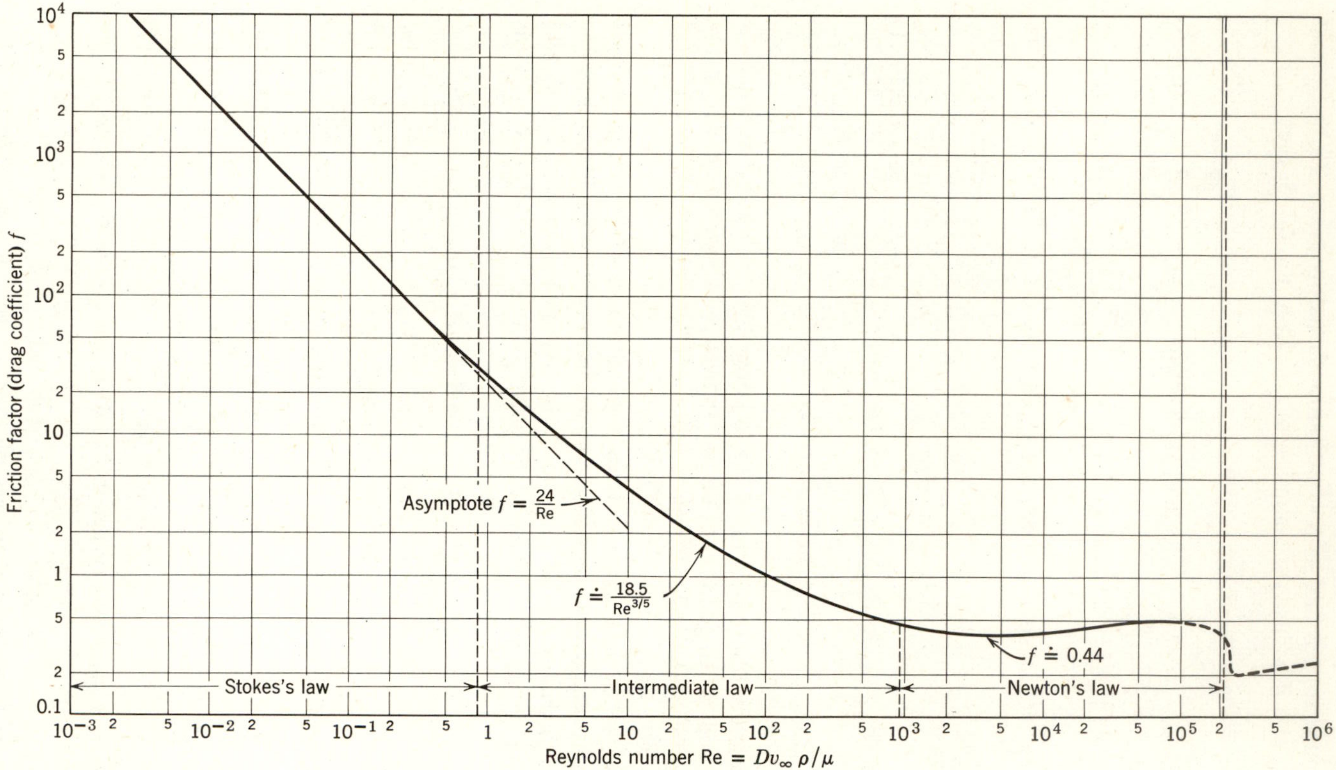

Fig. 6.3–1. Friction factor (or drag coefficient) for spheres moving relative to a fluid with a velocity v_∞. See definition of f in Eq. 6.1–5. [Curve taken from C. E. Lapple, "Dust and Mist Collection," in *Chemical Engineers' Handbook* (ed. by J. H. Perry), McGraw-Hill, New York (1950), Third Edition, p. 1018.]

by using arguments similar to those in §6.2. Hence from the dimensional analysis of the partial differential equations describing the flow and from the definition of the friction factor we have obtained the result that f may be correlated as a function of Re alone.

Many experimental data have been taken for flow around spheres, so that a chart of f versus Re is available for smooth spheres. (See Fig. 6.3–1.) For this system there is no sharp transition from an unstable laminar flow curve to a stable turbulent flow curve as was indicated for tubes in Fig. 6.2–2 at $\text{Re} \doteq 2.1 \times 10^3$. In this system, as the flow rate increases, there is an increase in the amount of eddying behind the sphere. The kink in the curve at about $\text{Re} = 2 \times 10^5$ is associated with the shift of the boundary-layer separation zone from in front of the equator to in back of the equator of the sphere.[2]

We have purposely chosen to discuss the sphere immediately after the tube in order to emphasize the fact that various flow systems may behave quite differently. Several points of difference between the two systems are:

For tubes there is a rather well-defined laminar-turbulent transition at $\text{Re} \doteq 2 \times 10^3$.	For spheres the f-curve exhibits no well-defined laminar-turbulent transition.
For smooth tubes the only contribution to f is friction drag.	For spheres there are contributions to f owing to both friction drag and form drag.
For tubes there is no boundary layer separation.	For spheres there is a kink in f-curve associated with a shift in the separation zone.

The general shape of the curves in Figs. 6.2–2 and 6.3–1 should be carefully remembered.

For the *creeping flow region*, we already know that the drag force is given by *Stokes's law*, which is a consequence of an analytical solution of the equations of motion and continuity (with the term $\rho Dv/Dt$ omitted from the equation of motion given in Eq. 3.2–20). By rearranging Stokes's law (Eq. 2.6–14) in the form of Eq. 6.1–5, we get

$$F_k = \pi R^2 \cdot \frac{1}{2}\rho v_\infty^2 \cdot \frac{24}{\left(\dfrac{Dv_\infty \rho}{\mu}\right)} \tag{6.3–19}$$

Hence, for *creeping flow* around a sphere,

$$f = \frac{24}{\text{Re}} \qquad \text{Re} < 0.1 \tag{6.3–20}$$

This is the straight-line portion of the log f versus log Re curve.

[2] See H. Schlichting, *Boundary Layer Theory*, McGraw-Hill, New York (1955), pp. 34–35.

For higher values of the Reynolds number, it is very difficult to make purely theoretical calculations. Several investigators have managed to estimate f as far as Re = 10 but only with a considerable amount of effort. Hence the f-curve for Re > 0.1 is a result of experiment. Occasionally, simple analytical expressions for the higher Reynolds number regions are useful. For the *intermediate region,* we may write very approximately

$$f \doteq \frac{18.5}{\text{Re}^{3/5}} \qquad 2 < \text{Re} < 5 \times 10^2 \tag{6.3–21}$$

which indicates a lesser dependence on Re than in Stokes's law. This expression is less accurate than Stokes's law for Re < 2.

For higher Re, we see that the friction factor is approximately constant. This is known as the *Newton's law region,* for which

$$f \doteq 0.44 \qquad 5 \times 10^2 < \text{Re} < 2 \times 10^5 \tag{6.3–22}$$

In this region the drag force acting on the sphere is approximately proportional to the square of the velocity of the fluid moving past the sphere. Equation 6.3–22 is a useful approximation for making rapid estimates. (Note that Newton's "law" for the drag force on a sphere is not to be confused with Newton's law of viscosity or Newton's laws of motion.)

Many extensions of Fig. 6.3–1 have been made, but a systematic study would be beyond the scope of the text. Among the effects investigated are wall effects (see Problem 6.O), fall of droplets with internal circulation,[3] fall of particles in non-Newtonian fluids,[4] hindered settling (i.e., fall of clusters of particles which interfere with one another),[5] unsteady flow,[6] and nonspherical particles.[7,8]

Example 6.3–1. Determination of Diameter of a Falling Sphere

Glass spheres of density $\rho_{sph} = 2.62$ g cm^{-3} are allowed to fall through carbon tetrachloride ($\rho = 1.59$ g cm^{-3} and $\mu = 9.58$ millipoises) at 20° C in an experiment for studying reaction times in making time observations with stopwatches and more elaborate devices. What diameter should the spheres be in order to have a terminal velocity of about 65 cm sec^{-1}?

[3] H. Lamb, *Hydrodynamics*, Dover (1945), Sixth Edition pp. 600–601; S. Hu and R. C. Kintner, *A.I.Ch.E. Journal*, **1**, 42–48 (1955).

[4] J. C. Slattery, doctoral thesis, University of Wisconsin (1959).

[5] H. H. Steinour, *Ind. Eng. Chem.*, **36**, 618–624, 840–847, 900–901 (1947); see also C. E. Lapple, *Fluid and Particle Dynamics*, University of Delaware Press, Newark (1951), Chapter 13.

[6] R. R. Hughes and E. R. Gilliland, *Chem. Eng. Prog.*, **48**, 497–504 (1952).

[7] E. S. Pettyjohn and E. B. Christiansen, *Chem. Eng. Prog.*, **44**, 157 (1948).

[8] H. A. Becker, *Can. J. Chem. Eng.* **37**, 85–91 (1959).

Solution. To find the sphere diameter, we have to solve Eq. 6.1–7 for D. However, in this equation one has to know D in order to get f; and f is given by the solid curve in Fig. 6.3–1. A trial-and-error procedure can be used, taking $f = 0.44$ as a first guess.

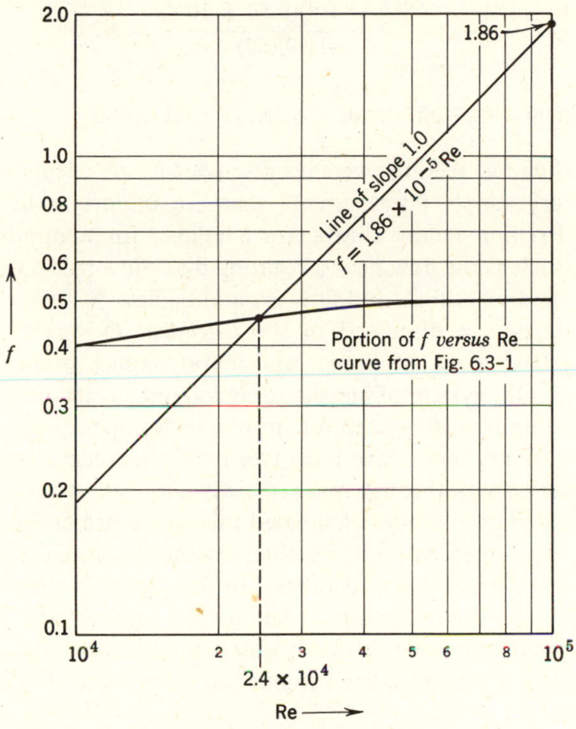

Fig. 6.3–2. Graphical procedure used in Example 6.3–1.

Alternatively, we can solve Eq. 6.1–7 for f and then note that f/Re is a quantity independent of D:

$$\frac{f}{\text{Re}} = \frac{4}{3} \frac{g\mu}{\rho v_\infty^3} \left(\frac{\rho_{\text{sph}} - \rho}{\rho} \right) \tag{6.3–23}$$

The quantity on the right side can be calculated with the foregoing data, and we can call it C. Hence we have two simultaneous equations to solve:

$$f = C\text{Re} \qquad \text{(from Eq. 6.3–23)} \tag{6.3–24}$$

$$f = f(\text{Re}) \qquad \text{(given in Fig. 6.3–1)} \tag{6.3–25}$$

Equation 6.3–24 is a straight line of slope unity on the log-log plot of f versus Re. For the problem at hand,

$$C = \frac{4}{3} \frac{(980)(9.58 \times 10^{-3})}{(1.59)(65)^3} \left(\frac{2.62 - 1.59}{1.59} \right) = 1.86 \times 10^{-5} \tag{6.3–26}$$

Hence at Re $= 10^5$, according to Eq. 6.3–24, $f = 1.86$. The line of slope 1 passing through $f = 1.86$ at Re $= 10^5$ is shown in Fig. 6.3–2. This line intersects the curve of Eq. 6.3–25 (i.e., the curve of Fig. 6.3–1) at Re $\equiv Dv_\infty\rho/\mu = 2.4 \times 10^4$. The sphere diameter is then found to be

$$D = \frac{\text{Re}\,\mu}{\rho v_\infty} = \frac{(2.4 \times 10^4)(9.58 \times 10^{-3})}{(1.59)(65)} = 2.2 \text{ cm}$$

§6.4 FRICTION FACTORS FOR PACKED COLUMNS

In the preceding two sections we have discussed at some length the correlation of two very simple flow systems that are important in engineering calculations. Friction factor charts are available for a number of other fluid motions, such as the flow near a rotating disk, flow past a cylinder, flow past a set of cylindrical tubes, and flow around baffles. Systems of chemical-engineering interest are discussed in the *Chemical Engineers' Handbook*,[1] and others for basic fluid mechanics and aerodynamics studies are treated by Schlichting.[2] One system of considerable interest in chemical engineering is the packed column, widely used for mass transfer operations.

Generally speaking, there have been two main theoretical approaches for studying pressure drops through packed beds. In one method the packed column is regarded as a bundle of tangled tubes of weird cross section; the theory is then developed by applying the previous results for single straight tubes to the collection of crooked tubes. In the second method the packed tower is visualized as a collection of submerged objects, and the pressure drop is calculated by summing up the resistances of the submerged particles.[3,4] The tube-bundle theories have been somewhat more successful, and we shall discuss them here.

The packing material may be spheres, cylinders, or various kinds of commercial packing for contacting apparatus.[5] It is assumed throughout the following discussion that the packing is everywhere uniform and that there is no "channeling" (in actual practice, channeling frequently occurs and the formulas given here are not valid). It is further assumed that the diameter of the packing is small in comparison with the diameter of the column in which the packing is contained and that the column diameter is constant.

[1] T. B. Drew, H. H. Dunkle, and R. P. Genereaux, Section 5 in *Chemical Engineers' Handbook* (J. H. Perry, Ed.), McGraw-Hill, New York (1950), Third Edition.

[2] H. Schlichting, *Boundary Layer Theory*, McGraw-Hill, New York (1955), pp. 75–80, 439, 445–447.

[3] H. C. Brinkman, *Appl. Sci. Research*, **A1**, 27–34, 81–86, 333–346 (1949).

[4] W. E. Ranz, *Chem. Eng. Prog.*, **48**, 247–253 (1952).

[5] See, for example, McCabe and Smith, p. 630 (Fig. 11.1); M. Leva, *Tower Packings and Packed-Tower Design*, U.S. Stoneware Co., Akron, Ohio (1953), Chapters 1 and 2.

We define the friction factor for the packed bed analogously to Eq. 6.1–4:

$$\frac{\mathscr{P}_0 - \mathscr{P}_L}{\frac{1}{2}\rho v_0^2} = \frac{L}{D_p} \cdot 4f \tag{6.4–1}$$

in which D_p is the particle diameter (defined presently) and v_0 is the "superficial velocity" (this is the average linear velocity the fluid would have in the column if no packing were present); L is the length of the packed column. We shall now estimate separately the friction factor for laminar flow and that for turbulent flow.

For *laminar flow* in circular tubes of radius R it was shown in §2.3 that

$$\langle v \rangle = \frac{(\mathscr{P}_0 - \mathscr{P}_L)R^2}{8\mu L} \tag{6.4–2}$$

We now imagine that a packed bed is just a tube of very complicated cross section with hydraulic radius R_h. (See Eq. 6.2–16.) The average flow velocity in the cross section available for flow is then

$$\langle v \rangle = \frac{(\mathscr{P}_0 - \mathscr{P}_L)R_h^2}{2\mu L} \tag{6.4–3}$$

The hydraulic radius may be expressed in terms of the "void fraction" ϵ and the wetted surface a per unit volume of bed in the following way:

$$\begin{aligned} R_h &= \left(\frac{\text{cross section available for flow}}{\text{wetted perimeter}}\right) \\ &= \left(\frac{\text{volume available for flow}}{\text{total wetted surface}}\right) \\ &= \frac{\left(\dfrac{\text{volume of voids}}{\text{volume of bed}}\right)}{\left(\dfrac{\text{wetted surface}}{\text{volume of bed}}\right)} = \frac{\epsilon}{a} \end{aligned} \tag{6.4–4}$$

The quantity a is related to the "specific surface" a_v (the total particle surface/the volume of the particles) by

$$a = a_v(1 - \epsilon) \tag{6.4–5}$$

The quantity a_v is in turn used to define the mean particle diameter D_p:

$$D_p = 6/a_v \tag{6.4–6}$$

This definition is chosen because, for spheres, Eq. 6.4–6 gives just D_p = diameter of sphere. Finally, we note that the average value of the velocity in the interstices $\langle v \rangle$ is not of general interest to the engineer but rather the superficial velocity v_0; these two velocities are related by $v_0 = \langle v \rangle \epsilon$.

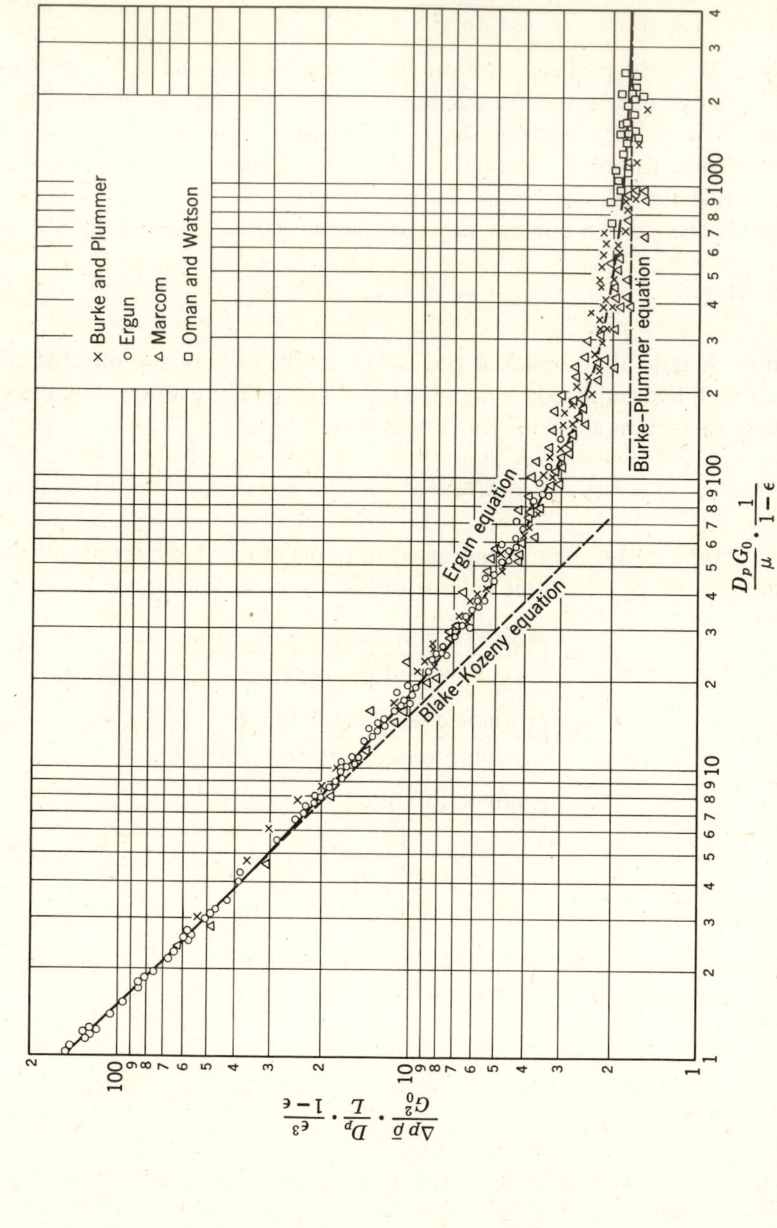

Fig. 6.4-1. Sketch showing the general behavior of the Ergun equation on a log-log plot. [S. Ergun, *Chem. Eng. Prog.*, **48**, 93 (1952).]

If we combine these definitions with the modified Hagen-Poiseuille formula in Eq. 6.4–3, we obtain

$$v_0 = \frac{(\mathscr{P}_0 - \mathscr{P}_L)R_h^2}{2\mu L}\epsilon = \frac{(\mathscr{P}_0 - \mathscr{P}_L)\epsilon^3}{2\mu La^2} = \frac{(\mathscr{P}_0 - \mathscr{P}_L)\epsilon^3}{2\mu La_v^2(1-\epsilon)^2}$$

$$= \frac{(\mathscr{P}_0 - \mathscr{P}_L)D_p^2}{2L(36\mu)}\frac{\epsilon^3}{(1-\epsilon)^2} \tag{6.4–7}$$

or finally

$$v_0 = \frac{(\mathscr{P}_0 - \mathscr{P}_L)}{L}\cdot\frac{D_p^2}{2(36\mu)}\frac{\epsilon^3}{(1-\epsilon)^2} \tag{6.4–8}$$

Now, in laminar flow the assumption of mean hydraulic radius frequently gives too large a throughput for a given pressure gradient; hence because of this assumption one would expect that the right side of Eq. 6.4–8 should be somewhat smaller. A second assumption implicitly made in the foregoing development is that the path of the fluid going through the bed is of length L; that is, it is the same as the length of the packed column. Actually, of course, the liquid traverses a very tortuous path, the length of which may be half again as long as the length L. Here, again, one would expect that the right side of Eq. 6.4–8 should be somewhat diminished.

Experimental measurements indicate that the theoretical formula can be improved if the 2 in the denominator on the right-hand side is changed to a value somewhere between 4 and 5. Analysis of a great deal of data has led to the value 25/6, which we accept here. Insertion of that value into Eq. 6.4–8 then gives

$$v_0 = \frac{(\mathscr{P}_0 - \mathscr{P}_L)}{L}\frac{D_p^2}{150\mu}\frac{\epsilon^3}{(1-\epsilon)^2} \tag{6.4–9}$$

which is the *Blake-Kozeny* equation. This result is generally good for void fractions less than 0.5 and is valid only in the laminar region given by $(D_pG_0/\mu)(1-\epsilon)^{-1} < 10$, where $G_0 = \rho v_0$. Note that the Blake-Kozeny equation corresponds to a bed friction factor of

$$f = \left(\frac{(1-\epsilon)^2}{\epsilon^3}\right)\frac{75}{D_pG_0/\mu} \tag{6.4–10}$$

This result is plotted in Fig. 6.4–1.

Exactly the same treatment can be repeated for *highly turbulent flow* in packed columns. We begin again with the expression for the friction-factor definition for flow in a circular tube. This time, however, we note that for highly turbulent flow in tubes with any appreciable roughness the friction factor becomes a function of the roughness only. Assuming that all packed beds have similar roughness characteristics, a unique friction factor f_0 may be used for turbulent flow. We now follow the procedure in §6.2 and make the same substitutions: $\langle v \rangle = v_0/\epsilon$; $D = 4R_h$; $R_h = \epsilon/a$; $a = a_v(1 - \epsilon)$, and finally $D_p = 6/a_v$. This leads to the following results:

$$\frac{\mathscr{P}_0 - \mathscr{P}_L}{L} = \frac{1}{D} \cdot \frac{1}{2} \rho \langle v \rangle^2 \cdot 4f_0 = 6f_0 \cdot \frac{1}{D_p} \cdot \frac{1}{2} \rho v_0^2 \cdot \frac{1 - \epsilon}{\epsilon^3} \tag{6.4–11}$$

Experimental data indicate that $6f_0 = 3.50$. Hence we obtain

$$\frac{\mathscr{P}_0 - \mathscr{P}_L}{L} = 3.50 \frac{1}{D_p} \frac{1}{2} \rho v_0^2 \frac{1 - \epsilon}{\epsilon^3} \tag{6.4–12}$$

which is the *Burke-Plummer* equation valid for $(D_p G_0/\mu)(1 - \epsilon)^{-1} > 1000$. This result corresponds to a friction factor given by

$$f = 0.875 \frac{1 - \epsilon}{\epsilon^3} \tag{6.4–13}$$

Note that the dependence on ϵ is different from that given for laminar flow.

When the Blake-Kozeny equation for laminar flow and the Burke-Plummer equation for turbulent flow are simply added together the result is

$$\frac{\mathscr{P}_0 - \mathscr{P}_L}{L} = \frac{150 \mu v_0}{D_p^2} \frac{(1 - \epsilon)^2}{\epsilon^3} + \frac{1.75 \rho v_0^2}{D_p} \frac{(1 - \epsilon)}{\epsilon^3} \tag{6.4–14}$$

This may be rewritten in terms of dimensionless groups:[6]

$$\left(\frac{(\mathscr{P}_0 - \mathscr{P}_L)\rho}{G_0^2}\right)\left(\frac{D_p}{L}\right)\left(\frac{\epsilon^3}{1 - \epsilon}\right) = 150 \frac{(1 - \epsilon)}{(D_p G_0/\mu)} + 1.75 \tag{6.4–15}$$

This is the *Ergun equation.* It has been applied with success to gases by using the density of the gas at the arithmetic average of the end pressures. For large pressure drops, however, it seems more reasonable to use Eq. 6.4–14, with the pressure gradient in differential form. Note that G_0 is a constant through the bed, whereas v_0 changes through the bed for a compressible fluid. The D_p used in this equation is that defined in Eq. 6.4–6.

Note also that for high rates of flow the first term on the right side drops out and the equation reduces to the Burke-Plummer equation. At low rates of flow the second term on the right side drops out and the Blake-Kozeny equation is obtained. In Fig. 6.4–1 the general behavior of these equations is sketched. It should be emphasized that the Ergun equation is but one of many that have been proposed for describing pressure drop across packed columns.

[6] S. Ergun, *Chem. Eng. Prog.*, **48,** 89–94 (1952).

QUESTIONS FOR DISCUSSION

1. How could the friction factor be defined for flow through an annulus and for transverse flow around a cylinder?

2. Use Eq. 6.1–7 to obtain expressions for the terminal falling velocity of a sphere for the three regions discussed in §6.3.

3. Does Eq. 6.1–7 have to be modified for spheres that are lighter than the fluid, and hence rise instead of fall? If so, how?

4. What precautions have to be taken in using formulas with friction factors taken from reference books and original sources?

5. How is the Hagen-Poiseuille law related to Fig. 6.2–2?

6. Verify Eqs. 6.2–13 and 15 by performing the missing manipulations.

7. Compare the calculation procedures when it is desired to use Fig. 6.2–2 to solve for (*a*) pressure drop, (*b*) average velocity, (*c*) tube diameter for given velocity and Δp, (*d*) tube diameter for given volumetric flow rate and Δp. Trial-and-error is to be avoided.

8. How are boundary conditions accounted for in the dimensional analysis applications in §§6.2 and 6.3?

9. Contrast the shapes of the f versus Re curves for tubes and spheres. What is the approximate dependence of f on Re in the various regions of these graphs?

10. Outline the physical arguments that lead to the establishment of the form of the Ergun equation (Eq. 6.4–15).

11. Why does the turbulent curve in Fig. 6.2–2 lie *above* rather than *below* the extension of the curve $f = 16/\text{Re}$?

12. What is the connection between the Blake-Kozeny equation (Eq. 6.4–9) and Darcy's law? (See Problem 4.J.)

13. How would the friction factor for *unsteady* tube flow (see Example 4.1–2) behave? Would Eqs. 6.1–3 and 4 be applicable?

14. Discuss the usefulness of the approximate relation $f = \sqrt{\frac{24}{\text{Re}}\left(\frac{24}{\text{Re}} + 4.5\right)}$ for flow around spheres.

15. Show from Eq. 6.2–3 that $f\,\text{Re}/2$ can be interpreted as the area average value of the dimensionless velocity gradient at the pipe wall.

16. Discuss the flow of water out of a $\frac{1}{2}$ in. rubber garden hose, which is attached to a house faucet with a pressure of 70 psig available.

17. Verify that the right sides of Eqs. 6.1–4 and 6.1–7 are dimensionless.

18. A baseball announcer says, "Because of the high humidity today, the baseball cannot go so far through the heavy humid air as it would on a dry day." Comment critically on this announcer's logic.

PROBLEMS

6.A$_1$ Pressure Drop Required for a Given Flow Rate with Fittings

What pressure drop is required in order to pump water at 20° C through a pipe 25 cm in diameter and length 1234 m at a rate of 1.97 m^3 sec^{-1}? The pipe is horizontal and contains four standard-radius 90° elbows and two 45° elbows. (A standard-radius 90° elbow is roughly equivalent to the resistance offered by a pipe of length 32 diameters; a 45° elbow, 15 diameters.)[1]

Answer: 4.63×10^3 psi

[1] An alternative method for calculating losses in fittings is given in §7.4.

6.B$_1$ Pressure Drop Required for a Given Flow Rate with Elevation Change

Water at 68° F is to be pumped through 95 ft of standard 3-in. pipe (internal diameter 3.068 in.) into an overhead reservoir, shown in Fig. 6.B. (*a*) What pressure is needed at the

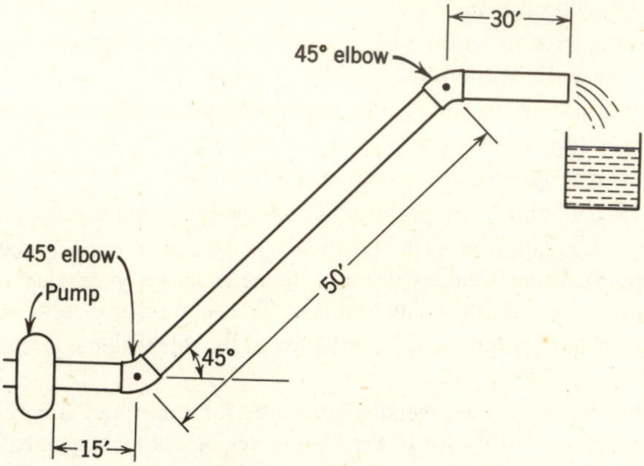

Fig. 6.B. Pipe flow system.

outlet of the pump to supply water to the overhead reservoir at a rate of 18 gal/min? (At 68° F the viscosity of water is 1.002 cp and the density is 0.9982 g ml^{-1}.) (*b*) What percentage of the pressure drop is needed for overcoming the pipe friction?

Answer: *a.* 15.2 psig

6.C$_1$ Flow Rate for a Given Pressure Drop

How many gal/hr of water at 68° F can be delivered through a 1320-ft length of smooth 6.00-in. i.d. pipe under a pressure difference of 0.25 psi? (*a*) Solve by Method *A* of Example 6.2–2; (*b*) Solve by Method *B* of Example 6.2–2; (*c*) Compare the value of $\langle v \rangle$ obtained here with that obtained in Problem 5.B, Part *e*. Assume that the pipe is "hydraulically smooth."

Answer: *a* and *b*: 4070 gal/hr

6.D$_1$ Motion of a Sphere in a Liquid

A hollow steel sphere, 5.00 mm in diameter, with a mass of 0.0500 g, is released in a column of liquid and attains a terminal velocity of 0.500 cm sec^{-1}. The liquid density is 0.900 gm cm^{-3}. The local acceleration of gravity is 980.7 cm sec^{-2}. The sphere is far enough from the containing walls so that their effect may be neglected.

a. Compute the drag force in dynes.
b. Compute the drag coefficient (friction factor).
c. Determine the viscosity of the liquid in centipoises.

Answer: *a.* 8.7 dynes
b. $f = 396$
c. $\mu = 370$ cp

6.E$_1$ Drag Calculations when Sphere Diameter is Unknown

a. How does one make drag calculations, using Fig. 6.3–1, if the sphere diameter is unknown? Show how a trial-and-error solution can be avoided.

b. Rework Problem 2.C, using Fig. 6.3–1.

c. Rework (*b*) when the gas velocity is 10 ft sec^{-1}.

6.F$_1$ Estimation of Void Fraction of a Packed Bed

A column of 146 in.^{2} cross section and 73 in. high is packed with spherical particles of diameter 2 mm. When a pressure difference of 158 psi is maintained across the bed, a 60 per cent aqueous sucrose solution at 20° C flows through the bed at a rate of 244 lb min^{-1}. At this temperature, the viscosity of the solution is $\mu = 56.5$ cp and its density is $\rho = 1.2865$ g cm^{-3}. What is the void fraction of the bed? Discuss the usefulness of this method of obtaining the void fraction. *Answer:* $\epsilon = 0.30$

6.G$_2$ Friction Factor for Flow around a Flat Plate

a. In §4.4 it is stated (just after Eq. 4.4–27) that the exact solution of the laminar boundary-layer equations leads to the following expression for the drag force on a flat plate of width W and length L, wetted on both sides:

$$F_k = 1.328\sqrt{\rho\mu L W^2 v_\infty^3} \tag{6.G–1}$$

This relation has been found to agree almost exactly with experimental measurements. Define a friction factor and a Reynolds number for the system and obtain an expression for f in terms of Re.

b. For *turbulent* flow around a flat plate an approximate[2] boundary-layer treatment based on the $\frac{1}{7}$-power velocity distribution yields, for the plate wetted on both sides,

$$F_k = 0.072\rho v_\infty^2 WL\left(\frac{Lv_\infty\rho}{\mu}\right)^{-1/5} \tag{6.G–2}$$

When 0.072 is replaced by 0.074, this relation describes the drag force within experimental error for $5 \times 10^5 < Lv_\infty\rho/\mu < 2 \times 10^7$. How does f depend on an appropriately defined Reynolds number in turbulent flow?

6.H$_2$ Friction Factor for Laminar Slit Flow

Use the results of Problem 2.E to show that for the laminar flow in a thin slit of width $2B$ the friction factor is $f = 12/\text{Re}$, if Re is defined as $2B\langle v\rangle\rho/\mu$. Compare this result for f with what one would obtain by the mean-hydraulic-radius approximation.

6.I$_2$ Friction Factor for a Rotating Disk

A thin circular disk with radius R is immersed in a large body of fluid with density ρ and viscosity μ. If a torque $\mathcal{T}$ is required to rotate the disk at an angular velocity Ω, then a friction factor f may be defined analogously to Eq. 6.1–1:

$$\frac{\mathcal{T}}{R} = AKf \tag{6.I–1}$$

where reasonable definitions for K and A are $K = \frac{1}{2}\rho R^2\Omega^2$ and $A = 2 \cdot \pi R^2$. A convenient way to define a Reynolds number for the system is $\text{Re} = R^2\Omega\rho/\mu$.

a. For *laminar* flow an exact boundary-layer calculation gives[3,4]

$$\mathcal{T} = 0.616\pi\rho R^4(\mu\Omega^3/\rho)^{1/2} \tag{6.I–2}$$

Re-express this result as a relation between f and Re.

b. For *turbulent* flow an approximate boundary-layer treatment[5] based on the $\frac{1}{7}$-power

[2] H. Schlichting, *Boundary Layer Theory*, McGraw-Hill, New York (1955), Chapter 21.
[3] T. von Kármán, *Z. angew. Math. Mech.*, **1**, 233–252 (1921).
[4] H. Schlichting, *Boundary Layer Theory*, McGraw-Hill, New York (1955), Chapter 5.
[5] H. Schlichting, *op. cit.*, Chapter 21.

velocity distribution leads to

$$\mathscr{T} = 0.073\rho\Omega^2R^5\left(\frac{\mu}{R^2\Omega\rho}\right)^{1/5} \tag{6.I-3}$$

Re-express this result as a relation between f and Re.

6.J$_2$ Friction Factor for Turbulent Flow in Smooth Tubes

Work through the derivation given in Eqs. 6.2–12 through 15 in which the *Blasius resistance law* is obtained from the $\frac{1}{7}$-power velocity profile for turbulent flow in smooth circular tubes. Then repeat this analysis for the logarithmic turbulent velocity profile:

$$v^+ = 2.5 \ln s^+ + 5.5 \tag{6.J-1}$$

in which $v^+ = \bar{v}/v_*$, $s^+ = sv_*\rho/\mu$, and $v_* = \sqrt{\tau_0/\rho}$. Eq. 6.J–1 is the same as Eq. 5.3–12, except that here we have used the numerical coefficients recommended by Schlichting.[6] Show that the foregoing velocity profile leads to this expression for the friction factor:

$$1/\sqrt{f} = 4.07 \log_{10} \mathrm{Re}\sqrt{f} - 0.60 \tag{6.J-2}$$

A good fit of the f versus Re curve from Re $= 2.1 \times 10^3$ to 5×10^6 is obtained if a slight modification in the numerical constants is made:

$$1/\sqrt{f} = 4.0 \log_{10} \mathrm{Re}\sqrt{f} - 0.40 \tag{6.J-3}$$

This is the rather famous *Prandtl* ("universal") *resistance law* for smooth circular tubes.

6.K$_2$ Friction Factor for Power-Law Non-Newtonian[7,8] Fluids in Tubes

For the non-Newtonian model defined in Eq. 1.2–3, derive an expression for the friction factor for laminar flow in tubes.

Answer: $f = \dfrac{16}{\mathrm{Re}_n'}$, where $\mathrm{Re}_n' = \dfrac{D^n\langle v\rangle^{2-n}\rho/m}{2^{n-3}\left(3 + \dfrac{1}{n}\right)^n}$

6.L$_3$ Inadequacy of Mean Hydraulic Radius for Laminar Flow

a. In §6.2 it is emphasized that the mean hydraulic radius should *not* be used for laminar flow. For laminar flow in an annulus with radii κR and R, use Eqs. 6.2–17 and 18 to derive an expression for the average velocity in terms of the pressure drop analogous to the exact expression in Eq. 2.4–15.

b. What is the per cent error in the result of (*a*) for $\kappa = \frac{1}{2}$? *Answer: b.* 47 per cent

6.M$_2$ Falling Sphere in Newton's Law Region

A sphere initially at rest at $z = 0$ falls (in the positive z-direction) under the influence of gravity. Conditions are such that within a short time interval the sphere is falling with a resisting force F_k proportional to the *square* of the velocity (i.e., Newton's drag law for spheres applies).

a. Find the distance z that the sphere has fallen as a function of time t.

b. What is the "terminal velocity" of the sphere? Assume that the density of the fluid is much less than the density of the sphere.

Answer: a. $z = (1/c^2g) \ln \cosh cgt$
where $c^2 = (3/8)(0.44)(\rho/\rho_{\mathrm{sph}})(1/gR)$
b. $1/c$

[6] H. Schlichting, *op. cit.*, pp. 368–369, 373–375.
[7] A. B. Metzner and J. C. Reed, *A.I.Ch.E. Journal*, **1**, 434–440 (1955).
[8] R. B. Bird, *A.I.Ch.E. Journal*, **2**, 428 (1956).

6.N$_3$ Power Input to an Agitated Tank

Show by dimensional analysis that the power imparted by a rotating impeller to an incompressible fluid in an agitated tank may be correlated, for any specific tank and impeller shape, by the expression

$$\left(\frac{P}{\rho N^3 D^5}\right) = \phi\left[\left(\frac{D^2 N \rho}{\mu}\right), \left(\frac{DN^2}{g}\right), (Nt)\right] \tag{6.N-1}$$

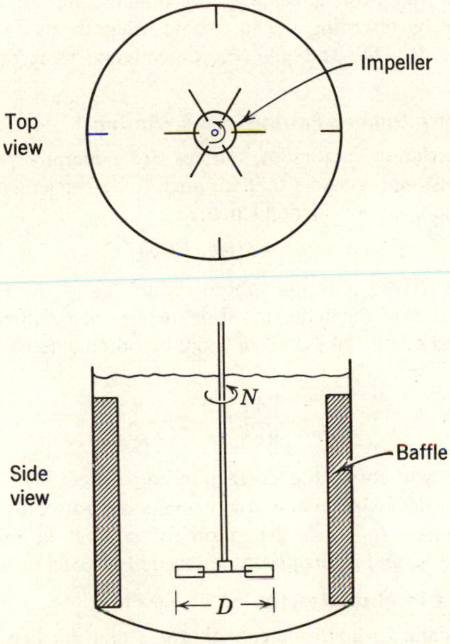

Fig. 6.N. Agitated tank with a six-bladed impeller and four vertical baffles.

where ϕ is a function whose form is to be determined experimentally.

Here P = power imparted by the impeller to the fluid (FLt^{-1})
N = rate of rotation of the impeller, (t^{-1})
D = impeller diameter (L)
g = gravitational acceleration (Lt^{-2})
ρ = liquid density (ML^{-3})
μ = liquid viscosity ($ML^{-1}t^{-1}$)
t = time since start of operation

Note: For the commonly used geometry in Fig. 6.N, the power is given by the sum of two integrals representing the contribution of friction drag of the cylindrical tank body and bottom and form drag of the radial baffles, respectively:

$$P = N \cdot \mathcal{T}$$

$$= N\left[\mu \iint_S R(\partial v_\theta / \partial n)_{\text{surf}}\, dS + \iint_A R p_{\text{surf}}\, dA\right] \tag{6.N-2}$$

Here $\mathscr{T}$ = the torque required to turn the impeller
= the resisting torque of the tank and baffle surfaces
S = total surface area of tank
A = surface area of the baffles, considered positive on the "upstream" side and negative on the "downstream" side
R = radial distance to any surface element dS or dA from the impeller axis of rotation
n = distance measured normally into the fluid from any element of tank surface, dS or dA

The desired solution may now be obtained by dimensional analysis of the equations of motion and continuity by rewriting the foregoing integrals in dimensionless form. Here it is convenient to use D, DN, and $\rho N^2 D^2$, respectively, as reference lengths, velocities, and pressures.

6.O$_3$ Wall Effects for a Sphere Falling in a Cylinder

a. Experiments on friction factors of spheres are generally performed in cylindrical tubes. Show by dimensional analysis that for such an arrangement the friction factor for the sphere will have the following dependence:

$$f = f(\text{Re}, R/R') \tag{6.O-1}$$

where $\text{Re} = 2Rv_\infty\rho/\mu$. (Here R is the sphere radius, v_∞ is the terminal velocity of the sphere, and R' is the radius of the cylinder.) For the *creeping-flow* region, it has been found empirically that the dependence of f on R/R' may be taken care of by the *Ladenburg-Faxén correction*:[9,10,11]

$$f = \frac{24}{\text{Re}}\left(1 + 2.1\,\frac{R}{R'}\right) \tag{6.O-2}$$

Little seems to be known about the corresponding corrections for the turbulent flow region.[11] Only recently have wall effects for droplets been investigated.[12]

b. Design an experiment to check the graph for spheres in Fig. 6.3-1. Select sphere sizes, cylinder dimensions, and appropriate materials for your experiments.

6.P$_4$ Friction Factors in Non-Newtonian Tube Flow

In Problem 6.K it is shown how one can obtain a relation between friction factor and Reynolds number for a power-law fluid flowing in a circular tube. This involves nothing more than a rewriting of the analytical solution for laminar tube flow. Use the expression for a power-law fluid given in Eq. 3.6–9 and the dimensional analysis method of §6.2 to determine the dimensionless groups on which f should depend.

6.Q$_3$ Non-Newtonian Flow through Porous Media

Use the line of reasoning in §6.4 to develop a relation analogous to Darcy's law (see Eq. 4.J–2), which is valid for the Ostwald-de Waele (power law) model given in §§1.2 and 3.6. Show that

$$v^n = -\frac{k}{\mu_{\text{eff}}}\nabla p \tag{6.Q-1}$$

[9] R. Ladenburg, *Ann. Physik* (4), **23,** 447 (1907); H. Faxén, dissertation, Uppsala (1921).
[10] An important (and neglected) work on particle movement in fluids is that of L. Schiller, *Handbuch der Experimentalphysik*, Vol. IV, Part 2, pp. 339–387; see also H. Falkenhagen, *op. cit.*, Vol. IV, Part 1, pp. 206–231.
[11] C. E. Lapple, *Fluid and Particle Dynamics*, University of Delaware Press, Newark (1951), Chapter 13.
[12] J. R. Strom and R. C. Kintner, *A.I.Ch.E. Journal*, **4,** 153–156 (1958).

in which μ_{eff} is given by

$$\mu_{\text{eff}} = m\left[\frac{2(25/12)^n\left(3 + \frac{1}{n}\right)^n 3^{n+1}}{150}\right]\frac{D_p^{1-n}\epsilon^{2(1-n)}}{(1-\epsilon)^{1-n}} \tag{6.Q-2}$$

Here m and n are the rheological parameters, and ϵ is the porosity of the porous medium as would be determined by letting a Newtonian fluid flow through the medium. Furthermore, k is the permeability of the medium as determined for Newtonian flow. Note that μ_{eff} does not have the units of viscosity.

6.R$_4$ Friction Factors for Turbulent Flow in Annuli

Extend the development in Problem 5.F to obtain an expression for the friction factor for turbulent flow in annuli.

CHAPTER 7

Macroscopic Balances for Isothermal Systems

In §§3.1, 2, and 3 the equations of change were presented for isothermal systems; these equations are statements of the laws of conservation of mass, conservation of momentum, and "lack of conservation" of mechanical energy, as applied to a "microscopic" volume element through which the fluid is flowing. In §§7.1, 2, and 3 we present analogous conservation statements for an arbitrary "macroscopic" flow system with one entrance and one exit for the fluid. (See Fig. 7.0–1.) Actually these macroscopic balances may be obtained by an integration of the equations of change[1] in Chapter 3. Here we adopt the simpler approach of deriving the macroscopic mass and momentum balances by paralleling the derivations in §§3.1 and 3.2. The macroscopic mechanical energy balance, on the other hand, is presented without derivation, inasmuch as simplified derivations seem to be misleading and a rigorous derivation is too lengthy to give here.[2] The subject matter of this chapter is extended to nonisothermal systems in Chapter 15 and to multicomponent systems in Chapter 22.

The macroscopic balances are used widely in the analysis of engineering

[1] R. B. Bird, *Chem. Eng. Sci.*, **6**, 123–131 (1957).

[2] The derivation for the special case of steady incompressible flow is, however, given in Example 7.3–1.

flow systems. The balances are used by discarding terms that are felt to be negligible for any given problem. Deciding what terms can be neglected requires a certain amount of intuition and, in some cases, some experimental observations of the flow behavior. The examples in §§7.4, 7.5, and 7.6 should be useful to introduce the reader to methods of problem solving.

This chapter, as the last in the series on momentum transport, makes use of many of the concepts introduced in the preceding six chapters—viscosity,

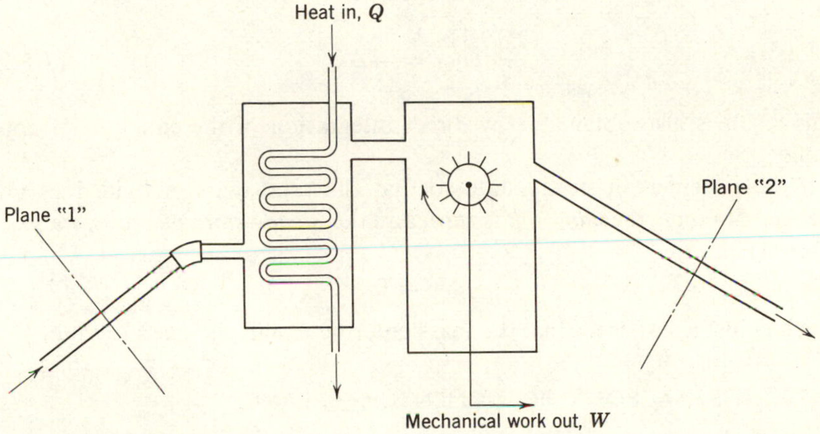

Fig. 7.0–1. Macroscopic flow system with fluid entering at plane "1" and leaving at plane "2." It may be necessary to add heat at a rate Q to maintain the system temperature constant; and the rate of doing mechanical work on the surroundings is W. Note that the sign conventions for Q and W are the same as those used in most thermodynamics texts.

velocity, profiles, laminar and turbulent flow, and friction factors. Hence this chapter provides the reader with an opportunity for reviewing while simultaneously introducing him to some useful systematic methods of solving a large class of engineering flow problems.

§7.1 THE MACROSCOPIC MASS BALANCE

We consider the system in Fig. 7.0–1 with a single fluid entrance (at plane "1" with cross section S_1) and a single exit (at plane "2" with cross section S_2). These reference planes are chosen to be perpendicular to the tube walls. In this and the following sections we make two assumptions that for most problems are not very restrictive: (*a*) at the planes "1" and "2" the time-smoothed velocity is parallel to the conduit walls; (*b*) at planes "1" and "2" the density ρ and other physical properties do not vary across the cross section.

We now apply the statement of conservation of mass in Eq. 3.1–1 to the system in Fig. 7.0–1 to obtain

$$\frac{d}{dt} m_{\text{tot}} = \rho_1 \langle \bar{v}_1 \rangle S_1 - \rho_2 \langle \bar{v}_2 \rangle S_2 \tag{7.1–1}$$

Here m_{tot} is the total mass[1] of fluid contained between planes "1" and "2." We now introduce the symbol $w = \rho\langle \bar{v} \rangle S$ for the mass rate of flow, and the notation Δw for $w_2 - w_1$ (exit value minus entrance value). The *unsteady-state macroscopic mass balance* then becomes

$$\frac{d}{dt} m_{\text{tot}} = -\Delta w \tag{7.1–2}$$

This result is also obtainable by direct integration of the equation of continuity.

If the system is in steady state so that the total mass of fluid does not change with time, then dm_{tot}/dt is zero, and the *steady-state macroscopic mass balance* is

$$\Delta w = 0 \tag{7.1–3}$$

which is just a statement that the mass entering equals the mass leaving.

§7.2 THE MACROSCOPIC MOMENTUM BALANCE

Once again we consider the system in Fig. 7.0–1 and make the same two assumptions as before. Application of the statement of conservation of momentum in Eq. 3.2–1 to this system leads to the following vector equation:

$$\begin{aligned} \frac{d}{dt} \boldsymbol{P}_{\text{tot}} = {} & \rho_1 \langle \bar{v}_1{}^2 \rangle \boldsymbol{S}_1 - \rho_2 \langle \bar{v}_2{}^2 \rangle \boldsymbol{S}_2 \\ & + \{p_1 \boldsymbol{S}_1 - p_2 \boldsymbol{S}_2\} \\ & - \{\boldsymbol{F}\} + \{m_{\text{tot}} \boldsymbol{g}\} \end{aligned} \tag{7.2–1}$$

Here $\boldsymbol{P}_{\text{tot}}$ is the total momentum[1] of the fluid in the system. The first two terms[2] on the right side represent the rate of momentum influx and efflux by virtue of the bulk-fluid motion; we have neglected the influx and efflux of momentum associated with the molecular and turbulent momentum flux components τ_{xx}, τ_{xy}, etc., on the ground that these contributions are comparatively

[1] That is, $m_{\text{tot}} = \int \rho \, dV$, the integral being taken over the volume of the fluid in the system.

[1] The total momentum is $\boldsymbol{P}_{\text{tot}} = \int \rho \bar{\boldsymbol{v}} \, dV$.

[2] By $\langle \bar{v}^n \rangle$ we mean, for example, that for circular tubes

$$\langle \bar{v}^n \rangle = \int_0^{2\pi} \int_0^R \bar{v}^n r \, dr \, d\theta \bigg/ \int_0^{2\pi} \int_0^R r \, dr \, d\theta$$

much smaller. The terms contained in the braces represent the various forces acting on the fluid in the system: the pressure forces acting at the ends of the system, the net force $-\boldsymbol{F}$ of the solid surfaces on the fluid, and the force of gravity $m_{tot}\boldsymbol{g}$ on the total mass of fluid. Keep in mind that $\boldsymbol{F}$ is the *force of the fluid on the solid* and is made up of the sum of all viscous and pressure forces.

In Eq. 7.2–1 we have written $\boldsymbol{S}_1$ and $\boldsymbol{S}_2$ in boldface to indicate that they are vectors. These are vectors with magnitudes S_1 and S_2, respectively, and with the direction of the prevailing time-smoothed velocity at sections "1" and "2." Corresponding to these vectors, we shall also introduce $\boldsymbol{w}_1 = \rho_1\langle\bar{v}_1\rangle\boldsymbol{S}_1$ and $\boldsymbol{w}_2 = \rho_2\langle\bar{v}_2\rangle\boldsymbol{S}_2$, which are defined similarly.

Using this new notation, we now rewrite Eq. 7.2–1 so that the *unsteady-state momentum balance* is

$$\frac{d}{dt}\boldsymbol{P}_{tot} = -\Delta\left(\frac{\langle\bar{v}^2\rangle}{\langle\bar{v}\rangle}\boldsymbol{w} + p\boldsymbol{S}\right) - \boldsymbol{F} + m_{tot}\boldsymbol{g} \tag{7.2–2}$$

This result is also obtainable by integration of the equation of motion in the form given in Eq. 3.2-8.

If the fluid system is at steady state, then $d\boldsymbol{P}_{tot}/dt = 0$, and Eq. 7.2–2 becomes the *steady-state momentum balance*:

$$\boldsymbol{F} = -\Delta\left(\frac{\langle\bar{v}^2\rangle}{\langle\bar{v}\rangle}\boldsymbol{w} + p\boldsymbol{S}\right) + m_{tot}\boldsymbol{g} \tag{7.2–3}$$

This result is useful for computing the forces acting on surfaces of pieces of equipment, such as turbine blades and pipe bends; it is also useful, along with the other macroscopic balances, in describing the operation of steam ejectors and other devices involving the mixing of fluid streams.

The ratio $\langle\bar{v}^2\rangle/\langle\bar{v}\rangle$ can be evaluated if the velocity profiles are known experimentally or can be calculated by the methods developed in Chapters 2, 3, and 5. For most engineering calculations involving *turbulent flow*, in which the velocity profile is nearly flat, it is permissible to replace $\langle\bar{v}^2\rangle/\langle\bar{v}\rangle$ by $\langle\bar{v}\rangle$. (See Problem 7.H.)

§7.3 THE MACROSCOPIC MECHANICAL ENERGY BALANCE (BERNOULLI EQUATION)

In §3.3 we showed how we could take the momentum conservation equation from §3.2 and transform it into an equation describing the various ways in which different forms of mechanical energy are interconverted and sometimes "lost" by irreversible conversion to thermal energy. It is, in general, not possible to perform an analogous manipulation on Eq. 7.2–2. On the other hand, it is possible[1] to integrate Eq. 3.3–2 over the volume

[1] R. B. Bird, *Chem. Eng. Sci.*, **6**, 123–131 (1957). In this development it is assumed that $\partial\hat{\Phi}/\partial t = 0$ at every point in the system.

of the system in Fig. 7.0–1 to obtain the following *unsteady-state macroscopic mechanical energy balance* for isothermal flow:

$$\frac{d}{dt}(K_{tot} + \Phi_{tot} + A_{tot}) = -\Delta\left[\left(\frac{1}{2}\frac{\langle \bar{v}^3 \rangle}{\langle \bar{v} \rangle} + \hat{\Phi} + \hat{G}\right)w\right] - W - E_v \quad (7.3\text{–}1)^2$$

Here K_{tot}, Φ_{tot}, and A_{tot} are, respectively, the total kinetic energy, potential energy, and (Helmholtz) free energy[3] in the flow system; W is the rate at which the system performs mechanical work on its surroundings, and E_v is the "friction loss," that is, the rate at which mechanical energy is irreversibly converted to thermal energy. The quantity $\hat{\Phi}$ is the potential energy per unit mass, and $\hat{G} = \hat{H} - T\hat{S}$ is the free enthalpy (or Gibbs free energy) per unit mass.

For systems in which the flow behavior is time-independent, the time derivative on the left side of Eq. 7.3–1 vanishes. Then, since $w_1 = w_2 = w$, we can divide the equation through by the mass rate of flow, introducing the notation $\hat{W} = W/w$ and $\hat{E}_v = E_v/w$. Furthermore, $\Delta\hat{G}$ may be replaced, for isothermal systems, by $\int_{p_1}^{p_2} (1/\rho)\, dp$. Hence we finally get the *steady-state macroscopic mechanical energy balance*:

$$\Delta \frac{1}{2}\frac{\langle \bar{v}^3 \rangle}{\langle \bar{v} \rangle} + \Delta\hat{\Phi} + \int_{p_1}^{p_2} \frac{1}{\rho}\, dp + \hat{W} + \hat{E}_v = 0 \quad (7.3\text{–}2)$$

This relation has sometimes been called the *Bernoulli equation*, although historically this nomenclature has been reserved for the corresponding equation for frictionless fluids with $\hat{W} = 0$ and $\hat{E}_v = 0$. Equation 7.3–2 has widespread applications, and the beginner will do well to familiarize himself thoroughly with this important relation.

Let us now see how various terms in Eq. 7.3–2 can be modified. The ratio $\langle \bar{v}^3 \rangle / \langle \bar{v} \rangle$ appearing in Eq. 7.3–2 can be calculated by the methods in Chapters 2, 3, and 5 or from experimental velocity profiles; but in most engineering problems dealing with *turbulent* flow—that is, with approximately flat velocity profiles—$\langle \bar{v}^3 \rangle / \langle \bar{v} \rangle$ may to a very good approximation be replaced by $\langle \bar{v} \rangle^2$. (See Problem 7.H.)

[2] For flow of a fluid of constant ρ, Eq. 7.3–1 becomes

$$\frac{d}{dt}(K_{tot} + \Phi_{tot}) = -\Delta\left[\left(\frac{1}{2}\frac{\langle \bar{v}^3 \rangle}{\langle \bar{v} \rangle} + \hat{\Phi} + \frac{p}{\rho}\right)w\right] - W - E_v \quad (7.3\text{–}1a)$$

[3] That is, $K_{tot} = \int \frac{1}{2}\rho\bar{v}^2\, dV$, $\Phi_{tot} = \int \rho\hat{\Phi}\, dV$, and $A_{tot} = \int \rho\hat{A}\, dV$, where all integrals are taken over the volume of the flow system; here $\hat{A} = \hat{U} - T\hat{S}$. Actually Eq. 7.3–1 is not completely correct for turbulent flow systems. The total kinetic energy should really be written $\int \frac{1}{2}\rho(\bar{v}^2 + \overline{v'^2})\, dV$ and the kinetic energy terms on the right side should be modified to account for the transport of energy by turbulent fluctuations. Because little is known about these effects and because they are generally small, we have contented ourselves with the simpler, approximate form given above.

In the usual situations, in which the gravitational acceleration is constant throughout the system, $\Delta\hat{\Phi}$ may be replaced by $g\Delta h$, where h_1 and h_2 are the elevations at planes "1" and "2," respectively.

The integral appearing in Eq. 7.3–2 can be evaluated if the equation of state is known (for isothermal situations, this means knowing ρ as a function of p, or *vice versa*). The following are two special cases:

isothermal ideal gas:
$$\int_{p_1}^{p_2} \frac{1}{\rho}\,dp = \int_{p_1}^{p_2} \frac{RT}{Mp}\,dp = \frac{RT}{M}\ln\frac{p_2}{p_1} \tag{7.3–3}$$

incompressible fluid:
$$\int_{p_1}^{p_2} \frac{1}{\rho}\,dp = \frac{1}{\rho}(p_2 - p_1) \tag{7.3–4}$$

These two limiting cases are frequently assumed in engineering work.

Example 7.3–1. Derivation of Mechanical Energy Balance for Steady Incompressible Flow

Integrate Eq. 3.3–2 to derive the mechanical energy balance for the flow in a system such as Fig. 7.0–1, except that there are no moving solid parts (hence no possibility for work being done by the fluid on the surroundings). Take the fluid to be incompressible and consider only the situation that the time-smoothed behavior is steady. Use the Gauss divergence theorem to transform volume integrals to surface integrals:

$$\int_V (\nabla \cdot \boldsymbol{A})\,dV = \int_{S_w} A_n\,dS + \int_{S_1} A_n\,dS + \int_{S_2} A_n\,dS \tag{7.3–5}$$

in which V is the volume between "1" and "2" containing the fluid. It is convenient to split up the surface of V into the wetted solid surface S_w and the cross-sectional areas at "1" and "2." Note that A_n is the *outward* normal component of $\boldsymbol{A}$.

Solution. For an incompressible fluid, Eq. 3.3–2 becomes (with $\boldsymbol{g} = -\nabla\hat{\Phi}$ and the signs changed)

$$\underset{(a)}{-\frac{\partial}{\partial t}(\tfrac{1}{2}\rho v^2)} = \underset{(b)}{(\nabla \cdot \tfrac{1}{2}\rho v^2 \boldsymbol{v})} + \underset{(c)}{(\nabla \cdot p\boldsymbol{v})} + \underset{(d)}{(\nabla \cdot [\boldsymbol{\tau} \cdot \boldsymbol{v}])}$$
$$\underset{(e)}{-\,(\boldsymbol{\tau} : \nabla \boldsymbol{v})} + \underset{(f)}{\rho(\boldsymbol{v} \cdot \nabla\hat{\Phi})} \tag{7.3–6}$$

This equation is valid for turbulent as well as laminar flow. Now we integrate, term by term, over the volume of the entire flow system:

$$(a) \qquad -\int_V \frac{\partial}{\partial t}(\tfrac{1}{2}\rho v^2)\,dV = -\frac{d}{dt}\int_V \tfrac{1}{2}\rho v^2\,dV = -\frac{d}{dt}K_{\text{tot}} = 0 \tag{7.3–7}$$

This integral vanishes because the total kinetic energy within the system does not change with time.

(b)
$$\int_V (\nabla \cdot \tfrac{1}{2}\rho v^2 \boldsymbol{v})\, dV = \int_{S_w+S_1+S_2} \tfrac{1}{2}\rho v^2 v_n\, dS$$
$$= -\tfrac{1}{2}\rho\langle v_1^3\rangle S_1 + \tfrac{1}{2}\rho\langle v_2^3\rangle S_2 \qquad (7.3\text{–}8)$$

The integral over S_w is zero because $\boldsymbol{v}$ is zero on the wetted surfaces. The one term acquires a minus sign, since the flow velocity is inwardly directed—that is, opposed to the outward normal.

(c)
$$\int_V (\nabla \cdot p\boldsymbol{v})\, dV = \int_{S_w+S_1+S_2} p v_n\, dS = -p_1\langle v_1\rangle S_1 + p_2\langle v_2\rangle S_2 \qquad (7.3\text{–}9)$$

(d)
$$\int_V (\nabla \cdot [\boldsymbol{\tau} \cdot \boldsymbol{v}])\, dV = \int_{S_w+S_1+S_2} [\boldsymbol{\tau} \cdot \boldsymbol{v}]_n\, dS \doteq 0 \qquad (7.3\text{–}10)$$

Here the integral on S_w is identically zero; the integrals on S_1 and S_2 represent the work being done by viscous forces to push fluid into or out of the system and this contribution can be safely neglected.

(e)
$$-\int_V (\boldsymbol{\tau} : \nabla \boldsymbol{v})\, dV = E_v \qquad (7.3\text{–}11)$$

No further manipulations are performed on this integral; it represents the total rate of irreversible conversion of mechanical to internal energy.

(f)
$$\rho\int_V (\boldsymbol{v} \cdot \nabla\hat{\Phi})\, dV = \rho\int_V [(\nabla \cdot \hat{\Phi}\boldsymbol{v}) - \hat{\Phi}(\nabla \cdot \boldsymbol{v})]\, dV$$
$$= \rho\int_{S_w+S_1+S_2} \hat{\Phi} v_n\, dS = -\rho\hat{\Phi}_1\langle v_1\rangle S_1 + \rho\hat{\Phi}_2\langle v_2\rangle S_2 \qquad (7.3\text{–}12)$$

Here use has been made of the fact that $(\nabla \cdot \boldsymbol{v}) = 0$ for a fluid of constant ρ.

When the foregoing results are substituted into the integrated version of Eq. 7.3–6, we get

$$0 = \Delta\tfrac{1}{2}\rho\langle v^3\rangle S + \Delta p\langle v\rangle S + E_v + \Delta\rho\hat{\Phi}\langle v\rangle S \qquad (7.3\text{–}13)$$

Division by $w = \rho\langle v\rangle S$, which is a constant for steady flow, gives

$$0 = \Delta\frac{1}{2}\frac{\langle v^3\rangle}{\langle v\rangle} + \frac{1}{\rho}\Delta p + \hat{E}_v + \Delta\hat{\Phi} \qquad (7.3\text{–}14)$$

This is a special case of Eq. 7.3–2. The $\hat{W}$ term can be obtained by allowing for the inclusion of moving solid surfaces, which permit mechanical energy to be transmitted to the surroundings.[1] (See Problem 7.R.)

§7.4 ESTIMATION OF THE FRICTION LOSS

Because of the great utility of Eq. 7.3–2 in quantitative calculations, considerable effort has been expended to develop methods for estimating the friction loss $\hat{E}_v$ in various parts of a flow system. Of course, $\hat{E}_v$ can be obtained for any system experimentally by measuring all the other terms in Eq. 7.3–2. More often than not $\hat{E}_v$ is estimated, and Eq. 7.3–2 is then used to find some other quantity, such as the work required for pumping a fluid or the exit flow velocity when the work is known. Hence this section is devoted to the theory behind the various estimation methods for $\hat{E}_v$.

In the derivation of Eq. 7.3–1 from integration of Eq. 3.3–2 the term E_v is found to be (see Eq. 7.3–11):

$$E_v = -\int (\boldsymbol{\tau} : \nabla \boldsymbol{v})\, dV \tag{7.4–1}$$

That is, E_v is just the integral of the local rate of dissipation of mechanical energy over the volume of the entire flow system. In simple laminar flow systems E_v can be calculated directly from this integral. In the usual complex flow situations encountered in engineering systems, evaluation of the friction loss integral is out of the question.

Dimensional analysis can, of course, be used to establish the general form of E_v. For example, for incompressible Newtonian flow, Eq. 7.4–1 becomes

$$E_v = +\mu \int \Phi_v \, dV \tag{7.4–2}$$

in which Φ_v is the dissipation function (with dimensions (velocity)^2/(length)^2) defined in Table 3.4–8. If we make the right side of Eq. 7.4–2 dimensionless by using a characteristic velocity v_0 and a characteristic length l_0, we get

$$E_v = (\rho v_0^3\, l_0^2\,)(\mu/l_0 v_0 \rho) \int \Phi_v{}^*\, dV^* \tag{7.4–3}$$

in which $\Phi_v{}^* = (l_0/v_0)^2 \Phi_v$ and $dV^* = d(V/l_0{}^3)$. If we make use of the dimensional arguments of §§3.7 and 6.2, we shall see that the integral in Eq. 7.4–3 will depend only on the various dimensionless groups in the equations of change and on various geometrical factors that enter in the concomitant boundary conditions. Hence, if the only significant dimensionless group is a Reynolds number, Re $= l_0 v_0 \rho/\mu$, then Eq. 7.4–3 has the general form

$$E_v = \rho v_0{}^3 l_0{}^2 \times \begin{pmatrix}\text{a dimensionless function of Re}\\ \text{and various geometrical ratios}\end{pmatrix} \tag{7.4–4}$$

In steady-state flow we work with the quantity $\hat{E}_v = E_v/w$, in which $w = \rho\langle\bar{v}\rangle S$ is the mass rate of flow passing *any* cross section. If we select the reference velocity v_0 to be $\langle\bar{v}\rangle$ and the reference length l_0 to be $\sqrt{S}$, then

$$\hat{E}_v = \tfrac{1}{2}\langle\bar{v}\rangle^2 e_v \tag{7.4–5}$$

in which e_v, the *friction loss factor*, is a function of a Reynolds number and pertinent dimensionless geometrical ratios. The factor $\frac{1}{2}$ has been introduced in keeping with the form of several related equations.

Let us now consider the friction loss in a straight conduit in order to appreciate the connection between the *friction-loss factor* e_v and the *friction factor* f, and between the *friction loss* $\hat{E}_v$ and the *force of the fluid on the solid* F. We consider only the special case of steady flow of a fluid of constant ρ in straight conduits of arbitrary but constant cross section S and length L. If the fluid is flowing under the influence of a pressure gradient and gravity (and in the direction of gravity), then Eqs. 7.2–3 and 7.3–2 become

(momentum) $$F = (p_1 - p_2)S + (\rho SL)g \tag{7.4–6}$$

(mech. energy) $$\hat{E}_v = \frac{1}{\rho}(p_1 - p_2) + gL \tag{7.4–7}$$

Multiplication of the second of these by ρS and subtracting gives

$$\hat{E}_v = \frac{F}{\rho S} \tag{7.4–8}$$

If, in addition, the fluid is in turbulent flow, then the expression for F in terms of the mean hydraulic radius R_h may be used (see Eq. 6.2–16) to get

$$\hat{E}_v = \tfrac{1}{2}\langle \bar{v}\rangle^2 \frac{L}{R_h} f \tag{7.4–9}$$

in which f is the friction factor discussed in Chapter 6. Equation 7.4–9 is of the form of Eq. 7.4–5. Comparison of these two equations shows that $e_v = (L/R_h)f$.

For straight lengths of pipe, then, Eq. 7.4–9 gives the friction loss $\hat{E}_v$. If in the flow system there are various "obstacles", such as fittings, sudden changes in diameter, valves, or flow-measuring devices, additional contributions to $\hat{E}_v$ must be included. These additional resistances may be written in the form of Eq. 7.4–5, with e_v determined by one of two methods: (*a*) simultaneous solution of the macroscopic balances or (*b*) experimental measurement. Some rough values of e_v are tabulated in Table 7.4–1 for the convention that $\langle \bar{v}\rangle$ is the average velocity *downstream* from the disturbance; these values are for turbulent flow for which the Reynolds number dependence is not too important.

On the basis of the foregoing discussion, we now rewrite Eq. 7.3–2 in the *approximate* form frequently used for turbulent flow calculations in a system composed of various kinds of piping and additional resistances:

$$\Delta\tfrac{1}{2}\langle \bar{v}\rangle^2 + g\,\Delta h + \int_{p_1}^{p_2} \frac{1}{\rho}\,dp + \hat{W} + \underbrace{\Sigma_i\left(\tfrac{1}{2}\langle \bar{v}\rangle^2 \frac{L}{R_h} f\right)_i}_{\text{sum on all sections of straight conduits}} + \underbrace{\Sigma_i(\tfrac{1}{2}\langle \bar{v}\rangle^2 e_v)_i}_{\text{sum on all fittings, valves, meters, etc.}} = 0 \tag{7.4–10}$$

Here R_h is the mean hydraulic radius defined in Eq. 6.2–16, f is the friction factor defined in Eq. 6.1–4, and e_v is the friction loss factor given in Table 7.4–1. Note that the $\langle \bar{v} \rangle$'s in the first terms refer to the average velocities at the planes "1" and "2"; the $\langle \bar{v} \rangle$ in the first sum indicates the average velocity in the ith pipe segment; and the $\langle \bar{v} \rangle$ in the second sum is the average flow velocity *downstream* from the ith fitting, valve, or other obstacle.

TABLE 7.4–1

BRIEF SUMMARY OF FRICTION LOSS FACTORS FOR USE WITH EQ. 7.4–10.

(Approximate Values for Turbulent Flow)[a]

Obstacles	e_v
Sudden Changes in Cross-Sectional Area[b]	
Rounded entrance to pipe	0.05
Sudden contraction	$0.45(1 - \beta)$
Sudden expansion[c]	$\left(\frac{1}{\beta} - 1\right)^2$
Orifice (sharp-edged)	$2.7(1 - \beta)(1 - \beta^2)\frac{1}{\beta^2}$
Fittings and Valves	
90° elbows (rounded)	0.4–0.9
90° elbows (square)	1.3–1.9
45° elbows	0.3–0.4
Globe valve (open)	6–10
Gate valve (open)	0.2

[a] Taken from H. Kramers, *Physische Transportverschijnselen*, Technische Hogeschool, Delft, Holland (1958), pp. 53–54.

[b] β = (smaller cross sectional area)/(larger cross sectional area)

[c] See derivation from the macroscopic balances in Example 7.5–1. When $\beta = 0$, $\hat{E}_v = \frac{1}{2}\langle \bar{v} \rangle^2$ where $\langle \bar{v} \rangle$ is the velocity *upstream* from the enlargement.

Example 7.4–1. Power Requirements for Pipe-Line Flow

What is the horsepower needed to pump the water in the system shown in Fig. 7.4–1? Water ($\rho = 62.4\ \text{lb}_m\ \text{ft}^{-3}$; $\mu = 1.0$ cp) is to be delivered to the upper tank at a rate of 12 ft^3 min^{-1}. All of the piping is 4-in. internal diameter smooth circular pipe.

Solution. The average velocity in the pipe is

$$\langle v \rangle = \frac{Q}{\pi R^2} = \frac{(12/60)}{\pi(1/6)^2} = 2.30\ \text{ft sec}^{-1}$$

and the Reynolds number is

$$\text{Re} = \frac{D\langle v\rangle\rho}{\mu} = \frac{(1/3)(2.30)(62.4)}{(1.0 \times 6.72 \times 10^{-4})} = 7.11 \times 10^4$$

Hence the flow is *turbulent*.

The contribution to $\hat{E}_v$ from the various lengths of pipe will be

$$\Sigma_i\left(\tfrac{1}{2}\langle v\rangle^2 \frac{L}{D} 4f\right)_i = \frac{2\langle v\rangle^2 f}{D}\Sigma_i L_i$$

$$= \frac{2(2.30)^2(0.0049)}{(1/3)}(5 + 300 + 120 + 80 + 40)$$

$$= (0.156)(545) = 85 \text{ ft}^2 \text{ sec}^{-2}$$

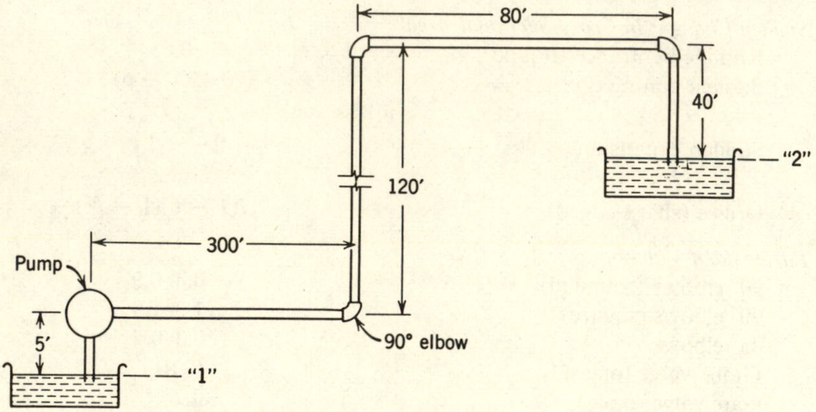

Fig. 7.4-1. Pipe-line flow with friction losses because of fittings.

The contribution to $\hat{E}_v$ from the sudden contraction, the three 90° elbows, and the sudden expansion (see Table 7.4–1) will be

$$\Sigma_i(\tfrac{1}{2}\langle v\rangle^2 e_v)_i = \tfrac{1}{2}(2.30)^2(0.45 + 3 \cdot \tfrac{1}{2} + 1)$$

$$= 8 \text{ ft}^2 \text{ sec}^{-2}$$

Then from Eq. 7.4–10 we get

$$0 + (32.2)(125 - 40) + 0 + \hat{W} + 85 + 8 = 0$$

Solving for $\hat{W}$, we get

$$\hat{W} = -2740 - 85 - 8 \doteq -2830 \text{ ft}^2 \text{ sec}^{-2}$$

This is the work (per unit mass) done *by* the fluid *in* the pump. Hence the pump does $+2830$ ft^2 sec^{-2} or $2830/32.2 = +88$ ft lb$_f$/lb$_m$ of work on the fluid passing through the system. The mass rate of flow is

$$w = Q\rho = (12/60)(62.4) = 12.5 \text{ lb}_m \text{ sec}^{-1}$$

Consequently

$$-W = -w\hat{W} = (12.5)(88) = 1100 \text{ ft lb}_f \text{ sec}^{-1}$$
$$= 2 \text{ hp}.$$

Hence 2 hp is the power delivered by the pump.

§7.5 USE OF THE MACROSCOPIC BALANCES TO SET UP STEADY FLOW PROBLEMS

In §3.5 it was shown how to set up the differential equations to calculate the velocity and pressure profiles for isothermal flow systems by simplifying the equations of change. In this section we do the same thing with the macroscopic balances in order to set up the algebraic equations describing the entrance and exit properties of the fluid in the system pictured in Fig. 7.0–1. This section is restricted to *steady* isothermal systems.

For each problem we write down a special form of each of the three macroscopic balances in §§7.1, 2, 3. By keeping track of the discarded terms, we automatically have a complete listing of the assumptions inherent in the final result. All of the illustrative examples given here will be for incompressible liquids; problems for compressible fluids frequently involve appreciable heat effects, hence are postponed to Chapter 15.

Example 7.5–1. Pressure Rise and Friction Loss in a Sudden Enlargement

An incompressible fluid flows turbulently in a circular tube of cross-sectional area S_1, which empties into a larger tube of cross-sectional area S_2, as shown in Fig. 7.5–1. Use the macroscopic balances to get an expression for the pressure change between "1" and "2" and for the friction loss associated with the sudden expansion. For the sake of brevity, let $v_1 = \langle \bar{v}_1 \rangle$ and $v_2 = \langle \bar{v}_2 \rangle$. The velocity profiles at "1" and "2" may be assumed flat.

Solution.

a. Mass Balance. For steady flow from Eq. 7.1–3,

$$w_1 = w_2 \quad \text{or} \quad \rho_1 v_1 S_1 = \rho_2 v_2 S_2 \tag{7.5–1}$$

For a fluid of constant ρ, this becomes (with $\beta = S_1/S_2$):

$$\frac{v_1}{v_2} = \frac{1}{\beta} \tag{7.5–2}$$

b. Momentum Balance. The component of the momentum balance (Eq. 7.2–3) in the direction of flow is

$$F = w_1 v_1 - w_2 v_2 + p_1 S_1 - p_2 S_2 \tag{7.5–3}$$

The force F is made up of two parts: the viscous force on the cylindrical surfaces parallel to the direction of flow and the pressure force on the "washer-shaped" surface just to the right of "1" and perpendicular to the flow axis. The former contribution we neglect, and the latter we take to be $p_1(S_2 - S_1)$ by assuming that the pressure on the washer-shaped surface is the same as that at "1." We then get

$$-p_1(S_2 - S_1) = \rho v_2 S_2(v_1 - v_2) + p_1 S_1 - p_2 S_2 \tag{7.5–4}$$

Solving for the pressure difference gives

$$p_2 - p_1 = \rho v_2(v_1 - v_2) \tag{7.5–5}$$

or, in terms of the downstream velocity,

$$p_2 - p_1 = \rho v_2^2\left(\frac{1}{\beta} - 1\right) \tag{7.5–6}$$

Note that the momentum balance predicts (correctly) a *rise* in the pressure.

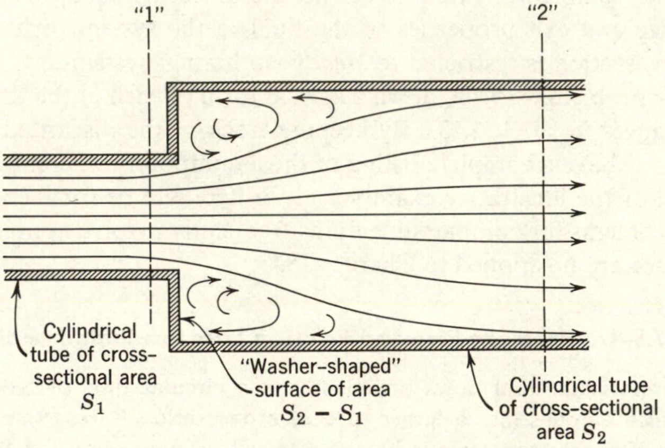

Fig. 7.5–1. Flow in a sudden enlargement.

c. Mechanical Energy Balance. Equation 7.3–2 for this system is

$$\tfrac{1}{2}(v_2^2 - v_1^2) + \frac{1}{\rho}(p_2 - p_1) + \hat{E}_v = 0 \tag{7.5–7}$$

Insertion of Eq. 7.5–6 for the pressure rise then gives after some simplification

$$\hat{E}_v = \tfrac{1}{2}v_2^2\left(\frac{1}{\beta} - 1\right)^2 \tag{7.5–8}$$

which is an entry in Table 7.4–1. This example has shown how the macroscopic balances can be used to obtain an expression for the friction loss in a simple obstacle in a flow system. At low Reynolds numbers the assumption of flat velocity profiles can lead to appreciable error. (See Problem 7.I.)

Example 7.5–2. Performance of a Liquid-Liquid Ejector

A schematic diagram of a liquid-liquid ejector device is shown in Fig. 7.5–2. It is desired to analyze the mixing of the two streams (of the same fluid) by means of the over-all balances. At plane "1" two fluid streams merge, one with velocity v_0, with cross-sectional area $\frac{1}{3}S_1$, and the other with velocity $v_0/2$ and cross-sectional area $\frac{2}{3}S_1$. Plane "2" is chosen far enough downstream that the two streams have mixed and the velocity is almost uniform at v_2. The flow is turbulent and the velocity profiles are assumed to be completely flat. In the following analysis $\boldsymbol{F}$ is neglected.

Solution.

a. Mass Balance. At steady state

$$w_1 = w_2 \quad \text{or} \quad \rho_1\langle\bar{v}_1\rangle S_1 = \rho_2\langle\bar{v}_2\rangle S_2 \tag{7.5–9}$$

For an incompressible fluid $\rho_1 = \rho_2 = \rho$; from the geometry $S_1 = S_2 = S$, hence

$$\langle\bar{v}_1\rangle = \langle\bar{v}_2\rangle = \tfrac{1}{3}v_0 + \tfrac{2}{3}(\tfrac{1}{2}v_0) = \tfrac{2}{3}v_0 \tag{7.5–10}$$

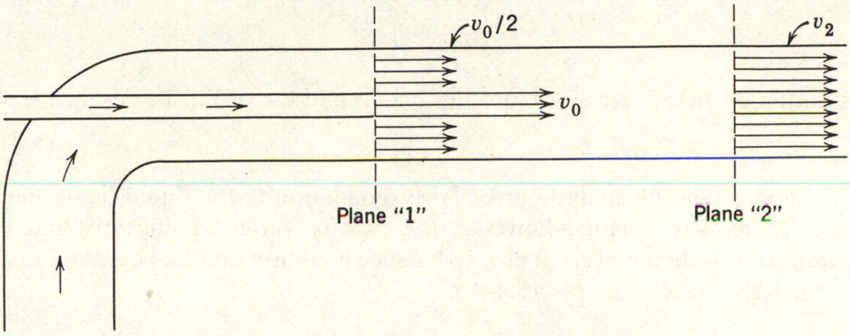

Fig. 7.5-2. Liquid-liquid ejector pump.

b. Momentum Balance. When $\boldsymbol{F}$ is neglected, the component of the momentum balance in the direction of the flow axis is

$$w_1\frac{\langle\bar{v}_1^2\rangle}{\langle\bar{v}_1\rangle} - w_2\frac{\langle\bar{v}_2^2\rangle}{\langle\bar{v}_2\rangle} + p_1S_1 - p_2S_2 = 0 \tag{7.5–11}$$

The ratios of averages are (assuming flat profiles)

$$\frac{\langle\bar{v}_2^2\rangle}{\langle\bar{v}_2\rangle} = \langle\bar{v}_2\rangle = \tfrac{2}{3}v_0 \tag{7.5–12}$$

$$\frac{\langle\bar{v}_1^2\rangle}{\langle\bar{v}_1\rangle} = \frac{\frac{1}{3}\,v_0^2 + \frac{2}{3}\left(\frac{v_0}{2}\right)^2}{\frac{1}{3}\,v_0 + \frac{2}{3}\left(\frac{v_0}{2}\right)} = \tfrac{3}{4}v_0 \tag{7.5–13}$$

Substituting these expressions into Eq. 7.5–11 and utilizing the result from (*a*) that $w_1 = w_2 = \frac{2}{3}\rho v_0 S$, we get, after some rearrangement,

$$p_2 - p_1 = \tfrac{1}{18}\rho v_0^2 \tag{7.5–14}$$

This is the expression for the pressure *increase* resulting from the mixing of the two streams.

c. Mechanical Energy Balance. When $\hat{W}$ and $\Delta\hat{\Phi}$ are omitted, Eq. 7.3–2 becomes, for an incompressible fluid,

$$\frac{1}{2}\frac{\langle \bar{v}_2^3 \rangle}{\langle \bar{v}_2 \rangle} - \frac{1}{2}\frac{\langle \bar{v}_1^3 \rangle}{\langle \bar{v}_1 \rangle} + \frac{1}{\rho}(p_2 - p_1) + \hat{E}_v = 0 \tag{7.5–15}$$

The ratios of averages are computed as before:

$$\frac{1}{2}\frac{\langle \bar{v}_2^3 \rangle}{\langle \bar{v}_2 \rangle} = \frac{1}{2}\langle \bar{v}_2 \rangle^2 = \frac{1}{2}\left(\frac{2}{3}v_0\right)^2 = \tfrac{2}{9}v_0^2 \tag{7.5–16}$$

$$\frac{1}{2}\frac{\langle \bar{v}_1^3 \rangle}{\langle \bar{v}_1 \rangle} = \frac{1}{2}\frac{\frac{1}{3}v_0^3 + \frac{2}{3}\left(\frac{v_0}{2}\right)^3}{\frac{1}{3}v_0 + \frac{2}{3}\left(\frac{v_0}{2}\right)} = \tfrac{5}{16}v_0^2 \tag{7.5–17}$$

Substitution of these expressions into Eq. 7.5–15 and use of Eq. 7.5–14 then lead to

$$\hat{E}_v = \tfrac{5}{144}v_0^2 \tag{7.5–18}$$

The foregoing type of analysis gives fairly good results for liquid-liquid ejector pumps. In gas-gas ejectors, however, the density varies significantly and it is necessary to include the over-all energy-balance equation and an equation of state in the analysis. (See Example 15.4–4.)

Example 7.5–3. Thrust on a Pipe Bend

Water at 95° C is flowing at a rate of 2.0 ft^3 sec^{-1} around a 60° bend, in which there is a contraction from 4 to 3 in. internal diameter. (See Fig. 7.5–3.) Compute the force exerted on the bend if the pressure at the downstream end is 1.1 atm.

Solution. The Reynolds number for the flow in the 3-in. pipe is (with $\rho = 0.962$ g cm^{-3} and $\mu = 0.299$ cp)

$$\begin{aligned} \text{Re} &= \frac{D\langle v \rangle \rho}{\mu} = \frac{4Q\rho}{\pi D \mu} \\ &= \frac{4(2.0 \times 2.83 \times 10^4)(0.962)}{\pi(3 \times 2.54)(0.00299)} \\ &= 3.0 \times 10^6 \end{aligned}$$

Hence the flow is highly turbulent. We therefore assume that the velocity profiles are almost flat, and we abbreviate $\langle \bar{v}_1 \rangle$ by v_1, $\langle \bar{v}_1^2 \rangle$ by v_1^2, etc. The three balances are applied as follows:

a. Mass Balance. For steady-state flow, $w_1 = w_2$; and when the density of the fluid does not change the mass balance gives

$$\frac{v_1}{v_2} = \frac{S_2}{S_1} \equiv \beta \tag{7.5–19}$$

in which β is the ratio of the smaller to the larger cross section.

b. Mechanical Energy Balance. For steady incompressible flow, Eq. 7.3–2 becomes, for this problem ($\hat{W}$ and $\Delta\hat{\Phi}$ being discarded),

$$\tfrac{1}{2}v_2^2 - \tfrac{1}{2}v_1^2 + \frac{1}{\rho}(p_2 - p_1) + \hat{E}_v = 0 \tag{7.5–20}$$

According to Table 7.4–1 and Eq. 7.4–5, we can take the friction loss as approximately $\frac{1}{2}v_2^2 \cdot \frac{2}{5}$ or $\frac{1}{5}v_2^2$. Inserting this into Eq. 7.5–20 and using the mass balance, we get

$$p_1 - p_2 = \rho v_2^2(\tfrac{1}{2} - \tfrac{1}{2}\beta^2 + \tfrac{1}{5}) \tag{7.5–21}$$

This is the pressure drop through the bend in terms of the known velocity v_2 and the known geometrical factor β.

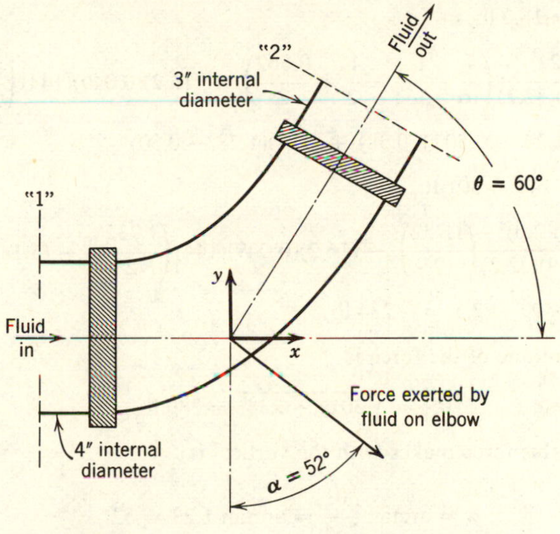

Fig. 7.5-3. Reaction force at a reducing bend in a pipe.

c. Momentum Balance. We now have to consider both the x- and y-component of the momentum balance; for the x-component, we have

$$\begin{aligned} F_x &= w_{1x}v_1 - w_{2x}v_2 + p_1S_{1x} - p_2S_{2x} \\ &= (\rho v_1 S_1)v_1 - (\rho v_2 S_2 \cos\theta)v_2 + p_1S_1 - p_2S_2\cos\theta \\ &= \rho v_2^2 S_2(\beta - \cos\theta) + (p_1 - p_2)S_1 + p_2(S_1 - S_2\cos\theta) \end{aligned} \tag{7.5–22}$$

Insertion of Eq. 7.5–21 then gives

$$\begin{aligned} F_x &= \rho v_2^2 S_2(\beta - \cos\theta) + \rho v_2^2 S_2 \beta^{-1}(\tfrac{7}{10} - \tfrac{1}{2}\beta^2) + p_2S_2(\beta^{-1} - \cos\theta) \\ &= \rho Q^2 S_2^{-1}(\tfrac{7}{10}\beta^{-1} - \cos\theta + \tfrac{1}{2}\beta) + p_2S_2(\beta^{-1} - \cos\theta) \end{aligned} \tag{7.5–23}$$

For the y-component, we have, with $w_{1y} = 0$ and $S_{1y} = 0$,

$$\begin{aligned} F_y &= -w_{2y}v_2 - p_2S_{2y} - m_{\text{tot}}g \\ &= -\rho Q^2 S_2^{-1} \sin\theta - p_2 S_2 \sin\theta - \pi R^2 L \rho g \end{aligned} \tag{7.5–24}$$

in which R and L are the radius and length of a roughly equivalent cylinder.

We have thus far obtained the components of the reaction force in terms of known quantities; it remains to insert the numerical values:

$$\begin{aligned} \rho &= 60 \text{ lb}_m \text{ ft}^{-3} & S_2 &= \pi/64 = 0.049 \text{ ft}^2 \\ Q &= 2.0 \text{ ft}^3 \text{ sec}^{-1} & \beta &= S_2/S_1 = 3^2/4^2 = 0.562 \\ \cos\theta &= \tfrac{1}{2} & R &\doteq (\tfrac{1}{8}) \text{ ft} \\ \sin\theta &= \sqrt{3}/2 & L &\doteq (\tfrac{5}{6}) \text{ ft} \\ p_2 &= 16.2 \text{ lb}_f \text{ in}^{-2} \end{aligned}$$

$$\begin{aligned} F_x &= \frac{(60)(2.0)^2}{(0.049)(32.2)}\left(\frac{7}{10}\frac{1}{0.562} - \frac{1}{2} + \frac{0.562}{2}\right) + (16.2)(0.049)(144)\left(\frac{1}{0.562} - \frac{1}{2}\right) \\ &= (152)(1.24 - 0.50 + 0.28) + (114)(1.78 - 0.50) \\ &= 155 + 146 = 301 \text{ lb}_f \end{aligned} \tag{7.5–25}$$

$$\begin{aligned} F_y &= -\frac{(60)(2.0)^2}{(0.049)(32.2)}\left(\frac{1.732}{2}\right) - (16.2)(0.049)(144)\left(\frac{1.732}{2}\right) - (3.142)(\tfrac{1}{8})^2(\tfrac{5}{6})(60) \\ &= -132 - 99 - 2.5 = -234 \text{ lb}_f \end{aligned} \tag{7.5–26}$$

Hence the magnitude of the force is

$$|F| = \sqrt{301^2 + 234^2} = 380 \text{ lb}_f \tag{7.5–27}$$

The angle that this force makes with the vertical is

$$\alpha = \arctan\frac{301}{234} = \arctan 1.29 = 52° \tag{7.5–28}$$

In looking back over the calculation, we see that all the effects we have included are important, with the possible exception of the 2.5 lb_f force associated with the mass of the fluid.

Example 7.5–4. Isothermal Flow of a Liquid through an Orifice

A common method for determining the mass rate of flow through a pipe is that in which the pressure drop is measured across some "obstacle" in the pipe. In an orifice the obstacle is a thin plate with a hole in the middle; as shown in Fig. 7.5–4. Here we derive a formula for the flow rate for the orifice, the other flow meters being handled similarly. Apply the mass and mechanical energy balances to the region between planes "1" and "2," located at the two pressure taps.

Solution.

a. Mass Balance. For a fluid of constant density with a system for which $S_1 = S_2 = S$, the mass balance leads to

$$\langle v_1 \rangle = \langle v_2 \rangle = \langle v \rangle \tag{7.5–29}$$

b. Mechanical Energy Balance. For a fluid of constant ρ with no work effects and no potential energy changes, we have

$$\frac{\langle v_2 \rangle^2}{2\alpha_2} - \frac{\langle v_1 \rangle^2}{2\alpha_1} + \frac{p_2 - p_1}{\rho} + \frac{1}{2}\langle v_2 \rangle^2 e_v = 0 \tag{7.5–30}$$

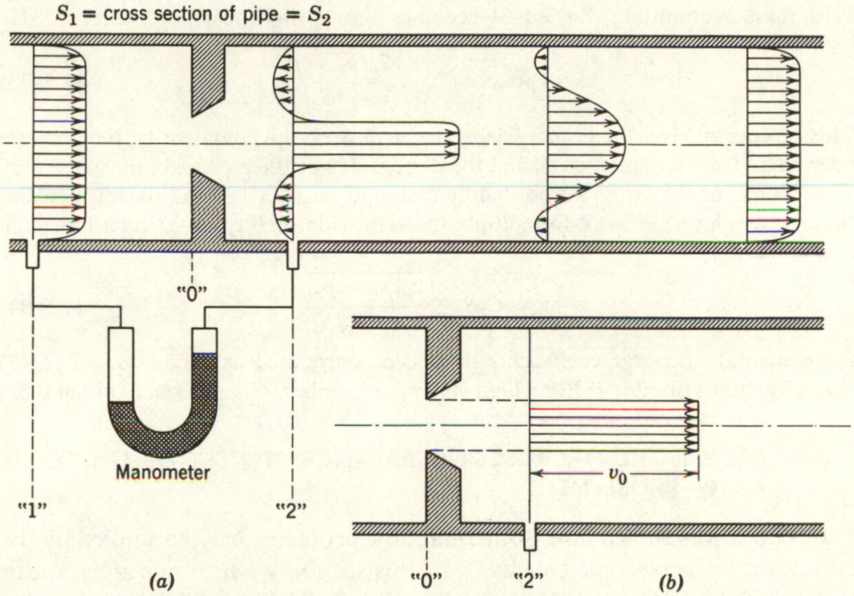

Fig. 7.5–4. (*a*) A sharp-edged orifice, showing the approximate velocity profiles at several planes near the orifice plate. Note that the fluid jet emerging from the hole is somewhat smaller than the hole itself; in highly turbulent flow this jet necks down to a minimum cross section at the *vena contracta*. Note that there is some back flow near the wall. The extent of this necking down can be estimated from ideal flow theory to be C_c = "contraction coefficient" = $(S_{vena\ contracta}/S_0) = \pi/(\pi + 2) = 0.611$. [See H. Lamb, *Hydrodynamics*, Dover, First American Edition (1945), p. 99.] (*b*) Approximate velocity profile at "2" used to calculate α_2.

Here the α's are factors introduced to account for the difference between $\langle v \rangle^2$ and $\langle v^3 \rangle / \langle v \rangle$. By combining Eqs. 7.5–29 and 30, we can get an expression for $\langle v \rangle$; when the latter is multiplied by ρS, we get

$$w = \rho \langle v \rangle S = \rho S \sqrt{\frac{2(p_1 - p_2)/\rho}{\dfrac{1}{\alpha_2} - \dfrac{1}{\alpha_1} + e_v}} \tag{7.5–31}$$

This gives the mass rate of flow in terms of the pressure drop. Because of the complicated flow patterns, α_1, α_2, and e_v may be obtained only approximately:

i. It is assumed that e_v is zero.

ii. It is assumed that the velocity profile at "1" is flat so that $\alpha_1 = 1$.

iii. It is assumed that the velocity profile at "2" is given by the approximate profile in Fig. 7.5–4b, so that $v_0 = \langle v \rangle (S/S_0)$ and

$$\frac{1}{\alpha_2} = \frac{\langle v_2^3 \rangle}{\langle v_2 \rangle^3} = \frac{S_0[\langle v \rangle (S/S_0)]^3 + (S - S_0)(0)}{S} \cdot \frac{1}{\langle v \rangle^3}$$

$$= \left(\frac{S}{S_0}\right)^2 \tag{7.5–32}$$

With these assumptions, Eq. 7.5–31 becomes after a slight rearrangement

$$w = \rho \langle v \rangle S = S_0 \sqrt{\frac{2\rho(p_1 - p_2)}{1 - (S_0/S)^2}} \tag{7.5–33}$$

This expression gives the primary dependence of w on the fluid density, the pressure drop (read from a manometer), and the dimensions of the pipe and orifice opening. To account for the errors introduced by assumptions (i), (ii), and (iii), it is conventional in engineering work to multiply the right side of Eq. 7.5–33 by a *discharge coefficient* C_d:

$$w = C_d S_0 \sqrt{\frac{2\rho(p_1 - p_2)}{1 - (S_0/S)^2}} \tag{7.5–34}$$

Experimental discharge coefficients have been correlated as a function of (S_0/S) and a Reynolds number.[1] For a high Reynolds number, C_d approaches about 0.61.

§7.6 USE OF THE MACROSCOPIC BALANCES TO SET UP UNSTEADY FLOW PROBLEMS

In §7.5 it was shown how isothermal flow problems may be studied by the steady-state macroscopic balances. In this section we turn our attention to the use of the unsteady balances in Eqs. 7.1–2, 7.2–2, and 7.3–1 to describe the time-dependent behavior of macroscopic systems. Although the engineering literature is full of examples of applications of the steady-state balances, only a limited amount of attention has been paid to analysis based on the unsteady-state balances. Two unsteady flow problems are presented here to show how to set up and solve problems of this kind. Further illustrations of the use of the unsteady macroscopic balances may be found in Problems 7.L and 7.P.

Example 7.6–1. Efflux Time for Flow from a Funnel

A conical funnel with dimensions shown in Fig. 7.6–1 is initially filled with a liquid. The fluid is allowed to drain out by gravity. Determine the efflux time.

[1] G. L. Tuve and R. E. Sprenkle, *Instruments*, **6**, 201–206, **8**, 202–205, 225, 232–234 (1935); J. H. Perry, *Chemical Engineers Handbook*, McGraw-Hill, New York (1950), p. 405.

Solution. From the geometry of the problem, it is evident that

$$\frac{r_0}{z_0} = \frac{r}{z} = \frac{\bar{r}}{\bar{z}} = \frac{r_2}{z_2} \tag{7.6–1}$$

Also, the corresponding fluid velocities are expressible by means of mass balances, in terms of the elevation of the liquid surface z:

$$v(\bar{z}) = \text{velocity at any plane in liquid} = -\frac{z^2}{\bar{z}^2}\frac{dz}{dt} \tag{7.6–2}$$

$$v_2 = \text{efflux velocity} = -\frac{z^2}{z_2^{\,2}}\frac{dz}{dt} \tag{7.6–3}$$

a. Mechanical Energy Balance. We define our system to be the fluid above the plane $z = z_2$ at any time t. In setting up the mechanical energy balance, we specify

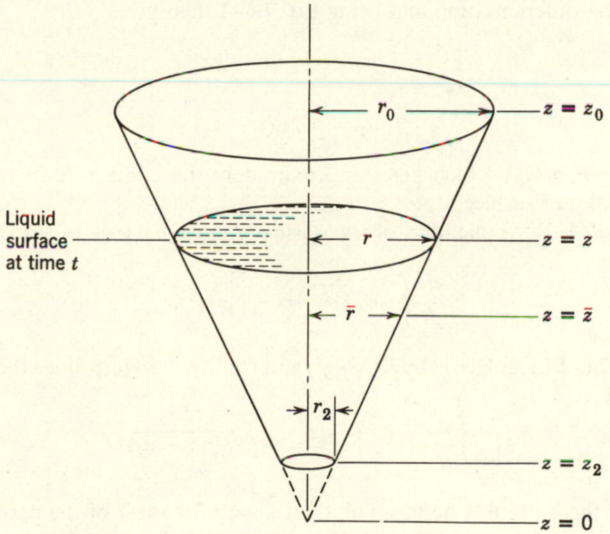

Fig. 7.6–1. Conical container from which a fluid is allowed to drain. The quantity r is the radius of the liquid surface at height z; $\bar{r}$ is the radius of the cone at any arbitrary height $\bar{z}$.

the following: (i) There is no plane "1" for this system; that is, there is no "inlet stream", and w_1 is therefore zero. (ii) Plane "2" is taken to be the outlet at $z = z_2$, with mass flow rate w_2. (iii) The free liquid surface is regarded as a moving surface capable of exchanging mechanical energy with the surroundings. Since the free surface and the emerging jet are open to the atmosphere, the pressure p is the same at $z = z$ and $z = z_2$. We wish to analyze the system by using the constant-density equation given in Eq. 7.3–1a:

$$\frac{d}{dt}(K_{\text{tot}} + \Phi_{\text{tot}}) = -\Delta[(\tfrac{1}{2}v^2 + gz)w] - \Delta(p\langle v\rangle S) - W - E_v \tag{7.6–4}$$

The rate at which the surroundings do work on the system is the force on the free surface multiplied by the velocity of the movement of the surface; that is, $-W = p(\pi r^2)(-dz/dt)$. The term $-\Delta(p\langle v\rangle S)$ becomes $-pv_2S_2 = -p(\pi r_2^2)(z^2/z_2^2)(-dz/dt)$. This contribution exactly cancels $-W$, as can be seen by using Eqs. 7.6–1 and 7.6–3. The viscous dissipation E_v is very small and is here taken to be zero. In addition, the kinetic energy term is neglected[1] on the left side. Equation 7.6–4 then becomes

$$\frac{d}{dt}\int_{z_2}^{z} (\rho g\bar{z})\pi\bar{r}^2\, d\bar{z} = -\tfrac{1}{2}v_2^2 w_2 - gz_2 w_2 \tag{7.6–5}$$

Dividing by $w_2 = \rho v_2 \pi r_2^2$ and using Eq. 7.6–1 to replace $\bar{r}^2$ by $\bar{z}^2(r_2^2/z_2^2)$, we get

$$\frac{g}{v_2 z_2^2}\frac{d}{dt}\left(\frac{z^4 - z_2^4}{4}\right) = -\tfrac{1}{2}v_2^2 - gz_2 \tag{7.6–6}$$

Performing the differentiation and using Eq. 7.6–3 then gives

$$-gz = -\tfrac{1}{2}v_2^2 - gz_2 \tag{7.6–7}$$

whence

$$v_2 = \sqrt{2g(z - z_2)} \tag{7.6–8}$$

This is *Torricelli's law*, which gives approximately the efflux rate in terms of the height of the liquid surface.

b. Mass Balance. Application of the unsteady mass balance of Eq. 7.1–2 gives

$$\frac{d}{dt}\int_{z_2}^{z} \rho\pi\bar{r}^2\, d\bar{z} = -\rho v_2 \pi r_2^2 \tag{7.6–9}$$

By using Eq. 7.6–1 to replace $\bar{r}^2$ by $\bar{z}^2(r_2^2/z_2^2)$ and Eq. 7.6–8 to introduce the expression for v_2, we get

$$\left(\frac{1}{z_2}\right)^2 \frac{d}{dt}\frac{1}{3}(z^3 - z_2^3) = -\sqrt{2g(z - z_2)} \tag{7.6–10}$$

If the hole at the bottom is quite small, then $z_2 \ll z$ for most of the period of efflux and $2g(z - z_2)$ can be approximated by $2gz$. Hence Eq. 7.6–10 (after performing the time differentiation) becomes

$$z^{3/2}\frac{dz}{dt} = -z_2^2\sqrt{2g} \tag{7.6–11}$$

Integration with the initial condition that $z = z_0$ at $t = 0$ gives

$$z_0^{5/2} - z^{5/2} = \tfrac{5}{2}z_2^2\sqrt{2g}\, t \tag{7.6–12}$$

The total efflux time is given by setting $z = z_2 \approx 0$:

$$t_{\text{efflux}} = \frac{1}{5}\left(\frac{z_0}{z_2}\right)^2\sqrt{\frac{2z_0}{g}} \tag{7.6–13}$$

[1] See Problem 7.P for an example in which the kinetic energy term is included.

This result has been reached after making quite a few assumptions. The neglect of the kinetic energy term on the left side is particularly serious when the fluid level is low. Furthermore, the entire treatment has assumed no vortex motion in the draining funnel.

Example 7.6–2. Oscillations of a Damped Manometer

The fluid in a U-tube manometer, initially at rest, is set in motion by suddenly imposing a pressure difference $p_a - p_b$. Determine the differential equation for the motion of the manometer fluid, assuming isothermal, incompressible flow. Obtain

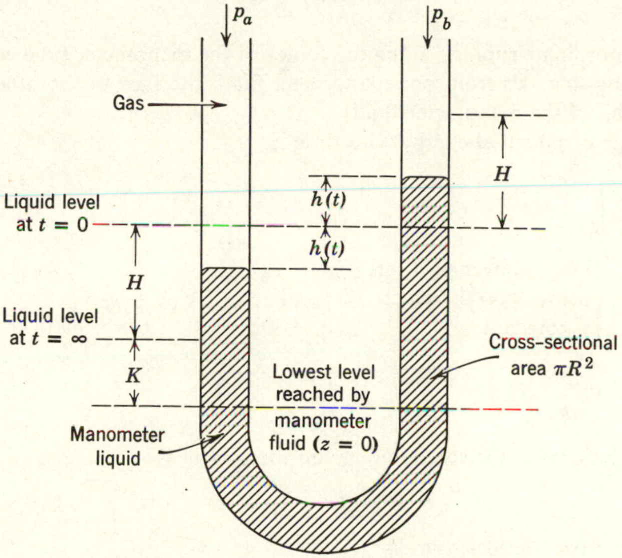

Fig. 7.6–2. Damped oscillation of a manometer fluid.

an expression for the tube radius for which "critical damping" occurs. Neglect the density of the gas above the manometer liquid. The notation is summarized in Fig. 7.6–2.

Solution. As the system we select the manometric fluid. We recognize that there are no planes "1" and "2" in this system through which fluid enters and leaves. The free liquid surfaces are capable of performing work on the surroundings. We apply the mechanical energy balance in the form given in Eq. 7.3–1*a*. Both w_1 and w_2 are zero, so that the only terms on the right side are $-W$ and $-E_v$.

In evaluating the kinetic energy term and the E_v term, we assume a parabolic profile of the form

$$v(r, t) = 2\langle v \rangle \left[1 - \left(\frac{r}{R}\right)^2\right] \tag{7.6–14}$$

where $\langle v \rangle$ is a function of time defined as positive when the flow is from left to right. Hence

$$\begin{aligned}\frac{d}{dt}K_{\text{tot}} &= \frac{d}{dt}\int_0^L\int_0^R \tfrac{1}{2}\rho v^2 \cdot 2\pi r\,dr\,dl \\ &= \pi\rho LR^2\frac{d}{dt}\int_0^1 v^2\frac{r}{R}\,d\frac{r}{R} \\ &= 4\pi\rho LR^2\int_0^1 (1-\xi^2)^2\xi\,d\xi\,\frac{d}{dt}\langle v\rangle^2 \\ &= \tfrac{4}{3}\rho LS\langle v\rangle\frac{d}{dt}\langle v\rangle \qquad (7.6\text{–}15)\end{aligned}$$

Here l is a coordinate running along the center of the manometer tube, and L is the distance along this axis from one manometer fluid interface to the other (i.e., the total "length" of the manometer fluid).

The change of potential energy with time is

$$\begin{aligned}\frac{d}{dt}\Phi_{\text{tot}} &= \frac{d}{dt}\int_0^L\int_0^R \rho gz\,2\pi r\,dr\,dl \\ &= \frac{d}{dt}\left[\begin{Bmatrix}\text{integral over portion} \\ \text{below } z=0,\text{ which} \\ \text{is constant}\end{Bmatrix} + \rho gS\int_0^{K+H-h} z\,dz + \rho gS\int_0^{K+H+h} z\,dz\right] \\ &= 2\rho gSh\frac{dh}{dt} \qquad (7.6\text{–}16)\end{aligned}$$

The net work done by the surroundings on the system is

$$-W = (p_a - p_b)S\langle v\rangle \qquad (7.6\text{–}17)$$

and finally the friction loss term is

$$\begin{aligned}E_v &= -\int_0^L\int_0^R (\boldsymbol{\tau}:\nabla \boldsymbol{v})2\pi r\,dr\,dl \\ &= 2\pi L\mu\int_0^R \left(\frac{dv_z}{dr}\right)^2 r\,dr \\ &= 8\pi L\mu\langle v\rangle^2\int_0^1 (-2\xi)^2\xi\,d\xi \\ &= 8LS\mu\langle v\rangle^2/R^2 \qquad (7.6\text{–}18)\end{aligned}$$

Substitution of these terms into the mechanical energy balance gives (with $\langle v\rangle = dh/dt$)

$$\frac{d^2h}{dt^2} + \left(\frac{6\mu}{R^2\rho}\right)\frac{dh}{dt} + 2\left(\frac{3}{4}\frac{g}{L}\right)h = \frac{3}{4}\left(\frac{p_a - p_b}{\rho L}\right) \qquad (7.6\text{–}19)$$

which is to be solved with the initial conditions that at $t = 0$, $h = 0$ and $dh/dt = 0$.

Equation 7.6–19 is a nonhomogeneous equation. It can easily be made homogeneous by the introduction of a new variable k defined by

$$k = 2h - \frac{p_a - p_b}{\rho g} \tag{7.6–20}$$

Then the equation for the motion of the manometer fluid is

$$\frac{d^2k}{dt^2} + \left(\frac{6\mu}{R^2\rho}\right)\frac{dk}{dt} + \left(\frac{3}{2}\frac{g}{L}\right)k = 0 \tag{7.6–21}$$

This is a second-order differential equation, which arises in describing the motion of a mass connected to a spring and dashpot, or the current in an *RLC*-circuit.

A possible solution to Eq. 7.6–21 is $k = e^{mt}$. When this function is substituted into the equation, we get two admissible values:

$$m_{\pm} = \tfrac{1}{2}[-(6\mu/R^2\rho) \pm \sqrt{(6\mu/R^2\rho)^2 - (6g/L)}] \tag{7.6–22}$$

and the solution has either of the following forms:

$$k = C_+e^{m_+t} + C_-e^{m_-t} \qquad m_+ \neq m_- \tag{7.6–23}$$

$$k = C_1e^{mt} + C_2te^{mt} \qquad m_+ = m_- = m \tag{7.6–24}$$

the constants being determined from the initial conditions. The type of motion which the manometer fluid undergoes depends upon the value of the discriminant in Eq. 7.6–22:

a. If $(6\mu/R^2\rho)^2 > (6g/L)$, then the system is "overdamped," and the fluid moves slowly to its final position.

b. If $(6\mu/R^2\rho)^2 < (6g/L)$, then the system is "underdamped" and oscillates about its final position, the amplitude of the oscillations becoming smaller and smaller.

c. If $(6\mu/R^2\rho)^2 = (6g/L)$, then the system is "critically damped," and the fluid moves to its final position in the most rapid monotone fashion.

The tube radius for critical damping is then

$$R_{cr} = \left(\frac{6\mu^2L}{g\rho^2}\right)^{1/4} \tag{7.6–25}$$

If the tube radius R is greater than R_{cr}, then an oscillatory motion occurs.

QUESTIONS FOR DISCUSSION

1. How are the macroscopic balances related to the equations of change in Chapter 3?
2. How is $\langle \bar{v}^n \rangle$ defined for axial flow in a cylindrical annulus?

3. In Eq. 7.2–3 verify that the equation is dimensionally consistent and explain the meaning of $\boldsymbol{w}$, $\boldsymbol{S}$, and $\boldsymbol{g}$ for some specific flow system.

4. Verify that Eq. 7.3–2 is dimensionally consistent and summarize ways of estimating $\hat{E}_v$.

5. How would one evaluate the integral in Eq. 7.3–2 for a van der Waals gas?

6. Evaluate E_v and $\hat{E}_v$ for laminar flow in a circular tube by using Eq. 7.4–1; for that case $-(\boldsymbol{\tau}:\nabla \boldsymbol{v}) = \mu(dv_z/dr)^2$.

7. What is the relation between the *friction factor* f and the *friction loss factor* e_v for incompressible flow in straight channels?

8. How is Eq. 7.4–10 used to calculate power requirements for pumping at a given flow rate and to calculate the flow rate at a given power input?

9. In Example 7.4–1 what would be the error in the final result if the estimation of $\hat{E}_v$ were off by a factor of 2? Under what circumstances would such an error be more serious?

10. In Example 7.5–3 how would the results be affected if p_2 were 11 atm instead of 1.1 atm?

11. Explain how an orifice meter works. What are some advantages and disadvantages of orifice meters?

12. Would the orifice meter in Fig. 7.5–4 function properly in a space ship?

13. Compare the derivation of Torricelli's law in Example 7.6–1 with that in other textbooks, most of which base their derivation on a "quasi steady-state" approach. Explain this term.

14. Complete the analysis in Example 7.6–2 by computing the height of the fluid as a function of time. Evaluate the radius for critical damping for a realistic system.

15. What would happen if 5′ were replaced by 50′ in Fig. 7.4–1?

16. Does Eq. 7.4–8 apply to incompressible non-Newtonian flow?

PROBLEMS

7.A$_1$ Pressure Rise in a Sudden Expansion

An aqueous salt solution is flowing in the sudden enlargement shown in Fig. 7.5–1 at a rate of 450 gal/min. The inside diameter of the smaller pipe is 5 in. and that of the larger pipe is 9 in. What is the pressure rise in pounds per square inch if the density of the solution is 63 $\text{lb}_m\ \text{ft}^{-3}$?
Answer: 0.157 psi

7.B$_1$ Compressible Gas Flow in Cylindrical Pipes

Gaseous nitrogen is in isothermal turbulent flow at 25° C through a straight length of horizontal pipe with 3-in. inside diameter at a rate of 0.28 $\text{lb}_m\ \text{sec}^{-1}$. The absolute pressures at inlet and outlet are 2 atm and 1 atm, respectively. From the macroscopic mass and mechanical energy balances, evaluate $\hat{E}_v$, the "friction loss," in Btu lb_m^{-1}, assuming ideal gas behavior.
Answer: 26.3 Btu lb_m^{-1}

7.C$_1$ Incompressible Flow in an Annulus

Water at 60° F is being delivered from a pump through a coaxial annular conduit 20.3 ft long at a rate of 240 U.S. gal/min. The inner and outer diameters are 3 in. and 7 in., respectively. The inlet is 5 ft lower than the outlet. Determine the horsepower needed to pump the water. Use the mean hydraulic radius to solve the problem. Assume that the pressures at the pump inlet and annular outlet are the same.
Answer: 0.31 hp

7.D$_1$ Force on a U-bend in a Pipe

Water at 68° F (ρ = 62.4 lb$_m$ ft^{-3} and μ = 1.0 cp) is flowing in turbulent flow in a U-shaped pipe bend at 3 ft^3 sec^{-1}. (See Fig. 7.D.) What is the horizontal force exerted by the water on the U-bend? *Answer:* 903 lb$_f$

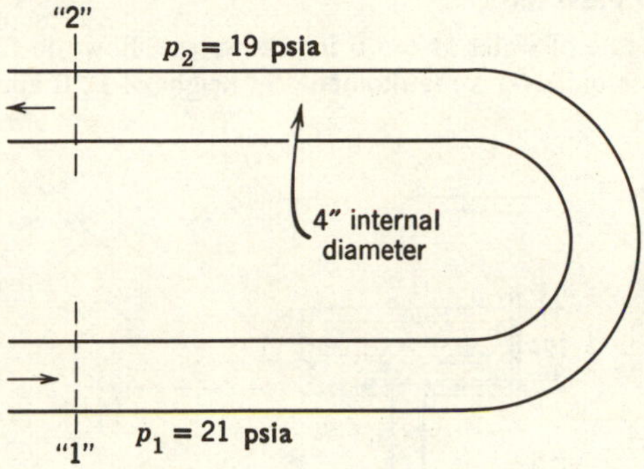

Fig. 7.D. Flow in a U-bend; both arms of bend are at the same elevation.

7.E$_1$ Disintegration of Wood Chips

In the manufacture of paper pulp the cellulose fibers of wood chips are freed from the lignin binder by heating in alkaline solutions under pressure in large cylindrical tanks called digesters. At the end of the "cooking" period a small port in one end of the digester is

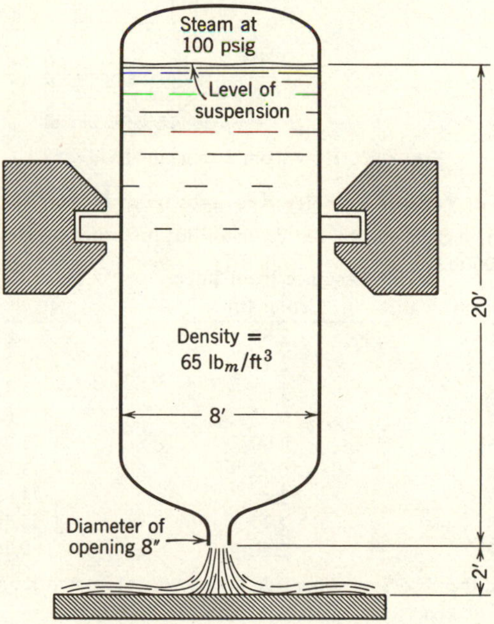

Fig. 7.E. Pulp digester.

opened, and the slurry of softened wood chips is allowed to blow against an impact plate to complete breakup of the chips and separation of the fibers. Estimate the velocity of the discharging stream and the force on the impact plate for the conditions shown in Fig. 7.E at the moment the discharge begins. Frictional effects inside the digester, and the small kinetic energy of the fluid inside the tank, may be neglected.

Answer: 2810 $lb_m\ sec^{-1}$; 10,900 lb_f

7.F$_1$ Calculation of Flow Rate

Calculate the flow rate of water at 68° F in the system shown in Fig. 7.F. The tank is so arranged that the liquid level is maintained at a height of 12 ft above the outlet.

Answer: 1930 gal min^{-1}

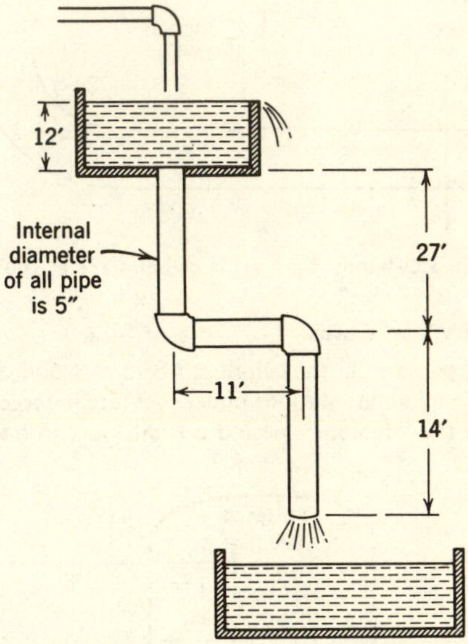

Fig. 7.F. Flow from a constant-head tank.

7.G$_1$ Evaluation of Various Velocity Averages from Pitot Tube Data

Following are some experimental data[1] for a Pitot tube traverse for the flow of water in a pipe of radius 3.06 in.

Position	Distance from Tube Center (in.)	Local Velocity (ft sec^{-1})
1	2.80	7.85
2	2.17	10.39
3	1.43	11.31
4	0.72	11.66
5	0.00	11.79
6	0.72	11.70
7	1.43	11.47
8	2.17	11.10
9	2.80	9.26

Plot these data and then use Simpson's rule for numerical integration to compute $\langle \bar{v} \rangle / \bar{v}_{max}$, $\langle \bar{v}^2 \rangle / \bar{v}^2_{max}$, and $\langle \bar{v}^3 \rangle / \bar{v}^3_{max}$.

[1] B. Bird, C. E. thesis, University of Wisconsin (1915).

7.H₂ Velocity Averages in Turbulent Flow

In §§7.2 and 7.3 we have pointed out that for turbulent flow it is customary to replace (*a*) $\langle\bar{v}^2\rangle/\langle\bar{v}\rangle$ by $\langle\bar{v}\rangle$ and (*b*) $\langle\bar{v}^3\rangle/\langle\bar{v}\rangle$ by $\langle\bar{v}\rangle^2$. What is the per cent error introduced by such an approximation for tube flow? Use the $\frac{1}{7}$-power law (see §5.1) to estimate the error.

Answer: a. 2 per cent, *b.* 6 per cent

7.I₂ Multiple Discharge into a Common Conduit[2]

Extend Example 7.5–1 in such a way that the (incompressible) fluid may flow out of several tubes into a larger tube with a net increase in cross section, as shown in Fig. 7.I.

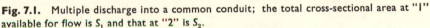

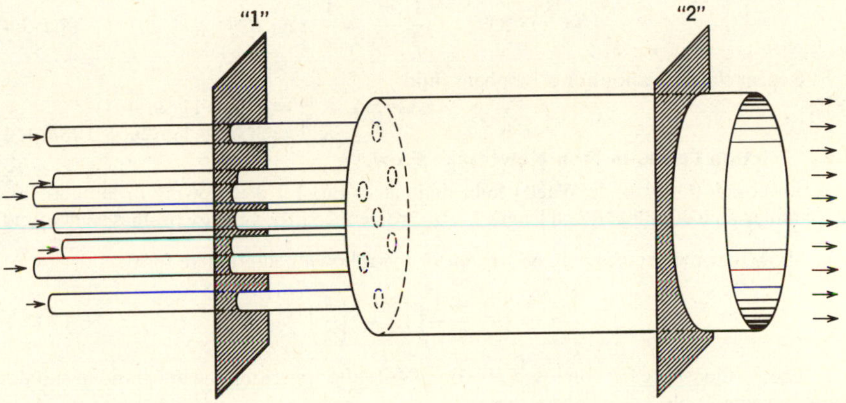

Fig. 7.I. Multiple discharge into a common conduit; the total cross-sectional area at "1" available for flow is S_1 and that at "2" is S_2.

Such systems are important in heat exchangers of certain types, for which the expansion and contraction losses account for an appreciable fraction of the over-all pressure drop. The flow in the small tubes and the large tube may each be either laminar or turbulent.

a. Show that the macroscopic balances lead to the following expressions for the pressure rise and the friction loss:

$$p_2 - p_1 = \rho\langle v_1\rangle^2(\beta K_1^{(2)} - \beta^2 K_2^{(2)}) \tag{7.I–1}$$

$$\hat{E}_v = \tfrac{1}{2}K_1^{(3)}\langle v_1\rangle^2\left[1 - 2\beta\frac{K_1^{(2)}}{K_1^{(3)}} + \beta^2\left(2\frac{K_2^{(2)}}{K_1^{(3)}} - \frac{K_2^{(3)}}{K_1^{(3)}}\right)\right] \tag{7.I–2}$$

in which we have introduced the dimensionless ratios:

$$K_i^{(j)} = \frac{\langle v_i^j\rangle}{\langle v_i\rangle^j}; \qquad \beta = \frac{S_1}{S_2} = \frac{\langle v_2\rangle}{\langle v_1\rangle} \tag{7.I–3,4}$$

b. Show that for laminar flow in circular tubes $K_1^{(2)} = \frac{4}{3}$ and $K_1^{(3)} = 2$.

c. Discuss the friction loss behavior when $\beta = 0$ and when $\beta \approx 1$.

d. Discuss the various possibilities with regard to laminar and turbulent flow.

[2] W. M. Kays, *Trans. ASME*, **72**, 1067 (1950).

7.J₂ Evaluation of Velocity Averages for Laminar Flow of Non-Newtonian Fluids in Circular Tubes

a. Evaluate the quantities (defined in Problem 7.I)

$$K^{(2)} = \frac{\langle v^2 \rangle}{\langle v \rangle^2}; \qquad K^{(3)} = \frac{\langle v^3 \rangle}{\langle v \rangle^3} \tag{7.J-1,2}$$

for the flow of a power-law fluid (Eq. 1.2–3) in long circular tubes when the flow is laminar. Note that the $K^{(j)}$ enable one to rewrite the macroscopic mass balances in such a way that the ratios of average velocities are written as the momentum or kinetic energy of the stream multiplied by a correction factor:

$$\frac{\langle v^2 \rangle}{\langle v \rangle} = \langle v \rangle \cdot K^{(2)}; \qquad \frac{1}{2}\frac{\langle v^3 \rangle}{\langle v \rangle} = \frac{1}{2}\langle v \rangle^2 \cdot K^{(3)} \tag{7.J-3,4}$$

b. Repeat the derivation for a Bingham fluid.

Answer: a. $K^{(2)} = (3n + 1)/(2n + 1)$
$K^{(3)} = 3(3n + 1)^2/(2n + 1)(5n + 3)$

7.K₂ Friction Losses in Non-Newtonian Flow

A power-law (Ostwald-de Waele) fluid defined in Eq. 1.2–3 is flowing in laminar flow in a circular pipe of radius R and length L. In Problem 2.H the velocity profiles were found.

a. Show that the profiles may be written in the following alternative form:

$$\frac{v_z}{\langle v_z \rangle} = \left(\frac{3n + 1}{n + 1}\right)\left(1 - \left(\frac{r}{R}\right)^{\frac{n+1}{n}}\right) \tag{7.K-1}$$

b. Then evaluate the friction loss E_v by Eq. 7.4–1, after rewriting the integrand in suitable form by using Table 3.4–8. Show that

$$E_v = \frac{2\pi m L \langle v \rangle^{n+1}}{R^{n-1}}\left(3 + \frac{1}{n}\right)^n \tag{7.K-2}$$

c. Next show that the result in Eq. 7.4–8 can be used to get this same result; the calculation of the friction factor for this system is performed in Problem 6.K.

7.L₂ Inventory Variations in a Gas Reservoir

A natural gas reservoir is to be supplied from a pipeline at a steady rate of w_1 lb hr^{-1}. The fuel demand from the reservoir, w_2, varies approximately as follows with a 24-hour period:

$$w_2 = A + B \cos \omega t \tag{7.L-1}$$

where ωt is a dimensionless time measured from the time of peak demand (approximately 6 A.M.).

a. Determine the maximum, minimum, and average values of w_2 for a 24-hour period in terms of A and B.

b. Determine the required value of w_1 in terms of A and B.

c. Let $m = m_0$ at $t = 0$ and integrate the unsteady mass balance with this boundary condition to obtain m as a function of time.

d. If $A = 5000$ lb$_m$ hr^{-1}, $B = 2000$ lb$_m$ hr^{-1}, and $\rho = 0.044$ lb$_m$ ft^{-3} in the reservoir, determine the absolute minimum reservoir capacity in cubic feet to meet the demand without interruption. At what time of day must the reservoir be full to permit such operation?

e. Determine the minimum reservoir capacity in cubic feet required to permit maintaining at least a three-day reserve at all times.

Answer: d. 3.47×10^5 ft^3
e. 8.53×10^6 ft^3

7.M₃ Efflux Time for Tank with Exit Pipe

a. The tank and pipe in Fig. 7.M are initially filled with a liquid of density ρ and viscosity μ. Obtain an expression for the time required to drain the tank (but not the pipe). Use a quasi steady-state approach. Use the unsteady-state mass balance with a steady-state

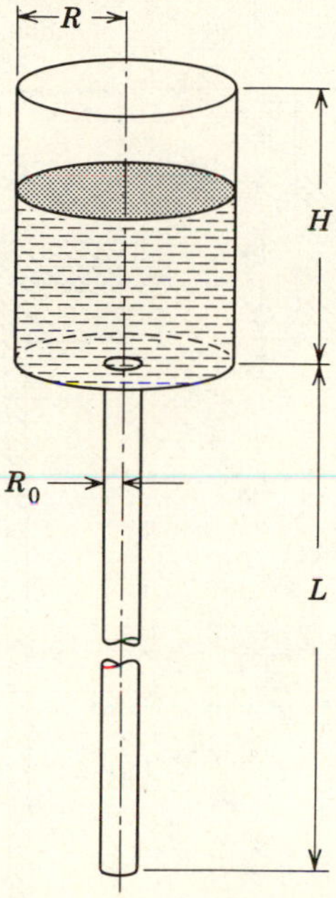

Fig. 7.M. Tank with long pipe attached; fluid surface and pipe exits are open to atmosphere.

mechanical energy balance. Neglect the entrance loss and assume that the flow in the tube is always laminar. Also neglect the kinetic energy of the emerging stream.

b. Rework for turbulent flow in pipe.

Answer: *a.* $t_{\text{efflux}} = \dfrac{8\mu L R^2}{\rho g R_0^4} \ln\left(1 + \dfrac{H}{L}\right)$

7.N₃ End Corrections in Tube Viscometers[3]

In analyzing tube-flow viscometric data to determine viscosity one compares pressure drop versus flow-rate data with the theoretical expression for pressure drop versus flow rate. The theoretical expression assumes that the flow is fully developed in the region between the two planes at which the pressure is measured. In an apparatus such as that sketched in Fig. 7.N the pressure is known at the tube exit ("2") and also above the fluid in the reservoir ("1"). However, in the entrance region of the tube the velocity profiles are not yet fully developed. Hence the theoretical expression relating the pressure drop and the volume rate of flow (e.g., the Hagen-Poiseuille law for Newtonian fluids) cannot be applied.

[3] A. G. Fredrickson [doctoral thesis, University of Wisconsin (1959)] developed and applied this method for viscometric studies of non-Newtonian fluids.

There is, however, a scheme whereby the Hagen-Poiseuille law can be used, provided that flow measurements are made in two tubes of different lengths, L_A and L_B; the shorter of the two tubes must be long enough so that the velocity profiles are fully developed at the exit, in which case the end section of the long tube of length $L_B - L_A$ will be a region of fully developed flow. (See Fig. 7.N.) If we knew the value of $\mathscr{P}_0 - \mathscr{P}_4$ for this region, we could then apply the Hagen-Poiseuille law of Eq. 2.3–19.

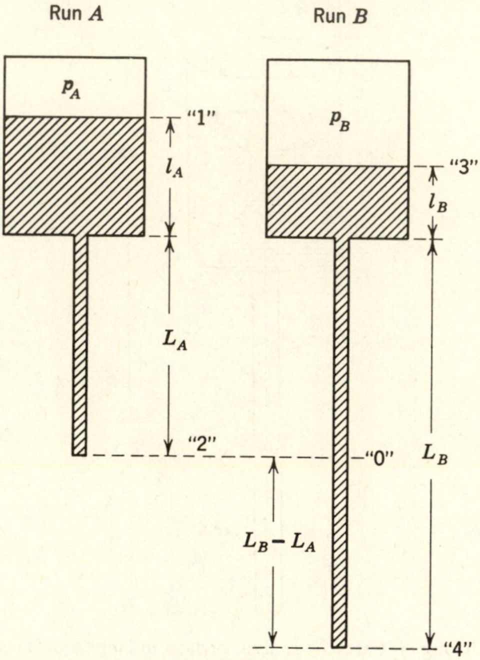

Fig. 7.N. Two tube-viscometer runs with the *same flow rate and the same outlet pressure.* Pressures p_A and p_B are maintained by an inert gas.

Show that proper combination of the mechanical energy balances written for the systems "1–2," "3–4," and "0–4" gives the following expression for $\mathscr{P}_0 - \mathscr{P}_4$ when each viscometer has the *same flow rate*:

$$\frac{\mathscr{P}_0 - \mathscr{P}_4}{L_B - L_A} = \frac{p_B - p_A}{L_B - L_A} + \rho g\left(1 + \frac{l_B - l_A}{L_B - L_A}\right) \tag{7.N–1}$$

where $\mathscr{P}_0 = p_0 + \rho g z_0$. Then $\mathscr{P}_0 - \mathscr{P}_4$ is the effective pressure difference that applies over the length $L_B - L_A$. Explain carefully how you would use Eq. 7.N–1 to analyze experimental measurements. Does Eq. 7.N–1 hold for tubes with cross section other than circular?

7.O$_3$ Relation between Friction Loss and Force on Annular Surfaces

Verify Eq. 7.4–8 for laminar flow of a Newtonian, incompressible fluid in a long annulus by using the known velocity profiles in §2.4. The length of the annulus is L, and the inner and outer radii are κR and R.

7.P$_4$ Acceleration Effects in Unsteady Flow from a Tank

An open cylinder of height H and radius R is initially entirely filled with a liquid. At time $t = 0$ the liquid is allowed to drain out through a small convergent nozzle of radius R_0 at the bottom of the tank. (See Fig. 7.P.)

a. Find the efflux time first by assuming Torricelli's law to describe the relation between efflux velocity and instantaneous height of fluid.

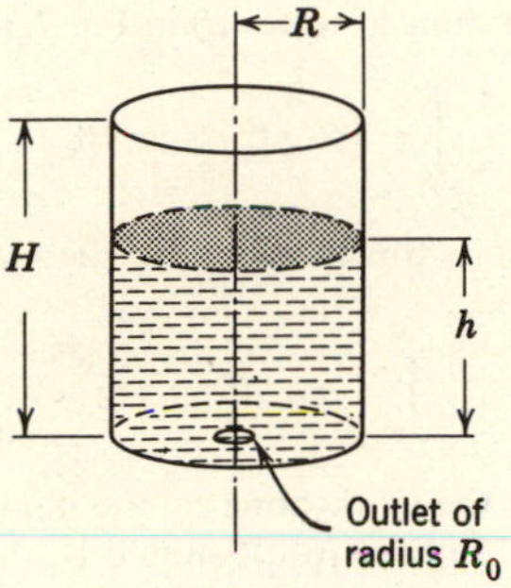

Fig. 7.P. Flow out of a cylindrical tank.

b. Find the efflux time using the unsteady mass balance and the unsteady mechanical energy balance. Express the result as the product of the efflux time in (*a*) and a function of N, where $N = (R/R_0)^4$.

c. How important is the correction factor obtained in (*b*)?

Answer: b. $t_{\text{efflux}} = 2(R/R_0)^2\sqrt{H/2g} \cdot \phi(N)$

where $\phi(N) = (\tfrac{1}{2})\sqrt{(N-2)/N}\int_0^1 (\eta - \eta^{N-1})^{-1/2}\, d\eta$

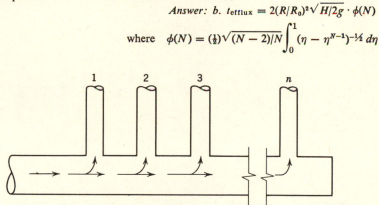

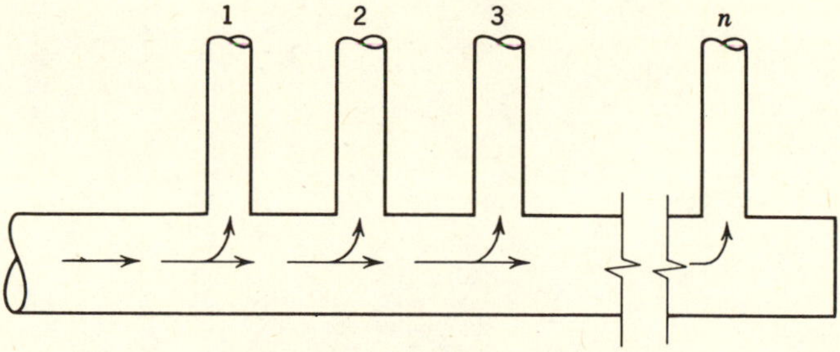

Fig. 7.Q. Flow in a manifold.

7.Q$_4$ Flow Distributions in Manifolds

A fluid flows in a pipe, capped at one end; the fluid leaves through n ports, as shown in Fig. 7.Q. The diameters of the ports are all the same but are different from that of the main pipe. Obtain a difference equation for the velocity in any given segment of the main pipe. Such a difference equation has been solved by a high-speed computer.[1]

[1] A. Acrivos, B. D. Babcock, and R. L. Pigford, *Chem. Eng. Sci.*, **10**, 112–124 (1959).

7.R$_4$ Derivation of Mechanical Energy Balance for Unsteady Incompressible Flow

It is desired to extend Example 7.3–1 to include unsteady flow in a system with moving parts. If there are moving solid surfaces in the system, then, even though the total volume of the system may remain constant, the shape changes, hence removing $\partial/\partial t$ from inside the integral sign in Eq. 7.3–6 is no longer permissible. We must then make use of the three-dimensional version of Leibnitz' formula for differentiating an integral[1,2] given in §A.5.

In applying the Gauss divergence theorem, the surface should be broken up into S_{wf} (the fixed wetted solid surface), S_{wm} (the moving wetted solid surface), and the mathematical surfaces S_1 and S_2. Then integration of term (*d*) in Eq. 7.3–6 will give a term

$$\int_{S_{wm}} [\boldsymbol{\tau} \cdot \boldsymbol{v}]_n \, dS = W_v \tag{7.R-1}$$

which is the work done by viscous forces acting on the moving surfaces. Also, term (*c*) will yield an additional term:

$$\int_{S_{wm}} p v_n \, dS = W_p \tag{7.R-2}$$

These two terms add up to give the work done by the fluid on its surroundings, as transmitted by mechanical devices. The final result should be that given in Eq. 7.3–1*a*.

[1] R. B. Bird, *Chem. Eng. Sci.*, **6**, 123–131 (1957).
[2] L. M. Grossman, *Am. J. Phys.*, **25**, 257–261 (1957).

PART II

ENERGY TRANSPORT

CHAPTER 8

Thermal Conductivity and the Mechanism of Energy Transport

The thermal conductivity k is a property that arises in most heat transfer problems. Its importance in energy transport parallels that of viscosity in momentum transport. We begin in §8.1 by stating Fourier's law of heat conduction, which defines the thermal conductivity of a gas, liquid, or solid. In §8.2 we summarize the dependence of thermal conductivities of fluids on temperature and pressure, as given by correlations based on the principle of corresponding states. In §§8.3 and 8.4 the thermal conductivities of gases and liquids are discussed from a molecular point of view. §8.5 deals with the thermal conductivities of solids and their relationship to electrical conductivity, including the famous law of Wiedemann, Franz, and Lorenz for metals.

Thermal conductivity data are rapidly accumulating in the literature, and, when reliable experimental data are available, they should be used. In the absence of reliable data, the estimation methods given in this chapter may prove useful.

In this chapter we consider energy transport by the mechanism of *conduction* only. In later chapters we consider energy transport by *convection*, *radiation*, and *diffusion*.

§8.1 FOURIER'S LAW OF HEAT CONDUCTION

Consider a slab of solid material of area A between two large parallel plates a distance Y apart. We imagine that initially (for time $t < 0$) the solid material is at a temperature T_0 throughout. At time $t = 0$ the lower plate is suddenly brought to a slightly higher temperature T_1 and maintained at that temperature. As time proceeds, the temperature profile in the slab changes,

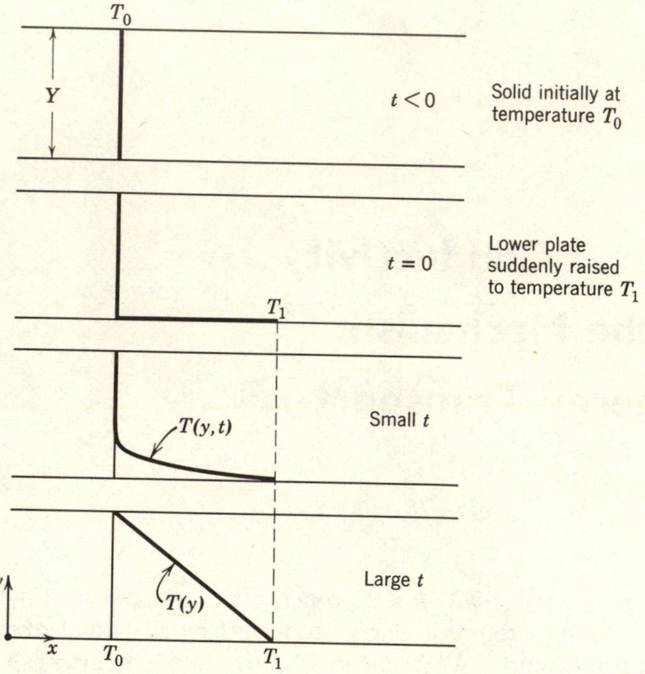

Fig. 8.1-1. Build up to steady-state temperature profile for a solid slab between two plates; see Fig. 1.1-1 for analogous situation for momentum transport.

and ultimately a linear steady-state temperature distribution is attained. (See Fig. 8.1-1.) When this steady-state condition has been reached, a constant rate of heat flow Q through the slab is required to maintain the temperature difference $\Delta T = T_1 - T_0$. It is found then that for sufficiently small values of ΔT the following relation holds:

$$\frac{Q}{A} = k\frac{\Delta T}{Y} \tag{8.1-1}$$

That is, the heat flow per unit area is proportional to the temperature decrease in the distance Y; the constant of proportionality k is the *thermal conductivity* of the slab.

Equation 8.1–1 is also valid if a liquid or gas is placed between the plates, provided that suitable precautions are taken to eliminate convection and radiation. Equation 8.1–1 therefore describes the heat-conduction process in solids, liquids, and gases. Convection and radiation are treated in later chapters.

In the analytic treatments that follow it will be more useful to work with Eq. 8.1–1 in differential form; that is, we shall use the limiting form of this equation as the slab thickness Y approaches zero. The local heat flow per unit area (heat flux) in the positive y-direction is designated q_y. In this notation Eq. 8.1–1 becomes[1]

$$q_y = -k \frac{dT}{dy} \tag{8.1–2}$$

This equation is the one-dimensional form of *Fourier's law of heat conduction*, valid when $T = T(y)$. It states that the heat flux by conduction is proportional to the temperature gradient, or, to put it somewhat pictorially, heat "slides downhill on the temperature versus distance graph."

In an isotropic[2] medium in which the temperature varies in all three directions we can write an equation like Eq. 8.1–2 for each of the coordinate directions:

$$q_x = -k \frac{\partial T}{\partial x}, \qquad q_y = -k \frac{\partial T}{\partial y}, \qquad q_z = -k \frac{\partial T}{\partial z} \tag{8.1–3,4,5}$$

These three relations are the components of the single vector equation

$$\boldsymbol{q} = -k\boldsymbol{\nabla} T \tag{8.1–6}$$

which is the three-dimensional form of Fourier's law. It states that the heat flux *vector* $\boldsymbol{q}$ is proportional to the temperature gradient[3] $\boldsymbol{\nabla} T$ and is oppositely directed. Thus in an isotropic medium heat flows by conduction in the direction of steepest temperature descent. In a moving fluid $\boldsymbol{q}$ represents the flux of thermal energy relative to the local fluid velocity.

The reader will have noticed by this time that there is a striking similarity between Eq. 8.1–2 for one-dimensional energy flux and Eq. 1.1–2 for one-dimensional momentum flux. In both cases the flux is proportional to the

[1] We shall have use for three kinds of heat quantities in the following chapters: the *heat flow* Q (Btu hr^{-1}), the *heat flux* $\boldsymbol{q}$ (Btu hr^{-1} ft^{-2}), and the *heat source strength* S (Btu hr^{-1} ft^{-3}).

[2] By isotropic we mean that the coefficient k in Eqs. 8.1–3, 4, and 5 has the same value in all three directions. The assumption of isotropy is satisfactory for fluids and for most homogeneous solids; the principal nonisotropic materials are single noncubic crystals and fibrous or laminated solids. (See wood in Table 8.1–4.)

[3] The vector quantity $\boldsymbol{\nabla} T$ is read "gradient of T" or "del T," and in some books it is written as grad T. For a discussion of the gradient of a scalar field, see G. B. Thomas, *Analytic Geometry and Calculus*, Addison-Wesley (1953), Second Edition pp. 497–500. See Table 10.2–1 for the components of Eq. 8.1–6 in curvilinear coordinates.

negative of the gradient of a macroscopic variable, and the coefficients of proportionality are physical properties dependent on the substance and on the local temperature and pressure. For more complicated situations in which the temperature and velocity vary in all three directions, however, we notice that Eqs. 8.1–3, 4, and 5 for the energy flux are simpler than Eqs. 3.2–11 through 16 for the momentum flux. This difference in form arises because energy is a scalar quantity but momentum is a vector quantity. As a result, the energy flux is a vector with three components, whereas the momentum flux is a tensor with nine components. We can thus anticipate that problems of momentum transport and energy transport will not be mathematically analogous except in certain geometrically simple situations.

In addition to the thermal conductivity k, defined by Eq. 8.1–2, a quantity known as the *thermal diffusivity*, α, is widely used in the heat-transfer literature; it is defined as[4]

$$\alpha = \frac{k}{\rho \hat{C}_p} \tag{8.1-7}$$

With regard to the units of the quantities defined above, it is found that there are several useful systems. The commonest way to assign units in the cgs system is as follows:

$$\begin{aligned} q_y &[=] \text{ cal cm}^{-2}\text{ sec}^{-1} \\ T &[=] \text{ °K} \\ y &[=] \text{ cm} \end{aligned} \tag{8.1-8}$$

hence from Eq. 8.1–2

$$\begin{aligned} k &[=] \text{ cal cm}^{-1}\text{ sec}^{-1}\text{ (° K)}^{-1} \\ \alpha &[=] \text{ cm}^2\text{ sec}^{-1} \end{aligned}$$

Sometimes experimental data are given in terms of other units simply related to the foregoing, such as kcal m^{-1} hr^{-1} $(\text{° K})^{-1}$.

The analogous set of units in the English system is

$$\begin{aligned} q_y &[=] \text{ Btu ft}^{-2}\text{ hr}^{-1} \\ T &[=] \text{ °R} \\ y &[=] \text{ ft} \end{aligned} \tag{8.1-9}$$

hence

$$\begin{aligned} k &[=] \text{ Btu ft}^{-1}\text{ hr}^{-1}\text{ (° R)}^{-1} \\ \alpha &[=] \text{ ft}^2\text{ hr}^{-1} \end{aligned}$$

In working problems involving interconversion of mechanical and thermal energy, Fourier's law is sometimes written

$$q_y = -J_c k \frac{dT}{dy} \tag{8.1-10}$$

[4] $\hat{C}_p$ is the heat capacity at constant pressure *per unit mass*; $\tilde{C}_p$ is the analogous quantity *per mole*.

in which

$$
\begin{aligned}
q_y &[=] \text{erg cm}^{-2}\text{ sec}^{-1} && \text{or ft-lb}_f\text{ ft}^{-2}\text{ hr}^{-1} \\
k &[=] \text{cal cm}^{-1}\text{ sec}^{-1}\,(^\circ\text{ K})^{-1} && \text{or Btu ft}^{-1}\text{ hr}^{-1}\,(^\circ\text{ R})^{-1} \\
J_c &[=] \text{erg cal}^{-1} && \text{or ft-lb}_f\text{ Btu}^{-1} \\
T &[=]\ ^\circ\text{ K} && \text{or }^\circ\text{ R} \\
y &[=] \text{cm} && \text{or ft}
\end{aligned}
\tag{8.1-11}
$$

It will be noted that q_y/J_c in Eq. 8.1–10 is the same as q_y in Eq. 8.1–2, so that the two methods of writing Fourier's law are equivalent. We shall use the simpler Eq. 8.1–2. Because the thermal and mechanical systems of units are both in common use, the engineer should get in the habit of making frequent checks for dimensional consistency. Appendix C will be found helpful for converting units as needed.

Some experimental values of the thermal conductivity of gases, liquids, and solids are given in Tables 8.1–1, 8.1–2, 8.1–3, and 8.1–4. Prediction of thermal conductivities is discussed in subsequent sections.

TABLE 8.1–1

EXPERIMENTAL VALUES OF THERMAL CONDUCTIVITY OF SOME COMMON GASES AT 1 ATM PRESSURE[a]

$k \times 10^7$ cal cm^{-1} sec^{-1} (° K)$^{-1}$

Gas	100° K	200° K	300° K
H_2	1625	3064	4227
O_2	216	438	635
CO_2	—	227	398
CH_4	254	522	819
NO	—	425	619

[a] Taken from J. O. Hirschfelder, C. F. Curtiss, and R. B. Bird, *Molecular Theory of Gases and Liquids*, Table 8.4–10, Wiley, New York (1954).

Example 8.1–1. Measurement of Thermal Conductivity

A plastic panel of area $A = 1$ ft^2 and thickness $Y = 0.252$ in. was found to conduct heat at a rate of 3.0 watts at steady state with temperatures of $T_0 = 24.00°$ C and $T_1 = 26.00°$ C on the two main surfaces. What is the thermal conductivity of the plastic in cal sec^{-1} cm^{-1} (° K)$^{-1}$ at 25° C?

Solution. First convert units with the aid of Appendix C:

$$
\begin{aligned}
A &= 144 \text{ in.}^2 \times (2.54)^2 = 929 \text{ cm}^2 \\
Y &= 0.252 \text{ in.} \times 2.54 = 0.640 \text{ cm} \\
Q &= 3.0 \text{ watts} = 3.0 \text{ joules sec}^{-1} \\
&\quad 3.0 \text{ joules sec}^{-1} \times 0.23901 = 0.717 \text{ cal sec}^{-1} \\
\Delta T &= 26.00 - 24.00 = 2.00^\circ \text{ K}
\end{aligned}
$$

TABLE 8.1-2

EXPERIMENTAL VALUES OF THERMAL CONDUCTIVITY OF SOME LIQUIDS AT ATMOSPHERIC PRESSURE[a]

Substance	Temperature T (° C)	Thermal Conductivity k (cal sec^{-1} cm^{-1} (° K)$^{-1}$)
Benzene	22.5	0.000378
	60	0.000363
Carbon tetrachloride	20	0.000247
Ether	30	0.000328
Ethyl alcohol	20	0.000400
Glycerol	20	0.000703
Water	20	0.00143
	60	0.00156
	100	0.00160

[a] Taken from *Handbook of Chemistry and Physics*, Thirty-ninth Edition, Chemical Rubber Publishing Co., Cleveland, Ohio (1957), pp. 2257–2259. Corrected data for water are given here by courtesy of the publishers.

TABLE 8.1-3

EXPERIMENTAL VALUES OF THERMAL CONDUCTIVITIES OF LIQUID METALS AT ATMOSPHERIC PRESSURE[a]

Metal	Temperature T (° C)	Thermal Conductivity k (cal sec^{-1} cm^{-1} (° K)$^{-1}$)
Al	700	0.247
	790	0.290
Cd	355	0.106
	435	0.119
Pb	330	0.039
	500	0.037
	700	0.036
Hg	0	0.0196
	120	0.0261
	222	0.0303
K	200	0.1073
	400	0.0956
	600	0.0846
Na	100	0.2055
	300	0.1809
	500	0.1596
Na-K alloy	100	0.0617
56% Na by wt.	300	0.0648
44% K by wt.	500	0.0675

[a] Data taken from the *Reactor Handbook*, Vol. 2, Atomic Energy Commission, AECD-3646, U.S. Government Printing Office, Washington, D.C. (May, 1955), pp. 258 *et seq.*

TABLE 8.1–4

EXPERIMENTAL VALUES OF THERMAL CONDUCTIVITIES OF SOME SOLIDS[a]

Substance	Temperature T (° C)	Thermal Conductivity k (cal sec^{-1} cm^{-1} (° K)$^{-1}$)
Aluminum	100	0.492
	300	0.64
	600	1.01
Cadmium	0	0.220
	100	0.216
Copper	18	0.918
	100	0.908
Steel	18	0.112
	100	0.107
Tin	0	0.1528
	100	0.143
Brick (common red)	—	0.0015
Concrete (stone)	—	0.0022
Earth's crust (av.)	—	0.004
Glass (soda)	200	0.0017
Graphite	—	0.012
Sand (dry)	—	0.00093
Wood (fir)		
parallel to axis	—	0.00030
perpendicular to axis	—	0.00009

[a] Data taken from the *Reactor Handbook*, Vol. 2, Atomic Energy Commission, AECD–3646, U.S. Government Printing Office, Washington, D.C. (May, 1955), pp. 1766 *et seq.*

Substitution in Eq. 8.1–1 then gives

$$k = \frac{QY}{A\,\Delta T} = \frac{0.717 \times 0.640}{929 \times 2.00} \frac{(\text{cal sec}^{-1})(\text{cm})}{(\text{cm}^2)(^\circ\,\text{K})}$$
$$= 2.47 \times 10^{-4}\ \text{cal sec}^{-1}\,\text{cm}^{-1}\,(^\circ\,\text{K})^{-1}$$

For ΔT as small as this, it is usually reasonable to assume that the value of k applies at the average temperature $(T_1 + T_0)/2$, which in this case is 25° C. See Problems 9.F and 9.J for methods of allowing for variation of k with T.

§8.2 TEMPERATURE AND PRESSURE DEPENDENCE OF THERMAL CONDUCTIVITY IN GASES AND LIQUIDS

The scarcity of reliable thermal conductivity data for fluids frequently makes it necessary to estimate k from other data on the given substance. We present here two correlations to aid in such estimation and to illustrate how

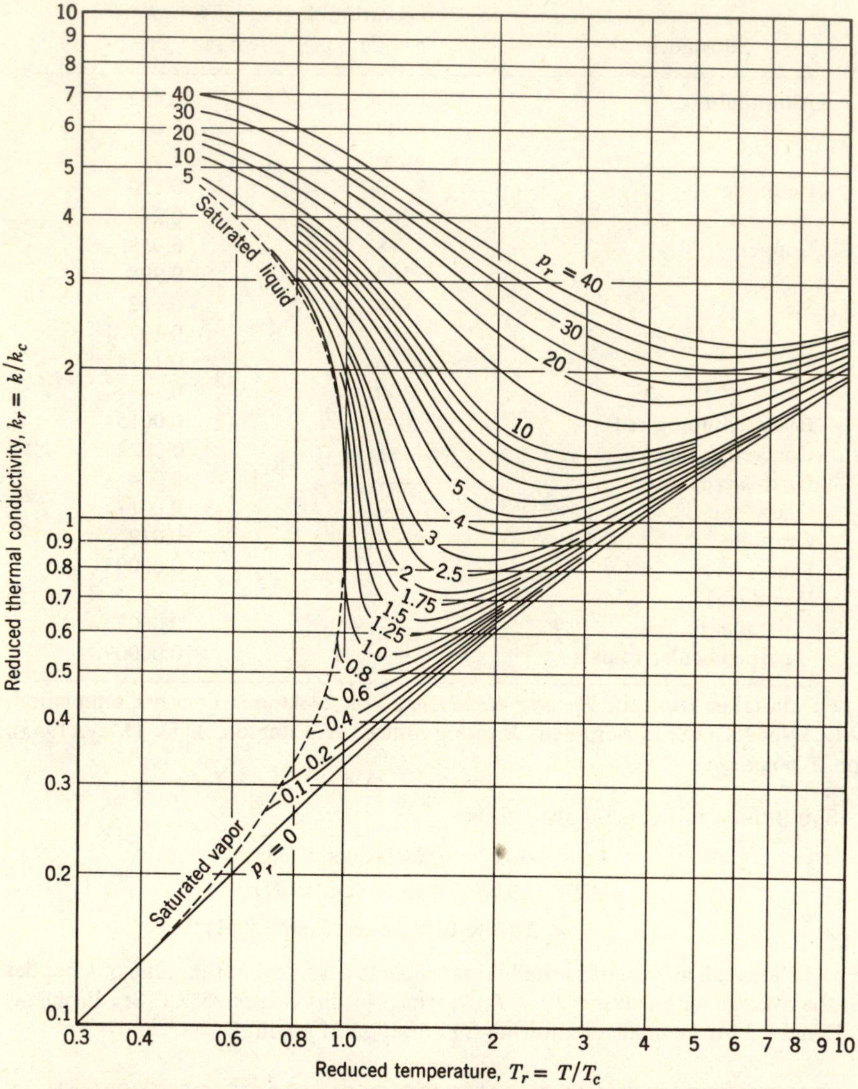

Fig. 8.2–1. Reduced thermal conductivity for monatomic substances as a function of reduced temperature and pressure. [E. J. Owens and G. Thodos, *A.I.Ch.E. Journal*, **3**, 454–461 (1957). A large-scale version of this chart is given in O. A. Hougen, K. M. Watson, and R. A. Ragatz, *Chemical Process Principles Charts*, Wiley, New York (1960). Second Edition.]

the thermal conductivity may vary with pressure and temperature for pure fluids. These corresponding-states correlations parallel the two presented for viscosity in Figs. 1.3–1 and 1.3–2.

In Fig. 8.2–1 a plot is given of the reduced thermal conductivity $k_r = k/k_c$, which is the thermal conductivity at a given temperature and pressure divided by the thermal conductivity at the critical point. This quantity is plotted as a function of the reduced temperature $T_r = T/T_c$ and the reduced pressure $p_r = p/p_c$. Figure 8.2–1 is analogous to the reduced viscosity plot given in Fig. 1.3–1; it was developed for monatomic substances but may be used as an approximation for polyatomic substances.

It will be noted in Fig. 8.2–1 that the thermal conductivity of a gas approaches a limiting function of T at low pressures; for most gases this limit is essentially reached at 1 atm pressure. The thermal conductivities of *gases* at low density *increase* with increasing temperature, whereas the thermal conductivities of most *liquids decrease* with increasing temperature. The correlation is less reliable in the liquid region; polar or associated liquids, like water, may exhibit a maximum in the curve of k versus T.

Experimental values of k_c are seldom available. However, k_c may be estimated in one of two ways: (*a*) given k at a known temperature and pressure, preferably close to the conditions at which k is to be predicted, one can read k_r from the chart and compute $k_c = k/k_r$; or (*b*) one can estimate a value of k in the low-density region by the methods given in §8.3 and then proceed as in (*a*). Values of k_c for a number of gases obtained by Method *a* are given in Appendix B, Table B.1.

Figure 8.2–2 is a plot of $k^{\#} = k/k^0$, which is the thermal conductivity at a given pressure and temperature divided by the thermal conductivity at atmospheric pressure and the same temperature. This chart is used exactly as one uses the corresponding viscosity chart in Fig. 1.3–2. One may find k^0 from experimental data or estimate it from the theory given in the next section. It should be emphasized that these charts are based on a limited amount of data; this is especially true of Fig. 8.2–1. Neither correlation is very accurate for polyatomic substances. (See Example 8.2–1.)

For mixtures, one might estimate the thermal conductivity by methods analogous to those described in §1.3. For best results, one should estimate k^0 by the method given in §8.3 for gas mixtures at low pressure and then apply a pressure correction based on Fig. 8.2–2. Very little is known about the accuracy of pseudocritical procedures as applied to thermal conductivity, largely because there are so few data on mixtures at elevated pressures.

Example 8.2–1. Effect of Pressure on Thermal Conductivity

Estimate the thermal conductivity of ethane at 153° F and 191.9 atm from the atmospheric value[1] $k^0 = 0.0159$ Btu hr^{-1} ft^{-1} °F^{-1} at this temperature.

[1] J. M. Lenoir, W. A. Junk, and E. W. Comings, *Chem. Eng. Progr.*, **49**, 539–542 (1949).

Solution. The simplest procedure is to use Fig. 8.2–2. We first compute the required coordinates, using critical properties from Table B.1:

$$T_r = \frac{153 + 460}{1.8 \times 305.4} = 1.115$$

$$p_r = \frac{191.9}{48.2} = 3.98$$

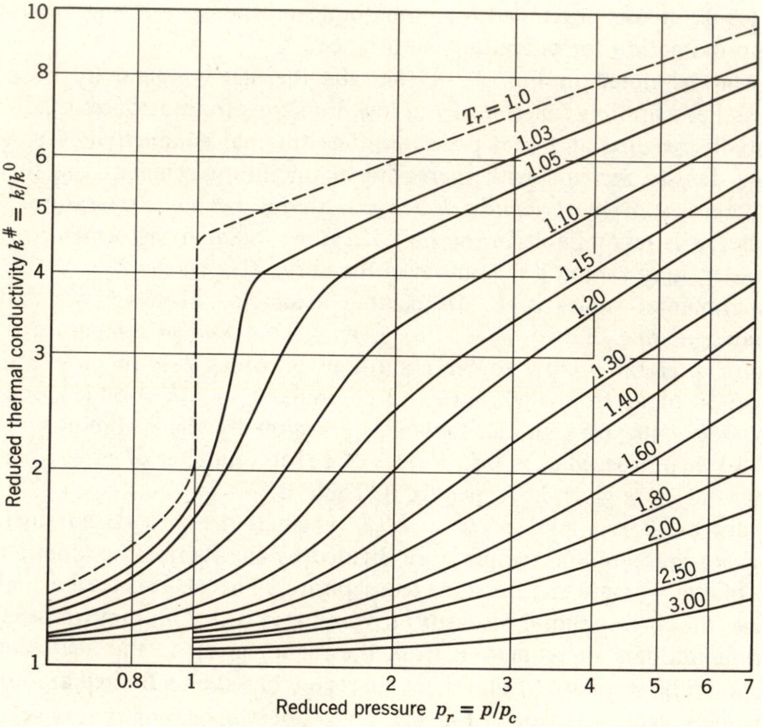

Fig. 8.2–2. Reduced thermal conductivity $k^{\#} = k/k^0$ as a function of reduced temperature and reduced pressure [J. M. Lenoir, W. A. Junk, and E. W. Comings, *Chem. Eng. Prog.*, **49**, 539 (1953).]

From Fig. 8.2–2, we find $k^{\#} = 4.7$, hence

$$k = k^{\#}k^0 = 4.7 \times 0.0159$$
$$= 0.075 \text{ Btu hr}^{-1}\text{ ft}^{-1}\,{}^\circ\text{F}^{-1}$$

An observed value of 0.0453 Btu hr^{-1} ft^{-1} °F^{-1} has been reported.[1] This poor agreement shows that one should not rely heavily on this correlation for polyatomic substances or for conditions near the critical point.

Alternate Solution. This problem may also be solved by use of Fig. 8.2–1. At the conditions of measurement of k^0 ($p_r = 0$, $T_r = 1.115$), we read $k_r = 0.36$. Then

$$k_c = \frac{k}{k_r} = \frac{0.0159}{0.36}$$

$$= 0.0442 \text{ Btu hr}^{-1}\text{ ft}^{-1}\,{}^\circ\text{F}^{-1}$$

At the conditions of prediction ($p_r = 3.98$, $T_r = 1.115$) we read $k_r = 2.07$. The predicted thermal conductivity is then

$$k = k_r k_c = 2.07 \times 0.0442$$

$$= 0.0914 \text{ Btu hr}^{-1}\text{ ft}^{-1}\,{}^\circ\text{F}^{-1}$$

This result is somewhat less accurate than that obtained from Fig. 8.2–2.

§8.3 THEORY OF THERMAL CONDUCTIVITY OF GASES AT LOW DENSITY

The thermal conductivities of dilute *monatomic* gases are well understood and can be accurately predicted by kinetic theory. The theory for *polyatomic* gases is only partially developed, but some convenient and rough approximations are available. Here again, as in §1.4, we present a simplified derivation to illustrate the transport mechanisms and then give the more accurate results of modern kinetic theory.

It is convenient to begin by calculating the thermal conductivity of a monatomic gas at low density. As in §1.4, we consider the molecules to be rigid, nonattracting spheres of mass m and diameter d. The gas as a whole is assumed to be at rest ($\boldsymbol{v} = 0$), but molecular motion has to be taken into account.

As before, we shall use the following results of kinetic theory for a rigid-sphere dilute gas in which the temperature, pressure, and velocity gradients are small:

$$\bar{u} = \sqrt{\frac{8\kappa T}{\pi m}} = \text{mean molecular speed} \tag{8.3–1}$$

$$Z = \tfrac{1}{4} n \bar{u} = \text{wall collision frequency per unit area} \tag{8.3–2}$$

$$\lambda = \frac{1}{\sqrt{2}\pi d^2 n} = \text{mean free path} \tag{8.3–3}$$

The molecules reaching any plane in the gas have, on an average, had their last collision at a distance a from the plane, where

$$a = \tfrac{2}{3}\lambda \tag{8.3–4}$$

In these equations κ is the Boltzmann constant and n is the number of molecules per unit volume.

The only form of energy that can be exchanged in collision by smooth rigid spheres is translational energy; the mean translational energy per molecule under equilibrium conditions (see Problem 1.I) is

$$\tfrac{1}{2}m\overline{u^2} = \tfrac{3}{2}\kappa T \tag{8.3–5}$$

For such a gas, the heat capacity per mole at constant volume is

$$\tilde{C}_v = \tilde{N}\frac{d}{dT}\left(\frac{1}{2}m\overline{u^2}\right) = \frac{3}{2}R \tag{8.3–6}$$

in which R is the gas constant. Equation 8.3–6 is satisfactory for monatomic gases up to temperatures of several thousand degrees.

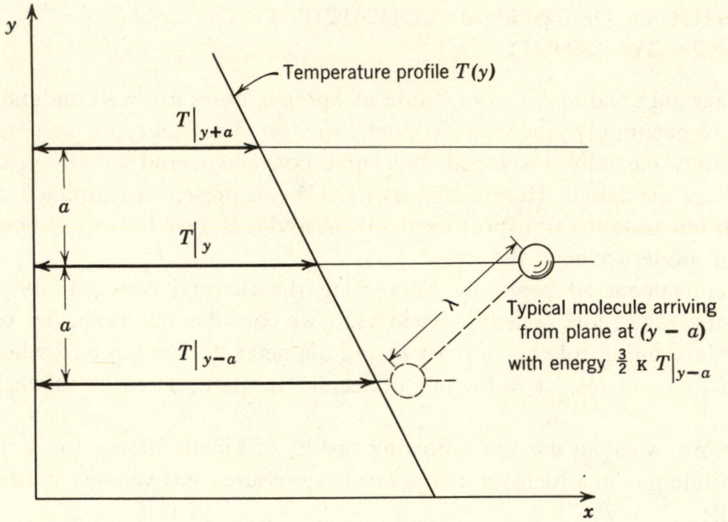

Fig. 8.3–1. Molecular transport of energy from plane at $(y - a)$ to plane at y.

To determine the thermal conductivity, we consider the behavior of the gas under a temperature gradient dT/dy. (See Fig. 8.3–1.) We assume that Eqs. 8.3–1 through 6 remain valid in this nonequilibrium situation, except that $m\overline{u^2}/2$ in Eq. 8.3–5 is taken as the average kinetic energy for molecules that had their last collision in a region of temperature T. The heat flux q_y across any plane of constant y is found by summing the kinetic energies of the molecules that cross the plane per unit time in the positive y-direction and subtracting the kinetic energies of the equal number that cross in the negative y-direction:

$$q_y = Z\tfrac{1}{2}m\overline{u^2}|_{y-a} - Z\tfrac{1}{2}m\overline{u^2}|_{y+a} \tag{8.3–7}$$

$$= \tfrac{3}{2}\kappa Z(T|_{y-a} - T|_{y+a}) \tag{8.3–8}$$

Equations 8.3–7 and 8 are based on the assumptions that all molecules have velocities representative of the region of their last collision and that the temperature profile $T(y)$ is essentially linear for a distance of several mean free paths. In view of the latter assumption, we may write

$$T|_{y-a} = T|_y - \frac{2}{3}\lambda\frac{dT}{dy}$$
$$T|_{y+a} = T|_y + \frac{2}{3}\lambda\frac{dT}{dy} \qquad (8.3\text{–}9)$$

By combining Eqs. 8.3–2, 8, and 9, we get

$$q_y = -\tfrac{1}{2}n\kappa\bar{u}\lambda\frac{dT}{dy} \qquad (8.3\text{–}10)$$

This corresponds to Fourier's law of heat conduction (Eq. 8.1–2), with the thermal conductivity given by

$$k = \tfrac{1}{2}n\kappa\bar{u}\lambda = \tfrac{1}{3}\rho\hat{C}_v\bar{u}\lambda \qquad \text{monatomic} \qquad (8.3\text{–}11)$$

in which $\rho = nm$ is the mass density of the gas.

Evaluation of $\bar{u}$ and λ from Eqs. 8.3–1 and 3 gives

$$k = \frac{1}{d^2}\sqrt{\frac{\kappa^3T}{\pi^3m}} \qquad \text{monatomic} \qquad (8.3\text{–}12)$$

which represents the thermal conductivity of a dilute gas composed of rigid spheres. This equation predicts that k is independent of pressure; Figs. 8.2–1 and 2 show that this prediction is in good agreement with data up to about 10 atm for most gases. The predicted temperature dependence is too weak, just as it was for viscosity, but it is qualitatively correct.

For a more accurate treatment of the monatomic gas, we turn again to the rigorous Chapman-Enskog theory discussed in §1.4. The Chapman-Enskog formula[1] for the thermal conductivity of a monatomic gas at low density at temperature T (° K) is

$$k = 1.9891 \times 10^{-4}\frac{\sqrt{T/M}}{\sigma^2\Omega_k} \qquad \text{monatomic} \qquad (8.3\text{–}13)$$

in which k [=] cal cm^{-1} sec^{-1} (° K)$^{-1}$, σ [=] Å, and Ω_k is identical with the collision function Ω_μ that appeared in the theory of viscosity. Values of $\Omega_k = \Omega_\mu$ are given for the Lennard-Jones intermolecular potential model in Appendix B, Table B.2, as functions of the dimensionless temperature

[1] J. O. Hirschfelder, C. F. Curtiss, and R. B. Bird, *Molecular Theory of Gases and Liquids*, Wiley, New York (1954), p. 534.

TABLE 8.3–1

PREDICTED AND OBSERVED VALUES OF $\frac{\hat{C}_p\mu}{k}$ FOR GASES AT ATMOSPHERIC PRESSURE[a]

Gas	T (° K)	$\frac{\hat{C}_p\mu}{k}$ by Eq. 8.3–16	$\frac{\hat{C}_p\mu}{k}$ from observed values of $\hat{C}_p$, μ, and k
Ne[b]	273.2	0.667	0.66
Ar[b]	273.2	0.667	0.67
H_2	90.6	0.68	0.68
	273.2	0.73	0.70
	673.2	0.74	0.65
N_2	273.2	0.74	0.73
O_2	273.2	0.74	0.74
Air	273.2	0.74	0.73
CO	273.2	0.74	0.76
NO	273.2	0.74	0.77
Cl_2	273.2	0.76	0.76
H_2O	373.2	0.77	0.94
	673.2	0.78	0.90
CO_2	273.2	0.78	0.78
SO_2	273.2	0.79	0.86
NH_3	273.2	0.77	0.85
C_2H_4	273.2	0.80	0.80
C_2H_6	273.2	0.83	0.77
$CHCl_3$	273.2	0.86	0.78
CCl_4	273.2	0.89	0.81

[a] Calculated from values given by M. Jakob, *Heat Transfer*, Wiley, New York (1949), pp. 75–76.

[b] J. O. Hirschfelder, C. F. Curtiss, and R. B. Bird, *Molecular Theory of Gases and Liquids*, Wiley, New York (1954), p. 16.

$\kappa T/\epsilon$. Equation 8.3–13 and Table B.2 have been found remarkably accurate and permit one to predict thermal conductivities of *monatomic* gases with the same numerical values of σ and ϵ that are used in viscosity calculations. (See Table B.1.)

Equation 8.3–13 is very similar to the corresponding viscosity formula, Eq. 1.4–18; the relation between k and μ is

$$k = \frac{15}{4}\frac{R}{M}\mu = \frac{5}{2}\hat{C}_v\mu \qquad \text{monatomic} \qquad (8.3\text{–}14)$$

The simplified rigid-sphere theory (see Eqs. 1.4–8 and 8.3–11) gives $k = \hat{C}_v\mu$ and is thus in error by a factor of 2.5. The discrepancy is not surprising in view of the many approximations that were made in the simple treatment.

So far we have discussed only *monatomic* gases; we now consider an approximate generalization of the foregoing results to *polyatomic* gases.[2] This generalization is needed because polyatomic molecules have rotational and vibrational energy in addition to the kinetic energy of translation of the center of mass, and all of these forms of energy may be exchanged in collision.

A simple semiempirical method of handling energy exchange in polyatomic gases was developed by Eucken.[2] His equation for thermal conductivity of a polyatomic gas at low density is

$$k = \left(\hat{C}_p + \frac{5}{4}\frac{R}{M}\right)\mu \qquad \text{polyatomic (Eucken)} \qquad (8.3\text{–}15)$$

This equation includes the monatomic formula, Eq. 8.3–14, as a special case, because $\hat{C}_p = \frac{5}{2}(R/M)$ for monatomic gases. Hirschfelder[3] has obtained an equation similar to Eq. 8.3–15 on theoretical grounds.

Equation 8.3–15 gives a simple method of estimating the Prandtl number, $\text{Pr} = \hat{C}_p\mu/k$, a dimensionless quantity of importance in convective heat transfer:

$$\text{Pr} = \frac{\hat{C}_p\mu}{k} = \frac{\tilde{C}_p}{\tilde{C}_p + 1.25R} \qquad \text{polyatomic} \qquad (8.3\text{–}16)$$

This is the Eucken formula for the Prandtl number of a nonpolar polyatomic gas at low density.

In Table 8.3–1 values of $\hat{C}_p\mu/k$ given by Eq. 8.3–16 are compared with those computed from selected data on $\hat{C}_p$, μ, and k. Since $\hat{C}_p$ and μ are accurately known for the gases shown, the comparisons are a direct measure of the agreement between observed and predicted values of k. The agreement is good for the monatomic and diatomic gases (H_2 excepted) but less satisfactory for the more complex gases. The greatest deviation is 20 per cent for steam at 100° C. Clearly, for polyatomic gases, experimental data should be used when available rather than the Eucken rule.

A more detailed empirical method of predicting k for polyatomic gases has been presented by Bromley.[4] A more satisfying method based on kinetic theory is now available for polyatomic and polar gases.[5]

The thermal conductivities of gas mixtures at low density may be estimated

[2] A. Eucken, *Physik. Z.*, **14,** 324–332 (1913).

[3] J. O. Hirschfelder, *J. Chem. Phys.*, **26,** 274–281, 282–285 (1957).

[4] L. A. Bromley, *University of California Radiation Lab. Report UCRL*-1852, Berkeley, California (1952); R. C. Reid and T. K. Sherwood, *The Properties of Gases and Liquids*, McGraw-Hill, New York (1958), pp. 228–231.

[5] E. A. Mason and L. Monchick, *J. Chem. Phys.*, **35** (1961).

by a method[6] analogous to that previously given for viscosity (see Eqs. 1.4–19 and 20):

$$k_{\text{mix}} = \sum_{i=1}^{n} \frac{x_i k_i}{\sum_{j=1}^{n} x_j \Phi_{ij}} \tag{8.3–17}$$

The x_i are mole fractions, and the k_i are the thermal conductivities of the pure components. The coefficients Φ_{ij} are identical with those that appeared in the viscosity equation:

$$\Phi_{ij} = \frac{1}{\sqrt{8}}\left(1 + \frac{M_i}{M_j}\right)^{-1/2}\left[1 + \left(\frac{\mu_i}{\mu_j}\right)^{1/2}\left(\frac{M_j}{M_i}\right)^{1/4}\right]^2 \tag{8.3–18}$$

All values of μ and k in these equations are low-density values at the given temperature. If viscosity data are not available, they may be estimated from k and $\hat{C}_p$ by means of Eq. 8.3–15. Comparisons with experimental data indicate an average deviation of about 4 per cent for mixtures[6] containing nonpolar polyatomic gases, including CH_4, O_2, N_2, C_2H_2, and CO.

Example 8.3–1. Computation of the Thermal Conductivity of a Monatomic Gas at Low Density

Compute the thermal conductivity of neon at 1 atm and 373.2° K.

Solution. The Lennard-Jones constants for neon, from Table B.1, are σ = 2.789 Å and ϵ/κ = 35.7° K; the molecular weight M is 20.183. Then, at 373.2° K, $\kappa T/\epsilon = 373.2/35.7 = 10.45$. From Table B.2, we find $\Omega_k = \Omega_\mu = 0.821$. Substitution in Eq. 8.3–13 gives

$$k = 1.9891 \times 10^{-4} \frac{\sqrt{T/M}}{\sigma^2 \Omega_k}$$

$$= 1.9891 \times 10^{-4} \frac{\sqrt{373.2/20.183}}{(2.789)^2(0.821)}$$

$$= 1.338 \times 10^{-4} \text{ cal cm}^{-1} \text{ sec}^{-1} \text{ °K}^{-1}$$

A measured value of 1.35×10^{-4} cal cm^{-1} sec^{-1} °K^{-1} has been reported at these conditions.[7]

Example 8.3–2. Estimation of the Thermal Conductivity of a Polyatomic Gas at Low Density

Estimate the thermal conductivity of molecular oxygen at 300° K and low pressure.

[6] E. A. Mason and S. C. Saxena, *The Physics of Fluids*, **1**, 361–369 (1958). This method is an approximation to the more accurate method given by J. O. Hirschfelder, *J. Chem. Phys.*, **26**, 274–281, 282–285 (1957). With Professor E. A. Mason's approval we have omitted the empirical factor 1.065 in the Φ_{ij} (for $i \neq j$).

[7] W. G. Kannuluik and E. H. Carman, *Proc. Phys. Soc. (London)*, **65B**, 701–704 (1952).

Solution. The molecular weight of O_2 is 32.000; its heat capacity $\tilde{C}_p$ at 300° K and low pressure is 7.019 cal g-mole^{-1} ° K^{-1}; its Lennard-Jones parameters, from Table B.1, are $\sigma = 3.433$ Å and $\epsilon/\kappa = 113°$ K. At 300° K, then, $\kappa T/\epsilon = 300/113 = 2.655$. From Table B.2, we find $\Omega_\mu = 1.074$. The viscosity, from Eq. 1.4–18, is

$$\mu = 2.6693 \times 10^{-5} \frac{\sqrt{MT}}{\sigma^2 \Omega_\mu}$$

$$= 2.6693 \times 10^{-5} \frac{\sqrt{(32.00)(300)}}{(3.433)^2(1.074)}$$

$$= 2.065 \times 10^{-4} \text{ g cm}^{-1} \text{ sec}^{-1}$$

The Eucken approximation to the thermal conductivity, from Equation 8.3–15, is

$$k = \left(\tilde{C}_p + \frac{5}{4} R\right) \frac{\mu}{M}$$

$$= (7.019 + 2.484) \frac{2.065 \times 10^{-4}}{32.000}$$

$$= 6.14 \times 10^{-5} \text{ cal cm}^{-1} \text{ sec}^{-1} \text{ °K}^{-1}$$

This estimate compares favorably with the measured value of 6.35×10^{-5} cal cm^{-1} sec^{-1} ° K^{-1} given in Table 8.1–1.

Example 8.3–3. Prediction of the Thermal Conductivity of a Gas Mixture at Low Density

Predict the thermal conductivity of the following gas mixture at 1 atm and 293° K from the given data on the pure components at 1 atm and 293° K:

Species	i	Mole Fraction x_i	Molecular Weight M_i	$\mu_i \times 10^7$ (g cm^{-1} sec^{-1})	$k_i \times 10^7$ (cal sec^{-1} cm^{-1} ° K^{-1})
CO_2	1	0.133	44.010	1462	383
O_2	2	0.039	32.000	2031	612
N_2	3	0.828	28.016	1754	627

Solution. Use Eqs. 8.3–17 and 18. To shorten the work, we note that the Φ_{ij} for this gas mixture at these conditions have already been computed in the viscosity calculation of Example 1.4–2. In that example we evaluated the following summations, which also appear in Eq. 8.3–17:

	$i = 1$	$i = 2$	$i = 3$
$\sum_{j=1}^{3} x_j \Phi_{ij}$	0.763	1.057	1.049

Substitution in Eq. 8.3–17 gives

$$k_{\text{mix}} = \sum_{i=1}^{3} \frac{x_i k_i}{\sum_{j=1}^{3} x_j \Phi_{ij}}$$

$$= \frac{(0.133)(383)(10^{-7})}{0.763} + \frac{(0.039)(612)(10^{-7})}{1.057} + \frac{(0.828)(627)(10^{-7})}{1.049}$$

$$= 584 \times 10^{-7} \text{ cal sec}^{-1} \text{ cm}^{-1} \, ^\circ\text{K}^{-1}$$

No data are available for comparison at these conditions.

§8.4 THEORY OF THERMAL CONDUCTIVITY OF LIQUIDS

A simple theory of energy transport in pure liquids was proposed by Bridgman[1] in 1923. He assumed the molecules to be arranged in a cubic lattice, with a center-to-center spacing $(\tilde{V}/\tilde{N})^{1/3}$, in which $\tilde{V}/\tilde{N}$ is the volume per molecule. He further assumed energy to be transferred from one lattice plane to the next at the sonic velocity v_s for the given fluid. The development here is based on a reinterpretation of Eq. 8.3–11 of the rigid-sphere gas theory:

$$k = \tfrac{1}{3}\rho \hat{C}_v \bar{u} \lambda = \rho \hat{C}_v \overline{|u_y|} a \qquad (8.4\text{–}1)$$

The heat capacity of a monatomic liquid at constant volume is about the same as for a solid at high temperature, so that $\rho \hat{C}_v \doteq 3(\tilde{N}/\tilde{V})\kappa$. The mean molecular speed in the y-direction, $\overline{|u_y|}$, is replaced by the sonic velocity v_s. The distance a that energy travels per single collision is taken to be the lattice spacing $(\tilde{V}/\tilde{N})^{1/3}$. By making these substitutions in Eq. 8.4–1, one obtains

$$k = 3\left(\frac{\tilde{N}}{\tilde{V}}\right)^{2/3} \kappa v_s \qquad (8.4\text{–}2)$$

which is Bridgman's equation. Experimental data show good agreement with Eq. 8.4–2 even for polyatomic liquids, but the numerical coefficient of 3.0 is somewhat too high. Very good agreement is obtained[2] if the coefficient is taken as 2.80:

$$k = 2.80\left(\frac{\tilde{N}}{\tilde{V}}\right)^{2/3} \kappa v_s \qquad (8.4\text{–}3)$$

This equation is limited to densities well above the critical density because of the assumption that each molecule oscillates in a "cage" formed by its

[1] P. W. Bridgman, *Proc. Am. Acad. Arts. Sci.*, **59**, 141–169 (1923). Bridgman's equation is often misquoted because he gave it in terms of a little-known gas constant equal to $3\kappa/2$.

[2] Equation 8.4–3 gives approximate agreement with a formula derived by R. E. Powell, W. E. Roseveare, and H. Eyring, *Ind. Eng. Chem.*, **33**, 430 (1941).

nearest neighbors. The success of this equation for polyatomic fluids seems to imply that the energy transfer in collisions of polyatomic molecules is incomplete, since the molecular heat capacity 3κ used here is less than the heat capacities of polyatomic liquids at room temperature and above.

The velocity of low-frequency sound, v_s (see Problem 10.Q), is given by the relation

$$v_s = \sqrt{\frac{C_p}{C_v}\left(\frac{\partial p}{\partial \rho}\right)_T} \tag{8.4–4}$$

The quantity $(\partial p/\partial \rho)_T$ is readily obtainable from isothermal compressibility measurements, or from an equation of state or compressibility correlation,[3] and C_p/C_v is very nearly unity for liquids except near the critical point.

Empirical correlations of thermal conductivity for liquids are summarized elsewhere.[4,5]

Example 8.4–1. Prediction of the Thermal Conductivity of a Liquid

The density of liquid CCl_4 at 20° C and 1 atm is 1.595 g cm^{-3}, and its compressibility $\rho^{-1}(\partial\rho/\partial p)_T$ is 90.7×10^{-6} atm^{-1}. What is its thermal conductivity?

Solution. First compute

$$\left(\frac{\partial p}{\partial \rho}\right)_T = \frac{1}{\rho[\rho^{-1}(\partial\rho/\partial p)_T]} = \frac{1}{(1.595)(90.7 \times 10^{-6})}$$

$$= 6.91 \times 10^3 \text{ atm cm}^3 \text{ g}^{-1} = 7.00 \times 10^9 \text{ cm}^2 \text{ sec}^{-2}$$

Assuming $C_p/C_v = 1.0$ and substituting into Eq. 8.4–4, we get

$$v_s = \sqrt{(1.0)(7.00 \times 10^9)} = 8.37 \times 10^4 \text{ cm sec}^{-1}$$

The molar volume is $\tilde{V} = M/\rho = 153.84/1.595 = 96.5$ cm^3 g-mole^{-1}. Substitution of these values in Eq. 8.4–3 gives

$$k = 2.80\left(\frac{\tilde{N}}{\tilde{V}}\right)^{2/3} \kappa v_s$$

$$= 2.80\left(\frac{6.023 \times 10^{23}}{0.965 \times 10^2}\right)^{2/3}(1.3805 \times 10^{-16})(8.37 \times 10^4)$$

$$= 1.10 \times 10^4 \,(\text{cm}^{-2})(\text{erg °K}^{-1})(\text{cm sec}^{-1})$$

$$= 2.62 \times 10^{-4} \text{ cal sec}^{-1} \text{ cm}^{-1} \text{ °K}^{-1}$$

The observed value (Table 8.1–2) is 2.47×10^{-4} cal sec^{-1} cm^{-1} °K^{-1}.

[3] O. A. Hougen, K. M. Watson, and R. A. Ragatz, *Chemical Process Principles*, Second Edition, Wiley, New York (1959), Part II.

[4] R. C. Reid and T. K. Sherwood, *The Properties of Gases and Liquids*, McGraw-Hill, New York (1958), Chapter 7.

[5] L. Riedel, *Chemie-Ing.-Techn.*, **27**, 209–213 (1955).

§8.5 THERMAL CONDUCTIVITY OF SOLIDS

Thermal conductivities of solids have to be established experimentally because they depend on many factors that are difficult to measure or predict. In porous solids, for example, the thermal conductivity is strongly dependent on the void fraction, the pore size, and the fluid contained in the pores; in crystalline materials the phase and crystallite size are important; in amorphous solids the degree of molecular orientation has considerable effect. A detailed discussion of thermal conductivities of solids is given by Jakob.[1]

In general, metals are better heat conductors than nonmetals, and crystalline materials conduct heat more readily than amorphous materials. Dry porous solids are very poor heat conductors and are therefore excellent for thermal insulation. The conductivities of most pure metals decrease with increasing temperature, whereas the conductivities of nonmetals increase; alloys show intermediate behavior. Perhaps the most useful of these rules of thumb is that thermal and electrical conductivity go hand in hand.

For pure metals, the thermal conductivity k and the electrical conductivity k_e are related approximately[2,3] as follows:

$$\frac{k}{k_e T} = L = \text{constant} \tag{8.5-1}$$

This is the famous equation of Wiedemann, Franz, and Lorenz. The Lorenz number, L, is about 22 to 29 $\times$ 10^{-9} volts^2 $(°\text{K})^{-2}$ for pure metals at 0° C and changes but little with temperature above 0° C, increases of 10 to 20 per cent/1000° C being typical.[1] At very low temperatures (−269.4° C for mercury) metals become superconductors of electricity, but not of heat, and L thus varies strongly with T near the superconduction region. Equation 8.5–1 is of limited utility for alloys, since L varies strongly with composition and, in some cases, with T.

The success of Eq. 8.5–1 for pure metals is due to the fact that free electrons are the major heat carriers in pure metals. The equation is not suitable for nonmetals, in which the concentration of free electrons is so low that energy transmission by molecular motion predominates.

QUESTIONS FOR DISCUSSION

1. What is the order of magnitude of the thermal conductivity for gases, liquids, and solids at room temperature and 1 atm pressure?
2. Would you expect wood to have the same thermal conductivity in all three directions?
3. How is thermal diffusivity defined? What are its units?

[1] M. Jakob, *Heat Transfer*, Vol. I, Wiley, New York (1949), Chapter 6.
[2] L. Lorenz, *Poggendorff's Annalen*, **147**, 429 (1872).
[3] G. Wiedemann and R. Franz, *Ann. Phys. Chemie*, **89**, 530 (1853).

4. Compare Fourier's law of heat conduction with Newton's law of viscosity.

5. Roughly, how does the thermal conductivity depend on the size of the constituent molecules in a gas at low density and in a liquid?

6. How are the thermal conductivities and viscosities of gases related?

7. What is the physical significance of the Eucken formula?

8. Compare the temperature dependence of k for gases, liquids, and solids.

9. Qualitatively, compare the thermal conductivities of Ne^{20} and Ne^{22} in the gaseous state at low density.

10. Show that Eq. 8.4–3 is dimensionally consistent.

11. Verify the units of the Lorenz number L.

12. What are the limitations on the relation $\tilde{C}_p - \tilde{C}_v = R$?

PROBLEMS

8.A$_1$ Prediction of Thermal Conductivities of Gases at Low Density

a. Compute the thermal conductivity of argon at 100° C and atmospheric pressure, using the Chapman-Enskog theory and the Lennard-Jones constants derived from viscosity data. Compare your result with the observed value[1] of 506×10^{-7} cal cm^{-1} sec^{-1} ° C^{-1}.

b. Compute the thermal conductivities of nitric oxide (NO) and methane (CH_4) at 300° K and atmospheric pressure from the following data for these conditions:

	$\mu \times 10^7$, g cm^{-1} sec^{-1}	$\tilde{C}_p$, cal g-mole^{-1} ° K^{-1}
NO	1929	7.15
CH_4	1116	8.55

Compare your results with the experimental values given in Table 8.1–1.

8.B$_1$ Computation of the Prandtl Number for Gases at Low Density

a. Estimate the Prandtl number, $\text{Pr} = \hat{C}_p\mu/k$, at 1 atm and 300° K for each of the gases listed below by the Eucken method, using only the heat-capacity data.

b. Compute the Prandtl number for each gas directly from the tabulated values of $\tilde{C}_p$, μ, and k and compare with the results of (*a*). All properties are given at low pressure and 300° K.

Gas	$\tilde{C}_p$ (cal g-mole^{-1} ° K^{-1})[a]	$\mu \times 10^7$ (g cm^{-1} sec^{-1})[b]	$k \times 10^7$ (cal cm^{-1} sec^{-1} ° K^{-1})[c]
He	4.968	1987	3540
Ar	4.968	2270	421
H_2	6.895	896	4250
Air	6.973	1851	602
CO_2	8.894	1495	383
H_2O	8.026	959	426 (extrapolated)

[a] *Selected Values of Thermodynamic Properties*, National Bureau of Standards (1947).
[b] J. O. Hirschfelder, R. B. Bird, and E. L. Spotz, *Chem. Revs.*, **44**, 205–231 (1949).
[c] *Lange's Handbook of Chemistry*, McGraw-Hill, New York (1956), Ninth Edition, p. 1544. (The values given for this problem were interpolated assuming $\log k = a + b \log T$.)

[1] W. G. Kannuluik and E. H. Carman, *Proc. Phys. Soc.* (*London*), **65B**, 701–704 (1952).

8.C$_1$ Prediction of the Thermal Conductivity of a Dense Gas

Predict the thermal conductivity of methane (CH_4) at 110.4 atm and 127° F by the following two methods:

a. Use Fig. 8.2–1. Obtain the needed critical properties from Appendix B.

b. Predict the thermal conductivity at 127° F and low pressure by the Eucken method and then apply a pressure correction based on Fig. 8.2–2. The observed value[2] is 0.0282 Btu hr^{-1} ft^{-1} ° F^{-1}.
Answer: a. 0.0294 Btu hr^{-1} ft^{-1} ° F^{-1}.

8.D$_1$ Prediction of the Thermal Conductivity of a Gas Mixture

Predict the thermal conductivity of a mixture containing 20 mole per cent CO_2 and 80 mole per cent H_2 at 1 atm and 300° K. Use the data of Problem 8.B for your calculations.
Answer: 2850×10^{-7} cal sec^{-1} cm^{-1} (° K)$^{-1}$.

8.E$_1$ Prediction of the Thermal Conductivity of a Pure Liquid

Predict the thermal conductivity of water at 40° C and 40 megabars pressure (1 megabar = 10^6 dynes cm^{-2}). The isothermal compressibility, $\rho^{-1}(\partial\rho/\partial p)_T$, is 38×10^{-6} megabar^{-1} and the density is 0.9938 g cm^{-3}. Assume $C_p = C_v$. *Answer:* 0.375 Btu hr^{-1} ft^{-1} (° F)$^{-1}$.

8.F$_2$ Calculation of Molecular Diameters from Transport Properties

a. Determine the molecular diameter d for argon from Eq. 1.4–9 and the viscosity given in Problem 8.B.

b. Determine the molecular diameter of argon from Eq. 8.3–12 and the thermal conductivity value given in Problem 8.B. Compare this value with that obtained from viscosity.

c. Calculate and compare the values of σ, the Lennard-Jones collision diameter, from the same data, using $\epsilon/\kappa = 124$° K.

d. What advantage of the Chapman-Enskog theory is shown by the foregoing results?

Answer: a. 2.95 Å
b. 1.88 Å
c. 3.415 Å from Eq. 1.4–18
3.425 Å from Eq. 8.3–13

[2] J. M. Lenoir, W. A. Junk, and E. W. Comings, *Chem. Eng. Prog.*, **49**, 539–542 (1953).

CHAPTER 9

Temperature Distributions in Solids and in Laminar Flow

In Chapter 2 it is shown how certain simple viscous flow problems are solved by a procedure consisting of two steps: (*a*) a momentum balance is made over a thin slab or shell perpendicular to the direction of momentum transport, and this shell balance leads to a first-order differential equation, which may be solved to get the momentum-flux distributions; (*b*) then into this expression for momentum flux Newton's law of viscosity is inserted, and this results in a first-order differential equation for the fluid velocity as a function of the distance. The integration constants appearing are evaluated by use of boundary conditions which specify the velocity or momentum flux at the bounding surfaces.

In this chapter we show how a number of heat-conduction problems are solved by an analogous procedure: (*a*) an energy balance is made over a thin slab or shell perpendicular to the direction of heat flow, and this balance leads to a first-order differential equation from which the heat-flux distribution is obtained; (*b*) then into this expression for the heat flux Fourier's law of heat conduction is inserted, and this gives a first-order differential equation for the temperature as a function of position. The integration constants that appear here are determined by use of boundary conditions which specify the temperature or heat flux at the bounding surfaces.

It should be apparent from the similar wording of the foregoing two paragraphs that the mathematical methods used in this chapter are the same as those introduced in Chapter 2—only the notation and nomenclature are different.

After a brief introduction to the shell energy balance in §9.1, we give an analysis of the heat conduction in a series of uncomplicated systems. It is true that these illustrations are somewhat idealized; nevertheless, the results of these problems find application in numerous standard engineering design methods. The systems discussed were chosen to introduce the beginner to a number of important physical concepts associated with the heat-transfer field. In addition, they should familiarize the reader with a wide assortment of boundary conditions and illustrate setting up problems in cartesian, cylindrical, and spherical coordinates. In §§9.2–9.5 four kinds of *heat sources* are investigated: electrical, nuclear, viscous, and chemical. In §§9.6 and 9.7 two topics with rather widespread application are covered, namely, *heat flow through composite walls* and *heat transfer from fins*. Finally, in §§9.8 and 9.9, we analyze two limiting cases of heat transfer in moving fluids: *forced convection* and *free* (or *natural*) *convection*. The study of this group of topics paves the way for the general equations given in Chapter 10.

§9.1 SHELL ENERGY BALANCES: BOUNDARY CONDITIONS

The systems described in the sections that follow are discussed in terms of shell energy balances. One selects a slab or shell, the surfaces of which are normal to the direction of heat conduction, and then one writes for this system a statement of the law of conservation of energy. For steady-state (time-independent) conditions, we write for the purposes of this chapter

$$\begin{Bmatrix}\text{rate of}\\ \text{thermal}\\ \text{energy in}\end{Bmatrix} - \begin{Bmatrix}\text{rate of}\\ \text{thermal}\\ \text{energy out}\end{Bmatrix} + \begin{Bmatrix}\text{rate of thermal}\\ \text{energy production}\end{Bmatrix} = 0 \qquad (9.1\text{–}1)$$

Thermal energy may enter or leave the system by the mechanism of heat conduction according to Fourier's law. Thermal energy may also enter or leave the system by virtue of the over-all fluid motion; this type of transport is sometimes referred to as *convective transport*, and the energy entering and leaving in this way is commonly called the *sensible heat* in and out. Thermal energy may be "produced" by the degradation of electrical energy, by the slowing down of neutrons and nuclear fragments liberated in the fission process, by the degradation of mechanical energy (viscous dissipation), and by conversion of chemical energy into heat.

It must be emphasized that Eq. 9.1–1 is only a restricted form of the energy balance, for no mention is made of kinetic energy, potential energy, or work. Nevertheless, the statement in Eq. 9.1–1 will be useful for setting up and solving a number of steady-state heat-conduction problems in *solids* and *incompressible fluids*. We use Eq. 9.1–1 in this chapter without justifying it and postpone any attempts at rigorous interpretations until the following chapter.

When Eq. 9.1–1 has been written for a system consisting of a thin slab or shell, the thickness of the slab or shell may be allowed to approach zero. This procedure leads us ultimately to a differential equation for the temperature distribution. When the differential equation is integrated, constants of integration appear. These are evaluated by the use of boundary conditions. The commonest types of boundary conditions are the following:

a. The temperature at a surface may be specified, e.g., $T = T_0$.

b. The heat flux at a surface may be given, e.g., $q = q_0$ (this is tantamount to specifying the temperature gradient).

c. At a solid-fluid interface the heat flux may be related to the difference between the temperature at the interface and that in the fluid, thus:

$$q = h(T - T_{\text{fluid}}) \tag{9.1–2}$$

This relation is referred to as "Newton's law of cooling"; it is not really a law but rather a defining equation for h, which is called a "heat-transfer coefficient."

d. At solid-solid interfaces the continuity of temperature and heat flux may be specified.

All four types of boundary conditions are encountered in this chapter. The foregoing list will probably mean more to the reader after he has seen how these boundary conditions are applied. The heat-transfer coefficient, h, introduced in Eq. 9.1–2, is assumed to be known in problems discussed in this chapter; the dependence of h on the physical properties of the fluid and the flow variables is discussed in Chapter 13.

§9.2 HEAT CONDUCTION WITH AN ELECTRICAL HEAT SOURCE

The first system we consider is an electric wire of circular cross section with radius R and electrical conductivity k_e ohm^{-1} cm^{-1}. Through this wire there is an electric current with current density I amps cm^{-2}. The transmission of an electric current is an irreversible process, and some electrical energy is converted into heat (thermal energy). The rate of heat production per unit volume is given by the expression

$$S_e = \frac{I^2}{k_e} \tag{9.2–1}$$

The quantity S_e is the heat source owing to electrical dissipation. We assume here that the temperature rise in the wire is not so large that the temperature dependence of either the thermal or electrical conductivity need be considered. The surface of the wire is maintained at temperature T_0. We now show how one can determine the radial temperature distribution within the heated wire.

For the energy balance we select as the system a cylindrical shell of thickness Δr and length L. (See Fig. 9.2–1.) The various contributions to the energy balance are

rate of thermal energy in across cylindrical surface at r	$(2\pi rL)(q_r\|_r)$	(9.2–2)
rate of thermal energy out across cylindrical surface at $r + \Delta r$	$(2\pi(r + \Delta r)L)(q_r\|_{r+\Delta r})$	(9.2–3)
rate of production of thermal energy by electrical dissipation	$(2\pi r\,\Delta rL)S_e$	(9.2–4)

The notation q_r means "flux of energy in the r-direction," and $|_r$ means "evaluated at r." Note that we take "in" and "out" to be in the positive r-direction.

We now substitute these three expressions into Eq. 9.1–1. Division by $2\pi L\,\Delta r$ and taking the limit as Δr goes to zero gives

$$\left\{\lim_{\Delta r\to 0}\frac{(rq_r)|_{r+\Delta r} - (rq_r)|_r}{\Delta r}\right\} = S_e r \tag{9.2–5}$$

The expression within braces is just the first derivative of rq_r with respect to r, so that Eq. 9.2–5 becomes

$$\frac{d}{dr}(rq_r) = S_e r \tag{9.2–6}$$

This is a first-order ordinary differential equation for the energy flux, which may be integrated to give

$$q_r = \frac{S_e r}{2} + \frac{C_1}{r} \tag{9.2–7}$$

The integration constant C_1 must be zero because of the boundary condition

B.C. 1: at $r = 0$ q_r is not infinite (9.2–8)

Hence the final expression for the energy flux distribution is

$$\boxed{q_r = \frac{S_e r}{2}} \tag{9.2–9}$$

This states that the heat flux increases linearly with r.

We now substitute Fourier's law (see Eq. 8.1–2) in the form $q_r = -k(dT/dr)$ into Eq. 9.2–9 to obtain

$$-k\frac{dT}{dr} = \frac{S_e r}{2} \tag{9.2–10}$$

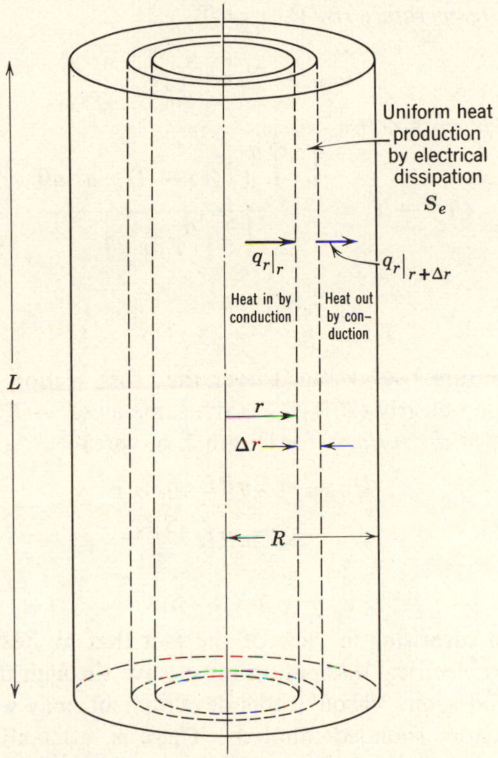

Fig. 9.2–1. Cylindrical shell over which energy balance is made in order to get temperature distribution in an electrically heated wire.

When k is assumed to be constant, this first-order differential equation may be integrated to give

$$T = -\frac{S_e r^2}{4k} + C_2 \tag{9.2–11}$$

The integration constant C_2 is determined from

B.C. 2: at $r = R$ $T = T_0$ (9.2–12)

Hence C_2 is found to be $T_0 + (S_e R^2/4k)$ and Eq. 9.2–11 becomes

$$\boxed{T - T_0 = \frac{S_e R^2}{4k}\left[1 - \left(\frac{r}{R}\right)^2\right]} \tag{9.2–13}$$

Equation 9.2–13 gives the temperature rise as a parabolic function of the distance r from the wire axis.

Once the temperature and energy flux distributions are known, various information about the system may be obtained:

(i) *Maximum temperature rise* (at $r = 0$)

$$T_{\max} - T_0 = \frac{S_e R^2}{4k} \tag{9.2–14}$$

(ii) *Average temperature rise*

$$\langle T \rangle - T_0 = \frac{\int_0^{2\pi}\int_0^R (T(r) - T_0) r\, dr\, d\theta}{\int_0^{2\pi}\int_0^R r\, dr\, d\theta} = \frac{S_e R^2}{8k} \tag{9.2–15}$$

Thus the temperature rise averaged over the cross section is just one half the maximum rise; clearly $\langle T \rangle - T_0$ is the same as $\langle T - T_0 \rangle$.

(iii) *Heat flow at the surface* (for length L of wire)

$$\begin{aligned} Q|_{r=R} &= 2\pi RL \cdot q_r|_{r=R} \\ &= 2\pi RL \cdot \frac{S_e R}{2} \\ &= \pi R^2 L \cdot S_e \end{aligned} \tag{9.2–16}$$

This result is not surprising in view of the fact that at steady state all the heat produced by electrical dissipation must leave through the wall.

The reader, while going through this development, may well have had the feeling that the story sounded familiar. There is, after all, a pronounced similarity between the heated-wire problem and the problem of viscous flow through a circular tube; only the notation is different, as shown in the following table:

	Viscous Tube Flow	Heated Wire
First integration gives distribution of	τ_{rz}	q_r
Second integration gives distribution of	v_z	$T - T_0$
Boundary condition at $r = 0$ is	τ_{rz} = finite	q_r = finite
Boundary condition at $r = R$ is	$v_z = 0$	$T - T_0 = 0$
Transport property	μ	k
Source term	$(\mathcal{P}_0 - \mathcal{P}_L)/L$	S_e
Assumptions	μ = constant	k, k_e = constant

That is, when proper quantities are chosen, the differential equations *and* the boundary conditions for the two problems are identical, and the physical processes are said to be "analogous." Not all problems in momentum transport have analogies in energy and mass transport. However, when such analogies can be found, they may be useful in predicting the behavior of systems. For example, the reader should have no trouble in finding a heat conduction analog for the viscous flow in an annulus and in writing down the solution at once.

Examples of heat-conduction problems in the electrical industry are legion.[1] The minimizing of temperature rises inside electrical machines prolongs insulation life. One example is the use of internally liquid-cooled stator conductors in very large (500,000 kw) AC generators. To illustrate further the nature of electrical heating, we give two examples concerning the temperature rise in wires: the first indicates the order of magnitude of the heating effect, and the second shows how variations in the thermal and electrical conductivity can be accounted for.

Example 9.2-1. Voltage Required for a Given Temperature Rise in a Wire Heated by an Electric Current

A copper wire has a radius of 2 mm and a length of 5 m. For what voltage drop would the temperature rise at the wire axis be 10° C, if the surface temperature of the wire is 20° C?

Solution. Combination of Eq. 9.2-14 and Eq. 9.2-1 gives

$$T_{\max} - T_0 = \frac{I^2R^2}{4kk_e} \tag{9.2-17}$$

The current density is related to the voltage drop E over a length L:

$$I = k_e\frac{E}{L} \tag{9.2-18}$$

Hence

$$T_{\max} - T_0 = \left(\frac{E^2R^2}{4L^2}\right)\left(\frac{k_e}{k}\right) \tag{9.2-19}$$

from which

$$E = 2\left(\frac{L}{R}\right)\sqrt{\frac{k}{k_eT_0}}\sqrt{T_0(T_{\max} - T_0)} \tag{9.2-20}$$

For copper, the Lorenz number[2] k/k_eT_0 is 2.23×10^{-8} volt^2 °K^{-2}. Therefore, the voltage drop needed to cause a 10° C temperature rise is

$$\begin{aligned} E &= 2\left(\frac{5000\text{ mm}}{2\text{ mm}}\right)\sqrt{2.23 \times 10^{-8}}\left(\frac{\text{volt}}{°\text{K}}\right)\sqrt{(293)(10)}\ (°\text{ K}) \\ &= (5000)(1.49 \times 10^{-4})(54.1) \\ &= 40\text{ volts} \end{aligned} \tag{9.2-21}$$

[1] M. Jakob, *Heat Transfer*, Vol. 1, Wiley, New York (1949), Chapter 10, pp. 167–199.
[2] See §8.5 for the interrelation of k and k_e.

Example 9.2–2. Heating of an Electric Wire with Temperature-Dependent Electrical and Thermal Conductivity[3]

Find the temperature distribution in an electrically heated wire if the thermal and electrical conductivities vary with position because of the change in temperature throughout the metal. Assume that k and k_e may be represented as functions of the temperature in the following way:

$$\frac{k}{k_0} = 1 - \alpha_1\Theta - \alpha_2\Theta^2 - \cdots \tag{9.2–22}$$

$$\frac{k_e}{k_{e0}} = 1 - \beta_1\Theta - \beta_2\Theta^2 - \cdots \tag{9.2–23}$$

Here k_0 and k_{e0} are the values of the conductivities at temperature T_0, and Θ is a dimensionless temperature rise defined by $\Theta = (T - T_0)/T_0$. Relations of this kind may always be used over moderate temperature ranges.

Solution. Because of the temperature gradient in the wire, the electrical conductivity is a function of position, $k_e(r)$. Therefore, the current density is also a function of r:

$$I(r) = k_e(r)\frac{E}{L} \tag{9.2–24}$$

Hence the electrical heat source S_e has the following dependence on the radial coordinate:

$$S_e(r) = k_e(r)\frac{E^2}{L^2} \tag{9.2–25}$$

according to Eq. 9.2–1. When this result is substituted into Eq. 9.2–6 and when Fourier's law is used to write q_r as $-k(r)\,dT/dr$, we get

$$-\frac{1}{r}\frac{d}{dr}\left(rk(r)\frac{dT}{dr}\right) = k_e(r)\frac{E^2}{L^2} \tag{9.2–26}$$

We now introduce these dimensionless quantities:

$$\Theta = \frac{T - T_0}{T_0} = \text{dimensionless temperature} \tag{9.2–27}$$

$$\xi = \frac{r}{R} = \text{dimensionless radial coordinate} \tag{9.2–28}$$

$$B = \frac{k_{e0}R^2E^2}{k_0L^2T_0} = \text{dimensionless heat source} \tag{9.2–29}$$

Multiplication of Eq. 9.2–26 by R^2/k_0T_0 then gives

$$-\frac{1}{\xi}\frac{d}{d\xi}\left(\frac{k}{k_0}\xi\frac{d\Theta}{d\xi}\right) = B\frac{k_e}{k_{e0}} \tag{9.2–30}$$

[3] The solution presented here was given by Professor L. J. F. Broer, of the Technische Hogeschool in Delft, Holland (private communication to RBB dated 20 August 1958).

Substitution of Eqs. 9.2–22 and 23 into Eq. 9.2–30 gives

$$-\frac{1}{\xi}\frac{d}{d\xi}\left((1 - \alpha_1\Theta - \alpha_2\Theta^2 - \cdots)\xi\frac{d\Theta}{d\xi}\right) = B(1 - \beta_1\Theta - \beta_2\Theta^2 - \cdots) \quad (9.2\text{–}31)$$

This is the equation that is to be solved for the dimensionless temperature distribution.

We begin by noting that if all the α_i and β_i were zero (i.e., constant k and k_e) then Eq. 9.2–31 would simplify to

$$-\frac{1}{\xi}\frac{d}{d\xi}\left(\xi\frac{d\Theta}{d\xi}\right) = B \quad (9.2\text{–}32)$$

which has as its solution for Θ = finite at $\xi = 0$ and $\Theta = 0$ at $\xi = 1$:

$$\Theta = \frac{B}{4}(1 - \xi^2) \quad (9.2\text{–}33)$$

This is just Eq. 9.2–13 in dimensionless notation.

Now we know that Eq. 9.2–31 will have the solution in Eq. 9.2–33 for small values of B, that is for weak heat sources. For stronger heat sources, we presume that the temperature distribution can be expressed as a power series in the reduced heat-source strength B:

$$\Theta = \frac{B}{4}(1 - \xi^2)(1 + B\Theta_1 + B^2\Theta_2 + \cdots) \quad (9.2\text{–}34)$$

where the Θ_n are functions of ξ but not of B. Substitution of Eq. 9.2–34 into Eq. 9.2–31 and equating like powers of B gives a set of ordinary differential equations to determine $\Theta_1, \Theta_2, \ldots$. These may be solved with the boundary conditions that Θ_i = finite at $\xi = 0$ and $\Theta_i = 0$ at $\xi = 1$. Thereby one obtains

$$\Theta = \frac{B}{4}(1 - \xi^2)\left[1 + B\left(\frac{\alpha_1}{8}(1 - \xi^2) - \frac{\beta_1}{16}(3 - \xi^2)\right) + O(B^2)\right] \quad (9.2\text{–}35)$$

where $O(B^2)$ means "terms of the order of B^2 and higher."

For materials that are described by the Wiedemann-Franz-Lorenz law (see §8.5), the ratio k/k_eT is a constant (independent of temperature). Hence

$$\frac{k}{k_eT} = \frac{k_0}{k_{e0}T_0} \quad (9.2\text{–}36)$$

Combination of this with Eqs. 9.2–22 and 23 gives

$$(1 - \alpha_1\Theta - \alpha_2\Theta^2 - \cdots) = (1 - \beta_1\Theta - \beta_2\Theta^2 - \cdots)(1 + \Theta) \quad (9.2\text{–}37)$$

Equating equal powers of Θ gives relations among the α_i and β_i:

$$\alpha_1 = \beta_1 - 1 \quad (9.2\text{–}38)$$

$$\alpha_2 = \beta_2 + \beta_1 \quad (9.2\text{–}39)$$

These relations may be substituted into Eq. 9.2–35 to effect a simplification:

$$\Theta = \frac{B}{4}(1 - \xi^2)\left[1 - \frac{B}{16}\left((\beta_1 + 2) + (\beta_1 - 2)\xi^2\right) + O(B^2)\right] \quad (9.2\text{–}40)$$

§9.3 HEAT CONDUCTION WITH A NUCLEAR HEAT SOURCE

We consider a nuclear fuel element of spherical form, as shown in Fig. 9.3–1. It consists of a sphere of fissionable material with radius $R^{(F)}$, surrounded by a spherical shell of aluminum "cladding" with outer radius $R^{(C)}$. Inside the fuel element fission fragments are produced which have very

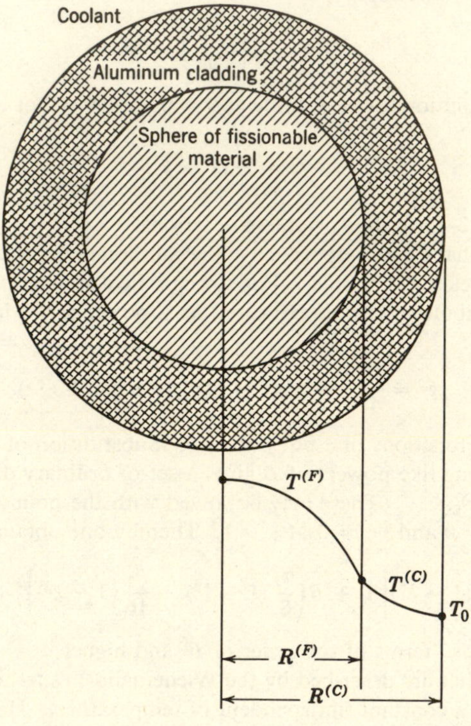

Fig. 9.3–1. Temperature distribution in a spherical nuclear fuel assembly.

high kinetic energies. Collisions between these fragments and the atoms of the fissionable material provide the major source of thermal energy in the reactor. Such a volume source of thermal energy resulting from nuclear fission we call S_n (cal cm^{-3} sec^{-1}). This source will not be uniform throughout the sphere of fissionable material; it will be the smallest at the center of the sphere. For the purpose of this problem, we assume that the source can be approximated by a simple parabolic function:

$$S_n = S_{n0}\left[1 + b\left(\frac{r}{R^{(F)}}\right)^2\right] \tag{9.3–1}$$

Here S_{n0} is the volume rate of heat production at the center of the sphere, and b is a dimensionless constant between 0 and 1.

We begin by making a thermal energy balance over a spherical shell of thickness Δr within the sphere of fissionable material:

thermal energy in at r
$$q_r^{(F)}|_r \cdot 4\pi r^2 \tag{9.3-2}$$

thermal energy out at $r + \Delta r$
$$q_r^{(F)}|_{r+\Delta r} \cdot 4\pi(r + \Delta r)^2 \tag{9.3-3}$$

thermal energy produced
$$S_n \cdot 4\pi r^2 \Delta r \tag{9.3-4}$$

Substitution of these terms into the thermal energy balance of Eq. 9.1-1 gives in the limit as $\Delta r \to 0$

$$\lim_{\Delta r \to 0} \frac{(r^2 q_r^{(F)})|_{r+\Delta r} - (r^2 q_r^{(F)})|_r}{\Delta r} = S_n r^2 \tag{9.3-5}$$

whence

$$\frac{d}{dr}(r^2 q_r^{(F)}) = S_{n0} r^2 \left[1 + b\left(\frac{r}{R^{(F)}}\right)^2\right] \tag{9.3-6}$$

The differential equation for the heat flux $q_r^{(C)}$ in the cladding is of the same form as Eq. 9.3-6, except that there is no source term:

$$\frac{d}{dr}(r^2 q_r^{(C)}) = 0 \tag{9.3-7}$$

Integration of these two equations gives

$$q_r^{(F)} = S_{n0}\left(\frac{r}{3} + \frac{b}{R^{(F)2}} \frac{r^3}{5}\right) + \frac{C_1^{(F)}}{r^2} \tag{9.3-8}$$

$$q_r^{(C)} = \frac{C_1^{(C)}}{r^2} \tag{9.3-9}$$

in which $C_1^{(F)}$ and $C_1^{(C)}$ are integration constants. These are evaluated by means of the boundary conditions:

B.C. 1: at $r = 0$, $q_r^{(F)}$ is not infinite (9.3-10)

B.C. 2: at $r = R^{(F)}$, $q_r^{(F)} = q_r^{(C)}$ (9.3-11)

Evaluation of the constants then leads to

$$\boxed{q_r^{(F)} = S_{n0}\left(\frac{r}{3} + \frac{b}{R^{(F)2}} \frac{r^3}{5}\right)} \tag{9.3-12}$$

$$\boxed{q_r^{(C)} = S_{n0} R^{(F)3}\left(\frac{1}{3} + \frac{b}{5}\right)\frac{1}{r^2}} \tag{9.3-13}$$

These are the heat-flux distributions in the fissionable sphere and in the spherical shell cladding.

Into these distributions we now substitute Fourier's law of heat conduction:

$$-k^{(F)}\frac{dT^{(F)}}{dr} = S_{n0}\left(\frac{r}{3} + \frac{b}{R^{(F)2}}\frac{r^3}{5}\right) \tag{9.3–14}$$

$$-k^{(C)}\frac{dT^{(C)}}{dr} = S_{n0}R^{(F)3}\left(\frac{1}{3} + \frac{b}{5}\right)\frac{1}{r^2} \tag{9.3–15}$$

These equations may be integrated for constant $k^{(C)}$ and $k^{(F)}$ to give

$$T^{(F)} = -\frac{S_{n0}}{k^{(F)}}\left(\frac{r^2}{6} + \frac{b}{R^{(F)2}}\frac{r^4}{20}\right) + C_2^{(F)} \tag{9.3–16}$$

$$T^{(C)} = +\frac{S_{n0}}{k^{(C)}}R^{(F)3}\left(\frac{1}{3} + \frac{b}{5}\right)\frac{1}{r} + C_2^{(C)} \tag{9.3–17}$$

The integration constants $C_2^{(F)}$ and $C_2^{(C)}$ are determined from the following boundary conditions:

B.C. 3: at $r = R^{(F)}$, $T^{(F)} = T^{(C)}$ (9.3–18)

B.C. 4: at $r = R^{(C)}$, $T^{(C)} = T_0$ (9.3–19)

where T_0 is the known temperature at the outside of the cladding. The final expressions for the temperature profiles are

$$T^{(F)} - T_0 = \frac{S_{n0}R^{(F)2}}{6k^{(F)}}\left\{\left[1 - \left(\frac{r}{R^{(F)}}\right)^2\right] + \frac{3}{10}b\left[1 - \left(\frac{r}{R^{(F)}}\right)^4\right]\right\} + \frac{S_{n0}R^{(F)2}}{3k^{(C)}}\left(1 + \frac{3}{5}b\right)\left(1 - \frac{R^{(F)}}{R^{(C)}}\right) \tag{9.3–20}$$

$$T^{(C)} - T_0 = \frac{S_{n0}R^{(F)2}}{3k^{(C)}}\left(1 + \frac{3}{5}b\right)\left(\frac{R^{(F)}}{r} - \frac{R^{(F)}}{R^{(C)}}\right) \tag{9.3–21}$$

Clearly one can find the maximum temperature in the sphere of fissionable material by setting $r = 0$ in Eq. 9.3–20. This is a quantity one might well want to know when making estimates of structural deterioration.

§9.4 HEAT CONDUCTION WITH A VISCOUS HEAT SOURCE

We consider the flow of an incompressible Newtonian fluid between two coaxial cylinders, as shown in Fig. 9.4–1. As the outer cylinder rotates, each cylindrical shell of fluid rubs against an adjacent shell of fluid. This rubbing together of adjacent layers of fluid produces heat; that is, mechanical energy is steadily degraded into thermal energy. The volume heat source resulting from this "viscous dissipation" we designate by S_v. Its magnitude depends on the local velocity gradient; the more rapidly two adjacent layers move with respect to one another, the greater will be the viscous dissipation heating. The surfaces of the inner and outer cylinders are maintained at $T = T_0$ and $T = T_b$, respectively. Clearly T will be a function of r alone.

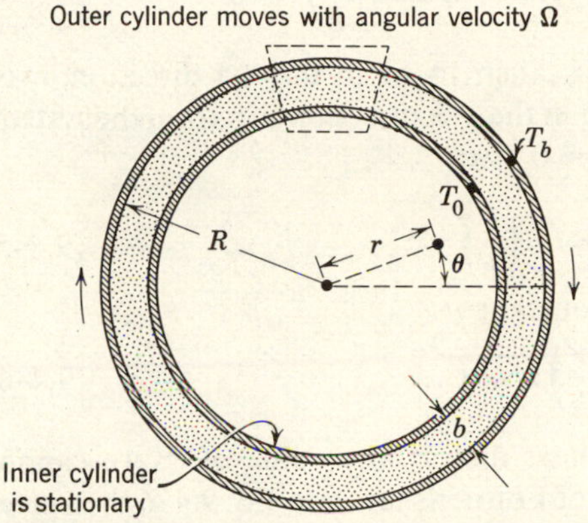

Fig. 9.4–1. Flow between cylinders with viscous heat generation. That part of the system enclosed within the dotted lines is shown in idealized form in Fig. 9.4–2.

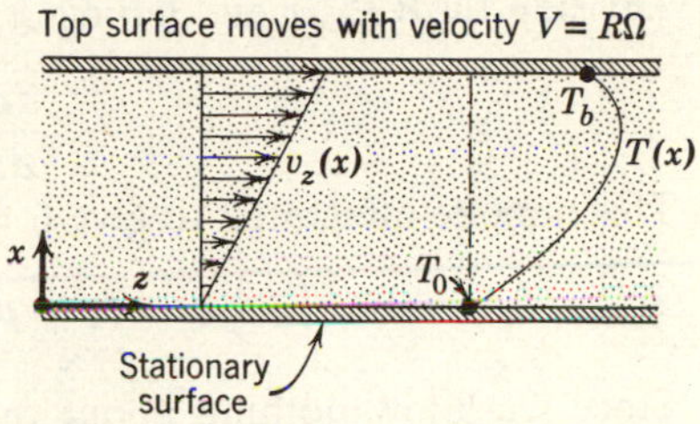

Fig. 9.4–2. Idealization of portion of flow system in Fig. 9.4–1, in which curvature of the containing surfaces is neglected.

If the slit width b is small with respect to the radius R of the outer cylinder, then the problem may be solved approximately by using the somewhat simplified system shown in Fig. 9.4–2; that is, we ignore curvature effects and solve the problem in cartesian coordinates.[1] For this modified problem, the viscous heat source is given by

$$S_v = -\tau_{xz}\left(\frac{dv_z}{dx}\right) = \mu\left(\frac{dv_z}{dx}\right)^2 \tag{9.4–1}$$

The reader is asked to accept this expression at this point; the general expression for the heat of viscous dissipation is given in Chapter 10. For the steady laminar flow of a fluid with constant viscosity in a slit, as shown in Fig. 9.4–2, the velocity profile is linear:

$$v_z = \left(\frac{x}{b}\right) V \tag{9.4–2}$$

[1] In Problems 2.F and 2.P it is shown how annulus problems behave as flat slit problems in the limit as the spacing between the cylinders goes to zero. The problem treated here is solved in cylindrical coordinates in Example 10.5–1.

so that the rate of viscous heat production per unit volume is

$$S_v = \mu\left(\frac{V}{b}\right)^2 \tag{9.4–3}$$

We are now ready to insert this quantity into an energy balance.

A thermal energy balance over a shell of thickness Δx, width W, and length L gives for the steady state

$$WLq_x|_x - WLq_x|_{x+\Delta x} + WL\,\Delta x\mu\left(\frac{V}{b}\right)^2 = 0 \tag{9.4–4}$$

Note that both "in" and "out" are taken to be in the $+x$-direction, even though in this problem heat is flowing in the $-x$-direction in part of the system. Division by $WL\,\Delta x$ and letting $\Delta x \to 0$ gives

$$\frac{dq_x}{dx} = \mu\left(\frac{V}{b}\right)^2 \tag{9.4–5}$$

This may be integrated to give for constant μ

$$q_x = \mu\left(\frac{V}{b}\right)^2 x + C_1 \tag{9.4–6}$$

Since we know nothing about the heat flux at any value of x, we cannot determine C_1 at this stage. Insertion of Fourier's law into Eq. 9.4–6 then gives

$$-k\frac{dT}{dx} = \mu\left(\frac{V}{b}\right)^2 x + C_1 \tag{9.4–7}$$

in which k is the thermal conductivity of the fluid. Equation 9.4–7 may be integrated with respect to x to give (for constant k)

$$T = -\left(\frac{\mu}{k}\right)\left(\frac{V}{b}\right)^2\frac{x^2}{2} - \frac{C_1}{k}x + C_2 \tag{9.4–8}$$

The two integration constants C_1 and C_2 are determined from the boundary conditions:

B.C. 1: at $x = 0,\quad T = T_0$ (9.4–9)

B.C. 2: at $x = b,\quad T = T_b$ (9.4–10)

When the constants are thus determined and substituted back into Eq. 9.4–8, one obtains

$$\boxed{\frac{T - T_0}{T_b - T_0} = \left(\frac{x}{b}\right) + \frac{1}{2}\,\mathrm{Br}\left(\frac{x}{b}\right)\left[1 - \left(\frac{x}{b}\right)\right]} \tag{9.4–11}$$

Here $\mathrm{Br} = [\mu V^2/k(T_b - T_0)]$ is the "Brinkman number," which is a measure of the extent to which viscous heating is important[2] relative to the

[2] Named after Professor H. C. Brinkman, who solved the problem of flow in a circular tube with viscous heat effects [*Appl. Sci. Research*, **A2**, 120–124 (1951)]; see §11.2.

heat flow resulting from the impressed temperature difference $(T_b - T_0)$. If Br > 2, there is a maximum temperature at a position intermediate between the two walls.

In *most* flow problems viscous heating is *not* important. The viscous heating effect is important in several problems encountered in engineering work, however, in which large velocity changes occur over short distances: (*a*) flow of a lubricant between fast-moving parts, (*b*) flow of plastics through dies in high-speed extrusion, and (*c*) flow of air in the boundary layer near an earth satellite or a rocket (the re-entry problem). The first two of these problems are generally further complicated because many lubricants and molten plastics are non-Newtonian. An example of non-Newtonian flow with heat generation is given in Problem 9.K; an example of heat generation with temperature-dependent viscosity and thermal conductivity may be found in Problem 9.O.

§9.5 HEAT CONDUCTION WITH CHEMICAL HEAT SOURCE

A chemical reaction is being carried out in a fixed-bed flow reactor as shown in Fig. 9.5–1. The reactor extends from $z = -\infty$ to $z = +\infty$ and is divided into three zones. The reaction zone $(0 < z < L)$ is packed with catalyst pellets, and the entrance and exit zones are packed with pellets that are physically similar but noncatalytic. The radial velocity gradients are neglected (that is, "plug flow" is assumed). In addition, the reactor wall is well insulated, so that the temperature can be considered essentially independent of r. We want to find the steady-state axial temperature distribution $T(z)$ when the fluid enters at $z = -\infty$ with a uniform temperature T_1 and a superficial velocity $v_1 = w/(\pi R^2 \rho_1)$. (See §6.4.)

Problems of this type can be solved by assuming that the axial heat conduction follows Fourier's law, with an "effective thermal conductivity" for the packed bed. The axial and radial effective conductivities $k_{z,\text{eff}}$ and $k_{r,\text{eff}}$ of packed beds have been studied extensively because of their importance in the theory of catalytic reactors. In general, $k_{z,\text{eff}}$ and $k_{r,\text{eff}}$ have quite different values.

In a chemical reaction thermal energy is produced or consumed when the atoms of the reactant molecules rearrange to form the products. The volume rate of thermal energy production by chemical reactions, S_c, is in general a complicated function of pressure, temperature, composition, and catalyst activity. For simplicity, we represent S_c here as a function of temperature only and assume a linear temperature dependence:

$$S_c = S_{c1}\left(\frac{T - T^\circ}{T_1 - T^\circ}\right) \tag{9.5–1}$$

Here T is the local temperature in the catalyst bed (assumed equal for catalyst and fluid[1]), and S_{c1} and T° are empirical constants for the given reactor inlet conditions. More realistic expressions for S_c can be handled by numerical methods.

For the shell balance we select a disk of radius R and thickness Δz in the catalyst zone (see Fig. 9.5–1), and we choose Δz to be much larger than the

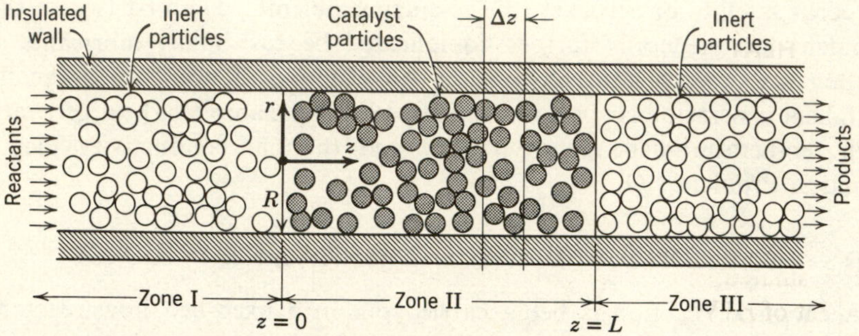

Fig. 9.5–1. Fixed-bed axial-flow reactor. Reactants enter at $z = -\infty$ and products leave at $z = +\infty$. The reaction zone extends from $z = 0$ to $z = L$.

catalyst particle dimensions. Then, at steady state, the thermal energy balance contains the following terms:

thermal energy in by conduction at z	$\pi R^2 q_z\|_z$	(9.5–2)
thermal energy out by conduction at $z + \Delta z$	$\pi R^2 q_z\|_{z+\Delta z}$	(9.5–3)
thermal energy in by flow at z	$\pi R^2 \rho_1 v_1 \hat{C}_p (T - T_0)\|_z$	(9.5–4)
thermal energy out by flow at $z + \Delta z$	$\pi R^2 \rho_1 v_1 \hat{C}_p (T - T_0)\|_{z+\Delta z}$	(9.5–5)
thermal energy produced	$(\pi R^2 \Delta z) S_c$	(9.5–6)

Here $\hat{C}_p$ has been assumed constant for the reacting mixture, and the mass flow rate $\pi R^2 \rho_1 v_1$ has been expressed in terms of the entrance conditions. Note that the thermal energy of the flowing fluid is expressed relative to a datum temperature T_0; such a datum temperature has to be chosen because energy content cannot be expressed absolutely but only relatively.

[1] The temperature difference between catalyst and fluid is not always negligible. (See Problems 21.I,J.)

We now substitute these terms into Eq. 9.1–1 and divide by $\pi R^2 \Delta z$. The result is

$$\frac{q_z|_{z+\Delta z} - q_z|_z}{\Delta z} + \rho_1 v_1 \hat{C}_p \frac{T|_{z+\Delta z} - T|_z}{\Delta z} = S_c \tag{9.5–7}$$

Next we take the limit of Eq. 9.5–7 as $\Delta z \to 0$. Strictly speaking, this operation is not "legal," since we are not dealing with a continuum. Nevertheless, we perform this limiting process with the understanding that the resulting equation describes, not point values, but rather average values of q_z, T, and S_c for entire flow cross sections of constant z. This gives

$$\frac{dq_z}{dz} + \rho_1 v_1 \hat{C}_p \frac{dT}{dz} = S_c \tag{9.5–8}$$

Now we insert Fourier's law into Eq. 9.5–8, once again treating the particulate system as a continuum. Assuming the effective axial conductivity $k_{z,\text{eff}}$ to be constant, we get

$$-k_{z,\text{eff}} \frac{d^2T}{dz^2} + \rho_1 v_1 \hat{C}_p \frac{dT}{dz} = S_c \tag{9.5–9}$$

Equation 9.5–9 is also valid in Zones I and III if S_c is set equal to zero. The resulting differential equations for the temperature distribution in the three zones of Fig. 9.5–1 are

Zone I ($z < 0$): $$-k_{z,\text{eff}} \frac{d^2T^{\text{I}}}{dz^2} + \rho_1 v_1 \hat{C}_p \frac{dT^{\text{I}}}{dz} = 0 \tag{9.5–10}$$

Zone II ($0 < z < L$):

$$-k_{z,\text{eff}} \frac{d^2T^{\text{II}}}{dz^2} + \rho_1 v_1 \hat{C}_p \frac{dT^{\text{II}}}{dz} = S_{c1}\left(\frac{T^{\text{II}} - T^\circ}{T_1 - T^\circ}\right) \tag{9.5–11}$$

Zone III ($z > L$): $$-k_{z,\text{eff}} \frac{d^2T^{\text{III}}}{dz^2} + \rho_1 v_1 \hat{C}_p \frac{dT^{\text{III}}}{dz} = 0 \tag{9.5–12}$$

Here we have used the same value of $k_{z,\text{eff}}$ for all three zones. These three second-order differential equations are to be solved with the following six boundary conditions:

$$\text{at } z = -\infty, \quad T^{\text{I}} = T_1 \tag{9.5–13}$$

$$\text{at } z = 0, \quad T^{\text{I}} = T^{\text{II}} \tag{9.5–14}$$

$$\text{at } z = 0, \quad k_{z,\text{eff}} \frac{dT^{\text{I}}}{dz} = k_{z,\text{eff}} \frac{dT^{\text{II}}}{dz} \tag{9.5–15}$$

$$\text{at } z = L, \quad T^{\text{II}} = T^{\text{III}} \tag{9.5–16}$$

$$\text{at } z = L, \quad k_{z,\text{eff}} \frac{dT^{\text{II}}}{dz} = k_{z,\text{eff}} \frac{dT^{\text{III}}}{dz} \tag{9.5–17}$$

$$\text{at } z = \infty, \quad T^{\text{III}} = \text{finite} \tag{9.5–18}$$

Equations 9.5–14 through 17 express the continuity of temperature and heat flux between zones, and Eq. 9.5–18 is added on physical grounds.

The solution is facilitated by introducing dimensionless quantities:

$$Z = \frac{z}{L} \tag{9.5–19}$$

$$\Theta = \frac{T - T^\circ}{T_1 - T^\circ} \tag{9.5–20}$$

$$B = \frac{\rho_1 v_1 \hat{C}_p L}{k_{z,\text{eff}}} \tag{9.5–21}$$

$$N = \frac{S_{c1} L}{\rho_1 v_1 \hat{C}_p (T_1 - T^\circ)} \tag{9.5–22}$$

Then the differential equations 9.5–10, 11, and 12 become

$$-\frac{1}{B}\frac{d^2\Theta^{\text{I}}}{dZ^2} + \frac{d\Theta^{\text{I}}}{dZ} = 0 \tag{9.5–23}$$

$$-\frac{1}{B}\frac{d^2\Theta^{\text{II}}}{dZ^2} + \frac{d\Theta^{\text{II}}}{dZ} = N\Theta^{\text{II}} \tag{9.5–24}$$

$$-\frac{1}{B}\frac{d^2\Theta^{\text{III}}}{dZ^2} + \frac{d\Theta^{\text{III}}}{dZ} = 0 \tag{9.5–25}$$

The solutions of these linear, homogeneous differential equations are

$$\Theta^{\text{I}} = c_1 + c_2 e^{BZ} \tag{9.5–26}$$

$$\Theta^{\text{II}} = c_3 e^{m_3 Z} + c_4 e^{m_4 Z} \qquad (\text{for } m_3 \neq m_4) \tag{9.5–27}$$

$$\Theta^{\text{III}} = c_5 + c_6 e^{BZ} \tag{9.5–28}$$

in which

$$m_3 = \tfrac{1}{2}B(1 - \sqrt{1 - (4N/B)}) \tag{9.5–29}$$

$$m_4 = \tfrac{1}{2}B(1 + \sqrt{1 - (4N/B)}) \tag{9.5–30}$$

Application of the boundary condition in Eq. 9.5–18 gives $c_6 = 0$; hence the temperature is constant in Zone III. From Eq. 9.5–13 one finds that c_1 is unity. Application of the four remaining boundary conditions gives the following results for the temperature profile in the three zones (when $1 - (4N/B) > 0$):

$$\Theta^{\text{I}} = 1 + \left(\frac{m_3 m_4 (e^{m_4} - e^{m_3})}{m_4^2 e^{m_4} - m_3^2 e^{m_3}}\right) e^{(m_3 + m_4)Z} \tag{9.5–31}$$

$$\Theta^{\text{II}} = \left(\frac{m_4 e^{m_4} e^{m_3 Z} - m_3 e^{m_3} e^{m_4 Z}}{m_4^2 e^{m_4} - m_3^2 e^{m_3}}\right)(m_3 + m_4) \tag{9.5–32}$$

$$\Theta^{\text{III}} = \left(\frac{m_4^2 - m_3^2}{m_4^2 e^{m_4} - m_3^2 e^{m_3}}\right) e^{m_3 + m_4} \tag{9.5–33}$$

Some sample temperature profiles based on these equations are given in Fig. 9.5–2. Note that the reacting mixture may be appreciably preheated or precooled by axial conduction before it enters the reaction zone; this effect deserves careful attention in the interpretation of catalytic experiments. Note also that the slope of the temperature profile always approaches zero at the exit of the catalyst bed, as required by Eqs. 9.5–17 and 33.

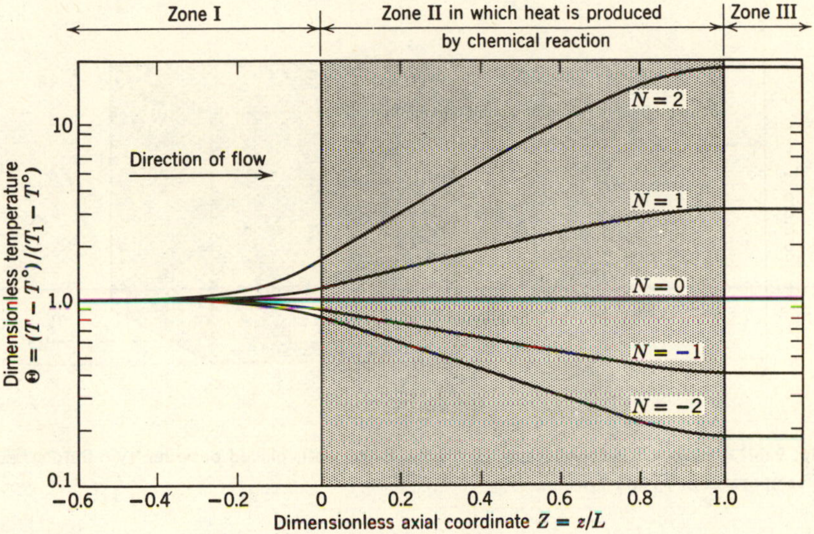

Fig. 9.5–2. Predicted temperature profiles in fixed-bed axial-flow reactor for $B = 8$ and various values of N.

It should be kept in mind that Eq. 9.5–1 is reasonable only when the composition change is small. Such a situation occurs for short reactors, for high flow rates, or for slow reactions—that is, when N is small. In the foregoing development it has been assumed that the reacting system does attain a steady state, if started under suitable initial conditions. This assumption can be tested only by experiment or by analyzing the unsteady-state equations corresponding to Eqs. 9.5–23, 24, and 25. For the system at hand, steady state is not attainable for very large N/B because the flow velocity is insufficient to counteract the rate of heat generation in Zone II and the preheating of the reactant stream in Zone I.

§9.6 HEAT CONDUCTION THROUGH COMPOSITE WALLS: ADDITION OF RESISTANCES

In industrial heat transfer problems one is frequently concerned with conduction through walls made up of layers of various materials each with its own characteristic thermal conductivity. In this section we show how the various resistances to heat transfer are combined into a total resistance.

In Fig. 9.6–1 is shown a composite wall made up of three materials of different thicknesses, $x_1 - x_0$, $x_2 - x_1$, and $x_3 - x_2$ and different thermal conductivities k^{01}, k^{12}, and k^{23}. At $x = x_0$, substance "01" is in contact with a fluid with ambient temperature T_a, and at $x = x_3$ substance "23" is in contact with a fluid at temperature T_b. The heat transfer at the boundaries

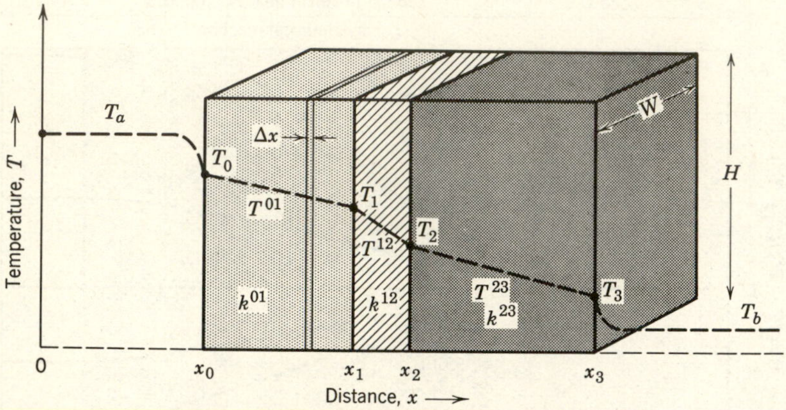

Fig. 9.6–1. Heat conduction through a composite wall, placed between two fluid streams of temperatures T_a and T_b.

$x = x_0$ and $x = x_3$ is given by Newton's "law of cooling" with heat transfer coefficients h_0 and h_3, respectively. The temperature profile is sketched in Fig. 9.6–1; one can make such a sketch before one starts the problem—and a sketch is frequently very helpful in a problem setup.

First we derive the differential equation for heat conduction in region "01." A balance on the slab of volume $WH\,\Delta x$ gives

$$q_x^{01}|_x WH - q_x^{01}|_{x+\Delta x} WH = 0 \tag{9.6–1}$$

or, after division by $WH\,\Delta x$ and taking the limit as Δx approaches zero,

$$\frac{dq_x^{01}}{dx} = 0 \tag{9.6–2}$$

Integration of this equation gives

$$q_x^{01} = q_0 \qquad \text{(a constant)} \tag{9.6–3}$$

The constant q_0 is the heat flux at the plane $x = x_0$. On physical grounds, we know that *at steady state* the heat flux in all three regions will be the same. Hence

$$q_x^{01} = q_x^{12} = q_x^{23} = q_0 \tag{9.6–4}$$

We also know that

$$q_x{}^{01} = -k^{01}\frac{dT^{01}}{dx} \tag{9.6–5}$$

with similar relations for $q_x{}^{12}$ and $q_x{}^{23}$. Combination of these relations with Eq. 9.6–4 then gives

$$-k^{01}\frac{dT^{01}}{dx} = q_0 \tag{9.6–6}$$

$$-k^{12}\frac{dT^{12}}{dx} = q_0 \tag{9.6–7}$$

$$-k^{23}\frac{dT^{23}}{dx} = q_0 \tag{9.6–8}$$

Integration of these equations then gives for constant k^{01}, k^{12}, and k^{23}:

$$T_0 - T_1 = -q_0\left(\frac{x_0 - x_1}{k^{01}}\right) \tag{9.6–9}$$

$$T_1 - T_2 = -q_0\left(\frac{x_1 - x_2}{k^{12}}\right) \tag{9.6–10}$$

$$T_2 - T_3 = -q_0\left(\frac{x_2 - x_3}{k^{23}}\right) \tag{9.6–11}$$

In addition we have the two statements regarding the heat transfer at the surfaces:

$$T_a - T_0 = \frac{q_0}{h_0} \tag{9.6–12}$$

$$T_3 - T_b = \frac{q_0}{h_3} \tag{9.6–13}$$

Addition of all five of these equations gives

$$T_a - T_b = q_0\left(\frac{1}{h_0} + \frac{x_1 - x_0}{k^{01}} + \frac{x_2 - x_1}{k^{12}} + \frac{x_3 - x_2}{k^{23}} + \frac{1}{h_3}\right) \tag{9.6–14}$$

or

$$q_0 = \frac{T_a - T_b}{\left(\dfrac{1}{h_0} + \displaystyle\sum_{i=1}^{3}\frac{x_i - x_{i-1}}{k^{\,i-1,i}} + \dfrac{1}{h_3}\right)} \tag{9.6–15}$$

Sometimes this result is rewritten in a form reminiscent of Newton's law of cooling

$$q_0 = U(T_a - T_b) \qquad \text{or} \qquad Q_0 = U(WH)(T_a - T_b) \tag{9.6–16}$$

and the quantity U (called the "over-all heat transfer coefficient") is given by

$$U = \left(\frac{1}{h_0} + \sum_{i=1}^{3}\frac{x_i - x_{i-1}}{k^{i-1,i}} + \frac{1}{h_3}\right)^{-1} \tag{9.6–17}$$

This result can clearly be generalized to include more layers in the composite wall by replacing "3" in the upper limit of the sum by any integer "n" representing the number of layers in the wall and by replacing h_3 by h_n. Equation 9.6–15 is useful for calculating the heat transfer rate through a composite wall separating two fluid streams, when the heat transfer coefficients and the thermal conductivities are known. The calculation of heat transfer coefficients is discussed in a later chapter.

One further comment needs to be made concerning the foregoing development. It has been tacitly assumed that the various layers are tightly fitted together with no intervening "air spaces." Clearly, if the layers touch each other only at several points, the resistance to heat transfer will be appreciably increased.

Example 9.6–1. Composite Cylindrical Walls

Perform a development similar to that above for a composite cylindrical wall, such as that shown in Fig. 9.6–2.

Solution. A thermal energy balance on a shell of volume $2\pi rL\,\Delta r$ for region "01" is

$$q_r^{01}\big|_r \cdot 2\pi rL - q_r^{01}\big|_{r+\Delta r} \cdot 2\pi(r + \Delta r)L = 0 \tag{9.6–18}$$

Division by $2\pi L\,\Delta r$ and taking the limit as $\Delta r \to 0$ gives

$$\frac{d}{dr}(rq_r^{01}) = 0 \tag{9.6–19}$$

Integration of this equation gives for constant k^{01}, k^{12}, and k^{23}

$$rq_r^{01} = r_0 q_0 \qquad \text{(a constant)} \tag{9.6–20}$$

in which r_0 is the inner radius of region "01" and q_0 is the heat flux there. Hence for steady state we write, analogous to Eqs. 9.6–6, 7, and 8,

$$-k^{01} r \frac{dT^{01}}{dr} = r_0 q_0 \tag{9.6–21}$$

$$-k^{12} r \frac{dT^{12}}{dr} = r_0 q_0 \tag{9.6–22}$$

$$-k^{23} r \frac{dT^{23}}{dr} = r_0 q_0 \tag{9.6–23}$$

Integration of these equations gives for constant k^{01}, k^{12}, and k^{23}

$$T_0 - T_1 = r_0 q_0 \left(\frac{\ln r_1/r_0}{k^{01}}\right) \tag{9.6–24}$$

$$T_1 - T_2 = r_0 q_0 \left(\frac{\ln r_2/r_1}{k^{12}}\right) \tag{9.6–25}$$

$$T_2 - T_3 = r_0 q_0 \left(\frac{\ln r_3/r_2}{k^{23}}\right) \tag{9.6–26}$$

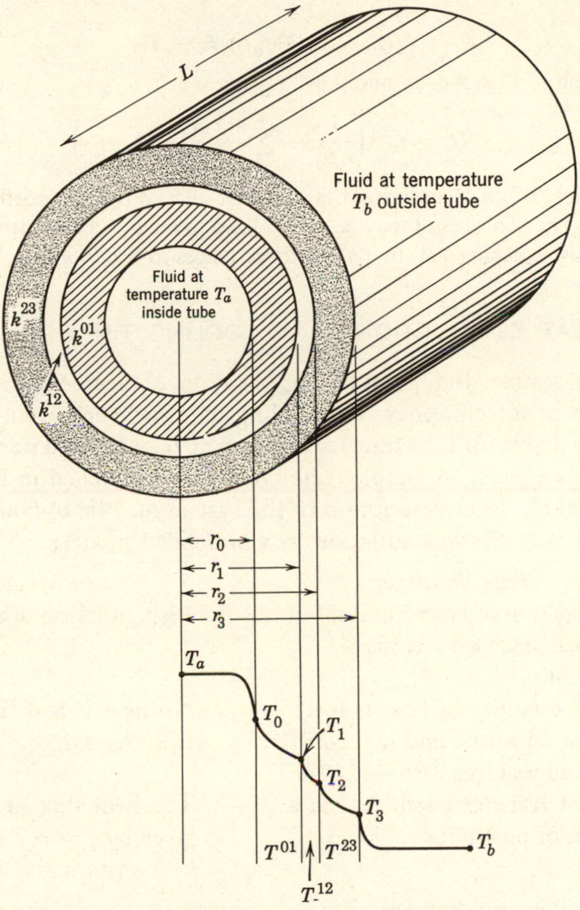

Fig. 9.6–2. Heat conduction through a laminated tube with fluid at temperature T_a inside and fluid at temperature T_b outside.

At the two fluid-solid interfaces we have the requirements that

$$T_a - T_0 = \frac{q_0}{h_0} \tag{9.6–27}$$

$$T_3 - T_b = \frac{q_3}{h_3} = \frac{q_0}{h_3}\frac{r_0}{r_3} \tag{9.6–28}$$

Addition of all five equations gives an expression for $T_a - T_b$, which may be solved for q_0 to give

$$Q_0 = 2\pi L r_0 q_0 = \frac{2\pi L(T_a - T_b)}{\left(\dfrac{1}{r_0 h_0} + \dfrac{\ln r_1/r_0}{k^{01}} + \dfrac{\ln r_2/r_1}{k^{12}} + \dfrac{\ln r_3/r_2}{k^{23}} + \dfrac{1}{r_3 h_3}\right)} \tag{9.6–29}$$

We now define an "over-all heat transfer coefficient based on the inner surface" U_0:

$$Q_0 = U_0(2\pi r_0 L)(T_a - T_b) \tag{9.6–30}$$

Combination of Eqs. 9.6–29 and 9.6–30 gives

$$U_0 = r_0^{-1}\left(\frac{1}{r_0 h_0} + \sum_{i=1}^{3} \frac{\ln r_i/r_{i-1}}{k^{i-1,i}} + \frac{1}{r_3 h_3}\right)^{-1} \tag{9.6–31}$$

The subscript "0" on U_0 indicates that the over-all heat transfer coefficient is referred to the radius r_0. This result may be generalized to include a cylindrical tube made of n laminae by replacing "3" by "n" in three places in Eq. 9.6–31.

§9.7 HEAT CONDUCTION IN A COOLING FIN[1]

Another simple, but practical, application of heat conduction is in the calculation of the efficiency of a cooling fin. Such fins are used to increase the area available for heat transfer between metal walls and poorly conducting fluids such as gases. A simple rectangular fin is sketched in Fig. 9.7–1.

A reasonably good description of the system may be obtained by approximating the true physical situation by a simplified model:

True Situation	*Model*
1. T is a function of both x and z, but the dependence on z is more important.	1. T is a function of z alone.
2. A small quantity of heat is lost from the fin at the end (area $2BW$) and the edges (area $2BL + 2BL$).	2. No heat is lost from the end or from the edges.
3. The heat transfer coefficient is a function of position.	3. The heat flux at the surface is given by $q = h(T - T_a)$, in which h is constant and $T = T(z)$.

A thermal energy balance on a segment Δz of the bar gives

$$q_z|_z \cdot 2BW - q_z|_{z+\Delta z} \cdot 2BW - h(2W\,\Delta z)(T - T_a) = 0 \tag{9.7–1}$$

Division by $2BW\,\Delta z$ and taking the limit as Δz approaches zero gives

$$-\frac{dq_z}{dz} = \frac{h}{B}(T - T_a) \tag{9.7–2}$$

Insertion of Fourier's law ($q_z = -k\,dT/dz$) in which k is the thermal conductivity of the metal gives for constant k

$$\frac{d^2T}{dz^2} = \frac{h}{kB}(T - T_a) \tag{9.7–3}$$

[1] For a further discussion of fins, see, for example, M. Jakob, *Heat Transfer*, Wiley, New York (1949), Vol. I, Chapter 11; see also E. R. G. Eckert and R. M. Drake, Jr., *Heat and Mass Transfer*, McGraw-Hill, New York (1959), Second Edition, Chapter 3.

This equation is to be solved with the boundary conditions

B.C. 1: at $z = 0$ $\quad T = T_w$ (9.7–4)

B.C. 2: at $z = L$ $\quad \dfrac{dT}{dz} = 0$ (9.7–5)

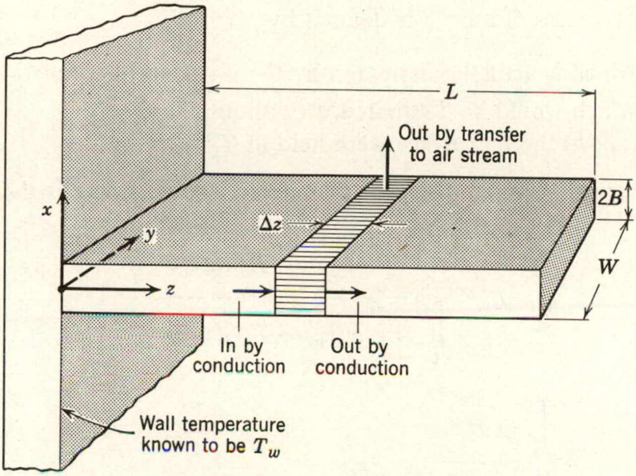

Fig. 9.7–1. A simple cooling fin with $B \ll L$.

We can now introduce the following dimensionless quantities:

$$\Theta = \frac{T - T_a}{T_w - T_a} = \text{dimensionless temperature} \tag{9.7–6}$$

$$\zeta = \frac{z}{L} = \text{dimensionless distance} \tag{9.7–7}$$

$$N = \sqrt{\frac{hL^2}{kB}} = \text{dimensionless heat transfer coefficient} \tag{9.7–8}$$

This problem may be restated:

$$\frac{d^2\Theta}{d\zeta^2} = N^2\Theta \quad \text{with} \quad \Theta|_{\zeta=0} = 1 \quad \text{and} \quad \left.\frac{d\Theta}{d\zeta}\right|_{\zeta=1} = 0 \tag{9.7–9,10,11}$$

Equation 9.7–9 may be integrated to give hyperbolic functions. When the two integration constants have been determined, we get

$$\Theta = \cosh N\zeta - (\tanh N) \sinh N\zeta \tag{9.7–12}$$

This may be rearranged to give

$$\boxed{\Theta = \frac{\cosh N(1 - \zeta)}{\cosh N}} \tag{9.7-13}$$

It should be emphasized that this expression is reasonable only if the heat lost at the edges is negligible.

The "effectiveness of a fin"[2] is defined by

$$\eta = \frac{(\text{heat which is actually dissipated by the fin surface})}{\left[\begin{array}{c}\text{heat which would be dissipated if (without change}\\ \text{in } h\text{) the fin surface were held at } T_w\end{array}\right]} \tag{9.7-14}$$

The theoretical value of η for the problem considered here is then

$$\eta = \frac{\int_0^W \int_0^L h(T - T_a)\,dz\,dy}{\int_0^W \int_0^L h(T_w - T_a)\,dz\,dy} \tag{9.7-15}$$

or

$$\eta = \frac{\int_0^1 \Theta\,d\zeta}{\int_0^1 d\zeta}$$

$$= \frac{1}{\cosh N}\left(-\frac{1}{N}\sinh N(1 - \zeta)\right)\Bigg|_0^1 = \frac{\tanh N}{N} \tag{9.7-16}$$

in which N is the dimensionless quantity defined in Eq. 9.7–8.

Example 9.7–1. Error in Thermocouple Measurement

In Fig. 9.7–2 a thermocouple is shown in a cylindrical well inserted into a gas stream. Estimate the true temperature of the gas stream if

T_1 = 500° F = temperature indicated by thermocouple
T_w = 350° F = wall temperature
h = 120 Btu hr^{-1} ft^{-2} (° F)$^{-1}$
k = 60 Btu hr^{-1} ft^{-1} (° F)$^{-1}$
B = 0.08 in. = thickness of well wall
L = 0.2 ft = length of well

Solution. The thermocouple well wall of thickness B is in contact with the gas stream on one side only, and the tube thickness is small compared with the diameter. Hence the temperature distribution along this wall will be about the same as that

[2] M. Jakob, *loc. cit.*, p. 235.

along a bar of thickness $2B$, in contact with the gas stream on both sides. According to Eq. 9.7–13, the temperature at the end of the well (that registered by the thermocouple) is

$$\frac{T_1 - T_a}{T_w - T_a} = \frac{\cosh 0}{\cosh N} = \frac{1}{\cosh \sqrt{hL^2/kB}}$$
$$= \frac{1}{\cosh \sqrt{(120)(0.2)^2/(60)(0.08/12)}}$$
$$= \frac{1}{\cosh 2\sqrt{3}} = \frac{1}{16.0} \qquad (9.7\text{–}17)$$

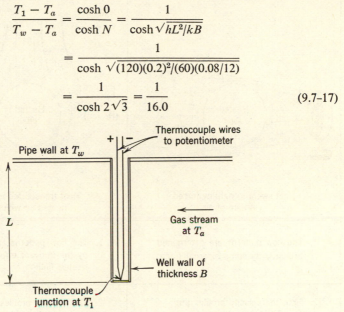

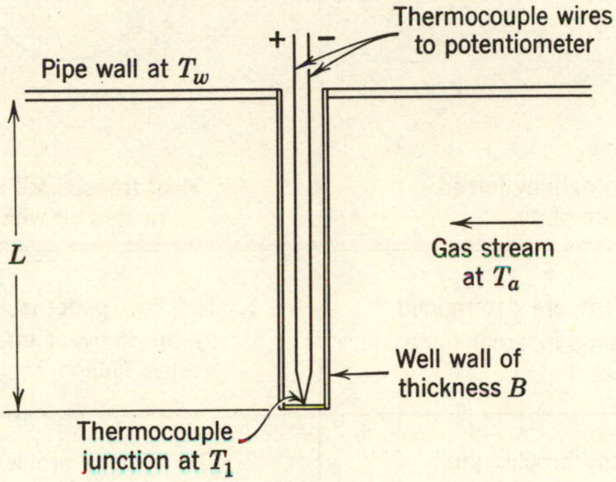

Fig. 9.7–2. Thermocouple in cylindrical well.

Hence the actual ambient gas temperature is obtained by solving this equation for T_a:

$$\frac{500 - T_a}{350 - T_a} = \frac{1}{16.0}$$

whence

$$T_a = 510° \text{ F}$$

Therefore, the reading is 10° F too low. (Had the variation of temperature in the x-direction been taken into account and the corresponding two-directional heat-flow problem been solved, an answer of 511° F would have been obtained.)

This example has indicated one kind of error that may be made in thermometry. Frequently a simple analysis, such as the foregoing, can be used to estimate the magnitude of the errors involved.[3]

§9.8 FORCED CONVECTION

In the preceding sections the emphasis has been placed on heat conduction in solids. In this and the following section we study two limiting types of heat transfer in fluids: *forced convection* and *free convection*. The salient

[3] For a further discussion, see M. Jakob, *loc. cit.*, Vol. II, Chapter 33, pp. 147–201.

Forced Convection Heat Transfer	Free Convection Heat Transfer
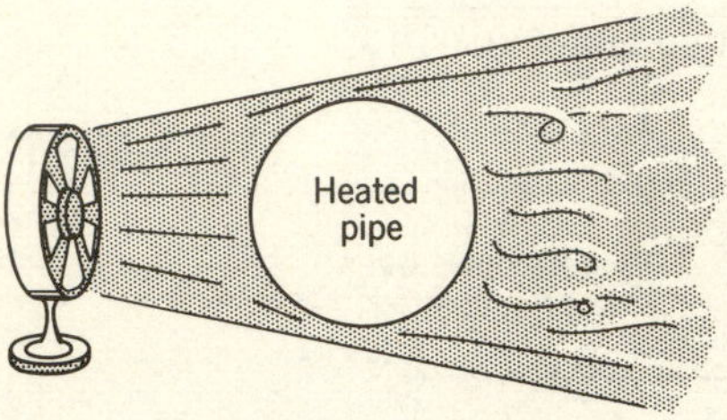	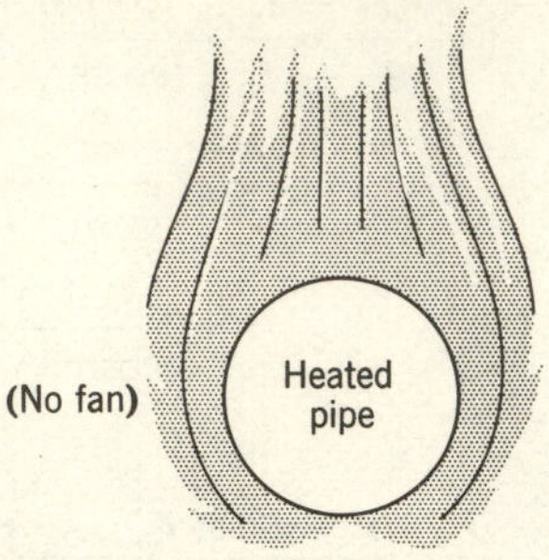
Heat swept to right by forced stream of air	Heat transported upward by heated air which rises
1. The flow patterns are determined primarily by some external force	1. The flow patterns are determined by the buoyant effect of the heated fluid
2. First, the velocity profiles are found; then they are used to find the temperature profiles (usual procedure for fluids with constant physical properties)	2. The velocity profiles and temperature profiles are intimately connected
3. The Nusselt number depends on Reynolds number and Prandtl number. (See Chapter 13)	3. The Nusselt number depends on Grashof number and Prandtl number. (See Chapter 13)

Fig. 9.8–1. Comparison of forced and free convection in nonisothermal systems.

differences between these two kinds of behavior are set forth in Fig. 9.8–1. Most industrial heat transfer problems have in the past been arbitrarily put into either one or the other of these two limiting categories; in some problems, however, both effects must be reckoned with.

In this section we consider a problem in steady-state forced convection, a limiting case of which is simple enough to be solved analytically.[1,2] A

[1] A. Eagle and R. M. Ferguson, *Proc. Roy. Soc.* (*London*), **A127**, 540–566 (1930).

[2] S. Goldstein, *Modern Developments in Fluid Dynamics*, Oxford University Press (1938), Vol. II, p. 622.

viscous fluid with constant physical properties (ρ, μ, k, $\hat{C}_p$) is in laminar flow in a circular tube of radius R. For $z < 0$ the fluid temperature is uniform at T_0; For $z > 0$ there is a constant wall heat flux q_1. Such a condition exists, for example, when a pipe is wrapped uniformly with an electrical heating coil.

As indicated in Fig. 9.8–1, the first step in solving a forced-convection heat transfer problem is the calculation of the velocity profiles in the system. We have seen in §2.3 how this may be accomplished for tube flow by application of the method of the shell momentum balance. We know that the velocity distribution so obtained, at least far enough downstream from the inlet so that the entrance length has been exceeded, is

$$v_z = v_{z,\max}\left[1 - \left(\frac{r}{R}\right)^2\right] \qquad (9.8\text{–}1)$$

in which $v_{z,\max}$ is given as $(\mathscr{P}_0 - \mathscr{P}_L)R^2/4\mu L$.

To get the temperature distribution, we have to make a thermal energy balance over a ring-shaped element such as that shown in Fig. 9.8–2, inasmuch as T is clearly a function of r and z. Energy enters and leaves this ring by thermal conduction in both r- and z-directions (solid arrows). Also, energy will enter and leave by fluid entering and leaving the ring, the fluid carrying with it a certain amount of "sensible heat." Hence we have the following contributions to the energy balance:

energy in by conduction at r	$q_r\|_r \cdot 2\pi r\,\Delta z$	(9.8–2)
energy out by conduction at $r + \Delta r$	$q_r\|_{r+\Delta r} \cdot 2\pi(r + \Delta r)\,\Delta z$	(9.8–3)
energy in by conduction at z	$q_z\|_z \cdot 2\pi r\,\Delta r$	(9.8–4)
energy out by conduction at $z + \Delta z$	$q_z\|_{z+\Delta z} \cdot 2\pi r\,\Delta r$	(9.8–5)
energy in with flowing fluid at z	$\rho\hat{C}_p v_z(T - T_0)\|_z \cdot 2\pi r\,\Delta r$	(9.8–6)
energy out with flowing fluid at $z + \Delta z$	$\rho\hat{C}_p v_z(T - T_0)\|_{z+\Delta z} \cdot 2\pi r\,\Delta r$	(9.8–7)

These last two terms represent the heat content of the entering and leaving streams with respect to the datum temperature T_0; such a datum temperature has to be chosen because energies cannot be expressed absolutely but only relatively. The choice of datum plane is arbitrary; we shall see presently that it does not appear in the differential equation.

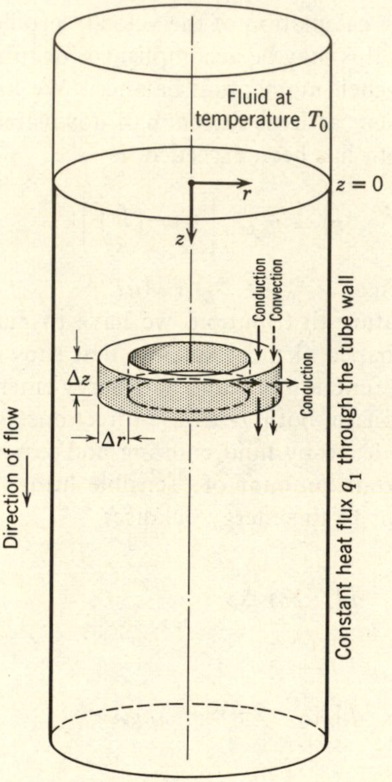

Fig. 9.8–2. Annular ring over which energy balance is made to get differential equations for the temperature distribution in laminar tube flow.

When energy inputs are equated to energy outputs, one obtains the statement of the energy balance on the annular ring in Fig. 9.8–2. Division by $2\pi\ \Delta r\ \Delta z$ then gives

$$\frac{(rq_r)|_{r+\Delta r} - (rq_r)|_r}{\Delta r} + r\,\frac{q_z|_{z+\Delta z} - q_z|_z}{\Delta z} + r\rho\hat{C}_p v_z\,\frac{T|_{z+\Delta z} - T|_z}{\Delta z} = 0 \qquad (9.8\text{–}8)$$

Now Δr and Δz are allowed to approach zero to give

$$\rho\hat{C}_p v_z\,\frac{\partial T}{\partial z} = -\frac{1}{r}\frac{\partial}{\partial r}(rq_r) - \frac{\partial q_z}{\partial z} \qquad (9.8\text{–}9)$$

Into this equation we introduce the velocity distribution of Eq. 9.8–1 and Fourier's law for the heat conduction in both r- and z-directions:

$$q_r = -k\frac{\partial T}{\partial r}; \qquad q_z = -k\frac{\partial T}{\partial z} \tag{9.8–10}$$

Then we get the following partial differential equation:

$$\rho\hat{C}_p v_{\max}\left[1 - \left(\frac{r}{R}\right)^2\right]\frac{\partial T}{\partial z} = k\left[\frac{1}{r}\frac{\partial}{\partial r}\left(r\frac{\partial T}{\partial r}\right) + \frac{\partial^2 T}{\partial z^2}\right] \tag{9.8–11}$$

Usually the heat conduction in the z-direction (the term containing $\partial^2 T/\partial z^2$) is small in comparison with the convective transfer term (the term containing $\partial T/\partial z$) with the result that $\partial^2 T/\partial z^2$ may be dropped from Eq. 9.8–11. One situation in which this is not permissible is in the slow flow of substances with high thermal conductivity, such as liquid metals. When the z-direction conductive term is omitted we finally obtain

$$\rho\hat{C}_p v_{\max}\left[1 - \left(\frac{r}{R}\right)^2\right]\frac{\partial T}{\partial z} = k\frac{1}{r}\frac{\partial}{\partial r}\left(r\frac{\partial T}{\partial r}\right) \tag{9.8–12}$$

This is a partial differential equation, which, when solved, gives the temperature as a function of both r and z in the tube. The boundary conditions are

B.C. 1: at $r = 0$, $T = \text{finite}$ (9.8–13)

B.C. 2: at $r = R$, $-k\dfrac{\partial T}{\partial r} = q_1$ (a constant) (9.8–14)

B.C. 3: at $z = 0$, $T = T_0$ (for all r) (9.8–15)

To simplify manipulations, we introduce the following dimensionless quantities:

$$\Theta = \frac{T - T_0}{q_1 R/k}; \qquad \xi = \frac{r}{R}; \qquad \zeta = \frac{zk}{\rho\hat{C}_p v_{z,\max} R^2} \tag{9.8–16,17,18}$$

The equation for the temperature distribution then becomes

$$(1 - \xi^2)\frac{\partial\Theta}{\partial\zeta} = \frac{1}{\xi}\frac{\partial}{\partial\xi}\left(\xi\frac{\partial\Theta}{\partial\xi}\right) \tag{9.8–19}$$

with boundary conditions

B.C. 1: at $\xi = 0$, $\Theta = \text{finite}$ (9.8–20)

B.C. 2: at $\xi = 1$, $-\dfrac{\partial\Theta}{\partial\xi} = 1$ (9.8–21)

B.C. 3: at $\zeta = 0$, $\Theta = 0$ (9.8–22)

The partial differential equation in Eq. 9.8–19 has been solved for these boundary conditions,[3] but we do not give the complete solution here.

It is, however, instructive to obtain the limiting form of the solution to

[3] R. Siegel, E. M. Sparrow, and T. M. Hallman, *Appl. Sci. Research*, **A7**, 386–392 (1958).

Eq. 9.8–19 for large ζ. After the fluid is far downstream from the beginning of the heated section, one expects intuitively that the constant heat flux through the wall will result in a rise in the fluid temperature that is linear in ζ. One further expects that the shape of the radial temperature profiles will ultimately not undergo further change with increasing ζ. Hence a solution of the following form seems reasonable for large ζ.

$$\Theta = C_0\zeta + \Psi(\xi) \tag{9.8–23}$$

in which C_0 is a constant to be determined presently.

The function in Eq. 9.8–23 is clearly not the complete solution to the problem. Although that function does allow the partial differential equation and also boundary conditions "1" and "2" to be satisfied, it is apparent that boundary condition "3" cannot be satisfied. Hence we replace the latter by

B.C. 3′: $$-2\pi R z q_1 = \int_0^{2\pi}\int_0^R \rho \hat{C}_p (T - T_0) v_z r\, dr\, d\theta \tag{9.8–24}$$

or

$$-\zeta = \int_0^1 \Theta(\xi, \zeta)(1 - \xi^2)\xi\, d\xi \tag{9.8–25}$$

This condition states that the heat entering through the walls is the same as the difference between the heat transported through the cross section at $\zeta = \zeta$ and that at $\zeta = 0$.

Substitution of Eq. 9.8–23 into Eq. 9.8–19 then gives the following ordinary differential equation for Ψ:

$$\frac{1}{\xi}\frac{d}{d\xi}\left(\xi \frac{d\Psi}{d\xi}\right) = C_0(1 - \xi^2) \tag{9.8–26}$$

This equation may be integrated twice, with respect to ξ, and the result substituted into Eq. 9.8–23 to give

$$\Theta = C_0\zeta + C_0\left\{\frac{\xi^2}{4} - \frac{\xi^4}{16}\right\} + C_1 \ln \xi + C_2 \tag{9.8–27}$$

in which C_1 and C_2 are constants of integration. These constants of integration are determined from the boundary conditions given above:

B.C. 1: $$C_1 = 0 \tag{9.8–28}$$

B.C. 2: $$C_0 = -4 \tag{9.8–29}$$

B.C. 3′: $$C_2 = +\tfrac{7}{24} \tag{9.8–30}$$

Substitution of these expressions into Eq. 9.8–27 gives finally[4]

$$\boxed{\Theta = -4\zeta - \xi^2 + \tfrac{1}{4}\xi^4 + \tfrac{7}{24}} \tag{9.8–31}$$

[4] A solution for the temperature distribution for very small values of ζ is to be found in Problem 9.R.

This result gives the temperature as a function of the dimensionless radial coordinate ξ and axial coordinate ζ. This expression is exact in the limit as $\zeta \rightarrow \infty$; for $\zeta = 0.1$, it predicts the local value of Θ to within about 2 per cent.

Once the temperature distribution is known, one can get various derived quantities. There are two kinds of average temperatures commonly used in connection with the flow of fluids with essentially constant ρ and $\hat{C}_p$:

$$\langle T \rangle = \frac{\int_0^{2\pi}\int_0^R T(r) r\,dr\,d\theta}{\int_0^{2\pi}\int_0^R r\,dr\,d\theta} \tag{9.8-32}$$

$$\frac{\langle v_z T \rangle}{\langle v_z \rangle} = \frac{\int_0^{2\pi}\int_0^R v_z(r) T(r) r\,dr\,d\theta}{\int_0^{2\pi}\int_0^R v_z(r) r\,dr\,d\theta} = T_b \tag{9.8-33}$$

Both averages are functions of z. The quantity $\langle T \rangle$ is just the arithmetic average of temperatures at any cross section. The "bulk temperature" T_b is the temperature one would measure if the tube were chopped off at z and if the fluid issuing forth were collected in a container and thoroughly mixed (hence this average temperature is sometimes referred to as the "cup-mixing temperature" or the "flow-average temperature").

Before leaving this section, we point out that the dimensionless axial coordinate ζ appearing above may be rewritten as

$$\zeta = \left[\frac{z}{R}\right]\left[\frac{\mu}{D\langle v_z \rangle \rho}\right]\left[\frac{k}{\mu \hat{C}_p}\right] = \left[\frac{z}{R}\right]\frac{1}{\text{Re Pr}} \tag{9.8-34}$$

in which D is the tube diameter, Re is the Reynolds number used in Part I, and $\text{Pr} = \hat{C}_p \mu / k$ is the Prandtl number. We shall find in Chapter 10 that the Reynolds and Prandtl numbers can generally be expected to appear in forced convection problems. This point is capitalized on in Chapter 13 in connection with correlation of heat transfer coefficients.

§9.9 FREE CONVECTION

In §9.8 an example of forced convection was studied. In this section we turn our attention to a free convection problem, namely the flow between two parallel walls of different temperatures. A fluid with density ρ and viscosity μ is placed between two vertical walls a distance $2b$ apart, as shown in Fig. 9.9-1. The heated wall at $y = -b$ is maintained at a temperature T_2 and the cooled wall at $y = +b$ is maintained at a temperature T_1. Because of the temperature gradient, the fluid near the hot wall rises and that near the cold wall descends. It is assumed that the system is so constructed that the volume rate of flow in the upward moving stream is the same as that in the downward moving stream.

If the plates are very long in the z-direction, then the temperature will be a

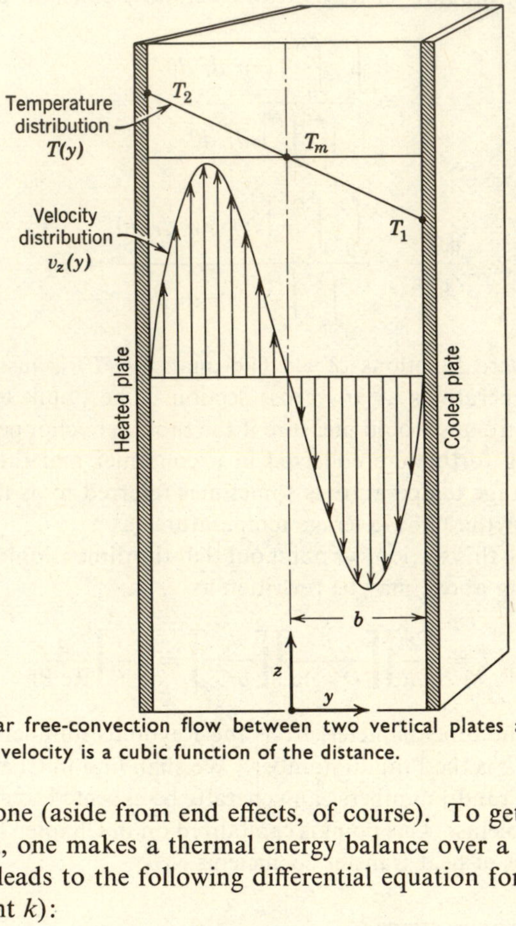

Fig. 9.9–1. Laminar free-convection flow between two vertical plates at two different temperatures; the velocity is a cubic function of the distance.

function of y alone (aside from end effects, of course). To get the temperature distribution, one makes a thermal energy balance over a shell of thickness Δy, which leads to the following differential equation for the temperature (for constant k):

$$k \frac{d^2T}{dy^2} = 0 \tag{9.9–1}$$

This equation is to be solved with the boundary conditions:

$$\text{B.C. 1:} \qquad \text{at} \quad y = -b, \qquad T = T_2 \tag{9.9–2}$$

$$\text{B.C. 2:} \qquad \text{at} \quad y = +b, \qquad T = T_1 \tag{9.9–3}$$

The solution is easily found to be

$$\boxed{T = T_m - \frac{1}{2}\Delta T\left(\frac{y}{b}\right)} \tag{9.9–4}$$

in which $\Delta T = T_2 - T_1$ is the difference between the wall temperatures, and $T_m = \frac{1}{2}(T_1 + T_2)$ is their arithmetic mean.

By making a momentum balance over the same slab of thickness Δy, one arrives at a differential equation for the velocity distribution:

$$\mu \frac{d^2 v_z}{dy^2} = \frac{dp}{dz} + \rho g \tag{9.9–5}$$

Here the viscosity μ has been assumed to be constant.

We now expand ρ in a Taylor series in T about some reference temperature $\bar{T}$ (as yet unspecified):

$$\begin{aligned} \rho &= \rho|_{\bar{T}} + \left.\frac{\partial \rho}{\partial T}\right|_{\bar{T}} (T - \bar{T}) + \cdots \\ &= \bar{\rho} - \bar{\rho}\bar{\beta}(T - \bar{T}) + \cdots \end{aligned} \tag{9.9–6}$$

Here we have introduced $\bar{\rho}$, the density at temperature $\bar{T}$, and the coefficient of volume expansion[1] $\bar{\beta}$, also evaluated at $\bar{T}$. Equation 9.9–6 might be regarded as a "Taylor-made" equation of state. Substitution of the first two terms of the Taylor series into Eq. 9.9–5 gives

$$\mu \frac{d^2 v_z}{dy^2} = \frac{dp}{dz} + \bar{\rho} g - \bar{\rho}\bar{\beta} g(T - \bar{T}) \tag{9.9–7}$$

If the pressure gradient in the system is due solely to the weight of the fluid in the slit, then $dp/dz = -\bar{\rho} g$ and the equation of motion becomes

$$\mu \frac{d^2 v_z}{dy^2} = -\bar{\rho}\bar{\beta} g(T - \bar{T}) \tag{9.9–8}$$

The physical meaning of this equation is that the viscous forces are just balanced by the buoyancy forces.

We now insert the expression for the temperature distribution (see Eq. 9.9–4) into Eq. 9.9–8 to get

$$\mu \frac{d^2 v_z}{dy^2} = -\bar{\rho}\bar{\beta} g\left[(T_m - \bar{T}) - \frac{1}{2}\Delta T\left(\frac{y}{b}\right)\right] \tag{9.9–9}$$

[1] The coefficient of volume expansion is defined by

$$\beta = \frac{1}{V}\left(\frac{\partial V}{\partial T}\right)_p = \frac{1}{(1/\rho)}\left(\frac{\partial (1/\rho)}{\partial T}\right)_p = -\frac{1}{\rho}\left(\frac{\partial \rho}{\partial T}\right)_p \tag{9.9–6a}$$

This equation is to be solved with the boundary conditions

B.C. 1: at $y = -b, \quad v_z = 0$ (9.9–10)

B.C. 2: at $y = +b, \quad v_z = 0$ (9.9–11)

The solution is

$$v_z = \frac{\bar{\rho}\bar{\beta}gb^2\,\Delta T}{12\mu}[\eta^3 - A\eta^2 - \eta + A] \tag{9.9–12}$$

in which $A = 6(T_m - \bar{T})/\Delta T$ and $\eta = y/b$.

Now we require that the net volume flow in the z-direction be zero; that is

$$\int_{-1}^{+1} v_z\,d\eta = 0 \tag{9.9–13}$$

Substitution of Eq. 9.9–12 into this expression gives

$$-\tfrac{2}{3}A + 4A = 0 \tag{9.9–14}$$

whence $A = 0$ or $\bar{T} = T_m$. Hence the final expression for the velocity distribution is

$$\boxed{v_z = \frac{\bar{\rho}\bar{\beta}gb^2\,\Delta T}{12\mu}(\eta^3 - \eta)} \tag{9.9–15}$$

We have seen how the velocity distribution owes its existence to the buoyancy forces resulting from the temperature inequalities in the system. The velocity distribution shown in Eq. 9.9–15 is given graphically in Fig. 9.9–1. It is this sort of velocity distribution that occurs in the operation of a Clusius-Dickel column for separating isotopes or organic liquid mixtures by the combined effects of thermal diffusion and free convection.[2]

Before leaving this section, we note that Eq. 9.9–15 may be written in terms of a dimensionless velocity $\phi = bv_z\bar{\rho}/\mu$ and the dimensionless length $\eta = y/b$ thus:

$$\phi = \tfrac{1}{12}\,\text{Gr}\,(\eta^3 - \eta) \tag{9.9–16}$$

in which Gr is the Grashof number:

$$\text{Gr} = \left[\frac{\bar{\rho}^2\bar{\beta}gb^3\,\Delta T}{\mu^2}\right] = \left[\frac{\bar{\rho}gb^3\,\Delta\rho}{\mu^2}\right] \tag{9.9–17}$$

Here $\Delta\rho = \rho_1 - \rho_2$, and the second form of the Grashof number given here is obtained from the first form by Eq. 9.9–6. The Grashof number arises in problems on free convection, as is shown generally in Chapter 10. The use of the Grashof number in experimental heat transfer coefficient correlations is discussed in Chapter 13.

[2] See Chapter 18 for a brief discussion of thermal diffusion.

QUESTIONS FOR DISCUSSION

1. Compare and contrast the four systems discussed in §§9.2–9.5, which involve heat sources.

2. What is the momentum-transport analog of a heat source?

3. To what problem in electrical circuits is the addition of thermal resistances analogous?

4. What is the coefficient of volume expansion β for an ideal gas? What is the corresponding expression for the Grashof number?

5. Give examples of various kinds of convection heat transfer in meteorology.

6. How would B.C. 2 in Eq. 9.2–12 be changed if one were told that heat is being lost by Newton's law of cooling to a surrounding gas of temperature T_g?

7. Explain in detail the averaging operations in Eq. 9.2–15, 9.8–32 and 33.

8. What might be some consequences of large temperature gradients produced by viscous heat effects in plastics extrusion and in lubrication?

9. At steady state the temperature profiles in a laminated system appear thus:

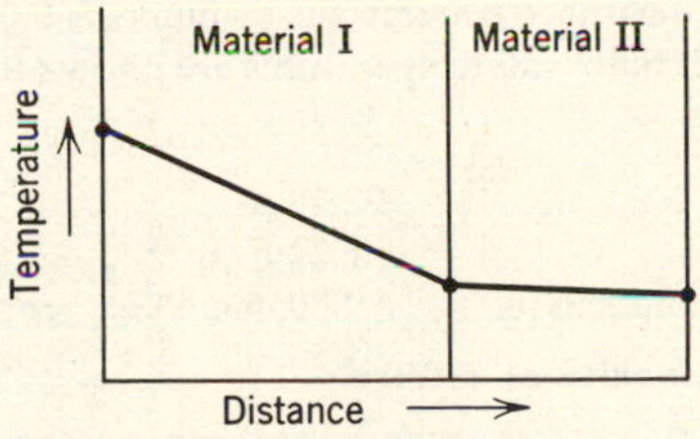

Which material has the higher thermal conductivity?

10. Is Eq. 9.5–1 consistent with the Arrhenius equation of chemical kinetics?

11. In Eq. 9.6–30 why is it necessary to specify that U_0 is "based on the inner surface"?

12. Show that the Grashof, Brinkman, and Prandtl numbers are dimensionless.

13. What is the average velocity in the upward moving stream in Fig. 9.9–1?

14. In §9.4 how (qualitatively) do the results have to be modified if the viscosity changes noticeably because of the temperature rise?

15. In Eq. 9.4–11 what happens when $T_b = T_0$?

16. What is the limiting value of η in Eq. 9.7–16 when the fin becomes very short? Is this reasonable?

17. How might the effect studied in §9.4 be important as an error in viscometric measurements?

PROBLEMS

9.A$_1$ Heat Loss from an Insulated Pipe

A standard Schedule 40 two-inch steel pipe (inside diameter 2.067 in. and wall thickness 0.154 in.) carrying steam is lagged (i.e., insulated) with 2 in. of 85 per cent magnesia covered in turn with 2 in. of cork. Estimate the heat loss per hour per foot of pipe if the inner surface of the pipe is at 250° F and the outer surface of the cork is at 90° F. The thermal conductivities of the substances concerned are

steel	26.1 Btu hr^{-1} ft^{-1} ° F^{-1}
85 per cent magnesia	0.04
cork	0.03

Answer: 24 Btu hr^{-1} ft^{-1}

9.B_1 Heat Loss from a Rectangular Fin

Calculate the heat loss from a rectangular fin (see Fig. 9.7–1) for the following conditions:

air temperature	350° F
wall temperature	500° F
thermal conductivity of fin	60 Btu hr^{-1} ft^{-1} ° F^{-1}
heat transfer coefficient	120 Btu hr^{-1} ft^{-2} ° F^{-1}
length of fin	0.2 ft
width of fin	1.0 ft
thickness of fin	0.16 in.

Answer: 2080 Btu hr^{-1}

9.C_1 Maximum Temperature in a Lubricant

An oil is acting as a lubricant for a pair of cylindrical surfaces such as those shown in Fig. 9.4–1. The angular velocity of the outer cylinder is 7908 rpm. The outer cylinder has a radius of 5.06 cm, and the clearance between the cylinders is 0.027 cm. What is the maximum temperature in the oil if both wall temperatures are known to be 158° F? The physical properties of the oil are

viscosity	92.3 cp
density	1.22 g cm^{-3}
thermal conductivity	0.0055 cal sec^{-1} cm^{-1} ° C^{-1}

Answer: 174° F

9.D_1 Current-Carrying Capacity of a Wire

A copper wire 0.040 in. in diameter is insulated uniformly with plastic to an outer diameter of 0.12 in. and is exposed to surroundings at 100° F. The heat transfer coefficient from the outer surface of the plastic to the surroundings is 1.5 Btu hr^{-1} ft^{-2} ° F^{-1}. What is the maximum steady current, in amperes, that this wire can carry without heating any part of the plastic above its operating limit of 200° F? The thermal and electrical conductivities may be assumed constant at the values listed below.

	k, Btu hr^{-1} ft^{-1} ° F^{-1}	k_e, ohm^{-1} cm^{-1}
copper	220	5.1×10^5
plastic	0.20	0.0

Answer: 13.7 amp

9.E_1 Free Convection Velocity

What is the *average* velocity in the upward-moving stream in the system described in Fig. 9.9–1 for air flowing under the following conditions?

pressure	1 atm
temperature of the heated wall	100° C
temperature of the cooled wall	20° C
spacing between the walls	0.6 cm

Answer: 2.3 cm sec^{-1}

9.F_2 Evaporation Loss from an Oxygen Tank

a. Liquefied gases are sometimes stored in well-insulated spherical containers vented to the atmosphere. Develop an expression for the steady-state heat transfer rate through the walls of such a container; call the radii of the inner and outer walls r_0 and r_1. Let it be assumed that the temperatures T_0 and T_1 (at $r = r_0$ and $r = r_1$) are known. Assume that the thermal conductivity of the insulation varies linearly with the temperature according to the relation

$$k = k_0 + (k_1 - k_0)\left(\frac{T - T_0}{T_1 - T_0}\right)$$

What is the physical significance of the constants k_0 and k_1?

b. Estimate the rate of evaporation of liquid oxygen from a spherical container of 6 ft inside diameter covered with 1 ft of asbestos insulation. The following information is available:

temperature at inner surface of insulation	$-183°$ C
temperature at outer surface of insulation	$0°$ C
boiling point of O_2	$-183°$ C
heat of vaporization of O_2	1636 cal g-mole^{-1}
thermal conductivity of insulation at $0°$ C	0.090 Btu hr^{-1} ft^{-1} °F^{-1}
thermal conductivity of insulation at $-183°$ C	0.072 Btu hr^{-1} ft^{-1} °F^{-1}

Answer: a. $Q_0 = 4\pi r_0 r_1\left(\frac{k_0 + k_1}{2}\right)\left(\frac{T_1 - T_0}{r_1 - r_0}\right)$

b. 19.8 kg hr^{-1}

9.G$_2$ Alternative Methods of Setting Up the Heated Wire Problem

a. Show that Eq. 9.2–9 can be derived by making a thermal energy balance on a *cylinder* of radius r and length L in the heated wire (rather than on a *cylindrical shell* as in §9.2).

b. Substitute Fourier's law into Eq. 9.2–6 and obtain a second-order equation for the temperature. What are the boundary conditions to be placed on the equation? Integrate the equation with these boundary conditions and obtain Eq. 9.2–13.

9.H$_2$ Heat Conduction from a Sphere to a Stagnant Fluid

A heated sphere of radius R is suspended in a large, motionless body of fluid. It is desired to study the heat conduction in the fluid surrounding the sphere. It is assumed in this problem that free convection effects (see §9.9) can be neglected.

a. Set up the differential equation describing the temperature T in the surrounding fluid as a function of r, the distance from the center of the sphere. The thermal conductivity of the fluid k is constant.

b. Integrate the differential equation and use the boundary conditions

B.C. 1: at $r = R$, $T = T_R$

B.C. 2: at $r = \infty$, $T = T_\infty$

to determine the constants of integration.

c. From the temperature profile, obtain an expression for the heat flux at the surface. Equate this result to the heat flux written as "Newton's law of cooling" and show that a dimensionless heat transfer coefficient (known as the "Nusselt number") is given by

$$\text{Nu} = \frac{hD}{k} = 2 \tag{9.H-1}$$

in which D is the sphere diameter. This is a well-known result, which is the limiting value of Nu for heat transfer from spheres at low Reynolds or Grashof numbers (see Chapters 13 and 21), e.g., for small spheres.

9.I$_2$ Heat Conduction in a Nuclear Fuel Rod Assembly

Consider a long nuclear fuel rod, which is surrounded by an annular layer of aluminum "cladding," as shown in Fig. 9.I. Within the fuel rod heat is produced by fission; this heat source is dependent on position, with a source strength varying approximately as

$$S_n = S_{n0}\left[1 + b\left(\frac{r}{R_F}\right)^2\right] \tag{9.I-1}$$

Here S_{n0} is the heat per unit volume per unit time produced at $r = 0$, and r is the distance from the axis of the fuel rod. Calculate the maximum temperature in the fuel rod, if the

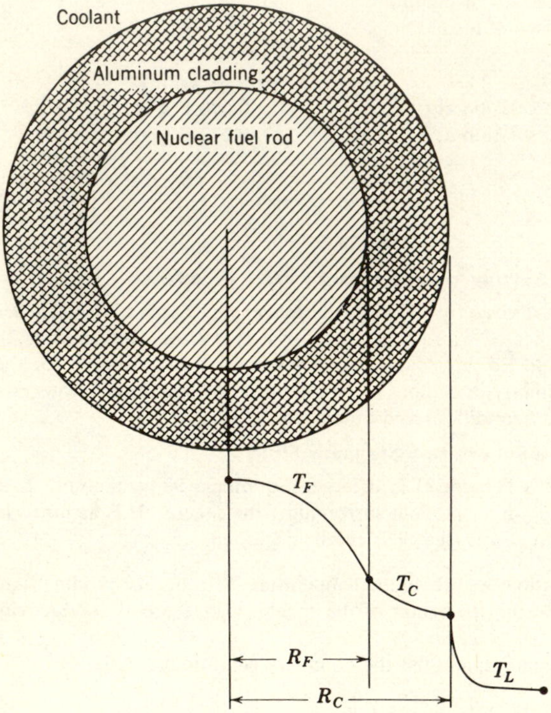

Fig. 9.I. Temperature distribution in a fuel rod assembly.

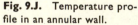

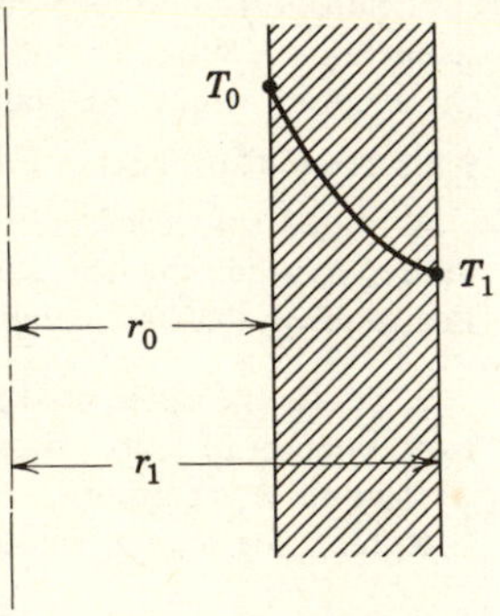

Fig. 9.J. Temperature profile in an annular wall.

outer surface of the cladding is in contact with a liquid coolant at temperature T_L, the heat transfer coefficient at the cladding-coolant interface being h_L. The thermal conductivities of fuel rod and cladding are k_F and k_C.

Answer: $$(T_{F,\max} - T_L) = \frac{S_{n0}R_F^2}{4k_F}\left(1 + \frac{b}{4}\right) + \frac{S_{n0}R_F^2}{2k_C}\left(1 + \frac{b}{2}\right)\left(\frac{k_C}{R_C h_L} + \ln\frac{R_C}{R_F}\right)$$

9.J$_2$ Heat Conduction in an Annulus

a. Heat is flowing through an annular wall of inside radius r_0 and outside radius r_1. The thermal conductivity varies linearly with temperature from k_0 at T_0 to k_1 at T_1. Develop an expression for the heat flow through the wall at $r = r_0$. (See Fig. 9.J.).

b. Show how the expression in (*a*) can be simplified when $(r_1 - r_0)$ is very small. Interpret the result physically.

Answer: *a.* $Q_0 = 2\pi L(T_0 - T_1)\left(\frac{k_0 + k_1}{2}\right)\left(\ln \frac{r_1}{r_0}\right)^{-1}$

b. $Q_0 = 2\pi r_0 L\left(\frac{k_0 + k_1}{2}\right)\left(\frac{T_0 - T_1}{r_1 - r_0}\right)$

9.K₂ Heat Generation in a Non-Newtonian Fluid

Rework the problem discussed in §9.4 for an Ostwald-de Waele non-Newtonian fluid (see §1.2). Show that the temperature distribution is the same as that in Eq. 9.4–11 with Br replaced by Br_n where

$$\text{Br}_n = \left[\frac{mV^{n+1}}{kb^{n-1}(T_b - T_0)}\right] \tag{9.K-1}$$

9.L₂ Calculation of Insulation Thickness for a Furnace Wall

A furnace wall consists of three layers, as shown in Fig. 9.L: first, a layer of heat resistant or refractory brick; second, a layer of insulating brick; and, finally, a steel plate for mechanical protection, which is ¼ in. thick.

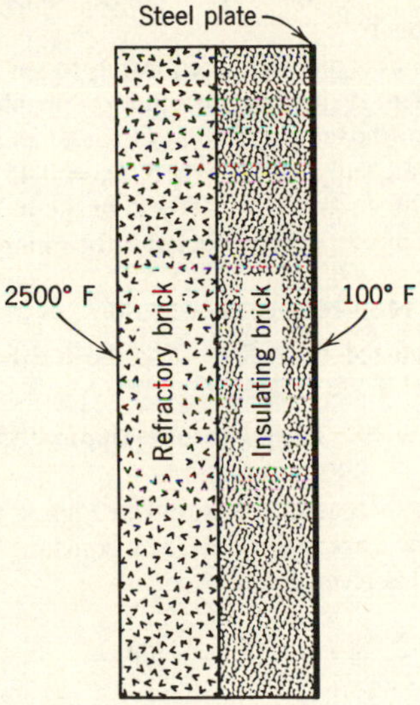

Fig. 9.L. A composite furnace wall.

Calculate the thickness of each layer of brick to give minimum total wall thickness if the heat loss through the wall is to be 5000 Btu ft^{-2} hr^{-1}, assuming that the layers are in good thermal contact. The following information is available:

Material	Maximum Service Temperature	Thermal Conductivity (Btu hr^{-1} ft^{-1} °F^{-1}) at 100° F	at 2000° F
Refractory brick	2600° F	1.8	3.6
Insulating brick	2000° F	0.9	1.8
Steel	—	26.1	—

Answer: refractory brick, 0.39 ft
insulating brick, 0.51 ft

9.M$_2$ Radial Temperature Gradients in an Annular Chemical Reactor

A catalytic reaction is being carried out at constant pressure in a packed bed in an annular space between two coaxial cylinders with inner radius R_0 and outer radius R_1. Such a configuration occurs when temperatures are measured with a centered thermowell and is in addition useful for controlling temperature gradients if a thin annulus is used. The entire inner wall is at uniform temperature T_0, and it can be assumed that there is almost no heat transfer through this surface. The reaction releases heat at a uniform volumetric rate S_c throughout the reactor. The effective thermal conductivity k_{eff} of the reactor contents is to be treated as a constant throughout.

a. By a shell thermal energy balance, derive a second-order differential equation that describes the temperature profiles, assuming that the temperature gradients in the axial direction can be neglected. What boundary conditions must be used?

b. Rewrite the differential equation and boundary conditions in terms of the dimensionless length and dimensionless temperature defined here:

$$\xi = \frac{r}{R_0}; \qquad \Theta = \frac{T - T_0}{S_c R_0^2/4k_{\text{eff}}} \tag{9.M-1,2}$$

Why are these logical choices?

c. Integrate the dimensionless differential equation to get the radial temperature profile. To what viscous flow problem is this heat conduction problem analogous?

d. Develop expressions for the temperature at the outer wall and the average temperature.

e. Calculate the outer wall temperature when $R_0 = 0.45$ in., $R_1 = 0.50$ in., $k_{\text{eff}} = 0.3$ Btu hr^{-1} ft^{-1} °F^{-1}, $T_0 = 900°$ F, and $S_c = 480$ cal hr^{-1} cm^{-3}.

f. How would the results of part (*e*) be affected if the inner and outer radii were doubled?

Answer: e. 898.8° F

9.N$_2$ Heat Transfer to a Non-Newtonian Fluid

Extend the problem discussed in §9.8 to describe the heat transfer to a fluid that is described by the Ellis model in Eq. 1.2–5.

9.O$_3$ Viscous Heating with Temperature-Dependent Viscosity and Thermal Conductivity

Consider a flow situation such as that pictured in Figs. 9.4–1 and 2. Both the stationary surface and the moving surface are maintained at a constant temperature T_0. The temperature-dependence of k and μ is given by

$$\frac{k}{k_0} = 1 + \alpha_1\Theta + \alpha_2\Theta^2 + \cdots \tag{9.O-1}$$

$$\frac{\mu_0}{\mu} = \frac{\varphi}{\varphi_0} = 1 + \beta_1\Theta + \beta_2\Theta^2 + \cdots \tag{9.O-2}$$

in which φ is the fluidity and subscript 0 means "evaluated at T_0"; the dimensionless temperature Θ is defined by $\Theta = (T - T_0)/T_0$.

a. Show that the differential equations describing the viscous flow and heat conduction may be written in the following form:

$$\frac{d}{d\xi}\left(\frac{\mu}{\mu_0}\frac{d\phi}{d\xi}\right) = 0 \tag{9.O-3}$$

$$\frac{d}{d\xi}\left(\frac{k}{k_0}\frac{d\Theta}{d\xi}\right) + B\frac{\mu}{\mu_0}\left(\frac{d\phi}{d\xi}\right)^2 = 0 \tag{9.O-4}$$

in which $\phi = v_z/V$, $\xi = x/b$, and $B = \mu_0 V^2/k_0 T_0$.

b. The equation for the reduced velocity distribution may be integrated once to give $d\phi/d\xi = C_1 \cdot (\varphi/\varphi_0)$. This expression is then substituted into the heat balance equation to get

$$\frac{d}{d\xi}\left((1 + \alpha_1\Theta + \alpha_2\Theta^2 + \cdots)\frac{d\Theta}{d\xi}\right) + BC_1^2(1 + \beta_1\Theta + \beta_2\Theta^2 + \cdots) = 0 \quad (9.O\text{–}5)$$

in which C_1 is an integration constant from the ϕ-equation. Obtain the first two terms of a solution in the form

$$\Theta = B\Theta_1 + B^2\Theta_2 + B^3\Theta_3 + \cdots \quad (9.O\text{–}6)$$

$$\phi = \phi_0 + B\phi_1 + B^2\phi_2 + \cdots \quad (9.O\text{–}7)$$

in which the ϕ_i and Θ_i are functions of ξ but do not depend on B.

Hint: Example 9.2–2 suggests the technique to be used. It is further suggested that C_1 be expanded thus:

$$C_1 = C_{10} + BC_{11} + B^2C_{12} + \cdots \quad (9.O\text{–}8)$$

Answer: $\phi = \xi - \frac{1}{12}B\beta_1(\xi - 3\xi^2 + 2\xi^3) + \cdots$

$$\Theta = \tfrac{1}{2}B(\xi - \xi^2) - \tfrac{1}{8}B^2\alpha_1(\xi^2 - 2\xi^3 + \xi^4) - \tfrac{1}{24}B^2\beta_1(\xi - 2\xi^2 + 2\xi^3 - \xi^4) + \cdots$$

9.P$_3$ Forced Convection Heat Transfer for Non-Newtonian Flow in Tubes—Short Contact Times

A non-Newtonian fluid which can be described by the model in Eq. 1.2–5 is flowing in the system shown in Fig. 9.P. It is desired to find the temperature profiles and the wall heat flux for the special case of "short contact times." This means that the heat does not "penetrate" very far into the fluid.

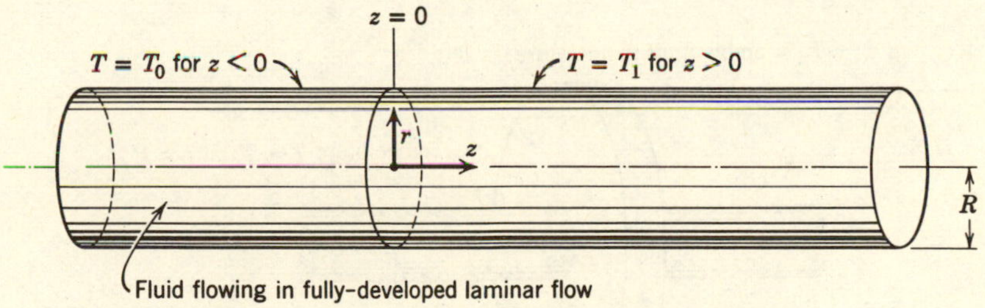

Fig. 9.P. Pipe with sudden change in wall temperature.

a. Obtain the velocity profile.

b. Establish by a thermal energy balance the differential equation for $T(r, z)$ by the method described in §9.8. Then set $s = R - r$ and discard any terms that are not important in the vicinity of the wall. Show that the equation becomes

$$\rho\hat{C}_p(\varphi_0\tau_R + \varphi_1\tau_R^{\alpha})s\frac{\partial T}{\partial z} = k\frac{\partial^2 T}{\partial s^2} \quad (9.P\text{–}1)$$

in which τ_R is the momentum flux at $r = R$. Rewrite Eq. 9.P–1 in terms of the following dimensionless variables:

$$\Theta = \frac{T - T_0}{T_1 - T_0}; \quad \zeta = \frac{z}{R}; \quad \sigma = \frac{s}{R}; \quad N = \frac{\rho\hat{C}_pR^2}{k}(\varphi_0\tau_R + \varphi_1\tau_R^{\alpha}) \quad (9.P\text{–}2,3,4,5)$$

c. Then show that if a solution of the following form is assumed

$$\Theta = f(\eta) \qquad \text{where } \eta = \left(\frac{N\sigma^3}{9\zeta}\right)^{1/3} \tag{9.P-6,7}$$

the partial differential equation from (*b*) is transformed into the ordinary differential equation

$$f'' + 3\eta^2 f' = 0 \tag{9.P-8}$$

in which the primes indicate differentiation with respect to η. What are the boundary conditions that go with Eq. 9.P–8? Solve Eq. 9.P–8 and get

$$f = [\Gamma(\tfrac{4}{3})]^{-1} \int_{\eta}^{\infty} e^{-\eta^3}\, d\eta \tag{9.P-9}$$

in which $\Gamma(\frac{4}{3})$ is the "gamma function" evaluated at $\frac{4}{3}$.

d. From the temperature distribution in (*c*), evaluate the wall heat flux as a function of the distance down the tube.

e. Integrate the result in (*d*) to obtain the total heat flow through the pipe surface between $z = 0$ and $z = L$.

Answer: e. $Q = 4\pi Rk(T_1 - T_0)\left(\frac{N}{9}\right)^{1/3}\left(\frac{L}{R}\right)^{2/3}[\Gamma(\frac{7}{3})]^{-1}$

9.Q_4 Heat Loss From a Circular Fin

a. Obtain the temperature profile $T(r)$ for a circular fin of thickness B on a pipe with outside wall temperature T_0. Make the same assumptions that were made in the study of the rectangular fin in §9.7. (See Fig. 9.Q.)

b. Derive an expression for the total heat loss from the fin.

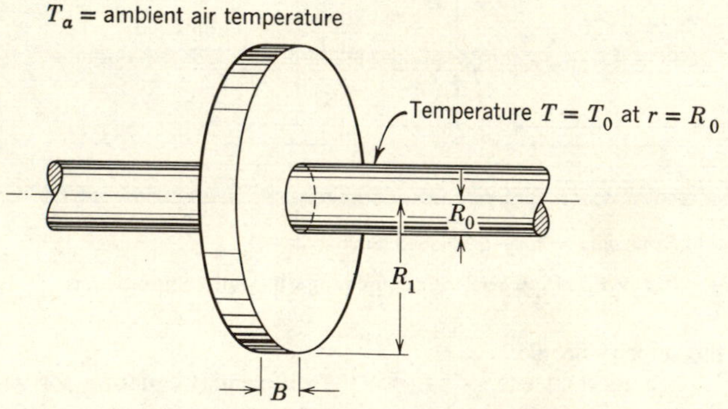

Fig. 9.Q. Circular fin on a heated pipe.

9.R_4 Forced Convection Heat Transfer for Non-Newtonian Flow in Tubes—Short Contact Times with Constant Wall Heat Flux[1]

In Problem 9.N the temperature profiles were calculated for a three-constant, non-Newtonian fluid flowing in a circular tube with constant wall flux. The calculation was patterned after the development in §9.8, hence the result is an asymptotic expression for the temperature profile valid only for distances far down the tube.

[1] R. B. Bird, *Chemie-Ing. Techn.*, **31**, 569–572 (1959).

In this problem we obtain an asymptotic solution in the entrance region to the heated tube. We imagine the fluid flowing in fully developed laminar flow at temperature $T = T_0$ in the region $z < 0$. In the region $z > 0$ there is a constant heat flux at the wall q_1. It is desired to find T as a function of r and z for small values of z. We use a method similar to that in Problem 9.P.

a. Show that the partial differential equation describing the heat transport is

$$\frac{\rho \hat{C}_p v_0}{R} s \frac{\partial T}{\partial z} = k \frac{\partial^2 T}{\partial s^2} \tag{9.R–1}$$

in which $v_0 = (\varphi_0 \tau_R + \varphi_1 \tau_R^{\alpha})R$. What is the physical significance of v_0?

b. Make use of Fourier's law to get the following equation for the heat flux:

$$\frac{\rho \hat{C}_p v_0}{R} \frac{\partial q_s}{\partial z} = k \frac{\partial}{\partial s}\left(\frac{1}{s} \frac{\partial q_s}{\partial s}\right) \tag{9.R–2}$$

c. Rewrite the equation in (*b*) in terms of the following dimensionless variables:

$$Q = \frac{q_s}{q_1}; \qquad \sigma = \frac{s}{R}; \qquad \zeta = \frac{z}{R}; \qquad \text{Pé}_0 = \frac{\rho \hat{C}_p R v_0}{k} \tag{9.R–3,4,5,6}$$

where Pé_0 is the Péclet number based on the velocity v_0 and the tube radius R. What are the boundary conditions?

d. Solve the partial differential equation in (*c*) by the method of combination of variables by anticipating the fact that Q depends upon χ only where χ is

$$\chi = \frac{\sigma}{\sqrt[3]{9\zeta/\text{Pé}_0}} \tag{9.R–7}$$

Show that this leads to the following ordinary differential equation for $Q(\chi)$:

$$\chi \frac{d^2 Q}{d\chi^2} + (3\chi^3 - 1)\frac{dQ}{d\chi} = 0 \tag{9.R–8}$$

What are the two boundary conditions?

e. Solve the equation in (*d*) and then use it to get the temperature profiles.

$$\textit{Answer:}\ \Theta = \frac{T - T_0}{q_1 R/k} = \sqrt[3]{\frac{9\zeta}{\text{Pé}_0}}\left\{\frac{e^{-\chi^3}}{\Gamma(\frac{2}{3})} - \chi\left[1 - \frac{\Gamma(\frac{2}{3}, \chi^3)}{\Gamma(\frac{2}{3})}\right]\right\}$$

where $\Gamma(\frac{2}{3}, \chi^3)$ is an incomplete gamma function.

CHAPTER 10

The Equations of Change for Nonisothermal Systems

In Chapter 9 we determined temperature profiles in several simple heat-transfer systems by the use of shell energy balances, and we showed how these profiles could be used to obtain other information, such as average temperatures and energy fluxes. For those situations in which heat transfer is accompanied by fluid flow, we found it necessary to use the equations of continuity and motion in addition to an energy balance. The developments of Chapter 9 should have given the reader some understanding of differential energy balances and their application to simple heat-transfer problems. In this chapter we generalize the shell energy balance to obtain *the equation of energy*, a partial differential equation that describes the transport of energy in a homogeneous fluid or solid.

This chapter is also closely related to Chapter 3, in which the reader was introduced to the equation of continuity (conservation of mass), the equation of motion (conservation of momentum), and the equation of mechanical energy (derivable from the equation of motion). The addition of the equation of energy (conservation of energy *or* the first law of thermodynamics) to this set permits extension of the preceding analyses to nonisothermal systems.

We begin in §10.1 by deriving the energy equation. Then we subtract from it the *mechanical* energy equation to get the more commonly used *thermal* energy equation. In §10.2 the thermal energy equation is given in rectangular, cylindrical, and spherical coordinates.

Although our main interest in this chapter is in the various energy equations, we find it useful in §10.3 to introduce an approximation to the equation of motion that is convenient for solving problems in free convection.

Because the equations of change for nonisothermal systems may be written in a variety of useful forms, we find it desirable to devote §10.4 to a summary for handy reference. In §10.5 we show how these tabulated equations can be used to set up heat-transfer problems by discarding unneeded terms.

Finally, in §10.6, we extend the dimensional analysis of §3.7 to obtain the key dimensionless groups that arise in heat-transfer problems.

§10.1 THE EQUATIONS OF ENERGY

Just as in §3.2 we generalized the shell momentum balances of Chapter 2 to obtain the equation of motion, we wish here to generalize the shell energy balances of Chapter 9 to obtain the *equations of energy*. To do this, we start as before with a stationary volume element through which a pure fluid is flowing; we then write the law of conservation of energy for the fluid contained within this volume element at any given time:

$$\left\{\begin{matrix}\text{rate of}\\ \text{accumulation}\\ \text{of internal}\\ \text{and kinetic}\\ \text{energy}\end{matrix}\right\} = \left\{\begin{matrix}\text{rate of}\\ \text{internal and}\\ \text{kinetic energy}\\ \text{in}\\ \text{by convection}\end{matrix}\right\} - \left\{\begin{matrix}\text{rate of}\\ \text{internal and}\\ \text{kinetic energy}\\ \text{out}\\ \text{by convection}\end{matrix}\right\}$$

$$+ \left\{\begin{matrix}\text{net rate of}\\ \text{heat addition}\\ \text{by conduction}\end{matrix}\right\} - \left\{\begin{matrix}\text{net rate of}\\ \text{work done by}\\ \text{system on}\\ \text{surroundings}\end{matrix}\right\} \quad (10.1\text{–}1)$$

This is the first law of thermodynamics written for an "open," unsteady-state system. Actually this statement of the first law is not complete in that other forms of energy and energy transport, such as nuclear, radiative, and electromagnetic, are not included. Equation 10.1–1 does represent a generalization over Eq. 9.1–1 because work and kinetic energy effects are included; also unsteady behavior is allowed. In Eq. 10.1–1 *kinetic energy* is interpreted as the energy associated with the observable fluid motion (i.e. $\frac{1}{2}\rho v^2$, on a per-unit-volume basis). By *internal energy* we understand the energy associated with the random translational and internal motions of the molecules plus the energy of interaction between the molecules; that is, the internal energy is dependent on the local temperature and density of the fluid. The *potential energy* of the fluid as a whole does not appear explicitly in

Eq. 10.1–1, inasmuch as we have chosen to include it in the work term; this will be discussed presently.

We now write Eq. 10.1–1 explicitly for the fixed volume element $\Delta x\, \Delta y\, \Delta z$. (See Fig. 3.2–1.) The rate of accumulation of internal and kinetic energy within $\Delta x\, \Delta y\, \Delta z$ is

$$\Delta x\, \Delta y\, \Delta z \frac{\partial}{\partial t}(\rho \hat{U} + \tfrac{1}{2}\rho v^2) \tag{10.1–2}$$

Here $\hat{U}$ is the internal energy per unit mass of the fluid in the element, and v is the magnitude of the local fluid velocity.

The net rate of convection of internal and kinetic energy into the element is

$$\begin{aligned} &\Delta y\, \Delta z\{v_x(\rho \hat{U} + \tfrac{1}{2}\rho v^2)|_x - v_x(\rho \hat{U} + \tfrac{1}{2}\rho v^2)|_{x+\Delta x}\} \\ &+ \Delta x\, \Delta z\{v_y(\rho \hat{U} + \tfrac{1}{2}\rho v^2)|_y - v_y(\rho \hat{U} + \tfrac{1}{2}\rho v^2)|_{y+\Delta y}\} \\ &+ \Delta x\, \Delta y\{v_z(\rho \hat{U} + \tfrac{1}{2}\rho v^2)|_z - v_z(\rho \hat{U} + \tfrac{1}{2}\rho v^2)|_{z+\Delta z}\} \end{aligned} \tag{10.1–3}$$

The net rate of energy input by conduction is

$$\begin{aligned} \Delta y\, \Delta z\{q_x|_x - q_x|_{x+\Delta x}\} + \Delta x\, \Delta z\{q_y|_y - q_y|_{y+\Delta y}\} \\ + \Delta x\, \Delta y\{q_z|_z - q_z|_{z+\Delta z}\} \end{aligned} \tag{10.1–4}$$

Here q_x, q_y, and q_z are the x-, y-, and z-components of the heat flux vector $\boldsymbol{q}$.

The work done by the fluid element against its surroundings consists of two parts: the work against the volume forces (e.g., gravity) and the work against the surface forces (i.e., pressure and viscous forces). Recall that (work) = (force) × (distance in the direction of the force) and that (rate of doing work) = (force) × (velocity in the direction of the force). The rate of doing work against the three components of the gravitational force per unit mass $\boldsymbol{g}$ is

$$-\rho\, \Delta x\, \Delta y\, \Delta z(v_x g_x + v_y g_y + v_z g_z) \tag{10.1–5}$$

The minus sign arises because work is done against gravity when $\boldsymbol{v}$ and $\boldsymbol{g}$ are opposed. The rate of doing work against the static pressure p at the six faces of $\Delta x\, \Delta y\, \Delta z$ is

$$\begin{aligned} \Delta y\, \Delta z\{(pv_x)|_{x+\Delta x} - (pv_x)|_x\} + \Delta x\, \Delta z\{(pv_y)|_{y+\Delta y} - (pv_y)|_y\} \\ + \Delta x\, \Delta y\{(pv_z)|_{z+\Delta z} - (pv_z)|_z\} \end{aligned} \tag{10.1–6}$$

Similarly, the rate of doing work against the viscous forces is

$$\begin{aligned} &\Delta y\, \Delta z\{(\tau_{xx}v_x + \tau_{xy}v_y + \tau_{xz}v_z)|_{x+\Delta x} - (\tau_{xx}v_x + \tau_{xy}v_y + \tau_{xz}v_z)|_x\} \\ &+ \Delta x\, \Delta z\{(\tau_{yx}v_x + \tau_{yy}v_y + \tau_{yz}v_z)|_{y+\Delta y} - (\tau_{yx}v_x + \tau_{yy}v_y + \tau_{yz}v_z)|_y\} \\ &+ \Delta x\, \Delta y\{(\tau_{zx}v_x + \tau_{zy}v_y + \tau_{zz}v_z)|_{z+\Delta z} - (\tau_{zx}v_x + \tau_{zy}v_y + \tau_{zz}v_z)|_z\} \end{aligned} \tag{10.1–7}$$

We now substitute the foregoing expressions into Eq. 10.1–1 and divide the entire equation by $\Delta x\, \Delta y\, \Delta z$. Taking the limit of the resulting expression as

Δx, Δy, and Δz approach zero, we obtain one form of the equation of energy:

$$\frac{\partial}{\partial t}(\rho\hat{U} + \tfrac{1}{2}\rho v^2)$$
$$= -\left(\frac{\partial}{\partial x} v_x(\rho\hat{U} + \tfrac{1}{2}\rho v^2) + \frac{\partial}{\partial y} v_y(\rho\hat{U} + \tfrac{1}{2}\rho v^2) + \frac{\partial}{\partial z} v_z(\rho\hat{U} + \tfrac{1}{2}\rho v^2)\right)$$
$$- \left(\frac{\partial q_x}{\partial x} + \frac{\partial q_y}{\partial y} + \frac{\partial q_z}{\partial z}\right) + \rho(v_x g_x + v_y g_y + v_z g_z)$$
$$- \left(\frac{\partial}{\partial x} p v_x + \frac{\partial}{\partial y} p v_y + \frac{\partial}{\partial z} p v_z\right)$$
$$- \left(\frac{\partial}{\partial x}(\tau_{xx}v_x + \tau_{xy}v_y + \tau_{xz}v_z) + \frac{\partial}{\partial y}(\tau_{yx}v_x + \tau_{yy}v_y + \tau_{yz}v_z)\right.$$
$$\left. + \frac{\partial}{\partial z}(\tau_{zx}v_x + \tau_{zy}v_y + \tau_{zz}v_z)\right) \qquad (10.1\text{–}8)$$

This equation may be written more compactly in vector-tensor notation as[1]

$$\underbrace{\frac{\partial}{\partial t}\rho(\hat{U} + \tfrac{1}{2}v^2)}_{\text{rate of gain of energy per unit volume}} = \underbrace{-(\nabla \cdot \rho \boldsymbol{v}(\hat{U} + \tfrac{1}{2}v^2))}_{\text{rate of energy input per unit volume by convection}} \quad \underbrace{-(\nabla \cdot \boldsymbol{q})}_{\text{rate of energy input per unit volume by conduction}}$$
$$\underbrace{+\rho(\boldsymbol{v} \cdot \boldsymbol{g})}_{\text{rate of work done on fluid per unit volume by gravitational forces}} \quad \underbrace{-(\nabla \cdot p\boldsymbol{v})}_{\text{rate of work done on fluid per unit volume by pressure forces}}$$
$$\underbrace{-(\nabla \cdot [\boldsymbol{\tau} \cdot \boldsymbol{v}])}_{\text{rate of work done on fluid per unit volume by viscous forces}} \qquad (10.1\text{–}9)$$

We now find it convenient here to rearrange the equation of energy with the aid of the equations of continuity and motion. Note that a similar rearrangement was made in going from one form of the equation of motion in Eq. 3.2–5 to the form given in Eq. 3.2–9 with the aid of the equation of continuity.

We start by carrying out the indicated differentiations of the left side of Eq. 10.1–9 and of the convective contribution to obtain on rearrangement

$$\rho\left[\frac{\partial}{\partial t}(\hat{U} + \tfrac{1}{2}v^2) + (\boldsymbol{v} \cdot \nabla(\hat{U} + \tfrac{1}{2}v^2))\right] + (\hat{U} + \tfrac{1}{2}v^2)\left[\frac{\partial \rho}{\partial t} + (\nabla \cdot \rho\boldsymbol{v})\right]$$
$$= -(\nabla \cdot \boldsymbol{q}) + \rho(\boldsymbol{v} \cdot \boldsymbol{g}) - (\nabla \cdot p\boldsymbol{v}) - (\nabla \cdot [\boldsymbol{\tau} \cdot \boldsymbol{v}]) \quad (10.1\text{–}10)$$

The first term on the left side of Eq. 10.1–10 is the local fluid density times the

[1] The development that follows is primarily devoted to mathematical juggling, in which use is made of standard mathematical and thermodynamic formulas. The reader may prefer, on first reading at least, to skip at once to Eq. 10.1–19, which follows with *no* additional assumptions from Eq. 10.1–9.

substantial derivative of $(\hat{U} + \frac{1}{2}v^2)$. The second can be seen, with the aid of the equation of continuity (Eq. 3.1–4), to be zero. We may then write

$$\rho \frac{D}{Dt}(\hat{U} + \tfrac{1}{2}v^2) = -(\nabla \cdot \boldsymbol{q}) + \rho(\boldsymbol{v} \cdot \boldsymbol{g}) - (\nabla \cdot p\boldsymbol{v}) - (\nabla \cdot [\boldsymbol{\tau} \cdot \boldsymbol{v}]) \tag{10.1–11}$$

Note that we have now obtained two forms of the equation of energy corresponding to the two forms obtained earlier for the equations of continuity (Eqs. 3.1–4 and 6) and motion (Eqs. 3.2–8 and 10): Eq. 10.1–9 describes the energy interchanges in a fluid from the standpoint of a stationary observer, and Eq. 10.1–11 describes these interchanges as seen by an observer moving with the fluid.

Equation 10.1–11 is an equation of change for the sum of $\hat{U}$ and $\frac{1}{2}v^2$. The reader may recall that an equation of change for $\frac{1}{2}v^2$ was derived in §3.3; that equation may be rewritten as

$$\rho \frac{D}{Dt}(\tfrac{1}{2}v^2) = p(\nabla \cdot \boldsymbol{v}) - (\nabla \cdot p\boldsymbol{v}) + \rho(\boldsymbol{v} \cdot \boldsymbol{g}) - (\nabla \cdot [\boldsymbol{\tau} \cdot \boldsymbol{v}]) + (\boldsymbol{\tau} : \nabla \boldsymbol{v}) \tag{10.1–12}$$

Subtraction of Eq. 10.1–12 from Eq. 10.1–11 yields an equation of change for $\hat{U}$

$$\rho \frac{D\hat{U}}{Dt} = -(\nabla \cdot \boldsymbol{q}) - p(\nabla \cdot \boldsymbol{v}) - (\boldsymbol{\tau} : \nabla \boldsymbol{v}) \tag{10.1–13}$$

$\rho \frac{D\hat{U}}{Dt}$	$-(\nabla \cdot \boldsymbol{q})$	$-p(\nabla \cdot \boldsymbol{v})$	$-(\boldsymbol{\tau} : \nabla \boldsymbol{v})$
rate of gain of internal energy per unit volume	rate of internal energy input by conduction per unit volume	*reversible* rate of internal energy increase per unit volume by compression	*irreversible* rate of internal energy increase per unit volume by viscous dissipation

By analogy with the equation of mechanical energy, Eq. 10.1–13 might be called the *equation of thermal energy*. It is generally more convenient than the complete equation of energy for heat-transfer work.

It is of particular interest at this point to compare the equations of mechanical and thermal energy, Eqs. 10.1–12 and 10.1–13. Note that only the terms $p(\nabla \cdot \boldsymbol{v})$ and $(\boldsymbol{\tau} : \nabla \boldsymbol{v})$ are common to both equations and that they appear with opposite signs in the two equations. Therefore, these terms describe the interconversion of mechanical and thermal energy. The term $p(\nabla \cdot \boldsymbol{v})$ can be either positive or negative, depending upon whether the fluid is expanding or contracting; hence it represents a *reversible* mode of interchange. The term $(-\boldsymbol{\tau} : \nabla \boldsymbol{v})$, on the other hand, is always positive (see Eq. 3.3–3) and therefore represents an *irreversible* degradation of mechanical to thermal energy.

Up to this point, no mention has been made of the potential energy, $\hat{\Phi}$. If the external force $\boldsymbol{g}$ is expressible in terms of the gradient of a scalar function (i.e., $\boldsymbol{g} = -\nabla\hat{\Phi}$), then

$$\rho(\boldsymbol{v} \cdot \boldsymbol{g}) = -\rho(\boldsymbol{v} \cdot \nabla\hat{\Phi}) = -\rho \frac{D\hat{\Phi}}{Dt} + \rho \frac{\partial\hat{\Phi}}{\partial t} \tag{10.1–14}$$

If $\hat{\Phi}$ is time-independent, then the last term vanishes, and Eq. 10.1–11 may be transformed into

$$\rho \frac{D}{Dt}(\hat{U} + \hat{\Phi} + \tfrac{1}{2}v^2) = -(\nabla \cdot \boldsymbol{q}) - (\nabla \cdot p\boldsymbol{v}) - (\nabla \cdot [\boldsymbol{\tau} \cdot \boldsymbol{v}]) \qquad (10.1\text{–}15)$$

This is an equation of change for $\hat{E} = \hat{U} + \hat{\Phi} + \frac{1}{2}v^2$, which we term the *total energy*. For terrestrial problems involving no body forces other than gravity, $\hat{\Phi}$ is time-independent and Eq. 10.1–15 holds.

For most engineering applications, it is convenient to have the equation of thermal energy (Eq. 10.1–13) in terms of the fluid temperature and heat capacity rather than the internal energy. We may rewrite the equation in these terms by recognizing that internal energy $\hat{U}$ may be considered as a function of $\hat{V}$ and T, so that

$$\begin{aligned} d\hat{U} &= \left(\frac{\partial \hat{U}}{\partial \hat{V}}\right)_T d\hat{V} + \left(\frac{\partial \hat{U}}{\partial T}\right)_{\hat{V}} dT \\ &= \left[-p + T\left(\frac{\partial p}{\partial T}\right)_{\hat{V}}\right] d\hat{V} + \hat{C}_v\, dT \end{aligned} \qquad (10.1\text{–}16)$$

Here $\hat{C}_v$ is the heat capacity of the fluid at constant volume, per unit mass. Then ρ times the substantial derivative of $\hat{U}$ becomes

$$\rho \frac{D\hat{U}}{Dt} = \left[-p + T\left(\frac{\partial p}{\partial T}\right)_{\hat{V}}\right] \rho \frac{D\hat{V}}{Dt} + \rho \hat{C}_v \frac{DT}{Dt} \qquad (10.1\text{–}17)$$

The term $\rho(D\hat{V}/Dt)$ may be transformed as follows with the aid of the equation of continuity:

$$\rho \frac{D\hat{V}}{Dt} = \rho \frac{D}{Dt}\left(\frac{1}{\rho}\right) = -\frac{1}{\rho}\frac{D\rho}{Dt} = (\nabla \cdot \boldsymbol{v}) \qquad (10.1\text{–}18)$$

By substituting Eq. 10.1–18 into Eq. 10.1–17 and Eq. 10.1–17 into Eq. 10.1–13, we obtain

$$\rho \hat{C}_v \frac{DT}{Dt} = -(\nabla \cdot \boldsymbol{q}) - T\left(\frac{\partial p}{\partial T}\right)_{\hat{V}} (\nabla \cdot \boldsymbol{v}) - (\boldsymbol{\tau} : \nabla \boldsymbol{v}) \qquad (10.1\text{–}19)$$

This is the *equation of energy* in terms of the fluid temperature T. It is as general as Eq. 10.1–9, but it is in more useful form for calculating temperature profiles. The remainder of this section is devoted to the simplified forms of this equation that find widest application in practical work.

The first simplification of Eq. 10.1–19 is obtained by expressing $\boldsymbol{q}$ in terms of temperature gradients (see Eqs. 8.1–3 through 5) and $\boldsymbol{\tau}$ in terms of velocity gradients (see Eqs. 3.2–11 through 16). Then, for a Newtonian fluid with constant thermal conductivity, Eq. 10.1–19 becomes

$$\rho \hat{C}_v \frac{DT}{Dt} = k\nabla^2 T - T\left(\frac{\partial p}{\partial T}\right)_{\rho} (\nabla \cdot \boldsymbol{v}) + \mu \Phi_v \qquad (10.1\text{–}20)$$

This equation states that the temperature of a moving fluid element changes because of (*a*) heat conduction, (*b*) expansion effects, and (*c*) viscous heating. The quantity Φ_v is known as the dissipation function and its form in rectangular coordinates is

$$\Phi_v = 2\left[\left(\frac{\partial v_x}{\partial x}\right)^2 + \left(\frac{\partial v_y}{\partial y}\right)^2 + \left(\frac{\partial v_z}{\partial z}\right)^2\right] + \left(\frac{\partial v_y}{\partial x} + \frac{\partial v_x}{\partial y}\right)^2 + \left(\frac{\partial v_z}{\partial y} + \frac{\partial v_y}{\partial z}\right)^2 + \left(\frac{\partial v_x}{\partial z} + \frac{\partial v_z}{\partial x}\right)^2 - \frac{2}{3}\left(\frac{\partial v_x}{\partial x} + \frac{\partial v_y}{\partial y} + \frac{\partial v_z}{\partial z}\right)^2 \quad (10.1\text{–}21)$$

Equation 10.1–20 may also be used for generalized Newtonian fluids (see Eq. 3.6–2) by substituting η for μ.

Four more simplifications of the energy equation are so widely used that they deserve special mention here. In each we shall omit the viscous dissipation term, which is needed only in special situations. We shall also assume *constant thermal conductivity*.

(i) For an *ideal gas*, $(\partial p/\partial T)_{\hat{V}} = p/T$, and we obtain

(ideal gas) $$\rho \hat{C}_v \frac{DT}{Dt} = k\nabla^2 T - p(\nabla \cdot \boldsymbol{v}) \quad (10.1\text{–}22)$$

(ii) For a *fluid at constant pressure*, Eq. 10.1–16 may be modified to give

$$d\hat{U} = -p\, d\hat{V} + \hat{C}_p\, dT \quad (10.1\text{–}23)$$

The development between Eqs. 10.1–16 and 10.1–20 may be paralleled for this case to give (neglecting viscous dissipation):

(fluid at constant pressure) $$\rho \hat{C}_p \frac{DT}{Dt} = k\nabla^2 T \quad (10.1\text{–}24)$$

(iii) For fluids with ρ *independent of* T, Eq. 10.1–20 (or Eq. J of Table 10.4–1) becomes

(ρ independent of T) $$\rho \hat{C}_p \frac{DT}{Dt} = k\nabla^2 T \quad (10.1\text{–}25)$$

Note that Eqs. 10.1–24 and 25 are the same but that they describe entirely different physical situations.

(iv) For *solids*, the density may usually be considered constant and we may set $\boldsymbol{v} = 0$:

(solid) $$\rho \hat{C}_p \frac{\partial T}{\partial t} = k\nabla^2 T \quad (10.1\text{–}26)$$

Equations 10.1–22 through 26 are used as the starting point for most of our subsequent discussions of heat transfer.

The reader will recall that in Chapter 9 various "source terms" were included in the shell thermal-energy balances. In the foregoing development the *viscous dissipation* source S_v was found to be $-(\boldsymbol{\tau}:\nabla \boldsymbol{v})$; this term appeared automatically because the mechanical energy and work terms were all correctly accounted for. The development in this section did not lead to a *chemical* source S_c because we considered only pure fluids; we shall find that in the treatment of mixtures in Chapter 18 such a term arises naturally. Similarly, the foregoing treatment does not lead to the *nuclear* source S_n or the *electrical* source S_e because these forms of energy have not been accounted for.[2] In order to describe temperature profiles resulting from chemical, nuclear, and electrical heat generation, one may add the appropriate source term to the right side of Eqs. 10.1–20 to 26.

§10.2 THE ENERGY EQUATION IN CURVILINEAR COORDINATES

In this section we express Eqs. 8.1–6 and 10.1–19 and 20 in terms of rectangular, cylindrical, and spherical coordinates. Equations 10.1–22, 24, 25, and 26 are not listed in Table 10.2–2; they may be easily obtained in curvilinear coordinates by analogy with Eq. 10.1–20.

The advantage of using curvilinear coordinates pointed out in §3.4 exists here as well. It is a needless expenditure of effort to set up problems in curvilinear coordinates by means of shell balances. Rather, one should always start with the general equations of change and discard unnecessary terms.

§10.3 THE EQUATIONS OF MOTION FOR FORCED AND FREE CONVECTION IN NONISOTHERMAL FLOW

As pointed out at the beginning of this chapter, the equation of motion developed in §3.2 is valid for nonisothermal flow. In using it, of course, we must now consider ρ and μ to be functions of temperature as well as of

TABLE 10.2–1

COMPONENTS OF THE ENERGY FLUX q

Rectangular	Cylindrical	Spherical
$q_x = -k\dfrac{\partial T}{\partial x}$ (A)	$q_r = -k\dfrac{\partial T}{\partial r}$ (D)	$q_r = -k\dfrac{\partial T}{\partial r}$ (G)
$q_y = -k\dfrac{\partial T}{\partial y}$ (B)	$q_\theta = -k\dfrac{1}{r}\dfrac{\partial T}{\partial \theta}$ (E)	$q_\theta = -k\dfrac{1}{r}\dfrac{\partial T}{\partial \theta}$ (H)
$q_z = -k\dfrac{\partial T}{\partial z}$ (C)	$q_z = -k\dfrac{\partial T}{\partial z}$ (F)	$q_\phi = -k\dfrac{1}{r\sin\theta}\dfrac{\partial T}{\partial \phi}$ (I)

[2] B.-T. Chu, *Physics of Fluids*, **2**, 473–484 (1959).

TABLE 10.2–2
THE EQUATION OF ENERGY IN TERMS OF ENERGY AND MOMENTUM FLUXES
(Eq. 10.1–19)

Rectangular coordinates:

$$\rho \hat{C}_v \left(\frac{\partial T}{\partial t} + v_x \frac{\partial T}{\partial x} + v_y \frac{\partial T}{\partial y} + v_z \frac{\partial T}{\partial z} \right) = -\left[\frac{\partial q_x}{\partial x} + \frac{\partial q_y}{\partial y} + \frac{\partial q_z}{\partial z} \right]$$

$$- T \left(\frac{\partial p}{\partial T} \right)_\rho \left(\frac{\partial v_x}{\partial x} + \frac{\partial v_y}{\partial y} + \frac{\partial v_z}{\partial z} \right) - \left\{ \tau_{xx} \frac{\partial v_x}{\partial x} + \tau_{yy} \frac{\partial v_y}{\partial y} + \tau_{zz} \frac{\partial v_z}{\partial z} \right\}$$

$$- \left\{ \tau_{xy} \left(\frac{\partial v_x}{\partial y} + \frac{\partial v_y}{\partial x} \right) + \tau_{xz} \left(\frac{\partial v_x}{\partial z} + \frac{\partial v_z}{\partial x} \right) + \tau_{yz} \left(\frac{\partial v_y}{\partial z} + \frac{\partial v_z}{\partial y} \right) \right\} \quad (A)$$

Cylindrical coordinates:

$$\rho \hat{C}_v \left(\frac{\partial T}{\partial t} + v_r \frac{\partial T}{\partial r} + \frac{v_\theta}{r} \frac{\partial T}{\partial \theta} + v_z \frac{\partial T}{\partial z} \right) = -\left[\frac{1}{r} \frac{\partial}{\partial r} (r q_r) + \frac{1}{r} \frac{\partial q_\theta}{\partial \theta} + \frac{\partial q_z}{\partial z} \right]$$

$$- T \left(\frac{\partial p}{\partial T} \right)_\rho \left(\frac{1}{r} \frac{\partial}{\partial r} (r v_r) + \frac{1}{r} \frac{\partial v_\theta}{\partial \theta} + \frac{\partial v_z}{\partial z} \right) - \left\{ \tau_{rr} \frac{\partial v_r}{\partial r} + \tau_{\theta\theta} \frac{1}{r} \left(\frac{\partial v_\theta}{\partial \theta} + v_r \right) \right.$$

$$\left. + \tau_{zz} \frac{\partial v_z}{\partial z} \right\} - \left\{ \tau_{r\theta} \left[r \frac{\partial}{\partial r} \left(\frac{v_\theta}{r} \right) + \frac{1}{r} \frac{\partial v_r}{\partial \theta} \right] + \tau_{rz} \left(\frac{\partial v_z}{\partial r} + \frac{\partial v_r}{\partial z} \right) \right.$$

$$\left. + \tau_{\theta z} \left(\frac{1}{r} \frac{\partial v_z}{\partial \theta} + \frac{\partial v_\theta}{\partial z} \right) \right\} \quad (B)$$

Spherical coordinates:

$$\rho \hat{C}_v \left(\frac{\partial T}{\partial t} + v_r \frac{\partial T}{\partial r} + \frac{v_\theta}{r} \frac{\partial T}{\partial \theta} + \frac{v_\phi}{r \sin \theta} \frac{\partial T}{\partial \phi} \right) = -\left[\frac{1}{r^2} \frac{\partial}{\partial r} (r^2 q_r) \right.$$

$$\left. + \frac{1}{r \sin \theta} \frac{\partial}{\partial \theta} (q_\theta \sin \theta) + \frac{1}{r \sin \theta} \frac{\partial q_\phi}{\partial \phi} \right] - T \left(\frac{\partial p}{\partial T} \right)_\rho \left(\frac{1}{r^2} \frac{\partial}{\partial r} (r^2 v_r) \right.$$

$$\left. + \frac{1}{r \sin \theta} \frac{\partial}{\partial \theta} (v_\theta \sin \theta) + \frac{1}{r \sin \theta} \frac{\partial v_\phi}{\partial \phi} \right) - \left\{ \tau_{rr} \frac{\partial v_r}{\partial r} + \tau_{\theta\theta} \left(\frac{1}{r} \frac{\partial v_\theta}{\partial \theta} + \frac{v_r}{r} \right) \right.$$

$$\left. + \tau_{\phi\phi} \left(\frac{1}{r \sin \theta} \frac{\partial v_\phi}{\partial \phi} + \frac{v_r}{r} + \frac{v_\theta \cot \theta}{r} \right) \right\} - \left\{ \tau_{r\theta} \left(\frac{\partial v_\theta}{\partial r} + \frac{1}{r} \frac{\partial v_r}{\partial \theta} - \frac{v_\theta}{r} \right) \right.$$

$$\left. + \tau_{r\phi} \left(\frac{\partial v_\phi}{\partial r} + \frac{1}{r \sin \theta} \frac{\partial v_r}{\partial \phi} - \frac{v_\phi}{r} \right) + \tau_{\theta\phi} \left(\frac{1}{r} \frac{\partial v_\phi}{\partial \theta} + \frac{1}{r \sin \theta} \frac{\partial v_\theta}{\partial \phi} - \frac{\cot \theta}{r} v_\phi \right) \right\}$$

$$(C)$$

Note: The terms contained in braces { } are associated with viscous dissipation and may usually be neglected, except for systems with large velocity gradients.

TABLE 10.2–3
THE EQUATION OF ENERGY IN TERMS OF THE TRANSPORT PROPERTIES
(*for Newtonian fluids of constant ρ and k*)
(Eq. 10.1–25 with viscous dissipation terms included)

Rectangular coordinates:

$$\rho\hat{C}_p\left(\frac{\partial T}{\partial t} + v_x\frac{\partial T}{\partial x} + v_y\frac{\partial T}{\partial y} + v_z\frac{\partial T}{\partial z}\right) = k\left[\frac{\partial^2 T}{\partial x^2} + \frac{\partial^2 T}{\partial y^2} + \frac{\partial^2 T}{\partial z^2}\right]$$

$$+ 2\mu\left\{\left(\frac{\partial v_x}{\partial x}\right)^2 + \left(\frac{\partial v_y}{\partial y}\right)^2 + \left(\frac{\partial v_z}{\partial z}\right)^2\right\} + \mu\left\{\left(\frac{\partial v_x}{\partial y} + \frac{\partial v_y}{\partial x}\right)^2\right.$$

$$\left. + \left(\frac{\partial v_x}{\partial z} + \frac{\partial v_z}{\partial x}\right)^2 + \left(\frac{\partial v_y}{\partial z} + \frac{\partial v_z}{\partial y}\right)^2\right\} \qquad (A)$$

Cylindrical coordinates:

$$\rho\hat{C}_p\left(\frac{\partial T}{\partial t} + v_r\frac{\partial T}{\partial r} + \frac{v_\theta}{r}\frac{\partial T}{\partial \theta} + v_z\frac{\partial T}{\partial z}\right) = k\left[\frac{1}{r}\frac{\partial}{\partial r}\left(r\frac{\partial T}{\partial r}\right) + \frac{1}{r^2}\frac{\partial^2 T}{\partial \theta^2} + \frac{\partial^2 T}{\partial z^2}\right]$$

$$+ 2\mu\left\{\left(\frac{\partial v_r}{\partial r}\right)^2 + \left[\frac{1}{r}\left(\frac{\partial v_\theta}{\partial \theta} + v_r\right)\right]^2 + \left(\frac{\partial v_z}{\partial z}\right)^2\right\} + \mu\left\{\left(\frac{\partial v_\theta}{\partial z} + \frac{1}{r}\frac{\partial v_z}{\partial \theta}\right)^2\right.$$

$$\left. + \left(\frac{\partial v_z}{\partial r} + \frac{\partial v_r}{\partial z}\right)^2 + \left[\frac{1}{r}\frac{\partial v_r}{\partial \theta} + r\frac{\partial}{\partial r}\left(\frac{v_\theta}{r}\right)\right]^2\right\} \qquad (B)$$

Spherical coordinates:

$$\rho\hat{C}_p\left(\frac{\partial T}{\partial t} + v_r\frac{\partial T}{\partial r} + \frac{v_\theta}{r}\frac{\partial T}{\partial \theta} + \frac{v_\phi}{r\sin\theta}\frac{\partial T}{\partial \phi}\right) = k\left[\frac{1}{r^2}\frac{\partial}{\partial r}\left(r^2\frac{\partial T}{\partial r}\right)\right.$$

$$\left. + \frac{1}{r^2\sin\theta}\frac{\partial}{\partial \theta}\left(\sin\theta\frac{\partial T}{\partial \theta}\right) + \frac{1}{r^2\sin^2\theta}\frac{\partial^2 T}{\partial \phi^2}\right] + 2\mu\left\{\left(\frac{\partial v_r}{\partial r}\right)^2\right.$$

$$\left. + \left(\frac{1}{r}\frac{\partial v_\theta}{\partial \theta} + \frac{v_r}{r}\right)^2 + \left(\frac{1}{r\sin\theta}\frac{\partial v_\phi}{\partial \phi} + \frac{v_r}{r} + \frac{v_\theta\cot\theta}{r}\right)^2\right\}$$

$$+ \mu\left\{\left[r\frac{\partial}{\partial r}\left(\frac{v_\theta}{r}\right) + \frac{1}{r}\frac{\partial v_r}{\partial \theta}\right]^2 + \left[\frac{1}{r\sin\theta}\frac{\partial v_r}{\partial \phi} + r\frac{\partial}{\partial r}\left(\frac{v_\phi}{r}\right)\right]^2\right.$$

$$\left. + \left[\frac{\sin\theta}{r}\frac{\partial}{\partial \theta}\left(\frac{v_\phi}{\sin\theta}\right) + \frac{1}{r\sin\theta}\frac{\partial v_\theta}{\partial \phi}\right]^2\right\} \qquad (C)$$

Note: The terms contained in braces { } are associated with viscous dissipation and may usually be neglected, except for systems with large velocity gradients.

pressure. In forced-convection problems it is customary to use the equation of motion in one of the forms given in Chapter 3. In free convection, on the other hand, the temperature dependence of ρ is of critical importance, and it is convenient to modify the equation of motion to account automatically for buoyancy effects. The desirability of such modification has already been shown in §9.9.

Let us begin by considering a free-convection system in which the fluid temperature varies about some mean value $\bar{T}$. If all of the fluid were at $\bar{T}$ and if the fluid were not moving, the pressure gradient in the system would be given by the equation of motion with $\boldsymbol{v} = 0$:

$$\nabla p = \bar{\rho} \boldsymbol{g} \tag{10.3–1}$$

in which $\bar{\rho}$ is the fluid density at $\bar{T}$ and the local pressure. If the velocity gradients result entirely from temperature inequalities, the fluid motion is usually quite slow, and Eq. 10.3–1 may be assumed to be a reasonably good approximation of the pressure gradient even in the moving fluid. By making this assumption, we may express the equation of motion as

$$\rho \frac{D\boldsymbol{v}}{Dt} = -\bar{\rho}\boldsymbol{g} - [\nabla \cdot \boldsymbol{\tau}] + \rho \boldsymbol{g} \tag{10.3–2}$$

This equation may be simplified by replacing ρ by $\bar{\rho}$ on the left side and ρ by $\bar{\rho} - \bar{\rho}\bar{\beta}(T - \bar{T})$ in the term $\rho\boldsymbol{g}$ (see §9.9), so that

$$\underset{\substack{\text{mass per unit}\\ \text{volume} \times\\ \text{acceleration}}}{\bar{\rho} \frac{D\boldsymbol{v}}{Dt}} = \underset{\substack{\text{viscous forces}\\ \text{per unit volume}}}{-[\nabla \cdot \boldsymbol{\tau}]} \quad \underset{\substack{\text{buoyant force}\\ \text{per unit volume}}}{- \bar{\rho}\bar{\beta}\boldsymbol{g}(T - \bar{T})} \tag{10.3–3}$$

This is the equation of motion used for setting up problems in free convection when a mean temperature $\bar{T}$ may be defined. Note that it is an approximate equation, limited to low fluid velocities and small temperature variations. Throughout the remainder of this book in all discussions of free convection we follow the usual convention of omitting the bars on ρ and β.

We have spoken here as if a nonisothermal system would be either in a state of forced or free convection; actually, in engineering applications this arbitrary distinction is usually made for convenience. It must be kept in mind, however, that forced and free convection represent two limiting conditions, one in which buoyancy forces are negligible, and the other in which the effects of pressure and gravitational forces can be expressed entirely in terms of buoyancy forces. The transition between forced and free convection is gradual,[1] and there is a large and ill-defined region in which system behavior cannot be reliably predicted from forced- and free-convection analyses alone. This intermediate kind of behavior appears to occur in many industrially important situations.

[1] See, for example, A. Acrivos, *A.I.Ch.E. Journal*, **4**, 285–289 (1958).

§10.4 SUMMARY OF THE EQUATIONS OF CHANGE

In Chapter 3 and in the first three sections of this chapter we have derived the three equations of change for pure fluids, and we have presented each of them in a number of different ways. At this point a summary of these equations seems most appropriate—not only to provide a convenient "dictionary," but also to show the orderly way in which discussion of energy and momentum transport in homogeneous fluids can be organized. Such a summary is provided by Tables 10.4–1 and 2, in which vector-tensor notation is used for brevity. Whenever possible, reference is made to the first appearance of each equation in the text. Some of the entries in these tables have not previously been introduced, however, and the reader may find it interesting to verify them.

We start in Table 10.4–1 with the equations of change in terms of $\boldsymbol{\tau}$ and $\boldsymbol{q}$, both for a stationary observer and for an observer moving with the fluid. All of the forms of the energy equation are exactly equivalent to Eq. 10.1–9, except for Eqs. *D* and *N*.

In Table 10.4–2 we summarize the equations of change for the special case of constant ρ, μ, and k. In spite of these restrictions, these equations have been widely used for analysis of heat-transfer problems.

For some problems one can obtain useful information from the equations of Table 10.4–1 with $\boldsymbol{\tau}$ and $\boldsymbol{q}$ set equal to zero—that is, by neglecting the molecular transport phenomena entirely. Deciding whether or not these assumptions are valid requires some experience and judgment. Examples of situations in which $\boldsymbol{\tau}$ and $\boldsymbol{q}$ may be neglected are (*a*) adiabatic flow processes in systems designed to minimize frictional effects, such as venturi meters and turbines, and (*b*) high-speed flow around submerged objects. In neither of these examples would one be able to obtain information about the state of the fluid near the solid-fluid boundaries, but the description of the flow at a distance from the boundaries would be reasonably reliable.

§10.5 USE OF EQUATIONS OF CHANGE TO SET UP STEADY-STATE HEAT TRANSFER PROBLEMS

In Chapter 3 we found it quicker and safer to set up fluid-flow problems by simplification of the equations of continuity and motion than to make a differential momentum balance for each new situation. Similarly, we shall find here that the best method of setting up heat-transfer problems is by simplification of the equations of continuity, motion, and energy. To illustrate the procedures, we shall work out several examples, most of which involve only ordinary differential equations. Some more complex situations of engineering interest, which lead to partial differential equations, are discussed in Chapter 11.

TABLE 10.4–1

EQUATIONS OF CHANGE FOR PURE FLUIDS IN TERMS OF THE FLUXES

Eq.	Special Form	In Terms of D/Dt		Comments
Cont.	—	$\frac{D\rho}{Dt} = -\rho(\nabla \cdot \boldsymbol{v})$	3.1–6 (*A*)	For $D\rho/Dt = 0$ simplifies to $(\nabla \cdot \boldsymbol{v}) = 0$
Motion	Forced convection	$\rho \frac{D\boldsymbol{v}}{Dt} = -\nabla p - [\nabla \cdot \boldsymbol{\tau}] + \rho \boldsymbol{g}$	3.2–10 (*B*)	For $\boldsymbol{\tau} = 0$ this becomes Euler's equation
	Free convection	$\rho \frac{D\boldsymbol{v}}{Dt} = -[\nabla \cdot \boldsymbol{\tau}] - \rho\beta \boldsymbol{g}(T - \bar{T})$	10.3–3 (*C*)	Approximate
Energy	In terms of $\hat{E} = \hat{U} + \hat{K} + \hat{\Phi}$	$\rho \frac{D\hat{E}}{Dt} = -(\nabla \cdot \boldsymbol{q}) - (\nabla \cdot p\boldsymbol{v}) - (\nabla \cdot [\boldsymbol{\tau} \cdot \boldsymbol{v}])$	10.1–15 (*D*)	Exact only for $\hat{\Phi}$ time independent
	In terms of $\hat{U} + \hat{K}$	$\rho \frac{D(\hat{U} + \hat{K})}{Dt} = -(\nabla \cdot \boldsymbol{q}) - (\nabla \cdot p\boldsymbol{v}) - (\nabla \cdot [\boldsymbol{\tau} \cdot \boldsymbol{v}]) + \rho(\boldsymbol{v} \cdot \boldsymbol{g})$	10.1–11 (*E*)	
	In terms of $\hat{K} = \frac{1}{2}v^2$	$\rho \frac{D\hat{K}}{Dt} = -(\boldsymbol{v} \cdot \nabla p) - (\boldsymbol{v} \cdot [\nabla \cdot \boldsymbol{\tau}]) + \rho(\boldsymbol{v} \cdot \boldsymbol{g})$	3.3–1 (*F*)	
	In terms of $\hat{U}$	$\rho \frac{D\hat{U}}{Dt} = -(\nabla \cdot \boldsymbol{q}) - p(\nabla \cdot \boldsymbol{v}) - (\boldsymbol{\tau} : \nabla \boldsymbol{v})$	10.1–13 (*G*)	Term containing p is zero for $D\rho/Dt = 0$
	In terms of $\hat{H}$	$\rho \frac{D\hat{H}}{Dt} = -(\nabla \cdot \boldsymbol{q}) - (\boldsymbol{\tau} : \nabla \boldsymbol{v}) + \frac{Dp}{Dt}$	(*H*)	
	In terms of $\hat{C}_v$ and T	$\rho \hat{C}_v \frac{DT}{Dt} = -(\nabla \cdot \boldsymbol{q}) - T\left(\frac{\partial p}{\partial T}\right)_\rho (\nabla \cdot \boldsymbol{v}) - (\boldsymbol{\tau} : \nabla \boldsymbol{v})$	10.1–19 (*I*)	For an ideal gas $T(\partial p/\partial T)_\rho = p$
	In terms of $\hat{C}_p$ and T	$\rho \hat{C}_p \frac{DT}{Dt} = -(\nabla \cdot \boldsymbol{q}) + \left(\frac{\partial \ln \hat{V}}{\partial \ln T}\right)_p \frac{Dp}{Dt} - (\boldsymbol{\tau} : \nabla \boldsymbol{v})$	(*J*)	For an ideal gas $(\partial \ln \hat{V}/\partial \ln T)_p = 1$

Table 10.4-1 (*Continued*)

Eq.	Special Form	In Terms of $\partial/\partial t$	
Cont.	—	$\frac{\partial}{\partial t}\rho = -(\nabla \cdot \rho \boldsymbol{v})$	3.1–4 (*K*)
Motion	Forced convection	$\frac{\partial}{\partial t}\rho \boldsymbol{v} = -[\nabla \cdot \rho \boldsymbol{v}\boldsymbol{v}] - \nabla p - [\nabla \cdot \boldsymbol{\tau}] + \rho \boldsymbol{g}$	3.2–8 (*L*)
	Free convection	$\frac{\partial}{\partial t}\rho \boldsymbol{v} = -[\nabla \cdot \rho \boldsymbol{v}\boldsymbol{v}] - [\nabla \cdot \boldsymbol{\tau}] - \rho\beta \boldsymbol{g}(T - \bar{T})$ (Approximate)	(*M*)
Energy	In terms of $\hat{E} = \hat{U} + \hat{K} + \hat{\Phi}$	$\frac{\partial}{\partial t}\rho \hat{E} = -(\nabla \cdot \rho \hat{E}\boldsymbol{v}) - (\nabla \cdot \boldsymbol{q}) - (\nabla \cdot p\boldsymbol{v}) - (\nabla \cdot [\boldsymbol{\tau} \cdot \boldsymbol{v}])$	(*N*)
	In terms of $\hat{U} + \hat{K}$	$\frac{\partial}{\partial t}\rho(\hat{U} + \hat{K}) = -(\nabla \cdot \rho(\hat{U} + \hat{K})\boldsymbol{v}) - (\nabla \cdot \boldsymbol{q}) - (\nabla \cdot p\boldsymbol{v}) - (\nabla \cdot [\boldsymbol{\tau} \cdot \boldsymbol{v}]) + \rho(\boldsymbol{v} \cdot \boldsymbol{g})$	10.1–9 (*O*)
	In terms of $\hat{K} = \frac{1}{2}v^2$	$\frac{\partial}{\partial t}\rho \hat{K} = -(\nabla \cdot \rho \hat{K}\boldsymbol{v}) - (\boldsymbol{v} \cdot \nabla p) - (\boldsymbol{v} \cdot [\nabla \cdot \boldsymbol{\tau}]) + \rho(\boldsymbol{v} \cdot \boldsymbol{g})$	3.3–2 (*P*)
	In terms of $\hat{U}$	$\frac{\partial}{\partial t}\rho \hat{U} = -(\nabla \cdot \rho \hat{U}\boldsymbol{v}) - (\nabla \cdot \boldsymbol{q}) - p(\nabla \cdot \boldsymbol{v}) - (\boldsymbol{\tau} : \nabla \boldsymbol{v})$	(*Q*)
	In terms of $\hat{H}$	$\frac{\partial}{\partial t}\rho \hat{H} = -(\nabla \cdot \rho \hat{H}\boldsymbol{v}) - (\nabla \cdot \boldsymbol{q}) - (\boldsymbol{\tau} : \nabla \boldsymbol{v}) + \frac{Dp}{Dt}$	(*R*)
	In terms of $\hat{C}_v$ and T	$\frac{\partial}{\partial t}\rho \hat{C}_v T = -(\nabla \cdot \rho \hat{C}_v T\boldsymbol{v}) - (\nabla \cdot \boldsymbol{q}) - T\left(\frac{\partial p}{\partial T}\right)_\rho (\nabla \cdot \boldsymbol{v}) - (\boldsymbol{\tau} : \nabla \boldsymbol{v}) + \rho T \frac{D\hat{C}_v}{Dt}$	(*S*)
	In terms of $\hat{C}_p$ and T	$\frac{\partial}{\partial t}\rho \hat{C}_p T = -(\nabla \cdot \rho \hat{C}_p T\boldsymbol{v}) - (\nabla \cdot \boldsymbol{q}) - (\boldsymbol{\tau} : \nabla \boldsymbol{v}) + \left(\frac{\partial \ln \hat{V}}{\partial \ln T}\right)_p \frac{Dp}{Dt} + \rho T \frac{D\hat{C}_p}{Dt}$	(*T*)

TABLE 10.4–2

EQUATIONS OF CHANGE FOR PURE FLUIDS OF CONSTANT ρ, μ, AND k

Eq.	Special Form	Equation in Symbolic Form		Coordinate System: Rect.	Cyl.	Sph.
Cont.	—	$(\nabla \cdot \boldsymbol{v}) = 0$	(A)	Table 3.4–1 Eq. A	Table 3.4–1 Eq. B	Table 3.4–1 Eq. C
Motion	Forced convection	$\rho \dfrac{D\boldsymbol{v}}{Dt} = -\nabla p + \mu \nabla^2 \boldsymbol{v} + \rho \boldsymbol{g}$	(B)	Table 3.4–2 Eqs. D, E, F	Table 3.4–3 Eqs. D, E, F	Table 3.4–4 Eqs. D, E, F
	Free convection	$\rho \dfrac{D\boldsymbol{v}}{Dt} = \mu \nabla^2 \boldsymbol{v} - \rho \beta \boldsymbol{g}(T - \bar{T})$	(C)	—	—	—
Energy	In terms of $\hat{U}$	$\rho \dfrac{D\hat{U}}{Dt} = k \nabla^2 T + \mu \Phi_v$	(D)	—	—	—
	In terms of $\hat{C}_p$ and T	$\rho \hat{C}_p \dfrac{DT}{Dt} = k \nabla^2 T + \mu \Phi_v$	(E)	Table 10.2–3 Eq. A	Table 10.2–3 Eq. B	Table 10.2–3 Eq. C

Before taking up these examples, we shall show how one simplifies the equations of change to derive Eq. 9.8–11, the differential equation for laminar forced convection in a long round tube. We start by noting that $v_r = v_\theta = q_\theta = 0$. We further assume constancy of all physical properties and neglect viscous dissipation. The equations of change in cylindrical coordinates for this system then become

$$\text{(continuity)} \qquad \frac{\partial v_z}{\partial z} = 0 \qquad (10.5\text{–}1)$$

$$\text{(motion)} \qquad \rho v_z \frac{\partial v_z}{\partial z} = -\frac{\partial \mathscr{P}}{\partial z} + \mu\left[\frac{1}{r}\frac{\partial}{\partial r}\left(r\frac{\partial v_z}{\partial r}\right) + \frac{\partial^2 v_z}{\partial z^2}\right] \qquad (10.5\text{–}2)$$

$$\text{(energy)} \qquad \rho \hat{C}_p v_z \frac{\partial T}{\partial z} = +k\left[\frac{1}{r}\frac{\partial}{\partial r}\left(r\frac{\partial T}{\partial r}\right) + \frac{\partial^2 T}{\partial z^2}\right] \qquad (10.5\text{–}3)$$

Use of the equation of continuity permits us to discard the left side of the equation of motion and the last term on the right side. We may then integrate the equation of motion directly to obtain the parabolic velocity profile obtained in §2.3 and substitute the resulting velocity distribution into Eq. 10.5–3, thereby obtaining Eq. 9.8–11. Similarly, the other problems of Chapter 9 can be set up easily by simplifying the equations of change.

Example 10.5–1. Tangential Flow in an Annulus with Viscous Heat Generation

Determine the temperature distribution in an incompressible Newtonian fluid held between two coaxial cylinders, the outer one of which is rotating at a steady angular velocity Ω_o. (See §9.4 and Example 3.5–1.) Use the nomenclature of Example 3.5–1 and assume κ is fairly small so that the curvature of the fluid streamlines must be considered. Consider that the wetted surfaces of the outer and inner cylinders are at temperatures T_1 and T_κ, respectively; assume steady laminar flow, and neglect the temperature dependence of μ, ρ, and k. This example is an illustration of a forced-convection problem: the equation of motion is solved first to get the velocity profiles; the resulting velocity distribution is inserted into the energy equation; then the latter is solved to get the temperature profiles. This problem is of interest in connection with heat effects in viscometry.

Solution. Since we are neglecting the effect of temperature on physical properties, we may use the velocity distribution of Example 3.5–1:

$$v_\theta = \Omega_o R \frac{\left(\dfrac{r}{\kappa R} - \dfrac{\kappa R}{r}\right)}{\left(\dfrac{1}{\kappa} - \kappa\right)}; \qquad v_r = v_z = 0 \qquad (10.5\text{–}4)$$

The energy equation may be obtained from Eq. *B* of Table 10.2–3:

$$0 = k\frac{1}{r}\frac{d}{dr}\left(r\frac{dT}{dr}\right) + \mu\left[r\frac{d}{dr}\left(\frac{v_\theta}{r}\right)\right]^2 \qquad (10.5\text{–}5)$$

We now substitute Eq. 10.5–4 into 10.5–5 and perform the indicated differentiation to obtain a single differential equation for the desired temperature distribution:

$$0 = k\frac{1}{r}\frac{d}{dr}\left(r\frac{dT}{dr}\right) + \frac{4\mu\Omega_o^2\kappa^4R^4}{(1-\kappa^2)^2}\frac{1}{r^4} \tag{10.5–6}$$

This is the differential equation for the temperature distribution. It may be rewritten in terms of dimensionless quantities by putting

$$\xi = \frac{r}{R} = \text{dimensionless radial coordinate} \tag{10.5–7}$$

$$\Theta = \frac{T - T_\kappa}{T_1 - T_\kappa} = \text{dimensionless temperature} \tag{10.5–8}$$

$$N = \frac{\mu\Omega_o^2R^2}{k(T_1 - T_\kappa)}\cdot\frac{\kappa^4}{(1-\kappa^2)^2} = \text{Br}\frac{\kappa^4}{(1-\kappa^2)^2} \tag{10.5–9}$$

in which Br is the Brinkman number for the system. (See §9.4.) Then Eq. 10.5–6 becomes

$$\frac{1}{\xi}\frac{d}{d\xi}\left(\xi\frac{d\Theta}{d\xi}\right) = -4N\frac{1}{\xi^4} \tag{10.5–10}$$

Integration twice with respect to ξ gives

$$\Theta = -N\frac{1}{\xi^2} + C_1\ln\xi + C_2 \tag{10.5–11}$$

The constants of integration C_1 and C_2 are determined by the boundary conditions:

B.C. 1: at $\xi = \kappa$, $\Theta = 0$ (10.5–12)

B.C. 2: at $\xi = 1$, $\Theta = 1$ (10.5–13)

The final expression for the reduced temperature distribution is then

$$\Theta = \left((N+1) - \frac{N}{\xi^2}\right) - \left((N+1) - \frac{N}{\kappa^2}\right)\frac{\ln\xi}{\ln\kappa} \tag{10.5–14}$$

When $N = 0$, we obtain the temperature distribution for a motionless cylindrical shell of thickness $R(1-\kappa)$ with inner and outer wall temperatures T_κ and T_1, respectively. If N is large enough, there may actually be a maximum in the temperature distribution located at

$$\xi = \left(\frac{-2N\ln\kappa}{N/\kappa^2 - (N+1)}\right)^{1/2} \tag{10.5–15}$$

with the temperature at this point being greater than either T_κ or T_1.

Example 10.5–2. Steady Flow of a Nonisothermal Film

A liquid is falling in steady laminar flow over an inclined plane surface, as shown in Figs. 2.2–1 and 2. Conditions are such that the free liquid surface is maintained at $T = T_0$ and the solid surface at $x = \delta$ is maintained at $T = T_\delta$. The corresponding fluid viscosities are μ_0 and μ_δ, respectively, and the density and thermal conductivity may be assumed constant. Develop expressions for the velocity distribution as a function of fluid properties. Assume that the temperature dependence of viscosity may be expressed by an equation of the form

$$\mu = Ae^{B/T} \tag{10.5-16}$$

as suggested by the Eyring theory discussed in §1.5, in which A and B are empirical constants. End effects and viscous heating may be neglected.

In this example the energy equation is solved first to get the temperature profile; the temperature profile is used to determine the change of viscosity with distance; then the equation of motion may be solved to give the velocity distribution.

Solution. We begin with the equation of energy, which here reduces to

$$\frac{d^2T}{dx^2} = 0 \tag{10.5-17}$$

This equation may be immediately integrated by making use of the known terminal temperatures to give the temperature distribution in the film:

$$\frac{T - T_0}{T_\delta - T_0} = \frac{x}{\delta} \tag{10.5-18}$$

With the aid of Eq. 10.5-18 and the known terminal viscosities, we may now determine the viscosity as a function of position in the film. When the temperature difference across the film is moderate we may write:[1]

$$\frac{\mu}{\mu_0} = \left(\frac{\mu_\delta}{\mu_0}\right)^{x/\delta} \tag{10.5-19}$$

This expression is equivalent to Eq. 2.2-22, with $-\alpha = \ln (\mu_\delta/\mu_0)$. We see then that the apparently arbitrary choice of the viscosity distribution in Example 2.2-2 was in fact reasonable and conforms well to the observed behavior of typical liquids.

[1] According to Eq. 1.5-12, μ/μ_0 should have the form

$$\frac{\mu}{\mu_0} = \exp\left[B\left(\frac{1}{T} - \frac{1}{T_0}\right)\right] \tag{10.5-19a}$$

in which B is a constant. Combination of Eq. 10.5-19*a* with Eq. 10.5-18 gives

$$\frac{\mu}{\mu_0} = \exp\left(\frac{B}{TT_0}(T_0 - T_\delta)\frac{x}{\delta}\right)$$
$$\approx \exp\left(\frac{B}{T_\delta T_0}(T_0 - T_\delta)\frac{x}{\delta}\right) \tag{10.5-19b}$$

the second expression being a good approximation if the temperature change is small. Insertion of Eq. 10.5-19*a*, written for $T = T_\delta$, into Eq. 10.5-19*b* gives

$$\frac{\mu}{\mu_0} = \exp\left(\frac{x}{\delta}\ln\frac{\mu_\delta}{\mu_0}\right) = \left(\frac{\mu_\delta}{\mu_0}\right)^{x/\delta} \tag{10.5-19c}$$

We may now write the desired expression for velocity distribution from the results already obtained in Example 2.2–2:

$$v_z = \left(\frac{\rho g \cos\beta}{\mu_0}\right)\left(\frac{\delta}{\ln \mu_\delta/\mu_0}\right)^2 \left[\frac{1 + (x/\delta)\ln(\mu_\delta/\mu_0)}{(\mu_\delta/\mu_0)^{x/\delta}} - \frac{1 + \ln(\mu_\delta/\mu_0)}{(\mu_\delta/\mu_0)}\right] \tag{10.5–20}$$

Example 10.5–3. Transpiration Cooling

Consider two concentric porous spherical shells of radii κR and R, as shown in Fig. 10.5–1. The inner surface of the outer one is at $T = T_1$, and the outer surface of the inner one is to be maintained at a lower temperature, T_κ. Dry air at

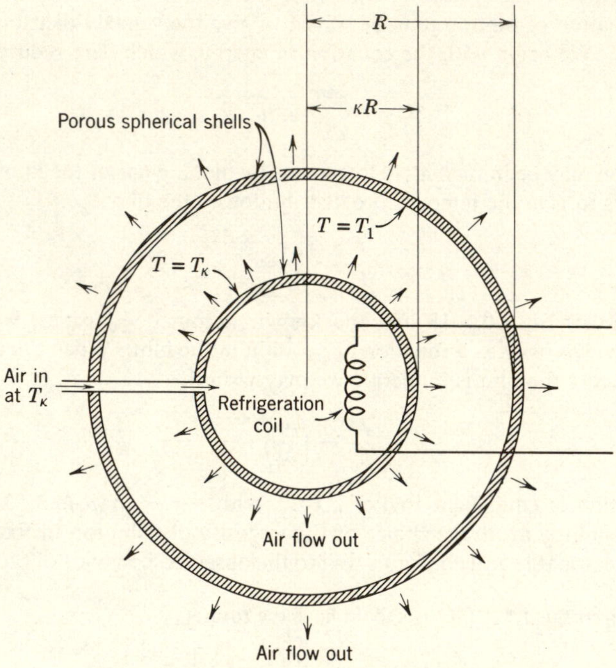

Fig. 10.5–1. Transpiration cooling. The inner sphere is being cooled by means of the refrigeration coil to maintain its temperature at T_κ. When air is blown outward, as shown, less refrigeration is required.

temperature $T = T_\kappa$ is blown outward radially from the inner shell into the intervening space and out through the outer shell. Develop an expression for the required rate of heat removal from the inner sphere as a function of the mass rate of flow of gas. Assume steady laminar flow, and low gas velocity.

In this example the equations of continuity and energy are solved to get the temperature distribution.

Solution. For this system $v_\theta = v_\phi = 0$, and the equation of *continuity* becomes

$$\frac{1}{r^2}\frac{d}{dr}(r^2 \rho v_r) = 0 \tag{10.5–21}$$

This equation may be integrated to give

$$r^2\rho v_r = \text{const.} = \frac{w_r}{4\pi} \tag{10.5-22}$$

Here w_r is the radial mass rate of flow of gas.

The equation of *motion* may be solved to get the pressure rise between inner and outer spheres. For low gas rates, however, this information would be of little interest.

The equation of *energy* for the system is

$$\rho\hat{C}_p v_r \frac{dT}{dr} = k \frac{1}{r^2}\frac{d}{dr}\left(r^2 \frac{dT}{dr}\right) \tag{10.5-23}$$

Here we have used Eq. 10.1–24 for a pure fluid at constant pressure with constant thermal conductivity; we have been guided by Eq. *C* of Table 10.2–3 to write the equation in spherical coordinates. Equation 10.5–22 is used next to eliminate v_r from Eq. 10.5–23; thereby we obtain a differential equation to describe the temperature distribution $T(r)$ in the space between shells:

$$\frac{dT}{dr} = \frac{4\pi k}{w_r\hat{C}_p}\frac{d}{dr}\left(r^2\frac{dT}{dr}\right) \tag{10.5-24}$$

By integrating twice and making use of the known terminal temperatures we find

$$\frac{T - T_1}{T_\kappa - T_1} = \frac{e^{-R_0/r} - e^{-R_0/R}}{e^{-R_0/\kappa R} - e^{-R_0/R}} \tag{10.5-25}$$

in which $R_0 = w_r\hat{C}_p/4\pi k$. Note that for small R_0 Eq. 10.5–25 reduces to

$$\frac{T - T_1}{T_\kappa - T_1} = \frac{1/r - 1/R}{1/\kappa R - 1/R} \tag{10.5-26}$$

If, in addition, R is infinite and κR is finite, then near the inner-sphere surface

$$\frac{T - T_\infty}{T_\kappa - T_\infty} = \frac{\kappa R}{r} \tag{10.5-27}$$

Equation 10.5–27 has already been obtained in Problem 9.H.

The rate of heat flow to the inner sphere in the negative r-direction (i.e., the rate of heat removal by the refrigerant) is

$$Q = -4\pi\kappa^2R^2 q_r|_{r=\kappa R} \tag{10.5-28}$$

By substituting the expression for q_r from Eq. *G*, Table 10.2–1, into Eq. 10.5–28, we get

$$Q = +4\pi k\kappa^2R^2 \left.\frac{dT}{dr}\right|_{r=\kappa R} \tag{10.5-29}$$

We now evaluate dT/dr at the surface $r = \kappa R$ with the aid of Eq. 10.5–25 to obtain the desired relation for the heat removal rate:

$$Q = 4\pi k R_0(T_1 - T_\kappa)/(e^{(R_0/\kappa R)(1-\kappa)} - 1) \tag{10.5-30}$$

When $R_0 = 0$, we get the rate of heat removal at zero gas rate:

$$Q_0 = 4\pi k\kappa R(T_1 - T_\kappa)/(1 - \kappa) \tag{10.5-31}$$

Then the "effectiveness," ϵ, of the transpiration of gas, defined as $\epsilon = (Q_0 - Q)/Q_0$, is

$$\epsilon = 1 - \frac{\phi}{e^\phi - 1} \tag{10.5-32}$$

Here $\phi = R_0(1 - \kappa)/\kappa R = w_r\hat{C}_p(1 - \kappa)/4\pi k\kappa R$ is the dimensionless transpiration rate. Equation 10.5-32 is shown graphically in Fig. 10.5-2. For low values of ϕ, the effectiveness ϵ is about $\phi/2$.

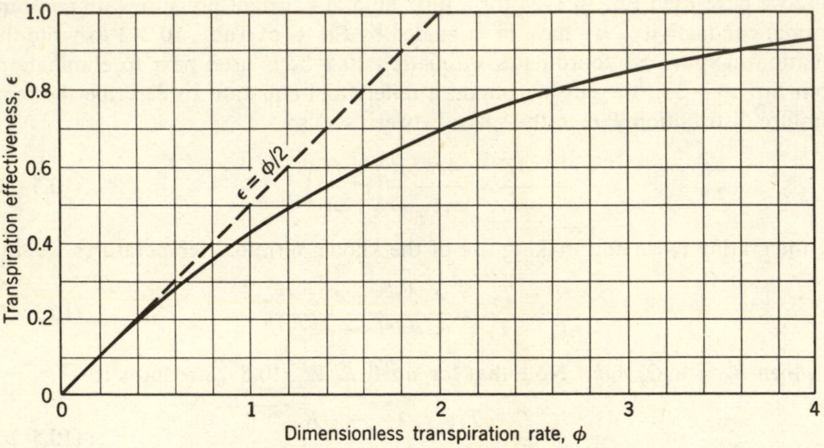

Fig. 10.5-2. Effect of transpiration cooling.

It can be seen that transpiration can be a very effective method for reducing heat-transfer rates. Transpiration cooling has been suggested for cooling rocket nose cones during re-entry to the atmosphere. This effect may be important in situations in which heat and mass transfer occur simultaneously. (See §21.5.)

Example 10.5-4. Free Convection Heat Transfer from a Vertical Plate

A flat plate heated to a temperature T_0 is suspended in a large body of fluid, which is at temperature T_1. In the neighborhood of the heated plate the fluid rises because of the buoyancy force. (See Fig. 10.5-3.) From the equations of change, deduce the dependence of the heat loss on the system variables. The physical properties of the fluid may be considered constant, except that the free-convection form of the equation of motion should be used.

Solution. Let the height of the heated plate be H, and let the dimension in the x-direction be very large so that both v_y and v_z may be considered as functions of y and z alone. We shall assume that the velocity component v_y is quite small—that is, that the fluid moves almost directly upward. Hence we eliminate the y-component of the equation of motion from our analysis and write the equations of continuity, motion, and energy as follows:

$$\text{(continuity)} \qquad \frac{\partial v_y}{\partial y} + \frac{\partial v_z}{\partial z} = 0 \tag{10.5–33}$$

$$\text{(motion)} \qquad \rho\left(v_y \frac{\partial}{\partial y} + v_z \frac{\partial}{\partial z}\right) v_z = \mu\left(\frac{\partial^2 v_z}{\partial y^2} + \underline{\frac{\partial^2 v_z}{\partial z^2}}\right) + \rho g \beta (T - T_1) \tag{10.5–34}$$

$$\text{(energy)} \qquad \rho \hat{C}_p\left(v_y \frac{\partial}{\partial y} + v_z \frac{\partial}{\partial z}\right)(T - T_1) = k\left(\frac{\partial^2}{\partial y^2} + \underline{\frac{\partial^2}{\partial z^2}}\right)(T - T_1) \tag{10.5–35}$$

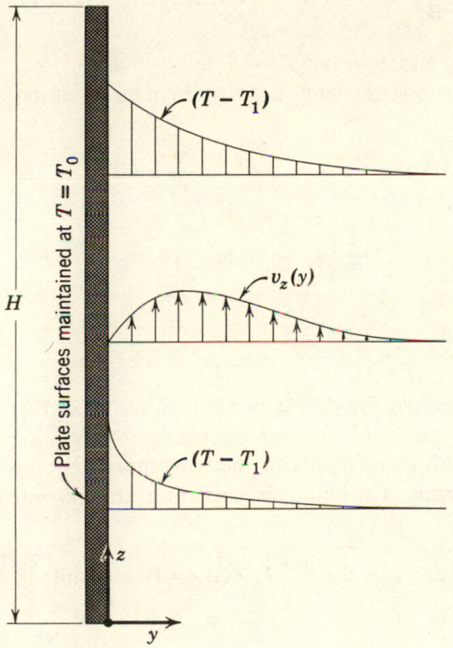

Fig. 10.5–3. Temperature and velocity profiles in the neighborhood of a heated vertical plate.

The concomitant boundary conditions are

B.C. 1: at $y = 0$, $v_y = v_z = 0$, and $T = T_0$ (10.5–36)

B.C. 2: at $y = \infty$, $v_z = 0$, and $T = T_1$ (10.5–37)

B.C. 3: at $z = -\infty$, $v_y = v_z = 0$, and $T = T_1$ (10.5–38)

Note that the temperature rise appears in the equation of motion and that the velocity distribution appears in the equation of energy. Thus the equations are "coupled." Analytic solutions of such coupled, nonlinear differential equations are very difficult, and we content ourselves with a dimensional analysis approach.

It is convenient to introduce the following dimensionless variables:

$$\Theta = (T - T_1)/(T_0 - T_1) = \text{dimensionless temperature} \quad (10.5\text{–}39)$$

$$\zeta = z/H = \text{dimensionless vertical distance} \quad (10.5\text{–}40)$$

$$\eta = \left(\frac{B}{\mu\alpha H}\right)^{1/4} y = \text{dimensionless horizontal distance} \quad (10.5\text{–}41)$$

$$\phi_z = \left(\frac{\mu}{B\alpha H}\right)^{1/2} v_z = \text{dimensionless vertical velocity} \quad (10.5\text{–}42)$$

$$\phi_y = \left(\frac{\mu H}{\alpha^3 B}\right)^{1/4} v_y = \text{dimensionless horizontal velocity} \quad (10.5\text{–}43)$$

where $\alpha = k/\rho\hat{C}_p$ and $B = \rho g\beta(T_0 - T_1)$.

When the equations of change are written in terms of these dimensional variables, we get

(continuity)
$$\frac{\partial\phi_y}{\partial\eta} + \frac{\partial\phi_z}{\partial\zeta} = 0 \quad (10.5\text{–}44)$$

(motion)
$$\frac{1}{\text{Pr}}\left(\phi_y \frac{\partial}{\partial\eta} + \phi_z \frac{\partial}{\partial\zeta}\right)\phi_z = \frac{\partial^2\phi_z}{\partial\eta^2} + \Theta \quad (10.5\text{–}45)$$

(energy)
$$\phi_y \frac{\partial\Theta}{\partial\eta} + \phi_z \frac{\partial\Theta}{\partial\zeta} = \frac{\partial^2\Theta}{\partial\eta^2} \quad (10.5\text{–}46)$$

Here we have omitted the dotted-underlined terms in Eqs. 10.5–34 and 35 on the ground that momentum and energy transport by molecular processes in the z-direction is small compared with the corresponding convective terms on the left side of the equations. The boundary conditions that go with the foregoing equations are

B.C. 1: at $\eta = 0$, $\phi_y = \phi_z = 0$, and $\Theta = 1$ (10.5–47)

B.C. 2: at $\eta = \infty$, $\phi_y = \phi_z = 0$, and $\Theta = 0$ (10.5–48)

B.C. 3: at $\zeta = -\infty$, $\phi_y = \phi_z = 0$, and $\Theta = 0$ (10.5–49)

One can see immediately from these equations and boundary conditions that the dimensionless velocity components ϕ_y and ϕ_z and the dimensionless temperature Θ will depend on η and ζ and also on the parameter Pr, the Prandtl number. Inasmuch as the flow is usually very slow in free convection, the terms in which Pr appears will generally be rather small (setting them exactly equal to zero would correspond to the "creeping flow assumption"). Hence we conclude that the dependence on the Prandtl number will be slight.

The average heat flux from the plate may be written as

$$q_{\text{avg}} = +\frac{k}{H}\int_0^H -\left.\frac{\partial T}{\partial y}\right|_{y=0} dz \quad (10.5\text{–}50)$$

The integral appearing here may be rewritten in terms of the dimensionless parameters introduced above:

$$q_{\text{avg}} = +k(T_0 - T_1)\left(\frac{B}{\mu\alpha H}\right)^{1/4} \cdot \int_0^1 - \frac{\partial\Theta}{\partial\eta}\bigg|_{\eta=0} d\zeta$$
$$= +k(T_0 - T_1)\left(\frac{B}{\mu\alpha H}\right)^{1/4} \cdot C$$
$$= C \cdot \frac{k}{H}(T_0 - T_1)(\text{Gr Pr})^{1/4} \tag{10.5-51}$$

Because Θ is a function of η, ζ, and Pr, the derivative $\partial\Theta/\partial\eta$ depends also on η, ζ, and Pr. Then $\partial\Theta/\partial\eta$, evaluated at $\eta = 0$, depends only on ζ and Pr. The definite integral over ζ then leaves us with a dimensionless function of Pr. From the preceding remarks, we can infer that this function, called C, will be very nearly a constant—that is, a weak function of the Prandtl number.

The foregoing analysis has shown that even without solving the equations we can predict that the average heat flux is proportional to the $\frac{5}{4}$-power of $(T_0 - T_1)$ and inversely proportional to the $\frac{1}{4}$-power of H. Both predictions have been borne out by experiment. The only thing we could not do by this analysis was to determine the value of C as a function of the Prandtl number.

In order to determine C, we have to resort either to experimental measurements or to the solution of Eqs. 10.5-44 through 46. About 80 years ago, Lorenz[2] obtained an approximate solution to these equations and thereby found a value of 0.548 for C. Later, more refined calculations[3] gave the following dependence of C on Pr:

Pr	0.73 (air)	10	10^2	10^3
C	0.517	0.612	0.652	0.653

These values of C are in nearly exact agreement with the best experimental measurements available in the laminar flow range (i.e., for Gr Pr $< 10^9$).

Example 10.5-5. One-Dimensional Compressible Flow. Velocity, Temperature, and Pressure Gradients in a Stationary Shock Wave

Consider the adiabatic flow of an ideal gas through a "nonequilibrium" region in a duct of constant cross section, as shown in Fig. 10.5-4. Assume for the moment a flat velocity profile in the duct. In the nonequilibrium region the velocity, temperature, and pressure change from upstream values v_1, T_1, and p_1 to stable downstream values v_2, T_2, and p_2. There is no change of velocity, temperature, or pressure outside this region. It can be shown[4] that under the proper conditions almost all

[2] L. Lorenz, Wiedemann's *Ann. Physik.*, **13**, 582–606 (1881).

[3] E. Schmidt and W. Beckmann, *Techn. Mech. Thermodynam.*, **1**, 341, 391 (1930); see E. M. Sparrow and J. L. Gregg, *Trans. ASME*, **80**, 379–386 (1958).

[4] For more detailed discussions of shock phenomena, the reader is referred to H. W. Liepmann and A. Roshko, *Elements of Gasdynamics*, Wiley, New York (1957); also M. Morduchow and P. A. Libby, *J. Aeronaut. Sci.*, **16**, 674–84 (1949); R. von Mises, *ibid.*, **17**, 551–554 (1950); G. S. S. Ludford, *ibid.*, **18**, 830–834 (1951).

of the changes of gas condition take place in an extremely thin region called a "shock wave," as shown in Fig. 10.5–5. Under these circumstances, the assumptions of a constant-area duct and flat velocity profile are no longer needed: one-dimensional solutions of the equations of change always apply *locally* to conditions in the immediate region of the "shock," provided only that it is normal to the streamline in question.

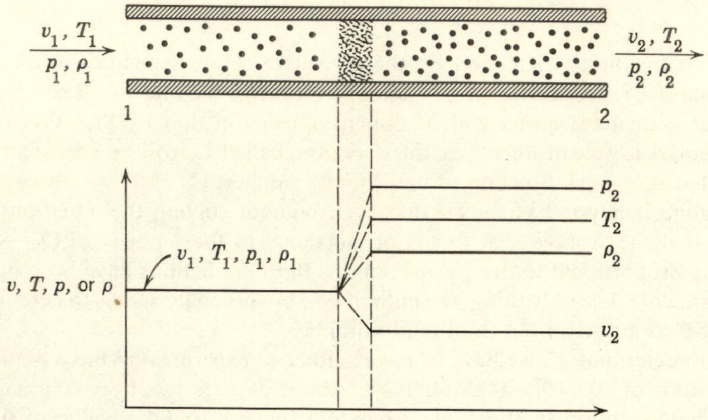

Fig. 10.5–4. Velocity, temperature, pressure, and density changes in a shock wave.

Use the three equations of change to determine the conditions under which a shock wave is possible and to determine the velocity, temperature, and pressure distributions in such a shock wave. Assume steady-state, one-dimensional flow of an ideal gas and ignore changes of μ, k, and $\hat{C}_p$ with temperature and pressure.

Solution. The equations of change for the system under consideration are

(continuity)
$$\frac{d}{dx}\rho v_x = 0 \tag{10.5–52}$$

(motion)
$$\rho v_x \frac{dv_x}{dx} = -\frac{dp}{dx} + \frac{4}{3}\frac{d}{dx}\left(\mu \frac{dv_x}{dx}\right) \tag{10.5–53}$$

(energy)
$$\rho \hat{C}_p v_x \frac{dT}{dx} = \frac{d}{dx}\left(k\frac{dT}{dx}\right) + v_x \frac{dp}{dx} + \frac{4}{3}\mu\left(\frac{dv_x}{dx}\right)^2 \tag{10.5–54}$$

The energy equation here is Eq. *J* of Table 10.4–1, written for steady state and in which $(\partial \ln \hat{V}/\partial \ln T)_p = 1$ for an ideal gas.

The equation of *continuity* may be integrated to give

$$\rho v_x = \rho_1 v_1 \tag{10.5–55}$$

in which ρ_1 and v_1 are quantities evaluated a short distance upstream from the shock.

In the *energy* equation we eliminate ρv_x by use of Eq. 10.5–55 and dp/dx by use of the equation of motion to get

$$\hat{C}_p \frac{dT}{dx} + v_x \frac{dv_x}{dx} = \frac{4}{3} \frac{\mu}{\rho_1 v_1} \frac{d}{dx}\left(v_x \frac{dv_x}{dx}\right) + \frac{1}{\rho_1 v_1} \frac{d}{dx}\left(k \frac{dT}{dx}\right) \tag{10.5–56}$$

This equation may be integrated once to give on rearrangement

$$\hat{C}_p T + \tfrac{1}{2}v_x^2 = \frac{k}{\hat{C}_p \rho_1 v_1}\left[\frac{4}{3} \frac{\hat{C}_p \mu}{k} \frac{d}{dx}\left(\frac{v_x^2}{2}\right) + \hat{C}_p \frac{dT}{dx}\right] + C_{\text{I}} \tag{10.5–57}$$

in which C_{I} is an integration constant. Note that Eq. 10.5–57 is exact for one-dimensional flow of an ideal gas. Further integration must in general be carried out by numerical means. However, if we assume that the Prandtl number $\hat{C}_p\mu/k$ is $\frac{3}{4}$—a fairly reasonable assumption for most gases[5]—we may then complete the integration of the energy equation to obtain

$$\tfrac{1}{2}v_x^2 + \hat{C}_p T = C_{\text{I}} + C_{\text{II}} \exp\left(\frac{\rho_1 v_1 \hat{C}_p x}{k}\right) \tag{10.5–58}$$

We now evaluate the constants of integration. First of all we must set C_{II} equal to zero, since the sum $\hat{C}_p T + \frac{1}{2}v_x^2$ cannot increase without limit in the $+x$-direction. We evaluate C_{I} from the upstream conditions so that $C_{\text{I}} = \frac{1}{2}v_1^2 + \hat{C}_p T_1$; hence the integrated energy equation is

$$\tfrac{1}{2}v_x^2 + \hat{C}_p T = \tfrac{1}{2}v_1^2 + \hat{C}_p T_1 \tag{10.5–59}$$

We next substitute the integrated continuity equation into the equation of *motion* and integrate once to obtain

$$\rho_1 v_1 v_x = -p + \frac{4}{3}\mu \frac{dv_x}{dx} + C_{\text{III}} \tag{10.5–60}$$

Evaluation of C_{III} from upstream conditions (where $dv_x/dx = 0$) gives $C_{\text{III}} = p_1 + \rho_1 v_1^2 = \rho_1(v_1^2 + RT_1/M)$. With the aid of the ideal gas law and the integrated energy equation, we may eliminate p from Eq. 10.5–60 to obtain a relation containing only v_x and x as variables:

$$\frac{4}{3} \frac{\mu}{\rho_1 v_1} v_x \frac{dv_x}{dx} - \frac{\gamma + 1}{2\gamma} v_x^2 + \frac{C_{\text{III}}}{\rho_1 v_1} v_x = \frac{\gamma - 1}{\gamma} C_{\text{I}} \tag{10.5–61}$$

Here γ is the ratio of heat capacities $\hat{C}_p/\hat{C}_v$.

This equation may be written in terms of dimensionless variables to give, after considerable rearrangement,

$$\phi \frac{d\phi}{d\xi} = \beta \text{Ma}_1(\phi - 1)(\phi - \alpha) \tag{10.5–62}$$

[5] Equation 8.3–14, with $\hat{C}_p/\hat{C}_v = \frac{5}{3}$, indicates that the Prandtl number of an ideal, monatomic gas should be two-thirds. It is found experimentally that for most polyatomic gases at low density the Prandtl number is between about 0.65 and 0.85, with an average quite close to 0.75. See, for example, Table 8.3–1.

The dimensionless variables are defined by

$$\phi = \frac{v_x}{v_1} = \text{dimensionless velocity} \tag{10.5–63}$$

$$\xi = \frac{x}{\lambda} = \text{dimensionless distance} \tag{10.5–64}$$

$$\text{Ma}_1 = v_1\sqrt{\frac{M}{\gamma RT_1}} = \text{Mach number}^6 \text{ at upstream conditions} \tag{10.5–65}$$

$$\alpha = \frac{\gamma - 1}{\gamma + 1} + \frac{2}{\gamma + 1}\frac{1}{\text{Ma}_1^2} \tag{10.5–66}$$

$$\beta = \tfrac{9}{8}(\gamma + 1)\sqrt{\pi/8\gamma} \tag{10.5–67}$$

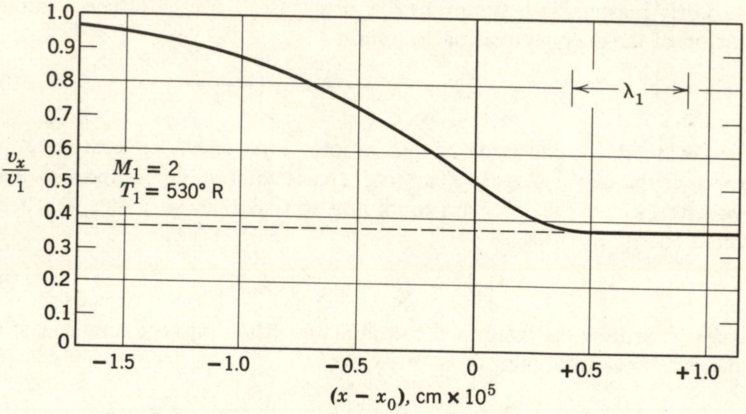

Fig. 10.5–5. Velocity distribution in a stationary shock wave.

The reference length λ is the mean free path defined in Eq. 1.4–3 (with d^2 eliminated by use of Eq. 1.4–9):

$$\lambda = 3\frac{\mu_1}{\rho_1}\sqrt{\frac{\pi M}{8RT_1}} \tag{10.5–68}$$

We may integrate Eq. 10.5–62 to obtain

$$\frac{1 - \phi}{(\phi - \alpha)^\alpha} = \exp\left[\beta(1 - \alpha)\text{Ma}_1(\xi - \xi_0)\right] \tag{10.5–69}$$

This equation is the dimensionless velocity distribution, which contains an integration constant $\xi_0 = x_0/\lambda$. The temperature and pressure distributions may be determined from this equation and Eqs. 10.5–59 and 60. Since ϕ must approach unity as $\xi \to -\infty$, the constant α is less than 1. This can be true only if $\text{Ma}_1 > 1$, that is, if the upstream flow is supersonic. It can also be seen that for very large positive ξ the reduced velocity ϕ approaches α.

[6] The Mach number can be shown to be the ratio of v_1 to sonic velocity at T_1. See Problems 10.L and 10.Q.

It can be seen from Fig. 10.5–5 that shock waves are indeed very thin. Von Mises[4] has extended the foregoing result to gases with Prandtl numbers other than $\frac{3}{4}$, and Morduchow and Libby[4] have taken into account temperature variation of viscosity.

The tendency of gas in supersonic flow to "degrade" spontaneously to subsonic flow is important in wind tunnels and in the design of high-velocity systems, e.g., in turbines and rocket engines. Note that the changes taking place in shock waves are irreversible and that since the velocity gradients are so very steep a considerable amount of mechanical energy is dissipated.

Example 10.5–6. Adiabatic Frictionless Processes in an Ideal Gas

Develop equations for the relationship of local pressure to density or temperature in a stream of ideal gas in which the momentum flux τ and the heat flux q are negligible.

Solution. For the conditions assumed, the equation of energy (Eq. *J* in Table 10.4–1) may be written

$$\rho \hat{C}_p \frac{DT}{Dt} = \left(\frac{\partial \ln \hat{V}}{\partial \ln T}\right)_p \cdot \frac{Dp}{Dt} \tag{10.5–70}$$

But, for an ideal gas $(\partial \ln \hat{V}/\partial \ln T)_p = 1$, and we may write for any element of the moving fluid

$$\frac{dp}{dT} = \rho \hat{C}_p \tag{10.5–71}$$

Again making use of the ideal gas law, we may write

$$\frac{dp}{dT} = \frac{p\hat{C}_p M}{RT} \tag{10.5–72}$$

in which M is the molecular weight of the gas. Defining $\gamma = C_p/C_v$, we may then write

$$\frac{d \ln p}{d \ln T} = \frac{\gamma}{\gamma - 1} \tag{10.5–73}$$

This may be integrated for constant γ to give

$$p^{(\gamma-1)/\gamma}\, T^{-1} = \text{constant} \tag{10.5–74}$$

Evaluation of T in terms of ρ and p gives the alternate solution

$$p\rho^{-\gamma} = \text{constant} \tag{10.5–75}$$

It may be noted from the conditions of this derivation that the same solution is obtained for flow or nonflow processes and for steady-state or unsteady-state processes; furthermore, no assumption is necessary regarding the work performed by the gas. The result is therefore quite useful and is worth remembering. It should, however, be kept in mind that the neglect of q and τ is equivalent to assuming adiabatic, reversible behavior and that Eqs. 10.5–74 and 75 therefore represent the limiting case of *isentropic* flow. In any real adiabatic system the temperature increase on compression will be greater than that calculated from Eq. 10.5–74.

However, Eqs. 10.5–74 and 75 are useful as limiting cases in determining the effectiveness of compressors and blowers. In addition, these equations give quite good descriptions of the p-T and p-ρ relationships in sound waves.[7]

§10.6 DIMENSIONAL ANALYSIS OF THE EQUATIONS OF CHANGE

Now that we have presented the equations of change for nonisothermal systems and have shown how they may be used to solve a variety of heat-transport problems, we discuss briefly the dimensional analysis of these equations. This discussion parallels that given in §3.7 and is intended to provide an introduction to Chapter 13. For simplicity, we again restrict ourselves to systems of constant physical properties. With this restriction we may write

$$\text{(continuity)} \qquad (\nabla \cdot \boldsymbol{v}) = 0 \tag{10.6–1}$$

$$\text{(motion)} \qquad \rho \frac{D\boldsymbol{v}}{Dt} = \mu \nabla^2 \boldsymbol{v} + \begin{cases} -\nabla p + \rho \boldsymbol{g} & \text{(forced convection)} \\ -\rho \beta \boldsymbol{g}(T - T_0) & \text{(free convection)} \end{cases} \tag{10.6–2}$$

$$\text{(energy)} \qquad \rho \hat{C}_p \frac{DT}{Dt} = k \nabla^2 T + \mu \Phi_v \tag{10.6–3}$$

Note that we provided for inclusion of either the forced- or the free-convection "momentum sources" in Eq. 10.6–2. We shall consider forced and free convection separately as it is convenient to define dimensionless variables in a slightly different way for each. The T_0 appearing in Eq. 10.6–2 is the ambient temperature.

For *forced convection*, we define the following dimensionless variables:

$$\boldsymbol{v}^* = \frac{\boldsymbol{v}}{V} = \text{dimensionless velocity} \tag{10.6–4}$$

$$p^* = \frac{p - p_0}{\rho V^2} = \text{dimensionless pressure} \tag{10.6–5}$$

$$t^* = \frac{tV}{D} = \text{dimensionless time} \tag{10.6–6}$$

$$T^* = \frac{T - T_0}{T_1 - T_0} = \text{dimensionless temperature} \tag{10.6–7}$$

$$x^*, y^*, z^* = \frac{x}{D}, \frac{y}{D}, \frac{z}{D} = \text{dimensionless coordinates} \tag{10.6–8}$$

Here V, D, and $(T_1 - T_0)$ represent, respectively, any convenient characteristic velocity, length, and temperature difference in the system. If we write

[7] See, for example, Problem 10.Q.

our three equations of change for forced convection in terms of these dimensionless variables we obtain, on rearrangement,

$$\text{(continuity)} \qquad (\nabla^* \cdot \boldsymbol{v}^*) = 0 \tag{10.6–9}$$

$$\text{(motion)} \qquad \frac{D\boldsymbol{v}^*}{Dt^*} = \frac{1}{\text{Re}} \nabla^{*2}\boldsymbol{v}^* - \nabla^* p^* + \frac{1}{\text{Fr}}\frac{\boldsymbol{g}}{g} \tag{10.6–10}$$

$$\text{(energy)} \qquad \frac{DT^*}{Dt^*} = \frac{1}{\text{Re Pr}} \nabla^{*2}T^* + \frac{\text{Br}}{\text{Re Pr}} \Phi_v^* \tag{10.6–11}$$

Here Φ_v^* is the dissipation function written in terms of v^*, x^*, y^*, and z^*. Note that there are four dimensionless groups appearing in these three

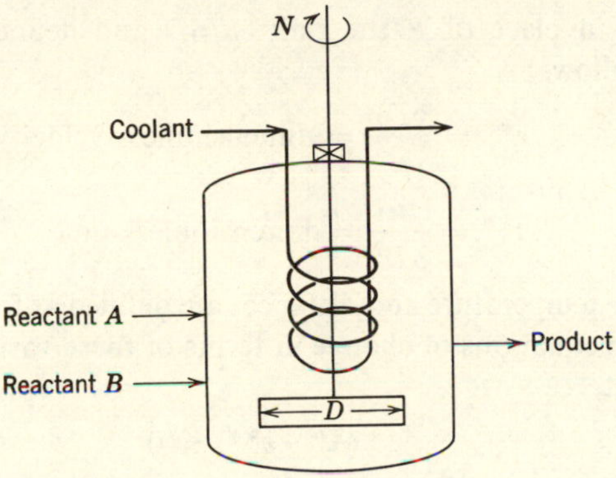

Fig. 10.6–1. Continuous blender.

equations: $\text{Re} = [DV\rho/\mu]$, $\text{Fr} = [V^2/gD]$, the Prandtl number, $\text{Pr} = [\hat{C}_p\mu/k]$, and the Brinkman number,[1] $[\mu V^2/k(T_1 - T_0)]$.

Example 10.6–1. Forced-Convection Heat Transfer in an Agitated Tank

Two fluid streams are being continuously blended in fixed proportions in a completely filled agitated reactor, as shown in Fig. 10.6–1. Under the conditions of operation, heat must be removed through the cooling coil at a rate Q proportional to the flow rate of the exit stream. Estimate the effect of tank diameter on permissible mass flow rates if the solution and coil-surface temperatures, T_1 and T_0, respectively, and the Reynolds number $(D^2N\rho/\mu)$ are to be maintained constant. Assume geometric similarity and constant fluid properties; assume further that the stream flow rates have negligible effect on the flow pattern. Neglect heat loss from the tank wall.

[1] In many texts the ratio (Br/Pr) $= V^2/\hat{C}_p(T_1 - T_0)$ is given as a single dimensionless group called the Eckert number, Ec. Clearly the choice of dimensionless groups is arbitrary. The Eckert number is closely related to the Mach number. (See Problem 10.L.)

Solution. The required rate of heat removal at the coil surface is

$$Q = +k \int_A \left.\frac{\partial T}{\partial n}\right|_{\text{surface}} dA \tag{10.6–12}$$

in which A is the total coil-surface area and n is the distance measured outward from and normal to the coil surface. In terms of the dimensionless variables $A^* = A/D^2$, $n^* = n/D$, and $T^* = (T - T_0)/(T_1 - T_0)$, we may write Eq. 10.6–12 as

$$Q = +k(T_1 - T_0)D \int_{A^*} \left.\frac{\partial T^*}{\partial n^*}\right|_{\text{surface}} dA^* \tag{10.6–13}$$

We may write $(\partial T^*/\partial n^*)|_{\text{surface}}$ as a function of Re, Pr, and the boundary conditions. Here, then, the value of the integral in Eq. 10.6–13 is constant. Then Q, and therefore the permissible mass throughput rate, varies directly with the diameter.

For *free convection*, there is no readily available reference velocity. We therefore use in place of V the ratio $\mu/\rho D$ and define our dimensionless variables as follows:

$$v^{**} = \frac{vD\rho}{\mu} = \text{dimensionless velocity} \tag{10.6–14}$$

$$t^{**} = \frac{t\mu}{\rho D^2} = \text{dimensionless time} \tag{10.6–15}$$

Dimensionless temperature and distances are defined as for forced convection. By writing the equations of change in terms of these variabies, we obtain, on rearrangement,

(continuity)
$$(\nabla^* \cdot v^{**}) = 0 \tag{10.6–16}$$

(motion)
$$\frac{Dv^{**}}{Dt^{**}} = \nabla^{*2} v^{**} - T^* \,\text{Gr}\, \frac{g}{g} \tag{10.6–17}$$

(energy)
$$\frac{DT^*}{Dt^{**}} = \frac{1}{\text{Pr}} \nabla^{*2} T^* \tag{10.6–18}$$

We see then that for free convection only the two dimensionless groups, Pr and the Grashof number, $\text{Gr} = [g\rho^2\beta(T_1 - T_0)D^3/\mu^2]$, appear in the equations of change. Note that we have omitted the viscous dissipation term in the equation of energy, that term being clearly unimportant for free convection.

Example 10.6–2. Surface Temperature of an Electric Heating Coil

An electric heating coil of diameter D is being designed to keep a large tank of liquid above its freezing point. It is desired to predict the temperature that will be reached by the coil surface as a function of the heating rate Q and the tank surface temperature T_0. This prediction is to be made on the basis of experiments with a smaller geometrically similar apparatus filled with the same liquid.

Outline a suitable experimental procedure for making the desired prediction. Temperature dependence of the physical properties other than ρ may be neglected, and the entire heating coil surface may be assumed to be at the same temperature, T_1.

Solution. We may write for the total energy input rate through the coil surface

$$Q = -k \int_A \left. \frac{\partial T}{\partial r} \right|_s dA \tag{10.6–19}$$

Here A is the surface area of the coil, the subscript s refers to conditions in the fluid immediately adjacent to the coil surface, and r is the distance measured outward from and normal to the coil surface. By rewriting Eq. 10.6–19 in terms of dimensionless variables, we get

$$\frac{Q}{k(T_1 - T_0)D} = -\int_{A^*} \left. \frac{\partial T^*}{\partial r^*} \right|_{s^*} dA^* \tag{10.6–20}$$

$$= \psi\left[\left(\frac{\hat{C}_p \mu}{k}\right), \left(\frac{\rho^2 \beta (T_1 - T_0) g D^3}{\mu^2}\right)\right] \tag{10.6–21}$$

Here

$$T^* = \frac{T - T_0}{T_1 - T_0} \tag{10.6–22}$$

$$r^* = \frac{r}{D} \tag{10.6–23}$$

$$A^* = \frac{A}{D^2} \tag{10.6–24}$$

$$\psi = \text{some function of Gr and Pr to be determined} \tag{10.6–25}$$

Since our two systems are to be geometrically similar, the upper limit of integration in Eq. 10.6–20, A/D^2, will be the same in both and need not be included in the function ψ. Similarly, if we write our boundary conditions for temperature, velocity, and pressure at the coil and tank surface, we will obtain only size ratios that will be identical in the two systems. We therefore omit this step.

We now note that the desired quantity $(T_1 - T_0)$ appears on both sides of Eq. 10.6–21. To avoid this inconvenient situation, we multiply both sides by the Grashof number to get

$$\left(\frac{D^2 Q \rho^2 \beta g}{k \mu^2}\right) = \text{Gr} \cdot \psi(\text{Gr}, \text{Pr}) \tag{10.6–26}$$

In principle, we may solve Eq. 10.6–26 for Gr and obtain an explicit expression for $(T_1 - T_0)$. Since we are neglecting the temperature dependence of physical properties, we may consider the Prandtl number constant for the given fluid and write

$$(T_1 - T_0) = \left(\frac{\mu^2}{\rho^2 \beta g D^3}\right) \cdot \phi\left(\frac{D^2 Q \rho^2 \beta g}{k \mu^2}\right) \tag{10.6–27}$$

Here ϕ is some function of $(D^2 Q \rho^2 \beta g / k \mu^2)$ to be determined experimentally. We may then construct a plot of Eq. 10.6–27 from experimental measurements of T_1, T_0, D, and the physical properties of the fluid and use this plot to predict the behavior of the larger system.

In fact, since we are permitted to neglect the temperature dependence of the fluid properties, we may go even further. If we maintain the ratio of Q in the two systems inversely proportional to the square of the ratio of the diameters, then the corresponding ratio of $(T_1 - T_0)$ will be inversely proportional to the cube of the ratio of the diameters.

We have already seen in Chapter 9 how several of the dimensionless groups used in the foregoing examples appear when the differential equations describing specific systems are put in dimensionless form. All we have done here is to show that these same dimensionless groups arise in terms of the general partial differential equations for nonisothermal processes. We shall make use of this discussion in later chapters, in which heat transfer coefficient correlations are given.

It is sometimes useful to think of the dimensionless groups as ratios of various forces or effects in the system:[2]

$$\text{Re} = \frac{\rho V^2/D}{\mu V/D^2} = \frac{\text{inertial forces}}{\text{viscous forces}}$$

$$\text{Fr} = \frac{\rho V^2/D}{\rho g} = \frac{\text{inertial forces}}{\text{gravity forces}}$$

$$\frac{\text{Gr}}{\text{Re}^2} = \frac{\rho\beta g(T_0 - T_1)}{\rho V^2/D} = \frac{\text{buoyancy forces}}{\text{inertial forces}}$$

$$\text{Pr Re} = \frac{\rho\hat{C}_p V(T_0 - T_1)/D}{k(T_0 - T_1)/D^2} = \frac{\text{heat transport by convection}}{\text{heat transport by conduction}}$$

$$\text{Br} = \frac{\mu(V/D)^2}{k(T_0 - T_1)/D^2} = \frac{\text{heat production by viscous dissipation}}{\text{heat transport by conduction}}$$

Hence a low value of the Reynolds number means that viscous forces are large in comparison with inertial forces. A low value of the Brinkman number means that any heat produced by viscous dissipation can be transported away by heat conduction. When Gr/Re^2 is large, then buoyancy forces are important in determining flow patterns.

QUESTIONS FOR DISCUSSION

1. Define energy, potential energy, kinetic energy, and internal energy.

2. Why in the derivation of Eq. 10.1–9 is the rate of doing work given by the product of pressure (or viscous stress), area, and velocity?

3. How are the "energy equation" (Eq. 10.1–9), the "equation of mechanical energy" (Eq. 10.1–12), and the "equation of thermal energy" (Eq. 10.1–13) related? Are Eqs. 10.1–12 and 13 general?

[2] The inertial terms in the equation of motion are $\rho[\boldsymbol{v}\cdot\nabla\boldsymbol{v}]$ and the viscous terms are $\mu\nabla^2\boldsymbol{v}$. To get "typical" values of these quantities, replace the various variables by the characteristic "yardsticks" used in obtaining the dimensionless variables. Hence $\rho[\boldsymbol{v}\cdot\nabla\boldsymbol{v}] \rightarrow \rho(V^2/D)$ and $\mu\nabla^2\boldsymbol{v} \rightarrow \mu(V/D^2)$.

4. Is thermal energy ever completely convertible to mechanical energy?

5. What is the difference between Eqs. 10.1–24 and 25? State the assumptions involved in obtaining each from Eq. 10.1–20.

6. Distinguish between free- and forced-convection heat transfer. Is such a distinction always possible?

7. What are the assumptions needed to obtain Eq. 10.3–3 from Eq. 10.3–2? Explain how a Taylor series is used in obtaining Eq. 10.3–3.

8. Is the equation of motion developed in Chapter 3 applicable to nonisothermal systems?

9. When, if ever, can the equation of motion be completely solved for a nonisothermal system without detailed knowledge of the thermal behavior of the system? Give examples.

10. When, if ever, can the equation of energy be completely and exactly solved without detailed knowledge of the flow behavior of the system? Give examples.

11. In Example 10.5–1 how would the temperature dependence of μ distort the velocity profile from that used to calculate temperature profiles? How would it distort the temperature profile? How could you calculate the true temperature and velocity profiles?

12. If a rocket nose cone were made of a porous material and a volatile liquid were forced slowly through the pores during re-entry to the atmosphere, how would the cone surface temperature be affected and why?

13. What is the physical significance of R_0 in Example 10.5–3?

14. In Example 10.5–4 do we really ignore the temperature dependence of ρ? Explain.

15. We found in Example 10.5–5 that velocity must decrease as one proceeds downstream through a shock wave. Explain in terms of the equation of mechanical energy.

16. Is the development in Example 10.5–5 restricted to fluid with a flat velocity profile, to flow in a duct of constant cross section, or to laminar flow?

17. Is the restriction to fluids with a Pr of $\frac{3}{4}$ in Example 10.5–5 a severe limitation? Would the results of this example be applicable to liquids?

18. How might you make use of the results of Example 10.5–6 in designing a gas compressor or a gas turbine?

19. How does one know how to select the dimensionless variables used in §10.6?

20. What would happen if V were chosen as $\mu/\rho D$ for *forced* convection?

21. Would one ever need *more* dimensionless ratios than are obtained for forced convection in §10.6?

22. How are $\hat{C}_p$ and $\hat{C}_v$ defined? What are their units? Is $\tilde{C}_p - \tilde{C}_v$ always equal to R? Does $\hat{C}_p/\hat{C}_v$ for an ideal gas change with temperature or with pressure?

23. Check the units in Eq. 10.1–9.

24. Describe the term $\bar{\rho}\bar{\beta}g(T - \bar{T})$ in Eq. 10.3–3 in terms of Archimedes' principle.

25. In Table 10.4–1 show how one may obtain the $\partial/\partial t$-form of any of the energy equations from the D/Dt-form.

26. Show that Eq. 10.5–14 satisfies its differential equation and boundary conditions.

PROBLEMS

10.A$_1$ Temperature in a Friction Bearing

Calculate the maximum temperature in the friction bearing of Problem 3.A$_1$, assuming the thermal conductivity of the lubricant to be 4×10^{-4} cal sec^{-1} cm^{-1} (° C)$^{-1}$, metal temperature 200° C, and the rate of rotation to be 4000 rpm. *Answer: ca.* 225° C

10.B$_1$ Viscosity Variation and Velocity Gradients in a Nonisothermal Film

Water is falling over a vertical wall in a film 0.1 mm thick. The water temperature is 100° C at the free liquid surface and 80° C at the wall surface.

a. Show that the maximum fractional deviation between viscosities predicted by Eqs. 10.5–19 and 10.5–19*a* occurs when $T = \sqrt{T_0 T_\delta}$.

b. Calculate the maximum fractional deviation for the conditions given.

Answer: 0.5 per cent

10.C$_1$ Transpiration Cooling

a. Calculate the temperature distribution between the two shells of Example 10.5–3 for radial mass flow rates of zero and of 10^{-5} g sec^{-1} for the following conditions:

$R = 500$ microns	$T_R = 300°$ C	$k = 6.13 \times 10^{-5}$ cal cm^{-1} sec^{-1} ° C^{-1}
$\kappa R = 100$ microns	$T_\kappa = 100°$ C	$\hat{C}_p = 0.25$ cal g^{-1} ° C^{-1}

b. Compare the rates of heat conduction to the surface at κR in the presence and absence of convection.

Answer: $\epsilon = 0.124$

10.D$_1$ Free Convection From a Vertical Surface

A small heating panel consists essentially of a flat, vertical, rectangular surface 30 cm high and 50 cm wide. Estimate the total rate of heat loss from one side of this panel *by free convection* if the panel surface is at 150° F and the surrounding air is at 70° F and 1 atm.

Use the expressions of Lorenz and of Schmidt and Beckmann and compare the results of these two calculations.

Answer: 8.1 cal sec^{-1} by Lorenz expression
7.6 cal sec^{-1} by Schmidt and Beckmann expression

10.E$_1$ Velocity, Temperature, and Pressure Changes in a Shock Wave

Air at 1 atm and 70° F is flowing at an upstream Mach number of 2 across a stationary shock wave.

a. Calculate the initial velocity of the air.
b. Calculate the velocity, temperature, and pressure downstream from the shock wave
c. Calculate the velocity profile in the shock wave.
d. Calculate the changes of internal and kinetic energy across the shock wave.

Assume that γ is constant at 1.4 and that $\hat{C}_p = 0.24$ Btu lb$_m^{-1}$ ° R^{-1}

Answers: *a.* 2250 ft sec^{-1}
b. 844 ft sec^{-1}, 888° R, 4.48 atm
d. $\Delta\hat{U} = +61.4$ Btu lb$_m^{-1}$
$\Delta\hat{K} = -86.9$ Btu lb$_m^{-1}$

10.F$_1$ Adiabatic Frictionless Compression of an Ideal Gas

Calculate the temperature attained by compressing air, initially at 100° F and 1 atm, to $\frac{1}{10}$ of its initial volume. $\hat{C}_p/\hat{C}_v$ is to be assumed constant at 1.40, and frictionless adiabatic compression may be assumed. The resulting temperature is the approximate temperature just before ignition in an automotive engine with a 10/1 compression ratio.

Answer: 950° F

10.G$_2$ Use of the Energy Equation to Set up Problems

Verify the following equations in Chapter 9, using the appropriate forms of the energy equation from §10.2: (*a*) 9.2–6, (*b*) 9.3–6, (*c*) 9.6–2, (*d*) 9.7–3.

10.H₂ Viscous Heating in Laminar Slit Flow

Derive an expression for the temperature distribution $T(x)$ in a viscous fluid in steady laminar flow between large flat parallel plates, as shown in Fig. 2.E. Both plates are maintained at constant temperature T_0. Take into account explicitly the heat generated by viscous dissipation. Neglect the temperature dependence of μ and k.

$$\text{Answer: } T - T_0 = \frac{1}{3}\frac{\mu v_{\max}^2}{k}\left[1 - \left(\frac{x}{B}\right)^4\right]$$

10.I₂ Velocity Distribution in a Nonisothermal Film

Show that Eq. 10.5–20 meets the following requirements:

a. At $x = \delta$, $v_z = 0$.

b. At $x = 0$, $\partial v_z / \partial x = 0$.

c. $\lim_{\mu_\delta \to \mu_0} \{v_z\} = (\rho g \delta^2 \cos \beta / 2\mu_0)[1 - (x/\delta)^2]$.

10.J₂ Transpiration Cooling in a Planar System

Two large flat porous horizontal plates are separated by a relatively small distance L. The upper plate at $y = L$ is at $T = T_L$ and the lower one at $y = 0$ is to be maintained at a lower temperature T_0. To reduce the amount of heat that must be removed from the lower plate, an ideal gas at T_0 is blown upward through both plates at a steady rate. Develop an expression for the temperature distribution and the amount of heat q_0 that must be removed from the cold plate per unit area as a function of the fluid properties and flow rate of gas.

$$\text{Answer: } \frac{T - T_L}{T_0 - T_L} = \frac{e^{y\phi/L} - e^{\phi}}{1 - e^{\phi}}; \qquad \phi = \frac{L\rho \hat{C}_p v_y}{k}$$

$$q_0 = \frac{k(T_L - T_0)}{L}\left(\frac{\phi}{e^{\phi} - 1}\right)$$

10.K₂ Use of Transpiration to Reduce Evaporation Losses

It is proposed to reduce the rate of evaporation of liquefied oxygen in small containers by taking advantage of transpiration. To do this, the liquid is to be stored in a spherical container surrounded by a spherical shell of porous insulating material, as shown in Fig. 10.K. A thin gas space is to be left between the container and insulation, and the opening

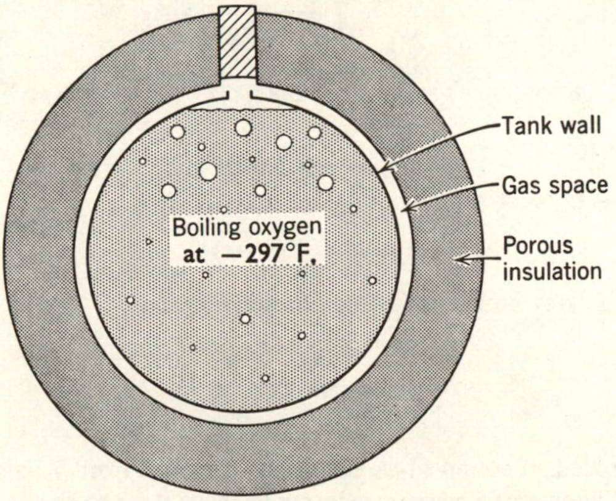

Fig. 10.K. Use of transpiration to reduce evaporation rate.

in the insulation is to be stoppered. In operation, the evaporating oxygen is to leave the container proper, move through the gas space, and then flow uniformly out through the porous insulation.

Calculate the rate of heat gain and evaporation loss from a tank 1 ft in diameter covered with a shell of insulation 6 in. thick under the following conditions, with and without transpiration.

Temperature of liquid oxygen:	$-297°$ F
Temperature of outer surface of insulation:	$+30°$ F
Effective thermal conductivity of insulation:	0.02 Btu hr^{-1} ft^{-1} $°F^{-1}$
Average heat capacity of O_2 flowing through insulation:	0.22 Btu lb^{-1} $°F^{-1}$
Heat of vaporization of oxygen:	91.7 Btu lb^{-1}

Neglect the thermal resistance of the liquid oxygen, container wall, and gas space, and neglect heat losses through the stopper. Assume the particles of insulation to be in thermal equilibrium with the surrounding gas. *Answers:* 82 Btu hr^{-1} without transpiration
61 Btu hr^{-1} with transpiration

10.L$_2$ Frictionless Processes in an Ideal Gas

a. Show that the equation of energy balance for an ideal gas in steady-state flow with no viscous dissipation

$$\rho \hat{C}_v \left(v_x \frac{\partial}{\partial x} + v_y \frac{\partial}{\partial y} + v_z \frac{\partial}{\partial z} \right) T = k\nabla^2 T - p(\nabla \cdot v) \quad \text{(10.L–1)}$$

may be rewritten as

$$\rho \hat{C}_p \left(v_x \frac{\partial}{\partial x} + v_y \frac{\partial}{\partial y} + v_z \frac{\partial}{\partial z} \right) T = k\nabla^2 T + \left(v_x \frac{\partial}{\partial x} + v_y \frac{\partial}{\partial y} + v_z \frac{\partial}{\partial z} \right) p \quad \text{(10.L–2)}$$

b. Rewrite the equation obtained in (*a*) in dimensionless form, using a characteristic length D, a characteristic velocity V, and a characteristic temperature T_0. Let $v^* = v/V$, $T^* = (T - T_0)/T_0$, and $p^* = p/\rho V^2$. Show that the reduced temperature equation contains in addition to these dimensionless variables the dimensionless groups: $\text{Re} = DV\rho/\mu$, $\text{Pr} = \hat{C}_p\mu/k$, and $\text{Ec} = V^2/\hat{C}_p T_0$ (known as the *Eckert number*).

c. The speed of sound v_s is related to the p-V-T behavior of the medium through which the sound is propagated:

$$v_s = \sqrt{\gamma(\partial p/\partial \rho)_T} \quad \text{(10.L–3)}$$

in which $\gamma = C_p/C_v$. Show that for an ideal gas

$$v_s = \sqrt{\gamma RT/M} \quad \text{(10.L–4)}$$

and that (since $\tilde{C}_p - \tilde{C}_v = R$ for such a gas)

$$v_s = \sqrt{\hat{C}_v T(\gamma - 1)} \quad \text{(10.L–5)}$$

d. Show that the Eckert number may also be written as

$$\text{Ec} = (\gamma - 1)\frac{V^2}{v_{s0}^2} = (\gamma - 1)\text{Ma}^2 \quad \text{(10.L–6)}$$

in which v_{s0} is the speed of sound in an ideal gas at temperature T_0 and Ma is the *Mach number*, the ratio of some characteristic velocity in the system to the velocity of sound. If the Mach number is small compared with unity, the flow can be considered as incompressible.

10.M_2 Dimensional Analysis of Forced-Convection Heat Transfer in an Agitated Tank

A fluid of constant ρ, μ, $\hat{C}_p$, and k is being heated in a completely filled, steam-jacketed, and agitated tank. It is desired to use the tank to predict the rate of temperature rise in a geometrically similar tank of twice the linear dimensions and at a specified agitator speed. The same fluid is to be used in the two cases, and the jacket surface temperature T_1 and initial bulk fluid temperature T_0 are to be the same in both. What should the ratio of agitator speeds be and what will the ratio of rates of temperature increase be?

Answer: The angular velocity of the small agitator should be four times that of the larger. If this ratio is used, the time required for any portion of the large tank to reach a given temperature will be four times that in the smaller.

10.N_2 Equivalence of Different Forms of the Energy Equation

a. Show how Eq. H of Table 10.4–1 may be obtained from Eq. G.

b. Show how Eq. T of Table 10.4–1 may be obtained from Eq. J.

10.O_3 Tangential Annular Flow of a Highly Viscous Fluid

Show that Eq. 10.5–14 reduces to Eq. 9.4–11 as κ approaches unity.

10.P_3 Dimensional Analysis for Combined Free- and Forced-Convection Heat Transfer

It is desired to estimate the rate of heat loss from the meteorological installation shown in Fig. 10.P from an experiment on a small geometrically similar model of $\frac{1}{5}$ the linear

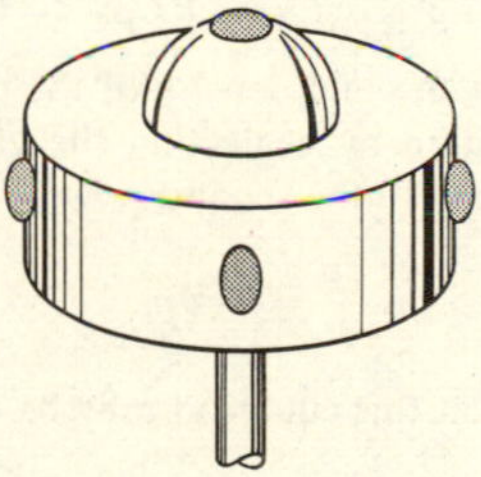

Fig. 10.P. Meteorological installation.

dimensions. The desired surface temperature of the installation and the expected air temperature and wind velocity and direction are known. Under these expected conditions, both free and forced convection are probably important, and it therefore appears reasonable that the reduced velocity distribution will depend upon both the Reynolds and Grashof numbers. It is desired to use air in the model experiment and to maintain similarity between the model and full-scale apparatus by varying air pressure and velocity. The temperature dependence of fluid properties may be ignored.

a. What pressure and relative air velocity should be used in the model experiment to maintain dynamic similarity?

b. What will be the relative heat flux from the model?

c. Why would it not be practical to maintain dynamic similarity by varying ΔT?

Answers: *a.* For equal temperatures in the two systems, the pressure in the model study should be 11.2 times the pressure prevailing at the proposed installation. The velocity in the model experiment should then be 0.446 times the expected wind velocity.

b. The surface flux from the model will be 5 times that from the full-scale apparatus.

10.Q$_3$ Speed of Propagation of Sound Waves

Sound waves may be considered as harmonic compression waves of very small amplitude traveling through a compressible fluid. The velocity of propagation of such waves may be estimated by assuming that both $\boldsymbol{\tau}$ and $\boldsymbol{q}$ are zero and that the velocity of the fluid, $\boldsymbol{v}$, is small.[1]

a. Show that by neglecting $\boldsymbol{\tau}$ and $\boldsymbol{q}$ Eq. J of Table 10.4–1 may be written to give the p-T relation in a sound wave as

$$\frac{dp}{dT} = \rho\hat{C}_p\left(\frac{\partial \ln T}{\partial \ln \hat{V}}\right)_p \tag{10.Q–1}$$

Use this expression to show that

$$\begin{aligned}\frac{dp}{d\rho} &= \left(\frac{\partial p}{\partial \rho}\right)_T\left[\frac{\rho\hat{C}_p(\partial \ln T/\partial \ln \hat{V})_p}{\rho\hat{C}_p(\partial \ln T/\partial \ln \hat{V})_p - (\partial p/\partial T)_\rho}\right]\\ &= \gamma\left(\frac{\partial p}{\partial \rho}\right)_T\end{aligned} \tag{10.Q–2}$$

Show that for the specific case of an ideal gas,

$$\frac{dp}{d\rho} = \left(\frac{\tilde{C}_p}{\tilde{C}_p - R}\right)\frac{p_0}{\rho_0} \tag{10.Q–3}$$

Here p_0 and ρ_0 are the average pressure and density of the fluid.

b. Linearize the equation of motion by neglecting the viscous shear and gravitational terms and all terms involving velocity to the second power, to obtain

$$\rho_0\frac{\partial \boldsymbol{v}}{\partial t} = -\nabla p \tag{10.Q–4}$$

Justify this simplification. Show that this equation may be rewritten as

$$\rho_0\frac{\partial \boldsymbol{v}}{\partial t} = -v_s^2\nabla\rho \tag{10.Q–5}$$

where

$$v_s^2 = \gamma\left(\frac{\partial p}{\partial \rho}\right)_T \tag{10.Q–6}$$

c. Linearize the equation of continuity by neglecting the term $(\boldsymbol{v}\cdot\nabla\rho)$, and justify this simplification. Differentiate the linearized equation with respect to time and combine your result with that of (*b*) to obtain

$$\frac{\partial^2\rho}{\partial t^2} = v_s^2\,\nabla^2\rho \tag{10.Q–7}$$

[1] These assumptions are appropriate only for infinitesimal low-frequency disturbances. Large disturbances, e.g., shock waves, propagate at much higher velocities—see for example, J. O. Hirschfelder, C. F. Curtiss, and R. B. Bird, *Molecular Theory of Gases and Liquids*, Wiley, New York (1954), p. 728.

d. Show that a solution of this equation is $\rho = \rho_0\left[1 + \phi_0 \sin\left(\frac{2\pi}{\lambda}(z - v_s t)\right)\right]$. This solution represents a harmonic wave of wavelength λ and amplitude $\rho_0\phi_0$ traveling in the z-direction at a speed v_s. The speed of sound is then v_s and is independent of amplitude and wavelength—provided that the amplitude is small enough to satisfy the foregoing limiting assumptions.

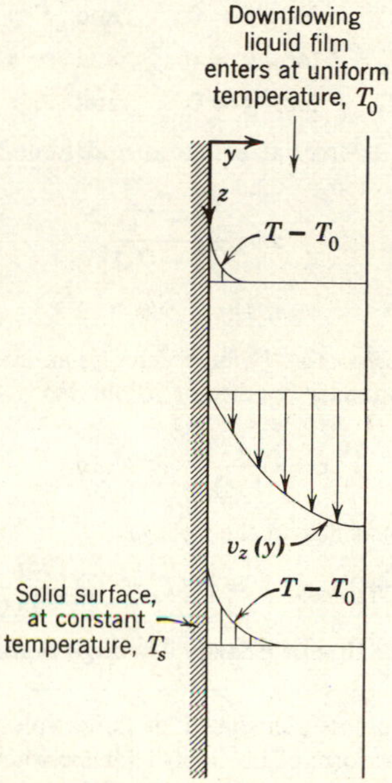

Fig. 10.R. Heat transfer to a falling liquid film.

10.R$_3$ Heat Transfer from a Wall to a Falling Film: Short Contact Times[2]

A cold liquid film flowing down a vertical or inclined solid wall, as shown in Fig. 10.R, has a considerable cooling effect on the solid surface. Estimate the rate of heat transfer from the wall to the fluid for such short contact times that the fluid temperature changes appreciably only in the immediate vicinity of the wall.

a. Show that $v_z = v_{\max}[2y/\delta - (y/\delta)^2]$, in which $v_{\max} = \delta^2\rho g/2\mu$ and therefore that near the wall

$$v_z \doteq \frac{\rho g \delta}{\mu} y \tag{10.R-1}$$

b. Show that the energy equation reduces to

$$\rho \hat{C}_p v_z \frac{\partial T}{\partial z} \doteq k \frac{\partial^2 T}{\partial y^2} \tag{10.R-2}$$

[2] See R. L. Pigford, *C. E. P. Symposium Series*, No. 17, **51**, 79–92 (1955).

(What simplifying assumptions are required to obtain this result?) Combine the foregoing results to obtain the approximate differential equation

$$y \frac{\partial T}{\partial z} = \beta \frac{\partial^2 T}{\partial y^2} \tag{10.R-3}$$

in which $\beta = \mu k / \rho^2 \hat{C}_p g \delta$.

c. Show that for short contact times we may write as boundary conditions

B.C. 1:	$T = T_0$	for $z = 0$	and	$y > 0$	(10.R-4)
B.C. 2:	$T = T_0$	for $y = \infty$	and	z finite	(10.R-5)
B.C. 3:	$T = T_s$	for $y = 0$	and	$z > 0$	(10.R-6)

d. Rewrite the foregoing differential equation and boundary conditions in terms of the reduced variables

$$\Theta = \frac{(T - T_0)}{(T_s - T_0)} \tag{10.R-7}$$

$$\eta = y / \sqrt[3]{9\beta z} \tag{10.R-8}$$

Integrate once to obtain $d\Theta/d\eta = Ce^{-\eta^3}$, in which C is an integration constant. Integrate a second time and use the boundary conditions to obtain

$$\Theta = \frac{1}{\Gamma(\frac{4}{3})} \int_\eta^\infty e^{-\eta^3} \, d\eta \tag{10.R-9}$$

e. Show that the *average* heat flux to the fluid is

$$(q_y)_{\text{avg}}|_{y=0} = \tfrac{3}{2} k (T_s - T_0) \frac{(9\beta L)^{-1/3}}{\Gamma(\frac{4}{3})} \tag{10.R-10}$$

10.S₄ Derivation of the Equation of Energy Through Application of the Divergence Theorem

Consider any stationary volume element G of finite volume and surface V_G and S_G. Write an energy balance of the form of Eq. 10.1–1 for the whole element. Since conditions cannot be considered uniform over the element, it will be necessary to integrate the rate of energy gain and work per unit volume against body forces over V_G and the rate of energy convection and work done against surface forces over S_G. Now convert all surface integrals into volume integrals by the use of the divergence theorem and show that the resulting expression can be reduced to Eq. 10.1–9.

10.T₄ The Equation of Change for Entropy of a Pure Fluid

a. Consider a fixed volume element $\Delta x \, \Delta y \, \Delta z$ and write an "entropy balance" for this system. Let s_x, s_y, and s_z be the components of an "entropy flux vector" $\boldsymbol{s}$, which is measured with respect to the fluid velocity $\boldsymbol{v}$. Further, let the "rate of entropy production" per unit volume be designated by g. Show that when the volume $\Delta x \, \Delta y \, \Delta z$ is allowed to approach zero one finally obtains

$$\rho \frac{D\hat{S}}{Dt} = -(\nabla \cdot \boldsymbol{s}) + g \tag{10.T-1}$$

where $\hat{S}$ is the entropy per unit mass.

b. If one assumes that the thermodynamic quantities can be locally defined in a non-equilibrium situation, then $\hat{U}$ can be related to $\hat{S}$ and $\hat{V}$ according to the thermodynamic relation $d\hat{U} = T\,d\hat{S} - p\,d\hat{V}$. Combine this relation with Eq. 10.1–13 to get

$$\rho \frac{D\hat{S}}{Dt} = -\frac{1}{T}(\nabla \cdot \boldsymbol{q}) - \frac{1}{T}(\boldsymbol{\tau} : \nabla \boldsymbol{v}) \tag{10.T-2}$$

c. The local entropy flux is equal to the local energy flux divided by the local temperature—that is, $\boldsymbol{s} = \boldsymbol{q}/T$. Once this relation between $\boldsymbol{s}$ and $\boldsymbol{q}$ is recognized, we can compare Eqs. 10.T–1 and 2 to get the following expression for the rate of entropy production per unit volume:

$$g = -\frac{1}{T^2}(\boldsymbol{q} \cdot \nabla T) - \frac{1}{T}(\boldsymbol{\tau} : \nabla \boldsymbol{v}) \qquad (10.\text{T–3})$$

On the right side the first term represents the rate of entropy production associated with energy transport and the second term is the rate of entropy production associated with momentum transport. Equation 10.T–3 is the starting point for the thermodynamic study of the irreversible processes in a pure fluid.[3]

d. What conclusions can be drawn when Newton's law of viscosity and Fourier's law of heat conduction are inserted into Eq. 10.T–3?

[3] J. G. Kirkwood and B. L. Crawford, Jr., *J. Phys. Chem.*, **56**, 1048–1051 (1952); R. B. Bird, C. F. Curtiss, and J. O. Hirschfelder, *Chem. Engr. Prog. Symp. Series, No. 16*, **51**, 69–85 (1955).

CHAPTER 11

Temperature Distributions with More than One Independent Variable

In Chapter 9 it was shown how a number of simple heat-flow problems can be solved by means of shell energy balances. In Chapter 10 the energy equation for flow systems was developed, which in principle describes heat-transfer processes in more complex situations. Just to illustrate the usefulness of the energy equation, we presented in §10.5 a series of examples, most of which required no knowledge of partial differential equations. In this chapter we discuss several classes of heat-transfer problems which involve more than one independent variable: unsteady heat conduction in solids, steady heat conduction in viscous flow, steady two-dimensional heat conduction in solids, and heat flow in laminar boundary layers. These topics roughly parallel those given in Chapter 4 both in physical processes and mathematical techniques.

§11.1 UNSTEADY HEAT CONDUCTION IN SOLIDS

For solids, the energy equation of Eq. 10.1–19, after insertion of Fourier's law of heat conduction, becomes

$$\rho \hat{C}_p \frac{\partial T}{\partial t} = (\nabla \cdot k \nabla T) \tag{11.1–1}$$

If the thermal conductivity is independent of the temperature or position, then Eq. 11.1–1 becomes

$$\frac{\partial T}{\partial t} = \alpha \nabla^2 T \tag{11.1-2}$$

in which $\alpha = k/\rho \hat{C}_p$ is the thermal diffusivity of the solid. Equation 11.1–2 is one of the most worked-over equations of theoretical physics. The treatise of Carslaw and Jaeger[1] is devoted entirely to methods of solution of this equation; their book should be familiar to all engineers and applied scientists because of its extensive tabulation of solutions to Eq. 11.1–2 for an enormous number of boundary and initial conditions. Many frequently encountered heat-conduction problems may be solved simply by looking up the solution in Carslaw and Jaeger's reference work.

In this section we begin by giving two of the very simplest unsteady-state solutions to Eq. 11.1–2 to introduce beginners to the subject. These solutions illustrate the method of combination of variables and the method of separation of variables, which were also used in §4.1. Then we give one example of a problem solved by means of Laplace transform, a technique that is of great importance in solving unsteady-state problems. Readers desiring more elaborate examples will have no trouble finding them in Carslaw and Jaeger.

Example 11.1–1. Heating of a Semi-Infinite Slab[2]

A solid body occupying the space from $y = 0$ to $y = \infty$ is initially at temperature T_0. At time $t = 0$, the surface at $y = 0$ is suddenly raised to temperature T_1 and maintained at that temperature for $t > 0$. Find the time-dependent temperature profiles $T(y, t)$.

Solution. For this problem, Eq. 11.1–2 becomes

$$\frac{\partial \Theta}{\partial t} = \alpha \frac{\partial^2 \Theta}{\partial y^2} \tag{11.1-3}$$

in which we have introduced a dimensionless temperature $\Theta = (T - T_0)/(T_1 - T_0)$. With this dimensionless temperature, the initial and boundary conditions assume this simple form:

I.C.:	at $t \leqslant 0$,	$\Theta = 0$	for all y	(11.1–4)
B.C. 1:	at $y = 0$,	$\Theta = 1$	for all $t > 0$	(11.1–5)
B.C. 2:	at $y = \infty$,	$\Theta = 0$	for all $t > 0$	(11.1–6)

This problem is mathematically analogous to that formulated in Eqs. 4.1–1 through

[1] H. S. Carslaw and J. C. Jaeger, *Conduction of Heat in Solids*, Oxford University Press, (1959), Second Edition.

[2] See K. T. Yang, *J. Appl. Mechanics*, **25**, 146–147 (1958) for a solution with variable thermal conductivity.

4; hence the solution in Eq. 4.1–13 can be taken over directly by appropriate change in notation.

$$\Theta = 1 - \frac{2}{\sqrt{\pi}} \int_0^{y/\sqrt{4\alpha t}} e^{-\eta^2}\, d\eta \tag{11.1–7}$$

or

$$\frac{T - T_0}{T_1 - T_0} = 1 - \operatorname{erf} \frac{y}{\sqrt{4\alpha t}} \tag{11.1–8}$$

The graphical solution in Fig. 4.1–2 for $n = 1$ describes the temperature profiles when the ordinate is labeled $(T - T_0)/(T_1 - T_0)$ and the abscissa, $y/\sqrt{4\alpha t}$.

Because the error function reaches a value of 0.99 when the argument is about 2, the "thermal penetration thickness" δ_T is

$$\delta_T \doteq 4\sqrt{\alpha t} \tag{11.1–9}$$

That is, for distances $y > \delta_T$, the temperature has changed by less than 1 per cent of the difference $(T_1 - T_0)$. If it is necessary to calculate the temperature in a slab of finite thickness, the solution in Eq. 11.1–8 will be a good approximation when δ_T is small with respect to the slab thickness. When δ_T is of the order of magnitude of the slab thickness, then the series solution of Example 11.1–2 has to be used.

The wall heat flux can be calculated from Eq. 11.1–8:

$$q_y\big|_{y=0} = -k \left.\frac{\partial T}{\partial y}\right|_{y=0} = \frac{k}{\sqrt{\pi \alpha t}} (T_1 - T_0) \tag{11.1–10}$$

Hence the penetration thickness varies as $t^{1/2}$ and the wall heat flux as $t^{-1/2}$.

Example 11.1–2. Heating of a Finite Slab

A slab occupying the space between $y = -b$ and $y = +b$ is initially at temperature T_0. At time $t = 0$ the surfaces at $y = \pm b$ are suddenly raised to T_1 and maintained there. Find $T(y, t)$.

Solution. For this problem we introduce the following dimensionless quantities:

$$\Theta = \frac{T_1 - T}{T_1 - T_0} = \begin{array}{l}\text{dimensionless temperature} \\ \text{(different from } \Theta \text{ in Example 11.1–1)}\end{array} \tag{11.1–11}$$

$$\eta = \frac{y}{b} = \text{dimensionless length} \tag{11.1–12}$$

$$\tau = \frac{\alpha t}{b^2} = \text{dimensionless time} \tag{11.1–13}$$

Experience teaches us that it is convenient to introduce such dimensionless quantities so that the differential equations and boundary conditions assume a simpler form:

$$\frac{\partial \Theta}{\partial \tau} = \frac{\partial^2 \Theta}{\partial \eta^2} \tag{11.1–14}$$

I.C.: at $\tau = 0$, $\Theta = 1$ (11.1–15)

B.C.'s 1 and 2: at $\eta = \pm 1$, $\Theta = 0$ (11.1–16)

We present here the classical solution by the method of separation of variables. We anticipate that a solution of the following product form can be found:

$$\Theta(\eta, \tau) = f(\eta)g(\tau) \tag{11.1–17}$$

Substitution of this trial function into Eq. 11.1–14 gives, after division by fg,

$$\frac{1}{g}\frac{dg}{d\tau} = \frac{1}{f}\frac{d^2f}{d\eta^2} \tag{11.1–18}$$

The left side is a function of τ alone, and the right side is a function of η alone. This can be true only if both sides equal a constant, which we call $-c^2$. The problem then becomes one of solving the two ordinary differential equations:

$$\frac{dg}{d\tau} = -c^2 g \tag{11.1–19}$$

$$\frac{d^2f}{d\eta^2} = -c^2 f \tag{11.1–20}$$

These may be integrated to give

$$g = A \exp(-c^2\tau) \tag{11.1–21}$$

$$f = B \sin c\eta + C \cos c\eta \tag{11.1–22}$$

in which A, B, and C are constants.

We note that Θ, hence f, must be even functions of η by virtue of the symmetry of the problem. Therefore we must set B equal to zero. Use of either one of the boundary conditions gives

$$C \cos c = 0 \tag{11.1–23}$$

Clearly, C cannot be zero because this choice would lead to a physically inadmissable solution. Consequently, we are forced to let

$$c = (n + \tfrac{1}{2})\pi \qquad n = 0, \pm 1, \pm 2, \cdots \pm \infty \tag{11.1–24}$$

The foregoing choices for C and c lead then to the fact that

$$\Theta_n = A_n C_n e^{-(n+\frac{1}{2})^2\pi^2\tau} \cos (n + \tfrac{1}{2})\pi\eta \tag{11.1–25}$$

is an admissable solution. The subscripts n remind us that A and C may be different for each n. The most general solution of this form is obtained by adding the solutions of the form of Eq. 11.1–25 for all integral n from $n = -\infty$ to $n = +\infty$:

$$\Theta = \sum_{n=0}^{\infty} D_n e^{-(n+\frac{1}{2})^2\pi^2\tau} \cos (n + \tfrac{1}{2})\pi\eta \tag{11.1–26}$$

in which $D_n = A_n C_n + A_{-(n+1)} C_{-(n+1)}$.

The set of D_n are now determined by using the initial condition, which states that

$$1 = \sum_{n=0}^{\infty} D_n \cos (n + \tfrac{1}{2})\pi\eta \tag{11.1–27}$$

Multiplication by $\cos (m + \frac{1}{2})\pi\eta \, d\eta$ and integrating from $\eta = -1$ to $\eta = +1$ gives

$$\int_{-1}^{+1} \cos (m + \tfrac{1}{2})\pi\eta \, d\eta = \sum_{n=0}^{\infty} D_n \int_{-1}^{+1} \cos (m + \tfrac{1}{2})\pi\eta \cos (n + \tfrac{1}{2})\pi\eta \, d\eta \tag{11.1–28}$$

When the indicated integrations are performed, all the integrals in the summation on the right are identically zero, with the exception of the single term in which $n = m$. Hence we get

$$\left.\frac{\sin (m + \frac{1}{2})\pi\eta}{(m + \frac{1}{2})\pi}\right|_{\eta=-1}^{\eta=+1} = D_m \left.\left(\frac{\frac{1}{2}(m + \frac{1}{2})\pi\eta + \frac{1}{4}\sin (m + \frac{1}{2})2\pi\eta}{(m + \frac{1}{2})\pi}\right)\right|_{\eta=-1}^{\eta=+1} \tag{11.1-29}$$

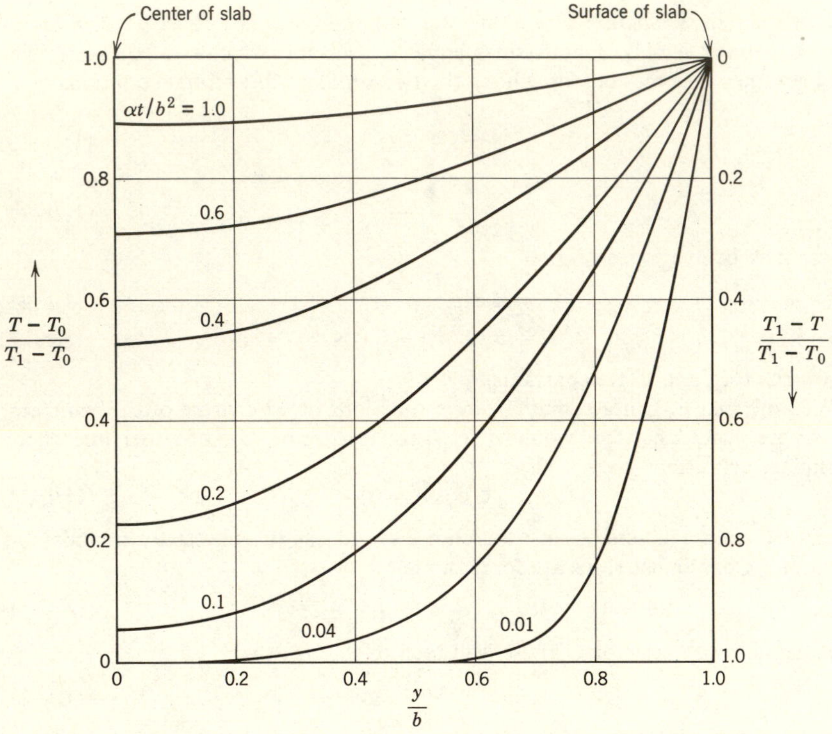

Fig. 11.1-1. Temperature profiles for unsteady-state heat conduction in a slab of finite thickness. [H. S. Carslaw and J. C. Jaeger, *Conduction of Heat in Solids*, Oxford University Press (1959), p. 101.]

Inserting the limits and solving for D_m gives

$$D_m = \frac{2(-1)^m}{(m + \frac{1}{2})\pi} \tag{11.1-30}$$

Substitution of this expression into Eq. 11.1-26 and rewriting in terms of the original variables gives

$$\frac{T_1 - T}{T_1 - T_0} = 2\sum_{n=0}^{\infty} \frac{(-1)^n}{(n + \frac{1}{2})\pi} e^{-(n+\frac{1}{2})^2\pi^2\alpha t/b^2} \cos (n + \tfrac{1}{2}) \frac{\pi y}{b} \tag{11.1-31}$$

The solutions to many unsteady-state heat-conduction problems come out in the form of infinite series, such as that obtained here. Such a series converges rapidly for large dimensionless times $\alpha t/b^2$. For very short times the convergence is very slow, and in the limit as $\alpha t/b^2 \to 0$, the solution in Eq. 11.1–31 may be shown (by use of Laplace transform) to approach that given in Eq. 11.1–8. Although Eq. 11.1–31 is unwieldy for some practical calculations, a graphical presentation, such as that in Fig. 11.1–1, is easy to use. (See Problem 11.C.)

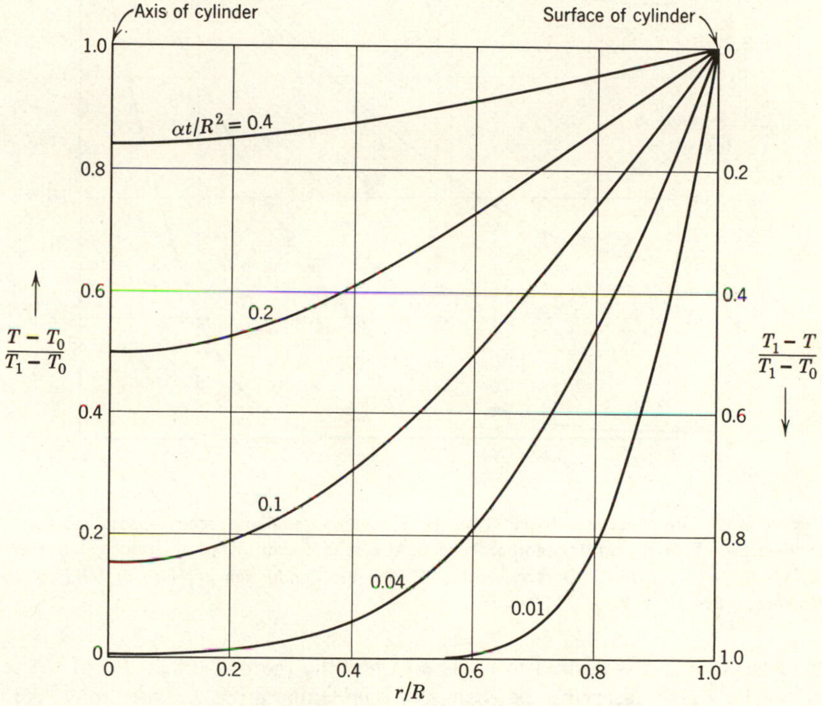

Fig. 11.1–2. Temperature profiles for unsteady-state heat conduction in a cylinder. T_0 is the initial temperature of the cylinder, and T_1 is the temperature at the surface for $t > 0$. [H. S. Carslaw and J. C. Jaeger, *Conduction of Heat in Solids*, Oxford University Press (1959), p. 200.]

Results analogous to Fig. 11.1–1 are given for infinite cylinders and spheres in Figs. 11.1–2 and 3. These charts can be used to build up the solutions for the analogous heat conduction problems in rectangular parallelopipeds and finite cylinders. (See Problem 11.L.)

Example 11.1–3. Cooling of a Sphere in Contact with a Well-Stirred Fluid

A homogeneous sphere of solid material, initially at a uniform temperature T_1, is suddenly immersed in a volume V_f of well-stirred fluid of temperature T_0 in an

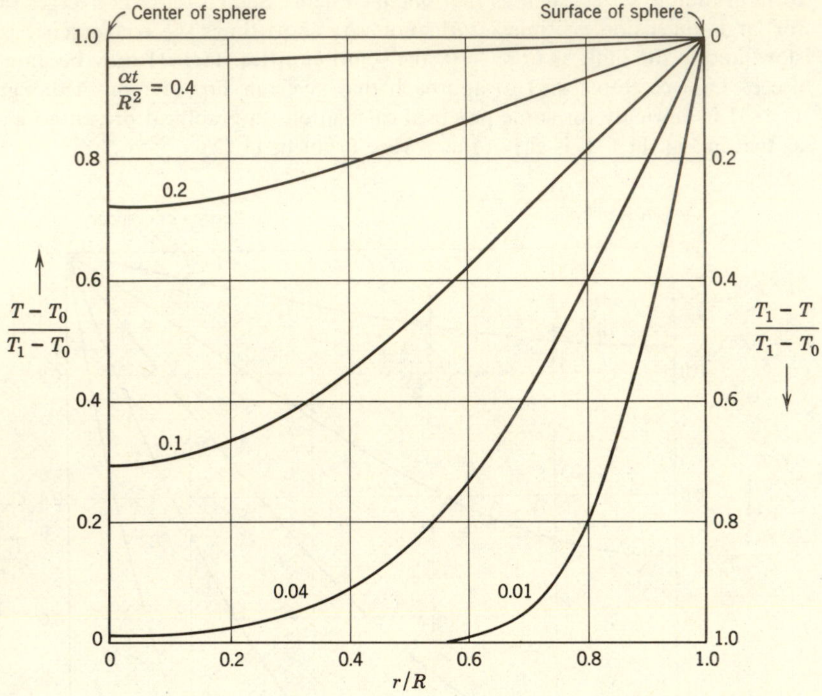

Fig. 11.1-3. Temperature distribution in a sphere resulting from unsteady-state heat conduction. T_0 is the initial temperature of the sphere, and T_1 is the temperature at the surface for $t > 0$. [H. S. Carslaw and J. C. Jaeger, *Conduction of Heat in Solids*, Oxford University Press (1959), p. 234.]

insulated tank. It is desired to be able to find the thermal diffusivity of the solid $\alpha_s = k_s/\rho_s \hat{C}_{ps}$ by observing the change of *fluid* temperature T_f with time. We use the following dimensionless variables:

$$\Theta_s = \frac{T_1 - T_s}{T_1 - T_0} = \text{dimensionless solid temperature (a function of } \xi \text{ and } \tau) \tag{11.1-32}$$

$$\Theta_f = \frac{T_1 - T_f}{T_1 - T_0} = \text{dimensionless fluid temperature (a function of } \tau \text{ alone—this assumes "perfect stirring")} \tag{11.1-33}$$

$$\xi = \frac{r}{R} = \text{dimensionless radial coordinate} \tag{11.1-34}$$

$$\tau = \frac{\alpha_s t}{R^2} = \text{dimensionless time} \tag{11.1-35}$$

Solution. The reader may verify that the problem stated in dimensionless variables is

Solid		Fluid	
$\dfrac{\partial \Theta_s}{\partial \tau} = \dfrac{1}{\xi^2}\dfrac{\partial}{\partial \xi}\left(\xi^2 \dfrac{\partial \Theta_s}{\partial \xi}\right)$	(11.1–36)	$\dfrac{d\Theta_f}{d\tau} = -\dfrac{3}{B}\dfrac{\partial \Theta_s}{\partial \xi}\bigg\rvert_{\xi=1}$	(11.1–40)
At $\tau = 0$, $\Theta_s = 0$	(11.1–37)	At $\tau = 0$, $\Theta_f = 1$	(11.1–41)
At $\xi = 1$, $\Theta_s = \Theta_f$	(11.1–38)		
At $\xi = 0$, Θ_s = finite	(11.1–39)		

in which B is a dimensionless quantity $V_f \rho_f \hat{C}_{pf}/V_s \rho_s \hat{C}_{ps}$, the V's representing the volume of the fluid and of the solid. We now take the Laplace transform[3,4] of the foregoing equations and their boundary conditions:

Solid		Fluid	
$p\overline{\Theta}_s = \dfrac{1}{\xi^2}\dfrac{d}{d\xi}\left(\xi^2 \dfrac{d\overline{\Theta}_s}{d\xi}\right)$	(11.1–42)	$p\overline{\Theta}_f - 1 = -\dfrac{3}{B}\dfrac{d\overline{\Theta}_s}{d\xi}\bigg\rvert_{\xi=1}$	(11.1–45)
At $\xi = 1$, $\overline{\Theta}_s = \overline{\Theta}_f$	(11.1–43)		
At $\xi = 0$, $\overline{\Theta}_s$ = finite	(11.1–44)		

The solution to Eq. 11.1–42 is

$$\overline{\Theta}_s = \frac{C}{\xi}\sinh\sqrt{p}\,\xi + \frac{D}{\xi}\cosh\sqrt{p}\,\xi \tag{11.1–46}$$

But $D = 0$, according to Eq. 11.1–44. Substitution of this result into Eq. 11.1–45 then gives

$$\overline{\Theta}_f = \frac{1}{p} + 3\frac{C}{Bp}(\sinh\sqrt{p} - \sqrt{p}\cosh\sqrt{p}) \tag{11.1–47}$$

Next, we insert the last two results into the transformed boundary condition in Eq. 11.1–43 in order to determine C. This gives us finally

$$\overline{\Theta}_f = \frac{1}{p} + 3\left\{\frac{1 - (1/\sqrt{p})\tanh\sqrt{p}}{(3 - Bp)\sqrt{p}\tanh\sqrt{p} - 3p}\right\} \tag{11.1–48}$$

[3] Carslaw and Jaeger, *op. cit.*, Chapter XII; we use the definition

$$\mathscr{L}\{f(\tau)\} = \bar{f}(p) = \int_0^\infty f(\tau)\exp(-p\tau)\,d\tau$$

[4] R. V. Churchill, *Modern Operational Mathematics in Engineering*, McGraw-Hill, New York (1944).

We now take the inverse Laplace transform

$$\Theta_f = 1 + 3\mathscr{L}^{-1}\left\{\frac{1 - (1/\sqrt{p}) \tanh \sqrt{p}}{(3 - Bp)\sqrt{p} \tanh \sqrt{p} - 3p}\right\} \tag{11.1-49}$$

The last term has a denominator with a double root at $p = 0$ and a set of roots $\sqrt{p_k} = ib_k$, in which $k = 1, 2, 3, \ldots \infty$ and $i = \sqrt{-1}$; the b_k are the nonzero

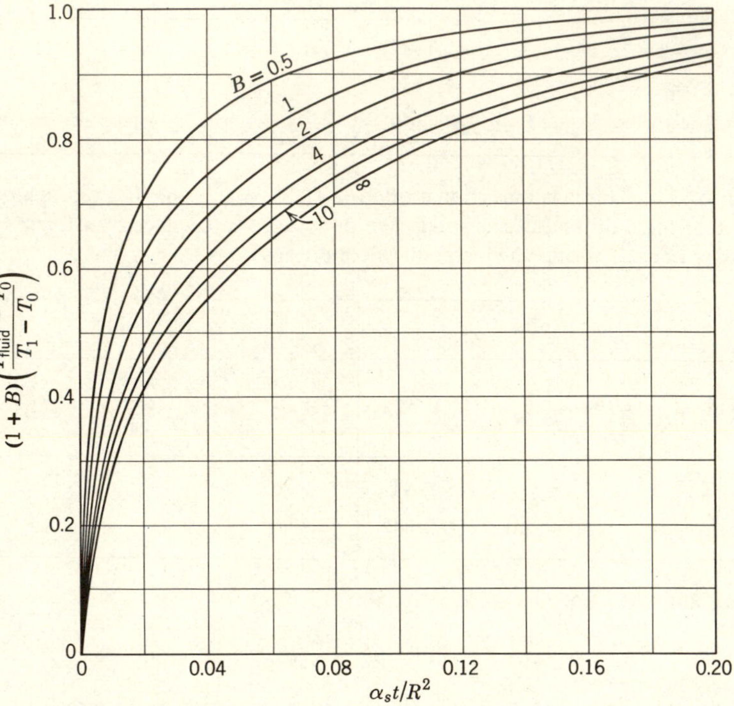

Fig. 11.1-4. Variation of fluid temperature with time when a sphere of radius R at temperature T_1 is placed in a well-stirred fluid at temperature T_0. [Carslaw and Jaeger, *Conduction of Heat in Solids*, Oxford University Press (1959), p. 206.]

roots of $\tan b = 3b/(3 + Bb^2)$. The Heaviside partial fractions expansion theorem for repeated roots[5] may be used to complete the inversion. The result is[6]

$$\Theta_f = \frac{B}{1 + B} + 6B \sum_{k=1}^{\infty} \frac{e^{-b_k^2 \tau}}{B^2 b_k^2 + 9(1 + B)} \tag{11.1-50}$$

The only place in which α_s appears in this expression is in the dimensionless time $\tau = \alpha_s t/R^2$. Equation 11.1-50 is shown graphically in Fig. 11.1-4. Note that the Laplace transform technique has enabled us to get the temperature history of the fluid without obtaining the temperature profiles in the solid.

[5] R. V. Churchill, *op. cit.*, pp. 44–50.
[6] Carslaw and Jaeger, *op. cit.*, pp. 240–241.

§11.2 STEADY HEAT CONDUCTION IN LAMINAR FLOW OF A VISCOUS FLUID

In Chapters 9 and 10 we distinguished between free and forced convection. To get the temperature profiles for forced convection in laminar flow, the following procedure is used, provided that the temperature dependence of viscosity can be neglected: first the equation of motion is solved to get the velocity distribution, which is then substituted into the energy equation to give a second order partial differential equation for the temperature distribution.

Not many forced-convection heat-flow problems have been solved analytically. We should at least mention the classical *Graetz-Nusselt problem*, which describes the temperature profiles in tube flow where the wall temperature suddenly changes at some plane through the tube from one fixed value to another fixed value (that is, the wall temperature profile is a step-function). This problem is discussed in numerous textbooks and review articles; hence we do not give it here.[1,2,3,4] The problem has, in addition, been extended to non-Newtonian flow.[5,6] A further problem of interest is tube flow with viscous heating effects, sometimes referred to as the *Brinkman problem*.[7,8,9]

In this section we give a further discussion of the problem treated in §9.8, namely the determination of temperature profiles in laminar flow in a circular tube with constant heat flux at the wall. In that section we obtained the partial differential equation given in Eq. 9.8–19. Then we found an asymptotic solution valid for large distances down the tube from the start of the heated zone. Here, we give the complete solution to the partial differential equation and also an asymptotic solution for short distances. That is, the system shown in Fig. 9.8–2 is discussed from three viewpoints in this text:

a. Solution of partial differential equation by the method of separation of variables (Example 11.2–1).

b. Asymptotic solution for short distances down the tube by the method of combination of variables (Example 11.2–2).

c. Asymptotic solution for large distances down the tube (§9.8).

[1] M. Jakob, *Heat Transfer*, Wiley, New York (1949), Vol. I, pp. 451–464.

[2] U. Grigull, *Wärmeübertragung*, Springer, Berlin (1955), Third Edition,pp. 179–185.

[3] K. Yamagata, *Memoirs of the Faculty of Engineering*, Kyushu Imperial University, Vol. VIII, No. 6, Fukuoka, Japan (1940).

[4] M. Tribus and J. Klein, Chapter in *Heat Transfer—A Symposium*, Eng. Res. Inst., University of Michigan (1953), pp. 211–235.

[5] B. C. Lyche and R. B. Bird, *Chem. Eng. Sci.*, **6**, 35–41 (1956).

[6] J. Schenk and J. van Laar, *Appl. Sci. Research*, **A7**, 449–462 (1958).

[7] H. C. Brinkman, *Appl. Sci. Research*, **A2**, 120–124 (1951).

[8] R. B. Bird, *SPE Journal*, **11**, 35–40 (1955), extension of the Brinkman problem to non-Newtonian flow.

[9] H. L. Toor, *Ind. Eng. Chem.*, **48**, 922–926 (1956), extension of the Brinkman problem to include heat effects associated with compressibility.

Example 11.2–1. Laminar Tube Flow with Constant Heat Flux at Wall

Solve Eq. 9.8–19 with the initial and boundary conditions given in Eqs. 9.8–20, 21, and 22.

Solution. The complete solution for the temperature profile will be of the form

$$\Theta(\xi, \zeta) = \Theta_\infty(\xi, \zeta) - \Theta_d(\xi, \zeta) \tag{11.2–1}$$

in which $\Theta_\infty(\xi, \zeta)$ is the asymptotic solution given in Eq. 9.8–31 and Θ_d is a function that will be damped out exponentially with ζ. The function Θ_d must satisfy Eq. 9.8–19 and also the boundary conditions

$$\text{B.C. 1:} \qquad \text{at } \xi = 0, \qquad \frac{\partial \Theta_d}{\partial \xi} = 0 \tag{11.2–2}$$

$$\text{B.C. 2:} \qquad \text{at } \xi = 1, \qquad \frac{\partial \Theta_d}{\partial \xi} = 0 \tag{11.2–3}$$

$$\text{B.C. 3:} \qquad \text{at } \zeta = 0, \qquad \Theta_d = \Theta_\infty(\xi, 0) \tag{11.2–4}$$

We anticipate that $\Theta_d(\xi, \zeta)$ will be of the form

$$\Theta_d(\xi, \zeta) = \Xi(\xi) Z(\zeta) \tag{11.2–5}$$

and thereby separate Eq. 9.8–19 into two ordinary differential equations:

$$\frac{dZ}{d\zeta} = -c^2 Z \tag{11.2–6}$$

$$\frac{1}{\xi}\frac{d}{d\xi}\left(\xi \frac{d\Xi}{d\xi}\right) + c^2(1 - \xi^2)\Xi = 0 \tag{11.2–7}$$

in which $-c^2$ is the separation constant. Since the boundary conditions on Ξ are $d\Xi/d\xi = 0$ at $\xi = 0, 1$, we have a Sturm-Liouville problem. We know that there will be an infinite number of eigenvalues c_i and eigenfunctions Ξ_i and that the final solution must be of the form

$$\Theta = \Theta_\infty(\xi, \zeta) - \sum_{i=1}^{\infty} B_i e^{-c_i^2 \zeta} \Xi_i(\xi) \tag{11.2–8}$$

where

$$B_i = \frac{\displaystyle\int_0^1 \Theta_\infty(\xi, 0)\Xi_i(\xi)(1 - \xi^2)\xi \, d\xi}{\displaystyle\int_0^1 [\Xi_i(\xi)]^2(1 - \xi^2)\xi \, d\xi} \tag{11.2–9}$$

The problem is thus reduced to that of finding the eigenfunctions Ξ_i by solving Eq. 11.2–7 and getting the c_i by applying the boundary condition at $\xi = 1$. This procedure, which requires considerable numerical effort, has been carried out up to $i = 7$ for this problem.[10]

[10] R. Siegel, E. M. Sparrow, and T. M. Hallman, *Appl. Sci. Research*, A7, 386–392 (1958).

Example 11.2–2. Laminar Tube Flow with Constant Heat Flux at Wall: Asymptotic Solution for Small Distances

Note that the sum in Eq. 11.2–8 converges rapidly for large z but slowly for small z. Develop an expression for $T(r, z)$ that is useful for small z values.

Solution. For small z, the heat penetration is restricted to a thin shell near the wall, so that the following three approximations lead to results that are accurate in the limit as $z \to 0$:

a. Curvature effects may be neglected and the problem treated as though the wall were flat; call the distance from the wall $s = R - r$.

b. The fluid may be regarded as extending from the (flat) heat-transfer surface, $s = 0$, to $s = \infty$.

c. The velocity profile may be regarded as linear, with a slope given by the true slope at the wall: $v_z(s) = v_0 s/R$, where $v_0 = (p_0 - p_L)R^2/2\mu L$.

The energy equation for this simplified model then becomes

$$v_0\left(\frac{s}{R}\right)\frac{\partial T}{\partial z} = \alpha\frac{\partial^2 T}{\partial s^2} \tag{11.2–10}$$

Actually, it is easier to work with the corresponding equation for q_s obtained by dividing Eq. 11.2–10 by s and then differentiating with respect to s:

$$v_0\left(\frac{1}{R}\right)\frac{\partial q_s}{\partial z} = \alpha\frac{\partial}{\partial s}\left(\frac{1}{s}\frac{\partial q_s}{\partial s}\right) \tag{11.2–11}$$

It is more convenient to work in terms of dimensionless variables:

$$\psi = \frac{q_s}{q_1}; \qquad \eta = \frac{s}{R}; \qquad \lambda = \frac{z\alpha}{v_0 R^2} \tag{11.2–12}$$

so that Eq. 11.2–11 becomes

$$\frac{\partial\psi}{\partial\lambda} = \frac{\partial}{\partial\eta}\left(\frac{1}{\eta}\frac{\partial\psi}{\partial\eta}\right) \tag{11.2–13}$$

with boundary conditions

B.C. 1: $\quad$ at $\lambda = 0, \quad \psi = 0 \qquad$ (11.2–14)

B.C. 2: $\quad$ at $\eta = 0, \quad \psi = 1 \qquad$ (11.2–15)

B.C. 3: $\quad$ at $\eta = \infty, \quad \psi = 0 \qquad$ (11.2–16)

This problem can be solved by the method of combination of variables (see Examples 4.1–1 and 11.1–1) by using the new independent variable:

$$\chi = \frac{\eta}{\sqrt[3]{9\lambda}} \tag{11.2–17}$$

Then the heat flux equation becomes

$$\chi\psi'' + (3\chi^3 - 1)\psi' = 0 \tag{11.2–18}$$

where the accents mean differentiation with respect to χ. The boundary conditions are the following: at $\chi = 0$, $\psi = 1$ and at $\chi = \infty$, $\psi = 0$. The solution is

$$\psi = \frac{\int_\chi^\infty \chi e^{-\chi^3}\,d\chi}{\int_0^\infty \chi e^{-\chi^3}\,d\chi} = \frac{3}{\Gamma(\frac{2}{3})}\int_\chi^\infty \chi e^{-\chi^3}\,d\chi \qquad (11.2\text{–}19)$$

The temperature profile may then be obtained by integration:

$$\int_T^{T_0} dT = -\frac{1}{k}\int_s^\infty q_s\,ds \qquad (11.2\text{–}20)$$

or in dimensionless quantities

$$\Theta = \frac{T - T_0}{q_1 R/k} = \sqrt[3]{9\lambda}\int_\chi^\infty \psi\,d\chi \qquad (11.2\text{–}21)$$

Insertion of the foregoing expression for ψ into this result gives

$$\Theta = \sqrt[3]{9\lambda}\left[\frac{e^{-\chi^3}}{\Gamma(\frac{2}{3})} - \chi\left\{1 - \frac{\Gamma(\frac{2}{3}, \chi^3)}{\Gamma(\frac{2}{3})}\right\}\right] \qquad (11.2\text{–}22)$$

Here $\Gamma(\frac{2}{3})$ is a "(complete) gamma function," and $\Gamma(\frac{2}{3}, \chi^3)$ is an "incomplete gamma function." Similar results have been given for the constant wall temperature problem[11,12] and for the analogous mass-transfer problem;[13] application to non-Newtonian flow has also been discussed. (See Problem 9.R.)

In order to compare the results of this example with those of Example 11.2–1, we mention that $\eta = 1 - \xi$, $\lambda = \zeta/2$, and the dimensionless temperatures Θ are identical.

§11.3 STEADY TWO-DIMENSIONAL POTENTIAL FLOW OF HEAT IN SOLIDS

In §4.3 it was pointed out that there is a class of inviscid, steady flow problems that can be solved by the methods of conformal mapping. These are problems for which the potential function and the stream function are both described by the two-dimensional Laplace equation. From Eq. 11.1–2 it is easy to see that the steady-state two-dimensional heat conduction in solids of constant thermal conductivity is also described by the two-dimensional Laplace equation:

$$\frac{\partial^2 T}{\partial x^2} + \frac{\partial^2 T}{\partial y^2} = 0 \qquad (11.3\text{–}1)$$

We now make use of the fact that *any* analytic function $w(z) = S(x, y) + iT(x, y)$ provides two scalar functions S and T, which are solutions of Eq. 11.3–1; curves of S = constant may be interpreted as lines of heat flow, and curves of T = constant are the corresponding isothermals for *some* heat-flow problem. These two sets of curves are orthogonal—that is, they intersect at

[11] M. A. Lévêque, *Ann. mines*, **13**, 201 (1928).
[12] R. L. Pigford, *C.E.P. Symposium Series*, No. 17, Vol. 51, 79–92 (1955).
[13] H. Kramers and P. J. Kreyger, *Chem. Eng. Sci.*, **6**, 42–48 (1956).

right angles. Furthermore the components of the heat flux vector at any point are given by

$$ik\frac{dw}{dz} = q_x - iq_y \tag{11.3-2}$$

Given an analytic function, it is easy to find heat-flow problems that are described by it; but the inverse process of finding an analytic function suitable for a given heat-flow problem is generally very difficult. Some methods for this are available, but they are outside the scope of this book.[1,2]

For every function $w(z)$, two heat flow nets are obtained by interchanging lines of constant S and lines of constant T. Furthermore, two additional nets are obtained by working with the inverse function $z(w)$, just as is the case for ideal fluid flow.

Note that potential fluid flow and potential heat flow are mathematically similar, the two-dimensional flow nets in both cases being described by analytic functions. Physically, however, there are certain important differences. The fluid flow nets described in §4.3 are for a fluid with no viscosity, and therefore one cannot use them to calculate the momentum transport (friction drag) at solid surfaces. The heat flow nets described in this section are for solids that have a finite thermal conductivity, and therefore one can calculate the heat transport at all surfaces. Moreover, in §4.3 the velocity profiles do *not* satisfy the Laplace equation, whereas in this section the temperature profiles *do* satisfy Laplace's equation. The reader desirous of learning more about analogous physical processes described by Laplace's equation can find an interesting summary in the monograph of Sneddon.[3]

Here we give just one example to provide a glimpse of the method using analytic functions; further examples may be found in the references cited.

Example 11.3-1. Temperature Distribution in a Wall

Consider a wall of thickness b extending from 0 to ∞ in the y-direction and from $-\infty$ to $+\infty$ in the z-direction. (See Fig. 11.3-1.) The surfaces at $x = \pm b/2$ are held at temperature T_0, whereas the bottom of the wall at the surface $y = 0$ is at T_1. Show that the imaginary part of the function[4]

$$w(z) = \frac{1}{\pi}\ln\left(\frac{(\sin \pi z/b) - 1}{(\sin \pi z/b) + 1}\right) \tag{11.3-3}$$

gives the steady temperature distribution $\Theta(x, y) = (T - T_0)/(T_1 - T_0)$.

[1] H. S. Carslaw and J. C. Jaeger, *Conduction of Heat in Solids*, Oxford University Press (1959), Second Edition, Chapter XVI.

[2] R. V. Churchill, *Introduction to Complex Variables and Applications*, McGraw-Hill, New York (1948), Chapter IX.

[3] I. N. Sneddon, *Elements of Partial Differential Equations*, McGraw-Hill, New York (1957), Chapter 4.

[4] R. V. Churchill, *op. cit.*, pp. 149–152; there it is shown how one arrives at Eq. 11.3-3 with the help of a table of conformal mappings.

Solution. The imaginary part of $w(z)$ in Eq. 11.3–3 is

$$\Theta(x, y) = \frac{2}{\pi} \arctan \left(\frac{\cos \pi x/b}{\sinh \pi y/b} \right) \tag{11.3–4}$$

in which the arctangent has the range 0 to $\pi/2$. When $x = \pm b/2$, Eq. 11.3–4 gives $\Theta = 0$; and when $y = 0$, Eq. 11.3–4 gives $\Theta = (2/\pi) \arctan \infty = 1$. From Eq. 11.3–4 it is easy to get the heat flux through the base of the wall:

$$q_y|_{y=0} = -k \left.\frac{\partial T}{\partial y}\right|_{y=0} = \left(\frac{2k}{b}\right)(T_1 - T_0)\left(\sec \frac{\pi x}{b}\right) \tag{11.3–5}$$

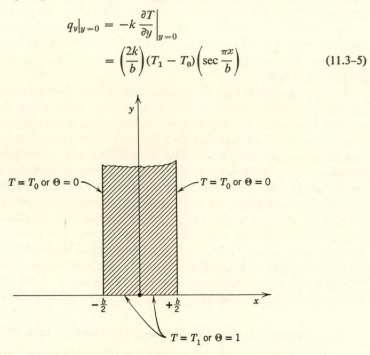

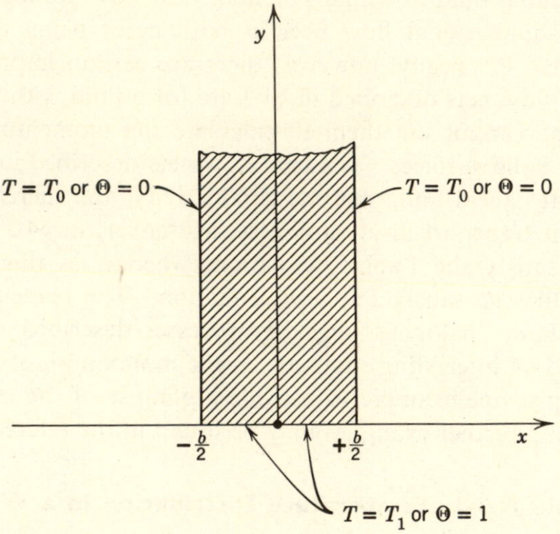

Fig. 11.3–1. Steady two-dimensional temperature distribution in a wall.

§11.4 BOUNDARY-LAYER THEORY[1,2,3]

If the temperature profiles in a system are flat except in the neighborhood of a surface, then the energy transport can be regarded as occurring in a "boundary layer." In the problem discussed in Example 11.1–1, for example, for short times the heat has penetrated only a short distance into the solid; this problem can be worked by assuming a thermal boundary-layer thickness

[1] H. Schlichting, *Boundary-Layer Theory*, McGraw-Hill, New York (1955), pp. 263–279.

[2] S. Goldstein, *Modern Developments in Fluid Dynamics*, Oxford University Press (1938), Vol. II, pp. 610–616, 623–645.

[3] E. R. G. Eckert and R. M. Drake, Jr., *Heat and Mass Transfer*, McGraw-Hill, New York (1959), Second Edition, pp. 131–153, 167–178.

δ_T, which is a function of time, and by assuming that the temperature profiles within this boundary layer are similar in time. The details of this procedure are already given in Example 4.4–1.

In this section we give a boundary-layer solution of the steady-state heat transfer to a fluid of constant properties which is flowing past a heated flat plate. This example is thus an extension of the isothermal problem presented in Example 4.4–2.

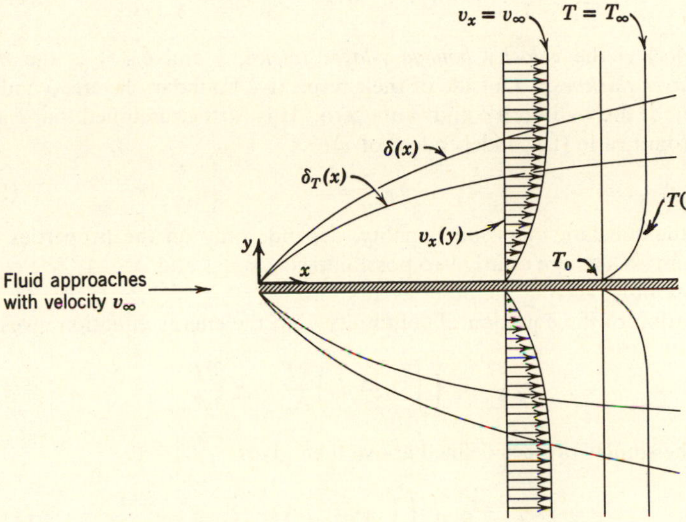

Fig. 11.4–1. Boundary-layer development for flow along a heated flat plate, showing the thermal boundary layer for $\Delta < 1$. Surface of plate is at temperature T_0.

Example 11.4–1. Heat Transfer in Forced Convection Laminar Flow along a Heated Flat Plate

Obtain the temperature profiles near a flat plate immersed in a viscous fluid, as shown in Fig. 11.4–1. The wetted surface of the heated plate is maintained at T_0 and the temperature of the approaching fluid is T_∞.

Solution. The boundary-layer equations for the flow of a fluid with constant ρ, μ, k, and $\hat{C}_p$ past a flat plate are

(continuity)
$$\frac{\partial v_x}{\partial x} + \frac{\partial v_y}{\partial y} = 0 \tag{11.4–1}$$

(motion)
$$v_x \frac{\partial v_x}{\partial x} + v_y \frac{\partial v_x}{\partial y} = \nu \frac{\partial^2 v_x}{\partial y^2} \tag{11.4–2}$$

(energy)
$$v_x \frac{\partial T}{\partial x} + v_y \frac{\partial T}{\partial y} = \alpha \frac{\partial^2 T}{\partial y^2} \tag{11.4–3}$$

The first two equations were discussed in Example 4.4–2. The third equation is written without the viscous dissipation term. A boundary-layer solution for these equations is obtained by assuming similar velocity profiles and similar temperature profiles:

$$\frac{v_x}{v_\infty} = \phi(\eta) \qquad \text{where} \qquad \eta = \frac{y}{\delta(x)} \tag{11.4–4,5}$$

$$\frac{T_0 - T}{T_0 - T_\infty} = \Theta(\eta_T) \qquad \text{where} \qquad \eta_T = \frac{y}{\delta_T(x)} \tag{11.4–6,7}$$

in which $\delta(x)$ is the *velocity boundary-layer thickness* and $\delta_T(x)$ is the *thermal boundary-layer thickness*. Outside of their respective boundary layers, ϕ and Θ are both unity; at the wall both ϕ and Θ are zero. It is further assumed that δ and δ_T bear a constant ratio (i.e., independent of x):

$$\delta_T = \delta\Delta \tag{11.4–8}$$

where Δ, the constant of proportionality, depends only on the properties of the fluid medium. There are clearly two possibilities: $\Delta \leqslant 1$ and $\Delta \geqslant 1$. We consider the first case only, leaving the other as an exercise.

Combination of the equation of continuity with the energy equation gives

$$v_x \frac{\partial T}{\partial x} - \left(\int_0^y \frac{\partial v_x}{\partial x}\, dy\right) \frac{\partial T}{\partial y} = \alpha \frac{\partial^2 T}{\partial y^2} \tag{11.4–9}$$

Inserting the similar profiles defined above then gives

$$-v_\infty \phi \Theta' \eta_T \frac{\delta'_T}{\delta_T} + v_\infty \left(\int_0^\eta \phi' \eta \delta'\, d\eta\right) \Theta' \frac{1}{\delta_T} = \alpha \Theta'' \frac{1}{\delta_T^2} \tag{11.4–10}$$

in which $\delta' = d\delta/dx$, $\delta_T' = d\delta_T/dx$, $\phi' = d\phi/d\eta$, and $\Theta' = d\Theta/d\eta_T$. We now perform the following operations: multiply Eq. 11.4–10 by δ_T^2; let $\eta = \eta_T\Delta$ and $\delta_T = \delta\Delta$; integrate over the thermal boundary layer (which lies inside the momentum boundary layer if $\Delta < 1$). This gives

$$(E - D)\delta \frac{d\delta}{dx} = \frac{\alpha}{v_\infty} F \tag{11.4–11}$$

in which

$$D = \Delta^2 \int_0^1 \phi(\eta_T\Delta)\Theta' \eta_T\, d\eta_T \tag{11.4–12}$$

$$E = \Delta \int_0^1 \left(\int_0^{\eta_T\Delta} \phi' \eta\, d\eta\right) \Theta'\, d\eta_T \tag{11.4–13}$$

$$F = \int_0^1 \Theta''\, d\eta_T \tag{11.4–14}$$

But from Eq. 4.4–19 we know that

$$(B - A)\delta \frac{d\delta}{dx} = \frac{\nu}{v_\infty} C \tag{11.4–15}$$

Combination of Eqs. 11.4–11 and 15 gives Δ implicitly as a function of $\Pr = \nu/\alpha$:

$$\left(\frac{E - D}{F}\right) = \left(\frac{B - A}{C}\right)\frac{1}{\Pr} \tag{11.4–16}$$

Once Δ is known, then from Eqs. 11.4–8 and 4.4–23 we can get $\delta_T(x)$. When $\delta_T(x)$ is known, we know the temperature profile from Eq. 11.4–6. As an illustration of the method, we assume the following profiles:[4]

$$\phi = 2\eta - 2\eta^3 + \eta^4 \tag{11.4–17}$$

$$\Theta = 2\eta_T - 2\eta_T^{\,3} + \eta_T^{\,4} \tag{11.4–18}$$

For these profiles, we have to evaluate *A*, *B*, and *C* (Eqs. 4.4–20, 21, and 22) and *D*, *E*, and *F* (Eqs. 11.4–12, 13, and 14). When this is done, we get from Eq. 4.4–23

$$\delta = \sqrt{\frac{1260}{37}\frac{\nu x}{v_\infty}} \tag{11.4–19}$$

and from Eq. 11.4–16

$$\frac{1}{15}\Delta^3 - \frac{3}{280}\Delta^5 + \frac{1}{360}\Delta^6 = \frac{37}{630}\frac{1}{\Pr} \qquad (\Delta < 1) \tag{11.4–20}$$

This sixth-order algebraic equation has to be solved for Δ as a function of Pr. When this is done, the result may be curve-fitted within 5 per cent by the relation[1]

$$\Delta = \Pr^{-1/3} \qquad (\Delta < 1) \tag{11.4–21}$$

The temperature profile is given explicitly by

$$\frac{T_0 - T}{T_0 - T_\infty} = 2\left(\frac{y}{\delta\Delta}\right) - 2\left(\frac{y}{\delta\Delta}\right)^3 + \left(\frac{y}{\delta\Delta}\right)^4 \tag{11.4–22}$$

in which δ is given by Eq. 11.4–19 and Δ is given by Eq. 11.4–21. This solution is valid for the laminar flow region that extends from the leading edge downstream to a position x_{crit} where $x_{\text{crit}} v_\infty \rho/\mu \geqslant 10^5$.

Actually Eq. 11.4–9 can be integrated formally to give

$$\frac{1}{\rho \hat{C}_p} q_y\bigg|_{y=0} = \frac{d}{dx}\left[\delta_3 v_\infty (T_0 - T_\infty)\right] \tag{11.4–23}$$

where

$$\delta_3 = \int_0^\infty \frac{v_x}{v_\infty}\left(1 - \frac{T_0 - T}{T_0 - T_\infty}\right) dy \tag{11.4–24}$$

is the *energy thickness*, defined analogously to the *displacement thickness* δ_1 and the *momentum thickness* δ_2 in Eqs. 4.4–30 and 31.

Just as for momentum transport, there are for energy transport two types of boundary-layer approaches: the "exact solutions,"[5] obtained by solving

[4] H. Schlichting, *Boundary-Layer Theory*, McGraw-Hill, New York (1955), pp. 270–271

[5] See §19.3.

Eqs. 11.4–1, 2, and 3 analytically or numerically, and the "approximate solutions," obtained by assuming a form for the profiles in Eq. 11.4–24 and using Eq. 11.4–23 (this is the same as what we did in Example 4.4–1).

The methods described above have been the subject of considerable research. Although we have illustrated the approximate method for laminar flow, it should be mentioned that similar calculations have also been made for turbulent flow.

QUESTIONS FOR DISCUSSION

1. How should Eq. 11.1–2 be modified if there is a heat source within the solid? Give examples in which unsteady-state conduction with heat sources occurs in electrical and nuclear installations.

2. Discuss the use of Carslaw and Jaeger's *Heat Conduction in Solids* as a tool for solving heat-transfer problems. Give some concrete examples.

3. Verify the result in Eq. 11.1–10.

4. In Example 11.1–2, what would have happened if Θ had been chosen to be $(T - T_1)/(T_0 - T_1)$? Could $\partial\Theta/\partial\eta = 0$ at $\eta = 0$ have been used as one of the boundary conditions? Could $+c^2$ have been chosen as the separation constant? Where did the expression for D_n come from?

5. What kinds of heat-conduction problems can be solved by Laplace transform? What kinds cannot?

6. Compare the Graetz-Nusselt problem with that discussed in Examples 11.2–1 and 2.

7. When does the heat flux satisfy the same equation as the temperature (see Eqs. 11.2–10 and 11)?

8. Suggest some heat-conduction problems that are described by the analytic functions used in Examples 4.3–1 and 2.

9. Justify Eq. 11.3–2 by using the fact that $dw/dz = (\partial S/\partial x) + i(\partial T/\partial x) = (\partial T/\partial y) - i(\partial S/\partial y)$ and the Cauchy-Riemann equations.

10. By means of a carefully labeled diagram, show what is meant by the two cases $\Delta \leqslant 1$ and $\Delta \geqslant 1$ in Example 11.4–1. What is the difference between these two cases as far as the mathematical treatment is concerned?

11. What is the set of equations, analogous to Eq. 11.4–1, 2, and 3, for describing the *free*-convection heat transfer along a vertical flat plate?

12. What kinds of boundary and initial conditions are needed in describing unsteady heat conduction in solid laminated systems?

13. Discuss and compare various special methods for solving unsteady heat conduction in solids, such as electrical analogs, difference equations, relaxation methods, and variational methods.

PROBLEMS

11.A$_1$ Unsteady-State Heat Conduction in an Iron Sphere

An iron sphere of 1-in. diameter has the following physical properties: $\rho = 436$ lb$_m$ ft^{-3}, $k = 30$ Btu hr^{-1} ft^{-1} °F^{-1}, $\hat{C}_p = 0.12$ Btu lb$_m^{-1}$ °F^{-1}. The sphere is at a temperature of 70° F.

a. What is the numerical value of the thermal diffusivity of the sphere?

b. If the sphere is suddenly plunged into a large body of fluid of temperature 270° F, how much time is required for the center of the sphere to attain a temperature of 128° F?

c. A sphere of the same size and same initial temperature but of another material requires twice as long for its center to reach 128° F. What is its thermal diffusivity?

d. The chart used in the solution of (*b*) and (*c*) was prepared from the solution to a partial differential equation. What is that differential equation?

Answer: *a.* 0.574 $ft^2\ hr^{-1}$
b. 1.1 sec
c. 0.287 $ft^2\ hr^{-1}$

11.B_1 Comparison of the Two Slab Solutions for Short Times

What error is made in using Eq. 11.1–8 (based on a semi-infinite slab) instead of Eq. 11.1–31 (based on a slab of finite thickness) when $\alpha t/b^2 = 0.01$ and for a position 0.9 of the way from the mid-plane to the slab surface? Use the graphically presented solutions for the comparison. *Answer:* 4 per cent

11.C_1 Bonding with Thermoplastic Adhesive[1]

It is desired to bond together two sheets of a solid material, each of thickness $b = 0.77$ cm. This is done by using a thin layer of thermoplastic material, which fuses and forms a good

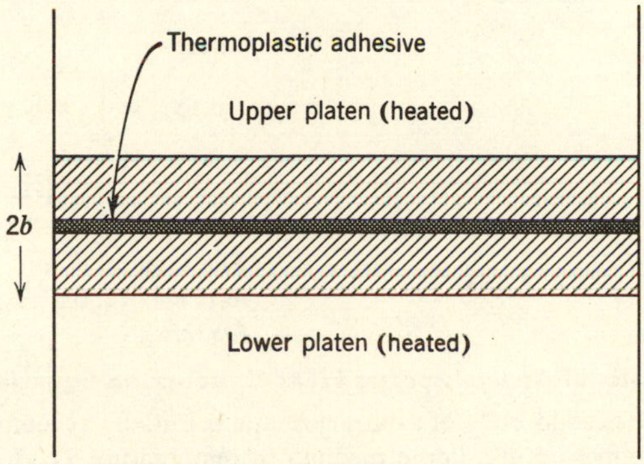

Fig. 11.C. Two sheets of solid material with a thin layer of adhesive.

bond at 160° C. The two plates are inserted in a press, as shown in Fig. 11.C, with both platens of the press maintained at a constant temperature of 220° C. How long will the sheets have to be held in the press if they are initially at 20° C? The solid sheets have a thermal diffusivity of $4.2 \times 10^{-3}\ cm^2\ sec^{-1}$. *Answer:* 85 sec

11.D_1 Quenching of a Steel Billet

A cylindrical steel billet 1 ft in diameter and 3 ft long, initially at 1000° F, is quenched in oil. Assume that the surface of the billet is at 200° F throughout the quenching. The steel has the following properties, which may be assumed independent of temperature:

$$k = 25\ \text{Btu hr}^{-1}\ \text{ft}^{-1}\ {}^\circ\text{F}^{-1}$$
$$\rho = 7.7\ \text{g cm}^{-3}$$
$$\hat{C}_p = 0.12\ \text{cal g}^{-1}\ {}^\circ\text{C}^{-1}$$

[1] This problem is based on Example 10 of J. M. McKelvey, *Heat Transfer and Thermodynamics*, Chapter 2 of *Processing of Thermoplastic Materials* (E. C. Bernhardt, Ed.), Reinhold, New York (1959), p. 93.

Estimate the temperature of the hottest point in the billet after five minutes of quenching. Neglect end effects, that is, make the calculation for a cylinder of the given diameter but of infinite length. *Answer:* 750° F

11.E$_2$ Temperature in a Slab with Heat Production

The slab of thermal conductivity k in Example 11.1–2 is initially at temperature T_0. For time $t > 0$ there is a uniform volume production of heat S_0 within the slab.

a. Obtain an expression for the dimensionless temperature $k(T - T_0)/S_0 b^2$ as a function of $\eta = y/b$ and $\tau = \alpha t/b^2$ by looking up the solution in Carslaw and Jaeger's *Heat Conduction in Solids* [Oxford University Press (1959), Second Edition].

b. What is the maximum temperature reached at the center of the slab?

c. What is the time that must elapse before 90 per cent of the temperature rise occurs?

Answer: *c.* $t \doteq b^2/\alpha$

11.F$_2$ Heating of a Semi-Infinite Slab: Constant Wall Flux

A solid occupying the space from $y = 0$ to $y = \infty$ is initially at the temperature T_0. At time $t = 0$ there is instituted a heat flux q_0 through the surface at $y = 0$ and this flux is maintained. Find the time-dependent profiles $T(y, t)$.

Hint: First write Eq. 11.1–2 for heat conduction in the y-direction; then differentiate it with respect to y and get an equation for q_y, which can be solved exactly as Eq. 11.1–3 in Example 11.1–1.

Answer: $$T - T_0 = \frac{q_0}{k}\sqrt{4\alpha t}\ \mathrm{ierfc}\ \frac{y}{\sqrt{4\alpha t}}$$

where

$$\mathrm{ierfc}\ x = \int_x^\infty \mathrm{erfc}\ x\, dx = \int_x^\infty (1 - \mathrm{erf}\ x)\, dx$$

The integral complementary error function ierfc x is tabulated in Carslaw and Jaeger (*op. cit.*).

11.G$_2$ Dimensional Analysis of the Heat-Conduction Equation

A homogeneous solid body of arbitrary shape is initially at temperature T_0 throughout. At $t = 0$ it is immersed in a liquid medium of temperature T_1. Let L be a characteristic length in the solid. Show that dimensional analysis predicts that

$$\Theta = \Theta(\xi, \eta, \zeta, \tau, \text{and geometrical ratios}) \qquad (11.G\text{–}1)$$

where $\Theta = (T - T_0)/(T_1 - T_0)$, $\xi = x/L$, $\eta = y/L$, $\zeta = z/L$, and $\tau = \alpha t/L^2$. Relate this result to the graphs given in §11.1.

11.H$_2$ Time Table for Roasting Turkey

a. A typical timetable for roasting turkey at 350° F is[2]

Mass of Turkey (lb_m)	Time Required per Unit Mass (min/lb_m)
6–10	20–25
10–16	18–20
18–25	15–18

[2] *Woman's Home Companion Cook Book*, Garden City Publishing Co., 1946.

It is interesting to compare this empirically determined cooking schedule with the results of Problem 11.G. If two geometrically similar turkeys at initial temperature T_0 are cooked at a given surface temperature T_1 to the same dimensionless temperature distribution $\Theta(\xi, \eta, \zeta)$, then the dimensionless time $\tau = \alpha t/L^2$ will have to be the same for both turkeys. Show that the latter implies that

$$(\text{cooking time}) \times (\text{mass of turkey})^{-2/3} = \text{constant} \qquad (11.H\text{–}1)$$

and make a numerical comparison of this equation with the table.

b. What assumptions were made in (*a*)?

11.I$_2$ Mean Temperature in Slab

From the result in Eq. 11.1–31, obtain the mean temperature in the slab as a function of time.

Answer: $$(T_1 - T_{\text{avg}})/(T_1 - T_0) = 2 \sum_{n=0}^{\infty} \frac{1}{(n + \frac{1}{2})^2\pi^2} \exp\left(-(n + \tfrac{1}{2})^2 \frac{\pi^2\alpha t}{b^2}\right)$$

11.J$_3$ Forced Convection Heat Transfer from Flat Plate (thermal boundary layer *outside* momentum boundary layer)

Obtain the result analogous to Eq. 11.4–20 for $\Delta \geqslant 1$.

Answer: $$\frac{3}{20}\Delta^2 - \frac{3}{20}\Delta + \frac{1}{15} - \frac{3}{280}\frac{1}{\Delta^2} + \frac{1}{360}\frac{1}{\Delta^3} = \frac{37}{630}\frac{1}{\text{Pr}}$$

11.K$_3$ Product Solutions for Unsteady Heat Conduction in Solids[3]

a. In Example 11.1–2 the unsteady-state heat conduction equation is solved for a slab of thickness $2b$. Show that the solution to Eq. 11.1–2 for the analogous problem for a rectangular block of finite dimensions may be written as the product of three solutions such as those in Eq. 11.1–31:

$$\frac{T_1 - T(x, y, z)}{T_1 - T_0} = 8 \sum_{m=0}^{\infty} \sum_{n=0}^{\infty} \sum_{p=0}^{\infty} \frac{(-1)^{m+n+p}}{(m + \frac{1}{2})(n + \frac{1}{2})(p + \frac{1}{2})\pi^3}$$
$$\times \exp\left[-\left(\frac{(m + \frac{1}{2})^2}{a^2} + \frac{(n + \frac{1}{2})^2}{b^2} + \frac{(p + \frac{1}{2})^2}{c^2}\right)\pi^2\alpha t\right]$$
$$\times \cos\,(m + \tfrac{1}{2})\frac{\pi x}{a} \cos\,(n + \tfrac{1}{2})\frac{\pi y}{b} \cos\,(p + \tfrac{1}{2})\frac{\pi z}{c}$$

b. Prove a similar result for cylinders of finite length; then rework Problem 11.D, not assuming the cylinder to be infinite.

11.L$_3$ Periodic Heating of the Earth's Crust

The temperature of the surface of the earth undergoes a periodic variation by virtue of the diurnal fluctuations in the radiation received. Take the earth's surface to be a plane and find an expression for $(T - T_{\text{avg}})/(T_{\max} - T_{\text{avg}})$ as a function of z (the distance into the earth's crust), t (the time measured after $T = T_{\max}$ at $z = 0$), and ω (the frequency of the diurnal variations, assumed to be sinusoidal).

Hint: Begin by postulating a solution of the form $(T - T_{\text{avg}})/(T_{\max} - T_{\text{avg}}) = f(z) \cdot \exp i\omega t$, where $i = \sqrt{-1}$, and then take the real part of the final answer.

Answer: $$\frac{T - T_{\text{avg}}}{T_{\max} - T_{\text{avg}}} = \exp\left(-\sqrt{\frac{\omega}{2\alpha}}\,z\right) \cos\left(\omega t - \sqrt{\frac{\omega}{2\alpha}}\,z\right)$$

[3] See, for example, H. S. Carslaw and J. C. Jaeger, *Conduction of Heat in Solids*, Oxford University Press (1959), Second Edition pp. 33–35.

11.M₃ Heating of Semi-Infinite Slab with Variable Thermal Conductivity

Rework Example 11.1–1 for the case that the thermal conductivity varies with temperature according to

$$\frac{k}{k_0} = 1 + \beta\left(\frac{T - T_0}{T_1 - T_0}\right) \tag{11.M–1}$$

in which k_0 is the thermal conductivity at T_0 and β is a constant. Use a boundary-layer method to obtain the temperature profile.

11.N₄ Unsteady Heat Conduction in Cylinders and Spheres

Derive the equations used to plot the graphs in Figs. 11.1–2 and 3. Verify the results by looking up the solutions in Carslaw and Jaeger.

11.O₄ Solution of Heat Conduction Problems in Solids by a Variational Method

a. Show that the steady-state heat conduction equation with a position-dependent heat source $S(x, y, z)$ and constant thermal conductivity k is equivalent to the variational problem that the integral

$$I = \iiint_V [(\nabla T \cdot \nabla T) - (2/k)ST]\, dx\, dy\, dz \tag{11.O–1}$$

be an extremum. Here V is the volume in which heat conduction is occurring; and the temperature is assumed to be specified on the surface of V.

b. Resolve Problem 9.J (with constant k) by using the foregoing variational method. Select as a trial temperature profile

$$\frac{T - T_0}{T_1 - T_0} = -a\left(\frac{r - r_0}{r_1 - r_0}\right) + (1 + a)\left(\frac{r - r_0}{r_1 - r_0}\right)^2 \tag{11.O–2}$$

in which a is an unspecified parameter. Why is this a reasonable choice? Evaluate I from Eq. 11.O–1 and then set dI/da equal to zero and find a. This then gives the value of a which makes a function of the form of Eq. 11.O–2 do the best possible job of simulating the exact solution.

c. Compare the value of the heat flux at the outer surface as obtained by (*b*) with that obtained by the exact solution for several values of (r_1/r_0).

d. Rework the problem in Example 11.3–1 by a variational method.

Answer: *b.* $a = -2r_1/(r_0 + r_1)$

c. $$\frac{q_1 r_1}{k(T_1 - T_0)} = -\frac{2(r_1/r_0)}{\left(\frac{r_1}{r_0}\right)^2 - 1}$$

CHAPTER 12

Temperature Distributions in Turbulent Flow

In the preceding chapters we have shown how to obtain temperature distributions in solids and in fluids in laminar motion. In this chapter we have tried to summarize the little that is known about the calculation of temperature profiles in turbulent flow.

The discussion here parallels that of Chapter 5. We begin by time-smoothing the energy balance equation. This operation gives rise to the turbulent energy flux $\bar{q}^{(t)}$, a quantity that cannot be predicted a priori. About all we can do at present is to indicate several empirical expressions for the turbulent energy flux which have been used in the technical literature. To illustrate the use of these expressions, we present a brief analysis of the turbulent heat transport in a tube.

At the end of the chapter we say a few words about the double temperature correlations in homogeneous, isotropic turbulence. This treatment is similar to that given in connection with the derivation of the von Kármán-Howarth equation.

§12.1 TEMPERATURE FLUCTUATIONS AND THE TIME-SMOOTHED TEMPERATURE

Before presenting the analytical description of turbulent heat transport, we discuss briefly the physical picture with regard to the heat transport in tube flow. As in §5.1, it is convenient to employ a model in which the tube is subdivided into three rather vaguely defined regions: the turbulent core, the buffer zone, and the viscous sublayer near the wall.

In the turbulent core thermal energy is transported very quickly from place to place by virtue of the frenzied eddy activity. This mechanism for rapid transit of thermal energy results in the fact that the time-smoothed temperature varies very little throughout the turbulent core. Close to the

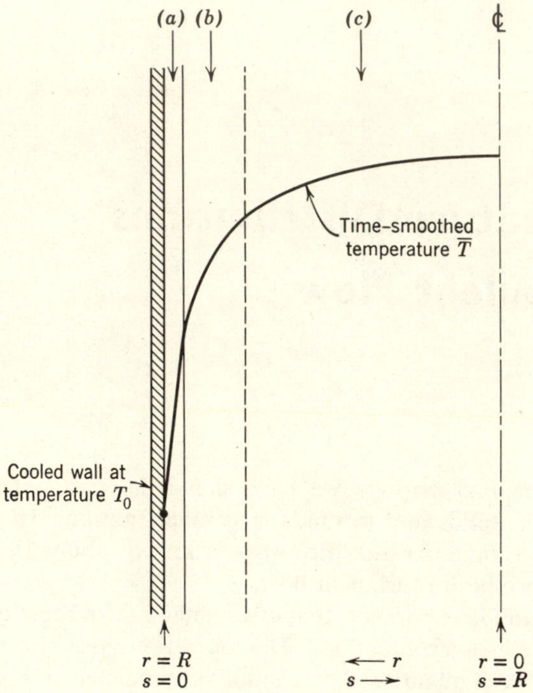

Fig. 12.1–1. Hot fluid flowing in a pipe with cooled walls with $\mathrm{Re} \gg 2.1 \times 10^3$. Sketch shows how temperature changes in (*a*) viscous zone, (*b*) buffer zone, and (*c*) turbulent core.

wall, on the other hand, eddy activity becomes negligible, and it is convenient to visualize a laminar region wherein energy is transported by heat conduction alone—a slow process in comparison with eddy transport. Hence one expects that there will be a large temperature drop through the thin viscous zone. In the buffer zone one anticipates a situation somewhat intermediate between that in the turbulent core and the viscous sublayer; in the buffer region energy transport both by conduction and by eddies is appreciable. A sketch of a typical temperature profile is shown in Fig. 12.1–1.

Although this picture is an idealized one, it has nevertheless provided a useful model in the development of the theory of turbulent heat transfer.

A rather complete account of the use of this model in the study of the analogies between heat and momentum transport has been given by Goldstein.[1]

Thus far we have been discussing the appearance of the time-smoothed temperature distribution. This is the temperature profile that one would measure by means of a traverse with a mercury thermometer inserted in the fluid stream. The use of more delicate measuring devices indicates that at any given point in a turbulent stream the temperature T is a wildly oscillating function of time, fluctuating about the time-smoothed value of the temperature $\bar{T}$. Hence we write

$$T = \bar{T} + T' \tag{12.1-1}$$

in which T' is the temperature fluctuation; this fluctuation may occasionally amount to as much as 5 to 10 per cent of the total temperature difference across the system. Clearly $\overline{T'} = 0$ by virtue of the definition of T'. But the quantities $\overline{v_x'T'}$, $\overline{v_y'T'}$, $\overline{v_z'T'}$ will not be zero because of the "correlation" between the velocity and temperature fluctuations at any one point.

§12.2 TIME-SMOOTHING THE ENERGY EQUATION[1,2]

In §5.2 it was shown how the equations of continuity and motion for an incompressible fluid may be "time-smoothed." In the time-smoothing operation additional terms arose in the equation of motion—the Reynolds stresses—describing the eddy transport of momentum. In this section we add to the previous results the time-smoothed energy equation, which contains additional terms describing the eddy energy transport. The development is restricted to a fluid of constant ρ, $\hat{C}_p$, μ, and k.

We begin with the energy equation in the form

$$\rho\hat{C}_p \frac{\partial T}{\partial t} = -\left(\frac{\partial}{\partial x}\rho\hat{C}_p v_x T + \frac{\partial}{\partial y}\rho\hat{C}_p v_y T + \frac{\partial}{\partial z}\rho\hat{C}_p v_z T\right)$$
$$+ k\left(\frac{\partial^2 T}{\partial x^2} + \frac{\partial^2 T}{\partial y^2} + \frac{\partial^2 T}{\partial z^2}\right)$$
$$+ \mu\left[2\left(\frac{\partial v_x}{\partial x}\right)^2 + \left(\frac{\partial v_x}{\partial y}\right)^2 + 2\left(\frac{\partial v_x}{\partial y}\frac{\partial v_y}{\partial x}\right) + \cdots\right] \tag{12.2-1}$$

in which only a few sample terms in the viscous dissipation function Φ_v have been written down.

[1] S. Goldstein, *Modern Developments in Fluid Dynamics*, Oxford University Press (1938), Vol. II, pp. 649–660.

[1] S. Goldstein, *Modern Developments in Fluid Dynamics*, Oxford University Press (1938), Vol. II, pp. 646–648.

[2] H. B. Squire, *Heat Transfer*, Chapter 14 in *Modern Developments in Fluid Dynamics: High Speed Flow* (L. Howarth, Ed.), Oxford University Press (1953).

In Eq. 12.2–1 we replace T by $\bar{T} + T'$, v_x by $\bar{v}_x + v_x'$, etc. Then the equation is time-averaged to give

$$\rho\hat{C}_p \frac{\partial \bar{T}}{\partial t} = -\left(\frac{\partial}{\partial x}\rho\hat{C}_p\bar{v}_x\bar{T} + \frac{\partial}{\partial y}\rho\hat{C}_p\bar{v}_y\bar{T} + \frac{\partial}{\partial z}\rho\hat{C}_p\bar{v}_z\bar{T}\right)$$

$$- \underline{\left(\frac{\partial}{\partial x}\rho\hat{C}_p\overline{v_x'T'} + \frac{\partial}{\partial y}\rho\hat{C}_p\overline{v_y'T'} + \frac{\partial}{\partial z}\rho\hat{C}_p\overline{v_z'T'}\right)}$$

$$+ k\left(\frac{\partial^2 \bar{T}}{\partial x^2} + \frac{\partial^2 \bar{T}}{\partial y^2} + \frac{\partial^2 \bar{T}}{\partial z^2}\right)$$

$$+ \mu\left[2\left(\frac{\partial \bar{v}_x}{\partial x}\right)^2 + \left(\frac{\partial \bar{v}_x}{\partial y}\right)^2 + 2\left(\frac{\partial \bar{v}_x}{\partial y}\frac{\partial \bar{v}_y}{\partial x}\right) + \cdots\right]$$

$$\underline{+ \mu\left[2\overline{\frac{\partial v_x'}{\partial x}\frac{\partial v_x'}{\partial x}} + \overline{\frac{\partial v_x'}{\partial y}\frac{\partial v_x'}{\partial y}} + 2\overline{\frac{\partial v_x'}{\partial y}\frac{\partial v_y'}{\partial x}} + \cdots\right]} \quad (12.2\text{–}2)$$

Comparison of this equation with the preceding one shows that the time-smoothed equation is the same as the original equation, except that there are new terms—indicated by dashed underlines—that are concerned with the turbulent eddy processes. We are thus led to the definition of the turbulent energy flux $\bar{q}^{(t)}$ with components,

$$\bar{q}_x^{(t)} = \rho\hat{C}_p\overline{v_x'T'}; \qquad \bar{q}_y^{(t)} = \rho\hat{C}_p\overline{v_y'T'}; \qquad \bar{q}_z^{(t)} = \rho\hat{C}_p\overline{v_z'T'} \quad (12.2\text{–}3)$$

and the turbulent energy dissipation function $\bar{\Phi}_v^{(t)}$,

$$\bar{\Phi}_v^{(t)} = \sum_{i=1}^{3}\sum_{j=1}^{3}\overline{\left(\frac{\partial v_i'}{\partial x_j}\frac{\partial v_i'}{\partial x_j} + \frac{\partial v_i'}{\partial x_j}\frac{\partial v_j'}{\partial x_i}\right)} \quad (12.2\text{–}4)$$

The similarity between the components of $\bar{q}^{(t)}$ in Eq. 12.2–3 and those of $\bar{\boldsymbol{\tau}}^{(t)}$ in Eqs. 5.2–5 and 6 should be noted.

By way of summary, we list all three of the equations of change for the turbulent flow of fluids with constant ρ, $\hat{C}_p$, μ, and k:

time-smoothed equation of continuity $$(\nabla \cdot \bar{\boldsymbol{v}}) = 0 \quad (12.2\text{–}5)$$

time-smoothed equation of motion $$\rho\frac{D\bar{\boldsymbol{v}}}{Dt} = -\nabla\bar{p} - [\nabla\cdot\bar{\boldsymbol{\tau}}^{(l)}] - \underline{[\nabla\cdot\bar{\boldsymbol{\tau}}^{(t)}]} + \rho\boldsymbol{g} \quad (12.2\text{–}6)$$

time-smoothed equation of energy $$\rho\hat{C}_p\frac{D\bar{T}}{Dt} = -(\nabla\cdot\bar{\boldsymbol{q}}^{(l)}) - \underline{(\nabla\cdot\bar{\boldsymbol{q}}^{(t)})} + \mu\bar{\Phi}_v^{(l)} + \underline{\mu\bar{\Phi}_v^{(t)}} \quad (12.2\text{–}7)$$

Here $\bar{q}^{(l)} = -k\nabla\bar{T}$, and $\bar{\Phi}_v^{(l)}$ is the viscous dissipation function with v_i replaced by $\bar{v}_i$. It is further understood that D/Dt in these equations is written with the time-smoothed velocity $\bar{v}$ in it.

In discussing turbulent heat flow problems, it has been customary to drop the viscous dissipation terms. Then, in order to set up turbulent heat-transfer problems, one sets up the problem as for laminar flow, except that q is replaced by $\bar{q}^{(l)} + \bar{q}^{(t)}$ and time-smoothed $\bar{v}$ and $\bar{T}$ are used throughout.

§12.3 SEMIEMPIRICAL EXPRESSIONS FOR THE TURBULENT ENERGY FLUX

In the preceding section it has been shown that the time-smoothing of the energy equation gives rise to a turbulent energy flux $\bar{q}^{(t)}$. In order to solve the energy equation to get time-smoothed temperature profiles, some relation between $\bar{q}^{(t)}$ and $\bar{T}$ has to be postulated. We summarize here some of the more widely used empirical expressions.

The Eddy Thermal Conductivity

By analogy with the Fourier law of heat conduction, one may write

$$\bar{q}_y^{(t)} = -k^{(t)}\frac{d\bar{T}}{dy} \tag{12.3-1}$$

the quantity $k^{(t)}$ being called the "turbulent coefficient of thermal conductivity" or "eddy conductivity." It should be kept in mind that $k^{(t)}$ is not a physical property characteristic of a given fluid but depends on position, direction, and the nature of the turbulent flow.

The eddy kinematic viscosity $\nu^{(t)} = \mu^{(t)}/\rho$ and the eddy thermal diffusivity $\alpha^{(t)} = k^{(t)}/\rho\hat{C}_p$ have the same dimensions. These quantities may be compared for turbulent momentum and energy transport. The ratio $\nu^{(t)}/\alpha^{(t)}$ is of the order of unity, values in the turbulence literature varying from 0.5 to 1.0. For air flow in conduits,[1] apparently $\nu^{(t)}/\alpha^{(t)}$ varies from 0.7 to 0.9, whereas for flow in jets[2] (i.e., in "free turbulence") the value is more nearly 0.5.

Mixing Length Expressions of Prandtl and Taylor

According to Prandtl's theory, momentum and energy are transferred in turbulent flow by the same mechanism. Hence, by analogy with Eq. 5.3-2,

$$\bar{q}_y^{(t)} = -\rho\hat{C}_p l^2\left|\frac{d\bar{v}_x}{dy}\right|\frac{d\bar{T}}{dy} \tag{12.3-2}$$

in which l is the Prandtl mixing length. Note that this is tantamount to

[1] J. B. Opfell and B. H. Sage, *Advances in Chemical Engineering*, Vol. I, Academic Press, New York (1956), p. 259.

[2] H. Schlichting, *Boundary-Layer Theory*, McGraw-Hill, New York (1955), p. 504.

saying that $\nu^{(t)}/\alpha^{(t)} = 1$. The vorticity-transport theory of Taylor[3] differs from the Prandtl theory in that the mixing length in the expression $\bar{\tau}^{(t)}$ is different from that in the expression for $\bar{q}^{(t)}$. In fact, Taylor's theory predicts that $\nu^{(t)}/\alpha^{(t)} = 0.5$. It is for this reason that Taylor's theory is said to be preferable for flow in jets and wakes, whereas Prandtl's theory is preferable for flow in conduits.[4] In this case Prandtl recommends setting $l = \kappa_1 y$, in which y is the distance from the wall.

Expression Based on the von Kármán Similarity Hypothesis

Insertion of the von Kármán expression for mixing length into Eq. 12.3–2 gives

$$\bar{q}_y^{(t)} = -\rho \hat{C}_p \kappa_2^2 \left| \frac{(d\bar{v}_x/dy)^3}{(d^2\bar{v}_x/dy^2)^2} \right| \frac{d\bar{T}}{dy} \qquad (12.3\text{–}3)$$

in which κ_2 is the same constant as that appearing in Eq. 5.3–3. This also assumes that $\nu^{(t)}/\alpha^{(t)} = 1$.

Deissler's Empirical Formula for the Region near the Wall

In order to describe the transport process near solid surfaces (where the Prandtl and von Kármán expressions are not valid), Deissler suggested the following empirical expression:

$$\bar{q}_y^{(t)} = -\rho \hat{C}_p n^2 \bar{v}_x y(1 - \exp\{-n^2 \bar{v}_x y/\nu\}) \frac{d\bar{T}}{dy} \qquad (12.3\text{–}4)$$

The n appearing here is the same as that employed in Eq. 5.3–4. Here again it is assumed that the mechanisms for momentum and energy transport are the same, so that $\nu^{(t)}/\alpha^{(t)} = 1$. In the following illustrative example we discuss turbulent heat transfer in tubes in terms of the foregoing empirical expressions.

Example 12.3–1. Temperature Profiles in Steady Turbulent Flow in Smooth Circular Tubes

A fluid is flowing in turbulent flow in a smooth circular pipe of diameter $D = 2R$, at a uniform temperature T_1. (See Fig. 12.3–1.) Beginning at $z = 0$, there is a cooling device that withdraws heat from the tube at a constant heat flux q_0. At some distance downstream from the start of this constant wall heat flux, the radial temperature profiles will have become stabilized; that is, $\bar{T}(r, z) - \bar{T}(0, z)$ is a function of r alone, so that the constancy of the wall flux implies that

$$\bar{T}(r, z) = Az + \psi(r) \qquad (12.3\text{–}5)$$

in which A is a constant. (For the analogous development in laminar flow see

[3] G. I. Taylor, *Proc. Roy. Soc.* (*London*), **A135,** 685–702 (1932); *Phil. Trans.*, **A215,** 1 (1915).

[4] H. Schlichting, *op. cit.*, p. 391.

§9.8.) We wish to derive some approximate expressions for the turbulent temperature profiles by using the results of Example 5.3–1 and the expressions for the turbulent heat flux in this section. We follow here the development given by Deissler,[5] which leads to results that seem to be particularly useful for Pr > 1. The discussion consists of two parts: (*a*) heat transport in the turbulent core and (*b*) heat transport near the wall.

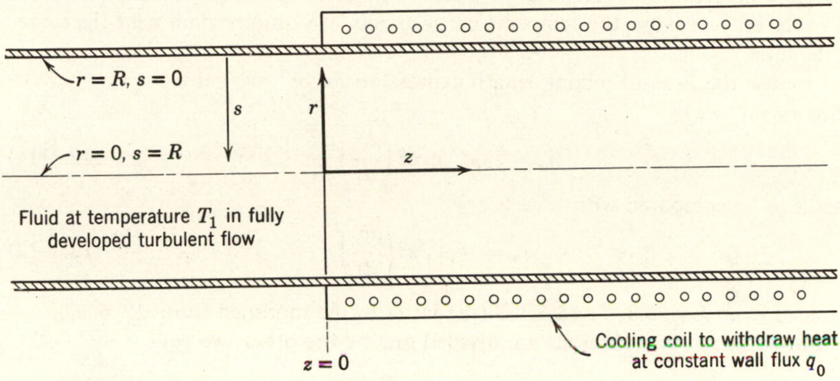

Fig. 12.3–1. Turbulent tube flow with constant heat flux at wall.

Solution. We begin with Eq. 12.2–7 and discard the viscous dissipation terms and the unsteady term in the left side; this gives

$$\rho \hat{C}_p(\bar{v} \cdot \nabla \bar{T}) = -(\nabla \cdot \bar{q}^{(l)}) - (\nabla \cdot \bar{q}^{(t)}) \tag{12.3–6}$$

or in component form in cylindrical coordinates

$$\rho \hat{C}_p\left(\bar{v}_r \frac{\partial \bar{T}}{\partial r} + \frac{\bar{v}_\theta}{r}\frac{\partial \bar{T}}{\partial \theta} + \bar{v}_z \frac{\partial \bar{T}}{\partial z}\right) = -\left(\frac{1}{r}\frac{\partial}{\partial r} r(\bar{q}_r^{(l)} + \bar{q}_r^{(t)}) + \frac{1}{r}\frac{\partial}{\partial \theta}(\bar{q}_\theta^{(l)} + \bar{q}_\theta^{(t)}) + \frac{\partial}{\partial z}(\bar{q}_z^{(l)} + \bar{q}_z^{(t)})\right) \tag{12.3–7}$$

The first two terms on the left vanish because there is no mean flow in the radial or tangential directions. The $\partial/\partial\theta$ term on the right is zero because of cylindrical symmetry, and the $\partial/\partial z$ term is zero because of the linear dependence of $\bar{T}$ on z. Hence we are left with

$$\rho \hat{C}_p \bar{v}_z A = -\frac{1}{r}\frac{d}{dr}[r(\bar{q}_r^{(l)} + \bar{q}_r^{(t)})] \tag{12.3–8}$$

Integration over r from the wall to any arbitrary position gives

$$\frac{1}{r}\int_r^R \rho \hat{C}_p \bar{v}_z A r\, dr = -q_0 + (\bar{q}_r^{(t)} + \bar{q}_r^{(l)}) \tag{12.3–9}$$

in which q_0 is the wall heat flux.

[5] R. G. Deissler, *NACA Report 1210* (1955).

a. For the *turbulent core*, we set $\bar{q}_r^{(l)} = 0$ in Eq. 12.3–9. In addition, we neglect the integral on the left side, which then gives

$$\bar{q}_r^{(t)} = q_0 \tag{12.3–10}$$

Note that the assumption that $\bar{q}_r^{(t)}$ is a constant for all r in the turbulent core is to some extent analogous to the assumption $\bar{\tau}_{rz}^{(t)} = \tau_0$ for all r in the Prandtl development in connection with Eq. 5.3–8. Such an apparently unsound assumption can be made only because the temperature gradients are unimportant near the center of the tube.

If we use the Prandtl mixing length expression for $\bar{q}_r^{(t)}$, we get from Eq. 12.3–10 (with $s = R - r$)

$$q_0 = +\rho \hat{C}_p \kappa_1{}^2 s^2 \left(\frac{d\bar{v}_z}{ds}\right)\frac{d\bar{T}}{ds} \tag{12.3–11}$$

This is to be compared with

$$\tau_0 = +\rho \kappa_1{}^2 s^2 \left(\frac{d\bar{v}_z}{ds}\right)^2 \tag{12.3–12}$$

obtained from Eqs. 5.3–5 and 5.3–7 (the latter in the modified form $\bar{\tau}_{rz}^{(t)} = \tau_0$).

If, now, these two equations are divided one by the other, we get

$$\frac{q_0}{\tau_0} = \frac{\hat{C}_p\, d\bar{T}}{d\bar{v}_z} \tag{12.3–13}$$

which may be integrated from a reference surface $s = s_1$ to an arbitrary surface $s = s$:

$$\frac{\hat{C}_p(\bar{T} - \bar{T}_1)}{q_0} = \frac{\bar{v}_z - \bar{v}_{z,1}}{\tau_0} \tag{12.3–14}$$

In Deissler's development[5] the surface $s = s_1$ is taken to be the surface separating the buffer zone and the turbulent core.[6] Equation 12.3–14 simply states that

[6] If the surface $s = s_1$ is taken to be the wetted surface of the tube (i.e., $s_1 = 0$), then $v_{z,1} = 0$ and $\bar{T}_1 = T_0$, in which T_0 is the wall temperature, which varies with z. If both sides are multiplied by $\rho\bar{v}_z$ and the entire equation is flow-averaged, we get

$$\frac{\rho\hat{C}_p\langle\bar{v}_z\rangle\left[\dfrac{\langle\bar{v}_z\bar{T}\rangle}{\langle\bar{v}_z\rangle} - T_0\right]}{q_0} = \frac{\rho\langle\bar{v}_z{}^2\rangle}{\tau_0} \tag{12.3–14a}$$

This is a statement of the *Reynolds analogy* between heat and momentum transfer. Equation 12.3–14*a* states that the ratio of the transport of energy downstream to the transport of energy across the solid fluid interface is equal to a similar ratio for momentum transport. If $\langle\bar{v}_z{}^2\rangle$ is replaced by $\langle\bar{v}_z\rangle^2$ (an error of not more than 5 per cent for tubes) and if a heat-transfer coefficient (see Chapter 13) is defined on the basis of the "cup-mixing" temperature difference in Eq. 12.3–14*a*, then Eq. 12.3–14*a* is the same as

$$\frac{h}{\rho\hat{C}_p\langle\bar{v}_z\rangle} = \frac{f}{2} \quad \text{or} \quad \frac{\text{Nu}}{\text{Re Pr}} = \frac{f}{2} \tag{12.3–14b}$$

which is known to be very good for Pr = 1. For Pr ≠ 1, the simplest relation of the form of Eq. 12.3–14*b* is the Chilton-Colburn relation (Eq. 13.2–18).

within the turbulent core the velocity and temperature profiles are similar; this agrees roughly with experiment.

It is convenient to introduce dimensionless quantities that contain $v_* = \sqrt{\tau_0/\rho}$ and q_0:

$$v^+ = \frac{\bar{v}_z}{v_*}; \qquad T^+ = \frac{\rho \hat{C}_p v_* (\bar{T} - T_0)}{q_0}; \qquad s^+ = \frac{s v_* \rho}{\mu} \qquad (12.3\text{–}15,16,17)$$

Then Eq. 12.3–14 becomes

$$T^+ - T_1^+ = v^+ - v_1^+ \qquad (s^+ \geqslant 26) \qquad (12.3\text{–}18)$$

The velocity difference $v^+ - v_1^+$ has already been given as a logarithmic function of s^+ in Eq. 5.3–11. Hence from Eq. 12.3–18 one deduces that the temperature profile in the turbulent core will also be a logarithmic function:

$$T^+ - T_1^+ = \frac{1}{\kappa_1} \ln \frac{s^+}{s_1^+} \qquad (s^+ \geqslant 26) \qquad (12.3\text{–}19)$$

in which $s_1^+ = 26$ is the boundary between the buffer zone and the turbulent core.

b. For the *region near the wall*, we cannot neglect $\bar{q}_r^{(l)}$ in Eq. 12.3–9. However, setting the integral on the left side equal to zero is, for this region, a very good approximation. Hence we use as our starting point

$$\bar{q}_r^{(l)} + \bar{q}_r^{(t)} = q_0 \qquad (12.3\text{–}20)$$

Insertion of Fourier's law for $\bar{q}_r^{(l)}$ and Deissler's expression (Eq. 12.3–4) for $\bar{q}_r^{(t)}$ into Eq. 12.3–20 then gives

$$q_0 = +k \frac{d\bar{T}}{ds} + \rho \hat{C}_p n^2 \bar{v}_z s (1 - \exp\{-n^2 \bar{v}_z s/\nu\}) \frac{d\bar{T}}{ds} \qquad (12.3\text{–}21)$$

This equation may be integrated with respect to s, and the result in terms of dimensionless quantities is

$$T^+ = \int_0^{s^+} \frac{ds^+}{(1/\text{Pr}) + n^2 v^+ s^+ (1 - \exp\{-n^2 v^+ s^+\})} \qquad (s^+ \leqslant 26) \qquad (12.3\text{–}22)$$

in which v^+ is given by Eq. 5.3–15.

The temperature profiles from Eqs. 12.3–19 and 12.3–22 are plotted in Fig. 12.3–2. There one sees how T^+ depends on the radial position in the tube and on the Prandtl number. From these curves Deissler has calculated the curves to show how the Nusselt number depends on the Reynolds and Prandtl numbers—that is, an improved version of Eq. 12.3–14*b*. (See Chapter 13.) Deissler has further generalized the foregoing treatment to fluids with variable physical properties.[5] Only very limited research has been done on the calculation of axial and radial dependence of the temperature in turbulent tube flow. Complete numerical solutions of

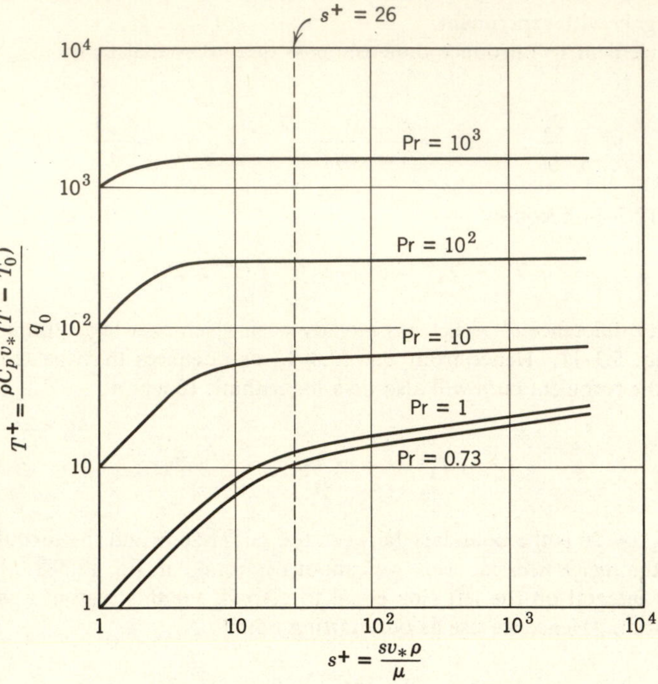

Fig. 12.3–2. Theoretically calculated temperature profiles for heat transfer to fluids flowing in the turbulent range in smooth circular tubes; Pr = 0.73 is the curve for air. [R. G. Deissler, *NACA Report* 1210 (1955), p. 4.] Note that these curves have positive slope throughout, hence predict that the temperature profile at the center of the tube is slightly pointed. Clearly this anomaly is important only at low Prandtl numbers.

Eq. 12.3–6, using empirical expressions for $\bar{q}_r^{(t)}$ and the velocity distribution, have been obtained for both uniform wall temperature[7] and uniform wall flux.[8]

§12.4 THE DOUBLE TEMPERATURE CORRELATION AND ITS PROPAGATION: THE CORRSIN EQUATION

In §5.4 we devoted considerable space to the study of decay of homogeneous, isotropic turbulence. This led us to the von Kármán-Howarth equation, which describes the way in which the double velocity correlation $\overline{v_i' v_j''}$ is propagated, and to the Taylor decay law for the intensity of the turbulence.

[7] H. L. Beckers, *Appl. Sci. Research*, **A6**, 147–190 (1956).

[8] E. M. Sparrow, T. M. Hallman, and R. Siegel, *Appl. Sci. Research*, **A7**, 37–52 (1957); this article contains a helpful comparison of the results of several other investigators. For distances past the thermal entrance region, their solution gives heat transfer rates that differ by only a few per cent from those obtained by Deissler.[5]

In this section we summarize very briefly the more recent studies on the decay of isotropic temperature fluctuations behind a heated grid.[1,2]

The three dimensionless correlation functions considered here are

$$V_i = \frac{\overline{T'v_i''}}{\sqrt{\overline{T'^2}}\sqrt{\overline{v_i'^2}}} = \text{the } i\text{th component of the temperature-velocity correlation vector} \tag{12.4–1}$$

$$W = \frac{\overline{T'T''}}{\overline{T'^2}} = \text{double temperature correlation scalar} \tag{12.4–2}$$

$$Q_i = \frac{\overline{v_i'T'T''}}{\sqrt{\overline{v_i'^2}}\,\overline{T'^2}} = i\text{th component of the triple velocity-temperature correlation vector} \tag{12.4–3}$$

Here, as before, singly primed quantities refer to fluctuations at point B' and doubly primed quantities refer to fluctuations at point B''. (See Figs. 5.4–1 and 2.) These correlations can be written in simpler form when homogeneity and isotropy are assumed:

a. By the same arguments used in §5.4 to prove that $P_i = 0$, we can show that

$$V_i = 0 \qquad (\text{for } i = 1, 2, 3) \tag{12.4–4}$$

b. The correlation W can be written as a scalar function of the distance r between B' and B'':

$$W = w(r, t) = 1 + w''(0, t)\frac{r^2}{2!} + \cdots \tag{12.4–5}$$

It is further known that w is an even function of r, as indicated. Here the double prime indicates the second derivative with respect to r.

c. Because Q_i must transform as a vector, it may be shown that

$$Q_i = q(r, t)\frac{\xi_i}{r} = \left(q'''(0, t)\frac{r^3}{3!} + \cdots\right)\frac{\xi_i}{r} \tag{12.4–6}$$

in which $q(r, t)$ is an odd function of r beginning with r^3, as indicated. Here the triple prime indicates the third derivative with respect to r.

An equation relating $w(r, t)$ and $q(r, t)$ may be obtained from the energy equation written in terms of fluctuations:

$$\frac{\partial T'}{\partial t} + \sum_{j=1}^{3} v_j'\frac{\partial T'}{\partial x_j} = \alpha\sum_{j=1}^{3}\frac{\partial^2 T'}{\partial x_j^2} \tag{12.4–7}$$

[1] S. Corrsin, *J. Aeronaut. Sci.*, **18,** 417–423 (1951).

[2] R. R. Mills, Jr., A. L. Kistler, V. O'Brien, and S. Corrsin, *NACA Tech. Note 4288*, August 1958.

This equation is valid at B' for $\bar{v} = 0$ and homogeneous turbulence. At B'' we have

$$\frac{\partial T''}{\partial t} + \sum_{j=1}^{3} v_j'' \frac{\partial T''}{\partial x_j} = \alpha \sum_{j=1}^{3} \frac{\partial^2 T''}{\partial x_j^2} \tag{12.4–8}$$

Multiplication of Eq. 12.4–7 by T'' and Eq. 12.4–8 by T', time-smoothing, and adding the two resulting equations gives

$$\frac{\partial}{\partial t} \overline{T'T''} + \sum_{j=1}^{3} \frac{\partial}{\partial \xi_j} (\overline{v_j''T'T''} - \overline{v_j'T'T''}) = 2\alpha \sum_{j=1}^{3} \frac{\partial^2}{\partial \xi_j^2} \overline{T'T''} \tag{12.4–9}$$

Introduction of the correlation coefficients leads finally to the following relation between $w(r, t)$ and $q(r, t)$:

$$\frac{\partial}{\partial t} (\overline{T'^2}w) - 2\overline{T'^2}\sqrt{\overline{v_i'^2}}\left(\frac{\partial q}{\partial r} + 2\frac{q}{r}\right) = 2\alpha\overline{T'^2}\left(\frac{\partial^2 w}{\partial r^2} + \frac{2}{r}\frac{\partial w}{\partial r}\right) \tag{12.4–10}$$

This is the *Corrsin equation* for the propagation of the double temperature correlation.[1] It is the heat-transport analog of the von Kármán-Howarth equation. This equation has been experimentally verified by measurement of the various correlations appearing therein.[2] Some solutions to Eq. 12.4–10 for various simple cases have been worked out by Corrsin;[1] moreover, he has discussed thermal fluctuations from the spectral viewpoint.[3] Measurements of $w(r, t)$ have shown it to be roughly equal to the function $f(r, t)$ in Chapter 5.

Example 12.4–1. Decay Equation for the Double Temperature Correlation[1]

Set $r = 0$ in the Corrsin equation and obtain an equation for the change of $\overline{T'^2}$ with time.

Solution. Because of the properties of $w(r, t)$ and $q(r, t)$ in Eqs. 12.4–5 and 6, setting $r = 0$ gives at once

$$\frac{d}{dt} \overline{T'^2} = 6\alpha\overline{T'^2}w''(0, t) \tag{12.4–11}$$

If we now define a "thermal microscale" by

$$\lambda_T^2 = \frac{-1}{w''(0, t)} \tag{12.4–12}$$

then

$$\frac{d}{dt} \overline{T'^2} = -6\alpha\left(\frac{\overline{T'^2}}{\lambda_T^2}\right) \tag{12.4–13}$$

This equation, obtained by Corrsin,[1] is analogous to the Taylor decay law in Eq. 5.4–39.

QUESTIONS FOR DISCUSSION

1. By means of a sketch, define T, $\bar{T}$, and T'.
2. Explain why $\overline{v_x'} = 0$ and $\overline{T'} = 0$, whereas the quantity $\overline{v_x'T'}$ is not equal to zero.

[3] S. Corrsin, *J. Appl. Phys.*, **22**, 469–473 (1951).

3. Compare eddy thermal conductivity and eddy viscosity as to definition, order of magnitude, and dependence upon physical properties and nature of the flow.

4. Verify that Eq. 12.3–2 is dimensionally consistent.

5. Summarize the assumptions made in the derivation of temperature profiles in Fig. 12.3–2.

6. Discuss possible ways of justifying the assumption made just before Eq. 12.3–10 and the analogous assumption made in connection with Eq. 5.3–8.

7. What is the Reynolds analogy and what does it mean?

8. How can Fig. 12.3–2 be used to plot $T(r)$ for flow in a tube and to get the wall heat flux?

9. What does the Corrsin equation describe?

10. To what extent can a priori calculations of turbulent temperature profiles be made?

PROBLEMS

12.A$_1$ Temperature Profile in Turbulent Tube Flow

a. Use the definition of v_* following Eq. 6.2–12 to show that

$$s^+ = \frac{1}{2^{3/2}} \text{Re} \sqrt{f}\left(1 - \frac{r}{R}\right) \tag{12.A–1}$$

$$T^+ = \frac{1}{2^{1/2}} \text{Re Pr} \sqrt{f}\left(\frac{k(\bar{T} - T_0)}{q_0 D}\right) \tag{12.A–2}$$

Do these relations depend on any particular turbulent flow model?

b. Use Fig. 12.3–2 to plot $k(\bar{T} - T_0)/q_0 D$ versus r/R for Pr = 10^2 and Re = 10^6.

12.B$_2$ Asymptotic Expression for Cup-Mixing Temperature at Very High Prandtl Numbers[1]

At very high Prandtl numbers (e.g., for fluids of low thermal conductivity) the principal thermal resistance in pipe flow is that in the laminar sublayer. Hence the temperature change takes place in the immediate neighborhood of the wall, where it is known (cf. §5.3) that $v^+ = s^+$. Hence set v^+ equal to s^+ in Eq. 12.3–22, and then expand the exponential, discarding all terms beyond the second. Next, perform the indicated integration and set s^+ equal to ∞. Show that the result is

$$T^+(\infty) = \frac{\pi}{2^{3/2} n} \text{Pr}^{3/4} \tag{12.B–1}$$

This may be interpreted[1] as the dimensionless cup-mixing temperature of the fluid. Explain.

12.C$_3$ Derivation of Radial Temperature Profiles for Turbulent Flow with Constant Wall Heat Flux[2]

a. Consider the problem posed in Example 12.3–1. Use Eq. 5.1–3 for the velocity profile and show that that profile leads to the following expression:

$$\nu + \nu^{(t)} = 7\left(\frac{v_*^2 R}{v_{\max}}\right)\left(\frac{r}{R}\right)\left(1 - \frac{r}{R}\right)^{6/7} \tag{12.C–1}$$

where $v_* = \sqrt{\tau_0/\rho}$.

[1] R. G. Deissler, *NACA Report 1210* (1955), p. 6.

[2] A somewhat similar problem is discussed by M. Jakob, *Heat Transfer*, Wiley, New York (1949), Vol. 1, pp. 476 et seq.

b. Assume that $\nu + \nu^{(t)} = \alpha + \alpha^{(t)}$; assume further that

$$\Theta = \left(\frac{T - T_{\text{wall}}}{q_0 R^2/k}\right) = A\zeta + \Psi(\xi) \tag{12.C–2}$$

in which q_0 is the (outwardly-directed) wall heat flux, $\xi = r/R$, $\zeta = (z/R)(v_*/v_{\text{max}})^2$, and A is a constant. Then show that Ψ satisfies the following differential equation:

$$(1 - \xi)^{1/7} A = \frac{7}{\xi}\frac{d}{d\xi}\left(\xi^2 (1 - \xi)^{6/7} \frac{d\Psi}{d\xi}\right) \tag{12.C–3}$$

c. Show that the solution to the differential equation in (*b*) is

$$\Psi = -\tfrac{49}{120} A(\beta + \tfrac{1}{4}\beta^8 - \tfrac{5}{21}\beta^9 + \tfrac{1}{5}\beta^{15} - \tfrac{11}{56}\beta^{16} + \cdots) \tag{12.C–4}$$

in which $\beta = (1 - \xi)^{1/7}$.

The more realistic Deissler velocity distribution has been used[3] to solve this problem and also the problem of the development of the temperature profiles in the entry-length region. High-speed computers were used.

[3] E. M. Sparrow, T. M. Hallman, and R. Siegel, *Appl. Sci. Research*, **A7**, 37–52 (1957).

CHAPTER 13

Interphase Transport in Nonisothermal Systems

Thus far we have seen how energy balances may be set up for various simple problems and how these balances lead to ordinary or partial differential equations from which temperature profiles may be calculated. We have also seen that the energy balance over an arbitrary differential fluid element leads to a partial differential equation—the energy equation—which may be used to set up more complex problems. It was seen in Chapter 12 that the time-smoothed energy equation, together with empirical expressions for the turbulent heat flux $\bar{q}^{(t)}$, provides a useful basis for summarizing and extrapolating temperature-profile measurements in turbulent systems. Hence by this time the reader should have a fairly good idea of the meaning of the equations of change for nonisothermal flow systems and their range of applicability.

It should be apparent that all of the problems discussed have pertained to systems of rather simple geometry and furthermore that most of these problems have contained assumptions, such as the assumption that the viscosity is temperature-independent or that the fluid density is constant. This does not mean that the solutions we have thus far obtained are unimportant—quite the contrary, in fact. Many of the solutions have been used in design work, under conditions in which the assumptions are reasonably valid. Many of these solutions are useful for order-of-magnitude calculations. In addition, a study of these simpler systems provides the stepping stone to the discussion of more complex problems.

In this chapter we turn to some of the problems in which it is necessary or convenient to use a less detailed analysis. In such problems the usual engineering approach is to formulate energy balances over larger regions, as discussed in Chapter 15. In preparation for the discussion of the macroscopic energy balance, we present in this chapter a series of representative correlations for predicting rates of heat flow across solid-fluid boundaries; the radiative contribution to this heat flow is considered in Chapter 14.

The reader will recall that in Chapter 6 complex momentum transfer problems were described by combining experimental knowledge with the results of dimensional analysis; the results were given as plots of the friction factor versus the Reynolds number for various geometries. In this chapter we extend this procedure to heat-transfer problems. From the dimensional analysis of the equations of change for nonisothermal flow, plus experimental data on heat-transfer rates, one can arrive at dimensionless presentations of the rates of forced- and free-convection heat transfer. These correlations are given in terms of a dimensionless heat flux, which is known as the Nusselt number for heat transfer, designated by Nu.

The correlations for nonisothermal systems are more complicated in many ways than those encountered in Chapter 6. The number of dimensionless groups is larger, the possible variations in boundary conditions are more numerous, and the temperature dependence of the physical properties now has to be dealt with. In addition, the phenomena of free convection, condensation, and boiling are encountered in nonisothermal systems. Whenever possible, we refer back to analogous topics in Chapter 6 and indicate the modifications necessary to apply those correlations to nonisothermal systems.

We have limited our discussion here to a small number of heat-transfer formulas to introduce the reader to the subject without attempting to be encyclopedic. Wider selections of formulas may be found in a number of heat-transfer texts,[1,2,3,4,5] and in the recent research literature.

§13.1 THE DEFINITION OF THE HEAT-TRANSFER COEFFICIENT

Let us consider a flow system with the fluid flowing either in a conduit or around a solid object. Suppose that the solid surface is warmer than the

[1] H. Gröber, S. Erk, and U. Grigull, *Die Grundgesetze der Wärmeübertragung*, Springer, Berlin (1955), Third Edition.

[2] M. Jakob, *Heat Transfer*, Wiley, New York, Vol. I (1949), Vol. II (1957).

[3] W. H. McAdams, *Heat Transmission*, McGraw-Hill, New York (1954), Third Edition.

[4] E. R. G. Eckert and R. M. Drake, *Heat and Mass Transfer*, McGraw-Hill, New York (1959), Second Edition.

[5] M. Jakob and G. A. Hawkins, *Elements of Heat Transfer*, Wiley, New York (1957), Third Edition.

fluid so that heat is being transferred from the solid to the fluid. Then the rate of heat flow across the solid-fluid interface would be expected to depend on the area of the interface and the temperature drop between fluid and solid. Hence we write

$$Q = hA\,\Delta T \tag{13.1-1}$$

in which Q is the heat flow into the fluid (Btu hr^{-1}), A is a characteristic area, ΔT is a characteristic temperature difference, and the proportionality factor h is known as the *heat-transfer coefficient*. Clearly the same definition can be used when the fluid is cooled. Equation 13.1-1 is sometimes called "Newton's law of cooling"; this is misleading, since it is not really a "law" at all—rather it is the defining equation for h. Note that h is not defined for a specific situation until A and ΔT are stipulated. Let us now consider the usual definitions of h for two types of flow geometry.

As an example of *flow in conduits*, consider a fluid flowing in a circular tube of diameter D (see Fig. 13.1-1), in which there is a heated wall section of length L and varying inside surface temperature $T_0(z)$. Suppose that the bulk temperature T_b of the fluid[1] increases from T_{b1} to T_{b2} in the heated section. Then there are three conventional definitions of heat-transfer coefficients for the fluid in the entire heated section:

$$Q = h_1(\pi DL)(T_{01} - T_{b1}) \tag{13.1-2}$$

$$Q = h_a(\pi DL)\left(\frac{(T_{01} - T_{b1}) + (T_{02} - T_{b2})}{2}\right) \tag{13.1-3}$$

$$Q = h_{\text{ln}}(\pi DL)\left[\frac{(T_{01} - T_{b1}) - (T_{02} - T_{b2})}{\ln\,[(T_{01} - T_{b1})/(T_{02} - T_{b2})]}\right] \tag{13.1-4}$$

Note that h_1 is based on the initial temperature difference $(T_0 - T_b)_1$, h_a is based on the arithmetic mean of the terminal temperature differences, $(T_0 - T_b)_a$, and h_{ln} is based on the corresponding logarithmic mean temperature difference $(T_0 - T_b)_{\text{ln}}$. The coefficient h_{ln} is preferable for most calculations because it is less dependent on L/D than the other two;[2] however, it is not universally used. In using heat-transfer results from the literature, one should make a special note of the definitions of the heat-transfer coefficients.

If the wall-temperature distribution is initially unknown or if the fluid properties change appreciably along the pipe, it is difficult to predict the heat-transfer coefficients defined above. Under these conditions, it is customary to rewrite Eq. 13.1-1 in the differential form:

$$dQ = h_{\text{loc}}(\pi D\,dz)(T_0 - T_b) \tag{13.1-5}$$

[1] For the definition of bulk temperature for constant ρ and $\hat{C}_p$, see Eq. 9.8-33.

[2] It is useful to note, however, that if $\Delta T_2/\Delta T_1$ is between 0.5 and 2.0 then ΔT_a may be substituted for ΔT_{ln}, or h_a for h_{ln}, with a maximum error of 4 per cent. This degree of accuracy is adequate for most heat-transfer calculations.

Here dQ is the heat added to the fluid in the distance dz along the pipe, $(T_0 - T_b)$ is the local temperature difference, and h_{loc} is the *local heat-transfer coefficient*.[3] This equation is widely used in engineering design.

As an example of *flow around submerged objects*, consider a fluid flowing about a sphere of radius R, whose surface temperature is everywhere maintained at T_0. Suppose that the fluid approaches the sphere with a uniform

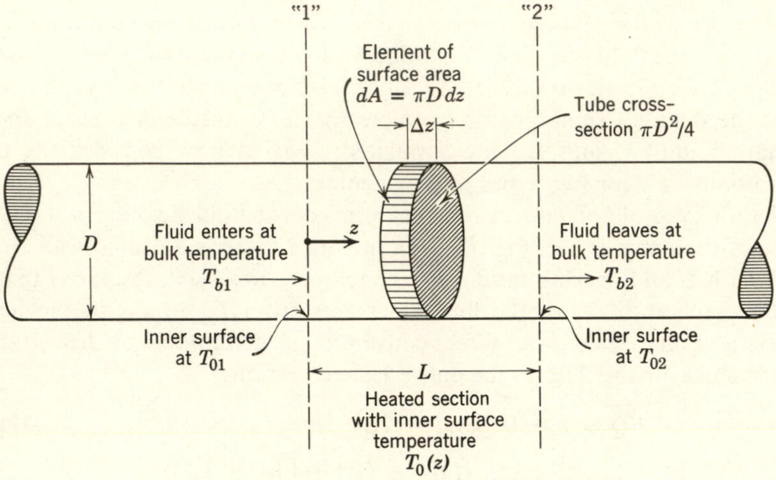

Fig. 13.1-1. Heat transfer in a circular tube.

temperature T_∞, which is different from T_0. Then we may define a *mean* heat-transfer coefficient for the entire surface of the sphere by the relation

$$Q = h_m(4\pi R^2)(T_0 - T_\infty) \tag{13.1-6}$$

Note that the choice of characteristic area for the sphere in this case is different from that in Eq. 6.1-5.

A local coefficient can also be defined for submerged objects by analogy with Eq. 13.1-5:

$$dQ = h_{loc}(dA)(T_0 - T_\infty) \tag{13.1-7}$$

This coefficient is more informative than h_m because it shows how the heat flux is distributed over the surface; however, most experimenters report only h_m, which is more easily measured.

Let us emphasize that the definitions of A and ΔT must be made clear before h is defined. Keep in mind, too, that h is not a constant characteristic of the fluid medium. On the contrary, the heat-transfer coefficient depends in a

[3] Note that h_{loc} and $(\Delta T)_{loc}$ are not defined until one specifies the form of the element of area. In Eq. 13.1-5 we have set $dA = \pi D\,dz$, which means that h_{loc} and T_0 are the mean values for the shaded area dA in Fig. 13.1-1.

complicated way on many variables, including the fluid properties (k, μ, ρ, $\hat{C}_p$), the system geometry, the flow velocity, the value of the characteristic temperature difference, and the surface temperature distribution. The remainder of this chapter is concerned with predicting the dependence of h on these quantities. Usually this is done by collecting experimental data and performing correlations on the basis of dimensional analysis; however, direct calculations of h from the equations of change are available for a number of simple laminar flow systems, and in turbulent flow semitheoretical calculations of h have proven useful. Some sample values of h are given in Table 13.1–1.

TABLE 13.1–1[a]

ORDER OF MAGNITUDE OF THE HEAT TRANSFER COEFFICIENT h

Situation	h(kcal m^{-2} hr^{-1} °C^{-1})	h(Btu ft^{-2} hr^{-1} °F^{-1})
Free convection		
Gases	3–20	1–4
Liquids	100–600	20–120
Boiling water	1000–20000	200–4000
Forced convection		
Gases	10–100	2–20
Viscous fluids	50–500	10–100
Water	500–10000	100–2000
Condensing vapors	1000–100000	200–20000

[a] Taken from H. Gröber, S. Erk, and U. Grigull, *Wärmeübertragung*, Springer, Berlin (1955), Third Edition, p. 158. When given h in kcal m^{-2} hr^{-1} °C^{-1}, multiply by 0.204 to get h in Btu ft^{-2} hr^{-1} °F^{-1}.

We have seen in §9.6 that in the calculation of heat-transfer rates between two fluid streams separated by one or more solid layers it is convenient to utilize an *over-all heat-transfer coefficient*, U_0, which expresses the combined effect of the whole series of resistances through which the transferred heat must flow. We give here a more precise definition of U_0 and show how to calculate it in the special case of heat exchange between two coaxial streams with bulk temperatures T_h and T_c, separated by a cylindrical tube of inside diameter D_0 and outside diameter D_1:

$$dQ = U_0(\pi D_0\, dz)(T_h - T_c) \tag{13.1–8}$$

$$\frac{1}{D_0 U_0} = \left(\frac{1}{D_0 h_0} + \frac{\ln D_1/D_0}{2k^{01}} + \frac{1}{D_1 h_1}\right)_{\text{loc}} \tag{13.1–9}$$

Note that we have defined U_0 as a local coefficient; this is the definition implied in most design procedures. (See Example 15.4–2.)

Note that Eqs. 13.1–8 and 9 are restricted to heat flow through thermal resistances connected in *series*. Under some conditions there will be an appreciable *parallel* heat flux at one or both surfaces by radiation, and these equations will require modification. (See Example 14.5–1.)

To illustrate the physical significance of heat-transfer coefficients and

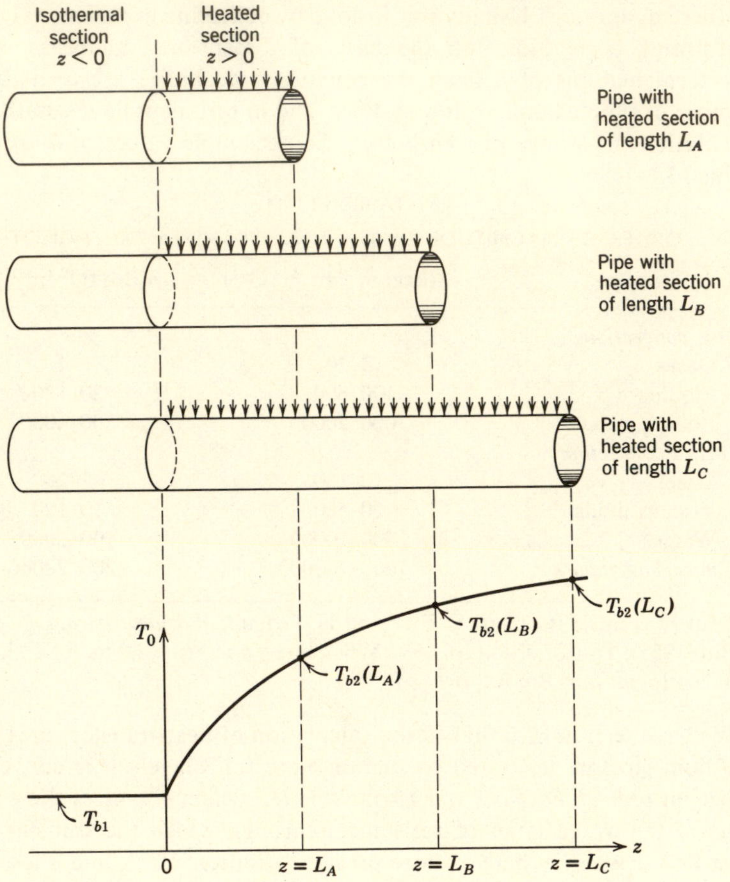

Fig. 13.1–2. Series of experiments to obtain heat-transfer coefficients.

illustrate one method of measuring them, we conclude this section with an analysis of a hypothetical set of heat-transfer data.

Example 13.1–1. Calculation of Heat-Transfer Coefficients from Experimental Data

A series of steady-state experiments on the heating of air in tubes is depicted in Fig. 13.1–2. In the first experiment air at $T_{b1} = 200.0°$ F is flowing in a 0.5-in. i.d. tube with fully developed laminar velocity profile in the isothermal pipe section for $z < 0$. At $z = 0$ the wall temperature is suddenly increased to $T_0 = 212.0°$ F and maintained there for the remaining tube length L_A. At $z = L_A$ the fluid issues into a mixing chamber in which the cup-mixing (or "bulk") temperature T_{b2} is measured. Similar experiments are done with tubes of different lengths, L_B, L_C, etc., with the following results:

Experiment	A	B	C	D	E	F	G
L (in.)	1.5	3.0	6.0	12.0	24.0	48.0	96.0
T_{b2} (° F)	201.4	202.2	203.1	204.6	206.6	209.0	211.0

The air flow rate w is 3.0 lb_m hr^{-1} in all of the experiments. Calculate h_1, h_a, h_{ln}, and the exit value of h_{loc} as functions of the reduced length L/D.

Solution. An energy balance over the heated length L in any experiment gives, for steady-state and constant $\hat{C}_p$,

$$w\hat{C}_p(T_{b2} - T_{b1}) = Q \tag{13.1-10}$$

Combining this with Eq. 13.1–2 gives

$$w\hat{C}_p(T_{b2} - T_{b1}) = h_1\pi DL(T_0 - T_{b1}) \tag{13.1-11}$$

from which

$$h_1 = \frac{w\hat{C}_p}{\pi DL}\left(\frac{T_{b2} - T_{b1}}{T_0 - T_{b1}}\right) \tag{13.1-12}$$

Analogous substitutions of Eqs. 13.1–3 and 4 into the energy balance give

$$h_a = \frac{w\hat{C}_p}{\pi DL}\frac{(T_{b2} - T_{b1})}{(T_0 - T_b)_a} \tag{13.1-13}$$

$$h_{\text{ln}} = \frac{w\hat{C}_p}{\pi DL}\frac{(T_{b2} - T_{b1})}{(T_0 - T_b)_{\text{ln}}} \tag{13.1-14}$$

To evaluate h_{loc}, we have to use the foregoing data to construct a continuous curve $T_b(z)$ to represent the change in bulk temperature with z in the longest (96-in.) tube. Then Eq. 13.1–10 becomes

$$w\hat{C}_p(T_b(z) - T_{b1}) = Q(z) \tag{13.1-15}$$

By differentiating this expression with respect to z and combining the result with Eq. 13.1–5, we get

$$w\hat{C}_p\frac{dT_b}{dz} = h_{\text{loc}}\pi D(T_0 - T_b) \tag{13.1-16}$$

or

$$h_{\text{loc}} = \frac{w\hat{C}_p}{\pi D}\left(\frac{dT_b}{dz}\right)\left(\frac{1}{T_0 - T_b}\right) \tag{13.1-17}$$

and for constant T_0 this becomes

$$h_{\text{loc}} = -\frac{w\hat{C}_p}{\pi D}\frac{d\ln(T_0 - T_b)}{dz} \tag{13.1-18}$$

The derivative in Eq. 13.1–18 is conveniently determined from a plot of ln $(T_0 - T_b)$ versus z. Because a differentiation is involved, it is difficult to determine h_{loc} precisely.

The calculated results are shown in Fig. 13.1–3. Note that all of the coefficients decrease with increasing L/D but that h_{loc} and h_{ln} are less variable than the others and approach a common asymptote with increasing L/D. Somewhat similar

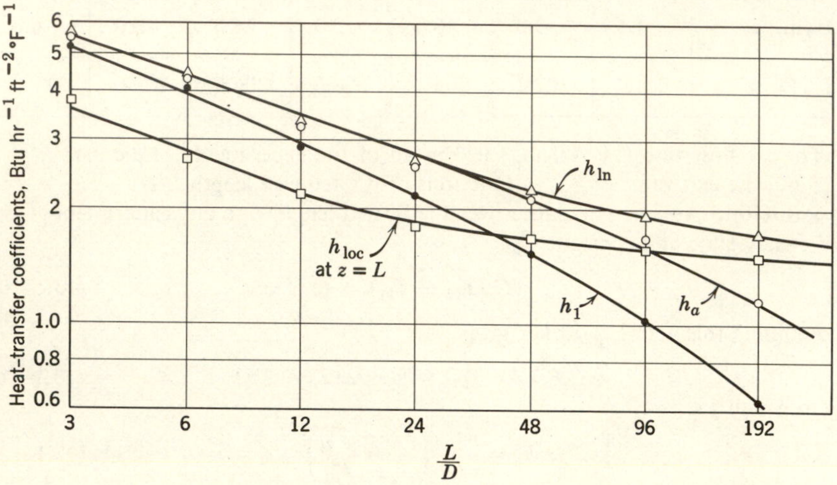

Fig. 13.1–3. Calculated heat-transfer coefficients for Example 13.1–1.

behavior is observed in turbulent flow with constant wall temperature, except that h_{loc} approaches the asymptote much more rapidly. (See Fig. 13.2–1.)

§13.2 HEAT-TRANSFER COEFFICIENTS FOR FORCED CONVECTION IN TUBES

The subject of heating or cooling of fluids inside tubes is of considerable practical importance. There is, consequently, a voluminous literature on the subject, and many design equations of varying degrees of generality are available. Our purpose here is simply to introduce the reader to this subject and to point out a few well-known results.

We first extend the dimensional analysis given in §10.6 in order to predict the form of the correlations for the heat-transfer coefficient h in forced convection. Consider the steady flow of a fluid in the pipe shown in Fig. 13.1–1. We assume that the velocity distribution at plane "1" is known and that the surface temperature T_0 in the heated section from $z = 0$ to $z = L$ is constant. We further assume that the physical properties ρ, μ, $\hat{C}_p$, and k are constants. Later in this section we consider the effects of the temperature dependence of μ, which are significant under some conditions.

For a tube of radius R and length L, the total heat flow into the fluid at the pipe wall is

$$Q = \int_0^L \int_0^{2\pi} \left(+k \frac{\partial T}{\partial r}\right)\Big|_{r=R} R\,d\theta\,dz \tag{13.2–1}$$

this statement being valid for laminar or turbulent flow.[1] The + sign appears here because the heat is added to the system in the $-r$-direction. If we equate the expressions for Q given in Eqs. 13.1–2 and 13.2–1 and solve for h_1, we get

$$h_1 = \frac{1}{\pi DL(T_0 - T_{b1})} \int_0^L \int_0^{2\pi} \left(+k \frac{\partial T}{\partial r}\right)\Big|_{r=R} R\,d\theta\,dz \tag{13.2–2}$$

Next we introduce the dimensionless quantities $r^* = r/D$, $z^* = z/D$, $T^* = (T - T_0)/(T_{b1} - T_0)$, and the dimensionless *Nusselt number* $\text{Nu}_1 = h_1D/k$. Multiplication of Eq. 13.2–2 by D/k and introduction of the dimensionless quantities then gives

$$\text{Nu}_1 = \frac{1}{2\pi L/D} \int_0^{L/D} \int_0^{2\pi} \left(-\frac{\partial T^*}{\partial r^*}\right)\Big|_{r^*=\frac{1}{2}} d\theta\,dz^* \tag{13.2–3}$$

Hence the Nusselt number Nu_1 is basically a dimensionless temperature gradient averaged over the heat-transfer surface.

This dimensionless temperature gradient can, in principle, be evaluated by differentiating the expression for T^* obtained by solving Eqs. 10.6–9, 10, and 11 with the boundary conditions

$$\boldsymbol{v}^* = \text{a given function of } r^* \text{ and } \theta \text{ at } z^* = 0 \tag{13.2–4}$$

$$\boldsymbol{v}^* = 0 \qquad \text{at } r^* = \tfrac{1}{2} \tag{13.2–5}$$

$$p^* = 0 \qquad \text{at } z^* = 0, \quad r^* = 0 \tag{13.2–6}$$

$$T^* = 1 \qquad \text{at } z^* = 0 \tag{13.2–7}$$

$$T^* = 0 \qquad \text{at } r^* = \tfrac{1}{2} \tag{13.2–8}$$

From the dimensionless groups that appear in the differential equations[2] (Eqs. 10.6–9, 10, and 11) and the fact that no new dimensionless groups appear in the boundary conditions, one can conclude that the dimensionless time-smoothed temperature will have the following dependence (cf. Eqs. 6.2–7 and 8):

$$T^* = T^*(r^*, \theta, z^*; \text{Re}, \text{Pr}, \text{Br}) \tag{13.2–9}$$

Substitution of this relation into Eq. 13.2–3 gives:

$$\text{Nu}_1 = \text{Nu}_1(\text{Re}, \text{Pr}, \text{Br}, L/D) \tag{13.2–10}$$

[1] All temperatures, pressures, and velocities are understood to be time-smoothed in the following development; for brevity the overlines are omitted.

[2] The Froude number has no effect here, since there is no free liquid surface; that is, the pipe is running full.

In most practical situations the viscous dissipation effects are small so that the Brinkman group can be neglected. Equation 13.2–10 then simplifies to

$$\text{Nu}_1 = \text{Nu}_1(\text{Re}, \text{Pr}, L/D) \tag{13.2–11}$$

Hence dimensional analysis has told us that for forced convection in tubes with constant wall temperature, the heat-transfer coefficient h_1 may be correlated[3] in terms of a dimensionless group Nu_1, which depends upon the Reynolds number, the Prandtl number, and the geometric factor L/D. It can correspondingly be shown (see Problem 13.K) that

$$\text{Nu}_a = \text{Nu}_a(\text{Re}, \text{Pr}, L/D) \tag{13.2–12}$$

$$\text{Nu}_{\text{ln}} = \text{Nu}_{\text{ln}}(\text{Re}, \text{Pr}, L/D) \tag{13.2–13}$$

$$\text{Nu}_{\text{loc}} = \text{Nu}_{\text{loc}}(\text{Re}, \text{Pr}, z/D) \tag{13.2–14}$$

in which $\text{Nu}_a = h_a D/k$, $\text{Nu}_{\text{ln}} = h_{\text{ln}} D/k$, and $\text{Nu}_{\text{loc}} = h_{\text{loc}} D/k$.

Thus far we have assumed that the physical properties are constants over the temperature range encountered in the system. For large temperature differences, this assumption leads to serious errors, the variation in μ and ρ being most important. The temperature dependence of the viscosity may be handled approximately[4] by including the group μ_b/μ_0, in which μ_b is the viscosity at some average value T_b of the fluid bulk temperature and μ_0 is the viscosity at the temperature of the solid surface. Hence some correlations of the Nusselt numbers defined above are of the form

$$\text{Nu} = \text{Nu}(\text{Re}, \text{Pr}, L/D, \mu_b/\mu_0) \tag{13.2–15}$$

which seems to have been first used by Sieder and Tate.[5] If, in addition, the

[3] This result is also given by the Buckingham Π-theorem (cf. footnote in connection with Eq. 6.2–9). One seeks a relation expressing h in terms of the physical properties ρ, μ, $\hat{C}_p$, k, the geometrical quantities D, L, and the flow velocity $\langle v \rangle$. There are four fundamental units involved (mass m, length l, time t, and temperature T). Hence the Π-theorem correctly predicts that there will be four interrelated dimensionless groups, though it fails to reveal the assumptions made in choosing the variables.

[4] One can arrive at a group of this sort by inserting in the equations of change a temperature-dependent coefficient of viscosity, described by a Taylor expansion of μ as a function of T:

$$\mu = \mu_0 + \left(\frac{\partial \mu}{\partial T}\right)_{T=T_0} (T - T_0) + \cdots \tag{13.2–15a}$$

We now approximate the differential quotient by a difference quotient and rearrange so as to obtain a dimensionless viscosity:

$$\mu \doteq \mu_0 + \left(\frac{\mu_b - \mu_0}{T_b - T_0}\right)(T - T_0) \quad \text{or} \quad \frac{\mu}{\mu_0} \doteq 1 + \left(\frac{\mu_b}{\mu_0} - 1\right)\left(\frac{T - T_0}{T_b - T_0}\right) \tag{13.2–15b,c}$$

The group μ_b/μ_0 then appears naturally in the equations of change.

[5] E. N. Sieder and G. E. Tate, *Ind. Eng. Chem.*, **28**, 1429–1435 (1936).

density ρ varies significantly, then some free convection may occur. This effect has been accounted for in correlations by adding the Grashof number to the foregoing set of dimensionless groups. Discussion of this effect is outside the scope of this text.[6]

The dimensional-analysis results just given are of great utility in empirical heat-transfer studies. For example, Eq. 13.2–15 shows that although h depends upon *eight* physical quantities (D, $\langle v \rangle$, ρ, μ_b, μ_0, $\hat{C}_p$, k, L) this dependence can be expressed by giving Nu as a function of only *four* dimensionless groups (Re, Pr, L/D, μ_b/μ_0). This reduction in the number of independent variables to be studied greatly reduces the number of tests required; for example, to study all combinations of eight independent variables for ten values of each variable would take 10^8 tests, whereas for four independent variables 10^4 tests would suffice.

Correlations Obtained from Dimensional Analysis

Let us now consider a few of the better-known correlations that have been given for heat transfer in pipes. The correlation of Sieder and Tate[5] for developed flow in smooth tubes with nearly constant wall temperature is plotted in Fig. 13.2–1. This correlation is consistent with Eq. 13.2–15 and gives a convenient over-all picture of heat transfer in pipe flow. Note that all physical properties are evaluated at $(T_{b1} + T_{b2})/2$, except μ_0, which is evaluated at $(T_{01} + T_{02})/2$. The Reynolds number employed here, $\text{Re}_b = DG/\mu_b = Dw/S\mu_b$, is convenient because it has been found empirically[7] that transition to turbulence usually begins at about $\text{Re}_b = 2100$, even when μ varies appreciably in the radial direction. The quantity $G = w/S = \langle \rho v \rangle$ is known as the "mass velocity."

For *highly turbulent flow*, the curves for $L/D > 10$ converge to a single line. For $\text{Re}_b > 20{,}000$, this line follows the equation

$$\frac{h_{\ln} D}{k_b} = 0.026\left(\frac{DG}{\mu_b}\right)^{0.8}\left(\frac{\hat{C}_p\mu}{k}\right)^{1/3}\left(\frac{\mu_b}{\mu_0}\right)^{0.14} \tag{13.2–16}$$

which reproduces available experimental data within about ± 20 per cent in the range $\text{Re}_b = 10^4$ to 10^5, $\text{Pr}_b = 0.6$ to 100, and $L/D > 10$.

For *laminar flow*, the lines are given by the equation

$$\frac{h_{\ln} D}{k_b} = 1.86(\text{Re}_b \, \text{Pr}_b \, D/L)^{1/3}(\mu_b/\mu_0)^{0.14} \tag{13.2–17}$$

which is an empirical modification[5] of the Graetz theoretical solution[8] for heat

[6] A detailed treatment of laminar pipe flow with temperature-dependent ρ and μ has been given by R. L. Pigford, *C.E.P. Symposium Series No. 17*, **51**, 79–92 (1955).

[7] A. P. Colburn, *Trans. A.I.Ch.E.*, **29**, 174–210 (1933).

[8] L. Graetz, *Ann. d. Physik*, **25**, 337–357 (1885); J. Lévêque, *Ann. mines*, **13**, 201, 305, 381 (1928). These solutions are discussed at length by M. Jakob, *Heat Transfer*, Vol. I, Wiley, New York (1949), pp. 451–464.

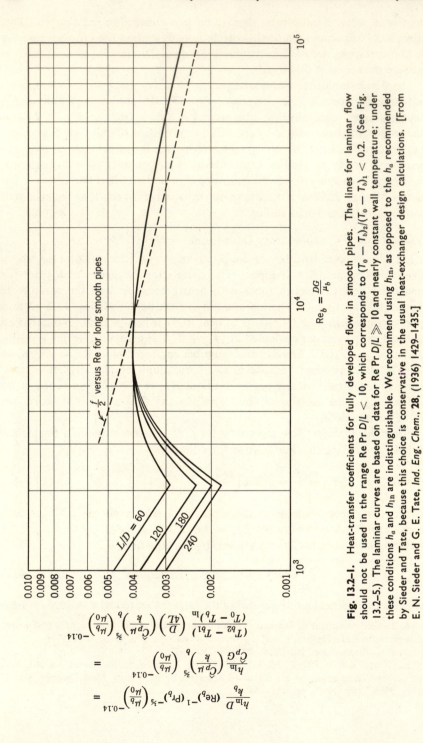

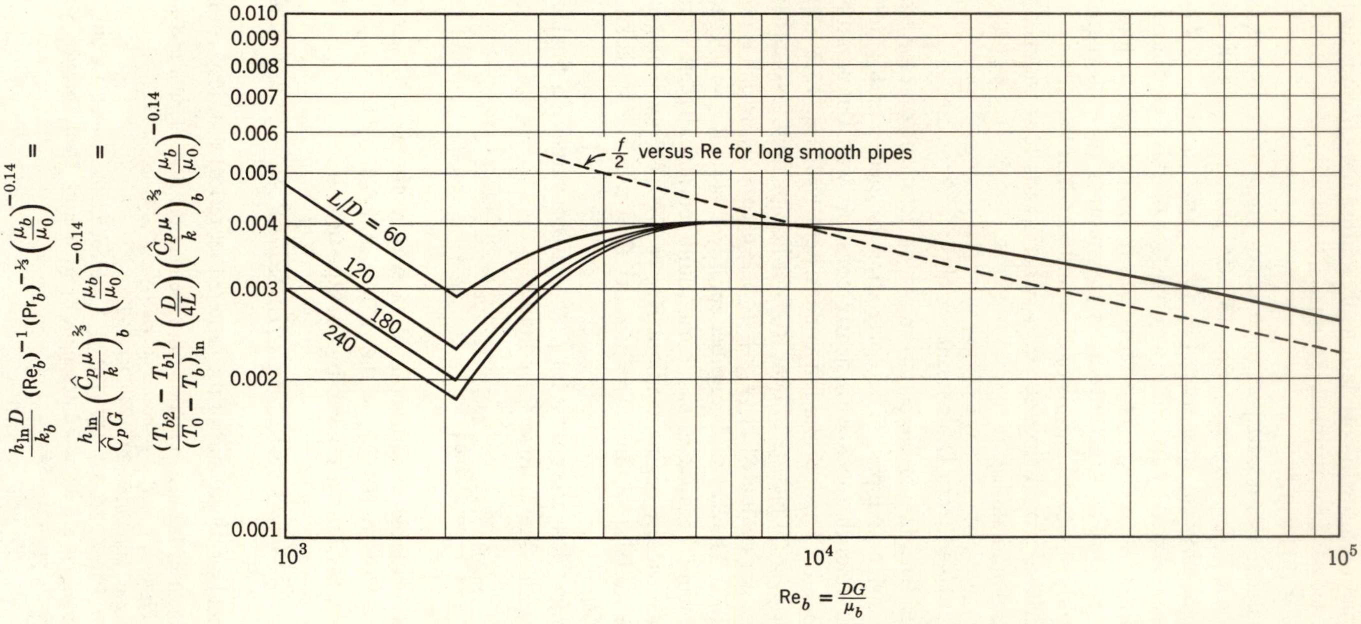

Fig. 13.2–1. Heat-transfer coefficients for fully developed flow in smooth pipes. The lines for laminar flow should not be used in the range Re Pr $D/L < 10$, which corresponds to $(T_0 - T_b)_2/(T_0 - T_b)_1 < 0.2$. (See Fig. 13.2–5.) The laminar curves are based on data for Re Pr $D/L \gg 10$ and nearly constant wall temperature; under these conditions h_a and $h_{\ln}$ are indistinguishable. We recommend using $h_{\ln}$, as opposed to the h_a recommended by Sieder and Tate, because this choice is conservative in the usual heat-exchanger design calculations. [From E. N. Sieder and G. E. Tate, *Ind. Eng. Chem.*, **28**, (1936) 1429–1435.]

transfer in fully developed laminar flow with constant wall temperature. (See Eq. 13.2–23.) Equation 13.2–17 is good within about 20 per cent for Re Pr $D/L > 10$; at lower flow rates it underestimates h_{ln} considerably.

The *transition region* extending roughly from $\text{Re}_b = 2100$ to 10,000 is not well understood and is usually avoided in design if possible. The curves in this region are supported by data[5] but are less reliable than the rest of the plot.

The general characteristics of the curves in Fig. 13.2–1 deserve careful study. Note that for a heated section of given L and D, and a fluid of given physical properties, the ordinate is proportional to the dimensionless temperature rise $(T_{b2} - T_{b1})/(T_0 - T_b)_{\text{ln}}$ of the fluid passing through. Under these conditions, as one increases the flow rate, hence the Reynolds number, the exit fluid temperature will first decrease (until Re_b reaches about 2100), then increase (until Re_b reaches about 8000), and finally decrease again. Note also that the influence of L/D on h_{ln} is marked in laminar flow but becomes insignificant for $\text{Re}_b > 8000$ in the range of L/D shown here.

Figure 13.2–1 somewhat resembles the friction-factor plot, Fig. 6.2–2, although the physical situation is quite different. In the highly turbulent range ($\text{Re}_b > 10{,}000$) the heat-transfer ordinate agrees approximately with $f/2$ for long smooth pipes. This fact was pointed out by Colburn,[7] who proposed the following empirical analogy for $\text{Re}_b > 10{,}000$ in long smooth pipes:

$$j_H = \frac{f}{2} \tag{13.2–18}$$

in which

$$j_H = \frac{h_{\text{ln}}}{\hat{C}_p G}\left(\frac{\hat{C}_p \mu}{k}\right)_f^{2/3} \tag{13.2–19}$$

and $f/2$ is obtainable from Fig. 6.2–2. The Reynolds number for this purpose is calculated as DG/μ_f. In this correlation the subscript f refers to properties evaluated at the "film temperature" $(T_b + T_0)/2$, in which T_b and T_0 are averages of the terminal values; the $\hat{C}_p$ in the denominator is evaluated at T_b. If the physical properties are constant throughout the tube, then Fig. 13.2–1 becomes a plot of j_H versus the Reynolds number; clearly, the analogy 13.2–18 is not valid below $\text{Re} = 10{,}000$. The analogy breaks down completely for rough tubes in fully turbulent flow because f is affected more by roughness than is j_H.

One final remark concerning the use of Fig. 13.2–1 has to do with the application to conduits of noncircular cross section. For highly turbulent flow, one may use[9] the mean hydraulic radius defined in Eq. 6.2–16. In order to apply this empiricism, one simply replaces D by $4R_h$ everywhere in the Nusselt and Reynolds numbers.

[9] W. H. McAdams, *Heat Transmission*, McGraw-Hill, New York (1954), Third Edition, pp. 241–243.

The simple correlation presented in Fig. 13.2–1 is satisfactory for many design problems, but the representation of laminar and turbulent flow by a single grid of lines is an over-simplification. The correlations that follow deal only with one or the other of these flow regimes.[10]

Correlations Obtained from Temperature Profiles

For *highly turbulent flow with constant fluid properties* (Re > 10,000) and Prandtl numbers above 0.5, the Deissler correlation[11] (Fig. 13.2–2) is highly

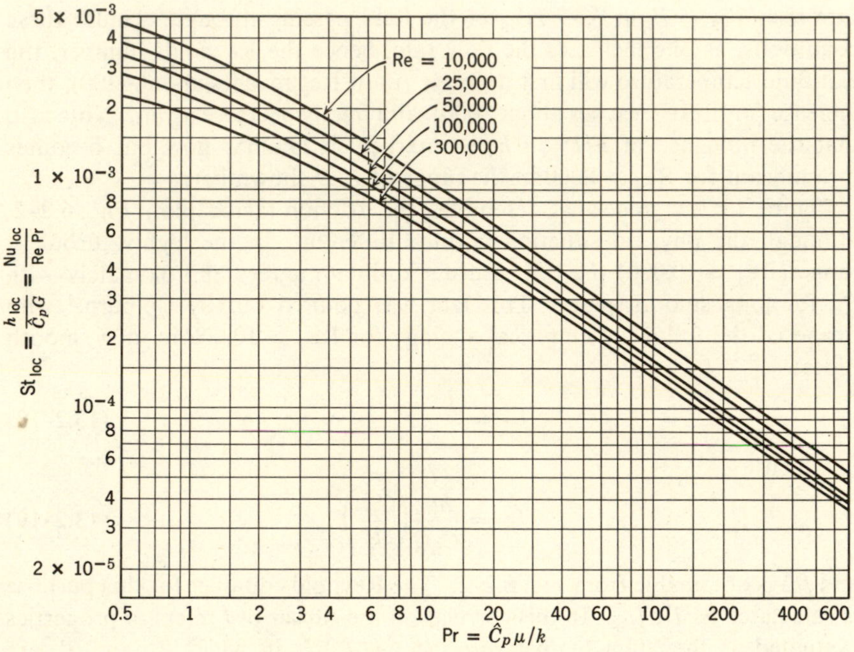

Fig. 13.2–2. Predicted asymptotic values of the local Stanton number for turbulent flow in smooth pipes with constant heat flux. [R. G. Deissler, *NACA Report* 1210 (1955).]

regarded. The ordinate, $h/\hat{C}_p G = \mathrm{Nu}\,\mathrm{Re}^{-1}\,\mathrm{Pr}^{-1}$, is known as the Stanton number, St. This plot is based on the semiempirical velocity and temperature profiles of Figs. 5.3–1 and 12.3–2; it holds for constant wall heat flux q_0 and fully developed temperature and velocity profiles. In this range of Reynolds and Prandtl numbers the temperature profile develops very quickly; thus, as shown in Fig. 13.2–3, the asymptotic Nusselt numbers hold with good

[10] The material immediately following may be omitted on first reading. The reader interested in more detailed correlations may continue; others should skip directly to Example 13.2–1.

[11] R. G. Deissler, *NACA Report 1210* (1955); Deissler has, in addition, computed h_{loc} and f, taking into account the variation in the physical properties.

accuracy, except for a short "thermal entrance region" that may be neglected in most problems.[12] This rapid thermal response of the fluid also permits the use of Fig. 13.2–2 with good accuracy for other thermal boundary conditions, such as constant or smoothly varying surface temperature T_0, and justifies the corresponding application of local coefficients estimated from Fig. 13.2–1.

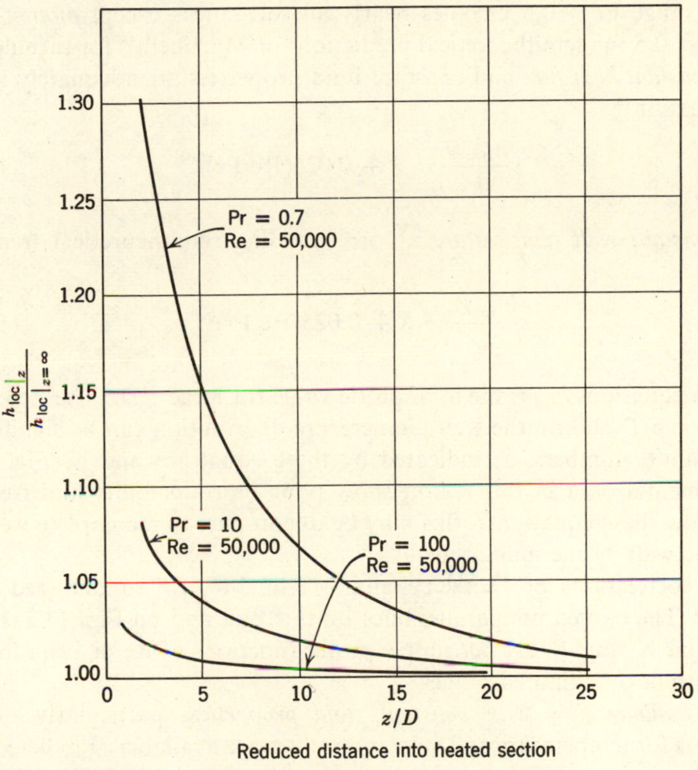

Fig. 13.2–3. Predicted variation of heat-transfer coefficients in the thermal entrance region for developed flow and constant heat flux. [E. M. Sparrow, T. M. Hallman, and R. Siegel, *Appl. Sci. Research* **A7**, 37–52 (1957).]

The Deissler correlation can be accurately represented for Pr > 200 by the asymptotic formula

$$\frac{h_{loc}}{\hat{C}_p G} = 0.0789 \frac{\sqrt{f}}{\text{Pr}^{3/4}} \tag{13.2–20}$$

Note that for constant fluid properties the turbulent-flow correlations of

[12] Other semiempirical calculations of turbulent heat transfer in the thermal entrance region have been made by H. L. Beckers, *Appl. Sci. Research*, **A6**, 147–190 (1956); C. A. Sleicher and M. Tribus, *1956 Heat Transfer and Fluid Mechanics Institute Preprints*, Stanford University Press, pp. 59–78; and R. G. Deissler, *NACA Report 1210* (1955).

Colburn and of Sieder and Tate predict $h/\hat{C}_p G \doteq (f/2)\text{Pr}^{-2/3}$ in which $h = h_{\text{loc}}$ or h_{ln}; this difference in form is probably too large to be explained by the difference in boundary conditions (constant wall temperature versus constant wall heat flux). Experimental data indicate that the Deissler asymptote in Eq. 13.2–20 is somewhat more reliable than either Eq. 13.2–16 or 18.

The range $\text{Pr} > 0.5$ includes nearly all substances except *molten metals*; for $\text{Pr} < 0.5$ the semitheoretical predictions of Martinelli[13] for turbulent flow with *constant heat flux* and constant fluid properties are adequately fitted by the equation[14]

$$\frac{h_{\text{loc}} D}{k} = 7 + 0.025(\text{Re Pr})^{0.8} \tag{13.2–21}$$

For *constant wall temperature* a corresponding semitheoretical treatment[15] gives

$$\frac{h_{\text{loc}} D}{k} = 5 + 0.025(\text{Re Pr})^{0.8} \tag{13.2–22}$$

In both equations h_{loc} is the asymptotic value for large z/D. The dependence of h_{loc} on z/D and on the wall-temperature distribution can be significant at low Prandtl numbers, as indicated by these equations and by Fig. 13.2–3. Experimental data in this region show poor reproducibility and frequently fall below these equations; this may be due in part to incomplete wetting of the tube wall by the molten metal.

The correlations of Deissler[11] and Martinelli[13] are summarized in Fig. 13.2–4. The curved nonparallel lines on this plot and on Fig. 13.2–2 clearly show that Nu and St are *not* simple power functions of Re or Pr in turbulent flow, except over limited ranges.

For *laminar flow with constant fluid properties*, particularly extensive solutions for temperature and velocity profiles are available. Predicted values of Nu_{ln} are shown in Fig. 13.2–5 for various boundary conditions. Note that in laminar flow the dependence of Nu on L/D and on surface-temperature distribution is significant in contrast to the minor dependence shown in turbulent flow. (See Fig. 13.2–1.) All of the curves approach asymptotes corresponding to fully developed flow as $L/D \to \infty$; for constant T_0 the asymptote is $\text{Nu}_{\text{ln}} = 3.657$, and for constant $T_0 - T_b$ it is $\text{Nu}_{\text{ln}} = 48/11 = 4.364$. Curve A, an extension of the calculations of Graetz,[8] is probably the earliest-known solution of a forced convection problem by the equations of change.

For small values of L, or high flow rates, the Graetz solution reduces

[13] R. C. Martinelli, *Trans. ASME,* **69,** 947–959 (1947).
[14] R. N. Lyon, *Chem. Eng. Prog.,* **47,** 75–79 (1951).
[15] R. A. Seban and T. T. Shimazaki, *Trans. ASME,* **73,** 803–809 (1951).

asymptotically to a simpler form given by Lévêque.[8] Lévêque's solution gives

$$\mathrm{Nu_{ln}} = 1.62(\mathrm{Re\ Pr}\ D/L)^{1/3}$$
$$= 1.75(w\hat{C}_p/kL)^{1/3} \qquad (13.2\text{–}23)^{16}$$

This equation is the basis for the semiempirical Eq. 13.2–17. Note that the empirically determined coefficient in Eq. 13.2–17 does not agree with the

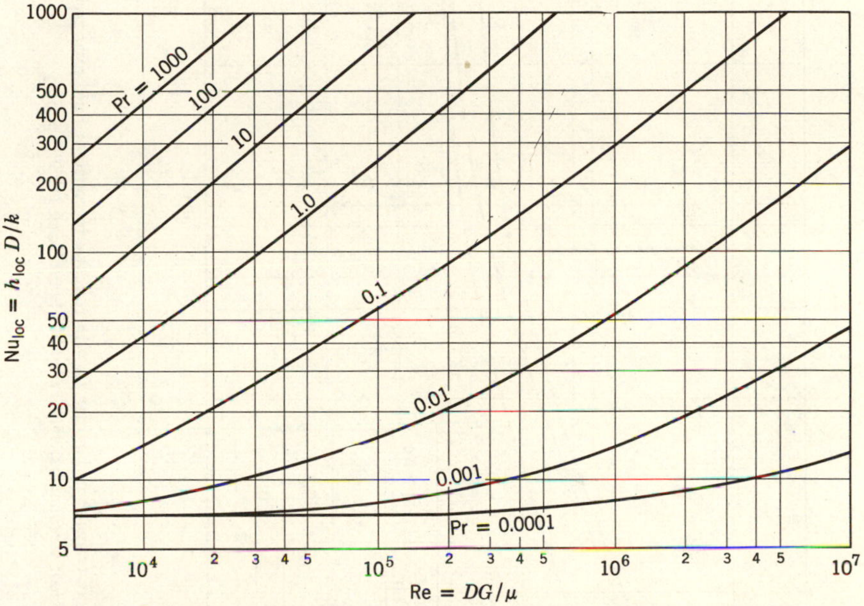

Fig. 13.2–4. Predicted Nusselt numbers for developed turbulent flow with constant wall heat flux. [Results for Pr $\geqslant$ 1 are from R. G. Deissler, *NACA Report* 1210 (1955); others are from R. C. Martinelli, *Trans. ASME*, **69**, 947–959 (1947).]

exact solution for constant properties; the discrepancy has been attributed to small free-convection effects caused by the high temperature differences employed in the experiments.

Example 13.2–1. Design of a Tubular Heater

Air at 70° F and 1 atm is to be pumped through a long, straight 2-in. i.d. pipe at a rate of 70 $\mathrm{lb}_m\,\mathrm{hr}^{-1}$. A section of the pipe is to be heated to an inside wall temperature of 250° F to raise the air temperature to 230° F. What heated length L is required?

[16] The quantity $w\hat{C}_p/kL = (\pi/4)\mathrm{Re\ Pr}\ D/L$ in this equation is known as the Graetz number, Gz. The coefficient 1.75 in this equation is sometimes misquoted as 2.0.

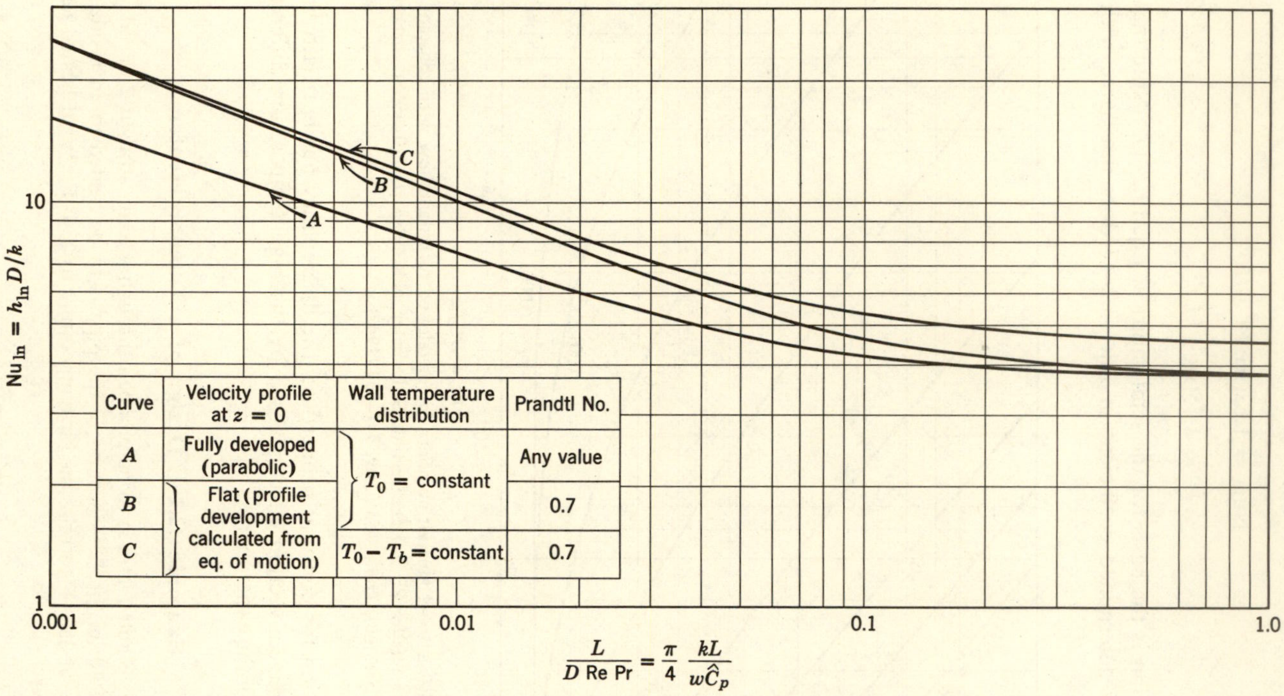

Fig. 13.2-5. Theoretical solutions for heat transfer in laminar flow in tubes with constant fluid properties. [Adapted from W. M. Kays and A. L. London, *Compact Heat Exchangers*, McGraw-Hill, New York (1958), p. 50.]

$$\frac{L}{D} \frac{1}{\mathrm{Re\ Pr}} = \frac{\pi}{4} \frac{kL}{w\hat{C}_p}$$

Curve	Velocity profile at $z = 0$	Wall temperature distribution	Prandtl No.
A	Fully developed (parabolic)	$T_0 = $ constant	Any value
B	Flat (profile development calculated from eq. of motion)	$T_0 = $ constant	0.7
C	Flat (profile development calculated from eq. of motion)	$T_0 - T_b = $ constant	0.7

$\mathrm{Nu_{ln}} = h_{\mathrm{ln}} D/k$

Solution. The required physical properties of air are

	At average T_b of 150° F	At T_0 of 250° F
μ, $lb_m\ ft^{-1}\ hr^{-1}$	0.0490	0.055
$\hat{C}_p$, $Btu\ lb_m^{-1}\ °F^{-1}$	0.241	—
k, $Btu\ hr^{-1}\ ft^{-1}\ °F^{-1}$	0.0169	—
$\Pr = \dfrac{\hat{C}_p\mu}{k}$	0.698	—

The Reynolds number, evaluated at $\frac{1}{2}(T_{b1} + T_{b2}) = \frac{1}{2}(70 + 230) = 150°$ F is

$$\text{Re}_b = \frac{Dw}{S\mu} = \frac{4w}{\pi D\mu} = \frac{(4)(70)}{(\pi)(\frac{2}{12})(0.0490)} = 1.09 \times 10^4 \tag{13.2-24}$$

From Fig. 13.2-1 we obtain

$$\frac{(T_{b2} - T_{b1})}{(T_0 - T_b)_{\ln}} \frac{D}{4L}\left(\frac{\hat{C}_p\mu}{k}\right)_b^{2/3}\left(\frac{\mu_b}{\mu_0}\right)^{-0.14} = 0.0039 \tag{13.2-25}$$

from which

$$\begin{aligned}\frac{L}{D} &= \frac{1}{4(0.0039)}\frac{(T_{b2} - T_{b1})}{(T_0 - T_b)_{\ln}}\left(\frac{\hat{C}_p\mu}{k}\right)_b^{2/3}\left(\frac{\mu_b}{\mu_0}\right)^{-0.14} \\ &= \frac{1}{4(0.0039)}\frac{(230 - 70)}{72.8}(0.698)^{2/3}\left(\frac{0.0490}{0.055}\right)^{-0.14} \\ &= 113\end{aligned} \tag{13.2-26}$$

Hence the required length is

$$L = 113D = (113)(\tfrac{2}{12}) = 19 \text{ ft} \tag{13.2-27}$$

If Re_b had been much smaller, it would have been necessary to estimate L/D before reading Fig. 13.2-1, thus initiating a trial-and-error process.

Note that in this problem we did not have to calculate h. Numerical evaluation of h is desirable, however, in more complicated problems such as heat exchange between two fluids with an intervening wall.

§13.3 HEAT-TRANSFER COEFFICIENTS FOR FORCED CONVECTION AROUND SUBMERGED OBJECTS

Another topic of industrial importance is the transfer of heat to or from an object around which a fluid is flowing. The object concerned may be relatively simple, such as a single cylinder or sphere, or it may be more complex, such as a "tube bundle" made up of a set of cylindrical tubes with a stream of gas or liquid flowing between them. We consider here only a few selected correlations for simple systems to introduce the reader to the subject; many additional correlations may be found in the references cited earlier.

In the following correlations the heat-transfer coefficient h_m is defined for the total surface of the submerged object, as in Eq. 13.1–6. The correlations are for a uniform surface temperature T_0; we use the subscript ∞ for conditions in the approaching stream and f for properties at the so-called "film temperature" $T_f = (T_0 + T_\infty)/2$.

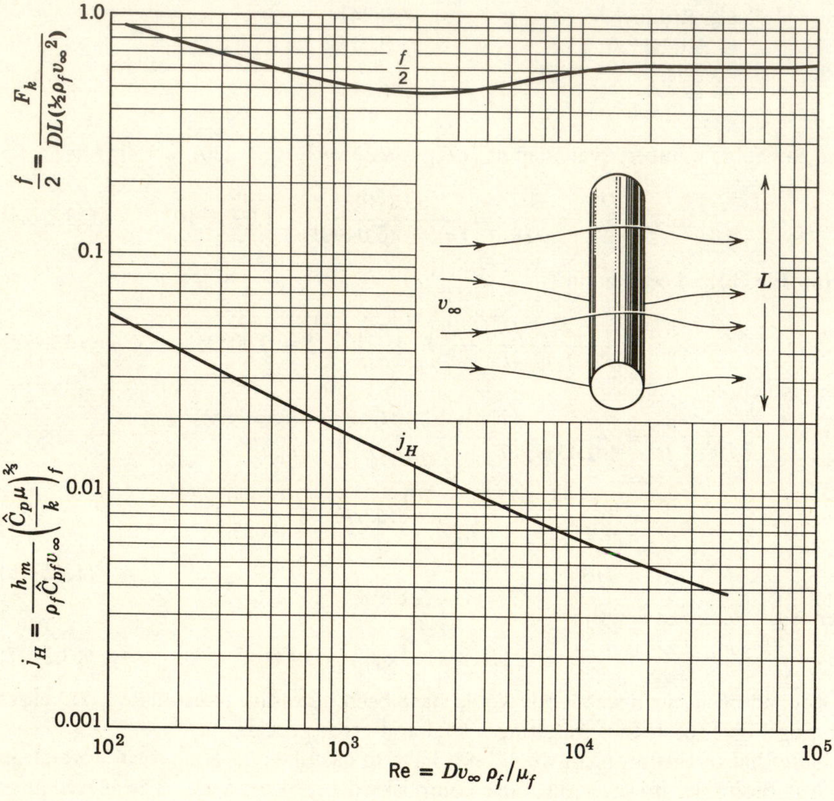

Fig. 13.3–1. Heat and momentum transfer between a long cylinder and a transverse stream. [j_H from T. K. Sherwood and R. L. Pigford, *Absorption and Extraction*, McGraw-Hill, New York (1952), Second Edition, p. 70; $f/2$ from H. Schlichting, *Boundary-Layer Theory*, Pergamon, New York (1955), p. 16.]

In Fig. 13.3–1, $j_H = \text{Nu Re}^{-1} \text{Pr}^{-1/3}$ is plotted versus Re for a long *cylinder* with its axis perpendicular to an approaching unbounded stream of velocity v_∞. The physical properties of the fluid are evaluated at T_f, as recommended by Douglas and Churchill[1] from a study of data on flow of gases across cylinders with large temperature differences. Also shown in Fig. 13.3–1 is a plot of $f/2$ versus Re, to emphasize that $j_H < f/2$ for this flow system, as is

[1] W. J. M. Douglas and S. W. Churchill, *Chem. Eng. Prog. Symposium Series*, No. 18, **52**, 23–28 (1956).

usually the case in flows with curved streamlines; the discrepancy is mainly due to form drag (see §6.3), which has no counterpart in heat transfer. These results are useful in connection with hot wire anemometry (see Problem 13.J) and heat transfer to bundles of parallel tubes.

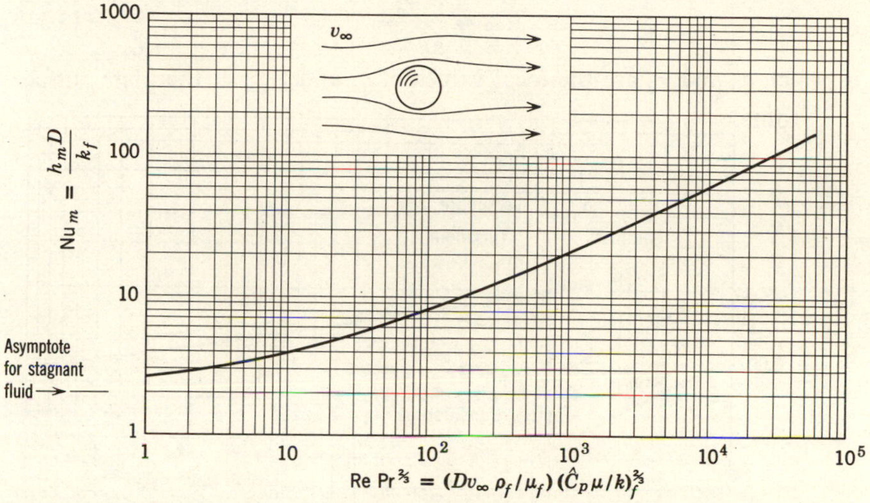

Fig. 13.3–2. Forced-convection heat transfer from a single sphere. [W. E. Ranz and W. R. Marshall, Jr., *Chem. Eng. Prog.* **48,** 141–146, 173–180 (1952).]

In Fig. 13.3–2, Nu is given as a function of Re and Pr for the total surface of a *sphere* in an infinite fluid. The relation plotted is[2]

$$\frac{h_m D}{k_f} = 2.0 + 0.60\left(\frac{Dv_\infty \rho_f}{\mu_f}\right)^{1/2}\left(\frac{\hat{C}_p \mu}{k}\right)_f^{1/3} \tag{13.3–1}$$

This equation predicts Nu = 2 for a motionless fluid; this asymptote is theoretically derived (see Problem 9.H) and is confirmed by experiments at low Reynolds and Grashof numbers (see §13.5). This result finds application in many processes that involve sprays of bubbles or droplets.

In Fig. 13.3–3 results are shown for tangential flow over an isothermal semi-infinite *flat plate* in an infinite fluid medium. The flow system and coordinates are shown in Fig. 11.4–1. In this flow system the Colburn analogy between heat transfer and fluid friction (see Eq. 13.2–18) holds very well because the quantity $p + \rho gh$ is essentially constant throughout the system and there is no form drag. The simple analogies that result when $p + \rho gh$

[2] W. E. Ranz and W. R. Marshall, Jr., *Chem. Eng. Prog.*, **48**, 141–146, 173–180 (1952). N. Frössling, *Gerlands Beitr. Geophys.*, **52**, 170 (1938), first gave a correlation of this form, with a coefficient of 0.552 instead of 0.60 in the last term. The properties are evaluated at T_f in Eq. 13.3–1 by analogy with the results for cylinders; this choice remains to be tested experimentally.

is a constant are discussed in §19.3. The coefficients h_{loc} and f_{loc} are defined as follows for any point on the plate surface:

$$h_{\text{loc}} = \frac{q_0}{T_0 - T_\infty} \tag{13.3-2}$$

$$f_{\text{loc}} = \frac{\tau_0}{\frac{1}{2}\rho_f v_\infty^{\,2}} \tag{13.3-3}$$

in which q_0 and τ_0 are the local values of q_y and $|\tau_{yx}|$ at the plate surface.

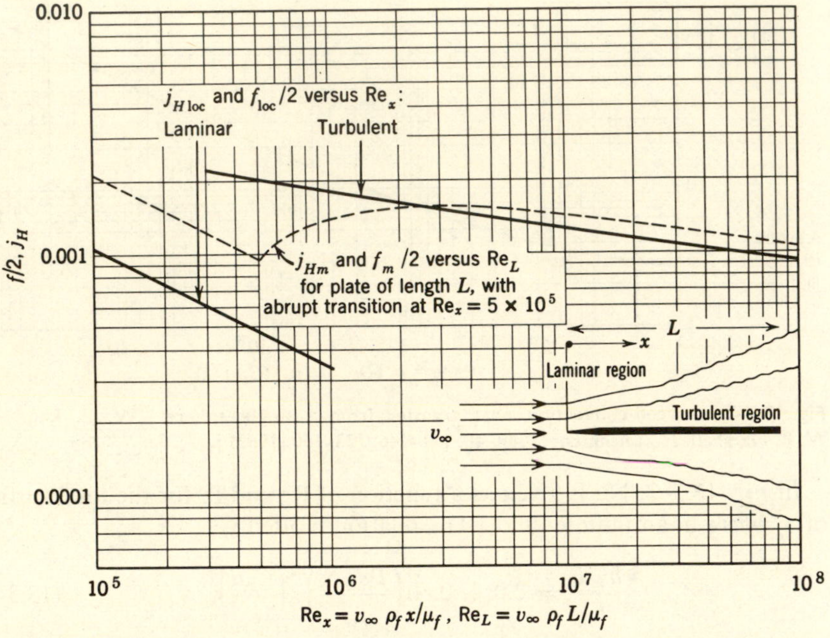

Fig. 13.3-3. Transfer coefficients for a smooth flat plate in tangential flow. [Adapted from H. Schlichting, *Boundary-Layer Theory*, Pergamon Press, New York (1955), pp. 438–439.]

In the laminar region which usually exists near the leading edge of the plate, the following theoretically derived expressions hold:

$$j_{H\text{loc}} \doteq \frac{f_{\text{loc}}}{2} \tag{13.3-4}$$

$$\frac{f_{\text{loc}}}{2} = 0.332\left(\frac{v_\infty \rho_f x}{\mu_f}\right)^{-\frac{1}{2}} \tag{13.3-5}$$

These equations were originally derived for fluids with constant physical properties;[3] in practice, they are usually written as shown in terms of the

[3] The result for $f/2$ in the case of constant fluid properties was derived by H. Blasius, *Z. Math. Phys.*, **56,** 1 (1908); the evaluation of j_H for constant fluid properties follows from the results of E. Pohlhausen, *Z. angew. Math. Mech.*, **1,** 115–121 (1921).

fluid properties at temperature T_f, a procedure that works very well for gases.[4] The analogy given in Eq. 13.3–4 is accurate within 2 per cent for $\text{Pr} > 0.6$ but becomes inaccurate at lower Prandtl numbers. For highly turbulent flow, the analogy still holds with fair accuracy, with f_{loc} given by the empirical curve in Fig. 13.3–3. The transition between laminar and turbulent flow resembles that shown for pipes in Fig. 13.2–1, but the limits of the transition region are harder to predict. For smooth, sharp-edged flat plates in an isothermal flow the transition usually begins at a *length* Reynolds number, $\text{Re}_x = v_\infty x \rho_f / \mu_f$, of 100,000 to 300,000 and is almost complete at a 50 per cent higher Reynolds number.

In general, the prediction of total heat-transfer rates from submerged objects becomes difficult for $\text{Re} > 10^5$ because of the uncertainty in the location of the transition region. Extrapolation of the correlations for spheres and cylinders into this region should be avoided because of the abrupt change in flow behavior at $\text{Re} \doteq 10^5$. (See Fig. 6.3–1.)

§13.4 HEAT-TRANSFER COEFFICIENTS FOR FORCED CONVECTION THROUGH PACKED BEDS

Heat-transfer coefficients between solid and fluid in a packed bed are generally defined as "local" values representative of a cross section through the bed by the following modification of Eq. 13.1–5:

$$dQ = h_{\text{loc}}(aS\,dz)(T_0 - T_b) \tag{13.4-1}$$

in which $S\,dz$ is the bed volume (solid plus fluid) between two cross sections a distance dz apart in the flow direction, and a is the solid particle surface area per unit bed volume.

A large amount of experimental information on heat and mass transfer in packed beds has been analyzed to arrive at the following empirical correlation:[1]

$$j_H = 0.91\ \text{Re}^{-0.51}\,\psi \qquad (\text{Re} < 50) \tag{13.4-2}$$

$$j_H = 0.61\ \text{Re}^{-0.41}\,\psi \qquad (\text{Re} > 50) \tag{13.4-3}$$

Here the Colburn j_H factor and the Reynolds number are defined by

$$j_H = \frac{h_{\text{loc}}}{\hat{C}_{pb}G_0}\left(\frac{\hat{C}_p\mu}{k}\right)_f^{2/3} \tag{13.4-4}$$

$$\text{Re} = \frac{G_0}{a\mu_f\psi} \tag{13.4-5}$$

[4] E. R. G. Eckert, *Trans. ASME*, **56**, 1273–1283 (1956).

[1] F. Yoshida, D. Ramaswami, and O. A. Hougen, *A.I.Ch.E. Journal*, **8**, 5–11 (1962). See also O. A. Hougen, K. M. Watson, and R. A. Ragatz, *CPP Charts*, Wiley, New York (1960). To obtain the quantities h and j_H in these references, divide our values by ψ.

In these equations the subscript f denotes properties evaluated at the "film temperature" $T_f = \frac{1}{2}(T_0 + T_b)$, and $G_0 = w/S$ is the superficial mass velocity. (See §6.4.) The quantity ψ is an empirical coefficient that depends on the particle shape. A few sample values of ψ are given in Table 13.4–1. This

TABLE 13.4–1
PARTICLE SHAPE FACTORS FOR PACKED-BED CORRELATIONS[a]

Particle Shape	ψ
Spheres	1.00
Cylinders	0.91
Flakes	0.86
Raschig rings	0.79
Partition rings	0.67
Berl saddles	0.80

[a] B. Gamson, *Chem. Engrg. Progr.*, **47**, 19–28 (1951).

correlation is useful in the design of fixed-bed systems such as driers, catalytic reactors, and pebble-bed heat reservoirs.

§13.5 HEAT-TRANSFER COEFFICIENTS FOR FREE CONVECTION

In §13.2 at the beginning of the discussion on forced convection we determined the functional dependence of Nu by dimensional analysis of the equations of change and the boundary conditions for the problem at hand. For free convection we can make a similar analysis, except that the equation of motion has to be written for variable ρ, and this gives rise to a "buoyant force" term. (See §§10.3, 10.6, and Example 10.5–4.) By an analysis such as that leading up to Eq. 13.2–10, one finds that the Nusselt number for free-convection heat transfer for an object submerged in an infinite fluid is of the form

$$\text{Nu}_m = \text{Nu}(\text{Gr}, \text{Pr}) \tag{13.5-1}$$

in which Nu_m is based on the heat-transfer coefficient h_m for the total surface of the submerged object. There may, of course, be additional dimensionless ratios describing the geometry of the system.

The following correlations are useful for estimating heat losses from various structures and engineering equipment. In each case it is assumed that the temperature is constant at T_0 on the surface and at T_∞ far from the surface; the characteristic temperature difference ΔT is $|T_0 - T_\infty|$. Fluid properties are to be evaluated at $(T_0 + T_\infty)/2$. The quantity β, which appears in the Grashof number, is just $1/T_f$ for ideal gases.

For a *single sphere* of diameter D in a large body of fluid the semiempirical relation[1]

$$\frac{h_m D}{k} = 2 + 0.60\left(\frac{D^3\rho^2 g\beta\,\Delta T}{\mu^2}\right)^{1/4}\left(\frac{\hat{C}_p\mu}{k}\right)^{1/3} \qquad (13.5\text{–}2)$$

agrees well with available data for $Gr^{1/4}\,Pr^{1/3} < 200$. Note that this equation gives $Nu_m = 2$ when the fluid is motionless, as does Eq. 13.3–1, in which free convection is neglected.

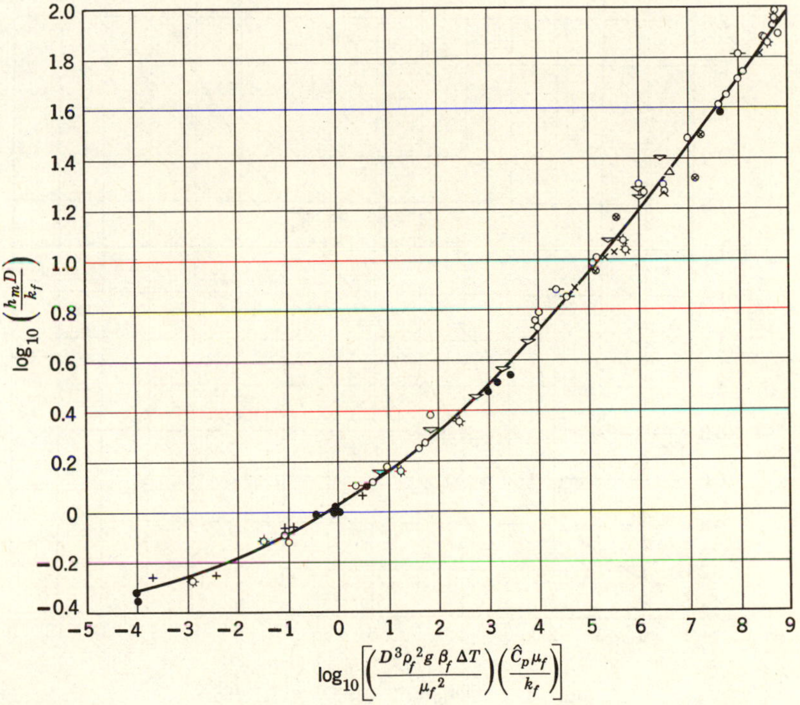

Fig. 13.5–1. Free convection from long horizontal cylinders to various fluids. [W. H. McAdams, *Heat Transmission*, McGraw-Hill, New York (1954), Third Edition, p. 176.]

For a long *horizontal cylinder* in an infinite fluid, Fig. 13.5–1 is recommended.[2] For Gr Pr > 10^4, this plot is closely represented by the equation

$$Nu_m = 0.525(Gr\,Pr)^{1/4} \qquad (13.5\text{–}3)$$

This correlation is based on extensive data for gases and liquids with Pr > 0.6.

[1] W. E. Ranz and W. R. Marshall, Jr., *Chem. Eng. Prog.*, **48**, 141–146, 173–180 (1952).
[2] W. H. McAdams, *Heat Transmission*, McGraw-Hill, New York (1954), Third Edition, pp. 172–176.

Experiments with thin, heated *vertical plates* of various heights, L, suspended in air are correlated[2] in Fig. 13.5–2. Note that the Nusselt and Grashof numbers here are based on the characteristic length L. The linear portion of the curve is closely approximated by the equation

$$\text{Nu}_m = 0.59(\text{Gr Pr})^{1/4} \qquad 10^4 < \text{Gr Pr} < 10^9 \tag{13.5–4}$$

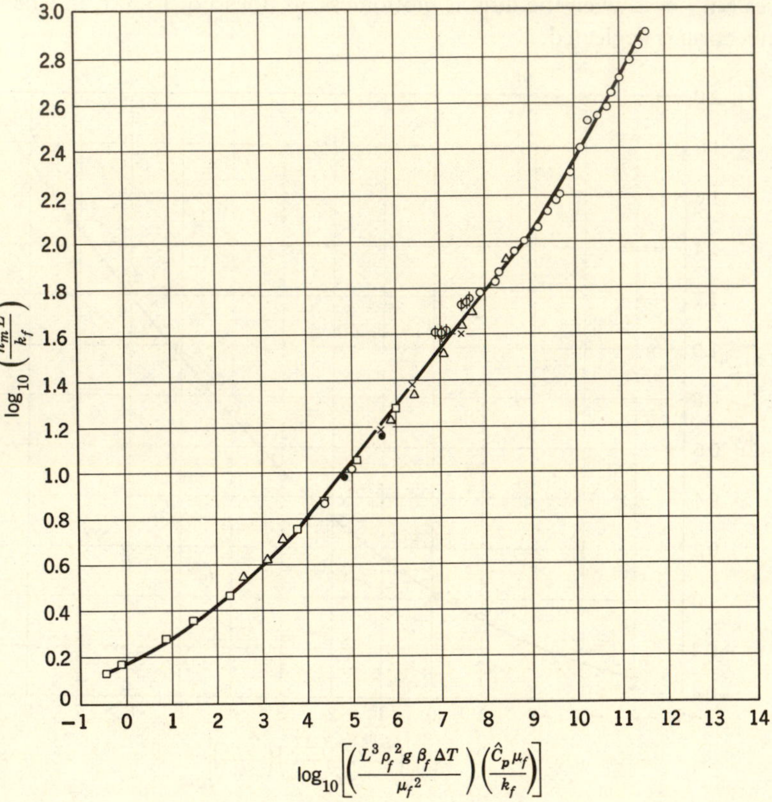

Fig. 13.5–2. Free convection from vertical plates, based on data for air. [W. H. McAdams, *Heat Transmission*, McGraw-Hill, New York (1954), Third Edition, p. 173.]

which is in close agreement with the results of boundary-layer theory cited in Example 10.5–4. For Gr Pr > 10^9, the flow is turbulent.

Example 13.5–1. Heat Loss by Free Convection from a Horizontal Pipe

Estimate the rate of heat loss by free convection from a unit length of a long horizontal pipe, 6 in. in outside diameter, if the outer surface temperature is 100° F and the surrounding air is at 1 atm and 80° F.

Solution. The properties of air at 1 atm and a film temperature of 90° F (550° R) are

$$\mu = 0.0190 \text{ cp} = 0.0460 \text{ lb}_m \text{ ft}^{-1} \text{ hr}^{-1}$$
$$\rho = 0.0723 \text{ lb}_m \text{ ft}^{-3}$$
$$\hat{C}_p = 0.241 \text{ Btu lb}_m^{-1} \text{ °R}^{-1}$$
$$k = 0.0152 \text{ Btu hr}^{-1} \text{ ft}^{-1} \text{ °R}^{-1}$$
$$\beta = 1/T_f = \tfrac{1}{550} \text{ °R}^{-1}$$

Other pertinent values are $D = 0.5$ ft, $\Delta T = 20°$ R, and $g = 4.17 \times 10^8$ ft hr^{-2}. From these data we obtain

$$\text{Gr Pr} = \left(\frac{(0.5)^3(0.0723)^2(4.17 \times 10^8)(\frac{20}{550})}{0.0460}\right)\left(\frac{0.241}{0.0152}\right)$$
$$= 3.4 \times 10^6 \tag{13.5-5}$$

From Fig. 13.5-1 or Eq. 13.5-3 we obtain $\text{Nu}_m = 22.5$; hence

$$h_m = \text{Nu}_m \frac{k}{D} = \frac{(22.5)(0.0152)}{0.5} = 0.68 \text{ Btu hr}^{-1} \text{ ft}^{-2} \text{ °F}^{-1} \tag{13.5-6}$$

The rate of heat loss per unit length of pipe is then

$$\frac{Q}{L} = h_m A \frac{\Delta T}{L} = h_m \pi D \, \Delta T$$
$$= (0.68)(\pi)(0.5)(20) = 21 \text{ Btu hr}^{-1} \text{ ft}^{-1} \tag{13.5-7}$$

This is the heat loss by convection only; radiation is considered in Example 14.5-2.

§13.6 HEAT-TRANSFER COEFFICIENTS FOR CONDENSATION OF PURE VAPORS ON SOLID SURFACES

Condensation of a pure vapor on a solid surface is a particularly complicated heat-transfer process because it involves two flowing fluid phases: the vapor and the condensate. Condensation occurs industrially in many types of equipment; for simplicity, we consider here only the common cases of condensation of a slowly moving vapor on the outside of horizontal or vertical tubes or vertical flat walls.

The condensation process on a vertical wall is illustrated schematically in Fig. 13.6-1. Vapor flows over the condensing surface and is moved toward it by the small pressure gradient near the liquid surface.[1] Some of the molecules from the vapor phase strike the liquid surface and bounce off; others

[1] Note that there occur small but abrupt changes in pressure and temperature at the interface. These discontinuities are essential to the condensation process but are generally of negligible magnitude in engineering calculations for pure fluids. For mixtures, they may be more important. See R. W. Schrage, *Interphase Mass Transfer*, Columbia University Press (1953).

penetrate the surface and give off their latent heat of condensation. The heat thus released must then flow through the condensate to the wall, thence to the coolant on the other side of the wall. At the same time, the condensate must drain from the surface by gravity flow.

The condensate on the wall is normally the sole important resistance to heat transfer on the condensing side of the wall. If the solid surface is clean,

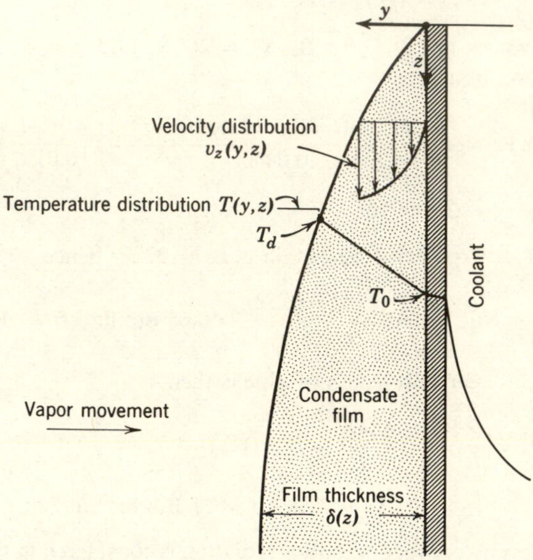

Fig. 13.6–1. Film condensation on a vertical surface (interfacial temperature discontinuity exaggerated).

the condensate will usually form a continuous film over the surface, but if traces of certain impurities are present, *e.g.*, fatty acids in a steam condenser, the condensate will form in droplets. "Dropwise condensation"[2] gives much higher rates of heat transfer than "film condensation," but is difficult to maintain, so that it is common practice to assume film condensation in condenser design. The correlations which follow apply only to film condensation.

The usual definition of h_m for condensation of a pure vapor on a solid surface of area A and constant temperature T_0 is

$$Q = h_m A(T_d - T_0) \tag{13.6–1}$$

in which Q is the rate of heat flow into the solid surface and T_d is the *dew point* of the vapor approaching the tube surface, that is, the temperature at which the vapor would condense if cooled slowly at the prevailing pressure.

[2] Dropwise condensation is discussed at length by McAdams, *Heat Transmission*, McGraw-Hill, New York (1954), Third Edition, pp. 347–351, and by H. Gröber, S. Erk, and U. Grigull, *Wärmeübertragung*, Springer, Berlin (1955), Third Edition, pp. 303–310.

This temperature is essentially that of the liquid at the liquid-gas interface; hence h_m may be regarded as a heat-transfer coefficient for the liquid film.

Expressions for h_m have been derived[3] for *laminar nonrippling* condensate flow by approximate solution of the equations of energy and motion for a falling liquid film. (See Problem 13.N.) For film condensation on a *horizontal tube* of outside diameter D, length L, and constant surface temperature T_0, the result of Nusselt may be written

$$h_m = 0.954\left(\frac{k_f^3\rho_f^2 gL}{\mu_f w}\right)^{1/3} \qquad (13.6\text{–}2)$$

in which w/L is the mass rate of condensation per unit length of tube. For moderate temperature differences, Eq. 13.6–2 may be rewritten with the aid of an energy balance on the condensate film to give

$$h_m = 0.725\left(\frac{k_f^3\rho_f^2 g\,\Delta\hat{H}_{\text{vap}}}{\mu_f D(T_d - T_0)}\right)^{1/4} \qquad (13.6\text{–}3)$$

Equations 13.6–2 and 3 have been experimentally confirmed within ± 10 per cent for single horizontal tubes. They also seem to give satisfactory results for bundles of horizontal tubes,[4] in spite of the complications introduced by condensate dripping from tube to tube.

For film condensation on *vertical tubes* or *vertical walls*, the theoretical results corresponding to Eqs. 13.6–2 and 3 are[3]

$$h_m = \frac{4}{3}\left(\frac{k_f^3\rho_f^2 g}{3\mu_f\Gamma}\right)^{1/3} \qquad (13.6\text{–}4)$$

and

$$h_m = \tfrac{2}{3}\sqrt{2}\left(\frac{k_f^3\rho_f^2 g\,\Delta\hat{H}_{\text{vap}}}{\mu_f L(T_d - T_0)}\right)^{1/4} \qquad (13.6\text{–}5)$$

The quantity Γ in Eq. 13.6–4 is the total rate of condensate flow from the bottom of the condensing surface per unit width of that surface; for a vertical tube, $\Gamma = w/\pi D$, where w is the total mass rate of condensation on the tube. For *short* vertical tubes ($L < 0.5$ ft), the experimental values of h_m confirm the theory well, but the measured values for *long* vertical tubes ($L > 8$ ft) may exceed the theory for given ΔT by as much as 70 per cent; the discrepancy is attributed to ripples that attain greatest amplitude on long vertical tubes.[5]

We now turn our attention to the empirical expressions for *turbulent* condensate flow. Turbulent flow begins, on *vertical tubes or walls*, at a

[3] W. Nusselt, *Z. Ver. dtsch. Ing.*, **60**, 541–546, 569–575 (1916).

[4] B. E. Short and H. E. Brown, *Proc. General Disc. Heat Transfer*, London, pp. 27–31 (1951).

[5] W. H. McAdams, *loc. cit.*, p. 333.

Reynolds number $\text{Re} = \Gamma/\mu_f$ of about 350; for higher Reynolds numbers the following empirical formula has been proposed:[6]

$$h_m = 0.003\left(\frac{k_f^3\rho_f^2 g(T_d - T_0)L}{\mu_f^3\,\Delta\hat{H}_{\text{vap}}}\right)^{1/2} \qquad (13.6\text{–}6)$$

Equation 13.6–6 is equivalent, for small ΔT, to the formula

$$h_m = 0.021\left(\frac{k_f^3\rho_f^2 g\Gamma}{\mu_f^3}\right)^{1/3} \qquad (13.6\text{–}7)$$

For convenience in calculation, and to show the degree of agreement with data, Eqs. 13.6–4, 5, 6, and 7 are summarized in Fig. 13.6–2. Somewhat better agreement could have been obtained by using a family of lines in the turbulent range to represent the effect of Prandtl number; however, in view of the scattering of data, a single line is adequate.

Turbulent condensate flow is very difficult to obtain on horizontal tubes, unless very large diameters or high temperature differences are employed. Equations 13.6–2 and 3 are believed satisfactory up to the estimated transition Reynolds number, $\text{Re} = w_T/L\mu_f$, of about 1000, where w_T is the *total* condensate flow leaving a given tube, including condensate from the tubes above.[7]

The inverse process of vaporization of a pure fluid is considerably more complicated than condensation. We do not attempt to discuss heat transfer to boiling liquids here, but refer the reader to some reviews on this fast-developing field of research.[8,9]

Example 13.6–1. Condensation of Steam on a Vertical Surface

A boiling liquid flowing in a vertical tube is being heated by condensation of steam on the outside of the tube. The steam-heated tube section is 10 ft high and 2 in. in outside diameter. If saturated steam is used, what steam temperature is required to supply 92,000 Btu hr^{-1} of heat to the tube at a tube-surface temperature of 200° F? Assume film condensation.

Solution The fluid properties depend on the unknown temperature T_d. Assuming $T_d = 200°$ F, one obtains

$$\begin{aligned}\Delta\hat{H}_{\text{vap}} &= 978 \text{ Btu lb}_m^{-1}\\ k_f &= 0.393 \text{ Btu hr}^{-1}\text{ ft}^{-1}\text{ °F}^{-1}\\ \rho_f &= 60.1 \text{ lb}_m\text{ ft}^{-3}\\ \mu_f &= 0.738 \text{ lb}_m\text{ ft}^{-1}\text{ hr}^{-1}\end{aligned}$$

[6] U. Grigull, *Forsch. IngWes.*, **13**, 49–57 (1942); *Z. Ver. dtsch. Ing.*, **86**, 444–445 (1942).
[7] W. H. McAdams, *loc. cit.*, p. 338–339.
[8] W. H. McAdams, *loc. cit.*, Chapter 14.
[9] J. J. Westwater, *Advances in Chemical Engineering*, Academic Press, New York, Vol. I, Chapter I (1956); Vol. II, Chapter I (1958).

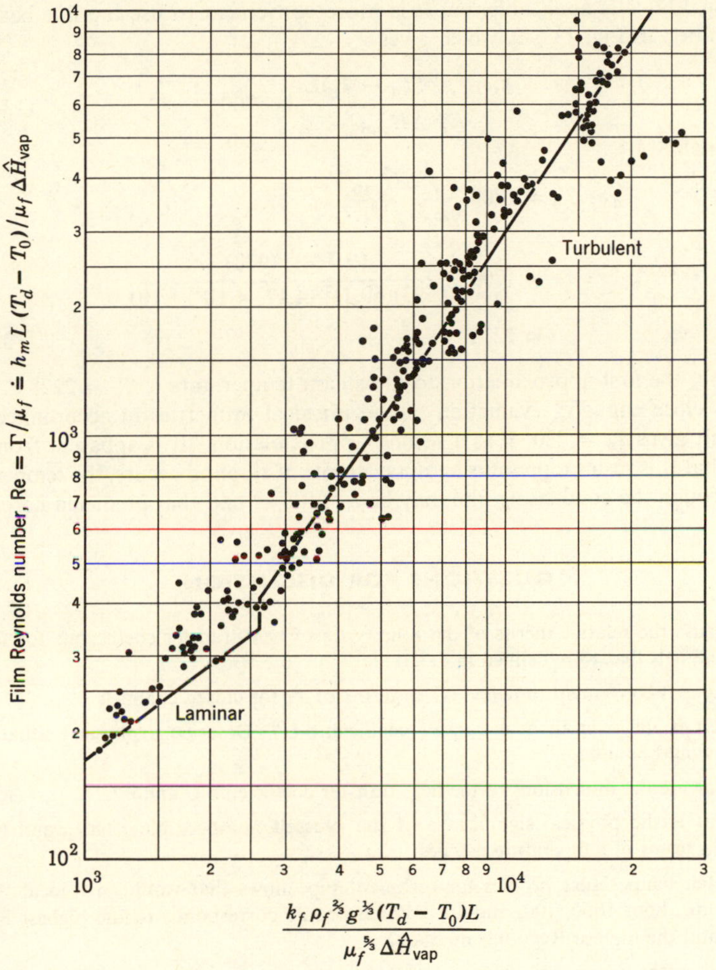

Fig. 13.6–2. Correlation of heat-transfer data for film condensation of pure vapors on vertical surfaces. [U. Grigull, *Die Grundgesetze der Wärmeübertragung*, Springer, Berlin (1955), Third Edition, p. 296.]

Assuming that the steam gives up only latent heat (the assumption $T_d = T_0 = 200°$ F implies this), an energy balance around the tube gives

$$Q = w\,\Delta\hat{H}_{\text{vap}} = \pi D\Gamma\,\Delta\hat{H}_{\text{vap}} \tag{13.6–8}$$

in which Q is the heat flow into the tube wall. The film Reynolds number is

$$\frac{\Gamma}{\mu} = \frac{Q}{\pi D\mu\,\Delta\hat{H}_{\text{vap}}} = \frac{92{,}000}{\pi(\frac{2}{12})(0.738)(978)} = 244 \tag{13.6–9}$$

Reading Fig. 13.6–2 at this value of the ordinate, we find that the flow is laminar. Equation 13.6–4 is applicable, but it is more convenient to use the line based on this equation in Fig. 13.6–2, which gives

$$\frac{k_f \rho_f^{2/3} g^{1/3} (T_d - T_0) L}{\mu_f^{5/3} \Delta H_{\text{vap}}} = 1700 \tag{13.6–10}$$

from which

$$\begin{aligned} T_d - T_0 &= 1700 \frac{\mu_f^{5/3} \Delta \hat{H}_{\text{vap}}}{k_f \rho_f^{2/3} g^{1/3} L} \\ &= 1700 \frac{(0.738)^{5/3}(978)}{(0.393)(60.1)^{2/3}(4.17 \times 10^8)^{1/3}(10)} \\ &= 22^\circ \text{ F} \end{aligned} \tag{13.6–11}$$

Therefore, the first approximation to the steam temperature is $T_d = 222^\circ$ F. This result is close enough; evaluation of the physical properties in accordance with this result gives $T_d = 220^\circ$ F as a second approximation. It is apparent from Fig. 13.6–2 that this result represents an upper limit; if rippling occurs, the temperature drop through the condensate film may be as little as half that predicted here.

QUESTIONS FOR DISCUSSION

1. Discuss the relative merits of the four types of heat-transfer coefficients for flow in closed channels that were defined in §13.1.

2. Why is it convenient to define ΔT in terms of T_b for closed channels?

3. What is the over-all heat-transfer coefficient U? Describe a physical situation in which it would be used.

4. What are the dimensions of the heat-transfer coefficients U and h?

5. What is the physical significance of the Nusselt numbers Nu_1, Nu_{ln}, and Nu_{loc}? Explain in terms of temperature profiles.

6. Sketch temperature profiles for turbulent pipe flows that would give local Nusselt numbers of about 1000, 100, and 10. Which profile corresponds to the highest Prandtl number and the highest Reynolds number?

7. What is the physical significance of the Stanton number, $\text{St} = \text{Nu}/\text{Re Pr}$, for heating a fluid in a pipe section of length L and diameter D? Interpret the significance of the local Stanton number in this situation.

8. In Fig. 13.2–3 it is clear that the Nusselt number decreases for increasing values of L/D. Explain this behavior in terms of the physical processes occurring near the entrance of the heat-exchange section.

9. Name some treatises that should be consulted for detailed information and literature references on heat-transfer coefficient correlations.

10. How does dimensional analysis help in the planning and analysis of heat-transfer experiments?

11. What dimensionless groups may arise in problems of forced convection heat transfer, in free convection, and in simultaneous forced and free convection? Give an example of the latter situation.

12. What additional dimensionless group or groups would one expect to arise in heat-transfer correlations for high-speed liquid flow and high-speed compressible gas flow?

13. Most empirical correlations of heat-transfer coefficients make no allowance for the dependence of the local heat flux on the surface-temperature distribution. Give an example from this chapter to illustrate the quantitative effect of different thermal boundary conditions on the local heat-transfer coefficient.

14. What is the Colburn j_H factor? Under what conditions does a useful relation exist between j_H and f? Is this relation empirical or theoretically justified?

15. Why is the use of j_H not convenient for forced convection around a sphere? Illustrate by sketching a plot of j_H versus Reynolds number.

16. Describe two methods that have been used in heat-transfer correlations to allow for the temperature dependence of the physical properties.

17. What do free convection and condensation have in common?

18. Show how Eq. 13.6–5 may be obtained from Eq. 13.6–4.

19. In what position (horizontal or vertical) should a condenser tube of outside diameter 1 in. and length 100 in. be placed to condense the maximum amount of steam for a given surface temperature? What are the relative amounts condensed in the two positions?

20. For what fluids can the Sieder and Tate plot and the Deissler plot *not* be used? What correlations are available for such fluids?

21. Show that the three ordinate quantities in Fig. 13.2–1 are equivalent.

PROBLEMS

13.A$_1$ Average Heat-Transfer Coefficients

Ten thousand pounds per hour of an oil with a specific heat of 0.6 are being heated from 100 to 200° F in a simple heat exchanger, as shown in Fig. 13.A. The oil is flowing

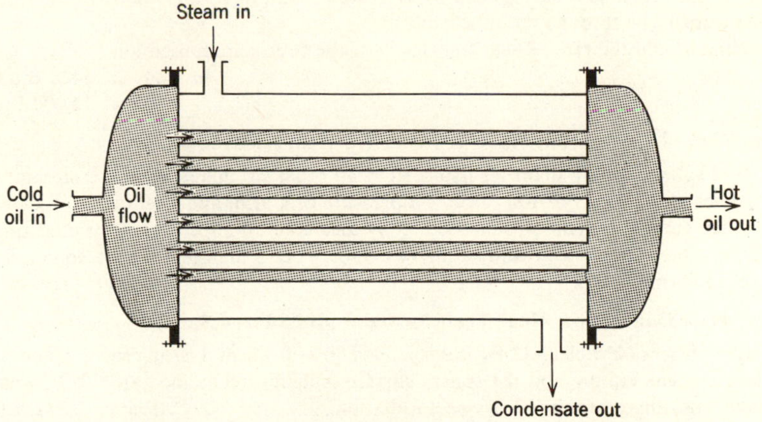

Fig. 13.A. A single-pass "shell-and-tube" heat exchanger.

through the tubes, which are copper, 1 in. in outside diameter, with 0.065-in. walls. The combined length of the tubes is 300 ft. The required heat is supplied by condensation of saturated steam at 15.0 psia on the outside of the tubes.

Calculate h_1, h_a, and h_{ln} for the oil, assuming that the inside surfaces of the tubes are at the saturation temperature of the steam, 213° F.

Answers: 78, 139, 190 Btu hr^{-1} ft^{-2} ° F^{-1}

13.B$_1$ Heat Transfer in Laminar Tube Flow

One hundred pounds per hour of oil at 100° F are flowing through a 1-in. i.d. copper tube, 20 ft long. The inside surface of the tube is maintained at 215° F by condensing steam on the outside surface. Fully developed flow may be assumed throughout the length of the tube, and the physical properties of the oil may be considered constant at the following values: $\hat{C}_p = 0.49$ Btu lb$_m^{-1}$ (° F)$^{-1}$, $\rho = 55$ lb$_m$ ft^{-3}, $\mu = 1.42$ lb$_m$ hr^{-1} ft^{-1}, $k = 0.0825$ Btu hr^{-1} ft^{-1} (° F)$^{-1}$.

a. Calculate Re.
b. Calculate Pr.
c. Calculate the exit temperature of the oil.

Answers: *a.* 1075
b. 8.44
c. 155° F

13.C$_1$ Effect of Flow Rate on Exit Temperature from a Heat Exchanger

a. Repeat parts (*a*) and (*c*) of Problem 13.B for oil flow rates of 200, 400, 800, 1600, and 3200 lb hr^{-1}.

b. Calculate the total heat flow through the tube wall for each of the foregoing oil flow rates.

13.D$_1$ Local Heat-Transfer Coefficient for Turbulent Forced Convection in a Tube

Water is flowing in a 2-in. i.d. tube at a mass flow rate $w = 15{,}000$ lb$_m$ hr^{-1}. The inner wall temperature at some point along the tube is 160° F, and the bulk fluid temperature at that point is 60° F.

What is the local heat flux q_r at the pipe wall? Assume h_{loc} has attained a constant asymptotic value.

Answer: -6.25×10^4 Btu hr^{-1} ft^{-2}

13.E$_1$ Heat Transfer from Condensing Vapors

a. The outer surface of a vertical tube 1 in. in outside diameter and 1 ft long is maintained at 190° F. If this tube is surrounded by saturated steam at 1 atm, what will the total rate of heat transfer be through the tube wall?

b. What would the rate of heat transfer be if the tube were horizontal?

Answers: *a.* 8400 Btu hr^{-1}
b. 12,000 Btu hr^{-1}

13.F$_1$ Forced-Convection Heat Transfer from an Isolated Sphere

A solid sphere 1 in. in diameter is placed in an otherwise undisturbed air stream, which approaches at a velocity of 100 ft sec^{-1}, a pressure of 1 atm, and a temperature of 100° F. The sphere surface is maintained at 200° F by means of an imbedded electric heating coil. What must the rate of electrical heating in cal sec^{-1} be to maintain the stated conditions? Neglect radiation.

Answer: 3.35 cal sec^{-1}

13.G$_1$ Free-Convection Heat Transfer from an Isolated Sphere

If the sphere of Problem 13.F$_1$ is suspended in still air at 1 atm pressure and 100° F ambient air temperature and the sphere surface is again maintained at 200° F, what rate of electric heating is required? Neglect radiation.

Answer: 0.332 cal sec^{-1}

13.H$_2$ Limiting Local Nusselt Number for Laminar Pipe Flow with Constant Heat Flux

Equation 9.8–31 gives the asymptotic temperature distribution for heating a fluid of constant ρ, μ, $\hat{C}_p$, and k in a long tube with constant heat flux at the wall. Use this temperature profile to show that the limiting Nusselt number for these conditions is $\frac{48}{11}$.

13.I$_2$ Local Over-all Heat-Transfer Coefficients

In Problem 13.A the thermal resistances of the condensed steam film and tube wall were neglected. Justify this neglect by calculating the actual inner-surface temperature of the tubes at that cross-section in the exchanger at which the oil temperature is 150° F. You may assume that for the oil h_{loc} is constant throughout the exchanger at 190 Btu hr^{-1} ft^{-2} ° F^{-1}. The tubes are horizontal.

13.J$_2$ The Hot-Wire Anemometer[1]

A hot-wire anemometer is essentially a fine wire, usually made of platinum, which is heated electrically and exposed to a flowing fluid. Its temperature, which is a function of the fluid temperature, fluid velocity, and the rate of heating, may be determined by measuring its electrical resistance.

a. A straight cylindrical wire 0.5 in. long and 0.01 in. in diameter is exposed to a stream of air at 70° F flowing past the wire at 100 ft sec^{-1}. What must the rate of energy input be in watts to maintain the wire surface at 600° F? Neglect radiation and heat conduction along the wire.

b. It has been reported[2] that for a given fluid and wire at given fluid and wire temperatures (hence a given wire resistance)

$$i^2 = B\sqrt{v_\infty} + C \tag{13.J-1}$$

in which i is the current required to maintain the desired temperature, and v_∞ is the velocity of the approaching fluid. How well does this equation agree with the predictions of Fig. 13.3-1 for the fluid and wire of (*a*) over a fluid velocity range of 100 to 300 ft sec^{-1}? What is the significance of the constant C in Eq. 13.J-1?

13.K$_2$ Extension of Dimensional Analysis

Consider the flow system described in the first paragraph of §13.2, for which dimensional analysis has already given the dimensionless velocity profile (Eq. 6.2-7) and temperature profile (Eq. 13.2-9).

a. Use Eqs. 6.2-7 and 13.2-9 and the definition of cup-mixing temperature to show that

$$\frac{T_{b2} - T_{b1}}{T_0 - T_{b1}} = \text{a function of Re, Pr, } L/D \tag{13.K-1}$$

b. Use the result just obtained and the definitions of the heat transfer coefficients to derive Eqs. 13.2-12, 13, and 14, neglecting axial heat conduction.

13.L$_2$ Relation between h_{loc} and h_{ln}

In many industrial tubular heat exchangers (see Example 15.4-2) the tube-surface temperature T_0 varies linearly with the bulk fluid temperature T_b. For this common situation h_{loc} and h_{ln} may be simply related:

a. Starting with Eq. 13.1-5, show that

$$h_{loc}\pi D\, dz(T_b - T_0) = -\left(\frac{\pi}{4}\right) D^2\rho C_p\langle v\rangle\, dT_b \tag{13.L-1}$$

and therefore that

$$\int_0^L h_{loc}\, dz = \tfrac{1}{4}\rho C_p\langle v\rangle D \frac{T_b(L) - T_b(0)}{(T_0 - T_b)_{ln}} \tag{13.L-2}$$

[1] See, for example, M. Jakob, *Heat Transfer*, Vol. II, Wiley, New York (1957), pp. 196–198.

[2] L. V. King, *Phil. Trans. Roy. Soc.* (*London*), **A214**, 373 (1914).

b. Combine the foregoing result with Eq. 13.1–4 to show that

$$h_{\ln} = \int_0^1 h_{\text{loc}}\, d\frac{z}{L} \tag{13.L–3}$$

in which L is the total tube length, and therefore that (if $(\partial h_{\text{loc}}/\partial L)_z = 0$):

$$h_{\text{loc}}\big|_{z=L} = h_{\ln}\big|_{z=L} + \frac{dh_{\ln}}{d\ln L} \tag{13.L–4}$$

13.M$_2$ Heat Loss by Free Convection from a Pipe

In Example 13.5–1 would the heat loss be higher or lower if the pipe-surface temperature were 200° F and the air temperature were 180° F?

13.N$_3$ The Nusselt Expression for Condensing Vapor Heat-Transfer Coefficients

Consider a laminar film of condensate flowing down a vertical wall (see Fig. 13.6–1) and assume that this liquid film constitutes the sole heat-transfer resistance on the vapor side of the wall. Further assume (i) that the shear stress between liquid and vapor may be neglected, (ii) that physical properties in the film may be evaluated at the arithmetic mean of vapor and cooling-surface temperatures and that the cooling-surface temperature may be assumed constant, (iii) that acceleration of fluid elements in the film may be neglected in comparison with gravitational and viscous forces, (iv) that sensible heat changes in the condensate film are unimportant compared to the latent heat transferred through it, and (v) that the heat flux is very nearly normal to the wall surface.

a. Recall from §2.2 that the average velocity of a film of constant thickness δ is

$$\langle v_z \rangle = \frac{\rho g \delta^2}{3\mu} \tag{13.N–1}$$

Assume that this relation is valid for any value of z.

b. Write the energy equation for the film, neglecting film curvature and convection; show that the heat flux through the film toward the cold surface is

$$-q_y = k\left(\frac{T_d - T_0}{\delta}\right) \tag{13.N–2}$$

c. As the film proceeds down the wall, it picks up additional material by the condensation process. In this process heat is liberated to the extent of $\Delta\hat{H}_{\text{vap}}$ per unit mass of material that undergoes the change in state. Show that equating the heat liberation by condensation with the heat flowing through the film in a segment dz of the film leads to

$$\rho\, \Delta\hat{H}_{\text{vap}}\, d(\langle v \rangle \delta) = k\left(\frac{T_d - T_0}{\delta}\right) dz \tag{13.N–3}$$

d. Insert the expression for the average velocity from (*a*) into Eq. 13.N–3 and integrate from $z = 0$ to $z = z$ to obtain

$$\delta(z) = \left(\frac{4k(T_d - T_0)\mu z}{\rho^2 g\, \Delta\hat{H}_{\text{vap}}}\right)^{1/4} \tag{13.N–4}$$

e. Use the definition of the heat-transfer coefficient and the result in (*d*) to obtain Eq. 13.6–5.

f. Show that Eqs. 13.6–4 and 5 are equivalent for the conditions of the problem.

13.O$_3$ Heat-Transfer Correlations for Agitated Tanks

A liquid of essentially constant physical properties is being continuously heated by passage through an agitated tank, as shown in Fig. 13.O. Heat is supplied by condensation of steam on the outer wall of the tank. The thermal resistance of the condensate film and tank wall may be considered small compared to that of the fluid in the tank, and the unjacketed portion of the tank may be assumed to be well insulated. The rate of liquid flow through the tank has a negligible effect on the flow pattern in the tank.

Develop the general form of a dimensionless heat-transfer correlation for the tank corresponding to the correlation for pipe flow in §13.2. Choose as reference length, D,

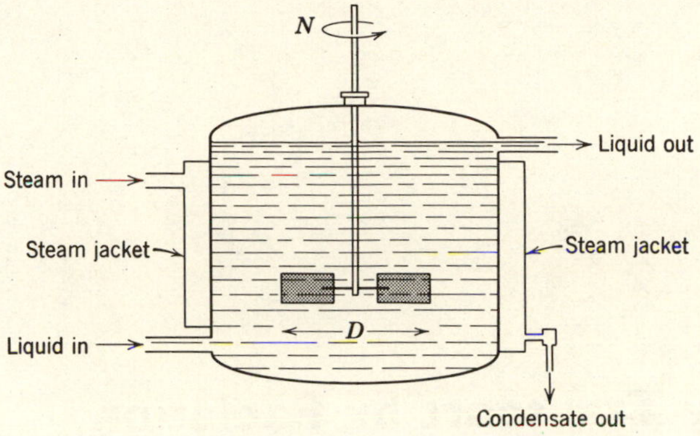

Fig. 13.O. Continuous heating of a liquid in an agitated tank.

the impeller diameter; as reference velocity, ND, in which N is the rate of shaft rotation in revolutions per unit time; and as reference pressure, $\rho(ND)^2$, in which ρ is the fluid density.

13.P_3 Analytical Calculation of Nusselt Numbers

Extend the results of the following examples and problems to obtain theoretical calculations for Nusselt numbers: Problem 9.P, Problem 9.R, Example 11.2–1, Example 11.2–2, Example 11.4–1, and Problem 12.B.

CHAPTER 14

Energy Transport by Radiation

In the preceding chapters the transport of energy by conduction and by convection has been discussed. Both modes of transport rely on the existence of a material medium. For heat conduction to occur, there must be temperature inequalities at neighboring points in the material medium. For heat convection to occur, there must be a fluid that is free to move and transport energy with it. In this chapter we turn our attention to a third mechanism for energy transport, namely, radiation. This is basically an electromagnetic mechanism, which allows energy to be transported with the speed of light through regions of space that are devoid of any matter. The rate of radiative energy transport between two "black" bodies in a vacuum is proportional to the difference of the fourth powers of their absolute temperatures. This mechanism is quantitatively very different from the three transport processes considered elsewhere in this book: momentum transport (in Newtonian fluids) proportional to a velocity gradient, energy transport (by conduction) proportional to a temperature gradient, and mass transport (by ordinary diffusion) proportional to a concentration gradient. Because of the uniqueness of radiation as a means of transport and because of the importance of radiation heat transfer in industrial calculations, we have devoted a separate chapter to an introduction to this subject.

A thorough understanding of the physics of radiative transport requires the use of several different disciplines:[1,2] electromagnetic theory is needed to describe the essentially wavelike nature of radiation, in particular the energy and pressure associated with electromagnetic waves; thermodynamics is useful for obtaining some relations between the "bulk properties" of an enclosure containing radiation; quantum mechanics is necessary in order to describe in detail the atomic and molecular processes that occur when radiation is produced within matter and when it is absorbed by matter; and statistical mechanics is needed in order to describe the way in which the energy of radiation is distributed over the wavelength spectrum. All we have attempted to do in the elementary discussion given here is to define certain key quantities and to set forth the important results of theory and experiment; we then show how some of these results can be used to compute the rate of heat transfer by radiant processes in simple systems.

In §14.1 we introduce the reader to some basic concepts and definitions. Then in §14.2 some of the principal physical results regarding black-body radiation are given. In the following section, §14.3, the rate of heat exchange between two black bodies is discussed; this section introduces no new physical principles, the basic problems being those of geometry. Next, §14.4 is devoted to an extension of the preceding section to nonblack surfaces. Finally, in the last section, there is a brief discussion of radiation processes in absorbing media.

§14.1 THE SPECTRUM OF ELECTROMAGNETIC RADIATION

When a solid body is heated, for example by an electric coil, the surface of the solid emits radiation of wavelength primarily in the range 0.1 to 10 microns. Such radiation is usually referred to as *thermal radiation*. A quantitative description of the atomic and molecular mechanisms by which the radiation is produced is given by quantum mechanics, and is hence outside the scope of this discussion. A qualitative description may, however, be given as follows: When energy is supplied to a solid body, some of the constituent molecules and atoms are raised to "excited states." There is a tendency for the atoms or molecules to return spontaneously to lower energy states. When this occurs, energy is emitted in the form of electromagnetic radiation. Because the emitted radiation may result from changes in the electronic, vibrational, and rotational states of the atoms and molecules, the radiation will be distributed over a range of wavelengths.

Actually, thermal radiation represents just a small part of the total spectrum of electromagnetic radiation. In Fig. 14.1-1 a diagram, which represents the various kinds of radiation, is shown with a rough indication of the kinds of

[1] M. Planck, *Theory of Heat*, Macmillan, London (1932), Parts III and IV.

[2] W. Heitler, *Quantum Theory of Radiation*, Oxford University Press (1944), Second Edition.

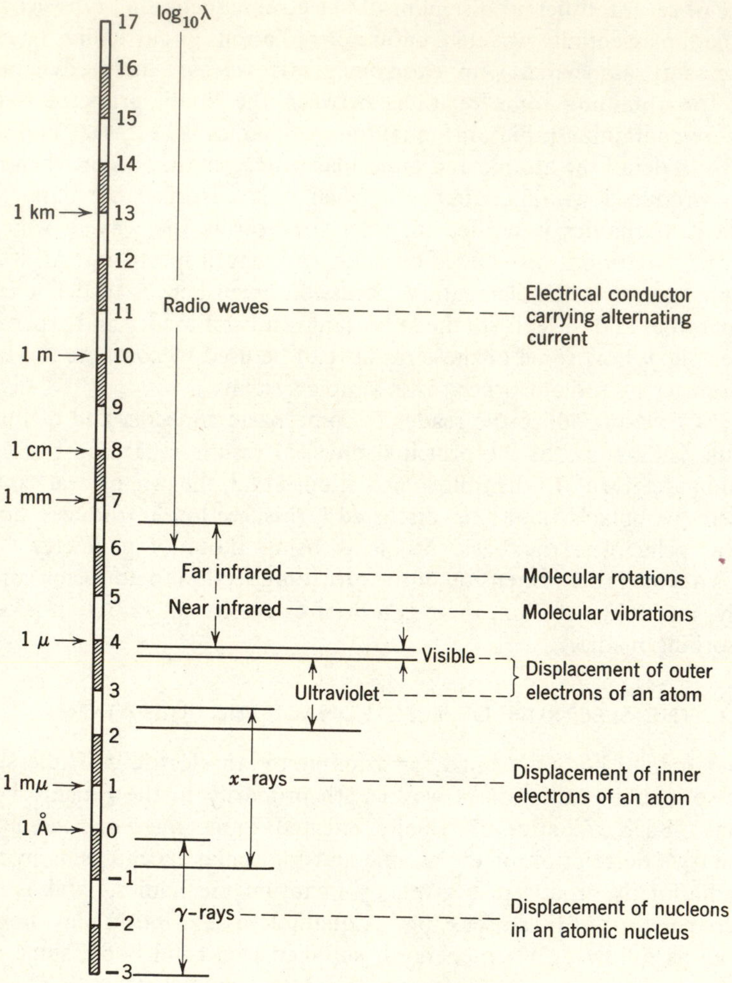

Fig. 14.1-1. The spectrum of electromagnetic radiation, showing roughly the mechanisms by which various types of radiation are produced: 1 Å = 1 Ångstrom unit = 10^{-8} cm; 1 μ = 1 micron = 10^{-4} cm.

mechanisms that are responsible for the radiation. The various kinds of radiation are distinguished one from the other only by the range of wavelengths they include. In a vacuum all these forms of radiant energy travel with the speed of light c. The wavelength λ, characterizing an electromagnetic wave, is then related to its frequency ν by the equation

$$\lambda = \frac{c}{\nu} \tag{14.1-1}$$

in which c has the value 2.9979×10^{10} cm sec^{-1}. In the visible part of the spectrum the various wavelengths are associated with the "color" of the light.

For some purposes, it is convenient to think of electromagnetic radiation from a corpuscular point of view. In that case we associate with an electromagnetic wave of frequency ν a *photon*, which is a "particle" with charge zero and mass zero and which has an energy ϵ given by

$$\epsilon = h\nu \tag{14.1–2}$$

Here h is Planck's constant, which has a value of 6.624×10^{-27} erg sec. From these two equations and the information in Fig. 14.1–1, we see that decreasing the wavelength of electromagnetic radiation corresponds to increasing the energy of the corresponding photons. This fact ties in with the various mechanisms that produce the radiation. For example, relatively small energies are released when a molecule decreases its speed of rotation; the corresponding radiation is generally in the infrared. On the other hand, relatively large energies are released when an atomic nucleus goes from a high energy state to a lower one; here the corresponding radiation is either gamma or x-radiation. The foregoing statements also make it seem reasonable that the radiant energy emitted from heated objects will tend towards shorter wavelengths (higher energy photons) as the temperature of the body is raised.

Thus far we have sketched the phenomenon of the *emission* of radiant energy or photons when a molecular or atomic system goes from a high to a low energy state. The reverse process, which is known as *absorption*, occurs when the addition of radiant energy to a molecular or atomic system causes the system to go from a low to a high energy state. The latter process is then what occurs when radiant energy impinges upon a solid surface and causes it to be heated up.

§14.2 ABSORPTION AND EMISSION AT SOLID SURFACES

Having introduced the reader to the concepts of absorption and emission in terms of the atomic picture, we now proceed to the discussion of the same processes from a macroscopic viewpoint. We restrict the discussion here to opaque solids.

Radiation impinging upon the surface of an opaque solid is either absorbed or reflected. The fraction of the incident radiation that is absorbed is called the *absorptivity* and is given the symbol a; similarly the fraction of the incident radiation with frequency ν that is absorbed is given the symbol a_ν. That is, a and a_ν are defined by

$$a = \frac{q^{(a)}}{q^{(i)}}; \qquad a_\nu = \frac{q_\nu^{(a)}}{q_\nu^{(i)}} \tag{14.2–1,2}$$

in which $q_\nu^{(a)}\,d\nu$ and $q_\nu^{(i)}\,d\nu$ are the absorbed and incident radiation per unit area per unit time in the frequency range ν to $\nu + d\nu$. For any *real body*, a_ν will be less than unity and will vary considerably with the frequency. A hypothetical body for which a_ν is a constant (less than unity) over the entire frequency range and at all temperatures is referred to as a *gray body*; that is, a gray body always absorbs the same fraction of the incident radiation at all frequencies. A limiting case of the gray body is that for which $a_\nu = 1$ for all frequencies and all temperatures; this limiting behavior describes what is known as a *black body*.

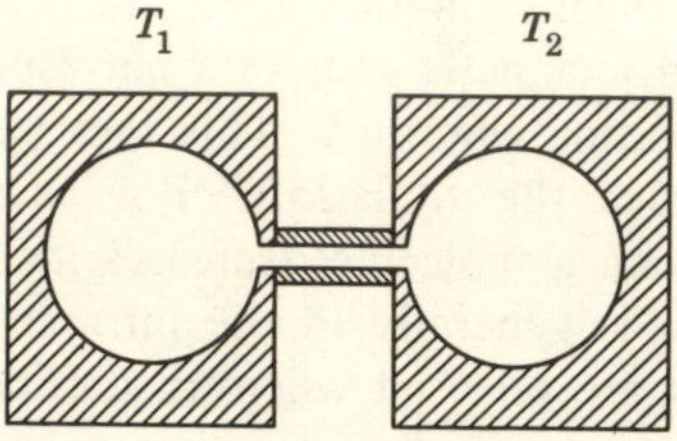

Fig. 14.2–1. Thought experiment for proof that cavity radiation is independent of the wall materials.

All solid surfaces emit radiant energy. The total radiant energy emitted per unit area per unit time is designated by $q^{(e)}$ and that emitted in the frequency range ν to $\nu + d\nu$ is called $q_\nu^{(e)}\,d\nu$. The corresponding rates of energy emission from a black body are given the symbols $q_b^{(e)}$ and $q_{b\nu}^{(e)}\,d\nu$. In terms of these quantities, the *emissivity* for the total radiant-energy emission as well as that for a given frequency are defined by

$$e = \frac{q^{(e)}}{q_b^{(e)}}; \qquad e_\nu = \frac{q_\nu^{(e)}}{q_{b\nu}^{(e)}} \tag{14.2–3,4}$$

The emissivity is also a quantity less than unity for real, nonfluorescing surfaces and equal to unity for black bodies. At any given temperature the radiant energy emitted by a black body represents an upper limit to the radiant energy emitted by real, nonfluorescing surfaces.

We now consider the radiation within an evacuated enclosure or "cavity" with isothermal walls. We imagine that the entire system is at equilibrium. Under these conditions, there is no net flux of energy across the inteıface between the solid and the cavity. It can be shown that the radiation in such a cavity is independent of the nature of the walls and dependent solely on the temperature of the walls of the cavity: we connect two cavities, the walls of which are at the same temperature but are made of two different materials, as shown in Fig. 14.2–1. If the radiation intensities in the two cavities were different, there would be a net transport of radiant energy from one cavity to the other. Because such a flux would violate the second law of thermodynamics, the radiation intensities in the two cavities must be equal, regardless of the composition of the cavity surfaces. Furthermore, it can be shown that the radiation is uniform and unpolarized throughout the cavity. This *cavity radiation* plays an important role in the development of Planck's law. We designate the intensity of the radiation as $q^{(\text{cav})}$; this is the radiant energy that would impinge on a solid surface of unit area placed anywhere within the cavity.

We now perform two additional thought experiments. In the first we put into a cavity a small black body at the same temperature as the walls of the cavity. There will be no net interchange of energy between the black body and the cavity walls. Hence the energy impinging on the black-body surface will just equal the energy emitted by the black body:

$$q^{(\text{cav})} = q_b^{(e)} \tag{14.2–5}$$

From this result, we draw the important conclusion that the radiation emitted by a black body is the same as the equilibrium radiation intensity within a cavity at the same temperature. In the second thought experiment we put a small nonblack body into the cavity, once again specifying that its temperature be the same as that of the walls of the cavity. There is no net heat exchange between the nonblack body and the cavity walls. Hence we can state that the energy absorbed by the nonblack body will be the same as that radiating from it:

$$aq^{(\text{cav})} = q^{(e)} \tag{14.2–6}$$

Comparison of Eqs. 14.2–5 and 6 leads to the result

$$a = \frac{q^{(e)}}{q_b^{(e)}} \tag{14.2–7}$$

Note, however, that the definition of emissivity e in Eq. 14.2–3 allows us to conclude that

$$e = a \tag{14.2–8}$$

This is *Kirchhoff's law*, which states that at a given temperature the emissivity and absorptivity of any solid surface are the same when the radiation is in equilibrium with the solid surface. It can be shown that Eq. 14.2–8 is also valid for each wavelength separately:

$$e_\nu = a_\nu \tag{14.2–9}$$

In Table 14.2–1 are some values of the total emissivity e for some solids. Actually, e depends also on frequency and on the angle of emission, but the averaged values given there have found widespread use. The tabulated values are (with a few exceptions) for emission normal to the surface, but they may be used for hemispheric emissivity, particularly for rough surfaces. Unoxidized, clean, metallic surfaces have very low emissivities, whereas most nonmetals and metallic oxides have emissivities above 0.8 at room temperature or higher. Note that emissivity increases with increasing temperature for nearly all materials.

TABLE 14.2–1

THE TOTAL EMISSIVITIES OF VARIOUS SURFACES FOR PERPENDICULAR EMISSION[a]

	T(°R)	e	T(°R)	e
Aluminum				
Highly polished, 98.3% pure	900	0.039	1530	0.057
Oxidized at 1110° F	850	0.11	1570	0.19
Al-coated roofing	560	0.216		
Copper				
Highly polished, electrolytic	636	0.018		
Oxidized at 1110° F	850	0.57	1570	0.57
Iron				
Highly polished, electrolytic	810	0.052	900	0.064
Completely rusted	527	0.685		
Cast iron, polished	852	0.21		
Cast iron, oxidized at 1100° F	850	0.64	1570	0.78
Asbestos paper	560	0.93	1160	0.945
Brick				
Red, rough	530	0.93		
Silica, unglazed, rough	2292	0.80		
Silica, glazed, rough	2472	0.85		
Lampblack, 0.003 in. or thicker	560	0.945	1160	0.945
Paints				
Black shiny lacquer on iron	536	0.875		
White lacquer	560	0.80	660	0.95
Oil paints, 16 colors	672	0.92–0.96		
Aluminum paints, varying age and lacquer content	672	0.27–0.67		
Refractories, 40 different				
Poor radiators	1570	0.65–0.70	2290	0.75
Good radiators	1570	0.80–0.85	2290	0.85–0.90
Water, liquid, thick layer[b]	492	0.95	672	0.963

[a] Selected values from the table compiled by H. C. Hottel for W. H. McAdams, *Heat Transmission*, McGraw-Hill, New York (1942), Second Edition, pp. 393–396.

[b] Calculated from spectroscopic data.

We have indicated that the radiant energy emitted by a black body is an upper limit to the radiant energy emitted by real surfaces and that this energy is a function of the temperature. It has been shown experimentally that the total emitted energy from a black surface is

$$q_b^{(e)} = \sigma T^4 \tag{14.2–10}$$

in which T is the absolute temperature. This is known as the *Stefan-Boltzmann law*. The Stefan-Boltzmann constant σ has been found to have the value 0.1712×10^{-8} Btu hr^{-1} ft^{-2} °R^{-4} or 1.355×10^{-12} cal sec^{-1} cm^{-2} °K^{-4}. In the next section we indicate two routes by which this important formula has been confirmed theoretically. For nonblack bodies at temperature T the radiant energy emitted is

$$q^{(e)} = e\sigma T^4 \tag{14.2–11}$$

in which e must be evaluated at temperature T. The use of Eqs. 14.2–10 and 11 to calculate radiant-heat-transfer rates between heated surfaces is discussed in §§14.4 and 5.

We have mentioned above that the Stefan-Boltzmann constant has been experimentally determined. This implies that we have a true black body at our disposal. Solids with perfectly black surfaces do not exist. However, we can get an excellent approximation to a black surface by piercing a very small hole in an isothermal cavity. The hole itself is then very nearly a black surface. The extent to which this is a good approximation may be seen from the following relation, which gives the effective emissivity of the hole e_{hole} in a rough-walled enclosure in terms of the actual emissivity e of the cavity walls and the fraction f of the total internal cavity area that is cut away by the hole:

$$e_{\text{hole}} \doteq \frac{e}{e + f(1 - e)} \tag{14.2–12}$$

If $e = 0.8$ and $f = 0.001$, then e_{hole} has the value 0.99975. Therefore, 99.975 per cent of the radiation that falls on the hole will be absorbed. The radiation that emerges from the hole will then be very nearly black-body radiation.

§14.3 PLANCK'S DISTRIBUTION LAW, WIEN'S DISPLACEMENT LAW, AND THE STEFAN-BOLTZMANN LAW[1,2,3]

The Stefan-Boltzmann law may be deduced from thermodynamics, provided that certain results of the electromagnetic field theory are known. Specifically, it can be shown that for cavity radiation the energy density (that is, the energy per unit volume) within the cavity is

$$u^{(r)} = \frac{4}{c} q_b^{(e)} \tag{14.3–1}$$

Since the radiant energy emitted by a black body depends on temperature alone, the energy density $u^{(r)}$ must also be a function of temperature only.

[1] J. de Boer, Chapter VII, *Textbook of Physics* (R. Kronig, Ed.), Scheltema and Holkema, Amsterdam (1951), Third Edition.

[2] G. Joos, *Theoretical Physics*, Hafner, New York (1939).

[3] M. Planck, *Vorlesungen über die Theorie der Wärmestrahlung*, Barth, Leipzig (1923), Fifth Edition.

It can be shown further that the electromagnetic radiation exerts a pressure $p^{(r)}$ on the walls of the cavity given by

$$p^{(r)} = \tfrac{1}{3}u^{(r)} \tag{14.3-2}$$

It should be noted that the foregoing results for cavity radiation can also be obtained by considering that the cavity is filled with a gas made up of photons, each endowed with energy $h\nu$ and momentum $h\nu/c$. We now apply the thermodynamic formula

$$\left(\frac{\partial U}{\partial V}\right)_T = T\left(\frac{\partial p}{\partial T}\right)_V - p \tag{14.3-3}$$

to the photon gas or radiation in the cavity. Insertion of $U^{(r)} = Vu^{(r)}$ and $p^{(r)} = (\frac{1}{3})u^{(r)}$ into this relation gives the following ordinary differential equation for $u^{(r)}(T)$:

$$u^{(r)} = \frac{T}{3}\frac{du^{(r)}}{dT} - \frac{u^{(r)}}{3} \tag{14.3-4}$$

This equation may be integrated to give

$$u^{(r)} = bT^4 \tag{14.3-5}$$

in which b is a constant of integration. Combination of this result with Eq. 14.3-1 gives the radiant energy emitted from the surface of a black body per unit area per unit time:

$$q_b^{(e)} = \frac{c}{4}\,u^{(r)} = \frac{cb}{4}\,T^4 = \sigma T^4 \tag{14.3-6}$$

which is just the Stefan-Boltzmann law. Note that the thermodynamic development does not predict the value of σ.

A second way of deducing the Stefan-Boltzmann law is by the integration of the *Planck distribution law.*[3] This famous equation gives the radiated energy flux $q_{b\lambda}^{(e)}\,d\lambda$ in the wavelength range λ to $\lambda + d\lambda$ from a black surface:

$$q_{b\lambda}^{(e)} = \frac{2\pi c^2 h}{\lambda^5}\,\frac{1}{e^{ch/\lambda \kappa T} - 1} \tag{14.3-7}$$

in which h is Planck's constant. This result may be derived by applying quantum statistics to a photon gas in a cavity, the photons being regarded as obeying the Bose-Einstein statistics.[4,5] Actually, it was Planck's successful attempt to explain the frequency distribution of black-body radiation that led to the development of the quantum theory and the quantum statistics in the first few years of the twentieth century. The Planck distribution is

[4] J. E. Mayer and M. G. Mayer, *Statistical Mechanics*, Wiley, New York (1940), pp. 363–374.

[5] R. C. Tolman, *Principles of Statistical Mechanics*, Oxford University Press (1938), pp. 380–383.

shown in Fig. 14.3–1; it correctly predicts the entire energy versus wavelength curve and the shift of the maximum towards shorter wavelengths at higher temperatures. When Eq. 14.3–7 is integrated over all wavelengths, we get

$$\begin{aligned} q_b^{(e)} &= \int_0^\infty q_{b\lambda}^{(e)}\, d\lambda \\ &= 2\pi c^2 h \int_0^\infty \frac{\lambda^{-5}}{e^{ch/\lambda\kappa T} - 1}\, d\lambda \\ &= \frac{2\pi\kappa^4 T^4}{c^2 h^3} \int_0^\infty \frac{x^3}{e^x - 1}\, dx \\ &= \frac{2\pi\kappa^4 T^4}{c^2 h^3}\left(6 \sum_{n=1}^{\infty} \frac{1}{n^4}\right) \\ &= \frac{2\pi\kappa^4 T^4}{c^2 h^3}\left(\frac{\pi^4}{15}\right) \\ &= \left(\frac{2}{15}\frac{\pi^5\kappa^4}{c^2 h^3}\right) T^4 \end{aligned} \tag{14.3–8}$$

In the above integration we changed the variable of integration from λ to $x = ch/\lambda\kappa T$. The integration over x is performed by expanding $1/(e^x - 1)$ in e^x and integrating term by term. The quantum statistical approach thus gives the details of the spectral distribution of the radiation. It also gives the value of the constant σ in terms of the universal constants c, h, and κ; that is, $\sigma = (2\pi^5\kappa^4/15c^2h^3) = 0.1712 \times 10^{-8}$ Btu hr^{-1} ft^{-2} °R^{-4}. This value is about $1\frac{1}{2}$ per cent below the value found by direct radiation measurements, but it is believed to be more reliable.

In addition to obtaining the Stefan-Boltzmann law from the Planck distribution, we can get an important relation pertaining to the maximum in the Planck distribution. First we rewrite Eq. 14.3–7 in terms of x and then set $dq_{b\lambda}^{(e)}/dx = 0$. This gives the following equation for $x_{\max}$, which is the value of x for which the Planck distribution shows a maximum:

$$x_{\max} = 5(1 - e^{-x_{\max}}) \tag{14.3–9}$$

The solution to this equation may be found numerically to be $x_{\max} = 4.9651$. Hence at a given temperature T

$$\lambda_{\max} T = \frac{ch}{\kappa x_{\max}} \tag{14.3–10}$$

By inserting the values of the universal constants and the value for $x_{\max}$, we get

$$\lambda_{\max} T = 0.2884 \text{ cm °K} \tag{14.3–11}$$

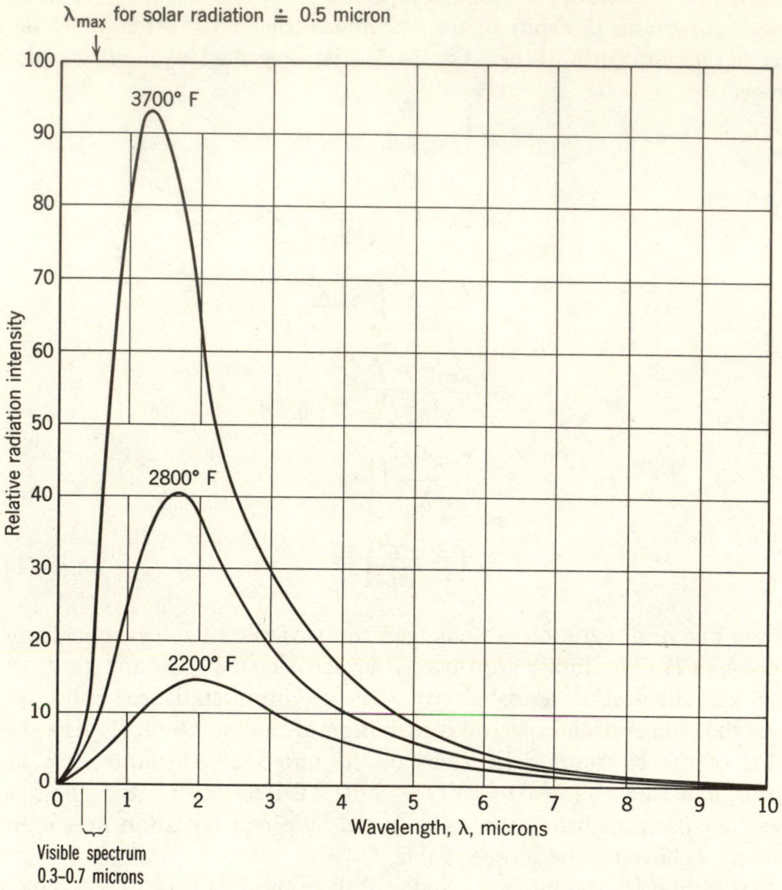

Fig. 14.3–1. The spectrum of equilibrium radiation as given by Planck's law. [M. Planck, Verh. der deutschen physik. Gesell., **2,** 202, 237 (1900); Ann. der Physik, **4,** 553 (1901).

This result is known as *Wien's displacement law*. It is useful primarily for estimating the temperature of remote objects. The law predicts, in agreement with experience, that the apparent color of radiation shifts from red (long wavelengths) towards blue (short wavelengths) as the temperature increases.

Finally we may reinterpret some of our previous remarks in terms of the Planck distribution law. In Fig. 14.3–2 we have sketched three curves: the Planck distribution curve for a hypothetical black body, the distribution curve for a hypothetical gray body, and a distribution curve for some real body. It is thus clear that when we use total emissivity values, such as those in Table 14.2–1, we are just accounting empirically for the deviations from Planck's law over the entire spectrum.

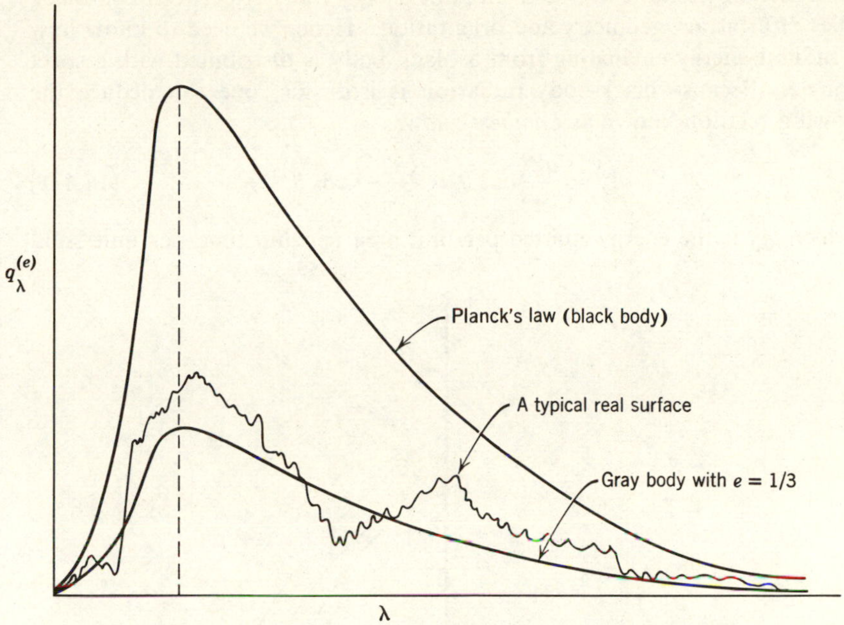

Fig. 14.3–2. Comparison of black, gray, and real surfaces.

Example 14.3–1. Temperature and Radiant-Energy Emission of the Sun

For approximate calculations, the sun may be considered a black body, emitting radiation with maximum intensity at $\lambda = 0.5$ microns (5000 Å). On this basis, estimate (*a*) the surface temperature of the sun and (*b*) the emitted heat flux at the surface of the sun.

Solution.

a. From Wien's displacement law,

$$T = \frac{0.2884}{\lambda_{\max}} = \frac{(0.2884 \text{ cm } ^\circ\text{K})}{(0.5 \times 10^{-4} \text{ cm})} = 5760^\circ \text{ K} = 10{,}400^\circ \text{ R}$$

b. From the Stefan-Boltzmann law,

$$\begin{aligned} q_b^{(e)} &= \sigma T^4 \\ &= (0.1712 \times 10^{-8})(10{,}400)^4 \\ &= 2.0 \times 10^7 \text{ Btu hr}^{-1} \text{ ft}^{-2} \end{aligned}$$

§14.4 DIRECT RADIATION BETWEEN BLACK BODIES IN VACUO AT DIFFERENT TEMPERATURES

In the preceding sections we have given the Stefan-Boltzmann law, which describes the total radiant-energy emission from a perfectly black surface.

In this section we want to discuss radiant-energy transfer between two black bodies of arbitrary geometry and orientation. Hence we need to know how the radiant energy emanating from a black body is distributed with respect to angle. Because black-body radiation is isotropic, one can deduce the following relation known as *Lambert's law:*

$$q_{b\theta}^{(e)} = \frac{q_b^{(e)}}{\pi} \cos\theta = \frac{(\sigma T^4)}{\pi} \cos\theta \tag{14.4–1}$$

in which $q_{b\theta}^{(e)}$ is the energy emitted per unit area per unit time per unit solid

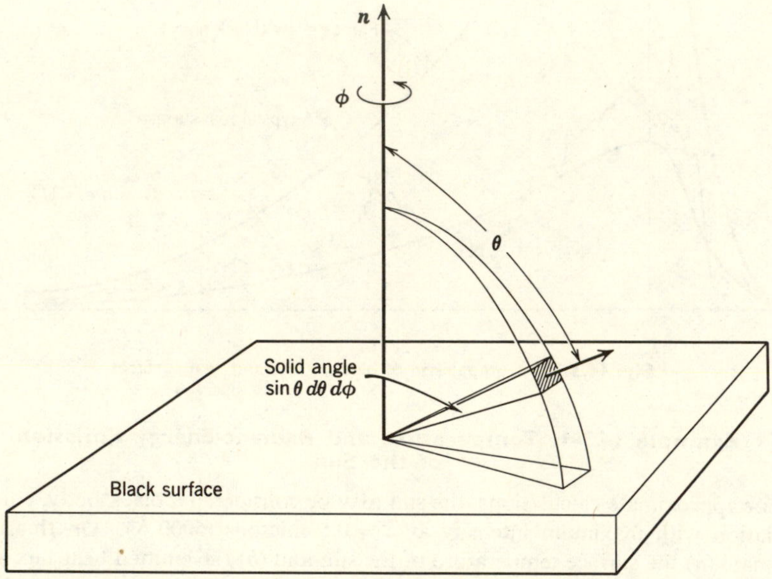

Fig. 14.4–1. Radiation at an angle θ from the normal to the surface into a solid angle $\sin\theta\, d\theta\, d\phi$.

angle in a direction θ (see Fig. 14.4–1). The energy emitted through the shaded solid angle is then $q_{b\theta}^{(e)} \sin\theta\, d\theta\, d\phi$ per unit area of black solid surface. Integration of the foregoing expression for $q_{b\theta}^{(e)}$ over the entire hemisphere gives the known total energy emissions:

$$\int_0^{2\pi}\int_0^{\pi/2} q_{b\theta}^{(e)} \sin\theta\, d\theta\, d\phi = \frac{\sigma T^4}{\pi}\int_0^{2\pi}\int_0^{\pi/2} \cos\theta \sin\theta\, d\theta\, d\phi$$

$$= \sigma T^4 = q_b^{(e)} \tag{14.4–2}$$

This calculation explains why the factor of π^{-1} was included in Eq. 14.4–1.

We are now in a position to get the net heat-transfer rate from body "1" to body "2," where these are black bodies of any arbitrary shape and orientation. (See Fig. 14.4–2.) We do this by getting the net heat-transfer rate between a pair of surface elements dA_1 and dA_2 that "see" each other and then integrate over all such possible pairs of areas. First we join the elements dA_1 and dA_2 by a straight line r_{12}, which makes an angle θ_1 with the normal to dA_1 and an angle θ_2 with the normal to dA_2.

We begin by writing down an expression for the energy radiated from dA_1 into a solid angle $\sin\theta_1\, d\theta_1\, d\phi_1$ about r_{12}. We choose this solid angle large enough that dA_2 will lie entirely within the "beam." (See Fig. 14.4–2.)

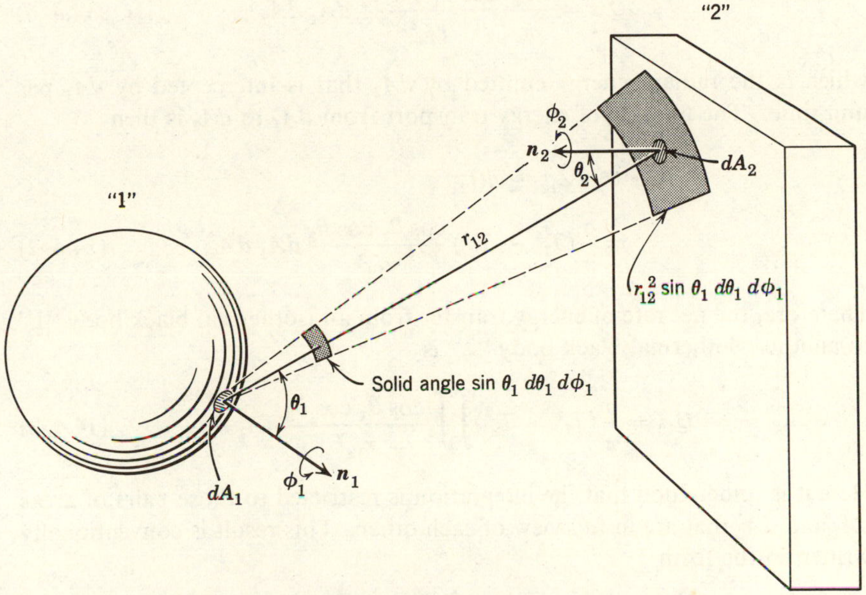

Fig. 14.4–2. Radiant interchange between two black bodies.

According to Lambert's cosine law, the energy radiated in a unit time will be

$$\left(\frac{\sigma T_1^4}{\pi}\cos\theta_1\right) dA_1 \sin\theta_1\, d\theta_1\, d\phi_1 \tag{14.4–3}$$

Of the energy leaving dA_1 at an angle θ_1, only that fraction will be intercepted by dA_2 which is given by the following ratio:

$$\frac{\begin{pmatrix}\text{area of } dA_2 \text{ projected onto a}\\ \text{plane perpendicular to } r_{12}\end{pmatrix}}{\begin{pmatrix}\text{area formed by the intersection}\\ \text{of the solid angle } \sin\theta_1\, d\theta_1\, d\phi_1\\ \text{with a sphere of radius } r_{12} \text{ with}\\ \text{center at } dA_1\end{pmatrix}} = \frac{dA_2 \cdot \cos\theta_2}{r_{12}^2 \sin\theta_1\, d\theta_1\, d\phi_1} \tag{14.4–4}$$

Multiplication of these last two expressions then gives

$$dQ_{\overrightarrow{12}} = \frac{\sigma T_1^4}{\pi} \frac{\cos \theta_1 \cos \theta_2}{r_{12}^2} dA_1 \, dA_2 \tag{14.4-5}$$

This is the radiant energy emitted by dA_1 and intercepted by dA_2 per unit time. In a similar way we can write

$$dQ_{\overrightarrow{21}} = \frac{\sigma T_2^4}{\pi} \frac{\cos \theta_1 \cos \theta_2}{r_{12}^2} dA_1 \, dA_2 \tag{14.4-6}$$

which is the radiant energy emitted by dA_2 that is intercepted by dA_1 per unit time. The net rate of energy transport from dA_1 to dA_2 is then

$$\begin{aligned} dQ_{12} &= dQ_{\overrightarrow{12}} - dQ_{\overrightarrow{21}} \\ &= \frac{\sigma}{\pi}(T_1^4 - T_2^4) \frac{\cos \theta_1 \cos \theta_2}{r_{12}^2} dA_1 \, dA_2 \end{aligned} \tag{14.4-7}$$

Therefore, the net rate of energy transfer from an isothermal black body "1" to another isothermal black body "2" is

$$Q_{12} = \frac{\sigma}{\pi}(T_1^4 - T_2^4) \iint \frac{\cos \theta_1 \cos \theta_2}{r_{12}^2} dA_1 \, dA_2 \tag{14.4-8}$$

Here it is understood that the integration is restricted to those pairs of areas dA_1 and dA_2 that are in full view of each other. This result is conventionally written in the form

$$Q_{12} = A_1 F_{12} \sigma(T_1^4 - T_2^4) = A_2 F_{21} \sigma(T_1^4 - T_2^4) \tag{14.4-9}$$

Here A_1 and A_2 are usually chosen to be the total areas of bodies "1" and "2," and the dimensionless *view factors*[1] F_{12} and F_{21} are given by

$$F_{12} = \frac{1}{\pi A_1} \iint \frac{\cos \theta_1 \cos \theta_2}{r_{12}^2} dA_1 \, dA_2 \tag{14.4-10}$$

$$F_{21} = \frac{1}{\pi A_2} \iint \frac{\cos \theta_1 \cos \theta_2}{r_{12}^2} dA_1 \, dA_2 \tag{14.4-11}$$

Clearly, $A_1 F_{12} = A_2 F_{21}$. The view factor F_{12} represents the fraction of radiation leaving A_1 that is directly intercepted by A_2.

The actual calculation of view factors is a difficult problem, except for some very simple situations. In Figs. 14.4–3 and 4 some view factors for direct radiation are shown, which have been calculated by Hottel and his

[1] Also called *angle factors*.

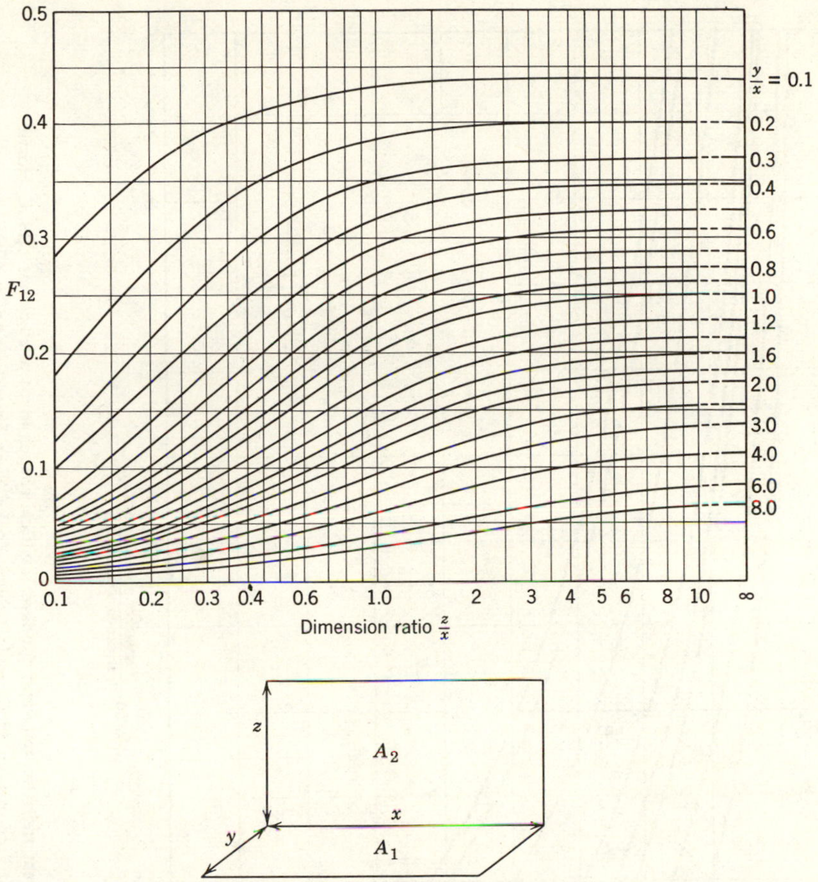

Fig. 14.4-3. View factors for direct radiation between adjacent rectangles in perpendicular planes. [H. C. Hottel, Chapter 3 in W. H. McAdams *Heat Transmission*, McGraw-Hill, New York (1954), p. 68.]

collaborators.[2,3,4] Once these charts are available, the calculation of energy interchanges by Eq. 14.4-9 is easy.

In the above development we have assumed that Lambert's law and the Stefan-Boltzmann law may be used to describe the nonequilibrium radiative transport process, in spite of the fact that they are strictly valid only for radiative equilibrium. The errors thus introduced do not seem to have been

[2] H. C. Hottel, Radiant Heat Transmission, *Chemical Engineers' Handbook*, McGraw-Hill, New York (1950), Third Edition, pp. 483–498.

[3] H. C. Hottel, Radiant Heat Transmission, in W. H. McAdams, *Heat Transmission*, McGraw-Hill, New York (1954), Third Edition, Chapter 4.

[4] M. Jakob, *Heat Transfer*, Wiley, New York (1957), Vol. II, Chapter 31.

Curves for F	Shape	Curves for $\bar{F}$
1	Disks	5
2	Squares	6
3	Rectangles, $\frac{L}{W} = 2$	7
4	Rectangles, $\frac{L}{W} = \infty$	8

Ratio, $\dfrac{\text{Diameter or shorter side}}{\text{Distance between planes}}$

View factor F_{12} or $\bar{F}_{12}$

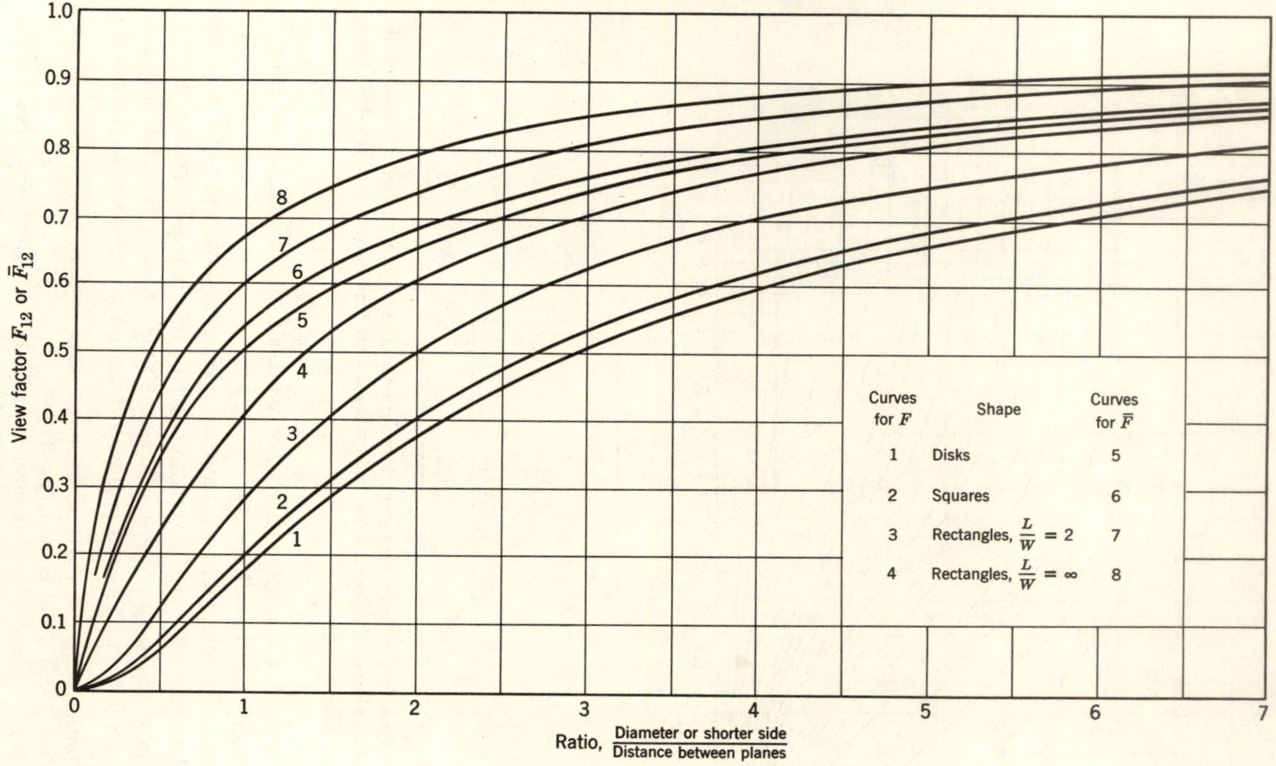

Fig. 14.4-4. View factors for direct radiation between opposed identical shapes in parallel planes. [H. C. Hottel, Chapter 3 in W. H. McAdams *Heat Transmission*, McGraw-Hill, New York (1954), Third Edition, p. 69.]

studied thoroughly, but apparently the formulas derived give a good quantitative description.

Thus far we have concerned ourselves with systems containing two black bodies, "1" and "2," only. We now wish to consider a set of black surfaces "1," "2," ..., "n," which form the walls of a complete enclosure. The surfaces are maintained at temperatures $T_1, T_2, \ldots, T_n$, respectively. The net heat flow from any surface "i" to the other $n - 1$ surfaces is then

$$Q_{is} = \sigma A_i \sum_{j=1}^{n} F_{ij}(T_i^4 - T_j^4) \qquad i = 1, 2, \cdots, n \qquad (14.4\text{–}12)$$

or

$$Q_{is} = \sigma A_i \left\{ T_i^4 - \sum_{j=1}^{n} F_{ij} T_j^4 \right\} \qquad (14.4\text{–}13)$$

In writing the second form, we have made use of the relation

$$\sum_{j=1}^{n} F_{ij} = 1 \qquad (14.4\text{–}14)$$

The sums in Eqs. 14.4–13 and 14 include the term F_{ii}, which is zero for any object that intercepts none of its own rays. The set of n equations given in Eq. 14.4–12 (or Eq. 14.4–13) may be solved to get temperatures or heat flows according to the data available.

A simultaneous solution of Eqs. 14.4–13 of special interest is that for which $Q_{3s}, Q_{4s}, \ldots, Q_{ns} = 0$; surfaces such as 3, 4, ... n are here called "adiabatic". In this situation one can eliminate the temperatures of all surfaces except "1" and "2" from the heat-flow calculation and obtain a direct solution for the net heat flow from surface "1" to surface "2":

$$Q_{12} = \sigma A_1 \bar{F}_{12}(T_1{}^4 - T_2{}^4) = \sigma A_2 \bar{F}_{21}(T_1{}^4 - T_2{}^4) \qquad (14.4\text{–}15)$$

Values of $\bar{F}_{12}$ for use in this equation are given in Fig. 14.4–4. These values apply only when the adiabatic walls are formed from line elements perpendicular to "1" and "2."

The use of these view factors F and $\bar{F}$ greatly simplifies calculations of black-body radiation when the temperatures of surfaces "1" and "2" are known to be uniform. The reader desirous of further information on radiative exchange in enclosures is referred to the literature.[4]

Example 14.4–1. Estimation of the Solar Constant

The radiant heat flux entering the earth's atmosphere from the sun has been termed the "solar constant" and is important in solar energy utilization as well as in meteorology. Designate the sun as body 1 and the earth as body 2 and use the following data to calculate the solar constant: $D_1 = 8.60 \times 10^5$ miles; $r_{12} = 9.29 \times 10^7$ miles; $q_{b1}^{(e)} = 2.0 \times 10^7$ Btu hr^{-1} ft^{-2} (from Example 14.3–1).

Solution. In the terminology of Eq. 14.4–5 and Fig. 14.4–5

$$\text{solar constant} = \frac{dQ_{\overrightarrow{12}}}{\cos\theta_2\, dA_2} = \frac{\sigma T_1^4}{\pi r_{12}^2}\int \cos\theta_1\, dA_1$$

$$= \frac{\sigma T_1^4}{\pi}\frac{\pi D_1^2}{4r_{12}^2} = \frac{q_{b1}^{(e)} D_1^2}{4r_{12}^2}$$

$$= \frac{2.0 \times 10^7}{4}\left(\frac{8.60 \times 10^5 \text{ miles}}{9.29 \times 10^7 \text{ miles}}\right)^2$$

$$= 430 \text{ Btu hr}^{-1}\text{ ft}^{-2} \qquad (14.4\text{–}16)$$

This is in satisfactory agreement with other estimates that have been made. The removal of r_{12}^2 from the integrand is permissible here because the distance r_{12}

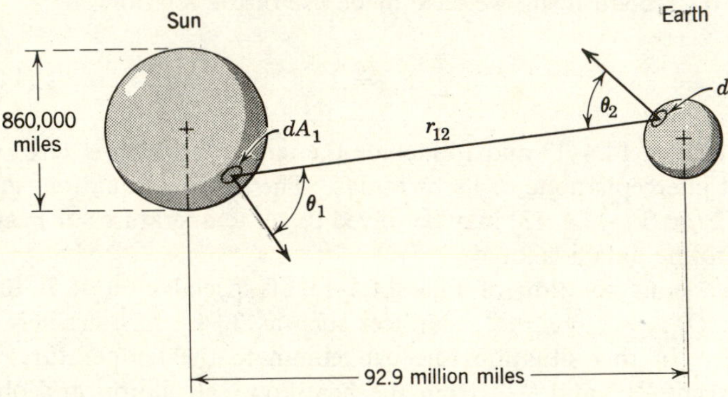

Fig. 14.4–5. Estimation of the solar constant.

varies by less than 0.5 per cent over the visible surface of the sun. The remaining integral $\int \cos\theta_1\, dA_1$ is the projected area of the sun as seen from the earth, or very nearly $\pi D_1^2/4$.

Example 14.4–2. Radiant Heat Transfer between Disks

Two black disks of diameter 2 ft are placed directly opposite one another at a distance of 4 ft. Disk "1" is maintained at 2000° R, and disk "2" at 1000° R. Calculate the heat flow between the two disks (*a*) when no other surfaces are present and (*b*) when the two disks are connected by an adiabatic right-cylindrical black surface.

Solution

a. From Eq. 14.4–9 and curve 1 of Fig. 14.4–4,

$$Q_{12} = A_1 F_{12}\sigma(T_1^4 - T_2^4)$$
$$= (\pi)(0.06)(0.1712 \times 10^{-8})[(2000)^4 - (1000)^4]$$
$$= 4.83 \times 10^3 \text{ Btu hr}^{-1}$$

b. From Eq. 14.4–15 and curve 5 of Fig. 14.4–4,

$$\begin{aligned} Q_{12} &= A_1\bar{F}_{12}\sigma(T_1^4 - T_2^4) \\ &= (\pi)(0.34)(0.1712 \times 10^{-8})[(2000)^4 - (1000)^4] \\ &= 27.4 \times 10^3 \text{ Btu hr}^{-1} \end{aligned}$$

§14.5 RADIATION BETWEEN NONBLACK BODIES AT DIFFERENT TEMPERATURES

In principle, radiation between nonblack surfaces can be treated by differential analysis of emitted rays and their successive reflected components. For nearly black surfaces this is feasible, as only one or two reflections need be considered. For highly reflecting surfaces, however, the analysis is complicated, and the distributions of emitted and reflected rays with respect to angle and wavelength are not usually known with enough accuracy to justify a detailed calculation.

A reasonably accurate treatment is possible for a small convex surface in a large, nearly isothermal enclosure (i.e., a "cavity"), such as a steam pipe in a room with walls at constant temperature. The rate of energy emission from a nonblack surface "1" to the surrounding enclosure "2" is given by

$$Q_{\overrightarrow{12}} = e_1 A_1 \sigma T_1^4 \tag{14.5–1}$$

and the rate of energy absorption from the surroundings by surface "1" is

$$Q_{\overrightarrow{21}} = a_1 A_1 \sigma T_2^4 \tag{14.5–2}$$

Here we have made use of the fact that the radiation impinging on surface "1" is very nearly cavity radiation or black-body radiation corresponding to temperature T_2. Since A_1 is convex, it intercepts none of its own rays; hence F_{12} has been set equal to unity. The net radiation rate from A_1 to the surroundings is therefore

$$Q_{12} = \sigma A_1(e_1 T_1^4 - a_1 T_2^4) \tag{14.5–3}$$

In Eq. 14.5–3 e_1 is the value of emissivity of surface "1" at T_1. The absorptivity a_1 is usually *estimated* as the value of e of surface "1" at T_2.

We next consider an enclosure consisting of n gray, diffuse-reflecting surfaces at different temperatures $T_1, T_2 \ldots T_n$. If all but two of the surfaces are adiabatic (i.e., $Q_{3s} = Q_{4s} = \ldots = 0$), the heat flow between surfaces "1" and "2" is given approximately by the following extension of Eq. 14.4–15:[1,2]

$$Q_{12} = A_1\mathscr{F}_{12}\sigma(T_1^4 - T_2^4) = A_2\mathscr{F}_{21}\sigma(T_1^4 - T_2^4) \tag{14.5–4}$$

in which

$$\frac{1}{A_1\mathscr{F}_{12}} = \frac{1}{A_1\bar{F}_{12}} + \frac{1}{A_1}\left(\frac{1}{e_1} - 1\right) + \frac{1}{A_2}\left(\frac{1}{e_2} - 1\right) \tag{14.5–5}$$

[1] H. Ç. Hottel, in W. H. McAdams, *Heat Transmission*, McGraw-Hill, New York (1954), Chapter 4.

[2] M. Jakob, *Heat Transfer*, Wiley, New York (1957), Vol. II, Chapter 31.

Equations 14.5–4 and 5 constitute an exact solution if A_1 and A_2 are parallel infinite planes, coaxial infinite cylinders, or concentric spheres, and no other surfaces are in view; it is accurate for other geometries if the enclosure looks the same when viewed from any point on A_1, and similarly for A_2. The equations also become exact as the emissivities of all surfaces in the system

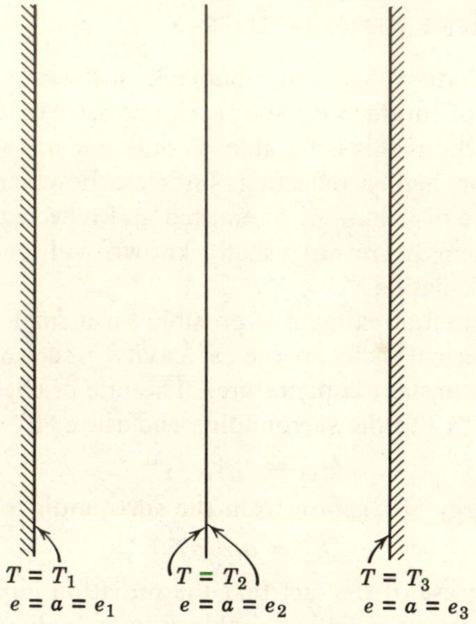

Fig. 14.5–1. Radiation shield.

approach unity. The quantity $\bar{F}_{12}$ is given for some simple geometries in Fig. 14.4–4.

Example 14.5–1. Radiation Shields

Develop an expression for the reduction in radiant heat transfer between two infinite parallel gray planes having the same area, A, when a thin parallel gray sheet of very high thermal conductivity is placed between them.

Solution. A radiation balance may be drawn between each pair of adjacent surfaces by using the nomenclature of Fig. 14.5–1 and by summing up the energies absorbed from any given primary beam on repeated reflection. For example, in Fig. 14.5–2 we show schematically the way in which energy leaving surface "1" is absorbed. A similar picture may be drawn for the energy originally leaving body "2." When these results are combined, we get

$$Q_{12} = A\sigma(T_1^4 - T_2^4)[e_1e_2 + e_1e_2(1 - e_2)(1 - e_1) + e_1e_2(1 - e_2)^2(1 - e_1)^2 + \cdots]$$

$$= A\sigma e_1e_2(T_1^4 - T_2^4)\sum_{i=0}^{\infty}[(1 - e_2)(1 - e_1)]^i \qquad (14.5\text{–}6)$$

It may be shown that the summation reduces to

$$\sum_{i=0}^{\infty} [(1 - e_2)(1 - e_1)]^i = \frac{1}{1 - (1 - e_2)(1 - e_1)} \tag{14.5-7}$$

so that

$$Q_{12} = \frac{A\sigma(T_1^4 - T_2^4)}{\left(\dfrac{1}{e_1} + \dfrac{1}{e_2} - 1\right)} \tag{14.5-8}$$

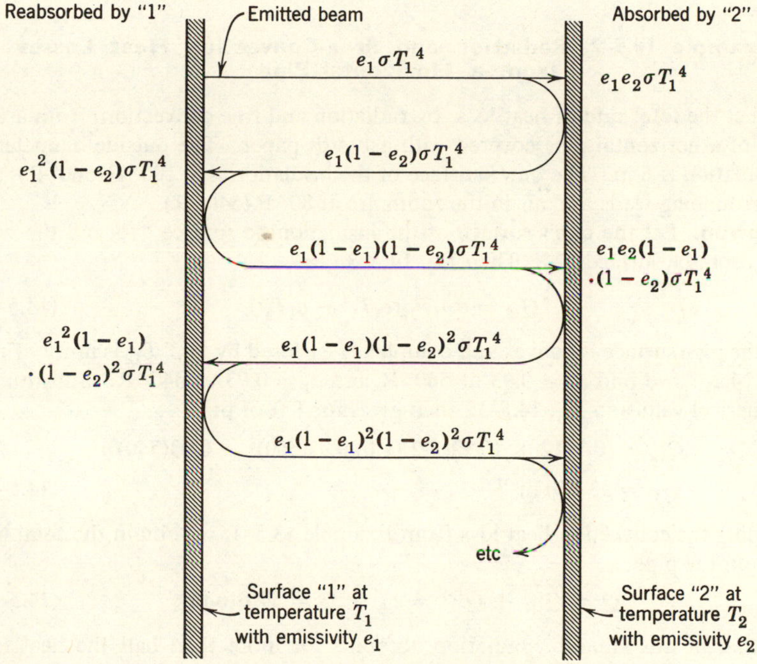

Fig. 14.5-2. Schematic diagram showing what happens to radiation emitted from surface "1."

(Note that this is a special case of Eq. 14.5-4.) Similarly

$$Q_{23} = \frac{A\sigma(T_2^4 - T_3^4)}{\left(\dfrac{1}{e_2} + \dfrac{1}{e_3} - 1\right)} \tag{14.5-9}$$

The temperature of the radiation shield T_2 may now be eliminated from these equations to give

$$Q_{12} = \frac{A\sigma(T_1^4 - T_3^4)}{\left(\dfrac{1}{e_1} + \dfrac{1}{e_2} - 1\right) + \left(\dfrac{1}{e_2} + \dfrac{1}{e_3} - 1\right)} \tag{14.5-10}$$

The ratio of radiant energy transfer with a shield to that without one is

$$\frac{Q_{12}}{Q_{13}} = \frac{\left(\dfrac{1}{e_1} + \dfrac{1}{e_3} - 1\right)}{\left(\dfrac{1}{e_1} + \dfrac{1}{e_2} - 1\right) + \left(\dfrac{1}{e_2} + \dfrac{1}{e_3} - 1\right)} \tag{14.5–11}$$

If all three surfaces have the same emissivity, this fraction is one half. Multiple shields are even more effective (see Problem 14.H) and are frequently used for high-temperature insulation or for thermocouple shielding.

Example 14.5–2. Radiation and Free-Convection Heat Losses from a Horizontal Pipe

Predict the total rate of heat loss, by radiation and free convection, from a unit length of a horizontal pipe covered with asbestos paper. The outside diameter of the insulation is 6 in. The outer surface of the insulation is at 100° F (560° R), and the surrounding walls and air in the room are at 80° F (540° R).

Solution. Let the outer surface of the insulation be surface "1" and the walls of the room be surface "2." Then Eq. 14.5–3 gives

$$Q_{12} = \sigma A_1 F_{12}(e_1 T_1^4 - a_1 T_2^4) \tag{14.5–12}$$

Since the pipe surface is convex and completely enclosed by "2," F_{12} is unity. From Table 14.2–1, we find $e_1 = 0.93$ at 560° R and $a_1 = 0.93$ at 540° R. Substitution of numerical values in Eq. 14.5–12 then gives for 1 ft of pipe

$$\begin{aligned} Q_{12} &= (0.1712 \times 10^{-8})(\pi/2)(1.0)[0.93(560)^4 - 0.93(540)^4] \\ &= 33 \text{ Btu hr}^{-1} \end{aligned} \tag{14.5–13}$$

By adding the convection heat loss from Example 13.5–1, we obtain the total heat loss from the pipe:

$$Q = Q^{(c)} + Q^{(r)} = 21 + 33 = 54 \text{ Btu hr}^{-1} \tag{14.5–14}$$

Note that in this situation radiation accounts for more than half the heat loss. If the fluid is *not* transparent, the convection and radiation processes are not independent, and $Q^{(c)}$ and $Q^{(r)}$ cannot be added directly. The calculation given here is almost always suitable for gases; for most problems involving liquids $Q^{(r)}$ can be neglected.

Example 14.5–3. Combined Radiation and Convection

A body directly exposed to the night sky will be cooled below ambient temperature because of radiation to outer space. This effect can be used to freeze water in shallow trays well insulated from the ground. Estimate the maximum air temperature for which freezing is possible, neglecting evaporation.

Solution. As a first approximation, the following may be assumed:

a. All heat received by the water is by free convection from the surrounding air, which is assumed to be quiescent.

b. The heat effect of evaporation or condensation of water is not significant.
c. Steady state has been achieved.
d. The pan of water is square in cross section.
e. Back radiation from the atmosphere may be neglected.

The maximum permitted air temperature at the water surface is $T_1 = 492°$ R. The rate of *heat loss by radiation* is

$$Q^{(r)} = e_1 A_1 \sigma T_1^4 = (0.95)(0.1712 \times 10^{-8})(492)^4 L^2$$
$$= 95L^2 \text{ Btu hr}^{-1}\text{ ft}^{-2} \qquad (14.5\text{–}15)$$

in which L is the length of one edge of the pan.

To get the *heat gain by convection*, we use the relation

$$Q^{(\text{conv})} = hL^2(T_{\text{air}} - T_{\text{water}}) \qquad (14.5\text{–}16)$$

in which h is the heat transfer coefficient for free convection. For cooling atmospheric air at ordinary temperatures by a horizontal square surface facing upward, the heat-transfer coefficient is given by[3]

$$h = 0.2(T_{\text{air}} - T_{\text{water}})^{1/4} \qquad (14.5\text{–}17)$$

in which h is expressed in Btu hr^{-1} ft^{-2} ° F^{-1} and the temperature is given in degrees Rankine.

When the foregoing expressions for heat loss by radiation and heat gain by free convection are equated, we get

$$0.2L^2(T_{\text{air}} - 492)^{5/4} = 95L^2 \qquad (14.5\text{–}18)$$

From this we find that the maximum ambient air temperature is 630° R or 170° F. Except under desert conditions, back radiation and moisture condensation from the surrounding air greatly lower the required air temperature.

§14.6 RADIANT ENERGY TRANSPORT IN ABSORBING MEDIA[1]

The methods given in the preceding sections are applicable only to materials that are completely transparent or completely opaque. In order to describe energy transport in nontransparent media, we write differential equations for the local rate of change of energy as viewed both from the material and the radiation standpoint. That is, we regard a material medium traversed by electromagnetic radiation as two coexisting "phases": a "material phase," consisting of all the mass in the system, and a "photon phase," consisting of the electromagnetic radiation.

In Chapter 10 we have already given an energy balance equation for a system containing no radiation. Here we extend Eq. Q of Table 10.4–1 for

[3] W. H. McAdams, in *Chemical Engineers' Handbook* (J. H. Perry, Ed.), McGraw-Hill, New York (1950), Third Edition, p. 474.

[1] An extensive treatment of the theory of radiation in absorbing media is given by S. Chandrasekhar, *Radiative Transfer*, Oxford University Press (1950).

the material phase to take into account the energy that is being interchanged with the photon phase by emission and absorption processes:

$$\frac{\partial}{\partial t}\rho\hat{U} = -(\nabla \cdot \rho\hat{U}\boldsymbol{v}) - (\nabla \cdot \boldsymbol{q}) - p(\nabla \cdot \boldsymbol{v}) - (\boldsymbol{\tau} : \nabla\boldsymbol{v}) - (\mathscr{E} - \mathscr{A}) \tag{14.6–1}$$

Here we have introduced $\mathscr{E}$ and $\mathscr{A}$, which are the local rates of photon emission and absorption per unit volume, respectively. That is, $\mathscr{E}$ represents

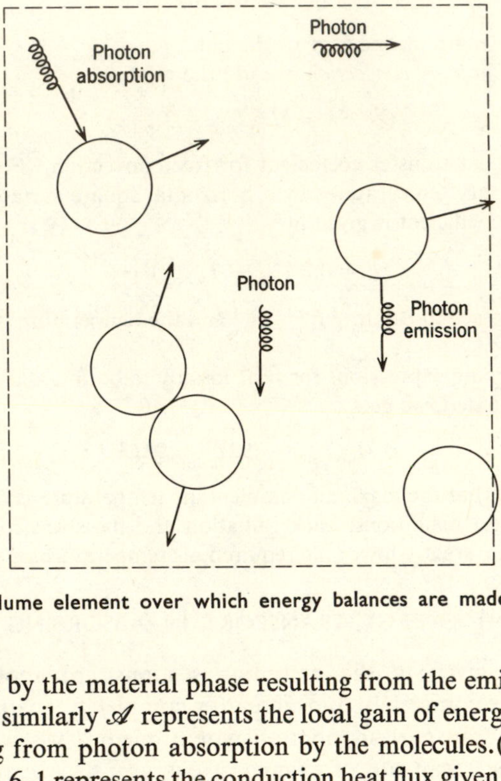

Fig. 14.6–1. Volume element over which energy balances are made; circles represent molecules.

the energy lost by the material phase resulting from the emission of photons by molecules; similarly $\mathscr{A}$ represents the local gain of energy by the material phase resulting from photon absorption by the molecules. (See Fig. 14.6–1.) The $\boldsymbol{q}$ in Eq. 14.6–1 represents the conduction heat flux given by Fourier's law.

For the "photon phase," we may write an equation describing the local rate of change of radiant-energy density $u^{(r)}$:

$$\frac{\partial}{\partial t}u^{(r)} = -(\nabla \cdot \boldsymbol{q}^{(r)}) + (\mathscr{E} - \mathscr{A}) \tag{14.6–2}$$

in which $\boldsymbol{q}^{(r)}$ is the radiant-energy flux. This equation is obtained by writing a radiant energy balance on a small unit of volume fixed in space. Note that there is no convective term in Eq. 14.6–2, since the photons move independently of the local material velocity. Note further that the term $(\mathscr{E} - \mathscr{A})$ appears with opposite signs in Eqs. 14.6–1 and 2, indicating that a net gain of radiant energy occurs at the expense of molecular energy. Equation 14.6–2 can also be written for the radiant energy within a frequency range ν to $\nu + d\nu$:

$$\frac{\partial}{\partial t} u_\nu^{(r)} = -(\nabla \cdot \mathbf{q}_\nu^{(r)}) + (\mathscr{E}_\nu - \mathscr{A}_\nu) \qquad (14.6\text{–}3)$$

This expression is obtained by differentiating Eq. 14.6–2 with respect to ν.

For the purpose of simplifying the discussion, we consider a steady-state nonflow system in which the radiation travels only in the $+z$-direction. Such a system can be closely approximated by passing a collimated light beam through a solution at temperatures sufficiently low that the emission by the solution is unimportant.[2] These are the conditions commonly encountered in spectrophotometry. For such a system Eqs. 14.6–1 and 2 become

$$0 = -\frac{d}{dz} q_z + \mathscr{A} \qquad (14.6\text{–}4)$$

$$0 = -\frac{d}{dz} q_z^{(r)} - \mathscr{A} \qquad (14.6\text{–}5)$$

In order to use these equations, we need information about the volumetric absorption rate $\mathscr{A}$. For a unidirectional beam a conventional expression is

$$\mathscr{A} = m_a q^{(r)} \qquad (14.6\text{–}6)$$

in which m_a is known as the "extinction coefficient." Basically, this just states that the probability for photon absorption is proportional to the concentration of photons.

Example 14.6–1. Absorption of a Monochromatic Radiant Beam

A monochromatic radiant beam of frequency ν, focused parallel to the z-axis, passes through an absorbing fluid. The local rate of energy absorption is given by $m_{a\nu} q_\nu^{(r)}$, in which $m_{a\nu}$ is the "extinction coefficient" for radiation of frequency ν. Determine the distribution of the radiant flux $q_\nu^{(r)}$ in the system.

Solution. We neglect refraction and scattering of the incident beam. Also, we assume that the liquid is cooled so that re-radiation can be neglected. Then Eq. 14.6–5 becomes for steady state

$$0 = -\frac{d}{dz} q_\nu^{(r)} - m_{a\nu} q_\nu^{(r)} \qquad (14.6\text{–}7)$$

Integration with respect to z gives

$$q_\nu^{(r)} = q_\nu^{(r)}\big|_{z=0} \exp(-m_{a\nu} z) \qquad (14.6\text{–}8)$$

This is *Lambert's law* of absorption widely used in spectrometry. For any given pure material, $m_{a\nu}$ depends in a characteristic way on ν. The shape of the absorption spectrum is therefore a very useful tool for qualitative analysis.

[2] Note that it would be necessary to consider radiation in all directions if emission were important.

QUESTIONS FOR DISCUSSION

1. What is the Stefan-Boltzmann law and under what conditions does it apply? Give an example of a physical system in which these conditions can be realized. How is it related to Planck's law?

2. What is a black body? Do any such bodies exist? Why is the concept of a black body useful?

3. What is a gray body? Why is the concept of such a body useful? Can a black body be considered gray?

4. In specular (mirrorlike) reflection the angle of incidence equals the angle of reflection. How are these angles related for diffuse reflection?

5. What is the physical significance of the view factor F_{12} and of the factor $\bar{F}_{12}$.

6. Under what conditions is the effect of geometry on radiant-heat interchange between two surfaces completely expressible in terms of F_{12}? In terms of $\bar{F}_{12}$?

7. Which of the equations in this chapter show that the apparent brightness of a black body with a uniform surface temperature is independent of the position (distance and direction) from which it is viewed?

8. Is it possible for a very thick layer of luminous gas to radiate more intensely than a black body at the same temperature?

9. What is the relation analogous to Eq. 14.3–2 for an ideal monatomic gas?

10. What are the units of $q^{(e)}$, $q_\nu^{(e)}$, $q_\lambda^{(e)}$?

11. Verify the numerical value and units of σ given after Eq. 14.3–8 by computing σ from the universal constants.

12. What is the Wien displacement law? Lambert's law?

PROBLEMS

14.A$_1$ Approximation of a Black Body by a Hole in a Sphere

A thin sphere of copper, with its internal surface highly oxidized, has a diameter of 6 in. How small a hole must be made in the sphere to make an opening that will have an absorptivity of 0.99? *Answer:* Radius = 0.70 in.

14.B$_1$ Efficiency of a Solar Engine

A device for utilizing solar energy, developed by Abbot,[1] consists of a parabolic mirror that focuses the impinging sunlight onto a Pyrex tube containing a high-boiling, nearly black liquid. This liquid is circulated to a heat exchanger in which the heat energy is transferred to super-heated water at 25 atm pressure. Steam may be withdrawn and used to run an engine. The most efficient design requires a mirror 10 ft in diameter to generate 2 hp when the axis of the mirror is pointed directly at the sun. What is the over-all efficiency of the device? *Answer:* 15 per cent

14.C$_1$ Radiant Heating Requirement

A shed is rectangular in shape, with the floor 15 by 30 ft and the roof $7\frac{1}{2}$ ft above the floor. The floor is heated by hot water running through coils. On cold winter days the walls and roof are at about $-10°$ F. At what rate must heat be supplied through the floor in order to maintain the floor temperature at 75° F? (Assume that all the surfaces of the system are black.)

[1] C. G. Abbot, in *Solar Energy Research* (F. Daniels and J. A. Duffie, Eds.), University of Wisconsin Press, Madison, (1955), pp. 91–95; see also U.S. Patent No. 2,460,482 (Feb. 1, 1945).

14.D_1 Steady-State Temperature of a Roof

Estimate the maximum temperature attained by a level roof at 45° north latitude on June 21 in clear weather. Radiation from sources other than the sun may be neglected, and a convection-heat-transfer coefficient of 2.0 Btu hr^{-1} ft^{-2} ° F^{-1} may be assumed. A maximum temperature of 100° F may be assumed for the surrounding air.

a. Solve for the case of a perfectly black roof.

b. Solve for aluminum-coated roof, with an absorptivity of 0.3 for solar radiation and an emissivity of 0.07 at the temperature of the roof.

The solar constant of Example 14.4–1 may be used, and the absorption and scattering of the sun's rays by the atmosphere are to be neglected.

14.E_1 Radiation Errors in Temperature Measurements

The temperature of an air stream in a duct is being measured by means of a thermocouple. The thermocouple wires and junction are cylindrical, 0.05 in. in diameter, and extend across the duct perpendicular to the flow with the junction in the center of the duct. Estimate the temperature of the gas stream from the following data obtained under steady conditions:

Thermocouple junction temperature	= 500° F
Duct wall temperature	= 300° F
Emissivity of thermocouple wires	= 0.8
Convection heat transfer coefficient from wire to air	= 50 Btu hr^{-1} ft^{-2} ° F^{-1}

The wall temperature is constant at the value given for 20 duct diameters upstream and downstream of the thermocouple installation. The thermocouple leads are positioned so that the effect of conduction losses on the junction temperature may be neglected.

14.F_2 Mean Temperature for Effective Emissivity

Show that if the emissivity increases linearly with the temperature Eq. 14.5–3 may be written as

$$Q_{12} = e_1^\circ \sigma A_1(T_1^4 - T_2^4) \tag{14.F-1}$$

in which e_1° is the emissivity of surface "1" evaluated at a reference temperature T° given by

$$T^\circ = \frac{T_1^5 - T_2^5}{T_1^4 - T_2^4} \tag{14.F-2}$$

14.G_2 Radiation Across an Annulus

Develop an expression for the rate of radiant heat transfer between two long, gray, coaxial cylinders by considering multiple reflections, as in Example 14.5–1. Then show that your answer satisfies Eq. 14.5–5.

$$\textit{Answer:}\quad Q_{12} = \frac{\sigma(T_1^4 - T_2^4)}{\left[\dfrac{1}{A_1 e_1} + \dfrac{1}{A_2}\left(\dfrac{1}{e_2} - 1\right)\right]}$$

where A_1 is the surface area of the inner cylinder

14.H_2 Multiple Radiation Shields

a. Develop an equation for the rate of radiant heat transfer through a series of n very thin, flat, parallel metal sheets, each having a different emissivity e, when the first sheet is at temperature T_1 and the nth sheet is at temperature T_n. Give your result in terms of the $\mathscr{F}$ factors for the successive pairs of planes. Edge effects and conduction across the air gap between the sheets are to be neglected.

b. Determine the ratio of the radiant heat transfer rate for n identical sheets to that for two identical sheets.

c. Compare your result for three sheets with that obtained in Example 14.5–1.

The marked reduction in heat-transfer rates produced by a number of radiation shields in series has led to the use of multiple layers of metal foils for high-temperature insulation.

$14.I_2$ Radiation and Conduction through Absorbing Media

Consider a glass slab extending from $z = 0$ to $z = \delta$ and of infinite extent in the x- and y-directions. The temperatures of the surfaces at $z = 0$ and $z = \delta$ are maintained at T_0 and T_δ, respectively. Incident on the face at $z = 0$ is a uniform monochromatic radiant beam in the z-direction of intensity $q_0^{(r)}$. Emission within the slab, and incident radiation in the $-z$-direction, can be neglected.

a. Determine the temperature distribution in the slab, assuming m_a and k to be constants.

b. How does the distribution of the conduction energy flux q_z depend on m_a?

$14.J_3$ Cooling of a Black Body in Vacuo

A thin black body of very high thermal conductivity has a volume V, surface area A, density ρ, and heat capacity $\hat{C}_p$. At time $t = 0$ this body at temperature T_1 is placed in a black enclosure, the walls of which are maintained permanently at temperature $T_2 (T_2 < T_1)$. Derive an expression for the temperature T of the black body as a function of time.

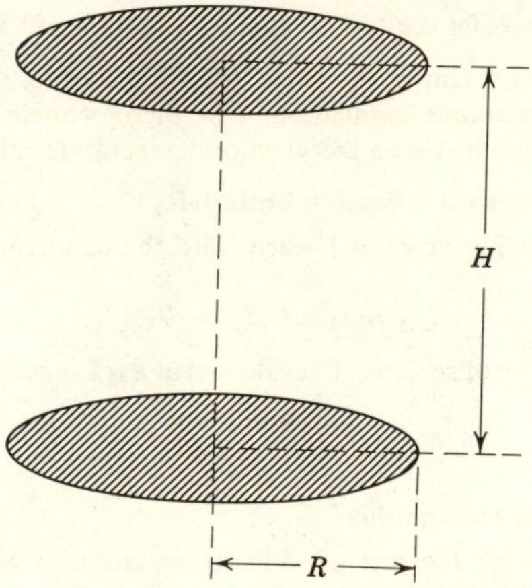

Fig. 14.K. Two perfectly black disks.

$14.K_3$ Integration of the View-Factor Integral for a Specific Problem

Two identical, perfectly black disks of radius R are placed a distance H apart, as shown in Fig. 14.K. Integrate the view-factor integrals for this case and show that

$$F_{12} = F_{21} = \frac{1 + 2B^2 - \sqrt{1 + 4B^2}}{2B^2} \tag{14.K-1}$$

in which $B = R/H$; this result was derived by Christiansen.[2]

[2] C. Christiansen, *Wiedemann's Ann.*, **19**, 267 (1883); see also M. Jakob, *Heat Transfer*, Wiley, New York (1957), Vol. II, p. 14.

14.L_4 Heat Loss from Wire Carrying an Electric Current[3]

An electrically heated wire of length L loses heat to the surroundings by radiative heat transfer. If the ends of the wire are maintained at constant temperature T_0, obtain an expression for the axial variation in wire temperature. The wire can be considered to be radiating into a black enclosure at temperature T_0.

[3] H. S. Carslaw and J. C. Jaeger, *Conduction of Heat in Solids*, Oxford University Press (1959), Second Edition, pp. 154–156.

CHAPTER 15

Macroscopic Balances for Nonisothermal Systems

In Chapter 7 we discussed the macroscopic mass balance, momentum balance, and mechanical energy balance. The discussion there was restricted to systems at constant temperature. Actually, this restriction is somewhat artificial, inasmuch as real flow systems always have some mechanical energy being converted into thermal energy. What we really assumed in Chapter 7 is that any heat so produced is either too small to change the fluid properties or is immediately conducted away through the walls of the system containing the fluid. In this chapter we generalize the previous results to describe the over-all behavior of nonisothermal macroscopic flow systems.

For a nonisothermal system, there are four macroscopic balances that describe the relations between the inlet and outlet conditions of the stream:

a. The macroscopic *mass* balance given in Eq. 7.1–2, obtained by integrating the equation of continuity over the volume of the flow system (or derived as indicated in §7.1.)

b. The macroscopic *momentum* balance, given in Eq. 7.2–2, obtained by integrating the equation of motion over the volume of the flow system (or derived as indicated in §7.2.)

c. The macroscopic *energy* balance, given in Eq. 15.1–3, obtained by integrating the energy equation over the volume of the flow system (or derived as indicated in §15.1).

d. The macroscopic *mechanical energy* "balance," special cases of which are given in Eqs. 15.2–1 and 2, obtained by forming the dot product of the local stream velocity with the equation of motion and integrating over the volume of the flow system.

These four relations comprise the conservation statements for mass, momentum, and energy (the last of these being the first law of thermodynamics) and a fourth relation, which is in essence a statement of the restrictions imposed by the second law of thermodynamics; the term "mechanical energy balance" is really somewhat of a misnomer, although it is true that the equation accounts for the various forms of mechanical energy in the system.

The macroscopic mass and momentum balances for nonisothermal systems are the same as those for isothermal systems. Hence there is no need to discuss them further here. We begin with the macroscopic energy balance in §15.1 and then turn to the mechanical energy "balance" in §15.2. Then, after a short summary of all the macroscopic balances in §15.3, we devote the remaining two sections to steady-state and unsteady-state applications. The latter are important in connection with the theory of process control.

§15.1 THE MACROSCOPIC ENERGY BALANCE

We consider the system sketched in Fig. 7.0–1, and assume, as we did in Chapter 7, that the time-smoothed velocities at "1" and "2" are parallel to the conduit walls and that the fluid properties do not vary across the cross section.

We now apply the statement of conservation of energy to this system. We could use the conservation statement of Eq. 10.1–1 and include the work that the fluid does against gravity in the last term. It is more usual, however, to introduce the concept of potential energy (see Eq. 10.1–14) and exclude the work against gravity from the last term of Eq. 10.1–1. In other words, we prefer to base our development on the energy conservation statement embodied in Eq. *N* of Table 10.4–1, which in words is stated

$$\begin{Bmatrix} \text{rate of accumulation} \\ \text{of internal, kinetic,} \\ \text{and potential energy} \end{Bmatrix} = \begin{Bmatrix} \text{rate of internal,} \\ \text{kinetic, and} \\ \text{potential energy} \\ \text{in} \\ \text{by convection} \end{Bmatrix} - \begin{Bmatrix} \text{rate of internal,} \\ \text{kinetic, and} \\ \text{potential energy} \\ \text{out} \\ \text{by convection} \end{Bmatrix}$$

$$+ \begin{Bmatrix} \text{net rate of heat} \\ \text{addition } \textit{to} \text{ system} \\ \text{from surroundings} \end{Bmatrix} - \begin{Bmatrix} \text{net rate of work} \\ \text{done } \textit{by} \text{ system} \\ \text{on surroundings} \end{Bmatrix} \qquad (15.1\text{–}1)$$

This statement includes the tacit assumption that the external force (i.e., gravity) is not changing with time and that it is derivable from a potential energy; otherwise, the statement in Eq. 10.1–1 should be used.

We now apply Eq. 15.1–1 to the macroscopic flow system, including in the last term the work W transferred by means of moving parts (e.g. a turbine or compressor) and the work needed to introduce new material at "1" and remove material at "2"; in Q we include all thermal energy entering the fluid via the solid surfaces of the system and by conduction or radiation in the fluid at the inlet and outlet:

$$\begin{aligned}\frac{d}{dt}(U_{\text{tot}} + K_{\text{tot}} + \Phi_{\text{tot}}) = {} & \rho_1\langle\bar{v}_1\rangle\hat{U}_1S_1 - \rho_2\langle\bar{v}_2\rangle\hat{U}_2S_2 \\ & + \tfrac{1}{2}\rho_1\langle\bar{v}_1^{\,3}\rangle S_1 - \tfrac{1}{2}\rho_2\langle\bar{v}_2^{\,3}\rangle S_2 \\ & + \rho_1\langle\bar{v}_1\rangle\hat{\Phi}_1S_1 - \rho_2\langle\bar{v}_2\rangle\hat{\Phi}_2S_2 \\ & + Q - W \\ & + p_1\langle\bar{v}_1\rangle S_1 - p_2\langle\bar{v}_2\rangle S_2 \end{aligned} \tag{15.1–2}$$

Here U_{tot}, K_{tot}, and Φ_{tot} are the total internal, kinetic, and potential energies in the system.[1] This equation may be written in more compact form by introducing the mass rates of flow $w_1 = \rho_1\langle\bar{v}_1\rangle S_1$ and $w_2 = \rho_2\langle\bar{v}_2\rangle S_2$ and the total energy $E_{\text{tot}} = U_{\text{tot}} + K_{\text{tot}} + \Phi_{\text{tot}}$; we then have for the *unsteady-state macroscopic energy balance*

$$\frac{d}{dt}E_{\text{tot}} = -\Delta\left[\left(\hat{U} + p\hat{V} + \frac{1}{2}\frac{\langle\bar{v}^3\rangle}{\langle\bar{v}\rangle} + \hat{\Phi}\right)w\right] + Q - W \tag{15.1–3}$$

This is just the statement of the first law of thermodynamics as applied to a flow system. Note that our sign convention for Q and W is consistent with that used by most thermodynamicists. Equation 15.1–3 can also be obtained by integrating the energy balance equation (Eq. N of Table 10.4–1) over the volume of the flow system. (See Problem 15.O.)

If the flow system is operating at steady state, then dE_{tot}/dt is zero, and $w_1 = w_2 = w$. Equation 15.1–3 may then be divided through by w to give the *steady-state macroscopic energy balance*

$$\Delta\left(\hat{U} + p\hat{V} + \frac{1}{2}\frac{\langle\bar{v}^3\rangle}{\langle\bar{v}\rangle} + \hat{\Phi}\right) = \hat{Q} - \hat{W} \tag{15.1–4}$$

Here $\hat{Q}$ is the heat added per unit mass of fluid flowing through the system, and $\hat{W}$ is the amount of work done by a unit mass of fluid in traversing the system from "1" to "2."

[1] These quantities are defined as

$$U_{\text{tot}} = \int_V \rho\hat{U}\,dV; \qquad K_{\text{tot}} = \int_V \tfrac{1}{2}\rho v^2\,dV; \qquad \Phi_{\text{tot}} = \int_V \rho\hat{\Phi}\,dV \tag{15.1–2a,b,c}$$

The quantity $\hat{U} + p\hat{V}$, which appears in the foregoing equations, is called the enthalpy (per unit mass) and is given the symbol $\hat{H}$. Two limiting forms of $\Delta\hat{H}$ are useful for making engineering calculations:

ideal gases:
($p\tilde{V} = RT$ and $\tilde{C}_p - \tilde{C}_v = R$)

$$\Delta\hat{H} = \int_{T_1}^{T_2} \hat{C}_p \, dT = \frac{R}{M} \int_{T_1}^{T_2} \frac{\gamma}{\gamma - 1} \, dT \tag{15.1–5}$$

incompressible liquids:
(ρ = constant and $\tilde{C}_p = \tilde{C}_v$)

$$\Delta\hat{H} = \int_{T_1}^{T_2} \hat{C}_p \, dT + \frac{1}{\rho}(p_2 - p_1) \tag{15.1–6}$$

In order to complete the evaluation of $\Delta\hat{H}$ in Eqs. 15.1–5, 6, one has to have information regarding the dependence of $\hat{C}_p$ on temperature.

For many chemical plant systems, kinetic energy, potential energy, and work effects are small, with the result that a large class of systems of interest in engineering are described by the relation

$$\Delta\hat{H} = \hat{Q} \tag{15.1–7}$$

Problems of this type are discussed extensively in chemical engineering courses on material and energy balances.[2]

With regard to the terms containing $\frac{1}{2}\langle\bar{v}^3\rangle/\langle\bar{v}\rangle$, the same statements made in §7.3 (just after Eq. 7.3–2) apply here as well. As long as the kinetic energy terms are small, and particularly if the fluid is in turbulent flow, it is an excellent approximation to write $\frac{1}{2}\langle\bar{v}\rangle^2$ instead of the exact expression.

The term $\Delta\hat{\Phi}$ is generally replaced by $g\,\Delta h$, in which Δh represents the difference in elevation between planes "1" and "2."

A careful distinction has to be made between the quantities on the two sides of Eq. 15.1–4. Those on the left depend solely on the conditions at the reference planes "1" and "2"; that is, they are "point functions." Those on the right, on the other hand, depend on the thermodynamic path followed by the fluid. The evaluation of $\hat{Q}$ and $\hat{W}$ therefore presents special difficulties. For those problems in which $\hat{Q}$ is known or $\hat{Q}$ is the only unknown, Eq. 15.1–4 is directly usable. For those problems in which $\hat{Q}$ has to be calculated in terms of the local heat-transfer coefficient and local temperature differences, it is more convenient to write Eq. 15.1–4 for an incremental volume, thereby generating a differential equation. This equation is then integrated between planes "1" and "2"; this procedure yields some information regarding the temperature distribution in the direction of flow, as indicated in Example 15.4–2. Whereas heat is usually added over extended surfaces in a system, the performance of work is generally localized. Hence it is customary to regard work as being done at a specific cross section of flow.

[2] O. A. Hougen, K. M. Watson, and R. A. Ragatz, *Chemical Process Principles*, Wiley, New York (1954), Part I, pp. 247, *et seq.*

§15.2 THE MACROSCOPIC MECHANICAL ENERGY BALANCE (Bernoulli Equation)

When Eq. 3.3–2 is integrated over the volume of the flow system in Fig. 7.0–1, one obtains an equation that describes the mechanical energy changes between the inlet and outlet of the system. There are two important limiting forms of this result: *isothermal flow* (already given as Eq. 7.3–1) and *isentropic flow.* We give both here as special forms of the *unsteady-state mechanical energy-balance*:[1]

isothermal flow:

$$\frac{d}{dt}(K_{\text{tot}} + \Phi_{\text{tot}} + A_{\text{tot}}) = -\Delta\left[\left(\frac{1}{2}\frac{\langle \bar{v}^3 \rangle}{\langle \bar{v} \rangle} + \hat{\Phi} + \hat{G}\right)w\right] - W - E_v \tag{15.2–1}$$

isentropic flow:

$$\frac{d}{dt}(K_{\text{tot}} + \Phi_{\text{tot}} + U_{\text{tot}}) = -\Delta\left[\left(\frac{1}{2}\frac{\langle \bar{v}^3 \rangle}{\langle \bar{v} \rangle} + \hat{\Phi} + \hat{H}\right)w\right] - W - E_v \tag{15.2–2}$$

in which E_v is the integral in Eq. 7.4–1, which represents the total rate of irreversible conversion of mechanical to internal energy. Many physical processes are either approximately isothermal or isentropic, so that these limiting equations are useful. It is instructive to note that Eqs. 15.1–3 and 15.2–2 are almost the same. In fact, by combining them we find that $-Q = E_v$ for an isentropic process; that is, the rate of thermal-energy production by viscous dissipation is equal to the rate of heat removal through the walls of the system. Note that the restriction $-Q = E_v$ does *not* imply isothermal behavior.

For those systems in which the flow behavior is independent of the time, both of the foregoing relations as well as the more general equation from which they are derived may be simplified to a single relation, which is the *steady-state mechanical energy balance*:

$$\Delta\frac{1}{2}\frac{\langle \bar{v}^3 \rangle}{\langle \bar{v} \rangle} + \Delta\hat{\Phi} + \int_1^2 \frac{1}{\rho}\,dp + \hat{W} + \hat{E}_v = 0 \tag{15.2–3}$$

This result, sometimes called the *Bernoulli equation,* has already been discussed for isothermal systems in §7.3. The only terms that need further discussion here are the integral and $\hat{E}_v$.

The integral must be evaluated along a representative "streamline" in the system. To do this, one must know the equation of state $\rho = \rho(p, T)$ and also how T changes with p along the streamline. In Fig. 15.2–1 the surface $\hat{V}(p, T)$ for an ideal gas is shown. In the pT-plane there is shown a curve

[1] R. B. Bird, *Chem. Eng. Sci.*, **6**, 123–131 (1957).

beginning at p_1, T_1 (the inlet stream conditions) and ending at p_2, T_2 (the outlet conditions). The curve in the pT-plane indicates the succession of states through which the gas passes in going from the initial state to the final state. The integral $\int_1^2 (1/\rho)\,dp$ or $\int_1^2 \hat{V}\,dp$ is then the projection of the shaded area in Fig. 15.2–1 on the $p\hat{V}$-plane. It should be clear that the value of this integral

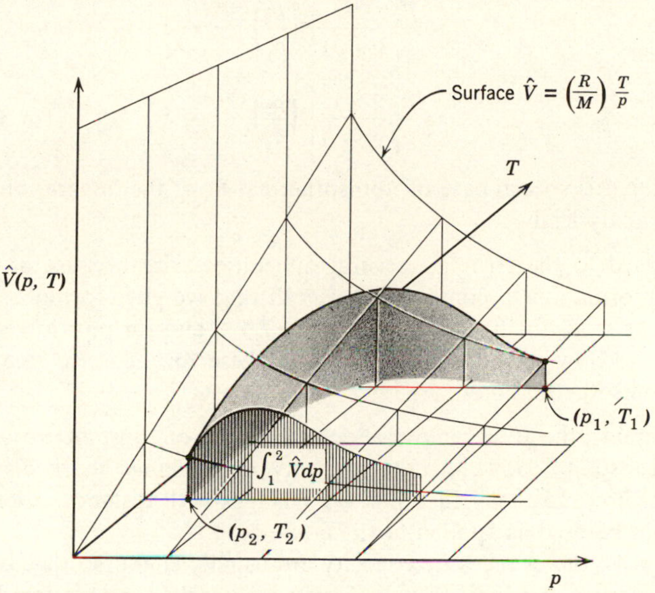

Fig. 15.2–1. Graphical representation of the integral in Eq. 15.2–3. The ruled area is $\int_1^2 \hat{V}\,dp$. Note that the value of this integral is *negative* here because we are integrating from right to left.

changes as the "thermodynamic path" of the process from 1 to 2 is altered. If one knows the path and the equation of state, then one can compute $\int_1^2 \hat{V}\,dp$.

The integral may easily be evaluated for two special cases:

a. For *isothermal systems*, the integral is easily evaluated by prescribing the isothermal equation of state—that is, by giving a relation for ρ as a function of p. For example, for ideal gases $\rho = pM/RT$ and

$$\int_1^2 \frac{1}{\rho}\,dp = \frac{RT}{M}\int_{p_1}^{p_2} \frac{1}{p}\,dp = \frac{RT}{M}\ln\frac{p_2}{p_1} \qquad \text{(ideal gases)} \qquad (15.2\text{–}4)$$

And for incompressible liquids ρ is constant so that

$$\int_1^2 \frac{1}{\rho}\,dp = \frac{1}{\rho}(p_2 - p_1) \qquad \text{(incompressible liquids)} \qquad (15.2\text{–}5)$$

b. For frictionless *adiabatic flow of ideal gases* with constant heat capacity, p and ρ are related according to the relation $p\rho^{-\gamma}$ = constant, in which $\gamma = \hat{C}_p/\hat{C}_v$, as shown in Example 10.5–6. Then the integral becomes

$$\int_1^2 \frac{1}{\rho}\,dp = \frac{p_1^{1/\gamma}}{\rho_1}\int_{p_1}^{p_2}\frac{dp}{p^{1/\gamma}}$$

$$= \frac{p_1}{\rho_1}\frac{\gamma}{\gamma - 1}\left[\left(\frac{p_2}{p_1}\right)^{\frac{\gamma-1}{\gamma}} - 1\right]$$

$$= \frac{p_1}{\rho_1}\frac{\gamma}{\gamma - 1}\left[\left(\frac{\rho_2}{\rho_1}\right)^{\gamma-1} - 1\right] \qquad (15.2\text{–}6)$$

Hence, for this special case of nonisothermal flow, the integration can be performed analytically.

With regard to the term $\hat{E}_v$, several difficulties arise because of lack of sufficient information to compute it. Recall that we gave formulas for the contributions to $\hat{E}_v$ for fittings losses in Eq. 7.4–5 and for pipe losses in Eq. 7.4–9. Let us now consider $\hat{E}_v$ in terms of these formulas for two special kinds of nonisothermal flow:

a. For liquids, the average flow velocity in a tube of constant cross section is nearly constant. However, the viscosity may change markedly in the direction of flow, so that f in Eq. 7.4–9 changes with distance. Hence Eq. 7.4–9 cannot be used as applied to the entire pipe.

b. For gases, the changes in viscosity are usually slight, so that the local Reynolds number and local friction factor are nearly constant for ducts of constant cross section. However, the average velocity may change considerably along the duct as a result of changes in density. Hence Eq. 7.4–9 cannot be used as applied to the entire duct.

Therefore, for systems in which μ or ρ are changing significantly in the direction of flow, we rewrite Eq. 15.2–3 on an incremental basis and in this way generate a differential equation. In such a differential balance one can take into account the change in local friction factor or local average velocity. In this way one can describe nonisothermal flow in conduits. (See Example 15.4–3.)

The energy losses in fittings can be obtained from Eq. 7.4–5 for liquids. For gases, the same formula may be used, provided the flow is subsonic.

§15.3 SUMMARY OF THE MACROSCOPIC BALANCES FOR PURE FLUIDS

In describing the relations between the input and output stream conditions of a flow system, it is important to use all four of the relations discussed above—the statements of the laws of conservation of mass, momentum, and energy, and the Bernoulli equation. In addition, one needs information about the thermal equation of state [$\rho = \rho(p, T)$] and the caloric equation of state [usually in the form $\hat{C}_p = \hat{C}_p(p, T)$]. The four macroscopic balances are summarized in Table 15.3–1 for ready reference.

§15.4 USE OF THE MACROSCOPIC BALANCES FOR SOLVING STEADY-STATE PROBLEMS

In §10.5 it was shown how to find a wide variety of steady-state temperature profiles by solving simplified forms of the general equations of change. Here, in this section, we show how to use simplified forms of the general macroscopic balances to set up the relations that describe the entrance and exit conditions of a steady-state flow system. By keeping track of the discarded terms, we know exactly what restrictions have to be placed on the derived results.

In the following examples the utility of macroscopic balances is demonstrated and the meaning of the various terms in them is illustrated. In Examples 15.4–2 and 3 we make use of the macroscopic balances as applied to a system between control surfaces that are infinitesimally separated.

Example 15.4–1. The Cooling of an Ideal Gas

Two hundred pounds per hour of dry air enter the inner tube of the heat exchanger shown in Fig. 15.4–1 at 300° F and 30 psia and at a velocity of 100 ft sec^{-1}. The air leaves the exchanger at 0° F and 15 psia and 10 ft above the exchanger entrance.

Calculate the rate of energy removal across the tube wall. Assume turbulent flow and ideal-gas behavior, and use the following expression for the heat capacity of air:

$$\tilde{C}_p = 6.39 + (9.8 \times 10^{-4})T - (8.18 \times 10^{-8})T^2 \tag{15.4–1}$$

where $\tilde{C}_p$ is in Btu lb-mole^{-1} ° R^{-1} and T is in ° R.

Solution. For this system, the macroscopic energy balance, Eq. 15.1–4, becomes

$$(\hat{H}_2 - \hat{H}_1) + \tfrac{1}{2}(v_2^2 - v_1^2) + (\hat{\Phi}_2 - \hat{\Phi}_1) = \hat{Q} \tag{15.4–2}$$

For ideal gases, we may express enthalpy according to Eq. 15.1–5, and we may express velocity in terms of temperature and pressure with the aid of the macroscopic mass balance $\rho_1 v_1 = \rho_2 v_2$ and the ideal gas law $p = \rho RT/M$. We may therefore rewrite Eq. 15.4–2 as

$$\frac{1}{M}\int_1^2 \tilde{C}_p\, dT + \frac{1}{2}v_1^2\left[\left(\frac{p_1 T_2}{p_2 T_1}\right)^2 - 1\right] + (\hat{\Phi}_2 - \hat{\Phi}_1) = \hat{Q} \tag{15.4–3}$$

The indicated integration can be performed after inserting Eq. 15.4–1 for $\tilde{C}_p$.

TABLE 15.3–1

SUMMARY OF THE MACROSCOPIC BALANCES FOR NONISOTHERMAL FLOW SYSTEMS CONTAINING A SINGLE CHEMICAL SPECIES

Balance	Special Form	Unsteady State	Steady State
Mass	—	$\frac{dm_{tot}}{dt} = -\Delta w$ (A)	$\Delta w = 0$ (F)
Momentum	—	$\frac{d\boldsymbol{P}_{tot}}{dt} = -\Delta\left(\frac{\langle v^2\rangle}{\langle v\rangle}\boldsymbol{w} + p\boldsymbol{S}\right) - \boldsymbol{F} + m_{tot}\boldsymbol{g}$ (B)	$\boldsymbol{F} = -\Delta\left(\frac{\langle v^2\rangle}{\langle v\rangle}\boldsymbol{w} + p\boldsymbol{S}\right) + m_{tot}\boldsymbol{g}$ (G)
Energy	—	$\frac{dE_{tot}}{dt} = -\Delta\left[\left(\hat{U} + p\hat{V} + \frac{1}{2}\frac{\langle v^3\rangle}{\langle v\rangle} + \hat{\Phi}\right)w\right] + Q - W$ (C)	$\Delta\left(\hat{U} + p\hat{V} + \frac{1}{2}\frac{\langle v^3\rangle}{\langle v\rangle} + \hat{\Phi}\right) = \hat{Q} - \hat{W}$ (H)
Mechanical Energy	Isothermal	$\frac{d}{dt}(K_{tot} + \Phi_{tot} + A_{tot}) = -\Delta\left[\left(\frac{1}{2}\frac{\langle v^3\rangle}{\langle v\rangle} + \hat{\Phi} + \hat{G}\right)w\right] - W - E_v$ (D)	$\Delta\left(\frac{1}{2}\frac{\langle v^3\rangle}{\langle v\rangle} + \hat{\Phi} + \hat{G}\right) + \hat{W} + \hat{E}_v = 0$ (I)
	Isentropic	$\frac{d}{dt}(K_{tot} + \Phi_{tot} + U_{tot}) = -\Delta\left[\left(\frac{1}{2}\frac{\langle v^3\rangle}{\langle v\rangle} + \hat{\Phi} + \hat{H}\right)w\right] - W - E_v$ (E)	$\Delta\left(\frac{1}{2}\frac{\langle v^3\rangle}{\langle v\rangle} + \hat{\Phi} + \hat{H}\right) + \hat{W} + \hat{E}_v = 0$ (J)

Note: Special forms of $\Delta\hat{H} = \Delta(\hat{U} + p\hat{V})$ are given in Eqs. 15.1–5 and 6. Special forms of $\int_1^2 (1/\rho)\,dp$ are given in Eqs. 15.2–4, 5 and 6.

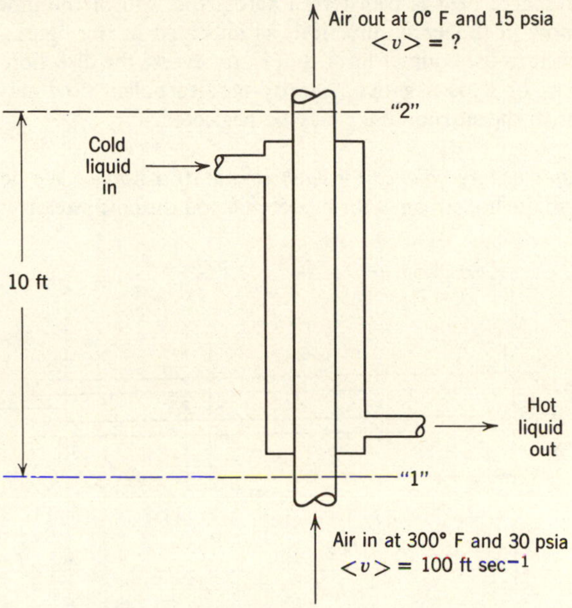

Fig. 15.4–1. Cooling of air in a heat exchanger.

By substituting numerical values into Eq. 15.4–3, we obtain the heat removal per pound of fluid passing through the exchanger:

$$
\begin{aligned}
-\hat{Q} &= \tfrac{1}{29}[(6.39)(300) + \tfrac{1}{2}(9.8 \times 10^{-4})(5.78 - 2.12)(10^5) \\
&\quad -\tfrac{1}{3}(8.18 \times 10^{-8})(4.39 - 0.976)(10^8)] \\
&\quad + \frac{1}{2}\left(\frac{10^4}{(32.2)(778)}\right)[1 - (1.21)^2] - \left(\frac{10}{778}\right) \\
&= 72.0 - 0.093 - 0.0128 \\
&= 71.9 \text{ Btu lb}^{-1} \qquad (15.4\text{–}4)
\end{aligned}
$$

The rate of heat removal is then

$$-\hat{Q}w = 14{,}400 \text{ Btu hr}^{-1} \qquad (15.4\text{–}5)$$

Note in Eq. 15.4–4 that the kinetic and potential energy contributions are negligible in comparison with the change in enthalpy.

Example 15.4–2. Parallel- or Counter-Flow Heat Exchangers

It is desired to describe the performance of the simple double-pipe heat exchanger shown in Fig. 15.4–2 in terms of the heat-transfer coefficients of the two streams and the thermal resistance of the pipe wall. The exchanger consists essentially of two coaxial pipes with one fluid stream flowing through the inner pipe and another in the annular space; heat is transferred across the wall of the inner pipe. Both streams may flow in the same direction, as indicated in the figure, but normally it is more efficient to use counter flow, that is, to reverse the direction of one stream so that either w_h or w_c is negative. Steady-state turbulent flow may be assumed, and heat losses to the surroundings may be neglected.

Solution.

a. Macroscopic energy balance for each stream as a whole. We designate quantities referring to the hot stream with subscript h and the cold stream with subscript c.

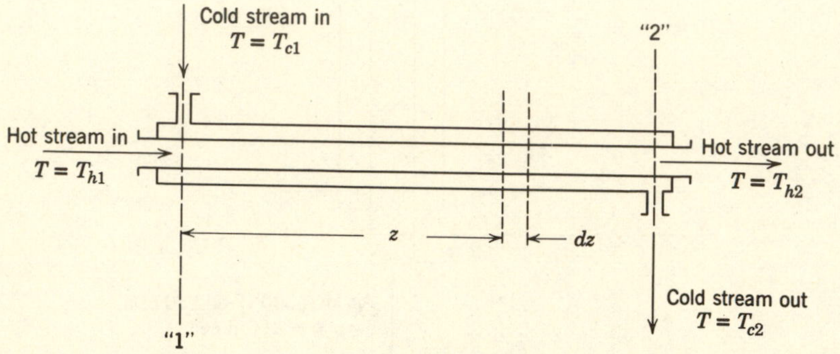

Fig. 15.4–2. A double-pipe heat exchanger.

The steady-state energy balance in Eq. 15.1–4 becomes (for negligible changes in kinetic and potential energy)

$$w_h \, \Delta\hat{H}_h = Q_h \tag{15.4–6}$$

$$w_c \, \Delta\hat{H}_c = Q_c \tag{15.4–7}$$

Because there is no heat loss to the surroundings, $Q_h = -Q_c$. For incompressible liquids flowing under reasonable pressure drop or for ideal gases, Eqs. 15.1–5 and 6 give for constant $\hat{C}_p$ the relation $\Delta\hat{H} = \hat{C}_p \, \Delta T$. Hence Eqs. 15.4–6 and 7 can be rewritten as

$$w_h\hat{C}_{ph}(T_{h2} - T_{h1}) = Q_h \tag{15.4–8}$$

$$w_c\hat{C}_{pc}(T_{c2} - T_{c1}) = Q_c \tag{15.4–9}$$

b. Macroscopic energy balance applied in differential form. Similar application of Eq. 15.1–4 to a segment dz of the hot stream in the heat exchanger (see Fig. 15.4–2) gives

$$w_h\hat{C}_{ph} \, dT_h = dQ_h \tag{15.4–10}$$

According to Eq. 9.6–30 (or Eq. 13.1–8), dQ_h can be replaced by $U_0(2\pi r_0 \, dz) \times (T_c - T_h)$, in which U_0 is the over-all heat-transfer coefficient given by Eq. 9.6–31 (or Eq. 13.1–9). Hence, for the hot stream, we have

$$\frac{dT_h}{T_c - T_h} = U_0 \frac{2\pi r_0 \, dz}{w_h\hat{C}_{ph}} \tag{15.4–11}$$

The corresponding equation for the cold stream is clearly

$$-\frac{dT_c}{T_c - T_h} = U_0 \frac{2\pi r_0\, dz}{w_c \hat{C}_{pc}} \tag{15.4–12}$$

By adding Eqs. 15.4–11 and 12, we obtain an expression for the temperature difference between the two fluids at any cross section in the exchanger:

$$-\frac{d(T_h - T_c)}{(T_h - T_c)} = U_0\left(\frac{1}{w_h \hat{C}_{ph}} + \frac{1}{w_c \hat{C}_{pc}}\right)(2\pi r_0\, dz) \tag{15.4–13}$$

By assuming that U_0 is independent[1] of z and integrating both sides of this equation between planes 1 and 2, we get

$$\ln\left(\frac{T_{h1} - T_{c1}}{T_{h2} - T_{c2}}\right) = U_0\left(\frac{1}{w_h \hat{C}_{ph}} + \frac{1}{w_c \hat{C}_{pc}}\right)(2\pi r_0 L) \tag{15.4–14}$$

This expression relates the temperature change of each stream with the stream rates and exchanger dimensions, and it can thus be used to describe the performance of the exchanger. However, it is conventional to rearrange Eq. 15.4–14 by taking advantage of the steady-state energy balances in Eqs. 15.4–8 and 9. We solve each of these relations for $w\hat{C}_p$ and substitute the results into Eq. 15.4–14 to obtain

$$Q_c = U_0(2\pi r_0 L)\left\{\frac{(T_{h2} - T_{c2}) - (T_{h1} - T_{c1})}{\ln\left(\dfrac{T_{h2} - T_{c2}}{T_{h1} - T_{c1}}\right)}\right\} \tag{15.4–15}$$

or

$$Q_c = U_0 A_0 (T_h - T_c)_{\ln} \tag{15.4–16}$$

Here A_0 is the total outer surface area of the inner tube and $(T_h - T_c)_{\ln}$ is the "logarithmic mean temperature difference" between the two streams. Equations 15.4–15 and 16 describe the rate of heat exchange between the two streams and find wide application in engineering practice. Note that the stream rates do not appear explicitly in these equations, which are valid for both parallel flow and counter flow. (See Problem 15.A.)

From Eqs. 15.4–11 and 12 we can also get the temperature as a function of z if desired.

Example 15.4–3. Power Requirement for Pumping a Compressible Fluid Through a Long Pipe

A natural gas, which may be considered pure methane, is to be pumped through a long smooth pipeline with a 2-ft inside diameter. The gas enters the line at 100

[1] The case of variable coefficients is treated briefly in Problem 15.E. Considerable care must be used in applying the results of this example to laminar flow for which the variation of U_0 with position may be quite large.

psia at a velocity of 40 ft sec^{-1} and at the ambient temperature of 70° F. Pumping stations are provided every 10 miles along the line, and at each of these stations the gas is recompressed and cooled to its original temperature and pressure. (See Fig. 15.4–3.) Estimate the power that must be expended on the gas by each pumping station, assuming ideal gas behavior, flat velocity profiles, and negligible changes in elevation.

Solution. We find it convenient to consider the pipe and compressor separately. First we apply Eq. 15.2–3 in differential form to a length dz of the pipe. We then integrate this equation between planes "1" and "2" to obtain the unknown pressure

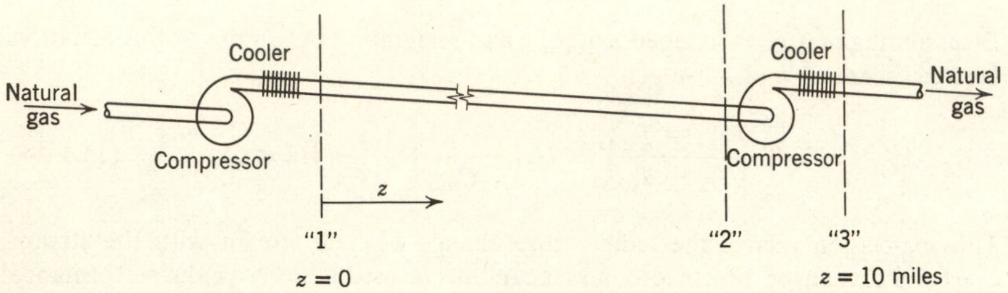

Fig. 15.4–3. Pumping a compressible fluid through a pipeline.

p_2. Once we know p_2, we may apply Eq. 15.2–3 to the system between planes "2" and "3" and obtain the work done by the pump.

a. Flow through the pipe. For this portion of the system, Eq. 15.2–3 may be expressed as

$$v\,dv + \frac{1}{\rho}\,dp + \frac{2v^2 f}{D}\,dz = 0 \tag{15.4–17}$$

Since the pipe is quite long, we assume the flow to be isothermal at 70° F. We may then eliminate both v and ρ from Eq. 15.4–17 by use of the assumed equation of state, $p = \rho RT/M$, and the macroscopic mass balance, which may be written $\rho v = \rho_1 v_1$. With ρ and v written in terms of pressure, Eq. 15.4–17 becomes

$$-\frac{1}{p}\,dp + \frac{RT_1}{M}\frac{p\,dp}{(p_1 v_1)^2} + \frac{2f\,dz}{D} = 0 \tag{15.4–18}$$

It was shown in §1.3 that the viscosity of ideal gases is unaffected by pressure. It then follows that the Reynolds number of the gas, Re $= Dw/S\mu$, hence the friction factor f, must be constant. We may then integrate Eq. 15.4–18 to obtain

$$-\ln\frac{p_2}{p_1} + \frac{1}{2}\left[\left(\frac{p_2}{p_1}\right)^2 - 1\right]\frac{RT_1}{Mv_1^2} + \frac{2fL}{D} = 0 \tag{15.4–19}$$

This equation gives p_2 in terms of quantities that are already known, except for f which can easily be calculated: the kinematic viscosity of methane at 100 psi and 70° F is about 2.61×10^{-5} ft^2 sec^{-1}, and therefore Re $= (2.00 \text{ ft})(40 \text{ ft sec}^{-1})/(2.61 \times 10^{-5} \text{ ft}^2 \text{ sec}^{-1}) = 3.07 \times 10^6$. The friction factor can then be estimated as 0.0025 (Fig. 6.2–2). By substituting numerical values into Eq. 15.4–19, we obtain

$$-\ln\frac{p_2}{p_1} + \frac{1}{2}\left[\left(\frac{p_2}{p_1}\right)^2 - 1\right]\frac{(1545)(530)(32.2)}{(16.04)(1600)} + \frac{(2)(0.0025)(52{,}800)}{(2.00)} = 0 \qquad (15.4\text{–}20)$$

or

$$-\ln\frac{p_2}{p_1} + 514\left[\left(\frac{p_2}{p_1}\right)^2 - 1\right] + 132 = 0 \qquad (15.4\text{–}21)$$

By solving this equation with $p_1 = 100$ psia, we obtain $p_2 = 86$ psia. We are now ready to apply the mechanical energy balance to the compressor.

b. Flow through the compressor. Here we may put Eq. 15.2–3 in the form

$$-\hat{W} = \tfrac{1}{2}(v_3^2 - v_2^2) + \int_2^3 \frac{1}{\rho}\,dp + \hat{E}_v \qquad (15.4\text{–}22)$$

To evaluate the integral in this equation, we assume[2] that the compression is adiabatic and neglect $\hat{E}_v$ between planes "2" and "3." We may use the results of Example 10.5–6 to rewrite Eq. 15.4–22 as

$$-\hat{W}' = \tfrac{1}{2}(v_3^2 - v_2^2) + \frac{p_2^{1/\gamma}}{\rho_2}\int_{p_2}^{p_3} p^{-1/\gamma}\,dp$$

$$= \frac{v_1^2}{2}\left[1 - \left(\frac{p_1}{p_2}\right)^2\right] + \frac{RT_2}{M}\frac{\gamma}{\gamma - 1}\left[\left(\frac{p_1}{p_2}\right)^{\frac{\gamma-1}{\gamma}} - 1\right] \qquad (15.4\text{–}23)$$

in which $-\hat{W}'$ is the *calculated* energy required of the compressor. By substituting numerical values into Eq. 15.4–23, we get

$$-\hat{W}' = \frac{1600}{2 \times 32.2}[1-(1.163)^2] + \frac{(1545)(530)}{16}\frac{1.3}{0.3}\left[1.163^{\left(\frac{0.3}{1.3}\right)} - 1\right]$$

$$= -9 + 7834 = 7825 \text{ ft lb}_f\text{ lb}_m^{-1} \qquad (15.4\text{–}24)$$

The power required to compress the fluid is

$$-w\hat{W}' = -\frac{\pi D^2}{4}\frac{p_1 M}{RT_1}v_1\hat{W}'$$

$$= \pi\frac{(100)(16.04)(40)}{(10.73)(530)}(7825) \text{ ft lb}_f\text{ sec}^{-1}$$

$$= 277{,}000 \text{ ft lb}_f\text{ sec}^{-1} \text{ or } 504 \text{ hp} \qquad (15.4\text{–}25)$$

[2] These assumptions are conventional in the design of compressor-cooler combinations. Note that the energy required to run the compressor is greater than the calculated work, $-\hat{W}'$, by (i) $\hat{E}_v$ between "2" and "3," (ii) mechanical losses in the compressor itself, and (iii) errors in the assumed p–ρ path. Normally the energy required at the pump shaft is at least 15 to 20 per cent greater than $-\hat{W}'$.

The power required for adiabatic flow in both pipeline and compressor is about 4850 hp. (See Problem 15.B.)

Example 15.4–4. Mixing of Two Ideal-Gas Streams

Two steady turbulent streams of the same ideal gas flowing at different velocities, temperatures, and pressures are mixed as shown in Fig. 15.4–4. Calculate the velocity, temperature, and pressure of the resulting stream.

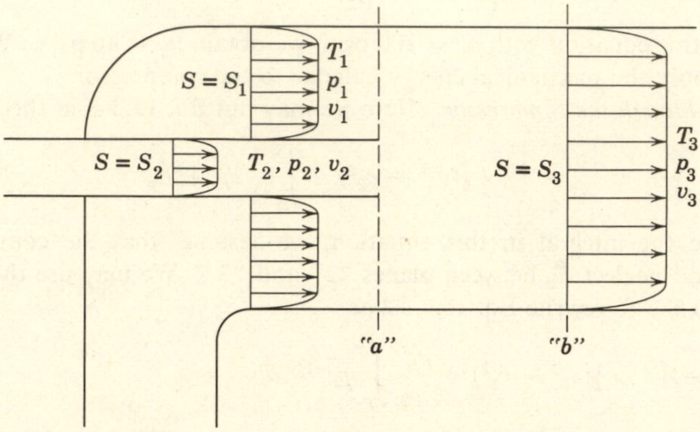

Fig. 15.4–4. Mixing of two ideal-gas streams.

Solution. The behavior here is more complex than that in Example 7.5–2 (incompressible isothermal fluids), because here changes in density and temperature may be important. We shall therefore need to use the steady-state macroscopic energy balance, Eq. 15.1–4, and the ideal gas equation of state, in addition to the mass and momentum balances. With these exceptions, we proceed as in Example 7.5–2.

We choose reference planes at cross sections at which the streams first begin to mix (plane "a") and far enough downstream that complete mixing has taken place (plane "b"). As in Example 7.5–2, we assume flat velocity profiles, no changes in potential energy, and negligible shear stresses on the pipe wall. In addition, we neglect changes in the heat capacity of the fluid and assume adiabatic operation. We may now write

(mass) $$w_3 = w_1 + w_2 \tag{15.4–26}$$

(momentum) $$w_3 v_3 + p_3 S_3 = w_1 v_1 + p_1 S_1 + w_2 v_2 + p_2 S_2 \tag{15.4–27}$$

(energy) $$w_3[\hat{C}_p(T_3 - T^\circ) + \tfrac{1}{2}v_3^2] = w_1[\hat{C}_p(T_1 - T^\circ) + \tfrac{1}{2}v_1^2] + w_2[\hat{C}_p(T_2 - T^\circ) + \tfrac{1}{2}v_2^2] \tag{15.4–28}$$

(equation of state) $$p_3 = \frac{\rho_3 R T_3}{M} \tag{15.4–28a}$$

In these equations the unknown quantities are $w_3 = \rho_3 v_3 S_3$, p_3, T_3, and ρ_3; and $T°$ is the reference temperature for enthalpy. By multiplying the mass balance by $\hat{C}_p T°$ and subtracting the result from Eq. 15.4–28, we get

$$\text{(energy)} \qquad w_3(\hat{C}_p T_3 + \tfrac{1}{2}v_3^2) = w_1(\hat{C}_p T_1 + \tfrac{1}{2}v_1^2) + w_2(\hat{C}_p T_2 + \tfrac{1}{2}v_2^2) \qquad (15.4\text{–}29)$$

The right sides of Eqs. 15.4–26, 27, and 29 are known; we designate them by w, P, and E, respectively.

We now eliminate the pressure p_3 and the cross-sectional area S_3 from Eq. 15.4–27 by use of the ideal-gas law:

$$\text{(momentum)} \qquad v_3 + \frac{RT_3}{Mv_3} = \frac{P}{w} \qquad (15.4\text{–}30)$$

Next, the elimination of T_3 between Eqs. 15.4–29 and 30 gives

$$v_3^2 - \left[2\left(\frac{\gamma}{\gamma+1}\right)\frac{P}{w}\right]v_3 + 2\left(\frac{\gamma-1}{\gamma+1}\right)\frac{E}{w} = 0 \qquad (15.4\text{–}31)$$

in which $\gamma = \hat{C}_p/\hat{C}_v$. We thus have a quadratic equation for v_3 in terms of the known quantities w, P, E, and γ. The solutions are

$$v_3 = \frac{P}{w}\left(\frac{\gamma}{\gamma+1}\right)\left[1 \pm \sqrt{1 - 2\left(\frac{\gamma^2-1}{\gamma^2}\right)\frac{wE}{P^2}}\right] \qquad (15.4\text{–}32)$$

It can be shown (see Problem 15.G) that when the quantity in brackets is unity the velocity of the final stream is sonic. Therefore, in general one of the solutions for v_3 is supersonic and one is subsonic. Only the lower (subsonic) solution can be obtained in the turbulent mixing process under consideration, since, as shown in Example 10.5–5, supersonic flow is unstable in the presence of any significant disturbance.

Once the velocity v_3 is known, the pressure and the temperature may be calculated from Eqs. 15.4–27 and 30.

It is interesting to note that $\hat{E}_v$ cannot be calculated exactly from Eq. 15.2–3, since the thermodynamic paths followed by the fluid streams are not known.

Example 15.4–5. Flow of Compressible Fluids through Head Meters

Extend the development of Example 7.5–4 to the flow of compressible fluids through orifice meters and venturi tubes.

Solution. We begin, as in Example 7.5–4, by writing down the steady-state mass and mechanical energy balances between reference planes "1" and "2" of the two flow meters shown in Fig. 15.4–5. For compressible fluids, these may be expressed as

$$\text{(mass)} \qquad w = \rho_1\langle v_1\rangle S_1 = \rho_2\langle v_2\rangle S_2 \qquad (15.4\text{–}33)$$

$$\text{(mechanical energy)} \qquad \frac{\langle v_2\rangle^2}{2\alpha_2} - \frac{\langle v_1\rangle^2}{2\alpha_1} + \int_1^2 \frac{1}{\rho}\,dp + \tfrac{1}{2}\langle v_2\rangle^2 e_v = 0 \qquad (15.4\text{–}34)$$

We next eliminate $\langle v_1 \rangle$ and $\langle v_2 \rangle$ between these equations to obtain an expression for the mass flow rate:

$$w = \rho_2 S_2 \sqrt{\frac{-2\alpha_2 \int_1^2 (1/\rho)\, dp}{1 - \frac{\alpha_2}{\alpha_1}\left(\frac{\rho_2 S_2}{\rho_1 S_1}\right)^2 + \alpha_2 e_v}} \tag{15.4-35}$$

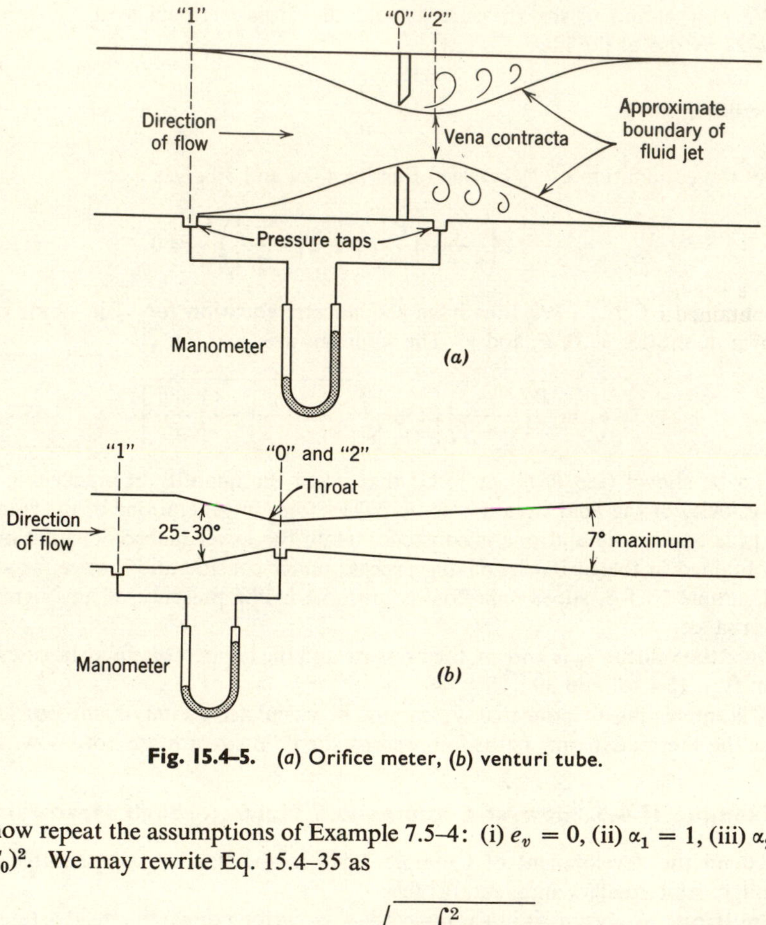

Fig. 15.4-5. (*a*) Orifice meter, (*b*) venturi tube.

We now repeat the assumptions of Example 7.5-4: (i) $e_v = 0$, (ii) $\alpha_1 = 1$, (iii) $\alpha_2 = (S_2/S_0)^2$. We may rewrite Eq. 15.4-35 as

$$w = C_d \rho_2 S_0 \sqrt{\frac{-2\int_1^2 (1/\rho)\, dp}{1 - \left(\frac{\rho_2 S_0}{\rho_1 S_1}\right)^2}} \tag{15.4-36}$$

The discharge coefficient C_d is used in Eq. 15.4-36 to permit correction of this expression for errors introduced by the three assumptions; C_d must be determined experimentally. For venturi meters, it is convenient to put plane "2" at the point of minimum cross section of the meter so that $S_2 = S_0$. Then α_2 is very nearly unity, and it has been found experimentally that C_d is almost the same for compressible and incompressible fluids, that is, about 0.98 for well-designed venturi meters. For orifice meters, the degree of contraction of the fluid stream at plane "2" is somewhat less than for incompressible fluids, especially at high flow rates, and a different discharge coefficient[3] is required.

In order to use Eq. 15.4–36, the fluid density must be known as a function of pressure; that is, one must know both the path of the expansion and the equation of state of the fluid. In most cases the assumption of frictionless adiabatic behavior appears to be acceptable. For ideal gases, one may write $p\rho^{-\gamma}$ = constant, where $\gamma = \hat{C}_p/\hat{C}_v$ (see Example 10.5–6); then Eq. 15.4–36 becomes

$$\begin{bmatrix}\text{ideal gas,}\\ \text{frictionless}\\ \text{adiabatic flow}\end{bmatrix} \qquad w = C_d\rho_2 S_0 \sqrt{\frac{2\left(\dfrac{p_1}{\rho_1}\right)\left(\dfrac{\gamma}{\gamma-1}\right)\left[1-\left(\dfrac{p_2}{p_1}\right)^{\frac{\gamma-1}{\gamma}}\right]}{1-\left(\dfrac{S_0}{S_1}\right)^2\left(\dfrac{p_2}{p_1}\right)^{\frac{2}{\gamma}}}} \tag{15.4–37}$$

This formula expresses the mass flow rate as a function of measurable quantities and the discharge coefficient; values of the latter may be found in engineering handbooks.[3]

§15.5 USE OF THE MACROSCOPIC BALANCES FOR SOLVING UNSTEADY-STATE PROBLEMS

Steady-state problems have been studied in §15.4. In this section we turn to the description of the transient behavior of nonisothermal flow systems. Particular emphasis is placed on time-dependent heat-transfer phenomena, which are described by the unsteady-state macroscopic energy balance. Such problems are of importance in estimating the time required for various industrial heating operations and also in connection with process control and instrumentation.

Example 15.5–1. Heating of a Liquid in an Agitated Tank[1]

A cylindrical tank capable of holding 1000 ft^3 of liquid is equipped with an agitator having sufficient power to keep the liquid contents at uniform temperature. (See Fig. 15.5–1.) Heat is transferred to the contents by means of a coil arranged in such a way that the area available for heat transfer is proportional to the quantity of liquid in the tank. This heating coil consists of 10 turns, 4 ft in diameter, of 1-in. o.d. tubing. Water at 20° C is fed into this tank at a rate of 20 lb min^{-1}, starting with no water in the tank at time $t = 0$. Steam at 105° C is contained

[3] See, for example, J. H. Perry, *Chemical Engineers' Handbook*, McGraw-Hill, New York (1950), Third Edition, pp. 402–403.

[1] This problem is taken in modified form from W. R. Marshall, Jr., and R. L. Pigford, *Applications of Differential Equations to Chemical Engineering Problems*, University of Delaware Press, Newark (1947), pp. 16–18.

within the heating coil and the over-all heat-transfer coefficient is 100 Btu hr^{-1} ft^{-2} ° F^{-1}. What is the temperature of the water when the tank is filled?

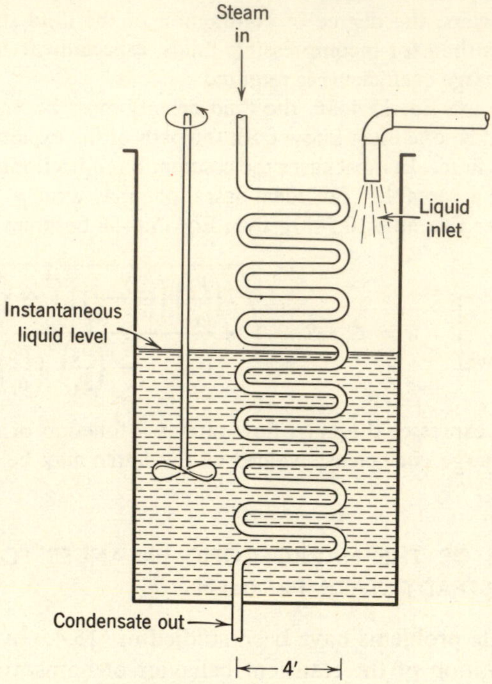

Fig. 15.5-1. Heating a liquid in a tank with the liquid level variable.

Solution. Problems of this kind are best solved in terms of symbols, with the numerical values being inserted at the last possible moment. Let us introduce the following notation:

A_0 = total area available for heat transfer
$A(t)$ = instantaneous heat-transfer area
V_0 = volume of liquid in filled tank
$V(t)$ = instantaneous volume of liquid in tank
T = instantaneous temperature of liquid
T_1 = temperature of inlet liquid
T_0 = temperature of liquid in filled tank
T_s = temperature of steam
U_0 = over-all heat-transfer coefficient
t = time measured from start of water flow
$t_0 = \rho V_0/w$ = time required to fill tank
$\hat{C}_p$ = heat capacity per unit mass
w = mass rate of flow of water into the tank
ρ = density of liquid

We shall now formulate the problem by making the following assumptions:

a. Steam temperature remains constant in coil.

b. Density and heat capacity of water do not change very much with change in temperature.

c. Because the fluid is approximately incompressible, $\hat{C}_p = \hat{C}_v$.

d. Agitator maintains uniform temperature throughout.

e. Heat-transfer coefficient is independent of position and time.

f. The walls of the tank are adiabatic.

We choose the fluid within the tank as our system. Equation 15.1–3 describes the way in which the energy of the system changes with time. The E_{tot} on the left side comprises internal, kinetic, and potential energy, the last two of which may be neglected in this problem. On the right side we can omit the work term W as well as the kinetic and potential energy terms (which will be small in comparison with other terms). Inasmuch as there is no outlet stream, $w_2 = 0$. Hence, for the system at hand,

$$\frac{d}{dt} U_{tot} = w\hat{H}_1 + Q \tag{15.5–1}$$

which states that the internal energy of the system is increasing because of the addition of fluid with enthalpy $\hat{H}_1$ and because of the addition of heat Q through the steam coil.

In view of the fact that U_{tot} and $\hat{H}_1$ cannot be given absolutely, we now select the inlet temperature T_1 as the thermal datum plane. Then $\hat{H}_1 = 0$ and $U_{tot} = \rho\hat{C}_v V(T - T_1) \doteq \rho\hat{C}_p V(T - T_1)$. Furthermore, the rate of heat addition to the system Q is given by $Q = U_0 A(T_s - T)$. Hence Eq. 15.5–1 becomes

$$\rho\hat{C}_p \frac{d}{dt} V(T - T_1) = U_0 A(T_s - T) \tag{15.5–2}$$

Now the instantaneous area and volume are simply related to the total available area and volume according to the relations

$$V(t) = \frac{wt}{\rho}; \qquad A(t) = \frac{wt}{\rho V_0} A_0 \tag{15.5–3,4}$$

Substitution of these expressions in Eq. 15.5–2 gives

$$U_0 A_0 \frac{wt}{\rho V_0}(T_s - T) = w\hat{C}_p(T - T_1) + w\hat{C}_p t\left(\frac{d(T - T_1)}{dt}\right) \tag{15.5–5}$$

which is to be solved with the initial condition that at $t = 0$, $T = T_1$.

Equation 15.5–5, along with the initial condition, is the analytical statement of the problem consistent with the assumptions listed above. One could proceed to solve the differential equation Eq. 15.5–5 as stated, but it is preferable to make use of the reduced variables:

$$\Theta = \frac{T - T_1}{T_s - T_1} = \text{reduced temperature} \tag{15.5–6}$$

$$\tau = \frac{t}{t_0} = \frac{wt}{\rho V_0} = \text{reduced time} \tag{15.5–7}$$

The reduced temperature represents the fractional approach of the liquid temperature to the steam temperature. The reduced time represents time relative to that required to fill the tank completely. In order to rewrite Eq. 15.5–5 in terms of the foregoing dimensionless variables, the entire equation is multiplied through by $[1/w\hat{C}_p(T_s - T_1)]$ to obtain

$$\left(\frac{U_0A_0}{w\hat{C}_p}\right)\left(\frac{wt}{\rho V_0}\right)\left(1 - \frac{T - T_1}{T_s - T_1}\right) = \left(\frac{T - T_1}{T_s - T_1}\right) + t\frac{d}{dt}\left(\frac{T - T_1}{T_s - T_1}\right) \tag{15.5–8}$$

or

$$N\tau(1 - \Theta) = \Theta + \tau\frac{d\Theta}{d\tau} \tag{15.5–9}$$

in which $N = U_0A_0/w\hat{C}_p$ is a dimensionless group appearing in the differential equation. The equation may be further simplified by noting that if the last term is multiplied by N/N then the reduced time variable always appears in the combination $N\tau$, which combination is designated by η. The final analytical statement of the problem is

differential equation:
$$\frac{d\Theta}{d\eta} + \left(1 + \frac{1}{\eta}\right)\Theta = 1 \tag{15.5–10}$$

initial condition:
$$\Theta = 0 \quad \text{at} \quad \eta = 0 \tag{15.5–11}$$

From the statement of the problem in dimensionless terms, we see at once that Θ will be a function of η alone. Equation 15.5–10 is simpler to handle than Eq. 15.5–5 written in dimensional quantities. Equation 15.5–10 is a first-order linear differential equation. Its solution is

$$\Theta = 1 - \frac{1}{\eta} + \frac{C}{\eta \exp \eta} \tag{15.5–12}$$

The constant of integration may be determined from the initial condition by multiplying Eq. 15.5–12 by η and letting $\eta = 0$; thereby, it is shown that $C = 1$, so that the final solution of the differential equation is

$$\Theta = 1 - \frac{1 - \exp(-\eta)}{\eta} \tag{15.5–13}$$

This function is shown graphically in Fig. 15.5–2. Finally, the temperature of the liquid in the tank, when it has been filled, is given by Eq. 15.5–13 when $\tau = 1$ (or $\eta = N$); in terms of the original variables, this result is

$$\frac{T_0 - T_1}{T_s - T_1} = 1 - \frac{1 - \exp(-U_0A_0/w\hat{C}_p)}{(U_0A_0/w\hat{C}_p)} \tag{15.5–14}$$

Numerical calculation. It is easy to show that for the data given in the problem the dimensionless group $N = U_0A_0/w\hat{C}_p$ has a value of 2.74. For this value, Eq. 15.5–14 gives $(T_0 - T_1)/(T_s - T_1) = 0.659$, whence $T_0 = 76°$ C.

Example 15.5–2. Operation of a Simple Temperature Controller

It is desired to control the temperature of a liquid flowing through the well-insulated agitated tank shown in Fig. 15.5–3. The volume V_l of liquid in the tank and the mass rate of liquid flow w are constant, but the temperature at which liquid

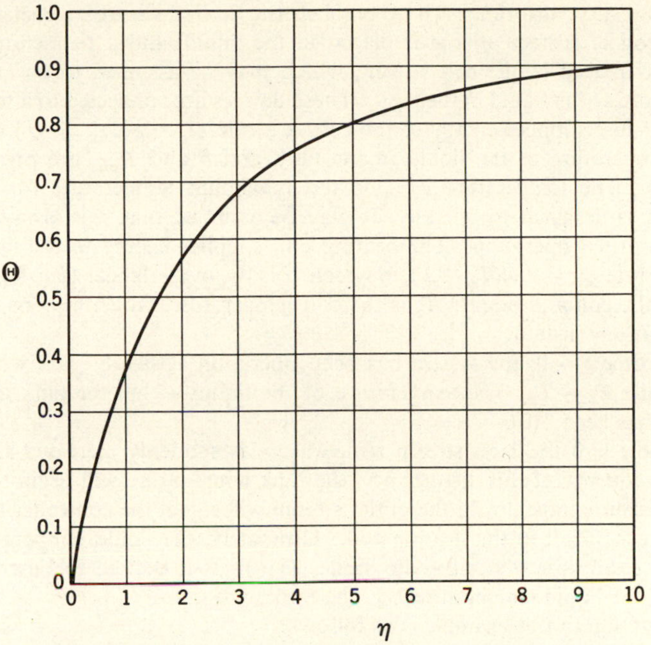

Fig. 15.5–2. Plot of dimensionless temperature versus $\eta = N_T$ according to Eq. 15.5–13. [W. R. Marshall, Jr. and R. L. Pigford, *Application of Differential Equations to Chemical Engineering*, University of Delaware Press, Newark (1947), p. 18.]

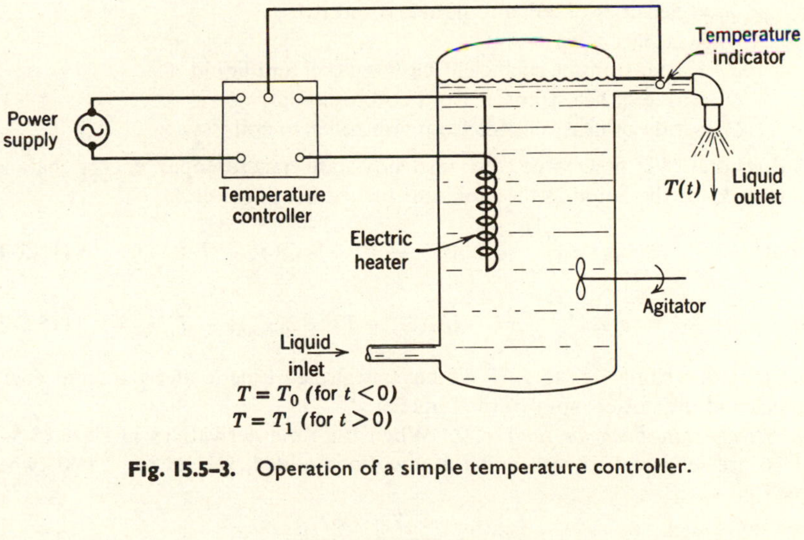

Fig. 15.5–3. Operation of a simple temperature controller.

enters may vary with time. To accomplish the desired control, a metallic electric heating coil of surface area A is placed in the liquid, and a temperature-sensing element is placed in the exit stream, which may be assumed to be at the same temperature as the liquid in the tank. These devices are connected to a temperature controller that supplies energy to the coil at a rate $Q_e = b(T_{max} - T)$, in which T is the temperature of the liquid in the tank and b and T_{max} are predetermined constants. The temperature T_{max} is the maximum temperature for which the controller is designed to operate. It may be assumed that T is always less than T_{max} in normal operation. The heating coil supplies energy to the liquid in the tank at a rate $Q_c = UA(T_c - T)$, in which U is the over-all heat-transfer coefficient between the coil and liquid; T_c is the coil temperature, which may be considered uniform at any instant.

Up to time $t = 0$, the system has been operating at steady state with a liquid temperature $T_i = T_0$. The temperature of the liquid within the tank and leaving the tank has been $T(0)$.

At time $t = 0$ the inlet stream temperature is suddenly increased to $T_i = T_1$. As a consequence of this disturbance, the tank temperature will begin to rise, and the temperature indicator in the outlet stream will signal the controller to decrease the power supplied to the heating coil. Ultimately, the liquid temperature in the tank will attain a new steady-state value, $T(\infty)$. It is desired to describe the behavior of the liquid temperature T in the tank with time.

The notation in this example is as follows:

$T(t)$ = instantaneous liquid temperature in the tank and outgoing stream
T_c = instantaneous coil temperature
T_i = instantaneous inlet temperature
$\hat{C}_c, \hat{C}_l$ = heat capacity of coil and liquid, respectively
V_c, V_l = volumes of coil and liquid, respectively
ρ_c, ρ_l = densities of coil and liquid, respectively
A = surface area of coil
U = heat-transfer coefficient between coil and liquid
Q_c = rate of heat transfer from coil to liquid
Q_e = rate of heat transfer from controller to coil

Solution. We first write the unsteady-state macroscopic energy balance, Eq. 15.1–3, for the liquid in the tank and for the heating coil:

$$\text{(liquid)} \qquad \rho_l V_l \hat{C}_l \frac{dT}{dt} = UA(T_c - T) + w\hat{C}_l(T_i - T) \qquad (15.5\text{–}15)$$

$$\text{(coil)} \qquad \rho_c V_c \hat{C}_c \frac{dT_c}{dt} = -UA(T_c - T) + b(T_{max} - T) \qquad (15.5\text{–}16)$$

Note that in writing Eq. 15.5–15 we have neglected kinetic and potential energy changes and the power input of the impeller.

a. Steady-state behavior for $t < 0$. When the time derivatives in Eqs. 15.5–15 and 16 are set equal to zero and the equations added, we get for $t < 0$ (where $T_i = T_0$)

$$T(0) = \frac{w\hat{C}_l T_0 + bT_{max}}{w\hat{C}_l + b} \qquad (15.5\text{–}17)$$

The steady-state form of Eq. 15.5–16 gives

$$T_c(0) = \frac{bT_{\max}}{UA} + \left(1 - \frac{b}{UA}\right)T(0) \tag{15.5–18}$$

b. Steady-state behavior for $t \gg 0$. When similar operations are performed with $T_i = T_1$, we get for $t \gg 0$

$$T(\infty) = \frac{w\hat{C}_l T_1 + bT_{\max}}{w\hat{C}_l + b} \tag{15.5–19}$$

and

$$T_c(\infty) = \frac{bT_{\max}}{UA} + \left(1 - \frac{b}{UA}\right)T(\infty) \tag{15.5–20}$$

c. Unsteady-state behavior for $t > 0$. It is convenient to define dimensionless variables using the steady-state quantities for $t < 0$ and $t \gg 0$.

$$\Theta = \frac{T - T(\infty)}{T(0) - T(\infty)} = \text{dimensionless liquid temperature} \tag{15.5–21}$$

$$\Theta_c = \frac{T_c - T_c(\infty)}{T_c(0) - T_c(\infty)} = \text{dimensionless coil temperature} \tag{15.5–22}$$

$$\tau = \frac{UAt}{\rho_l V_l \hat{C}_l} = \text{dimensionless time} \tag{15.5–23}$$

In addition, we define these dimensionless parameters:

$$R = \frac{\rho_l V_l \hat{C}_l}{\rho_c V_c \hat{C}_c} = \text{ratio of thermal capacities} \tag{15.5–24}$$

$$F = \frac{w\hat{C}_l}{UA} = \text{flow rate parameter} \tag{15.5–25}$$

$$B = \frac{b}{UA} = \text{controller parameter} \tag{15.5–26}$$

In terms of these quantities, the unsteady-state balances in Eqs. 15.5–15 and 16 become after considerable rearrangement

$$\frac{d\Theta}{d\tau} = \Theta_c(1 - B) - \Theta(1 + F) \tag{15.5–27}$$

$$\frac{1}{R}\frac{d\Theta_c}{d\tau} = -\Theta_c + \Theta \tag{15.5–28}$$

Elimination of Θ_c between this pair of equations gives a single second-order, linear, ordinary differential equation for liquid temperature as a function of time:

$$\frac{d^2\Theta}{d\tau^2} + (1 + F + R)\frac{d\Theta}{d\tau} + R(B + F)\Theta = 0 \tag{15.5–29}$$

It is of interest to note that this is the same differential equation as that obtained for the damped manometer, Eq. 7.6–21. The general solution is then either of the form of Eq. 7.6–23 or 24:

$$\Theta = C_+ e^{m_+\tau} + C_- e^{m_-\tau} \qquad (m_+ \neq m_-) \tag{15.5–30}$$

or

$$\Theta = C_1 e^{m\tau} + C_2 \tau e^{m\tau} \qquad (m_+ = m_- = m) \tag{15.5–31}$$

where

$$m_\pm = \tfrac{1}{2}[-(1 + F + R) \pm \sqrt{(1 + F + R)^2 - 4R(B + F)}] \tag{15.5–32}$$

The temperature changes of the fluid are then similar to the fluctuations of the damped manometer of Example 7.6–2. That is, the fluid temperature may approach

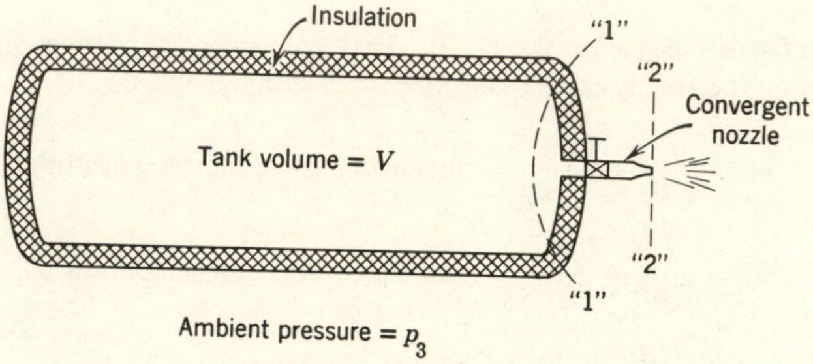

Fig. 15.5–4. Free batch expansion of a compressible fluid.

its final value either with or without oscillation about $T(\infty)$. The system parameters appear in the time variable τ and in the constants B, F, and R. Therefore, numerical calculations are necessary in order to assess the relative merits of over- and under-damping for this system.

Example 15.5–3. Free Batch Expansion of a Compressible Fluid

A fluid initially at $T = T_0, p = p_0$, and $\rho = \rho_0$ is discharged from a large stationary insulated tank through a small convergent nozzle, as shown in Fig. 15.5–4. Show how the fractional remaining mass of fluid in the tank ρ/ρ_0 may be determined as a function of time. Develop working equations for the specific case of an ideal gas.

Solution. For convenience, we consider the tank to be divided into two parts, separated by surface "1." (See Fig. 15.5–4.) We assume surface "1" is near enough to the tank exit that essentially all of the fluid mass is to its left but far enough from the exit that the fluid velocity through surface "1" is negligible. We further assume that the average fluid properties to the left of "1" are identical with those at surface "1." We now consider the behavior of these two parts of the system separately.

a. The bulk of the fluid in the tank. For the region to the left of "1," the unsteady-state energy balance, Eq. 15.1–3, becomes

$$\frac{d}{dt}[\rho_1 V(\hat{U}_1 + \hat{\Phi}_1)] = -w_1\left(\hat{U}_1 + \frac{p_1}{\rho_1} + \hat{\Phi}_1\right) \tag{15.5–33}$$

in which V is the total volume in the system being considered and w_1 is the mass rate of flow from the system. In writing this equation, we have neglected the kinetic energy of the fluid.

The mass balance for that portion of the system to the left of "1" is

$$\frac{d}{dt}(\rho_1 V) = -w_1 \tag{15.5–34}$$

Combination of the mass and energy balances then gives

$$\rho_1\left(\frac{d\hat{U}_1}{dt} + \frac{d\hat{\Phi}_1}{dt}\right) = \frac{p_1}{\rho_1}\frac{d\rho_1}{dt} \tag{15.5–35}$$

For a stationary system under the influence of no other external forces than gravity, $d\hat{\Phi}_1/dt = 0$, and we may write

$$\frac{d\hat{U}_1}{d\rho_1} = \frac{p_1}{\rho_1^{\,2}} \tag{15.5–36}$$

This equation may be combined with the thermal and caloric equations of state for the fluid to obtain $p_1(\rho_1)$ and $T_1(\rho_1)$. We find, then, that the condition of the fluid in the tank depends only upon the degree to which the tank has been emptied and not upon the rate of discharge. For the special case of an ideal gas with constant $\hat{C}_v$, in which $d\hat{U} = \hat{C}_v\, dT$ and $p = \rho RT/M$, we may integrate Eq. 15.5–36 to get

$$p_1\rho_1^{\,-\gamma} = p_0\rho_0^{\,-\gamma} \tag{15.5–37}$$

in which $\gamma = C_p/C_v$.

b. Discharge of fluid through the nozzle. We will assume here for the sake of simplicity that the flow between surfaces "1" and "2" is both frictionless and adiabatic. The behavior of this portion of the system is then described by the isentropic unsteady-state mechanical energy balance, Eq. 15.2–2. However, for relatively small tank openings, as considered here, (dE_{tot}/dt) is small and w_1 is about equal to w_2. We therefore assume "quasi-steady-state" behavior and use the Bernoulli equation to provide an approximate description of our system:

$$\tfrac{1}{2}v_2^{\,2} + \int_1^2 \frac{1}{\rho}\, dp = 0 \tag{15.5–38}$$

For the specific case of ideal gases, we may use the results of Example 15.4–5 to write an expression for the instantaneous discharge rate:

$$w_2 = -V\frac{d\rho_1}{dt} = S_2\sqrt{2p_1\rho_1\left(\frac{\gamma}{\gamma-1}\right)\left[\left(\frac{p_2}{p_1}\right)^{(2/\gamma)} - \left(\frac{p_2}{p_1}\right)^{\left(\frac{\gamma+1}{\gamma}\right)}\right]} \tag{15.5–39}$$

in which S_2 is the area of the nozzle opening. We may now use Eq. 15.5–37 to eliminate p_1 from Eq. 15.5–39 and thereby obtain an expression for the mass fraction of the original fluid, ρ_1/ρ_0, remaining at any time t:

$$t = \frac{V/S_2}{\sqrt{2\left(\frac{\gamma}{\gamma-1}\right)\left(\frac{p_0}{\rho_0}\right)}}\int_{\frac{\rho_1}{\rho_0}}^{1} \frac{d\eta}{\sqrt{\left(\frac{p_2}{p_0}\right)^{2/\gamma}\eta^{\gamma-1} - \left(\frac{p_2}{p_0}\right)^{1+(1/\gamma)}}} \tag{15.5–40}$$

Here p_2, the pressure at the nozzle opening, must be carefully evaluated. At low flow rates the pressure p_2 at the nozzle opening is equal to the ambient pressure, just as for an incompressible fluid. (See Problem 15.C.) However, examination of Eq. 15.5–39 shows that as the ambient pressure is reduced the calculated mass rate of flow reaches a maximum at a critical pressure ratio:

$$r = \left(\frac{p_2}{p_1}\right)_{\text{crit}} = \left(\frac{2}{\gamma + 1}\right)^{\frac{\gamma}{\gamma-1}} \tag{15.5–41}$$

For air ($\gamma = 1.4$), this critical pressure ratio is 0.53. If the ambient pressure is further reduced, the pressure just inside the nozzle will remain at the value of p_2 calculated from Eq. 15.5–41, and the mass flow rate will become independent of ambient pressure. It may easily be shown that the discharge rate under these conditions is

$$w_{\max} = S_2 \sqrt{\gamma p_1 \rho_1 \left(\frac{2}{\gamma + 1}\right)^{\frac{\gamma+1}{\gamma-1}}} \tag{15.5–42}$$

Then, for $p_3/p_1 < r$, we may write Eq. 15.5–40 more simply as

$$t = \frac{V/S_2}{\sqrt{\gamma\left(\frac{p_0}{\rho_0}\right)\left(\frac{2}{\gamma+1}\right)^{\frac{\gamma+1}{\gamma-1}}}} \int_{\frac{\rho_1}{\rho_0}}^{1} \frac{d\eta}{\eta^{(\gamma+1)/2}} \tag{15.5–43}$$

or

$$t = \frac{V/S_2}{\sqrt{\gamma\left(\frac{p_0}{\rho_0}\right)\left(\frac{2}{\gamma+1}\right)^{\frac{\gamma+1}{\gamma-1}}}} \left(\frac{2}{\gamma - 1}\right)\left[\left(\frac{\rho_1}{\rho_0}\right)^{\frac{1-\gamma}{2}} - 1\right] \quad (p_3/p_1 < r) \tag{15.5–44}$$

If p_3/p_1 is initially less than r, both Eqs. 15.5–44 and 40 will be useful in calculating the total discharge time.

QUESTIONS FOR DISCUSSION

1. Give the physical significance of each term in the four macroscopic balances.
2. How are the three equations of change related to the four macroscopic balances?
3. Does each of the four terms between parentheses in Eq. 15.1–3 represent a form of energy? Explain.
4. Verify the expressions for $\Delta\hat{H}$ in Eqs. 15.1–5 and 6.
5. What is $\Delta\hat{H}$ for a low-density monatomic gas?
6. Compare the kinds of average velocities that appear in the four macroscopic balances.
7. Discuss the evaluation of $\int(1/\rho)\,dp$ in terms of Fig. 15.2–1.
8. What information can be obtained from Eq. 15.2–3 about a fluid at rest?
9. What information is obtained by combining Eqs. 15.1–4 and 15.2–3 for an incompressible fluid?

PROBLEMS

15.A_1 Rates of Heat Transfer in a Double-Pipe Exchanger

a. Hot oil entering the heat exchanger in Example 15.4–2 at surface "2" is to be cooled by water entering at surface "1"; that is, the exchanger is being operated in *counterflow.* Compute the required exchanger area A if the heat-transfer coefficient U is 200 Btu hr^{-1} ft^{-2} ° F^{-1} and the fluid streams have the following properties:

	Mass Flow Rate lb_m hr^{-1}	Heat Capacity Btu lb_m^{-1} ° F^{-1}	Temperature Entering	Temperature Leaving
Oil	10,000	0.60	200° F	100° F
Water	5,000	1.00	60° F	—

b. Repeat the calculation of part (*a*) if $U_1 = 50$ and $U_2 = 350$ Btu hr^{-1} ft^{-2} ° F^{-1}. Assume that U varies linearly with water temperature, and use the result of Problem 15.E.

c. What is the minimum amount of water that can be used in parts (*a*) and (*b*) to obtain the desired oil temperature change? What is the minimum amount of water that can be used in parallel flow?

d. Calculate the required exchanger area for parallel flow operation if the mass rate of flow of water is 15,500 lb_m hr^{-1} and U is constant at 200 Btu hr^{-1} ft^{-2} ° F^{-1}.

Answer: *a.* 104 ft^2
b. 122 ft^2
c. 4290 lb_m hr^{-1}, 15,000 lb_m hr^{-1}
d. ca. 101 ft^2

15.B_1 Adiabatic Flow of Natural Gas in a Pipeline

Recalculate the power requirement $-w\hat{W}'$ in Example 15.4–3 if the flow in the pipeline is adiabatic rather than isothermal:

a. Use the result of Problem 15.F*d* to determine the density of the gas at plane "2."
b. Use your answer to part (*a*) with the result of Problem 15.F*e* to obtain p_2.
c. Calculate the power requirement, as in Example 15.4–3.

Answer: *a.* 0.25 lb_m ft^{-3}
b. 86 psia
c. 504 hp

15.C_1 Mixing of Two Ideal-Gas Streams

a. Calculate the resulting velocity, temperature, and pressure when the following two air streams are mixed in an apparatus such as that described in Example 15.4–4. The heat capacity $\tilde{C}_p$, of air may be considered constant at 7.0 Btu lb-$mole^{-1}$ ° F^{-1}.

	w(lb hr^{-1})	v(ft sec^{-1})	T(° F)	p(atm)
Stream 1:	1000	1000	80	1.00
Stream 2:	10,000	100	80	1.00
Answer:	11,000	ca. 110	88	1.00

b What would the calculated velocity be if the flow were treated as isothermal and incompressible?

c. Estimate $\hat{E}_v$ for this operation, basing your calculation on the results of part (*b*).

Answer: *a.* In table above
b. 109 ft sec^{-1}
c. 1.4 × 10^3 ft lb_f lb_m^{-1}

15.D$_1$ Flow through a Venturi Tube

A venturi tube with a throat 3 in. in diameter is placed in a circular pipe 1 ft in diameter carrying dry air. The discharge coefficient C_d of the meter is 0.98. Calculate the mass flow rate of air in the pipe if the air enters the venturi at 70° F and 1 atm and the throat pressure is 0.75 atm:

a. Assuming adiabatic frictionless flow and $\gamma = 1.4$.
b. Assuming isothermal flow.
c. Assuming incompressible flow at entering density.

Answer: *a.* 2.07 $\text{lb}_m\ \text{sec}^{-1}$
b. 1.96 $\text{lb}_m\ \text{sec}^{-1}$
c. 2.43 $\text{lb}_m\ \text{sec}^{-1}$

15.E$_2$ Performance of a Double-Pipe Heat Exchanger with Variable Over-all Heat-Transfer Coefficient

Develop an expression for the amount of heat transferred in an exchanger of the type discussed in Example 15.4–2 if the overall heat transfer coefficient U varies linearly with the temperature of either stream:

a. Show that $(T_h - T_c)$ is a linear function of both T_h and T_c so that $(U - U_1)/(U_2 - U_1) = (\Delta T - \Delta T_1)/(\Delta T_2 - \Delta T_1)$, in which $\Delta T = T_h - T_c$ and the subscripts 1 and 2 refer to conditions at control surfaces "1" and "2."

b. Substitute the foregoing result for $(T_h - T_c)$ in Eq. 15.4–13 and integrate the resulting equation over the length of the exchanger. Use your answer to show that

$$Q = A\,\frac{(U_1\Delta T_2 - U_2\,\Delta T_1)}{\ln\,(U_1\,\Delta T_2/U_2\,\Delta T_1)} \tag{15.E–1}$$

This expression was first suggested by A. P. Colburn, *Ind. Eng. Chem.*, **25**, 873 (1933).

15.F$_2$ Steady-State Flow of Ideal Gases in Ducts of Constant Cross Section

a. Show that for horizontal flow of any fluid in a duct of constant cross section the Bernoulli equation, Eq. 15.2–3, may be written as

$$v\,dv + \hat{V}\,dp + \tfrac{1}{2}v^2\,de_v = 0 \tag{15.F–1}$$

in which $de_v = (4f/D)\,dL$. Assume flat velocity profiles.

b. Show that Eq. 15.F–1 may be rewritten to give

$$v\,dv + d(p\hat{V}) - p\hat{V}\frac{d\hat{V}}{\hat{V}} + \tfrac{1}{2}v^2\,de_v = 0 \tag{15.F–2}$$

c. Show that for isothermal flow of ideal gases Eq. 15.F–2 may be written as

$$de_v = \frac{2RT}{M}\frac{dv}{v^3} - 2\frac{dv}{v} \tag{15.F–3}$$

Integrate this expression between any two pipe cross sections "1" and "2" enclosing a total pipe length L and rearrange your answer to obtain

$$G = \frac{v_1}{\hat{V}_1} = \left\{\frac{(p_1/\hat{V}_1)[1 - (p_2/p_1)^2]}{e_v - \ln\,(p_2/p_1)^2}\right\}^{1/2} \quad \text{(isothermal flow of ideal gases)} \tag{15.F–4}$$

Show for any given values of e_v and conditions at "1" that G reaches its maximum possible value at a critical value of $(p_2/p_1)^2 = r$ defined by $e_v - \ln r = (1 - r)/r$. (See also Problem 15.H.)

d. Use the steady-state macroscopic energy balance, Eq. 15.1–4, to show that for adiabatic flow of ideal gases with constant $\hat{C}_v$ in ducts of constant cross section

$$p\hat{V} + \left(\frac{\gamma - 1}{\gamma}\right)\frac{1}{2}v^2 = \text{constant} \tag{15.F–5}$$

Use this expression and the result of part (*b*) to obtain

$$\frac{\gamma + 1}{\gamma}\frac{dv}{v} - \left(2p_1\hat{V}_1 + \frac{\gamma - 1}{\gamma}v_1^2\right)\frac{dv}{v^3} = -de_v \tag{15.F–6}$$

Integrate this equation between two pipe cross sections "1" and "2" enclosing a dimensionless resistance e_v. Assume constant γ in this integration. Rearrange the result with the aid of the macroscopic mass balance to obtain a relation for the mass velocity of gas:

$$G = \frac{v_1}{\hat{V}_1} = \left\{\frac{p_1/\hat{V}_1}{\left[\dfrac{e_v - \left(\dfrac{\gamma + 1}{2\gamma}\right)\ln(\hat{V}_1/\hat{V}_2)^2}{1 - (\hat{V}_1/\hat{V}_2)^2} - \dfrac{\gamma - 1}{2\gamma}\right]}\right\}^{1/2} \quad \begin{array}{l}\text{(adiabatic}\\ \text{flow of ideal}\\ \text{gases)}\end{array} \tag{15.F–7}$$

e. Show through use of the macroscopic energy and mass balances that for horizontal adiabatic flow of ideal gases of constant γ

$$\frac{p_2}{p_1} = \left(\frac{\hat{V}_1}{\hat{V}_2}\right)\left\{\frac{[1 - (\hat{V}_2/\hat{V}_1)^2]G^2\hat{V}_1}{p_1}\left(\frac{\gamma - 1}{2\gamma}\right) + 1\right\} \tag{15.F–8}$$

This equation can be combined with Eq. 15.F–7 to show that, as for isothermal flow, there is a critical pressure ratio (p_2/p_1) corresponding to the maximum possible mass flow rate.

15.G$_2$ The Mach Number in the Mixing of Two Fluid Streams

a. Show that when $(wE/P^2)[(\gamma^2 - 1)/\gamma^2]$ is equal to $\frac{1}{2}$ in Eq. 15.4–32, the Mach number of the final stream is unity. Note that the Mach number, Ma, which is the ratio of local fluid velocity to the velocity of sound at the local conditions, may be written for an ideal gas as $v/v_s = v/\sqrt{\gamma RT/M}$. (See Problem 10.L.)

b. Show how the results of Example 15.4–4 may be used to predict the behavior of a gas passing through a sudden enlargement of duct cross section.

15.H$_2$ Limiting Discharge Rates for Venturi Meters

a. Starting with Eq. 15.4–37, show that as the throat pressure in a Venturi meter is reduced the mass rate of flow reaches a maximum when the ratio $r = (p_2/p_1)$ of throat pressure to entrance pressure is defined by the expression

$$\frac{\gamma + 1}{r^{2/\gamma}} - \frac{2}{r^{\left(\frac{\gamma+1}{\gamma}\right)}} - \frac{\gamma - 1}{(S_1/S_0)^2} = 0 \tag{15.H–1}$$

b. Show that for $S_1 \gg S_0$ the mass flow rate under these limiting conditions is

$$w = C_d p_1 S_0 \sqrt{\frac{\gamma M}{RT_1}\left(\frac{2}{\gamma + 1}\right)^{\left(\frac{\gamma+1}{\gamma-1}\right)}} \tag{15.H–2}$$

c. Obtain expressions corresponding to the foregoing for isothermal flow.

15.I$_2$ Flow of a Compressible Fluid through a Convergent-Divergent Nozzle

In many applications, such as steam turbines or rockets, hot compressed gases are expanded through nozzles of the kind shown in Fig. 15.I in order to convert the gas enthalpy into kinetic energy. This operation is in many ways similar to the flow of gases through orifices. Here, however, the purpose of the expansion is to produce power, for example, by the impingement of the fast-moving exit fluid on a turbine blade or by direct thrust as in a rocket engine.

To explain the behavior of such systems and to justify the general shape of the nozzle described, follow the path of expansion of an ideal gas. Assume that the gas is initially in a very large reservoir at essentially zero velocity and that it expands through an adiabatic

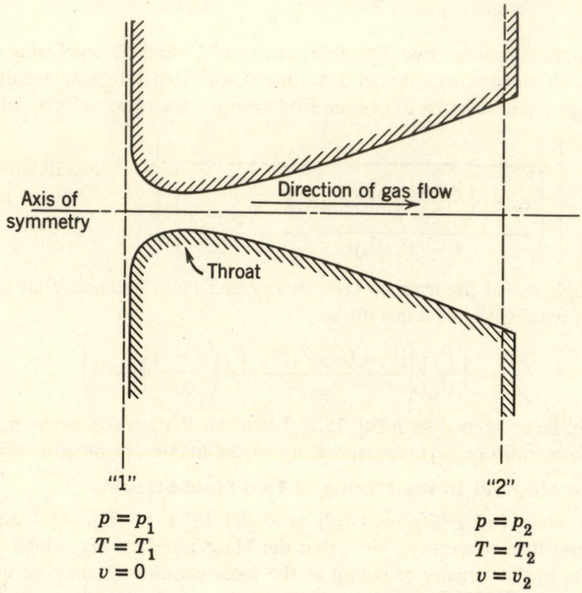

Fig. 15.I. Schematic cross section of a convergent-divergent nozzle.

frictionless nozzle to zero pressure. Further assume flat velocity profiles and neglect changes in elevation.

a. Show, by writing the Bernoulli equation between planes "1" and "2" of Fig. 15.I, that

$$\tfrac{1}{2}v_2^{\,2} = \frac{RT_1}{M}\frac{\gamma}{\gamma-1}\left[1 - \left(\frac{p_2}{p_1}\right)^{\left(\frac{\gamma-1}{\gamma}\right)}\right] \tag{15.I–1}$$

b. Show, by use of the ideal-gas law, the steady-state macroscopic mass balance, and Eq. 15.I–1, that the cross section S of the expanding stream goes through a minimum at a critical pressure:

$$p_{2,\,\mathrm{crit}} = p_1\left(\frac{2}{\gamma+1}\right)^{\left(\frac{\gamma}{\gamma-1}\right)} \tag{15.I–2}$$

c. Show that the Mach number of the fluid at this minimum cross section is unity. (See Problem 10.L.) How do the results of parts (*a*) and (*b*) above compare with those of Problem 15.H?

d. Calculate fluid velocity v, fluid temperature T, and stream cross section S as functions of the local pressure p for the discharge of 10 lb-moles of air per second from 560° R and 10 atm to zero pressure. Discuss the significance of your results.

Answer:

p, atm	10	9	8	7	6	5.3	5	4	3	2	1	0
v, ft sec^{-1}	0	442	638	795	1020	1050	1092	1237	1390	1560	1790	2575
T, ° R	562	546	525	506	485	468	461	433	399	354	292	0
S, ft^2	∞	0.99	0.745	0.656	0.620	0.612	0.620	0.635	0.693	0.860	1.182	∞

15.J$_3$ Parallel-Counterflow Heat Exchangers

In the heat exchanger shown in Fig. 15.J the "tube fluid," fluid A, enters and leaves at the same end of the exchanger, whereas the "shell fluid," fluid B, always moves in the same direction; there is thus both parallel flow and counterflow in the same apparatus. This flow

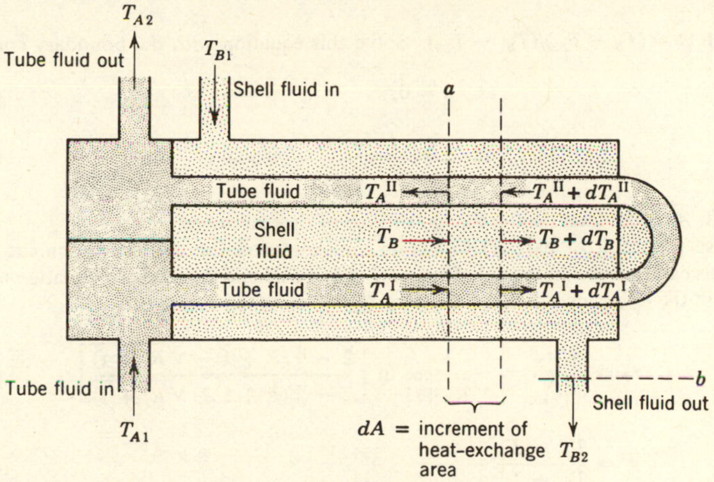

Fig. 15.J. A 1–2 parallel-counterflow heat exchanger.

arrangement is one of the simplest examples of "mixed flow," often used in practice to reduce exchanger length.[1] The behavior of this kind of equipment may be simply analyzed by making the following assumptions: (i) steady-state conditions exist; (ii) the over-all heat-transfer coefficient U and the heat capacities of the two fluids are constant; (iii) the shell-fluid temperature T_B is constant over any cross section perpendicular to the direction of flow; and (iv) there is an equal amount of heating area in each "pass," that is, for tube fluid streams I and II in the figure.

a. Show by an energy balance over the portion of the system between planes "*a*" and "*b*" that

$$(T_B - T_{B2}) = R(T_A^{II} - T_A^{I}) \qquad \text{where} \qquad R = |w_A\hat{C}_{pA}/w_B\hat{C}_{pB}|$$

[1] See D. Q. Kern, *Process Heat Transfer*, McGraw-Hill, New York, pp. 127–189 (1950); J. H. Perry, *Chemical Engineers' Handbook*, McGraw-Hill, New York (1950), Third Edition, pp. 464–465.

b. Show that over a differential section of the exchanger, including a *total* heat exchange surface dA,

$$\frac{dT_A^{\mathrm{I}}}{d\alpha} = \frac{1}{2}(T_B - T_A^{\mathrm{I}}) \tag{15.J–1}$$

$$\frac{dT_A^{\mathrm{II}}}{d\alpha} = \frac{1}{2}(T_A^{\mathrm{II}} - T_B) \tag{15.J–2}$$

$$\frac{1}{R}\frac{dT_B}{d\alpha} = -[T_B - \tfrac{1}{2}(T_A^{\mathrm{I}} + T_A^{\mathrm{II}})] \tag{15.J–3}$$

in which $d\alpha = (U/w_A\hat{C}_{pA})\,dA$, and w_A and $\hat{C}_{pA}$ are defined as in Example 15.4–2.

c. Show that T_A^{I} and T_A^{II} can be eliminated between these three equations to obtain a differential equation for the temperature distribution of the shell fluid:

$$\frac{d^2\Theta}{d\alpha^2} + \frac{R\,d\Theta}{d\alpha} - \frac{\Theta}{4} = 0 \tag{15.J–4}$$

in which $\Theta = (T_B - T_{B2})/(T_{B1} - T_{B2})$. Solve this equation with the boundary conditions

$$\text{at} \quad \alpha = 0, \qquad \Theta = 1 \tag{15.J–5}$$

$$\text{at} \quad \alpha = \frac{UA_T}{w_A\hat{C}_{pA}}, \qquad \Theta = 0 \tag{15.J–6}$$

in which A_T is the total heat-exchange surface of the exchanger.

d. Use the result of part (*c*) to obtain an expression for $dT_B/d\alpha$. Eliminate $dT_B/d\alpha$ from this expression with the aid of Eq. 15.J–3 and evaluate the resulting equation at $\alpha = 0$ to obtain the following relation for the performance of the exchanger:

$$\alpha_T = \frac{UA_T}{w_A\hat{C}_{pA}} = \frac{1}{\sqrt{R^2+1}} \ln\left[\frac{2 - \Psi(R + 1 - \sqrt{R^2+1})}{2 - \Psi(R + 1 + \sqrt{R^2+1})}\right] \tag{15.J–7}$$

in which

$$\Psi = \frac{T_{A2} - T_{A1}}{T_{B1} - T_{A1}} \tag{15.J–8}$$

e. Use this result to obtain the following expression for the rate of heat transfer in the exchanger:

$$Q = UA(\Delta T)_{\ln} \cdot Y \tag{15.J–9}$$

in which

$$(\Delta T)_{\ln} = \frac{[(T_{B1} - T_{A2}) - (T_{B2} - T_{A1})]}{\ln\left[\dfrac{(T_{B1} - T_{A2})}{(T_{B2} - T_{A1})}\right]} \tag{15.J–10}$$

$$Y = \frac{\sqrt{R^2+1}\,\ln[(1 - \Psi)/(1 - R\Psi)]}{(R - 1)\ln\left[\dfrac{2 - \Psi(R + 1 - \sqrt{R^2+1})}{2 - \Psi(R + 1 + \sqrt{R^2+1})}\right]} \tag{15.J–11}$$

Note that Y represents the ratio of the heat transferred in the 1-2-parallel-counterflow exchanger shown to that transferred in a true counterflow exchanger of the same area and terminal fluid temperatures. Values of $Y(R, \Psi)$ are given graphically in Perry (*op. cit.*) and in most unit operations texts. It may be seen that Y is always less than unity.

15.K_3 Discharge of Air from a Large Tank

It is desired to withdraw 5 lb_m sec^{-1} of air from a large storage tank through an equivalent length of 55 ft of new steel pipe 2.067 in. in inside diameter. The air undergoes a sudden contraction on entering the pipe, and the accompanying contraction loss is not included in the equivalent length of the pipe. Can the desired flow rate be obtained if the air in the tank is at 150 psig and 70° F and the pressure at the downstream end of the pipe is 50 psig? (The effect of the sudden contraction may be estimated with reasonable accuracy[2] by considering the entrance to consist of an ideal nozzle converging to a cross section equal to that of the pipe followed by a section of pipe with $e_v = 0.5$. (See Table 7.4–1.) The behavior of the nozzle can be determined from Eq. 15.4–37 by assuming the cross-sectional area S_1 to be infinite and C_a to be unity.)

Answer: Yes. The calculated discharge rate is about 6 lb_m sec^{-1} if isothermal flow is assumed (see Problem 15.F) and about 6.3 lb_m sec^{-1} for adiabatic flow. The actual rate should be between these limits for an ambient temperature of 70° F.

15.L_3 Stagnation Temperature

A "total temperature probe," as shown in Fig. 15.L, is inserted in a steady stream of an ideal gas at a temperature T_1 and moving with a velocity v_1. Part of the moving gas enters the open end of the probe and is decelerated to nearly zero velocity before slowly leaking

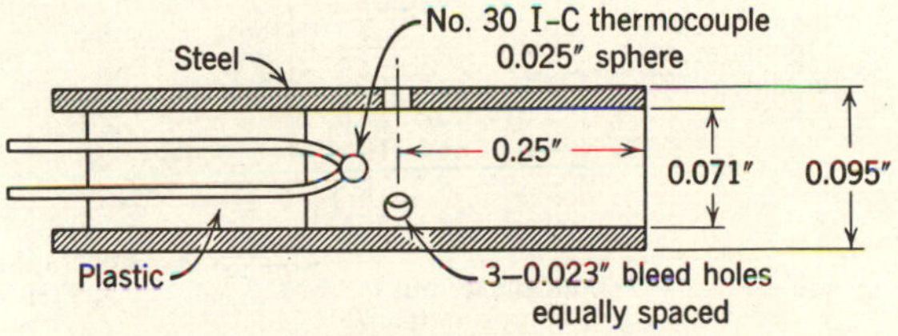

Fig. 15.L. Total temperature probe. [H. C. Hottel and A. Kalitinsky, *J. Appl. Mech.* **12,** A25 (1945).]

out of the bleed holes; this deceleration results in a temperature rise, which is measured by the thermocouple. Since the deceleration is rapid it is nearly adiabatic.

a. Develop an expression for the temperature registered by the thermocouple in terms of v_1 and T_1 by use of the steady-state macroscopic energy balance, Eq. 15.1–4. Use as your system a representative stream of fluid entering the probe. Draw reference plane "1" far enough upstream that conditions may be assumed unaffected by the probe, and reference plane "2" in the probe itself. Assume zero velocity at plane "2," neglect radiation, and neglect conduction of heat from the fluid as it passes between the reference planes.

b. What is the function of the bleed holes?

Answer: a. $(T_2 - T_1) = v_1^2/2\hat{C}_p$. Temperature rises within about 2 per cent of those given by this expression may be obtained with well-designed probes.

15.M_3 Continuous Heating of a Slurry in an Agitated Tank

A slurry is being heated by pumping it through a well-stirred heating tank as shown in Fig. 15.M. The inlet temperature of the slurry is T_i, and the temperature of the outer surface

[2] See C. E. Lapple, *Trans. Am. Inst. Chem. Engrs.*, **39**, 385 (1943).

of the steam coil is T_s. In addition, we define the following symbols:

V = volume of slurry in tank
ρ = density of slurry
w = mass rate of flow of slurry through tank
$\hat{C}_p$ = heat capacity of slurry
U = over-all heat-transfer coefficient of heating coil
A = total heat-transfer surface of coil
t = time since start of heating
$T(t)$ = temperature of slurry in tank

The stirring is sufficiently good that the fluid temperature in the tank is uniform and the same as the outlet fluid temperature.

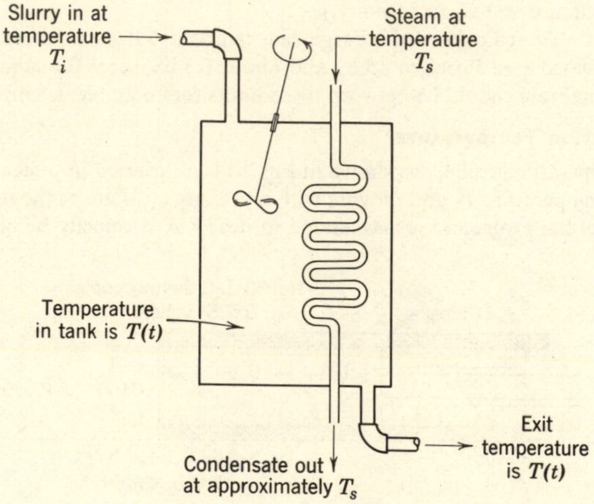

Fig. 15.M. Heating of a slurry in a stirred tank.

a. By means of an energy balance, show that the slurry temperature is described by the differential equation

$$\frac{dT}{dt} = \left(\frac{UA}{\rho \hat{C}_p V}\right)(T_s - T) - \left(\frac{w}{\rho V}\right)(T - T_i) \tag{15.M-1}$$

b. Rewrite this differential equation in terms of the dimensionless variables:

$$\tau = \frac{wt}{\rho V} \tag{15.M-2}$$

$$\Theta = \frac{T - T_\infty}{T_i - T_\infty} \tag{15.M-3}$$

where

$$T_\infty = \frac{\left(\frac{UA}{w\hat{C}_p}\right)T_s + T_i}{1 + \left(\frac{UA}{w\hat{C}_p}\right)} \tag{15.M-4}$$

What is the physical significance of τ, Θ, and T_∞?

c. Solve the dimensionless equation obtained in (*b*) for the boundary condition that at $t = 0$, $T = T_i$.

d. Check the solution. Are the differential equation and the initial condition satisfied? How does the system behave at large time? Is this limiting behavior in agreement with your intuitive expectations?

e. How is the temperature at infinite time affected by flow rate? Is this reasonable?

$$\textit{Answer: c.}\quad \frac{T - T_\infty}{T_i - T_\infty} = \exp\left[-\left(\frac{UA}{\rho V\hat{C}_p} + \frac{w}{\rho V}\right)t\right]$$

15.N$_4$ Discharge of a Gas from a Moving Tank

Equation 15.5–36 in Example 15.5–3 was obtained by setting $d\hat{\Phi}/dt$ equal to zero, a procedure justified only because the tank considered was stationary. It is nevertheless true that Eq. 15.5–36 is correct for moving tanks as well; this statement may be proved as follows:

a. Consider a tank such as that pictured in Fig. 15.5–4 but moving at a velocity $\boldsymbol{v}$ which is much larger than the relative velocity of fluid and tank in the region to the left of surface "1." Show that for this region of the tank the macroscopic momentum balance becomes

$$-\left(\boldsymbol{F} + \int_{S_1} p_1\, d\boldsymbol{S}_1\right) = m_{\text{tot}}\left(\frac{d\boldsymbol{v}}{dt} - \boldsymbol{g}\right) \tag{15.N-1}$$

in which the fluid velocity is assumed to be uniform and equal to $\boldsymbol{v}$. Then take the dot product of both sides of Eq. 15.N–1 with $\boldsymbol{v}$ to obtain

$$-W = m_{\text{tot}}\left(\frac{d\hat{K}}{dt} + \frac{d\hat{\Phi}}{dt}\right) \tag{15.N-2}$$

where $\partial\hat{\Phi}/\partial t$ is neglected.

b. Substitute this result into the macroscopic energy balance and continue as in Example 15.5–3.

15.O$_4$ Derivation of the Macroscopic Energy Balance

Integrate Eq. N of Table 10.4–1 over the flow system sketched in Fig. 7.0–1 to obtain Eq. 15.1–3. Note that the three-dimensional Leibnitz formula (see Appendix A, §A.5) must be used to perform this integration for the left side of the equation. On the right side use the Gauss divergence theorem to transform volume integrals to surface integrals.

15.P$_4$ An Alternate Form of the Macroscopic Energy Balance

Repeat Problem 15.O for Eq. O of Table 10.4–1 and show that a macroscopic energy balance may be written in the form

$$\frac{d}{dt}(U_{\text{tot}} + \Phi_{\text{tot}} + K_{\text{tot}}) = -\Delta[(\hat{U} + \hat{K} + \hat{\Phi})w] + Q - W + \int_V \rho\frac{\partial\hat{\Phi}}{\partial t}\, dV \tag{15.P-1}$$

Note: You may find Eq. 10.1–14 useful. Note also that for $\partial\hat{\Phi}/\partial t = 0$ this expression reduces to Eq. 15.1–3.

PART III

MASS TRANSPORT

CHAPTER 16

Diffusivity and the Mechanisms of Mass Transport

In Chapter 1 we began by stating Newton's law of viscosity, and in Chapter 8 we began with Fourier's law of heat conduction. In this chapter we present Fick's law of diffusion, which describes the movement of one species, say *A*, through a binary mixture of *A* and *B* because of the concentration gradient of *A*.

The movement of a chemical species from a region of high concentration to a region of low concentration can be observed with the naked eye by dropping a small crystal of potassium permanganate into a beaker of water. The $KMnO_4$ begins to dissolve in the water, and very near the crystal there is a dark purple concentrated solution of $KMnO_4$. Because of the concentration gradient that is established, the $KMnO_4$ diffuses away from the crystal. The progress of the diffusion can then be followed by observing the growth of the purple region—dark purple where the $KMnO_4$ concentration is high and light purple where it is low.

Diffusion is more complicated than viscous flow or heat conduction because here for the first time we have to deal with mixtures. In a diffusing mixture the velocities of the individual species are different, and there are several useful ways of averaging the velocities of the species to get a local velocity for the mixture. It is necessary to choose such a local velocity before the rates of diffusion can be defined. Hence it is appropriate that in §16.1 we discuss in some detail the definitions of concentrations, velocities, and fluxes. Although the discussion in §16.1 involves no new physical principles, a mastery of these definitions is essential for study of the later topics.

In §16.2 Fick's law of diffusion is stated, and the binary diffusivity $\mathscr{D}_{AB}$ is thereby defined. In §16.3 the temperature and pressure dependence of $\mathscr{D}_{AB}$ are discussed. In §16.4 and §16.5 the diffusivities of binary gas and liquid mixtures are considered from a molecular point of view.

The present lack of diffusivity data on most mixtures of interest makes it necessary to use estimated or extrapolated values of $\mathscr{D}_{AB}$ in most calculations. One should, of course, endeavor to use measured values, since they are usually (but not always) more reliable.

We have indicated that diffusion of A in a binary system of A and B occurs because of a concentration gradient of A. This phenomenon is sometimes called *ordinary diffusion* to distinguish it from *pressure diffusion* (motion of A resulting from a pressure gradient), *thermal diffusion* (motion of A resulting from a thermal gradient), and *forced diffusion* (motion of A because of unequal external forces on A and B). These additional effects, and also diffusion in multicomponent systems, are discussed briefly in Chapter 18.

Note: In this and the following chapters we adopt a fixed rule on notation. When considering *two-component systems*, we label the species A and B. When dealing with *multicomponent systems*, we label specific species 1, 2, 3, etc., or in general discussions we may use a dummy index such as i, j, or k to stand for the various species in turn. Formulas that are limited to binary systems are thus distinguished by the appearance of A and B as subscripts.

§16.1 DEFINITIONS OF CONCENTRATIONS, VELOCITIES, AND MASS FLUXES[1]

In a multicomponent system the concentrations of the various species may be expressed in numerous ways. We limit our discussion to four: the *mass concentration*, ρ_i, which is the mass of species i per unit volume of solution; the *molar concentration*, $c_i = \rho_i/M_i$, which is the number of moles of species i per unit volume of solution; the *mass fraction*, $\omega_i = \rho_i/\rho$, which is the mass concentration of species i divided by the total mass density of the solution; and the *mole fraction*, $x_i = c_i/c$, which is the molar concentration of species i divided by the total molar density of the solution. (By the word "solution" we mean a one-phase gaseous, liquid, or solid mixture.) In Table 16.1–1 a summary is given of these concentration units and their interrelation for binary systems.

[1] R. B. Bird, Theory of Diffusion, in *Advances in Chemical Engineering*, Vol. I, Academic Press, New York (1956), pp. 170, *et seq.*

In a diffusing mixture the various chemical species are moving at different velocities. Let $\boldsymbol{v}_i$ denote the velocity of the *i*th species with respect to stationary coordinate axes.[2] Then, for a mixture of *n* species, the local *mass average velocity* $\boldsymbol{v}$ is defined as

$$\boldsymbol{v} = \frac{\sum_{i=1}^{n} \rho_i \boldsymbol{v}_i}{\sum_{i=1}^{n} \rho_i} \tag{16.1-1}$$

Note that $\rho\boldsymbol{v}$ is the local rate at which mass passes through a unit cross section placed perpendicular to the velocity $\boldsymbol{v}$. This is the local velocity one would measure by means of a pitot tube and corresponds to $\boldsymbol{v}$ as used in the preceding chapters for pure fluids. Similarly, one may define a local *molar average velocity* $\boldsymbol{v}^\star$ as

$$\boldsymbol{v}^\star = \frac{\sum_{i=1}^{n} c_i \boldsymbol{v}_i}{\sum_{i=1}^{n} c_i} \tag{16.1-2}$$

Note that $c\boldsymbol{v}^\star$ is the local rate at which moles pass through a unit cross section placed perpendicular to the velocity $\boldsymbol{v}^\star$. We shall find extensive use for both of these average velocities. Still other average velocities are sometimes used, such as the *volume average velocity* $\boldsymbol{v}^\blacksquare$. (See Problem 16.K.)

In flow systems one is frequently interested in the velocity of a given species with respect to $\boldsymbol{v}$ or $\boldsymbol{v}^\star$ rather than with respect to stationary coordinates. This leads to the definition of the "diffusion velocities":

$$\boldsymbol{v}_i - \boldsymbol{v} = \text{diffusion velocity of } i \text{ with respect to } \boldsymbol{v} \tag{16.1-3}$$

$$\boldsymbol{v}_i - \boldsymbol{v}^\star = \text{diffusion velocity of } i \text{ with respect to } \boldsymbol{v}^\star \tag{16.1-4}$$

These diffusion velocities indicate the motion of component *i* relative to the local motion of the fluid stream.

The notation for velocities in binary systems is summarized in Table 16.1–2, together with some useful relations among the various velocities.

It is important that the reader thoroughly understand the meaning of the various kinds of velocities. Therefore, let us examine a specific system and represent its behavior pictorially. In Fig. 16.1–1 we consider what happens when liquid *A* evaporates and diffuses upward through a long tube initially filled with vapor *B*. As *A* evaporates, it pushes gas *B* upward. However, there is not a clear dividing line between pure *A* vapor and pure *B* vapor, as suggested in Fig. 16.1–1*a*; instead, the upward displacement of *B* is

[2] By "velocity" we do not mean the velocity of an individual molecule of species *i*; rather we mean the sum of the velocities of the molecules of species *i* within a small volume element divided by the number of such molecules.

TABLE 16.1–1

NOTATION FOR CONCENTRATIONS IN BINARY SYSTEMS

Basic definitions	$\rho = \rho_A + \rho_B$ = mass density of solution (g/cm³)	(*A*)		
	$\rho_A = c_A M_A$ = mass concentration of *A* (g of *A*/cm³ of solution)	(*B*)		
	$\omega_A = \dfrac{\rho_A}{\rho}$ = mass fraction of *A*	(*C*)		
	$c = c_A + c_B$ = molar density of solution (g-moles/cm³)	(*D*)		
	$c_A = \dfrac{\rho_A}{M_A}$ = molar concentration of *A* (g-moles of *A*/cm³ of solution)	(*E*)		
	$x_A = \dfrac{c_A}{c}$ = mole fraction of *A*	(*F*)		
	$M = \dfrac{\rho}{c}$ = number-mean molecular weight of mixture	(*G*)		
Additional relations, for reference only	$x_A + x_B = 1$	(*H*)	$\omega_A + \omega_B = 1$	(*I*)
	$x_A M_A + x_B M_B = M$	(*J*)	$\dfrac{\omega_A}{M_A} + \dfrac{\omega_B}{M_B} = \dfrac{1}{M}$	(*K*)
	$x_A = \dfrac{\dfrac{\omega_A}{M_A}}{\dfrac{\omega_A}{M_A} + \dfrac{\omega_B}{M_B}}$	(*L*)	$\omega_A = \dfrac{x_A M_A}{x_A M_A + x_B M_B}$	(*M*)
	$dx_A = \dfrac{d\omega_A}{M_A M_B \left(\dfrac{\omega_A}{M_A} + \dfrac{\omega_B}{M_B}\right)^2}$	(*N*)	$d\omega_A = \dfrac{M_A M_B \, dx_A}{(x_A M_A + x_B M_B)^2}$	(*O*)

TABLE 16.1–2

NOTATION FOR VELOCITIES IN BINARY SYSTEMS

Basic definitions	$\boldsymbol{v}_A$ = velocity of species *A* relative to stationary coordinates	(*A*)
	$\boldsymbol{v}_A - \boldsymbol{v}$ = diffusion velocity of species *A* relative to $\boldsymbol{v}$	(*B*)
	$\boldsymbol{v}_A - \boldsymbol{v}^\star$ = diffusion velocity of species *A* relative to $\boldsymbol{v}^\star$	(*C*)
	$\boldsymbol{v}$ = mass average velocity $= (1/\rho)(\rho_A \boldsymbol{v}_A + \rho_B \boldsymbol{v}_B) = \omega_A \boldsymbol{v}_A + \omega_B \boldsymbol{v}_B$	(*D*)
	$\boldsymbol{v}^\star$ = molar average velocity $= (1/c)(c_A \boldsymbol{v}_A + c_B \boldsymbol{v}_B) = x_A \boldsymbol{v}_A + x_B \boldsymbol{v}_B$	(*E*)
Additional relations	$\boldsymbol{v} - \boldsymbol{v}^\star = \omega_A(\boldsymbol{v}_A - \boldsymbol{v}^\star) + \omega_B(\boldsymbol{v}_B - \boldsymbol{v}^\star)$	(*F*)
	$\boldsymbol{v}^\star - \boldsymbol{v} = x_A(\boldsymbol{v}_A - \boldsymbol{v}) + x_B(\boldsymbol{v}_B - \boldsymbol{v})$	(*G*)

TABLE 16.1–3

MASS AND MOLAR FLUXES IN BINARY SYSTEMS

	Quantity	With Respect to Stationary Axes	With Respect to $\boldsymbol{v}$	With Respect to $\boldsymbol{v}^\star$
Basic definitions	Velocity of species A (cm sec^{-1})	$\boldsymbol{v}_A$ (A)	$\boldsymbol{v}_A - \boldsymbol{v}$ (B)	$\boldsymbol{v}_A - \boldsymbol{v}^\star$ (C)
	Mass flux of species A (g cm^{-2} sec^{-1})	$\boldsymbol{n}_A = \rho_A \boldsymbol{v}_A$ (D)	$\boldsymbol{j}_A = \rho_A(\boldsymbol{v}_A - \boldsymbol{v})$ (E)	$\boldsymbol{j}_A^\star = \rho_A(\boldsymbol{v}_A - \boldsymbol{v}^\star)$ (F)
	Molar flux of species A (g-moles cm^{-2} sec^{-1})	$\boldsymbol{N}_A = c_A \boldsymbol{v}_A$ (G)	$\boldsymbol{J}_A = c_A(\boldsymbol{v}_A - \boldsymbol{v})$ (H)	$\boldsymbol{J}_A^\star = c_A(\boldsymbol{v}_A - \boldsymbol{v}^\star)$ (I)
Relations among the fluxes, for reference only	Sum of mass fluxes (g cm^{-2} sec^{-1})	$\boldsymbol{n}_A + \boldsymbol{n}_B = \rho \boldsymbol{v}$ (J)	$\boldsymbol{j}_A + \boldsymbol{j}_B = 0$ (K)	$\boldsymbol{j}_A^\star + \boldsymbol{j}_B^\star = \rho(\boldsymbol{v} - \boldsymbol{v}^\star)$ (L)
	Sum of molar fluxes (g-moles cm^{-2} sec^{-1})	$\boldsymbol{N}_A + \boldsymbol{N}_B = c\boldsymbol{v}^\star$ (M)	$\boldsymbol{J}_A + \boldsymbol{J}_B = c(\boldsymbol{v}^\star - \boldsymbol{v})$ (N)	$\boldsymbol{J}_A^\star + \boldsymbol{J}_B^\star = 0$ (O)
	Fluxes in terms of $\boldsymbol{n}_A$ and $\boldsymbol{n}_B$	$\boldsymbol{N}_A = \dfrac{\boldsymbol{n}_A}{M_A}$ (P)	$\boldsymbol{j}_A = \boldsymbol{n}_A - \omega_A(\boldsymbol{n}_A + \boldsymbol{n}_B)$ (Q)	$\boldsymbol{j}_A^\star = \boldsymbol{n}_A - x_A\left(\boldsymbol{n}_A + \dfrac{M_A}{M_B}\boldsymbol{n}_B\right)$ (R)
	Fluxes in terms of $\boldsymbol{N}_A$ and $\boldsymbol{N}_B$	$\boldsymbol{n}_A = \boldsymbol{N}_A M_A$ (S)	$\boldsymbol{J}_A = \boldsymbol{N}_A - \omega_A\left(\boldsymbol{N}_A + \dfrac{M_B}{M_A}\boldsymbol{N}_B\right)$ (T)	$\boldsymbol{J}_A^\star = \boldsymbol{N}_A - x_A(\boldsymbol{N}_A + \boldsymbol{N}_B)$ (U)
	Fluxes in terms of $\boldsymbol{j}_A$ and $\boldsymbol{v}$	$\boldsymbol{n}_A = \boldsymbol{j}_A + \rho_A \boldsymbol{v}$ (V)	$\boldsymbol{J}_A = \dfrac{\boldsymbol{j}_A}{M_A}$ (W)	$\boldsymbol{j}_A^\star = \dfrac{M}{M_B}\boldsymbol{j}_A$ (X)
	Fluxes in terms of $\boldsymbol{J}_A^\star$ and $\boldsymbol{v}^\star$	$\boldsymbol{N}_A = \boldsymbol{J}_A^\star + c_A \boldsymbol{v}^\star$ (Y)	$\boldsymbol{J}_A = \dfrac{M_B}{M}\boldsymbol{J}_A^\star$ (Z)	$\boldsymbol{j}_A^\star = \boldsymbol{J}_A^\star M_A$ (AA)

accompanied by an intermingling of the two vapors, as shown in Fig. 16.1–1*b*. Thus at any point in the column A is moving upward more rapidly than the over-all average motion and B is moving less rapidly, because of diffusion. In Fig. 16.1–1*b* the various velocity vectors are shown for the situation that $x_A = \frac{1}{6}$, $v^\star = 12$, $v_A - v^\star = 3$, and $M_A = 5M_B$. The reader should verify the other velocity vectors.

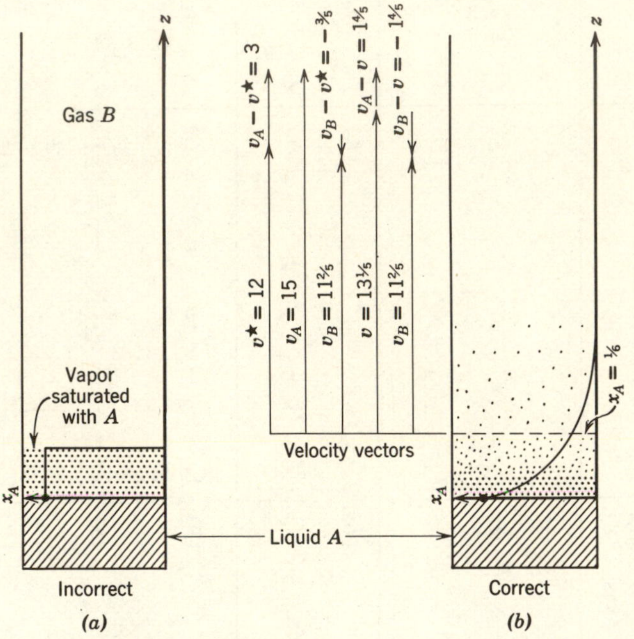

Fig. 16.1–1. Concentration distribution at a specific time in a diffusing system.

Now that concentrations and velocities have been discussed, we are in a position to define mass and molar fluxes. The mass (or molar) flux of species i is a vector quantity denoting the mass (or moles) of species i that passes through a unit area per unit time. The motion may be referred to stationary coordinates, to the local mass average velocity $\boldsymbol{v}$, or to the local molar average velocity $\boldsymbol{v}^\star$. Thus the mass and molar fluxes *relative to stationary coordinates* are

$$\boldsymbol{n}_i = \rho_i \boldsymbol{v}_i \qquad \text{mass} \tag{16.1–5}$$

$$\boldsymbol{N}_i = c_i \boldsymbol{v}_i \qquad \text{molar} \tag{16.1–6}$$

The mass and molar fluxes *relative to the mass-average velocity* $\boldsymbol{v}$ are

$$\boldsymbol{j}_i = \rho_i(\boldsymbol{v}_i - \boldsymbol{v}) \qquad \text{mass} \tag{16.1–7}$$

$$\boldsymbol{J}_i = c_i(\boldsymbol{v}_i - \boldsymbol{v}) \qquad \text{molar} \tag{16.1–8}$$

and the mass and molar fluxes *relative to the molar-average velocity* $\boldsymbol{v}^\star$ are

$$\boldsymbol{j}_i^\star = \rho_i(\boldsymbol{v}_i - \boldsymbol{v}^\star) \qquad \text{mass} \tag{16.1-9}$$

$$\boldsymbol{J}_i^\star = c_i(\boldsymbol{v}_i - \boldsymbol{v}^\star) \qquad \text{molar} \tag{16.1-10}$$

The notation for fluxes in binary systems is summarized in Table 16.1–3, together with some relations among the various fluxes. It should be emphasized that the definition of a mass flux is not complete until both *units* and *reference frame* have been given.

Mathematically speaking, any *one* of the foregoing six flux notations is adequate for all diffusion problems, but each has certain advantages and all have appeared in the literature. The flux $\boldsymbol{N}_i$ (and to a lesser extent $\boldsymbol{n}_i$) has been used in engineering because in process calculations it is usually desirable to refer to a coordinate system fixed in the equipment. The fluxes $\boldsymbol{j}_i$ and $\boldsymbol{J}_i^\star$ are the usual measures of diffusion rates and are useful in formulating the equations of change for multicomponent systems. The fluxes $\boldsymbol{J}_i$ and $\boldsymbol{j}_i^\star$ are seldom used but are included here for the sake of completeness. We use $\boldsymbol{N}_i$ in most subsequent discussions.

Example 16.1–1. Relations Among the Molar Fluxes

a. How are the fluxes $\boldsymbol{J}_i^\star$ and $\boldsymbol{N}_i$ related in an *n*-component system?

b. Show that the sum of the fluxes $\boldsymbol{J}_i^\star$ is zero.

Solution. *a.* First we combine the definitions of $\boldsymbol{v}^\star$ and $\boldsymbol{J}_i^\star$ (Eqs. 16.1–2 and 10) to get

$$\begin{aligned}\boldsymbol{J}_i^\star &= c_i(\boldsymbol{v}_i - \boldsymbol{v}^\star)\\ &= c_i\boldsymbol{v}_i - \frac{c_i}{c}\sum_{j=1}^{n} c_j\boldsymbol{v}_j\end{aligned} \tag{16.1-11}$$

Application of the definitions of $\boldsymbol{N}_i$ and x_i then gives

$$\boldsymbol{J}_i^\star = \boldsymbol{N}_i - x_i\sum_{j=1}^{n}\boldsymbol{N}_j \tag{16.1-12}$$

From this result it can be seen that the molar diffusion flux $\boldsymbol{J}_i^\star$ is simply the difference between the molar flux $\boldsymbol{N}_i$ and the rate of bodily transport (bulk flow) of species *i* due to the local molar flux of the mixture. Equation 16.1–10 may be similarly interpreted.

b. Summation of Eq. 16.1–12 from $i = 1$ to $i = n$ gives

$$\sum_{i=1}^{n}\boldsymbol{J}_i^\star = 0 \tag{16.1-13}$$

which shows that the sum of the *molar* diffusion fluxes relative to the *molar* average velocity is zero in any mixture. Thus in a binary mixture the diffusion fluxes $\boldsymbol{J}_A^\star$ and $\boldsymbol{J}_B^\star$ are of equal magnitude and are oppositely directed:

$$\boldsymbol{J}_A^\star = -\boldsymbol{J}_B^\star \tag{16.1-14}$$

An application of this result is given in Problem 16.I.

§16.2 FICK'S LAW OF DIFFUSION

In Eq. 1.1–2 the viscosity μ is defined as the proportionality factor between momentum flux and velocity gradient (Newton's law of viscosity). In Eq. 8.1–6 the thermal conductivity k is defined as the proportionality factor between heat flux and temperature gradient (Fourier's law of heat conduction).

TABLE 16.2–1

EQUIVALENT FORMS OF FICK'S FIRST LAW OF BINARY DIFFUSION

Flux	Gradient	Form of Fick's First Law	
$\boldsymbol{n}_A$	$\nabla\omega_A$	$\boldsymbol{n}_A - \omega_A(\boldsymbol{n}_A + \boldsymbol{n}_B) = -\rho\mathscr{D}_{AB}\nabla\omega_A$	(A)
$\boldsymbol{N}_A$	∇x_A	$\boldsymbol{N}_A - x_A(\boldsymbol{N}_A + \boldsymbol{N}_B) = -c\mathscr{D}_{AB}\nabla x_A$	(B)
$\boldsymbol{j}_A$	$\nabla\omega_A$	$\boldsymbol{j}_A = -\rho\mathscr{D}_{AB}\nabla\omega_A$	(C)
$\boldsymbol{J}_A^\star$	∇x_A	$\boldsymbol{J}_A^\star = -c\mathscr{D}_{AB}\nabla x_A$	(D)
$\boldsymbol{j}_A$	∇x_A	$\boldsymbol{j}_A = -\left(\frac{c^2}{\rho}\right)M_AM_B\mathscr{D}_{AB}\nabla x_A$	(E)
$\boldsymbol{J}_A^\star$	$\nabla\omega_A$	$\boldsymbol{J}_A^\star = -\left(\frac{\rho^2}{cM_AM_B}\right)\mathscr{D}_{AB}\nabla\omega_A$	(F)
$c(\boldsymbol{v}_A - \boldsymbol{v}_B)$	∇x_A	$c(\boldsymbol{v}_A - \boldsymbol{v}_B) = -\frac{c\mathscr{D}_{AB}}{x_Ax_B}\nabla x_A$	(G)

Now we define the mass diffusivity $\mathscr{D}_{AB} = \mathscr{D}_{BA}$ in a binary system in an analogous fashion:

$$\boldsymbol{J}_A^\star = -c\mathscr{D}_{AB}\nabla x_A \tag{16.2–1}$$

This is *Fick's first law of diffusion*,[1] written in terms of the molar diffusion flux $\boldsymbol{J}_A^\star$. This equation states that species A diffuses (moves relative to the mixture) in the direction of decreasing mole fraction of A, just as heat flows by conduction in the direction of decreasing temperature.

A number of other mathematically equivalent statements of Fick's first law have appeared in the literature, and some of them are summarized in Table 16.2–1 for reference only. The diffusivity $\mathscr{D}_{AB}$ is *identical* in all these equations. Of special importance in the following chapters is the form of Fick's first law in terms of $\boldsymbol{N}_A$, the molar flux relative to stationary coordinates:

$$\boldsymbol{N}_A = x_A(\boldsymbol{N}_A + \boldsymbol{N}_B) - c\mathscr{D}_{AB}\nabla x_A \tag{16.2–2}$$

This equation shows that the diffusion flux $\boldsymbol{N}_A$ relative to stationary coordinates is the *resultant* of two *vector* quantities: the vector $x_A(\boldsymbol{N}_A + \boldsymbol{N}_B)$,

[1] Temperature gradients, pressure gradients, and external forces also contribute to the diffusion flux, although their effects are usually minor. More complete expressions for the diffusion flux are given in §18.4.

which is the molar flux of A resulting from the bulk motion of the fluid, and the vector $\boldsymbol{J}_A^\star = -c\mathscr{D}_{AB}\nabla x_A$, which is the molar flux of A resulting from the diffusion superimposed on the bulk flow. Thus in Fig. 16.1–1*b* the bulk flow and diffusion terms in Eq. 16.2–2 are in the same direction for species A (because A is diffusing with the current) and are opposed for species B (because B is diffusing against the current).

The units of the mass diffusivity $\mathscr{D}_{AB}$ are cm^2 sec^{-1} or ft^2 hr^{-1}. Note that the kinematic viscosity ν and the thermal diffusivity α also have the same units. The way in which these three quantities are analogous can be seen from the following equations for the fluxes of mass, momentum, and energy in one-dimensional systems:

$$j_{Ay} = -\mathscr{D}_{AB}\frac{d}{dy}(\rho_A) \qquad \text{(Fick's law}^1 \text{ for constant } \rho) \qquad (16.2\text{–}3)$$

$$\tau_{yx} = -\nu\frac{d}{dy}(\rho v_x) \qquad \text{(Newton's law for constant } \rho) \qquad (16.2\text{–}4)$$

$$q_y = -\alpha\frac{d}{dy}(\rho\hat{C}_pT) \qquad \text{(Fourier's law for constant } \rho\hat{C}_p) \qquad (16.2\text{–}5)$$

These equations state, respectively, that (*a*) mass transport occurs because of a gradient in mass concentration, (*b*) momentum transport occurs because of a gradient in momentum concentration, and (*c*) energy transport occurs because of a gradient in energy concentration. These analogies do not apply in two- and three-dimensional problems, however, because $\boldsymbol{\tau}$ is a tensor quantity with nine components, whereas $\boldsymbol{j}_A$ and $\boldsymbol{q}$ are vectors with three components.

In Tables 16.2–2, 3, and 4 some values of $\mathscr{D}_{AB}$ are given for a few gas,

TABLE 16.2–2

EXPERIMENTAL DIFFUSIVITIES OF SOME DILUTE GAS PAIRS[a]

Gas Pair	Temperature (° K)	$\mathscr{D}_{AB}$ (cm^2 sec^{-1})
CO_2—N_2O	273.2	0.096
CO_2—CO	273.2	0.139
CO_2—N_2	273.2	0.144
	288.2	0.158
	298.2	0.165
Ar—O_2	293.2	0.20
H_2—SF_6	298.2	0.420
H_2—CH_4	298.2	0.726

[a] This table is abstracted from J. O. Hirschfelder, C. F. Curtiss, and R. B. Bird, *Molecular Theory of Gases and Liquids*, Wiley, New York (1954), p. 579. The values given are for 1 atm pressure.

TABLE 16.2-3

EXPERIMENTAL DIFFUSIVITIES IN THE LIQUID STATE[a]

A	B	T (° C)	x_A	$\mathscr{D}_{AB} \times 10^5$ (cm² sec⁻¹)
Chlorobenzene	Bromobenzene	10.01	0.0332	1.007
			0.2642	1.069
			0.5122	1.146
			0.7617	1.226
			0.9652	1.291
		39.97	0.0332	1.584
			0.2642	1.691
			0.5122	1.806
			0.7617	1.902
			0.9652	1.996
Ethanol	Water	25	0.05	1.13
			0.275	0.41
			0.50	0.90
			0.70	1.40
			0.95	2.20
Water	*n*-Butanol	30	0.131	1.24
			0.222	0.920
			0.358	0.560
			0.454	0.437
			0.524	0.267

[a] This table is abstracted from a review article by P. A. Johnson and A. L. Babb, Liquid Diffusion in Non-Electrolytes, *Chem. Revs.*, **56**, 387–453 (1956); in this article a summary of experimental diffusion coefficients for liquid systems is given, as well as a survey of methods of measurement. Another excellent review article is that of L. J. Gosting, Measurement and Interpretation of Diffusion Coefficients of Proteins, *Advances in Protein Chemistry*, Vol. XI, Academic Press, New York (1956).

liquid, and solid systems. Diffusivities of gases at low density are almost composition independent, increase with the temperature, and vary inversely with pressure. Liquid and solid diffusivities are strongly concentration-dependent and generally increase with temperature. In the next three sections we summarize the available means for estimating diffusivities.

§16.3 TEMPERATURE AND PRESSURE DEPENDENCE OF MASS DIFFUSIVITY

The mass diffusivity $\mathscr{D}_{AB}$ for a binary system is a function of temperature, pressure, and composition, whereas the viscosity μ and thermal conductivity

TABLE 16.2–4

EXPERIMENTAL DIFFUSIVITIES IN THE SOLID STATE[a]

System	T (° C)	Diffusivity, $\mathscr{D}_{AB}$ ($cm^2\ sec^{-1}$)
He in SiO_2	20	$2.4 - 5.5 \times 10^{-10}$
He in pyrex	20	4.5×10^{-11}
	500	2×10^{-8}
H_2 in SiO_2	500	$0.6 - 2.1 \times 10^{-8}$
H_2 in Ni	85	1.16×10^{-8}
	165	10.5×10^{-8}
Bi in Pb	20	1.1×10^{-16}
Hg in Pb	20	2.5×10^{-15}
Sb in Ag	20	3.5×10^{-21}
Al in Cu	20	1.3×10^{-30}
Cd in Cu	20	2.7×10^{-15}

[a] Values taken from R. M. Barrer, *Diffusion in and through Solids*, Macmillan, New York (1941), pp. 141, 222, and 275.

k for a pure fluid are functions only of temperature and pressure. The data available on $\mathscr{D}_{AB}$ for most binary mixtures are, moreover, quite limited in range and accuracy. The available correlations of $\mathscr{D}_{AB}$ are of limited scope and are based more on theory than on experiment.

For binary gas mixtures at low pressure, $\mathscr{D}_{AB}$ is inversely proportional to the pressure, increases with increasing temperature, and is almost independent of composition for a given gas-pair. The following equation for estimation of $\mathscr{D}_{AB}$ at *low pressures* has been developed[1] from a combination of kinetic theory and corresponding-states arguments:

$$\frac{p\mathscr{D}_{AB}}{(p_{cA}p_{cB})^{1/3}(T_{cA}T_{cB})^{5/12}\left(\dfrac{1}{M_A}+\dfrac{1}{M_B}\right)^{1/2}} = a\left(\frac{T}{\sqrt{T_{cA}T_{cB}}}\right)^b \tag{16.3-1}$$

in which $\mathscr{D}_{AB}$ [=] $cm^2\ sec^{-1}$, p [=] atm, and T [=] ° K. Analysis of experimental data gave the following values of the constants a and b:

For nonpolar gas-pairs:

$$a = 2.745 \times 10^{-4}$$
$$b = 1.823$$

For H_2O with a nonpolar gas:

$$a = 3.640 \times 10^{-4}$$
$$b = 2.334$$

[1] J. C. Slattery and R. B. Bird, *A.I.Ch.E. Journal*, **4**, 137–142 (1958).

Equation 16.3–1 agrees with experimental data at atmospheric pressure within about 8 per cent.[1,2] If gases A and B are nonpolar and their Lennard-Jones parameters are available, the kinetic theory method given in §16.4 will usually give better accuracy.

At high pressures $\mathscr{D}_{AB}$ no longer varies inversely with the pressure. Very little is known about this pressure dependence, except in the limiting case of self-diffusion (interdiffusion of identical molecules), which can be closely

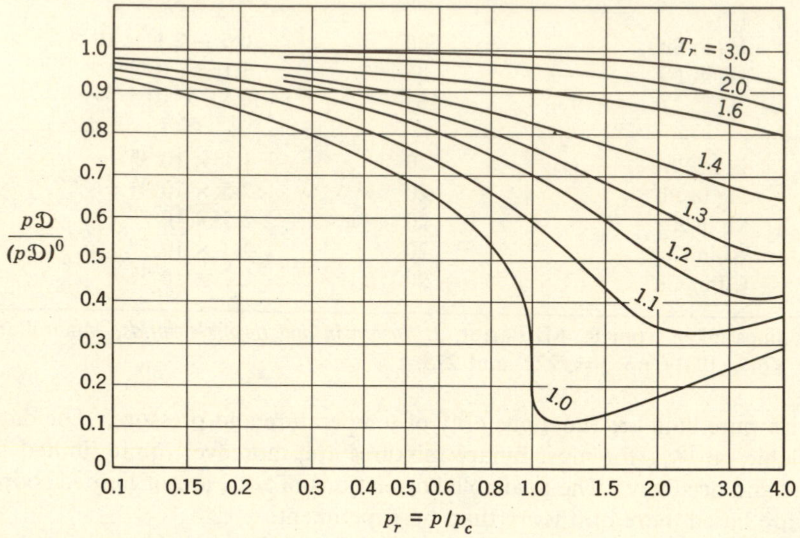

Fig. 16.3–1. Generalized chart for self-diffusivities of dense gases [J. C. Slattery, M.S. thesis, University of Wisconsin (1955)]; this chart is based on very few data.

approached experimentally by using isotopic tracers. The pressure dependence of the self-diffusivity, $\mathscr{D}_{AA}$, may be estimated from Fig. 16.3–1; this plot is based on the Enskog kinetic theory of dense gases and on some fragmentary data.[1] The ordinate of the plot is $p\mathscr{D}_{AA}/(p\mathscr{D}_{AA})^0$, the ratio of the pressure-diffusivity product at pressure p and temperature T to the pressure-diffusivity product at the same temperature but at low pressure. This ratio is plotted as a function of the reduced temperature $T_r = T/T_c$ and the reduced pressure $p_r = p/p_c$. The chart is analogous to Figs. 1.3–2 and 8.2–2; its use is illustrated below.

In the absence of other information, it has been suggested that Fig. 16.3–1 may be used to estimate $\mathscr{D}_{AB}$ if p_c and T_c are replaced by the pseudocritical values p_c' and T_c' given in Eqs. 1.3–3 and 4. Very few data are available to test this procedure, and therefore it has to be regarded as provisional.

[2] Detailed comparisons of this and other estimation methods with experimental data are given by R. C. Reid and T. K. Sherwood, *The Properties of Gases and Liquids*, McGraw-Hill, New York (1958), pp. 267–276.

Example 16.3–1. Estimation of Mass Diffusivity at Low Density

Estimate $\mathscr{D}_{AB}$ for the system argon-oxygen at 293.2° K and 1 atm total pressure.

Solution. The properties needed for Eq. 16.3–1 (see Table B–1) are

	M	T_c	p_c
A (argon)	39.94	151.2	48.0
B (oxygen)	32.00	154.4	49.7

Therefore

$$(p_{cA}p_{cB})^{1/3} = (48.0 \times 49.7)^{1/3} = 13.36$$

$$(T_{cA}T_{cB})^{5/12} = (151.2 \times 154.4)^{5/12} = 66.08$$

$$\left(\frac{1}{M_A} + \frac{1}{M_B}\right)^{1/2} = \left(\frac{1}{39.94} + \frac{1}{32.00}\right)^{1/2} = 0.2372$$

$$a\left(\frac{T}{\sqrt{T_{cA}T_{cB}}}\right)^b = 2.745 \times 10^{-4}\left(\frac{293.2}{\sqrt{151.2 \times 154.4}}\right)^{1.823}$$

$$= 9.01 \times 10^{-4}$$

Substitution of these numerical values into Eq. 16.3–1 gives

$$(1.0)\mathscr{D}_{AB} = (9.01 \times 10^{-4})(13.36)(66.08)(0.2372)$$

$$\mathscr{D}_{AB} = 0.189 \text{ cm}^2 \text{ sec}^{-1}$$

The measured value (Table 16.2–2) is 0.20 cm^2 sec^{-1}. See Example 16.4–1 for a method based on kinetic theory.

Example 16.3–2. Estimation of Mass Diffusivity at High Density

Estimate $\mathscr{D}_{AB}$ for a mixture of 80 mole per cent methane and 20 mole per cent ethane at 2000 psia and 104° F (136 atm and 313° K). The experimental value of $(p\mathscr{D}_{AB})^0$ at 293° K is 0.163 atm cm^2 sec^{-1}.

Solution. First correct the given value of $(p\mathscr{D}_{AB})^0$ to the desired temperature, using Eq. 16.3–1 with the nonpolar constants:

$$(p\mathscr{D}_{AB})^0 \text{ at } 313^\circ \text{ K} = 0.163\left(\frac{313}{293}\right)^{1.823}$$

$$= 0.184 \text{ atm cm}^2 \text{ sec}^{-1}$$

The critical properties from Table B–1 in the Appendix are

methane:	$T_c = 190.7^\circ$ K,	$p_c = 45.8$ atm
ethane:	$T_c = 305.4^\circ$ K,	$p_c = 48.2$ atm

The pseudocritical properties according to Eqs. 1.3–4 and 5 are

$$p_c' = \sum_{i=1}^{n} x_i p_{ci} = (0.80 \times 45.8) + (0.20 \times 48.2)$$

$$= 46.3 \text{ atm}$$

$$T_c' = \sum_{i=1}^{n} x_i T_{ci} = (0.80 \times 190.7) + (0.20 \times 305.4)$$

$$= 213.6^\circ \text{ K}$$

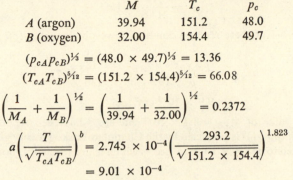

The pseudo-reduced pressure and temperature are

$$\frac{p}{p_c'} = \frac{136}{46.3} = 2.94$$

$$\frac{T}{T_c'} = \frac{313}{213.6} = 1.47$$

From Fig. 16.3–1 at these reduced conditions we obtain $p\mathscr{D}_{AB}/(p\mathscr{D}_{AB})^0 \doteq 0.73$. Then

$$\mathscr{D}_{AB} = \frac{(p\mathscr{D}_{AB})^0}{p} \frac{p\mathscr{D}_{AB}}{(p\mathscr{D}_{AB})^0} = \frac{0.184}{136}(0.73)$$
$$= 9.9 \times 10^{-4} \text{ cm}^2 \text{ sec}^{-1}$$

An observed value of $\mathscr{D}_{AB} = 8.4 \times 10^{-4}$ cm^2 sec^{-1} has been reported at these conditions.[3] If $(p\mathscr{D}_{AB})^0$ is predicted by the Chapman-Enskog theory given in §16.4, $\mathscr{D}_{AB}$ is predicted to be 8.8×10^{-4} cm^2 sec^{-1}, which is unexpectedly good agreement.

§16.4 THEORY OF ORDINARY DIFFUSION IN GASES AT LOW DENSITY

The mass diffusivity $\mathscr{D}_{AB}$ for binary mixtures of nonpolar gases is predictable within about 5 per cent by kinetic theory. Again, as in §§1.4 and 8.3, we begin with a simplified derivation to illustrate the mechanisms involved and then present the more accurate results of the Chapman-Enskog theory.

Consider a large body of gas containing two molecular species A and A^*, both species having the same mass m_A and the same size and shape. Such a pair is closely approximated by two heavy molecules containing different isotopes, such as $U^{235}F_6$ and $U^{238}F_6$. We wish to determine the mass diffusivity $\mathscr{D}_{AA^*}$ in terms of the molecular properties on the assumption that the molecules are rigid spheres of diameter d_A.

As before, we shall use the following results of kinetic theory for a pure[1] rigid-sphere gas at low density in which the temperature, pressure, and velocity gradients are small:

$$\bar{u} = \sqrt{\frac{8\kappa T}{\pi m}} = \text{mean molecular speed relative to } \boldsymbol{v} \text{ or } \boldsymbol{v}^* \quad (16.4\text{–}1)$$

$$Z = \tfrac{1}{4}n\bar{u} = \text{wall collision frequency per unit area in stationary gas} \quad (16.4\text{–}2)$$

$$\lambda = \frac{1}{\sqrt{2}\,\pi d^2 n} = \text{mean free path} \quad (16.4\text{–}3)$$

[3] V. J. Berry, Jr. and R. C. Koeller, Diffusion in Compressed Binary Gaseous Systems, *Preprint 45*, A.I.Ch.E. 40th National Meeting, 1959.

[1] The use of the expressions for a pure gas is permissible here because the mechanical properties of A and A^* are the same.

The molecules reaching any plane in the gas have, on the average, had their last collision at a distance a from the plane, where

$$a = \tfrac{2}{3}\lambda \tag{16.4–4}$$

To determine the diffusivity $\mathscr{D}_{AA^*}$, we consider the motion of species A in the y-direction under a concentration gradient dx_A/dy (see Fig. 16.4–1), when the mixture moves at a finite velocity $v_y^\star$ throughout. The temperature T and total molar concentration c are assumed constant. We assume that

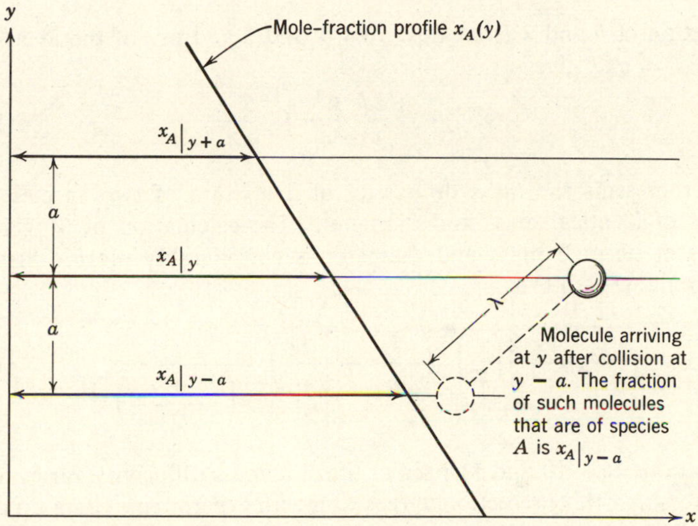

Fig. 16.4–1. Molecular transport of species A from plane at $(y - a)$ to plane at y.

Eqs. 16.4–1, 2, 3, and 4 remain valid in this nonequilibrium situation. The molar flux, N_{Ay}, of species A across any plane of constant y is found by counting the molecules of A that cross unit area of the plane in the positive y-direction and subtracting the number that cross in the negative y-direction. Thus

$$N_{Ay} = \frac{1}{\tilde{N}}\{nx_Av_y^\star|_y + \tfrac{1}{4}nx_A\bar{u}|_{y-a} - \tfrac{1}{4}nx_A\bar{u}|_{y+a}\} \tag{16.4–5}$$

In writing the last two terms, it is assumed that the concentration profile $x_A(y)$ is essentially linear for a distance of several mean free paths. As a further consequence of this assumption we may write

$$x_A|_{y-a} = x_A|_y - \frac{2}{3}\lambda\frac{dx_A}{dy} \tag{16.4–6}$$

$$x_A|_{y+a} = x_A|_y + \frac{2}{3}\lambda\frac{dx_A}{dy} \tag{16.4–7}$$

By combining Eqs. 16.4–5, 6, and 7 and noting that $cv^\star = N_A + N_{A^*}$ (see Table 16.1–3, Eq. *M*), we get

$$N_{Ay} = x_A(N_{Ay} + N_{A^*y}) - \frac{1}{3} c\bar{u}\lambda \frac{dx_A}{dy} \tag{16.4–8}$$

This equation corresponds to the y-component of Fick's law (Eq. 16.2–2) with the following approximate value for $\mathscr{D}_{AB}$:

$$\mathscr{D}_{AB} = \tfrac{1}{3}\bar{u}\lambda \tag{16.4–9}$$

Evaluation of $\bar{u}$ and λ from Eqs. 16.4–1 and 3 and use of the ideal-gas law $p = cRT = n\kappa T$ gives

$$\mathscr{D}_{AA^*} = \frac{2}{3}\left(\frac{\kappa^3}{\pi^3 m_A}\right)^{1/2} \frac{T^{3/2}}{p\, d_A^2} \tag{16.4–10}$$

which represents the mass diffusivity of a mixture of two species of rigid spheres of identical mass and diameter. The calculation of $\mathscr{D}_{AB}$ for rigid spheres of unequal mass and diameter is considerably more difficult; the corresponding result[2] is

$$\mathscr{D}_{AB} = \frac{2}{3}\left(\frac{\kappa^3}{\pi^3}\right)^{1/2} \left(\frac{1}{2m_A} + \frac{1}{2m_B}\right)^{1/2} \frac{T^{3/2}}{p\left(\dfrac{d_A + d_B}{2}\right)^2} \tag{16.4–11}$$

Equations 16.4–10 and 11 predict that the mass diffusivity varies inversely with pressure; this prediction agrees well with experimental data up to about 10 atm for many gas mixtures. (See Fig. 16.3–1.) The predicted temperature dependence is too weak, just as it was for viscosity and thermal conductivity, but it is qualitatively correct.

The foregoing development is illustrative only; for accurate results, the Chapman-Enskog kinetic theory should be used. The Chapman-Enskog formulas for viscosity and thermal conductivity were given in §§1.4 and 8.3; the corresponding formula for $c\mathscr{D}_{AB}$ for the gaseous state at low density is:[3,4]

$$c\mathscr{D}_{AB} = 2.2646 \times 10^{-5} \frac{\sqrt{T\left(\dfrac{1}{M_A} + \dfrac{1}{M_B}\right)}}{\sigma_{AB}^2\, \Omega_{\mathscr{D},AB}} \tag{16.4–12}$$

[2] A similar result is given by R. D. Present, *Kinetic Theory of Gases*, McGraw-Hill, New York (1958), p. 55.

[3] S. Chapman and T. G. Cowling, *Mathematical Theory of Non-Uniform Gases*, Second Edition, Cambridge University Press (1951), Chapters 10 and 14.

[4] J. O. Hirschfelder, C. F. Curtiss, and R. B. Bird, *Molecular Theory of Gases and Liquids*, Wiley, New York (1954), p. 539.

If we approximate c by the ideal-gas law $c = p/RT$, this becomes

$$\mathscr{D}_{AB} = 0.0018583 \frac{\sqrt{T^3\left(\frac{1}{M_A} + \frac{1}{M_B}\right)}}{p\sigma_{AB}{}^2\,\Omega_{\mathscr{D},AB}} \tag{16.4–13}$$

in which $\mathscr{D}_{AB}$ [=] $cm^2\ sec^{-1}$, c [=] g-moles cm^{-3}, T [=] °K, p [=] atm, σ_{AB} [=] Ångström units, and $\Omega_{\mathscr{D},AB}$ is a dimensionless function of the temperature and of the intermolecular potential field for one molecule of A and one of B. It is convenient to approximate this potential field by the Lennard-Jones function:

$$\varphi_{AB}(r) = 4\epsilon_{AB}\left[\left(\frac{\sigma_{AB}}{r}\right)^{12} - \left(\frac{\sigma_{AB}}{r}\right)^{6}\right] \tag{16.4–14}$$

Table B–2 gives $\Omega_{\mathscr{D},AB}$ for this potential energy model as a function of $\kappa T/\epsilon_{AB}$. From these results one can compute that $\mathscr{D}_{AB}$ increases roughly as the 2.0 power of T at low temperatures and as the 1.65 power of T at very high temperatures. (Compare Eq. 16.3–1.) For rigid spheres, $\Omega_{\mathscr{D},AB}$ would be unity at all temperatures and a result analogous to Eq. 16.4–11 would be obtained.

The Lennard-Jones parameters σ_{AB} and ϵ_{AB} could, in principle, be determined directly from accurate measurements of $\mathscr{D}_{AB}$ over a wide range of temperature; with somewhat greater difficulty, other measured properties of mixtures of A and B could be used. Suitable data are, however, extremely rare; therefore, one usually has to estimate σ_{AB} and ϵ_{AB}. It turns out that fairly satisfactory estimates can be made for nonpolar, nonreacting[5] molecule pairs by combining the Lennard-Jones parameters of species A and B empirically:[6]

$$\sigma_{AB} = \tfrac{1}{2}(\sigma_A + \sigma_B) \tag{16.4–15}$$

$$\epsilon_{AB} = \sqrt{\epsilon_A\epsilon_B} \tag{16.4–16}$$

In this way it is possible to predict measured values of $\mathscr{D}_{AB}$ within an average deviation of about 6 per cent by use of viscosity data on the pure species A and B or within about 10 per cent if the Lennard-Jones parameters for A and B are estimated from boiling-point data by means of Eqs. 1.4–14 and 15.[7]

For isotopic pairs, $\sigma_{AA^*} = \sigma_A = \sigma_{A^*}$ and $\epsilon_{AA^*} = \epsilon_A = \epsilon_{A^*}$; that is, the intermolecular force fields for the various pairs A–A, A^*–A^*, and A–A^*

[5] By nonreacting we mean here that A and B, in a *bimolecular* collision, do not react. Such a collision, of course, rules out the presence of any catalytic agent, so that molecule-pairs that react only catalytically are nonreacting pairs in the present discussion.

[6] J. O. Hirschfelder, R. B. Bird, and E. L. Spotz, *Chem. Revs.*, **44**, 205–231 (1949).

[7] R. C. Reid and T. K. Sherwood, *The Properties of Gases and Liquids*, McGraw-Hill, New York (1958), pp. 274–5.

are identical and the parameters σ_A and ϵ_A may be obtained from viscosity data on pure A. If, in addition, M_A is large, Eq. 16.4–12 simplifies to

$$c\mathscr{D}_{AA^*} = 3.2027 \times 10^{-5} \frac{\sqrt{T/M_A}}{\sigma_A^2 \Omega_{\mathscr{D},A}} \qquad (16.4\text{–}17)$$

An approximate equation for $\mathscr{D}_{AA^*}$ was derived earlier. (Compare Eq. 16.4–10.)

Comparison of Eqs. 16.4–17 and 1.4–18 shows that the self-diffusivity $\mathscr{D}_{AA^*}$ and the viscosity μ are related for heavy isotopic gas-pairs at low density:

$$\frac{\mu}{\rho\mathscr{D}_{AA^*}} = \frac{\nu}{\mathscr{D}_{AA^*}} = \frac{5}{6}\frac{\Omega_{\mathscr{D},A}}{\Omega_\mu} \qquad (16.4\text{–}18)$$

in which $\Omega_{\mathscr{D},A}$ and Ω_μ are nearly equal functions of $\kappa T/\epsilon_A$. (See Table B–2.) Thus the kinematic viscosity ν and the *self-diffusivity* $\mathscr{D}_{AA^*}$ are of the same order of magnitude *for gases at low density*. The relation between ν and the *binary diffusivity* $\mathscr{D}_{AB}$ is not so simple because $\nu = \mu/\rho$ may vary considerably with composition, but the ratio $\mu/\rho\mathscr{D}_{AB}$ lies between 0.2 and 5.0 for most gas-pairs. The quantity $\mu/\rho\mathscr{D}_{AB} = \nu/\mathscr{D}_{AB}$ is known as the *Schmidt number*; we shall encounter this quantity repeatedly in problems of diffusion in flow systems just as we encountered the Prandtl number $\hat{C}_p\mu/k = \nu/\alpha$ in problems of heat conduction in flow systems.

Equations 16.4–12, 13, 17, and 18 were derived for monatomic nonpolar gases but have been found to work well for polyatomic nonpolar gases as well. In addition, these equations may be used to predict $\mathscr{D}_{AB}$ for inter-diffusion of a polar gas and a nonpolar gas[8] by using combining laws different from Eqs. 16.4–15 and 16.

Example 16.4–1. Computation of Mass Diffusivity at Low Density

Predict the value of $\mathscr{D}_{AB}$ for mixtures of argon (A) and oxygen (B) at 293.2° K and 1 atm total pressure.

Solution. From Table B–1 we obtain the following constants:

$$M_A = 39.944; \qquad \sigma_A = 3.418\ \text{Å}; \qquad \frac{\epsilon_A}{\kappa} = 124°\ \text{K}$$

$$M_B = 32.000; \qquad \sigma_B = 3.433\ \text{Å}; \qquad \frac{\epsilon_B}{\kappa} = 113°\ \text{K}$$

The parameters σ_{AB} and ϵ_{AB}/κ for collisions of argon with oxygen may be estimated by means of Eqs. 16.4–15 and 16:

$$\sigma_{AB} = (\tfrac{1}{2})(3.418 + 3.433) = 3.426\ \text{Å}$$

$$\frac{\epsilon_{AB}}{\kappa} = \sqrt{(124)(113)} = 118.5°\ \text{K}$$

[8] J. O. Hirschfelder, C. F. Curtiss, and R. B. Bird, *loc. cit.*, §8.6*b*.

This gives $\kappa T/\epsilon_{AB} = 293.2/118.5 = 2.47$, and from Table B–2 we then obtain $\Omega_{\mathscr{D},AB} = 1.003$. Substitution of the foregoing values into Eq. 16.4–13 gives

$$\mathscr{D}_{AB} = 0.0018583 \frac{\sqrt{T^3\left(\frac{1}{M_A} + \frac{1}{M_B}\right)}}{p\sigma_{AB}^2\,\Omega_{\mathscr{D},AB}}$$

$$= 0.0018583 \frac{\sqrt{(293.2)^3\left(\frac{1}{39.944} + \frac{1}{32.00}\right)}}{(1.00)(3.426)^2(1.003)}$$

$$= 0.188 \text{ cm}^2 \text{ sec}^{-1}$$

This compares favorably with the measured value of 0.20 cm² sec⁻¹ given in Table 16.2–2.

§16.5 THEORIES OF ORDINARY DIFFUSION IN LIQUIDS

In the absence of rigorous theory for diffusion in liquids, there are two rough theories that have been useful in making order of magnitude of calculations: the *hydrodynamical theory* and the *Eyring theory*. (We have already given some results of the theories of Eyring and co-workers in connection with the calculation of viscosity and thermal conductivity of liquids.)

The *hydrodynamical theory* takes as its starting point the Nernst-Einstein equation,[1] which states that the diffusivity of a single particle or solute molecule of A through a stationary medium B is

$$\mathscr{D}_{AB} = \kappa T \frac{u_A}{F_A} \tag{16.5–1}$$

in which u_A/F_A is the "mobility" of the particle of A (that is, the steady-state velocity attained by the particle under the action of a unit force). A relation between force and velocity for a rigid sphere moving in "creeping flow" (that is, $\text{Re} \ll 1$) may be obtained from hydrodynamics. If the possibility of "slip" at the sphere-fluid interface is taken into account, then[2]

$$F_A = 6\pi\mu_B u_A R_A \left(\frac{2\mu_B + R_A\beta_{AB}}{3\mu_B + R_A\beta_{AB}}\right) \tag{16.5–2}$$

in which μ_B is the viscosity of the pure solvent, R_A is the radius of the diffusing particle, and β_{AB} is the *coefficient of sliding friction*. It has been pointed out[3] that there are two limiting cases of interest:

[1] F. Daniels and R. A. Alberty, *Physical Chemistry*, Wiley, New York (1955), p. 650. See also Eq. 18.4–14*a*.

[2] H. Lamb, *Hydrodynamics*, Dover, New York, §337.

[3] R. Fürth, in *Handbuch der physikalischen und technischen Mechanik*, Barth, Leipzig (1931), Vol. 7, p. 635.

a. If there is no tendency for the fluid to slip at the surface of the diffusing particle ($\beta_{AB} = \infty$), then Eq. 16.5–2 becomes Stokes's law. (See §2.6.)

$$F_A = 6\pi\mu_B u_A R_A \tag{16.5–3}$$

and substitution into Eq. 16.5–1 gives

$$\frac{\mathscr{D}_{AB}\mu_B}{\kappa T} = \frac{1}{6\pi R_A} \tag{16.5–4}$$

which is usually called the *Stokes-Einstein equation.* Equation 16.5–4 has been shown to be fairly good for describing the diffusion of *large spherical particles* or *large spherical molecules,* under which conditions the solvent appears to the diffusing species as a continuum.[4]

b. If there is no tendency for the fluid to stick at the surface of the diffusing particle ($\beta_{AB} = 0$), Eq. 16.5–2 becomes

$$F_A = 4\pi\mu_B u_A R_A \tag{16.5–5}$$

and Eq. 16.5–1 becomes

$$\frac{\mathscr{D}_{AB}\mu_B}{\kappa T} = \frac{1}{4\pi R_A} \tag{16.5–6}$$

If the molecules are all alike (that is, *self-diffusion*) and if they can be assumed to be arranged in a cubic lattice with all molecules just touching, then $2R_A$ may be set equal to $(\tilde{V}_A/\tilde{N})^{1/3}$ and

$$\frac{\mathscr{D}_{AA}\mu_A}{\kappa T} = \frac{1}{2\pi}\left(\frac{\tilde{N}}{\tilde{V}_A}\right)^{1/3} \tag{16.5–7}$$

It has been shown[5] that Eq. 16.5–7 predicts the self-diffusion data for a number of liquids within about ± 12 per cent; that comparison includes polar substances, associated substances, liquid metals, and molten sulfur.

It is thus seen that the simple hydrodynamical approach gives expressions for the diffusion coefficient for spherical molecules in dilute solution and also for the coefficient of self-diffusion. The theory further predicts that there should be a variation of $\mathscr{D}_{AB}$ with the size of the diffusing species. The hydrodynamic theory suggests that the shape of the diffusing species may well be important, since the friction factor increases by a factor of about 2 as the length-to-width ratio of a body goes from 1 to 10.

The *Eyring rate theory*[6] attempts to explain the transport phenomena on the basis of a simple model for the liquid state. It is assumed in this theory that there is some unimolecular rate process in terms of which the diffusion

[4] W. Sutherland, *Phil. Mag.*, **9**, 781–785 (1905).

[5] J. C. M. Li and P. Chang, *J. Chem. Phys.*, **23**, 518–520 (1955).

[6] S. Glasstone, K. J. Laidler, and H. Eyring, *Theory of Rate Processes*, McGraw-Hill, New York (1941), Chapter IX.

process can be described, and it is further assumed that in this process there is some configuration that can be identified as the *activated state*. The Eyring theory of reaction rates is applied to this elementary process.

On the assumption of a cubic lattice configuration, this theory gives the following relation between the self-diffusion coefficient and the coefficient of viscosity:

$$\frac{\mathscr{D}_{AA}\mu_A}{\kappa T} = \left(\frac{\tilde{N}}{\tilde{V}_A}\right)^{1/3} \tag{16.5–8}$$

This relation differs by a factor of 2π from Eq. 16.5–7, the expression for the self-diffusion coefficient according to the hydrodynamic theory; but Eq. 16.5–7 has been shown to fit the experimental data better than Eq. 16.5–8. Recently Li and Chang[7] have suggested a modification of the Eyring theory in order to account for the discrepancy.

Because of the approximate nature of the foregoing theories, a number of *empirical relations* have been proposed,[8] one of which we mention briefly.

Wilke[9] has developed a correlation for diffusion coefficients on the basis of the Stokes-Einstein equation. His results may be summarized by the following approximate analytical relation,[10] which gives the diffusion coefficient in cm^2 sec^{-1} for small concentrations of A in B:

$$\mathscr{D}_{AB} = 7.4 \times 10^{-8} \frac{(\psi_B M_B)^{1/2} T}{\mu \tilde{V}_A^{0.6}} \tag{16.5–9}$$

Here $\tilde{V}_A$ is the molar volume of the solute A in cm^3 g-mole^{-1} as liquid at its normal boiling point, μ is the viscosity of the solution in centipoises, ψ_B is an "association parameter" for the solvent B, and T is the absolute temperature in ° K. Recommended values of ψ_B are 2.6 for water, 1.9 for methanol, 1.5 for ethanol, and 1.0 for benzene, ether, heptane, and other unassociated solvents. This equation is good only for dilute solutions of nondissociating solutes; for such solutions it is usually good within ± 10 per cent.

If the reader has by now concluded that little is known about the prediction of dense gas and liquid diffusivities, he is correct. There is an urgent need for experimental measurements, both for their own value and for development of future theories.

Example 16.5–1. Estimation of Mass Diffusivity for a Binary Liquid Mixture

Estimate $\mathscr{D}_{AB}$ for a dilute solution of TNT (2,4,6-trinitrotoluene) in benzene at 15° C.

[7] Li and Chang, *loc. cit.*

[8] See, for example, R. C. Reid and T. K. Sherwood, *The Properties of Gases and Liquids*, McGraw-Hill, New York (1958), pp. 284–298.

[9] C. R. Wilke, *Chem. Eng. Prog.*, **45**, 218–224 (1949).

[10] C. R. Wilke and P. Chang, *A.I.Ch.E. Journal*, **1**, 264–270, 1955.

Solution. Use the equation of Wilke and Chang, taking TNT as component A and benzene as component B. The required data are

$$\begin{aligned} \mu &= 0.705 \text{ cp (for solution considered as pure benzene)} \\ \tilde{V}_A &= 140 \text{ cc g-mole}^{-1} \text{ for TNT}^{11} \\ \psi_B &= 1.0 \text{ for benzene} \\ M_B &= 78.11 \text{ for benzene} \end{aligned}$$

Substitution in Eq. 16.5–9 gives

$$\begin{aligned} \mathscr{D}_{AB} &= 7.4 \times 10^{-8} \frac{(1.0 \times 78.11)^{1/2}(273 + 15)}{0.705(140)^{0.6}} \\ &= 1.38 \times 10^{-5} \text{ cm}^2 \text{ sec}^{-1} \end{aligned}$$

This result may be compared with a measured value of 1.39×10^{-5} cm^2 sec^{-1}.

QUESTIONS FOR DISCUSSION

1. What is diffusion? What factors may cause diffusion to occur?

2. When is mass fraction equal to mole fraction?

3. Compare the mass average velocity $\boldsymbol{v}$ in Eq. 16.1–1 with the average velocity defined in §2.2.

4. Compare the significance of the fluxes $\boldsymbol{N}_i$ and $\boldsymbol{J}_i^{\star}$. Compare the significance of the fluxes $\boldsymbol{n}_i$ and $\boldsymbol{j}_i$.

5. Why is it necessary to specify a reference velocity in reporting data on rates of diffusion?

6. Compare Fick's law of diffusion with Newton's law of viscosity and Fourier's law of thermal conductivity. To what extent are these three relations analogous?

7. Give the special forms of Fick's law for the flux $\boldsymbol{N}_A$ that apply to (*a*) equimolar counterdiffusion ($\boldsymbol{N}_A = -\boldsymbol{N}_B$), (*b*) diffusion of A through stagnant B ($\boldsymbol{N}_B = 0$).

8. What are the dimensions of $\mathscr{D}_{AB}$, ν, and α?

9. Compare the order of magnitude of $\mathscr{D}_{AB}$ and of $c\mathscr{D}_{AB}$ for gases and liquids at 1 atm.

10. What is the difference in behavior of the diffusivity-concentration curve for ethanol-water and that for n-butanol-water in Table 16.2–3?

11. By means of a small calculation, verify the statement concerning the temperature dependence of $\mathscr{D}_{AB}$ just after Eq. 16.4–14.

12. What empirical method is available for predicting the intermolecular force field between a pair of dissimilar nonpolar molecules?

13. What bulk physical properties are related by the Stokes-Einstein equation? Verify that the equation is dimensionally consistent.

14. Compare the relationship between $\mathscr{D}_{AB}$ and μ for gases with that for liquids.

15. What percentage of error in the predicted value of $\mathscr{D}_{AB}$ from Eq. 16.4–13 would result from a 5 per cent error in σ_{AB}? In T? In ϵ_{AB}/κ?

16. In binary diffusion, does the statement that the mole fraction gradient ∇x_A is not zero imply that the molar flux $\boldsymbol{N}_A$ is not zero? Explain.

[11] Estimated by the atomic-volume method of Schroeder as quoted by R. C. Reid and T. K. Sherwood, *The Properties of Gases and Liquids*, McGraw-Hill, New York (1958), p. 51.

PROBLEMS

16.A$_1$ Prediction of Mass Diffusivity for a Gas-Pair at Low Density

Predict $\mathscr{D}_{AB}$ for the methane-ethane system at 104° F and 1 atm, using the Chapman-Enskog theory. Compare your result with the experimental value given in Example 16.3–2.

Answer: 0.166 cm² sec⁻¹

16.B$_1$ Prediction of Mass Diffusivity at Low Density from Critical Properties

Predict $\mathscr{D}_{AB}$ for the situation of Problem 16.A by the following two methods:

a. The Slattery equation, Eq. 16.3–1.

b. The Chapman-Enskog theoretical equation, Eq. 16.4–13, using critical pressures and temperatures to estimate the Lennard-Jones parameters. (See Eqs. 1.4–11 and 13.)

Answers: *a.* 0.172 $cm^2\ sec^{-1}$
b. 0.156 $cm^2\ sec^{-1}$

16.C$_1$ Correction of Mass Diffusivity for Temperature at Low Density

A value of $\mathscr{D}_{AB} = 0.151$ $cm^2\ sec^{-1}$ has been reported[1] for the system CO_2-air at 293° K and 1 atm. Extrapolate $\mathscr{D}_{AB}$ to 1500° K by the following methods:

a. Eq. 16.3–1.
b. Eq. 16.4–11.
c. Eq. 16.4–13.

Answers in $cm^2\ sec^{-1}$: *a.* 2.96
b. 1.75
c. 2.49
Observed:[1] 2.45 $cm^2\ sec^{-1}$

16.D$_1$ Prediction of Mass Diffusivity for a Gas Mixture at High Density

Predict $\mathscr{D}_{AB}$ for an equimolar mixture of CO_2 and N_2 at 288.2° K and 40 atm:

a. Use the value of $\mathscr{D}_{AB}$ at 1 atm from Table 16.2–2.
b. Use the Chapman-Enskog theory to predict $(p\mathscr{D}_{AB})^{\circ}$.

Answers: *a.* 0.0033 $cm^2\ sec^{-1}$
b. 0.0030 $cm^2\ sec^{-1}$

16.E$_1$ Estimation of Mass Diffusivity for a Binary Liquid Mixture

Estimate $\mathscr{D}_{AB}$ for acetic acid in dilute aqueous solution at 12.5° C by using Eq. 16.5–9. The density of acetic acid at its normal boiling point is 0.937 g cm^{-3}. (An experimental value of $0.91 \pm 0.04 \times 10^{-5}$ $cm^2\ sec^{-1}$ has been reported.[2])

16.F$_1$ Correction of Mass Diffusivity for Temperature for a Binary Liquid Mixture

The value of $\mathscr{D}_{AB}$ for a dilute solution of methanol in water at 15° C is 1.28×10^{-5} $cm^2\ sec^{-1}$. Estimate $\mathscr{D}_{AB}$ for the same solution at 100° C, using the Wilke-Chang equation.

Answer: 6.7×10^{-5} $cm^2\ sec^{-1}$

16.G$_1$ Correction of Mass Diffusivity for Temperature for a Dense Gas Mixture

The observed value of $\mathscr{D}_{AB}$ for a mixture of 80 mole per cent methane and 20 mole per cent ethane at 104° F and 2000 psia is 8.4×10^{-4} $cm^2\ sec^{-1}$. Predict $\mathscr{D}_{AB}$ for the same mixture at 2000 psia and 171° F:

a. Use Eq. 16.3–1 and Fig. 16.3–1.
b. Use Eq. 16.4–13 and Fig. 16.3–1.

Answer: *a.* 1.19×10^{-3} $cm^2\ sec^{-1}$

[1] Ts. M. Klibanova. V. V. Pomerantsev, and D. A. Frank-Kamenetskii, *J. Tech. Phys.* (*U.S.S.R.*), **12,** 14–30 (1942), as quoted by C. R. Wilke and C. Y. Lee, *Ind. Eng. Chem.*, **47,** 1253 (1955).

[2] *International Critical Tables*, McGraw-Hill, New York (1929) Vol. V, p. 70.

16.H$_2$ Relations Among the Fluxes and Concentrations in a Binary System

Prove Eqs. *K*, *L*, and *Z* in Table 16.1–3, and *O* in Table 16.1–1, using only the definitions of the concentrations, velocities, and fluxes.

16.I$_2$ Equivalence of Various Forms of Fick's Law

a. Prove that all of the forms of Fick's first law given in Table 16.2–1 are equivalent, that is, that the value of $\mathscr{D}_{AB}$ is identical in all seven equations. Any relations given in §16.1 may be used.

b. Use Eq. 16.2–1 and the relations tabulated in §16.1 to prove that $\mathscr{D}_{AB} = \mathscr{D}_{BA}$.

16.J$_3$ Determination of Collision Parameters from Diffusivity Data

a. Use the following data[3] for the system CO_2-air at 1 atm pressure to determine σ_{AB} and ϵ_{AB}/κ directly, that is, without using the values of σ and ϵ/κ for the separate species:

T, °K	293	400	600	800	1000
$\mathscr{D}_{AB}$, cm^2 sec^{-1}	0.151	0.273	0.555	0.915	1.32

It is suggested that the data be plotted as log $(T^{3/2}/\mathscr{D}_{AB})$ versus log T on a thin sheet of graph paper and that a plot of $\Omega_{\mathscr{D},AB}$ vs. $\kappa T/\epsilon_{AB}$ be made on a separate sheet to the same fineness of scale. Then one can superimpose the data plot on the $\Omega_{\mathscr{D}}$-function plot and from the scales of the two overlapping plots determine the numerical ratios $T/(\kappa T/\epsilon_{AB})$ and $(T^{3/2}/\mathscr{D}_{AB})/\Omega_{\mathscr{D},AB}$. These two ratios, together with Eq. 16.4–12, may be used to solve for the two parameters σ_{AB} and ϵ_{AB}/κ.

b. Estimate σ_{AB} and ϵ_{AB}/κ by the procedure of §16.4.

16.K$_3$ Mass Flux with Respect to Volume-Average Velocity

Let the *volume-average velocity* be defined by

$$\boldsymbol{v}^{\blacksquare} = \Sigma_i \rho_i \boldsymbol{v}_i (\bar{V}_i/M_i) \tag{16.K–1}$$

in which $\bar{V}_i$ is the partial molal volume of component i and M_i is its molecular weight. Let the mass flux with respect to the volume average velocity then be

$$\boldsymbol{j}_i^{\blacksquare} = \rho_i(\boldsymbol{v}_i - \boldsymbol{v}^{\blacksquare}) \tag{16.K–2}$$

Show that for a binary system composed of species A and B

$$\boldsymbol{j}_A^{\blacksquare} = \boldsymbol{j}_A \rho(\bar{V}_B/M_B) \tag{16.K–3}$$

and that Fick's first law assumes the form

$$\boldsymbol{j}_A^{\blacksquare} = -\mathscr{D}_{AB}\nabla\rho_A \quad \text{for constant } T \text{ and } p. \tag{16.K–4}$$

[3] From Footnote 1, Problem 16.C.

CHAPTER 17

Concentration Distributions in Solids and in Laminar Flow

In Chapter 2 it was shown how a number of simple viscous flow problems can be set up and solved by making a shell *momentum balance*. In Chapter 9 it was further shown how heat-conduction problems can be handled by means of a shell *energy balance*. In this chapter we show how elementary diffusion problems may be formulated by a shell *mass balance*. The procedure used here is essentially the same as that used previously:

a. A mass balance is made over a thin shell perpendicular to the direction of mass transport, and this shell balance leads to a first-order differential equation, which may be solved to get the mass-flux distribution.

b. Into this expression we insert the relation between mass flux and concentration gradient, and this results in a second-order differential equation for the concentration profile. The integration constants that appear in the resulting expression are determined by the boundary conditions that specify the concentration or the mass flux at the bounding surfaces.

In Chapter 16 it was pointed out that several kinds of mass fluxes are in common use. For the sake of simplicity, we shall use the flux N_A exclusively, that is, the number of moles of A that go through a unit area in unit time, the unit area being fixed in space. We shall relate the molar flux to the concentration gradient (see Eq. 16.2–2) by

$$\underset{\left\{\substack{\text{flux with respect}\\ \text{to fixed axes}}\right\}}{N_{Az}} = \underset{\left\{\substack{\text{flux resulting}\\ \text{from diffusion}}\right\}}{-c\mathscr{D}_{AB}\frac{\partial x_A}{\partial z}} + \underset{\left\{\substack{\text{flux resulting}\\ \text{from total molar bulk flow}}\right\}}{x_A(N_{Az} + N_{Bz})} \tag{17.0–1}$$

for two-component systems. Keep in mind that N_{Az} is the z-component of the vector $\boldsymbol{N}_A$. Before Eq. 17.0–1 can be used, we shall have to eliminate N_{Bz}. This can be done only if something is known beforehand about the ratio N_{Bz}/N_{Az}. In each of the problems discussed in this chapter we begin by specifying this ratio on physical grounds.

In this chapter we study diffusion in both *nonreacting* and *reacting* systems. When chemical reactions occur, we distinguish between two types: *homogeneous*, in which the chemical change occurs in the entire volume of the fluid, and *heterogeneous*, in which the chemical change takes place only in a restricted region in the system, such as at the surface of a catalyst. Not only is the physical picture different for homogeneous and heterogeneous reactions, but there is also a difference in the way in which the two reactions are described. The rate of production by a homogeneous reaction appears in a source term in the differential equation obtained from the shell mass balance, just as the thermal source term appears in the shell energy balance. The rate of production by a heterogeneous reaction, on the other hand, appears not in the differential equation but rather in the boundary condition at the surface on which the reaction occurs.

In order to set up problems involving chemical reactions, some information has to be available about the rate at which the various chemical species appear or disappear by reaction. This brings us then to the subject of *chemical kinetics*, that branch of physical chemistry that deals with the mechanisms of chemical reactions and the rates at which they occur.[1] In this chapter we assume that the mechanisms of the reactions are known and that the reaction rates are describable by means of simple functions of the concentrations of the reacting species. Mention needs to be made at this point of the notation to be used for the chemical rate constants. For homogeneous reactions, the volume rate of production of species A may be given by an expression of the form

$$R_A = k_n'''c_A{}^n \tag{17.0–2}$$

in which R_A [=] moles cm^{-3} sec^{-1} and c_A [=] moles cm^{-3}. The index n indicates the "order" of the reaction;[2] for a first-order reaction, k_1''' [=] sec^{-1}.

[1] For an introduction to chemical kinetics, see F. Daniels and R. A. Alberty, *Physical Chemistry*, Wiley, New York (1955), Ch. 13. A more mathematical discussion is given by E. A. Moelwyn-Hughes, *Physical Chemistry*, Pergamon Press, New York (1957).

[2] Not all rate expressions are of the simple form of Eq. 17.0–2. The reaction rate may depend in a more complex way on the concentrations of all species present. Similar remarks hold for Eq. 17.0–3.

For heterogeneous reactions, the rate of reaction at the catalytic surface may be specified by a relation of the form

$$N_{Az}|_{\text{surface}} = k_n''c_A^n|_{\text{surface}} \tag{17.0-3}$$

in which N_{Az} [=] moles cm^{-2} sec^{-1} and c_A [=] moles cm^{-3}. Here k_1'' [=] cm sec^{-1}. Note that $'''$ indicates a rate constant related to a volume source and $''$ indicates a rate constant related to a surface source.

We begin in §17.1 with a statement of the shell mass balance and the kinds of boundary conditions that may arise in solving diffusion problems. In §17.2 a discussion of diffusion in a stagnant film is given, this topic being necessary to the understanding of the film theories of the diffusional operations in chemical engineering. Then, in §§17.3 and 17.4, we give some elementary examples of diffusion with chemical reaction—both homogeneous and heterogeneous; it is hoped that these examples will illustrate the role that diffusion plays in chemical kinetics and the important fact that there is generally a difference between the rate of the chemical reaction and the rate of the combined diffusion-reaction process. In §17.5 we turn our attention to the subject of forced-convection mass transfer—that is, diffusion superimposed on a flow field; we could also have included for the sake of completeness a section on free-convection mass transfer, which would have paralleled the discussion on free-convection heat transfer in Chapter 9. Finally, in the last section, §17.6, we discuss diffusion in porous catalysts.

§17.1 SHELL MASS BALANCES: BOUNDARY CONDITIONS

The diffusion problems in this chapter are solved by making mass balances for a specific chemical species over a thin shell of solid or fluid. Having selected an appropriate system, the law of conservation of mass is written in the form

$$\left\{\begin{matrix}\text{rate of}\\ \text{mass of}\\ A \text{ in}\end{matrix}\right\} - \left\{\begin{matrix}\text{rate of}\\ \text{mass of}\\ A \text{ out}\end{matrix}\right\} + \left\{\begin{matrix}\text{rate of production}\\ \text{of mass of } A \text{ by}\\ \text{homogeneous chemical}\\ \text{reaction}\end{matrix}\right\} = 0 \tag{17.1-1}$$

The conservation statement may, of course, also be expressed in terms of moles. The chemical species A may enter or leave the system by means of diffusion and by virtue of the over-all motion of the fluid. Also species A may be produced or destroyed by homogeneous chemical reactions.

After a balance is made on a shell of finite thickness, according to Eq. 17.1-1, we then let the dimensions of the system become infinitesimally small. As a result of this process, a differential equation is generated, the solution of which gives the distribution of species A in the system.

When the differential equation has been integrated, constants of integration appear, which have to be evaluated by use of boundary conditions.

The boundary conditions used are very similar to those used in energy transfer. (See §9.1.)

a. The concentration at a surface can be specified; for example, $x_A = x_{A0}$.

b. The mass flux at a surface can be specified (if the ratio N_A/N_B is known, this is tantamount to giving the concentration gradient); for example, $N_A = N_{A0}$.

c. If diffusion is occurring in a solid, it may happen that at the solid surface substance A is lost to a surrounding fluid stream according to the relation

$$N_{A0} = k_c(c_{A0} - c_{Af}) \tag{17.1-2}$$

in which N_{A0} is the mass flux at the surface, c_{A0} is the surface concentration, c_{Af} is the concentration in the fluid stream, and the proportionality constant k_c is a "mass-transfer coefficient"; the methods of correlating mass-transfer coefficients are discussed in Chapter 21. Equation 17.1–2 is analogous to "Newton's law of cooling" defined in Eq. 9.1–2.

d. The rate of chemical reaction at the surface can be specified. For example, if substance A disappears at a surface by a first-order chemical reaction, $N_{A0} = k_1''c_A$; that is, the rate of disappearance at a surface is proportional to the surface concentration, the proportionality constant k_1'' being a first-order chemical rate constant.

In the following sections we shall see how these various boundary conditions are used.

§17.2 DIFFUSION THROUGH A STAGNANT GAS FILM

Consider the diffusion system shown in Fig. 17.2–1. Liquid A is evaporating into gas B, and we imagine that there is some device which maintains the liquid level at $z = z_1$. Right at the liquid-gas interface the gas-phase concentration of A, expressed as mole fraction, is x_{A1}. This is taken to be the gas-phase concentration of A corresponding to equilibrium[1] with the liquid at the interface; that is, x_{A1} is the vapor pressure of A divided by the total pressure, $p_A^{(\text{vap})}/p$, provided that A and B form an ideal gas mixture. We further assume that the solubility of B in liquid A is negligible.

At the top of the tube (at $z = z_2$) a stream of gas mixture A-B of concentration x_{A2} flows past slowly; thereby the mole fraction of A at the top of the column is maintained at x_{A2}. The entire system is presumed to be held

[1] The assumption of equilibrium at the interface has been subjected to experimental test by L. N. Tung and H. G. Drickamer, *J. Chem. Phys.*, **20**, 6–12 (1952), and by R. E. Emmert and R. L. Pigford, *Chem. Eng. Prog.*, **50**, 87–93 (1954). A kinetic theory approach was made by R. W. Schrage, *A Theoretical Study of Interphase Mass Transfer*, Columbia University Press, New York (1953). These studies seem to indicate that only at very high mass-transfer rates is there a significant departure from equilibrium at the interface.

at constant temperature and pressure. Gases A and B are assumed to be ideal.

When this evaporating system attains a steady state, there is a net motion of A away from the evaporating surface and the vapor B is stationary. Hence

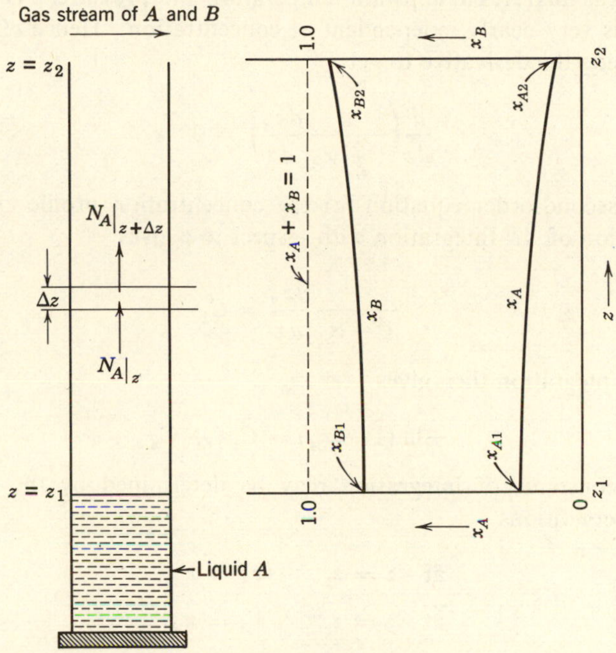

Fig. 17.2–1. Diffusion of A through B in steady state with B not in motion. Graph shows how concentration profile for B is distorted because of diffusion of A.

we can use the expression for N_{Az} given in Eq. 17.0–1 with $N_{Bz} = 0$. Solving for N_{Az}, we get

$$N_{Az} = -\frac{c\mathscr{D}_{AB}}{1 - x_A}\frac{dx_A}{dz} \tag{17.2–1}$$

A mass balance over an incremental column height Δz (see Fig. 17.2–1) states that at steady state

$$SN_{Az}|_z - SN_{Az}|_{z+\Delta z} = 0 \tag{17.2–2}$$

in which S is the cross-sectional area of the column. Division by $S\,\Delta z$ and taking the limit as Δz approaches zero gives

$$-\frac{dN_{Az}}{dz} = 0 \tag{17.2–3}$$

Substitution of Eq. 17.2–1 into Eq. 17.2–3 gives

$$\frac{d}{dz}\left(\frac{c\mathscr{D}_{AB}}{1-x_A}\frac{dx_A}{dz}\right) = 0 \tag{17.2–4}$$

For ideal-gas mixtures at constant temperature and pressure, c is a constant, and $\mathscr{D}_{AB}$ is very nearly independent of concentration. Hence $c\mathscr{D}_{AB}$ can be taken outside the derivative to get

$$\frac{d}{dz}\left(\frac{1}{1-x_A}\frac{dx_A}{dz}\right) = 0 \tag{17.2–5}$$

This is a second-order equation for the concentration profile expressed as mole fraction of A. Integration with respect to z gives

$$\frac{1}{1-x_A}\frac{dx_A}{dz} = C_1 \tag{17.2–6}$$

A second integration then gives

$$-\ln(1-x_A) = C_1 z + C_2 \tag{17.2–7}$$

The two constants of integration may be determined by the use of the boundary conditions

B.C. 1: at $z = z_1, \quad x_A = x_{A1}$ (17.2–8)

B.C. 2: at $z = z_2, \quad x_A = x_{A2}$ (17.2–9)

When the constants so obtained[2] are substituted into Eq. 17.2–7, the following expressions for the concentration profiles are obtained:

$$\boxed{\left(\frac{1-x_A}{1-x_{A1}}\right) = \left(\frac{1-x_{A2}}{1-x_{A1}}\right)^{\frac{z-z_1}{z_2-z_1}}} \tag{17.2–10}$$

or

$$\boxed{\left(\frac{x_B}{x_{B1}}\right) = \left(\frac{x_{B2}}{x_{B1}}\right)^{\frac{z-z_1}{z_2-z_1}}} \tag{17.2–11}$$

[2] The constants are

$$C_1 = \frac{\ln x_{B1} - \ln x_{B2}}{z_2 - z_1} = \frac{\ln(x_{B1}/x_{B2})}{z_2 - z_1}; \qquad C_2 = \frac{z_1 \ln x_{B2} - z_2 \ln x_{B1}}{z_2 - z_1}$$

To get Eq. 17.2–10, first rewrite C_2 thus:

$$C_2 = \frac{z_1 \ln x_{B2} - (z_2 - z_1)\ln x_{B1} - z_1 \ln x_{B1}}{z_2 - z_1} = -\frac{z_1}{z_2 - z_1}\ln\frac{x_{B1}}{x_{B2}} - \ln x_{B1}$$

These concentration distributions are depicted in Fig. 17.2–1. An examination of these curves shows that the slope dx_A/dz is not constant with respect to z, although the molar flux N_{Az} is.

Although the concentration profiles are helpful in picturing the diffusion process, in engineering calculations it is usually the average concentration or the mass flux at some surface that is of interest. For example, the average concentration of B in the region between $z = z_1$ and $z = z_2$ is

$$\frac{x_{B,\text{avg}}}{x_{B1}} = \frac{\int_{z_1}^{z_2} (x_B/x_{B1})\, dz}{\int_{z_1}^{z_2} dz} = \frac{\int_0^1 (x_{B2}/x_{B1})^{\zeta}\, d\zeta}{\int_0^1 d\zeta} = \left. \frac{(x_{B2}/x_{B1})^{\zeta}}{\ln (x_{B2}/x_{B1})} \right|_0^1 \tag{17.2–12}$$

whence

$$x_{B,\text{avg}} = \frac{x_{B2} - x_{B1}}{\ln (x_{B2}/x_{B1})} \tag{17.2–13}$$

That is, the average value of x_B is the logarithmic mean of the terminal values, $(x_B)_{\text{ln}}$. In the foregoing ζ is a reduced length $(z - z_1)/(z_2 - z_1)$.

The rate of mass transfer at the liquid-gas interface—that is, the rate of evaporation—is obtained by using Eq. 17.2–1:

$$N_{Az}\big|_{z=z_1} = - \frac{c\mathscr{D}_{AB}}{1 - x_{A1}} \left. \frac{dx_A}{dz} \right|_{z=z_1} = + \frac{c\mathscr{D}_{AB}}{x_{B1}} \left. \frac{dx_B}{dz} \right|_{z=z_1} = \frac{c\mathscr{D}_{AB}}{(z_2 - z_1)} \ln \left(\frac{x_{B2}}{x_{B1}} \right) \tag{17.2–14}$$

Equations 17.2–13 and 17.2–14 may be combined to give an alternative expression for the mass-transfer rate:

$$N_{Az}\big|_{z=z_1} = \frac{c\mathscr{D}_{AB}}{(z_2 - z_1)(x_B)_{\text{ln}}} (x_{A1} - x_{A2}) \tag{17.2–15}$$

This expression shows how the rate of mass transfer is related to a characteristic concentration driving force $x_{A1} - x_{A2}$. Equations 17.2–14 and 15 may also be expressed in terms of total pressure and partial pressures:

$$N_{Az}\big|_{z=z_1} = \frac{(p\mathscr{D}_{AB}/RT)}{(z_2 - z_1)} \ln \frac{p_{B2}}{p_{B1}} = \frac{(p\mathscr{D}_{AB}/RT)}{(z_2 - z_1)(p_B)_{\text{ln}}} (p_{A1} - p_{A2}) \tag{17.2–15a}$$

in which $(p_B)_{\text{ln}}$ is the logarithmic mean of p_{B1} and p_{B2} defined analogously to $(x_B)_{\text{ln}}$ in Eq. 17.2–15.

The results of this section have been used for experimental determination of gas diffusivities.[3] Furthermore, these results find use in the "film theories"

[3] See, for example, C. Y. Lee and C. R. Wilke, *Ind. Eng. Chem.*, **46**, 2381–2387 (1954); T. K. Sherwood and R. L. Pigford, *Absorption and Extraction*, McGraw-Hill (1952).

of mass transfer. (See §21.5.) In Fig. 17.2–2 a solid or liquid surface is shown along which a gas is flowing. Near the surface is a slowly moving film through which A diffuses; z_1 is taken to be the solid-gas or liquid-gas interface, and z_2 is taken to be the outer limit of the gas "film" through which the diffusion occurs. In this "model" it is assumed that there is a sharp transition between a stagnant film and a well-mixed fluid in which concentration

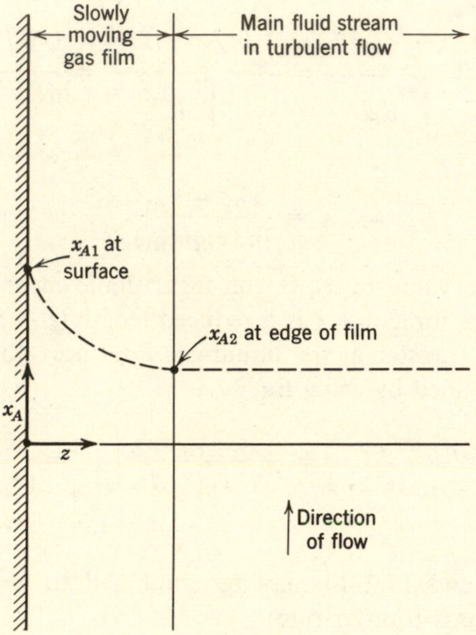

Fig. 17.2–2. Film model for mass transfer; component A is diffusing from the surface into the gas stream.

gradients are negligible. Although this model is physically unrealistic, it has nevertheless proven very useful as a basis for correlating mass-transfer coefficients in terms of a simple physical picture.

Example 17.2–1. Determination of Diffusivity

The diffusivity of the gas-pair oxygen-carbon tetrachloride is determined by observing the steady-state evaporation of CCl_4 into a tube containing O_2, as shown in Fig. 17.2–1. The distance between the CCl_4 liquid level and the top of the tube is $z_2 - z_1 = 17.1$ cm. The total pressure on the system is 755 mm Hg, and the temperature is 0° C. The vapor pressure of CCl_4 at that temperature is 33.0 mm Hg. The cross-sectional area of the diffusion tube is 0.82 cm². If it is found that 0.0208 cm³ of CCl_4 evaporate in a 10-hour period after steady state has been attained, what is the diffusivity of the gas-pair CCl_4—O_2?

Solution. First we get the molar flux N_{CCl_4} from the evaporation rate:

$$N_{CCl_4} = (0.0208 \text{ cm}^3)\left(1.59 \frac{\text{g}}{\text{cm}^3}\right)\left(\frac{1}{154} \frac{\text{g-mole}}{\text{g}}\right)\left(\frac{1}{0.82} \frac{1}{\text{cm}^2}\right)\left(\frac{1}{3.6 \times 10^4} \frac{1}{\text{sec}}\right)$$

$$= 7.26 \times 10^{-9} \frac{\text{g-mole}}{\text{cm}^2 \text{ sec}}$$

Then from Eq. 17.2–15*a* we get

$$\mathscr{D}_{CCl_4-O_2} = \frac{N_{CCl_4}(z_2 - z_1)RT}{p \ln (p_{O_2 2}/p_{O_2 1})}$$

$$= \frac{(7.26 \times 10^{-9})(17.1)(82.06)(273)}{(\frac{755}{760})(2.303 \log_{10} \frac{755}{722})}$$

$$= 0.0636 \text{ cm}^2 \text{ sec}^{-1}$$

This method of determining gas-phase diffusivities suffers from several defects. The most serious is probably the cooling effect associated with the evaporation. Furthermore, there may be free-convection effects if the vapor from the evaporating liquid is lighter than the gas above the liquid. Also, because of the meniscus, the length of the diffusion path will not be uniform across the tube.

Example 17.2–2. Diffusion Through a Nonisothermal Spherical Film

a. Derive expressions for diffusion through a spherical shell that are analogous to Eq. 17.2–10 (concentration profile) and Eq. 17.2–14 (molar flux); see Fig. 17.2–3 for specification of system.

b. Extend these results to describe the diffusion in a nonisothermal film in which the temperature changes with the distance according to

$$\left(\frac{T}{T_1}\right) = \left(\frac{r}{r_1}\right)^n \tag{17.2–16}$$

in which T_1 is the temperature at $r = r_1$. Assume as a *rough* approximation that $\mathscr{D}_{AB}$ varies as the $\frac{3}{2}$-power of the temperature:

$$\left(\frac{\mathscr{D}_{AB}}{\mathscr{D}_{AB,1}}\right) = \left(\frac{T}{T_1}\right)^{3/2} \tag{17.2–17}$$

in which $\mathscr{D}_{AB,1}$ is the diffusivity at $T = T_1$. Problems of this kind arise in connection with drying of droplets and diffusion through gas films near spherical catalyst pellets. (The temperature distribution in Eq. 17.2–16 has been chosen solely for mathematical simplicity. This example has been included to emphasize that in nonisothermal systems Eq. 17.0–1 is the correct starting point rather than $N_{Az} = -\mathscr{D}_{AB}(\partial c_A/\partial z) + x_A(N_{Az} + N_{Bz})$, as given in many engineering texts.)

Solution. *a.* A mass balance on a spherical shell leads to

$$\frac{d}{dr}(r^2 N_{Ar}) = 0 \tag{17.2–18}$$

Substitution of the expression for molar flux N_{Ar} from Eq. 16.2–2 with $N_{Br} = 0$ then gives

$$\frac{d}{dr}\left(r^2 \frac{c\mathscr{D}_{AB}}{1 - x_A} \frac{dx_A}{dr}\right) = 0 \tag{17.2–19}$$

For constant temperature the product $c\mathscr{D}_{AB}$ is constant, and Eq. 17.2–19 may be integrated to give the concentration distribution

$$\left(\frac{x_B}{x_{B_1}}\right) = \left(\frac{x_{B_2}}{x_{B_1}}\right)^{\frac{(1/r_1)-(1/r)}{(1/r_1)-(1/r_2)}} \tag{17.2–20}$$

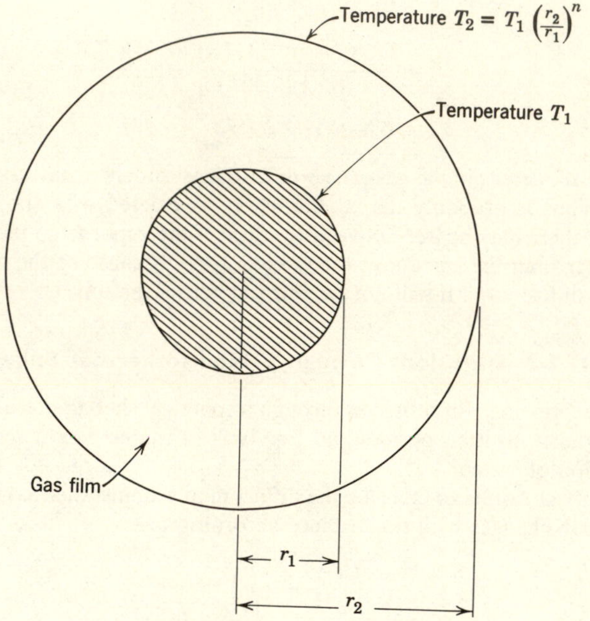

Fig. 17.2–3. Diffusion through a spherical film.

with $x_A = 1 - x_B$. From Eq. 17.2–20 one can then get

$$\mathscr{W}_A = 4\pi r_1^2 N_{Ar}\big|_{r=r_1} = \frac{4\pi c\mathscr{D}_{AB}}{(1/r_1) - (1/r_2)} \ln \frac{x_{B2}}{x_{B1}} \tag{17.2–21}$$

which is the molar flow of A across any spherical surface.

b. For the nonisothermal situation, combination of Eqs. 17.2–16 and 17 gives the variation of diffusivity with position:

$$\left(\frac{\mathscr{D}_{AB}}{\mathscr{D}_{AB,1}}\right) = \left(\frac{r}{r_1}\right)^{3n/2} \tag{17.2–22}$$

When this expression is inserted into Eq. 17.2–19 and c is set equal to p/RT, we get

$$\frac{d}{dr}\left(r^2 \frac{p\mathscr{D}_{AB,1}/RT_1}{1 - x_A}\left(\frac{r}{r_1}\right)^{n/2} \frac{dx_A}{dr}\right) = 0 \tag{17.2–23}$$

When this equation is integrated between $r = r_1$ and $r = r_2$, we can obtain (for $n \neq -2$)

$$\mathscr{W}_A = 4\pi r_1^2 N_{Ar}\big|_{r=r_1} = \frac{4\pi(p\mathscr{D}_{AB,1}/RT_1)[1 + (n/2)]}{(r_1^{-1-(n/2)} - r_2^{-1-(n/2)})r_1^{n/2}} \ln \frac{x_{B2}}{x_{B1}} \quad (17.2\text{–}24)$$

For $n = 0$, this result simplifies to that in Eq. 17.2–21.

§17.3 DIFFUSION WITH HETEROGENEOUS CHEMICAL REACTION

Consider a catalytic reactor, such as that shown in Fig. 17.3–1*a*, in which the dimerization reaction $2A \rightarrow A_2$ is being carried out. A system of this

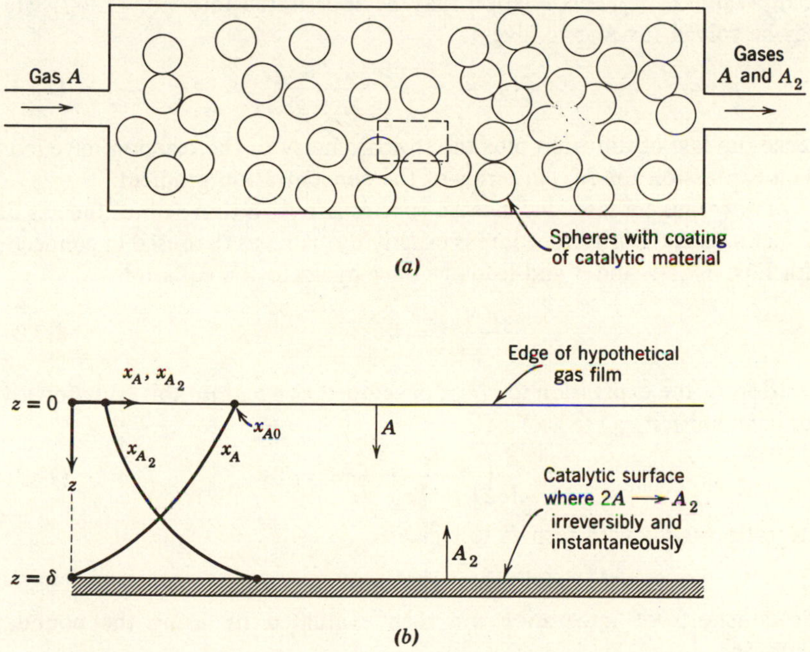

Fig. 17.3–1. (*a*) Schematic diagram of a catalytic reactor in which A is being converted into A_2. (*b*) Idealized picture (or "model") of the diffusion problem near a catalyst particle—that is, the enlargement of dotted region in (*a*).

degree of complexity cannot be described exactly by a theoretical development. Some information about the behavior of the system can, however, be obtained by analysis of a highly simplified model.

For example, we may imagine that each catalyst particle is surrounded by a stagnant gas film through which A has to diffuse in order to arrive at the catalytic surface. (See Fig. 17.3–1*b*.) At the catalyst surface we presume that the reaction $2A \rightarrow A_2$ occurs instantaneously and that the product A_2 then diffuses back out through the gas film to the main turbulent gas stream composed of A and A_2. We seek then an expression for the local rate of conversion from A to A_2 when the effective gas-film thickness δ and the main gas-stream compositions x_{A0} and x_{A_20} are known. We assume that the gas film is isothermal, although in many catalytic reactions the heat generated by the reaction cannot be neglected.

For the situation depicted in Fig. 17.3–1*b*, there is *one* mole of A_2 moving in the *minus* z-direction for every *two* moles of A moving in the *plus* z-direction. We know this from the stoichiometry of the reaction. Hence we know that at steady state

$$N_{A_2z} = -\tfrac{1}{2}N_{Az} \qquad (17.3\text{–}1)$$

at any value of z. This relation may be substituted into Eq. 17.0–1, which may be solved for N_{Az} to give

$$N_{Az} = -\frac{c\mathscr{D}_{AA_2}}{1 - \frac{1}{2}x_A}\frac{dx_A}{dz} \qquad (17.3\text{–}2)$$

Hence the law of diffusion plus the stoichiometry of the reaction have led us to an expression for N_{Az} in terms of the concentration gradient.

We now make a mass balance on species A over a thin slab of the gas film of thickness Δz. This procedure is exactly the same as that used in connection with Eqs. 17.2–2 and 3 and leads us once again to the equation

$$\frac{dN_{Az}}{dz} = 0 \qquad (17.3\text{–}3)$$

Insertion of the expression for N_{Az}, developed above, into this equation gives (for constant $c\mathscr{D}_{AA_2}$)

$$\frac{d}{dz}\left(\frac{1}{1 - \frac{1}{2}x_A}\frac{dx_A}{dz}\right) = 0 \qquad (17.3\text{–}4)$$

Integration twice with respect to z gives

$$-2\ln\left(1 - \tfrac{1}{2}x_A\right) = C_1 z + C_2 \qquad (17.3\text{–}5)$$

The constants of integration are then evaluated by using the boundary conditions

B. C. 1: at $z = 0, \quad x_A = x_{A0}$ (17.3–6)

B. C. 2: at $z = \delta, \quad x_A = 0$ (17.3–7)

The final result, after a bit of algebraic juggling, is

$$\boxed{\left(1 - \tfrac{1}{2}x_A\right) = \left(1 - \tfrac{1}{2}x_{A0}\right)^{1-(z/\delta)}} \qquad (17.3\text{–}8)$$

for the concentration profile in the gas film. Equation 17.3–2 may now be used to get the molar flux through the film:

$$N_{Az} = \frac{2c\mathscr{D}_{AA_2}}{\delta}\ln\left(\frac{1}{1 - \frac{1}{2}x_{A0}}\right) \qquad (17.3\text{–}9)$$

The quantity N_{Az} may also be interpreted as the local rate of dimerization per unit area of catalytic surface. This information can be combined with other information about the catalytic reactor sketched in Fig. 17.3–1*a* to obtain information about the over-all conversion rate in the entire reactor.

One point deserves to be emphasized. Although the chemical reaction occurs instantaneously at the catalytic surface, the conversion of A to A_2 proceeds at a finite rate because of the diffusion process which is "in series" with the reaction process. Hence we speak of the conversion of A to A_2 as being *diffusion-controlled.*

In the foregoing discussion we have assumed that the dimerization occurs instantaneously at the catalytic surface. In the illustrative example that follows we show how one can account for finite reaction rates at the catalytic surface.

Example 17.3–1. Diffusion with Slow Heterogeneous Reaction

Rework the problem just considered in §17.3 when the reaction $2A \rightarrow A_2$ is not instantaneous at the catalytic surface (at $z = \delta$). Instead, assume that the rate at which A disappears at the catalyst-coated surface is proportional to the concentration of A at that surface:

$$N_{Az} = k_1''c_A = ck_1''x_A \tag{17.3–10}$$

in which k_1'' is a rate constant.

Solution. We proceed exactly as before, except that B. C. 2 in Eq. 17.3–7 must be replaced by

$$\text{B. C. 2}': \qquad \text{at } z = \delta, \qquad x_A = \frac{N_{Az}}{ck_1''} \tag{17.3–11}$$

N_{Az} being, of course, a constant at steady state. The determination of the integration constants for B. C. 1 and B. C. 2′ leads to

$$(1 - \tfrac{1}{2}x_A) = \left(1 - \frac{1}{2}\frac{N_{Az}}{ck_1''}\right)^{z/\delta} (1 - \tfrac{1}{2}x_{A0})^{1-(z/\delta)} \tag{17.3–12}$$

Then we may calculate N_{Az} from Eq. 17.3–2 at $z = 0$ and get

$$N_{Az} = \frac{2c\mathscr{D}_{AA_2}}{\delta} \ln \left(\frac{1 - \frac{1}{2}(N_{Az}/ck_1'')}{1 - \frac{1}{2}x_{A0}}\right) \tag{17.3–13}$$

This is a transcendental equation for N_{Az} as a function of x_{A0}, k_1'', $c\mathscr{D}_{AA_2}$, and δ. When k_1'' is large, the logarithm of $1 - \frac{1}{2}(N_{Az}/ck_1'')$ may be expanded in a Taylor series and all terms discarded but the first. We then get

$$N_{Az} = \frac{2c\mathscr{D}_{AA_2}/\delta}{1 + (\mathscr{D}_{AA_2}/k_1''\delta)} \ln \left(\frac{1}{1 - \frac{1}{2}x_{A0}}\right) \tag{17.3–14}$$

Note once again that the rate of the *combined* reaction and diffusion process is obtained by the foregoing calculation. Note also that the dimensionless group $\mathscr{D}_{AA_2}/k_1''\delta$ describes the effect of the surface reaction rate on the over-all diffusion-reaction process.

§17.4 DIFFUSION WITH HOMOGENEOUS CHEMICAL REACTION

As the next illustration of setting up a mass balance, we consider the system shown in Fig. 17.4–1. Here gas A dissolves in liquid B and diffuses into the liquid phase. As it diffuses, A also undergoes an irreversible first-order chemical reaction: $A + B \rightarrow AB$. Hence the mass balance (input − output + production = 0) takes the form

$$N_{Az}\big|_z S - N_{Az}\big|_{z+\Delta z} S - k_1''' c_A S \,\Delta z = 0 \qquad (17.4\text{–}1)$$

in which k_1''' is a first-order rate constant for the chemical decomposition of

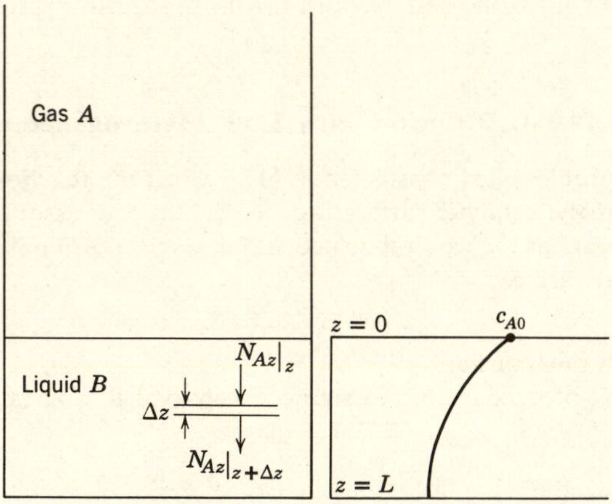

Fig. 17.4–1. Absorption with a homogeneous chemical reaction.

A, and S is the cross-sectional area of the liquid. The quantity $k_1''' c_A$ represents the moles of A disappearing per unit volume per unit time. Division of Eq. 17.4–1 by $S\Delta z$ and taking the limit as Δz goes to zero gives

$$\frac{dN_{Az}}{dz} + k_1''' c_A = 0 \qquad (17.4\text{–}2)$$

If A and AB are present in small concentrations, then we may to a good approximation write Eq. 17.0–1 as

$$N_{Az} = -\mathscr{D}_{AB} \frac{dc_A}{dz} \qquad (17.4\text{–}3)$$

Substitution of Eq. 17.4–3 into 17.4–2 gives

$$-\mathscr{D}_{AB} \frac{d^2 c_A}{dz^2} + k_1''' c_A = 0 \qquad (17.4\text{–}4)$$

This is to be solved with the boundary conditions

B. C. 1: at $z = 0$, $c_A = c_{A0}$ (17.4–5)

B. C. 2: at $z = L$, $N_{Az} = 0$ or $\dfrac{dc_A}{dz} = 0$ (17.4–6)

The first boundary condition states that the surface concentration is maintained at a fixed value c_{A0}. The second states that no A diffuses through the bottom of the container. The solution of Eq. 17.4–4, subject to these boundary conditions, gives

$$\frac{c_A}{c_{A0}} = \frac{\cosh b_1[1 - (z/L)]}{\cosh b_1} \tag{17.4–7}$$

in which $b_1 = \sqrt{k_1'''L^2/\mathscr{D}_{AB}}$. The concentration profile thus obtained is shown in Fig. 17.4–1.

From Eq. 17.4–7 one can obtain the average concentration of A in the liquid phase:

$$\begin{aligned}
\frac{c_{A,\text{avg}}}{c_{A0}} &= \frac{\int_0^L (c_A/c_{A0})\,dz}{\int_0^L dz} \\
&= \frac{\int_0^1 (\cosh b_1\zeta)/(\cosh b_1)\,d\zeta}{\int_0^1 d\zeta} \\
&= \frac{\sinh b_1\zeta}{b_1 \cosh b_1}\bigg|_0^1 = \frac{1}{b_1}\tanh b_1
\end{aligned} \tag{17.4–8}$$

In the foregoing ζ is used as an abbreviation for $[1 - (z/L)]$.

The molar flux of A at plane $z = 0$ can be computed from Eq. 17.4–7:

$$\begin{aligned}
N_{Az}|_{z=0} &= -\mathscr{D}_{AB}\frac{dc_A}{dz}\bigg|_{z=0} \\
&= -\mathscr{D}_{AB}c_{A0}\frac{\sinh b_1[1 - (z/L)]}{\cosh b_1}\left(-\frac{b_1}{L}\right)\bigg|_{z=0} \\
&= +\left(\frac{\mathscr{D}_{AB}c_{A0}}{L}\right) b_1 \tanh b_1
\end{aligned} \tag{17.4–9}$$

in which $b_1 = \sqrt{k_1'''L^2/\mathscr{D}_{AB}}$.

In this treatment it has been assumed that A is present in small concentration (see Eq. 17.4–3) and that the product of the chemical reaction AB does not interfere with the diffusion of A through B.

Example 17.4–1. Gas Absorption with Chemical Reaction in an Agitated Tank[1]

Estimate the effect of chemical reaction rate on the rate of gas absorption in an agitated tank. (See Fig. 17.4–2.) Consider the case in which the dissolved gas A undergoes an irreversible first-order reaction with the liquid B—that is, A disappears in the liquid phase at a rate proportional to the local concentration of A.

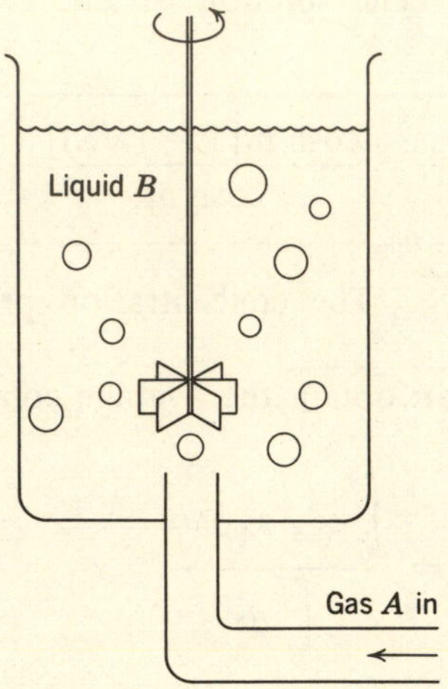

Fig. 17.4–2. Gas-absorption apparatus.

Solution. An exact analysis of this situation is not yet possible because of the complexity of the gas-absorption process. But a useful semiquantitative understanding can be obtained by the analysis of a relatively simple model. The model we employ embodies the following assumptions:

a. Each gas bubble is surrounded by a stagnant liquid film of thickness δ, which is small with respect to the bubble diameter.

b. A steady-state concentration is quickly established in the liquid film after the bubble is formed.

c. The gas A is only sparingly soluble in the liquid, so that we can neglect the bulk flow term in Eq. 17.0–1.

d. The bulk of the liquid outside the stagnant film is at concentration $c_{A\delta}$, which changes so slowly with respect to time that it can be considered to be constant.

The differential equation describing the diffusion with chemical reaction is the same as that in Eq. 17.4–4, but the boundary conditions are now

B.C. 1: at $z = 0, \quad c_A = c_{A0}$ (17.4–10)

B.C. 2: at $z = \delta, \quad c_A = c_{A\delta}$ (17.4–11)

The concentration c_{A0} is the interfacial concentration of A in the liquid phase,

[1] E. N. Lightfoot, *A.I.Ch.E. Journal*, **4**, 499–500 (1958).

which is assumed to be at equilibrium with the gas phase at the interface, and $c_{A\delta}$ is the concentration of A in the main body of the liquid. The solution of Eq. 17.4–4 with these boundary conditions is

$$\left(\frac{c_A}{c_{A0}}\right) = \frac{\Gamma \sinh b_1\zeta + \sinh b_1(1 - \zeta)}{\sinh b_1} \tag{17.4–12}$$

in which $\zeta = z/\delta$, $\Gamma = c_{A\delta}/c_{A0}$, and $b_1 = \sqrt{k_1''' \delta^2/\mathscr{D}_{AB}}$. The analogous solution

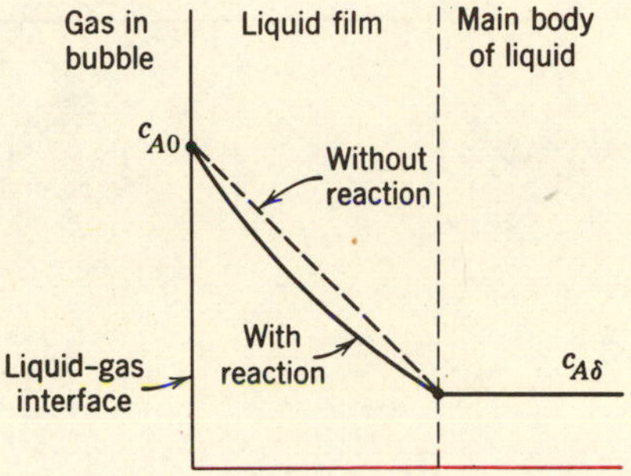

Fig. 17.4–3. Predicted concentration profile in liquid film near a bubble.

in the absence of chemical reaction (that is, $b_1 = 0$) is

$$\left(\frac{c_A}{c_{A0}}\right)_{\text{no reaction}} = \Gamma\zeta + (1 - \zeta) \tag{17.4–13}$$

The concentration profiles from Eqs. 17.4–12 and 17.4–13 are shown in Fig. 17.4–3. From these concentration profiles we can then get the mass fluxes for absorption with and without reaction, which are, at $z = 0$,

$$N_{Az}|_{z=0} = \left(-\mathscr{D}_{AB}\left.\frac{dc_A}{dz}\right|_{z=0}\right) = \frac{\mathscr{D}_{AB}c_{A0}}{\delta}\left(\frac{b_1 \cosh b_1 - b_1\Gamma}{\sinh b_1}\right) \tag{17.4–14}$$

$$(N_{Az})_{\text{no reaction}} = \left(-\mathscr{D}_{AB}\left.\frac{dc_A}{dz}\right|_{z=0}\right)_{\text{no reaction}} = \frac{\mathscr{D}_{AB}c_{A0}}{\delta}(1 - \Gamma) \tag{17.4–15}$$

It is common practice to estimate apparent film thicknesses *in the presence of chemical reaction* by using Eq. 17.4–15 and setting $\Gamma = 0$ in it. This is equivalent to assuming a value of unity for the ratio:

$$N^* = \frac{\begin{pmatrix}\text{absorption rate with}\\ \text{first-order reaction}\end{pmatrix}}{\begin{pmatrix}\text{absorption rate with no}\\ \text{reaction and with } c_{A\delta} = 0\end{pmatrix}} = \left(\frac{b_1}{\sinh b_1}\right)(\cosh b_1 - \Gamma) \tag{17.4–16}$$

To determine the conditions under which this assumption is valid, we need an expression for Γ. This is obtained by equating the mass of A crossing the outer surface of the film with the amount of A being removed by chemical reaction in the bulk of the liquid [here we make use of assumption (*d*) above]:

$$-A\mathscr{D}_{AB}\left.\frac{dc_A}{dz}\right|_{z=\delta} = Vk_1'''c_{A\delta} \tag{17.4-17}$$

in which A is the total bubble surface area and V is the volume of the main body

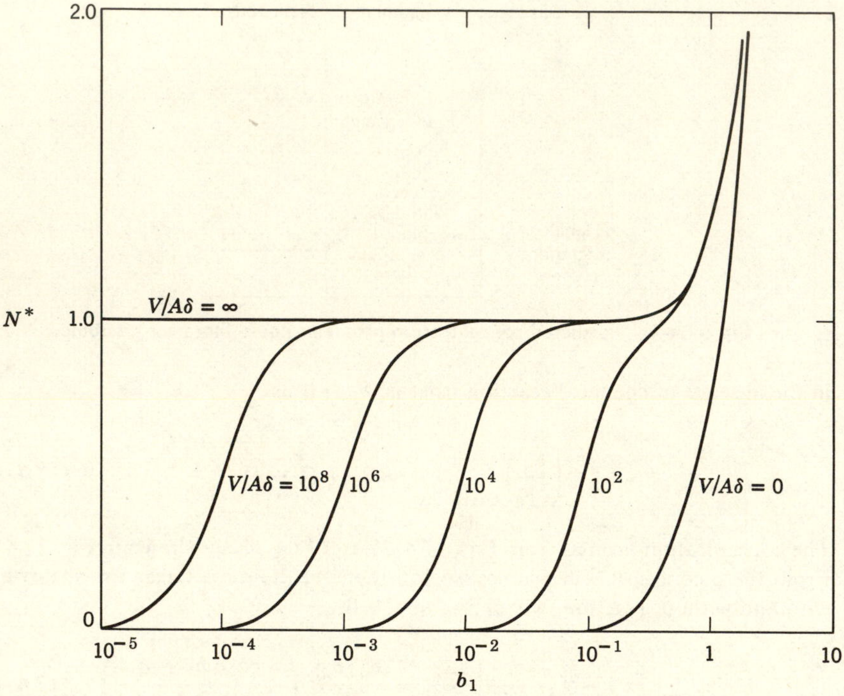

Fig. 17.4–4. Gas absorption accompanied by an irreversible first-order chemical reaction.

of liquid. Using Eq. 17.4–12, we evaluate $dc_A/dz\,|_{z=\delta}$ and $c_{A\delta}$ to get

$$\Gamma = \frac{1}{\cosh b_1 + b_1(V/A\delta)\sinh b_1} \tag{17.4-18}$$

We can then rewrite Eq. 17.4–16 as

$$N^* = \frac{b_1}{\sinh b_1}\left(\cosh b_1 - \frac{1}{\cosh b_1 + b_1(V/A\delta)\sinh b_1}\right) \tag{17.4-19}$$

This result is plotted in Fig. 17.4–4. It can be seen from this graph that over a fairly wide range of b_1 the ratio N^* is indeed very nearly unity, and Eq. 17.4–15 may be used to describe the absorption rate merely by setting $\Gamma = 0$. Similar simplification is possible for more realistic mass-transfer models.[1]

§17.5 DIFFUSION INTO A FALLING LIQUID FILM: FORCED-CONVECTION MASS TRANSFER

In this section we present an illustration of forced-convection mass transfer in which viscous flow and diffusion occur under such conditions that the

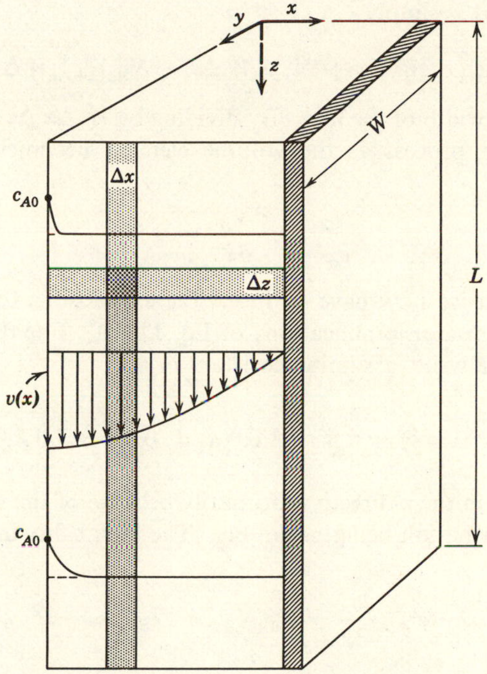

Fig. 17.5–1. Absorption into a falling film.

velocity field can be considered virtually unaffected by the diffusion. Specifically, we consider the absorption of gas A by a laminar falling film of liquid B. The material A is only slightly soluble in B, so that the viscosity of the liquids is not changed appreciably. We shall make the further restriction that the diffusion take place so slowly in the liquid film that A will not "penetrate" very far into B—that the penetration distance be small in comparison with the film thickness. The system is sketched in Fig. 17.5–1.

Let us now set up the differential equations describing the process. First of all we have to solve the momentum transfer problem to obtain the velocity profile $v_z(x)$ for the film; this has already been worked out in §2.2 in the absence of mass transfer at the fluid surface, and we know that the result is

$$v_z(x) = v_{\max}\left[1 - \left(\frac{x}{\delta}\right)^2\right] \tag{17.5–1}$$

provided that "end effects" are not considered.

Next we have to establish a mass balance on component A. We note that c_A will be changing both with x and with z. Hence as the element of volume over which we set up the mass balance we select the volume formed by the intersection of a slab of thickness Δz with a slab of thickness Δx. Then the mass balance on A is simply

$$N_{Az}|_z W\,\Delta x - N_{Az}|_{z+\Delta z} W\,\Delta x + N_{Ax}|_x W\,\Delta z - N_{Ax}|_{x+\Delta x} W\,\Delta z = 0 \tag{17.5–2}$$

in which W is the width of the film. By dividing by $W\,\Delta x\,\Delta z$ and performing the usual limiting process as the volume element becomes infinitesimally small, we get

$$\frac{\partial N_{Az}}{\partial z} + \frac{\partial N_{Ax}}{\partial x} = 0 \tag{17.5–3}$$

Into this equation we now have to insert the expressions for N_{Az} and N_{Ax} by making appropriate simplifications of Eq. 17.0–1. For the molar flux in the z-direction, we write, assuming constant c

$$N_{Az} = -\mathscr{D}_{AB}\frac{\partial c_A}{\partial z} + x_A(N_{Az} + N_{Bz}) \approx c_A v_z(x) \tag{17.5–4}$$

That is, A moves in the z-direction primarily because of the flow of the film, the diffusive contribution being negligible. The molar flux in the x-direction will be

$$N_{Ax} = -\mathscr{D}_{AB}\frac{\partial c_A}{\partial x} + x_A(N_{Ax} + N_{Bx}) \approx -\mathscr{D}_{AB}\frac{\partial c_A}{\partial x} \tag{17.5–5}$$

That is, in the x-direction A is transported primarily by diffusion, there being almost no convective transport because of the very slight solubility of A in B. Substitution of these expressions for N_{Ax} and N_{Az} into Eq. 17.5–3 gives

$$v_z\frac{\partial c_A}{\partial z} = \mathscr{D}_{AB}\frac{\partial^2 c_A}{\partial x^2} \tag{17.5–6}$$

which is the differential equation describing $c_A(x, z)$.

The forced-convection mass transfer is then described by Eqs. 17.5–1 and 17.5–6. When these two equations are combined, we get finally

$$v_{\max}\left[1 - \left(\frac{x}{\delta}\right)^2\right]\frac{\partial c_A}{\partial z} = \mathscr{D}_{AB}\frac{\partial^2 c_A}{\partial x^2} \tag{17.5–7}$$

This partial differential equation is to be solved with the following boundary conditions:

B. C. 1: at $z = 0, \quad c_A = 0$ (17.5-8)

B. C. 2: at $x = 0, \quad c_A = c_{A0}$ (17.5-9)

B. C. 3: at $x = \delta, \quad \dfrac{\partial c_A}{\partial x} = 0$ (17.5-10)

when the fluid film begins as pure B, and the interface concentration in the liquid is taken to be the solubility of A in B, here designated as c_{A0}. This problem has been solved by Pigford,[1] but we do not give that solution here. Instead, we solve to obtain only a limiting expression valid for "short contact times"—that is, small values of $L/v_{\max}$.

If, as is indicated in Fig. 17.5-1, the substance A has penetrated only a short distance into the film, then the A species for the most part has the impression that the film is moving throughout with a velocity equal to $v_{\max}$. Furthermore, if A does not penetrate very far, it does not "feel" the presence of the solid wall at $x = \delta$. Hence, if the film were of infinite thickness moving with the velocity $v_{\max}$, the diffusing material would not know the difference. This physical argument suggests then that we replace Eq. 17.5-7 and its boundary conditions by

$$v_{\max} \frac{\partial c_A}{\partial z} = \mathscr{D}_{AB} \frac{\partial^2 c_A}{\partial x^2} \tag{17.5-11}$$

B. C. 1: at $z = 0, \quad c_A = 0$ (17.5-12)

B. C. 2: at $x = 0, \quad c_A = c_{A0}$ (17.5-13)

B. C. 3: at $x = \infty, \quad c_A = 0$ (17.5-14)

The solution to Eq. 17.5-11 with these boundary conditions is[2]

$$\frac{c_A}{c_{A0}} = 1 - \frac{2}{\sqrt{\pi}} \int_0^{x/\sqrt{4\mathscr{D}_{AB}z/v_{\max}}} e^{-\xi^2}\, d\xi$$

$$= 1 - \operatorname{erf} \frac{x}{\sqrt{4\mathscr{D}_{AB}z/v_{\max}}}$$

or

$$\boxed{\frac{c_A}{c_{A0}} = \operatorname{erfc} \frac{x}{\sqrt{4\mathscr{D}_{AB}z/v_{\max}}}} \tag{17.5-15}$$

[1] R. L. Pigford, doctoral dissertation, University of Illinois (1941).

[2] The solution is worked out in detail in Example 4.1-1 by the "method of combination of variables."

Here erf y is the "error function" defined in §4.1 and erfc $y = 1 - \text{erf}\, y$ is the "complementary error function"; both are standard tabulated functions. Once the concentration profiles are known, the total mass-transfer rate may be found by either of the following two methods:

Integration of Mass Flux over Length of Film

The local mass flux at the surface $x = 0$ at a position z down the plate is

$$N_{Ax}(z)|_{x=0} = -\mathscr{D}_{AB} \frac{\partial c_A}{\partial x}\bigg|_{x=0} = c_{A0}\sqrt{\frac{\mathscr{D}_{AB} v_{\max}}{\pi z}} \qquad (17.5\text{–}16)$$

The total moles of A transferred per unit time from the gas to the liquid film is

$$\begin{aligned} \mathscr{W}_A &= \int_0^W \int_0^L N_{Ax}|_{x=0}\, dz\, dy \\ &= W c_{A0}\sqrt{\frac{\mathscr{D}_{AB} v_{\max}}{\pi}} \int_0^L z^{-1/2}\, dz \\ &= W L c_{A0}\sqrt{\frac{4\mathscr{D}_{AB} v_{\max}}{\pi L}} \end{aligned} \qquad (17.5\text{–}17)$$

Integration of Concentration Profile over (Infinite) Film Thickness at $z = L$

According to an over-all mass balance on the film, the total moles of A transferred per unit time across the gas-liquid interface must be the same as the total molar rate of flow of A across the plane $z = L$, which may be calculated by multiplying the volume rate of flow across the plane $z = L$ by the average concentration at that plane:

$$\begin{aligned} \mathscr{W}_A &= \lim_{\delta \to \infty} (W \delta v_{\max}) \left(\frac{1}{\delta} \int_0^\delta c_A|_{z=L}\, dx \right) \\ &= W v_{\max} \int_0^\infty c_A|_{z=L}\, dx \\ &= W v_{\max} c_{A0} \int_0^\infty \text{erfc}\, \frac{x}{\sqrt{4\mathscr{D}_{AB} L / v_{\max}}}\, dx \\ &= W v_{\max} c_{A0} \cdot \frac{2}{\sqrt{\pi}} \int_0^\infty \left(\int_{\frac{x}{\sqrt{4\mathscr{D}_{AB} L / v_{\max}}}}^\infty e^{-\xi^2}\, d\xi \right) dx \\ &= W v_{\max} c_{A0} \sqrt{\frac{4\mathscr{D}_{AB} L}{v_{\max}}} \cdot \frac{2}{\sqrt{\pi}} \cdot \int_0^\infty \left(\int_u^\infty e^{-\xi^2}\, d\xi \right) du \end{aligned} \qquad (17.5\text{–}18)$$

In the last line the new variable $u = x/\sqrt{4\mathscr{D}_{AB} L / v_{\max}}$ has been introduced.

At this juncture we change the order of integration[3] in the double integral, which procedure enables us to evaluate the double integral analytically and to obtain Eq. 17.5–17 thus:

$$\mathscr{W}_A = WLc_{A0}\sqrt{\frac{4\mathscr{D}_{AB}v_{\max}}{\pi L}} \cdot 2\int_0^\infty e^{-\xi^2}\left(\int_0^\xi du\right) d\xi$$

$$= WLc_{A0}\sqrt{\frac{4\mathscr{D}_{AB}v_{\max}}{\pi L}} \cdot 2\int_0^\infty e^{-\xi^2}\xi \, d\xi$$

$$= WLc_{A0}\sqrt{\frac{4\mathscr{D}_{AB}v_{\max}}{\pi L}} \tag{17.5–19}$$

From this development it is seen that the mass transfer rate is directly proportional to the square root of the diffusivity and inversely proportional to the square root of the "exposure time," $t_{\text{exp}} = L/v_{\max}$. This approach for studying gas absorption was apparently first proposed by Higbie.[4]

Example 17.5–1. Gas Absorption from Rising Bubbles

Estimate the rate at which gas bubbles of A are absorbed by liquid B as the gas bubbles rise at their terminal velocity v_t through a "clean" quiescent liquid.

Solution. For bubbles of moderate size rising in liquids with no surface-active agents, the gas in the bubble undergoes a toroidal circulation as shown in Fig. 17.5–2. As the gas circulates, it encounters fresh liquid at the top of the bubble. As the bubble rises, the liquid moves downward in relation to the bubble and leaves when it reaches the bottom of the bubble. The liquid near the interface is usually in laminar flow and appears to maintain its identity. Thus the liquid behaves much like the liquid at the surface of the falling film just discussed. The contact time is certainly short, so that the penetration of the dissolved gas is slight and the assumptions introduced in §17.5 are valid. Therefore, to a first approximation, we can use the result in Eq. 17.5–17 (or 19) to estimate the rate of gas absorption and the change in bubble size.

The average rate of mass transfer of A is

$$(N_A)_{\text{avg}} = \sqrt{\frac{4\mathscr{D}_{AB}}{\pi t_{\text{exp}}}}\, c_{A0} \tag{17.5–20}$$

in which c_{A0} is the solubility of gas A in liquid B. The exposure time t_{exp} is the time required for the liquid to slide along the bubble from top to bottom. This can be taken to be approximately[5] $t_{\text{exp}} \doteq D/v_t$, in which D is the bubble diameter and

[3] See, for example, G. B. Thomas, Jr., *Calculus and Analytic Geometry*, Addison-Wesley Reading, Mass. (1953), Chapter 15.

[4] R. Higbie, *Trans. A.I.Ch.E.*, **31**, 365–389 (1935).

[5] R. Higbie, *op. cit.*

v_t is the terminal velocity of the rising bubble. Hence the absorption rate through the bubble-liquid interface is

$$(N_A)_{\text{avg}} = \sqrt{\frac{4\mathscr{D}_{AB}v_t}{\pi D}}\,c_{A0} \tag{17.5–21}$$

This result has been substantiated[6] for gas bubbles of about 0.3–0.5 cm in diameter rising through carefully purified water; small amounts of surface-active agents cause a marked decrease in $(N_A)_{\text{avg}}$ because of the formation of a "skin" around

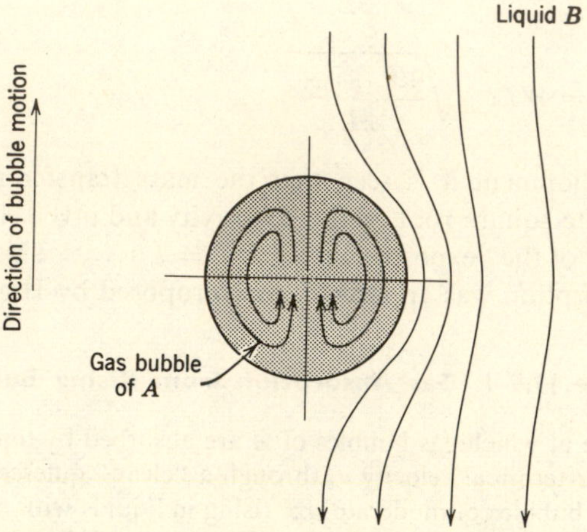

Fig. 17.5–2. Absorption from a gas bubble.

the bubble that effectively prevents circulation. A similar approach has been successfully used for predicting mass transfer rates during drop-formation at a capillary tip.[7]

§17.6 DIFFUSION AND CHEMICAL REACTION INSIDE A POROUS CATALYST: THE "EFFECTIVENESS FACTOR"[1,2,3]

Up to this point we have discussed diffusion in gases and liquids in systems of simple geometry. We now wish to apply the shell mass-balance method and Fick's first law to describe diffusion on the inside of a porous catalyst pellet. We make no attempt to describe the diffusion inside the tortuous void passages in the medium. Instead, we describe the "average" diffusion of the chemical species in terms of an effective diffusion coefficient.

[6] D. Hammerton and F. H. Garner, *Trans. Inst. Chem. Engrs. (London)*, **32**, 518 (1954).
[7] H. Groothuis and H. Kramers, *Chem. Eng. Sci.*, **4**, 17–25 (1955).
[1] E. W. Thiele, *Ind. Eng. Chem.*, **31**, 916–920 (1939).
[2] R. Aris, *Chem. Eng. Sci.*, **6**, 265–268 (1957).
[3] A. Wheeler, *Advances in Catalysis*, Academic Press, New York (1950), Vol. 3, pp. 250–326.

Specifically, we consider a spherical porous catalyst particle of radius R. (See Fig. 17.6–1.) This particle is in a catalytic reactor, where it is submerged in a gas stream containing the reactant A and the product B. In the neighborhood of the surface of the particular catalyst particle under consideration we presume that the concentration is c_{As} moles of A per unit volume. Species A diffuses through the tortuous passages in the catalyst and is converted to B on the catalytic surfaces. (See Fig. 17.6–2.)

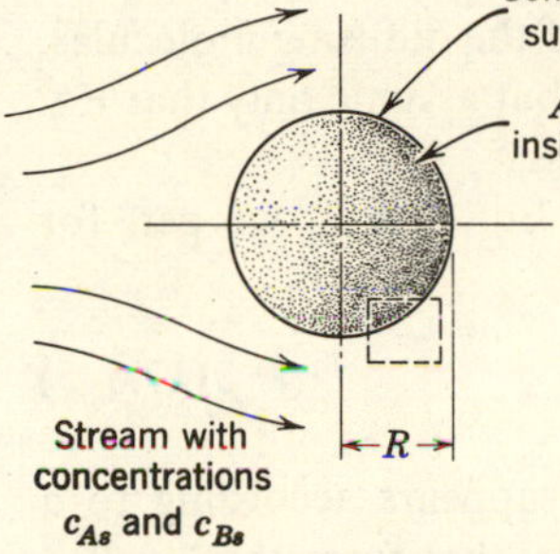

Fig. 17.6–1. Spherical catalyst particle which is porous. For magnified version of inset, see Fig. 17.6–2.

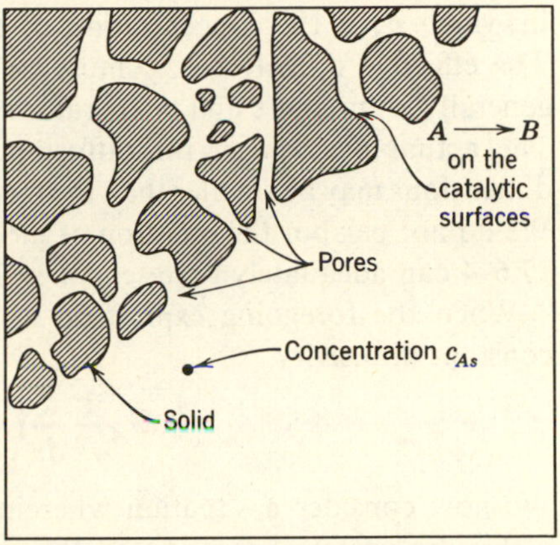

Fig. 17.6–2. Pores in the catalyst in which diffusion and chemical reaction occur.

We begin by making a mass balance for species A on a spherical shell of thickness Δr within a single catalyst particle:

$$N_{Ar}|_r \cdot 4\pi r^2 - N_{Ar}|_{r+\Delta r} \cdot 4\pi(r + \Delta r)^2 + R_A \cdot 4\pi r^2 \, \Delta r = 0 \quad (17.6\text{–}1)$$

Here $N_{Ar}|_r$ is the number of moles of A passing in the r-direction through an imaginary spherical surface at a distance r from the center of the sphere. The source term $R_A \cdot 4\pi r^2 \, \Delta r$ gives the number of moles of A being produced by chemical reaction in the shell of thickness Δr. Division by $4\pi \, \Delta r$ and letting $\Delta r \rightarrow 0$ gives

$$\lim_{\Delta r \to 0} \frac{(r^2 N_{Ar})|_{r+\Delta r} - (r^2 N_{Ar})|_r}{\Delta r} = r^2 R_A \quad (17.6\text{–}2)$$

or

$$\frac{d}{dr}(r^2 N_{Ar}) = r^2 R_A \quad (17.6\text{–}3)$$

Clearly, in view of the fact that the medium is granular rather than homogeneous, such a limiting process is in conflict with the physical picture. Consequently, in Eq. 17.6–3 the symbols N_{Ar} and R_A cannot be interpreted as quantities having a meaningful value at a point; rather we have to interpret them as quantities averaged over a small neighborhood of the point in question—a neighborhood small with respect to the dimension R but large with respect to the dimensions of the passages within the porous particle.

Now we *define* an "effective diffusivity" for species A in the porous medium by

$$N_{Ar} = -\mathscr{D}_A \frac{dc_A}{dr} \tag{17.6–4}$$

in which c_A is the concentration of the gas A contained within the pores. The effective diffusivity $\mathscr{D}_A$ must be measured experimentally; it depends generally on pressure and temperature and also on the catalyst pore structure. The actual mechanism for diffusion in pores is complex, since the pore dimensions may be smaller than the mean free path of the diffusing molecules. We do not belabor the question of mechanism here but assume only that Eq. 17.6–4 can adequately represent the diffusion process.

When the foregoing expression is inserted into Eq. 17.6–3, we get, for constant diffusivity,

$$-\mathscr{D}_A \frac{1}{r^2}\frac{d}{dr}\left(r^2 \frac{dc_A}{dr}\right) = R_A \tag{17.6–5}$$

We now consider a situation wherein species A disappears according to a first-order chemical reaction on the catalytic surfaces that form the "walls" of the winding passages. Let a be the available catalytic surface area per unit volume (of solids + voids). Then $R_A = -k_1''ac_A$ and Eq. 17.6–5 becomes

$$\mathscr{D}_A \frac{1}{r^2}\frac{d}{dr}\left(r^2 \frac{dc_A}{dr}\right) = k_1''ac_A \tag{17.6–6}$$

This equation is to be solved with the boundary conditions that $c_A = c_{As}$ at $r = R$ and that c_A is finite at $r = 0$. Differential equations in which the operator $(1/r^2)(d/dr)[r^2(d/dr)]$ appears can frequently be simplified by a change of variable of the type $c_A/c_{As} = f(r)/r$. The equation for $f(r)$ is

$$\frac{d^2f}{dr^2} = \left(\frac{k_1''a}{\mathscr{D}_A}\right)f \tag{17.6–7}$$

The general solution containing two integration constants is

$$\frac{c_A}{c_{As}} = \frac{C_1}{r}\cosh\sqrt{\frac{k_1''a}{\mathscr{D}_A}}\,r + \frac{C_2}{r}\sinh\sqrt{\frac{k_1''a}{\mathscr{D}_A}}\,r \tag{17.6–8}$$

Application of the boundary conditions gives finally

$$\boxed{\frac{c_A}{c_{As}} = \left(\frac{R}{r}\right)\frac{\sinh\sqrt{k_1''a/\mathscr{D}_A}\,r}{\sinh\sqrt{k_1''a/\mathscr{D}_A}\,R}} \tag{17.6–9}$$

In studies on chemical kinetics and catalysis one is frequently interested in the molar flux N_{As} or the molar flow $\mathcal{W}_{As}$ at the surface $r = R$:

$$\mathcal{W}_{As} = 4\pi R^2 N_{As} = -4\pi R^2 \mathscr{D}_A \left.\frac{dc_A}{dr}\right|_{r=R} \tag{17.6–10}$$

By making use of Eq. 17.6–9, we find that

$$\mathcal{W}_{As} = 4\pi R \mathscr{D}_A c_{As}\left(1 - \sqrt{\frac{k_1''a}{\mathscr{D}_A}}\, R \coth \sqrt{\frac{k_1''a}{\mathscr{D}_A}}\, R\right) \tag{17.6–11}$$

This result gives the rate of conversion (moles/sec) of A to B in a single catalyst particle of radius R in terms of the diffusive processes involved.

If the catalytically active surface were all exposed to the stream of concentration c_{As}, then the species A would not have to diffuse through the pores to a reaction site; the molar rate of conversion would then be given by the product of available surface and surface reaction rate:

$$\mathcal{W}_{A0} = (\tfrac{4}{3}\pi R^3)(a)(-k_1''c_{As}) \tag{17.6–12}$$

Division of Eq. 17.6–11 by Eq. 17.6–12 gives

$$\eta_A = \frac{3}{K^2}(K \coth K - 1) \tag{17.6–13}$$

in which $K = \sqrt{k_1''a/\mathscr{D}_A}R$, a dimensionless group. The quantity η_A is called the *effectiveness factor*.[1,2,4] It is the quantity by which $\mathcal{W}_{A0}$ has to be multiplied to account for the resistance of diffusion to the over-all conversion process.

For nonspherical catalyst particles, the foregoing results may be applied approximately by reinterpreting R. We note that for a sphere of radius R the ratio of volume to external surface is $R/3$. For nonspherical particles, we redefine R in Eq. 17.6–13 as

$$R_{\text{nonsph}} = 3\left(\frac{V_P}{S_P}\right) \tag{17.6–14}$$

in which V_P and S_P are the volume and external surface of a single catalyst particle. The conversion rate is then given approximately by

$$|\mathcal{W}_{As}| = V_P a k_1'' c_{As} \eta_A \tag{17.6–15}$$

where

$$\eta_A = \frac{1}{3\Lambda^2}(3\Lambda \coth 3\Lambda - 1) \tag{17.6–16}$$

in which $\Lambda = \sqrt{k_1''a/\mathscr{D}_A}(V_P/S_P)$.

The particular utility of the quantity Λ may be seen in Fig. 17.6–3. It is clear that when the exact theoretical expressions for η_A are plotted as a

[4] O. A. Hougen and K. M. Watson, *Chemical Process Principles*, Wiley, New York (1947), Part III, Chapter XIX. See also *CPP Charts* by O. A. Hougen, K. M. Watson, and R. A. Ragatz, Wiley, New York (1960), Fig. E.

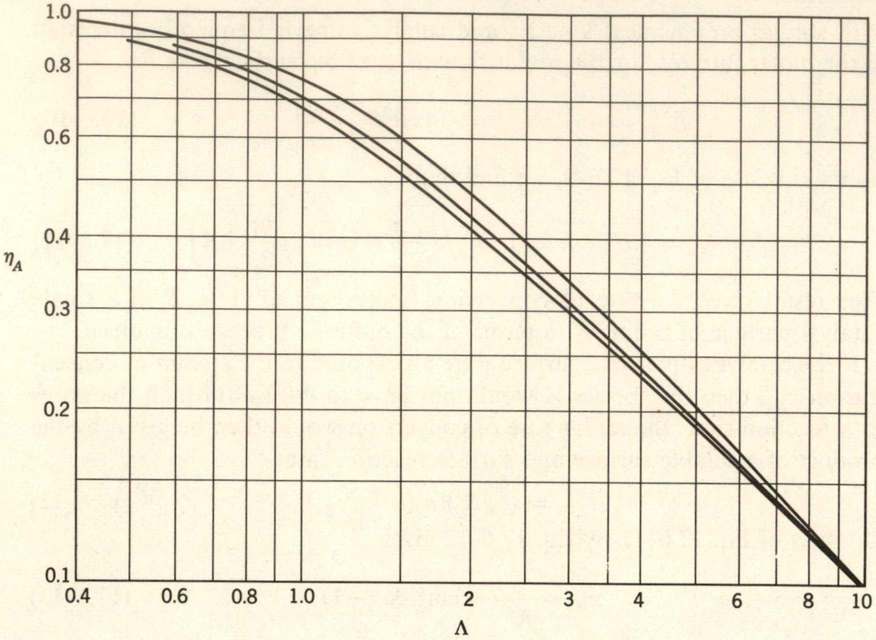

Fig. 17.6–3. Effectiveness factors for porous solid catalysts: top curve, flat particles; middle curve, long cylindrical particles; bottom curve, spherical particles [R. Aris, *Chem. Eng. Sci.*, **6**, 265 (1957).] An enlarged graph for spheres is given in O. A. Hougen, K. M. Watson, and R. A. Ragatz, *CPP Charts*, Wiley, New York (1960), Fig. E.

function of Λ the curves have common asymptotes for large and small Λ and do not differ from one another very much for intermediate values of Λ. Thus Fig. 17.6–3 provides a justification for the use of Eq. 17.6–16 to estimate η_A for nonspherical particles.

QUESTIONS FOR DISCUSSION

1. What physical information is used to eliminate N_{Bz} from Eq. 17.0–1?
2. Compare and contrast *homogeneous* and *heterogeneous* chemical reactions.
3. How would the results of §17.2 be affected if the net motion of the mixture were disregarded [i.e., if $(1 - x_A)$ in Eq. 17.2–4 were set equal to unity]?
4. How do N_{Az} and N_{Bz} vary with distance in §17.2?
5. In Eq. 17.2–9 how is x_{A2} determined?
6. Under what physical situations would you expect the model sketched in Fig. 17.2–1 to be realistic? unrealistic?
7. Discuss the term "diffusion-controlled chemical reaction."
8. Discuss the effect of chemical reaction on gas absorption.
9. Show how Eq. 17.3–14 is obtained from Eq. 17.3–13.

10. Discuss the physical significance of the approximations made in Eqs. 17.5–4 and 5.

11. Sketch the concentration profiles in Eq. 17.5–15 by looking up a table of the error function.

12. How do the results in §17.5 have to be modified if the film at $z = 0$ contains some dissolved A, say c_{Ai} moles per unit volume?

13. Sketch a graph of N_{Az} as a function of z for the problem described by Fig. 17.4–1.

14. How does the size of a soluble bubble change as it moves upward through a liquid?

15. In §17.2 there is a gradient in the concentration of B, yet the flux N_{Bz} is zero. Explain.

16. Use Fig. 17.6–3 to show that for very large Λ the rate of reaction for a given mass of catalyst is proportional to the external surface area of the catalyst particles.

17. How can Eq. 17.3–9 be obtained from Eq. 17.3–2 without getting the concentration profile?

PROBLEMS

17.A$_1$ Rate of Evaporation

Consider the system described in Fig. 17.2–1. What is the rate of evaporation (in g hr^{-1}) of CCl_3NO_2 (chloropicrin) into air (considered here as a pure substance) at 25° C?

Total pressure	770 mm Hg
Diffusivity	0.088 cm^2 sec^{-1}
Vapor pressure	23.81 mm Hg
Distance from liquid level to top of tube	11.14 cm
Density of chloropicrin	1.65 g cm^{-3}
Surface area of liquid exposed for evaporation	2.29 cm^2

Answer: 0.0139 g hr^{-1}

17.B$_1$ Error in Calculating Absorption Rate

What is the maximum possible error in computing the absorption rate from Eq. 17.5–17 if the solubility of A in B is known within ± 5 per cent and the diffusivity of A in B is known within ± 10 per cent? (Assume that the errors in the film dimensions and velocity are known to be of a lower order of magnitude.) *Answer:* ± 10 per cent

17.C$_1$ Rate of Absorption in a Falling Film

Chlorine is being absorbed from a gas in a small experimental wetted wall tower, as shown in Fig. 17.C. The absorbing fluid is water, which is moving with an average velocity of 17.7 cm sec^{-1}. What is the absorption rate in g-moles hr^{-1} if $\mathscr{D}_{Cl_2-H_2O} = 1.26 \times 10^{-5}$ cm^2 sec^{-1} in the liquid phase and if the saturation concentration of chlorine in water is 0.823 g Cl_2 per 100 g of water (these are the experimental values at 16° C). The dimensions of the column are given in Fig. 17.C.

Hint. Ignore the chemical reaction between Cl_2 and H_2O. *Answer:* 0.273 g-moles hr^{-1}

17.D$_2$ Diffusion Method for Separating Helium from Natural Gas

In a short note in *Scientific American*, Vol. 199, No. 1, (July 1958), p. 52, it is described how helium may be separated from natural gas by a method developed by K. B. McAfee of Bell Telephone Laboratories. The method is based on the fact that pyrex glass is almost impermeable to all gases but helium; for example, the diffusion coefficient of He through pyrex is about 25 times the diffusion coefficient of H_2 through pyrex—hydrogen being the closest "competitor" in the diffusion process. This method presumably offers possibility for more efficient and less costly separations than the previous method of low-temperature distillation.

Suppose a natural-gas mixture is contained in a pyrex tube with dimensions shown in Fig. 17.D. Obtain an expression for the rate at which helium will leak through the tube in terms of the diffusivity of helium in pyrex, the interfacial concentrations of the helium in the pyrex, and the dimensions of the tube.

Answer: $\mathcal{W}_{\text{He}} = 2\pi L \dfrac{\mathcal{D}_{\text{He-Pyr}}(c_{\text{He},1} - c_{\text{He},2})}{\ln (R_2/R_1)}$

17.E$_2$ Diffusion Through a Stagnant Film—Alternate Derivation

In §17.2 an expression for the evaporation rate was obtained in Eq. 17.2–14 by differentiating the concentration profile found a few lines before. Show that the same result may be obtained much more easily (without finding the concentration profile) in the following way. First note that at steady state N_{Az} will be a constant and N_{Bz} will be zero. Then integrate Eq. 17.0–1 directly between the limits of z_1 and z_2 to get Eq. 17.2–14.

17.F$_2$ Diffusion Through a Stagnant Liquid Film

In studying the rate of leaching of a substance A from solid particles by a solvent B, we may postulate that the rate-controlling step is the diffusion of A from the particle surface through a liquid film out into the main liquid stream. (See Fig. 17.F.) The solubility of A in B is c_{A0} g-moles/cm^3, and the concentration in the main stream—beyond the liquid film of thickness δ—is $c_{A\delta}$.

a. Obtain a differential equation for c_A as a function of z by making a mass balance on A over a thin slab of thickness Δz. Assume that $\mathcal{D}_{AB}$ is constant and that A is only slightly soluble in B. Neglect the curvature of the particle.

b. Show that in the absence of chemical reaction in the liquid phase the concentration profile is linear.

c. Show that the rate of leaching is then given by

$$N_{Az} = \frac{\mathcal{D}_{AB}(c_{A0} - c_{A\delta})}{\delta} \tag{17.F–1}$$

17.G$_2$ Diffusion from a Droplet into a Quiescent Gas

A droplet of substance A is suspended in a stream of gas B. The droplet radius is r_1. We postulate that there is a spherical stagnant gas film of radius r_2. (See Fig. 17.2–3.) The concentration of A in the gas phase is x_{A1} at $r = r_1$ and x_{A2} at $r = r_2$.

a. By a shell balance, show that for steady-state diffusion $r^2 N_{Ar}$ is a constant and set the constant equal to $r_1^2 N_{Ar1}$, the value at the droplet surface.

b. Show that Eq. 17.0–1 and the result in (*a*) lead to the following equation for x_A:

$$r_1^2 N_{Ar1} = -\frac{c\mathcal{D}_{AB}}{1 - x_A} r^2 \frac{dx_A}{dr} \tag{17.G–1}$$

c. Integrate this equation between the limits r_1 and r_2 and get

$$N_{Ar1} = \frac{c\mathcal{D}_{AB}}{r_2 - r_1}\left(\frac{r_2}{r_1}\right) \ln \frac{x_{B2}}{x_{B1}} \tag{17.G–2}$$

d. If a mass transfer coefficient k_p is defined by $N_{Ar1} = k_p(p_{A1} - p_{A2})$, show that when $r_2 \to \infty$

$$k_p = \frac{2c\mathcal{D}_{AB}/D}{(p_B)_{\ln}} \tag{17.G–3}$$

in which D is the droplet diameter. Discuss the significance of this result as applied to a droplet evaporating into a large body of gas that is not in motion.

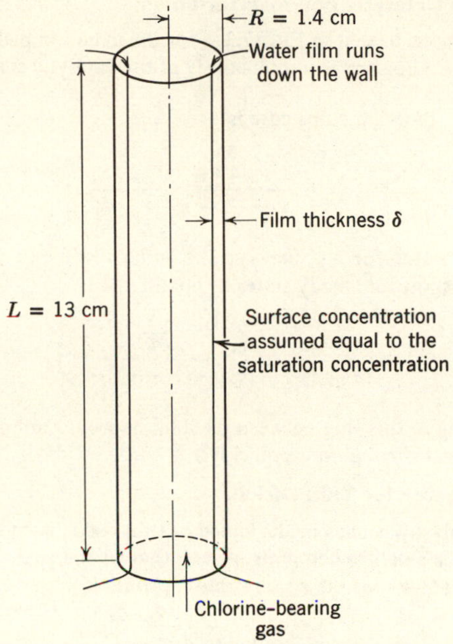

Fig. 17.C. Schematic drawing of a wetted-wall column.

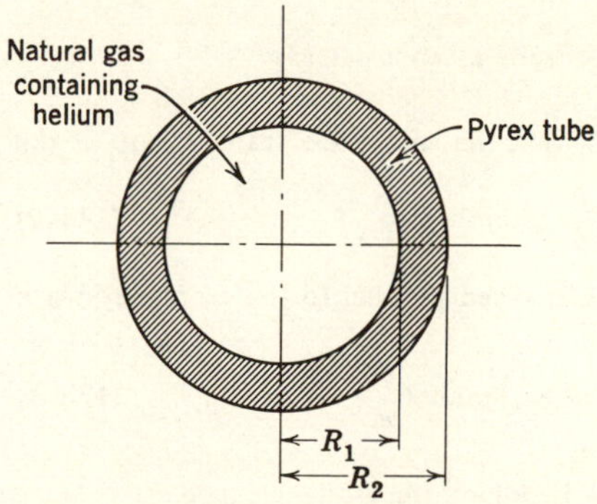

Fig. 17.D. Diffusion through pyrex tubing; length of tubing is L.

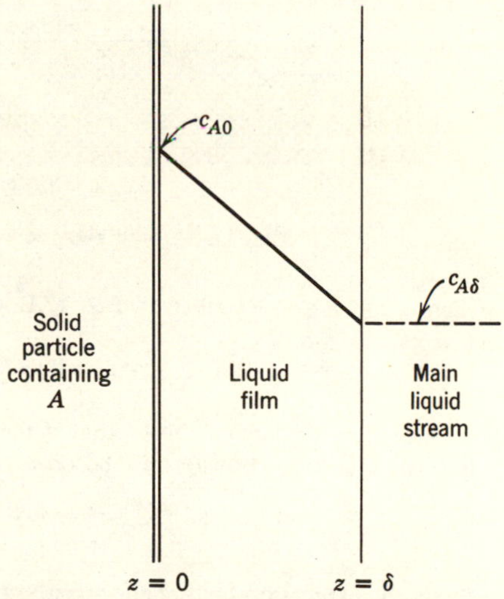

Fig. 17.F. Diffusion through a stagnant liquid film.

17.H₂ Diffusion with Catalytic Polymerization

Obtain a result analogous to that in Eq. 17.3–9 for the situation pictured in Fig. 17.3–1*b* when the reaction $nA \rightarrow A_n$ occurs instantaneously at the catalytic surface.

a. Show first that Eq. 17.0–1 for this case is

$$N_{Az} = -\frac{c\mathscr{D}_{AA_n}}{1 - (1 - n^{-1})x_A}\frac{dx_A}{dz} \tag{17.H–1}$$

b. Integrate this expression for N_{Az} between the limits $z = 0$ and $z = \delta$ (making use of the fact that N_{Az} is constant for steady state) to obtain

$$N_{Az} = \frac{nc\mathscr{D}_{AA_n}}{(n-1)\delta}\ln\left(\frac{1}{1 - (1 - n^{-1})x_{A0}}\right) \tag{17.H–2}$$

(Note that by integrating in this way between fixed limits we do not obtain the concentration profile, but we do get the flux more quickly.)

17.I₂ Effectiveness Factor for Thin Disks

Consider porous catalyst particles in the shape of thin disks, such that the surface area of the edge of the disk is small in comparison with that of the two circular faces. Apply the method of §17.6 to show that the concentration profile is

$$\frac{c_A}{c_{As}} = \frac{\cosh\sqrt{k_1''a/\mathscr{D}_A}\,z}{\cosh\sqrt{k_1''a/\mathscr{D}_A}\,b} \tag{17.I–1}$$

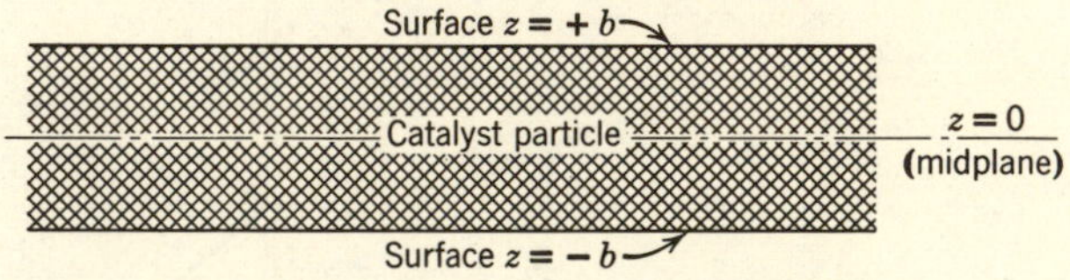

Fig. 17.I. Side view of a disk-shaped catalyst particle.

where z and b are described in Fig. 17.I. Show that the total mass transfer rate at the surfaces $z = \pm b$ is

$$|\mathscr{W}_A| = 2 \cdot \pi R^2 \cdot \mathscr{D}_A c_{As}\lambda \tanh \lambda b \tag{17.I–2}$$

in which $\lambda = \sqrt{k_1''a/\mathscr{D}_A}$. Show that if the disk is sliced parallel to the xy-plane into n slices the total mass-transfer rate becomes

$$|\mathscr{W}_A^{(n)}| = 2 \cdot \pi R^2 \cdot n \cdot \mathscr{D}_A c_{As}\lambda \tanh \lambda \frac{b}{n} \tag{17.I–3}$$

Obtain the expression for the effectiveness factor by taking the limit

$$\eta_A = \lim_{n\rightarrow\infty}\frac{|\mathscr{W}_A|}{|\mathscr{W}_A^{(n)}|} = \frac{\tanh \lambda b}{\lambda b} \tag{17.I–4}$$

Re-express this result in terms of the parameter Λ defined in §17.6.

17.J$_3$ Solid Dissolution into a Falling Film[1]

Liquid B is flowing in laminar motion down a vertical wall. For $z < 0$, the wall does not dissolve in the fluid, but for $0 < z < L$ the wall contains a species A that is slightly soluble in B. The film begins far enough up the wall so that v_z depends only on y for $z \geqslant 0$. (See Fig. 17.J.)

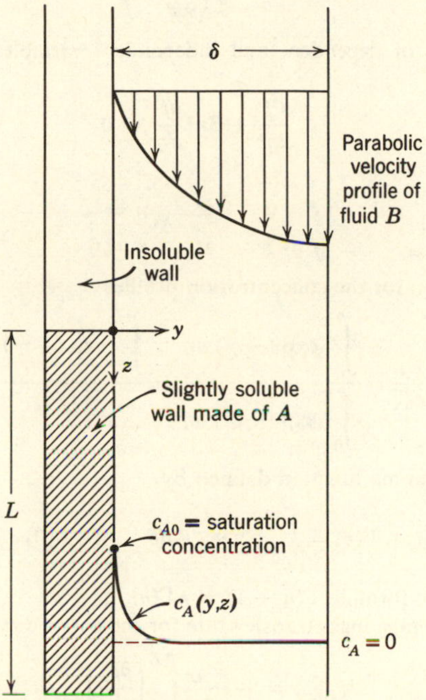

Fig. 17.J. Solid dissolution into a falling film, with fully developed parabolic velocity profile.

a. Clearly, Eq. 17.5–6 applies here. If the "contact time" $L/v_{\max}$ is small, then show that a good approximation to Eq. 17.5–6 is

$$ay \frac{\partial c_A}{\partial z} = \mathscr{D}_{AB} \frac{\partial^2 c_A}{\partial y^2} \tag{17.J–1}$$

in which $a = \rho g \delta / \mu$. Justify the use of the boundary conditions

B.C. 1: $c_A = 0$ at $z = 0$ (17.J–2)

B.C. 2: $c_A = 0$ at $y = \infty$ (17.J–3)

B.C. 3: $c_A = c_{A0}$ at $y = 0$ (17.J–4)

in which c_{A0} is the solubility of A in B.

[1] H. Kramers and P. J. Kreyger, *Chem. Eng. Sci.*, **6**, 42–48 (1956); see also R. L. Pigford, *Chem. Eng. Prog. Symposium Series No. 17*, Vol. 51, pp. 79–92 (1955) for the analogous heat-conduction problem.

b. Solve Eq. 17.J–1 by anticipating a solution of the form

$$\frac{c_A}{c_{A0}} = f(\eta) \tag{17.J–5}$$

in which η is a dimensionless variable:

$$\eta = y\left(\frac{a}{9\mathscr{D}_{AB}z}\right)^{1/3} \tag{17.J–6}$$

Show that this choice of dependent and independent variables allows Eq. 17.J–1 to be rewritten as

$$\frac{d^2f}{d\eta^2} + 3\eta^2\frac{df}{d\eta} = 0 \tag{17.J–7}$$

with the boundary conditions

B. C. 1: $f = 0 \quad \text{at} \quad \eta = \infty$ (17.J–8)

B. C. 2: $f = 1 \quad \text{at} \quad \eta = 0$ (17.J–9)

c. Solve this equation for the concentration profiles

$$\frac{c_A}{c_{A0}} = \frac{\int_\eta^\infty \exp(-\eta^3)\,d\eta}{\int_0^\infty \exp(-\eta^3)\,d\eta} = \frac{\int_\eta^\infty \exp(-\eta^3)\,d\eta}{\Gamma(\frac{4}{3})} \tag{17.J–10}$$

in which $\Gamma(\frac{4}{3})$ is the gamma function defined by

$$\Gamma(n) = \int_0^\infty \beta^{n-1}e^{-\beta}\,d\beta \qquad (n > 0) \tag{17.J–11}$$

which has the recursion formula $\Gamma(n + 1) = n\Gamma(n)$.

d. Show that the average mass transfer rate for the entire dissolving surface is

$$N_{A,\text{avg}} = -\frac{\mathscr{D}_{AB}}{L}\int_0^L \left(\frac{\partial c_A}{\partial y}\right)_{y=0} dz$$

$$= \frac{2\mathscr{D}_{AB}c_{A0}}{\Gamma(\frac{7}{3})}\sqrt[3]{\frac{a}{9\mathscr{D}_{AB}L}} \tag{17.J–12}$$

17.K$_3$ Diffusion from a Point Source in a Moving Stream

A stream of fluid (of chemical species B) in laminar motion has a uniform velocity v_0. At some point in the stream species A is injected in a small amount $\mathscr{W}_A$ moles sec^{-1}. This amount is assumed to be small enough so that the mass average velocity will not deviate appreciably from v_0. The system is pictured in Fig. 17.K. Species A is swept downstream (in the z-direction), and at the same time it diffuses both axially and radially.

a. Show that a mass balance on species A over the cylindrical ring in Fig. 17.K leads to the following partial differential equations if ρ and $\mathscr{D}_{AB}$ are assumed to be constant:

$$v_0\frac{\partial c_A}{\partial z} = \mathscr{D}_{AB}\left[\frac{1}{r}\frac{\partial}{\partial r}\left(r\frac{\partial c_A}{\partial r}\right) + \frac{\partial^2 c_A}{\partial z^2}\right] \tag{17.K–1}$$

b. Verify (lengthy!) that the solution

$$c_A = \frac{\mathscr{W}_A}{4\pi\mathscr{D}_{AB}s}\exp\left[(-v_0/2\mathscr{D}_{AB})(s - z)\right] \tag{17.K–2}$$

satisfies the differential equation above. Note that $s^2 = r^2 + z^2$.

c. Show that this solution also satisfies the following boundary conditions:

B. C. 1: $$c_A = 0 \quad \text{at} \quad s = \infty \tag{17.K-3}$$

B. C. 2: $$-4\pi s^2 \mathscr{D}_{AB} \left(\frac{\partial c_A}{\partial s} \right)_z = \mathscr{W}_A \quad \text{at} \quad s \to 0 \tag{17.K-4}$$

B. C. 3: $$\left(\frac{\partial c_A}{\partial r} \right)_z = 0 \quad \text{at} \quad r = 0 \tag{17.K-5}$$

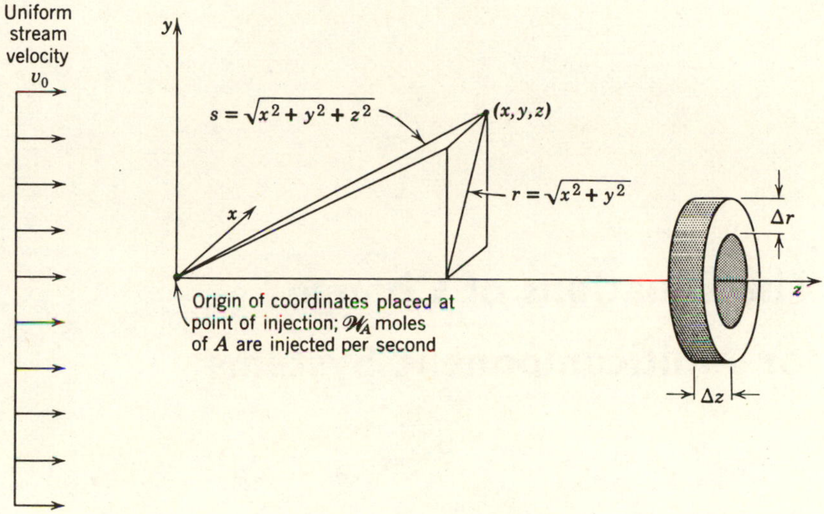

Fig. 17.K. Diffusion from a point source in a constant-velocity stream.

d. Show how data on c_A as a function of r and z for given v_0 and $\mathscr{W}_A$ may be plotted (when the foregoing solution applies) to give a straight line with slope $v_0/2\mathscr{D}_{AB}$ and intercept $\ln \mathscr{D}_{AB}$.

Note. One of the applications of this method is in the determination of "eddy diffusivities."[2]

17.L$_4$ Gas Absorption in a Falling Film with Chemical Reaction

Rework the problem discussed in §17.5 and described in Fig. 17.5-1 when gas A reacts with liquid B by a first-order irreversible chemical reaction in the liquid phase, with rate constant k_1'''. Specifically, find the expression for total absorption analogous to that given in Eq. 17.5-17. Show that the result for absorption with reaction properly simplifies to that for absorption without reaction.

Answer: $$\mathscr{W}_A = Wc_{A0}v_{\max}\sqrt{\mathscr{D}_{AB}/k_1'''}\left[\left(\frac{1}{2} + u\right) \operatorname{erf}\sqrt{u} + \sqrt{\frac{u}{\pi}}\, e^{-u}\right]$$

where $u = k_1'''L/v_{\max}$

17.M$_4$ Effectiveness Factors for Long Cylinders

Derive the expression for η_A analogous to Eq. 17.6-16 for cylinders. Neglect the diffusion through the ends of the cylinders.

Answer: $$\eta_A = \frac{I_1(2\Lambda)}{\Lambda I_0(2\Lambda)}$$

[2] T. K. Sherwood and R. L. Pigford, *Absorption and Extraction*, McGraw-Hill, New York (1952), Second Edition, p. 42.

CHAPTER 18

The Equations of Change for Multicomponent Systems

In Chapter 17 a number of problems in ordinary diffusion were formulated by making mass balances on one of the diffusing species. In this chapter we begin by making a mass balance over an arbitrary differential fluid element to establish the equations of continuity for the two chemical species in a binary fluid mixture. Then the insertion of the expressions for the mass flux gives the diffusion equations in a variety of forms. These diffusion equations can be used to set up any of the problems in Chapter 17 and more complicated ones as well.

After this introduction to the equations of ordinary diffusion for a binary mixture, we proceed to the exposition of the equations of change for a multicomponent mixture with chemical reactions and heat effects, which include an equation of continuity for each chemical species present, the equation of motion, and the equation of energy balance. These relations supply a full description of multicomponent flow systems and include the equations given in Chapters 3 and 10 as special cases. They are not normally used in their complete form given here; usually one uses them by discarding terms that are identically zero or physically negligible and by this procedure obtaining simpler equations for the special situation under consideration.

After the presentation of these general equations, we are then in a position to consider some of their special applications, such as the description of combined heat and mass transfer, thermal, pressure, and forced diffusion, and three-component ordinary diffusion. The equations of change also provide the basis for dimensional analysis, which is used later in connection with the determination of the functional form of the mass-transfer coefficient correlations.

§18.1 THE EQUATIONS OF CONTINUITY FOR A BINARY MIXTURE

In this section we apply the law of conservation of mass of species A to a volume element $\Delta x\, \Delta y\, \Delta z$ *fixed in space*, through which a binary mixture of A and B is flowing. (See Fig. 3.1–1.) Within this element, A may be produced by chemical reaction at a rate r_A (g cm^{-3} sec^{-1}). The various contributions to the mass balance are

time rate of change of mass of A in volume element	$\dfrac{\partial \rho_A}{\partial t} \Delta x\, \Delta y\, \Delta z$	(18.1–1)
input of A across face at x	$n_{Ax}\|_x\, \Delta y\, \Delta z$	(18.1–2)
output of A across face at $x + \Delta x$	$n_{Ax}\|_{x+\Delta x}\, \Delta y\, \Delta z$	(18.1–3)
rate of production of A by chemical reaction	$r_A\, \Delta x\, \Delta y\, \Delta z$	(18.1–4)

There are also input and output terms in the y- and z-directions. When the entire mass balance is written down and divided through by $\Delta x\, \Delta y\, \Delta z$, one obtains, after letting the size of the volume element decrease to zero,

$$\frac{\partial \rho_A}{\partial t} + \left(\frac{\partial n_{Ax}}{\partial x} + \frac{\partial n_{Ay}}{\partial y} + \frac{\partial n_{Az}}{\partial z}\right) = r_A \tag{18.1–5}$$

This is the *equation of continuity for component* A in a binary mixture. It describes the change of mass concentration of A with respect to time at a fixed point in space, this change resulting from motion of A and chemical reactions producing A. The quantities n_{Ax}, n_{Ay}, n_{Az} are the rectangular components of the mass flux vector $\boldsymbol{n}_A = \rho_A \boldsymbol{v}_A$ defined in Eq. 16.1–5. In vector notation Eq. 18.1–5 may be rewritten

$$\frac{\partial \rho_A}{\partial t} + (\nabla \cdot \boldsymbol{n}_A) = r_A \tag{18.1–6}$$

Similarly *the equation of continuity for component B* is

$$\frac{\partial \rho_B}{\partial t} + (\nabla \cdot \boldsymbol{n}_B) = r_B \tag{18.1–7}$$

Addition of these two equations gives

$$\frac{\partial \rho}{\partial t} + (\nabla \cdot \rho \boldsymbol{v}) = 0 \tag{18.1–8}$$

which is the *equation of continuity for the mixture.* It is the same as that for a pure fluid given in Eq. 3.1–4. In obtaining Eq. 18.1–8, we have made use of the relation $\boldsymbol{n}_A + \boldsymbol{n}_B = \rho \boldsymbol{v}$ (Eq. J in Table 16.1–3) and also the law of conservation of mass in the form $r_A + r_B = 0$. Finally, we note that Eq. 18.1–8 becomes

$$(\nabla \cdot \boldsymbol{v}) = 0 \tag{18.1–9}$$

for a fluid of *constant mass density* ρ.

The foregoing development could equally well have been made in terms of molar units. If R_A is the molar rate of production of A per unit volume, then the molar analog of Eq. 18.1–6 is

$$\frac{\partial c_A}{\partial t} + (\nabla \cdot \boldsymbol{N}_A) = R_A \tag{18.1–10}$$

Similarly, for component B, we have

$$\frac{\partial c_B}{\partial t} + (\nabla \cdot \boldsymbol{N}_B) = R_B \tag{18.1–11}$$

Addition of these two equations gives

$$\frac{\partial c}{\partial t} + (\nabla \cdot c\boldsymbol{v}^\star) = (R_A + R_B) \tag{18.1–12}$$

for the equation of continuity for the mixture. Here we have used the relation $\boldsymbol{N}_A + \boldsymbol{N}_B = c\boldsymbol{v}^\star$. However, since moles are not in general conserved, we cannot set $R_A + R_B$ equal to zero unless one mole of B is produced for every mole of A disappearing (or *vice versa*). Finally, we note that Eq. 18.1–12 becomes

$$(\nabla \cdot \boldsymbol{v}^\star) = \frac{1}{c}(R_A + R_B) \tag{18.1–13}$$

for a fluid of *constant molar density* c.

Equations 18.1–6 and 10 are not in useful form for obtaining concentration profiles. In order to get the equations generally used for describing binary diffusion, we replace the fluxes $\boldsymbol{n}_A$ and $\boldsymbol{N}_A$ by the appropriate expressions involving the concentration gradients. When Eq. *A* of Table 16.2–1 is substituted into Eq. 18.1–6 and when Eq. *B* of Table 16.2–1 is substituted into Eq. 18.1–10, we get the following completely equivalent binary diffusion equations:

$$\frac{\partial \rho_A}{\partial t} + (\nabla \cdot \rho_A \boldsymbol{v}) = (\nabla \cdot \rho \mathscr{D}_{AB} \nabla \omega_A) + r_A \tag{18.1–14}$$

$$\frac{\partial c_A}{\partial t} + (\nabla \cdot c_A \boldsymbol{v}^\star) = (\nabla \cdot c \mathscr{D}_{AB} \nabla x_A) + R_A \tag{18.1–15}$$

Either one of these equations describes the concentration profiles in a *binary* diffusing system. The only restriction is the absence of thermal, pressure, and forced diffusion. Equations 18.1–14 and 15 are valid for systems with variable total density (ρ or c) and variable diffusivity $\mathscr{D}_{AB}$.

Because Eqs. 18.1–14 and 15 are fairly general, they are also fairly unwieldy. In the analysis of diffusing systems one can often legitimately assume either constant mass density or constant molar density and thereby effect some simplification:

Assumption of Constant ρ and $\mathscr{D}_{AB}$

For this assumption, Eq. 18.1–14 becomes

$$\frac{\partial \rho_A}{\partial t} + \rho_A(\nabla \cdot \boldsymbol{v}) + (\boldsymbol{v} \cdot \nabla \rho_A) = \mathscr{D}_{AB} \nabla^2 \rho_A + r_A \tag{18.1–16}$$

But according to Eq. 18.1–9, $(\nabla \cdot \boldsymbol{v})$ is zero. When Eq. 18.1–16 is divided by M_A we get

$$\frac{\partial c_A}{\partial t} + (\boldsymbol{v} \cdot \nabla c_A) = \mathscr{D}_{AB} \nabla^2 c_A + R_A \tag{18.1–17}$$

This equation is usually used for diffusion in *dilute liquid solutions* at constant temperature and pressure. The left side can be written as Dc_A/Dt. Equation 18.1–17 is of the same form as Eq. 10.1–25 if $R_A = 0$; this similarity is the basis for the analogies that are frequently drawn between heat and mass transport in flowing fluids with constant ρ.

Assumption of Constant c and $\mathscr{D}_{AB}$

For this assumption, Eq. 18.1–15 becomes

$$\frac{\partial c_A}{\partial t} + c_A(\nabla \cdot \boldsymbol{v}^\star) + (\boldsymbol{v}^\star \cdot \nabla c_A) = \mathscr{D}_{AB} \nabla^2 c_A + R_A \tag{18.1–18}$$

But according to Eq. 18.1–13, $(\nabla \cdot \boldsymbol{v}^\star)$ can be replaced by $(1/c)(R_A + R_B)$. Hence Eq. 18.1–18 becomes[1]

$$\frac{\partial c_A}{\partial t} + (\boldsymbol{v}^\star \cdot \nabla c_A) = \mathscr{D}_{AB}\nabla^2 c_A + R_A - \frac{c_A}{c}(R_A + R_B) \qquad (18.1\text{–}19)$$

This equation is usually used for *low-density gases* at constant temperature and pressure. The left side of this equation *cannot* be written as Dc_A/Dt because of the appearance of $\boldsymbol{v}^\star$ instead of $\boldsymbol{v}$.

Assumption of Zero Velocity

There is one more simplified form of Eqs. 18.1–14 and 15 that must be mentioned. If there are no chemical changes occurring, then r_A, r_B, R_A, and R_B are all zero. If, in addition, $\boldsymbol{v}$ is zero in Eq. 18.1–17 or if $\boldsymbol{v}^\star$ is zero in Eq. 18.1–18, then we get

$$\frac{\partial c_A}{\partial t} = \mathscr{D}_{AB}\nabla^2 c_A \qquad (18.1\text{–}20)$$

which is called *Fick's second law of diffusion* or sometimes simply *the diffusion equation*. This equation is usually used for diffusion in *solids* or *stationary liquids* ($\boldsymbol{v} = 0$ in Eq. 18.1–17) and for *equimolar counter-diffusion* in gases[2] ($\boldsymbol{v}^\star = 0$ in Eq. 18.1–19). Note that Eq. 18.1–20 is similar to the *heat-conduction equation* given in Eq. 10.1–26; this similarity is the basis for the analogous treatment of many heat-conduction and diffusion problems in solids. Keep in mind that many hundreds of problems described by Fick's second law have been solved; the solutions may be found in the monographs of Crank[3] and of Carslaw and Jaeger.[4]

§18.2 THE EQUATION OF CONTINUITY OF A IN CURVILINEAR COORDINATES

In this section the most important equations given in §18.1 are summarized in rectangular, cylindrical, and spherical coordinates. They are tabulated for ready reference in setting up problems. In Table 18.2–1 we give the equation of continuity in terms of N_A, and in Table 18.2–2, the diffusion

[1] In the absence of chemical reactions, Eq. 18.1–19 can be written in terms of $\boldsymbol{v}$ rather than $\boldsymbol{v}^\star$ by using a different type of concentration, namely the logarithm of the mean molecular weight:

$$\frac{\partial}{\partial t}\ln M + (\boldsymbol{v} \cdot \nabla \ln M) = \mathscr{D}_{AB}\nabla^2 \ln M \qquad (18.1\text{–}19a)$$

in which $M = x_A M_A + x_B M_B$. This relation seems to have been first suggested by C. H. Bedingfield, Jr., and T. B. Drew, *Ind. Eng. Chem.*, **42,** 1164 (1950).

[2] By equimolar counter-diffusion, we mean that the total molar flux with respect to stationary coordinates is zero.

[3] J. Crank, *The Mathematics of Diffusion*, Oxford University Press (1956).

[4] H. S. Carslaw and J. C. Jaeger, *Heat Conduction in Solids*, Oxford University Press (1959), Second Edition.

TABLE 18.2–1

THE EQUATION OF CONTINUITY OF A IN VARIOUS COORDINATE SYSTEMS (Eq. 18.1–10)

Rectangular coordinates:

$$\frac{\partial c_A}{\partial t} + \left(\frac{\partial N_{Ax}}{\partial x} + \frac{\partial N_{Ay}}{\partial y} + \frac{\partial N_{Az}}{\partial z}\right) = R_A \tag{A}$$

Cylindrical coordinates:

$$\frac{\partial c_A}{\partial t} + \left(\frac{1}{r}\frac{\partial}{\partial r}(rN_{Ar}) + \frac{1}{r}\frac{\partial N_{A\theta}}{\partial \theta} + \frac{\partial N_{Az}}{\partial z}\right) = R_A \tag{B}$$

Spherical coordinates:

$$\frac{\partial c_A}{\partial t} + \left(\frac{1}{r^2}\frac{\partial}{\partial r}(r^2N_{Ar}) + \frac{1}{r\sin\theta}\frac{\partial}{\partial \theta}(N_{A\theta}\sin\theta) + \frac{1}{r\sin\theta}\frac{\partial N_{A\phi}}{\partial \phi}\right) = R_A \tag{C}$$

TABLE 18.2–2

THE EQUATION OF CONTINUITY OF A FOR CONSTANT ρ AND $\mathscr{D}_{AB}$ (Eq. 18.1–17)

Rectangular coordinates:

$$\frac{\partial c_A}{\partial t} + \left(v_x\frac{\partial c_A}{\partial x} + v_y\frac{\partial c_A}{\partial y} + v_z\frac{\partial c_A}{\partial z}\right) = \mathscr{D}_{AB}\left(\frac{\partial^2 c_A}{\partial x^2} + \frac{\partial^2 c_A}{\partial y^2} + \frac{\partial^2 c_A}{\partial z^2}\right) + R_A \tag{A}$$

Cylindrical coordinates:

$$\frac{\partial c_A}{\partial t} + \left(v_r\frac{\partial c_A}{\partial r} + v_\theta\frac{1}{r}\frac{\partial c_A}{\partial \theta} + v_z\frac{\partial c_A}{\partial z}\right)$$
$$= \mathscr{D}_{AB}\left(\frac{1}{r}\frac{\partial}{\partial r}\left(r\frac{\partial c_A}{\partial r}\right) + \frac{1}{r^2}\frac{\partial^2 c_A}{\partial \theta^2} + \frac{\partial^2 c_A}{\partial z^2}\right) + R_A \tag{B}$$

Spherical coordinates:

$$\frac{\partial c_A}{\partial t} + \left(v_r\frac{\partial c_A}{\partial r} + v_\theta\frac{1}{r}\frac{\partial c_A}{\partial \theta} + v_\phi\frac{1}{r\sin\theta}\frac{\partial c_A}{\partial \phi}\right)$$
$$= \mathscr{D}_{AB}\left(\frac{1}{r^2}\frac{\partial}{\partial r}\left(r^2\frac{\partial c_A}{\partial r}\right) + \frac{1}{r^2\sin\theta}\frac{\partial}{\partial \theta}\left(\sin\theta\frac{\partial c_A}{\partial \theta}\right) + \frac{1}{r^2\sin^2\theta}\frac{\partial^2 c_A}{\partial \phi^2}\right) + R_A \tag{C}$$

equation in the form of Eq. 18.1–17. Other equations of §18.1 may be written down by analogy. The notation for curvilinear coordinate systems is given in Fig. A.6–1. Note that the diffusion equations for solids can be obtained by setting the velocity components in Table 18.2–2 equal to zero.

§18.3 THE MULTICOMPONENT EQUATIONS OF CHANGE IN TERMS OF THE FLUXES

In Chapters 3 and 10 the equations of change were given for a pure fluid. Here we extend these discussions and give the equations of change for a nonisothermal multicomponent fluid of n chemical species.

(i) An *equation of continuity* for each chemical species present in the fluid:

$$\frac{D}{Dt}\rho_i = -\rho_i(\nabla \cdot v) - (\nabla \cdot j_i) + r_i \qquad i = 1, 2, \cdots, n \qquad (18.3\text{–}1)$$

Addition of all n equations of this kind gives the equation of continuity for the mixture, described in Eq. 18.1–8. Any one of the n equations above may be replaced by the equation of continuity for the mixture in any given problem.

(ii) The *equation of motion* for the mixture:

$$\rho\frac{Dv}{Dt} = -[\nabla \cdot \pi] + \sum_{i=1}^{n} \rho_i g_i \qquad (18.3\text{–}2)$$

Here we have introduced for the sake of brevity the pressure tensor $\pi = \tau + p\delta$, in which τ is the viscous part of the momentum flux (or shear stress tensor), p is the static pressure, and δ is the unit tensor. Note that Eq. 18.3–2 differs from the equation of motion for a pure fluid (Eq. 3.2–10) only in the last term, where ρg has been replaced by $\Sigma\rho_i g_i$. In that term account is taken of the fact that each chemical species present may be acted on by a different external force per unit mass g_i.

(iii) The *equation of energy* for the mixture:

$$\rho\frac{D}{Dt}\{\hat{U} + \tfrac{1}{2}v^2\} = -(\nabla \cdot q) - (\nabla \cdot [\pi \cdot v]) + \sum_{i=1}^{n}(n_i \cdot g_i) \qquad (18.3\text{–}3)$$

Note that this equation differs in appearance from the energy equation for a pure fluid (Eq. 10.1–11) only in the last term, where $(\rho v \cdot g)$ has been replaced by $\Sigma_i(n_i \cdot g_i)$. Here q is the multicomponent energy flux relative to the mass average velocity v, defined in §18.4. Emission and absorption of radiant energy are neglected. (See §14.6.)

The complete description of mass, momentum, and energy transport in multicomponent systems is contained in these equations. One also needs the thermal equation of state $p = p(\rho, T, x_i)$, the caloric equation of state $\hat{U} = \hat{U}(\rho, T, x_i)$, and information about the chemical kinetics. In addition, one generally needs explicit expressions for the fluxes $\boldsymbol{j}_i$, $\boldsymbol{\pi}$, and $\boldsymbol{q}$ in terms of the gradients and the transport coefficients. (See §18.4.) The transport coefficients also need to be known as functions of ρ, T, and composition.

It has been pointed out in preceding discussions regarding the equations of change that these equations may assume a myriad of different forms, depending on (*a*) whether D/Dt or $\partial/\partial t$ is used, (*b*) which reference frame is selected for the fluxes, (*c*) whether mass or molar units are used, and (*d*) how the various forms of energy are broken up.

For example, we may rewrite Eqs. 18.3–1, 2, and 3 in terms of $\partial/\partial t$:

$$\text{(continuity)} \quad \frac{\partial}{\partial t}\rho_i = -(\nabla \cdot \{\rho_i \boldsymbol{v} + \boldsymbol{j}_i\}) + r_i \qquad i = 1, 2, \cdots, n \tag{18.3–4}$$

$$\text{(motion)} \quad \frac{\partial}{\partial t}\rho \boldsymbol{v} = -[\nabla \cdot \{\rho \boldsymbol{v}\boldsymbol{v} + \boldsymbol{\pi}\}] + \sum_{i=1}^{n} \rho_i \boldsymbol{g}_i \tag{18.3–5}$$

$$\text{(energy)} \quad \frac{\partial}{\partial t}\rho\{\hat{U} + \tfrac{1}{2}v^2\} = -(\nabla \cdot \{\rho(\hat{U} + \tfrac{1}{2}v^2)\boldsymbol{v} + \boldsymbol{q} + [\boldsymbol{\pi} \cdot \boldsymbol{v}]\}) + \sum_{i=1}^{n}(\boldsymbol{n}_i \cdot \boldsymbol{g}_i) \tag{18.3–6}$$

From Table 16.1–3 we know that the quantity $\{\rho_i \boldsymbol{v} + \boldsymbol{j}_i\}$ in Eq. 18.3–4 is just $\boldsymbol{n}_i$, the mass flux with respect to a fixed coordinate system. Similar fluxes may be introduced for momentum and energy transport. We therefore define the following mass, momentum, and energy *fluxes with respect to a coordinate system fixed in space:*

$$\boldsymbol{n}_i = \rho\omega_i \boldsymbol{v} + \boldsymbol{j}_i \qquad i = 1, 2, \cdots, n \tag{18.3–7}$$

$$\boldsymbol{\phi} = \rho \boldsymbol{v}\boldsymbol{v} + \boldsymbol{\pi} \tag{18.3–8}$$

$$\boldsymbol{e} = \rho\{\hat{U} + \tfrac{1}{2}v^2\}\boldsymbol{v} + \boldsymbol{q} + [\boldsymbol{\pi} \cdot \boldsymbol{v}] \tag{18.3–9}$$

Equations 18.3–4, 5, and 6 may be rewritten as

$$\text{(continuity)} \quad \frac{\partial}{\partial t}\rho_i = -(\nabla \cdot \boldsymbol{n}_i) + r_i \qquad i = 1, 2, \cdots, n \tag{18.3–10}$$

$$\text{(motion)} \quad \frac{\partial}{\partial t}\rho \boldsymbol{v} = -[\nabla \cdot \boldsymbol{\phi}] + \sum_{i=1}^{n} \rho_i \boldsymbol{g}_i \tag{18.3–11}$$

$$\text{(energy)} \quad \frac{\partial}{\partial t}\rho\{\hat{U} + \tfrac{1}{2}v^2\} = -(\nabla \cdot \boldsymbol{e}) + \sum_{i=1}^{n}(\boldsymbol{n}_i \cdot \boldsymbol{g}_i) \tag{18.3–12}$$

For one-dimensional steady-state systems without chemical reactions or external forces the fluxes $\boldsymbol{n}_i$, $\boldsymbol{\phi}$, and $\boldsymbol{e}$ are constant. The fluxes $\boldsymbol{n}_i$ and $\boldsymbol{e}$ are particularly convenient for setting up problems involving combined heat and mass transfer. (See Example 18.5–1.)

TABLE 18.3–1

THE EQUATION OF ENERGY FOR MULTICOMPONENT SYSTEMS[a]

In Terms of $\hat{E} = \hat{U} + \hat{K} + \hat{\Phi}$: (exact only for $\partial\hat{\Phi}/\partial t = 0$)

$$\rho \frac{D\hat{E}}{Dt} = -(\nabla \cdot \boldsymbol{q}) - (\nabla \cdot [\boldsymbol{\pi} \cdot \boldsymbol{v}]) + \sum_{i=1}^{n} (\boldsymbol{j}_i \cdot \boldsymbol{g}_i) \qquad (A)$$

In Terms of $\hat{U} + \hat{K} = \hat{U} + \frac{1}{2}v^2$:

$$\rho \frac{D}{Dt}(\hat{U} + \hat{K}) = -(\nabla \cdot \boldsymbol{q}) - (\nabla \cdot [\boldsymbol{\pi} \cdot \boldsymbol{v}]) + \sum_{i=1}^{n} (\boldsymbol{n}_i \cdot \boldsymbol{g}_i) \qquad (B)$$

In Terms of $\hat{K} = \frac{1}{2}v^2$:

$$\rho \frac{D\hat{K}}{Dt} = -(\boldsymbol{v} \cdot [\nabla \cdot \boldsymbol{\pi}]) + \sum_{i=1}^{n} \rho_i(\boldsymbol{v} \cdot \boldsymbol{g}_i) \qquad (C)$$

In Terms of $\hat{U}$:

$$\rho \frac{D\hat{U}}{Dt} = -(\nabla \cdot \boldsymbol{q}) - (\boldsymbol{\pi} : \nabla \boldsymbol{v}) + \sum_{i=1}^{n} (\boldsymbol{j}_i \cdot \boldsymbol{g}_i) \qquad (D)$$

In Terms of $\hat{H}$:

$$\rho \frac{D\hat{H}}{Dt} = -(\nabla \cdot \boldsymbol{q}) + \frac{Dp}{Dt} - (\boldsymbol{\tau} : \nabla \boldsymbol{v}) + \sum_{i=1}^{n} (\boldsymbol{j}_i \cdot \boldsymbol{g}_i) \qquad (E)$$

In Terms of $\hat{C}_p$:

$$\rho \hat{C}_p \frac{DT}{Dt} = -(\nabla \cdot \boldsymbol{q}) - (\boldsymbol{\tau} : \nabla \boldsymbol{v}) + \sum_{i=1}^{n} (\boldsymbol{j}_i \cdot \boldsymbol{g}_i) + \left(\frac{\partial \ln \hat{V}}{\partial \ln T}\right)_{p,x_i} \frac{Dp}{Dt} + \sum_{i=1}^{n} \bar{H}_i[(\nabla \cdot \boldsymbol{J}_i) - R_i] \qquad (F)^{\text{b}}$$

In Terms of $\hat{C}_v$:

$$\rho \hat{C}_v \frac{DT}{Dt} = -(\nabla \cdot \boldsymbol{q}) - (\boldsymbol{\pi} : \nabla \boldsymbol{v}) + \sum_{i=1}^{n} (\boldsymbol{j}_i \cdot \boldsymbol{g}_i) + \left(p - T\left(\frac{\partial p}{\partial T}\right)_{\rho,x_i}\right)(\nabla \cdot \boldsymbol{v}) + \sum_{i=1}^{n} \left[\bar{U}_i + \left(p - T\frac{\partial p}{\partial T}\right)\bar{V}_i\right][(\nabla \cdot \boldsymbol{J}_i) - R_i] \qquad (G)$$

In Terms of $\bar{H}_i$:

$$\frac{\partial}{\partial t}\left(\sum_{i=1}^{n} c_i \bar{H}_i\right) + \left(\nabla \cdot \sum_{i=1}^{n} \boldsymbol{N}_i \bar{H}_i\right) = (\nabla \cdot k \nabla T) + \frac{Dp}{Dt} - (\boldsymbol{\tau} : \nabla \boldsymbol{v}) + \sum_{i=1}^{n} (\boldsymbol{j}_i \cdot \boldsymbol{g}_i) \qquad (H)^{\text{c}}$$

[a] See §18.4 for the definition of the heat fluxes in these equations.

[b] L. B. Rothfeld, personal communication.

[c] The usually unimportant Dufour energy flux $\boldsymbol{q}^{(x)}$ has been neglected here.

We now conclude this discussion with a few remarks about special forms of the equations of motion and energy. In §10.3 it was pointed out that the *equation of motion* as it is usually written is in suitable form for setting up forced-convection problems but that an alternate form (Eq. 10.3–3) is desirable for describing the limiting case of pure free convection resulting from temperature inequalities in the system. In systems with concentration inequalities, as well as temperature inequalities, we write the equation of motion as in Eq. 10.3–2 and use an equation of state formed by making a double Taylor series of ρ in T and x_A (for a two-component system):

$$\begin{aligned}\rho &= \bar{\rho} + \left.\frac{\partial \rho}{\partial T}\right|_{\bar{T},\bar{x}_A} (T - \bar{T}) + \left.\frac{\partial \rho}{\partial x_A}\right|_{\bar{T},\bar{x}_A} (x_A - \bar{x}_A) + \cdots \\ &= \bar{\rho} - \bar{\rho}\bar{\beta}(T - \bar{T}) - \bar{\rho}\bar{\zeta}(x_A - \bar{x}_A) + \cdots \qquad (18.3\text{–}13)\end{aligned}$$

in which $\zeta = -(1/\rho)(\partial\rho/\partial x_A)_T$ is a quantity (defined analogously to β) that indicates how much the density varies with composition. The equation of motion then becomes, for gravity as the only external force,

$$\rho \frac{Dv}{Dt} = -[\nabla \cdot \tau] - \bar{\rho}\bar{\beta}g(T - \bar{T}) - \bar{\rho}\bar{\zeta}g(x_A - \bar{x}_A) \qquad (18.3\text{–}14)$$

This equation reduces to Eq. 10.3–3 when there are no concentration differences within the system. The last term in Eq. 18.3–14 gives the "buoyancy force" resulting from deviations from the mean concentration $\bar{x}_A$.

Next we consider the *equation of energy*. We recall from the discussion for pure fluids in §10.1 that the energy equation may be written in many different ways—in terms of $\hat{U}$, $\hat{H}$, or T. The same is true for mixtures. In Eq. 18.3–3 we have given the energy equation in terms of $\hat{U} + \frac{1}{2}v^2$. Several other forms of the energy equation are presented in Table 18.3–1. Note that it is not necessary to add a term S_c to describe thermal energy released by homogeneous chemical reactions. This information is included implicitly in $\hat{H}$ and $\hat{U}$, and appears explicitly as $-\Sigma_i \bar{H}_i R_i$ or $-\Sigma_i \bar{U}_i R_i$ in Eqs. (F) and (G). Remember that in calculating $\hat{H}$ or $\hat{U}$, the energies of formation and mixing of the various species must be included. (See Example 22.5–1.)

§18.4 THE MULTICOMPONENT FLUXES IN TERMS OF THE TRANSPORT PROPERTIES

In §18.3 the equations of change for a multicomponent mixture are given in terms of the fluxes of mass, momentum, and energy. In order to obtain expressions for the profiles, we have to be able to replace the fluxes by expressions that contain the transport properties and the gradients of the concentration, velocity, and temperature. This procedure has been used before: in Chapter 3 the equation of motion was rewritten by inserting the expression for momentum flux in terms of velocity gradients; in Chapter 10 the energy equation was rewritten by inserting the expression for energy flux in terms of the temperature gradient; and in §18.1 the equation of continuity was rewritten by inserting an expression for the mass flux in terms of the concentration gradient.

Actually, the discussions we have had so far regarding mass fluxes and concentration gradients have been somewhat oversimplified. Certainly the most important contribution to the mass flux is that resulting from the concentration gradient. It is known, however, that even in an isothermal system there are actually three "mechanical driving forces" that tend to produce the movement of a species with respect to the mean fluid motion: (*a*) the concentration gradient, (*b*) the pressure gradient, and (*c*) external forces acting unequally on the various chemical species. In §16.2 and §18.1 the second and third of these "mechanical driving forces" have been neglected in order to simplify the discussions there. In a multicomponent system, then, we have fluxes of momentum, energy, and mass, each resulting from an associated driving force as indicated by the main diagonal in Fig. 18.4–1. But the story is not quite that simple. According to the thermodynamics of irreversible processes, there will be a contribution to each flux owing to each driving force in the system. This "coupling" can occur, however, only between flux-force pairs that are tensors of equal order or which differ in order by two. Consequently, in a multicomponent system (*a*) the momentum flux depends only upon the velocity gradients, (*b*) the energy flux depends both on the temperature gradient (heat conduction) and on the mechanical driving forces (the "diffusion-thermo effect" or "Dufour effect"), and (*c*) the mass flux depends both on the mechanical driving forces (ordinary, pressure, and forced diffusion) and on the temperature gradient (the "thermal-diffusion effect" or "Soret effect"). Furthermore, the Onsager reciprocal relations of the thermodynamics of irreversible processes give information as to the interrelation of the two coupled effects, the Dufour and the Soret. In order to describe the Soret effect, an additional transport property (i.e., in addition to viscosity, thermal conductivity, and diffusivity) had to be introduced, namely the "thermal diffusion ratio" or the "Soret coefficient," depending upon the exact definition. Because of the interconnection of the Soret and Dufour effects, as described by the Onsager relations, this one additional transport property will take care of the quantitative description of both phenomena. (See the nondiagonal entries in Fig. 18.4–1.)

It is hoped that these introductory remarks will give the beginner a glimpse of the insight thai thermodynamic arguments can offer in connection with coupled phenomena. Also, perhaps, the reader will be somewhat more appreciative of the set of general flux relations that we are about to give for multicomponent systems. Those desirous of exploring the connection between thermodynamics and the transport processes will find several suitable references in the literature.[1]

The expressions for the *momentum flux* $\boldsymbol{\tau}$ in Chapter 3 are valid for mixtures as well as for pure substances. For Newtonian fluids $\boldsymbol{\tau}$ is a second order tensor given by

$$\boldsymbol{\tau} = -\mu(\nabla \boldsymbol{v} + (\nabla \boldsymbol{v})^{\dagger}) + (\tfrac{2}{3}\mu - \kappa)(\nabla \cdot \boldsymbol{v})\boldsymbol{\delta} \qquad (18.4\text{–}1)$$

in which $\nabla \boldsymbol{v}$ is a dyadic product, $(\nabla \boldsymbol{v})^{\dagger}$ is the transpose of the dyadic $\nabla \boldsymbol{v}$ and $\boldsymbol{\delta}$ is the unit tensor. The components of $\boldsymbol{\tau}$ for $\kappa = 0$ are given in Tables

Driving forces / Fluxes	Velocity gradients	Temperature gradient	Concentration gradient Pressure gradient External force differences
Momentum (second order tensor)	Newton's law $[\mu, \kappa]$		
Energy (vector)		Fourier's law $[k]$	Dufour effect $[\mathfrak{D}_A^T]$
Mass (vector)		Soret effect $[\mathfrak{D}_A^T]$	Fick's law $[\mathfrak{D}_{AB}]$

Fig. 18.4–1. Schematic diagram showing roughly the relations between fluxes and driving forces in a binary system. The associated transport coefficients are shown within brackets.

3.4–5, 6, and 7. Equation 18.4–1 shows how the momentum flux is related to the velocity gradients at any point in the system. The μ and κ in Eq. 18.4–1 are the instantaneous local viscosity and bulk viscosity of the fluid mixture. Expressions for $\boldsymbol{\tau}$ in non-Newtonian fluids were discussed in §3.6.

The expression for the *energy flux* $\boldsymbol{q}$ given in Eq. 8.1–6 is valid only for heat conduction in pure substances. For mixtures there are, in addition to the conductive flux, contributions resulting from the interdiffusion of the

[1] Two articles dealing specifically with the expressions for the mass, momentum, and energy fluxes are J. G. Kirkwood and B. L. Crawford, Jr., *J. Phys. Chem.*, **56,** 1048–1051 (1952) and R. B. Bird, C. F. Curtiss, and J. O. Hirschfelder, *Chem. Eng. Prog. Symposium Series*, No. 16, **51,** 69–85 (1955). More general references to irreversible thermodynamics are S. R. de Groot, *Thermodynamics of Irreversible Processes*, North Holland Publishing Co., Amsterdam (1951), pp. 94–123 for transport phenomena; I. Prigogine, *Thermodynamics of Irreversible Processes*, Thomas, Springfield, Illinois (1955); K. G. Denbigh, *The Thermodynamics of the Steady State*, Methuen, London (1951), pp. 78–86. All of these references are based upon the original development of the concepts of irreversible thermodynamics of L. Onsager, *Physical Review*, **37,** I, 405–426 (1931); *ibid.*, **38,** II, 2265–2279 (1931).

various species present and the Dufour or diffusion-thermo effect. We may then write for the total energy flux relative to the mass average velocity

$$\boldsymbol{q} = \boldsymbol{q}^{(c)} + \boldsymbol{q}^{(d)} + \boldsymbol{q}^{(x)} \tag{18.4–2}$$

Here $\boldsymbol{q}^{(c)} = -k\nabla T$ is the conductive energy flux, as defined in §8.1, and k is the instantaneous local thermal conductivity of the mixture. The energy flux $\boldsymbol{q}^{(d)}$ caused by interdiffusion is defined for a fluid containing n species by the expression

$$\boldsymbol{q}^{(d)} = \sum_{i=1}^{n} \frac{\bar{H}_i}{M_i} \boldsymbol{j}_i = \sum_{i=1}^{n} \bar{H}_i \boldsymbol{J}_i \tag{18.4–3}$$

Here $\bar{H}_i$ is the partial molal enthalpy of the ith species. The Dufour energy flux $\boldsymbol{q}^{(x)}$ is quite complex in nature and is usually of minor importance; hence it is not further discussed here.[2] The radiant energy flux $\boldsymbol{q}^{(r)}$ may be handled separately as described in §14.6.

Frequently it is desirable to use the energy flux with respect to stationary coordinates, $\boldsymbol{e}$, rather than $\boldsymbol{q}$. By using the definition of $\boldsymbol{e}$, Eq. 18.3–9, and the foregoing expression for $\boldsymbol{q}$, we may write

$$\boldsymbol{e} = \boldsymbol{q}^{(c)} + \boldsymbol{q}^{(d)} + \boldsymbol{q}^{(x)} + [\boldsymbol{\pi} \cdot \boldsymbol{v}] + \rho\{\hat{U} + \tfrac{1}{2}v^2\}\boldsymbol{v} \tag{18.4–4}$$

When $\boldsymbol{q}^{(x)}$, $[\boldsymbol{\tau} \cdot \boldsymbol{v}]$, and $(\frac{1}{2}\rho v^2)\boldsymbol{v}$ are of negligible importance, we may approximate $\boldsymbol{e}$ as

$$\begin{aligned} \boldsymbol{e} &= -k\nabla T + \sum_{i=1}^{n} \bar{H}_i \boldsymbol{J}_i + p\boldsymbol{v} + \rho\hat{U}\boldsymbol{v} \\ &= -k\nabla T + \sum_{i=1}^{n} \bar{H}_i \boldsymbol{J}_i + \rho\hat{H}\boldsymbol{v} \\ &= -k\nabla T + \sum_{i=1}^{n} \bar{H}_i \boldsymbol{J}_i + \sum_{i=1}^{n} c_i \bar{H}_i \boldsymbol{v} \end{aligned} \tag{18.4–5}$$

With the help of Table 16.1–3, we rewrite Eq. 18.4–5 to get

$$\boldsymbol{e} = -k\nabla T + \sum_{i=1}^{n} \boldsymbol{N}_i \bar{H}_i \tag{18.4–6}$$

This approximate expression is the usual starting point for engineering studies on heat transfer with simultaneous mass transfer.[3]

[2] The explicit form of the Dufour-effect term in multicomponent gas mixtures has been discussed by J. O. Hirschfelder, C. F. Curtiss, and R. B. Bird, *Molecular Theory of Gases and Liquids*, Wiley, New York (1954), pp. 522, 705. The separation of $\boldsymbol{q}$ into the component parts above is not entirely clean-cut because of a convention adopted in the definition of k (see R. B. Bird, C. F. Curtiss, and J. O. Hirschfelder, *Chem. Eng. Prog. Symposium Series*, No. 16, **51**, 69–85 (1955), p. 77, Eqs. 2.15 and 2.16).

[3] See, for example, T. K. Sherwood and R. L. Pigford, *Absorption and Extraction*, McGraw-Hill, New York (1952), Second Edition, p. 96, Eq. 128–1/2. See also Example 18.5–1 and §21.5.

The expression for the *mass flux* $\mathbf{j}_i$ in a multicomponent system will, in accordance with our brief preliminary discussion, consist of three contributions associated with the mechanical driving forces and an additional contribution associated with the thermal driving force

$$\mathbf{j}_i = \mathbf{j}_i^{(x)} + \mathbf{j}_i^{(p)} + \mathbf{j}_i^{(g)} + \mathbf{j}_i^{(T)} \tag{18.4-7}$$

Here we have written the mass flux as the sum of terms describing ordinary (concentration) diffusion $\mathbf{j}_i^{(x)}$, pressure diffusion $\mathbf{j}_i^{(p)}$, forced diffusion $\mathbf{j}_i^{(g)}$, and thermal diffusion $\mathbf{j}_i^{(T)}$. The formulas for these mass flux contributions are

$$\mathbf{j}_i^{(x)} = \frac{c^2}{\rho RT} \sum_{j=1}^{n} M_i M_j D_{ij} \left[x_j \sum_{\substack{k=1 \\ k \neq j}}^{n} \left(\frac{\partial \bar{G}_j}{\partial x_k}\right)_{\substack{T,p,x_s \\ s \neq j,k}} \nabla x_k \right] \tag{18.4-8}$$

$$\mathbf{j}_i^{(p)} = \frac{c^2}{\rho RT} \sum_{j=1}^{n} M_i M_j D_{ij} \left[x_j M_j \left(\frac{\bar{V}_j}{M_j} - \frac{1}{\rho}\right) \nabla p \right] \tag{18.4-9}$$

$$\mathbf{j}_i^{(g)} = -\frac{c^2}{\rho RT} \sum_{j=1}^{n} M_i M_j D_{ij} \left[x_j M_j \left(\mathbf{g}_j - \sum_{k=1}^{n} \frac{\rho_k}{\rho} \mathbf{g}_k\right) \right] \tag{18.4-10}$$

$$\mathbf{j}_i^{(T)} = -D_i^T \nabla \ln T \tag{18.4-11}$$

In these equations $\bar{G}_j$ and $\bar{V}_j$ are the partial molal free enthalpy (Gibbs free energy) and volume, respectively. The D_{ij} are multicomponent diffusion coefficients, and the D_i^T are multicomponent thermal diffusion coefficients. The D_{ij} and D_i^T have the following properties:

$$D_{ii} = 0; \qquad \sum_{i=1}^{n} D_i^T = 0 \tag{18.4-12}$$

$$\sum_{i=1}^{n} \{M_i M_h D_{ih} - M_i M_k D_{ik}\} = 0 \tag{18.4-13}$$

For $n > 2$ the quantities D_{ij} and D_{ji} are not in general equal.

The *ordinary diffusion* contribution to the mass flux is seen to depend in a complicated way on the concentration gradients of all the substances present. The *pressure diffusion* term indicates that there may be a net movement of the ith species in a mixture if there is a pressure gradient imposed on the system; the tendency for a mixture to separate under a pressure gradient is very small, but use is made of this effect in centrifuge separations in which tremendous pressure gradients may be established. The *forced diffusion* term is of primary importance in ionic systems, in which the external force on an ion is equal to the product of the ionic charge and the local electric field strength; each ionic species may thus be under the influence of a different force. If gravity is the only external force, then all the $\mathbf{g}_j$ are the same and $\mathbf{j}_i^{(g)}$ vanishes identically. The *thermal diffusion* term describes the tendency for species to diffuse under the influence of a temperature gradient; this effect is quite small, but devices can be arranged to produce very steep temperature gradients so that separations of mixtures are effected.[4]

To give the reader a better feeling for the physical aspects of these general relations, we discuss two important limiting cases.

Binary Systems

For a two-component system of A and B the quantities D_{AB} and D_{BA} are equal, and Eqs. 18.4–7 through 11 become[5]

$$j_A = -j_B = -\left(\frac{c^2}{\rho RT}\right) M_A{}^2 M_B D_{AB} x_A \cdot \left[\left(\frac{\partial}{\partial x_A}\frac{\bar{G}_A}{M_A}\right)_{T,p} \nabla x_A - \frac{\rho_B}{\rho}(g_A - g_B) + \left(\frac{\bar{V}_A}{M_A} - \frac{1}{\rho}\right)\nabla p\right] - D_A{}^T \nabla \ln T \quad (18.4\text{–}14)$$

This equation may be written in an alternate form by using $(d\bar{G}_A)_{T,p} = RTd \ln a_A$ and by defining[6] a "thermal diffusion ratio" $k_T = (\rho/c^2 M_A M_B) \times (D_A{}^T/D_{AB})$:

$$j_A = -j_B = -\left(\frac{c^2}{\rho}\right) M_A M_B D_{AB} \cdot \left[\left(\frac{\partial \ln a_A}{\partial \ln x_A}\right)_{T,p} \nabla x_A - \frac{M_A \rho_B x_A}{\rho RT}(g_A - g_B) + \frac{M_A x_A}{RT}\left(\frac{\bar{V}_A}{M_A} - \frac{1}{\rho}\right)\nabla p + k_T \nabla \ln T\right] \quad (18.4\text{–}15)$$

This equation is the starting point for the study of ordinary, forced, pressure,

[4] Separations by thermal diffusion may be considerably enhanced by combining the effect with free convection in a system similar to that shown in Fig. 9.9–1. A column that makes use of these two effects is called a *Clusius-Dickel column*. Such columns have been used for isotope separations and for separating complex mixtures of very similar organic compounds. A very readable introduction to thermal diffusion is the small book by K. E. Grew and T. L. Ibbs, *Thermal Diffusion in Gases*, Cambridge University Press (1952). For applications to liquids, see A. L. Jones and R. W. Foreman, *Ind. Eng. Chem.*, **44,** 2249–2253, (1952), and A. L. Jones and E. C. Milberger, *Ind. Eng. Chem.*, **45,** 2689–2696 (1953). The mathematical theory of the thermal diffusion column is described in a classic article by R. C. Jones and W. H. Furry, *Rev. Mod. Phys.*, **18,** 151–224 (1946).

[5] Note that the quantity in the square brackets has the dimensions of force per unit mass; hence let us designate it by $-\hat{F}_A$. When the relations in Table 16.1–3 are used, one can show that Eq. 18.4–14 (omitting the thermal diffusion term) becomes

$$v_A - v^\star = \frac{D_{AB} M_A \hat{F}_A}{RT} = m_{AB} \tilde{F}_A \quad (18.4\text{–}14a)$$

where $m_{AB} = D_{AB}/RT$ is the "mobility." This relation was used in Eq. 16.5–1 in connection with the hydrodynamic theory of diffusion.

[6] Other defined quantities for binary systems are the *thermal diffusion factor* α and the *Soret coefficient* σ:

$$k_T = \alpha x_A x_B = \sigma x_A x_B T \quad (18.4\text{–}15a)$$

The quantity α is almost independent of concentration for gases; the quantity σ is generally used for liquids. Note that $D_A{}^T = -D_B{}^T$. Because k_T is defined in terms of $D_A{}^T$, when k_T is positive, component A moves to the colder region; when k_T is negative, component A moves to the warmer region.

and thermal diffusion in binary nonideal mixtures. (See Examples 18.5–2, 3, and 4.) Some sample values of k_T for gases and liquids are given in Table 18.4–1.

TABLE 18.4–1
EXPERIMENTAL THERMAL DIFFUSION RATIOS FOR LIQUIDS AND LOW-DENSITY GASES

Liquids[a]				Gases[b]			
Components $A-B$	T(°K)	x_A	k_T	Components $A-B$	T(°K)	x_A	k_T
$C_2H_2Cl_4$—n-C_6H_{14}	298	0.5	1.08	Ne—He	330	0.20	0.0531
$C_2H_4Br_2$—$C_2H_4Cl_2$	298	0.5	0.225			0.60	0.1004
$C_2H_2Cl_4$—CCl_4	298	0.5	0.060	N_2—H_2	264	0.294	0.0548
CBr_4—CCl_4	298	0.09	0.129			0.775	0.0663
CCl_4—CH_3OH	313	0.5	1.23	D_2—H_2	327	0.10	0.0145
CH_3OH—H_2O	313	0.5	−0.137			0.50	0.0432
cyclo-C_6H_{12}—C_6H_6	313	0.5	0.100			0.90	0.0166

[a] Abstracted from R. L. Saxton, E. L. Dougherty, and H. G. Drickamer, *J. Chem. Phys.*, **22,** 1166–1168 (1954); R. L. Saxton and H. G. Drickamer, *ibid.*, 1287–1288; L. J. Tichacek, W. S. Kmak, and H. G. Drickamer, *J. Phys. Chem.*, **60,** 660–665 (1956).

[b] Abstracted from tables given by J. O. Hirschfelder, C. F. Curtiss, and R. B. Bird, *Molecular Theory of Gases and Liquids*, Wiley, New York (1954), §8.4.

In considering ordinary diffusion only, we see that Eq. 18.4–15 simplifies to

$$j_A = -\left(\frac{c^2}{\rho}\right) M_A M_B D_{AB} \left(\frac{\partial \ln a_A}{\partial \ln x_A}\right)_{T,p} \nabla x_A \tag{18.4–16}$$

This should be compared with Eq. *E* in Table 16.2–1:

$$j_A = -\left(\frac{c^2}{\rho}\right) M_A M_B \mathscr{D}_{AB} \nabla x_A \tag{18.4–17}$$

Comparison of these two equations shows that D_{AB} and $\mathscr{D}_{AB}$ are identical for ideal solutions (i.e., activity proportional to mole fraction). In nonideal systems one may use either D_{AB} or $\mathscr{D}_{AB}$ and the corresponding equation above. Generally, experimentalists report diffusion coefficients as $\mathscr{D}_{AB}$, since this requires no activity-concentration data. Available data indicate, however, that D_{AB} is less concentration dependent than $\mathscr{D}_{AB}$ in the liquid phase. (See Fig. 18.4–2.)

Ordinary Diffusion in Multicomponent Gases at Low Density

For an ideal gas mixture, Eq. 18.4–8 becomes

$$j_i = \frac{c^2}{\rho} \sum_{j=1}^{n} M_i M_j D_{ij} \nabla x_j \qquad i = 1, 2, \cdots, n \tag{18.4–18}$$

For an n-component ideal-gas mixture the relation is known between the D_{ij} (the diffusivity of the pair i-j in a multicomponent mixture) and the $\mathscr{D}_{ij}$ (the diffusivity of the pair i-j in a binary mixture).[7] Because the D_{ij} are concentration dependent, Eq. 18.4–18 is inconvenient to use. It has been shown by Curtiss and Hirschfelder that Eqs. 18.4–18 may be "turned wrong-side out" to obtain

$$\nabla x_i = \sum_{j=1}^{n} \frac{c_i c_j}{c^2 \mathscr{D}_{ij}} (\boldsymbol{v}_j - \boldsymbol{v}_i) = \sum_{j=1}^{n} \frac{1}{c \mathscr{D}_{ij}} (x_i \boldsymbol{N}_j - x_j \boldsymbol{N}_i) \qquad (18.4\text{–}19)$$

These equations are known as the *Stefan-Maxwell equations*.[8] Note that it

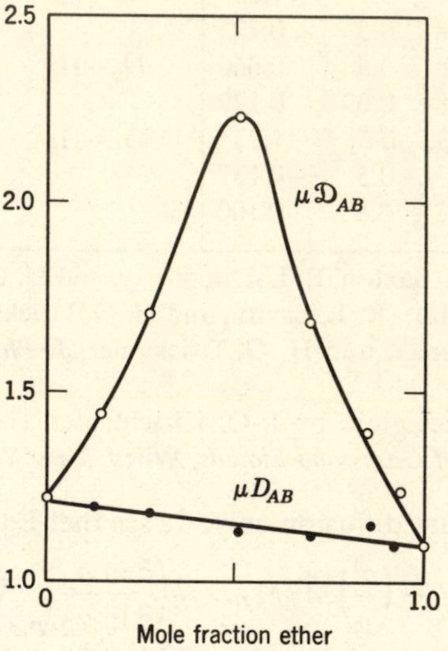

Fig. 18.4–2. Effect of activity on the product of viscosity and diffusivity for liquid mixtures of chloroform and ether. [From R. E. Powell, W. L. Roseveare, and Henry Eyring, *Ind. Eng. Chem.* **33**, 430–435 (1941).]

is the $\mathscr{D}_{ij}$ that appear here rather than the D_{ij} and that the $\mathscr{D}_{ij}$ are virtually independent of composition. (See Eq. 16.4–13.) This is the usual starting

[7] C. F. Curtiss and J. O. Hirschfelder, *J. Chem. Phys.*, **17**, 550–555 (1949). For a three-component system, the relations are of the form

$$D_{12} = \mathscr{D}_{12}\left\{1 + \frac{x_3[(M_3/M_2)\mathscr{D}_{13} - \mathscr{D}_{12}]}{x_1\mathscr{D}_{23} + x_2\mathscr{D}_{13} + x_3\mathscr{D}_{12}}\right\} \qquad (18.4\text{–}18a)$$

with similar relations for D_{21}, D_{23}, D_{32}, D_{13}, and D_{31}.

[8] The analogous set of equations, including pressure, thermal, and forced diffusion, has also been derived by Curtiss and Hirschfelder (*loc. cit.*); see also J. O. Hirschfelder, C. F. Curtiss, and R. B. Bird, *Molecular Theory of Gases and Liquids*, Wiley, New York (1954), Eq. 11.2–54, p. 718.

point for the calculation of ordinary diffusion in multicomponent gas mixtures. (See Example 18.5–5.)

For some calculations and for use in some correlations, it is convenient to define[9] an *effective binary diffusivity* $\mathscr{D}_{im}$ for the diffusion of i in a mixture. Recall that $\mathscr{D}_{AB}$ was defined by

$$N_A = -c\mathscr{D}_{AB}\nabla x_A + x_A(N_A + N_B) \tag{18.4–20}$$

and *define* $\mathscr{D}_{im}$ by this analogous relation:

$$N_i = -c\mathscr{D}_{im}\nabla x_i + x_i \sum_{j=1}^{n} N_j \tag{18.4–21}$$

By solving Eq. 18.4–21 for ∇x_i and equating the result to ∇x_i in the Stefan-Maxwell equations, we get immediately for collinear ∇x_i

$$\frac{1}{c\mathscr{D}_{im}} = \frac{\sum_{j=1}^{n} (1/c\mathscr{D}_{ij})(x_j N_i - x_i N_j)}{N_i - x_i \sum_{j=1}^{n} N_j} \tag{18.4–22}$$

In general the $\mathscr{D}_{im}$ are dependent on position. For situations in which this dependence is slight, we may generalize the binary diffusion formulas and mass-transfer coefficient correlations by simply replacing $\mathscr{D}_{AB}$ by $\mathscr{D}_{im}$. For some special kinds of diffusing systems, this formula (Eq. 18.4–22) for $\mathscr{D}_{im}$ becomes particularly simple:

a. For trace components 2, 3, . . . n in nearly pure species 1,

$$\mathscr{D}_{im} = \mathscr{D}_{i1} \tag{18.4–23}$$

b. For systems in which all the $\mathscr{D}_{ij}$ are the same

$$\mathscr{D}_{im} = \mathscr{D}_{ij} \tag{18.4–24}$$

c. For systems in which 2, 3, . . . n move with the same velocity (or are stationary),

$$\frac{1 - x_1}{\mathscr{D}_{1m}} = \sum_{j=2}^{n} \frac{x_j}{\mathscr{D}_{1j}} \tag{18.4–25}$$

In systems in which the variation of $\mathscr{D}_{im}$ is considerable, the assumption of linear variation with composition or distance has proven useful.[10] The

[9] The systematic use of an effective binary diffusivity seems first to have been suggested by O. A. Hougen and K. M. Watson, *Chemical Process Principles*, Vol. III, Wiley, New York (1947), pp. 977–979. Methods of evaluating $\mathscr{D}_{im}$ for special cases have been developed by C. R. Wilke, *Chem. Eng. Prog.*, **46**, 95–104 (1950) and W. E. Stewart, *NACA Tech. Note 3208* (1954).

[10] H. W. Hsu and R. B. Bird, *A.I.Ch.E. Journal* (in press).

$\mathscr{D}_{im}$ approach to solving multicomponent problems seems to give pretty good results for calculating mass-transfer rates but a less satisfactory quantitative description of concentration profiles.

Substitution of the expressions for the fluxes given in this section into the equations of change of §18.3 produces the general partial differential equations describing the flow of a multicomponent fluid mixture with heat transfer, mass transfer, and chemical reactions occurring. The word "general," of course, always has to be used with some caution, for one can frequently think up "more general" situations. In this case, the field of *magnetohydrodynamics* comes to mind; the equations describing multicomponent fluid mixtures with electromagnetic effects are the equations of change and Maxwell's equations of electromagnetic theory. This field is of interest in connection with astrophysical phenomena, ionized gas behavior, and plasma jets.[11,12,13] Another field not covered by our equations is that of *relativistic fluid mechanics*; this subject includes the relativistic effects that are important when the fluid velocity is near the velocity of light.[14]

§18.5 USE OF EQUATIONS OF CHANGE TO SET UP DIFFUSION PROBLEMS

All of the problems of Chapter 17, and more difficult ones as well, may be set up directly by means of the differential equations in this chapter. As examples we consider combined heat and mass transfer, thermal, pressure, and forced diffusion, and three-component ordinary diffusion.

Example 18.5–1. Simultaneous Heat and Mass Transfer

Develop expressions for the concentration profile $x_A(z)$ and temperature profile $T(z)$ for the system pictured in Fig. 18.5–1, given the concentrations and temperatures at both film boundaries ($z = 0, z = \delta$). Here a hot condensable vapor, A, is diffusing at steady state through a stagnant film of noncondensable gas, B, to a cold surface at $z = 0$ where A condenses. Assume that the gas behavior is ideal and that the pressure and the physical properties of the mixture are constant.[1] Neglect radiative heat transfer and thermal diffusion.

Solution. To determine the desired quantities, we must solve the equations of

[11] See T. G. Cowling, *Magnetohydrodynamics*, Interscience, New York (1957).

[12] L. Spitzer, Jr., *Physics of Fully Ionized Gases*, Interscience, New York (1956).

[13] B. T. Chu, *Physics of Fluids*, **2**, 473–484 (1959); in this paper the derivations of the equations of change are given for a pure, electrically conducting fluid. In the energy equation there are terms which account for the temperature rise resulting from both viscous dissipation and electrical dissipation (S_v and S_e in Chapter 9).

[14] L. D. Landau and E. M. Lifshitz, *Myekhanika Sploshnikh Sred*, Moscow (1954), Chapter XV, pp. 606–616.

[1] The simple system described here is often used as a model in psychrometric calculations. More general models are discussed in §§21.5, 6, 7.

continuity and energy for this system. These may be written from Eqs. 18.3–10 and 12 as

$$\text{(continuity of } A\text{)} \qquad \frac{dN_{Az}}{dz} = 0 \tag{18.5-1}$$

$$\text{(energy)} \qquad \frac{de_z}{dz} = 0 \tag{18.5-2}$$

Therefore both N_{Az} and e_z are constant through the film.

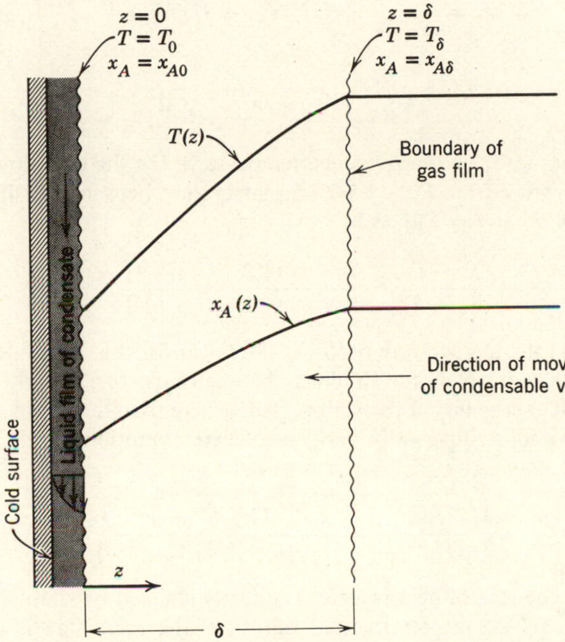

Fig. 18.5-1. Condensation of hot vapor A on a cold surface in the presence of a noncondensable gas B.

To determine the *concentration profile*, we need the mass flux for diffusion of A through stagnant B:

$$N_{Az} = -\frac{c\mathscr{D}_{AB}}{1 - x_A}\frac{dx_A}{dz} \tag{18.5-3}$$

Insertion of Eq. 18.5–3 into Eq. 18.5–1 and integration gives the concentration profiles

$$\left(\frac{1 - x_A}{1 - x_{A0}}\right) = \left(\frac{1 - x_{A\delta}}{1 - x_{A0}}\right)^{z/\delta} \tag{18.5-4}$$

This result was obtained in §17.2 for the isothermal problem. As before, the constant flux N_{Az} is

$$N_{Az} = \frac{c\mathscr{D}_{AB}}{\delta}\ln\frac{1 - x_{A\delta}}{1 - x_{A0}} \tag{18.5-5}$$

Note that N_{Az} is negative in Fig. 18.5–1 because A is condensing. The last two expressions may be combined to put the concentration profiles in an alternate form:

$$\frac{x_A - x_{A0}}{x_{A\delta} - x_{A0}} = \frac{1 - \exp\left[(N_{Az}/c\mathscr{D}_{AB})z\right]}{1 - \exp\left[(N_{Az}/c\mathscr{D}_{AB})\delta\right]} \tag{18.5–6}$$

To determine the *temperature profile*, we use the energy flux from Eq. 18.4–6 for an ideal gas:

$$e_z = -k\frac{dT}{dz} + (\tilde{H}_A N_{Az} + \tilde{H}_B N_{Bz})$$

$$= -k\frac{dT}{dz} + N_{Az}\tilde{C}_{pA}(T - T_0) \tag{18.5–7}$$

Here we have chosen T_0 as the reference temperature for the enthalpy. Insertion of this expression for e_z into Eq. 18.5–2 and integration between the limits $T = T_0$ at $z = 0$ and $T = T_\delta$ at $z = \delta$ gives

$$\frac{T - T_0}{T_\delta - T_0} = \frac{1 - \exp\left[(N_{Az}\tilde{C}_{pA}/k)z\right]}{1 - \exp\left[(N_{Az}\tilde{C}_{pA}/k)\delta\right]} \tag{18.5–8}$$

It can be seen that the temperature profile is not linear for this system except in the limit as $N_{Az}\tilde{C}_{pA}/k \rightarrow 0$. Note the similarity between Eqs. 18.5–6 and 8.

The conduction energy flux at the wall is greater here than in the absence of mass transfer. Thus, using a superscript zero to indicate conditions in the absence of mass transfer, we may write

$$\frac{-k(dT/dz)_{z=0}}{-k(dT/dz)^0_{z=0}} = \frac{-(N_{Az}\tilde{C}_{pA}/k)\delta}{1 - \exp\left[(N_{Az}\tilde{C}_{pA}/k)\delta\right]} \tag{18.5–9}$$

We see then that the rate of heat transfer is directly affected by simultaneous mass transfer, whereas (if we neglect thermal diffusion) the mass flux is not directly affected by simultaneous heat transfer. In many applications, for example in most psychrometric problems, $N_{Az}\tilde{C}_{pA}/k$ is small, and the right side of Eq. 18.5–9 is very nearly unity. (See Problem 18.A.) The relationships between heat and mass transfer are further discussed in Chapter 21.

Example 18.5–2. Thermal Diffusion

Consider two bulbs joined by an insulated tube of small diameter and filled with a mixture of ideal gases A and B as shown in Fig. 18.5–2. The bulbs are maintained at constant temperatures of T_1 and T_2, respectively, and the diameter of the insulated tube is small enough to eliminate convection currents substantially. Develop an expression for the steady-state difference in composition of the two bulbs.

Solution. According to Eq. 18.4–15, the temperature gradient in the system will cause a mass flux given by

$$j_{Az}^{(T)} = -\frac{c^2}{\rho} M_A M_B \mathscr{D}_{AB} \frac{k_T}{T}\frac{dT}{dz} \tag{18.5–10}$$

This mass flux will tend to establish a concentration gradient, which in turn will result in an opposed flux:

$$j_{Az}^{(x)} = -\frac{c^2}{\rho} M_A M_B \mathscr{D}_{AB} \frac{dx_A}{dz} \tag{18.5-11}$$

When steady state is reached, there will be no net mass flux and we may write

$$0 = j_{Az}^{(x)} + j_{Az}^{(T)} = -\frac{c^2}{\rho} M_A M_B \mathscr{D}_{AB} \left(\frac{dx_A}{dz} + \frac{k_T}{T} \frac{dT}{dz} \right) \tag{18.5-12}$$

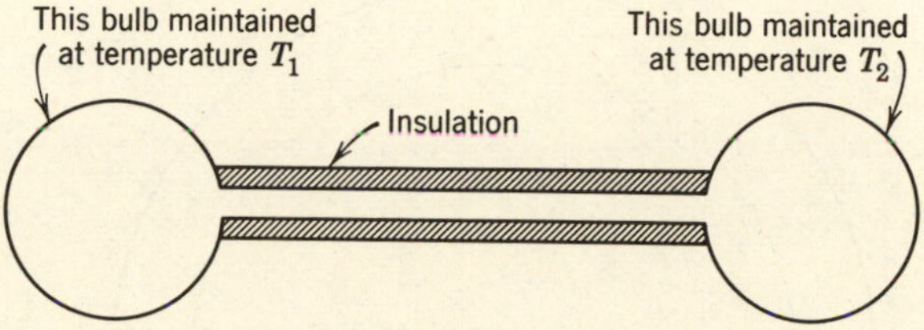

Fig. 18.5-2. Steady-state binary thermal diffusion in a two-bulb apparatus. The mixture of gases A and B tends to separate under the influence of the thermal gradient.

When k_T is positive, component A moves to the cold region; when it is negative, A moves to the hot region. We may write Eq. 18.5-12 as

$$\frac{dx_A}{dz} = -\frac{k_T}{T} \frac{dT}{dz} \tag{18.5-13}$$

In general, the degree of separation in an apparatus of the kind being considered is small. We may therefore ignore the effect of composition on k_T and integrate Eq. 18.5-13 to obtain

$$x_{A2} - x_{A1} = -\int_{T_1}^{T_2} \frac{k_T}{T} \, dT \tag{18.5-14}$$

Because the dependence of k_T on T is rather complicated, it is customary to assume k_T constant at some mean temperature T_m. Integration of Eq. 18.5-14 then gives approximately

$$x_{A2} - x_{A1} = -k_T \ln \frac{T_2}{T_1} \tag{18.5-15}$$

The recommended[2] mean temperature T_m is

$$T_m = \frac{T_1 T_2}{T_2 - T_1} \ln \frac{T_2}{T_1} \tag{18.5-16}$$

Equation 18.5-15 is useful for estimating the order of magnitude of thermal diffusion effects.

Example 18.5-3. Pressure Diffusion

A binary liquid solution is mounted in a cylindrical cell in a very high-speed centrifuge, as shown in Fig. 18.5-3. The length of the cell L may be considered

[2] H. Brown, *Phys. Rev.*, **58**, 661–662 (1940).

short with respect to the radius of rotation R_0, and the solution density may be considered a function of composition only. Determine the distribution of the two components at steady state in terms of their partial molal volumes, position in the cell, and the pressure gradient $dp/dz = -\rho g_\Omega \doteq -\rho\Omega^2 R_0$ in which Ω is the angular velocity of the centrifuge. Neglect changes in the partial molal volumes with composition and assume that the activity coefficients are constant over the range of compositions existing in the cell.

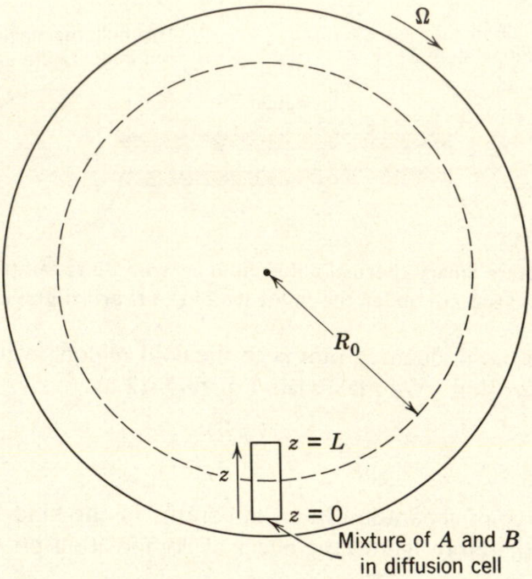

Fig. 18.5-3. Steady-state pressure diffusion in a centrifuge. The mixture of A and B tends to separate by virtue of the pressure gradient produced in the centrifuge.

Solution. At steady state the net mass flux j_{Az} is zero, and we may write Eq. 18.4–15 as

$$\frac{d}{dz}\ln x_A = -\left(\frac{g_\Omega}{RT}\right)(M_A - \rho\bar{V}_A) \tag{18.5–17}$$

This expression,[3] which may also be obtained by thermodynamic arguments, can be integrated to give

$$\left(\frac{x_A}{x_{A0}}\right)^{\bar{V}_B}\left(\frac{x_{B0}}{x_B}\right)^{\bar{V}_A} = \exp\left((\bar{V}_A M_B - \bar{V}_B M_A)\frac{g_\Omega z}{RT}\right) \tag{18.5–18}$$

Eq. 18.5–18 describes the steady-state concentration distribution for a binary system in a constant centrifugal field.

[3] See Problem 18.I for the details of this integration. A development of Eq. 18.5–17 by thermodynamic arguments is given by E. A. Guggenheim, *Thermodynamics*, North Holland Publishing Co., Amsterdam (1950), pp. 356–360. Note, however, that Guggenheim's final result is incorrect because of an algebraic error in the third line of Eq. 11.10–3.

Example 18.5–4. Forced Diffusion

A quiescent aqueous solution of a salt MX is situated between two flat parallel plates of metallic M, as shown in Fig. 18.5–4. A constant direct current is passed between the plates under such conditions that the only electrode reactions are dissolution of the anode and deposition of M on the cathode. Estimate the concentration profile of MX in the solution and the maximum possible current density.

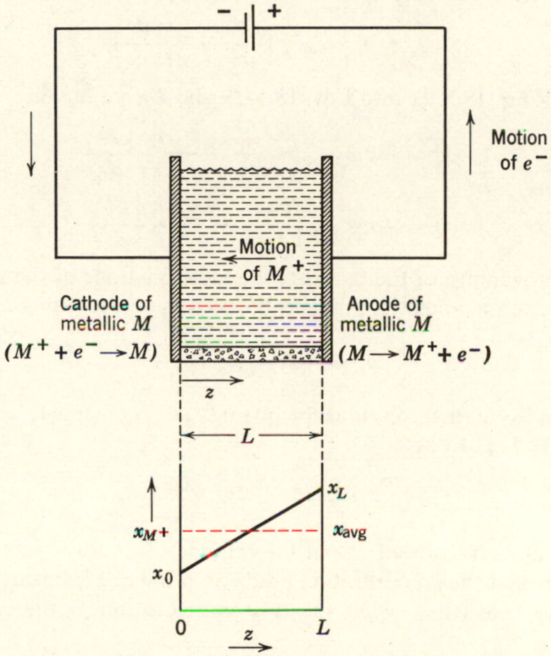

Fig. 18.5–4. Concentration polarization.

Assume the solutions to be quite dilute and ignore changes of temperature and physical properties.

Solution. We may consider the solution in this apparatus as a ternary mixture of water and the two ionic species M^+ and X^-. It can be seen from the electrode reactions that N_{M^+}, the magnitude of the molar flux of M^+, is proportional to the current density in the cell and that the fluxes of X^- and water are zero. It can be further seen that the local molar concentrations of M^+ and X^- must always be equal by the requirement of electrical neutrality. Finally, since the solution is dilute, we may consider that the fluxes N_i and $J_i^\star$ are equal.

We may immediately write

$$I = N_{M^+} \doteq J_{M^+}^{\star(x)} + J_{M^+}^{\star(g)} \tag{18.5–19}$$

$$0 = N_{X^-} \doteq J_{X^-}^{\star(x)} + J_{X^-}^{\star(g)} \tag{18.5–20}$$

Here I is the current density in the solution, expressed as equivalents per unit area per unit time. To express Eqs. 18.5–19 and 20 in terms of the transport coefficients, we must make some further assumptions. We assume that the M^+ and X^- ions are in such small concentration as to have no appreciable effect on one another. Then we consider each ion as diffusing in a binary system with water as the second component. We further assume the activity coefficients of the ions to be unity. Then $d\bar{G}_i = RT\,d\ln x_i$ and for the ions $g_{iz} = (\epsilon_i/m_i)(-d\phi/dz)$. Here ϕ is the local electrostatic potential, ϵ_i is the ionic charge, and m_i is the ionic mass. We may then write Eq. 18.4–15 for ionic species i diffusing through water (w) as

$$J_{iz}^{\star} = -c\mathscr{D}_{iw}\left(\frac{dx_i}{dz} + \frac{x_i\epsilon_i}{\kappa T}\frac{d\phi}{dz}\right) \tag{18.5–21}$$

By substituting Eq. 18.5–21 into Eqs. 18.5–19 and 20, we obtain

$$I_z = -c\mathscr{D}_{M^+w}\left(\frac{dx_{M^+}}{dz} + \frac{x_{M^+}|\epsilon_i|}{\kappa T}\frac{d\phi}{dz}\right) \tag{18.5–22}$$

$$0 = -c\mathscr{D}_{X^-w}\left(\frac{dx_{X^-}}{dz} - \frac{x_{X^-}|\epsilon_i|}{\kappa T}\frac{d\phi}{dz}\right) \tag{18.5–23}$$

We now take advantage of the fact that the mole fractions of the two ions are the same to eliminate the potential gradient between these two equations:

$$I_z = -2c\mathscr{D}_{M^+w}\frac{dx_{M^+}}{dz} \tag{18.5–24}$$

Since for dilute isothermal solutions the quantity $c\mathscr{D}_{M^+w}$ is nearly constant, we may integrate Eq. 18.5–24 to obtain

$$\frac{x_{M^+} - x_0}{z} = -\frac{I_z}{2c\mathscr{D}_{M^+w}} \tag{18.5–25}$$

Here x_0 is the mole fraction of M^+ at the cathode.

We find then that the concentration gradient in the cell is linear. The maximum current density is reached when the salt concentration at the cathode is zero; that is,

$$I_{\max} = \frac{4c\mathscr{D}_{M^+w}x_{\text{avg}}}{L} \tag{18.5–26}$$

Equations 18.5–25 and 26 provide a good qualitative description of "concentration polarization," even though their quantitative application is limited to quite dilute systems. For example, Eq. 18.5–26 shows that in an electroplating bath there is a limit to the rate at which metal may be deposited. Similarly, in a battery the diffusion of electrolyte limits the rate at which current can be drawn.

Example 18.5–5. Three-Component Ordinary Diffusion with Heterogeneous Chemical Reaction[4]

A gas A is diffusing at steady state in the z-direction through a film of stagnant B to a catalytic surface (see Fig. 18.5–5), where an irreversible reaction of the following form takes place:

$$nA \rightarrow A_n \tag{18.5–27}$$

[4] H. W. Hsu and R. B. Bird, *A.I.Ch.E. Journal* **6**, 516–524 (1960).

At the catalyst surface A is removed at a rate that may be expressed as

$$N_A = k_h'' x_A^h \qquad \text{at } z = \delta \tag{18.5-28}$$

Here k_h'' and h are constants assumed to be known. Develop an expression for the reaction rate in terms of the gas composition at $z = 0$. Assume constant temperature and pressure and ideal-gas behavior.

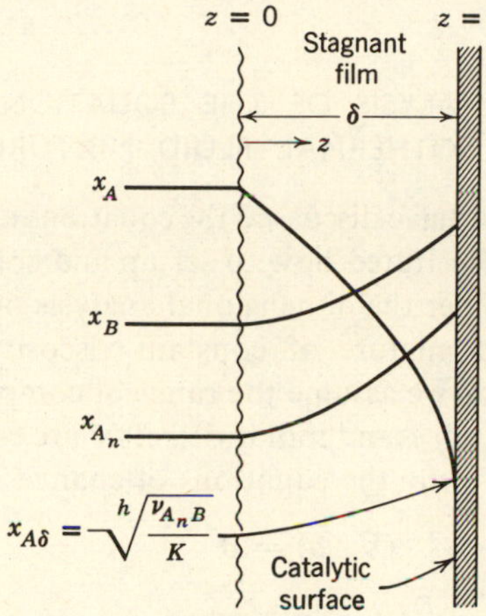

Fig. 18.5-5. Diffusion of A through stagnant B to form A_n at a catalytic surface.

Solution. From the stoichiometry of the problem, it is required that $N_{Az} = -nN_{A_nz}$. The Stefan-Maxwell equations for A and B are

$$\frac{1}{\nu_{A_nB}} \frac{dx_A}{d\zeta} = r_{A_n}\left(1 - \frac{1}{n}\right)x_A + (r_{A_n} - r_B)x_B - r_{A_n} \tag{18.5-29}$$

$$\frac{1}{\nu_{A_nB}} \frac{dx_B}{d\zeta} = \left(r_B - \frac{1}{n}\right)x_B \tag{18.5-30}$$

Here $\nu_{A_nB} = N_{Az}\delta/c\mathscr{D}_{A_nB}$, $r_B = \mathscr{D}_{A_nB}/\mathscr{D}_{AB}$, $r_{A_n} = \mathscr{D}_{A_nB}/\mathscr{D}_{AA_n}$, and $\zeta = z/\delta$. These equations may be integrated with the boundary conditions that $x_A = x_{A0}$ and $x_B = x_{B0}$ at $\zeta = 0$. We first integrate Eq. 18.5-30 to obtain

$$x_B = x_{B0} \exp{(R\nu_{A_nB}\zeta)} \tag{18.5-31}$$

Here $R = r_B - (1/n)$. We now substitute this expression into Eq. 18.5-29 and integrate the resulting first-order linear differential equation to get

$$x_A = (x_{A0} - N^{-1} + M^{-1}x_{B0}) \exp{(Nr_{A_n}\nu_{A_nB}\zeta)} + N^{-1} - M^{-1}x_{B0} \exp{(R\nu_{A_nB}\zeta)} \tag{18.5-32}$$

Here $N = (1 - n^{-1})$ and $M = 1 - n^{-1}[(1 - r_{A_n})/(r_B - r_{A_n})]$. Finally, we use

Eqs. 18.5–28 and 32 to obtain the desired relation[5] for the dimensionless reaction rate ν_{A_nB} as the following transcendental equation:

$$h\sqrt{\frac{\nu_{A_nB}}{K}} = (x_{A0} - N^{-1} + M^{-1}x_{B0})\exp(Nr_{A_n}\nu_{A_nB}) + N^{-1} - M^{-1}x_{B0}\exp(R\nu_{A_nB}) \quad (18.5\text{–}33)$$

in which $K = k_n''\delta/c\mathscr{D}_{A_nB}$. That is, Eq. 18.5–33 gives ν_{A_nB} as an implicit function of K, n, r_B, r_{A_n}, x_{A0}, and x_{B0}.

§18.6 DIMENSIONAL ANALYSIS OF THE EQUATIONS OF CHANGE FOR A BINARY ISOTHERMAL FLUID MIXTURE

Thus far in this chapter we have discussed the equations of change for a fluid mixture, and in §18.5 we illustrated how to set up and solve some problems in diffusion. Next we consider the dimensional analysis of the equations for an isothermal binary fluid mixture of constant viscosity μ and constant diffusivity $\mathscr{D}_{AB}$. In addition, we assume the range of composition to be small enough that both mass density ρ and molar density c are essentially constant. With these restrictions, we write the equations of change as

$$\text{(continuity)} \qquad (\nabla \cdot \boldsymbol{v}) = 0 \qquad (18.6\text{–}1)$$

$$\text{(continuity of } A\text{)} \qquad \frac{Dx_A}{Dt} = \mathscr{D}_{AB}\nabla^2 x_A \qquad (18.6\text{–}2)$$

$$\text{(motion)} \qquad \rho\frac{D\boldsymbol{v}}{Dt} = \mu\nabla^2\boldsymbol{v} + \begin{cases} -\nabla p + \rho\boldsymbol{g} & \textit{forced} \\ -\rho\zeta\boldsymbol{g}(x_A - x_{A0}) & \textit{free} \end{cases} \qquad (18.6\text{–}3)$$

in which x_{A0} is some reference concentration. In Eq. 18.6–3 we have written the equation of motion to include either the forced- or free-convection terms.

We first consider *forced convection* for which it is convenient to define the reduced quantities much as in §10.6:

$$\boldsymbol{v}^* = \frac{\boldsymbol{v}}{V} = \text{dimensionless velocity} \qquad (18.6\text{–}4)$$

$$p^* = \frac{p - p_0}{\rho V^2} = \text{dimensionless pressure} \qquad (18.6\text{–}5)$$

$$t^* = \frac{Vt}{D} = \text{dimensionless time} \qquad (18.6\text{–}6)$$

$$x_A^* = \frac{x_A - x_{A0}}{x_{A1} - x_{A0}} = \text{dimensionless concentration} \qquad (18.6\text{–}7)$$

[5] For the special cases, $n = 1$, $r_{A_n} = r_B$, and $(nr_B - 1) = r_{A_n}(n - 1)$, see Hsu and Bird, *op. cit.*

in which V, D, and $x_{A1} - x_{A0}$ are, respectively, a characteristic velocity, a characteristic linear dimension, and a characteristic concentration difference in the system. The equations of change may now be put into dimensionless form by multiplying Eq. 18.6–1 by D/V, Eq. 18.6–2 by $D/V(x_{A1} - x_{A0})$, and Eq. 18.6–3 by $D/\rho V^2$:

$$\text{(continuity)} \qquad (\nabla^* \cdot \boldsymbol{v}^*) = 0 \tag{18.6–8}$$

$$\text{(continuity of } A\text{)} \qquad \frac{Dx_A^*}{Dt^*} = \frac{1}{\text{Re Sc}} \nabla^{*2} x_A^* \tag{18.6–9}$$

$$\text{(motion)} \qquad \frac{D\boldsymbol{v}^*}{Dt^*} = -\nabla^* p^* + \frac{1}{\text{Re}} \nabla^{*2} \boldsymbol{v}^* + \frac{1}{\text{Fr}} \frac{\boldsymbol{g}}{g} \tag{18.6–10}$$

The dimensionless groups appearing are the Froude number, $\text{Fr} = V^2/gD$, the Reynolds number, $\text{Re} = DV\rho/\mu$, and the Schmidt number, $\text{Sc} = \mu/\rho\mathscr{D}_{AB} = \nu/\mathscr{D}_{AB}$. For isothermal mass transfer, the Schmidt number plays a role analogous to that of the Prandtl number in heat transfer.

In *free convection* we again proceed analogously to the development in §10.6. We define a reduced velocity and time as in Eqs. 10.6–14 and 15. Then, by multiplying Eq. 18.6–1 by $D^2\rho/\mu$, Eq. 18.6–2 by $\rho D^2/\mu(x_{A1} - x_{A0})$, and Eq. 18.6–3 by $D^3\rho/\mu^2$, we get the equations of change in the form

$$\text{(continuity)} \qquad (\nabla^* \cdot \boldsymbol{v}^{**}) = 0 \tag{18.6–11}$$

$$\text{(continuity of } A\text{)} \qquad \frac{Dx_A^*}{Dt^{**}} = \frac{1}{\text{Sc}} \nabla^{*2} x_A^* \tag{18.6–12}$$

$$\text{(motion)} \qquad \frac{D\boldsymbol{v}^{**}}{Dt^{**}} = \nabla^{*2} \boldsymbol{v}^{**} - x_A^* \,\text{Gr}_{AB} \frac{\boldsymbol{g}}{g} \tag{18.6–13}$$

in which $\text{Gr}_{AB} = \rho^2 \zeta g D^3 (x_{A1} - x_{A0})/\mu^2$ is the Grashof number for mass transfer. We see that Gr_{AB} differs from Gr only in that Gr_{AB} contains $\zeta(x_{A1} - x_{A0})$, whereas Gr contains $\beta(T_1 - T_0)$.

In Chapter 21 use is made of the groups Sc and Gr_{AB} in connection with the correlation of mass-transfer coefficients. Note that the foregoing treatment exactly parallels that for heat transfer and that, with the exception of the dissipation function, the corresponding equations in the two sections are the same. We may therefore expect to find many close analogies between heat and mass transfer. However, the extent and limitations of these analogies cannot be fully understood without examining the boundary conditions on p, $\boldsymbol{v}$, T, and x_A. The student will find it profitable to look for these analogies

—and also for differences between these two transfer processes—in the remaining chapters of the text.

Example 18.6-1. Blending of Miscible Fluids

Develop by the methods of dimensional analysis the general form of a correlation for the time required to blend two miscible fluids in an agitated tank. Assume that the two fluids and their mixtures have essentially the same physical properties.

Solution. It will be assumed here that the achievement of "equal degrees of blending" in any two operations means the achievement of the same reduced concentration pattern in each; that is, for any component A the concentration profile

$$x_A = x_A(r^*, \theta, z^*)$$

is the same. Here r^* and z^* are reduced position variables, r/D and z/D, respectively, in which D is the diameter of the impeller. It will also be convenient to define reduced time, velocity, and pressure, as tN, v/ND, and $p/\rho N^2D^2$, respectively, in which N is the rate of rotation of the impeller, revolutions per unit time, and ρ is the fluid density.

Then from Eqs. 18.6–9 and 10 it can be seen that $x_A(r^*, \theta, z^*)$ depends upon t^*, Re, Sc, Fr, geometry, and initial and boundary conditions. Here $\text{Re} = D^2N\rho/\mu$, $\text{Sc} = \mu/\rho\mathscr{D}_{AB}$, and $\text{Fr} = DN^2/g$. Important among the initial conditions are the relative amounts of the two fluids, the manner in which they are introduced to the tank, and the flow patterns at the start of mixing.

Frequently the number of independent dimensionless groups can be further reduced. For example, it has been experimentally observed that if the tank is baffled[1] no vortices of importance occur; that is, the fluid surface is effectively level. Under these circumstances, or in the absence of a free fluid surface, gravitational forces, hence the Froude number, are unimportant. At a high Reynolds number the effects of both Re and Sc can also be neglected[2] so that $x_A(r^*, \theta, z^*)$ depends only upon t^*; that is, the reduced mixing time is constant for any desired degree of blending. In other words, the number of revolutions of the impeller required to produce a given degree of mixing is fixed. For most impellers in common

[1] A common and effective baffling arrangement for vertical cylindrical tanks with axially mounted impellers is to mount four evenly spaced strips, with their flat surfaces in planes through the tank axis, along the tank wall, extending from the top to the bottom of the tank and at least two tenths of the distance to the tank center.

[2] This independence of mixing time upon Reynolds numbers can be seen intuitively from the fact that the term $(\nabla^{*2}v^*)/\text{Re}$ in Eq. 18.6–10 becomes small in relation to the acceleration terms at high Re. Such intuitive arguments are dangerous, however, and the effect of Re is always important in the immediate neighborhood of solid surfaces. Here the amount of mixing taking place in the immediate neighborhood of solid surfaces is small and can be neglected. For a more complete discussion of the effect of Reynolds number, see H. Schlichting, *Boundary Layer Theory*, McGraw-Hill, New York (1955), Chapter 4.

The independence of mixing time on Schmidt numbers can best be seen from the time-averaged equation of continuity in Chapter 20. At high Reynolds numbers the turbulent mass flux is much greater than that due to molecular diffusion except in the immediate neighborhood of the solid surfaces.

use, this situation occurs at Re greater than 10^4 to 2×10^4. At lower Re the Reynolds number has a pronounced effect. This behavior has been substantiated by a number of different investigators.[3]

QUESTIONS FOR DISCUSSION

1. Discuss the applicability of the equations of change as developed in Chapters 3 and 10 to the description of systems of more than one chemical species.
2. As the concentration of component A of a binary mixture becomes very small, both ρ and c become nearly constant; both Eqs. 18.1–17 and 19 are then valid. Comparison of these two equations suggests that $\boldsymbol{v} = \boldsymbol{v}^\star$. Is this true?
3. When is $(\nabla \cdot \boldsymbol{v}^\star)$ equal to zero?
4. Is the definition of $\boldsymbol{q}$ given in Chapter 8 suitable to describe combined heat and mass transfer in systems of interdiffusing pairs of isotopes? Geometric isomers? Optical isomers?
5. What does the term $\Sigma_i \bar{H}_i \boldsymbol{J}_i$ mean?
6. What are Fick's first and second laws? Under what circumstances are they applicable?
7. Show that forced diffusion does not occur when the same external force per unit mass acts on all species.
8. Define the Soret coefficient. What are its units?
9. Why should the connecting tube in Figure 18.5–2 be small in diameter? Need it be insulated?
10. The ratio of partial *specific* volumes of $PbNO_3$ and water is greater than the corresponding ratio for egg albumin and water. However, it is harder to separate the former by ultracentrifuging than the latter. Why?
11. What happens to the water in Example 18.5–4?
12. In gaseous diffusion systems air is usually treated as a single component. Is this reasonable?
13. A gas i is diffusing through a mixture of two heavy isotopes j and j^*. When can this system be considered as a binary one?
14. Compare the effects of mass transfer on simultaneous heat transfer with those of heat transfer on simultaneous mass transfer.
15. Compare the $\mathscr{D}_{AB}$ and D_{ij} used in this chapter.
16. What are the mass transfer analogs of the Grashof and Prandtl numbers?
17. In what ways are heat and mass transfer analogous and in what ways do they differ?
18. Why is the right side of Eq. 18.5–9 greater than unity?
19. How do forced and pressure diffusion differ?

PROBLEMS

18.A$_1$ Dehumidification of Air

Consider a system such as that pictured in Fig. 18.5–1 in which the vapor is H_2O and the stagnant gas is air. Assume the following conditions (which are representative in air conditioning): (i) at $z = \delta$, $T = 80°$ F, and $x_{H_2O} = 0.018$; (ii) at $z = 0$, $T = 50°$ F.

[3] E. A. Fox and V. E. Gex, *A.I.Ch.E. Journal*, **2**, 539–544 (1956); H. Kramers, G. M. Baars, and W. H. Knoll, *Chem. Eng. Sci.*, **2**, 35–42 (1955); J. G. van de Vusse, *Chem. Eng. Sci.*, **4**, 178–200, 209–220 (1955).

a. Calculate the ratio

$$\frac{-(N_{Az}\tilde{C}_{pA}/k)\delta}{1 - \exp(N_{Az}\tilde{C}_{pA}/k)\delta} \tag{18.A-1}$$

in Eq. 18.5-9, for $p = 1$ atm.

b. Compare $\boldsymbol{q}^{(c)}$ and $\boldsymbol{q}^{(d)}$ at $z = 0$. What is the significance of your answer?

Answer: a. 1.004

18.B₁ Thermal Diffusion

a. Estimate the steady-state separation of H_2 and D_2 achieved in the simple thermal diffusion apparatus shown in Fig. 18.5-2 under the following conditions: T_1 is 200° K, T_2 is 600° K, the mole fraction of deuterium in the charge is 0.10, and the effective average k_T is 0.0166.

b. At what temperature should this average k_T have been evaluated?

Answers: a. x_{H_2} is higher by 0.0183 in hot bulb
b. 330° K

18.C₁ Ultracentrifuging of Proteins

Estimate the steady-state concentration profile when a typical albumin solution is subjected to a centrifugal field of 50,000 times the force of gravity under the following conditions:

Cell length = 1.0 cm
Molecular weight of albumin = 45,000
Apparent density of albumin in solution = $M_A/\bar{V}_A = 1.34$ g/cm³
Mole fraction of albumin at $z = 0$, $x_{A0} = 5 \times 10^{-6}$
Apparent density of water in the solution = 1.00 g/cm³
Temperature = 75° F

Answer: $x_A = 5 \times 10^{-6} \exp(-22.7z)$; z is in centimeters

18.D₁ Electrode Polarization

The effective diffusivity of Ag^+ in dilute aqueous solutions at 20° C is about 10^{-5} cm² sec⁻¹. Estimate the limiting current density in amperes/cm² for the apparatus shown in Fig. 18.5-4 if the solution is tenth-normal in the absence of any applied voltage and the thickness of the solution is 0.10 cm.

Answer: 3.86×10^{-3} amperes/cm²

18.E₁ Effective Binary Diffusivities in a Multicomponent Gaseous Mixture

Compute $c\mathscr{D}_{im}$ for each species in the surface-catalyzed gas-phase hydrogenation of benzene, assuming that only the following reaction occurs:

$$C_6H_6 + 3H_2 \rightarrow \text{cyclo-}C_6H_{12} \tag{18.E-1}$$

The calculation is to be made for one point in the reactor, near the catalyst surface, where conditions are $T = 500°$ K, $p = 10$ atm, $x_1(C_6H_6) = 0.10$, $x_2(H_2) = 0.80$, $x_3(C_6H_{12}) = 0.05$, $x_4(CH_4) = 0.05$. CH_4 does not react in this system.

The binary $c\mathscr{D}_{ij}$ in g-mole cm⁻¹ sec⁻¹ are

$$\begin{aligned}
c\mathscr{D}_{12} &= c\mathscr{D}_{21} = 24.2 \times 10^{-6}\\
c\mathscr{D}_{13} &= c\mathscr{D}_{31} = 1.95 \times 10^{-6}\\
c\mathscr{D}_{14} &= c\mathscr{D}_{41} = 6.21 \times 10^{-6}\\
c\mathscr{D}_{23} &= c\mathscr{D}_{32} = 20.7 \times 10^{-6}\\
c\mathscr{D}_{24} &= c\mathscr{D}_{42} = 41.3 \times 10^{-6}\\
c\mathscr{D}_{34} &= c\mathscr{D}_{43} = 5.45 \times 10^{-6}
\end{aligned}$$

Answer:

Component:	C_6H_6	H_2	C_6H_{12}	CH_4
$c\mathscr{D}_{im}$, g-mole cm⁻¹ sec⁻¹ × 10⁶:	6.63	20.8	8.71	60.

18.F$_2$ Setting Up Diffusion Problems

Show how the general equations of Chapter 18 can be used to set up the following problems of Chapter 17: (i) 17.D; (ii) 17.E; (iii) 17.F; (iv) 17.G; (v) 17.H; (vi) 17.J; (vii) 17.K; (viii) 17.L.

18.G$_2$ Alternative Forms of the Equation of Continuity

Show that Eq. 18.1–6 may be written in these forms:

$$a.\ \frac{\partial \rho_A}{\partial t} + (\nabla \cdot \rho_A v) + (\nabla \cdot j_A) = r_A \tag{18.G-1}$$

$$b.\ \frac{\partial c_A}{\partial t} + (\nabla \cdot c_A v^\star) + (\nabla \cdot J_A^\star) = R_A \tag{18.G-2}$$

Discuss the significance of each term in these equations.

18.H$_2$ Simplification of the Multicomponent Mass Flux Expressions for Use in a Binary System

Show that Eqs. 18.4–7 through 11 may be combined to yield Eq. 18.4–14 for the special case of a binary mixture. (It will probably be convenient to take advantage of the relation $j_A + j_B = 0$.)

18.I$_2$ Pressure Diffusion

a. Combine Eq. 18.5–17 with the analogous equation for species B to obtain

$$\frac{g\Omega}{RT}(M_B\bar{V}_A - M_A\bar{V}_B)\,dz = -\bar{V}_A\frac{dx_B}{x_B} + \bar{V}_B\frac{dx_A}{x_A} \tag{18.I-1}$$

Integrate this expression to obtain Eq. 18.5–18.

b. Extend Example 18.5–3 to the case in which L cannot be considered short with respect to R_0. Note that for this case $dp/dr = \rho\Omega^2 r$, in which r is distance measured from the axis of rotation of the centrifuge.

c. Simplify Eq. 18.5–18 for the case in which the mole fraction of one component is negligible. (See Problem 18.C.)

18.J$_2$ Mobility

a. Estimate the total force required to move one equivalent of silver ions through a solution at 1 cm sec^{-1} at 25° C if the diffusivity of silver in the solution is 10^{-5} cm^2 sec^{-1}.

b. Comparable forces are clearly required to move uncharged species. What provides the motive force for such materials in ordinary diffusion?

Answer: a. 2.53×10^9 kilograms force

18.K$_3$ Binary and Ternary Diffusion in Air

Air is conventionally treated as a single component in evaluation of transport properties. Thus low-temperature viscosity data for air have been nicely fitted by using the monatomic-gas viscosity formula (Eq. 1.4–18) with the following Lennard-Jones parameters:

$$\sigma = 3.617\ \text{Å}, \qquad \frac{\epsilon}{\kappa} = 97.0°\ \text{K}$$

whereas, for O_2,

$$\sigma = 3.433\ \text{Å}, \qquad \frac{\epsilon}{\kappa} = 113.0°\ \text{K}$$

and, for N_2,

$$\sigma = 3.681\ \text{Å}, \qquad \frac{\epsilon}{\kappa} = 91.5°\ \text{K}$$

a. Compute $c\mathscr{D}_{AB}$ for the diffusion of methane in air at 300° K, treating air as one component with the Lennard-Jones parameters given above.

b. Compute $c\mathscr{D}_{Am}$ for methane in air at 300° K, taking air to be 21 mole per cent oxygen and 79 mole per cent nitrogen. Assume that the nitrogen and oxygen move at the same velocity.

Note that the two methods are in good agreement, as is usually to be expected. In cases in which the relative concentrations or fluxes of nitrogen and oxygen vary appreciably they should be treated as separate components.

18.L$_3$ Diffusivity in an Aqueous Solution of a Single Salt

When a single salt $(M^{p+})_a(X^{a-})_p$ diffuses through water in the absence of electric current, the molar fluxes $J_i^\star$ of the two ionic species are related by the requirement of electrical neutrality. Show by a procedure analogous to that used in Example 18.5–4 that, in dilute solution, this requirement may be used to determine the following relation between the diffusivity of the salt in water, $\mathscr{D}_{sw}$, and the diffusivities of the anion and cation, $\mathscr{D}_{Xw}$ and $\mathscr{D}_{Mw}$, respectively:

$$\mathscr{D}_{sw} = \frac{\mathscr{D}_{Mw}\mathscr{D}_{Xw}(a + p)}{p\mathscr{D}_{Mw} + a\mathscr{D}_{Xw}} \tag{18.L-1}$$

This result was first obtained by Nernst in 1888.[1]

18.M$_3$ The Bedingfield-Drew Equation of Continuity

Show how to get Eq. 18.1–19*a* from Eq. 18.1–19.

Note: Solution is lengthy.

18.N$_3$ Condensation of Mixed Vapors[2]

Chloroform (C) and benzene (B) vapors are condensing on a cold surface from an equimolar mixture at 1 atm. (See Fig. 18.N.) The temperature T_c of the cold surface can be varied. The more volatile species (C) accumulates near the cold surface, so that the condensation process is retarded by the presence of a gas "film" in which both the temperature and the composition depend on the distance z from the liquid surface. The film thickness δ is 0.1 mm, and the heat-transfer coefficient h of the condensate film is 200 Btu hr^{-1} ft^{-2} °F^{-1}. The ratio of benzene to chloroform in the condensate can be considered as equal to the ratio of rates of condensation and the following data may be used:

$\mathscr{D}_{BC} = 0.050$ cm^2 sec^{-1}; $\tilde{C}_{pC} = 16.5$ Btu/lb-mole °F; $\tilde{C}_{pB} = 22.8$ Btu/lb-mole °F

Vapor-liquid equilibrium data for this system at 1 atm are as follows:

x_C (vapor)	0	0.10	0.20	0.30	0.40	0.50	0.60	0.70	0.80	0.90	1.00
x_C (liquid)	0	0.08	0.15	0.22	0.29	0.36	0.44	0.54	0.66	0.79	1.00
Saturation temperature, °C	80.6	79.8	79.0	78.2	77.3	76.4	75.3	74.0	71.9	68.9	61.4

The molar heat of vaporization $\tilde{\lambda}_{\text{mix}}$ and thermal conductivity k_{mix} of chloroform-benzene mixtures may be considered constant at 12,800 Btu/lb-mole and 0.007 Btu/hr ft °F, respectively. The effect of temperature on physical properties may be neglected.

[1] W. Nernst, *Z. physik. Chem.*, **2**, 613–637 (1888).

[2] For a discussion of the condensation of mixed vapors, see A. P. Colburn and T. B. Drew, *Trans. Am. Inst. Chem. Engrs.*, **33**, 197–212 (1937).

a. Show that the total molar rate of condensation is

$$-(N_{Bz} + N_{Cz}) = \frac{c\mathscr{D}_{BC}}{\delta} \ln \left\{ \frac{x_{B0} - x_B^{(l)}}{x_{B\delta} - x_B^{(l)}} \right\} = \frac{c\mathscr{D}_{BC}}{\delta} \ln \left\{ \frac{x_{C0} - x_C^{(l)}}{x_{C\delta} - x_C^{(l)}} \right\} \qquad (18.N\text{-}1)$$

Here the subscripts 0 and δ refer to conditions at $z = 0$ and $z = \delta$. The quantity $x_A^{(l)}$ is the mole fraction of component A in the condensate.

b. Show that the energy flux through the *condensate film* at $z = 0$ is

$$-q_z\big|_{z=0} = (T_\delta - T_0)\left(\frac{k_{\text{mix}}}{\delta}\right)\left(\frac{\beta}{1 - e^{-\beta}}\right) - \tilde{\lambda}_{\text{mix}}\,(N_{Bz} + N_{Cz}) \qquad (18.N\text{-}2)$$

Here $\beta = -(N_{Bz} + N_{Cz})\tilde{C}_{p,\text{mix}}\delta/k_{\text{mix}}$ and $\tilde{C}_{p,\text{mix}} = x_B^{(l)}\tilde{C}_{pB} + x_C^{(l)}\tilde{C}_{pC}$.

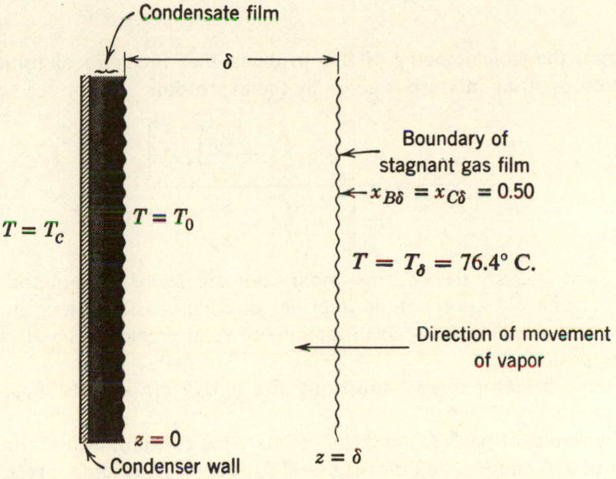

Fig. 18.N. Condensation of mixed vapors.

c. Estimate the rates of condensation and the composition of the condensate for several condenser wall temperatures. Note that the calculation will be simplified by first choosing a condensate composition. (i) This choice permits direct calculation of $N_{Bz} + N_{Cz}$ from the result of part (*a*). (ii) The energy flux through the condensate film may then be calculated from the result of part (*b*). (iii) The required wall temperature may then be estimated by an energy balance through the condensate film. In making this energy balance, assume that all energy transport through the condensate film occurs by conduction according to the expression $q_z = h(T_c - T_0)$.

Sample answer: If T_c is 86° F, the rate of condensation will be 1.26 lb-moles hr^{-1} ft^{-2} and the condensate will be 44 mole per cent chloroform.

d. When, if ever, will the condensate be in equilibrium with the saturated vapor?

e. What is the (approximate) minimum temperature that can be maintained at the condensate-vapor interface and what will the (exact) composition of the condensate be under these limiting conditions? Explain the significance of your answers.

18.O$_3$ Constant-Evaporating Mixtures

A mixture of toluene and ethanol is evaporating through a one-dimensional stagnant nitrogen film of thickness δ to a stream of pure nitrogen. The entire system is maintained at 60° F and at constant pressure. The ratio of the binary diffusivities of toluene and ethanol in nitrogen ($\mathscr{D}_{TN_2}/\mathscr{D}_{EN_2}$) is about 0.695, and the vapor-liquid equilibrium data for the toluene-ethanol system at 60° F are

Mole fraction toluene in liquid	0.096	0.155	0.233	0.274	0.375
Mole fraction toluene in vapor	0.147	0.198	0.242	0.256	0.277
Total pressure, mm Hg	388	397	397	395	390

It is desired to determine the composition of the constant-evaporating mixture of toluene and ethanol (i.e., the liquid composition for which the ratio of evaporation rates is the ratio of concentrations in the liquid) under a variety of conditions. Assume ideal gas behavior.

a. Show from the stoichiometry of this problem that the ratio of toluene to ethanol in a constant-evaporating mixture is given by the expression

$$\frac{\mathscr{D}_{Em}}{\mathscr{D}_{Tm}} = \frac{\ln\left[1 - \left(1 + \dfrac{x_E}{x_T}\right)y_T^*\right]}{\ln\left[1 - \left(1 + \dfrac{x_T}{x_E}\right)y_E^*\right]} \tag{18.O-1}$$

Here the $\mathscr{D}_{Em}$ and $\mathscr{D}_{Tm}$ are the effective binary diffusivities of ethanol and toluene in the mixture and y_E^* and y_T^* are the mole fractions of ethanol and toluene in the gas at the interface. (Note that y_T^* and y_E^* are functions of total pressure as well as temperature and liquid composition.)

b. Show that it is reasonable to substitute the binary diffusivities $\mathscr{D}_{EN_2}$ and $\mathscr{D}_{TN_2}$ for $\mathscr{D}_{Em}$ and $\mathscr{D}_{Tm}$.

c. Using the results of parts (*a*) and (*b*), estimate the constant evaporating mixtures for total pressure of 396 mm Hg, 760 mm Hg, and for very high external pressure.

Answers: a. 25 per cent toluene at 396 mm Hg
b. 17 per cent toluene at one atm[3]
c. 13 per cent toluene at high pressure

18.P₃ Concentration Polarization in the Presence of an Indifferent Electrolyte

Consider that the apparatus described in Example 18.5-4 contains both $Ag^+NO_3^-$ at an average concentration 10^{-6} *N* and $K^+NO_3^-$ at an average concentration of 0.1 *N*. A voltage just sufficient to cause the silver ion concentration at the cathode to drop essentially to zero is impressed across the cell. It may again be assumed that no electrode reactions take place except the dissolution and plating out of metallic silver; that is, the transport number of the silver ion is unity.

a. Calculate the concentration gradients of the three ionic species present, assuming, as in Example 18.5-4, that they diffuse independently of one another and that bulk flow is negligible. *Answer:* See Fig. 18.P.
All concentration gradients are linear.

[3] G. W. Bennett and W. A. Wright, *Ind. Eng. Chem.*, **28**, 646–648 (1936) obtained an experimental constant-evaporating composition of 20 per cent toluene. However, because of defects in experimental technique, their results should be intermediate between the azeotropic composition, 25 per cent toluene, and the true constant-evaporating composition. The calculated value of 17 per cent thus appears quite reasonable.

b. Show that the effect of the potential gradient on the movement of the silver ion through the solution is negligible. The purpose of the indifferent, that is, nondischarging, KNO_3 is to reduce the potential gradient so that silver ions are moving almost solely by ordinary diffusion. Systems of the type pictured here, but of more complicated geometry, are widely used for polarographic analysis and for the measurement of mass-transfer coefficients.

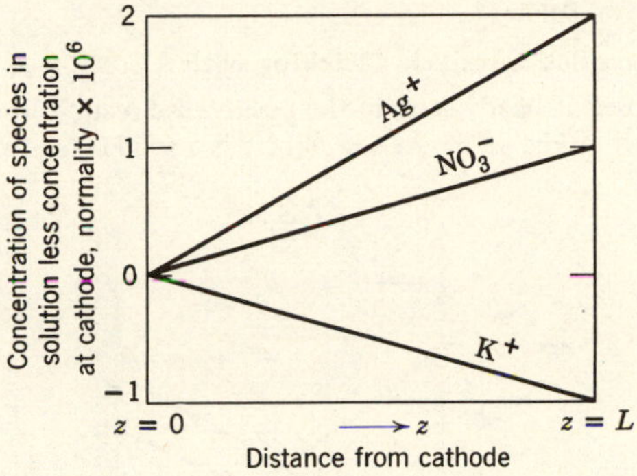

Fig. 18.P. Concentration profiles for a polarized cell containing an excess of indifferent electrolyte.

c. It is frequently stated that in systems such as this most of the current is "carried by the nondischarging species." Is this a reasonable statement?

18.Q$_3$ Diffusion of *A* through a Stagnant Film of *B* and *C* of Known Average Composition

Gas A is diffusing in the positive z-direction at steady state through a flat stagnant film of thickness δ, which contains nonmoving gases B and C. The average mole fraction of gas B in the film on an A-free basis is $x_B{}^\circ$. The system is isothermal.

a. Show that the Stefan-Maxwell equation, Eq. 18.4–19, may be written for gases B and C as $(dx_B/d\zeta) = \nu_{AB}x_B$ and $(dx_C/d\zeta) = \nu_{AC}x_C$. Here $\zeta = z/\delta$ is the fractional distance through the film and $\nu_{Ai} = \delta N_{Az}/c\mathscr{D}_{Ai}$.

b. Integrate these two differential equations with the aid of the boundary conditions that at $z = 0$, $x_A = x_{A0}$, and at $z = \delta$, $x_A = x_{A\delta}$ to obtain

$$x_B = \left(\frac{(1 - x_{A0}) \exp \nu_{AC} - (1 - x_{A\delta})}{\exp \nu_{AC} - \exp \nu_{AB}}\right) \exp \frac{zN_{Az}}{c\mathscr{D}_{AB}} \tag{18.Q-1}$$

$$x_C = \left(\frac{(1 - x_{A0}) \exp \nu_{AB} - (1 - x_{A\delta})}{\exp \nu_{AB} - \exp \nu_{AC}}\right) \exp \frac{zN_{Az}}{c\mathscr{D}_{AC}} \tag{18.Q-2}$$

c. We may write $x_B{}^\circ = \int_0^1 x_B \, d\zeta \Big/ \int_0^1 (x_B + x_C)\, d\zeta$. Substitute the results of part (*b*) into this expression and integrate to obtain

$$x_B{}^\circ = \left[1 - \frac{1}{r}\left(\frac{\exp r\nu_{AB} - 1}{\exp r\nu_{AB} - Q}\right)\left(\frac{\exp \nu_{AB} - Q}{\exp \nu_{AB} - 1}\right)\right]^{-1} \tag{18.Q-3}$$

Here $r = \mathscr{D}_{AB}/\mathscr{D}_{AC}$ and $Q = (1 - x_{A\delta})/(1 - x_{A0})$. Note that the desired flux N_{Az} is contained in ν_{AB}.

18.R$_4$ The Equations of Mechanical and Thermal Energy for Multicomponent Systems

a. Obtain an expression for the rate of change of kinetic energy per unit mass in multicomponent systems, Eq. *C* of Table 18.3–1, by taking the dot product of $\boldsymbol{v}$ and the equation of motion, Eq. 18.3–2.

b. Subtract the result of part (*a*) from Eq. (*B*) of Table 18.3–1 to obtain the equation of internal energy, Eq. *D* of Table 18.3–1.

18.S$_4$ Diffusion-Controlled Catalytic Cracking with $A \rightarrow pP + qQ$

An ideal gas A diffuses at steady state in the positive z-direction through a flat gas film of thickness δ, as shown in Fig. 18.S. At $z = \delta$ there is a solid catalytic surface at which A

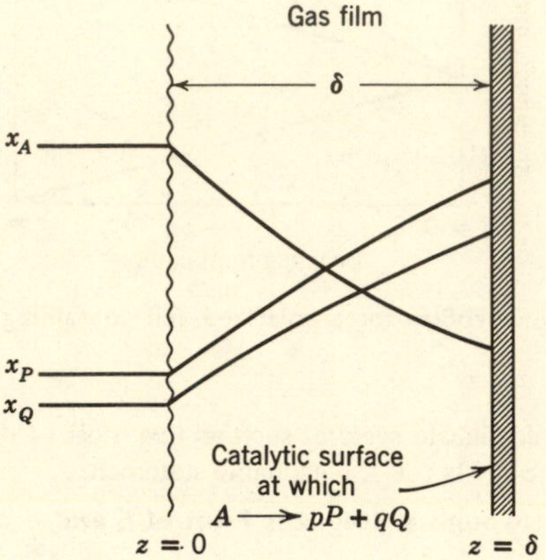

Fig. 18.S. Diffusion-controlled catalytic cracking.

undergoes an irreversible nth-order reaction: $A \rightarrow pP + qQ$. The reaction rate may be expressed as $N_A|_{z=\delta} = k_n''x_A^n$. At $z = 0$, $x_P = x_{P0}$ and $x_Q = x_{Q0}$; both of these mole fractions are assumed to be known.

a. Show that the Stefan-Maxwell equations for components P and Q are

$$\frac{1}{\nu_{PQ}}\frac{dx_P}{d\zeta} = A_Px_P + B_Px_Q + C_P \tag{18.S–1}$$

$$\frac{1}{\nu_{PQ}}\frac{dx_Q}{d\zeta} = A_Qx_Q + B_Qx_P + C_Q \tag{18.S–2}$$

Here $\nu_{PQ} = N_A\delta/c\mathscr{D}_{PQ}$, $A_P = (1 - p)r_P - q$, $B_P = (1 - r_P)p$, $C_P = pr_P$. The quantities A_Q, B_Q, and C_Q may be obtained by everywhere replacing P by Q and p by q. In these expressions $r_P = \mathscr{D}_{PQ}/\mathscr{D}_{AP}$ and $r_Q = \mathscr{D}_{PQ}/\mathscr{D}_{AQ}$.

b. Take the Laplace transform of Eqs. 18.S–1 and 2 with respect to $\nu_{PQ}\zeta$ to obtain the following algebraic expressions for the transformed mole fractions $\bar{x}_P(s)$ and $\bar{x}_Q(s)$:

$$\bar{x}_P = \frac{x_{P0} + C_Ps^{-1} + B_P\bar{x}_Q}{s - A_P} \tag{18.S–3}$$

$$\bar{x}_Q = \frac{x_{Q0} + C_Qs^{-1} + B_Q\bar{x}_P}{s - A_Q} \tag{18.S–4}$$

c. Solve these equations simultaneously to get

$$\bar{x}_P = \frac{x_{P0}s^2 + (B_P x_{Q0} - A_Q x_{P0} + C_P)s + (B_P C_Q - C_P A_Q)}{s^2 - (A_P + A_Q)s + (A_P A_Q - B_P B_Q)} \frac{1}{s} \tag{18.S-5}$$

d. Take the inverse transform of $\bar{x}_p$ to obtain

$$\begin{aligned} x_P = \frac{1}{\alpha_+ - \alpha_-}\Bigg[& x_{P0}(\alpha_+ e^{\alpha_+ \nu_{PQ}\zeta} - \alpha_- e^{\alpha_- \nu_{PQ}\zeta}) \\ & + (B_P x_{Q0} - A_Q x_{P0} + C_P)(e^{\alpha_+ \nu_{PQ}\zeta} - e^{\alpha_- \nu_{PQ}\zeta}) \\ & + (B_P C_Q - C_P A_Q)\left(\frac{e^{\alpha_+ \nu_{PQ}\zeta} - 1}{\alpha_+} - \frac{e^{\alpha_- \nu_{PQ}\zeta} - 1}{\alpha_-}\right)\Bigg] \end{aligned} \tag{18.S-6}$$

Here $\alpha_\pm = \frac{1}{2}[(A_P + A_Q) \pm \sqrt{(A_P - A_Q)^2 + 4B_P B_Q}]$.

e. Is this solution valid for the following special cases? (i) $\alpha_+ = \alpha_-$, (ii) $\alpha_+ \neq 0$ and $\alpha_- = 0$.

CHAPTER 19

Concentration Distributions with More Than One Independent Variable

The diffusion problems discussed in the preceding two chapters yield concentration distributions that for the most part are functions of but one independent variable. In this chapter we use the general equations of Chapter 18 to set up and solve some diffusion problems that involve two independent variables—either position and time, or two position variables.

A large number of diffusion problems can be solved by simply looking up the solutions to the analogous problems in heat conduction. When the differential equations *and* the boundary and initial conditions for the diffusion process are of exactly the same form as those for the heat-conduction process, then the solution may be taken over with appropriate change in notation. In Table 19.0–1 the three main heat-transport equations used in Chapter 11 are shown along with their mass-transport analogs. Many solutions to the nonflow equations may be found in the monographs of Carslaw and Jaeger[1] and of Crank.[2] Further analogies in other fields of physics are summarized

[1] H. S. Carslaw and J. C. Jaeger, *Heat Conduction in Solids*, Oxford University Press, (1959), Second Edition.

[2] J. Crank, *The Mathematics of Diffusion*, Oxford University Press (1956).

TABLE 19.0–1

ANALOGIES BETWEEN SPECIAL FORMS OF THE HEAT CONDUCTION AND DIFFUSION EQUATIONS

Process	Unsteady-state Nonflow	Steady-state Flow	Steady-state Nonflow
Solution Given in	§11.1—Exact solutions §11.4—Boundary-layer solutions	§11.2—Exact solutions §11.4—Boundary-layer solutions	§11.3—Exact solutions in two dimensions by analytic functions
Heat Conduction: Equations	$\dfrac{\partial T}{\partial t} = \alpha \nabla^2 T$	$(\boldsymbol{v} \cdot \nabla T) = \alpha \nabla^2 T$	$\nabla^2 T = 0$
Heat Conduction: Applications	Heat conduction in solids	Heat conduction in laminar incompressible flow	Steady heat conduction in solids
Heat Conduction: Assumptions	1. k = constant 2. $\boldsymbol{v} = 0$	1. k, ρ = constants 2. No viscous dissipation 3. Steady state	1. k = constant 2. $\boldsymbol{v} = 0$ 3. Steady state
Diffusion: Equations	$\dfrac{\partial c_A}{\partial t} = \mathscr{D}_{AB} \nabla^2 c_A$	$(\boldsymbol{v} \cdot \nabla c_A) = \mathscr{D}_{AB} \nabla^2 c_A$	$\nabla^2 c_A = 0$
Diffusion: Applications	Diffusion of traces of A through B	Diffusion in laminar flow (dilute solutions of A in B)	Steady diffusion in solids
Diffusion: Assumptions	1. $\mathscr{D}_{AB}, \rho$ = constants 2. $\boldsymbol{v} = 0$ 3. No chemical reactions	1. $\mathscr{D}_{AB}, \rho$ = constants 2. Steady state 3. No chemical reactions	1. $\mathscr{D}_{AB}, \rho$ = constants 2. Steady state 3. No chemical reactions 4. $\boldsymbol{v} = 0$
Diffusion: Applications	OR Equimolar counter-diffusion in low density gases		
Diffusion: Assumptions	1. $\mathscr{D}_{AB}, c$ = constants 2. $\boldsymbol{v}^\star = 0$ 3. No chemical reactions		

by Sneddon.[3] In addition, certain classes of neutron-diffusion problems can be solved by utilizing the analogy with heat conduction.[4]

Because the diffusion problems described by the equations in Table 19.0–1 are the same as those discussed in Chapter 11, we do not discuss them further here. Instead, we select problems involving diffusion with chemical reactions, diffusion with nonzero molar average velocity, diffusion with more than two components, and forced convection with high mass-transfer rates; these problems illustrate effects peculiar to mass-transfer systems. Therefore, we do not follow the same topical outline as we did in Chapters 4 and 11. In §19.1 we discuss several unsteady-state diffusion problems, and in §19.2 two diffusion problems are examined by approximate boundary-layer methods of the von Kármán type. In §19.3 we present exact solutions of the Prandtl boundary-layer equations for simultaneous heat, mass, and momentum transfer in laminar flow along a flat plate; this section is of special importance because it illustrates conditions of exact analogy between heat, mass, and momentum transfer, and also shows the boundary layer behavior at high mass-transfer rates.

§19.1 UNSTEADY DIFFUSION

In this section we present three illustrative examples. The first example shows how the method of combination of variables is used to describe the diffusion-controlled evaporation process. The second example concerns Danckwerts' method for solving diffusion problems with first-order irreversible reactions occurring. The third concerns diffusion with a moving reaction zone.

Example 19.1–1. Unsteady-State Evaporation

It is desired to derive an equation for the rate at which a liquid A evaporates into a vapor B in a tube of infinite length. (See the system sketched in Fig. 16.1–1b.) The liquid level is maintained at position $z = 0$ at all times. The entire system is maintained at constant temperature and pressure, and the vapors A and B are assumed to form an ideal-gas mixture; hence the molar density c is constant throughout the gas phase. It is further assumed that B is insoluble in A.

Solution. The equations of continuity for components A and B are

$$\frac{\partial c_A}{\partial t} = -\frac{\partial N_{Az}}{\partial z}; \qquad \frac{\partial c_B}{\partial t} = -\frac{\partial N_{Bz}}{\partial z} \tag{19.1–1,2}$$

Addition of these two equations gives

$$\frac{\partial c}{\partial t} = -\frac{\partial}{\partial z}(N_{Az} + N_{Bz}) \tag{19.1–3}$$

[3] I. N. Sneddon, *Elements of Partial Differential Equations*, McGraw-Hill, New York (1957), Chapter 4 (Laplace's equation) and Chapter 6 (diffusion equation).

[4] For an introduction to neutron diffusion, see, for example R. L. Murray, *Nuclear Reactor Physics*, Prentice-Hall, Englewood Cliffs, N.J. (1957), Chapter 3.

Inasmuch as c is constant, we conclude that $\partial c/\partial t = 0$ and $N_{Az} + N_{Bz}$ is a function of time alone. But we know that at $z = 0$ there is no motion of B, so that $N_{B0} = 0$; at the same position, the flux of A is obtained from Eq. 16.2–2 as

$$N_{A0} = -\frac{c\mathscr{D}_{AB}}{1 - x_{A0}} \left.\frac{\partial x_A}{\partial z}\right|_{z=0} \tag{19.1–4}$$

in which the subscripts 0 indicate quantities evaluated at the plane $z = 0$. Hence for all z

$$N_{Az} + N_{Bz} = -\frac{c\mathscr{D}_{AB}}{1 - x_{A0}} \left.\frac{\partial x_A}{\partial z}\right|_{z=0} \tag{19.1–5}$$

and the flux of A for all z is given according to Eq. 16.2–2 as

$$N_{Az} = -c\mathscr{D}_{AB}\frac{\partial x_A}{\partial z} - x_A\left(\frac{c\mathscr{D}_{AB}}{1 - x_{A0}}\right) \left.\frac{\partial x_A}{\partial z}\right|_{z=0} \tag{19.1–6}$$

Substitution of Eq. 19.1–6 into Eq. 19.1–1 gives for constant $\mathscr{D}_{AB}$

$$\frac{\partial x_A}{\partial t} = \mathscr{D}_{AB}\frac{\partial^2 x_A}{\partial z^2} + \frac{\mathscr{D}_{AB}}{1 - x_{A0}} \left.\frac{\partial x_A}{\partial z}\right|_{z=0} \frac{\partial x_A}{\partial z} \tag{19.1–7}$$

This equation is to be solved with the initial and boundary conditions:

I. C.: at $t = 0$, $x_A = 0$ (19.1–8)

B. C. 1: at $z = 0$, $x_A = x_{A0}$ (19.1–9)

B. C. 2: at $z = \infty$, $x_A = 0$ (19.1–10)

Here x_{A0} is the equilibrium gas-phase concentration;[1] for an ideal-gas mixture this is just the vapor pressure of pure A divided by the total pressure.

Because there is no characteristic length in the system and because $x_A = 0$ both at $t = 0$ and $z = \infty$, the use of the method of combination of variables suggests itself. We introduce then a dimensionless concentration X and a dimensionless distance Z defined thus:

$$X = \frac{x_A}{x_{A0}}; \qquad Z = \frac{z}{\sqrt{4\mathscr{D}_{AB}t}} \tag{19.1–11}$$

We anticipate that X is a function of Z alone for any x_{A0}, and that enables us to rewrite Eq. 19.1–7 thus:

$$X'' + 2(Z - \varphi)X' = 0 \tag{19.1–12}$$

in which primes indicate derivatives with respect to Z and[2]

$$\varphi(x_{A0}) = -\frac{1}{2}\frac{x_{A0}}{(1 - x_{A0})} \left.\frac{dX}{dZ}\right|_{z=0} \tag{19.1–13}$$

[2] Physically, φ is a dimensionless molar average velocity

$$\varphi = v^{\star}\sqrt{\frac{t}{\mathscr{D}_{AB}}} \tag{19.1–13a}$$

[1] See Footnote 1 in §17.2 regarding the appropriateness of this assumption of equilibrium at the interface.

Equation 19.1–12 is to be solved with boundary conditions

B. C. 1: at $Z = 0, \quad X = 1$ (19.1–14)

B. C. 2: at $Z = \infty, \quad X = 0$ (19.1–15)

Inasmuch as φ does not depend on Z, Eq. 19.1–12 can be integrated easily to give the following reduced concentration profiles:[3]

$$X = \frac{1 - \text{erf}\,(Z - \varphi)}{1 + \text{erf}\,\varphi} \tag{19.1–16}$$

From Eqs. 19.1–13 and 16 we get φ as a function of x_{A0}; actually, it is considerably easier to express the relation by giving x_{A0} as a function of φ:

$$x_{A0} = \frac{1}{1 + [\sqrt{\pi}(1 + \text{erf}\,\varphi)\varphi \exp \varphi^2]^{-1}} \tag{19.1–17}$$

A small table of $\varphi(x_{A0})$ is given in Table 19.1–1, and the concentration profiles are sketched in Fig. 19.1–1.

Once the concentration profiles have been determined, we can calculate the volume rate of production of vapor of A from a surface of area S. If V_A is the volume of A produced by a time t after evaporation has started, then

$$\frac{dV_A}{dt} = \frac{N_{A0}S}{c} = S\varphi\sqrt{\frac{\mathscr{D}_{AB}}{t}} \tag{19.1–18}$$

Integration with respect to t gives

$$V_A = S\varphi\sqrt{4\mathscr{D}_{AB}t} \tag{19.1–19}$$

in which φ is a function of x_{A0} given in Table 19.1–1.

TABLE 19.1–1[a]

TABLE OF φ AND $\psi^{(2)}$ AS A FUNCTION OF x_{A0}

x_{A0}	φ	$\psi^{(2)} = \varphi\sqrt{\pi}/x_{A0}$
0	0.0000	1.000
$\frac{1}{4}$	0.1562	1.108
$\frac{1}{2}$	0.3578	1.268
$\frac{3}{4}$	0.6618	1.564
1	∞	∞

[a] J. H. Arnold, *Trans. A.I.Ch.E.*, **40**, 361–378 (1944).

We are now in a position to assess the importance of including the convective transport of species A up the tube. If Fick's second law (Eq. 18.1–20) had been used, then for V_A we would have obtained

$$V_A^{(\text{Fick})} = Sx_{A0}\sqrt{\frac{4\mathscr{D}_{AB}t}{\pi}} \tag{19.1–20}$$

[3] This solution was given by J. H. Arnold, *Trans. A.I.Ch.E.*, **40**, 361–378 (1944).

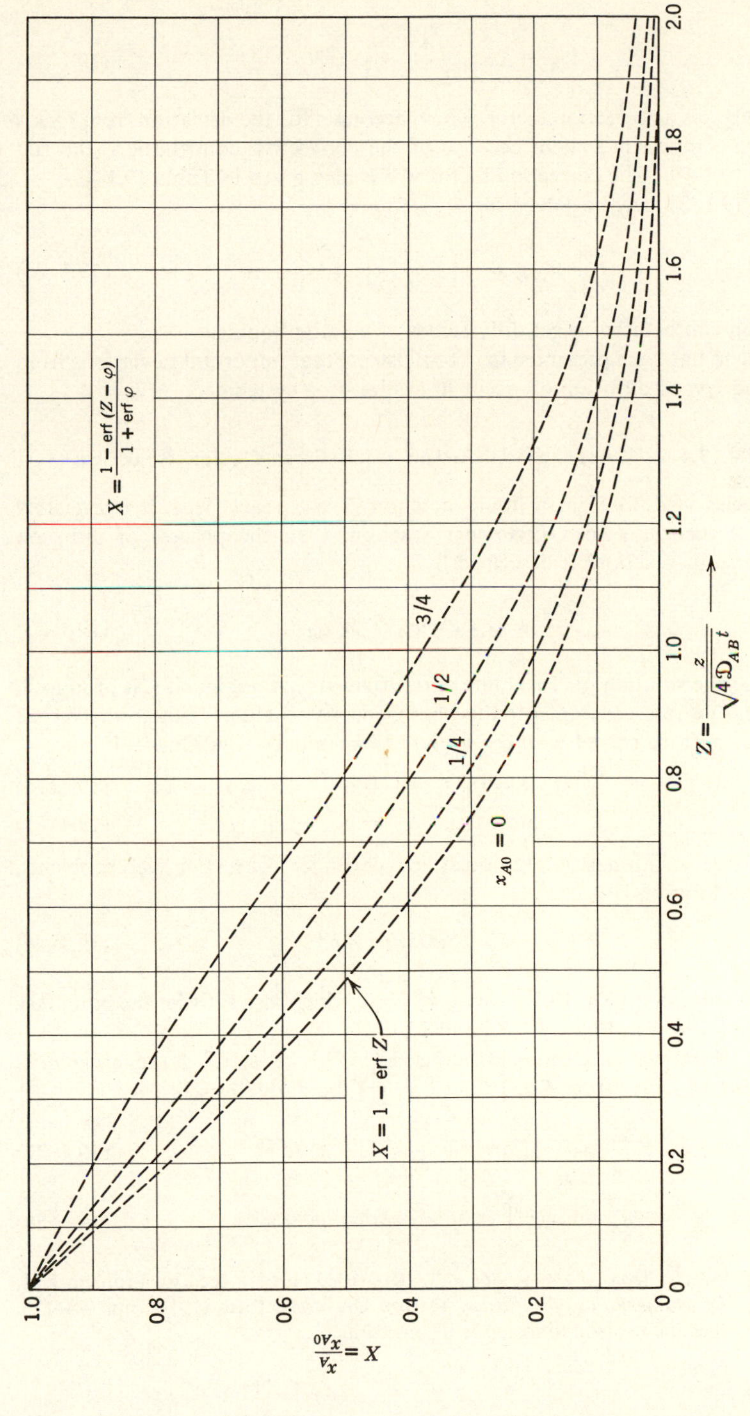

Fig. 19.1-1. Concentration profiles in an evaporating system, showing how the deviation from Fick's second law increases with the volatility of liquid A.

We can now rewrite Eq. 19.1–19 in the somewhat more convenient form:

$$V_A = Sx_{A0}\sqrt{\frac{4\mathscr{D}_{AB}t}{\pi}} \cdot \psi^{(2)} \tag{19.1–21}$$

$\psi^{(2)} = \varphi\sqrt{\pi}/x_{A0}$ is a correction factor, which accounts for the deviation from Fick's law in a two-component system because of the convective contribution (the last term in Eq. 19.1–7). The correction factor $\psi^{(2)}$ is also given in Table 19.1–1.

Equation 19.1–21 may be solved for $\mathscr{D}_{AB}$ to give

$$\mathscr{D}_{AB} = \frac{\pi}{t}\left(\frac{V_A}{2Sx_{A0}\psi^{(2)}}\right)^2 \tag{19.1–22}$$

This equation can be used to get diffusivities of volatile liquids.

This problem has been presented here to illustrate that important deviations from Fick's second law of diffusion do occur in evaporating systems.

Example 19.1–2. Unsteady Diffusion with First-Order Reaction[4,5]

When species A diffuses in a liquid medium B and reacts with it irreversibly ($A + B \rightarrow C$) according to a first-order reaction, then the process of diffusion and homogeneous reaction[6] is described by

$$\frac{\partial c_A}{\partial t} = \mathscr{D}_{AB}\nabla^2 c_A + k_1''' c_A \tag{19.1–23}$$

provided that the solution of A is dilute and provided that not much C is produced. Here $-k_1'''$ is the rate constant for the homogeneous reaction. Equation 19.1–23 in some cases is to be solved with the initial and boundary conditions:

I. C.: at $t = 0$, $c_A = 0$ (19.1–24)

B. C.: at surfaces, $c_A = c_{As}$ (19.1–25)

Here c_{As} may vary with position but must be independent of t. For such problems, show that the solution is[4]

$$c_A = -k_1'''\int_0^t f e^{k_1''' t'} dt' + f e^{k_1''' t} \tag{19.1–26}$$

where f is the solution to Eq. 19.1–23 with $k_1''' = 0$, which also satisfies the boundary and initial conditions in Eqs. 19.1–24 and 25.

Solution. First we test to see whether Eq. 19.1–26 satisfies the *differential equation* (Eq. 19.1–23); from Eq. 19.1–26 we get the derivatives

$$\frac{\partial c_A}{\partial t} = -k_1''' f e^{k_1''' t} + \frac{\partial f}{\partial t} e^{k_1''' t} + k_1''' f e^{k_1''' t} \tag{19.1–27}$$

$$\nabla^2 c_A = -k_1'''\int_0^t (\nabla^2 f) e^{k_1''' t'} dt' + (\nabla^2 f) e^{k_1''' t} \tag{19.1–28}$$

[4] P. V. Danckwerts, *Trans. Faraday Soc.*, **47**, 1014–1023 (1951); see also Problem 19.J.
[5] J. Crank, *The Mathematics of Diffusion*, Oxford University Press (1956), pp. 124–125.
[6] For some cases the neutron diffusion is of this form.

Substitution of Eqs. 19.1–26, 27, and 28 into Eq. 19.1–23 and making use of the fact that $\mathscr{D}_{AB}\nabla^2 f = \partial f/\partial t$ shows that Eq. 19.1–23 is indeed satisfied.

The *initial condition* is satisfied by Eq. 19.1–26, since

$$c_A(t=0) = -k_1''' \int_0^0 f e^{k_1''' t'} dt' + f(t=0)e^{k_1'''\cdot 0} = 0 \qquad (19.1\text{–}29)$$

because $f(t=0)$ has been defined to be zero.

The *boundary condition* is also satisfied, since on the surface $f = c_{As}$:

$$\begin{aligned} c_A \text{ (surface)} &= -k_1''' \int_0^t c_{As} e^{k_1''' t'} dt' + c_{As} e^{k_1''' t} \\ &= -k_1''' c_{As} \frac{e^{k_1''' t'}}{k_1'''}\Bigg|_0^t + c_{As} e^{k_1''' t} \\ &= c_{As} \end{aligned} \qquad (19.1\text{–}30)$$

Example 19.1–3. Gas Absorption with Rapid Chemical Reaction[7]

Gaseous A is absorbed by solvent S, the latter containing solute B. The gas-liquid interface is taken to be the plane $z = 0$. Species A reacts with B according to the reaction $A + B \rightarrow AB$ in an instantaneous irreversible reaction. It may be assumed that Fick's second law adequately describes the diffusion processes, since A, B, and AB are present in S in low concentrations. Obtain expressions for the concentration profiles.

Solution. Because of the instantaneous reaction of A and B, there will be a plane parallel to the liquid-vapor interface at a distance z' from it, which separates the region containing no A from that containing no B. The distance z' is a function of t, since the boundary between A and B retreats as B is used up in the chemical reaction.

The differential equations to be solved are

$$\frac{\partial c_A}{\partial t} = \mathscr{D}_{AS} \frac{\partial^2 c_A}{\partial z^2} \qquad 0 < z \leqslant z'(t) \qquad (19.1\text{–}31)$$

$$\frac{\partial c_B}{\partial t} = \mathscr{D}_{BS} \frac{\partial^2 c_B}{\partial z^2} \qquad z'(t) \leqslant z < \infty \qquad (19.1\text{–}32)$$

We now try solutions of the form

$$\frac{c_A}{c_{A0}} = a_1 + a_2 \operatorname{erf} \frac{z}{\sqrt{4\mathscr{D}_{AS}t}} \qquad (19.1\text{–}33)$$

$$\frac{c_B}{c_{B0}} = b_1 + b_2 \operatorname{erf} \frac{z}{\sqrt{4\mathscr{D}_{BS}t}} \qquad (19.1\text{–}34)$$

in which c_{A0} is the interfacial liquid-phase concentration of A and c_{B0} is the original

[7] T. K. Sherwood and R. L. Pigford, *Absorption and Extraction*, McGraw-Hill, New York (1952), pp. 332–337. Other problems of diffusion with chemical reaction are discussed in Chapter IX of this reference. See H. S. Carslaw and J. C. Jaeger, *Heat Conduction in Solids*, Oxford University Press, Second Edition (1959), for heat-conduction problems with moving boundaries associated with phase changes.

concentration of B in S. It can be shown a posteriori that the foregoing postulated forms of the solutions are appropriate.

The movement of the boundary with time must now be ascertained. We note that at the reaction surface Eq. 19.1–33 must reduce to

$$c_A(z', t) = 0 \tag{19.1–35}$$

From the perfect differential

$$dc_A = 0 = \left(\frac{\partial c_A}{\partial z'}\right)_t dz' + \left(\frac{\partial c_A}{\partial t}\right)_{z'} dt \tag{19.1–36}$$

one can solve and get

$$\frac{dz'}{dt} = -\frac{(\partial c_A/\partial t)_{z'}}{(\partial c_A/\partial z')_t} \tag{19.1–37}$$

Substitution of Eq. 19.1–33 (written in terms of z') into Eq. 19.1–37 provides a differential equation for z' in terms of t:

$$\frac{dz'}{dt} = \frac{z'}{2t} \tag{19.1–38}$$

which may be integrated to give

$$z' = \sqrt{4\alpha t} \tag{19.1–39}$$

in which α is an integration constant. This simple dependence of z' on t could also have been deduced by setting $c_A = 0$ and $z = z'$ in Eq. 19.1–33.

We have a total of five constants a_1, a_2, b_1, b_2, and α to determine. The following five initial and boundary conditions are used:

I. C.: at $t = 0$, $c_B = c_{B0}$ (19.1–40)

B. C. 1: at $z = 0$, $c_A = c_{A0}$ (19.1–41)

B. C. 2: at $z = z'(t)$, $c_A = 0$ (19.1–42)

B. C. 3: at $z = z'(t)$, $c_B = 0$ (19.1–43)

B. C. 4: at $z = z'(t)$, $-\mathscr{D}_{AS}\dfrac{\partial c_A}{\partial z} = +\mathscr{D}_{BS}\dfrac{\partial c_B}{\partial z}$ (19.1–44)

The last boundary condition is the stoichiometric requirement that one mole of A consume one mole of B. The five conditions allow us to get the five integration constants. The constant α is given implicitly by

$$1 - \operatorname{erf}\sqrt{\frac{\alpha}{\mathscr{D}_{BS}}} = \frac{c_{B0}}{c_{A0}}\sqrt{\frac{\mathscr{D}_{BS}}{\mathscr{D}_{AS}}}\operatorname{erf}\sqrt{\frac{\alpha}{\mathscr{D}_{AS}}}\exp\left\{\frac{\alpha}{\mathscr{D}_{AS}} - \frac{\alpha}{\mathscr{D}_{BS}}\right\} \tag{19.1–45}$$

The other constants are

$$a_1 = 1 \tag{19.1–46}$$

$$a_2 = -\left(\operatorname{erf}\sqrt{\frac{\alpha}{\mathscr{D}_{AS}}}\right)^{-1} \tag{19.1–47}$$

$$b_1 = 1 - \left(1 - \operatorname{erf}\sqrt{\frac{\alpha}{\mathscr{D}_{BS}}}\right)^{-1} \tag{19.1–48}$$

$$b_2 = +\left(1 - \operatorname{erf}\sqrt{\frac{\alpha}{\mathscr{D}_{BS}}}\right)^{-1} \tag{19.1–49}$$

Hence we first solve Eq. 19.1–45 for α (for given values of c_{B0}/c_{A0} and $\sqrt{\mathscr{D}_{BS}/\mathscr{D}_{AS}}$) and then insert this value of α into the subsequent relations to get a_2, b_1, b_2. In Fig. 19.1–2 a sample set of concentration profiles, as given by Sherwood and Pigford,[7] is shown. From this figure we can get quantitative information concerning the movement of the reaction zone.

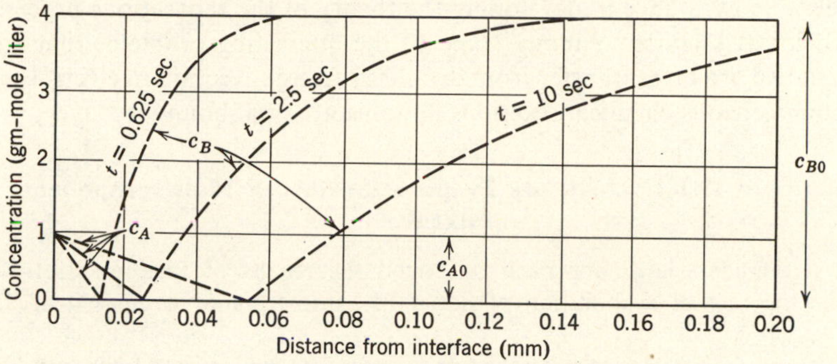

Fig. 19.1–2. Gas absorption with rapid chemical reaction. Concentration profiles according to Eqs. 19.1–33 and 19.1–34 with constants determined by Eqs. 19.1–45, 46, 47, 48, and 49. The calculation has been made for $\mathscr{D}_{AS} = 3.9 \times 10^{-5}$ ft² hr⁻¹ and $\mathscr{D}_{BS} = 1.95 \times 10^{-5}$ ft² hr⁻¹. [T. K. Sherwood and R. L. Pigford, *Absorption and Extraction*, McGraw-Hill, New York (1952), p. 336.]

From the concentration profiles we can calculate the rate of mass transfer at the interface:

$$N_A|_{z=0} = -\mathscr{D}_{AS} \left.\frac{\partial c_A}{\partial z}\right|_{z=0} = \frac{c_{A0}}{\operatorname{erf}\sqrt{\alpha/\mathscr{D}_{AS}}}\sqrt{\frac{\mathscr{D}_{AS}}{\pi t}} \tag{19.1–50}$$

The average rate of absorption up to time t is

$$N_{A,\text{avg}} = \frac{1}{t}\int_0^t N_A\,dt = 2\frac{c_{A0}}{\operatorname{erf}\sqrt{\alpha/\mathscr{D}_{AS}}}\sqrt{\frac{\mathscr{D}_{AS}}{\pi t}} \tag{19.1–51}$$

Hence the average rate up to time t is just twice the instantaneous rate.

§19.2 BOUNDARY-LAYER THEORY: VON KÁRMÁN APPROXIMATE METHOD

In this section we extend the boundary-layer theory discussions given earlier in §§4.4 and 11.4. Just as before, we consider two kinds of problems: unsteady transport, in which the boundary-layer thickness is a function of time, and steady transport, in which the boundary-layer thickness increases along a surface.

As illustrative examples, we have purposely selected two problems that have no analogs in heat transfer. In Example 19.2–1 we discuss the unsteady-state evaporation of a liquid into a multicomponent mixture; this leads to a discussion of the "sweep diffusion effect." In Example 19.2–2 we show how the concentration boundary-layer thickness depends on position and fluid properties in a mass-transfer system with a homogeneous reaction occurring. In the next section we consider the problem of calculating the velocity, temperature, and concentration profiles in laminar flow past a flat plate, with high mass-transfer rates at the plate surface.

The application of boundary-layer techniques to mass transfer has been of considerable assistance in developing the theory of the separations processes and chemical kinetics. Among some of the interesting problems that have been studied are mass transfer from droplets,[1] free convection in electrolysis,[2] and homogeneous chemical reactions in nonisothermal boundary layers.[3]

Example 19.2–1. Unsteady Evaporation into a Multicomponent Mixture[4]

Use a boundary-layer approach to extend the results of Example 19.1–1 to describe the rate of evaporation of liquid "1" into a vapor composed of "2," "3," . . . , "n".

Solution. (*a*) First we re-solve the two-component problem by a boundary-layer method. For our starting point we take an integrated form of Eq. 19.1–1 (the equation of continuity) and Eq. 19.1–4 (the expression for the flux at $z = 0$):

$$\frac{N_{A0}}{c} = \frac{d}{dt}\int_0^\infty x_A\,dz \tag{19.2–1}$$

$$\frac{N_{A0}}{c} = -\frac{\mathscr{D}_{AB}}{1 - x_{A0}}\left.\frac{\partial x_A}{\partial z}\right|_{z=0} \tag{19.2–2}$$

By equating these two expressions for N_{A0}/c, we get

$$\frac{d}{dt}\int_0^\infty x_A\,dz = -\frac{\mathscr{D}_{AB}}{1 - x_{A0}}\left.\frac{\partial x_A}{\partial z}\right|_{z=0} \tag{19.2–3}$$

This expression could also have been obtained by integrating Eq. 19.1–7 from $z = 0$ to $z = \infty$.

A concentration profile of the following form can be proposed:

$$\frac{x_A}{x_{A0}} = f\left(\frac{z}{\delta}\right) = f(\zeta) \qquad \text{for } \zeta \leqslant 1 \tag{19.2–4}$$

$$\frac{x_A}{x_{A0}} = 0 \qquad \text{for } \zeta \geqslant 1 \tag{19.2–5}$$

in which $\delta = \delta(t)$ is the concentration boundary-layer thickness. The function $f(\zeta)$ is required to have the following properties: $f(0) = 1, f(1) = 0$, and $f'(1) = 0$.

Substitution of the foregoing approximate profile into Eq. 19.2–3 gives

$$x_{A0}\left(\frac{d\delta}{dt}\right)\int_0^1 f\,d\zeta = \frac{(\mathscr{D}_{AB}x_{A0}/\delta)}{1 - x_{A0}}(-f_0') \tag{19.2–6}$$

[1] S. K. Friedlander, *A.I.Ch.E. Journal*, **3**, 43–49 (1957).
[2] C. R. Wilke, C. W. Tobias, and M. Eisenberg, *Chem. Eng. Prog.*, **49**, 663–674 (1953).
[3] F. E. Marble and T. C. Adamson, *Jet Propulsion*, **24**, 85–94 (1954).
[4] H-W. Hsu and R. B. Bird, *A.I.Ch.E. Journal*, **6**, 551–553 (1960).

Here f_0' is $df/d\zeta$ evaluated at $\zeta = 0$. This is an ordinary differential equation for δ as a function of t. When this is integrated with respect to time, using the initial condition that $\delta = 0$ at $t = 0$, we get

$$\delta = \sqrt{\frac{2\mathcal{D}_{AB}t(-f_0')}{(1 - x_{A0})\int_0^1 f\,d\zeta}} \tag{19.2-7}$$

Once δ is known, the evaporation rate is then found to be

$$\frac{dV_A}{dt} = \frac{N_{A0}S}{c} = \frac{(S\mathcal{D}_{AB}x_{A0}/\delta)}{1 - x_{A0}}(-f_0')$$

$$= Sx_{A0}\sqrt{\frac{\left(\int_0^1 f\,d\zeta\right)(-f_0')\mathcal{D}_{AB}}{2(1 - x_{A0})t}} \tag{19.2-8}$$

and the volume of A produced is

$$V_A = Sx_{A0}\sqrt{\frac{\left(\int_0^1 f\,d\zeta\right)(-f_0')\cdot 2\mathcal{D}_{AB}t}{(1 - x_{A0})}} \tag{19.2-9}$$

Comparison of Eq. 19.2-9 with Eq. 19.1-21 reveals that

$$\psi^{(2)}_{\text{bdry.lyr.}} = \sqrt{\frac{\left(\int_0^1 f\,d\zeta\right)(-f_0')\pi}{2(1 - x_{A0})}} \tag{19.2-10}$$

A very simple function $f(\zeta)$, which satisfies the restrictions mentioned above, is $f(\zeta) = (1 - \zeta)^2$. The reader may verify that for this function $\psi^{(2)}_{\text{bdry.lyr.}} = \sqrt{\pi/3(1 - x_{A0})}$; the $\psi^{(2)}$-values obtained thus are 1.02, 1.18, 1.44, 2.04, ∞ corresponding to the entries in the last column of Table 19.1-1. It is thus seen that the particular choice $f(\zeta) = (1 - \zeta)^2$ gives results that are progressively worse as the boiling point of the liquid is approached. This happens because the dependence of profile shape on x_{A0} was neglected. (See Fig. 19.1-1.)

b. For a multicomponent system we take as our starting point an integrated form of the equation of continuity for component "1" and the Stefan-Maxwell equations (see Eq. 18.4-19) for components "1," "2," . . . , "n" written for $z = 0$

$$\frac{N_{10}}{c} = \frac{d}{dt}\int_0^\infty x_1\,dz \tag{19.2-11}$$

$$\frac{N_{10}}{c} = -\left(\sum_{i=2}^{n}\frac{x_{i0}}{\mathcal{D}_{1i}}\right)^{-1}\left.\frac{\partial x_1}{\partial z}\right|_{z=0} \tag{19.2-12}$$

$$\frac{N_{10}}{c} = +\left(\frac{\mathcal{D}_{1j}}{x_{j0}}\right)\left.\frac{\partial x_j}{\partial z}\right|_{z=0} \qquad j = 2, 3, \ldots, n \tag{19.2-13}$$

in which the x_{i0} are the gas-phase mole fractions at the gas-liquid interface.

We now use the following approximate concentration profiles:

$$\frac{x_1}{x_{10}} = f_1\left(\frac{z}{\delta_1}\right) = f_1(\zeta_1) \qquad \text{for } \zeta_1 \leqslant 1 \tag{19.2–14}$$

$$\frac{x_j^\circ - x_j}{x_j^\circ - x_{j0}} = f_j\left(\frac{z}{\delta_j}\right) = f_j(\zeta_j) \qquad \text{for } \zeta_j \leqslant 1 \qquad j = 2, 3, \ldots, n \tag{19.2–15}$$

in which x_j° is the concentration of species j at $t = 0$ in the gas mixture. That is, we let each species have its own concentration boundary-layer thickness $\delta_j(t)$, and we specify that $f_j(0) = 1, f_j(1) = 0$, and $f_j'(1) = 0$ for all j.

When the function f_1 is introduced into Eqs. 19.2–11 and 12 and the two expressions for N_{10}/c are equated, we get a differential equation for $\delta_1(t)$. When that equation is solved, we get

$$\delta_1 = \sqrt{\frac{2(-f_{10}')t}{\int_0^1 f_1\, d\zeta_1}}\left(\sum_{i=2}^{n} \frac{x_{i0}}{\mathscr{D}_{1i}}\right)^{-1/2} \tag{19.2–16}$$

Expressions for $\delta_2, \delta_3 \ldots \delta_n$ can be derived by equating the right sides of Eqs. 19.2–11 and 13.

By the same procedure used in part (*a*), we can find the volume of species "1" produced:

$$\begin{aligned} V_1 &= Sx_{10}\sqrt{\left(\int_0^1 f_1\, d\zeta_1\right)(-f_{10}')2t}\left(\sum_{i=2}^{n} \frac{x_{i0}}{\mathscr{D}_{1i}}\right)^{-1/2} \\ &= Sx_{10}\sqrt{\frac{4\mathscr{D}_{12}t}{\pi}} \cdot \psi^{(n)}_{\text{bdry. lyr.}} \end{aligned} \tag{19.2–17}$$

which defines a correction factor analogous to $\psi^{(2)}_{\text{bdry. lyr.}}$ in Eq. 19.2–10.

If, now, we assume that f in part (*a*) is the same function as f_1 here, then combination of Eqs. 19.2–10 and 19.2–17 shows that

$$\frac{\psi^{(n)}_{\text{bdry. lyr.}}}{\psi^{(2)}_{\text{bdry. lyr.}}} = \left(\frac{\dfrac{1 - x_{10}}{\mathscr{D}_{12}}}{\dfrac{x_{20}}{\mathscr{D}_{12}} + \dfrac{x_{30}}{\mathscr{D}_{13}} + \cdots}\right)^{1/2} \tag{19.2–18}$$

If it is further postulated that the ratio $\psi^{(n)}/\psi^{(2)}$ computed by the boundary-layer method will not be very different from $\psi^{(n)}/\psi^{(2)}$ computed rigorously, then we may write:

$$V_1 = Sx_{10}\sqrt{\frac{4\mathscr{D}_{1,\text{eff}}t}{\pi}}\, \psi^{(2)} \tag{19.2–19}$$

in which $\psi^{(2)}$ is a function of x_{10}, as given in Table 19.1–1, and $\mathscr{D}_{1,\text{eff}}$ is an "effective diffusivity" defined by

$$\mathscr{D}_{1,\text{eff}} = \frac{1 - x_{10}}{\sum_{j=2}^{n} \dfrac{x_{j0}}{\mathscr{D}_{1j}}} \tag{19.2–20}$$

It should be noted that this is the same as $\mathscr{D}_{1m}$, defined in Eq. 18.4–21, provided that it is specified that $\mathscr{D}_{1m}$ be evaluated at the plane $z = 0$. We conclude, therefore, that the same expression for the evaporation rate is obtained by two procedures: (i) the boundary-layer treatment with f_1 set equal to the f of the binary problem, or (ii) extension of the binary result in Example 19.1–1 by replacing $\mathscr{D}_{AB}$ by $\mathscr{D}_{1m}$, the latter being evaluated at the liquid-vapor interface. Equation 19.2–19 has been tested experimentally for several evaporating systems and has been found to describe the evaporation rate within the error of experimental measurements.[5]

In the unsteady-state evaporation there is a tendency for the local ratios of the concentrations of the nonevaporating species to change with time. That is, a separation effect occurs. This is known as the *sweep diffusion effect*. To describe this effect quantitatively requires a knowledge of the profiles. Because the boundary-layer treatments give generally poorer results for profiles than for transfer rates, we make no attempt at a quantitative treatment of the sweep diffusion effect here. The use of sweep diffusion as a separation process has been studied by several groups of investigators.[6]

Example 19.2–2. Diffusion and Chemical Reaction in Isothermal Laminar Flow Along a Soluble Flat Plate

An appropriate mass-transfer analog to the problem discussed in Example 11.4–1 (*q.v.*) would be the flow along a flat plate that contains a species A slightly soluble in the fluid B. The concentration at the plate surface would be c_{A0}, the solubility of A in B, and the concentration of A far from the plate would be $c_{A\infty}$. In this example we let $c_{A\infty} = 0$ and break the analogy with Example 11.4–1 by letting A react with B by an nth order homogeneous chemical reaction, with rate constant k_n'''. It is desired to analyze the system by a boundary-layer method.

Solution. We assume that the concentration of dissolved A is small enough so that the physical properties ρ, μ, and $\mathscr{D}_{AB}$ are virtually constant throughout the fluid. The boundary-layer equations for this system (*cf.* Eqs. 11.4–1, 2, and 3) are

(continuity) $$\frac{\partial v_x}{\partial x} + \frac{\partial v_y}{\partial y} = 0 \qquad (19.2\text{–}21)$$

(motion) $$v_x \frac{\partial v_x}{\partial x} + v_y \frac{\partial v_x}{\partial y} = \nu \frac{\partial^2 v_x}{\partial y^2} \qquad (19.2\text{–}22)$$

(continuity of A) $$v_x \frac{\partial c_A}{\partial x} + v_y \frac{\partial c_A}{\partial y} = \mathscr{D}_{AB} \frac{\partial^2 c_A}{\partial y^2} - k_n''' c_A{}^n \qquad (19.2\text{–}23)$$

The first two equations may be used to get the velocity distribution $v_x(x, y)$, and this has already been done in Example 4.4–2. Then the latter has to be substituted into the equation for c_A, which may be combined with Eq. 19.2–21 to read

$$v_x \frac{\partial c_A}{\partial x} - \left(\int_0^y \frac{\partial v_x}{\partial x}\, dy\right) \frac{\partial c_A}{\partial y} = \mathscr{D}_{AB} \frac{\partial^2 c_A}{\partial y^2} - k_n''' c_A{}^n \qquad (19.2\text{–}24)$$

[5] D, F. Fairbanks and C. R. Wilke, *Ind. Eng. Chem.*, **42,** 471–475 (1950).

[6] M. Benedict and A. Boas, *Chem. Eng. Prog.*, **47,** 51–62, 111–122 (1951); M. T. Cichelli, W. E. Weatherford, Jr., and J. R. Bowman, *Chem. Eng. Prog.*, **47,** 63–74, 123–133 (1951).

Note that we have assumed that $v_y = 0$ at $y = 0$ in writing the last equation; at high rates of mass transfer that would be an improper assumption. Equation 19.2–24 may be integrated over y (*cf.* Eq. 11.4–23 for heat transfer and Eq. 4.4–29 for momentum transfer) to give

$$N_{Ay}\big|_{y=0} = \frac{d}{dx}\left[\delta_4 v_\infty (c_{A0} - c_{A\infty})\right] + k_n''' \int_0^\infty c_A{}^n \, dy \tag{19.2–25}$$

in which the *concentration thickness* δ_4 is

$$\delta_4 = \int_0^\infty \frac{v_x}{v_\infty}\left(1 - \frac{c_{A0} - c_A}{c_{A0} - c_{A\infty}}\right) dy \tag{19.2–26}$$

The integral for δ_4 has exactly the same form as that for δ_3 in Eq. 11.4–24.

For the present problem we set $c_{A\infty} = 0$ and introduce the dimensionless velocity $\phi = v_x/v_\infty$ and the dimensionless concentration $\Gamma = (c_{A0} - c_A)/c_{A0}$. Then Eq. 19.2–25 becomes

$$\frac{\mathscr{D}_{AB}}{v_\infty}\frac{\partial \Gamma}{\partial y}\bigg|_{y=0} = \frac{d}{dx}\int_0^\infty \phi(1 - \Gamma)\, dy + \frac{k_n''' c_{A0}^{n-1}}{v_\infty}\int_0^\infty (1 - \Gamma)^n \, dy \tag{19.2–27}$$

Now we introduce the following "similar" profiles:

$$\phi = \phi(\eta) \qquad \text{where} \quad \eta = \frac{y}{\delta(x)} \tag{19.2–28}$$

$$\Gamma = \Gamma(\eta_c) \qquad \text{where} \quad \eta_c = \frac{y}{\delta_c(x)} \tag{19.2–29}$$

That is, we have different boundary-layer thicknesses δ and δ_c for momentum and mass transfer, respectively. In order to relate this problem with that discussed in Example 11.4–1, we introduce the quantity $\Delta = \delta_c/\delta$, which in this case is a function of x because of the chemical reaction occurring. We restrict the discussion to $\Delta \leqslant 1$, for which the mass boundary layer lies entirely within the momentum boundary layer. Rewriting Eq. 19.2–27 in terms of the variables η and η_c gives

$$\frac{\mathscr{D}_{AB}}{v_\infty}\frac{\partial \Gamma}{\partial \eta_c}\bigg|_{\eta_c=0} \cdot \frac{1}{\delta(x)\,\Delta(x)} = \frac{d}{dx}\left\{\int_0^1 \phi(\eta_c \Delta)\,[1 - \Gamma(\eta_c)]\, d\eta_c \cdot \delta(x)\,\Delta(x)\right\} + \left\{\frac{k_n''' c_{A0}^{n-1}}{v_\infty}\int_0^1 [1 - \Gamma(\eta_c)]^n \, d\eta_c \cdot \delta(x)\,\Delta(x)\right\} \tag{19.2–30}$$

In order to get $\Delta(x)$, we now have to assume functional forms for $\phi(\eta)$ and $\Gamma(\eta_c)$. In order to minimize the algebra and still illustrate the method, we select very simple functions (clearly one can dream up more realistic functions):

$$\phi(\eta) = \begin{cases} \eta & \text{for} \quad \eta \leqslant 1 \\ 1 & \text{for} \quad \eta \geqslant 1 \end{cases} \tag{19.2–31}$$

$$\Gamma(\eta_c) = \begin{cases} \eta_c & \text{for} \quad \eta_c \leqslant 1 \\ 1 & \text{for} \quad \eta_c \geqslant 1 \end{cases} \tag{19.2–32}$$

The model is sketched in Fig. 19.2–1.

For this very simple choice, the reader may verify that Eqs. 4.4–29 and 31 give[7]

$$\delta(x) = \sqrt{12}\sqrt{\frac{\nu x}{v_\infty}} \tag{19.2–33}$$

Introducing into Eq. 19.2–30 the assumed forms for $\phi(\eta)$ and $\Gamma(\eta_c)$ gives after multiplication by $\delta\Delta$

$$\frac{\mathscr{D}_{AB}}{v_\infty} = \delta\Delta\frac{d}{dx}\left(\tfrac{1}{6}\delta\Delta^2\right) + \frac{k_n'''c_{A0}^{n-1}}{v_\infty(n+1)}\delta^2\Delta^2 \tag{19.2–34}$$

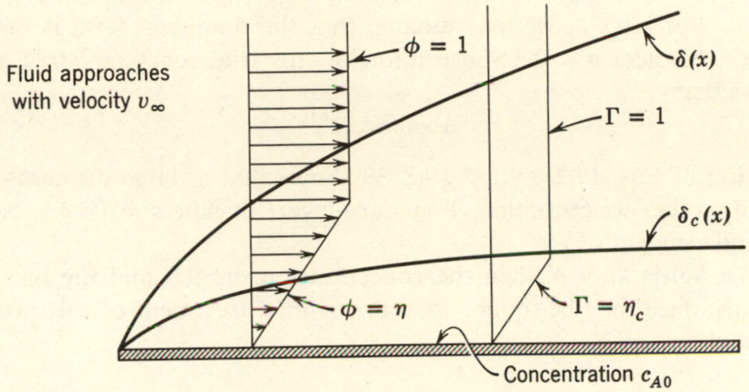

Fig. 19.2–1. Assumed velocity and concentration profiles for the laminar boundary layer with homogeneous chemical reaction.

Insertion of $\delta(x)$ from Eq. 19.2–33 and multiplication by v_∞/ν then gives

$$\frac{1}{\mathrm{Sc}} = \frac{4}{3}x\frac{d}{dx}\Delta^3 + \Delta^3 + \frac{12}{n+1}\left(\frac{k_n'''c_{A0}^{n-1}}{v_\infty}\right)x\Delta^2 \tag{19.2–35}$$

This is the differential equation for Δ, which is a function of the Schmidt number Sc and the distance x along the plate.

When there is *no reaction occurring*, k_n''' is zero and Eq. 19.2–35 becomes a linear first-order equation for Δ^3. When that equation is integrated, we get

$$\Delta^3 = \frac{1}{\mathrm{Sc}} + \frac{C}{x^{3/4}} \tag{19.2–36}$$

in which C is a constant of integration. Because Δ does not become infinite as x approaches zero, we obtain in the absence of chemical reaction (*cf.* Eq. 11.4–21)

$$\Delta = \mathrm{Sc}^{-1/3} \tag{19.2–37}$$

[7] Or, see H. Schlichting, *Boundary-Layer Theory*, McGraw-Hill, New York (1955), pp. 204–205. Presumably $\delta(x)$ can be calculated either from Eq. 4.4–29 or from Eq. 4.4–23. However, for the velocity profiles assumed in Eq. 19.2–31, the foregoing two methods of getting $\delta(x)$ are not equivalent because the assumed velocity profiles have a discontinuous first derivative at the outer edge of the boundary layer ($\eta = 1$). Similar remarks apply in the use of temperature or concentration profiles which have discontinuous first derivatives.

That is, when there is no reaction, the concentration and momentum boundary-layer thicknesses bear a constant ratio to one another, dependent only on the value of the Schmidt number.

When the *reaction is slow*, k_n''' is small and a series solution to Eq. 19.2–35 may be obtained:

$$\Delta = \text{Sc}^{-1/3}(1 + a_1\xi + a_2\xi^2 + \cdots) \qquad (19.2\text{–}38)$$

in which $\xi = [12/(n+1)](k_n''' c_{A0}^{n-1} x/v_\infty)$. Substitution of this expression into Eq. 19.2–35 gives $a_1 = -(\frac{1}{7})\text{Sc}^{1/3}$, $a_2 = (\frac{3}{539})\text{Sc}^{2/3}$, etc. Because a_1 is negative, the boundary-layer thickness is diminished by the chemical reaction.

When the *reaction is fast* (or when x is very large), a series solution in $1/\xi$ is more convenient. For large ξ, we may assume that the dominant term is of the form $\Delta = \text{const} \cdot \xi^n$ where $n < 0$. Substitution of this trial function into Eq. 19.2–35 then shows that

$$\Delta = (\text{Sc}\,\xi)^{-1/2} \qquad (19.2\text{–}39)$$

Combination of Eqs. 19.2–33 and 19.2–39 shows that at large distances from the leading edge the concentration boundary-layer thickness $\delta_c = \delta\Delta$ becomes a constant independent of ν.

Once $\Delta(x, \text{Sc})$ is known, then the concentration profiles and the mass transfer rate at the surface may be found. A more refined treatment of this problem has been given elsewhere.[8]

§19.3 BOUNDARY-LAYER THEORY: EXACT SOLUTIONS FOR SIMULTANEOUS HEAT, MASS, AND MOMENTUM TRANSFER

The boundary-layer developments we have discussed thus far are based on the von Kármán integral forms of the boundary-layer equations (Eqs. 4.4–29, 11.4–23, and 19.2–25), which are obtained from the Prandtl boundary-layer equations by integrating in the y-direction. To solve the von Kármán equations for the boundary-layer thicknesses δ, δ_T, and δ_c, one has to assume the shapes of the velocity, temperature, and concentration profiles. In this section we illustrate the more rigorous procedure of solving the Prandtl boundary-layer equations exactly;[1] this method gives greater accuracy and better illustrates the details of the transport mechanisms.

We are interested here in determining what happens when heat, mass, and momentum are simultaneously transferred across the boundary of a flowing fluid. This behavior is important in connection with many processes, such as

[8] See P. L. Chambré and J. D. Young, *Physics of Fluids*, **1**, 48–54 (1958). Catalytic surface reactions in boundary layers have been studied by P. L. Chambré and A. Acrivos, *J. Appl. Phys.*, **27**, 1322–1328 (1956).

[1] In the boundary-layer literature the term "exact solution" denotes a solution of the Prandtl boundary-layer equations as opposed to more approximate solutions. It should be noted that the Prandtl boundary-layer equations are actually asymptotic approximations to the equations of change and are accurate for two-dimensional laminar flow in the region $v_\infty x/\nu \gg 1$. (See H. Schlichting, *Boundary-Layer Theory*, Pergamon Press, New York (1955), Chapter VII.)

combustion of solid fuels, heterogeneous catalysis, and separation processes. The system chosen here is a simple one, but a study of the results can give valuable insight regarding the behavior of more complex systems.

Consider the flow system shown in Fig. 19.3–1. A thin, semi-infinite plate of volatile solid A sublimes, under steady conditions, into an unbounded gaseous stream of A and B, which approaches the plate tangentially in the x-direction with velocity v_∞. Species B is present in the gas phase only. We wish to determine the profiles of velocity, temperature, and concentration in the boundary layer for the case of known, uniform temperature and gas composition along the plate surface.

For simplicity, we assume that there are no chemical reactions and no external forces other than gravity, and we neglect viscous dissipation, heats of mixing, and emission and absorption of radiant energy in the gaseous boundary layer. These assumptions are reasonable for many gaseous systems. In addition, we treat the physical properties ρ, μ, $\hat{C}_p$, k, c, and $\mathscr{D}_{AB}$ as constants.[2] With these simplifications, the boundary-layer equations for this system become

(continuity)
$$\frac{\partial v_x}{\partial x} + \frac{\partial v_y}{\partial y} = 0 \tag{19.3–1}$$

(motion)
$$v_x \frac{\partial v_x}{\partial x} + v_y \frac{\partial v_x}{\partial y} = \nu \frac{\partial^2 v_x}{\partial y^2} \tag{19.3–2}$$

(energy)
$$v_x \frac{\partial T}{\partial x} + v_y \frac{\partial T}{\partial y} = \alpha \frac{\partial^2 T}{\partial y^2} \tag{19.3–3}$$

(continuity of A)
$$v_x \frac{\partial x_A}{\partial x} + v_y \frac{\partial x_A}{\partial y} = \mathscr{D}_{AB} \frac{\partial^2 x_A}{\partial y^2} \tag{19.3–4}$$

The boundary conditions are

$$\text{at } x \leqslant 0 \quad or \quad y = \infty: \quad v_x = v_\infty, \quad T = T_\infty, \quad x_A = x_{A\infty} \tag{19.3–5,6,7}$$

$$\text{at } y = 0: \quad v_x = 0, \quad T = T_0, \quad x_A = x_{A0}, \quad N_B = 0 \tag{19.3–8,9,10,11}$$

The last equation follows from the assumptions that B exists only in the gas phase and that there are no chemical reactions.

The equation of continuity may be integrated to obtain v_y as follows:

$$v_y = v_{y0} - \int_0^y \frac{\partial v_x}{\partial x}\, dy \tag{19.3–12}$$

or

$$v_y = \frac{M_A N_{A0}}{\rho} - \int_0^y \frac{\partial v_x}{\partial x}\, dy \tag{19.3–13}$$

[2] Note that these assumptions imply that $M_A = M_B$ and $\hat{C}_{pA} = \hat{C}_{pB}$, inasmuch as x_A is a function of position. Approximate methods of applying the present solutions to systems with variable properties are discussed in §21.7. Variation in the physical properties can be treated accurately by the numerical integration method of H. Schuh, *Z. angew. Math. Mech.* **25–27**, 54–60 (1947); *NACA Tech Memo 1275* (1950).

in which the subscript 0 denotes conditions at $y = 0$. The second relation follows from Eqs. *J* and *S* in Table 16.1–3. Evaluation of N_{A0} from Fick's law (Eq. 16.2–2) gives

$$N_{A0} = -c\mathscr{D}_{AB} \left. \frac{\partial x_A}{\partial y} \right|_{y=0} + x_{A0}(N_{A0} + N_{B0}) \tag{19.3–14}$$

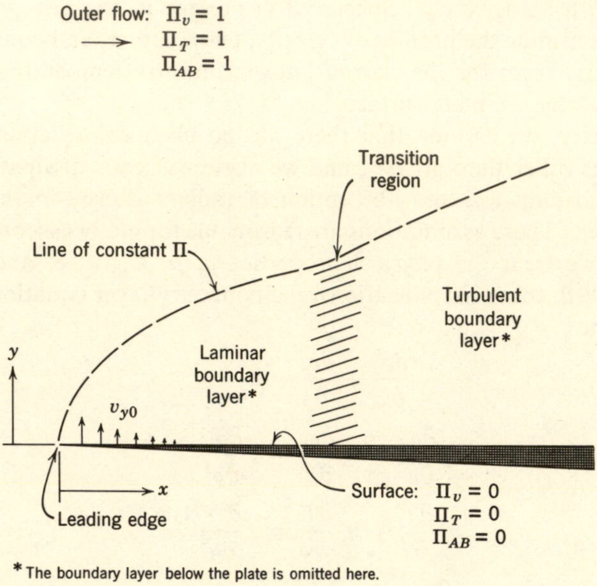

Fig. 19.3–1. Tangential flow along a sharp-edged semi-infinite flat plate with mass transfer into the stream. The laminar-turbulent transition usually occurs at a *length* Reynolds number ($v_\infty x/\nu$) on the order of 10^5.

By setting $N_{B0} = 0$ from Eq. 19.3–11 and combining the result with Eq. 19.3–13, we get

$$v_y = -\frac{M_A}{\rho} \frac{c\mathscr{D}_{AB}}{(1 - x_{A0})} \left. \frac{\partial x_A}{\partial y} \right|_{y=0} - \int_0^y \frac{\partial v_x}{\partial x}\, dy \tag{19.3–15}$$

This expression is to be used for v_y in Eqs. 19.3–2, 3, and 4. Because we have assumed that $M_A = M_B$, we may set $M_A c/\rho$ equal to unity.

We now define the dimensionless profile variables:

$$\Pi_v = \frac{v_x - v_{x0}}{v_{x\infty} - v_{x0}} = \frac{v_x}{v_\infty}; \quad \Pi_T = \frac{T - T_0}{T_\infty - T_0}; \quad \Pi_{AB} = \frac{x_A - x_{A0}}{x_{A\infty} - x_{A0}} \tag{19.3–16,17,18}$$

and the dimensionless physical property ratios

$$\Lambda_v = \frac{\nu}{\nu} = 1; \quad \Lambda_T = \frac{\nu}{\alpha} = \text{Pr}; \quad \Lambda_{AB} = \frac{\nu}{\mathscr{D}_{AB}} = \text{Sc} \tag{19.3–19,20,21}$$

With these definitions, the equations of motion, energy, and continuity of A take the common form

$$v_x \frac{\partial \Pi}{\partial x} - \left(\frac{\mathscr{D}_{AB}}{(1 - x_{A0})} \frac{\partial x_A}{\partial y}\bigg|_{y=0} + \int_0^y \frac{\partial v_x}{\partial x}\, dy \right) \frac{\partial \Pi}{\partial y} = \frac{\nu}{\Lambda} \frac{\partial^2 \Pi}{\partial y^2} \qquad (19.3\text{–}22)$$

and the boundary conditions on v_x, T, and x_A become

$$\text{at } x \leqslant 0 \quad or \quad y = \infty, \qquad \Pi = 1 \qquad (19.3\text{–}23)$$

$$\text{at } y = 0, \qquad \Pi = 0 \qquad (19.3\text{–}24)$$

The form of the boundary conditions on the profile variables Π, and the lack of a characteristic length in the flow system, suggest that the method of combination of variables may be used. Equation 4.4–26 suggests a combination of the form

$$\eta = \frac{y}{2} \sqrt{\frac{v_\infty}{\nu x}} \qquad (19.3\text{–}25)$$

By expressing the derivatives in Eq. 19.3–22 in terms of η, we obtain[3]

$$\Lambda \left[\frac{1}{\Lambda_{AB}} \left(\frac{x_{A0} - x_{A\infty}}{1 - x_{A0}} \right) \Pi'_{AB}(0) - \int_0^\eta 2\Pi_v \, d\eta \right] \Pi' = \Pi'' \qquad (19.3\text{–}26)$$

in which primes denote differentiation with respect to η. We see from this expression that the rate of mass transfer at the wall, expressed here in terms of the concentration gradient $\Pi'_{AB}(0)$, directly affects the three profiles Π_v, Π_T, and Π_{AB}.

For later convenience, we define the quantity

$$f = -K + \int_0^\eta 2\Pi_v \, d\eta \qquad (19.3\text{–}27)$$

The quantity K is a dimensionless mass flux at the wall, given here by

$$K = \frac{1}{\Lambda_{AB}} \left(\frac{x_{A0} - x_{A\infty}}{1 - x_{A0}} \right) \Pi'_{AB}(0) \qquad (19.3\text{–}28)$$

[3] It is instructive to compare the development at this point with that given in Example 19.1–1. The analogous quantities in the two developments are for small y:

Example 19.1–1	$1 - X$	Z	$-\frac{dX}{dZ}\Big\|_{Z=0}$	2φ	$\frac{x_{A0}}{1 - x_{A0}}$
This section	Π_{AB}	η	$\Pi'_{AB}(0)$	$K\Lambda_{AB}$	$\frac{x_{A0} - x_{A\infty}}{1 - x_{A0}}$

and is constant for the laminar boundary layer.[4] The boundary-layer equations of motion, energy, and continuity then become

$$-\Lambda f \Pi' = \Pi'' \tag{19.3–29}$$

The boundary conditions on Π are

$$\text{at } \eta = 0, \qquad \Pi = 0 \tag{19.3–30}$$

$$\text{at } \eta = \infty, \qquad \Pi = 1 \tag{19.3–31}$$

The given differential equations and their boundary conditions have now been expressed in terms of the single independent variable η, confirming the assumed combination of variables.

From Eqs. 19.3–27 through 31, we can compute the profiles of velocity, temperature, and concentration in the laminar boundary layer. We first compute the velocity profiles Π_v and the related function f for the desired value of K by solving Eq. 19.3–29 with $\Lambda = 1$; then the same equation can be directly integrated for any value of Λ to obtain the corresponding temperature or concentration profile:

$$\Pi(\eta, \Lambda, K) = \frac{\displaystyle\int_0^\eta \exp\left(-\Lambda \int_0^\eta f\, d\eta\right) d\eta}{\displaystyle\int_0^\infty \exp\left(-\Lambda \int_0^\eta f\, d\eta\right) d\eta} \tag{19.3–32}$$

Some profiles computed numerically from Eq. 19.3–32 are given in Fig. 19.3–2. Velocity profiles are given by the curves for $\Lambda = 1$; temperature and concentration profiles for various Prandtl and Schmidt numbers are given by the curves for the corresponding values of Λ. Note that the velocity, temperature, and concentration boundary layers get thicker when the dimensionless mass-transfer rate K is positive (as in evaporation) and thinner when K is negative (as in condensation).

The dimensionless gradients of velocity, temperature, and concentration at the wall, $\Pi'(0, \Lambda, K)$, are obtained from the derivative of Eq. 19.3–32:

$$\Pi'(0, \Lambda, K) = \frac{1}{\displaystyle\int_0^\infty \exp\left(-\Lambda \int_0^\eta f\, d\eta\right) d\eta} \tag{19.3–33}$$

A summary of values computed from this formula by numerical integration is given in Table 19.3–1.

[4] The dimensionless mass flux K is defined as

$$K = \frac{2v_{y0}}{v_\infty}\sqrt{\frac{v_\infty x}{\nu}} \tag{19.3–28a}$$

in which v_{y0} is the fluid velocity in the y-direction at the wall. This definition is applicable even when N_{A0} and N_{B0} are both nonzero. Note that v_{y0} varies as $1/\sqrt{x}$; this prediction, of course, is valid only in the region $v_\infty x/\nu \gg 1$ for which boundary-layer theory is valid.

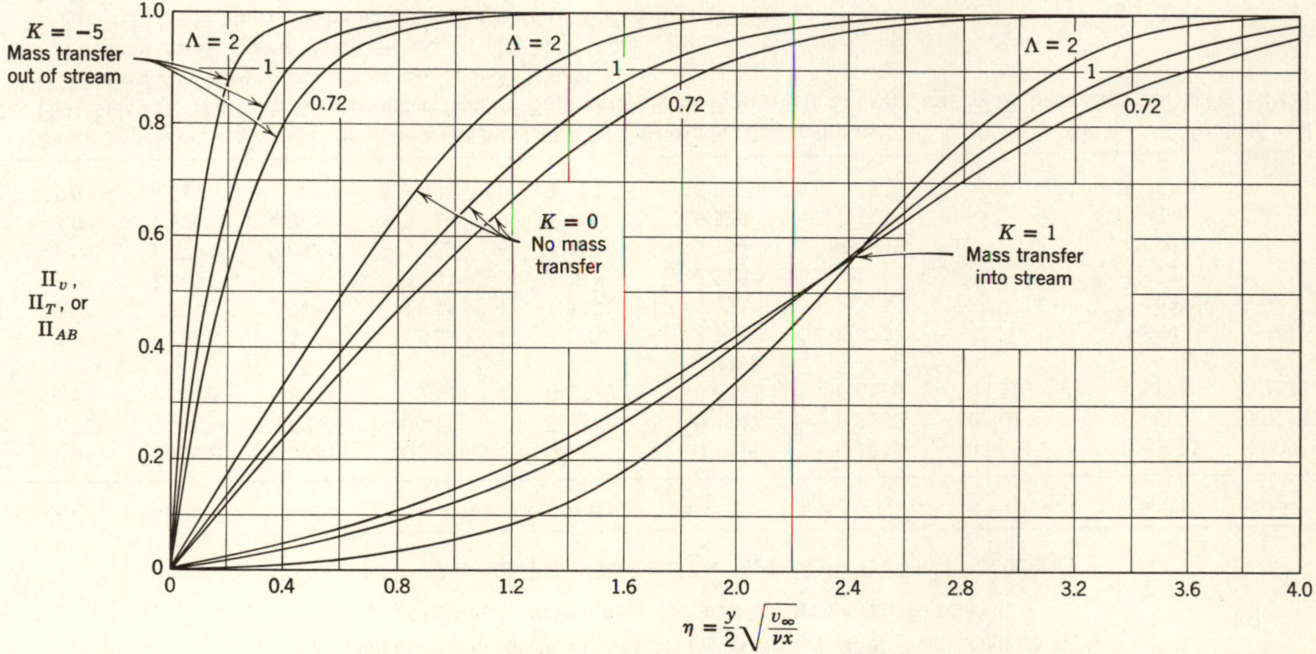

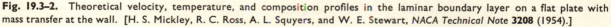

Fig. 19.3–2. Theoretical velocity, temperature, and composition profiles in the laminar boundary layer on a flat plate with mass transfer at the wall. [H. S. Mickley, R. C. Ross, A. L. Squyers, and W. E. Stewart, *NACA Technical Note* **3208** (1954).]

TABLE 19.3–1

DIMENSIONLESS GRADIENTS OF VELOCITY, TEMPERATURE, AND COMPOSITION CALCULATED FROM LAMINAR BOUNDARY-LAYER THEORY[a]

Dimensionless Gradient $2\Pi'(0, \Lambda, K)$ at the Flow Boundary

K	$\Lambda = 0.6$	$\Lambda = 0.7$	$\Lambda = 0.8$	$\Lambda = 0.9$	$\Lambda = 1.0$	$\Lambda = 1.1$	$\Lambda = 1.4$	$\Lambda = 2.0$	$\Lambda = 5.0$
1.0[e]	0.2321	0.2067	0.1831	0.1615	0.1419	0.1242	0.08219	0.03423	0.0002818
0.75[e]	0.4448	0.4290	0.4116	0.3933	0.3747	0.3560	0.3019	0.2098	0.02531
0.5[e]	0.6600	0.6644	0.6649	0.6626	0.6580	0.6516	0.6527	0.5587	0.2533
0.25[e]					0.9787				
0.0[c]	1.108	1.170	1.228	1.280	1.328[b]	1.374		1.689	
−0.5[d]	1.579	1.716	1.845	1.970	2.092	2.213		3.193	
−1.0[d]	2.069	2.288	2.501	2.709	2.916	3.121		4.892	
−1.5[d]	2.576	2.885	3.189	3.487	3.784	4.080		6.699	
−3.0[d]	4.174	4.765	5.354	5.943	6.529	7.115		12.41	
−5.0[d]	6.417	7.402	8.388	9.374	10.36	11.35		20.26	

[a] Based largely on the velocity profiles given by H. Schlichting and K. Bussmann, Schrift. d. deutsch. Akad. d. Luftfahrtforschung, *7B*, Heft 2 (1943); this table is reproduced from H. S. Mickley, R. C. Ross, A. L. Squyers, and W. E. Stewart, NACA TN 3208 (1954).

[b] H. Blasius, *Z. angew. Math. Phys.*, **56,** 1, 1–37 (1908).

[c] E. Pohlhausen, *Z. angew. Math. Mech.*, **1,** 115–121 (1921).

[d] J. A. Feyk, S.M. Thesis, Massachusetts Institute of Technology (1949).

[e] R. C. Ross, Private Communication (1950).

The fluxes of momentum, energy, and mass at the wall are given by the dimensionless expressions

$$\frac{\tau_{yx}|_{y=0}}{\rho v_\infty(0 - v_\infty)} = \frac{\Pi'(0, 1, K)}{2}\sqrt{\frac{\nu}{v_\infty x}} \tag{19.3–34}$$

$$\frac{q_y|_{y=0}}{\rho \hat{C}_p v_\infty(T_0 - T_\infty)} = \frac{\Pi'(0, \Lambda_T, K)}{2\Lambda_T}\sqrt{\frac{\nu}{v_\infty x}} \tag{19.3–35}$$

$$\frac{J_{Ay}^{\star}|_{y=0}}{c v_\infty(x_{A0} - x_{A\infty})} = \frac{\Pi'(0, \Lambda_{AB}, K)}{2\Lambda_{AB}}\sqrt{\frac{\nu}{v_\infty x}} \tag{19.3–36}$$

and the tabulated values of $\Pi'(0, \Lambda, K)$. We can thus compute the foregoing fluxes in terms of the dimensionless mass-transfer rate K. The solution for mass-transfer rate in terms of x_{A0} and $x_{A\infty}$ is illustrated in Example 19.3–1.

In a few limiting cases the profiles Π_T and Π_{AB} can be obtained analytically. For example, at small mass-transfer rates K and large values of Λ_T (the Prandtl number) or Λ_{AB} (the Schmidt number) the thermal or diffusional boundary layer is much thinner than the velocity boundary layer, and one needs to know only the velocity gradient at the wall to solve for the complete temperature and concentration profiles $\Pi(\eta, \Lambda, K)$. As we shall see, a small change in K has a marked effect on the temperature or concentration profile when Λ is large.

For small mass-transfer rates ($|K| \ll 1$), the velocity profile near the wall is

$$\Pi_v = \left(\frac{1.328}{2}\right)\eta \tag{19.3–37}$$

Here we have inserted the velocity gradient for $\eta = 0$ and $K = 0$, as given in Table 19.3–1. Application of Eqs. 19.3–27 and 32 then gives[5] for small K and $\Lambda \gg 1$

$$\Pi(\eta, \Lambda, K) \simeq \frac{\int_0^\eta \exp(\Lambda K\eta - 1.328\Lambda\eta^3/3!)\,d\eta}{\int_0^\infty \exp(\Lambda K\eta - 1.328\Lambda\eta^3/3!)\,d\eta} \tag{19.3–38}$$

and from Eq. 19.3–33 the corresponding temperature or concentration gradient at the wall is[6]

$$\Pi'(0, \Lambda, K) \simeq \frac{1}{\int_0^\infty \exp(\Lambda K\eta - 1.328\Lambda\eta^3/3!)\,d\eta} \tag{19.3–39}$$

[5] We use the symbol $\simeq$ to denote an asymptotic equality for large values of Λ.

[6] W. E. Stewart, Sc.D.Thesis, Massachusetts Institute of Technology (1951). See also, H. S. Mickley, R. C. Ross, A. L. Squyers, and W. E. Stewart, NACA TN 3208 (1954).

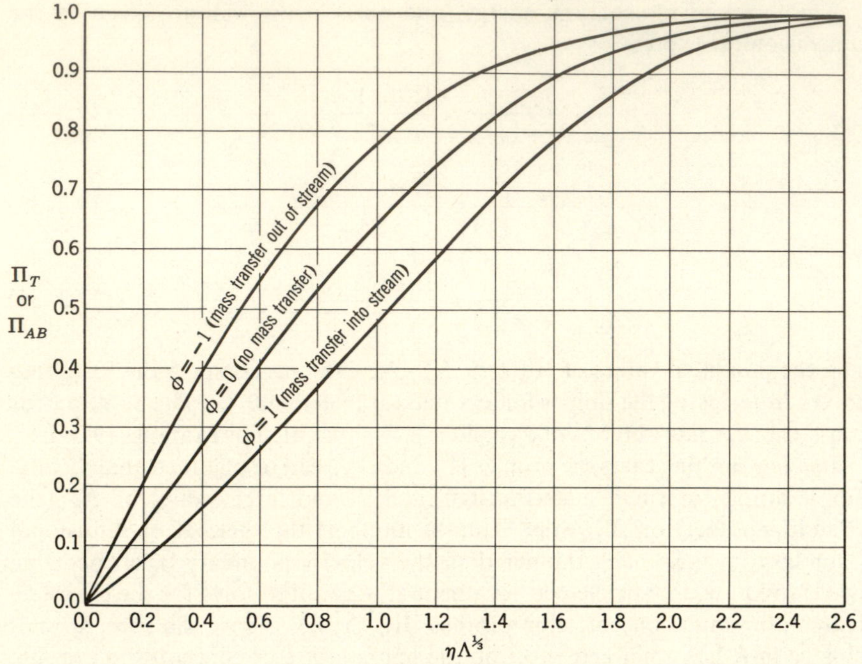

Fig. 19.3–3. Temperature or concentration profiles for high Prandtl or Schmidt number and various mass-transfer rates. The velocity profile used in these solutions is given by the line for $\Lambda = 1$ and $K = 0$ in Fig. 19.3–2.

In the following paragraphs we discuss the evaluation of the profiles and the gradients for small mass-transfer rates, K.

In the limit as $K \to 0$, the profiles take the form

$$\lim_{K\to 0} \Pi(\eta, \Lambda, K) \simeq \frac{\int_0^\eta \exp(-1.328\Lambda\eta^3/3!)\, d\eta}{\int_0^\infty \exp(-1.328\Lambda\eta^3/3!)\, d\eta} \tag{19.3–40}$$

$$= \frac{\Gamma(\frac{1}{3}, u)}{\Gamma(\frac{1}{3}, \infty)} \tag{19.3–41}$$

where

$$u = 1.328 \frac{\Lambda\eta^3}{3!} \tag{19.3–42}$$

and $\Gamma(x, u)$ is the incomplete gamma function.[7] This profile is plotted in Fig. 19.3–3; note that it has nearly the same shape as the profiles for finite

[7] E. Jahnke and F. Emde, *Tables of Functions*, Dover, New York (1945), Fourth Edition, p. 22. The function $\Gamma(x, \infty) = \Gamma(x)$ is the complete gamma function.

Λ and $K = 0$ in Fig. 19.3–2. From Eq. 19.3–39 the slope of the profile at the wall is

$$\lim_{K\to 0} \Pi'(0, \Lambda, K) \simeq \frac{1}{\displaystyle\int_0^\infty \exp(-1.328\Lambda\eta^3/3!)\, d\eta}$$
$$= \left[\frac{1}{3}\left(\frac{3!}{1.328\Lambda}\right)^{1/3} \Gamma\left(\frac{1}{3}\right)\right]^{-1}$$
$$\doteq 0.6774\Lambda^{1/3} \qquad (19.3\text{–}43)$$

This equation deviates by only 1 per cent, at $\Lambda = 2$, from the result obtained

TABLE 19.3–2

COMPARISON OF RESULTS FOR THE PROFILE GRADIENTS AT NEGLIGIBLE MASS-TRANSFER RATES

Λ	Predicted Values of $\Pi'(0, \Lambda, 0)$		
	Exact[a]	Asymptotic Solution $\Pi'(0, \Lambda, 0) = 0.6774\Lambda^{1/3}$	Pohlhausen Approximation $\Pi'(0, \Lambda, 0) = 0.664\Lambda^{1/3}$
0.6	0.552	0.571	0.560
0.8	0.614	0.629	0.616
1.0	0.664	0.677	0.664
2.0	0.845	0.853	0.837
7.0	1.29	1.296	1.27
10.0	1.46	1.459	1.43
15.0	1.67	1.671	1.64

[a] From the sources listed in Table 19.3–1.

by using the complete velocity profile, and the agreement becomes exact as $\Lambda \to \infty$. A detailed comparison is given in Table 19.3–2. Also shown are the results given by the equation

$$\Pi'(0, \Lambda, 0) \doteq 0.664\Lambda^{1/3} \qquad (19.3\text{–}44)$$

which was given by Pohlhausen[8] as a curve-fit of the exact values and is good within ± 2 per cent down to $\Lambda = 0.6$. Insertion of Pohlhausen's approximation in Eqs. 19.2–34, 35, and 36 gives the following convenient analogies for low mass-transfer rates:

$$\frac{\tau_{yx}|_{y=0}}{\rho v_\infty(0 - v_\infty)} = \frac{q_y|_{y=0}}{\rho \hat{C}_p v_\infty (T_0 - T_\infty)} (\mathrm{Pr})^{2/3} = \frac{J^\star_{Ay}|_{y=0}}{c v_\infty (x_{A0} - x_{A\infty})} (\mathrm{Sc})^{2/3}$$
$$= 0.332\,\mathrm{Re}^{-1/2} \qquad (19.3\text{–}45)$$

[8] E. Pohlhausen, *Z. angew. Math. Mech.*, **1**, 115–121 (1921).

in which $\text{Re} = v_\infty x/\nu$. This result justifies the Chilton-Colburn analogies (§§13.2 and 21.2) for this flow system for Pr or Sc greater than about $\frac{1}{2}$.

To extend the analytic solution to somewhat higher mass-transfer rates, the profiles are obtained from Eq. 19.3–38 by expanding exp $(\Lambda K\eta)$ in series and integrating term by term. The solution for any $K \ll 1$ is

$$\Pi(\eta, \Lambda, K) \simeq \frac{\sum_{m=0}^{\infty} c_m \phi^m \Gamma\left(\frac{m+1}{3}, u\right) \Big/ \Gamma\left(\frac{m+1}{3}, \infty\right)}{\sum_{m=0}^{\infty} c_m \phi^m} \tag{19.3–46}$$

Here $u = 1.328\Lambda\eta^3/3!$ as before, and the parameter ϕ is a dimensionless transfer rate:

$$\phi = \frac{K\Lambda}{\Pi'(0, \Lambda, 0)} \tag{19.3–47}$$

which, in view of Eq. 19.3–43 for large Λ, becomes

$$\phi \doteq \frac{K\Lambda^{2/3}}{0.6774} \tag{19.3–48}$$

The coefficients c_m are given by

$$c_m = \frac{1}{3}\frac{1}{m!}\Gamma\left(\frac{m+1}{3}\right)[\Gamma(\tfrac{4}{3})]^{-m-1} \tag{19.3–49}$$

TABLE 19.3–3

THE EFFECT OF MASS TRANSFER ON PROFILE GRADIENTS AT HIGH PRANDTL OR SCHMIDT NUMBER[a]

Mass flux Parameter $\lvert\phi\rvert$	$\dfrac{\Pi'(0, \Lambda, K)}{\Pi'(0, \Lambda, 0)}$ for positive ϕ (mass transfer into stream)	$\dfrac{\Pi'(0, \Lambda, K)}{\Pi'(0, \Lambda, 0)}$ for negative ϕ (mass transfer out of stream)
0.01	0.994	1.006
0.02	0.989	1.011
0.05	0.972	1.029
0.1	0.944	1.057
0.2	0.890	1.117
0.5	0.739	1.304
1.0	0.524	1.646
2.0	0.2300	2.421
3.0	0.0830	3.282
5.0	0.00613	5.140
∞	0	∞

[a] The values for $\lvert\phi\rvert < 5$ are taken from W. E. Stewart, Sc.D. Thesis, Massachusetts Institute of Technology (1951).

Some profiles computed from Eq. 19.3–46 are plotted in Fig. 19.3–3. The profiles are quite similar to those of Fig. 19.3–2 and show similar variations with mass-transfer rate. Note that the parameter ϕ, which determines the profile shapes, can have a large value in spite of the restriction $K \ll 1$ because Λ is assumed to be large. (See Eq. 19.3–48.) The gradients of these profiles at the wall are given by[6]

$$\frac{\Pi'(0, \Lambda, K)}{\Pi'(0, \Lambda, 0)} = \left(\sum_{m=0}^{\infty} c_m \phi^m\right)^{-1} \tag{19.3–50}$$

The ratio of gradients appearing in this equation gives a convenient measure of the effect of mass transfer on interphase transfer rates. Some values computed from this equation are given in Table 19.3–3; the results are further interpreted and applied to various mass-transfer problems in §21.7.

Example 19.3–1. Calculation of Mass-Transfer Rate

Give an expression for the local rate of sublimation from the plate shown in Fig. 19.3–1 into the laminar boundary layer. The mole fraction A next to the wall is $x_{A0} = 0.9$, the mole fraction A in the approaching stream is $x_{A\infty} = 0.01$, and the Schmidt number of the gas is constant at 2.0. M_A and M_B are equal, and the physical properties are constant.

Solution. Insertion of known quantities into Eq. 19.3–28 gives

$$\frac{2K}{\Pi_{AB}'(0, 2, K)} = \frac{0.9 - 0.01}{1 - 0.9} = 8.9 \tag{19.3–51}$$

To solve this equation for K, we evaluate the left-hand term as a function of K from Table 19.3–1:

K	$2\Pi'(0, 2, K)$	$2K/\Pi'(0, 2, K)$
1.00	0.03423	116.9
0.75	0.2098	14.30
0.50	0.5587	3.580
0.00	1.689	0.0000

From a plot of these values, we find that when $2K/\Pi'(0, 2, K) = 8.9$, $K = 0.65$. Therefore, the local mass flux is given by

$$K = \frac{2n_{A0}}{\rho v_\infty}\sqrt{\frac{v_\infty x \rho}{\mu}} = 0.65 \tag{19.3–52}$$

or

$$n_{A0} = 0.33\sqrt{\frac{\rho \mu v_\infty}{x}} \tag{19.3–53}$$

in the region $v_\infty x/\nu \gg 1$. In §21.7 this method of solution is simplified and extended to a variety of heat and mass-transfer problems.

QUESTIONS FOR DISCUSSION

1. To what extent can Table 19.0–1 be expanded to include analogies with momentum transport?

2. What phenomena in physics are described by the heat-conduction equation and the Laplace equation?

3. Prove that Eq. 19.1–16 satisfies the differential equation and boundary conditions in Eqs. 19.1–7 through 10.

4. What experimental difficulties can be anticipated in trying to measure gas-phase diffusivities by the system described in Example 19.1–1?

5. What is the utility of the result in Example 19.1–2? Suggest some problems that can be solved by this method.

6. Explain how one finds the rate of movement of the reaction zone in Example 19.1–3.

7. How do the results in Example 19.1–3 simplify if $\mathscr{D}_{AS} = \mathscr{D}_{BS}$?

8. From Eqs. 19.1–50 and 51, discuss the effect of reaction on absorption rates.

9. Obtain Eq. 19.2–3 by the alternate method suggested in the text.

10. Show quantitatively that the boundary-layer method used in Example 19.2–1 really does predict a sweep diffusion effect.

11. Show how Eq. 19.2–25 is derived.

12. Would you expect thermal diffusion ever to be important in a boundary-layer mass-transfer situation?

13. Why do the profiles in Fig. 19.3–2 and 3 depend on the mass-transfer rate? Explain in your own words.

14. Suggest some different physical situations for which the results of §19.3 could be used.

PROBLEMS

19.A$_1$ Estimation of Point Concentration in Binary Diffusional Evaporation

For the system in Example 19.1–1, calculate x_A at Z = 0.5 for a liquid A for which x_{A0} is 0.75. Compare your result with that given in Fig. 19.1–1. *Answer:* $x_A = 0.54$

19.B$_1$ Rate of Evaporation of *n*-Octane

At 20° C and 1 atm, how many grams of liquid *n*-octane will evaporate into N_2 in 24.5 hr in a system such as that studied in Example 19.1–1 if the area of the liquid surface is 1.29 cm^2? The vapor pressure of *n*-octane at 20° C is 10.45 mm. Hg. *Answer:* 6.71 mg

19.C$_2$ Boundary-Layer Results for Binary Diffusional Evaporation

Evaluate $\psi^{(2)}_{\text{bdry. lyr.}}$ in Example 19.2–1 by using the concentration profile $f(\zeta) = 1 - 2\zeta + 2\zeta^3 - \zeta^4$.

Answer: $\sqrt{\dfrac{3\pi}{10(1 - x_{A0})}}$

19.D$_3$ Absorption with Chemical Reaction in a Semi-Infinite Medium

Consider a semi-infinite medium extending from the plane boundary $x = 0$ to $x = \infty$. At time $t = 0$ substance A is brought into contact with this medium at the plane $x = 0$, the surface concentration being c_{A0} (for absorption of gas A by liquid B, c_{A0} would be the saturation concentration). A and B react in such a way that C is produced according to an irreversible first-order reaction, $A + B \rightarrow C$. It is assumed that A is present in such a small concentration that the equation describing the diffusional process is

$$\frac{\partial c_A}{\partial t} = \mathscr{D}_{AB} \frac{\partial^2 c_A}{\partial x^2} - k_1''' c_A \tag{19.D–1}$$

in which k_1''' is the first-order rate constant. This equation has been solved for the boundary and initial conditions:

I. C.: at $t = 0$, $c_A = 0$ (19.D–2)

B. C. 1: at $x = 0$, $c_A = c_{A0}$ (19.D–3)

B. C. 2: at $x = \infty$, $c_A = 0$ (19.D–4)

and the solution[1] is

$$\frac{c_A}{c_{A0}} = \frac{1}{2} \exp\left(-x\sqrt{\frac{k_1'''}{\mathscr{D}_{AB}}}\right) \operatorname{erfc}\left(\frac{x}{\sqrt{4\mathscr{D}_{AB}t}} - \sqrt{k_1'''t}\right) + \frac{1}{2} \exp\left(+x\sqrt{\frac{k_1'''}{\mathscr{D}_{AB}}}\right) \operatorname{erfc}\left(\frac{x}{\sqrt{4\mathscr{D}_{AB}t}} + \sqrt{k_1'''t}\right) \tag{19.D–5}$$

a. Prove that the foregoing solution satisfies the differential equation and boundary conditions.

b. Show that the molar flux of A at the interface $x = 0$ is

$$N_A\Big|_{x=0} = c_{A0}\sqrt{\mathscr{D}_{AB}k_1'''}\left(\operatorname{erf}\sqrt{k_1'''t} + \frac{e^{-k_1'''t}}{\sqrt{\pi k_1'''t}}\right) \tag{19.D–6}$$

c. Show further that the total moles of A absorbed from time $t = 0$ to $t = t_0$ is given by the expression

$$\mathscr{M}_A = c_{A0}\sqrt{\mathscr{D}_{AB}t_0}\left[\left(\sqrt{k_1'''t_0} + \frac{1}{2\sqrt{k_1'''t_0}}\right) \operatorname{erf}\sqrt{k_1'''t_0} + \frac{1}{\sqrt{\pi}} e^{-k_1'''t_0}\right] \tag{19.D–7}$$

d. Show that for large values of $k_1'''t_0$ the expression in (*c*) reduces asymptotically to

$$\mathscr{M}_A = c_{A0}\sqrt{\mathscr{D}_{AB}k_1'''}\left(t_0 + \frac{1}{2k_1'''}\right) \tag{19.D–8}$$

This equation[2] is good within 2 per cent for values of $k_1'''t_0 > 4$.

19.E$_3$ Unsteady Diffusion in a Solid with Chemical Reaction and Zero Surface Concentration

Show that the solution to Eq. 19.1–23 with

I. C.: at $t = 0$, $c_A = c_0(x, y, z)$ (19.E–1)

B. C.: at surface, $c_A = 0$ (19.E–2)

is

$$c_A(x, y, z, t) = g(x, y, z, t) \cdot e^{k_1'''t} \tag{19.E–3}$$

where $g(x, y, z, t)$ is the solution to Eq. 19.1–23 with $k_1''' = 0$, which satisfies the conditions in Eqs. 19.E–1 and 2.

[1] P. V. Danckwerts, *Trans. Faraday Soc.*, **46**, 300–304 (1950).

[2] R. A. T. O. Nijsing, *Absorptie van gassen in vloeistoffen, zonder en met chemische reactie*, Academisch Proefschrift, Delft (1957).

19.F₃ Simultaneous Heat, Mass, and Momentum Transfer: Alternate Boundary Conditions

Show that the dimensionless profiles $\Pi(\eta, \Lambda, K)$ still satisfy Eqs. 19.3–29, 30, and 31 (with f and K defined by Eqs. 19.3–27 and 28) for the following modifications of the system in §19.3–1. Obtain an expression for the constant K to replace Eq. 19.3–28 in each case.

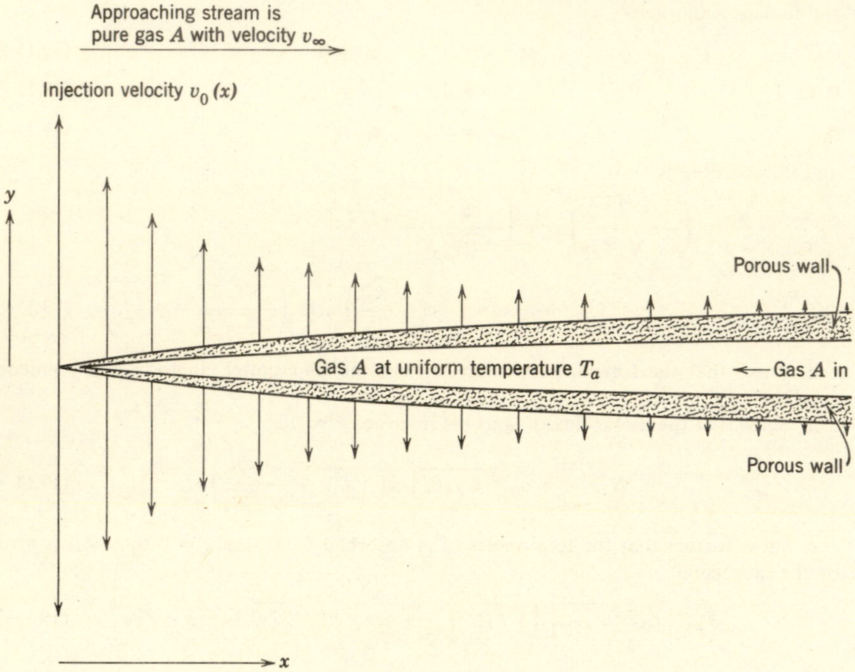

Fig. 19.F. Transpiration-cooled porous plate.

a. Same problem as stated in §19.3 except that the plate is a porous slab saturated with volatile species A and B and the ratio N_A/N_B is the same at all points on the surface.

b. The plate of A is replaced by a plate of nonvolatile solid C which reacts instantaneously with gas A to give 2 moles of gas B: $C + A \rightarrow 2B$. Assume $M_A = M_B$.

c. The plate is hollow and porous; it is transpiration-cooled by injection of cold gas A all along the plate, with a mass flux distribution such that the whole surface of the plate is at a uniform temperature T_0. The temperature of the gas is constant at T_a in the inner compartment. The approaching stream is pure A. Other assumptions remain as in §19.3. (See Fig. 19.F.)

Answers: $$a.\ K = \frac{1}{\Lambda_{AB}}\left(\frac{x_{A0} - x_{A\infty}}{\dfrac{N_{A0}}{N_{A0} + N_{B0}} - x_{A0}}\right)\Pi_{AB}'(0)$$

$$b.\ K = \frac{1}{\Lambda_{AB}}\left(\frac{x_{A0} - x_{A\infty}}{-1 - x_{A0}}\right)\Pi_{AB}'(0)$$

$$c.\ K = \frac{1}{\Lambda_{T}}\left(\frac{T_0 - T_\infty}{T_a - T_0}\right)\Pi_{T}'(0)$$

19.G₃ Solution of the Boundary-Layer Equations by Picard's Method

Equation 19.3–32 is the basis of a rapidly convergent iteration process for calculation of velocity profiles.[3,4] The procedure for any given K is simply to assume a trial velocity profile, insert it on the right-hand side of Eq. 19.3–32, and evaluate the integrals for $\Lambda = 1$; the resulting function $\Pi(\eta, 1, K)$ is a considerably closer approximation to the true velocity profile. This is an example of Picard's method[5] of successive integration, which has been widely used in boundary-layer problems.

a. Take $\Pi_v = A\eta$ as a first approximation and determine the second approximation to the velocity profile for $K = 0$ (no mass transfer) analytically. Determine A numerically by making the first and second approximations give the same velocity gradient at the wall. Compare your second approximation graphically with the exact solution.

b. Compute the limiting form of the velocity profiles for $K \to -\infty$. Note that in this case the result of the first iteration is exact, whatever finite trial profile Π_v is used.

c. Compute the temperature and concentration profiles for $K \to -\infty$. This is not an application of Picard's method, since Eq. 19.3–32 gives a direct solution when f is known.

Answer: *b* and *c*. $\Pi(\eta, \Lambda, K) = 1 - \exp(\Lambda K \eta)$

19.H₄ Critical Size of an Autocatalytic System

It is desired to use the result of Problem 19.E to discuss the critical size of a system in which an "autocatalytic reaction" is occurring. In such a system the reaction products increase the rate of reaction. If the ratio of system surface to system volume is large, then the reaction products tend to escape at the boundaries of the system. If the surface to volume ratio is small, however, the rate of escape may be less than the rate of formation, and the reaction rate will increase rapidly. For a system of given shape, there will be a critical size for which the rate of production just equals the rate of removal.

One example is that of nuclear fission; in a nuclear pile the rate of fission depends on the local neutron concentration. If neutrons are produced at a rate that exceeds the rate of escape by diffusion, the reaction is self-sustaining and a nuclear explosion results.

Similar behavior is also encountered in many chemical systems, although the behavior here is generally more complicated. An example is the thermal decomposition of acetylene gas, which is thermodynamically unstable according to the reaction

$$\mathrm{H{-}C{\equiv}C{-}H} \rightarrow \mathrm{H_2} + 2\mathrm{C} \qquad (19.H\text{-}1)$$

This reaction appears to proceed by a branched-chain, free-radical mechanism in which the free radicals behave qualitatively as the neutrons in the foregoing example, so that the decomposition is autocatalytic. However, the free radicals are effectively neutralized by contact with an iron surface, so that free-radical concentration is maintained near zero at such a surface. Acetylene gas can then be stored safely in iron pipe below a "critical" diameter, which is smaller the higher the pressure or temperature of the gas. If the pipe is too large, the formation of even one free radical is likely to cause a rapidly increasing rate of decomposition, which may result in a serious explosion.

a. Consider a system enclosed in a long cylinder in which the diffusion and reaction process is represented by

$$\frac{\partial c_A}{\partial t} = \mathscr{D}_{AB} \frac{1}{r} \frac{\partial}{\partial r}\left(r \frac{\partial c_A}{\partial r}\right) + k_1''' c_A \qquad (19.H\text{-}2)$$

[3] N. A. V. Piercy and J. H. Preston, *Phil. Mag.*, **21**, 995–1015 (1936).

[4] H. Schuh, *Z. angew. Math. Mech.* **25-27**, 54–60 (1947); *NACA Tech. Memo. 1275* (1950).

[5] See, for example, F. B. Hildebrand, *Advanced Calculus for Engineers*, Prentice-Hall, New York (1949), pp. 107–109.

with $c_A = 0$ at $r = R$ and $c_A = f(r)$ at $t = 0$, in which $f(r)$ is any reasonable function of r. Use the result of Problem 19.E to get a solution for c_A as a function of r and t.

b. Show that the critical radius for the system is

$$R_{\text{crit}} = \alpha_1 \sqrt{\frac{\mathscr{D}_{AB}}{k_1'''}} \tag{19.H-3}$$

in which α_1 is the first zero of the zero-order Bessel function J_0.

c. For a bare cylindrical nuclear reactor core[6] (0.2 per cent uranium, 39.8 per cent aluminum, and 60 per cent water by volume), the effective value of $k_1'''/\mathscr{D}_{AB}$ is $9 \times 10^{-3}\,\text{cm}^{-2}$. What is the critical radius? *Answer:* *c.* $R_{\text{crit}} = 25.3$ cm

19.I$_4$ Unsteady Diffusion in a Solid with Chemical Reaction and Mass Transfer at the Surface[7,8]

Show that Eq. 19.1–23 can be solved by using Eq. 19.1–26 when the boundary condition in Eq. 19.1–25 is replaced by

$$-(\boldsymbol{n} \cdot \mathscr{D}_{AB} \nabla c_A) = k_c(c_{As} - c_{Af}) \tag{19.I-1}$$

at the surfaces of the solid. Here $\boldsymbol{n}$ is the outwardly directed normal, c_{Af} is the concentration in the fluid surrounding the solid, and k_c is a mass-transfer coefficient.

19.J$_4$ Unsteady Diffusion with First-Order Reaction

Derive Eq. 19.1–26. Begin by taking the Laplace transform of Eq. 19.1–23 and the corresponding equation for f. Express the transform of c_A in terms of the transform of f; then, after some manipulations, perform the inversion to get Eq. 19.1–26.

19.K$_4$ Unsteady-State Interphase Diffusion

Consider the mass transfer of substance A between solvent I and solvent II, which are presumed to be entirely immiscible. It is further assumed that the concentration of A is sufficiently small that Fick's second law may be used to describe the diffusion in both regions. We then wish to solve the diffusion equations:

$$\frac{\partial c_{\text{I}}}{\partial t} = \mathscr{D}_{\text{I}} \frac{\partial^2 c_{\text{I}}}{\partial z^2} \qquad -\infty < z < 0 \tag{19.K-1}$$

$$\frac{\partial c_{\text{II}}}{\partial t} = \mathscr{D}_{\text{II}} \frac{\partial^2 c_{\text{II}}}{\partial z^2} \qquad 0 < z < +\infty \tag{19.K-2}$$

in which c_{I} is the concentration of A in phase I, and c_{II} is the concentration of A in the phase II. The diffusion coefficients $\mathscr{D}_{\text{I}}$ and $\mathscr{D}_{\text{II}}$ are the diffusion coefficients of A in phases I and II. The initial and boundary conditions are

$$\text{at } t = 0, \quad c_{\text{I}} = c_{\text{I}}^{\circ} \quad \text{for } -\infty < z < 0 \tag{19.K-3}$$

$$\text{at } t = 0, \quad c_{\text{II}} = c_{\text{II}}^{\circ} \quad \text{for } 0 < z < +\infty \tag{19.K-4}$$

$$\text{at } z = 0, \quad c_{\text{II}} = mc_{\text{I}} \quad \text{for all } t > 0 \tag{19.K-5}$$

$$\text{at } z = 0, \quad -\mathscr{D}_{\text{I}} \frac{\partial c_{\text{I}}}{\partial z} = -\mathscr{D}_{\text{II}} \frac{\partial c_{\text{II}}}{\partial z} \tag{19.K-6}$$

$$\text{at } z = -\infty, \quad c_{\text{I}} = c_{\text{I}}^{\circ} \tag{19.K-7}$$

$$\text{at } z = +\infty, \quad c_{\text{II}} = c_{\text{II}}^{\circ} \tag{19.K-8}$$

[6] R. L. Murray, *Nuclear Reactor Physics*, Prentice-Hall (1957), pp. 23, 30, 53.
[7] P. V. Danckwerts, *Trans. Faraday Soc.*, **47**, 1014–1023 (1951).
[8] J. Crank, *The Mathematics of Diffusion*, Oxford University Press (1956), pp. 124–125.

The first boundary condition at $z = 0$ is the statement of equilibrium at the interface, m being a "distribution coefficient" or "Henry law constant." The second boundary condition is a statement that the molar flux calculated at $z = 0-$ is the same at $z = 0+$; that is, there is no loss of A at the interface. Solve these simultaneous equations by means of Laplace transform[9] or other appropriate method to obtain these concentration profiles:

$$\frac{c_{\text{I}} - c_{\text{I}}^{\circ}}{c_{\text{II}}^{\circ} - mc_{\text{I}}^{\circ}} = \frac{1 + \operatorname{erf} z/\sqrt{4\mathscr{D}_{\text{I}}t}}{m + \sqrt{\mathscr{D}_{\text{I}}/\mathscr{D}_{\text{II}}}} \tag{19.K-9}$$

$$\frac{c_{\text{II}} - c_{\text{II}}^{\circ}}{c_{\text{I}}^{\circ} - (1/m)c_{\text{II}}^{\circ}} = \frac{1 - \operatorname{erf} z/\sqrt{4\mathscr{D}_{\text{II}}t}}{(1/m) + \sqrt{\mathscr{D}_{\text{II}}/\mathscr{D}_{\text{I}}}} \tag{19.K-10}$$

19.L$_4$ Measurement of Gas Diffusivities

A diffusion tube of length $2L$ and uniform cross section S has a partition at the center of the tube ($z = 0$), which may be removed at $t = 0$ to allow diffusion to occur. At each end ($z = \pm L$) the tube is joined to a reservoir of volume V containing stirrers to maintain a uniform concentration in the reservoir. Initially, the region $z > 0$ contains pure A and the region $z < 0$ contains pure B.

a. Show that at time t the concentration of A in the reservoirs is given by[10]

$$x_A = \tfrac{1}{2}\left[1 \pm \sum_{n=1}^{\infty} \frac{2N}{\gamma_n}(-1)^{n+1} \frac{\sqrt{\gamma_n^2 + N^2}}{\gamma_n^2 + N^2 + N} \exp\left(-\frac{\gamma_n^2 \mathscr{D}_{AB}t}{L^2}\right)\right] \tag{19.L-1}$$

in which γ_n is the nth root of $\gamma \tan \gamma = N$ and $N = SL/V$. Here the $\pm$ sign corresponds to the reservoirs attached at $\pm L$.

b. Make a numerical comparison of the results of Eq. 19.L-1 with the experimental measurements of Andrew.[11]

[9] See, for example, W. R. Marshall, Jr., and R. L. Pigford, *The Application of Differential Equations to Chemical Engineering Problems*, University of Delaware Press, Newark, Delaware (1947), pp. 134–136. Laplace transform methods were used by E. J. Scott, L. H. Tung, and H. G. Drickamer, *J. Chem. Phys.*, **19**, 1075–1078 (1951), in solving the same problem *without* the assumption of equilibrium at the interface.

[10] R. B. Bird, *Advances in Chemical Engineering*, Vol. I, Academic Press, New York (1956), p. 207.

[11] S. P. S. Andrew, *Chem. Eng. Sci.*, **4**, 269–272 (1955).

CHAPTER 20

Concentration Distributions in Turbulent Flow

In the preceding chapters we have derived the equations for diffusion in a fluid or solid, and we have shown how one can obtain expressions for the concentration distribution, provided the flow is not turbulent. Here we turn our attention to turbulent mass transport, a subject about which not a great deal can be said at the present time.

The discussion here is quite similar to that in Chapter 12. In fact, the treatment is kept quite short in view of the fact that much of the material is taken over by analogy. We restrict ourselves in this chapter to isothermal binary systems with approximately constant mass density ρ and diffusivity $\mathscr{D}_{AB}$. Therefore, the partial-differential equation describing the diffusion in a flowing fluid (Eq. 18.1–17) is of exactly the same form as that for heat transport in an incompressible fluid (Eq. 10.1–25) except for the inclusion of a chemical reaction term.

§20.1 CONCENTRATION FLUCTUATIONS AND THE TIME-SMOOTHED CONCENTRATION

The comments made in §12.1 regarding the physical picture can be taken over by analogy. We shall work here in terms of the molar concentration c_A. In a turbulent stream c_A will be a rapidly oscillating function of the time.

It is then convenient to replace c_A by the sum of a time-smoothed value $\bar{c}_A$ and a turbulent concentration fluctuation c_A':

$$c_A = \bar{c}_A + c_A' \tag{20.1-1}$$

analogously to Eq. 12.1-1 for temperature. By virtue of the definition of c_A', we see that $\overline{c_A'} = 0$. However, quantities such as $\overline{v_x'c_A'}$, $\overline{v_y'c_A'}$, and $\overline{v_z'c_A'}$ are not zero because the local fluctuations in concentration and velocity are not independent of one another.

The time-smoothed concentration profiles $\bar{c}_A(x, y, z, t)$ are those measured, for example, by the withdrawal of samples from the stream. In tube flow with mass transfer at the tube wall, one expects that the time-smoothed concentration $\bar{c}_A$ will vary only slightly in the turbulent core, where the eddy transport is predominant. In the slowly moving stream near the wall, on the other hand, the concentration $\bar{c}_A$ will be expected to change rapidly from the turbulent core value to the wall value in a short distance. The steep concentration gradient is then associated with the slow molecular diffusion process in the laminar sublayer, as opposed to the rapid eddy transport in the fully developed turbulent core.

§20.2 TIME-SMOOTHING THE EQUATION OF CONTINUITY OF A

We begin with the equation of continuity of substance A, which we presume to be disappearing by an nth-order chemical reaction.[1] (See Eq. 18.1-17.)

$$\frac{\partial c_A}{\partial t} = -\left(\frac{\partial}{\partial x} v_x c_A + \frac{\partial}{\partial y} v_y c_A + \frac{\partial}{\partial z} v_z c_A\right) + \mathscr{D}_{AB}\left(\frac{\partial^2 c_A}{\partial x^2} + \frac{\partial^2 c_A}{\partial y^2} + \frac{\partial^2 c_A}{\partial z^2}\right) - k_n''' c_A{}^n \tag{20.2-1}$$

Here k_n''' is the reaction-rate constant for the nth-order homogeneous reaction; it is presumed that k_n''' is independent of position. In subsequent equations we shall consider $n = 1$ and $n = 2$ just to emphasize the difference between a first-order and a sample higher-order reaction.

When c_A and v_i are replaced by $\bar{c}_A + c_A'$ and $\bar{v}_i + v_i'$, we obtain after time averaging

$$\frac{\partial \bar{c}_A}{\partial t} = -\left(\frac{\partial}{\partial x} \bar{v}_x \bar{c}_A + \frac{\partial}{\partial y} \bar{v}_y \bar{c}_A + \frac{\partial}{\partial z} \bar{v}_z \bar{c}_A\right) - \underline{\left(\frac{\partial}{\partial x} \overline{v_x' c_A'} + \frac{\partial}{\partial y} \overline{v_y' c_A'} + \frac{\partial}{\partial z} \overline{v_z' c_A'}\right)} + \mathscr{D}_{AB}\left(\frac{\partial^2 \bar{c}_A}{\partial x^2} + \frac{\partial^2 \bar{c}_A}{\partial y^2} + \frac{\partial^2 \bar{c}_A}{\partial z^2}\right) - \begin{cases} k_1''' \bar{c}_A & \text{or} \\ k_2'''(\bar{c}_A{}^2 + \underline{\overline{c_A'^2}}) \end{cases} \tag{20.2-2}$$

[1] S. Corrsin, *Physics of Fluids*, **1**, 42–47 (1958).

Comparison of this equation with Eq. 20.2–1 indicates that the time-smoothed equation differs only in the appearance of some extra terms (dashed underline). The terms containing $\overline{v_i'c_A'}$ describe the turbulent mass transport, and we designate the ith component of this flux by $\bar{J}_i^{(t)}$. We have now met the third of the turbulent fluxes, and we may summarize them thus:

turbulent momentum flux (tensor)
$$\bar{\tau}_{ij}^{(t)} = \rho\overline{v_i'v_j'} \tag{20.2–3}$$

turbulent energy flux (vector)
$$\bar{q}_i^{(t)} = \rho\hat{C}_p\overline{v_i'T'} \tag{20.2–4}$$

turbulent mass flux(vector)
$$\bar{J}_i^{(t)} = \overline{v_i'c_A'} \tag{20.2–5}$$

All of these are defined as fluxes with respect to the mass average velocity.

It is interesting to note that there is an essential difference between the behavior of chemical reactions of different orders. The first-order reaction is described by a linear differential equation, and the chemical reaction term has the same form in the time-smoothed equation as in the original equation. The second-order reaction is described by a nonlinear equation, and time-smoothing generates the extra term $-k_2'''\overline{c_A'^2}$. That is, for a first-order decomposition, the mean conversion rate is independent of the concentration fluctuations, whereas an explicit dependence on the concentration fluctuations is predicted for higher-order reactions.

By way of summary, we now list all three of the time-smoothed equations of change for turbulent flow of an isothermal, binary fluid mixture with constant ρ, $\mathscr{D}_{AB}$, and μ:

time-smoothed equation of continuity
$$(\nabla \cdot \bar{\boldsymbol{v}}) = 0 \tag{20.2–6}$$

time-smoothed equation of motion
$$\rho\frac{D\bar{\boldsymbol{v}}}{Dt} = -\nabla\bar{p} - [\nabla\cdot\bar{\boldsymbol{\tau}}^{(l)}] - \underline{[\nabla\cdot\bar{\boldsymbol{\tau}}^{(t)}]} + \rho\boldsymbol{g} \tag{20.2–7}$$

time-smoothed equation of continuity of A
$$\frac{D\bar{c}_A}{Dt} = -(\nabla\cdot\bar{\boldsymbol{J}}_A^{(l)}) - \underline{(\nabla\cdot\bar{\boldsymbol{J}}_A^{(t)})} - \begin{Bmatrix} k_1'''\bar{c}_A \quad \text{or} \\ k_2'''(\bar{c}_A^{\,2} + \underline{\overline{c_A'^2}}) \end{Bmatrix} \tag{20.2–8}$$

Here $\bar{\boldsymbol{J}}^{(l)} = -\mathscr{D}_{AB}\nabla\bar{c}_A$, and the D/Dt operator is understood to be written with the time-smoothed velocity $\bar{\boldsymbol{v}}$ in it.

§20.3 SEMIEMPIRICAL EXPRESSIONS FOR THE TURBULENT MASS FLUX

The turbulent mass flux $\bar{J}^{(t)}$ appearing in Eq. 20.2–8 has to be related to $\bar{c}_A$ or its gradient if Eq. 20.2–8 is to be solved. Just as for the turbulent momentum and energy fluxes, there are several semiempirical expressions available. Since they are all analogous to the expressions given in §12.3, we do little more than list them here:

The Eddy Diffusivity[1]

By analogy with Fick's law of diffusion, we write

$$\bar{J}_{Ay}^{(t)} = -\mathscr{D}_{AB}^{(t)} \frac{d\bar{c}_A}{dy} \tag{20.3–1}$$

in which $\mathscr{D}_{AB}^{(t)}$ is the "turbulent diffusivity" or "eddy diffusivity." Because of the analogy between heat and mass transfer, it is often assumed that $\mathscr{D}_{AB}^{(t)}/\alpha^{(t)} = 1$, in which $\alpha^{(t)}$ is the eddy thermal diffusivity. One also assumes that $\nu^{(t)}/\mathscr{D}_{AB}^{(t)}$ is between 0.5 and 1. (See §12.3.)

The Mixing Length Expressions of Prandtl and Taylor

By analogy with Eq. 12.3–2, we write

$$\bar{J}_{Ay}^{(t)} = -l^2 \left|\frac{d\bar{v}_x}{dy}\right| \frac{d\bar{c}_A}{dy} \tag{20.3–2}$$

Here l is the mixing length, which is usually set equal to $\kappa_1 y$ for flow in conduits, where y is the distance from the wall.

Expression Based on the von Kármán Similarity Hypothesis

By analogy with Eq. 12.3–3, we write

$$\bar{J}_{Ay}^{(t)} = -\kappa_2{}^2 \left|\frac{(d\bar{v}_x/dy)^3}{(d^2\bar{v}_x/dy^2)^2}\right| \frac{d\bar{c}_A}{dy} \tag{20.3–3}$$

in which κ_2 is the same constant appearing in Eqs. 5.3–3 and 12.3–3.

Deissler's Empirical Formula for the Region Near the Wall

By analogy with Eq. 12.3–4, Deissler proposed

$$\bar{J}_{Ay}^{(t)} = -n^2\bar{v}_x y\left[1 - \exp\left(\frac{-n^2\bar{v}_x y}{\nu}\right)\right]\frac{d\bar{c}_A}{dy} \tag{20.3–4}$$

The n here is the same as in Eqs. 5.3–4 and 12.3–4.

[1] For a summary of experimental eddy diffusivity measurements, see T. K. Sherwood and R. L. Pigford, *Absorption and Extraction*, McGraw-Hill, New York (1952), Chapter II, and J. B. Opfell and B. H. Sage, *Turbulence in Thermal and Material Transport*, chapter in *Advances in Chemical Engineering*, Vol. I, pp. 242–290 (1956).

Example 20.3–1. Concentration Profiles in Turbulent Flow in Smooth Circular Tubes

Use the analogy between heat and mass transfer to rewrite the key results in Example 12.3–1 for turbulent mass transfer in tubes with a constant wall mass flux.

Solution. First we note that without the chemical reaction terms Eq. 20.2–1 is of exactly the same form as Eq. 12.2–1 without the viscous dissipation terms. The solution obtained in Example 12.3–1 was for heat transfer with no viscous dissipation; hence, with appropriate notational changes, it will be valid for mass transfer without chemical reactions. We define a dimensionless concentration c_A^+ by

$$c_A^+ = \frac{v_*(\bar{c}_A - c_{A0})}{N_{A0}} \tag{20.3–5}$$

and immediately write down the concentration profiles by analogy with Eqs. 12.3–19 and 12.3–22:[2]

$$c_A^+ - c_{A1}^+ = \frac{1}{\kappa_1} \ln \frac{s^+}{s_1^+} \qquad s^+ \geqslant 26 \tag{20.3–6}$$

$$c_A^+ = \int_0^{s^+} \frac{ds^+}{\dfrac{1}{\mathrm{Sc}} + n^2v^+s^+[1 - \exp(-n^2v^+s^+)]} \qquad s^+ \leqslant 26 \tag{20.3–7}$$

in which v^+ is given by Eq. 5.3–15 and Sc, the Schmidt number, has replaced the Prandtl number. It is assumed in writing down an analogous solution that the mass-transfer rate at the wall N_{A0} is small enough so that the flow patterns are not disturbed; the effects of high mass-transfer rates are discussed in Chapter 21. The conclusion of the foregoing discussion is that we can use Fig. 12.3–2 for mass-transfer calculations, provided that T^+ is replaced by c_A^+ and Pr by Sc.

Example 20.3–2. Evaporation of Ammonia in a Wetted Wall Column

Aqueous NH_3 at 68° F flows in a thin film down the inside surface of a vertical cylindrical tube 2 in. in diameter. Air, saturated with water, flows up through the tube at a Reynolds number of 25,000. For dilute mixtures of NH_3 in air, the Schmidt number $\mathrm{Sc} = \mu_{\mathrm{air}}/\rho_{\mathrm{air}}\mathscr{D}_{\mathrm{air\text{-}NH_3}}$ is 0.61. The friction factor for the upward-flowing air stream may be taken as 0.007. Sketch the concentration profile in the column.

Solution. For our calculations, we need the following:

$$\nu_{\mathrm{air}} = \frac{\mu_{\mathrm{air}}}{\rho_{\mathrm{air}}} = 1.60 \times 10^{-4} \text{ ft}^2 \text{ sec}^{-1}$$

$$\langle \bar{v}_z \rangle = \frac{(25{,}000)(1.6 \times 10^{-4})}{(0.167)} = 24.0 \text{ ft sec}^{-1}$$

$$v^* = \sqrt{\frac{\tau_0}{\rho}} = \sqrt{\tfrac{1}{2} f \langle v_z \rangle^2} = \sqrt{2.02} = 1.42 \text{ ft sec}^{-1}$$

[2] These are the results of R. G. Deissler, *NACA Report 1210* (1955).

Hence at the center of the tube

$$s^+ = \frac{sv_*}{\nu} = \frac{(1/12)(1.42)}{(1.60 \times 10^{-4})} = 737$$

This is the highest value of s^+ encountered.

The NH_3 concentration profiles are then given by

$$c^+_{NH_3} = \int_0^{s^+} \frac{ds^+}{1.64 + 0.0154s^+v^+[1 - \exp(-0.0154s^+v^+)]} = \int_0^{s^+} X\,ds^+$$

$$(0 \leqslant s^+ \leqslant 26) \qquad (20.3\text{–}8)$$

$$c^+_{NH_3} = 9.3 + 2.78 \ln \frac{s^+}{26} = -0.3 + 2.78 \ln s^+ \qquad (26 \leqslant s^+ \leqslant 737) \qquad (20.3\text{–}9)$$

The computational results are given in Table 20.3–1.

TABLE 20.3–1

COMPUTATION OF THE CONCENTRATION PROFILES

v^+	s^+	X^{-1}	$X = \dfrac{dc^+_{NH_3}}{ds^+}$	$c^+_{NH_3}$
0	0	1.64 + 0	0.61	0
1	1	1.64 + 0.0002	0.61	0.61
1.5	1.5	1.64 + 0.001	0.61	0.92
2.0	2.0	1.64 + 0.004	0.61	1.2
3.0	3.1	1.64 + 0.02	0.60	1.9
5.0	5.2	1.64 + 0.13	0.57	3.1
7.5	8.6	1.64 + 0.66	0.44	4.9
10.0	15.0	1.64 + 2.10	0.27	7.0
13.0	26.0	1.64 + 5.19	0.15	9.3
	50			10.6
	100			12.5
	200			14.5
	400			16.4
	737			18.1

Values of X and of $c^+_{NH_3}/c^+_{NH_3}(\text{ctr})$ are plotted against s^+ in Figs. 20.3–1 and 20.3–2, respectively. It can be seen that the contribution of the turbulent core to the over-all resistance is appreciable and that the plotted concentration profile exhibits an abrupt change of slope at the center of the duct. This abrupt change does not actually occur but represents an inherent defect of the simplified flux distribution. The error introduced in using this expression is not very large in the present case but can become serious at lower Schmidt numbers. At higher Schmidt numbers, where almost all of the concentration change takes place in the buffer zone, this error is negligible.

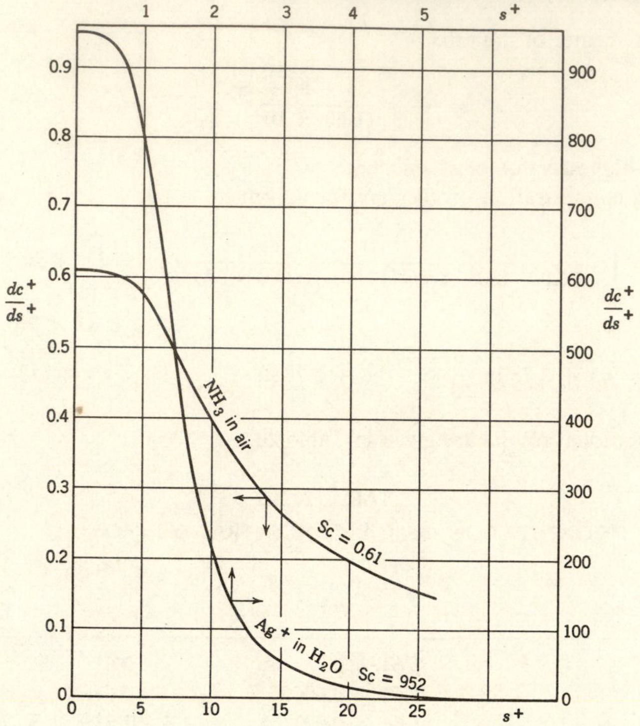

Fig. 20.3–1. Concentration gradients in turbulent tube flow for NH_3 evaporation; also shown are concentration gradients in electro-deposition of Ag at an electrode. (See Problem 20.B.)

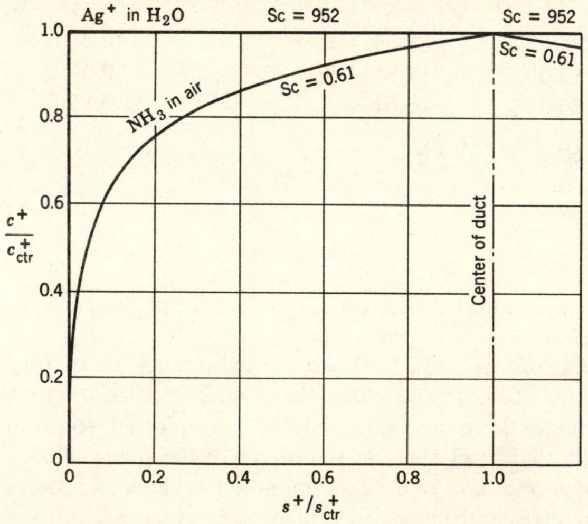

Fig. 20.3–2. Concentration profiles in turbulent tube flow.

§20.4 THE DOUBLE CONCENTRATION CORRELATION AND ITS PROPAGATION: THE CORRSIN EQUATION

The statistical approach to turbulent heat transfer has been extended to mass transfer with chemical reactions.[1] The treatment is identical with that given in §12.4, so that we give only the principal result here. First we define V_{Ai}, W_A, and Q_{Ai} as the correlations given in Eqs. 12.4–1, 2, and 3 with T' and T'' everywhere replaced by c_A' and c_A''. We further define w_A and q_A as the functions analogous to the w and q of Eqs. 12.4–5 and 6. The analog of Eq. 12.4–10 for isotropic, homogeneous turbulence with a first-order chemical reaction occurring is then

$$\frac{\partial}{\partial t}(\overline{c_A'^2}\, w_A) - 2\overline{c_A'^2}\sqrt{\overline{v'^2}}\left(\frac{\partial q_A}{\partial r} + 2\frac{q_A}{r}\right) = 2\mathscr{D}_{AB}\,\overline{c_A'^2}\left(\frac{\partial^2 w_A}{\partial r^2} + \frac{2}{r}\frac{\partial w_A}{\partial r}\right) - 2k_1'''\overline{c_A'^2}\, w_A \qquad (20.4\text{–}1)$$

which was derived by Corrsin.[1] This result differs from Eq. 12.4–10 only in the appearance of an additional term for the chemical reaction.

QUESTIONS FOR DISCUSSION

1. Discuss the similarities and differences between turbulent heat and mass transport.
2. Discuss the behavior of first- and higher-order chemical reactions in the time-smoothing of the equation of continuity for a given species. What are the consequences of this?
3. How are eddy diffusivity, eddy conductivity, and eddy viscosity interrelated?
4. What is the mass-transfer analog of the Reynolds analogy discussed in §12.3?
5. Interpret the curves in Fig. 20.3–1 and 2 in terms of the three hypothetical regions: turbulent core, buffer zone, and laminar sublayer.
6. How can eddy diffusivities be measured? On what do they depend?

PROBLEMS

20.A$_1$ Determination of Eddy Diffusivity

In Problem 17.K we gave the formula for the concentration profiles in diffusion from a point source in a moving stream. In isotropic highly turbulent flow Eq. 17.K–2 may be modified by replacing $\mathscr{D}_{AB}$ by $\mathscr{D}_{AB}^{(t)}$. Then the equation is useful for determining the eddy or turbulent diffusivity.

a. Show that if one plots $\ln sc_A$ versus $(s - z)$ the slope is $-v_0/2\mathscr{D}_{AB}^{(t)}$.

b. Use the data on diffusion of CO_2 from a point source in a turbulent air stream (see Fig. 20.A) to get $\mathscr{D}_{AB}^{(t)}$ for these conditions: diameter of pipe, 15.24 cm; $v_0 = 1512\ \text{cm sec}^{-1}$.

c. Compare the value of $\mathscr{D}_{AB}^{(t)}$ with the molecular diffusivity $\mathscr{D}_{AB}$ for the system CO_2-air.

Answer: $\mathscr{D}_{AB}^{(t)} = 19\ \text{cm}^2\ \text{sec}^{-1}$

[1] S. Corrsin, *Physics of Fluids*, **1**, 42–47 (1958).

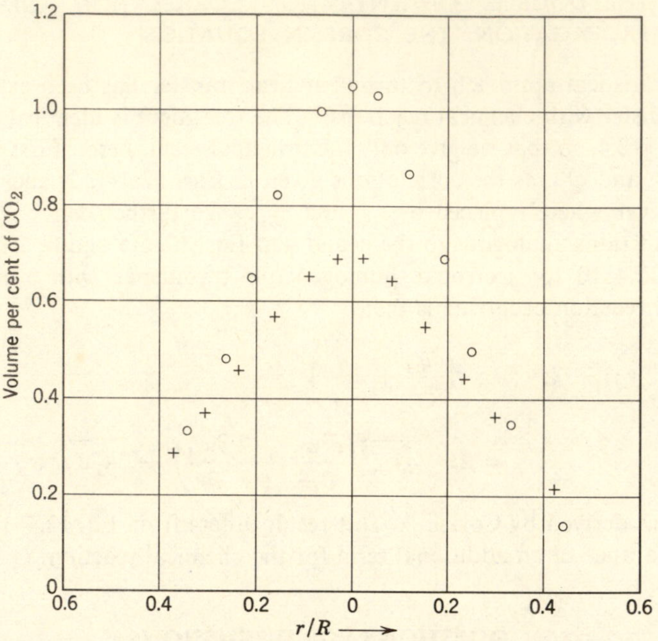

Fig. 20.A. Concentration traverse data for CO_2 injected into a turbulent air stream in a tube of diameter 15.24 cm. The Reynolds number for the flow is 119,000. The circles are concentrations at $z = 112.5$ cm, and the crosses are concentrations at $z = 152.7$ cm. [Experimental points taken from W. L. Towle and T. K. Sherwood, *Ind. Eng. Chem.*, **31**, 457–462 (1939).]

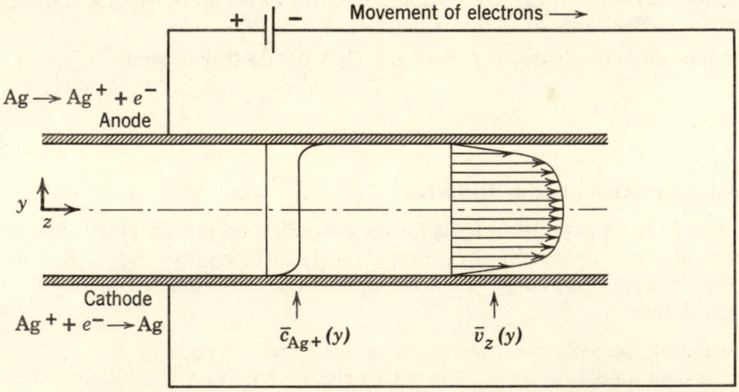

Fig. 20.B. Electrodeposition of Ag^+ from a turbulent stream flowing in the $+z$-direction between two parallel plates.

$20.B_3$ Deposition of Silver from a Turbulent Stream

An approximately $0.1N$ KNO_3 solution containing 1.00×10^{-6} g-eqt $AgNO_3$ per liter is flowing between parallel Ag plates, as shown in Fig. 20.B. A small voltage is applied across the plates to produce a deposition of Ag on the cathode (lower plate shown) and to polarize the circuit completely (i.e., to maintain the Ag^+ concentration at the cathode very nearly zero). Under the conditions given (see Problem 18.P), forced diffusion may be ignored and the Ag^+ may be considered moving to the cathode by ordinary and eddy diffusion only. Furthermore, this solution is dilute enough (see Example 18.5–4) so that the effects of the other ionic species on the diffusion of Ag^+ are negligible.

a. Calculate the Ag^+ concentration profile, assuming that (i) the effective binary diffusivity of Ag^+ through water is 1.06×10^{-5} cm² sec⁻¹ and the kinematic viscosity of water is 1.01×10^{-2} $cm^2\ sec^{-1}$; (ii) the Deissler-Prandtl velocity distribution for round tubes holds for "slit flow" as well if four times the hydraulic radius is substituted for tube diameter; (iii) the plates are 1.27 cm apart and $\sqrt{\tau_0/\rho}$ is 11.4 $cm\ sec^{-1}$.

b. Estimate the rate of deposition of Ag on the cathode, neglecting all other electrode reactions.

c. Does the method of calculation in part (*a*) predict a discontinuous slope for the concentration profile at the center plane of the system? Explain.

Answers: *a.* See Figs. 20.3–1 and 2
b. 6.7×10^{-12} $eqts\ sec^{-1}\ cm^{-2}$

CHAPTER 21

Interphase Transport in Multicomponent Systems

Correlations of interphase mass-transfer rates are used extensively in engineering analysis of processes such as distillation, absorption, extraction, drying, and heterogeneous chemical reactions. Such correlations are largely empirical because the processes are usually too complex to analyze in detail. However, the equations of change in Chapter 18 provide a sound basis for selection of dimensionless groups for correlation, and they are proving increasingly useful for analysis and extrapolation of experimental data.

This chapter is an extension of Chapters 6 and 13, in which we discussed correlations of momentum and heat transfer between a fluid and a solid boundary. Here we discuss the correlation of rates of interphase mass transfer and extend the results in Chapters 6 and 13 to mixtures.

In §21.1 we summarize the definitions for mass-transfer coefficients for binary mixtures. In §21.2 some simple rules are given for predicting mass-transfer coefficients on one side of a phase boundary at low mass-transfer rates by making use of analogies between heat and mass transfer. In §21.3 the calculation of fluid-fluid mass-transfer on both sides of the interface is discussed briefly in the framework of the Whitman two-film theory. It should be emphasized that the results in §§21.2 and 21.3 are restricted to binary systems and low mass-transfer rates. In §21.4, we define momentum-, heat-, and mass-transfer coefficients for binary systems at high mass-transfer rates, and in the following three sections we consider the prediction of these coefficients by film theory, penetration theory and boundary-layer theory. Finally, in §21.8, methods are given for estimating transfer coefficients in multicomponent systems.

We have limited this chapter to a few key topics on mass transfer and related correlations to introduce the reader to this rapidly developing subject. Further information is available in other mass-transfer texts.[1,2,3,4,5]

§21.1 DEFINITION OF BINARY MASS-TRANSFER COEFFICIENTS IN ONE PHASE

In this chapter we are concerned with the movement of material across an interface. The interface may be a true phase boundary, as in Fig. 21.1-1,

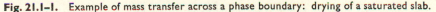

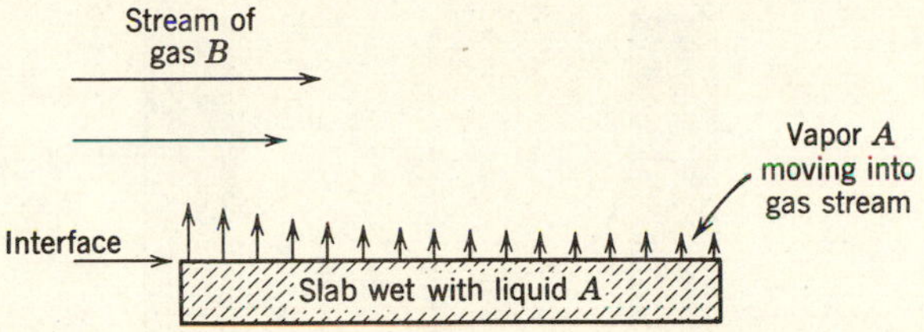

Fig. 21.1-1. Example of mass transfer across a phase boundary: drying of a saturated slab.

or a porous wall, as in Fig. 21.1–2. Mass transfer may occur in the presence of two species, as in Fig. 21.1–1, or in a pure fluid, as in Fig. 21.1–2. Most mass-transfer systems involve mixtures of varying composition, and in such applications mass-transfer coefficients provide a useful means of presenting experimental data for design.

Consider the flow of a stream of A and B along a mass-transfer surface $y = 0$ through which species A and B enter the stream at the local rates N_{A0} and N_{B0} moles/(unit area)(unit time). A typical system in Fig. 21.1–3 shows how the terms used here apply in a specific situation. Note that N_{A0} and N_{B0} are the components of the fluxes $\mathbf{N}_A$ and $\mathbf{N}_B$ at the interface measured *into* the phase of interest; in Fig. 21.1–3 this is the gas phase. The movement

[1] T. K. Sherwood and R. L. Pigford, *Absorption and Extraction*, McGraw-Hill, New York (1952).

[2] R. E. Treybal, *Mass Transfer Operations*, McGraw-Hill, New York (1952).

[3] A. P. Colburn and R. L. Pigford, "General Theory of Diffusional Operations," Section 8 in *Chemical Engineers' Handbook*, J. H. Perry (Ed.), McGraw-Hill, New York (1950), Third Edition.

[4] W. M. Ramm, *Absorptionsprozesse in der chemischen Industrie*, Verlag Technik Berlin (1952).

[5] W. Matz, *Die Thermodynamik des Wärme- und Stoffaustausches in der Verfahrenstechnik*," Steinkopff, Frankfurt-am-Main (1949).

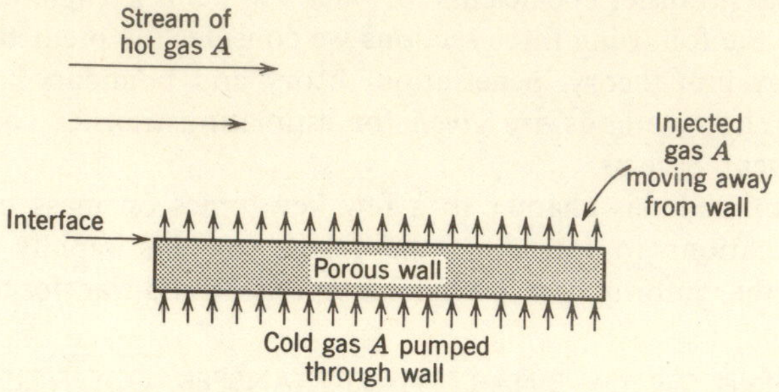

Fig. 21.1–2. Example of mass transfer through a porous wall: transpiration cooling

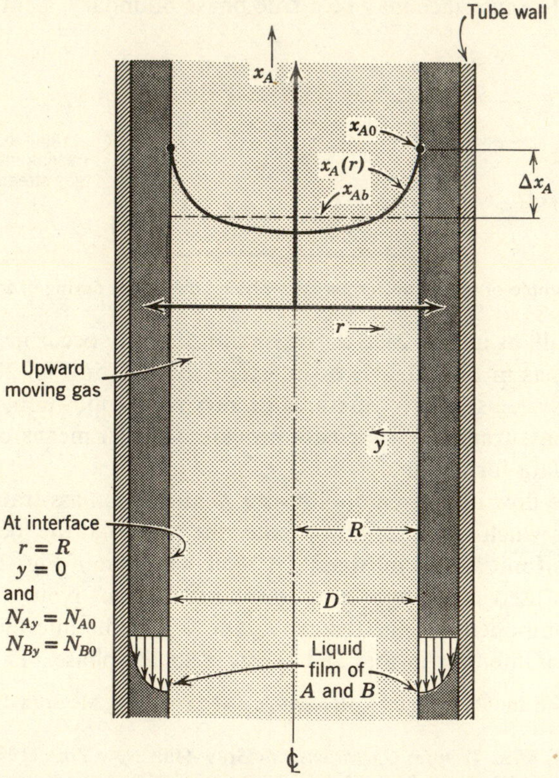

Fig. 21.1–3. Sketch showing mass transfer of *A* and *B* from a down-flowing liquid stream into an upward-flowing gas stream of *A* and *B* in a vertical cylindrical tube.

of each species through the surface may be treated as the sum of a diffusional contribution and a bulk-flow contribution (see Eq. 16.2–2). Now the diffusional contribution is proportional to the concentration gradient at the wall, hence should be roughly proportional to some characteristic composition difference, Δx_A, between the fluid at the surface and in the main stream; the bulk-flow contribution, on the other hand, can occur without any composition difference at all. (See Fig. 21.1–2, for example.) Therefore, it seems appropriate to define a mass-transfer coefficient, $k^{\bullet}_{x,\text{loc}}$, in terms of the rate of *diffusion* of either species normal to the interface:

$$J_{Ay}^{\star}|_{y=0} = -J_{By}^{\star}|_{y=0} = k^{\bullet}_{x,\text{loc}}\,\Delta x_A \tag{21.1–1}$$

In terms of the molar fluxes N_{A0} and N_{B0}, this gives

$$N_{A0} - x_{A0}(N_{A0} + N_{B0}) = k^{\bullet}_{x,\text{loc}}\,\Delta x_A \tag{21.1–2}$$

$$N_{B0} - x_{B0}(N_{A0} + N_{B0}) = k^{\bullet}_{x,\text{loc}}\,\Delta x_B \tag{21.1–3}$$

in which $x_{A0} = 1 - x_{B0}$ is the mole fraction of species A on the stream side of the interface. Note the parallelism between these equations and Fick's law (Eq. 16.2–2).

Equation 21.1–3 reduces to Eq. 21.1–2 on substitution of $1 - x_A$ for x_B; thus only one of the equations is needed, and the value of $k^{\bullet}_{x,\text{loc}}$ is the same whichever equation is used. We omit the equations corresponding to Eq. 21.1–3 in subsequent developments.

The black dot (●) on $k^{\bullet}_{x,\text{loc}}$ serves as a reminder that the mass-transfer coefficient itself depends on the mass-transfer rate. This effect arises from the distortion of the velocity and concentration profiles by the flow of A and B through the interface. The effect is small in many applications, and therefore we postpone detailed discussion of it to §§21.4 *et seq.*

In the limit of *small mass-transfer rates* N_{A0} *and* N_{B0}, the distortion of the velocity and concentration profiles by mass transfer may be neglected. It is convenient to define a mass-transfer coefficient for this limiting condition by the equation[1]

$$\lim_{\substack{N_{A0}\to 0\\ N_{B0}\to 0}} \left\{\frac{N_{A0} - x_{A0}(N_{A0} + N_{B0})}{\Delta x_A}\right\} = k_{x,\,\text{loc}} \tag{21.1–4}$$

[1] The mass-transfer coefficients defined here are not those used in the bulk of the published literature. The coefficient k_x, however, is similar to the coefficient F used by A. P. Colburn and T. B. Drew, *Trans. A.I.Ch.E.*, **33**, 197–212 (1937). In order to apply other published correlations in terms of the theory given here, one has to know how the various definitions are related. It is a common practice to define mass-transfer coefficients by expressions such as

$$N_{A0} = k_G\,\Delta p_A \tag{21.1–4a}$$

or

$$N_{A0} = k_L\,\Delta c_A \tag{21.1–4b}$$

which are used for gases and liquids, respectively, at any mass-transfer rate. Note that in

As shown in subsequent sections, it is permissible at moderate mass-transfer rates to insert $k_{x,\text{loc}}$ in place of $k^{\bullet}_{x,\text{loc}}$ in Eq. 21.1–2, that is, to neglect the correction of $k^{\bullet}_{x,\text{loc}}$ for mass-transfer rate. The difference between $k_{x,\text{loc}}$ and $k^{\bullet}_{x,\text{loc}}$ may also be shown by comparing the forms of the dimensionless correlations that will later be obtained for these quantities:

$$\frac{k_{x,\text{loc}}D}{c\mathscr{D}_{AB}} = \text{a function of Re, Sc, and geometry} \tag{21.1–5}$$

$$\frac{k^{\bullet}_{x,\text{loc}}D}{c\mathscr{D}_{AB}} = \text{a function of Re, Sc, } \frac{N_{A0} + N_{B0}}{k_{x,\text{loc}}}\text{, and geometry} \tag{21.1–6}$$

Most of the available forced-convection mass-transfer correlations can be put in the form of Eq. 21.1–5; methods of estimating the effect of the additional dimensionless group are considered in §§21.5, *et seq*.

For surfaces of finite area A, one can parallel Eq. 21.1–2 and define an average mass-transfer coefficient as follows:

$$\mathscr{W}_A^{(m)} - x_{A0}(\mathscr{W}_A^{(m)} + \mathscr{W}_B^{(m)}) = k^{\bullet}_x A\,\Delta x_A \tag{21.1–7}$$

or, for small mass-transfer rates, one can assume $k_x = k^{\bullet}_x$ and obtain

$$\mathscr{W}_A^{(m)} - x_{A0}(\mathscr{W}_A^{(m)} + \mathscr{W}_B^{(m)}) = k_x A\,\Delta x_A \tag{21.1–8}$$

in which $\mathscr{W}_A^{(m)}$ and $\mathscr{W}_B^{(m)}$ are the molar rates of addition of A and B to the stream over the entire surface.[2] The quantity x_{A0} is some characteristic composition of the fluid next to the interface, and Δx_A is a characteristic composition difference.

By comparing Eq. 21.1–8 with Eq. 13.1–1 for heat transfer, we see that the two equations are basically similar. Thus the diffusion rate $\mathscr{W}_A^{(m)} - x_{A0}(\mathscr{W}_A^{(m)} + \mathscr{W}_B^{(m)})$ corresponds to the heat-conduction rate Q, Δx_A

Eq. 21.1–2 we have subtracted the bulk-flow term so that $k^{\bullet}_x\,\Delta x_A$ is the rate of mass transfer at the interface by diffusion alone; in Eqs. 21.1–4*a*,*b*, however, the diffusion and bulk-flow contributions are combined. As a result, k_G and k_L show a more complicated dependence on concentration level and mass transfer rates than does $k^{\bullet}_x$.

For the frequently studied case in which $N_B = 0$, the relations between these coefficients are

$$k^{\bullet}_x = k_G p(1 - x_{A0}) = k_G p_{B0} \qquad \text{(constant } p\text{)} \tag{21.1–4c}$$

$$k^{\bullet}_x = k_L c(1 - x_{A0}) = k_L c_{B0} \qquad \text{(constant } c\text{)} \tag{21.1–4d}$$

For *small mass-transfer rates*, $k^{\bullet}_x$ may be replaced by k_x, and p_B and c_B become nearly constant in the given cross section. Then Eqs. 21.1–4*c*,*d* may be rewritten in terms of log-mean values of p_B and c_B computed from interfacial and main-stream conditions:

$$k_x = k_G p_{B,\text{ln}} \tag{21.1–4e}$$

$$k_x = k_L c_{B,\text{ln}} \tag{21.1–4f}$$

We have introduced $p_{B,\text{ln}}$ and $c_{B,\text{ln}}$ here to show how k_x may be determined from published correlations in which these quantities appear.

[2] The superscript (*m*) indicates a flow across a "mass-transfer surface." Although the notation is superfluous here, it is useful in §21.4 and in Chapter 22.

corresponds to ΔT, and k_x corresponds to h. Any of the equations for the various definitions of h in §13.1 can be taken over for mass transfer by analogous substitutions. For example, the mass-transfer analog of Eq. 13.1–2 is

$$\mathscr{W}_A^{(m)} - x_{A0}(\mathscr{W}_A^{(m)} + \mathscr{W}_B^{(m)}) = k_{x1}(\pi DL)(x_{A0} - x_{Ab1}) \quad (21.1\text{–}9)$$

This equation applies to *slow mass transfer* between a pipe wall and a fluid flowing in the pipe. The interfacial composition x_{A0} is here taken to be constant in the mass-transfer section of length L. The mass transfer might arise from vaporization, condensation, dissolution, crystallization, or catalytic reaction at the wall, to mention a few possibilities. Here x_{Ab1} is the bulk ("cup-mixing") composition of the flowing stream at the entrance to the mass-transfer section. Some more widely useful definitions, applicable to geometrically complicated systems, are given in the following paragraphs.

For *slow mass transfer* in a closed channel, with known interfacial area and composition, it is convenient to employ a local one-phase mass-transfer coefficient:

$$d\mathscr{W}_A^{(m)} = k_{x,\mathrm{loc}}(x_{A0} - x_{Ab})\, dA + x_{A0}(d\mathscr{W}_A^{(m)} + d\mathscr{W}_B^{(m)}) \quad (21.1\text{–}10)$$

or

$$N_{A0} = k_{x,\mathrm{loc}}(x_{A0} - x_{Ab}) + x_{A0}(N_{A0} + N_{B0}) \quad (21.1\text{–}11)$$

Here x_{A0} and x_{Ab} are the local interfacial and cup-mixing mole fractions of species A in the given phase, and $N_{A0} = d\mathscr{W}_A^{(m)}/dA$ and $N_{B0} = d\mathscr{W}_B^{(m)}/dA$ are the local molar fluxes *into* the given phase, in the element of area dA. Equation 21.1–10 corresponds to Eq. 13.1–5 for heat transfer to a fluid in a circular pipe, if one chooses $dA = \pi D\, dz$ and correspondingly works with average values of N_{A0} and x_{A0} for this ring-shaped element of area. (See Fig. 13.1–1.)

For mass transfer between a submerged object and a surrounding fluid, mass-transfer coefficients are usually reported for the entire surface area A. If the fluid concentration x_{A0} next to the surface is uniform, Eq. 21.1–7 is rewritten thus for *low mass-transfer rates*:

$$\mathscr{W}_A^{(m)} = k_{xm}A(x_{A0} - x_{A\infty}) + x_{A0}(\mathscr{W}_A^{(m)} + \mathscr{W}_B^{(m)}) \quad (21.1\text{–}12)$$

in which $x_{A\infty}$ is the uniform composition of the fluid approaching the object. This definition corresponds to Eq. 13.1–6 for heat transfer from a sphere to a surrounding fluid.

For enclosed systems of *unknown interfacial area*, such as sprays of bubbles or drops, Eqs. 21.1–10 and 11 require modification. One common procedure is to base the transfer coefficients on unit volume of the system; this leads to the following modification of Eq. 21.1–10 for *low mass-transfer rates*:

$$d\mathscr{W}_A^{(m)} = (k_x a)_{\mathrm{loc}}(x_{A0} - x_{Ab})\, dV + x_{A0}(d\mathscr{W}_A^{(m)} + d\mathscr{W}_B^{(m)}) \quad (21.1\text{–}13)$$

in which a_{loc} is the interfacial area per unit volume in the volume element dV. The combined quantity $(k_x a)_{\text{loc}}$ is measurable even when $k_{x,\text{loc}}$ and a_{loc} are not.

The calculation of the local interfacial composition x_{A0} is fairly straightforward if there is a pure solid or fluid on the other side of the interface. In that case, a good approximation to x_{A0} may be obtained by assuming that equilibrium exists at the interface, as in §17.2 and Example 19.1–1. The more general problem in which concentration gradients occur on both sides of the interface is treated in §21.3 and Example 22.5–2.

It may be noted that Eqs. 21.1–2 and 3 relate *one* concentration difference $(\Delta x_A = -\Delta x_B)$ to *two* mass-transfer rates. Thus somewhat more information is required to solve mass-transfer problems than the problems in Chapters 6 and 13. For example, it can be seen that specification of $k^{\bullet}_{x,\text{loc}}$, x_{A0}, and Δx_A in Eq. 21.1–2 does not, in general, determine N_{A0} or N_{B0} but only the rate of *diffusion* of A or B relative to the bulk flow across the interface; to complete the solution, a separate relation between N_{A0} and N_{B0} is needed. Examples of such relations are given in §17.2, where the relation $N_B = 0$ was obtained from solubility considerations, in §17.3, where the relation $N_{A_2z} = -\frac{1}{2}N_{Az}$ was obtained from the material balance (stoichiometry) of a chemical reaction, and in §17.4, where the relation $c_A(N_{Az} + N_{Bz}) \doteq 0$ was obtained from solubility considerations. Sometimes a momentum or energy balance may provide the needed relation, as in Problem 18.N.

§21.2 CORRELATIONS OF BINARY MASS-TRANSFER COEFFICIENTS IN ONE PHASE AT LOW MASS-TRANSFER RATES

The solutions of many mass-transfer problems at low mass-transfer rates can be obtained by analogy with corresponding problems in heat transfer, and the converse is also true. In this section, therefore, we do not present any new correlations but simply show how the results in Chapter 13 may be converted into the analogous mass-transfer correlations. Some of the results in Chapter 13 were, in fact, derived by analogy from mass-transfer data. To illustrate the background of these useful analogies and the conditions under which they apply, we discuss first the mass-transfer analog of the dimensional analysis given in §13.2.

Consider the steady isothermal flow of a liquid solution of A and B in the pipe shown in Fig. 21.2–1. We assume that the velocity distribution at plane "1" is known and that the fluid concentration is constant at x_{A1} in the region $z < 0$. From $z = 0$ to $z = L$, the pipe wall is coated with a solid solution of A and B, which dissolves slowly and maintains the liquid composition constant at x_{A0} along the dissolving surface. We further assume that the physical properties ρ, μ, c, and $\mathscr{D}_{AB}$ are constant.

The mass-transfer situation just described is mathematically analogous to the heat-transfer situation described at the beginning of §13.2. To emphasize this analogy, the equations for the two systems are presented together. Thus the rate of heat addition by conduction between "1" and "2" in Fig 13.1-1 and the molar rate of addition of species A by diffusion between

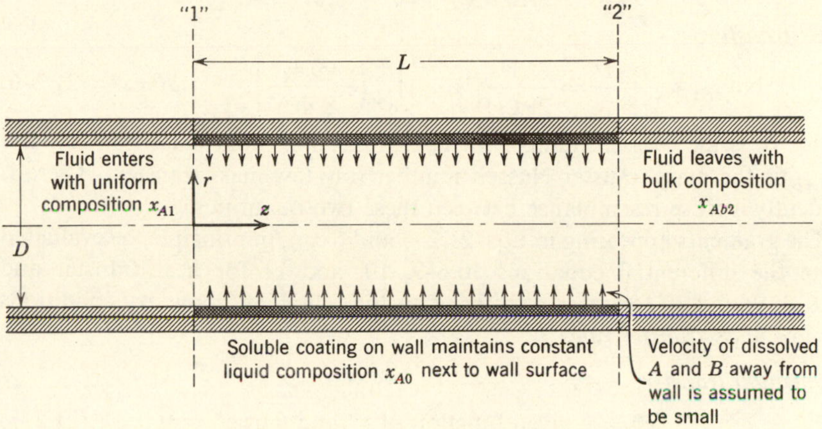

Fig. 21.2-1. Mass transfer in a pipe with a soluble wall.

"1" and "2" in Fig. 21.2-1 are given by the following expressions, valid for either laminar or turbulent flow:[1]

heat transfer:

$$Q = \int_0^L \int_0^{2\pi} \left(+k \left.\frac{\partial T}{\partial r}\right|_{r=R} \right) R\, d\theta\, dz \tag{21.2-1}$$

mass transfer:

$$\mathcal{W}_A^{(m)} - x_{A0}(\mathcal{W}_A^{(m)} + \mathcal{W}_B^{(m)}) = \int_0^L \int_0^{2\pi} \left(+c\mathcal{D}_{AB} \left.\frac{\partial x_A}{\partial r}\right|_{r=R} \right) R\, d\theta\, dz \tag{21.2-2}$$

Evaluating the left sides of these equations in terms of h_1 and k_{x1}, as defined in Eqs. 13.1-2 and 21.1-9, we get

heat transfer:

$$h_1 = \frac{1}{\pi DL(T_0 - T_1)} \int_0^L \int_0^{2\pi} \left(+k \left.\frac{\partial T}{\partial r}\right|_{r=R} \right) R\, d\theta\, dz \tag{21.2-3}$$

mass transfer:

$$k_{x1} = \frac{1}{\pi DL(x_{A0} - x_{A1})} \int_0^L \int_0^{2\pi} \left(+c\mathcal{D}_{AB} \left.\frac{\partial x_A}{\partial r}\right|_{r=R} \right) R\, d\theta\, dz \tag{21.2-4}$$

Next we introduce the dimensionless quantities $r^* = r/D$, $z^* = z/D$,

[1] If the flow is turbulent, it is necessary to regard all profiles as time-smoothed in the following development. For brevity we omit the overlines.

$T^* = (T - T_0)/(T_1 - T_0)$, and $x_A^* = (x_A - x_{A0})/(x_{A1} - x_{A0})$ and rearrange slightly to obtain

heat transfer:

$$\mathrm{Nu}_1 \equiv \frac{h_1 D}{k} = \frac{1}{2\pi L/D} \int_0^{L/D} \int_0^{2\pi} \left(- \left. \frac{\partial T^*}{\partial r^*} \right|_{r^* = \frac{1}{2}} \right) d\theta\, dz^* \qquad (21.2\text{–}5)$$

mass transfer:

$$\mathrm{Nu}_{AB1} \equiv \frac{k_{x1} D}{c\mathscr{D}_{AB}} = \frac{1}{2\pi L/D} \int_0^{L/D} \int_0^{2\pi} \left(- \left. \frac{\partial x_A^*}{\partial r^*} \right|_{r^* = \frac{1}{2}} \right) d\theta\, dz^* \qquad (21.2\text{–}6)$$

Here Nu is the Nusselt number for heat transfer without mass transfer, and Nu_{AB} is the mass-transfer Nusselt number for slow mass transfer. There is evidently a close resemblance between these two quantities.[2]

The gradients appearing in Eqs. 21.2–5 and 6 can, in principle, be evaluated from the differential equations 10.6–9, 10, and 11 for heat transfer and 18.6–8, 9, and 10 for mass transfer, with the following boundary conditions for *low mass-transfer rates:*[3]

momentum transfer:

$$\boldsymbol{v}^* = \text{a given function of } r^* \text{ and } \theta \text{ at } z^* = 0 \qquad (21.2\text{–}7)$$

$$\boldsymbol{v}^* = 0 \text{ at } r^* = \tfrac{1}{2} \qquad (21.2\text{–}8)$$

$$p^* = 0 \text{ at } z^* = r^* = 0 \qquad (21.2\text{–}9)$$

heat transfer:

$$T^* = 1 \text{ at } z^* = 0 \qquad (21.2\text{–}10)$$

$$T^* = 0 \text{ at } r^* = \tfrac{1}{2},\ z^* > 0 \qquad (21.2\text{–}11)$$

mass transfer:

$$x_A^* = 1 \text{ at } z^* = 0 \qquad (21.2\text{–}12)$$

$$x_A^* = 0 \text{ at } r^* = \tfrac{1}{2},\ z^* > 0 \qquad (21.2\text{–}13)$$

If we can neglect the dissipation term in Eq. 10.6–11 and if chemical reactions in the fluid can be neglected, as in Eq. 18.6–10, the differential equations for the heat-transfer system and the mass-transfer system are analogous; and by neglecting the velocity at the wall in mass transfer we have made the boundary conditions analogous. It follows then that the profiles of T^* and x_A^* are similar:

$$\begin{cases} T^* = \text{a function of } (r^*, \theta, z^*, \mathrm{Re}, \mathrm{Pr}) & (21.2\text{–}14) \\ x_A^* = \text{the same function of } (r^*, \theta, z^*, \mathrm{Re}, \mathrm{Sc}) & (21.2\text{–}15) \end{cases}$$

[2] These simple results illustrate why we chose to define k_{x1} in terms of the diffusion rate $\mathscr{W}_A^{0(m)} - x_{A0}(\mathscr{W}_A^{0(m)} + \mathscr{W}_B^{0(m)})$ rather than just $\mathscr{W}_A^{0(m)}$. The expression for Nu_{AB} in terms of the mass-transfer coefficient, and several later results, would be more complicated if we had made the latter choice.

[3] The assumption of zero fluid velocity at the wall (Eq. 21.2–8) restricts the analogies discussed here to low mass-transfer rates. Explicit allowance for the velocity of the fluid through the interface is considered in §§21.4, 5, 6, and 7.

That is, to obtain concentration profiles instead of temperature profiles one simply replaces the Prandtl number $\text{Pr} = \hat{C}_p\mu/k$ in Eq. 21.2–14 by the Schmidt number $\text{Sc} = \mu/\rho\mathscr{D}_{AB}$ and changes T^* to x_A^*.

Insertion of these expressions for the profiles into Eqs. 21.2–5 and 6 then gives

$$\begin{cases} \text{Nu}_1 = \text{a function of } (\text{Re}, \text{Pr}, L/D) & (21.2\text{–}16) \\ \text{Nu}_{AB1} = \text{the same function of } (\text{Re}, \text{Sc}, L/D) & (21.2\text{–}17) \end{cases}$$

The same functional similarity exists for Nu_a, Nu_{ln}, Nu_{loc}, and the corresponding forms of Nu_{AB}. This important analogy permits one to derive mass-transfer correlations from heat-transfer correlations for equivalent boundary conditions by merely substituting Nu_{AB} for Nu and Sc for Pr. The same can be done for any flow geometry and for laminar or turbulent flow. *Note* that in order to obtain this analogy one has to assume (i) constant physical properties, (ii) a small rate of mass transfer, (iii) no chemical reactions in the fluid, (iv) no viscous dissipation, (v) no emission or absorption of radiant energy, and (vi) no pressure diffusion, thermal diffusion or forced diffusion. Much of our subsequent discussion deals with means for getting around the restrictions (i) and (ii); the others are unimportant in many problems. Note further that assumption (i) implies that $\boldsymbol{v} = \boldsymbol{v}^\star$.

For free convection around submerged objects of any given shape one can similarly show that

$$\begin{cases} \text{Nu}_m = \text{a function of } (\text{Gr}, \text{Pr}) & (21.2\text{–}18) \\ \text{Nu}_{AB_m} = \text{the same function of } (\text{Gr}_{AB}, \text{Sc}) & (21.2\text{–}19) \end{cases}$$

in which Gr_{AB} is the Grashof number for binary diffusion. (See §18.6.) This analogy and that in Eqs. 21.2–16 and 17 have been confirmed experimentally for a number of flow systems.

A summary of analogous quantities for heat and mass transfer is given in Table 21.2–1. To convert any heat-transfer correlation into the corresponding mass-transfer correlation, one simply substitutes new dimensionless groups according to the table. This procedure may be used to determine concentration profiles from temperature profiles or mass-transfer coefficients from heat-transfer coefficients. Note that the boundary conditions should also be transformed to show when the correlation thus derived is applicable.

There is little to say about allowance for variable physical properties in mass-transfer systems. The effects of variable physical properties in isothermal mass-transfer problems should differ somewhat from those found for heat transfer in pure fluids because a different set of physical properties is involved and we are concerned with their concentration dependence rather than their temperature dependence.

As an illustration of the use of analogies, consider the correlation of heat

TABLE 21.2-1

ANALOGIES BETWEEN HEAT AND MASS TRANSFER AT LOW MASS-TRANSFER RATES

	Heat-Transfer Quantities	Binary Mass-Transfer Quantities
Profiles	T	x_A
Diffusivity	$\alpha = \dfrac{k}{\rho \hat{C}_p}$	$\mathscr{D}_{AB}$
Effect of profiles on density	$\beta = -\dfrac{1}{\rho}\left(\dfrac{\partial \rho}{\partial T}\right)_{p,x_A}$	$\zeta = -\dfrac{1}{\rho}\left(\dfrac{\partial \rho}{\partial x_A}\right)_{p,T}$
Flux	$q^{(c)}$	$J_A^{\star} = N_A - x_A(N_A + N_B)$
Transfer rate	Q	$\mathscr{W}_A^{(m)} - x_{A0}(\mathscr{W}_A^{(m)} + \mathscr{W}_B^{(m)})$
Transfer coefficient	$h = \dfrac{Q}{A\,\Delta T}$	$k_x = \dfrac{\mathscr{W}_A^{(m)} - x_{A0}(\mathscr{W}_A^{(m)} + \mathscr{W}_B^{(m)})}{A\,\Delta x_A}$
Dimensionless groups which are the same in both correlations	$\text{Re} = \dfrac{DV\rho}{\mu} = \dfrac{DG}{\mu}$ $\text{Fr} = \dfrac{V^2}{gD}$ $\dfrac{L}{D}$	$\text{Re} = \dfrac{DV\rho}{\mu} = \dfrac{DG}{\mu}$ $\text{Fr} = \dfrac{V^2}{gD}$ $\dfrac{L}{D}$
Basic dimensionless groups which are different	$\text{Nu} = \dfrac{hD}{k}$ $\text{Pr} = \dfrac{\hat{C}_p\mu}{k} = \dfrac{\nu}{\alpha}$ $\text{Gr} = \dfrac{D^3\rho^2 g\beta\,\Delta T}{\mu^2}$ $\text{St} = \dfrac{\text{Nu}}{\text{RePr}} = \dfrac{h}{\rho\hat{C}_p V}$	$\text{Nu}_{AB} = \dfrac{k_x D}{c\mathscr{D}_{AB}}$ $\text{Sc} = \dfrac{\mu}{\rho\mathscr{D}_{AB}} = \dfrac{\nu}{\mathscr{D}_{AB}}$ $\text{Gr}_{AB} = \dfrac{D^3\rho^2 g\zeta\,\Delta x_A}{\mu^2}$ $\text{St}_{AB} = \dfrac{\text{Nu}_{AB}}{\text{ReSc}} = \dfrac{k_x}{cV}$
Special combinations of dimensionless groups	$\text{Pé} = \text{RePr} = \dfrac{DV\rho\hat{C}_p}{k}$ $j_H = \text{NuRe}^{-1}\text{Pr}^{-1/3}$ $= \dfrac{h}{\rho\hat{C}_p V}\left(\dfrac{\hat{C}_p\mu}{k}\right)^{2/3}$	$\text{Pé}_{AB} = \text{ReSc} = \dfrac{DV}{\mathscr{D}_{AB}}$ $j_D = \text{Nu}_{AB}\text{Re}^{-1}\text{Sc}^{-1/3}$ $= \dfrac{k_x}{cV}\left(\dfrac{\mu}{\rho\mathscr{D}_{AB}}\right)^{2/3}$

and momentum transfer in Fig. 13.3–3 for tangential flow over an isothermal flat plate without mass transfer. This correlation is of the form

$$j_{H,\text{loc}} = \tfrac{1}{2}f_{\text{loc}} = \text{a function of Re} \qquad (21.2\text{–}20)$$

in which Re is the length Reynolds number $v_\infty \rho_f x/\mu_f$. Insertion of the corresponding quantity j_D for mass transfer gives the analogy

$$j_{D,\text{loc}} = j_{H,\text{loc}} = \tfrac{1}{2}f_{\text{loc}} = \text{a function of Re} \qquad (21.2\text{–}21)$$

or

$$\frac{k_{x,\text{loc}}}{c_f v_\infty}\left(\frac{\mu}{\rho \mathscr{D}_{AB}}\right)_f^{2/3} = \frac{h_{\text{loc}}}{\rho_f \hat{C}_{pf} v_\infty}\left(\frac{\hat{C}_p \mu}{k}\right)_f^{2/3} = \tfrac{1}{2}f_{\text{loc}} = \text{a function of Re} \qquad (21.2\text{–}22)$$

in which the subscript f denotes properties evaluated at the "film temperature" $T_f = \frac{1}{2}(T_0 + T_\infty)$ and the "film composition," $x_{Af} = \frac{1}{2}(x_{A0} + x_{A\infty})$. Since the heat-transfer correlation was given for uniform wall temperature T_0 and no mass transfer, the mass-transfer correlation is limited to uniform interface composition x_{A0} and low mass-transfer rates. This correlation of j_D, j_H, and f may also be applied to simultaneous heat, mass, and momentum transfer at low mass-transfer rates.

Equation 21.2–21 is one of the Chilton-Colburn analogies.[4] It agrees closely with the predictions of boundary-layer theory for the flat plate when Pr and Sc exceed 0.5 (see Eqs. 19.3–43, 44, and 45) and appears to be fairly good for turbulent flow. In flow around curved boundaries form-drag occurs (see §§2.6 and 6.3), so that $f/2$ may exceed j_H and j_D considerably. (See Fig. 13.3–1.) Even in straight conduits, the agreement between $f/2$ and the j-factors is only approximate. (See Fig. 13.2–1.) However, the more limited empirical analogy

$$j_H = j_D = \left\{\begin{array}{l}\text{a function of Re,}\\ \text{geometry, and}\\ \text{boundary conditions}\end{array}\right\} \qquad (21.2\text{–}23)$$

has proved useful for a number of flow situations, such as transverse flow over cylinders (Fig. 13.3–1), flow through packed beds (§13.4), and flow in pipes at high Reynolds numbers (Fig. 13.2–1). Equation 21.2–23 is the usual form of the Chilton-Colburn analogy.

As a second illustration, consider the heat-transfer correlation given in Eq. 13.3–1 for forced convection around a sphere of diameter D:

$$\frac{h_m D}{k_f} = 2.0 + 0.60\left(\frac{D v_\infty \rho_f}{\mu_f}\right)^{1/2}\left(\frac{\hat{C}_p \mu}{k}\right)_f^{1/3} \qquad (21.2\text{–}24)$$

Substitution of the corresponding mass-transfer quantities, according to Table 21.2–1, gives

$$\frac{k_{xm} D}{c_f \mathscr{D}_{ABf}} = 2.0 + 0.60\left(\frac{D v_\infty \rho_f}{\mu_f}\right)^{1/2}\left(\frac{\mu}{\rho \mathscr{D}_{AB}}\right)_f^{1/3} \qquad (21.2\text{–}25)$$

[4] T. H. Chilton and A. P. Colburn, *Ind. Eng. Chem.*, **26**, 1183 (1934).

Equations 21.2–24 and 25 hold for constant surface temperature and composition, respectively, and for small mass-transfer rates. They may also be applied to simultaneous heat and mass transfer under the same restrictions. Note that the Chilton-Colburn analogy, Eq. 21.2–23, does not hold for this flow system, except as an approximation when Nu and $\text{Nu}_{AB} \gg 2$.

Example 21.2–1. Evaporation of a Freely Falling Drop

A spherical drop of water, 0.05 cm in diameter, is falling at a velocity of 215 cm sec^{-1} through dry, still air at 1 atm pressure. Estimate the instantaneous rate of evaporation from the drop if the drop surface is at 70° F and the air is at 140° F. The vapor pressure of water at 70° F is 0.0247 atm. Assume pseudo-steady-state conditions.

Solution. Let water be species A and air be species B. The solubility of air in water may be neglected, so that $\mathcal{W}_B^{(m)} = 0$. Then, assuming that the evaporation rate is small,[5] we may apply Eq. 21.1–12 to the gas phase around the drop and obtain

$$\mathcal{W}_A^{(m)} = k_{xm}\pi D^2 \frac{x_{A0} - x_{A\infty}}{1 - x_{A0}} \tag{21.2-26}$$

and the mass-transfer coefficient k_{xm} may be predicted from Eq. 21.2–25 or Fig. 13.3–2.

The temperatures and compositions needed for this problem are

$$T_0 = 70^\circ \text{ F} \qquad T_\infty = 140^\circ \text{ F} \qquad T_f = \frac{T_0 + T_\infty}{2} = 105^\circ \text{ F}$$

$$x_{A0} = 0.0247 \qquad x_{A\infty} = 0 \qquad x_{Af} = \frac{x_{A0} + x_{A\infty}}{2} = 0.0124$$

In computing x_{A0}, we have assumed ideal gas behavior, equilibrium at the interface, and complete insolubility of air in water. The mean mole fraction, x_{Af}, of water in the gas is small enough that it can be neglected in evaluating the physical properties:

$$c_f = 3.88 \times 10^{-5} \text{ g-mole cm}^{-3}$$

$$\rho_f = 1.12 \times 10^{-3} \text{ g cm}^{-3}$$

$$\mu_f = 1.91 \times 10^{-4} \text{ g cm}^{-1} \text{ sec}^{-1} \text{ (from Table 1.1–1)}$$

$$\mathcal{D}_{ABf} = 0.292 \text{ cm}^2 \text{ sec}^{-1} \text{ (from Eq. 16.3–1)}$$

$$\left(\frac{\mu}{\rho \mathcal{D}_{AB}}\right)_f = 0.58$$

$$\frac{Dv_\infty \rho_f}{\mu_f} = \frac{(0.05)(215)(1.12 \times 10^{-3})}{1.91 \times 10^{-4}} = 63$$

[5] The validity of the assumption of a small mass-transfer rate, for the conditions given here, is shown in Example 21.5–2.

Insertion of these values in Eq. 21.2–25 gives

$$k_{xm} = \frac{c_f \mathscr{D}_{AB}}{D}\left[2 + 0.60\left(\frac{Dv_\infty \rho_f}{\mu_f}\right)^{1/2}\left(\frac{\mu}{\rho \mathscr{D}_{AB}}\right)_f^{1/3}\right]$$

$$= \frac{(3.88 \times 10^{-5})(0.292)}{0.05}[2 + 3.96]$$

$$= 1.35 \times 10^{-3} \text{ g-mole sec}^{-1}\text{ cm}^{-2} \tag{21.2–27}$$

From Eq. 21.2–26 the evaporation rate is then found to be

$$\mathscr{W}_A^{(m)} = (1.35 \times 10^{-3})(\pi)(0.05)^2\frac{0.0247 - 0}{1 - 0.0247}$$

$$= 2.70 \times 10^{-7} \text{ g-mole sec}^{-1}$$

This result corresponds to a decrease of 1.23×10^{-3} cm sec^{-1} in the drop diameter and indicates that a drop of this size will fall a considerable distance before it evaporates completely.

In this example, for simplicity, the velocity and surface temperature of the drop were given. In general, these conditions must be calculated from momentum and energy balances, as discussed in Problem 21.E.

Example 21.2–2. The Wet-and-Dry-Bulb Psychrometer

The cooling effect of evaporation from a wetted surface can be used to analyze certain simple gas mixtures. Consider, for example, the arrangement in Fig. 21.2–2, in which a mixture of condensable gas A and noncondensable gas B flows over a pair of long cylindrical thermometers. One thermometer bulb (the dry bulb) is left bare, and the other (the wet bulb) is covered with a wick saturated with liquid A. Fresh liquid A at the wet-bulb temperature continuously flows up the wick by capillary action from the reservoir below. Develop an equation for the composition of the gas stream in terms of the wet-bulb and dry-bulb thermometer readings.

Solution. For simplicity, we assume the fluid velocity to be high enough that the thermometer readings are unaffected by radiation and by heat conduction along the thermometer stems but not so high that dissipative heating becomes significant.[6] These assumptions are usually satisfactory for glass thermometers and gas velocities of 30 to 100 ft sec^{-1}. The dry-bulb temperature is then the same as the temperature T_∞ of the approaching gas, and the wet-bulb temperature is the same as the temperature T_0 of the outside of the wick.

An energy balance on a system that contains a length L of the wick (see Fig. 21.2–2) gives

$$\mathscr{W}_A^{(m)}(\bar{H}_{Ag0} - \tilde{H}_{A1}) = Q \tag{21.2–28}$$

in which Q is the heat flow to the wick in the gas phase at the interface and $\mathscr{W}_A^{(m)}$ is the evaporation rate. The quantity $\bar{H}_{Ag0}$ is the partial molal enthalpy of A in the

[6] A more detailed solution is considered in Problem 21.F.

gas phase at the interface. Neglecting radiation, and neglecting the effect of mass transfer on h, we may set $Q = h_m \pi DL(T_\infty - T_0)$. Also, neglecting heats of mixing in the gas phase, $\bar{H}_{Ag0} - \tilde{H}_{A1}$ may be replaced by the heat of vaporization

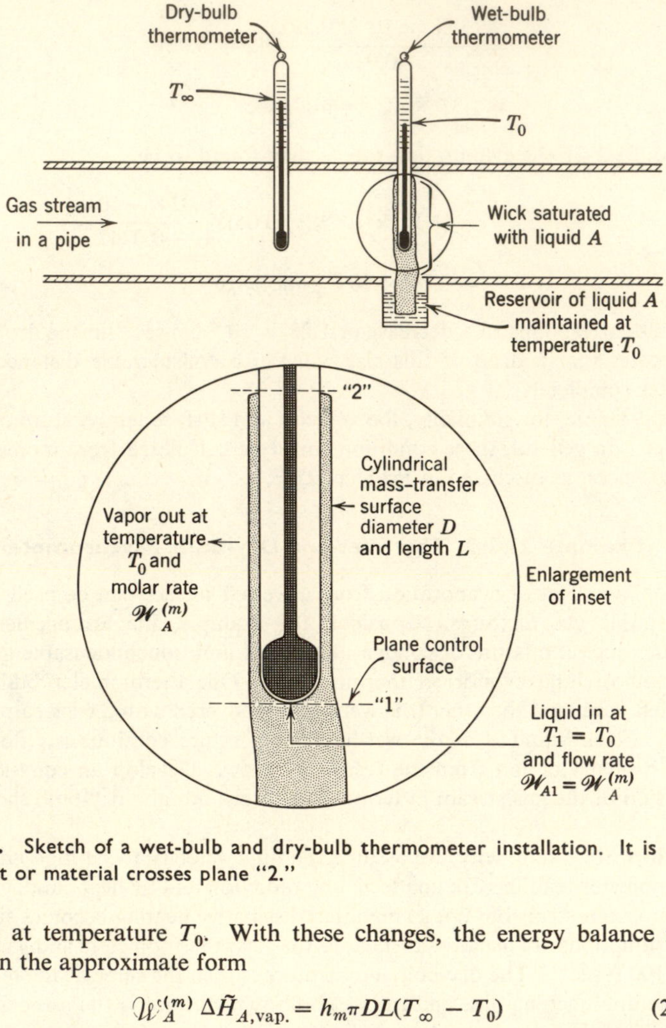

Fig. 21.2–2. Sketch of a wet-bulb and dry-bulb thermometer installation. It is assumed that no heat or material crosses plane "2."

of pure A at temperature T_0. With these changes, the energy balance may be rewritten in the approximate form

$$\mathcal{W}_A^{(m)} \Delta\tilde{H}_{A,\text{vap.}} = h_m \pi DL(T_\infty - T_0) \tag{21.2–29}$$

A second expression for $\mathcal{W}_A^{(m)}$, in terms of the interface and stream compositions, is obtained from Eq. 21.1–12:

$$\mathcal{W}_A^{(m)}(1 - x_{A0}) = k_{xm} \pi DL(x_{A0} - x_{A\infty}) \tag{21.2–30}$$

Combination of the last two equations gives

$$\frac{(x_{A0} - x_{A\infty})}{(T_\infty - T_0)(1 - x_{A0})} = \frac{h_m}{k_{xm} \Delta\tilde{H}_{A,\text{vap.}}} \tag{21.2–31}$$

Now the heat-transfer data for cylinders with constant surface temperature and no mass transfer are represented empirically as j_H versus Re in Fig. 13.3–1; the corresponding mass-transfer correlation for constant surface composition and slow mass transfer is obtained by setting

$$j_H = j_D \tag{21.2–32}$$

or

$$\frac{h_m}{\rho_f \hat{C}_{pf} v_\infty} \mathrm{Pr}_f^{2/3} = \frac{k_{xm}}{c_f v_\infty} \mathrm{Sc}_f^{2/3} \tag{21.2–33}$$

Rearranging this equation, and noting that $\rho \hat{C}_p = c\tilde{C}_p$, one gets the empirical formula[7]

$$\frac{h_m}{k_{xm}} = \tilde{C}_{pf} \left(\frac{\mathrm{Sc}}{\mathrm{Pr}}\right)_f^{2/3} \tag{21.2–34}$$

and with this substitution Eq. 21.2–31 becomes

$$\frac{(x_{A0} - x_{A\infty})}{(T_\infty - T_0)(1 - x_{A0})} = \frac{\tilde{C}_{pf}}{\Delta \tilde{H}_{A,\text{vap.}}} \left(\frac{\mathrm{Sc}}{\mathrm{Pr}}\right)_f^{2/3} \tag{21.2–35}$$

The interfacial gas composition x_{A0} can be accurately predicted, at low mass-transfer rates, by neglecting the heat- and mass-transfer resistance of the interface itself. (For further discussion see §21.3). One can then represent x_{A0} by the vapor-liquid equilibrium relationship:

$$x_{A0} = x_{A0}(T_0, p) \tag{21.2–36}$$

A relation of this kind will hold for given species A and B if the liquid is pure A as assumed above. A commonly used approximation of this relationship is

$$x_{A0} = \frac{p_{A,\text{vap.}}}{p} \tag{21.2–37}$$

in which $p_{A,\text{vap.}}$ is the vapor pressure of pure A at temperature T_0. This relation involves the added assumptions that the presence of B does not alter the partial pressure of A at the interface and that A and B form an ideal gaseous mixture.

Equation 21.2–36 or 37 can be combined with Eq. 21.2–35 to relate the approaching gas composition $x_{A\infty}$ to the thermometer readings T_0 and T_∞ for any system pressure. Thus, if an air-water mixture at 1 atm pressure gives a wet-bulb temperature T_0 of 70° F and a dry-bulb reading T_∞ of 140° F, the calculation proceeds as follows:

$$\begin{aligned}
x_{A0} &= 0.0247, \text{ from Eq. 21.2–37} \\
\tilde{C}_{pf} &= 6.98 \text{ Btu (lb-mole)}^{-1}{}^\circ\text{ F}^{-1} \text{ for air at } 105^\circ \text{ F} \\
\Delta \tilde{H}_{A,\text{vap.}} &= 18{,}900 \text{ Btu (lb-mole)}^{-1} \text{ at } 70^\circ \text{ F} \\
\mathrm{Sc}_f &= 0.58 \text{ (see Example 21.2–1)} \\
\mathrm{Pr}_f &= 0.74, \text{ from Eq. 8.3–16}
\end{aligned}$$

[7] A somewhat different equation, employing the 0.56 power of Sc, was recommended for measurements in air by C. H. Bedingfield and T. B. Drew, *Ind. Eng. Chem.*, **42**, 1164–1173 (1950).

Substitution in Eq. 21.2–35 gives

$$\frac{(0.0247 - x_{A\infty})}{(140 - 70)(1 - 0.0247)} = \frac{6.98}{18{,}900}\left(\frac{0.58}{0.74}\right)^{2/3}$$

from which the mole fraction of water in the approaching air is found to be

$$x_{A\infty} = 0.0033$$

In the foregoing calculations we have assumed $x_{Af} = 0$ as a first approximation; a second approximation could be made with the improved value $x_{Af} = \frac{1}{2}(0.0247 + 0.0033) = 0.0140$. The physical properties are not known accurately enough here to justify recalculation.

The calculated result is in fair agreement with published humidity charts,[8] which give $x_{A\infty} \doteq 0$ at the foregoing conditions.

§21.3 DEFINITION OF BINARY MASS-TRANSFER COEFFICIENTS IN TWO PHASES AT LOW MASS-TRANSFER RATES[1]

In most mass-transfer systems there is an interface with concentration gradients on both sides. Consider, as a simple example, the contacting of a gas mixture of A and B with a nonvolatile liquid C which absorbs only A; the concentration gradients in the two phases are sketched in Fig. 21.3–1. The symbol y_A is used for mole fraction A in the gas phase, and x_A is used for mole fraction A in the liquid phase.

There are, in general, three resistances to the movement of species A between phases: the gas phase, the liquid phase, and the interface itself. The interface resistance is thought to be negligible in most applications; exceptions do occur, for example, when traces of surface-active species concentrate at the interface or when the mass-transfer rate is exceedingly high. We assume here that the interface resistance is negligible and that any fluctuations in y_A and x_A are small, so that the time-smoothed gas and liquid compositions[2] lie essentially on the equilibrium curve:

$$y_{A0} = f(x_{A0}) \tag{21.3–1}$$

A sample equilibrium curve and interface composition are shown in Fig. 21.3–2.

The concentration difference across each phase can be expressed in terms of the local mass-transfer rate and the mass-transfer coefficient in that phase.

[8] O. A. Hougen, K. M. Watson, and R. A. Ragatz, *Chemical Process Principles*, Wiley, New York, Part I (1954), Second Edition,p. 120.

[1] This section parallels the original treatment by W. G. Whitman, *Chem. Met. Eng.*, **29**, 146–148 (1923), though the notation is different.

[2] All quantities are time-smoothed in the following development. For brevity, we omit the overlines.

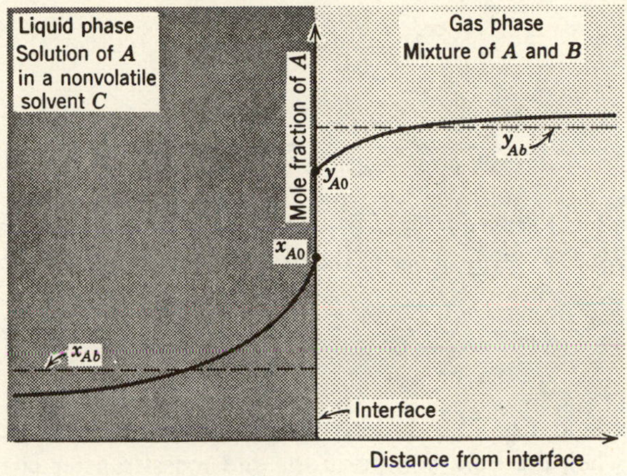

Fig. 21.3-1. Concentration profiles of A in the neighborhood of a gas-liquid interface.

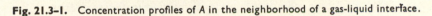

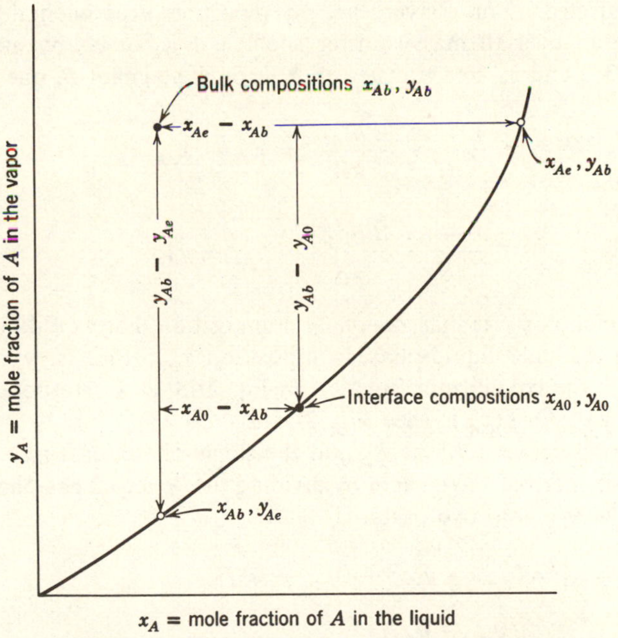

Fig. 21.3-2. Driving forces in interphase mass transfer.

For slow mass transfer of species A only, application of Eq. 21.1–10 to each phase gives

$$\frac{1 - x_{A0}}{k_x} \frac{d\mathcal{W}_{Al}^{\circ(m)}}{dA} = x_{A0} - x_{Ab} \tag{21.3–2}$$

$$\frac{1 - y_{A0}}{k_y} \frac{d\mathcal{W}_{Ag}^{\circ(m)}}{dA} = y_{A0} - y_{Ab} \tag{21.3–3}$$

and, since the local flux $(d\mathcal{W}_{Al}^{\circ(m)}/dA = N_{Al0})$ *into* the liquid equals the local flux $(-d\mathcal{W}_{Ag}^{\circ(m)}/dA = -N_{Ag0})$ *out of* the gas, these equations may be combined to give

$$\frac{y_{Ab} - y_{A0}}{x_{Ab} - x_{A0}} = -\frac{k_x/(1 - x_{A0})}{k_y/(1 - y_{A0})} \tag{21.3–4}$$

This equation and Eq. 21.3–1 determine the two unknown compositions y_{A0} and x_{A0} when the bulk compositions and mass-transfer coefficients are given. One can then compute the local mass-transfer rate from either Eq. 21.3–2 or 21.3–3.

The single-phase transfer coefficients are difficult to measure, except in experiments so designed that the concentration difference across one phase can be neglected. For convenience in reporting experimental data, one may define an over-all mass-transfer coefficient K_x or K_y by modification of Eqs. 21.3–2 and 3. For transfer of A through stagnant B, one may write

$$\frac{1 - x_{Ae}}{K_x} \frac{d\mathcal{W}_{Al}^{\circ(m)}}{dA} = x_{Ae} - x_{Ab} \tag{21.3–5}$$

or

$$\frac{1 - y_{Ae}}{K_y} \frac{d\mathcal{W}_{Ag}^{\circ(m)}}{dA} = y_{Ae} - y_{Ab} \tag{21.3–6}$$

The mole fraction y_{Ae} is that gas-phase composition that would be in equilibrium with the bulk liquid-phase composition x_{Ab}; that is, $y_{Ae} = f(x_{Ab})$, in which f is the equilibrium function in Eq. 21.3–1. Correspondingly, x_{Ae} is defined by $y_{Ab} = f(x_{Ae})$. (See Fig. 21.3–2.)

The relation between K_x or K_y and the single-phase mass-transfer coefficients is easily derived. We begin by dividing the "over-all gas-phase driving force," $y_{Ae} - y_{Ab}$, into two parts:

$$\begin{aligned}(y_{Ae} - y_{Ab}) &= (y_{Ae} - y_{A0}) + (y_{A0} - y_{Ab}) \\ &= \left(\frac{y_{Ae} - y_{A0}}{x_{Ab} - x_{A0}}\right)(x_{Ab} - x_{A0}) + (y_{A0} - y_{Ab}) \\ &= m_y(x_{Ab} - x_{A0}) + (y_{A0} - y_{Ab})\end{aligned} \tag{21.3–7}$$

Here m_y is the slope of a line joining the true interface point (y_{A0}, x_{A0}) to the fictitious one (y_{Ae}, x_{Ab}) used in computing the over-all gas-phase driving force. Insertion of Eqs. 21.3–2, 3, and 6 and division by $d\mathcal{W}_A^{(x)}/dA = -d\mathcal{W}_A^{(y)}/dA$ gives[3]

$$\frac{1 - y_{Ae}}{K_y} = \frac{m_y(1 - x_{A0})}{k_x} + \frac{1 - y_{A0}}{k_y} \qquad (21.3\text{–}8)$$

A similar development for $(x_{Ae} - x_{Ab})$ gives

$$\frac{1 - x_{Ae}}{K_x} = \frac{1 - x_{A0}}{k_x} + \frac{1 - y_{A0}}{m_x k_y} \qquad (21.3\text{–}9)$$

in which m_x is the slope of a straight line from (y_{A0}, x_{A0}) to (y_{Ab}, x_{Ae}). If the equilibrium line is straight with slope m, then $m_y = m_x = m$. It then follows that

$$\frac{1 - y_{Ae}}{K_y} = m\,\frac{1 - x_{Ae}}{K_x} \qquad (21.3\text{–}10)$$

which serves to emphasize that K_x and K_y convey the same basic information.

It is useful to know, in any given system, the relative resistances of the two phases to mass transfer. The terms in Eq. 21.3–8 or 9 represent the over-all, liquid-phase, and gas-phase resistances, respectively. The resistance offered by either phase decreases with increasing mass-transfer coefficient k_x or k_y, or increasing concentration of A, in that phase; thus experiments with nearly pure A in one phase are useful for determining one-phase transfer coefficients. If m is very large, that is, if A is nearly insoluble in C, the liquid-phase resistance tends to predominate, and conversely.

The method of derivation used here can be extended to the transfer of two or more components. (See Problem 21.G.)

Most of the available design data on mass transfer in two-fluid systems

[3] If a similar development is carried out in terms of k_G and k_L (see Eqs. 21.1–4*a* and 4*b*) and an over-all coefficient is defined by

$$N_{Ag0} = K_G(p_{Ae} - p_{Ab}) \qquad (21.3\text{–}8a)$$

then the corresponding result for K_G is

$$\frac{1}{K_G} = \frac{H}{k_L} + \frac{1}{k_G} \qquad (21.3\text{–}8b)$$

This equation applies when the equilibrium function Eq. 21.3–1 is approximated by the linear function $p_{A0} = Hc_{A0}$ (Henry's law). We prefer to use Eq. 21.3–8 because the mass-transfer coefficients appearing therein tend to vary less with concentration level; also because the use of k_x and k_y simplifies calculations when both A and B are transferred, as in distillation or partial condensation. (See Problem 21.G.)

are reported as over-all mass-transfer coefficients or related quantities. Extensive summaries are available in several references.[4,5,6]

Two-fluid mass-transfer systems offer many challenging problems: the flow behavior is complicated, the moving interface is virtually inaccessible to sampling, the interfacial area is usually unknown, and many of the practically important systems involve liquid-phase chemical reactions. A better basic understanding of these systems is needed.

§21.4 DEFINITION OF THE TRANSFER COEFFICIENTS FOR HIGH MASS-TRANSFER RATES

Interphase mass transfer generally involves a bulk flow of material through the interface. At *small mass-transfer rates*, this bulk flow is important only in calculating the fluxes of the different species across the boundary; one can then calculate momentum-, heat-, and mass-transfer rates at an interface of area A by the simple expressions given in Eqs. 6.1–1, 13.1–1, and 21.1–8:

$$F_{kz} = fA\tfrac{1}{2}\rho V^2 \tag{21.4–1}$$

$$Q = hA\,\Delta T \tag{21.4–2}$$

$$\mathcal{W}_A^{(m)} - x_{A0}(\mathcal{W}_A^{(m)} + \mathcal{W}_B^{(m)}) = k_x A\,\Delta x_A \tag{21.4–3}$$

Here F_{kz} is the force of the fluid on the interface in the main flow direction, Q is the rate of heat *conduction* into the fluid at the interface, and $\mathcal{W}_A^{(m)} - x_{A0}(\mathcal{W}_A^{(m)} + \mathcal{W}_B^{(m)})$ is the rate of *diffusion* of species A into the fluid at the interface.

For *high mass-transfer rates*, it is convenient to use analogous definitions but to add a superscript black dot (●) to indicate that the transfer coefficients now depend on the mass-transfer rate:

$$F_{kz} = f^{\bullet}A\tfrac{1}{2}\rho V^2 \tag{21.4–4}$$

$$Q = h^{\bullet}A\,\Delta T \tag{21.4–5}$$

$$\mathcal{W}_A^{(m)} - x_{A0}(\mathcal{W}_A^{(m)} + \mathcal{W}_B^{(m)}) = k_x^{\bullet} A\,\Delta x_A \tag{21.4–6}$$

The black dots will be employed only when the transfer coefficients have been corrected for mass-transfer rate. These corrections, which arise from the dependence of the velocity, temperature, and concentration profiles on the flow velocity through the interface, are the subject of §§21.5, 6, and 7. At low mass-transfer rates, one can insert the uncorrected transfer coefficients f, h, and k; these equations then reduce to Eqs. 21.4–1, 2, and 3.

[4] T. K. Sherwood and R. L. Pigford, *Absorption and Extraction*, McGraw-Hill, New York, Second Edition (1952).

[5] R. E. Treybal, *Mass Transfer Operations*, McGraw-Hill, New York (1955).

[6] *Chemical Engineers' Handbook*, J. H. Perry (Ed.), McGraw-Hill, New York (1950), Third Edition.

Note that the force F_{kz} and the heat-conduction rate Q represent only part of the momentum and energy flow through the surface. At high mass-transfer rates the additional transfer of momentum and energy by the fluid streaming through the interface may be significant. We designate these additional momentum and energy flows *into* the fluid by[1]

$$F_z^{(m)} = (\mathcal{W}_A^{(m)} M_A + \mathcal{W}_B^{(m)} M_B) v_{z0} \tag{21.4–7}$$

$$Q^{(m)} = (\mathcal{W}_A^{(m)} \bar{H}_{A0} + \mathcal{W}_B^{(m)} \bar{H}_{B0}) \tag{21.4–8}$$

in which the superscript (m) indicates that these terms arise only in the presence of mass transfer and the subscript 0 denotes conditions in the fluid next to the interface.

Addition of the bulk flow terms $F_z^{(m)}$, $Q^{(m)}$, and $x_{A0}(\mathcal{W}_A^{(m)} + \mathcal{W}_B^{(m)})$ to Eqs. 21.4–4, 5, and 6 gives

$$-F_{kz} + F_z^{(m)} = -f^\bullet A \tfrac{1}{2}\rho V^2 + (\mathcal{W}_A^{(m)} M_A + \mathcal{W}_B^{(m)} M_B) v_{z0} \tag{21.4–9}$$

$$Q + Q^{(m)} = h^\bullet A\,\Delta T + (\mathcal{W}_A^{(m)} \bar{H}_{A0} + \mathcal{W}_B^{(m)} \bar{H}_{B0}) \tag{21.4–10}$$

$$\mathcal{W}_A^{(m)} = k_x^\bullet A\,\Delta x_A + (\mathcal{W}_A^{(m)} + \mathcal{W}_B^{(m)}) x_{A0} \tag{21.4–11}$$

Here $-F_{kz} + F_z^{(m)}$ is the total rate of momentum addition to the fluid, $Q + Q^{(m)}$ is the rate of energy addition other than work,[1] and $\mathcal{W}_A^{(m)}$ is the molar rate of addition of species A, through the interface of area A.

The evaluation of $f^\bullet$, $h^\bullet$, and $k_x^\bullet$ is taken up in §§21.5, 6, and 7. Since most of the available information on transfer coefficients is limited to small mass-transfer rates, the emphasis is on evaluation of *correction factors* by which f, h, and k may be multiplied to obtain the corrected transfer coefficients:

$$\theta_v = \frac{f^\bullet}{f}, \qquad \theta_T = \frac{h^\bullet}{h}, \qquad \theta_{AB} = \frac{k_x^\bullet}{k_x} \tag{21.4–12,13,14}$$

The subscripts v, T, and AB are employed because these correction factors depend on the profiles of *velocity*, *temperature*, and *concentration of A or B*. These correction factors can be predicted for a few simple situations by solving the equations of change with allowance for the finite velocity through the interface. The model systems considered are necessarily idealized but may serve as useful approximations to the behavior of many practical systems.

[1] The expression for $Q^{(m)}$ has been simplified by omitting the potential and kinetic energy terms, which are usually unimportant. The complete expression for $Q^{(m)}$ is

$$Q^{(m)} = \mathcal{W}_A^{(m)} \bar{H}_{A0} + \mathcal{W}_B^{(m)} \bar{H}_{B0} + (\mathcal{W}_A^{(m)} M_A + \mathcal{W}_B^{(m)} M_B)\left(\hat{\Phi}_0 + \frac{v_0^2}{2}\right) \tag{21.4–8a}$$

if the potential energy and fluid mass-average velocity are constant at $\hat{\Phi}_0$ and v_0 along the interface.

§21.5 TRANSFER COEFFICIENTS AT HIGH MASS-TRANSFER RATES: FILM THEORY

In this section we use a simplified unidirectional transport model to predict the variation of the momentum-, heat-, and mass-transfer coefficients with mass-transfer rate. The solutions presented here were obtained by several different investigators.[1,2,3,4]

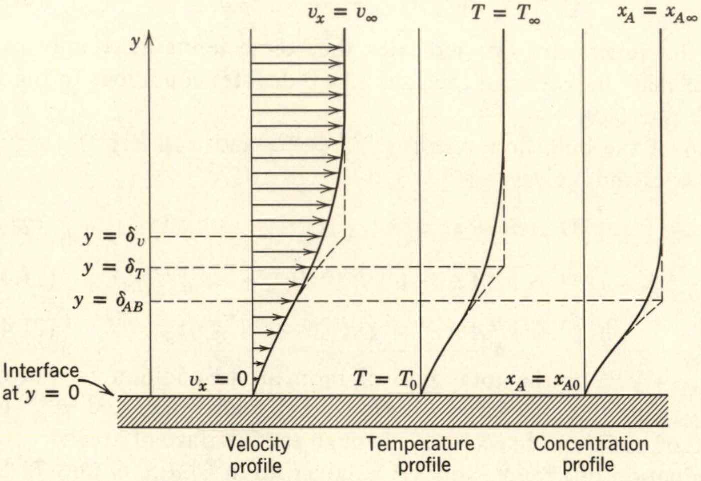

Fig. 21.5–1. Steady flow along a flat surface with rapid mass transfer into the stream. The unbroken curves represent the true profiles; the broken curves are predicted by the film theory.

Consider the steady-flow situation shown in Fig. 21.5–1, in which a binary fluid mixture flows horizontally along a plane surface and heat, mass, and momentum transfer occur simultaneously. We assume that the pressure p, velocity v_x, temperature T, and composition x_A depend only on the distance y from the wall and become constant when y exceeds the corresponding "film thicknesses" δ_v, δ_T, and δ_{AB}. The film thicknesses are assumed constant along the wall and independent of mass-transfer rate. The flow in the

[1] W. K. Lewis and K. C. Chang, *Trans. A.I.Ch.E.*, **21**, 127–136 (1928), applied the film theory to the interphase transfer of both species in a binary mixture at high mass-transfer rates.

[2] G. Ackermann, *Forschungsheft*, **382**, 1–16 (1937), applied the film theory to heat transfer in the presence of rapid mass transfer.

[3] A. P. Colburn and T. B. Drew, *Trans. A.I.Ch.E.*, **33**, 197–212 (1937), independently confirmed Ackermann's analysis and applied the results of Refs. 1 and 2 to the rapid condensation of mixed vapors.

[4] H. S. Mickley, R. C. Ross, A. L. Squyers and W. E. Stewart, *NACA Tech. Note 3208* (1954), summarized the results of Refs. 1, 2, and 3 in analogous form and extended the analogy to momentum transfer in constant-pressure systems.

film is assumed to be laminar, and the transport properties μ, k, and $c\mathscr{D}_{AB}$ are considered constant. Homogeneous chemical reactions, viscous dissipation, and radiant-energy emission or absorption in the fluid are neglected.

The following conditions in the system are considered to be known:

$$\text{at } y = 0, \qquad v_x = 0 \tag{21.5–1}$$

$$T = T_0 \tag{21.5–2}$$

$$x_A = x_{A0} \tag{21.5–3}$$

$$\frac{N_{Ay}}{N_{By}} = \frac{N_{A0}}{N_{B0}} \tag{21.5–4}$$

$$\text{at } y = \delta_v, \qquad v_x = v_\infty \tag{21.5–5}$$

$$\text{at } y = \delta_T, \qquad T = T_\infty \tag{21.5–6}$$

$$\text{at } y = \delta_{AB}, \qquad x_A = x_{A\infty} \tag{21.5–7}$$

That is, we know the terminal values of all three profiles and, in addition, the ratio of the molar fluxes at the wall. The velocity, temperature, and composition profiles, and the momentum, energy, and mass fluxes at the walls, are to be calculated. We will then show how to evaluate the film thicknesses so that the results can be applied to systems of practical interest.

The equations of change for this system are conveniently written in terms of the momentum, energy, and mass fluxes relative to stationary coordinates. From Eqs. 18.3–11 and 12 and 18.1–10 and the assumption that p, v_x, T, and x_A depend only on y, we get the following equations:

$$\text{(motion)} \qquad \frac{d\phi_{yx}}{dy} = 0 \qquad (y \leqslant \delta_v) \tag{21.5–8}$$

$$\text{(energy)} \qquad \frac{de_y}{dy} = 0 \qquad (y \leqslant \delta_T) \tag{21.5–9}$$

$$\text{(continuity)} \qquad \frac{dN_{Ay}}{dy} = 0 \qquad (y \leqslant \delta_{AB}) \tag{21.5–10}$$

$$\frac{dN_{By}}{dy} = 0 \qquad (y \leqslant \delta_{AB}) \tag{21.5–11}$$

Integration of these equations, and evaluation of ϕ_{yx}, e_y and N_{Ay} from Eqs. 18.3–8, 18.4–6, and 16.2–2, gives for the same ranges of y

$$\text{(motion)} \qquad \phi_{yx} = \rho v_y v_x - \mu \frac{dv_x}{dy} = \text{const.} \tag{21.5–12}$$

$$\text{(energy)} \qquad e_y = N_{Ay}\bar{H}_A + N_{By}\bar{H}_B - k\frac{dT}{dy} = \text{const.} \tag{21.5–13}$$

$$\text{(continuity)} \qquad N_{Ay} = x_A(N_{Ay} + N_{By}) - c\mathscr{D}_{AB}\frac{dx_A}{dy} = \text{const.} \tag{21.5–14}$$

$$N_{By} = \text{const.} \tag{21.5–15}$$

Hereafter we denote the constant fluxes N_{Ay} and N_{By} by N_{A0} and N_{B0}. The equation for N_{By} is then not needed further. In the event that δ_v or δ_T exceeds δ_{AB}, the fluxes N_{Ay} and N_{By} are still treated as constants throughout the range of Eqs. 21.5–8 and 21.5–9 for lack of any alternative information.

Expressing the constants in terms of the (yet unknown) fluxes at the wall gives[5]

$$\text{(motion)}\quad (v_x - 0)(N_{A0}M_A + N_{B0}M_B) - \mu\frac{dv_x}{dy} = \tau_{yx}|_{y=0} \equiv -\tau_0 \quad (21.5\text{–}16)$$

$$\text{(energy)}\quad (T - T_0)(N_{A0}\tilde{C}_{pA} + N_{B0}\tilde{C}_{pB}) - k\frac{dT}{dy} = q_y|_{y=0} \equiv q_0 \quad (21.5\text{–}17)$$

$$\text{(continuity)}\quad (x_A - x_{A0})(N_{A0} + N_{B0}) - c\mathscr{D}_{AB}\frac{dx_A}{dy}$$

$$= J_{Ay}^{\star}|_{y=0} \equiv N_{A0} - x_{A0}(N_{A0} + N_{B0}) \quad (21.5\text{–}18)$$

Here ρv_y has been evaluated from Table 16.1–3, and $\bar{H}_A$ and $\bar{H}_B$ have been evaluated on the assumptions of constant heat capacities and no heat of mixing. The usual convention $\tau_{yx}|_{y=0} = -\tau_0$ is adopted to make τ_0 positive. Note that in a given system Eqs. 21.5–16, 17, and 18 contain no variable quantities except v_x, T, x_A, and the independent variable y. Integration of Eqs. 21.5–16, 17, and 18, and application of the boundary conditions in Eqs. 21.5–1, 2, and 3, gives the following solutions for the profiles:

$$1 - \frac{(v_x - 0)(N_{A0}M_A + N_{B0}M_B)}{-\tau_0} = \exp\,(N_{A0}M_A + N_{B0}M_B)\frac{y}{\mu} \quad (21.5\text{–}19)$$

$$1 - \frac{(T - T_0)(N_{A0}\tilde{C}_{pA} + N_{B0}\tilde{C}_{pB})}{q_0} = \exp\,(N_{A0}\tilde{C}_{pA} + N_{B0}\tilde{C}_{pB})\frac{y}{k} \quad (21.5\text{–}20)$$

$$1 - \frac{(x_A - x_{A0})(N_{A0} + N_{B0})}{N_{A0} - x_{A0}(N_{A0} + N_{B0})} = \exp\,(N_{A0} + N_{B0})\frac{y}{c\mathscr{D}_{AB}} \quad (21.5\text{–}21)$$

Application of the boundary conditions in Eqs. 21.5–5, 6, and 7 gives the following equations for the fluxes τ_0, q_0, and $N_{A0} + N_{B0}$:

$$1 + \frac{(v_\infty - 0)(N_{A0}M_A + N_{B0}M_B)}{\tau_0} = \exp\,(N_{A0}M_A + N_{B0}M_B)\frac{\delta_v}{\mu} \quad (21.5\text{–}22)$$

$$1 + \frac{(T_0 - T_\infty)(N_{A0}\tilde{C}_{pA} + N_{B0}\tilde{C}_{pB})}{q_0} = \exp\,(N_{A0}\tilde{C}_{pA} + N_{B0}\tilde{C}_{pB})\frac{\delta_T}{k} \quad (21.5\text{–}23)$$

$$1 + \frac{(x_{A0} - x_{A\infty})(N_{A0} + N_{B0})}{N_{A0} - x_{A0}(N_{A0} + N_{B0})} = \exp\,(N_{A0} + N_{B0})\frac{\delta_{AB}}{c\mathscr{D}_{AB}} \quad (21.5\text{–}24)$$

These six equations give the fluxes and profiles. The results include those previously given for more restricted situations in §17.2 (diffusion through

[5] Note that τ_0, q_0, N_{A0}, and N_{B0} are employed here as convenient integration constants. Thus, after a second integration, we will have seven integration constants, which will be evaluated by means of Eqs. 21.5–1 through 7.

a stagnant gas film) and Example 18.5–1 (condensation of A in the presence of stagnant gas B). A numerical solution for given fluxes, physical properties, and film thicknesses would proceed as follows:

Knowing	*One can use Eq.*	*To compute*
$x_{A0}, x_{A\infty}, \dfrac{N_{A0}}{N_{B0}}$	21.5–24	N_{A0} and N_{B0}
$N_{A0}, N_{B0}, T_0, T_\infty$	21.5–23	q_0
N_{A0}, N_{B0}, v_∞	21.5–22	τ_0

Once the fluxes N_{A0}, N_{B0}, q_0, and τ_0 are computed, the profiles can be obtained directly from Eqs. 21.5–19, 20, and 21.

In order to apply these results to systems of practical interest, the film thicknesses have to be evaluated in terms of measurable quantities; the measurable quantities we select are the uncorrected transfer coefficients f_{loc}, h_{loc}, and $k_{x,\text{loc}}$. These coefficients are the limiting values, as N_{A0} and N_{B0} approach zero, of the corrected local transfer coefficients:

$$f^{\bullet}_{\text{loc}} = \frac{-\tau_0}{\frac{1}{2}\rho v_\infty(0 - v_\infty)} \tag{21.5–25}$$

$$h^{\bullet}_{\text{loc}} = \frac{q_0}{(T_0 - T_\infty)} \tag{21.5–26}$$

$$k^{\bullet}_{x,\text{loc}} = \frac{N_{A0} - x_{A0}(N_{A0} + N_{B0})}{(x_{A0} - x_{A\infty})} \tag{21.5–27}$$

Equations 21.5–22, 23, and 24 can now be rewritten in terms of these corrected local coefficients:

$$1 + \frac{N_{A0}M_A + N_{B0}M_B}{\frac{1}{2}\rho v_\infty f^{\bullet}_{\text{loc}}} = \exp\,(N_{A0}M_A + N_{B0}M_B)\frac{\delta_v}{\mu} \tag{21.5–28}$$

$$1 + \frac{N_{A0}\tilde{C}_{pA} + N_{B0}\tilde{C}_{pB}}{h^{\bullet}_{\text{loc}}} = \exp\,(N_{A0}\tilde{C}_{pA} + N_{B0}\tilde{C}_{pB})\frac{\delta_T}{k} \tag{21.5–29}$$

$$1 + \frac{N_{A0} + N_{B0}}{k^{\bullet}_{x,\text{loc}}} = \exp\,(N_{A0} + N_{B0})\frac{\delta_{AB}}{c\mathscr{D}_{AB}} \tag{21.5–30}$$

In the limit as N_{A0} and N_{B0} approach zero, these equations yield the following expressions for the film thicknesses:

$$\frac{1}{\frac{1}{2}\rho v_\infty f_{\text{loc}}} = \frac{\delta_v}{\mu} \tag{21.5–31}$$

$$\frac{1}{h_{\text{loc}}} = \frac{\delta_T}{k} \tag{21.5–32}$$

$$\frac{1}{k_{x,\text{loc}}} = \frac{\delta_{AB}}{c\mathscr{D}_{AB}} \tag{21.5–33}$$

These limiting values are easily found by expanding the right sides of Eqs. 21.5–28, 29, and 30 in Taylor series. Substitution of these expressions for the film thicknesses into Eqs. 21.5–22, 23, and 24, then gives

$$1 + \frac{(v_\infty - 0)(N_{A0}M_A + N_{B0}M_B)}{\tau_0} = \exp \frac{N_{A0}M_A + N_{B0}M_B}{\frac{1}{2}\rho v_\infty f_{\text{loc}}} \tag{21.5–34}$$

$$1 + \frac{(T_0 - T_\infty)(N_{A0}\tilde{C}_{pA} + N_{B0}\tilde{C}_{pB})}{q_0} = \exp \frac{N_{A0}\tilde{C}_{pA} + N_{B0}\tilde{C}_{pB}}{h_{\text{loc}}} \tag{21.5–35}$$

$$1 + \frac{x_{A0} - x_{A\infty}}{\dfrac{N_{A0}}{N_{A0} + N_{B0}} - x_{A0}} = \exp \frac{N_{A0} + N_{B0}}{k_{x,\text{loc}}} \tag{21.5–36}$$

These equations are the principal results of the film theory; they show how the shear stress, conductive energy flux, and diffusion flux at the wall depend on N_{A0} and N_{B0}. In this simple flow system the effects of mass transfer on the three flux relations are clearly analogous. Although Eqs. 21.5–34, 35, and 36 have been derived for laminar flow and constant physical properties, these results are also useful for turbulent flow and for variable physical properties. (See Problem 21.M.)

Most of the remainder of this section is devoted to defining the dimensionless quantities ϕ, R, θ, and Π, which are useful for summarizing the results of various mass transfer theories. We also present the foregoing results for the film theory in terms of these dimensionless quantities.

The dimensionless quantities on the right side of Eqs. 21.5–34, 35, and 36 vary directly with the mass-transfer rates and are therefore called the *rate factors*, ϕ:

$$\phi_v = \frac{N_{A0}M_A + N_{B0}M_B}{\frac{1}{2}\rho v_\infty f_{\text{loc}}} \tag{21.5–37}$$

$$\phi_T = \frac{N_{A0}\tilde{C}_{pA} + N_{B0}\tilde{C}_{pB}}{h_{\text{loc}}} \tag{21.5–38}$$

$$\phi_{AB} = \frac{N_{A0} + N_{B0}}{k_{x,\text{loc}}} \tag{21.5–39}$$

The dimensionless quantities on the left in Eqs. 21.5–34, 35, and 36 are called *flux ratios*, R:

$$R_v = \frac{N_{A0}M_A + N_{B0}M_B}{\frac{1}{2}\rho v_\infty f^{\bullet}_{\text{loc}}} = \frac{(N_{A0}M_A + N_{B0}M_B)(v_\infty - 0)}{\tau_0} \tag{21.5–40}$$

$$R_T = \frac{N_{A0}\tilde{C}_{pA} + N_{B0}\tilde{C}_{pB}}{h^{\bullet}_{\text{loc}}} = \frac{(N_{A0}\tilde{C}_{pA} + N_{B0}\tilde{C}_{pB})(T_0 - T_\infty)}{q_0} \tag{21.5–41}$$

$$R_{AB} = \frac{N_{A0} + N_{B0}}{k^{\bullet}_{x,\text{loc}}} = \frac{x_{A0} - x_{A\infty}}{\dfrac{N_{A0}}{N_{A0} + N_{B0}} - x_{A0}} \tag{21.5–42}$$

Note that R_v, R_T, and R_{AB} are ratios of momentum, energy, and molar fluxes by bulk flow to the fluxes by molecular transport at the interface. Equations 21.5–34, 35, and 36 may now be written more compactly as

$$R = e^{\phi} - 1 \qquad (21.5\text{–}43)$$

or

$$\phi = \ln(1 + R) \qquad (21.5\text{–}44)$$

Examples of calculation of heat-transfer and mass-transfer rates from these equations are given at the end of this section. The results for $h^{\bullet}$ and $k_x^{\bullet}$ are also valid for curved streamlines. (See Problems 21.K and L.)

The *correction factors*, θ, for the effect of mass transfer on the transfer coefficients are given by

$$\theta_v = \frac{f^{\bullet}}{f} = \frac{\phi_v}{R_v} \qquad (21.5\text{–}45)$$

$$\theta_T = \frac{h^{\bullet}}{h} = \frac{\phi_T}{R_T} \qquad (21.5\text{–}46)$$

$$\theta_{AB} = \frac{k_x^{\bullet}}{k_x} = \frac{\phi_{AB}}{R_{AB}} \qquad (21.5\text{–}47)$$

From Eqs. 21.5–43 and 44, the predicted values of these correction factors, according to the film theory, are

$$\theta = \frac{\phi}{e^{\phi} - 1} \qquad (21.5\text{–}48)$$

or

$$\theta = \frac{\ln(R + 1)}{R} \qquad (21.5\text{–}49)$$

These equations are convenient for examining the magnitude of the mass-transfer corrections, whereas Eqs. 21.5–43 and 44 give more direct results for the fluxes.

Equation 21.5–48 is plotted in Fig. 21.5–2. More detailed information is given in Figs. 21.7–1, 2, and 3, in which the results of the film theory and other theories are presented in working form. The results show that mass transfer of A and B *into* the stream makes ϕ_v, ϕ_T, and ϕ_{AB} positive and *decreases* the transfer coefficients, whereas mass transfer of A and B *out of* the stream makes ϕ_v, ϕ_T, and ϕ_{AB} negative and *increases* the transfer coefficients. If A and B are transferred in opposite directions, then the ϕ's may differ in sign, and, if so, the corresponding transfer coefficients will be corrected in opposite directions. (Note again that N_{A0} is *positive* when A is *added* to the phase under discussion and *negative* when A is *removed*; similarly for species B. The need for this sign convention is now apparent.)

The profiles obtainable from Eqs. 21.5–19 through 24 can be put in the dimensionless forms:

$$\Pi = \frac{e^{\phi\eta} - 1}{e^{\phi} - 1} \qquad (\eta \leqslant 1) \tag{21.5–50}$$

or

$$\Pi = \frac{(1 + R)^{\eta} - 1}{R} \qquad (\eta \leqslant 1) \tag{21.5–51}$$

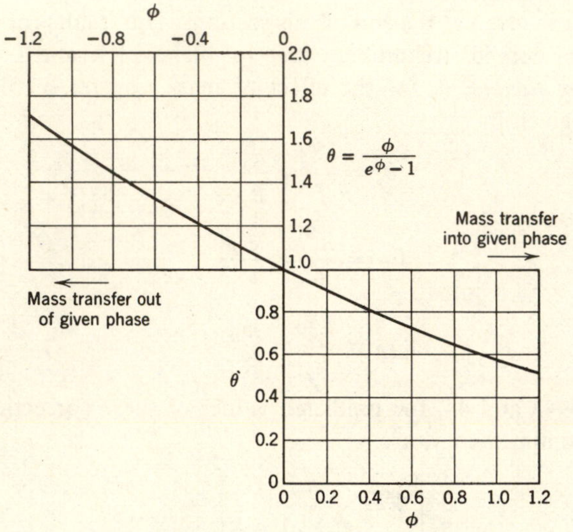

Fig. 21.5–2. The variation of the transfer coefficients with mass-transfer rate, as given by Eq. 21.5–48.

in which Π is a dimensionless velocity, temperature, or mole fraction

$$\Pi_v = \frac{v_x - 0}{v_\infty - 0}; \quad \Pi_T = \frac{T - T_0}{T_\infty - T_0}; \quad \Pi_{AB} = \frac{x_A - x_{A0}}{x_{A\infty} - x_{A0}} \tag{21.5–52,53,54}$$

and η is a dimensionless coordinate

$$\eta_v = \frac{y}{\delta_v}; \qquad \eta_T = \frac{y}{\delta_T}; \qquad \eta_{AB} = \frac{y}{\delta_{AB}} \tag{21.5–55,56,57}$$

In the limit of small mass-transfer rates, Eqs. 21.5–50 and 51 reduce to the linear form

$$\lim_{\substack{R \to 0 \\ \phi \to 0}} \Pi = \eta \tag{21.5–58}$$

Several special forms of these Π profiles were obtained in §17.2 (diffusion of A through stagnant B) and Example 18.5–1 (condensation of A in the presence of stagnant B).

A few sample profiles are plotted in Fig. 21.5–3. The effect of mass transfer on the profile shape is consistent with the effects on transfer coefficients already noted. The complete profiles are of only incidental interest here but do help to point out that the present development can be applied only approximately to most flow systems.

The foregoing derivations are stated in terms of the stream conditions v_∞, T_∞, and $x_{A\infty}$ at a great distance from the surface. For enclosed channels,

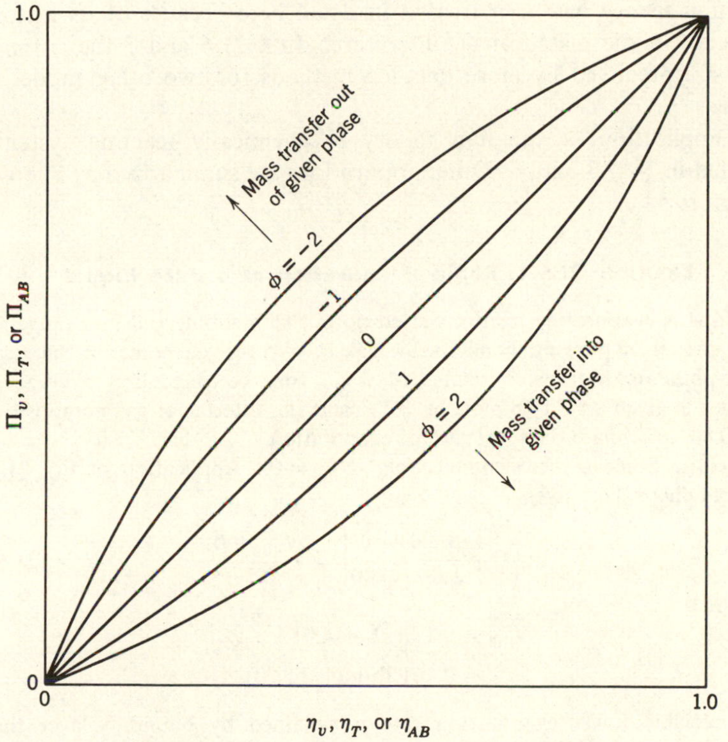

Fig. 21.5–3. Velocity, temperature, and concentration profiles in a laminar film, as calculated from Eq. 21.5–50.

the results are ordinarily applied by simply substituting the "bulk" stream conditions v_{xb}, T_b, and x_{Ab}. This substitution has the effect of chopping off each profile at that point at which the bulk value is reached and may thus underestimate the mass-transfer corrections somewhat; however, no better interpretation of the film theory is available at present.

The main limitation of the film theory lies in the assumed one-dimensional dependence of p, v_x, T, and x_A. The usefulness of this assumption is hard to assess in general, but it seems probable that the results will be at least qualitatively correct for flow in closed channels with fully developed profiles.

Since pressure gradient along the wall has been neglected, the results for $f^{\bullet}_{\mathrm{loc}}$ should not be applied in flows with form drag nor in flows in which the pressure gradient in the film parallel to the wall accounts for a significant fraction of the wall shear stress. The assumption that the film thicknesses are unaffected by mass transfer is open to question and can be tested only by experiment or by more realistic calculations. The assumptions of laminar flow and constant transport properties can be relaxed without materially complicating the results. (See Problem 21.M.)

The film theory has been treated in detail here because of its historical importance in the mass-transfer literature. In §§21.6 and 7 the correction factors are predicted by more detailed methods for two other model flow systems.

The application of the film theory to chemically reacting systems is illustrated in §§17.3 and 4. Other applications are summarized by Sherwood and Pigford.[6]

Example 21.5-1. Rapid Evaporation of a Pure Liquid

Liquid A is evaporating from a wetted porous slab submerged in a tangentially flowing stream of pure noncondensable gas B. At a given point on the surface, the gas-phase mass-transfer coefficient $k_{x,\mathrm{loc}}$ for the prevailing average fluid properties is given as 0.1 lb-mole hr^{-1} ft^{-2} and the interfacial gas composition is $x_{A0} = 0.80$. Estimate the local rate of evaporation.

Solution. Since B is noncondensable, $N_{B0} = 0$. Application of Eq. 21.5-36 to the gas phase then gives

$$\ln \left\{1 + \frac{0.80 - 0}{1.0 - 0.80}\right\} = \frac{N_{A0} + 0}{0.1} \tag{21.5-59}$$

from which

$$N_{A0} = 0.1 \ln \{1 + 4.0\}$$
$$= 0.161 \text{ lb-mole hr}^{-1} \text{ ft}^{-2}$$

An appreciably lower evaporation rate is obtained by boundary-layer theory. (See Example 21.7-1.)

It is interesting to compare the foregoing result with that obtained by using Eq. 21.1-2 with $k^{\bullet}_{x,\mathrm{loc}}$ replaced by $k_{x,\mathrm{loc}}$. This gives

$$N_{A0} - 0.80(N_{A0} + 0) = 0.1(0.80 - 0.00) \tag{21.5-60}$$

from which

$$N_{A0} = 0.40 \text{ lb-mole hr}^{-1} \text{ ft}^{-2}$$

This result is much too high and shows that the mass-transfer corrections are important in this system.

In this simple problem the quantities R_{AB} and ϕ_{AB} have appeared without

[6] T. K. Sherwood and R. L. Pigford, *Absorption and Extraction*, McGraw-Hill, New York (1952), Chapter IX.

being labeled as such (compare Eqs. 21.5–44 and 59). The symbols R and ϕ are helpful, however, in organizing the solution of more complicated problems. (See Examples 21.5–3 and 21.8–1.)

Example 21.5–2. Use of Correction Factors in Droplet Evaporation

Adjust the results of Example 21.2–1 for the prevailing mass-transfer rate by applying an appropriate correction factor.

Solution. In Example 21.2–1 the molar flux ratio R_{AB} at any point on the surface of the drop is

$$R_{AB} = \frac{x_{A0} - x_{A\infty}}{\dfrac{N_{A0}}{N_{A0} + N_{B0}} - x_{A0}} = \frac{0.0247}{1 - 0.0247} = 0.0253 \tag{21.5–61}$$

From Eq. 21.5–49 or Fig. 21.7–3, the predicted correction factor is $\theta_{AB} = 0.987$ at all points on the drop. Hence the corrected mass transfer rate is

$$\begin{aligned} \mathscr{W}_A^{(m)} &= \theta_{AB} k_{xm} \pi D^2 \frac{x_{A0} - x_{A\infty}}{1 - x_{A0}} \\ &= (0.987)(2.69 \times 10^{-7}) \\ &= 2.66 \times 10^{-7} \text{ g-mole sec}^{-1}\text{ cm}^{-2} \end{aligned} \tag{21.5–62}$$

This result agrees closely with that obtained in Example 21.2–1, indicating that the assumption of a small mass-transfer rate was satisfactory under the given conditions.

Example 21.5–3. Wet-Bulb Performance at High Mass-Transfer Rates

Extend the analysis in Example 21.2–2 to high mass-transfer rates.

Solution. By rewriting the energy balance, Eq. 21.2–29, for a finite mass-transfer rate and for any point on the wick, we obtain

$$N_{A0}\, \Delta\tilde{H}_{A,\text{vap.}} = h^{\bullet}_{\text{loc}}(T_\infty - T_0) \tag{21.5–63}$$

Multiplication of both sides by $\tilde{C}_{pA}/\Delta\tilde{H}_{A,\text{vap}} h^{\bullet}_{\text{loc}}$ gives, since $N_{B0} = 0$,

$$R_T = \frac{N_{A0}\tilde{C}_{pA}}{h^{\bullet}_{\text{loc}}} = \frac{\tilde{C}_{pA}(T_\infty - T_0)}{\Delta\tilde{H}_{A,\text{vap.}}} \tag{21.5–64}$$

The right-hand member of this equation is easily calculated if T_0, T_∞, and p are given.

Application of Eq. 21.5–44 for heat transfer with $N_{B0} = 0$ gives the following expression for the local evaporation rate:

$$N_{A0} = \frac{h_{\text{loc}}}{\tilde{C}_{pA}} \ln(1 + R_T) \tag{21.5–65}$$

A second expression for N_{A0} is obtained by applying Eq. 21.5–44 to diffusion of A or B and again setting $N_{B0} = 0$:

$$N_{A0} = k_{x,\text{loc}} \ln(1 + R_{AB}) \tag{21.5–66}$$

The two solutions for N_{A0} may be equated to give

$$\ln(1 + R_{AB}) = \frac{h_{\text{loc}}}{k_{x,\text{loc}}\tilde{C}_{pA}} \ln(1 + R_T) \tag{21.5–67}$$

Insertion of the expressions for R_{AB} and R_T for the present problem gives

$$\ln\left(1 + \frac{x_{A0} - x_{A\infty}}{1 - x_{A0}}\right) = \frac{h_{\text{loc}}}{k_{x,\text{loc}}\tilde{C}_{pA}} \ln\left(1 + \frac{\tilde{C}_{pA}(T_\infty - T_0)}{\Delta\tilde{H}_{A,\text{vap}}}\right) \quad (21.5\text{–}68)$$

This equation shows that x_{A0} and T_0 can be constant over the surface of the wick only if $h_{\text{loc}}/k_{x,\text{loc}}\tilde{C}_{pA}$ is constant and hence equal to $h_m/k_{xm}\tilde{C}_{pA}$. This constancy is assumed here for simplicity; such an assumption is particularly satisfactory for the system H_2O-air, in which Pr and Sc are nearly equal. With this change, the solution becomes

$$\ln\left(1 + \frac{x_{A0} - x_{A\infty}}{1 - x_{A0}}\right) = \frac{h_m}{k_{xm}\tilde{C}_{pA}} \ln\left(1 + \frac{\tilde{C}_{pA}(T_\infty - T_0)}{\Delta\tilde{H}_{A,\text{vap}}}\right) \quad (21.5\text{–}69)$$

and this solution reduces exactly to Eq. 21.2–31 at low mass-transfer rates.

For the numerical problem in Example 21.2–2, the following values apply:

$$x_{A0} = 0.0247$$

$$\tilde{C}_{pA} = 8.03 \text{ Btu (lb-mole)}^{-1}\ ^\circ\text{F}^{-1} \text{ for } H_2O \text{ vapor at } 105^\circ \text{ F}$$

$$\frac{h_m}{k_{xm}} = 5.93 \text{ Btu (lb-mole)}^{-1}\ ^\circ\text{F}^{-1} \text{ from Eq. 21.2–34}$$

$$\frac{\tilde{C}_{pA}(T_\infty - T_0)}{\Delta\tilde{H}_{A,\text{vap}}} = \frac{8.03(140 - 70)}{18{,}900} = 0.0297$$

Insertion of these values in Eq. 21.5–66 gives

$$\ln\left(1 + \frac{0.0247 - x_{A\infty}}{1 - 0.0247}\right) = \frac{5.93}{8.03} \ln(1.0297) = 0.0216$$

Solving this equation, we get

$$x_{A\infty} = 0.0034$$

This result differs only slightly from the value $x_{A\infty} = 0.0033$ obtained in Example 21.2–2 and justifies the neglect of mass-transfer corrections under the given conditions.

Numerical studies indicate that the simple Eq. 21.2–31 gives a close approximation to Eq. 21.5–69 under all foreseeable wet-bulb conditions. This is because Eqs. 21.2–29 and 30 overestimate the mass-transfer rate almost equally, and when these equations are combined the errors largely compensate.

§21.6 TRANSFER COEFFICIENTS AT HIGH MASS-TRANSFER RATES: PENETRATION THEORY

In this section we use the penetration theory to predict the dependence of the heat- and mass-transfer coefficients on the rate of mass transfer. This theory assumes a flat velocity profile near the interface and is thus applicable mainly to the liquid phase in gas-liquid systems. (See, for example, Figs. 17.5–1 and 2.) The theory was first applied to mass-transfer operations by Higbie,[1] and many variations of it have appeared in the literature. We first

[1] R. Higbie, *Trans. A.I.Ch.E.*, **31**, 365–389 (1935).

discuss a situation in which the theory is nearly exact and then discuss modifications for other flow systems.

Consider the contacting of a gas stream of A and B with a falling liquid film of A and B, as shown in Fig. 21.6–1. From §17.5, we know that, for laminar nonrippling flow and short times of exposure of the liquid surface, the liquid-phase diffusion process penetrates only part way into the falling film and is analogous to the unsteady-state diffusion in a quiescent infinite fluid.

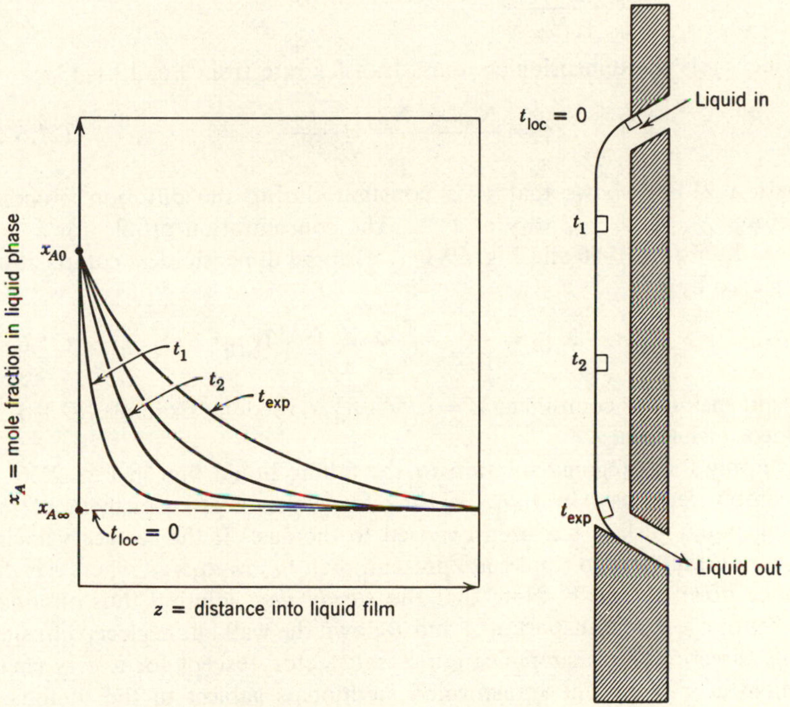

Fig. 21.6–1. Diffusion into a falling liquid film; t_{exp} is the total time of exposure of a typical element of volume near the surface.

The analysis in §17.5 is valid only for small mass-transfer rates and no chemical reaction; here we extend[2] the analysis to high mass-transfer rates by using the Arnold solution given in Example 19.1–1.

The analysis in §19.1 describes the unsteady-state diffusion that occurs when a large stationary mass of pure gas B is suddenly exposed along the plane surface $z = 0$ to a pure liquid A that does not dissolve B. Here we restate the solution in a more general form for systems in which *both* species cross the interface in any fixed ratio N_{A0}/N_{B0}. Both phases may initially

[2] W. E. Stewart, forthcoming publication.

contain A and B. The solution is written for the liquid or gaseous region $z > 0$, in which the initial composition is $x_{A\infty}$ and the interface composition, for $t > 0$, is constant at x_{A0}. The properties c and $\mathscr{D}_{AB}$ are assumed constant for $z \geq 0$.

The instantaneous mass-transfer rate is given implicitly by the following generalization of Eq. 19.1–17:

$$\frac{x_{A0} - x_{A\infty}}{\dfrac{N_{A0}}{N_{A0} + N_{B0}} - x_{A0}} = \sqrt{\pi}(1 + \text{erf } \varphi)\varphi \exp \varphi^2 \tag{21.6–1}$$

in which φ is the dimensionless mass-transfer rate from Eq. 19.1–13*a*:

$$\varphi = \frac{N_{A0} + N_{B0}}{c}\sqrt{\frac{t}{\mathscr{D}_{AB}}} \tag{21.6–2}$$

Equation 21.6–1 shows that φ is constant during the diffusion process; therefore, N_{A0} and N_{B0} vary as $t^{-1/2}$. The concentration profile for $z > 0$ is given by Eq. 19.1–16 and Fig. 19.1–1, with the dimensionless composition X replaced by

$$X_{AB} = \frac{x_A - x_{A\infty}}{x_{A0} - x_{A\infty}} = 1 - \Pi_{AB} \tag{21.6–3}$$

The dimensionless coordinate $Z = z/\sqrt{4\mathscr{D}_{AB}t}$ used in Example 19.1–1 is retained unchanged.

To apply the foregoing solution to the falling liquid film in Fig. 21.6–1, one simply replaces t by t_{loc}, the time for which a given element of the moving liquid surface has been exposed to the gas. If the surface velocity v_{max} is constant and no ripples are present, then $t_{\text{loc}} = x/v_{\text{max}}$, where x is the distance down the wall. Note that the *steady flow* solution thus obtained includes convective transport of A and B down the wall but neglects diffusion in this direction; this simplification is satisfactory except for a very small region near $x = 0$. The transformed solution is subject to the boundary conditions of constant x_{A0} and N_{A0}/N_{B0} along the interface and constant composition $x_{A\infty}$ in the interior of the falling film.

The local mass-transfer coefficient, $k^{\bullet}_{x,\text{loc}}$, at a given distance down the wetted wall is defined according to Eq. 21.1–2:

$$k^{\bullet}_{x,\text{loc}} = \frac{N_{A0} - x_{A0}(N_{A0} + N_{B0})}{x_{A0} - x_{A\infty}} \tag{21.6–4}$$

Here we have chosen $\Delta x_A = x_{A0} - x_{A\infty}$, in which $x_{A\infty}$ is the entering liquid composition.

Insertion of Eq. 21.6–4 into Eq. 21.6–1 gives

$$\frac{N_{A0} + N_{B0}}{k^{\bullet}_{x,\text{loc}}} = \sqrt{\pi}\varphi(1 + \text{erf } \varphi) \exp \varphi^2 \tag{21.6–5}$$

and in the limit of low mass-transfer rates this becomes

$$\left\{\lim_{\varphi\to 0} k^{\bullet}_{x,\text{loc}}\right\} = k_{x,\text{loc}} = \frac{N_{A0} + N_{B0}}{\varphi\sqrt{\pi}} \tag{21.6–6}$$

Evaluation of φ in this equation from Eq. 21.6–2 gives

$$k_{x,\text{loc}} = c\sqrt{\frac{\mathscr{D}_{AB}}{\pi t_{\text{loc}}}} \tag{21.6–7}$$

which corresponds to the result given by Higbie[1] for low mass-transfer rates.

The change in $k^{\bullet}_{x,\text{loc}}$ with mass-transfer rate is conveniently stated in terms of the same dimensionless quantities that appeared in the film theory:

$$R_{AB} = \frac{N_{A0} + N_{B0}}{k^{\bullet}_{x,\text{loc}}} = \frac{x_{A0} - x_{A\infty}}{\dfrac{N_{A0}}{N_{A0} + N_{B0}} - x_{A0}} \tag{21.6–8}$$

$$\phi_{AB} = \frac{N_{A0} + N_{B0}}{k_{x,\text{loc}}} \tag{21.6–9}$$

$$\theta_{AB} = \frac{k^{\bullet}_{x,\text{loc}}}{k_{x,\text{loc}}} \tag{21.6–10}$$

The flux ratio R_{AB} is the same as the left side of Eq. 21.6–1. The rate factor ϕ_{AB} can be related to the dimensionless flux φ by combining Eqs. 21.6–6 and 9:

$$\phi_{AB} = \varphi\sqrt{\pi} \tag{21.6–11}$$

Equation 21.6–1 can now be written in terms of R_{AB} and ϕ_{AB} to facilitate mass-transfer calculations:

$$R = \phi\left(1 + \text{erf}\,\frac{\phi}{\sqrt{\pi}}\right)\exp\frac{\phi^2}{\pi} \tag{21.6–12}$$

and the correction factor $\theta_{AB} = \phi_{AB}/R_{AB}$ is given by

$$\theta = \left(1 + \text{erf}\,\frac{\phi}{\sqrt{\pi}}\right)^{-1}\exp\left(-\frac{\phi^2}{\pi}\right) \tag{21.6–13}$$

These relations predict a stronger dependence of $k^{\bullet}_{x,\text{loc}}$ on mass-transfer rate than the corresponding results of the film theory, Eqs. 21.5–43 and 48. The results of both theories are plotted in Figs. 21.7–2 and 3.

Equations 21.6–12 and 13 can be carried over to heat transfer if k, ρ, and $\hat{C}_p$ are constant throughout the fluid and equal for both species. Approximate allowance for variable fluid properties may be made by defining R_T and ϕ_T as in Eqs. 21.5–38 and 41. The heat-transfer analog of Eq. 21.6–7 may also prove useful:

$$h_{\text{loc}} = \rho\hat{C}_p\sqrt{\frac{\alpha}{\pi t_{\text{loc}}}} \tag{21.6–14}$$

By combining this with Eq. 21.6–7, we obtain

$$h_{\text{loc}} = k_{x,\text{loc}}\tilde{C}_p \sqrt{\frac{\alpha}{\mathscr{D}_{AB}}} \tag{21.6–15}$$

Equation 21.6–15 provides a simple method of predicting heat-transfer coefficients at low mass-transfer rates from mass-transfer correlations. The corrected coefficient $h^{\bullet}$ may then be computed by Eq. 21.6–12 or 13 or from the plots in §21.7.

If ripples or turbulence are present in the falling film, one can still use the penetration theory in an approximate sense by assuming that various parts of the liquid surface are renewed with fresh material from time to time by some sort of mixing process. One then assumes that t_{loc} represents a sort of "average age" of the various surface elements arriving at a given location in the equipment over an extended period of time. If this interpretation is adopted, then t_{loc} may be *defined* by Eq. 21.6–7 or 14, and Eqs. 21.6–4 et seq. apply unchanged. This interpretation suggests that t_{loc} should depend on the flow conditions and system geometry but not on α or $\mathscr{D}_{AB}$.

The proportionality of k_x to $\mathscr{D}_{AB}^{1/2}$ under given flow conditions, predicted by Eq. 21.6–7, has been confirmed experimentally for the liquid phase in several gas-liquid mass-transfer systems, including short wetted-wall columns, packed columns, and liquids around gas bubbles in certain instances. (See Example 17.5–1.) Note that this dependence on $\mathscr{D}_{AB}$ applies only in systems in which the liquid-phase velocity is nearly constant throughout the region of diffusion; other velocity profiles give quite different results. (See Problem 17.J on dissolution in a laminar falling film.)

The penetration theory has also been applied to absorption with chemical reaction. A number of examples of this approach are given in the text by Sherwood and Pigford.[3] (See also Example 19.1–3 and Problem 19.D.)

§21.7 TRANSFER COEFFICIENTS AT HIGH MASS-TRANSFER RATES: BOUNDARY-LAYER THEORY

In this section we determine the effect of mass transfer on the transfer coefficients in two-dimensional laminar flow along a flat plate. (See Fig. 19.3–1.) This analysis differs from the film and penetration theories in that detailed allowance is made for the two-dimensional velocity profiles.

In §19.3 solutions are given for velocity, temperature, and composition profiles in simultaneous momentum, heat, and mass transfer in the laminar boundary layer on a flat plate. The analysis is given for a binary mixture with constant fluid properties and no dissipation or homogeneous chemical reaction. These solutions apply when the fluid temperature and composition are constant along the plate surface and when N_{A0}/N_{B0} is also constant.

[3] T. K. Sherwood and R. L. Pigford, *Absorption and Extraction*, McGraw-Hill, New York (1952), Chapter IX.

The resulting velocity, temperature, and composition profiles are given by the single function

$$\Pi = \Pi(\eta, \Lambda, K) \qquad (21.7\text{–}1)$$

which shows the dependence of the profiles Π_v, Π_T, and Π_{AB} on the position coordinate $\eta = (y/2)\sqrt{v_\infty/\nu x}$, the physical property group Λ (i.e., 1, Pr, or Sc), and the dimensionless mass flux at the wall, $K = (2v_{y0}/v_\infty)\sqrt{v_\infty x/\nu}$. Some sample profiles are plotted in Fig. 19.3–2. The derivation in §19.3 is given for $N_B = 0$; however, the results are valid for mass transfer of both A and B if K is evaluated as stated here. (See Problem 19.F.)

The local transfer coefficients for simultaneous momentum, heat, and mass transfer can be obtained directly from the wall-flux expressions given in Eqs. 19.3–34, 35, and 36; the following analogous equations result:

$$\frac{f^\bullet_{\text{loc}}}{2} = \frac{\Pi'(0, 1, K)}{2}\left(\frac{v_\infty x}{\nu}\right)^{-1/2} \qquad (21.7\text{–}2)$$

$$\frac{h^\bullet_{\text{loc}}}{\rho \hat{C}_p v_\infty} = \frac{\Pi'(0, \text{Pr}, K)}{2\ \text{Pr}}\left(\frac{v_\infty x}{\nu}\right)^{-1/2} \qquad (21.7\text{–}3)$$

$$\frac{k^\bullet_{x,\text{loc}}}{c v_\infty} = \frac{\Pi'(0, \text{Sc}, K)}{2\ \text{Sc}}\left(\frac{v_\infty x}{\nu}\right)^{-1/2} \qquad (21.7\text{–}4)$$

Numerical values of $2\Pi'(0, \Lambda, K)$ are given in Table 19.3–1.

For greater convenience of use, and for comparison with other theories of the mass-transfer corrections, we proceed to restate these results in terms of the quantities R, ϕ, and θ that appeared in the film and penetration theories. When all physical properties are constant in the mixture and equal for species A and B, Eqs. 21.5–40, 41, and 42 become

$$R_v = \frac{\rho v_{y0}}{\rho v_\infty f^\bullet/2}; \qquad R_T = \frac{\rho \hat{C}_p v_{y0}}{h^\bullet}; \qquad R_{AB} = \frac{c v_{y0}}{k^\bullet_x} \qquad (21.7\text{–}5,6,7)$$

Insertion of the boundary-layer solutions for $f^\bullet$, $h^\bullet$, and $k^\bullet_x$ gives

$$R_v = \frac{K}{\Pi'(0, 1, K)}; \qquad R_T = \frac{K\ \text{Pr}}{\Pi'(0, \text{Pr}, K)};$$

$$R_{AB} = \frac{K\ \text{Sc}}{\Pi'(0, \text{Sc}, K)} \qquad (21.7\text{–}8,9,10)$$

which can be summarized in the single equation

$$R = \frac{K\Lambda}{\Pi'(0, \Lambda, K)} \qquad (21.7\text{–}11)$$

To evaluate ϕ one simply replaces $f^\bullet$, $h^\bullet$, and $k^\bullet$ in Eqs. 21.7–5, 6, and 7 by the values for $K = 0$; the result is

$$\phi = \frac{K\Lambda}{\Pi'(0, \Lambda, 0)} \qquad (21.7\text{–}12)$$

and the mass-transfer corrections to f, h, and k are given by the following ratios of profile slopes at the wall:

$$\theta = \frac{\Pi'(0, \Lambda, K)}{\Pi'(0, \Lambda, 0)} \tag{21.7–13}$$

The predicted relations between R, ϕ, and θ are plotted in Figs. 21.7–1, 2, and 3. The notation for these figures is summarized in Table 21.7–1. The

TABLE 21.7–1

SUMMARY OF COORDINATES FOR MASS TRANSFER CORRECTION PLOTS

Process	Diffusivity Ratio, Λ	Flux Ratio, R	Rate Factor, ϕ	Correction Factor, $\theta = \frac{\phi}{R}$
Momentum transfer	1	$R_v = \dfrac{(N_{A0}M_A + N_{B0}M_B)(v_\infty - 0)}{\tau_0}$	$\phi_v = \dfrac{N_{A0}M_A + N_{B0}M_B}{\frac{1}{2}\rho v_\infty f_{\text{loc}}}$	$\theta_v = \dfrac{f^{\bullet}_{\text{loc}}}{f_{\text{loc}}}$
Heat transfer	$\dfrac{\hat{C}_p\mu}{k}$	$R_T = \dfrac{(N_{A0}\tilde{C}_{pA} + N_{B0}\tilde{C}_{pB})(T_0 - T_\infty)}{q_0}$	$\phi_T = \dfrac{N_{A0}\tilde{C}_{pA} + N_{B0}\tilde{C}_{pB}}{h_{\text{loc}}}$	$\theta_T = \dfrac{h^{\bullet}_{\text{loc}}}{h_{\text{loc}}}$
Diffusion of species A or B	$\dfrac{\mu}{\rho\mathscr{D}_{AB}}$	$R_{AB} = \dfrac{x_{A0} - x_{A\infty}}{\dfrac{N_{A0}}{N_{A0} + N_{B0}} - x_{A0}}$	$\phi_{AB} = \dfrac{N_{A0} + N_{B0}}{k_{x,\text{loc}}}$	$\theta_{AB} = \dfrac{k^{\bullet}_{x,\text{loc}}}{k_{x,\text{loc}}}$

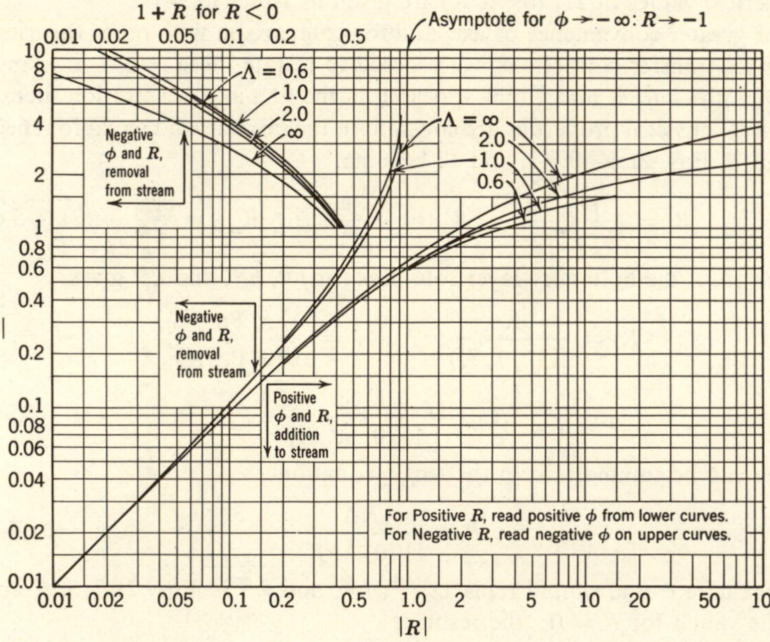

Fig. 21.7–1. Momentum, heat, and mass fluxes between a flat plate and a laminar boundary layer. [W. E. Stewart, Sc. D. Thesis, Massachusetts Institute of Technology (1950).]

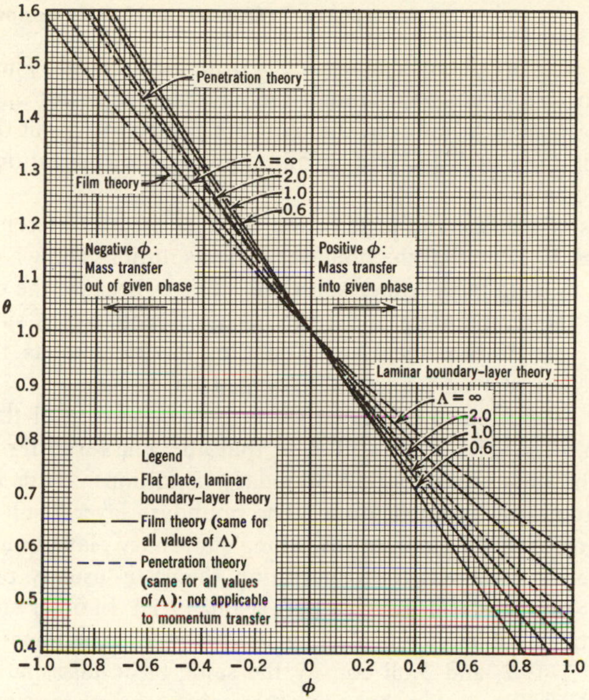

Fig. 21.7–2. The variation of the transfer coefficients with mass-transfer rate as predicted by various theories.

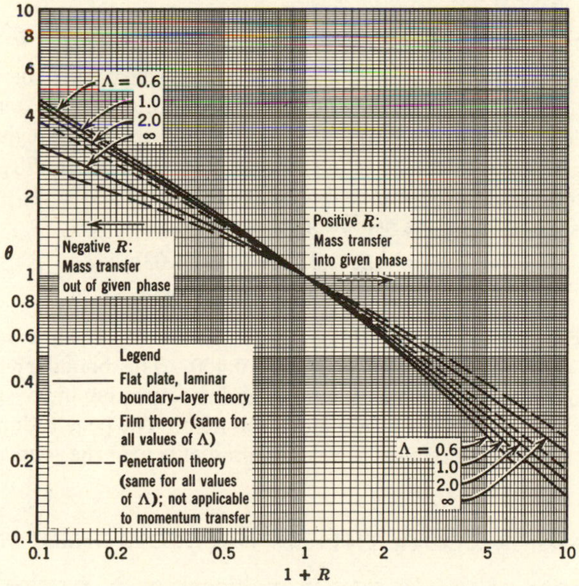

Fig. 21.7–3. The variation of the transfer coefficients with the flux ratio *R* as predicted by various theories.

values for $\Lambda = \infty$ are obtained analytically from Eq. 19.3–50. The boundary-layer solutions are strictly valid only if the physical properties ρ, μ, $\hat{C}_p$, k, c, and $\mathscr{D}_{AB}$ are constant; however, the plots are given in terms of the variable-property definitions of R and ϕ to permit approximate application to a wider range of problems.

The results of the film and penetration theories are also presented in Figs. 21.7–1, 2, and 3. Note that the boundary-layer solutions show a dependence on Λ that is missing in the other solutions; this is because the effect of the v_x-profile on the temperature and concentration profiles was suppressed in the film and penetration theories. The film theory predicts the smallest effect of mass transfer on the transfer coefficients.

The main conclusion to be drawn from the comparison of the three sets of results in Figs. 21.7–1, 2, and 3 is that the mass-transfer corrections depend on the flow geometry and boundary conditions. Thus each theory has a preferred range of application. The boundary layer results given here are clearly preferred for laminar flow along stationary plane boundaries and in pipe entrances, whereas the penetration theory is usually preferred for liquids at gas-liquid interfaces. (See §§17.5 and 19.1.) In other situations the choice must be based on judgment or, when possible, experimental evidence.

Figures 21.7–1, 2, and 3 all convey the same basic information. Figure 21.7–1 is the simplest to use, but the others give a better idea of the size of the corrections. An example is given below to illustrate conditions for which the corrections for mass transfer are significant.

Example 21.7–1. Rapid Evaporation from a Plane Surface

Rework Example 21.5–1, using the results of boundary-layer theory, given the additional information that Sc $= 0.6$ at the prevailing average temperature T_f and composition x_{Af}. Compare the result with those obtained previously.

Solution. As before, $R_{AB} = 4.00$. From Fig. 21.7–1, at $R_{AB} = 4.00$ and $\Lambda_{AB} = 0.6$, the result of laminar boundary-layer theory is $\phi_{AB} = 1.03$. By setting $N_{B0} = 0$ and using the definition of ϕ_{AB}, we get

$$\begin{aligned} N_{A0} &= k_{x,\text{loc}}\phi_{AB} = (0.1)(1.03) \\ &= 0.103 \text{ lb-mole hr}^{-1} \text{ ft}^{-2} \end{aligned} \tag{21.7–14}$$

The result according to the film theory is $N_{A0} = 0.161$, and the result if mass-transfer corrections are neglected is $N_{A0} = 0.400$. The boundary-layer solution should be accurate here if the flow is laminar and the variation in physical properties is not too great; the other results are much too high. The penetration theory is not applicable in this system because the velocity gradients near the solid surface cannot be neglected.

§21.8 TRANSFER COEFFICIENTS IN MULTICOMPONENT SYSTEMS

The prediction of transfer rates in multicomponent systems is a difficult problem and is currently a subject of intensive research. Two methods of attack have been used: (i) exact solutions of the multicomponent equations of change for simple systems and (ii) approximate generalizations of the correlations for one- and two-component systems. The two methods are complementary; exact solutions (see Example 18.5–5) have proven useful in developing and testing approximate methods. Here we discuss some generalized methods that are exact if the physical properties and effective diffusivities $\mathscr{D}_{im}$ are constant.

First let us consider a one-phase, n-component fluid stream in which the physical properties are constant and the mass-transfer rates of all species through the bounding surface are small. The velocity and temperature profiles in such a system are the same as for a pure fluid with the corresponding physical properties, and the momentum- and heat-transfer coefficients may be obtained directly from correlations such as those in Chapters 6 and 13. If, in addition, the effective diffusivity $\mathscr{D}_{im}$ defined in Eq. 18.4–21 is constant for a given species i, then species i behaves as if it were in a binary mixture of diffusivity $\mathscr{D}_{AB}$ equal to the prevailing $\mathscr{D}_{im}$. The results of §§21.1–3 can then be taken over with notational changes as illustrated below:

$$x_A \rightarrow x_i \tag{21.8-1}$$

$$\mathscr{D}_{AB} \rightarrow \mathscr{D}_{im} \tag{21.8-2}$$

$$N_{A0} - x_{A0}(N_{A0} + N_{B0}) \rightarrow N_{i0} - x_{i0}\Sigma_j N_{j0} \tag{21.8-3}$$

The multicomponent local mass-transfer coefficient for species i is thus defined by analogy with Eqs. 21.1–2 and 3:

$$N_{i0} - x_{i0}\Sigma_j N_{j0} = k^{\bullet}_{xi,\text{loc}}\, \Delta x_i \tag{21.8-4}$$

By writing Eq. 21.8–4 for all values of i, one obtains a set of n equations relating the fluxes N_{i0} to the interfacial and main-stream compositions. The use of this set of equations is illustrated in Example 21.8–1. At low mass-transfer rates, one can substitute $k_{xi,\text{loc}}$ for $k^{\bullet}_{xi,\text{loc}}$ with satisfactory accuracy.

For high mass-transfer rates and constant physical properties, the results of §§21.4–7 apply with analogous notational changes. The flux ratios R are generalized as follows for submerged objects in an infinite stream:

$$R_v = \frac{\Sigma_j N_{j0} M_j}{\frac{1}{2}\rho v_\infty f^{\bullet}_{\text{loc}}} = \frac{\Sigma_j N_{j0} M_j (v_\infty - 0)}{\tau_0} \tag{21.8-5}$$[1]

$$R_T = \frac{\Sigma_j N_{j0} \tilde{C}_{pj}}{h^{\bullet}_{\text{loc}}} = \frac{\Sigma_j N_{j0} \tilde{C}_{pj}(T_0 - T_\infty)}{q_0} \tag{21.8-6}$$

$$R_{im} = \frac{\Sigma_j N_{j0}}{k^{\bullet}_{xi,\text{loc}}} = \frac{x_{i0} - x_{i\infty}}{\dfrac{N_{i0}}{\Sigma_j N_{j0}} - x_{i0}} \tag{21.8-7}$$

[1] These two equations are not useful for blunt objects.

and the rate factors ϕ are generalized as follows:

$$\phi_v = \frac{\Sigma_j N_{j0} M_j}{\frac{1}{2}\rho v_\infty f_{\text{loc}}} \tag{21.8–8}$$[1]

$$\phi_T = \frac{\Sigma_j N_{j0} \tilde{C}_{pj}}{h_{\text{loc}}} \tag{21.8–9}$$

$$\phi_{im} = \frac{\Sigma_j N_{j0}}{k_{xi,\text{loc}}} \tag{21.8–10}$$

With these definitions, the relations between R, ϕ, and θ plotted in Figs. 21.7–1, 2, and 3 may be applied to systems of any number of components. These generalizations of the results for binary systems are exact if the physical properties, including the effective diffusivities $\mathscr{D}_{im}$, are constant.

In multicomponent systems the physical properties will generally vary from point to point. The effects of such property variations are quite complicated but can be handled, to a first approximation, by inserting average values of the fluid properties into the constant-property correlations. Tentatively, it is suggested that the fluid properties be evaluated at the mean of the interface and main-stream temperature and composition.

The approximation of using an average value of $\mathscr{D}_{im}$ for each species requires special comment. In many gaseous systems $\mathscr{D}_{im}$ is nearly constant (see discussion under Eq. 18.4–22), and this approximation works well. However, in liquids, or in systems containing species of widely different molecular weights, $\mathscr{D}_{im}$ may vary considerably and may even be negative over some range of composition. For such systems, the use of an average value of $\mathscr{D}_{im}$ may lead to serious errors, and better procedures are needed. Work along these lines is in progress at several institutions.

Example 21.8–1. Mass Transfer in a Fixed-Bed Catalytic Reactor

Benzene is being hydrogenated in a fixed-bed reactor at 30 atm total pressure. Estimate the gas composition next to the catalyst particles at a cross section where the flow-mean gas conditions are as follows (see Problem 18.E):

i	Species	x_{ib}	M_i	$(c\mathscr{D}_{im})_b$ g-mole cm^{-1} sec^{-1}	$(\mathrm{Sc}_{im})_b = (\mu/\rho\mathscr{D}_{im})_b$	R_i/R_3 Moles i produced per mole of cyclohexane produced
1	Benzene	0.100	78.11	6.63×10^{-6}	1.46	−1
2	Hydrogen	0.800	2.016	20.8×10^{-6}	0.466	−3
3	Cyclohexane	0.050	84.16	8.71×10^{-6}	1.11	1
4	Methane	0.050	16.04	$60. \times 10^{-6}$	0.162	0
	Total	1.000				−3

Other gas-stream conditions are $T_b = 500°K$, $\mu_b = 0.014$ centipoise, $M_b = \Sigma_i x_{ib} M_i = 14.43$ lb/lb-mole, and $G_0 = \langle \rho v \rangle = 600 \text{ lb}_m \text{ hr}^{-1} \text{ ft}^{-2}$, in which $\langle \rho v \rangle$ is based on the total bed cross section. The local rate of cyclohexane production is 1.5 lb-moles per hour per pound of catalyst. The catalyst particles are $\frac{1}{8} \times \frac{1}{8}$ in. cylinders with a particle density $\rho_p = 0.9 \text{ g cm}^{-3}$; the over-all bed density is $\rho_{\text{bed}} = 0.6 \text{ g cm}^{-3}$.

Solution. From the data given we can find the mass-transfer rates and then compute the unknown compositions, which will be assumed constant over the outer surface of a given catalyst particle.[2] To shorten the solution, the physical properties will be assumed constant.

The external particle surface per unit weight of catalyst for a cylindrical particle with $L = D = \frac{1}{8}$ in. is

$$a_m = \frac{2\dfrac{\pi D^2}{4} + \pi DL}{\dfrac{\pi D^2 L}{4}\rho_p} = \frac{\dfrac{2}{L} + \dfrac{4}{D}}{\rho_p} = \frac{\dfrac{2 \times 12}{1/8} + \dfrac{4 \times 12}{1/8}}{0.9 \times 62.4} = 10.2 \text{ ft}^2/\text{lb}_m \quad (21.8\text{–}11)$$

and the external particle surface per unit volume of bed is

$$a = a_m \rho_{\text{bed}} = (10.2)(0.6)(62.4) = 382 \text{ ft}^{-1} \quad (21.8\text{–}12)$$

The fluxes of the species into the gas stream, averaged over the outer particle surface, then become

$$N_{i0} = R_i/a = (1.5)(R_i/R_3)/(10.2) \quad (21.8\text{–}13)$$

Values computed from this equation are given in Table 21.8–1.

For constant surface compositions x_{i0}, we can predict the mass-transfer coefficients by using Eqs. 13.4–2 and 3 and setting $j_H = j_D$. In this multicomponent system j_D may be defined as

$$j_D = \frac{k_{xi}}{\langle cv \rangle}(\text{Sc}_{im})_f^{2/3} = \frac{k_{xi} M_b}{G_0}(\text{Sc}_{im})_f^{2/3} \quad (21.8\text{–}14)$$

in which k_{xi} is defined by setting $\Delta x_i = x_{i0} - x_{ib}$ in Eq. 21.8–4. The Reynolds number on which j_D depends is

$$\text{Re} = \frac{G_0}{a\psi\mu} = \frac{600}{(382)(0.91)(0.014 \times 2.42)} = 51 \quad (21.8\text{–}15)$$

in which $\psi = 0.91$ is the empirical shape factor given in §13.4 for packed beds of cylinders. From Eq. 13.4–3 and the analogy mentioned above, we get

$$j_D = j_H = 0.61 \text{ Re}^{-0.41}\psi = 0.111 \quad (21.8\text{–}16)$$

It should be noted that the same value of j_D applies to all species. Insertion of numerical values in Eq. 21.8–14 gives the local mass-transfer coefficients k_{xi}. (See Table 21.8–1.) The remainder of the calculations for computing the composition at the catalyst surface is summarized in the table. Because $\phi_{im} \ll 1$, the mass-transfer corrections are small and the film theory is adequate.

[2] A tentative method of allowing for variation in composition over the outer particle surface has been given by H. E. Hoelscher, *A.I.Ch.E. Journal*, **4**, 300–304 (1958).

The mole fractions at the catalyst surface differ considerably from those in the gas stream; there is also a considerable temperature difference, as shown in Problem 21.J. The reaction rate could probably be substantially increased by operating at a higher mass velocity; however, this question cannot be answered definitely unless one knows how the reaction rate varies with the temperature and composition at the catalyst surface.

TABLE 21.8–1

TABULATION OF CALCULATIONS FOR FIXED-BED CATALYTIC REACTOR[a]

	N_{i0}	k_{xi}	ϕ_{im}	R_{im}	x_{i0}
i	from Eq. 21.8–13	from Eq. 21.8–14	from Eq. 21.8–10	from Eq. 21.5–43	from Eq. 21.8–7[b]
1	−0.147	3.58	−0.123	−0.116	0.069
2	−0.441	7.66	−0.058	−0.056	0.788
3	0.147	4.30	−0.103	−0.098	0.092
4	0	15.5	−0.028	−0.028	0.051
$\sum_{i=1}^{4}$	−0.441	—	—	—	1.000

[a] N_{i0} and k_{xi} are given in lb-mole hr^{-1} ft^{-2}; the other quantities are dimensionless.
[b] With $x_{i\infty}$ replaced by x_{ib}.

Note that we have calculated x_{i0} independently for all four species. This is not essential, since any three mole fractions determine the fourth, but it is preferable to calculate all four independently as a check on the solution. Here the condition $\Sigma_j x_{j0} = 1$ is satisfied to three decimals; under other conditions, some adjustment of the calculated values may be necessary. Similar checking procedures can be devised in other multicomponent problems, as shown in Problem 21.H.

In general, for any n-component problem, there are only $n - 1$ independent equations in the set represented by Eq. 21.8–4. (See Problem 21.H.) Thus only $n - 1$ independent mass-transfer rates and compositions can be found from diffusional considerations, just as in binary systems.

QUESTIONS FOR DISCUSSION

1. What two types of interfaces are considered in this chapter? Give several examples of each type.
2. Can mass transfer occur in a system of constant composition?
3. Show that $k^{\bullet}_{x,\mathrm{loc}}$ has the same value in Eqs. 21.1–2 and 21.1–3.

4. Under what conditions can the analogies in Table 21.2–1 be applied?

5. What do Eqs. 21.2–16 and 17 predict, in general, regarding the functional dependence of j_H and j_D?

6. What assumptions have been made in deriving Eq. 21.3–4?

7. In what sense is Eq. 21.4–6 analogous to Eqs. 21.4–4 and 5?

8. Explain the physical significance of $F_z^{(m)}$ and of $Q^{(m)}$. What is the analogous quantity for mass transfer of species A in a binary system?

9. Compare the model systems used in §§21.5, 6, and 7.

10. How would the boundary conditions in §21.5 have to be modified to restrict the solution to low mass-transfer rates?

11. Show that ϕ_v, ϕ_T, and ϕ_{AB} may be regarded as dimensionless velocities through the interface. Identify the type of average velocity used in each of these quantities.

12. What is the physical significance of the flux ratios R_v, R_T, and R_{AB}?

13. In what approximate ranges of R and ϕ can one set $f^{\bullet} = f$, $h^{\bullet} = h$, or $k_x^{\bullet} = k_x$ if the predicted transfer coefficients are to be accurate within ± 5 per cent?

14. Compare the effects of mass transfer on the transfer coefficients with the effects of heat transfer on the transfer coefficients.

15. How are R and ϕ related at low mass-transfer rates? Show that Eqs. 21.5–43, 21.5–44, and 21.6–12 are correct in this limiting situation.

16. Could the penetration theory be used to predict rates of momentum transfer between phases? Explain.

17. In what way are Λ_v, Λ_T, and Λ_{AB} analogous?

18. Why does Λ appear in the results of boundary-layer theory but not in the film or penetration theories?

19. Under what conditions are the generalized procedures of §21.8 exact?

PROBLEMS

21.A$_1$ Prediction of Mass-Transfer Coefficients in Closed Channels

Estimate gas-phase mass-transfer coefficients for water evaporating into pure air at 2 atm, 25° C, and a mass flow rate of 1570 $\text{lb}_m\ \text{hr}^{-1}$, in the systems listed below. Use $\mathscr{D}_{AB} = 0.130\ \text{cm}^2\ \text{sec}^{-1}$.

a. A 6-in. i.d. vertical pipe with a falling film of water on the wall. Use the Gilliland correlation for gases in wetted-wall columns:

$$\frac{k_{x,\text{loc}}D}{c\mathscr{D}_{AB}} = 0.023\left(\frac{DG}{\mu}\right)^{0.83}\left(\frac{\mu}{\rho\mathscr{D}_{AB}}\right)^{0.44} \qquad \left(\frac{DG}{\mu} > 2000\right) \tag{21.A–1}$$

b. A 6 in. diameter packed bed of water-saturated spheres, with $a = 100\ \text{ft}^{-1}$.

21.B$_1$ Calculation of Gas Composition from Psychrometric Data

A stream of moist air has a wet-bulb temperature of 80° F and a dry-bulb temperature of 130° F, measured at 800 mm Hg total pressure and high air velocity. Compute the mole fraction of water vapor in the air stream. For simplicity, consider water (A) as a trace component in estimating the film properties. *Answer:* $x_{A\infty} = 0.0176$

21.C₁ Calculation of Air Temperature for Drying in a Fixed Bed

A shallow bed of water-saturated granular solids is to be dried by blowing dry air through it at 1.1 atm pressure and a superficial velocity of 15 ft sec^{-1}. What air temperature is required to keep the solids at a surface temperature of 60° F? Neglect radiation.

Answer: 112° F

21.D₁ Rate of Drying of Granular Solids in a Fixed Bed

Calculate the initial rate of water removal, in lb-moles hr^{-1} ft^{-3}, in the drying operation described in Problem 21.C, if the solids are flakes with $a = 180$ ft^{-1}. *Answer:* 22.7

21.E₂ Evaporation of a Freely Falling Drop

A drop of water, 1.00 mm in diameter, is falling freely through dry, still air at $p = 1$ atm and $T_\infty = 100°$ F. Assuming pseudo-steady-state behavior and a small mass-transfer rate, compute (*a*) the velocity of the falling drop; (*b*) the surface temperature of the drop; (*c*) the rate of change of the drop diameter in cm sec^{-1}. Assume that the "film properties" are those of dry air at 80° F.

Answers: a. 390 cm sec^{-1}
b. 54° F
c. 5.55 × 10^{-4}

21.F₃ Effect of Radiation on Psychrometric Measurements

Suppose that a wet-bulb and a dry-bulb thermometer are installed in a long duct with constant inside surface temperature T_s and that the gas velocity is small. Then the dry-bulb temperature T_{db} and the wet-bulb temperature T_{wb} should be corrected for radiation effects. Here again, for simplicity, we assume that the thermometers are installed so that heat conduction along the glass stems can be neglected.

a. Make an energy balance on unit area of the dry bulb to obtain an equation for the gas temperature T_∞ in terms of T_{db}, T_s, h_{db}, e_{db}, and a_{db} (the latter being the emissivity and the absorptivity of the dry bulb).

b. Make an energy balance on unit area of the wet bulb to obtain an expression for the evaporation rate.

c. Compute $x_{A\infty}$ for the pressure and thermometer readings of Example 21.2-2, with the additional data that $v_\infty = 15$ ft sec^{-1}, $T_s = 130°$ F, $e_{db} = a_{db} = e_{wb} = a_{wb} = 0.93$, diameter of dry bulb = 0.1 in., and diameter of wet bulb = 0.15 in. including wick.

Answer: c. $x_{A\infty} = 0.0021$

21.G₂ Transfer of *A* and *B* across a Fluid-Fluid Interface

In this problem the main results of §21.3 are to be extended to include transfer of *A* and *B* between two phases composed entirely of *A* and *B*. We assume that N_{A0}/N_{B0} is specified and that the bulk compositions and mass-transfer coefficients are known on both sides of the interface.

a. Extend Eqs. 21.3–2, 3, and 4, again using Eq. 21.1–10 to define the local mass-transfer coefficient in each phase.

b. Derive the relation between the one-phase coefficients k_x and k_y and the over-all coefficient K_y, defined by

$$d\mathcal{W}_{A0}^{(m)} - y_{Ae}(d\mathcal{W}_{A0}^{(m)} + d\mathcal{W}_{B0}^{(m)}) = K_y(y_{Ae} - y_{Ab})\, dA \qquad \text{(21.G–2)}$$

c. Give the special form of the result in (*b*) that applies when $N_{A0} = -N_{B0}$. This situation is known as equimolar counterdiffusion; the name is somewhat misleading, since the diffusion fluxes $J_A^\star$ and $J_B^\star$ are *always* equal and opposite in a binary system. (See Eq. 16.1–14.)

21.H₂ Interdependence of Mass-Transfer Coefficients in a Multicomponent System

As pointed out in §21.8, only $n - 1$ independent applications of mass-transfer coefficients can be made in an n-component mixture. Here we consider some checking procedures, which can be employed when the full set of n transfer coefficients is worked out.

a. First show from the definitions in §21.8 that

$$\Sigma_i k^{\bullet}_{xi,\text{loc}}\,\Delta x_i = 0 \tag{21.H-1}$$

$$\Sigma_i \frac{\Delta x_i}{R_{im}} = 0 \tag{21.H-2}$$

$$\phi_{im} k_{xi} = \text{const. for all } i \tag{21.H-3}$$

b. Use these identities to test the solution of Example 21.8-1.

c. Show how the foregoing identities simplify for binary systems.

21.I₂ Diffusion-Controlled Combustion in a Fixed Catalyst Bed

A fixed bed of catalyst is to be regenerated by passing nitrogen-diluted air through the bed to burn off carbonaceous deposits. The combustion reaction is assumed to occur in a single step, on the catalyst surface, and is represented by the empirical formula

$$\underset{\text{(gas)}}{12\,O_2} + \underset{\text{(solid)}}{C_{10}H_8} \rightarrow \underset{\text{(gas)}}{10\,CO_2} + \underset{\text{(gas)}}{4\,H_2O}$$

The heat of reaction is given by $\Delta\tilde{H}_1 = -100$ kcal per mole of O_2 (species 1) consumed at 1000° F. It is desired to know the maximum particle temperature that may occur when the local gas stream is at $T_b = 1000°$ F and contains $x_{b1} = 1.0$ mole per cent O_2, $x_{b2} =$ 99.0 mole per cent N_2.

a. Set up an equation for the maximum possible flux, N_{10}, of oxygen at the catalyst surface. The mass-transfer correction may be omitted, as R_{1m} is very small.

b. By means of an energy balance on a catalyst particle, obtain an expression for the temperature difference $T_0 - T_b$ in terms of N_{10}. To be consistent with (*a*), the mass-transfer correction should be omitted here also.

c. Combine the results of (*a*) and (*b*) with the analogy $j_H = j_D$ to obtain a relation between $(T_0 - T_b)$ and the gas-stream composition.

d. Compute the maximum possible value of T_0 for the conditions stated. Assume Sc = 0.8, Pr = 0.7, and $\tilde{C}_{pb} = 7.54$ cal g-mole^{-1} ° K^{-1}.

Answer: d. 1220° F

21.J₂ Catalyst Temperature in a Multicomponent Reaction

Estimate the catalyst temperature in the system of Example 21.8-1, if $\text{Pr}_f = 0.6$, $\tilde{C}_{pb} =$ 11.7 cal g-mole^{-1} ° K^{-1}, and the heat of reaction per mole of cyclohexane produced is $\Delta\tilde{H}_3 = -51$ kcal g-mole^{-1} at the catalyst surface conditions. Assume the other properties constant at the values previously given. *Answer:* 600° K

21.K₂ Film Theory for Spheres

Derive the results that correspond to Eqs. 21.5-20, 21, 23, and 24 for simultaneous heat and mass transfer in a system with spherical symmetry. That is, assume a spherical mass-transfer surface and assume that T and x_A depend only on the radial coordinate r. Show that Eqs. 21.5-35 and 36 apply unchanged.

What difficulties would be encountered if one tried to use the film theory to calculate the drag on a sphere?

21.L$_2$ Film Theory for Cylinders

Derive the results that correspond to Eqs. 21.5–20, 21, 23, and 24 for a system with cylindrical symmetry. That is, assume a cylindrical mass-transfer surface and assume that T and x_A depend only on r. Show that Eqs. 21.5–35 and 36 apply unchanged.

21.M$_3$ Film Theory with Variable Transport Properties

a. Show that in systems in which μ, k, and $c\mathscr{D}_{AB}$ vary as functions of y Eqs. 21.5–16, 17, and 18 may be integrated to give (for $y < \delta_v$, δ_T, or δ_{AB}, respectively)

$$1 - \frac{(v_x - 0)(N_{A0}M_A + N_{B0}M_B)}{-\tau_0} = \exp\,(N_{A0}M_A + N_{B0}M_B)\int_0^y \frac{dy}{\mu} \tag{21.M–1}$$

$$1 - \frac{(T - T_0)(N_{A0}\tilde{C}_{pA} + N_{B0}\tilde{C}_{pB})}{q_0} = \exp\,(N_{A0}\tilde{C}_{pA} + N_{B0}\tilde{C}_{pB})\int_0^y \frac{dy}{k} \tag{21.M–2}$$

$$1 - \frac{(x_A - x_{A0})(N_{A0} + N_{B0})}{N_{A0} - x_{A0}(N_{A0} + N_{B0})} = \exp\,(N_{A0} + N_{B0})\int_0^y \frac{dy}{c\mathscr{D}_{AB}} \tag{21.M–3}$$

Here again, if y in Eq. 21.M–1 or 2 exceeds δ_{AB}, the relations $N_A = N_{A0}$ and $N_B = N_{B0}$ have been extrapolated in performing the integration. By treating μ, k, and $c\mathscr{D}_{AB}$ as effective transport properties, one can use these equations to describe time-smoothed profiles in turbulent flow.

b. Make the corresponding changes in Eqs. 21.5–22 through 27 and 31 through 33 and show that Eqs. 21.5–34 to 49 remain correct. Thus one does not need to work with the integrals in calculating transfer rates if f_{loc}, h_{loc}, and $k_{x,\text{loc}}$ can be predicted.

c. Show that f_{loc}, h_{loc}, and $k_{x,\text{loc}}$ have to be evaluated in terms of the physical properties and flow regime (laminar or turbulent) that prevail at the conditions for which $f^{\bullet}_{\text{loc}}$, $h^{\bullet}_{\text{loc}}$, and $k^{\bullet}_{x,\text{loc}}$ are desired.

21.N$_3$ Unsteady-State One-Dimensional Diffusion: Generalization of Arnold Solution

A stationary, nonreacting fluid medium of uniform composition $x_{A\infty}$ and temperature T_∞ occupying the entire region of positive z has its surface $z = 0$ suddenly brought, at $t = 0$, to a new composition x_{A0} and temperature T_0, which are maintained for all time. The ratio N_{A0}/N_{B0} is also constant for $t > 0$. The physical properties are constant in the mixture and equal for both species.

a. Rework Example 19.1–1, with the necessary changes in boundary conditions, to obtain the dimensionless profile $X_{AB}(Z, \varphi)$ described under Eq. 21.6–2.

b. Verify the expression for the dimensionless mass flux φ given in Eq. 21.6–1.

c. Write out the energy equation and the boundary conditions on T for this system. Obtain, by analogy, the solutions for heat transfer that correspond to the results of (*a*) and (*b*).

CHAPTER 22

Macroscopic Balances for Multicomponent Systems

The applications of the laws of conservation of mass, momentum, and energy to engineering flow systems have been discussed in Chapter 7 (isothermal systems) and Chapter 15 (nonisothermal systems). In this chapter we continue the discussion by introducing three additional factors not taken into account heretofore: (*a*) the fluid in the system is composed of more than one chemical species; (*b*) chemical reactions may be occurring, with a concomitant change in the composition and a production or consumption of heat; and (*c*) mass may be entering the system through the bounding surfaces (that is, at surfaces other than planes "1" and "2"). Various mechanisms by which mass may enter or leave through the bounding surfaces of the system are portrayed in Fig. 22.0–1.

We begin by summarizing the macroscopic balances for the more generalized situation described above. Each of these balances will now contain one extra term, which accounts for the mass, momentum, or energy brought in across the bounding surface. The balances thus obtained are capable of describing industrial mass-transfer processes, such as absorption, extraction, ion-exchange, and selective adsorption. Entire treatises have been devoted to each of these topics; all we do here is to show how the material discussed in the preceding chapters paves the way for the study of the mass-transfer operations. The reader interested in pursuing the subject further will find several textbooks available.[1,2,3]

§22.1 THE MACROSCOPIC MASS BALANCES

The statement of the law of conservation of mass of the ith chemical species in a macroscopic flow system such as that described is

$$\frac{d}{dt} m_{i,\text{tot}} = -\Delta w_i + w_i^{(m)} + r_{i,\text{tot}} \qquad i = 1, 2, \cdots, n \qquad (22.1\text{–}1)$$

which is a generalization of Eq. 7.1–2. Here $m_{i,\text{tot}}$ is the instantaneous total mass of "i" in the system, and $\Delta w_i = w_{i2} - w_{i1} = \rho_{i2}\langle v_2\rangle S_2 - \rho_{i1}\langle v_1\rangle S_1$ is the difference between the mass rate of flow[1] of the ith species past planes "2" and "1." The quantity $w_i^{(m)}$ represents the mass rate of flow of the ith species across the bounding surfaces of the system by mass transfer; note that $w_i^{(m)}$ is positive when mass is *added* to the system, just as Q is taken to be positive when heat is *added* to the system. Finally, the symbol $r_{i,\text{tot}}$ stands for the rate of production of the ith species by homogeneous or heterogeneous reactions within the system.[2]

[1] T. K. Sherwood and R. L. Pigford, *Absorption and Extraction*, McGraw-Hill, New York (1952).

[2] R. E. Treybal, *Mass-Transfer Operations*, McGraw-Hill, New York (1955).

[3] W. Matz, *Die Thermodynamik des Wärme-und Stoffaustausches in der Verfahrenstechnik*, Steinkopff, Frankfurt-am-Main (1949).

[1] Recall that in Table 15.3–1 molecular and eddy transport of momentum and energy across surfaces "1" and "2" in the direction of flow have been neglected in comparison with bulk transport. The same is done here for the transport of species i. Really one should write $\Delta w_i = \Delta[\langle \rho_i v_i\rangle S]$. The use of this more accurate expression is recommended when velocity profiles are highly irregular, for example, in packed columns. Similar remarks apply to energy transfer. These effects are ordinarily neglected because there are usually not enough data to evaluate them.

[2] The quantities $m_{i,\text{tot}}$, $r_{i,\text{tot}}$, and $w_i^{(m)}$ may be expressed as integrals:

$$m_{i,\text{tot}} = \int_V \rho_i \, dV \qquad (22.1\text{–}1a)$$

$$r_{i,\text{tot}} = \int_V r_i \, dV \qquad (22.1\text{–}1b)$$

$$w_i^{(m)} = -\int_{S^{(m)}} (\boldsymbol{n} \cdot \rho_i \boldsymbol{v}_i) \, dS \qquad (22.1\text{–}1c)$$

in which $\boldsymbol{n}$ is the outwardly directed unit normal vector and $S^{(m)}$ is that portion of the bounding surface over which mass-transfer occurs. In defining $w_i^{(m)}$, we have assumed that the transfer surface $S^{(m)}$ is stationary. If this surface is in motion, then $\boldsymbol{v}_i$ must be replaced by the local fluid velocity relative to the surface.

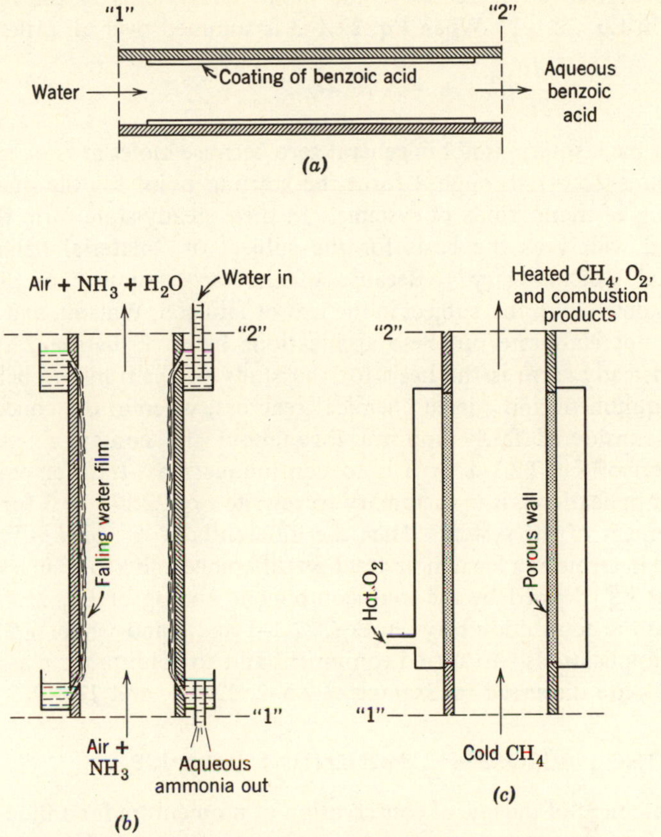

Fig. 22.0–1. Ways in which mass may enter or leave through bounding surfaces: (*a*) benzoic acid enters system by dissolution of the wall; (*b*) water vapor enters the system, defined as the gas phase, by evaporation, and ammonia vapor leaves by absorption; (*c*) oxygen enters the system by transpiration through a porous wall.

If all n equations of the form Eq. 22.1–1 are added together, we get

$$\frac{d}{dt} m_{\text{tot}} = -\Delta w + w^{(m)} \tag{22.1–2}$$

in which $w^{(m)} = \sum_{i=1}^{n} w_i^{(m)}$ and use has been made of the law of conservation of mass in the form $\sum_{i=1}^{n} r_{i,\text{tot}} = 0$.

Equation 22.1–1 may also be written in molar units:

$$\frac{d}{dt} \mathcal{M}_{i,\text{tot}} = -\Delta \mathcal{W}_i + \mathcal{W}_i^{(m)} + R_{i,\text{tot}} \qquad i = 1, 2, \cdots, n \tag{22.1–3}$$

Here the capital letters represent the molar equivalents of the lower-case symbols in Eq. 22.1–1. When Eq. 22.1–3 is summed over all i, the result is

$$\frac{d}{dt}\mathcal{M}_{\text{tot}} = -\Delta\mathcal{W} + \mathcal{W}^{(m)} + \sum_{i=1}^{n} R_{i,\text{tot}} \tag{22.1–4}$$

Note that the last term is not in general zero because moles are not conserved.

Equations 22.1–1 through 4 form the starting point for the quantitative description of many kinds of systems. In their steady-state form they have been used widely as the basis for the subject of "material balances" or "industrial stoichiometry." Because of the large number of illustrative examples devoted to this subject in the text of Hougen, Watson, and Ragatz,[3] we shall not elaborate on these applications here. Equation 22.1–1 or 3 in the unsteady form is the basis for the study of the transient behavior of mixing equipment and stirred chemical reactors, wherein the concentration may be considered fairly uniform throughout the container. A further application of Eq. 22.1–1 or 3 is to continuous mass-transfer equipment; for such applications, it is customary to rewrite Eq. 22.1–1 or 3 for a differential element of the system. Then the differential $dw_i^{(m)}$ or $d\mathcal{W}_i^{(m)}$ may be expressed in terms of a local mole-fraction difference and a local mass-transfer coefficient $k_{xi}^{\bullet}$, defined by the multicomponent analog of Eq. 21.4–6. This points out the connection between Eqs. 22.1–1 and 3 and Chapter 21. These last two applications—to stirred equipment and to continuous mass-transfer processes—are discussed in Examples 22.5–2, 22.6–1, and 22.6–2.

§22.2 THE MACROSCOPIC MOMENTUM BALANCE

The statement of the law of conservation of momentum for a fluid mixture, with gravity as the only external force acting on all species, is

$$\frac{d}{dt}\boldsymbol{P} = -\Delta\left(\frac{\langle v^2\rangle}{\langle v\rangle}w + p\boldsymbol{S}\right) + \boldsymbol{F}^{(m)} - \boldsymbol{F} + m_{\text{tot}}\boldsymbol{g} \tag{22.2–1}$$

This equation is the same as Eq. 7.2–2, save for the addition of the term $\boldsymbol{F}^{(m)}$, which is the net influx of momentum[1] into the system by mass transfer across the bounding surface. Actually, for most mass-transfer processes this additional term is so small that it can be neglected. The force $\boldsymbol{F}$ in Eq. 22.2–1 is to be calculated by using the friction factor $f^{\bullet}$ as defined in Eq. 21.4–4.

[3] O. A. Hougen, K. M. Watson, and R. A. Ragatz, *Material and Energy Balances*, Vol. 1, *Chemical Process Principles*, Wiley, New York (1954), Chapter 7.

[1] This term may be written as an integral:

$$\boldsymbol{F}^{(m)} = -\int_{S^{(m)}} (\boldsymbol{n}\cdot\rho\boldsymbol{v}\boldsymbol{v})\,dS \tag{22.2–1a}$$

in which $\boldsymbol{n}$ is the outwardly directed unit normal vector. As in §22.1, the transfer surface has been assumed stationary.

§22.3 THE MACROSCOPIC ENERGY BALANCE

For a fluid mixture the statement of the law of conservation of energy is written

$$\frac{d}{dt} E_{\text{tot}} = -\Delta\left[\left(\hat{U} + p\hat{V} + \frac{1}{2}\frac{\langle v^3 \rangle}{\langle v \rangle} + \hat{\Phi}\right)w\right] + Q^{(m)} + Q - W \quad (22.3\text{–}1)$$

This equation is the same as Eq. 15.1–3, except that an additional term $Q^{(m)}$ has been added.[1] This term accounts for the addition of energy to the system as a result of the mass transfer across the bounding surface. Note that Q is the rate of conductive heat transfer across the bounding surfaces. In the presence of mass transfer it must be calculated by using the heat-transfer coefficient $h^{\bullet}$ defined in Eq. 21.4–5. The term $Q^{(m)}$ may well be of considerable importance, particularly if material is entering through the bounding surface at a much higher or lower temperature than that of the fluid inside the flow system or if it reacts chemically in the system.

When chemical reactions are occurring, considerable heat may be released or absorbed. This heat of reaction is automatically taken into account in the calculation of the enthalpies of the entering and leaving streams. (See Example 22.5–1.)

In many applications, in which the energy-transfer rates across the surface are functions of position, it is more convenient to rewrite Eq. 22.3–1 over a differential portion of the flow system. Then the increment of heat added, dQ, is expressible in terms of $h^{\bullet}$, a local heat-transfer coefficient, corrected for the mass-transfer processes occurring. This procedure is similar to that illustrated in Example 15.4–2.

§22.4 THE MACROSCOPIC MECHANICAL ENERGY BALANCE

The mechanical energy balance, or Bernoulli equation, for a reacting fluid mixture with mass transfer across the bounding surfaces cannot be written in a simple form for the general situation. Hence we restrict the discussion here to fluids with *constant mass density* ρ. For such fluids, the mechanical energy balance is

$$\frac{d}{dt}(K_{\text{tot}} + \Phi_{\text{tot}}) = -\Delta\left[\left(\frac{1}{2}\frac{\langle v^3 \rangle}{\langle v \rangle} + \hat{\Phi} + \frac{p}{\rho}\right)w\right] + B^{(m)} - W - E_v \quad (22.4\text{–}1)$$

[1] This may be written as an integral:

$$Q^{(m)} = -\int_{S^{(m)}} \left(\boldsymbol{n} \cdot \sum_{i=1}^{n} c_i \boldsymbol{v}_i \bar{H}_i\right) dS \quad (22.3\text{–}1a)$$

in which $\boldsymbol{n}$ is the outwardly directed unit normal vector. The origin of this approximate expression may be seen by consulting Eq. 18.4–6. As in §22.1, the transfer surface has been assumed stationary.

This is the same as Eq. 7.3–1*a*, except for the addition of the term $B^{(m)}$, which represents the addition of energy to the flow system associated with mass-transfer processes; this term includes potential and kinetic energy added to the system and also work done against the system pressure in forcing a fluid through the walls of the system.[1] The steady-state form of the Bernoulli equation introduced for pure fluids (Eq. 7.3–2) is valid for reactive multicomponent systems if $B^{(m)}$ is zero. Its use under such circumstances is illustrated in Example 22.5–3.

§22.5 USE OF THE MACROSCOPIC BALANCES TO SOLVE STEADY-STATE PROBLEMS

In §18.5 it was shown how concentration and temperature profiles may be obtained for specific systems by solving simplified forms of the multicomponent equations of change. In this section we show how to use simplified forms of the multicomponent macroscopic balances to set up the relations between entrance and exit conditions in steady-state flow systems. By keeping track of the discarded terms, we know exactly what restrictions have to be placed on the derived results.

Example 22.5–1. Energy Balance for a Sulfur Dioxide Converter

Hot gases from a sulfur burner enter a converter, in which the sulfur dioxide present is to be oxidized catalytically to sulfur trioxide, according to the reaction $SO_2 + \frac{1}{2}O_2 \rightleftharpoons SO_3$. How much heat must be removed from the converter per hour to permit a 95 per cent conversion of the SO_2 for the conditions shown in Fig. 22.5–1? Assume that the converter is large enough for the components of the exit gas to be in thermodynamic equilibrium with one another; that is, the partial pressures of the exit gases are related by the equilibrium requirement

$$K_p = \frac{p_{SO_3}}{p_{SO_2} p_{O_2}^{1/2}} \tag{22.5–1}$$

Approximate values of K_p for this reaction are

T, °K	600	700	800	900
K_p, atm$^{-1/2}$	9500	880	69.5	9.8

[1] In terms of a surface integral, this term may be written

$$B^{(m)} = \int_{S^{(m)}} (\boldsymbol{n} \cdot \{\tfrac{1}{2}\rho v^2 + \rho\hat{\Phi} + p\}\boldsymbol{v})\, dS \tag{22.4–1a}$$

As in §22.1, the transfer surface has been assumed stationary.

Solution. It is convenient to break this problem up into two parts: first we use the mass balance and equilibrium expression to find the desired exit temperature, and then we use the energy balance to determine the required heat removal.

a. Determination[1] *of* T_2*:* We begin by writing the macroscopic mass balance,

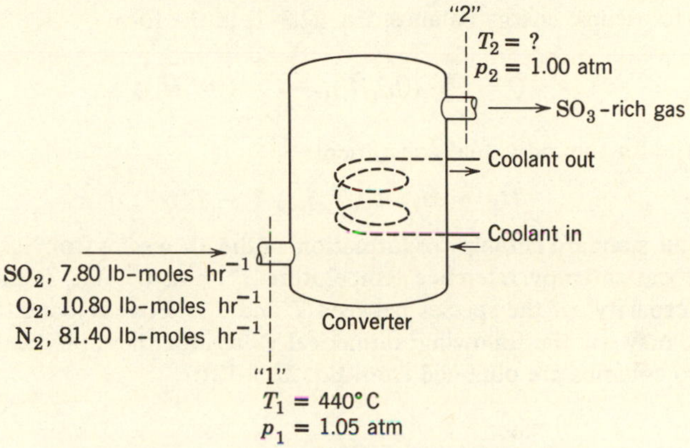

Fig. 22.5–1. Catalytic oxidation of sulfur dioxide.

Eq. 22.1–3, for the various constituents of the two streams in the form

$$\mathcal{W}_{i2} = \mathcal{W}_{i1} + R_{i,\text{tot}} \tag{22.5–2}$$

In addition, we take advantage of the two stoichiometric relations

$$R_{SO_2,\text{tot}} = -R_{SO_3,\text{tot}} \tag{22.5–3}$$

$$R_{O_2,\text{tot}} = \tfrac{1}{2}R_{SO_2,\text{tot}} \tag{22.5–4}$$

We thus obtain the desired molar flow rates through surface "2":

$$\mathcal{W}_{2,SO_2} = 7.80 - (0.95)(7.80) = 0.38 \text{ lb-mole hr}^{-1} \tag{22.5–5}$$

$$\mathcal{W}_{2,SO_3} = 0 + (0.95)(7.80) = 7.42 \text{ lb-moles hr}^{-1} \tag{22.5–6}$$

$$\mathcal{W}_{2,O_2} = 10.8 - \tfrac{1}{2}(0.95)(7.80) = 7.09 \text{ lb-moles hr}^{-1} \tag{22.5–7}$$

$$\mathcal{W}_{2,N_2} = \mathcal{W}_{1,N_2} = 81.40 \text{ lb-moles hr}^{-1} \tag{22.5–8}$$

$$\mathcal{W}_2 = 96.29 \text{ lb-moles hr}^{-1} \tag{22.5–9}$$

By substituting numerical values into the equilibrium expression Eq. 22.5–1, we obtain

$$K_p = \frac{(7.42/96.29)}{(0.38/96.29)(7.09/96.29)^{1/2}} = 72.0 \text{ atm}^{-1/2} \tag{22.5–10}$$

[1] See O. A. Hougen, K. M. Watson, and R. A. Ragatz, *Chemical Process Principles*, Part II, Wiley, New York (1959), Second Edition pp. 1017–1018.

This value of K_p corresponds to an exit temperature T_2 of about 510° C according to the given equilibrium data.

b. Calculation of the required heat removal. As indicated by the results of Example 15.4–1, changes in kinetic and potential energy may be neglected here in comparison with changes in enthalpy. In addition, for the conditions of this example, we may assume ideal-gas behavior. Then, for each constituent, $\tilde{H}_i = \tilde{H}_i(T)$. We may then write the macroscopic energy balance, Eq. 22.3–1, in the form

$$-Q = \sum_{i=1}^{n} (\mathcal{W}_i\tilde{H}_i)_1 - \sum_{i=1}^{n} (\mathcal{W}_i\tilde{H}_i)_2 \tag{22.5–11}$$

We may write for the individual constituents

$$\tilde{H}_i = \tilde{H}_i^\circ + (\tilde{C}_{pi})_{\text{avg}}(T - T^\circ) \tag{22.5–12}$$

Here $\tilde{H}_i^\circ$ is the standard enthalpy of formation[2] of the ith species from its constituent elements at the enthalpy reference temperature T°, and $(\tilde{C}_{pi})_{\text{avg}}$ is the enthalpy-mean heat capacity[2] of the species between T and T°. For the conditions of this problem we may use the following[2] numerical values for these physical properties (the last two columns are obtained from Eq. 22.5–12):

Component	$\tilde{H}_i^\circ$ (cal g-mole^{-1}) at 25° C	$(\tilde{C}_{pi})_{\text{avg}}$ [cal (g-mole)$^{-1}$ ° C^{-1}] from 25° C to 440° C	510° C	$(\mathcal{W}_i\tilde{H}_i)_1$ (Btu hr^{-1})	$(\mathcal{W}_i\tilde{H}_i)_2$ (Btu hr^{-1})
SO_2	−70,960	11.05	11.24	−931,900	−44,800
SO_3	−94,450	—	15.87	0	−1,158,700
O_2	0	7.45	7.53	60,100	46,600
N_2	0	7.12	7.17	433,000	509,500
				−438,800	−647,400

Substitution of the foregoing values into Eq. 22.5–11 gives the required rate of heat removal:

$$\begin{aligned} -Q &= -438{,}800 - (-647{,}400) \\ &= 208{,}600 \text{ Btu hr}^{-1} \end{aligned} \tag{22.5–13}$$

Example 22.5–2. Height of a Packed-Tower Absorber[3]

It is desired to remove a soluble gas A from a mixture of A and an insoluble gas G by contacting the mixture with a nonvolatile liquid solvent L in the apparatus shown in Fig. 22.5–2.

[2] See for example, O. A. Hougen, K. M. Watson, and R. A. Ragatz, Part I, *Chemical Process Principles*, Wiley, New York (1954), Second Edition pp. 257, 296.

[3] R. L. Pigford and A. P. Colburn, Section 10 in *Chemical Engineers' Handbook* (ed. J. H. Perry), McGraw-Hill, New York (1950), Third Edition.

The apparatus consists essentially of a vertical pipe filled with a randomly arranged packing of small rings of some chemically inert material. The liquid L is sprayed evenly over the top of the packing and trickles over the surfaces of these small rings. As it does so, it is intimately contacted with the gas mixture that is passing up the tower. This direct contacting between the two streams permits the transfer of A from the gas to the liquid.

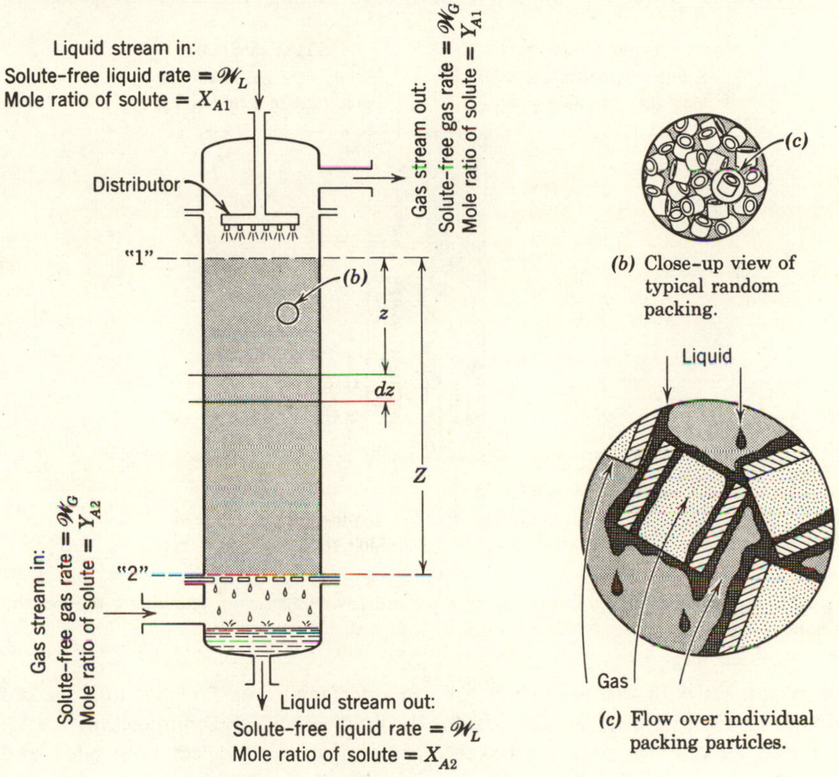

(a) Over-all cross section of column.

(b) Close-up view of typical random packing.

(c) Flow over individual packing particles.

Fig. 22.5-2. A packed column mass-transfer apparatus in which the descending phase is dispersed. Note that in this drawing $\mathscr{W}_G$ is negative; that is, the gas is flowing from "2" toward "1".

The gas and liquid streams enter the apparatus at molar rates of $-\mathscr{W}_G$ and $\mathscr{W}_L$, respectively, on an A-free basis. Note that the gas rate is negative because the gas stream is flowing from control surface "2" toward "1" in this problem. The molar ratio of A to G in the entering gas stream is $Y_{A2} = y_{A2}/(1 - y_{A2})$ and the molar ratio of A to L in the entering liquid stream is $X_{A1} = x_{A1}/(1 - x_{A1})$. Develop an expression for the tower height z required to reduce the molar ratio of A in the gas stream from Y_{A2} to Y_{A1}, in terms of the mass-transfer coefficients in the two streams and the foregoing stream rates and compositions.

Assume that the concentration of A is always small in both streams so that operation may be considered isothermal and so that the mass-transfer coefficients need not be corrected for mass-transfer rate. That is, the mass-transfer coefficients k_x and k_y defined in Eqs. 21.3–2 and 3 may be used.

Solution. Since the behavior of a packed tower is actually quite complex, we begin by setting up a simplified model. We consider the system to be equivalent to two streams flowing side-by-side without back mixing, as shown in Fig. 22.5–3,

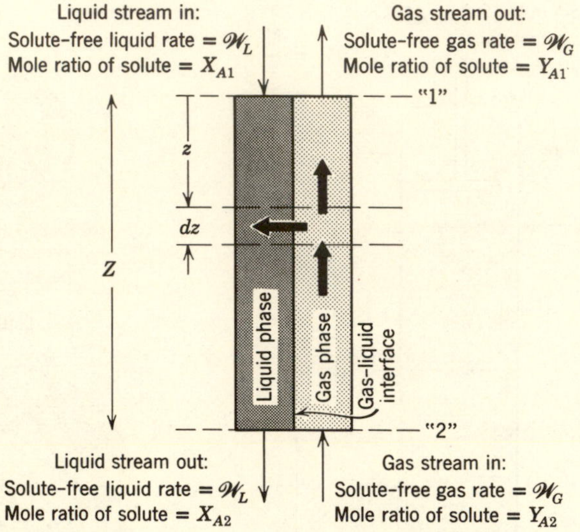

Fig. 22.5–3. Schematic representation of a packed-tower absorber, showing a differential element over which a mass balance is made.

and in contact with one another across an interfacial area "a" per unit packed volume (see Eq. 21.1–13). We also assume the fluid velocity and composition of each stream to be uniform over the tower cross section and neglect both eddy and molecular transport of solute in the direction of flow. Finally, we consider the concentration profiles in the direction of flow to be continuous curves not appreciably affected by the placement of individual packing particles. The model resulting from these simplifying assumptions is probably not a very satisfactory description of a packed tower; the neglect of back mixing and nonuniformity of fluid velocity are probably particularly serious. However, the presently available correlations for mass-transfer coefficients have been calculated on the basis of this model, which should therefore be employed when these correlations are used.

We are now in a position to develop an expression for column height, and we do this in two stages. First we use the over-all macroscopic mass balance to determine the exit liquid-phase composition and the relation between bulk compositions of the two phases at any point in the tower. We then use these results along with the differential form of the macroscopic mass balance to determine interfacial compositions and the required tower height.

a. Over-all macroscopic mass balances. For the solute A, we may write the macroscopic mass balance, Eq. 22.1–3, for each stream of the system between surfaces "1" and "2" as

(liquid stream) $$\mathcal{W}_{Al2} - \mathcal{W}_{Al1} = \mathcal{W}_{Al}^{(m)} \tag{22.5-14}$$

(gas stream) $$\mathcal{W}_{Ag2} - \mathcal{W}_{Ag1} = \mathcal{W}_{Ag}^{(m)} \tag{22.5-15}$$

Here the subscripts Al and Ag refer to the solute A in the liquid and gas streams, respectively. Clearly $\mathcal{W}_{Al}^{(m)} = -\mathcal{W}_{Ag}^{(m)}$, and therefore Eqs. 22.5–14 and 15 may be combined to give

$$\mathcal{W}_{Al2} - \mathcal{W}_{Al1} = -(\mathcal{W}_{Ag2} - \mathcal{W}_{Ag1}) \tag{22.5-16}$$

We now rewrite this equation in terms of the compositions of the entering and leaving streams by setting $\mathcal{W}_{Al2} = \mathcal{W}_L X_{A2}$, etc., and rearranging to get

$$X_{A2} = X_{A1} - (\mathcal{W}_G/\mathcal{W}_L)(Y_{A2} - Y_{A1}) \tag{22.5-17}$$

In this way we have found the concentration of A in the exit liquid stream.

By replacing surface "2" with a surface at a distance z down the column, we may use Eq. 22.5–17 to obtain an expression relating bulk stream compositions[3] at any point in the tower:

$$X_A = X_{A1} - (\mathcal{W}_G/\mathcal{W}_L)(Y_A - Y_{A1}) \tag{22.5-18}$$

Equation 22.5–18 (the "operating line") is shown in Fig. 22.5–4 along with the distribution equilibrium for the conditions of Problem 22.B.

b. Application of the macroscopic mass balances in differential form. We now apply Eq. 22.1–3 to a small length dz of the tower, first to estimate interfacial conditions and then to determine the tower height required to obtain a given separation.

(i) *Determination of interfacial conditions.* Because only A is transferred across the interface, we may use Eqs. 21.3–2 and 3 to express the rate of solute transfer in section dz for dilute solutions:

$$d\mathcal{W}_{Al}^{(m)} = (k_x a)\left(\frac{x_{A0} - x_A}{1 - x_{A0}}\right) S\, dz \tag{22.5-19}$$

$$d\mathcal{W}_{Ag}^{(m)} = (k_y a)\left(\frac{y_{A0} - y_A}{1 - y_{A0}}\right) S\, dz \tag{22.5-20}$$

where a = interfacial area per unit packed volume of tower

S = cross-sectional area of tower

x_A, y_A = bulk mole fraction of A in the liquid and gas phases, respectively

$$= \frac{X_A}{X_A + 1}, \qquad \frac{Y_A}{Y_A + 1}$$

x_{A0}, y_{A0} = interfacial mole fractions of A in the liquid and gas phases, respectively

For the dilute solutions being considered here, we may neglect X_A, x_A, Y_A, and

[3] For brevity in this section, the subscript b for bulk composition is omitted. The symbols X_A, Y_A, x_A, and y_A denote the bulk compositions.

y_A in comparison to unity. We next take advantage of the fact that $d\mathcal{W}_{Al}^{(m)} + d\mathcal{W}_{Ag}^{(m)} = 0$ and combine Eqs. 22.5–19 and 20 to obtain

$$\frac{Y_A - Y_{A0}}{X_A - X_{A0}} = -\left(\frac{k_x a}{k_y a}\right) \tag{22.5–21}$$

Equation 22.5–21 enables us to determine Y_{A0} as a function of Y_A. For any Y_A, one may locate X_A on the operating line (mass balance). One then draws a straight

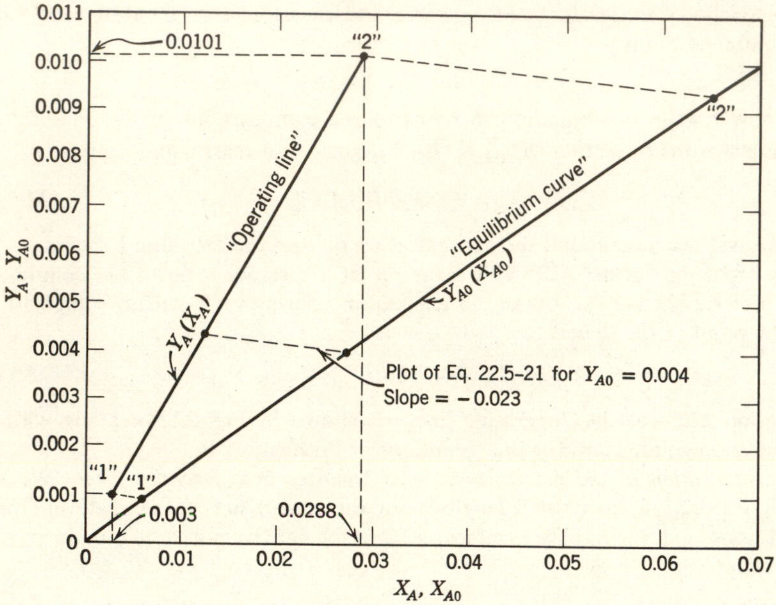

Fig. 22.5–4. Calculation of interfacial conditions in the absorption of cyclohexane from air in a packed column. (See Problem 22.B.)

line of slope $-(k_x a/k_y a)$ through the point (Y_A, X_A), as shown in Fig. 22.5–4. The intersection of this line with the equilibrium curve then gives the local interfacial compositions, Y_{A0} and X_{A0}. For concentrated solutions, it is preferable to start with a pair of interfacial compositions (y_{A0}, x_{A0}) and use Eq. 21.3–4.

(ii) *Determination of required column height.* Application of Eq. 22.1–1 to the gas stream in a volume $S\,dz$ of the tower gives

$$\mathcal{W}_G\, dY_A = d\mathcal{W}_{Ag}^{(m)} \tag{22.5–22}$$

We may combine this expression with Eq. 22.5–20 for the dilute solutions being considered to obtain

$$-\mathcal{W}_G\, dY_A = (k_y a)(Y_A - Y_{A0}) S\, dz \tag{22.5–23}$$

This equation may now be rearranged and integrated from $z = 0$ to $z = Z$:

$$Z = -\frac{\mathcal{W}_G}{S(k_y a)} \int_{Y_{A1}}^{Y_{A2}} \frac{dY_A}{Y_A - Y_{A0}} \tag{22.5–24}$$

Equation 22.5–24 is the desired expression for the column height required to effect the specified separation. Note that in writing Eq. 22.5–24 we have neglected variation of the mass-transfer coefficient k_x with composition; this is usually permissible only for dilute solutions.

In general, Eq. 22.5–24 must be integrated by numerical or graphical procedures. However, for dilute solutions, it may frequently be assumed that the operating and equilibrium lines, curves A and B, respectively, of Fig. 22.5–4, are straight. If, in addition, the ratio (k_x/k_y) is constant, then $(Y_A - Y_{A0})$ varies linearly with Y_A. We may then integrate Eq. 22.5–24 to obtain (see Problem 22.F)

$$Z = \frac{\mathcal{W}_G}{S(k_y a)} \frac{(Y_{A2} - Y_{A1})}{(Y_{A0} - Y_A)_{\ln}} \tag{22.5-25}$$

where

$$(Y_{A0} - Y_A)_{\ln} = \frac{(Y_{A0} - Y_A)_2 - (Y_{A0} - Y_A)_1}{\ln\left[\dfrac{(Y_{A0} - Y_A)_2}{(Y_{A0} - Y_A)_1}\right]} \tag{22.5-26}$$

Equation 22.5–25 may be rearranged to give

$$\mathcal{W}_{Ag}^{(m)} = \mathcal{W}_G(Y_{A2} - Y_{A1}) = (k_y a)ZS(Y_{A0} - Y_A)_{\ln} \tag{22.5-27}$$

Comparison of Eqs. 22.5–27 and 15.4–16 shows the close analogy between packed towers and simple heat exchangers. Expressions analogous to Eq. 22.5–27 but containing the over-all transfer coefficient K_y may also be obtained. (See Problem 22.F.) Again, we may use our final results, Eqs. 22.5–26 or 27, for either co-current or counter-current flow. It should be kept in mind, however, that the simplified model used to describe the packed tower is not so reliable as the corresponding one used for heat exchangers.

Example 22.5–3. Expansion of a Reactive Gas Mixture through a Frictionless Adiabatic Nozzle

An equimolar mixture of CO_2 and H_2 is confined at 1000° K and 1.50 atm in the large insulated pressure tank shown in Fig. 15.5–4. Under these conditions, the reaction

$$CO_2 + H_2 \rightleftharpoons CO + H_2O \tag{22.5-28}$$

may take place. After being stored in the tank long enough for this reaction to proceed to equilibrium, the gas is allowed to escape through the small convergent nozzle shown to the ambient pressure of 1 atm.

Estimate the temperature and velocity of the escaping gas through the nozzle throat (*a*) assuming no appreciable reaction takes place during passage of gas through the nozzle and (*b*) assuming instant attainment of thermodynamic equilibrium at all points in the nozzle. In each case, assume that the expansion is adiabatic and frictionless.

Solution. We begin by assuming quasi-steady-state operation, flat velocity profiles, and negligible changes in potential energy. We also assume constant heat capacities, ideal-gas behavior, and neglect diffusion in the direction of flow.

We may then write the macroscopic energy balance, Eq. 22.3–1, in the form

$$\tfrac{1}{2}v_2^2 = \hat{H}_1 - \hat{H}_2 \tag{22.5–29}$$

Here the subscripts "1" and "2" refer to conditions in the tank and the nozzle throat, respectively, and, as in Example 15.5–3, the fluid velocity in the tank is assumed to be zero.

We still have no way to determine the enthalpy change. We therefore combine Eq. 22.3–1 with the steady-state mechanical energy balance, Eq. 7.3–2, to obtain[4]

$$\frac{1}{\rho}\,dp = d\hat{H} \tag{22.5–30}$$

We may use this expression with the ideal-gas law and an expression for $\hat{H}(T)$ to evaluate $\hat{H}_2$, and therefore T_2 and v_2:

$$p = \frac{\rho RT}{\sum_{i=1}^{n} x_i M_i} \tag{22.5–31}$$

$$\hat{H} = \frac{\left\{\sum_{i=1}^{n} x_i[\tilde{H}_i^\circ + \tilde{C}_{pi}(T - T^\circ)]\right\}}{\sum_{i=1}^{n} x_i M_i} \tag{22.5–32}$$

Here the x_i are the mole fractions of the various components of the mixture at temperature T, and the $\tilde{H}_i^\circ$ are the molal enthalpies of the components at the enthalpy reference temperature T°. The evaluation of $\hat{H}$ is best discussed separately for the two parts of this problem.

Approximation A. Assumption of very slow chemical reaction.

Here the x_i are constant at the equilibrium values for 1000° K, and we may write

$$d\hat{H} = \left(\frac{\Sigma_i x_i \tilde{C}_{pi}}{\Sigma_i x_i M_i}\right) dT \tag{22.5–33}$$

We may therefore put Eq. 22.5–30 in the form

$$d \ln p = \left(\frac{\Sigma_i x_i \tilde{C}_{pi}}{R}\right) d \ln T \tag{22.5–34}$$

Since x_i and $\tilde{C}_{pi}$ are assumed constant, this equation may be integrated from p_1, T_1 to p_2, T_2 to give

$$T_2 = T_1\left(\frac{p_2}{p_1}\right)^{(R/\Sigma_i x_i \tilde{C}_{pi})} \tag{22.5–35}$$

We may now combine this expression with Eqs. 22.5–29 and 32 to obtain the desired expression for the gas velocity at surface 2:

$$v_2 = \left\{2T_1\left[1 - \left(\frac{p_2}{p_1}\right)^{(R/\Sigma_i x_i \tilde{C}_{pi})}\right] \frac{\Sigma_i x_i \tilde{C}_{pi}}{\Sigma_i x_i M_i}\right\}^{1/2} \tag{22.5–36}$$

[4] Note that Eq. 22.5–30 could also have been obtained directly from the energy equation, e.g., Eq. *E* of Table 18.3–1, by neglecting $\boldsymbol{\tau}$, $\boldsymbol{q}$, and the $\boldsymbol{j}_i$. Clearly, neither Eq. 22.3–1 nor Eq. 7.3–2 can be used directly in integrated form, since the thermodynamic path of the expansion is not yet known.

By substituting numerical values into Eqs. 22.5–35 and 36, we obtain (see Problem 22.A)

$$\begin{cases} T_2 = 920^\circ \text{ K} \\ v_2 = 1726 \text{ ft sec}^{-1} \end{cases}$$

It may be seen that this treatment is very similar to that presented in Example 15.5–3; it is also subject to the restriction that the throat velocity must be subsonic. That is, the pressure in the nozzle throat cannot fall below that fraction of p_1 required to produce sonic velocity at the throat. (See Eq. 15.I–2.) If the ambient pressure falls below this critical value of p_2, the throat pressure will remain at the critical value and there will be a shock wave beyond the nozzle exit.

Approximation B. Assumption of very rapid chemical reaction.

We may proceed here as in part *A*, except that the x_i must now be considered functions of temperature defined by the equilibrium relation:

$$\frac{(x_{H_2O})(x_{CO})}{(x_{H_2})(x_{CO_2})} = K_x(T) \tag{22.5–37}$$

and the stoichiometric relations

$$x_{H_2O} = x_{CO}; \quad x_{H_2} = x_{CO_2}; \quad \sum_{i=1}^{4} x_i = 1 \tag{22.5–38,39,40}$$

The quantity $K_x(T)$ in Eq. 22.5–37 is the known equilibrium constant for the reaction; it may be considered as a function only of temperature because of the assumed ideal-gas behavior and because the number of moles present is not affected by chemical reaction. Equations 22.5–38 and 39 follow from the stoichiometry of the reaction and the composition of the gas originally placed in the tank.

The expression for final temperature is now considerably more complex; for this reaction, where $\Sigma_i x_i M_i$ is constant, Eqs. 22.5–30 and 31 may be combined to give

$$R \ln \left(\frac{p_2}{p_1}\right) = \int_{T_1}^{T_2} \frac{d\tilde{H}}{dT} \, d \ln T \tag{22.5–41}$$

where

$$\frac{d\tilde{H}}{dT} = \sum_{i=1}^{4} [\tilde{H}_i^\circ + \tilde{C}_{pi}(T - T^\circ)] \frac{dx_i}{dT} + \sum_{i=1}^{4} \tilde{C}_{pi} x_i \tag{22.5–42}$$

In general, the integral in Eq. 22.5–41 must be evaluated numerically, since both the x_i and the dx_i/dT are complex functions of temperature governed by Eqs. 22.5–37 through 40. Once T_2 has been determined from Eq. 22.5–41, however, v_2 may be simply obtained by use of Eqs. 22.5–29 and 32. By substituting numerical values into these expressions, we obtain (see Problem 22.C)

$$\begin{cases} T_2 = 937^\circ \text{ K} \\ v_2 = 1752 \text{ ft sec}^{-1} \end{cases}$$

We find, then, that both the exit temperature and velocity from the nozzle are greater when chemical equilibrium is maintained throughout the expansion. This is because the equilibrium shifts with decreasing temperature in such a way as to release heat of reaction to the system. Such a release of energy will occur with decreasing temperature in any system at chemical equilibrium, regardless of the reactions involved; this is one consequence of the famous rule of Le Châtelier. In this case, the reaction is endothermic as written and the equilibrium constant decreases with falling temperature. As a result, CO and H_2O are partially reconverted to H_2 and CO_2 on expansion, with a corresponding release of energy.

It is interesting to note that in rocket engines the exhaust velocity, hence engine thrust, are also increased if rapid equilibration can be obtained, even though the combustion reactions encountered here are strongly exothermic. This is because the equilibrium constants for these reactions increase with falling temperature so that heat of reaction is again released on expansion. This principle has been suggested as a method for improving the thrust of rocket engines. The increase in thrust potentially obtainable in this way is quite large.

The foregoing example has been chosen for its simplicity. Note in particular that if a change in number of moles accompanies the chemical reaction the equilibrium constant, hence the enthalpy, are functions of pressure. In this case, which is quite common, the variables p and T implicit in Eq. 22.5–30 cannot be separated, and a step-by-step integration of this equation is required. Such integrations have been performed, for example, for the prediction of the behavior of supersonic wind tunnels[4] and rocket engines,[5] but the calculations involved are too tedious for presentation here.

§22.6 USE OF THE MACROSCOPIC BALANCES FOR SOLVING UNSTEADY-STATE PROBLEMS

In §22.5 steady-state problems have been discussed. Here we address ourselves to the transient behavior of multicomponent systems. Such behavior is important in a very large number of practical operations, for example, leaching and drying of solids, chromatographic separations, and chemical reactor operations. In many of these processes heats of reaction as well as mass transfer must be considered.

A complete discussion of these topics is outside the scope of this text, and we content ourselves here with two very simple examples. A more extensive discussion of the transient behavior of mass-transfer equipment has been given by Marshall and Pigford.[1]

Example 22.6–1. Start-Up of a Chemical Reactor

It is desired to produce a substance B from a raw material A in a chemical reactor of volume V equipped with a perfect stirrer. The formation of B is reversible, and the forward and reverse reactions may be considered first order. In

[4] J. G. Hall, "Dissociation Nonequlibrium in Hypersonic Nozzle Flow," *Reprint* 7, A.I.Ch.E. Meeting, Kansas City (1959).

[5] T. W. Reynolds and L. V. Baldwin, One-Dimensional Flow with Chemical Reaction in Nozzle Expansions, *Preprint* 6, A.I.Ch.E. Meeting, Kansas City (1959).

[1] W. R. Marshall, Jr., and R. L. Pigford, *The Application of Differential Equations to Chemical Engineering Problems*, University of Delaware Press, Newark, Delaware (1947).

addition, B undergoes an irreversible first-order decomposition to a third component C. The chemical reactions of interest may then be represented as

$$A \underset{k'''_{1A}}{\overset{k'''_{1B}}{\rightleftharpoons}} B \xrightarrow{k'''_{1C}} C \tag{22.6–1}$$

At zero time a solution of A at concentration c_{A0} is introduced to the initially empty reactor at a constant volumetric flow rate Q.

Develop an expression for the amount of B in the reactor when it is just filled to its capacity V, assuming that there is no B in the feed solution and neglecting changes of fluid properties.

Solution. We begin by writing the unsteady-state macroscopic mass balances for species A and B. In molar units these may be expressed as

$$\frac{d\mathcal{M}_A}{dt} = Qc_{A0} - k'''_{1B}\mathcal{M}_A + k'''_{1A}\mathcal{M}_B \tag{22.6–2}$$

$$\frac{d\mathcal{M}_B}{dt} = -(k'''_{1A} + k'''_{1C})\mathcal{M}_B + k'''_{1B}\mathcal{M}_A \tag{22.6–3}$$

We next eliminate $\mathcal{M}_A$ from Eq. 22.6–3. First we differentiate this equation with respect to time to get

$$\frac{d^2\mathcal{M}_B}{dt^2} = -(k'''_{1A} + k'''_{1C})\frac{d\mathcal{M}_B}{dt} + k'''_{1B}\frac{d\mathcal{M}_A}{dt} \tag{22.6–4}$$

We now substitute the right side of Eq. 22.6–2 for $d\mathcal{M}_A/dt$ and combine the resulting expression with Eq. 22.6–3 to obtain a second-order ordinary differential equation for $\mathcal{M}_B$ as a function of time:

$$\frac{d^2\mathcal{M}_B}{dt^2} + (k'''_{1A} + k'''_{1B} + k'''_{1C})\frac{d\mathcal{M}_B}{dt} + k'''_{1B}k'''_{1C}\mathcal{M}_B - k'''_{1B}Qc_{A0} = 0 \tag{22.6–5}$$

This equation is to be solved with the initial conditions:

I. C. 1: at $t = 0$, $\mathcal{M}_B = 0$ (22.6–6)

I. C. 2: at $t = 0$, $\dfrac{d\mathcal{M}_B}{dt} = 0$ (22.6–7)

By performing the indicated integration, we obtain

$$\mathcal{M}_B = \frac{Qc_{A0}}{k'''_{1C}}\left(1 + \frac{s_-}{s_+ - s_-}e^{s_+t} - \frac{s_+}{s_+ - s_-}e^{s_-t}\right) \tag{22.6–8}$$

where

$$2s_\pm = -(k'''_{1A} + k'''_{1B} + k'''_{1C}) \pm \sqrt{(k'''_{1A} + k'''_{1B} + k'''_{1C})^2 - 4k'''_{1B}k'''_{1C}} \tag{22.6–9}$$

Equations 22.6–8 and 9 give the total mass of B in the reactor as a function of time, up to the time at which the reactor is just completely filled. These expressions are very similar to the corresponding equations obtained for the damped manometer in Example 7.6–2 and the temperature controller in Example 15.5–2. It can be shown, however (see Problem 22.I), that s_+ and s_- are both real and negative and therefore that $\mathcal{M}_B$ cannot oscillate.

Example 22.6–2. Unsteady Operation of a Packed Column

There are many industrially important processes in which mass transfer takes place between a fluid and a granular solid: for example, recovery of organic vapors

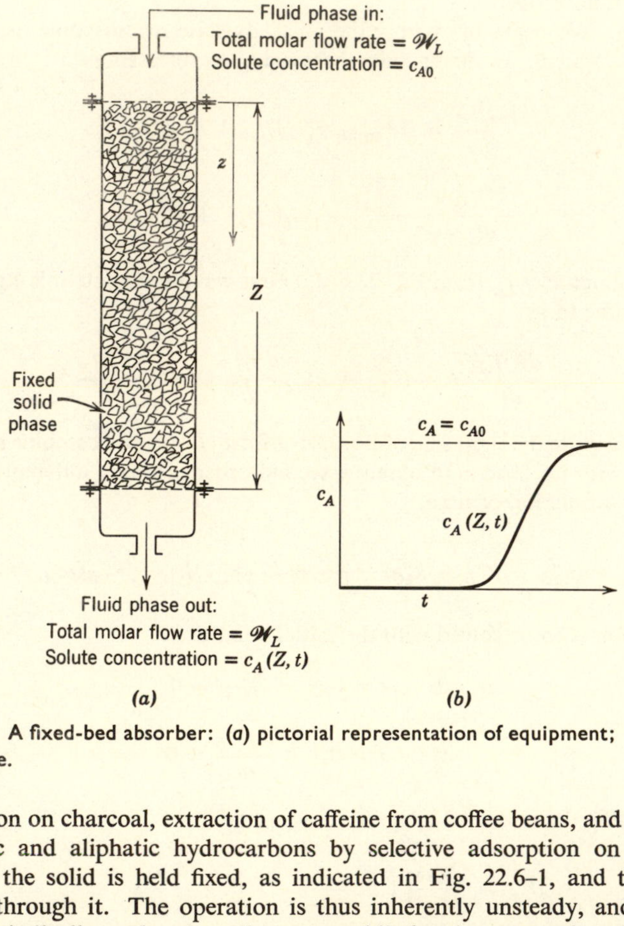

Fig. 22.6–1. A fixed-bed absorber: (*a*) pictorial representation of equipment; (*b*) a typical effluent curve.

by adsorption on charcoal, extraction of caffeine from coffee beans, and separation of aromatic and aliphatic hydrocarbons by selective adsorption on silica gel. Ordinarily, the solid is held fixed, as indicated in Fig. 22.6–1, and the fluid is percolated through it. The operation is thus inherently unsteady, and the solid must be periodically replaced or "regenerated," that is, returned to its original condition by heating or other treatment. To illustrate the behavior of such fixed-bed mass-transfer operations, we consider a physically simple case, the removal of a solute from a solution by passage through an adsorbent bed.

In this operation a solution containing a single solute A at mole fraction x_{A1}

in a solvent B is passed at a constant volumetric flow rate Q through a packed tower. The tower packing consists of a granular solid capable of adsorbing A from the solution. At the start of percolation, the interstices of the bed are completely filled with pure liquid B, and the solid is free of A. The percolating fluid displaces this solvent evenly so that the solution concentration of A is always uniform over any cross section. For simplicity, it will be assumed that the equilibrium concentration of A adsorbed on the solid is proportional to the local concentration of A in

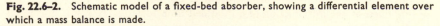

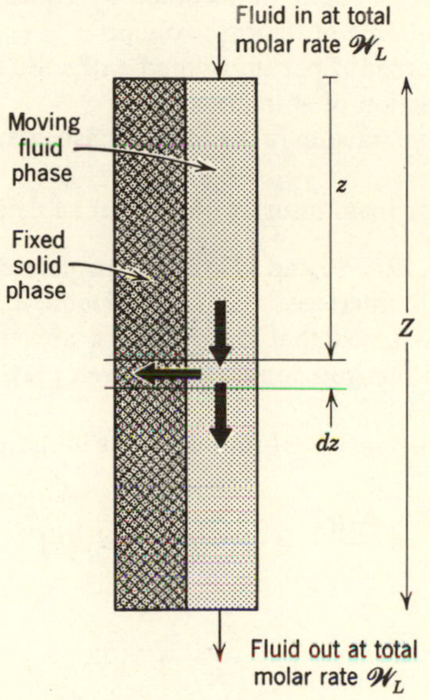

Fig. 22.6–2. Schematic model of a fixed-bed absorber, showing a differential element over which a mass balance is made.

solution. It will also be assumed that the concentration of A in the percolating solution is always small and that the resistance of the solid to mass transfer is negligible.

Develop an expression for the concentration of A in the column as a function of time and of distance down the column.

Solution. As for the gas absorber discussed in Example 22.5–2, it is convenient to think of the two phases as continuous and existing side by side as pictured in Fig. 22.6–2. We again define the contact area per unit packed volume of column as a. Now, however, one of the phases is stationary, and unsteady-state conditions prevail. Because of this unsteady behavior, the macroscopic mass balances are of limited usefulness in integrated form. We may use Eq. 22.1–3 and the assumption of dilute solutions to state that the molar rate of flow of solvent, $\mathscr{W}_B$, is essentially constant over the length of the column and the time of operation.

We now proceed to use Eq. 22.1–3 in differential form to write the continuity relationships for component A in each phase in a section of column of length dz.

For the *solid* in this section of column, we write

$$d\frac{\partial \mathcal{M}_A}{\partial t} = d\mathcal{W}_A^{(m)} \tag{22.6–10}$$

or

$$(1 - \epsilon)S\,dz\,\frac{\partial c_{As}}{\partial t} = (k_x a)(x_A - x_{A0})S\,dz \tag{22.6–11}$$

where ϵ = volume fraction of column occupied by liquid
S = cross-sectional area of (empty) column
c_{As} = moles of adsorbed A per unit volume of solid phase
x_A = bulk mole fraction of A in fluid phase
x_{A0} = interfacial mole fraction of A in fluid phase, assumed to be in equilibrium with c_{As}
k_x = the fluid-phase mass transfer coefficient as defined in Eq. 21.1–13

Note that in writing Eqs. 22.6–10 and 11 we have neglected convective mass transfer through the solution-solid interface; this is reasonable if x_{A0} is much smaller than unity. We have also assumed that the particles are small enough so that the concentration of the solution surrounding any given particle is essentially constant over the particle surface.

For the *fluid* in the differential column segment under consideration, Eq. 22.1–3 becomes

$$d\frac{\partial \mathcal{M}_A}{\partial t} = -d\mathcal{W}_A + d\mathcal{W}_A^{(m)} \tag{22.6–12}$$

or

$$\epsilon Sc\,\frac{\partial x_A}{\partial t} = -\mathcal{W}_B\,\frac{\partial x_A}{\partial z} - S(k_x a)(x_A - x_{A0}) \tag{22.6–13}$$

in which c is the total molar concentration of the liquid. Equation 22.6–13 may be simplified by the introduction of a modified time variable

$$t' = t - z\left(\frac{\epsilon Sc}{\mathcal{W}_B}\right) \tag{22.6–14}$$

It may be seen that for any point in the column t' is time measured from the instant that the percolating solvent "front" has reached the point in question. By introducing t' into Eq. 22.6–13 and rearranging the result, we get

$$\left(\frac{\partial x_A}{\partial z}\right)_{t'} = -\frac{(k_x a)S}{\mathcal{W}_B}(x_A - x_{A0}) \tag{22.6–15}$$

Similarly, we may rewrite Eq. 22.6–11 as

$$\left(\frac{\partial c_{As}}{\partial t'}\right)_z = \frac{(k_x a)}{(1 - \epsilon)}(x_A - x_{A0}) \tag{22.6–16}$$

Equations 22.6–15 and 16 combine the equations of continuity for each phase with our assumed mass-transfer rate expression. These two equations are to be solved simultaneously with the aid of the interphase equilibrium distribution, $x_{A0} = mc_{As}$, in which m is a constant. The boundary conditions are

B. C. 1: at $t' = 0, \quad c_{As} = 0$ for all $z > 0$ (22.6–17)

B. C. 2: at $z = 0, \quad x_A = x_{A1}$ for all $t' > 0$ (22.6–18)

Before solving these equations, it will be convenient to rewrite them in terms of the following dimensionless variables:

$$X = \frac{x_A}{x_{A1}} \tag{22.6–19}$$

$$Y = \frac{mc_{As}}{x_{A1}} \tag{22.6–20}$$

$$\zeta = \frac{zS(k_x a)}{\mathcal{W}_B} \tag{22.6–21}$$

$$\tau = \frac{mt'(k_x a)}{1 - \epsilon} \tag{22.6–22}$$

In terms of these variables, the differential equations and boundary conditions may be written

$$\frac{\partial X}{\partial \zeta} = -(X - Y) \tag{22.6–23}$$

$$\frac{\partial Y}{\partial \tau} = +(X - Y) \tag{22.6–24}$$

B. C. 1: at $\tau = 0, \quad Y = 0$ for all ζ (22.6–25)

B. C. 2: at $\zeta = 0, \quad X = 1$ for all τ (22.6–26)

The solution[1] to Eqs. 22.6–23 through 26 is

$$X = 1 - \int_0^\zeta e^{-(\tau+\zeta)} J_0(i\sqrt{4\tau\zeta})\, d\zeta \tag{22.6–27}$$

in which $J_0(ix)$ is a zero-order Bessel function of the first kind. This solution is presented graphically in several available references.[2]

QUESTIONS FOR DISCUSSION

1. Is it true that $\Sigma_i R_{i,\text{tot}} = 0$ for all systems?
2. Contrast $\boldsymbol{F}$ and $\boldsymbol{F}^{(m)}$.
3. Is the macroscopic energy balance in the form given by Eq. 22.3–1 correct for $\partial\hat{\Phi}/\partial t$ not equal to zero?

[1] This result was first obtained by A. Anzelius, *Z. angew. Math. u. Mech.*, **6**, 291–294 (1926), for the analogous problem in heat transfer. One method of obtaining this result is outlined in Problem 22.L. See also H. Bateman, *Partial Differential Equations of Mathematical Physics*, Dover, New York (1944), pp. 123–125.

[2] See, for example, O. A. Hougen and K. M. Watson, *Chemical Process Principles*, Wiley, New York (1947), First Edition, Part III, p. 1086. There y/y_0 corresponds to X, $b\tau$ corresponds to τ, and aZ corresponds to ζ.

4. Compare $Q^{(m)}$ with Q.

5. What is the physical significance of the term $B^{(m)}$ in the macroscopic mechanical energy balance?

6. In Example 22.5–2, use was made of the relation $\mathcal{W}_{Ag}^{(m)} + \mathcal{W}_{Al}^{(m)} = 0$. Would it also be true that $Q_g^{(m)} + Q_l^{(m)} = 0$?

7. How would you extend Example 15.4–3 to the pumping of a chemically reactive gas mixture?

8. Is the perfect mixing assumed in Example 22.6–1 desirable from the standpoint of maximizing the production of B? Or would it be preferable to have plug flow, that is, to have no mixing of fluid volume elements as they pass through the reactor? Would your answer be the same for the reaction

$$A_1 + A_2 \xrightarrow{k_2'''} B \qquad \text{where} \qquad R_B = k_2''' c_{A_1} c_{A_2}$$

9. Would it be possible to replace k_y with K_y in Example 22.6–2 to permit taking possible solid-phase resistance into account? If so, what difficulties might you encounter in predicting K_y?

10. Justify Eq. 22.6–7.

PROBLEMS

22.A₁ Expansion of a Gas Mixture: Very Slow Reaction Rate

Estimate the temperature and velocity of the water-gas mixture at the discharge end of the nozzle in Example 22.5–3 if the reaction rate is very slow. Use the following data: $\log_{10} K_x = -0.15$, $\tilde{C}_{pH_2} = 7.217$ Btu lb-mole^{-1} °F^{-1}, $\tilde{C}_{pCO_2} = 12.995$, $\tilde{C}_{pH_2O} = 9.861$, $\tilde{C}_{pCO} = 7.932$. Is the nozzle exit pressure equal to ambient pressure?

Answer: 920° K; 1726 ft sec^{-1}; yes, the nozzle flow is subsonic

22.B₁ Height of a Packed-Tower Absorber[1]

A packed tower of the type described in Example 22.5–2 is to be used to remove 90% of the cyclohexane from a cyclohexane-air mixture by absorption into nonvolatile light oil. The gas stream enters the bottom of the tower at a volumetric rate of 363 ft³ min^{-1}, at 30° C, and at 1.05 atm pressure. It contains 1 per cent cyclohexane by volume. The oil enters the top of the tower at a rate of 20 lb-mole hr^{-1}, also at 30° C, and it contains 0.3 per cent cyclohexane on a molar basis. The vapor pressure of cyclohexane at 30° C is 121 mm Hg, and solutions of it in the oil may be considered to follow Raoult's law.

a. Construct the operating line for the column.

b. Construct an equilibrium curve for the range of operation encountered here. Assume operation to be isothermal and isobaric.

c. Determine interfacial conditions at each end of the column.

d. Determine the required tower height using Eq. 22.5–24 if $k_x a = 0.32$ mole hr^{-1} ft^{-3}, $k_y a = 14.2$ mole hr^{-1} ft^{-3}, and the tower cross section S is 2.00 ft².

e. Repeat part (*d*), using Eq. 22.5–25.

Answer: d. ca. 62 ft
e. 60 ft

22.C₂ Expansion of a Gas Mixture: Very Fast Reaction Rate

Estimate the temperature and velocity of the water-gas mixture at the discharge end of the nozzle in Example 22.5–3 if the reaction rate may be considered infinitely fast. Use the data supplied in Problem 22.A and also the following: at 900° K $\log K_x = -0.34$;

[1] This problem is patterned after a similar development given by R. E. Treybal, *Mass-Transfer Operations*, McGraw-Hill, New York (1955), Chapter 8.

$\tilde{H}_{H_2} = +6340$ cal gm-mole^{-1}; $\tilde{H}_{H_2O}(g) = -49{,}378$; $\tilde{H}_{CO} = -16{,}636$; $\tilde{H}_{CO_2} = -83{,}242$, For simplicity, neglect the effect of temperature on heat capacity and assume log K_x to vary linearly with temperature between 900 and 1000° K. The following simplified procedure is recommended:

a. It may be seen in advance that T_2 will be higher than for slow reaction rates, hence greater than 920° K. (See Problem 22.A.) Show then that over the possible temperature range to be encountered $\tilde{H}$ varies very nearly linearly with temperature according to the expression $(d\tilde{H}/dT)_{\text{avg}} \doteq 12.40$ cal gm-mole^{-1} ° K^{-1}.

b. Substitute the foregoing result into Eq. 22.5–41 to show that $T_2 \doteq 937°$ K.

c. Calculate $\tilde{H}_1$ and $\tilde{H}_2$ and show by use of Eq. 22.5–29 that $v_2 \doteq 1750$ ft sec^{-1}.

22.D$_2$ Disposal of an Unstable Waste Material

A fluid stream of constant volumetric flow rate Q is to be discharged into a river. It contains a waste material A at concentration c_{A0}, which is unstable and decomposes at a rate proportional to its concentration according to the expression $-R_A = k_1'''c_A$. To reduce pollution, it is decided to pass the fluid through a holding tank of volume V before discharge to the river. At zero time the fluid begins to flow into the empty tank, which may be considered to have a perfect stirrer. No liquid leaves the tank until it is filled.

Develop an expression for the concentration of A in the tank and the effluent from the tank as a function of time by analogy with Problem 15.M.

22.E$_2$ Irreversible First-Order Reaction in a Continuous Reactor

A well-stirred reactor of volume V is initially completely filled with a solution of solute A in solvent S at concentration c_{A0}. At time $t = 0$, an identical solution of A in S is introduced at a constant volumetric flow rate Q. A small constant stream of dissolved catalyst is introduced at the same time, causing A to disappear according to an irreversible first-order reaction with a rate constant k_1''' sec^{-1}. The rate constant may be assumed independent of composition and time. Show that the concentration of A in the reactor at any time is

$$\frac{c_A}{c_{A0}} = \left(1 - \frac{Q}{Q + Vk_1'''}\right)\exp\left[-\left(\frac{Q + Vk_1'''}{V}\right)t\right] + \left(\frac{Q}{Q + Vk_1'''}\right) \qquad (22.\text{E}–1)$$

22.F$_2$ Effective Average Driving Forces in a Gas Absorber

Consider a packed-tower gas absorber of the type discussed in Example 22.5–2. Assume that the solute concentration is always low and that the equilibrium and operating lines are both very nearly straight. Under these conditions both k_xa and k_ya may be considered constant.

a. Show that $(Y_A - Y_{Ae})$ varies linearly with Y_A. Note that Y_A is the bulk mole ratio of A in the gas phase and Y_{Ae} is the equilibrium gas-phase mole ratio over liquid of the bulk composition, X_A. (See Fig. 21.3–2.)

b. Repeat part (*a*) for $(Y_A - Y_{A0})$.

c. Use the results of parts (*a*) and (*b*) to show that

$$\mathcal{W}_{Ag}^{0(m)} = (k_ya)ZS(Y_{A0} - Y_A)_{\ln} \qquad (22.\text{F}–1)$$

$$\mathcal{W}_{Ag}^{0(m)} = (K_ya)ZS(Y_{Ae} - Y_A)_{\ln} \qquad (22.\text{F}–2)$$

The over-all mass-transfer coefficient K_y is defined by Eq. 21.3–6. Note that this part of the problem may be solved simply by analogy with the development of Example 15.4–2.

22.G$_2$ Effect of Initial Solute Concentration on the Unsteady-State Operation of an Adsorption Column

Consider an adsorption column of the type described in Example 22.6–2. Assume the same conditions of operation except that at $t = 0$ the solid contains a uniformly distributed amount of solute at the initial concentration c_{Asi}. It may also be assumed that the liquid in the column is in equilibrium with the solid phase and therefore contains solute at the equilibrium mole fraction $x_{Ai} = x_{Ai}(c_{Asi})$.

Rewrite Eqs. 22.6–15 through 18 and the interphase equilibrium distribution in terms of the new variables $x_A' = x_A - x_{Ai}$ and $c_{As}' = c_{As} - c_{Asi}$. Show that Eqs. 22.6–23 through 26 are still valid if x_A' and c_{As}' are everywhere substituted for x_A and c_{As} in the definitions of X and Y.

22.H$_3$ Irreversible Second-Order Chemical Reaction in an Agitated Tank

Consider a system similar to that discussed in Problem 22.E, except that the solute A disappears according to a second-order reaction; that is $-R_{A,\text{tot}} = k_2''' V c_A^2$. Develop an expression for c_A as a function of time by the following procedure:

a. Write the macroscopic mass balance for the tank to obtain a differential equation describing c_A as a function of time.

b. Rewrite the differential equation, and the initial condition required to solve it, in terms of

$$u = c_A + \frac{Q}{2kV}\left[1 + \sqrt{1 + \frac{4k_2''' V c_{A0}}{Q}}\right] \tag{22.H-1}$$

The nonlinear differential equation obtained in this way is a "Bernoulli differential equation."

c. Now put $v = u^{-1}$ and integrate the resulting expressions. Put this integrated equation back into the original variable c_A and obtain the desired solution.

22.I$_3$ Startup of a Chemical Reactor

a. Integrate Eq. 22.6–5 with the aid of the given initial conditions to show that Eq. 22.6–8 correctly describes $\mathcal{M}_B$ as a function of time.

b. Show that s_+ and s_- in Eq. 22.6–9 are always real and negative. Suggested solution: show that

$$(k_{1A}''' + k_{1B}''' + k_{1C}''')^2 - 4k_{1B}'''k_{1C}''' = (k_{1A}''' - k_{1B}''' + k_{1C}''')^2 + 4k_{1A}'''k_{1B}''' \tag{22.I-1}$$

c. Obtain expressions for $\mathcal{M}_A$ and $\mathcal{M}_C$ as functions of time.

22.J$_3$ Analogy Between the Unsteady Operation of an Adsorption Column and Cross-Flow Heat Exchange[2]

An idealized cross-flow heat exchanger is shown in Fig. 22.J. In such an exchanger the two fluid streams flow at right angles to one another and the heat flux parallel to the wall is neglected. Here exchange of heat is clearly less than for a countercurrent exchanger of the same surface and over-all coefficient of heat transfer under otherwise identical conditions. The heat flow in these exchangers may be expressed for constant U_{loc} as

$$Q = U_{\text{loc}} A \, \Delta T_{\text{ln}} \Upsilon \tag{22.J-1}$$

Here Q is the total rate of heat transfer, A is the heat-transfer surface area, and $(\Delta T)_{\text{ln}}$ is the logarithmic mean of $(T_{h1} - T_{c1})$ and $(T_{h2} - T_{c2})$, as defined in the accompanying figure. (Note that T_{h2} and T_{c2} are the flow-averaged temperatures of the two exit streams.) We may then consider Υ as the ratio of heat transferred in cross flow to that which would be transferred in counter flow.

Use Eq. 22.6–27 to write an expression for Υ as a function of the stream rates, physical properties, A, and U_{loc}. Express the result in terms of definite integrals and assume $\hat{C}_{ph}$, $\hat{C}_{pc}$, and U_{loc} to be constant.

[2] See W. Nusselt, *Tech. Math. Therm.*, **1**, 417 (1930); D. M. Smith, *Engineering*, **138**, 479 (1934).

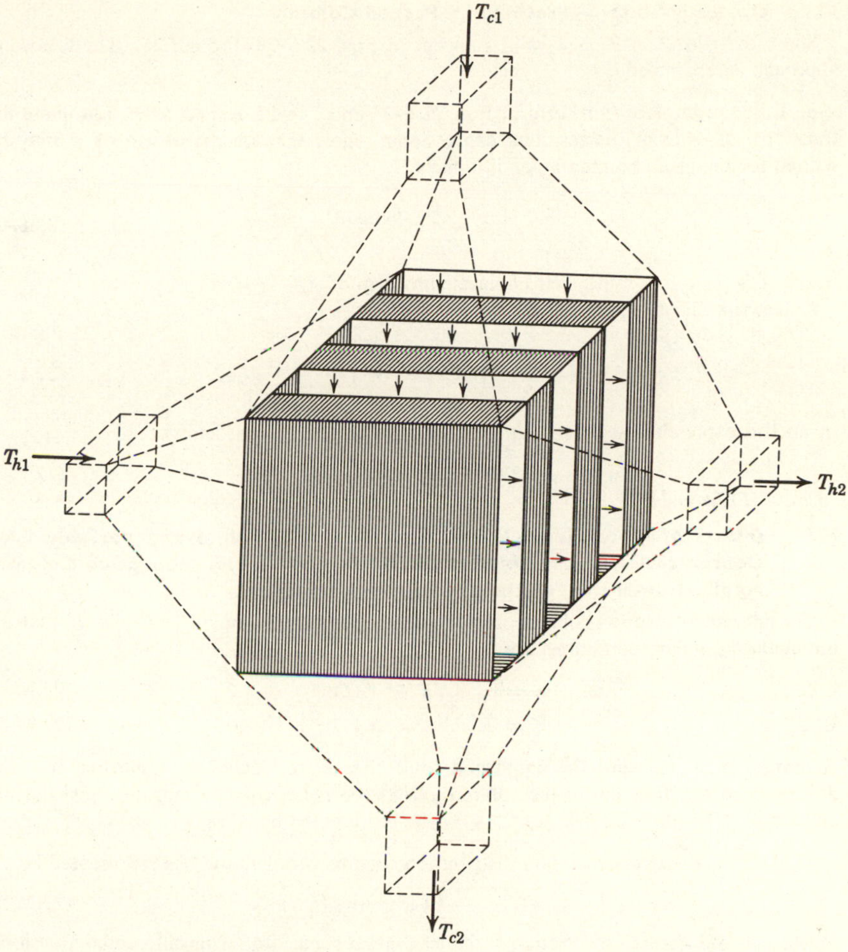

Fig. 22.J. Schematic representation of a "sandwich-type" cross-flow heat exchanger.

22.K$_4$ Imperfect Mixing in a Stirred Tank

A stirred tank has a capacity of V ft^3. Before time $t = 0$ the concentration of salt within the tank is c_i lb/ft^3. At time $t = 0$ pure water is run in at a rate of Q ft^3 min^{-1}, and brine is withdrawn from the bottom of the tank at the same volume rate. We assume negligible change in density of the fluid in the dilution process. How long will it take for the concentration to be reduced to some final value c_f?

Consider incomplete mixing, so that the average salt concentration within the tank, c, is not the same as the outlet concentration c_o. Assume that c and c_o may be related by the following simple function containing one parameter b, which is dependent on the stirrer speed and the geometry of the apparatus:

$$\frac{c_o - c}{c_i - c} = \exp -bt \tag{22.K-1}$$

22.L$_4$ Unsteady-State Operation of a Packed Column

Show that Eq. 22.6–27 is a valid solution of Eqs. 22.6–23 through 26. The following approach is recommended:

a. Take the Laplace transform of Eqs. 22.6–23 and 24 with respect to τ. Eliminate the transform of Y from the resulting expressions; show that the transform of X may be written for the given boundary conditions as

$$\bar{X} = \frac{1}{p} e^{-[p/(p+1)]\zeta} \tag{22.L–1}$$

where $\bar{X}$ is the non-p-multiplied Laplace transform of X.

b. Rewrite this expression in the form

$$\bar{X} = \frac{1}{p} - \int_0^\zeta e^{-\zeta}\left(\frac{1}{p+1}\right) e^{[\zeta/(p+1)]}\, d\zeta \tag{22.L–2}$$

Invert this expression to obtain Eq. 22.6–27 with the aid of the identity

$$\mathscr{L}\{e^{a\tau}F(\tau)\} = \bar{F}(p - a) \tag{22.L–3}$$

where $\mathscr{L}\{F(\tau)\} = \bar{F}(p)$.

22.M$_4$ Effects of Nonuniform Initial Distribution of Solute and Variable Feed Concentration on the Unsteady-State Behavior of an Adsorption Column: Application of the Principle of Superposition[3]

Consider an adsorption column of the type described in Example 22.6–2, except that the initial and boundary conditions are

I. C. at $\tau = 0,\quad Y = Y_0(\zeta);\quad \zeta > 0$ (22.M–1)

B. C. at $\zeta = 0,\quad X = X_0(\tau);\quad \tau > 0$ (22.M–2)

Assume that at zero time the liquid and solid phases are locally in equilibrium but that diffusivity in the direction of the column axis is zero. Let the concentration distribution developed in Example 22.6–2, Eq. 22.6–27, be represented by $X^\circ(\zeta, \tau)$.

a. Show that the concentration distribution for this problem may be represented by

$$X(\zeta, \tau) = U(\zeta, \tau) + V(\zeta, \tau) \tag{22.M–3}$$

in which U and V are concentration variables that satisfy the following differential equations, boundary conditions, and initial conditions:

$$\frac{\partial^2 U}{\partial\zeta\,\partial\tau} = -\left(\frac{\partial U}{\partial\zeta} + \frac{\partial U}{\partial\tau}\right) \tag{22.M–4}$$

I. C. at $\tau = 0,\quad U = Y_0(\zeta)$ (22.M–5)

B. C. at $\zeta = 0,\quad U = 0$ (22.M–6)

$$\frac{\partial^2 V}{\partial\zeta\,\partial\tau} = -\left(\frac{\partial V}{\partial\zeta} + \frac{\partial V}{\partial\tau}\right) \tag{22.M–7}$$

I. C. at $\tau = 0,\quad V = 0$ (22.M–8)

B. C. at $\zeta = 0,\quad V = X_0(\tau)$ (22.M–9)

[3] See, for example, H. S. Carslaw and J. C. Jaeger, *Conduction of Heat in Solids*, Oxford University Press (1959), Second Edition, pp. 30–32.

b. Show how U and V may be stated in terms of X° to obtain the general expression

$$X(\zeta, \tau) = \int_0^\tau X_0(p) \frac{\partial}{\partial \tau} X^\circ(\zeta, \tau - p)\, dp - \int_0^\zeta Y_0(\zeta - \lambda) \frac{\partial}{\partial \lambda} X^\circ(\lambda, \tau)\, d\lambda \quad (22.M\text{-}10)$$

Note that this result is not limited to the simple assumptions made in Example 22.6–2; it is correct provided only that the differential equations describing $X(\zeta, \tau)$ and $Y(\zeta, \tau)$ are linear with respect to X and Y and the interphase equilibrium distribution coefficient m is a constant.

c. Use the result of part (*b*) to describe the effluent composition from a column of length L for the following conditions, which are commonly encountered in practice:

I. C.	at	$t = 0,$	$c_{As} = c_{As0},$	a constant,	$(0 < z < L)$	(22.M–11)
B. C.	at	$z = 0,$	$x_A = x_{A1},$	a constant,	$(0 < t < t_0)$	(22.M–12)
			$x_A = 0,$		$(t > t_0)$	(22.M–13)

Express your answer in a form analogous to Eq. 22.6–27. What is the physical significance of this result?

22.N$_4$ Transient Behavior of *N* Chemical Reactors in Series[4]

Consider N identical chemical reactors of volume V connected in series, each equipped with a perfect stirrer. Initially, each tank is filled with pure solvent S. At zero time a solution of A in S is introduced to the first tank at a constant volumetric flow rate Q and at constant concentration $c_A(0)$. This solution also contains a small amount of a dissolved catalyst, introduced just prior to discharge into the first tank, which causes the following first-order reactions to take place:

$$A \underset{k'''_{1BA}}{\overset{k'''_{1AB}}{\rightleftharpoons}} B \underset{k'''_{1CB}}{\overset{k'''_{1BC}}{\rightleftharpoons}} C \quad (22.N\text{-}1)$$

These rate constants are assumed constant throughout the system. Let $h = Q/V$, the inverse of the "effective hold-up time" in each tank. It is desired to obtain an expression for $c_i(n)$, the concentration of component i in the nth tank at any time t.

22.O$_4$ Start-Up of a Chemical Reactor

Rework Example 22.6–1 by taking the Laplace transforms of Eqs. 22.6–2 and 3.

[4] This problem is taken from A. Acrivos and N. R. Amundson, *Ind. Eng. Chem.*, **47** 1533–1541 (1955).

Postface

On arriving at the end of the text, the reader should be conscious of the key role played by the *equations of change*, as developed in Chapters 3, 10, and 18. These equations provide the starting point for calculation of profiles, dimensional analysis, correlation of transfer rates, and development of the macroscopic balances. One or more of these results may be needed in any given engineering problem.

Needless to say in an introductory text no attempt can be made to go very far into special applications or special techniques. By way of conclusion we consider it appropriate to mention some of the areas that lie beyond the scope of this book.

We have confined ourselves largely to solutions of the equations of change obtainable by simple analytical procedures. It is anticipated that *numerical methods for solving transport phenomena problems* will find increasing use.

In this book we dealt mainly with the continuum aspects of the subject because this viewpoint is of more immediate use to the engineering student. It should be emphasized, however, that the *molecular theory of transport phenomena* complements the continuum theory in several ways: (*a*) the molecular theory can be used to derive the equations of change, (*b*) the molecular theory gives expressions for the transport properties in terms of intermolecular forces, and (*c*) the molecular approach is essential in understanding ultra-low-density gas phenomena.

Some material on *turbulent transport phenomena* has been included because of the practical importance of the subject, in spite of its present unsatisfactory state of development. It is hoped that this introduction will stimulate interest in the rapidly expanding literature of this field.

Certainly the *boundary-layer theory of transport phenomena* has developed well beyond the scope of this introductory text. The rapid growth of this subject was initiated by the aerodynamicists, but numerous recent applications have been made in other areas such as separations processes and applied chemical kinetics.

Of growing importance is the study of *transport phenomena in non-Newtonian flow*. Most large chemical industries are faced with non-Newtonian problems, and to date empirical procedures have been relied upon in design work. It is hoped that some of the non-Newtonian problems in this book will catalyze further interest in the fundamental approach to rheology.

Noteworthy advances have been made in the field of *transport phenomena in compressible flow*, which we have admittedly given less attention than it deserves. This field includes shock waves, sound propagation, supersonics, and aerothermochemistry. Readers interested in these subjects will find that special forms of the equations of change are generally taken as the starting point for the developments.

Some problems that involved *transport phenomena in chemically reacting systems* have been discussed. Usually, for the sake of simplification, we have taken the chemical kinetics to be of rather idealized form. Clearly, for many problems in combustion, detonation, and flame propagation, more realistic expressions for the reaction rates need to be known. Furthermore, much more needs to be known about the various physical properties of fluids containing free radicals and ions, particularly at elevated temperatures.

In this book we have restricted ourselves almost exclusively to systems in which electric and magnetic fields play no role. An important current area of interest is that of *transport phenomena in electrically conducting media*. For such problems we have to modify the equations of change to include electromagnetic forces in the equation of motion, electromagnetic energy terms in the equation of energy, and an additional equation of change for charge. In addition, we need the Maxwell equations of electromagnetism. The field formed by the union of these two subjects is called magnetohydrodynamics.

Numerous examples and problems that illustrate the *application of transport phenomena to engineering* have been given in the text. Most are relatively straightforward examples involving idealized situations. It is hoped that the coming years will see increased application of the principles of transport phenomena to more challenging problems.

R. B. B.
W. E. S.
E. N. L.

Summary of Vector and Tensor Notation

The physical quantities encountered in the theory of transport phenomena can be placed in the following categories: *scalars* such as temperature, energy, volume, and time; *vectors*, such as velocity, momentum, acceleration, and force; and second-order *tensors*, such as the shear-stress or momentum-flux tensor. We distinguish between these quantities by the following notation:

$$s = \text{scalar (lightface italic)}$$
$$\boldsymbol{v} = \text{vector (boldface italic)}$$
$$\boldsymbol{\tau} = \text{tensor (boldface Greek)}$$

For vectors and tensors, several different kinds of multiplication are possible. These operations are indicated by the use of three kinds of special multiplication signs to be defined later: the "single dot" $\cdot$, the "double dot" $:$, and the "cross" $\times$. The parentheses enclosing these special multiplications indicate the type of quantity produced by the multiplication:

$$(\quad) = \text{scalar}$$
$$[\quad] = \text{vector}$$
$$\{\quad\} = \text{tensor}$$

No special significance is attached to the kind of parentheses if the operation enclosed is addition or subtraction. Hence $(\boldsymbol{v} \cdot \boldsymbol{w})$ and $(\boldsymbol{\sigma} : \boldsymbol{\tau})$ are scalars,

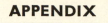

$[\boldsymbol{v} \times \boldsymbol{w}]$ and $[\boldsymbol{\tau} \cdot \boldsymbol{v}]$ are vectors, and $\{\boldsymbol{\sigma} \cdot \boldsymbol{\tau}\}$ is a tensor. On the other hand, $(\boldsymbol{v} - \boldsymbol{w})$ may be written $[\boldsymbol{v} - \boldsymbol{w}]$ or $\{\boldsymbol{v} - \boldsymbol{w}\}$ when convenient to do so. Actually, scalars can be regarded as zero-order tensors, vectors as first-order tensors. The multiplication signs may be interpreted thus:[1]

Multiplication Sign	Order of Result
None	Σ
$\times$	$\Sigma - 1$
$\cdot$	$\Sigma - 2$
$:$	$\Sigma - 4$

in which Σ represents the sum of the orders of the quantities being multiplied. For example $s\boldsymbol{\tau}$ is of the order $0 + 2 = 2$, $\boldsymbol{vw}$ is of the order $1 + 1 = 2$, $[\boldsymbol{v} \times \boldsymbol{w}]$ is of the order $1 + 1 - 1 = 1$, $(\boldsymbol{\sigma} : \boldsymbol{\tau})$ is of the order $2 + 2 - 4 = 0$, and $\{\boldsymbol{\sigma} \cdot \boldsymbol{\tau}\}$ is $2 + 2 - 2 = 2$.

The basic operations that can be performed on scalar quantities need not be elaborated on here. The laws for the algebra of scalars may be used to illustrate three terms that arise in the subsequent discussion of vector operations:

a. For the multiplication of two scalars, r and s, the order of multiplication is immaterial so that the *commutative* law is valid: $rs = sr$.

b. For the successive multiplication of three scalars q, r, and s, the order in which the multiplications are performed is immaterial, so that the *associative* law is valid: $(qr)s = q(rs)$.

c. For the multiplication of a scalar s by the sum of scalars p, q, and r, it is immaterial whether the addition or multiplication is performed first, so that the *distributive* law is valid: $s(p + q + r) = sp + sq + sr$.

These laws are not generally valid for the analogous vector and tensor operations described in the paragraphs to follow.

§A.1 VECTOR OPERATIONS FROM A GEOMETRICAL VIEWPOINT

In elementary physics courses one is introduced to vectors from a geometrical standpoint. In this section we extend this approach to include the operations of vector multiplication. In §A.2 a parallel analytic treatment is given.

Definition of a Vector and Its Magnitude

A vector $\boldsymbol{v}$ is defined as a quantity of a given magnitude and direction. The magnitude of the vector is designated by $|\boldsymbol{v}|$ or simply by the corresponding lightface symbol v. Two vectors $\boldsymbol{v}$ and $\boldsymbol{w}$ are equal when their

[1] This interpretation is consistent with the definitions in P. M. Morse and H. Feshbach, *Methods of Theoretical Physics*, McGraw-Hill, New York (1953), Chapter 1. In that book several additional operations, including some involving fourth-order tensors, are defined. Readers will be fascinated by the stereoscopic drawings that are used there.

magnitudes are equal and when they point in the same direction; they do not have to be collinear or have the same point of origin. If $\boldsymbol{v}$ and $\boldsymbol{w}$ have the same magnitude but point in opposite directions, then $\boldsymbol{v} = -\boldsymbol{w}$.

Addition and Subtraction of Vectors

The addition of two vectors can be accomplished by the familiar parallelo-

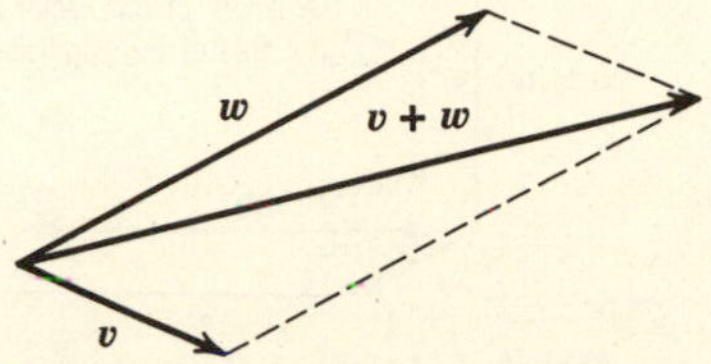

Fig. A.1–1. Parallelogram construction for adding two vectors.

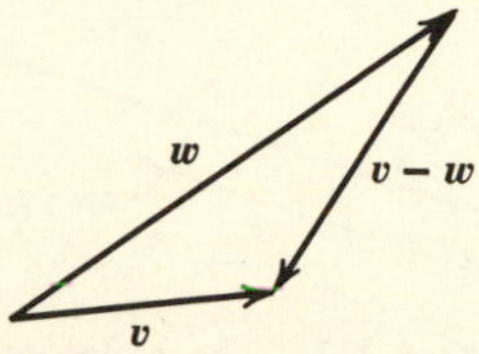

Fig. A.1–2. Construction for subtracting two vectors.

gram construction, as indicated by Fig. A.1–1. Vector addition obeys the following laws:

$$(\boldsymbol{v} + \boldsymbol{w}) = (\boldsymbol{w} + \boldsymbol{v}) \quad \text{(commutative)} \tag{A.1–1}$$

$$(\boldsymbol{v} + \boldsymbol{w}) + \boldsymbol{u} = \boldsymbol{v} + (\boldsymbol{w} + \boldsymbol{u}) \quad \text{(associative)} \tag{A.1–2}$$

Vector subtraction is performed by reversing the sign of one vector and adding; thus $\boldsymbol{v} - \boldsymbol{w} = \boldsymbol{v} + (-\boldsymbol{w})$. The geometrical construction for this is shown in Fig. A.1–2.

Multiplication of a Vector by a Scalar

When a vector is multiplied by a scalar, the magnitude of the vector is altered but its direction is not. The following laws are applicable:

$$s\boldsymbol{v} = \boldsymbol{v}s \quad \text{(commutative)} \tag{A.1–3}$$

$$r(s\boldsymbol{v}) = (rs)\boldsymbol{v} \quad \text{(associative)} \tag{A.1–4}$$

$$(q + r + s)\boldsymbol{v} = q\boldsymbol{v} + r\boldsymbol{v} + s\boldsymbol{v} \quad \text{(distributive)} \tag{A.1–5}$$

Scalar Product (or Dot Product) of Two Vectors

The scalar product of two vectors $\boldsymbol{v}$ and $\boldsymbol{w}$ is a scalar quantity defined by

$$(\boldsymbol{v} \cdot \boldsymbol{w}) = vw \cos \phi_{vw} \tag{A.1–6}$$

in which ϕ_{vw} is the angle (less than 180°) between the vectors $\boldsymbol{v}$ and $\boldsymbol{w}$. The scalar product is then the magnitude of $\boldsymbol{w}$ multiplied by the projection of $\boldsymbol{v}$ on $\boldsymbol{w}$, or vice versa. (See Fig. A.1–3.) Note that the scalar product of a vector with itself is just the square of the magnitude of the vector:

$$(\boldsymbol{v} \cdot \boldsymbol{v}) = |\boldsymbol{v}|^2 = v^2 \tag{A.1–7}$$

The rules governing scalar products are as follows:

(commutative) $$(\boldsymbol{u} \cdot \boldsymbol{v}) = (\boldsymbol{v} \cdot \boldsymbol{u}) \tag{A.1–8}$$

(not associative) $$(\boldsymbol{u} \cdot \boldsymbol{v})\boldsymbol{w} \neq \boldsymbol{u}(\boldsymbol{v} \cdot \boldsymbol{w}) \tag{A.1–9}$$

(distributive) $$(\boldsymbol{u} \cdot \{\boldsymbol{v} + \boldsymbol{w}\}) = (\boldsymbol{u} \cdot \boldsymbol{v}) + (\boldsymbol{u} \cdot \boldsymbol{w}) \tag{A.1–10}$$

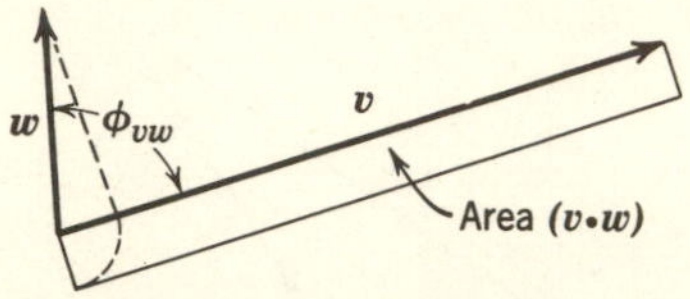

Fig. A.1–3. The scalar product of two vectors.

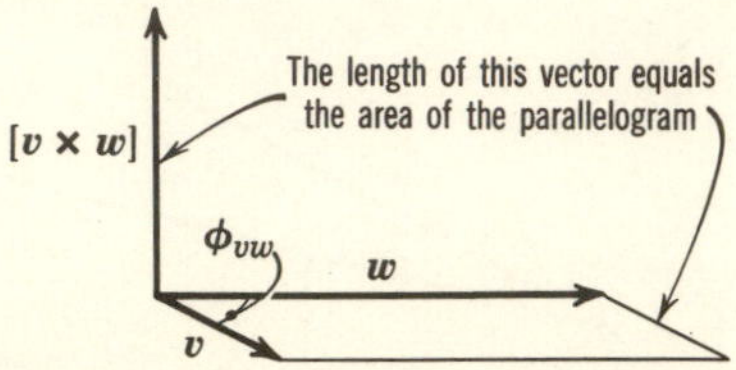

Fig. A.1–4. The vector product of two vectors.

Vector Product (or Cross Product) of Two Vectors

The vector product of two vectors $\boldsymbol{v}$ and $\boldsymbol{w}$ is a vector defined by

$$[\boldsymbol{v} \times \boldsymbol{w}] = \{vw \sin \phi_{vw}\}\boldsymbol{n}_{vw} \tag{A.1–11}$$

in which $\boldsymbol{n}_{vw}$ is a vector of unit length (a "unit vector") normal to the plane containing $\boldsymbol{v}$ and $\boldsymbol{w}$ and pointing in the direction that a right-handed screw will move if turned from $\boldsymbol{v}$ toward $\boldsymbol{w}$ by the shortest route. The vector product is illustrated in Fig. A.1–4. The magnitude of the vector product is just the area of the parallelogram defined by the vectors $\boldsymbol{v}$ and $\boldsymbol{w}$. It follows from the definition of the vector product that

$$[\boldsymbol{v} \times \boldsymbol{v}] = 0 \tag{A.1–12}$$

Note the following summary of laws governing the vector product operation:

(not commutative) $$[\boldsymbol{v} \times \boldsymbol{w}] = -[\boldsymbol{w} \times \boldsymbol{v}] \tag{A.1–13}$$

(not associative) $$[\boldsymbol{u} \times [\boldsymbol{v} \times \boldsymbol{w}]] \neq [[\boldsymbol{u} \times \boldsymbol{v}] \times \boldsymbol{w}] \tag{A.1–14}$$

(distributive) $$[\{\boldsymbol{u} + \boldsymbol{v}\} \times \boldsymbol{w}] = [\boldsymbol{u} \times \boldsymbol{w}] + [\boldsymbol{v} \times \boldsymbol{w}] \tag{A.1–15}$$

Multiple Products of Vectors

Somewhat more complicated are multiple products formed by combinations of the multiplication processes just described:

(*a*) $rs\boldsymbol{v}$ (*b*) $s(\boldsymbol{v} \cdot \boldsymbol{w})$ (*c*) $s[\boldsymbol{v} \times \boldsymbol{w}]$
(*d*) $(\boldsymbol{u} \cdot [\boldsymbol{v} \times \boldsymbol{w}])$ (*e*) $[\boldsymbol{u} \times [\boldsymbol{v} \times \boldsymbol{w}]]$ (*f*) $([\boldsymbol{u} \times \boldsymbol{v}] \cdot [\boldsymbol{w} \times \boldsymbol{z}])$
(g) $[[\boldsymbol{u} \times \boldsymbol{v}] \times [\boldsymbol{w} \times \boldsymbol{z}]]$

The geometrical interpretations of the first three of these are straightforward. The quantity $(\boldsymbol{u} \cdot [\boldsymbol{v} \times \boldsymbol{w}])$ can easily be shown to represent the volume of a parallelepiped defined by the vectors $\boldsymbol{u}$, $\boldsymbol{v}$, and $\boldsymbol{w}$.

§A.2 VECTOR OPERATIONS FROM AN ANALYTICAL VIEWPOINT

In this section a parallel analytical treatment is given to each of the topics presented geometrically in §A.1. In the discussion here we restrict ourselves to rectangular coordinates and label the axes as 1, 2, 3 corresponding to the usual notation of x, y, z.

Many formulas can be expressed more compactly in terms of the *Kronecker delta* δ_{ij} and the *alternating unit tensor* ϵ_{ijk}. These quantities are defined thus:

$$\begin{cases} \delta_{ij} = +1 & \text{if } i = j \\ \delta_{ij} = \;\;\, 0 & \text{if } i \neq j \end{cases} \qquad \begin{matrix} \text{(A.2–1)} \\ \text{(A.2–2)} \end{matrix}$$

$$\begin{cases} \epsilon_{ijk} = +1 & \text{if } ijk = 123, 231, \text{ or } 312 \\ \epsilon_{ijk} = -1 & \text{if } ijk = 321, 132, \text{ or } 213 \\ \epsilon_{ijk} = \;\;\, 0 & \text{if any two indices are alike} \end{cases} \qquad \begin{matrix} \text{(A.2–3)} \\ \text{(A.2–4)} \\ \text{(A.2–5)} \end{matrix}$$

Several relations involving these quantities are useful in proving some vector identities:

$$\Sigma_j \Sigma_k \epsilon_{ijk} \epsilon_{hjk} = 2\delta_{ih} \qquad \text{(A.2–6)}$$

$$\Sigma_k \epsilon_{ijk} \epsilon_{mnk} = \delta_{im}\delta_{jn} - \delta_{in}\delta_{jm} \qquad \text{(A.2–7)}$$

Note that a three-by-three determinant may be written in terms of the ϵ_{ijk}:

$$\begin{vmatrix} a_{11} & a_{12} & a_{13} \\ a_{21} & a_{22} & a_{23} \\ a_{31} & a_{32} & a_{33} \end{vmatrix} = \Sigma_i \Sigma_j \Sigma_k \epsilon_{ijk} a_{1i} a_{2j} a_{3k} \qquad \text{(A.2–8)}$$

The quantity ϵ_{ijk} thus selects the necessary terms that appear in the determinant and affixes the proper sign to each term.

Definition of a Vector and Its Magnitude: The Unit Vectors

A vector $\boldsymbol{v}$ can be completely described by giving the magnitudes of its projections v_1, v_2, v_3 on the coordinate axes 1, 2, 3. (See Fig. A.2–1.) Hence a vector may be represented analytically:

$$\boldsymbol{v} = \boldsymbol{\delta}_1 v_1 + \boldsymbol{\delta}_2 v_2 + \boldsymbol{\delta}_3 v_3 = \sum_{i=1}^{3} \boldsymbol{\delta}_i v_i \qquad \text{(A.2–9)}$$

Here $\boldsymbol{\delta}_1$, $\boldsymbol{\delta}_2$, $\boldsymbol{\delta}_3$ are the "unit vectors" (that is, vectors of unit length) in the direction of the 1, 2, 3 axes.[1,2] Throughout the discussion we restrict ourselves to right-handed coordinate axes.

[1] In most elementary texts the unit vectors are called $\boldsymbol{i}, \boldsymbol{j}, \boldsymbol{k}$. Note that we have violated our rule of notation here in using a boldface Greek symbol for a vector. We do this because the components of these vectors are given by the Kronecker delta. That is, the component of $\boldsymbol{\delta}_1$ in the 1-direction is δ_{11} or unity; the component of $\boldsymbol{\delta}_1$ in the 2-direction is δ_{12} or zero.

[2] We do not discuss *covariant* and *contravariant* vectors here; for an excellent introduction to the subject, see Morse and Feshbach, *op. cit.*, pp. 30–31. For more general theory, see A. J. McConnell, *Applications of Tensor Analysis*, Dover, New York (1957).

The magnitude of a vector is given by

$$|v| = v = \sqrt{v_1^2 + v_2^2 + v_3^2} = \sqrt{\Sigma_i v_i^2} \tag{A.2–10}$$

Two vectors $\boldsymbol{v}$ and $\boldsymbol{w}$ are equal if $v_1 = w_1$, $v_2 = w_2$, and $v_3 = w_3$. Also, $\boldsymbol{v} = -\boldsymbol{w}$, if $v_1 = -w_1$, etc.

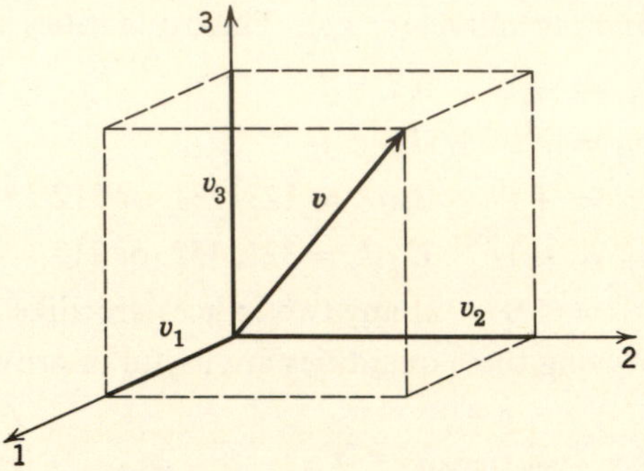

Fig. A.2–1. The projections of a vector on the coordinate axes 1, 2, and 3.

The unit vectors just introduced have the following properties, which are used in subsequent paragraphs:

$$\begin{cases} (\boldsymbol{\delta}_1 \cdot \boldsymbol{\delta}_1) = (\boldsymbol{\delta}_2 \cdot \boldsymbol{\delta}_2) = (\boldsymbol{\delta}_3 \cdot \boldsymbol{\delta}_3) = 1 & \text{(A.2–11)} \\ (\boldsymbol{\delta}_1 \cdot \boldsymbol{\delta}_2) = (\boldsymbol{\delta}_2 \cdot \boldsymbol{\delta}_3) = (\boldsymbol{\delta}_3 \cdot \boldsymbol{\delta}_1) = 0 & \text{(A.2–12)} \end{cases}$$

$$\begin{cases} [\boldsymbol{\delta}_1 \times \boldsymbol{\delta}_1] = [\boldsymbol{\delta}_2 \times \boldsymbol{\delta}_2] = [\boldsymbol{\delta}_3 \times \boldsymbol{\delta}_3] = 0 & \text{(A.2–13)} \\ [\boldsymbol{\delta}_1 \times \boldsymbol{\delta}_2] = \boldsymbol{\delta}_3; \quad [\boldsymbol{\delta}_2 \times \boldsymbol{\delta}_3] = \boldsymbol{\delta}_1; \quad [\boldsymbol{\delta}_3 \times \boldsymbol{\delta}_1] = \boldsymbol{\delta}_2 & \text{(A.2–14)} \\ [\boldsymbol{\delta}_2 \times \boldsymbol{\delta}_1] = -\boldsymbol{\delta}_3; \quad [\boldsymbol{\delta}_3 \times \boldsymbol{\delta}_2] = -\boldsymbol{\delta}_1; \quad [\boldsymbol{\delta}_1 \times \boldsymbol{\delta}_3] = -\boldsymbol{\delta}_2 & \text{(A.2–15)} \end{cases}$$

These relations follow from the geometrical definitions of dot and cross products given in §A.1. They may be summarized thus:

$$\boxed{(\boldsymbol{\delta}_i \cdot \boldsymbol{\delta}_j) = \delta_{ij}} \tag{A.2–16}$$

$$\boxed{[\boldsymbol{\delta}_i \times \boldsymbol{\delta}_j] = \sum_{k=1}^{3} \epsilon_{ijk} \boldsymbol{\delta}_k} \tag{A.2–17}$$

in which δ_{ij} is the Kronecker delta and ϵ_{ijk} is the alternating unit tensor defined in the introduction to this section. These two relations enable us to develop analytic expressions for all the common "dot" and "cross" operations. In the remainder of this section, and in the next section, all we do is to break all vectors up into components according to Eq. A.2–9 and then apply Eqs. A.2–16 and 17.

Addition and Subtraction of Vectors

The addition or subtraction of vectors $\boldsymbol{v}$ and $\boldsymbol{w}$ may be written in terms of components:

$$\boldsymbol{v} + \boldsymbol{w} = \Sigma_i \boldsymbol{\delta}_i v_i + \Sigma_i \boldsymbol{\delta}_i w_i = \Sigma_i \boldsymbol{\delta}_i (v_i + w_i) \tag{A.2–18}$$

Geometrically, this corresponds to adding up the projections of $\boldsymbol{v}$ and $\boldsymbol{w}$ on each individual axis and then constructing a vector with these new components. Three or more vectors may be added in exactly the same fashion.

Multiplication of a Vector by a Scalar

Multiplication of a vector by a scalar corresponds to multiplying each component of the vector by the scalar:

$$s\boldsymbol{v} = s\{\Sigma_i \boldsymbol{\delta}_i v_i\} = \Sigma_i \boldsymbol{\delta}_i \{s v_i\} \tag{A.2–19}$$

Scalar Product (or Dot Product) of Two Vectors

The scalar product of two vectors $\boldsymbol{v}$ and $\boldsymbol{w}$ is obtained by writing each vector in terms of components according to Eq. A.2–9 and then performing the scalar-product operations on the unit vectors, using Eq. A.2–16:

$$\begin{aligned}(\boldsymbol{v} \cdot \boldsymbol{w}) &= (\{\Sigma_i \boldsymbol{\delta}_i v_i\} \cdot \{\Sigma_j \boldsymbol{\delta}_j w_j\}) \\ &= \Sigma_i \Sigma_j (\boldsymbol{\delta}_i \cdot \boldsymbol{\delta}_j) v_i w_j \\ &= \Sigma_i \Sigma_j \delta_{ij} v_i w_j \\ &= \Sigma_i v_i w_i \end{aligned} \tag{A.2–20}$$

Hence the scalar product of two vectors is obtained by summing the products of the corresponding components of the two vectors.

Vector Product (or Cross Product) of Two Vectors

The vector product of two vectors $\boldsymbol{v}$ and $\boldsymbol{w}$ may be worked out by using Eqs. A.2–9 and 17:

$$\begin{aligned}[\boldsymbol{v} \times \boldsymbol{w}] &= [\{\Sigma_j \boldsymbol{\delta}_j v_j\} \times \{\Sigma_k \boldsymbol{\delta}_k w_k\}] \\ &= \Sigma_j \Sigma_k [\boldsymbol{\delta}_j \times \boldsymbol{\delta}_k] v_j w_k \\ &= \Sigma_i \Sigma_j \Sigma_k \epsilon_{ijk} \boldsymbol{\delta}_i v_j w_k \\ &= \begin{vmatrix} \boldsymbol{\delta}_1 & \boldsymbol{\delta}_2 & \boldsymbol{\delta}_3 \\ v_1 & v_2 & v_3 \\ w_1 & w_2 & w_3 \end{vmatrix}\end{aligned} \tag{A.2–21}$$

Here we have made use of Eq. A.2–8. Note that the ith component of $[\boldsymbol{v} \times \boldsymbol{w}]$ is $\Sigma_j \Sigma_k \epsilon_{ijk} v_j w_k$.

Multiple Vector Products

Expressions for the multiple products mentioned on p. 718 can be obtained by using the foregoing analytical expressions for the scalar and vector products. For example, the triple product $(\boldsymbol{u} \cdot [\boldsymbol{v} \times \boldsymbol{w}])$ may be written

$$(\boldsymbol{u} \cdot [\boldsymbol{v} \times \boldsymbol{w}]) = \Sigma_i u_i [\boldsymbol{v} \times \boldsymbol{w}]_i$$
$$= \Sigma_i \Sigma_j \Sigma_k \epsilon_{ijk} u_i v_j w_k \tag{A.2–22}$$

Then, from Eq. A.2–8, we obtain

$$(\boldsymbol{u} \cdot [\boldsymbol{v} \times \boldsymbol{w}]) = \begin{vmatrix} u_1 & u_2 & u_3 \\ v_1 & v_2 & v_3 \\ w_1 & w_2 & w_3 \end{vmatrix} \tag{A.2–23}$$

This is an alternative expression for the volume of a parallelepiped. Furthermore, the vanishing of the determinant is a necessary and sufficient condition that the vectors $\boldsymbol{u}$, $\boldsymbol{v}$, and $\boldsymbol{w}$ be coplanar.

Example A.2–1. Proof of a Vector Identity

The analytical expressions for dot and cross products may be used to prove vector identities; for example, let it be desired to verify the relation

$$[\boldsymbol{u} \times [\boldsymbol{v} \times \boldsymbol{w}]] = \boldsymbol{v}(\boldsymbol{u} \cdot \boldsymbol{w}) - \boldsymbol{w}(\boldsymbol{u} \cdot \boldsymbol{v}) \tag{A.2–24}$$

Solution. Let us rewrite the *i*-component of the expression on the left side in expanded form:

$$[\boldsymbol{u} \times [\boldsymbol{v} \times \boldsymbol{w}]]_i = \Sigma_j \Sigma_k \epsilon_{ijk} u_j [\boldsymbol{v} \times \boldsymbol{w}]_k$$
$$= \Sigma_j \Sigma_k \epsilon_{ijk} u_j \{\Sigma_l \Sigma_m \epsilon_{klm} v_l w_m\}$$
$$= \Sigma_j \Sigma_k \Sigma_l \Sigma_m \epsilon_{ijk} \epsilon_{klm} u_j v_l w_m$$
$$= \Sigma_j \Sigma_k \Sigma_l \Sigma_m \epsilon_{ijk} \epsilon_{lmk} u_j v_l w_m \tag{A.2–25}$$

Use may now be made of Eq. A.2–7 to complete the proof:

$$[\boldsymbol{u} \times [\boldsymbol{v} \times \boldsymbol{w}]]_i = \Sigma_j \Sigma_l \Sigma_m (\delta_{il}\delta_{jm} - \delta_{im}\delta_{jl}) u_j v_l w_m$$
$$= v_i \Sigma_j \Sigma_m \delta_{jm} u_j w_m - w_i \Sigma_j \Sigma_l \delta_{jl} u_j v_l$$
$$= v_i \Sigma_j u_j w_j - w_i \Sigma_j u_j v_j$$
$$= v_i(\boldsymbol{u} \cdot \boldsymbol{w}) - w_i(\boldsymbol{u} \cdot \boldsymbol{v}) \tag{A.2–26}$$

which is just the *i*-component of the right side of Eq. A.2–24. In a similar way one may verify such identities as

$$(\boldsymbol{u} \cdot [\boldsymbol{v} \times \boldsymbol{w}]) = (\boldsymbol{v} \cdot [\boldsymbol{w} \times \boldsymbol{u}]) \tag{A.2–27}$$

$$([\boldsymbol{u} \times \boldsymbol{v}] \cdot [\boldsymbol{w} \times \boldsymbol{z}]) = (\boldsymbol{u} \cdot \boldsymbol{w})(\boldsymbol{v} \cdot \boldsymbol{z}) - (\boldsymbol{u} \cdot \boldsymbol{z})(\boldsymbol{v} \cdot \boldsymbol{w}) \tag{A.2–28}$$

$$[[\boldsymbol{u} \times \boldsymbol{v}] \times [\boldsymbol{w} \times \boldsymbol{z}]] = ([\boldsymbol{u} \times \boldsymbol{v}] \cdot \boldsymbol{z})\boldsymbol{w} - ([\boldsymbol{u} \times \boldsymbol{v}] \cdot \boldsymbol{w})\boldsymbol{z} \tag{A.2–29}$$

§A.3 THE VECTOR DIFFERENTIAL OPERATIONS

The vector differential operator $\boldsymbol{\nabla}$, known as "nabla" or "del," is defined in rectangular coordinates as

$$\begin{aligned}\boldsymbol{\nabla} &= \boldsymbol{\delta}_1 \frac{\partial}{\partial x_1} + \boldsymbol{\delta}_2 \frac{\partial}{\partial x_2} + \boldsymbol{\delta}_3 \frac{\partial}{\partial x_3} \\ &= \Sigma_i \boldsymbol{\delta}_i \frac{\partial}{\partial x_i} \end{aligned} \tag{A.3-1}$$

in which the $\boldsymbol{\delta}_i$ are the unit vectors and the x_i are the variables associated with the 1, 2, 3 axes (that is, the x_i are the position coordinates normally referred to as x, y, z). The symbol $\boldsymbol{\nabla}$ is a vector-operator—it has components like a vector and it cannot stand alone but must operate on a scalar, vector, or tensor function. In this section we summarize the various uses of $\boldsymbol{\nabla}$ in its operations on scalars and vectors. Just as in §A.2, we break the vectors up into components and use Eqs. A.2–16 and 17.

The Gradient of a Scalar Field

If s is a scalar function of the variables x_1, x_2, x_3, then the operation of $\boldsymbol{\nabla}$ on s is

$$\begin{aligned}\boldsymbol{\nabla} s &= \boldsymbol{\delta}_1 \frac{\partial s}{\partial x_1} + \boldsymbol{\delta}_2 \frac{\partial s}{\partial x_2} + \boldsymbol{\delta}_3 \frac{\partial s}{\partial x_3} \\ &= \Sigma_i \boldsymbol{\delta}_i \frac{\partial s}{\partial x_i} \end{aligned} \tag{A.3-2}$$

The vector thus constructed from the derivatives of s is designated by $\boldsymbol{\nabla} s$ (or grad s) and is called the *gradient* of the scalar field s. The following properties of the gradient operation should be noted:

(not commutative) $$\boldsymbol{\nabla} s \neq s\boldsymbol{\nabla} \tag{A.3-3}$$

(not associative) $$(\boldsymbol{\nabla} r)s \neq \boldsymbol{\nabla}(rs) \tag{A.3-4}$$

(distributive) $$\boldsymbol{\nabla}(r + s) = \boldsymbol{\nabla} r + \boldsymbol{\nabla} s \tag{A.3-5}$$

The Divergence of a Vector Field

If the vector $\boldsymbol{v}$ is a function of the space variables x_1, x_2, x_3, then a scalar product may be formed with the operator $\boldsymbol{\nabla}$; in obtaining the final form, we use Eq. A.2–16:

$$\begin{aligned}(\boldsymbol{\nabla} \cdot \boldsymbol{v}) &= \left(\left\{\Sigma_i \boldsymbol{\delta}_i \frac{\partial}{\partial x_i}\right\} \cdot \{\Sigma_j \boldsymbol{\delta}_j v_j\}\right) \\ &= \Sigma_i \Sigma_j (\boldsymbol{\delta}_i \cdot \boldsymbol{\delta}_j) \frac{\partial}{\partial x_i} v_j \\ &= \Sigma_i \Sigma_j \delta_{ij} \frac{\partial}{\partial x_i} v_j = \Sigma_i \frac{\partial v_i}{\partial x_i} \end{aligned} \tag{A.3-6}$$

This collection of derivatives of the components of the vector $\boldsymbol{v}$ is called the *divergence* of $\boldsymbol{v}$ (sometimes abbreviated div $\boldsymbol{v}$). Some properties of the divergence operator should be noted:

(not commutative) $$(\boldsymbol{\nabla} \cdot \boldsymbol{v}) \neq (\boldsymbol{v} \cdot \boldsymbol{\nabla}) \tag{A.3–7}$$

(not associative) $$(\boldsymbol{\nabla} \cdot s\boldsymbol{v}) \neq (\boldsymbol{\nabla} s \cdot \boldsymbol{v}) \tag{A.3–8}$$

(distributive) $$(\boldsymbol{\nabla} \cdot \{\boldsymbol{v} + \boldsymbol{w}\}) = (\boldsymbol{\nabla} \cdot \boldsymbol{v}) + (\boldsymbol{\nabla} \cdot \boldsymbol{w}) \tag{A.3–9}$$

The Curl of a Vector Field

A cross product may also be formed between the $\boldsymbol{\nabla}$ operator and the vector $\boldsymbol{v}$, which is a function of the three space variables. This cross product may be simplified by using Eq. A.2–17 and written in a variety of forms:

$$\begin{aligned}
[\boldsymbol{\nabla} \times \boldsymbol{v}] &= \left[\left\{\Sigma_j \boldsymbol{\delta}_j \frac{\partial}{\partial x_j}\right\} \times \{\Sigma_k \boldsymbol{\delta}_k v_k\}\right] \\
&= \Sigma_j \Sigma_k [\boldsymbol{\delta}_j \times \boldsymbol{\delta}_k] \frac{\partial}{\partial x_j} v_k \\
&= \begin{vmatrix} \boldsymbol{\delta}_1 & \boldsymbol{\delta}_2 & \boldsymbol{\delta}_3 \\ \dfrac{\partial}{\partial x_1} & \dfrac{\partial}{\partial x_2} & \dfrac{\partial}{\partial x_3} \\ v_1 & v_2 & v_3 \end{vmatrix} \\
&= \boldsymbol{\delta}_1 \left\{\frac{\partial v_3}{\partial x_2} - \frac{\partial v_2}{\partial x_3}\right\} + \boldsymbol{\delta}_2 \left\{\frac{\partial v_1}{\partial x_3} - \frac{\partial v_3}{\partial x_1}\right\} + \boldsymbol{\delta}_3 \left\{\frac{\partial v_2}{\partial x_1} - \frac{\partial v_1}{\partial x_2}\right\}
\end{aligned} \tag{A.3–10}$$

The vector thus constructed is called the *curl* of $\boldsymbol{v}$. Alternate notations for $[\boldsymbol{\nabla} \times \boldsymbol{v}]$ are curl $\boldsymbol{v}$ and rot $\boldsymbol{v}$, the latter being common in the German literature. The curl operation, like the divergence, is distributive but not commutative or associative.

The Laplacian of a Scalar Field

If we take the divergence of the gradient of the scalar function s, we obtain

$$\begin{aligned}
(\boldsymbol{\nabla} \cdot \boldsymbol{\nabla} s) &= \left(\left\{\Sigma_i \boldsymbol{\delta}_i \frac{\partial}{\partial x_i}\right\} \cdot \left\{\Sigma_j \boldsymbol{\delta}_j \frac{\partial s}{\partial x_j}\right\}\right) \\
&= \Sigma_i \Sigma_j \delta_{ij} \frac{\partial}{\partial x_i} \frac{\partial s}{\partial x_j} \\
&= \left\{\Sigma_i \frac{\partial^2}{\partial x_i^2} s\right\}
\end{aligned} \tag{A.3–11}$$

The collection of differential operators which is operating on s in the last line is given the symbol $\boldsymbol{\nabla}^2$; hence in rectangular coordinates

$$\boldsymbol{\nabla}^2 = \frac{\partial^2}{\partial x_1^{\,2}} + \frac{\partial^2}{\partial x_2^{\,2}} + \frac{\partial^2}{\partial x_3^{\,2}} \tag{A.3–12}$$

This is called the *Laplacian* operator. (Some authors use the symbol $\boldsymbol{\Delta}$ for the Laplacian operator—this is particularly true in the older German literature.) The Laplacian operator has only the distributive property, just as the gradient, divergence, and curl.

The Laplacian of a Vector Field

In rectangular coordinates the operator $\boldsymbol{\nabla}^2$ may be applied to a vector field $\boldsymbol{v}$ by taking the derivatives with respect to the scalar components and adding vectorially:

$$\begin{aligned}\boldsymbol{\nabla}^2\boldsymbol{v} &= \boldsymbol{\nabla}^2\{\Sigma_i \boldsymbol{\delta}_i v_i\} \\ &= \boldsymbol{\delta}_1\boldsymbol{\nabla}^2 v_1 + \boldsymbol{\delta}_2\boldsymbol{\nabla}^2 v_2 + \boldsymbol{\delta}_3\boldsymbol{\nabla}^2 v_3\end{aligned} \tag{A.3–13}$$

Although this is correct in rectangular coordinates, it is not applicable in curvilinear coordinates.[1] It is preferable to *define* the Laplacian of a vector field thus:

$$\boldsymbol{\nabla}^2\boldsymbol{v} = \boldsymbol{\nabla}(\boldsymbol{\nabla}\cdot\boldsymbol{v}) - [\boldsymbol{\nabla}\times[\boldsymbol{\nabla}\times\boldsymbol{v}]] \tag{A.3–14}$$

In this way, the operation is reduced to a sequence of operations involving the gradient, divergence, and curl. The definition in Eq. A.3–14 gives the expression in Eq. A.3–13 for rectangular coordinates and, in addition, is easy to use to obtain $\boldsymbol{\nabla}^2\boldsymbol{v}$ in curvilinear coordinates. (See §A.7 for detailed expressions.)

The Substantial Derivative of a Scalar Field

At the beginning of Chapter 3 we introduced the *substantial derivative* operator:

$$\frac{D}{Dt} = \frac{\partial}{\partial t} + (\boldsymbol{v}\cdot\boldsymbol{\nabla}) \tag{A.3–15}$$

in which $\boldsymbol{v}$ is the local fluid velocity (or "mass average velocity" in fluid mixtures). When this operator operates on a scalar function s, we get

$$\begin{aligned}\frac{Ds}{Dt} &= \frac{\partial s}{\partial t} + (\boldsymbol{v}\cdot\boldsymbol{\nabla}s) \\ &= \frac{\partial s}{\partial t} + \Sigma_i v_i \frac{\partial s}{\partial x_i}\end{aligned} \tag{A.3–16}$$

[1] For an extensive discussion, see P. Moon and D. E. Spencer, *J. Franklin Inst.*, **256**, 551–558 (1953). An alternate formulation is to write $\boldsymbol{\nabla}^2\boldsymbol{v}$ as $[\boldsymbol{\nabla}\cdot\boldsymbol{\nabla}\boldsymbol{v}]$, that is, as the dot product of $\boldsymbol{\nabla}$ with the dyad $\boldsymbol{\nabla}\boldsymbol{v}$.

The Substantial Derivative of a Vector Field

The application of the substantial derivative to a vector can be performed formally for *rectangular* coordinates by applying the operator to each of the scalar components and adding vectorially:

$$\frac{D\boldsymbol{v}}{Dt} = \frac{\partial \boldsymbol{v}}{\partial t} + (\boldsymbol{v} \cdot \boldsymbol{\nabla})\boldsymbol{v}$$

$$= \Sigma_i \boldsymbol{\delta}_i \left\{ \frac{\partial v_i}{\partial t} + (\boldsymbol{v} \cdot \boldsymbol{\nabla}) v_i \right\} \qquad \text{(A.3–17)}$$

This is correct in rectangular coordinates but is not applicable in curvilinear coordinates; it is preferable to *define* the operation $(\boldsymbol{v} \cdot \boldsymbol{\nabla})\boldsymbol{v}$ thus:

$$(\boldsymbol{v} \cdot \boldsymbol{\nabla})\boldsymbol{v} = \tfrac{1}{2}\boldsymbol{\nabla}(\boldsymbol{v} \cdot \boldsymbol{v}) - [\boldsymbol{v} \times [\boldsymbol{\nabla} \times \boldsymbol{v}]] \qquad \text{(A.3–18)}$$

In this way, the operation is reduced to the more familiar gradient and curl operations.[2]

Differentiation of Products

The following relations may be verified by the definitions of the foregoing operations:

$$\boldsymbol{\nabla} rs = r\boldsymbol{\nabla} s + s\boldsymbol{\nabla} r \qquad \text{(A.3–19)}$$

$$(\boldsymbol{\nabla} \cdot s\boldsymbol{v}) = (\boldsymbol{\nabla} s \cdot \boldsymbol{v}) + s(\boldsymbol{\nabla} \cdot \boldsymbol{v}) \qquad \text{(A.3–20)}$$

$$(\boldsymbol{\nabla} \cdot [\boldsymbol{v} \times \boldsymbol{w}]) = (\boldsymbol{w} \cdot [\boldsymbol{\nabla} \times \boldsymbol{v}]) - (\boldsymbol{v} \cdot [\boldsymbol{\nabla} \times \boldsymbol{w}]) \qquad \text{(A.3–21)}$$

$$[\boldsymbol{\nabla} \times s\boldsymbol{v}] = [\boldsymbol{\nabla} s \times \boldsymbol{v}] + s[\boldsymbol{\nabla} \times \boldsymbol{v}] \qquad \text{(A.3–22)}$$

§A.4 SECOND-ORDER TENSORS[1]

In this section some of the operations needed in connection with tensors and dyads are given. These operations arise in the theory of transport phenomena, particularly in momentum transfer.

Definitions and Notation

In §A.2 it was pointed out that a vector $\boldsymbol{v}$ is specified by giving a set of components v_1, v_2, and v_3. Similarly, a second-order tensor $\boldsymbol{\tau}$ is specified by

[2] An alternate formulation is to write $(\boldsymbol{v} \cdot \boldsymbol{\nabla})\boldsymbol{v}$ as $[\boldsymbol{v} \cdot \boldsymbol{\nabla}\boldsymbol{v}]$, that is, as the dot product of $\boldsymbol{v}$ with the *dyad* $\boldsymbol{\nabla}\boldsymbol{v}$.

[1] A more extensive discussion is to be found in P. M. Morse and H. Feshbach, *Methods of Theoretical Physics*, McGraw-Hill, New York (1953), Vol. I, Chapter 1.

giving the nine components τ_{11}, τ_{12}, τ_{13}, τ_{22}, etc.[2] For the sake of "bookkeeping," these components may be written as

$$\boldsymbol{\tau} = \begin{pmatrix} \tau_{11} & \tau_{12} & \tau_{13} \\ \tau_{21} & \tau_{22} & \tau_{23} \\ \tau_{31} & \tau_{32} & \tau_{33} \end{pmatrix} \tag{A.4-1}$$

This array must not be confused with a determinant, which may be formed from such an array; the array is an ordered set of numbers, whereas the determinant is a certain sum of products of these numbers. The elements with both subscripts alike are called the *diagonal elements*, and the elements with dissimilar subscripts are referred to as the *nondiagonal elements*. If $\tau_{12} = \tau_{21}$, $\tau_{13} = \tau_{31}$, and $\tau_{23} = \tau_{32}$, then $\boldsymbol{\tau}$ is said to be a *symmetric tensor*. The *transpose* of a tensor $\boldsymbol{\tau}$ (designated by $\boldsymbol{\tau}^\dagger$) is that tensor formed by interchanging the subscripts on each of the elements:

$$\boldsymbol{\tau}^\dagger = \begin{pmatrix} \tau_{11} & \tau_{21} & \tau_{31} \\ \tau_{12} & \tau_{22} & \tau_{32} \\ \tau_{13} & \tau_{23} & \tau_{33} \end{pmatrix} \tag{A.4-2}$$

Clearly, when $\boldsymbol{\tau}$ is symmetric, then $\boldsymbol{\tau} = \boldsymbol{\tau}^\dagger$.

A dyadic product of two vectors $\boldsymbol{v}$ and $\boldsymbol{w}$ is a special form of second-order tensor, in which the elements of the array are products of the components of the vectors; the dyadic product $\boldsymbol{vw}$ is then

$$\boldsymbol{vw} = \begin{pmatrix} v_1w_1 & v_1w_2 & v_1w_3 \\ v_2w_1 & v_2w_2 & v_2w_3 \\ v_3w_1 & v_3w_2 & v_3w_3 \end{pmatrix} \tag{A.4-3}$$

The dyadic $\boldsymbol{vw}$ is, in general, different from the dyadic $\boldsymbol{wv}$. It should be emphasized that a dyadic product is written by placing the two vectors next to one another with no multiplication sign of any sort.

A *unit tensor* $\boldsymbol{\delta}$ is one whose diagonal components are unity and whose nondiagonal elements are zero:

$$\boldsymbol{\delta} = \begin{pmatrix} 1 & 0 & 0 \\ 0 & 1 & 0 \\ 0 & 0 & 1 \end{pmatrix} \tag{A.4-4}$$

[2] Not all arrays of nine components form a second-order tensor. The mathematical definition of a second-order tensor includes a statement as to how the components of the tensor transform under the coordinate transformations. The subject of coordinate transformations is dealt with briefly in §A.6.

The components of the unit tensor are δ_{ij}—that is, the Kronecker delta. Note further that the rows (or columns) of the unit tensor are the components of the three unit vectors $\boldsymbol{\delta}_1$, $\boldsymbol{\delta}_2$, $\boldsymbol{\delta}_3$.

It is now convenient to introduce a set of *unit dyads*.[3] These are just the dyadic products of unit vectors, $\boldsymbol{\delta}_m\boldsymbol{\delta}_n$, in which $m, n = 1, 2, 3$. The ijth component of $\boldsymbol{\delta}_m\boldsymbol{\delta}_n$ is $\delta_{mi}\delta_{nj}$, that is, the product of two Kronecker deltas. There are nine unit dyads in all:

$$\boldsymbol{\delta}_1\boldsymbol{\delta}_1 = \begin{pmatrix} 1 & 0 & 0 \\ 0 & 0 & 0 \\ 0 & 0 & 0 \end{pmatrix}; \quad \boldsymbol{\delta}_1\boldsymbol{\delta}_2 = \begin{pmatrix} 0 & 1 & 0 \\ 0 & 0 & 0 \\ 0 & 0 & 0 \end{pmatrix}; \quad \boldsymbol{\delta}_1\boldsymbol{\delta}_3 = \begin{pmatrix} 0 & 0 & 1 \\ 0 & 0 & 0 \\ 0 & 0 & 0 \end{pmatrix}$$

$$\boldsymbol{\delta}_2\boldsymbol{\delta}_1 = \begin{pmatrix} 0 & 0 & 0 \\ 1 & 0 & 0 \\ 0 & 0 & 0 \end{pmatrix}; \quad \text{etc.} \qquad \text{(A.4–5)}$$

In terms of these unit dyads, we may formally decompose a tensor or a dyadic product into its components analogously to the expression in Eq. A.2–9 for vectors:

$$\boldsymbol{\tau} = \Sigma_i\Sigma_j\boldsymbol{\delta}_i\boldsymbol{\delta}_j\tau_{ij} \qquad \text{(A.4–6)}$$

$$\boldsymbol{vw} = \Sigma_i\Sigma_j\boldsymbol{\delta}_i\boldsymbol{\delta}_j v_i w_j \qquad \text{(A.4–7)}$$

Now all we need to know is how the unit dyads may be multiplied with each other and with the unit vectors. There are four relations that play the same role as Eqs. A.2–16 and 17 do for vectors:

$(\boldsymbol{\delta}_i\boldsymbol{\delta}_j : \boldsymbol{\delta}_k\boldsymbol{\delta}_l) = \delta_{il}\delta_{jk}$	(A.4–8)
$[\boldsymbol{\delta}_i\boldsymbol{\delta}_j \cdot \boldsymbol{\delta}_k] = \boldsymbol{\delta}_i\delta_{jk}$	(A.4–9)
$[\boldsymbol{\delta}_i \cdot \boldsymbol{\delta}_j\boldsymbol{\delta}_k] = \delta_{ij}\boldsymbol{\delta}_k$	(A.4–10)
$\{\boldsymbol{\delta}_i\boldsymbol{\delta}_j \cdot \boldsymbol{\delta}_k\boldsymbol{\delta}_l\} = \delta_{jk}\boldsymbol{\delta}_i\boldsymbol{\delta}_l$	(A.4–11)

All of the tensor and dyadic formulas used in this book are simply derived from these four basic relations. They are easy to remember by the following mnemonic device: in the last three relations we simply take the dot product of the $\boldsymbol{\delta}$'s on either side of the dot; in the first relation we perform two such successive operations.

Addition of Tensors and Dyadic Products

Two tensors are added thus:

$$\begin{aligned} \boldsymbol{\sigma} + \boldsymbol{\tau} &= \Sigma_i\Sigma_j\boldsymbol{\delta}_i\boldsymbol{\delta}_j\sigma_{ij} + \Sigma_i\Sigma_j\boldsymbol{\delta}_i\boldsymbol{\delta}_j\tau_{ij} \\ &= \Sigma_i\Sigma_j\boldsymbol{\delta}_i\boldsymbol{\delta}_j(\sigma_{ij} + \tau_{ij}) \end{aligned} \qquad \text{(A.4–12)}$$

[3] See Morse and Feshbach, *op. cit.*, p. 55.

That is, the sum of two tensors is that tensor whose components are the sums of the corresponding components of the two tensors. The same is true for dyadic products.

Multiplication of a Tensor by a Scalar

Multiplication of a tensor by a scalar corresponds to multiplying each component of the tensor by the scalar:

$$\begin{aligned} s\boldsymbol{\tau} &= s\{\Sigma_i\Sigma_j\boldsymbol{\delta}_i\boldsymbol{\delta}_j\tau_{ij}\} \\ &= \Sigma_i\Sigma_j\boldsymbol{\delta}_i\boldsymbol{\delta}_j\{s\tau_{ij}\} \end{aligned} \tag{A.4–13}$$

The same is true for dyadic products.

The Scalar Product (or Double Dot Product) of Two Tensors

Two tensors may be multiplied according to the double-dot operation:

$$\begin{aligned} (\boldsymbol{\sigma}:\boldsymbol{\tau}) &= (\{\Sigma_i\Sigma_j\boldsymbol{\delta}_i\boldsymbol{\delta}_j\sigma_{ij}\}:\{\Sigma_k\Sigma_l\boldsymbol{\delta}_k\boldsymbol{\delta}_l\tau_{kl}\}) \\ &= \Sigma_i\Sigma_j\Sigma_k\Sigma_l(\boldsymbol{\delta}_i\boldsymbol{\delta}_j:\boldsymbol{\delta}_k\boldsymbol{\delta}_l)\sigma_{ij}\tau_{kl} \\ &= \Sigma_i\Sigma_j\Sigma_k\Sigma_l\delta_{il}\delta_{jk}\sigma_{ij}\tau_{kl} \\ &= \Sigma_i\Sigma_j\sigma_{ij}\tau_{ji} \end{aligned} \tag{A.4–14}$$

in which Eq. A.4–8 has been used. Similarly, we may show that

$$(\boldsymbol{\tau}:\boldsymbol{v}\boldsymbol{w}) = \Sigma_i\Sigma_j\tau_{ij}v_jw_i \tag{A.4–15}$$

$$(\boldsymbol{v}\boldsymbol{w}:\boldsymbol{x}\boldsymbol{y}) = \Sigma_i\Sigma_jv_iw_jx_jy_i \tag{A.4–16}$$

$$(\boldsymbol{\delta}:\boldsymbol{\tau}) = \Sigma_i\tau_{ii} \tag{A.4–17}$$

The Tensor Product (the Single Dot Product) of Two Tensors

Two tensors may also be multiplied according to the single dot operation:

$$\begin{aligned} \{\boldsymbol{\sigma}\cdot\boldsymbol{\tau}\} &= \{(\Sigma_i\Sigma_j\boldsymbol{\delta}_i\boldsymbol{\delta}_j\sigma_{ij})\cdot(\Sigma_k\Sigma_l\boldsymbol{\delta}_k\boldsymbol{\delta}_l\tau_{kl})\} \\ &= \Sigma_i\Sigma_j\Sigma_k\Sigma_l\{\boldsymbol{\delta}_i\boldsymbol{\delta}_j\cdot\boldsymbol{\delta}_k\boldsymbol{\delta}_l\}\sigma_{ij}\tau_{kl} \\ &= \Sigma_i\Sigma_j\Sigma_k\Sigma_l\delta_{jk}\boldsymbol{\delta}_i\boldsymbol{\delta}_l\sigma_{ij}\tau_{kl} \\ &= \Sigma_i\Sigma_l\boldsymbol{\delta}_i\boldsymbol{\delta}_l(\Sigma_j\sigma_{ij}\tau_{jl}) \end{aligned} \tag{A.4–18}$$

That is, the *il*-component of $\{\boldsymbol{\sigma}\cdot\boldsymbol{\tau}\}$ is $\Sigma_j\sigma_{ij}\tau_{jl}$. Similar operations may be performed with dyadic products.

The Vector Product (or Dot Product) of a Tensor with a Vector

When a tensor is dotted into a vector, we get a vector:

$$\begin{aligned}[\boldsymbol{\tau}\cdot\boldsymbol{v}] &= [\{\Sigma_i\Sigma_j\boldsymbol{\delta}_i\boldsymbol{\delta}_j\tau_{ij}\}\cdot\{\Sigma_k\boldsymbol{\delta}_k v_k\}] \\ &= \Sigma_i\Sigma_j\Sigma_k[\boldsymbol{\delta}_i\boldsymbol{\delta}_j\cdot\boldsymbol{\delta}_k]\tau_{ij}v_k \\ &= \Sigma_i\Sigma_j\Sigma_k\boldsymbol{\delta}_i\delta_{jk}\tau_{ij}v_k \\ &= \Sigma_i\boldsymbol{\delta}_i\{\Sigma_j\tau_{ij}v_j\}\end{aligned} \tag{A.4–19}$$

That is, the ith component of $[\boldsymbol{\tau}\cdot\boldsymbol{v}]$ is $\Sigma_j\tau_{ij}v_j$. Similarly, the ith component of $[\boldsymbol{v}\cdot\boldsymbol{\tau}]$ is $\Sigma_j v_j\tau_{ji}$. Clearly, $[\boldsymbol{\tau}\cdot\boldsymbol{v}] \neq [\boldsymbol{v}\cdot\boldsymbol{\tau}]$ unless $\boldsymbol{\tau}$ is symmetric.

Recall that when a vector $\boldsymbol{v}$ is multiplied by a scalar s the resultant vector $s\boldsymbol{v}$ points in the same direction as $\boldsymbol{v}$ but has a different length. But, when $\boldsymbol{\tau}$ is dotted into $\boldsymbol{v}$, the resultant vector $[\boldsymbol{\tau}\cdot\boldsymbol{v}]$ differs from $\boldsymbol{v}$ in *both* length and direction; that is, the tensor $\boldsymbol{\tau}$ "deflects" or "twists" the vector $\boldsymbol{v}$ to form a new vector pointing in a different direction.[4]

From the foregoing results, it is not difficult to prove the following identities:

$$[\boldsymbol{\delta}\cdot\boldsymbol{v}] = [\boldsymbol{v}\cdot\boldsymbol{\delta}] = \boldsymbol{v} \tag{A.4–20}$$

$$[\boldsymbol{uv}\cdot\boldsymbol{w}] = \boldsymbol{u}(\boldsymbol{v}\cdot\boldsymbol{w}) \tag{A.4–21}$$

$$[\boldsymbol{w}\cdot\boldsymbol{uv}] = (\boldsymbol{w}\cdot\boldsymbol{u})\boldsymbol{v} \tag{A.4–22}$$

$$(\boldsymbol{uv}:\boldsymbol{wz}) = (\boldsymbol{uw}:\boldsymbol{vz}) = (\boldsymbol{u}\cdot\boldsymbol{z})(\boldsymbol{v}\cdot\boldsymbol{w}) \tag{A.4–23}$$

$$(\boldsymbol{\tau}:\boldsymbol{uv}) = ([\boldsymbol{\tau}\cdot\boldsymbol{u}]\cdot\boldsymbol{v}) \tag{A.4–24}$$

$$(\boldsymbol{uv}:\boldsymbol{\tau}) = (\boldsymbol{u}\cdot[\boldsymbol{v}\cdot\boldsymbol{\tau}]) \tag{A.4–25}$$

Differential Operations Involving Tensors and Dyads

The differential operator $\boldsymbol{\nabla}$ may be used to operate on a tensor or a dyad:

$$\begin{aligned}[\boldsymbol{\nabla}\cdot\boldsymbol{\tau}] &= \left[\left\{\Sigma_i\boldsymbol{\delta}_i\frac{\partial}{\partial x_i}\right\}\cdot\{\Sigma_j\Sigma_k\boldsymbol{\delta}_j\boldsymbol{\delta}_k\tau_{jk}\}\right] \\ &= \Sigma_i\Sigma_j\Sigma_k[\boldsymbol{\delta}_i\cdot\boldsymbol{\delta}_j\boldsymbol{\delta}_k]\frac{\partial}{\partial x_i}\tau_{jk} \\ &= \Sigma_i\Sigma_j\Sigma_k\delta_{ij}\boldsymbol{\delta}_k\frac{\partial}{\partial x_i}\tau_{jk} \\ &= \Sigma_k\boldsymbol{\delta}_k\left\{\Sigma_i\frac{\partial}{\partial x_i}\tau_{ik}\right\}\end{aligned} \tag{A.4–26}$$

Hence the kth component of $[\boldsymbol{\nabla}\cdot\boldsymbol{\tau}]$ is $\Sigma_i\,\partial\tau_{ik}/\partial x_i$.

[4] For example, we cite the flow behavior of fluids in porous media. According to Darcy's law (*q.v.*), the flow in an isotropic medium is everywhere in the direction of the negative pressure gradient $-\boldsymbol{\nabla}p = \beta\boldsymbol{v}$, in which β is the quotient of the fluid viscosity and the medium permeability. If the medium is nonisotropic, then $-\boldsymbol{\nabla}p$ and $\boldsymbol{v}$ are no longer in the same direction, and Darcy's law has to be replaced by $-\boldsymbol{\nabla}p = [\boldsymbol{\beta}\cdot\boldsymbol{v}]$, in which $\boldsymbol{\beta}$ is a second-order tensor.

In addition, $\boldsymbol{\nabla}$ may appear in combination with a vector $\boldsymbol{v}$ in the dyadic product $\boldsymbol{\nabla v}$. Then we have

$$[\boldsymbol{w} \cdot \boldsymbol{\nabla v}] = \Sigma_i \Sigma_k \boldsymbol{\delta}_k w_i \frac{\partial}{\partial x_i} v_k \qquad \text{(A.4–27)}$$

$$(\boldsymbol{\tau} : \boldsymbol{\nabla v}) = \Sigma_i \Sigma_j \tau_{ij} \frac{\partial}{\partial x_j} v_i \qquad \text{(A.4–28)}$$

which may be worked out by similar methods.

The following identities may also be derived:

$$[\boldsymbol{\nabla} \cdot s\boldsymbol{\delta}] = \boldsymbol{\nabla} s \qquad \text{(A.4–29)}$$

$$[\boldsymbol{\nabla} \cdot \boldsymbol{vw}] = [\boldsymbol{v} \cdot \boldsymbol{\nabla w}] + \boldsymbol{w}(\boldsymbol{\nabla} \cdot \boldsymbol{v}) \qquad \text{(A.4–30)}$$

$$(s\boldsymbol{\delta} : \boldsymbol{\nabla v}) = s(\boldsymbol{\nabla} \cdot \boldsymbol{v}) \qquad \text{(A.4–31)}$$

Example A.4–1. Proof of a Tensor Identity

Prove that for symmetrical $\boldsymbol{\tau}$

$$(\boldsymbol{\tau} : \boldsymbol{\nabla v}) = (\boldsymbol{\nabla} \cdot [\boldsymbol{\tau} \cdot \boldsymbol{v}]) - (\boldsymbol{v} \cdot [\boldsymbol{\nabla} \cdot \boldsymbol{\tau}]) \qquad \text{(A.4–32)}$$

Solution. First we write out the right hand in terms of components:

$$(\boldsymbol{\nabla} \cdot [\boldsymbol{\tau} \cdot \boldsymbol{v}]) = \Sigma_i \frac{\partial}{\partial x_i} [\boldsymbol{\tau} \cdot \boldsymbol{v}]_i$$

$$= \Sigma_i \Sigma_j \frac{\partial}{\partial x_i} \tau_{ij} v_j \qquad \text{(A.4–33)}$$

$$(\boldsymbol{v} \cdot [\boldsymbol{\nabla} \cdot \boldsymbol{\tau}]) = \Sigma_j v_j [\boldsymbol{\nabla} \cdot \boldsymbol{\tau}]_j$$

$$= \Sigma_j \Sigma_i v_j \frac{\partial}{\partial x_i} \tau_{ij} \qquad \text{(A.4–34)}$$

Subtraction of Eq. A.4–34 from Eq. A.4–33 gives the expression for $(\boldsymbol{\tau} : \boldsymbol{\nabla v})$ in Eq. A.4–28.

§A.5 INTEGRAL THEOREMS FOR VECTORS AND TENSORS

For performing general proofs in continuum physics, several integral theorems are extremely useful.

The Gauss-Ostrogradskii Divergence Theorem

If V is a closed region in space surrounded by a surface S, then[1]

$$\iiint_V (\boldsymbol{\nabla} \cdot \boldsymbol{v})\, dV = \iint_S (\boldsymbol{n} \cdot \boldsymbol{v})\, dS \qquad \text{(A.5–1)}$$

[1] The right side may be written in a variety of ways:

$$\iint_S (\boldsymbol{v} \cdot \boldsymbol{n})\, dS \quad \text{or} \quad \iint_S v_n\, dS \quad \text{or} \quad \iint_S (\boldsymbol{v} \cdot \boldsymbol{dS})$$

in which $\boldsymbol{n}$ is the outwardly directed normal vector. This is known as the *divergence theorem* of Gauss and Ostrogradskii. Two closely allied theorems for scalars and tensors are

$$\iiint_V \nabla s \, dV = \iint_S \boldsymbol{n} s \, dS \tag{A.5–2}$$

$$\iiint_V [\nabla \cdot \boldsymbol{\tau}] \, dV = \iint_S [\boldsymbol{n} \cdot \boldsymbol{\tau}] \, dS \tag{A.5–3}$$ [2]

The last relation is also valid for dyadic products $\boldsymbol{vw}$.

The Stokes Curl Theorem

If S is a surface bounded by the closed curve C, then

$$\iint_S ([\nabla \times \boldsymbol{v}] \cdot \boldsymbol{n}) \, dS = \oint_C (\boldsymbol{v} \cdot \boldsymbol{t}) \, dC \tag{A.5–4}$$

in which $\boldsymbol{t}$ is a unit tangential vector in the direction of integration along C; $\boldsymbol{n}$ is the unit normal vector to S in the direction that a right-hand screw would move if its head were twisted in the direction of integration along C. A similar relation exists for tensors.[2]

The Leibnitz Formula for Differentiating a Triple Integral

Let V be a closed moving region in space surrounded by a surface S; let the velocity of any surface element be $\boldsymbol{v}_S$. Then, if $s(x, y, z, t)$ is a scalar function of position and time,

$$\frac{d}{dt} \iiint_V s \, dV = \iiint_V \frac{\partial s}{\partial t} \, dV + \iint_S s(\boldsymbol{v}_S \cdot \boldsymbol{n}) \, dS \tag{A.5–5}$$

This is an extension of the *Leibnitz formula* for differentiating an integral; keep in mind that $V = V(t)$ and $S = S(t)$.

If the integral is over a volume, the surface of which is moving with the local fluid velocity (so that $\boldsymbol{v}_S = \boldsymbol{v}$), then use of the equation of continuity leads to the additional useful result:

$$\frac{d}{dt} \iiint_V \rho s \, dV = \iiint_V \rho \frac{Ds}{Dt} \, dV \tag{A.5–6}$$

in which ρ is the fluid density.

[2] See P. M. Morse and H. Feshbach, *Methods of Theoretical Physics*, McGraw-Hill, New York (1953), p. 66.

§ A.6 VECTOR AND TENSOR COMPONENTS IN CURVILINEAR COORDINATES

Thus far we have considered only rectangular coordinates x, y, and z. Although formal derivations are usually made in rectangular coordinates, for working problems it is often more natural to use curvilinear coordinates. The two most common curvilinear coordinate systems are the *cylindrical* and the *spherical*. Here we discuss only these two systems, but the method can be applied to any *orthogonal* coordinate systems, that is, those in which the three families of coordinate surfaces are mutually perpendicular.

We are primarily interested in knowing how to write various differential operations, such as ∇s, $[\nabla \times \mathbf{v}]$, and $(\boldsymbol{\tau} : \nabla \mathbf{v})$ in curvilinear coordinates. It turns out that we can do this quite simply if we know, for the coordinate system being used, just two things:

a. The expression for ∇ in the curvilinear coordinates.
b. The spatial derivatives of the unit vectors in curvilinear coordinates.

Hence we focus our attention on these two points.

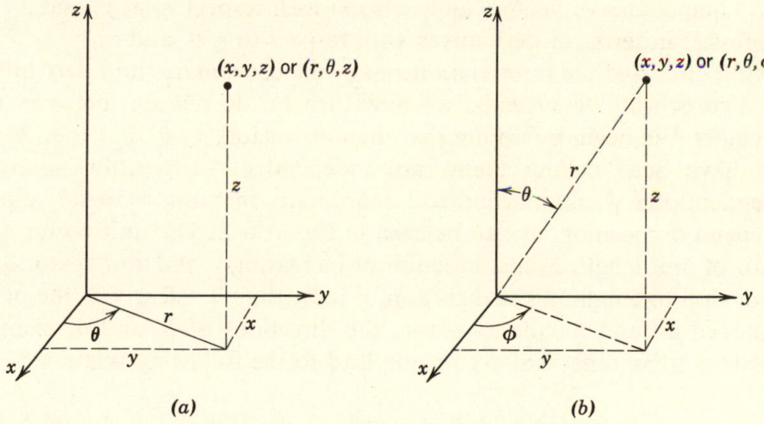

Fig. A.6-1. (a) Cylindrical coordinates. The ranges of the variables are $0 \leqslant r \leqslant \infty$, $0 \leqslant \theta < 2\pi$, $-\infty \leqslant z \leqslant \infty$. (b) Spherical coordinates. The ranges of the variables are $0 \leqslant r \leqslant \infty$, $0 \leqslant \theta \leqslant \pi$, and $0 \leqslant \phi < 2\pi$.

Cylindrical Coordinates

In cylindrical coordinates, instead of designating the coordinates of a point by x, y, z, we locate the point by giving the values of r, θ, z. These coordinates are shown in Fig. A.6–1a. They are related to the rectangular coordinates by

$$\begin{cases} x = r\cos\theta & \text{(A.6–1)} \\ y = r\sin\theta & \text{(A.6–2)} \\ z = z & \text{(A.6–3)} \end{cases} \qquad \begin{cases} r = +\sqrt{x^2+y^2} & \text{(A.6–4)} \\ \theta = \arctan(y/x) & \text{(A.6–5)} \\ z = z & \text{(A.6–6)} \end{cases}$$

To convert derivatives of scalars with respect to x,y,z into derivatives with respect to r,θ,z, the "chain rule" of partial differentiation[1] is used. The derivative operators are readily found to be related as follows:

$$\frac{\partial}{\partial x} = (\cos\theta)\frac{\partial}{\partial r} + \left(-\frac{\sin\theta}{r}\right)\frac{\partial}{\partial \theta} + (0)\frac{\partial}{\partial z} \qquad \text{(A.6–7)}$$

$$\frac{\partial}{\partial y} = (\sin\theta)\frac{\partial}{\partial r} + \left(\frac{\cos\theta}{r}\right)\frac{\partial}{\partial \theta} + (0)\frac{\partial}{\partial z} \qquad \text{(A.6–8)}$$

$$\frac{\partial}{\partial z} = (0)\frac{\partial}{\partial r} + (0)\frac{\partial}{\partial \theta} + (1)\frac{\partial}{\partial z} \qquad \text{(A.6–9)}$$

With these relations, derivatives of any scalar functions (including, of course, components of vectors and tensors) with respect to x, y, and z can be expressed in terms of derivatives with respect to r, θ, and z.

Having discussed the interrelationship of the coordinates and derivatives in the two coordinate systems, we now turn to the relation between the unit vectors. We begin by noting that the unit vectors $\boldsymbol{\delta}_x$, $\boldsymbol{\delta}_y$, $\boldsymbol{\delta}_z$ (or $\boldsymbol{\delta}_1$, $\boldsymbol{\delta}_2$, $\boldsymbol{\delta}_3$ as we have been calling them) are independent of position—that is, independent of x, y, z. In cylindrical coordinates the unit vectors $\boldsymbol{\delta}_r$ and $\boldsymbol{\delta}_\theta$ will depend on position, as can be seen in Fig. A.6–2. The unit vector $\boldsymbol{\delta}_r$ is a vector of unit length in the direction of increasing r; the unit vector $\boldsymbol{\delta}_\theta$ is a vector of unit length in the direction of increasing θ. Clearly as the point P is moved around on the xy-plane, the directions of $\boldsymbol{\delta}_r$ and $\boldsymbol{\delta}_\theta$ change. Elementary trigonometrical arguments lead to the following relations:

$$\boldsymbol{\delta}_r = (\cos\theta)\ \boldsymbol{\delta}_x + (\sin\theta)\ \boldsymbol{\delta}_y + (0)\ \boldsymbol{\delta}_z \qquad \text{(A.6–10)}$$

$$\boldsymbol{\delta}_\theta = (-\sin\theta)\ \boldsymbol{\delta}_x + (\cos\theta)\ \boldsymbol{\delta}_y + (0)\ \boldsymbol{\delta}_z \qquad \text{(A.6–11)}$$

$$\boldsymbol{\delta}_z = (0)\ \boldsymbol{\delta}_x + (0)\ \boldsymbol{\delta}_y + (1)\ \boldsymbol{\delta}_z \qquad \text{(A.6–12)}$$

[1] For example, for a scalar function $\phi(x,y,z) = \bar{\phi}(r,\theta,z)$

$$\left(\frac{\partial \phi}{\partial x}\right)_{y,z} = \left(\frac{\partial r}{\partial x}\right)_{y,z}\left(\frac{\partial \bar{\phi}}{\partial r}\right)_{\theta,z} + \left(\frac{\partial \theta}{\partial x}\right)_{y,z}\left(\frac{\partial \bar{\phi}}{\partial \theta}\right)_{r,z} + \left(\frac{\partial z}{\partial x}\right)_{y,z}\left(\frac{\partial \bar{\phi}}{\partial z}\right)_{r,\theta}$$

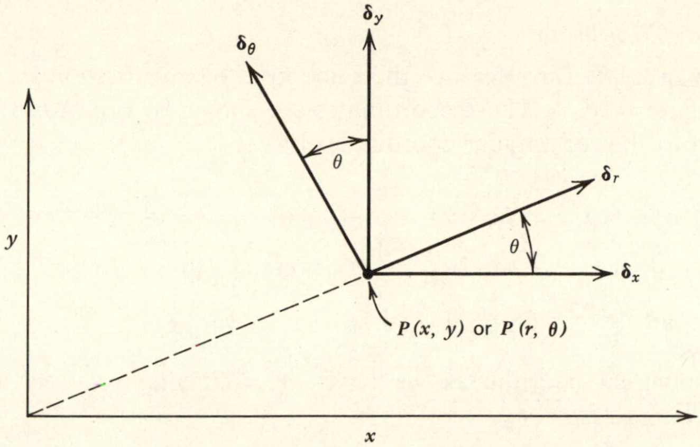

Fig. A.6-2. Unit vectors in rectangular and cylindrical coordinates.

These may be solved for δ_x, δ_y, and δ_z to give:

$$\delta_x = (\cos\theta)\ \delta_r + (-\sin\theta)\ \delta_\theta + (0)\ \delta_z \tag{A.6–13}$$

$$\delta_y = (\sin\theta)\ \delta_r + (\cos\theta)\ \delta_\theta + (0)\ \delta_z \tag{A.6–14}$$

$$\delta_z = (0)\ \delta_r + (0)\ \delta_\theta + (1)\ \delta_z \tag{A.6–15}$$

The utility of these two sets of relations will be made clear in the next section.

Vectors and tensors can be decomposed into components with respect to cylindrical coordinates just as with respect to rectangular coordinates, and the various dot and cross product operations (but *not* the differential operations!) performed as described in §A.2 and §A.4. For example:

$$(\mathbf{v}\cdot\mathbf{w}) = v_r w_r + v_\theta w_\theta + v_z w_z \tag{A.6–16}$$

$$\begin{aligned}[\mathbf{v}\times\mathbf{w}] = {} & \delta_r(v_\theta w_z - v_z w_\theta) + \delta_\theta(v_z w_r - v_r w_z) \\ & + \delta_z(v_r w_\theta - v_\theta w_r)\end{aligned} \tag{A.6–17}$$

$$\begin{aligned}\{\boldsymbol{\sigma}\cdot\boldsymbol{\tau}\} = {} & \delta_r\ \delta_r(\sigma_{rr}\tau_{rr} + \sigma_{r\theta}\tau_{\theta r} + \sigma_{rz}\tau_{zr}) \\ & + \delta_r\ \delta_\theta(\sigma_{rr}\tau_{r\theta} + \sigma_{r\theta}\tau_{\theta\theta} + \sigma_{rz}\tau_{z\theta}) \\ & + \delta_r\ \delta_z(\sigma_{rr}\tau_{rz} + \sigma_{r\theta}\tau_{\theta z} + \sigma_{rz}\tau_{zz}) \\ & + \text{etc.}\end{aligned} \tag{A.6–18}$$

Spherical Coordinates

We now tabulate for reference the same kind of information for spherical coordinates r, θ, ϕ. These coordinates are shown in Fig. A.6–1b. They are related to the rectangular coordinates by

$$\begin{cases} x = r\sin\theta\cos\phi & \text{(A.6–19)} \\ y = r\sin\theta\sin\phi & \text{(A.6–20)} \\ z = r\cos\theta & \text{(A.6–21)} \end{cases} \qquad \begin{cases} r = +\sqrt{x^2+y^2+z^2} & \text{(A.6–22)} \\ \theta = \arctan\left(\sqrt{x^2+y^2}/z\right) & \text{(A.6–23)} \\ \phi = \arctan(y/x) & \text{(A.6–24)} \end{cases}$$

For the spherical coordinates we have the following relations for the derivative operators:

$$\frac{\partial}{\partial x} = (\sin\theta\cos\phi)\frac{\partial}{\partial r} + \left(\frac{\cos\theta\cos\phi}{r}\right)\frac{\partial}{\partial\theta} + \left(-\frac{\sin\phi}{r\sin\theta}\right)\frac{\partial}{\partial\phi} \qquad \text{(A.6–25)}$$

$$\frac{\partial}{\partial y} = (\sin\theta\sin\phi)\frac{\partial}{\partial r} + \left(\frac{\cos\theta\sin\phi}{r}\right)\frac{\partial}{\partial\theta} + \left(\frac{\cos\phi}{r\sin\theta}\right)\frac{\partial}{\partial\phi} \qquad \text{(A.6–26)}$$

$$\frac{\partial}{\partial z} = (\cos\theta)\frac{\partial}{\partial r} + \left(-\frac{\sin\theta}{r}\right)\frac{\partial}{\partial\theta} + (0)\frac{\partial}{\partial\phi} \qquad \text{(A.6–27)}$$

The relations between the unit vectors are:

$$\boldsymbol{\delta}_r = (\sin\theta\cos\phi)\ \boldsymbol{\delta}_x + (\sin\theta\sin\phi)\ \boldsymbol{\delta}_y + (\cos\theta)\ \boldsymbol{\delta}_z \qquad \text{(A.6–28)}$$

$$\boldsymbol{\delta}_\theta = (\cos\theta\cos\phi)\ \boldsymbol{\delta}_x + (\cos\theta\sin\phi)\ \boldsymbol{\delta}_y + (-\sin\theta)\ \boldsymbol{\delta}_z \qquad \text{(A.6–29)}$$

$$\boldsymbol{\delta}_\phi = (-\sin\phi)\ \boldsymbol{\delta}_x + (\cos\phi)\ \boldsymbol{\delta}_y + (0)\ \boldsymbol{\delta}_z \qquad \text{(A.6–30)}$$

or

$$\boldsymbol{\delta}_x = (\sin\theta\cos\phi)\ \boldsymbol{\delta}_r + (\cos\theta\cos\phi)\ \boldsymbol{\delta}_\theta + (-\sin\phi)\ \boldsymbol{\delta}_\phi \qquad \text{(A.6–31)}$$

$$\boldsymbol{\delta}_y = (\sin\theta\sin\phi)\ \boldsymbol{\delta}_r + (\cos\theta\sin\phi)\ \boldsymbol{\delta}_\theta + (\cos\phi)\ \boldsymbol{\delta}_\phi \qquad \text{(A.6–32)}$$

$$\boldsymbol{\delta}_z = (\cos\theta)\ \boldsymbol{\delta}_r + (-\sin\theta)\ \boldsymbol{\delta}_\theta + (0)\ \boldsymbol{\delta}_\phi \qquad \text{(A.6–33)}$$

And, finally, some sample operations in spherical coordinates are:

$$(\boldsymbol{\sigma}:\boldsymbol{\tau}) = \sigma_{rr}\tau_{rr} + \sigma_{r\theta}\tau_{\theta r} + \sigma_{r\phi}\tau_{\phi r}$$
$$+ \sigma_{\theta r}\tau_{r\theta} + \sigma_{\theta\theta}\tau_{\theta\theta} + \sigma_{\theta\phi}\tau_{\phi\theta}$$
$$+ \sigma_{\phi r}\tau_{r\phi} + \sigma_{\phi\theta}\tau_{\theta\phi} + \sigma_{\phi\phi}\tau_{\phi\phi} \quad \text{(A.6–34)}$$

$$(\mathbf{u}\cdot[\mathbf{v}\times\mathbf{w}]) = \begin{vmatrix} u_r & u_\theta & u_\phi \\ v_r & v_\theta & v_\phi \\ w_r & w_\theta & w_\phi \end{vmatrix} \quad \text{(A.6–35)}$$

That is, the relations (not involving ∇!) given earlier can be written in terms of spherical components.

§ A.7 DIFFERENTIAL OPERATIONS IN CURVILINEAR COORDINATES

We now turn to the use of the ∇ operator in curvilinear coordinates. As in the preceding section, we restrict ourselves to cylindrical coordinates and spherical coordinates.

Cylindrical Coordinates

From Eqs. A.6–10, 11, and 12 we can obtain expressions for the spatial derivatives of the unit vectors $\boldsymbol{\delta}_r$, $\boldsymbol{\delta}_\theta$, and $\boldsymbol{\delta}_z$:

$$\frac{\partial}{\partial r}\boldsymbol{\delta}_r = 0 \qquad \frac{\partial}{\partial r}\boldsymbol{\delta}_\theta = 0 \qquad \frac{\partial}{\partial r}\boldsymbol{\delta}_z = 0 \quad \text{(A.7–1)}$$

$$\frac{\partial}{\partial \theta}\boldsymbol{\delta}_r = \boldsymbol{\delta}_\theta \qquad \frac{\partial}{\partial \theta}\boldsymbol{\delta}_\theta = -\boldsymbol{\delta}_r \qquad \frac{\partial}{\partial \theta}\boldsymbol{\delta}_z = 0 \quad \text{(A.7–2)}$$

$$\frac{\partial}{\partial z}\boldsymbol{\delta}_r = 0 \qquad \frac{\partial}{\partial z}\boldsymbol{\delta}_\theta = 0 \qquad \frac{\partial}{\partial z}\boldsymbol{\delta}_z = 0 \quad \text{(A.7–3)}$$

The reader would do well to interpret these derivatives geometrically by considering the way $\boldsymbol{\delta}_r$, $\boldsymbol{\delta}_\theta$, $\boldsymbol{\delta}_z$ change as the location of P is changed.

We now use the definition of the ∇ operator in Eq. A.3–1, the expressions in Eqs. A.6–13, 14, and 15, and the derivative operators in Eqs. A.6–7, 8, and 9 to obtain the formula for ∇ in cylindrical coordinates:

$$\nabla = \boldsymbol{\delta}_x\frac{\partial}{\partial x} + \boldsymbol{\delta}_y\frac{\partial}{\partial y} + \boldsymbol{\delta}_z\frac{\partial}{\partial z}$$
$$= (\boldsymbol{\delta}_r\cos\theta - \boldsymbol{\delta}_\theta\sin\theta)\left(\cos\theta\frac{\partial}{\partial r} - \frac{\sin\theta}{r}\frac{\partial}{\partial\theta}\right)$$
$$+ (\boldsymbol{\delta}_r\sin\theta + \boldsymbol{\delta}_\theta\cos\theta)\left(\sin\theta\frac{\partial}{\partial r} + \frac{\cos\theta}{r}\frac{\partial}{\partial\theta}\right) + \boldsymbol{\delta}_z\frac{\partial}{\partial z} \quad \text{(A.7–4)}$$

When this is multiplied, there is considerable simplification, and we get

$$\nabla = \boldsymbol{\delta}_r \frac{\partial}{\partial r} + \boldsymbol{\delta}_\theta \frac{1}{r}\frac{\partial}{\partial \theta} + \boldsymbol{\delta}_z \frac{\partial}{\partial z} \tag{A.7–5}$$

for *cylindrical* coordinates. This may be used for obtaining all differential operations in cylindrical coordinates, provided that Eqs. A.7–1, 2, and 3 are used to differentiate the unit vectors on which ∇ operates. This point will be made clear in the subsequent illustrative example.

Spherical Coordinates

The spatial derivatives of $\boldsymbol{\delta}_r$, $\boldsymbol{\delta}_\theta$, and $\boldsymbol{\delta}_\phi$ are given by differentiating Eqs. A.6–28, 29, and 30:

$$\frac{\partial}{\partial r}\boldsymbol{\delta}_r = 0 \qquad \frac{\partial}{\partial r}\boldsymbol{\delta}_\theta = 0 \qquad \frac{\partial}{\partial r}\boldsymbol{\delta}_\phi = 0 \tag{A.7–6}$$

$$\frac{\partial}{\partial \theta}\boldsymbol{\delta}_r = \boldsymbol{\delta}_\theta \qquad \frac{\partial}{\partial \theta}\boldsymbol{\delta}_\theta = -\boldsymbol{\delta}_r \qquad \frac{\partial}{\partial \theta}\boldsymbol{\delta}_\phi = 0 \tag{A.7–7}$$

$$\frac{\partial}{\partial \phi}\boldsymbol{\delta}_r = \boldsymbol{\delta}_\phi \sin\theta \qquad \frac{\partial}{\partial \phi}\boldsymbol{\delta}_\theta = \boldsymbol{\delta}_\phi \cos\theta \qquad \frac{\partial}{\partial \phi}\boldsymbol{\delta}_\phi = -\boldsymbol{\delta}_r \sin\theta - \boldsymbol{\delta}_\theta \cos\theta \tag{A.7–8}$$

The use of Eqs. A.6–31, 32, and 33 and Eqs. A.6–25, 26, and 27 gives the following expression for the ∇ operator in *spherical* coordinates:

$$\nabla = \boldsymbol{\delta}_r \frac{\partial}{\partial r} + \boldsymbol{\delta}_\theta \frac{1}{r}\frac{\partial}{\partial \theta} + \boldsymbol{\delta}_\phi \frac{1}{r\sin\theta}\frac{\partial}{\partial \phi} \tag{A.7–9}$$

This expression can be used to obtain differential operations in spherical coordinates, provided that Eqs. A.7–6, 7, and 8 are used concomitantly.

In Tables A.7–1, 2, and 3 we summarize the differential operations most commonly encountered in rectangular, cylindrical, and spherical coordinates. The curvilinear coordinate expressions given there are easily obtained by the method illustrated presently in two examples.

Once again we emphasize that one needs only two items of information in order to write the ∇-operations in orthogonal curvilinear coordinates: (1) the expression for ∇ and (2) the expressions for the spatial derivatives of the unit vectors. Thus far we have discussed the two most-used curvilinear coordinates. We now consider briefly the relations for any orthogonal coordinates q_i. Let the relation between Cartesian coordinates and the orthogonal coordinates be given by

$$\left.\begin{aligned} x_1 &= x_1(q_1, q_2, q_3) \\ x_2 &= x_2(q_1, q_2, q_3) \\ x_3 &= x_3(q_1, q_2, q_3) \end{aligned}\right\} \text{ or } \quad x_i = x_i(q_\alpha) \tag{A.7–10}$$

Then[1] the unit vectors in rectangular coordinates δ_i and those in curvilinear coordinates δ_α are related as follows:

$$\delta_\alpha = \Sigma_i \frac{1}{h_\alpha}\left(\frac{\partial x_i}{\partial q_\alpha}\right)\delta_i = \Sigma_i h_\alpha \left(\frac{\partial q_\alpha}{\partial x_i}\right)\delta_i \tag{A.7-11}$$

$$\delta_i = \Sigma_\alpha h_\alpha \left(\frac{\partial q_\alpha}{\partial x_i}\right)\delta_\alpha = \Sigma_\alpha \frac{1}{h_\alpha}\left(\frac{\partial x_i}{\partial q_\alpha}\right)\delta_\alpha \tag{A.7-12}$$

Consequently, we finally obtain:

$$\boxed{\frac{\partial \delta_\alpha}{\partial q_\beta} = \frac{\delta_\beta}{h_\alpha}\frac{\partial h_\beta}{\partial q_\alpha} - \delta_{\alpha\beta}\sum_{\gamma=1}^{3}\frac{\delta_\gamma}{h_\gamma}\frac{\partial h_\alpha}{\partial q_\gamma}} \tag{A.7-13}$$

and

$$\boxed{\nabla = \Sigma_\alpha \frac{\delta_\alpha}{h_\alpha}\frac{\partial}{\partial q_\alpha}} \tag{A.7-14}$$

where

$$h_\alpha^2 = \Sigma_i \left(\frac{\partial x_i}{\partial q_\alpha}\right)^2 = \left[\Sigma_i \left(\frac{\partial q_\alpha}{\partial x_i}\right)^2\right]^{-1} \tag{A.7-15}$$

The h_α are called "scale factors."

Example A.7-1. Differential Operations in Cylindrical Coordinates

Derive expressions for $(\nabla \cdot \mathbf{v})$ and $\nabla\mathbf{v}$ in cylindrical coordinates.

Solution

a. We begin by writing ∇ in cylindrical coordinates and decomposing $\mathbf{v}$ into its components:

$$(\nabla \cdot \mathbf{v}) = \left(\left\{\delta_r \frac{\partial}{\partial r} + \delta_\theta \frac{1}{r}\frac{\partial}{\partial \theta} + \delta_z \frac{\partial}{\partial z}\right\} \cdot \{\delta_r v_r + \delta_\theta v_\theta + \delta_z v_z\}\right) \tag{A.7-16}$$

Expanding, we get

$$\begin{aligned}(\nabla \cdot \mathbf{v}) &= \left(\delta_r \cdot \frac{\partial}{\partial r}\delta_r v_r\right) + \left(\delta_r \cdot \frac{\partial}{\partial r}\delta_\theta v_\theta\right) + \left(\delta_r \cdot \frac{\partial}{\partial r}\delta_z v_z\right)\\ &+ \left(\delta_\theta \cdot \frac{1}{r}\frac{\partial}{\partial \theta}\delta_r v_r\right) + \left(\delta_\theta \cdot \frac{1}{r}\frac{\partial}{\partial \theta}\delta_\theta v_\theta\right) + \left(\delta_\theta \cdot \frac{1}{r}\frac{\partial}{\partial \theta}\delta_z v_z\right)\\ &+ \left(\delta_z \cdot \frac{\partial}{\partial z}\delta_r v_r\right) + \left(\delta_z \cdot \frac{\partial}{\partial z}\delta_\theta v_\theta\right) + \left(\delta_z \cdot \frac{\partial}{\partial z}\delta_z v_z\right)\end{aligned} \tag{A.7-17}$$

[1]P. Morse and H. Feshbach, *Methods of Theoretical Physics*, McGraw-Hill, New York (1953), p. 36.

TABLE A.7–1

SUMMARY OF DIFFERENTIAL OPERATIONS INVOLVING THE ∇-OPERATOR IN RECTANGULAR COORDINATES[a] (x, y, z)

$$(\nabla \cdot \boldsymbol{v}) = \frac{\partial v_x}{\partial x} + \frac{\partial v_y}{\partial y} + \frac{\partial v_z}{\partial z} \tag{A}$$

$$(\nabla^2 s) = \frac{\partial^2 s}{\partial x^2} + \frac{\partial^2 s}{\partial y^2} + \frac{\partial^2 s}{\partial z^2} \tag{B}$$

$$\begin{aligned}(\boldsymbol{\tau} : \nabla \boldsymbol{v}) = {} & \tau_{xx}\left(\frac{\partial v_x}{\partial x}\right) + \tau_{yy}\left(\frac{\partial v_y}{\partial y}\right) + \tau_{zz}\left(\frac{\partial v_z}{\partial z}\right) + \tau_{xy}\left(\frac{\partial v_x}{\partial y} + \frac{\partial v_y}{\partial x}\right) \\ & + \tau_{yz}\left(\frac{\partial v_y}{\partial z} + \frac{\partial v_z}{\partial y}\right) + \tau_{zx}\left(\frac{\partial v_z}{\partial x} + \frac{\partial v_x}{\partial z}\right)\end{aligned} \tag{C}$$

$$[\nabla s]_x = \frac{\partial s}{\partial x} \tag{D}$$

$$[\nabla s]_y = \frac{\partial s}{\partial y} \tag{E}$$

$$[\nabla s]_z = \frac{\partial s}{\partial z} \tag{F}$$

$$[\nabla \times \boldsymbol{v}]_x = \frac{\partial v_z}{\partial y} - \frac{\partial v_y}{\partial z} \tag{G}$$

$$[\nabla \times \boldsymbol{v}]_y = \frac{\partial v_x}{\partial z} - \frac{\partial v_z}{\partial x} \tag{H}$$

$$[\nabla \times \boldsymbol{v}]_z = \frac{\partial v_y}{\partial x} - \frac{\partial v_x}{\partial y} \tag{I}$$

$$[\nabla \cdot \boldsymbol{\tau}]_x = \frac{\partial \tau_{xx}}{\partial x} + \frac{\partial \tau_{xy}}{\partial y} + \frac{\partial \tau_{xz}}{\partial z} \tag{J}$$

$$[\nabla \cdot \boldsymbol{\tau}]_y = \frac{\partial \tau_{xy}}{\partial x} + \frac{\partial \tau_{yy}}{\partial y} + \frac{\partial \tau_{yz}}{\partial z} \tag{K}$$

$$[\nabla \cdot \boldsymbol{\tau}]_z = \frac{\partial \tau_{xz}}{\partial x} + \frac{\partial \tau_{yz}}{\partial y} + \frac{\partial \tau_{zz}}{\partial z} \tag{L}$$

$$[\nabla^2 \boldsymbol{v}]_x = \frac{\partial^2 v_x}{\partial x^2} + \frac{\partial^2 v_x}{\partial y^2} + \frac{\partial^2 v_x}{\partial z^2} \tag{M}$$

$$[\nabla^2 \boldsymbol{v}]_y = \frac{\partial^2 v_y}{\partial x^2} + \frac{\partial^2 v_y}{\partial y^2} + \frac{\partial^2 v_y}{\partial z^2} \tag{N}$$

$$[\nabla^2 \boldsymbol{v}]_z = \frac{\partial^2 v_z}{\partial x^2} + \frac{\partial^2 v_z}{\partial y^2} + \frac{\partial^2 v_z}{\partial z^2} \tag{O}$$

$$[\boldsymbol{v} \cdot \nabla \boldsymbol{v}]_x = v_x \frac{\partial v_x}{\partial x} + v_y \frac{\partial v_x}{\partial y} + v_z \frac{\partial v_x}{\partial z} \tag{P}$$

$$[\boldsymbol{v} \cdot \nabla \boldsymbol{v}]_y = v_x \frac{\partial v_y}{\partial x} + v_y \frac{\partial v_y}{\partial y} + v_z \frac{\partial v_y}{\partial z} \tag{Q}$$

$$[\boldsymbol{v} \cdot \nabla \boldsymbol{v}]_z = v_x \frac{\partial v_z}{\partial x} + v_y \frac{\partial v_z}{\partial y} + v_z \frac{\partial v_z}{\partial z} \tag{R}$$

[a] Operations involving the tensor $\boldsymbol{\tau}$ are given for **symmetrical $\boldsymbol{\tau}$ only.**

TABLE A.7–2
SUMMARY OF DIFFERENTIAL OPERATIONS INVOLVING THE ∇-OPERATOR IN CYLINDRICAL COORDINATES[a] (r, θ, z)

$$(\nabla \cdot v) = \frac{1}{r}\frac{\partial}{\partial r}(rv_r) + \frac{1}{r}\frac{\partial v_\theta}{\partial \theta} + \frac{\partial v_z}{\partial z} \qquad (A)$$

$$(\nabla^2 s) = \frac{1}{r}\frac{\partial}{\partial r}\left(r\frac{\partial s}{\partial r}\right) + \frac{1}{r^2}\frac{\partial^2 s}{\partial \theta^2} + \frac{\partial^2 s}{\partial z^2} \qquad (B)$$

$$(\tau : \nabla v) = \tau_{rr}\left(\frac{\partial v_r}{\partial r}\right) + \tau_{\theta\theta}\left(\frac{1}{r}\frac{\partial v_\theta}{\partial \theta} + \frac{v_r}{r}\right) + \tau_{zz}\left(\frac{\partial v_z}{\partial z}\right) + \tau_{r\theta}\left[r\frac{\partial}{\partial r}\left(\frac{v_\theta}{r}\right) + \frac{1}{r}\frac{\partial v_r}{\partial \theta}\right] + \tau_{\theta z}\left(\frac{1}{r}\frac{\partial v_z}{\partial \theta} + \frac{\partial v_\theta}{\partial z}\right) + \tau_{rz}\left(\frac{\partial v_z}{\partial r} + \frac{\partial v_r}{\partial z}\right) \qquad (C)$$

$$[\nabla s]_r = \frac{\partial s}{\partial r} \qquad (D)$$

$$[\nabla s]_\theta = \frac{1}{r}\frac{\partial s}{\partial \theta} \qquad (E)$$

$$[\nabla s]_z = \frac{\partial s}{\partial z} \qquad (F)$$

$$[\nabla \times v]_r = \frac{1}{r}\frac{\partial v_z}{\partial \theta} - \frac{\partial v_\theta}{\partial z} \qquad (G)$$

$$[\nabla \times v]_\theta = \frac{\partial v_r}{\partial z} - \frac{\partial v_z}{\partial r} \qquad (H)$$

$$[\nabla \times v]_z = \frac{1}{r}\frac{\partial}{\partial r}(rv_\theta) - \frac{1}{r}\frac{\partial v_r}{\partial \theta} \qquad (I)$$

$$[\nabla \cdot \tau]_r = \frac{1}{r}\frac{\partial}{\partial r}(r\tau_{rr}) + \frac{1}{r}\frac{\partial}{\partial \theta}\tau_{r\theta} - \frac{1}{r}\tau_{\theta\theta} + \frac{\partial \tau_{rz}}{\partial z} \qquad (J)$$

$$[\nabla \cdot \tau]_\theta = \frac{1}{r}\frac{\partial \tau_{\theta\theta}}{\partial \theta} + \frac{\partial \tau_{r\theta}}{\partial r} + \frac{2}{r}\tau_{r\theta} + \frac{\partial \tau_{\theta z}}{\partial z} \qquad (K)$$

$$[\nabla \cdot \tau]_z = \frac{1}{r}\frac{\partial}{\partial r}(r\tau_{rz}) + \frac{1}{r}\frac{\partial \tau_{\theta z}}{\partial \theta} + \frac{\partial \tau_{zz}}{\partial z} \qquad (L)$$

$$[\nabla^2 v]_r = \frac{\partial}{\partial r}\left(\frac{1}{r}\frac{\partial}{\partial r}(rv_r)\right) + \frac{1}{r^2}\frac{\partial^2 v_r}{\partial \theta^2} - \frac{2}{r^2}\frac{\partial v_\theta}{\partial \theta} + \frac{\partial^2 v_r}{\partial z^2} \qquad (M)$$

$$[\nabla^2 v]_\theta = \frac{\partial}{\partial r}\left(\frac{1}{r}\frac{\partial}{\partial r}(rv_\theta)\right) + \frac{1}{r^2}\frac{\partial^2 v_\theta}{\partial \theta^2} + \frac{2}{r^2}\frac{\partial v_r}{\partial \theta} + \frac{\partial^2 v_\theta}{\partial z^2} \qquad (N)$$

$$[\nabla^2 v]_z = \frac{1}{r}\frac{\partial}{\partial r}\left(r\frac{\partial v_z}{\partial r}\right) + \frac{1}{r^2}\frac{\partial^2 v_z}{\partial \theta^2} + \frac{\partial^2 v_z}{\partial z^2} \qquad (O)$$

$$[v \cdot \nabla v]_r = v_r\frac{\partial v_r}{\partial r} + \frac{v_\theta}{r}\frac{\partial v_r}{\partial \theta} - \frac{v_\theta^2}{r} + v_z\frac{\partial v_r}{\partial z} \qquad (P)$$

$$[v \cdot \nabla v]_\theta = v_r\frac{\partial v_\theta}{\partial r} + \frac{v_\theta}{r}\frac{\partial v_\theta}{\partial \theta} + \frac{v_r v_\theta}{r} + v_z\frac{\partial v_\theta}{\partial z} \qquad (Q)$$

$$[v \cdot \nabla v]_z = v_r\frac{\partial v_z}{\partial r} + \frac{v_\theta}{r}\frac{\partial v_z}{\partial \theta} + v_z\frac{\partial v_z}{\partial z} \qquad (R)$$

[a] Operations involving the tensor τ are given for **symmetrical τ only.**

TABLE A.7–3

SUMMARY OF DIFFERENTIAL OPERATIONS INVOLVING THE ∇-OPERATOR IN SPHERICAL COORDINATES[a] (r, θ, ϕ)

$$(\nabla \cdot \boldsymbol{v}) = \frac{1}{r^2}\frac{\partial}{\partial r}(r^2 v_r) + \frac{1}{r\sin\theta}\frac{\partial}{\partial\theta}(v_\theta \sin\theta) + \frac{1}{r\sin\theta}\frac{\partial v_\phi}{\partial\phi} \qquad (A)$$

$$(\nabla^2 s) = \frac{1}{r^2}\frac{\partial}{\partial r}\left(r^2\frac{\partial s}{\partial r}\right) + \frac{1}{r^2\sin\theta}\frac{\partial}{\partial\theta}\left(\sin\theta\frac{\partial s}{\partial\theta}\right) + \frac{1}{r^2\sin^2\theta}\frac{\partial^2 s}{\partial\phi^2} \qquad (B)$$

$$\begin{aligned}(\boldsymbol{\tau} : \nabla \boldsymbol{v}) = {} & \tau_{rr}\left(\frac{\partial v_r}{\partial r}\right) + \tau_{\theta\theta}\left(\frac{1}{r}\frac{\partial v_\theta}{\partial\theta} + \frac{v_r}{r}\right) \\ & + \tau_{\phi\phi}\left(\frac{1}{r\sin\theta}\frac{\partial v_\phi}{\partial\phi} + \frac{v_r}{r} + \frac{v_\theta\cot\theta}{r}\right) \\ & + \tau_{r\theta}\left(\frac{\partial v_\theta}{\partial r} + \frac{1}{r}\frac{\partial v_r}{\partial\theta} - \frac{v_\theta}{r}\right) + \tau_{r\phi}\left(\frac{\partial v_\phi}{\partial r} + \frac{1}{r\sin\theta}\frac{\partial v_r}{\partial\phi} - \frac{v_\phi}{r}\right) \\ & + \tau_{\theta\phi}\left(\frac{1}{r}\frac{\partial v_\phi}{\partial\theta} + \frac{1}{r\sin\theta}\frac{\partial v_\theta}{\partial\phi} - \frac{\cot\theta}{r}v_\phi\right) \qquad (C)\end{aligned}$$

$$\begin{cases}[\nabla s]_r = \dfrac{\partial s}{\partial r} & (D) \\ [\nabla s]_\theta = \dfrac{1}{r}\dfrac{\partial s}{\partial\theta} & (E) \\ [\nabla s]_\phi = \dfrac{1}{r\sin\theta}\dfrac{\partial s}{\partial\phi} & (F)\end{cases}$$

$$\begin{cases}[\nabla \times \boldsymbol{v}]_r = \dfrac{1}{r\sin\theta}\dfrac{\partial}{\partial\theta}(v_\phi\sin\theta) - \dfrac{1}{r\sin\theta}\dfrac{\partial v_\theta}{\partial\phi} & (G) \\ [\nabla \times \boldsymbol{v}]_\theta = \dfrac{1}{r\sin\theta}\dfrac{\partial v_r}{\partial\phi} - \dfrac{1}{r}\dfrac{\partial}{\partial r}(r v_\phi) & (H) \\ [\nabla \times \boldsymbol{v}]_\phi = \dfrac{1}{r}\dfrac{\partial}{\partial r}(r v_\theta) - \dfrac{1}{r}\dfrac{\partial v_r}{\partial\theta} & (I)\end{cases}$$

TABLE A.7–3 (*continued*)

$$[\nabla \cdot \boldsymbol{\tau}]_r = \frac{1}{r^2}\frac{\partial}{\partial r}(r^2\tau_{rr}) + \frac{1}{r\sin\theta}\frac{\partial}{\partial\theta}(\tau_{r\theta}\sin\theta) + \frac{1}{r\sin\theta}\frac{\partial\tau_{r\phi}}{\partial\phi} - \frac{\tau_{\theta\theta}+\tau_{\phi\phi}}{r} \tag{J}$$

$$[\nabla \cdot \boldsymbol{\tau}]_\theta = \frac{1}{r^2}\frac{\partial}{\partial r}(r^2\tau_{r\theta}) + \frac{1}{r\sin\theta}\frac{\partial}{\partial\theta}(\tau_{\theta\theta}\sin\theta) + \frac{1}{r\sin\theta}\frac{\partial\tau_{\theta\phi}}{\partial\phi} + \frac{\tau_{r\theta}}{r} - \frac{\cot\theta}{r}\tau_{\phi\phi} \tag{K}$$

$$[\nabla \cdot \boldsymbol{\tau}]_\phi = \frac{1}{r^2}\frac{\partial}{\partial r}(r^2\tau_{r\phi}) + \frac{1}{r}\frac{\partial\tau_{\theta\phi}}{\partial\theta} + \frac{1}{r\sin\theta}\frac{\partial\tau_{\phi\phi}}{\partial\phi} + \frac{\tau_{r\phi}}{r} + \frac{2\cot\theta}{r}\tau_{\theta\phi} \tag{L}$$

$$[\nabla^2 \boldsymbol{v}]_r = \nabla^2 v_r - \frac{2v_r}{r^2} - \frac{2}{r^2}\frac{\partial v_\theta}{\partial\theta} - \frac{2v_\theta\cot\theta}{r^2} - \frac{2}{r^2\sin\theta}\frac{\partial v_\phi}{\partial\phi} \tag{M}$$

$$[\nabla^2 \boldsymbol{v}]_\theta = \nabla^2 v_\theta + \frac{2}{r^2}\frac{\partial v_r}{\partial\theta} - \frac{v_\theta}{r^2\sin^2\theta} - \frac{2\cos\theta}{r^2\sin^2\theta}\frac{\partial v_\phi}{\partial\phi} \tag{N}$$

$$[\nabla^2 \boldsymbol{v}]_\phi = \nabla^2 v_\phi - \frac{v_\phi}{r^2\sin^2\theta} + \frac{2}{r^2\sin\theta}\frac{\partial v_r}{\partial\phi} + \frac{2\cos\theta}{r^2\sin^2\theta}\frac{\partial v_\theta}{\partial\phi} \tag{O}$$

$$[\boldsymbol{v}\cdot\nabla \boldsymbol{v}]_r = v_r\frac{\partial v_r}{\partial r} + \frac{v_\theta}{r}\frac{\partial v_r}{\partial\theta} + \frac{v_\phi}{r\sin\theta}\frac{\partial v_r}{\partial\phi} - \frac{v_\theta^2+v_\phi^2}{r} \tag{P}$$

$$[\boldsymbol{v}\cdot\nabla \boldsymbol{v}]_\theta = v_r\frac{\partial v_\theta}{\partial r} + \frac{v_\theta}{r}\frac{\partial v_\theta}{\partial\theta} + \frac{v_\phi}{r\sin\theta}\frac{\partial v_\theta}{\partial\phi} + \frac{v_r v_\theta}{r} - \frac{v_\phi^2\cot\theta}{r} \tag{Q}$$

$$[\boldsymbol{v}\cdot\nabla \boldsymbol{v}]_\phi = v_r\frac{\partial v_\phi}{\partial r} + \frac{v_\theta}{r}\frac{\partial v_\phi}{\partial\theta} + \frac{v_\phi}{r\sin\theta}\frac{\partial v_\phi}{\partial\phi} + \frac{v_\phi v_r}{r} + \frac{v_\theta v_\phi\cot\theta}{r} \tag{R}$$

[a] Operations involving the tensor $\boldsymbol{\tau}$ are given for **symmetrical** $\boldsymbol{\tau}$ **only**.

We now use the relations given in Eqs. A.7–1, 2, and 3 to evaluate the derivatives of the unit vectors. This gives

$$(\nabla \cdot \mathbf{v}) = (\boldsymbol{\delta}_r \cdot \boldsymbol{\delta}_r)\frac{\partial v_r}{\partial r} + (\boldsymbol{\delta}_r \cdot \boldsymbol{\delta}_\theta)\frac{\partial v_\theta}{\partial r} + (\boldsymbol{\delta}_r \cdot \boldsymbol{\delta}_z)\frac{\partial v_z}{\partial r}$$

$$+ (\boldsymbol{\delta}_\theta \cdot \boldsymbol{\delta}_r)\frac{1}{r}\frac{\partial v_r}{\partial \theta} + (\boldsymbol{\delta}_\theta \cdot \boldsymbol{\delta}_\theta)\frac{1}{r}\frac{\partial v_\theta}{\partial \theta} + (\boldsymbol{\delta}_\theta \cdot \boldsymbol{\delta}_z)\frac{1}{r}\frac{\partial v_z}{\partial \theta}$$

$$+ \frac{v_r}{r}(\boldsymbol{\delta}_\theta \cdot \boldsymbol{\delta}_\theta) + \frac{v_\theta}{r}(\boldsymbol{\delta}_\theta \cdot \{-\boldsymbol{\delta}_r\})$$

$$+ (\boldsymbol{\delta}_z \cdot \boldsymbol{\delta}_r)\frac{\partial v_r}{\partial z} + (\boldsymbol{\delta}_z \cdot \boldsymbol{\delta}_\theta)\frac{\partial v_\theta}{\partial z} + (\boldsymbol{\delta}_z \cdot \boldsymbol{\delta}_z)\frac{\partial v_z}{\partial z} \qquad \text{(A.7–18)}$$

Since $(\boldsymbol{\delta}_r \cdot \boldsymbol{\delta}_r) = 1$, $(\boldsymbol{\delta}_r \cdot \boldsymbol{\delta}_\theta) = 0$, etc., the latter simplifies to

$$(\nabla \cdot \mathbf{v}) = \frac{\partial v_r}{\partial r} + \frac{1}{r}\frac{\partial v_\theta}{\partial \theta} + \frac{v_r}{r} + \frac{\partial v_z}{\partial z} \qquad \text{(A.7–19)}$$

which is the same as Eq. A of Table A.7–2.

b. Next we examine the dyadic product $\nabla \mathbf{v}$:

$$\nabla \mathbf{v} = \left\{ \boldsymbol{\delta}_r \frac{\partial}{\partial r} + \boldsymbol{\delta}_\theta \frac{1}{r}\frac{\partial}{\partial \theta} + \boldsymbol{\delta}_z \frac{\partial}{\partial z} \right\} \{ \boldsymbol{\delta}_r v_r + \boldsymbol{\delta}_\theta v_\theta + \boldsymbol{\delta}_z v_z \}$$

$$= \boldsymbol{\delta}_r \boldsymbol{\delta}_r \frac{\partial v_r}{\partial r} + \boldsymbol{\delta}_r \boldsymbol{\delta}_\theta \frac{\partial v_\theta}{\partial r} + \boldsymbol{\delta}_r \boldsymbol{\delta}_z \frac{\partial v_z}{\partial r}$$

$$+ \boldsymbol{\delta}_\theta \boldsymbol{\delta}_r \frac{1}{r}\frac{\partial v_r}{\partial \theta} + \boldsymbol{\delta}_\theta \boldsymbol{\delta}_\theta \frac{1}{r}\frac{\partial v_\theta}{\partial \theta} + \boldsymbol{\delta}_\theta \boldsymbol{\delta}_z \frac{1}{r}\frac{\partial v_z}{\partial \theta}$$

$$+ \boldsymbol{\delta}_\theta \boldsymbol{\delta}_\theta \frac{v_r}{r} - \boldsymbol{\delta}_\theta \boldsymbol{\delta}_r \frac{v_\theta}{r}$$

$$+ \boldsymbol{\delta}_z \boldsymbol{\delta}_r \frac{\partial v_r}{\partial z} + \boldsymbol{\delta}_z \boldsymbol{\delta}_\theta \frac{\partial v_\theta}{\partial z} + \boldsymbol{\delta}_z \boldsymbol{\delta}_z \frac{\partial v_z}{\partial z}$$

$$= \boldsymbol{\delta}_r \boldsymbol{\delta}_r \frac{\partial v_r}{\partial r} + \boldsymbol{\delta}_r \boldsymbol{\delta}_\theta \frac{\partial v_\theta}{\partial r} + \boldsymbol{\delta}_r \boldsymbol{\delta}_z \frac{\partial v_z}{\partial r}$$

$$+ \boldsymbol{\delta}_\theta \boldsymbol{\delta}_r \left(\frac{1}{r}\frac{\partial v_r}{\partial \theta} - \frac{v_\theta}{r} \right) + \boldsymbol{\delta}_\theta \boldsymbol{\delta}_\theta \left(\frac{1}{r}\frac{\partial v_\theta}{\partial \theta} + \frac{v_r}{r} \right) + \boldsymbol{\delta}_\theta \boldsymbol{\delta}_z \frac{1}{r}\frac{\partial v_z}{\partial \theta}$$

$$+ \boldsymbol{\delta}_z \boldsymbol{\delta}_r \frac{\partial v_r}{\partial z} + \boldsymbol{\delta}_z \boldsymbol{\delta}_\theta \frac{\partial v_\theta}{\partial z} + \boldsymbol{\delta}_z \boldsymbol{\delta}_z \frac{\partial v_z}{\partial z} \qquad \text{(A.7–20)}$$

Hence the *rr*-component is $\partial v_r / \partial r$, the $r\theta$-component is $\partial v_\theta / \partial r$, etc.

Example A.7-2. Differential Operations in Spherical Coordinates

Find the r-component of $[\nabla \cdot \tau]$ in spherical coordinates.

Solution Using Eq. A.7–9 we have:

$$[\nabla \cdot \tau]_r = \Bigg[\left\{ \delta_r \frac{\partial}{\partial r} + \delta_\theta \frac{1}{r}\frac{\partial}{\partial \theta} + \delta_\phi \frac{1}{r\sin\theta}\frac{\partial}{\partial \phi} \right\} \cdot \{ \delta_r\, \delta_r \tau_{rr} + \delta_r\, \delta_\theta \tau_{r\theta} + \delta_r\, \delta_\phi \tau_{r\phi} + \delta_\theta\, \delta_r \tau_{\theta r} + \delta_\theta\, \delta_\theta \tau_{\theta\theta} + \delta_\theta\, \delta_\phi \tau_{\theta\phi} + \delta_\phi\, \delta_r \tau_{\phi r} + \delta_\phi\, \delta_\theta \tau_{\phi\theta} + \delta_\phi\, \delta_\phi \tau_{\phi\phi} \} \Bigg]_r \tag{A.7–21}$$

We now use Eqs. A.7–6, 7, and 8 and Eq. A.4–10. Since we want only the r-component, we sort out only those terms that contribute to the coefficient of δ_r:

$$\left[\delta_r \frac{\partial}{\partial r} \cdot \delta_r\, \delta_r \tau_{rr} \right] = [\delta_r \cdot \delta_r\, \delta_r] \frac{\partial \tau_{rr}}{\partial r} = \delta_r \frac{\partial \tau_{rr}}{\partial r} \tag{A.7–22}$$

$$\left[\delta_\theta \frac{1}{r}\frac{\partial}{\partial \theta} \cdot \delta_\theta\, \delta_r \tau_{\theta r} \right] = [\delta_\theta \cdot \delta_\theta\, \delta_r] \frac{1}{r}\frac{\partial}{\partial \theta}\tau_{\theta r} + \text{other term} \tag{A.7–23}$$

$$\left[\delta_\phi \frac{1}{r\sin\theta}\frac{\partial}{\partial \phi} \cdot \delta_\phi\, \delta_r \tau_{\phi r} \right] = [\delta_\phi \cdot \delta_\phi\, \delta_r] \quad \frac{1}{r\sin\phi}\frac{\partial}{\partial \theta}\tau_{\phi r} + \text{other term} \tag{A.7–24}$$

$$\left[\delta_\theta \frac{1}{r}\frac{\partial}{\partial \theta} \cdot \delta_r\, \delta_r \tau_{rr} \right] = \frac{\tau_{rr}}{r}\left[\delta_\theta \cdot \left\{ \frac{\partial}{\partial \theta}\delta_r \right\} \delta_r \right] + \frac{\tau_{rr}}{r}\left[\delta_\theta \cdot \delta_r \left\{ \frac{\partial}{\partial \theta}\delta_r \right\} \right] = \frac{\tau_{rr}}{r}[\delta_\theta \cdot \delta_\theta\, \delta_r] = \delta_r \frac{\tau_{rr}}{r} \tag{A.7–25}$$

$$\left[\delta_\phi \frac{1}{r\sin\theta}\frac{\partial}{\partial \phi} \cdot \delta_r\, \delta_r \tau_{rr} \right] = \frac{\tau_{rr}}{r\sin\theta}\left[\delta_\phi \cdot \left\{ \frac{\partial}{\partial \phi}\delta_r \right\} \delta_r \right] = \frac{\tau_{rr}}{r\sin\theta}[\delta_\phi \cdot \delta_\phi \sin\theta\, \delta_r] = \delta_r \frac{\tau_{rr}}{r} \tag{A.7–26}$$

$$\left[\delta_\theta \frac{1}{r} \frac{\partial}{\partial \theta} \cdot \delta_\theta \, \delta_\theta \tau_{\theta\theta} \right] = \delta_r \left(-\frac{\tau_{\theta\theta}}{r} \right) + \text{other term} \qquad \text{(A.7–27)}$$

$$\left[\delta_\phi \frac{1}{r\sin\theta} \frac{\partial}{\partial \phi} \cdot \delta_\theta \, \delta_r \tau_{\theta r} \right] = \delta_r \frac{\tau_{\theta r} \cos\theta}{r \sin\theta} \qquad \text{(A.7–28)}$$

$$\left[\delta_\phi \frac{1}{r\sin\theta} \frac{\partial}{\partial \phi} \cdot \delta_\phi \, \delta_\phi \tau_{\phi\phi} \right] = \delta_r \left(\frac{-\tau_{\phi\phi}}{r} \right) + \text{other terms} \qquad \text{(A.7–29)}$$

Combining the results above we get

$$[\nabla \cdot \tau]_r = \frac{1}{r^2} \frac{\partial}{\partial r} (r^2 \tau_{rr}) + \frac{\tau_{\theta r}}{r} \cot\theta + \frac{1}{r} \frac{\partial}{\partial \theta} \tau_{\theta r} + \frac{1}{r \sin\theta} \frac{\partial \tau_{\phi r}}{\partial \phi} - \frac{\tau_{\theta\theta} + \tau_{\phi\phi}}{r} \qquad \text{(A.7–30)}$$

APPENDIX B

Tables for Prediction of Transport Properties

TABLE B–1

INTERMOLECULAR FORCE PARAMETERS AND CRITICAL PROPERTIES

Substance	Molecular Weight M	Lennard-Jones Parameters[a]		Critical Constants[b,c,d]				
		σ (Å)	ϵ/κ (° K)	T_c (° K)	p_c (atm)	$\tilde{V}_c$ (cm^3 g-mole^{-1})	μ_c (g cm^{-1} sec^{-1}) × 10^6	k_c (cal sec^{-1} cm^{-1} ° K^{-1}) × 10^6
Light elements:								
H_2	2.016	2.915	38.0	33.3	12.80	65.0	34.7	—
He	4.003	2.576	10.2	5.26	2.26	57.8	25.4	—
Noble gases:								
Ne	20.183	2.789	35.7	44.5	26.9	41.7	156.	79.2
Ar	39.944	3.418	124.	151.	48.0	75.2	264.	71.0
Kr	83.80	3.498	225.	209.4	54.3	92.2	396.	49.4
Xe	131.3	4.055	229.	289.8	58.0	118.8	490.	40.2
Simple polyatomic substances:								
Air	28.97[e]	3.617	97.0	132.[e]	36.4[e]	86.6[e]	193.	90.8
N_2	28.02	3.681	91.5	126.2	33.5	90.1	180.	86.8
O_2	32.00	3.433	113.	154.4	49.7	74.4	250.	105.3
O_3	48.00	—	—	268.	67.	89.4	—	—
CO	28.01	3.590	110.	133.	34.5	93.1	190.	86.5
CO_2	44.01	3.996	190.	304.2	72.9	94.0	343.	122.
NO	30.01	3.470	119.	180.	64.	57.	258.	118.2
N_2O	44.02	3.879	220.	309.7	71.7	96.3	332.	131.
SO_2	64.07	4.290	252.	430.7	77.8	122.	411.	98.6
F_2	38.00	3.653	112.	—	—	—	—	—
Cl_2	70.91	4.115	357.	417.	76.1	124.	420.	97.0
Br_2	159.83	4.268	520.	584.	102.	144.	—	—
I_2	253.82	4.982	550.	800.	—	—	—	—

Hydrocarbons:								
CH_4	16.04	3.822	137.	190.7	45.8	99.3	159.	158.0
C_2H_2	26.04	4.221	185.	309.5	61.6	113.	237.	—
C_2H_4	28.05	4.232	205.	282.4	50.0	124.	215.	—
C_2H_6	30.07	4.418	230.	305.4	48.2	148.	210.	203.0
C_3H_6	42.08	—	—	365.0	45.5	181.	233.	—
C_3H_8	44.09	5.061	254.	370.0	42.0	200.	228.	—
n-C_4H_{10}	58.12	—	—	425.2	37.5	255.	239.	—
i-C_4H_{10}	58.12	5.341	313.	408.1	36.0	263.	239.	—
n-C_5H_{12}	72.15	5.769	345.	469.8	33.3	311.	238.	—
n-C_6H_{14}	86.17	5.909	413.	507.9	29.9	368.	248.	—
n-C_7H_{16}	100.20	—	—	540.2	27.0	426.	254.	—
n-C_8H_{18}	114.22	7.451	320.	569.4	24.6	485.	259.	—
n-C_9H_{20}	128.25	—	—	595.0	22.5	543.	265.	—
Cyclohexane	84.16	6.093	324.	553.	40.0	308.	284.	—
C_6H_6	78.11	5.270	440.	562.6	48.6	260.	312.	—
Other organic compounds:								
CH_4	16.04	3.822	137.	190.7	45.8	99.3	159.	158.0
CH_3Cl	50.49	3.375	855.	416.3	65.9	143.	338.	—
CH_2Cl_2	84.94	4.759	406.	510.	60.	—	—	—
$CHCl_3$	119.39	5.430	327.	536.6	54.	240.	410.	—
CCl_4	153.84	5.881	327.	556.4	45.0	276.	413.	—
C_2N_2	52.04	4.38	339.	400.	59.	—	—	—
COS	60.08	4.13	335.	378.	61.	—	—	—
CS_2	76.14	4.438	488.	552.	78.	170.	404.	—

[a] Values of σ and ϵ/κ are from J. O. Hirschfelder, C. F. Curtiss, and R. B. Bird, *Molecular Theory of Gases and Liquids*, Wiley, New York (1954), pp. 1110–1112; also Addenda and Corrigenda, p. 11. The above values are computed from viscosity data and are applicable for temperatures above 100° K. Values for Kr are due to E. A. Mason, *J. Chem. Phys.*, **32**, 1832–1836 (1960).

[b] Values of T_c, p_c, and $\tilde{V}_c$ are from K. A. Kobe and R. E. Lynn, Jr., *Chem. Rev.*, **52**, 117–236 (1952), and Amer. Petroleum Inst. Research Proj. **44**, edited by F. D. Rossini, Carnegie Inst. of Technology (1952), with the exceptions noted.

[c] Values of μ_c are from O. A. Hougen and K. M. Watson, *Chemical Process Principles*, Volume III, Wiley, New York (1947), p. 873.

[d] Values of k_c are from E. J. Owens and G. Thodos, *A.I.Ch.E. Journal*, **3**, 454–461 (1957).

[e] For air, the molecular weight M and the pseudocritical properties T_c, p_c, and $\tilde{V}_c$ have been calculated from the average composition of dry air, as given in International Critical Tables, Vol. I, p. 393 (1926).

TABLE B–2

FUNCTIONS FOR PREDICTION OF TRANSPORT PROPERTIES OF GASES AT LOW DENSITIES[a]

$\kappa T/\epsilon$ or $\kappa T/\epsilon_{AB}$	$\Omega_\mu = \Omega_k$ (For viscosity and thermal conductivity)	$\Omega_{\mathscr{D},AB}$ (For mass diffusivity)	$\kappa T/\epsilon$ or $\kappa T/\epsilon_{AB}$	$\Omega_\mu = \Omega_k$ (For viscosity and thermal conductivity)	$\Omega_{\mathscr{D},AB}$ (For mass diffusivity)
			2.50	1.093	0.9996
0.30	2.785	2.662	2.60	1.081	0.9878
0.35	2.628	2.476	2.70	1.069	0.9770
0.40	2.492	2.318	2.80	1.058	0.9672
0.45	2.368	2.184	2.90	1.048	0.9576
0.50	2.257	2.066	3.00	1.039	0.9490
0.55	2.156	1.966	3.10	1.030	0.9406
0.60	2.065	1.877	3.20	1.022	0.9328
0.65	1.982	1.798	3.30	1.014	0.9256
0.70	1.908	1.729	3.40	1.007	0.9186
0.75	1.841	1.667	3.50	0.9999	0.9120
0.80	1.780	1.612	3.60	0.9932	0.9058
0.85	1.725	1.562	3.70	0.9870	0.8998
0.90	1.675	1.517	3.80	0.9811	0.8942
0.95	1.629	1.476	3.90	0.9755	0.8888
1.00	1.587	1.439	4.00	0.9700	0.8836
1.05	1.549	1.406	4.10	0.9649	0.8788
1.10	1.514	1.375	4.20	0.9600	0.8740
1.15	1.482	1.346	4.30	0.9553	0.8694
1.20	1.452	1.320	4.40	0.9507	0.8652
1.25	1.424	1.296	4.50	0.9464	0.8610
1.30	1.399	1.273	4.60	0.9422	0.8568
1.35	1.375	1.253	4.70	0.9382	0.8530
1.40	1.353	1.233	4.80	0.9343	0.8492
1.45	1.333	1.215	4.90	0.9305	0.8456
1.50	1.314	1.198	5.0	0.9269	0.8422
1.55	1.296	1.182	6.0	0.8963	0.8124
1.60	1.279	1.167	7.0	0.8727	0.7896
1.65	1.264	1.153	8.0	0.8538	0.7712
1.70	1.248	1.140	9.0	0.8379	0.7556
1.75	1.234	1.128	10.0	0.8242	0.7424
1.80	1.221	1.116	20.0	0.7432	0.6640
1.85	1.209	1.105	30.0	0.7005	0.6232
1.90	1.197	1.094	40.0	0.6718	0.5960
1.95	1.186	1.084	50.0	0.6504	0.5756
2.00	1.175	1.075	60.0	0.6335	0.5596
2.10	1.156	1.057	70.0	0.6194	0.5464
2.20	1.138	1.041	80.0	0.6076	0.5352
2.30	1.122	1.026	90.0	0.5973	0.5256
2.40	1.107	1.012	100.0	0.5882	0.5170

[a] Taken from J. O. Hirschfelder, R. B. Bird, and E. L. Spotz, *Chem. Revs.*, **44**, 205 (1949).

APPENDIX C

Constants and Conversion Factors

§C.1 MATHEMATICAL CONSTANTS

$$e = 2.71828\ldots$$
$$\ln 10 = 2.30259\ldots$$
$$\pi = 3.14159\ldots$$

§C.2 PHYSICAL CONSTANTS

Gas law constant

$$\begin{aligned} R &= 1.987_2 \text{ cal g-mole}^{-1}\ {}^\circ\text{K}^{-1} \\ &= 82.05_7 \text{ cm}^3 \text{ atm g-mole}^{-1}\ {}^\circ\text{K}^{-1} \\ &= 8.314_4 \times 10^7 \text{ g cm}^2 \text{ sec}^{-2} \text{ g-mole}^{-1}\ {}^\circ\text{K}^{-1} \\ &= 8.314_4 \times 10^3 \text{ kg m}^2 \text{ sec}^{-2} \text{ kg-mole}^{-1}\ {}^\circ\text{K}^{-1} \\ &= 4.968_6 \times 10^4 \text{ lb}_m \text{ ft}^2 \text{ sec}^{-2} \text{ lb-mole}^{-1}\ {}^\circ\text{R}^{-1} \\ &= 1.544_3 \times 10^3 \text{ ft lb}_f \text{ lb-mole}^{-1}\ {}^\circ\text{R}^{-1} \end{aligned}$$

Standard acceleration of gravity

$$\begin{aligned} g_0 &= 980.665 \text{ cm sec}^{-2} \\ &= 32.1740 \text{ ft sec}^{-2} \end{aligned}$$

Joule's constant (mechanical equivalent of heat)

$$\begin{aligned} J_c &= 4.1840 \times 10^7 \text{ erg cal}^{-1} \\ &= 778.16 \text{ ft lb}_f \text{ Btu}^{-1} \end{aligned}$$

Avogadro's number	$\tilde{N} = 6.02_3 \times 10^{23}$ molecules g-mole^{-1}
Boltzmann's constant	$\kappa = R/\tilde{N} = 1.380_5 \times 10^{-16}$ erg molecule^{-1} °K^{-1}
Faraday's constant	$\mathscr{F} = 9.652 \times 10^4$ abs-coulombs g-equivalent^{-1}
Planck's constant	$h = 6.62_4 \times 10^{-27}$ erg sec
Stefan-Boltzmann constant	$\sigma = 1.355 \times 10^{-12}$ cal sec^{-1} cm^{-2} °K^{-4}
	$= 0.1712 \times 10^{-8}$ Btu hr^{-1} ft^{-2} °R^{-4}
Electronic charge	$e = 1.602 \times 10^{-19}$ abs-coulomb
Speed of light	$c = 2.99793 \times 10^{10}$ cm sec^{-1}

§C.3 CONVERSION FACTORS

To convert any physical quantity from one set of units into another, multiply by the appropriate table entry. For example, suppose that p is given as 10 lb_f in^{-2} but is desired in units of poundals ft^{-2}. From Table C.3–2, the result is

$$p = 10 \times 4.6330 \times 10^3 = 4.6330 \times 10^4 \text{ poundals ft}^{-2} \quad \text{or} \quad lb_m \text{ ft}^{-1} \text{ sec}^{-2}$$

In addition to the extended tables, we give a few of the commonly used conversion factors here:

Given a quantity in these units	Multiply by	To get quantity in these units
Pounds	453.59	Grams
Kilograms	2.2046	Pounds
Inches	2.5400	Centimeters
Meters	39.370	Inches
Gallons (U.S.)	3.7853	Liters
Gallons (U.S.)	231.00	Cubic inches
Gallons (U.S.)	0.13368	Cubic feet
Cubic feet	28.316	Liters

TABLE C.3–1

CONVERSION FACTORS FOR QUANTITIES HAVING DIMENSIONS OF F OR MLt^{-2} (Force)

Given a quantity in these units ↓ / Multiply by table value to convert to these units→	g cm sec^{-2} (dynes)	kg m sec^{-2} (newtons)	lb_m ft sec^{-2} (poundals)	lb_f
g cm sec^{-2} (dynes)	1	10^{-5}	7.2330×10^{-5}	2.2481×10^{-6}
kg m sec^{-2} (newtons)	10^{5}	1	7.2330	2.2481×10^{-1}
lb_m ft sec^{-2} (poundals)	1.3826×10^{4}	1.3826×10^{-1}	1	3.1081×10^{-2}
lb_f	4.4482×10^{5}	4.4482	32.1740	1

TABLE C.3–2

CONVERSION FACTORS FOR QUANTITIES HAVING DIMENSIONS OF F/L^2 OR $ML^{-1}t^{-2}$

(Pressure, Momentum Flux)

Given a quantity in these units ↓ / Multiply by table value to convert to these units →	$g\ cm^{-1} sec^{-2}$ (dyne cm^{-2})	$kg\ m^{-1} sec^{-2}$ (newtons m^{-2})	$lb_m\ ft^{-1} sec^{-2}$ (poundals ft^{-2})	$lb_f\ ft^{-2}$	$lb_f\ in^{-2}$ (psia)[a]	Atmospheres (atm)	mm Hg	in. Hg
$g\ cm^{-1} sec^{-2}$	1	10^{-1}	6.7197×10^{-2}	2.0886×10^{-3}	1.4504×10^{-5}	9.8692×10^{-7}	7.5006×10^{-4}	2.9530×10^{-5}
$kg\ m^{-1} sec^{-2}$	10	1	6.7197×10^{-1}	2.0886×10^{-2}	1.4504×10^{-4}	9.8692×10^{-6}	7.5006×10^{-3}	2.9530×10^{-4}
$lb_m\ ft^{-1} sec^{-2}$	1.4882×10^{1}	1.4882	1	3.1081×10^{-2}	2.1584×10^{-4}	1.4687×10^{-5}	1.1162×10^{-2}	4.3945×10^{-4}
$lb_f\ ft^{-2}$	4.7880×10^{2}	4.7880×10^{1}	32.1740	1	6.9444×10^{-3}	4.7254×10^{-4}	3.5913×10^{-1}	1.4139×10^{-2}
$lb_f\ in^{-2}$	6.8947×10^{4}	6.8947×10^{3}	4.6330×10^{3}	144	1	6.8046×10^{-2}	5.1715×10^{1}	2.0360
Atmospheres	1.0133×10^{6}	1.0133×10^{5}	6.8087×10^{4}	2.1162×10^{3}	14.696	1	760	29.921
mm Hg	1.3332×10^{3}	1.3332×10^{2}	8.9588×10^{1}	2.7845	1.9337×10^{-2}	1.3158×10^{-3}	1	3.9370×10^{-2}
in. Hg	3.3864×10^{4}	3.3864×10^{3}	2.2756×10^{3}	7.0727×10^{1}	4.9116×10^{-1}	3.3421×10^{-2}	25.400	1

[a] This unit is preferably abbreviated psia (pounds per square inch absolute) or psig (pounds per square inch gage). Gage pressure is absolute pressure minus the prevailing barometric pressure.

TABLE C.3–3

CONVERSION FACTORS FOR QUANTITIES HAVING DIMENSIONS OF FL OR ML^2t^{-2}

(Work, Energy, Torque)

Given a quantity in these units ↓ / Multiply by table value to convert to these units →	$g\ cm^2\ sec^{-2}$ (ergs)	$kg\ m^2\ sec^{-2}$ (absolute joules)	$lb_m\ ft^2\ sec^{-2}$ (ft-poundals)	ft lb_f	cal	Btu	hp–hr	kw–hr
$g\ cm^2\ sec^{-2}$	1	10^{-7}	2.3730×10^{-6}	7.3756×10^{-8}	2.3901×10^{-8}	9.4783×10^{-11}	3.7251×10^{-14}	2.7778×10^{-14}
$kg\ m^2\ sec^{-2}$	10^7	1	2.3730×10^{1}	7.3756×10^{-1}	2.3901×10^{-1}	9.4783×10^{-4}	3.7251×10^{-7}	2.7778×10^{-7}
$lb_m\ ft^2\ sec^{-2}$	4.2140×10^{5}	4.2140×10^{-2}	1	3.1081×10^{-2}	1.0072×10^{-2}	3.9942×10^{-5}	1.5698×10^{-8}	1.1706×10^{-8}
ft lb_f	1.3558×10^{7}	1.3558	32.1740	1	3.2405×10^{-1}	1.2851×10^{-3}	5.0505×10^{-7}	3.7662×10^{-7}
Thermochemical calories[a]	4.1840×10^{7}	4.1840	9.9287×10^{1}	3.0860	1	3.9657×10^{-3}	1.5586×10^{-6}	1.1622×10^{-6}
British thermal units	1.0550×10^{10}	1.0550×10^{3}	2.5036×10^{4}	778.16	2.5216×10^{2}	1	3.9301×10^{-4}	2.9307×10^{-4}
Horsepower-hours	2.6845×10^{13}	2.6845×10^{6}	6.3705×10^{7}	1.9800×10^{6}	6.4162×10^{5}	2.5445×10^{3}	1	7.4570×10^{-1}
Absolute kilowatt-hours	3.6000×10^{13}	3.6000×10^{6}	8.5429×10^{7}	2.6552×10^{6}	8.6042×10^{5}	3.4122×10^{3}	1.3410	1

[a] This unit, abbreviated cal, is used in chemical thermodynamic tables. To convert quantities expressed in International Steam Table calories (abbreviated I.T. cal) to this unit, multiply by 1.000654.

TABLE C.3–4

CONVERSION FACTORS FOR QUANTITIES HAVING DIMENSIONS[a] OF $ML^{-1}t^{-1}$ OR FtL^{-2} OR MOLES $L^{-1}t^{-1}$

(Viscosity, Density times Diffusivity, Concentration times Diffusivity)

Given a quantity in these units ↓ / Multiply by table value to convert to these units→	$g\ cm^{-1}\ sec^{-1}$ (poises)	$kg\ m^{-1}\ sec^{-1}$	$lb_m\ ft^{-1}\ sec^{-1}$	$lb_f\ sec\ ft^{-2}$	Centipoises	$lb_m\ ft^{-1}\ hr^{-1}$
$g\ cm^{-1}\ sec^{-1}$	1	10^{-1}	6.7197×10^{-2}	2.0886×10^{-3}	10^{2}	2.4191×10^{2}
$kg\ m^{-1}\ sec^{-1}$	10	1	6.7197×10^{-1}	2.0886×10^{-2}	10^{3}	2.4191×10^{3}
$lb_m\ ft^{-1}\ sec^{-1}$	1.4882×10^{1}	1.4882	1	3.1081×10^{-2}	1.4882×10^{3}	3600
$lb_f\ sec\ ft^{-2}$	4.7880×10^{2}	4.7880×10^{1}	32.1740	1	4.7880×10^{4}	1.1583×10^{5}
Centipoises	10^{-2}	10^{-3}	6.7197×10^{-4}	2.0886×10^{-5}	1	2.4191
$lb_m\ ft^{-1}\ hr^{-1}$	4.1338×10^{-3}	4.1338×10^{-4}	2.7778×10^{-4}	8.6336×10^{-6}	4.1338×10^{-1}	1

[a] When moles appear in the given and desired units, the conversion factor is the same as for the corresponding mass units.

TABLE C.3–5

CONVERSION FACTORS FOR QUANTITIES HAVING DIMENSIONS OF $MLt^{-3}T^{-1}$ OR $Ft^{-1}T^{-1}$

(Thermal Conductivity)

Given a quantity in these units ↓ / Multiply by table value to convert to these units→	g cm sec^{-3} °K^{-1} (ergs sec^{-1} cm^{-1} °K^{-1})	kg m sec^{-3} °K^{-1} (watts m^{-1} °K^{-1})	lb$_m$ ft sec^{-3} °F^{-1}	lb$_f$ sec^{-1} °F^{-1}	cal sec^{-1} cm^{-1} °K^{-1}	Btu hr^{-1} ft^{-1} °F^{-1}
g cm sec^{-3} °K^{-1}	1	10^{-5}	4.0183×10^{-5}	1.2489×10^{-6}	2.3901×10^{-8}	5.7780×10^{-6}
kg m sec^{-3} °K^{-1}	10^{5}	1	4.0183	1.2489×10^{-1}	2.3901×10^{-3}	5.7780×10^{-1}
lb$_m$ ft sec^{-3} °F^{-1}	2.4886×10^{4}	2.4886×10^{-1}	1	3.1081×10^{-2}	5.9479×10^{-4}	1.4379×10^{-1}
lb$_f$ sec^{-1} °F^{-1}	8.0068×10^{5}	8.0068	3.2174×10^{1}	1	1.9137×10^{-2}	4.6263
cal sec^{-1} cm^{-1} °K^{-1}	4.1840×10^{7}	4.1840×10^{2}	1.6813×10^{3}	5.2256×10^{1}	1	2.4175×10^{2}
Btu hr^{-1} ft^{-1} °F^{-1}	1.7307×10^{5}	1.7307	6.9546	2.1616×10^{-1}	4.1365×10^{-3}	1

TABLE C.3–6

CONVERSION FACTORS FOR QUANTITIES HAVING DIMENSIONS OF L^2t^{-1}

(Momentum Diffusivity, Thermal Diffusivity, Molecular Diffusivity)

Given a quantity in these units ↓	Multiply by table value to convert to these units→	$cm^2\ sec^{-1}$	$m^2\ sec^{-1}$	$ft^2\ hr^{-1}$	Centistokes
$cm^2\ sec^{-1}$		1	10^{-4}	3.8750	10^2
$m^2\ sec^{-1}$		10^4	1	3.8750×10^4	10^6
$ft^2\ hr^{-1}$		2.5807×10^{-1}	2.5807×10^{-5}	1	2.5807×10^1
Centistokes		10^{-2}	10^{-6}	3.8750×10^{-2}	1

TABLE C.3–7

CONVERSION FACTORS FOR QUANTITIES HAVING DIMENSIONS OF $Mt^{-3}T^{-1}$ OR $FL^{-1}t^{-1}T^{-1}$

(Heat Transfer Coefficients)

Given a quantity in these units ↓ / Multiply by table value to convert to these units→	g sec^{-3} °K^{-1}	kg sec^{-3} °K^{-1} (watts m^{-2} °K^{-1})	lb_m sec^{-3} °F^{-1}	lb_f ft^{-1} sec^{-1} °F^{-1}	cal cm^{-2} sec^{-1} °K^{-1}	Watts cm^{-2} °K^{-1}	Btu ft^{-2} hr^{-1} °F^{-1}
g sec^{-3} °K^{-1}	1	10^{-3}	1.2248×10^{-3}	3.8068×10^{-5}	2.3901×10^{-8}	10^{-7}	1.7611×10^{-4}
kg sec^{-3} °K^{-1}	10^{3}	1	1.2248	3.8068×10^{-2}	2.3901×10^{-5}	10^{-4}	1.7611×10^{-1}
lb_m sec^{-3} °F^{-1}	8.1647×10^{2}	8.1647×10^{-1}	1	3.1081×10^{-2}	1.9514×10^{-5}	8.1647×10^{-5}	1.4379×10^{-1}
lb_f ft^{-1} sec^{-1} °F^{-1}	2.6269×10^{4}	2.6269×10^{1}	32.1740	1	6.2784×10^{-4}	2.6269×10^{-3}	4.6263
cal cm^{-2} sec^{-1} °K^{-1}	4.1840×10^{7}	4.1840×10^{4}	5.1245×10^{4}	1.5928×10^{3}	1	4.1840	7.3686×10^{3}
Watts cm^{-2} °K^{-1}	10^{7}	10^{4}	1.2248×10^{4}	3.8068×10^{2}	2.3901×10^{-1}	1	1.7611×10^{3}
Btu ft^{-2} hr^{-1} °F^{-1}	5.6782×10^{3}	5.6782	6.9546	2.1616×10^{-1}	1.3571×10^{-4}	5.6782×10^{-4}	1

TABLE C.3–8

CONVERSION FACTORS FOR QUANTITIES HAVING DIMENSIONS[a] OF $ML^{-2}t^{-1}$ OR MOLES $L^{-2}t^{-1}$ OR $FL^{-3}t$

(Mass Transfer Coefficients k_{xi})

Given a quantity in these units ↓ / Multiply by table value to convert to these units→	g cm^{-2} sec^{-1}	kg m^{-2} sec^{-1}	lb_m ft^{-2} sec^{-1}	lb_f ft^{-3} sec	lb_m ft^{-2} hr^{-1}
g cm^{-2} sec^{-1}	1	10^{1}	2.0482	6.3659×10^{-2}	7.3734×10^{3}
kg m^{-2} sec^{-1}	10^{-1}	1	2.0482×10^{-1}	6.3659×10^{-3}	7.3734×10^{2}
lb_m ft^{-2} sec^{-1}	4.8824×10^{-1}	4.8824	1	3.1081×10^{-2}	3600
lb_f ft^{-3} sec	1.5709×10^{1}	1.5709×10^{2}	32.1740	1	1.1583×10^{5}
lb_m ft^{-2} hr^{-1}	1.3562×10^{-4}	1.3562×10^{-3}	2.7778×10^{-4}	8.6336×10^{-6}	1

[a] When moles appear in the given and desired units, the conversion factor is the same as for the corresponding mass units.

Notation

Unless otherwise stated, numbers in parentheses after description refer to equations in which symbols are first used or thoroughly defined. Dimensions are given in terms of mass (M), length (L), time (t), and temperature (T). Boldface symbols are vectors or tensors; vector and tensor notations are discussed in Appendix A. Symbols that appear infrequently or in one section only are not listed.

A = area, L^2.
$A = U - TS$ = Helmholtz free energy or "work function" (7.3–1), ML^2/t^2.
a = absorptivity (14.2–1), dimensionless.
a = interfacial area per unit volume of bed (6.4–4), L^{-1}.
C_d = discharge coefficient (7.5–34), dimensionless.
$\hat{C}_p$ = heat capacity at constant pressure, per unit mass (8.1–7), L^2/t^2T.
$\hat{C}_v$ = heat capacity at constant volume, per unit mass (8.3–6), L^2/t^2T.
c = total molar concentration (§16.1), mols/L^3.
c = speed of light (14.1–1), L/t.
c_i = molar concentration of species i (16.1–2), mols/L^3.
D = characteristic length in dimensional analysis or diameter of sphere or cylinder, L.
D_p = particle diameter (6.4–6), L.
$\mathscr{D}_{AB}$ = binary diffusivity for system A-B (16.2–1), L^2/t.

D_{AB} = binary diffusivity for system A-B based on free energy driving force (18.4–14), L^2/t.

$D_A{}^T$ = binary thermal diffusion coefficient for A in system A-B (18.4–14), M/Lt.

$\mathscr{D}_{ij}$ = binary diffusivity for system i-j (18.4–19), L^2/t.

D_{ij} = multicomponent diffusivity of the pair i-j based on free energy driving force (18.4–8), L^2/t.

$D_i{}^T$ = multicomponent thermal diffusion coefficient for i in a mixture (18.4–11), M/Lt.

$\mathscr{D}_{im}$ = effective binary diffusivity of i in a multicomponent mixture (18.4–21), L^2/t.

d = molecular diameter (1.4–3), L.

$E = U + K + \Phi$ = total fluid energy (10.1–15), ML^2/t^2.

E_v = total rate of viscous dissipation of mechanical energy (7.3–1), ML^2/t^3.

e = 2.71828 . . .

e = emissivity (14.2–3), dimensionless.

$\mathbf{e}$ = total energy flux relative to stationary coordinates (18.3–9), M/t^3.

e_v = friction loss factor associated with viscous dissipation (7.4–5), dimensionless.

$\mathbf{F} = \mathbf{F}_n + \mathbf{F}_t = \mathbf{F}_k + \mathbf{F}_s$ = force of a fluid on an adjacent solid (§§2.6, 6.3), ML/t^2.

$\mathbf{F}^{(m)}$ = rate of addition of momentum to a macroscopic system by mass transfer through the boundaries (21.4–7), ML/t^2.

F_{12} = direct view factor (14.4–9), dimensionless.

$\bar{F}_{12}$ = indirect view factor (14.4–15), dimensionless.

$\mathscr{F}_{12}$ = combined emissivity and view factor (14.5–4), dimensionless.

f = friction factor or drag coefficient (6.1–1), dimensionless.

G = mass velocity (6.4–10), M/tL^2.

$G = H - TS$ = Gibbs free energy, or "free enthalpy" (7.3–1), ML^2/t^2.

$\mathbf{g}$ = gravitational acceleration (3.2–8), L/t^2.

$\mathbf{g}_i$ = total body force per unit mass of component i (18.3–2), L/t^2.

g_c = gravitational conversion factor (1.1–6), dimensionless.

$H = U + pV$ = enthalpy (15.1–5), ML^2/t^2.

h, h_1, h_{ln}, h_a, h_m = heat-transfer coefficients (9.1–2, §13.1), M/t^3T.

h = Planck's constant (14.1–2), ML^2/t.

h = elevation (2.3–10), L.

$i = \sqrt{-1}$ (4.3–12).

J_c = mechanical equivalent of heat (8.1–11), dimensionless.

$\boldsymbol{J}_i$ = molar flux of i relative to mass average velocity (16.1–8), moles/tL^2.

$\boldsymbol{J}_i^\star$ = molar flux of species i relative to the molar average velocity (16.1–10), moles/tL^2.

$\boldsymbol{j}_i, \boldsymbol{j}_i^{(x)}, \boldsymbol{j}_i^{(p)}, \boldsymbol{j}_i^{(g)}, \boldsymbol{j}_i^{(T)}$ = mass fluxes of i relative to mass average velocity (16.1–7, 18.4–7), M/tL^2.

$\boldsymbol{j}_i^\star$ = mass flux of species i relative to the molar average velocity (16.1–9), M/tL^2.

j_D = Chilton-Colburn j-factor for mass transfer (Table 21.2–1), dimensionless.

j_H = Chilton-Colburn j-factor for heat transfer (Eq. 13.2–19), dimensionless.

κ = Boltzmann constant (1.4–1), ML^2/t^2T.

K = kinetic energy (7.3–1), ML^2/t^2.

K = dimensionless boundary mass flux (19.3–28a).

k = thermal conductivity (8.1–1), ML/t^3T.

k_T = thermal diffusion ratio (18.4–15), dimensionless.

k_e = electrical conductivity (8.5–1), ohm^{-1} cm^{-1}.

k_n'' = heterogeneous chemical reaction rate constant (17.0–3), moles$^{1-n}/L^{2-3n}t$.

k_n''' = homogeneous chemical reaction rate constant (17.0–2), moles$^{1-n}/L^{3-3n}t$.

k_x = mass transfer coefficient in a binary system (21.1–8), moles/tL^2.

k_{xi} = mass transfer coefficient of species i in a multicomponent mixture (21.8–3), moles/tL^2.

L = length of tube or other characteristic length, L.

l = Prandtl mixing length (5.3–2), L.

M = molar mean molecular weight (Table 16.1–1), M/mole.

M_A = molecular weight of A, M/mole.

$\mathcal{M}$ = moles of material (22.1–4).

$\mathcal{M}_i$ = moles of component i (22.1–3).

m = parameter in Ostwald-de Waele model (1.2–3), dimensions depend on n ($q.v.$).

m = mass of a molecule (1.4–1), M.

m = mass of flow system (7.1–1), M.

m_i = mass of component i in flow system (22.1–1), M.

$\tilde{N}$ = Avogadro's number, (gm-mole)$^{-1}$.

N = rate of rotation of a shaft (3.7–25), t^{-1}.

$\boldsymbol{N}_i$ = molar flux with respect to stationary coordinates (16.1–6), moles/L^2t.

$\boldsymbol{n}_i$ = mass flux with respect to stationary coordinates (16.1–5), M/L^2t.

n = molecular concentration or number density (1.4–2), L^{-3}.

n = parameter in Ostwald-de Waele model (1.2–3), dimensionless.

n = Deissler's constant (5.3–4), dimensionless.

$\boldsymbol{P}$ = momentum (7.2–1), ML/t.

$\mathcal{P}$ = $p + \rho g h$ (for constant ρ and g) (2.3–10), M/Lt^2.

p = fluid pressure, M/Lt^2.

$p^{(r)}$ = radiant pressure (14.3–2), M/Lt^2.

Q = volumetric flow rate (2.3–19), L^3/t.

Q = rate of energy flow across a surface (8.1–1 and 15.1–2), ML^2/t^3.

$Q_{\overrightarrow{12}}$ = radiant energy flow from surface 1 to surface 2 (14.4–5), ML^2/t^3.

Q_{12} = net radiant energy interchange between surface 1 and surface 2 (14.4–8), ML^2/t^3.

$\boldsymbol{q}, \boldsymbol{q}^{(c)}, \boldsymbol{q}^{(d)}, \boldsymbol{q}^{(x)}$ = energy fluxes relative to mass average velocity (8.1–6 for pure fluids, 18.4–2 for mixtures), M/t^3.

R = gas constant, ML^2/t^2T mole.

R = radius of sphere or cylinder, L.

R_h = hydraulic radius (6.4–3), L.

R_A = molar rate of production of species A (18.1–10), moles$/tL^3$.

R_v, R_T, R_{AB} = flux ratios (21.5–40), dimensionless.

r = radial distance in both cylindrical and spherical coordinates, L.

r_A = mass rate of production of species A (18.1–5), M/tL^3.

S = cross-sectional area, L^2.

$\boldsymbol{S}$ = vector giving cross-sectional area and its orientation (7.2–1), L^2.

S_e, S_n, S_v, S_c = energy sources (9.2–1), M/Lt^3.

s = $R - r$ = distance into fluid from solid boundary in cylindrical coordinates (5.3–5), L.

s^+ = dimensionless distance into fluid in cylindrical coordinates (5.3–11).

T = *absolute* temperature, T.

t = time, t.

U = internal energy (10.1–2), ML^2/t^2.

U = over-all heat-transfer coefficient (9.6–16), M/t^3T.

$\bar{u}$ = arithmetic mean molecular speed (1.4–1), L/t.

V = characteristic speed in dimensional analysis (3.7–1), L/t.

V = volume, L^3.
$\boldsymbol{v}$ = mass average velocity (16.1–1), L/t.
$\boldsymbol{v}_i$ = velocity of species i (16.1–1), L/t.
v_s = speed of sound (8.4–2), L/t.
v_∞ = approach velocity (2.6–1), L/t.
v^+ = dimensionless velocity (5.3–10).
$v_* = \sqrt{\tau_0/\rho}$ = reference velocity (5.3–9), L/t.
$\boldsymbol{v}^\star$ = molar average velocity (16.1–2), L/t.
W = rate of doing work on surroundings (7.3–1), ML^2/t^3.
$\mathcal{W}$ = molar flow rate (22.1–4), moles$/t$.
$\mathcal{W}_A$ = molar flow of species A through a surface (17.2–21, 21.1–7), moles$/t$.
$\boldsymbol{w}$ = vector giving mass flow rate and its direction (7.2–2), M/t.
w = mass flow rate (7.1–2), M/t.
w_A = mass flow of species A through a surface (22.1–1), M/t.
x = rectangular coordinate, L.
x_i = mole fraction of species i (Table 16.1–1), dimensionless.
y = rectangular coordinate, L.
y_i = mole fraction of species i (21.3–1), dimensionless.
Z = wall collision frequency (1.4–2), $L^{-2}t^{-1}$.
z = rectangular coordinate, L.

$\alpha = k/\rho\hat{C}_p$ = thermal diffusivity (8.1–17), L^2/t.
β = thermal coefficient of volumetric expansion (9.9–6a), T^{-1}.
$\Gamma(x)$ = the gamma function of x.
Γ = mass flow rate of falling film per unit width of wetted wall (2.2–20), M/Lt.
$\gamma = \hat{C}_p/\hat{C}_v$ (10.5–75), dimensionless.
$\boldsymbol{\Delta}$ = rate of deformation tensor (3.6–1), t^{-1}.
$\Delta a = a_2 - a_1$, in which 1 and 2 refer to two control surfaces.
δ = falling film thickness (2.2–15), fictitious film thickness (17.4–11 and 21.5–5), penetration thickness (4.4–1), and boundary-layer thickness (4.4–15), L.
$\boldsymbol{\delta}$ = unit tensor (Appendix A), dimensionless.
$\boldsymbol{\delta}_i$ = unit vector (Appendix A), dimensionless.
ϵ, ϵ_{AB} = maximum attractive energy between two molecules (1.4–10, 16.4–16), ML^2/t^2.
ϵ = fractional void space (6.4–4), dimensionless.
ζ = concentration coefficient of volumetric expansion (18.3–13), dimensionless.

ζ = dimensionless position variable, variously defined.
η = dimensionless position variable, variously defined.
$\dot{\eta}$ = non-Newtonian viscosity (3.6–2), M/Lt.
Θ = dimensionless temperature, variously defined.
θ = angle in cylindrical or spherical coordinates (Fig. A.6–1), radians.
$\theta_v, \theta_T, \theta_{AB}$ = dimensionless correction factors (21.4–12, 13, and 14).
κ_1 = Prandtl's constant (5.3–5), dimensionless.
κ_2 = von Kármán's constant (5.3–3), dimensionless.
Λ = diffusivity ratios (19.3–19, 20, and 21), dimensionless.
λ = wavelength of electromagnetic radiation (14.1–1), L.
λ = molecular mean free path (1.4–3), L.
μ = viscosity (1.1–1, 3.2–11 through 16), M/Lt.
μ_0 = parameter in Bingham model (1.2–2*a*), M/Lt.
ν = frequency of electromagnetic radiation (14.1–1), t^{-1}.
$\nu = \mu/\rho$ = kinematic viscosity (1.1–3), L^2/t.
ξ = dimensionless position variable, variously defined.
Π = dimensionless profiles (19.3–16, 17, and 18).
π = 3.14159 . . .
$\boldsymbol{\pi}$ = pressure tensor (18.3–2), M/t^2L.
$\rho = nm$ = fluid density (1.1–3), M/L^3.
ρ_i = mass concentration of species *i* (16.1–1), M/L^3.
σ, σ_{AB} = collision diameter (1.4–10, 16.4–15), L.
σ = Stefan-Boltzmann constant (14.2–10), M/t^3T^4.
$\boldsymbol{\tau}$ = viscous stress tensor [1.1–2,3.2–8,3.2–3(components)], M/t^2L.
τ_0 = parameter in Bingham model (1.2–2*a*), M/t^2L.
τ_0 = magnitude of shear stress at fluid-solid interface (5.3–7), M/t^2L.
Φ = potential energy (7.3–1), ML^2/t^2.
Φ_v = viscous dissipation function (10.1–21), t^{-2}.
$\boldsymbol{\phi} = \rho \boldsymbol{v}\boldsymbol{v} + \boldsymbol{\pi}$ = total momentum flux relative to stationary coordinates (18.3–8), M/t^2L.
ϕ = angle in spherical coordinates (Fig. A.6–1), radians.
$\phi_v, \phi_T, \phi_{AB}$ = rate factors (21.5–37, 38, and 39), dimensionless.
ψ = stream function (4.2–1, and Table 4.2–1); dimensions depend on coordinate system.
Ω = angular velocity (3.5–11), radians/t.
Ω_μ = collision integral (1.4–18), dimensionless.
Ω_k = collision integral (8.3–13), dimensionless.
$\Omega_{\mathscr{D}}$ = collision integral (16.4–13), dimensionless.
ω_i = mass fraction of *i* (Table 16.1–1), dimensionless.

Overlines[1]

$\sim$	per mole
$\wedge$	per unit mass
$-$	partial molal (18.4–8)
$-$	time smoothed

Brackets

$\langle a \rangle$	average value of a over a flow cross section (7.2–1, 9.8–32).
[=]	has the dimensions of
( ), [ ], { }	for use in vector operations; see Appendix A.

Superscripts

#	reduced relative to 1 atm (§1.3).
0	value at 1 atm or other standard state (§1.3).
●	transfer coefficient corrected for finite mass-transfer rate (21.4–4, 5, and 6).
*	reduced with respect to some characteristic dimension (3.7–1).
′	deviation from time-smoothed value (5.1–6).
(t)	turbulent (5.2–5).
(l)	laminar (5.2–8).
(m)	quantity crossing a mass-transfer surface.

Subscripts

A, B	species in binary systems (§16.1).
a	arithmetic mean driving force or associated transfer coefficient (13.1–3).
b	bulk or "cup mixing" value for enclosed stream (9.8–33).
c	at critical point (§1.3).
i, j, k	species in multicomponent systems (§16.1).
ln	logarithmic mean driving force or associated transfer coefficient (13.1–4).
loc	local transfer coefficient (13.1–5).
m	mean transfer coefficient for a submerged object (13.1–6).
r	reduced relative to critical value (§1.3).
tot	total quantity in a macroscopic system (7.1–1).
0	quantity evaluated at a surface.
1, 2	quantity evaluated at cross sections "1" and "2" (7.1–1).
1	driving force at cross section "1" or associated transfer coefficient (13.1–2).

[1] Note that the markings $\wedge$, $\sim$, and $-$ (when used for partial molal quantities) serve to indicate the units of the quantities on which they appear. For example, H [=] ML^2/t^2, whereas $\hat{H}$ [=] L^2/t^2, $\tilde{H}$ [=] ML^2/t^2 mole, $\bar{H}_i$ [=] ML^2/t^2 mole.

Commonly Used Dimensionless Groups (see also Table 21.2–1).

Br = Brinkman number (9.4–11).
Fr = Froude number (3.7–15).
Gr = Grashof number for heat transfer (9.9–17).
Gr_{AB} = Grashof number for mass transfer (18.6–13).
Nu = Nusselt number for heat transfer (§13.2).
Nu_{AB} = Nusselt number for mass transfer (§21.2).
Pr = Prandtl number (8.3–16).
Re = Reynolds number (3.7–14).
Sc = Schmidt number (18.6–12).
St = Stanton number (Fig. 13.2–2).

Mathematical Operations

D/Dt = substantial derivative (3.0–2, A.3–15).

$$\operatorname{erf} x = \frac{2}{\sqrt{\pi}} \int_0^x e^{-t^2}\, dt$$ = the error function of x.

$\exp x = e^x$ = the exponential function of x.

$\ln x$ = the logarithm of x to the base e.

$\log_{10} x$ = the logarithm of x to the base 10.

$$\Gamma(x, u) = \int_0^u t^{x-1}\, e^{-t}\, dt$$ = the incomplete gamma function.

$$\Gamma(x) = \int_0^\infty t^{x-1} e^{-t}\, dt$$ = the (complete) gamma function.

∇ = the "del" or "nabla" operator (see Appendix A).

Author Index

Subject Index